HANDBOOK OF DOUBLE CONTAINMENT PIPING SYSTEMS

Other Handbooks of Interest from McGraw-Hill

Avallone and Baumeister • Mark's Standard Handbook for Mechanical Engineers
Haines and Wilson • HVAC Systems Design Handbook
Hicks • Standard Handbook of Engineering Calculations
Higgins • Maintenance Engineering Handbook
Hodson • Maynard's Industrial Engineering Handbook
Lingaiah • Machine Design Data Handbook
Nayyar • Piping Handbook
Rosaler • Standard Handbook of Plant Engineering
Shigley and Mischke • Standard Handbook of Machine Design
Wadsworth • Handbook of Statistical Methods for Engineers and Scientists
Walsh • Electromechanical Design Handbook
Walsh • McGraw-Hill Machining and Metalworking Handbook
Wang • Handbook of Air Conditioning and Refrigeration

HANDBOOK OF DOUBLE CONTAINMENT PIPING SYSTEMS

Christopher G. Ziu, P.E.

McGRAW-HILL, INC.
New York San Francisco Washington, D.C. Auckland Bogotá
Caracas Lisbon London Madrid Mexico City Milan
Montreal New Delhi San Juan Singapore
Sydney Tokyo Toronto

Library of Congress Cataloging-in-Publication Data

Ziu, Christopher.
    Handbook of double containment piping systems / Christopher G. Ziu.
      p.   cm.
    Includes index.
    ISBN 0-07-073012-1 (acid-free paper)
    1. Double wall piping—Handbooks, manuals, etc. I. Title.
TJ934.Z58   1995
621.8′672—dc20   95-19288
                                                                 CIP

Copyright © 1995 by Christopher G. Ziu. All rights reserved. Printed in the United States of America. Except as permitted under the United States Copyright Act of 1976, no part of this publication may be reproduced or distributed in any form or by any means, or stored in a data base or retrieval system, without the prior written permission of the publisher.

1 2 3 4 5 6 7 8 9 0   AGM/AGM   9 0 0 9 8 7 6 5

ISBN 0-07-073012-1

The sponsoring editor for this book was Robert W. Hauserman, the editing supervisor was Stephen M. Smith, and the production supervisor was Pamela A. Pelton.

Printed and bound by Quebecor/Martinsburg.

McGraw-Hill books are available at special quantity discounts to use as premiums and sales promotions, or for use in corporate training programs. For more information, please write to the Director of Special Sales, McGraw-Hill, Inc., 11 West 19th Street, New York, NY 10011. Or contact your local bookstore.

This book is printed on acid-free paper.

Information contained in this work has been obtained by McGraw-Hill, Inc., from sources believed to be reliable. However, neither McGraw-Hill nor its authors guarantee the accuracy or completeness of any information published herein and neither McGraw-Hill nor its authors shall be responsible for any errors, omissions, or damages arising out of use of this information. This work is published with the understanding that McGraw-Hill and its authors are supplying information but are not attempting to render engineering or other professional services. If such services are required, the assistance of an appropriate professional should be sought.

To my parents,

who taught me the value of an education, and have provided me with love, guidance and understanding, and a chance in life.

Chris Ziu

Contents

Preface, xi
Acknowledgments, xiii

1 Introduction 1

1.1 History of Double Containment Piping, 1
1.2 Ground Water Technology and Ground Water Contamination, 4
1.3 The Concept of Double Containment, 7
1.4 Leak Detection and Monitoring, 10
1.5 Design of Piping Systems, 11
1.6 Design of Double Containment Piping Systems, 14

2 Materials of Construction 19

2.1 Fundamentals of Metallic Corrosion, 19
2.2 Metallic Materials Used in Chemical Services, 32
2.3 Degradation of Nonmetallic Materials in Corrosive Media, 49
2.4 Nonmetallic Materials Used in Piping and Tanks, 55
2.5 Corrosion Testing Methods, 64
2.6 Sources of Corrosion Data, 68

3 System Selection 71

3.1 General Statements, 71
3.2 Classification of Material Combinations, 74

4 Fluid Dynamics and Sizing Analysis 89

4.1 General Fluid Considerations, 89
4.2 Flow in Pressure Pipes, 95
4.3 Pressure Flow in Annuli (Secondary Containment Piping Considerations), 118
4.4 Flow of Fluids in Chemical Process Sewer and DWV Piping, 125
4.5 Gravity Flow in Annuli and Venting of Annuli, 140

5 Design of Metallic Primary Components 151

5.1 Design Criteria and Conditions, 151
5.2 Pressure Design of Primary (Core) Piping Components, 161
5.3 Selecting Primary (Core) Piping Components for Fluid Services, 174
5.4 Selecting Joining Methods for Fluid Services, 178

5.5 Flexibility and Support, 181
5.6 Auxiliary Piping Systems, 190

6 Design of Nonmetallic Primary Components   193

6.1 Design Issues for Nonmetallic Primary Piping Components, 194
6.2 Design Criteria for Nonmetallic Primary Piping Components, 197
6.3 Pressure Design of Nonmetallic Primary Piping Components, 202
6.4 Selecting Nonmetallic Piping Components for Fluid Services, 211
6.5 Selecting Nonmetallic Piping Joining Methods for Primary Pipe Fluid Services, 212
6.6 Flexibility and Support of Nonmetallic Primary Piping, 213

7 Design of Secondary Containment Components (Metallic and Nonmetallic Materials),   219

7.1 Introduction, 219
7.2 Conditions and Criteria for Secondary Containment Components, 221
7.3 Design Criteria for Secondary Containment Components, 229
7.4 Pressure Design of Secondary Containment Piping Components, 232
7.5 Selecting Secondary Containment Piping Components and Joining Methods for Fluid Services, 244
7.6 Secondary Containment Piping Support, 245
7.7 Auxiliary Piping Systems, 245

8 Thermal Expansion Considerations   249

8.1 Thermal Expansion, 249
8.2 Axial Stress Calculations, 256
8.3 Simplified and Comprehensive Stress Analysis, 259
8.4 Stresses in Interconnecting Components, 264
8.5 Totally Restrained Double Containment Piping Systems, 268
8.6 Methods for Alleviation of Expansion and Contraction, 274
8.7 Use of Heat Transfer as a Controlling Medium, 303
8.8 Classification of Systems and Application of Compensation Techniques, 304

9 Structural Considerations   313

9.1 General, 313
9.2 Longitudinal Bending Analysis, 313
9.3 Support of Primary Piping and Nonburied Assemblies, 321
9.4 Burial Considerations, 336

10 Heat Transfer in Double Containment Piping   369

10.1 General, 369
10.2 Heat Transfer Theory Applied to Double Containment Pipe Assemblies, 392

11 Layout Concepts for Double Containment Piping   397

11.1 Pressure Piping Systems,   401
11.2 Layout Details for Double Containment Pressure Fittings,   446
11.3 Nonpressure System Layout and General Underground Considerations,   449
11.4 Layout Details of Double Containment Drainage Fittings,   463
11.5 Petroleum Station Double Containment Piping Layout,   469
11.6 Multiple-Primary-Pipe/Common Secondary Containment Pipe Systems,   473
11.7 Valves and Valve Layout for Double Containment Piping Systems,   479

12 Fabrication, Assembly, and Erection   497

12.1 Metallic Component Considerations,   497
12.2 Nonmetallic Component Considerations,   513
12.3 Double Containment Piping Installation Issues,   528

13 Inspection, Examination, and Testing   587

13.1 Metallic Component Considerations,   587
13.2 Nonmetallic Component Considerations,   600
13.3 Considerations for Double Containment Assemblies,   609

14 Associated Storage Tanks and Pressure Vessels   633

14.1 Overview,   633
14.2 Storage Tanks (Nonpressure Vessels),   634
14.3 Underground Storage Tanks,   635
14.4 Testing of Storage Tanks,   653
14.5 Double-Wall Storage Tanks and Other Alternatives,   653
14.6 Pressure Vessels,   667
14.7 Internally Lined Tanks and Vessels,   679
14.8 Associated Double Containment Piping,   680
14.9 Double Containment Piping Connections to Associated Tanks and Vessels,   682

15 Leak Detection   695

15.1 Introduction,   695
15.2 Interstitial (Annular) Monitoring of Double Containment Piping,   698
15.3 Internal Leak-Detection Methods,   730
15.4 External Leak-Detection Techniques,   736

16 Trenchless Reconstruction and Alternatives to Secondary Containment Piping   751

16.1 Trenchless Reconstruction Methods, 751
16.2 Impervious Barriers,   762
16.3 Concrete Encasement,   772
16.4 Concrete and Masonry RTRP Structures,   773
16.5 Cathodic Protection for Coated Single-Walled Piping and Tanks,   782

Appendix A, 793
Appendix B, 841
Appendix C, 847
Appendix D, 857
Appendix E, 861
Appendix F, 869
Index, 873

Preface

On December 23, 1988, a major law went into effect in the United States which has changed forever the way that industry views chemical and petroleum piping systems. This law, which was written as a revision to the Resource, Conservation and Recovery Act, requires that the majority of underground tanks and their associated pipes which are intended to store or transport petroleum fluids and/or chemical waste be provided with special means to prevent against leaks to the surrounding environment. With respect to piping systems, this now usually entails providing a separate outer pipe to house the inner pipe, so as to contain any leaks that may occur, and to provide a means to facilitate the detection of such leaks. We refer to such a system in this book as a double containment piping system.

To many, the thought of placing one pipe inside of another sounds like a relatively simple and straightforward task, and without much challenge involved. I can certainly be included among those who felt this way upon first considering this concept. However, as one gets involved in trying to design and implement such systems, one is soon overwhelmed with the enormous number of subtleties and complexities that arise. One can quickly be overwhelmed just by the task of even considering the sheer number of combinations of materials and systems types that exist for each and every design condition. The vast number of considerations and options that exist are quickly realized by examining the size of this handbook and the amount of information that is contained in it. This book reflects my ten years of study of and experience in dealing with this topic.

When I first started out on this project, I was aware of the challenge of incorporating the many types of material systems and their varying physical properties into one all-encompassing handbook. I have attempted to maintain a balance and neutrality with respect to the various materials considered. However, it is inevitable that some of my experiences with and preferences for certain materials have at times led to subjective coverage. It is my hope that the reader will bear this in mind, and keep an open mind with respect to material choices and system design options. It is my belief, and this is reflected in the text, that in any given application there is likely to be at least one suitable choice of metallic, RTRP (fiberglass), or thermoplastic material. The reader is encouraged to keep this in mind.

Chapter One presents a general introduction to the subject. It also provides information related to groundwater resources. Chapter Two covers material considerations, with particular emphasis on the three major classes of piping and tank materials: metallic, RTRP (fiberglass) and thermoplastic. The proper selection of materials is the first step in the process of designing a system. It is a step that can never be underestimated.

Chapter Three provides the reader with an understanding of when to use the various combinations of inner and outer material systems. It classifies system material combinations into 17 broad categories and attempts to point out major considerations and complications that each present. Chapter Four covers fluid dynamic considerations for both the inner and outer pipes. Fluid dynamics is the scientific basis by which piping systems are typically sized, at least initially. Ultimately the sizing of pipes involves economic factors, which is true for both primary and secondary containment pipe systems.

Chapters Five, Six and Seven present the considerations involved in designing each specific component of a primary piping system. The philosophy which is used in this text is clearly based upon that of the ASME B31.3 Chemical Plant and Petroleum Refinery Piping Code. It is the world's most widely used code for designing chemical and petroleum piping systems. It also governs the majority of double containment piping system applications in North America. The reader should be aware that not all systems are covered by the code. Other codes may apply (e.g., ASME BPV code, Section III, Div. 1, Subsections NB, NC and ND for Nuclear Piping, class 1, 2 and 3 components, respectively; ASME B31.1 for Power Piping; ASME B31.4 for "Liquid Transportation Systems for Hydrocarbons, Liquid Petroleum Gas, Anhydrous Ammonia, and Alcohols"), or the system may be outside the scope of a design code (e.g., certain gravity sewer systems, which may be governed by municipal building codes or plumbing codes).

Chapters Eight and Nine are important, covering in detail the topics of thermal expansion and structural criteria. The concepts here which apply to systems are interrelated to those concepts discussed in Chapters Five, Six and Seven for components. Chapter Ten discusses heat transfer theory, the concepts of which are applicable to double-walled piping systems of many types (e.g., jacketed process pipes, heat exchangers, insulated pipes, etc.).

Chapter Eleven provides the reader with an overview of the many layout options that exist for any given system design. The concepts discussed here are interrelated to those of several of the other chapters. In particular, thermal expansion calculations and considerations are affected anytime a layout change is made in a system. Chapters Twelve and Thirteen cover the fabrication, installation, inspection and testing considerations for systems of all types. Chapter Fourteen covers many details with respect to tanks, and the interacting of piping systems with tanks. This information is included since many double containment piping applications involve tanks.

Chapter Fifteen is very important. One of the main functions of a double containment piping system is to facilitate the detection of leaks. The early detection of leaks minimizes fluid loss, and allows for quick repair of the system and safe removal of leaked fluid. Leak detection options are presented and discussed for systems of all types, including piping and tanks of both single-wall and double-wall design.

Chapter Sixteen presents a discussion of some of the alternative means to provide secondary containment for piping and tanks. It also discusses some of the more novel approaches to repairing and retrofitting lines which are in place. It is in this area of system design that it is anticipated that many new approaches will be developed over the next few years.

It is my hope that this book will serve as the authoritative text on this subject for some time to come. It is expected that future editions of this text will be updated to include the many changes that are anticipated to occur in this continually developing area of technology.

Acknowledgments

There are many individuals and companies to whom I owe my gratitude for contributing in some way to the development of this text. Some were contributors in a direct manner (e.g., as editors), and others helped indirectly by contributing to the knowledge that I have gained over the years in piping, in nonmetallic materials (thermoplastic and RTRP), and in the art of double containment and leak detection. It is likely that I have left some names out inadvertently. However, I would like to offer my thanks to all of those individuals and organizations who lent support to me over the years by helping me to gain the knowledge that was necessary to write this text.

I owe many thanks to those who contributed greatly to my understanding of plastic piping and plastic materials, especially Jim Blumenkrantz, Roland Hall, and the late Gene Findley (each formerly of R&G Sloane Mfg. Co., Inc.); Karl Bohaty and Albert Lueghamer of AGRU; Dr. Julius Dohany, Dr. Steve Humphrey, Ed Bartoszek, and Roger Pecsok of AtoChem; Nestor Macquet and Roger Starr of Solvay; Dr. John Imbalzano and Pradip Khaladkar of DuPont; Bill Miller of Ausimont; Vince Bavaro of Himont; Carl Martin of Fibercast (whom I also owe many thanks for his contributions in educating me about restrained RTRP double containment system design); Ken Oswald and Bill McDonald of Smith Fiberglass Co.; Mike Luckenbill of Ameron Co.; Bob Davis, John Eddy and Nick Jones of the Ershigs div. of Chicago Bridge & Iron; numerous individuals at RPS/Abco; Terry McPherson of Eslon; Reijo Stoor of Chevron/Plexco; and Dagmar Ziegler of Wegener N.A. Special thanks go to my lifelong friends at Grewe Plastics, Renee, Norman, Allen and John Blum, for teaching me extensively about the nature of fabricating thermoplastic materials. I also owe an enormous amount of gratitude to Harvey Svetlick of Phillips Driscopipe, the "Grand Master" of the polyolefins, for providing countless answers to my many questions regarding HDPE piping systems.

I also would like to thank those individuals and organizations who contributed in a special way to my understanding of double containment and leak detection, especially Carl Martin of Fibercast, who is one of the great pioneers in double containment; Dennis McAtamney of Guardian, to whom I wish continued success in his endeavors; George Donovan (formerly of DuPont) and Carl Faller of DuPont; Cliff Smith, Lee Grulke, Ken McCoy, Adrienne Klopak, James Holmes and several others of Raychem Co.; Fati Elgendy of Permapipe; Wayne James of Simtech; Eben Williams (formerly of Juno Industries); Al Rubens of ACR; Dave Leger; Roger Berg of Harrington Industrial Plastics and Al Smith (formerly of Harrington); and Stan Lewis of A/A Manufacturing and Bud Lewis and Tim Robinson of Asahi/America, for allowing me an opportunity to develop my early understanding of the subject, and my former coworkers at Asahi/America Joe Durning and Pat Hollis. I owe a great deal of thanks to the individual who taught me more about the intricacies of fabricating double containment piping than anybody, namely Bob Silvia of Process Engineers & Constructors, who is a fellow ASME B31.3 Code Committee member. I would also like to thank Don Seccombe for providing me with a basic understanding of finite element analysis.

There are also many customers to whom I owe a special thanks in allowing me an opportunity to develop field experience and a first-hand working knowledge of the subject. Those who deserve a special mention include Chris Curtis, Jim Wheeler, Ralph Hutchins, Barry Goringe, Ray Gunner, Fred Erp, Don White, Fred Babarsky, Jeff West, Dale Hatcher, Luis Crespo, Adam Weissman, Robert Divjak, John Park, Bill Whorley, Bill Henderson, Doug Marshall, Ken Bruns, Dwight Hollingsworth, and Dan Loitz. I am certain that there are many other individuals who deserve to be mentioned, and I would like to express my thanks to all of those customers who believed enough in me to offer me the opportunity to gain experience.

There are also many individuals who contributed to the development of this text in some way and to whom I owe thanks. These include Laura Smith, Tom Berarducci and Cindy Poslusney, who contributed to the preparation of the manuscript. I also owe many thanks for the patience and work of many individuals at VCH who contributed in some way, in particular Charles Doering and Camille Pecoul. A special thank you goes out to Attilio Bisio for his contributions as an editor, for without his effort and guidance this text may never have gotten started. I am also forever grateful to the McGraw-Hill organization for having the foresight to go forward with this project, and in particular to Robert Hauserman for seeing this project through to completion.

I owe thanks to John Tverberg of Trent Tube, for educating me as to the science of corrosion of metallic materials and whose work and teachings I greatly admire. I would also like to offer thanks to Dr. Quy Truong for answering questions and providing input, and to Bruce Elliott for his editorial input on Chapters Five, Six and Seven. I would like to offer a special thank you to Bill Short of Pressure Systems Engineering and a fellow ASME B31.3 Code Committee member, who has been a big believer in me over the years, who provided me with additional insights to this subject matter, and for his editorial contribution to Chapters Five, Six and Seven. Many thanks also to Art Ivanick of AllCad for his artistry, including the cover concept, and various CAD contributions.

I owe a great deal of thanks to the individual who contributed perhaps more than any to my understanding of the subject matter, Chuck Becht of Becht Engineering, who is a fellow ASME B31.3 Code Committee member. Without his valuable editorial contributions to Chapters Five, Six, Seven, Eight and Nine, and my overall learning from his teachings, this text would not be nearly what it is today. Charles is widely recognized by his peers as a leading expert in piping stress analysis, and I am lucky to have enlisted his expertise.

There is no limit to the amount of thanks that are due to my parents for the work that went toward the creation of this text. I would like to express my gratitude to my father, whose brilliant CAD work was responsible for the creation of a majority of the many wonderful illustrations used in this text. Without his time and knowledge, all of which was donated to help me in this effort, this book would certainly not be as attractive and informative as it turned out to be. I am an admirer of his mechanical engineering knowledge and expertise as both a draftsman and machinist, which is a rare combination of skills to have. He is my idol, and will always serve as an inspiration to me. I also would like to express my gratitude to my mother for her love, patience and understanding in sacrificing much by allowing my father to dedicate his time as he did. I owe thanks as well to my brother Jim for his significant help by virtue of his computer skills, and for his generosity.

Acknowledgments

There are simply no words that I can find to express the thanks and love that are due to Abby, my wife, who contributed in so many ways to the creation and finalization of this text. First, I owe her thanks for the limitless support, love, patience and understanding she showed me during four years of effort. Second, I owe her thanks for the magnificent work she did in the typesetting and graphic design of the book. She is my inspiration and my strength.

Chapter One

Introduction

1.1 History of Double Containment Piping

The concept of placing a pipe within another pipe dates back to the early days of heat exchangers. Other related applications have also existed, including jacketed process piping systems and insulated utility and process piping equipped with external jacketing. Double-pipe heat exchangers were among the first types of heat exchangers used. The design of such systems has been considered as a relatively straightforward procedure in terms of heat exchanger design. Many of the design, construction, and fabrication techniques for double-pipe heat exchangers are the same as those required of double containment piping systems. The same is also true of jacketed process piping systems and insulated (and jacketed) pipes. Therefore, these related applications set some of the historical bases for successfully designing and implementing a double containment piping system. However, double containment piping systems tend to involve many unique and additional considerations in comparison to these systems. The main differences are due to the purpose of double containment piping, which is different from any of the other three applications mentioned; this is, to house a primary pipe for the purposes of preventing a corrosive, flammable, hazardous (including toxic and carcinogenic) or radioactive fluid from ever finding its way to the external atmosphere.

In accomplishing this purpose, it is best to design both the primary and secondary containment for mechanical integrity, while monitoring the containment (for possible breach of primary and secondary containment), and without adversely affecting flow capacity in the annulus (e.g., purging, flushing, draining, drying, and venting). As a result, these additional requirements of double containment piping result in many additional design and fabrication considerations beyond those of the related systems described in the preceding.

Up until the 1970s, dual arrangements of piping systems utilizing a carrier pipe with a secondary piping providing containment for purposes of environmental protection or safety were limited to highly specialized applications. These involved rare applications in the nuclear, gas, or chemical processing industry, where highly toxic or lethal chemicals or fluids were transported. The use of an outer jacket to maintain a positive seal around a primary carrier piping system and protect the primary piping was rare indeed until this time period.

In the 1970s, the world began focusing more attention on industrial pollution and its effect on our drinking water supply and air quality. In the United States, in particular, there were several incidents, such as the revealing of widespread pollution at Love Canal in upper New York State, that caused a public outcry and drew attention to the problem. In addition, there were several incidents of contamination caused by large electronic circuit manufacturers, electroplating manufacturers, chemical manufacturers, and similar industries that process hazardous chemicals. Also, many of the hundreds of thousands of underground storage tanks installed 20-50 years earlier had developed leakages due to soil corrosion. Contamination reports of drinking water supplies due to petroleum and their byproducts were becoming more and more frequent.

Several large U.S. electronics manufacturers began an unusual practice in the 1970s for their underground, and in some cases above-ground, hazardous chemical piping systems. Under the watchful eye of their respective local and state environmental agencies, they began placing their underground transportation piping within an outer jacket for the sole purpose of preventing any leakages from getting into the ground water supply. Part of the reason for doing so was due to inadequate performance on the part of ordinary single-wall piping components intended to handle these chemicals. Leakages from joints, failures of piping materials due to poor manufacture or installation practices, inappropriate material selection, and soil corrosion were some of the contributing factors. Thus began the modern day practice of placing underground piping systems within a secondary containment piping system for the sole purpose of protecting the local environment.

Unfortunately, many of these early systems met with complications that resulted in less-than-successful performance. The first attempts at dual arrangements met with frequent failure. This was primarily due to systems being designed with combinations of piping components whose geometries were not readily compatible. Also, there were technical issues that arose that had not been addressed in the design. This had to do with structural concerns (supporting and centering the inner system), differential thermal expansion issues, penetration sealing issues, and many others. Additionally, many of these systems were installed with poor fabrication techniques, resulting in installations that lacked inherent stability and were thus readily subject to premature failure. The lack of design consideration and installation deficiencies resulted in failures that manifested themselves in predictable ways. These included the separation of split outer pipe and split outer fittings, coupled with failed inner welds, thereby leading to a double failure. Other means of failure included premature failure of inner and outer pipes due to fatigue, excess strain, and many other reasons.

During this same time frame, the federal government of the United States was studying ways to protect against the failure of underground storage tanks and piping transportation systems. This resulted in revisions to the Resource Conservation and Recovery Act (RCRA) of 1976, which were enacted in 1984. Signed into law as the "Hazardous and Solid Waste Amendments of 1984" (U.S. Public Law 98-616, signed November 8, 1984), it extended and strengthened the Solid Waste Disposal Act, as originally amended by the Resource Conservation and Recovery Act of 1976. The amendments contained strict requirements and provisions for underground storage tank and piping systems.

For systems containing a hazardous fluid [according to the EPA as defined in the Code of Federal Regulations (CFR), Title 40, part 241] or petroleum-based product with 10% or more of its volume underground (including piping), the laws contained strict provisions.

The provisions gave the user a choice of providing secondary containment with leak detection or to use corrosion-resistant materials with frequent monitoring and tightness testing (petroleum-based products only). An alternative for petroleum piping applications involved the use of ordinary carbon steel (e.g., A-53 or A-106, Grade B) with some form of coating and cathodic protection added. In the case of the so-called hazardous chemicals, above-ground systems were regulated as well, unless the systems were to be inspected on a daily basis. The regulations applied to both new systems and existing systems. Existing systems were subjected to a retrofit requirement according to a time table, with the oldest systems being required to be replaced first. Thus, all existing systems in the United States (installed prior to December 1988) have been scheduled to be replaced within a 10-year time frame from the commencement of the regulations (by December 1998), by the strictest interpretations of the regulations, if they are not already in compliance. It is important to note that there are many aspects to the U.S. regulations (e.g., minimum volume requirements, regulated substances, leak detection requirements and options, etc.), and that the requirements and degree of enforcement vary according to individual state. Other countries beyond the United States are currently considering the development of similar regulations and are expected eventually to adopt them by the year 2000.

The 1984 amendments passed in the United States provide a framework by which individual states, territories, and possessions of the United States are mandated to draft state legislation, with their laws being at least as stringent as the U.S. Federal law. Enforcement is required at the individual state level, with the Federal Environmental Protection Agency overseeing each state program. Prior to the Federal laws being adopted in the United States, some of the individual states where ground water contamination has been a particularly bad problem, have adopted similar laws, prior to the Federal government requiring them to do so (e.g., California, New York, parts of Texas, and others).

A major incentive has existed since 1984 for inventive component suppliers to develop and market products that offer basic double containment features. Within a 5-year time frame from 1984, single-wall tank and pipe manufacturers recognized this need and began to develop and offer products that met the basic needs of the new regulations. What has become readily apparent, however, is that so-called "pre-engineered" double containment products are not enough. This is because each and every situation tends to be a unique one that requires some amount of custom design and fabrication. Requirements for custom design are even greater when piping or tank products are to be combined with some means of leak detection, as is true in the majority of applications. Therefore, it has became readily apparent that a coherent set of design and installation standards is needed in order to assist engineers and material suppliers in designing systems to work in a manner desired for each given application.

The idea for this text has grown from this recognized need. The book is an attempt to present a coherent set of design and installation standards. It is hoped that individual designers may apply these standards, together with a turnkey approach to each given potential design application. In doing so, a designer may solve many of the problems encountered early on, and result in high-quality, well-functioning double containment piping systems. In doing so, the possibility that piping and tank systems will contribute to future pollution of the earth and its atmosphere will be minimized to the greatest extent possible.

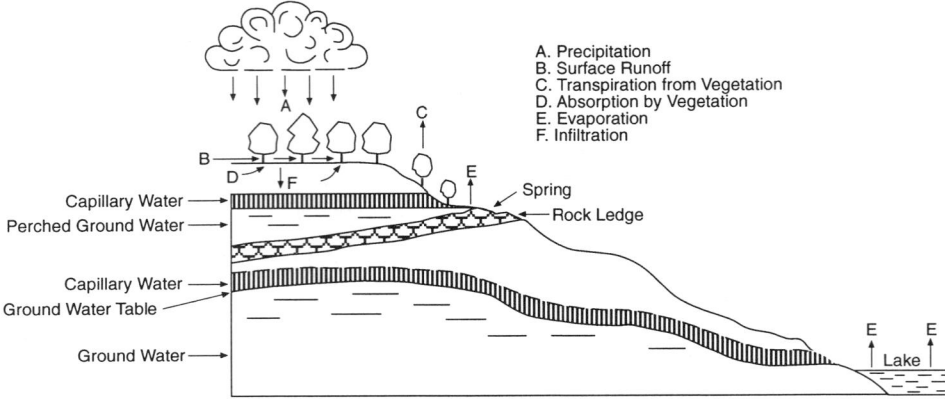

Figure 1.1. Normal hydrologic cycle. (Source: U.S. Army Corps of Engineers.)

1.2 Ground Water Technology and Ground Water Contamination

One of the main reasons to use a double containment piping system is to help preserve and protect the natural ground water supply from leaking chemicals. Therefore, it is important that the users and designers of piping systems gain a basic understanding of what ground water is and how the hydrologic cycle works. It is important that everyone involved in potential underground piping systems understands what it is, so that it may be realized exactly how important it is to prevent underground chemical leaks from occurring. Thus, an overview of ground water and the hydrologic cycle is essential in a handbook of this type to understand the full scope of double containment piping applications.

The term ground water is usually defined by geologists as any water found below the water table or in any geologic formation that is fully saturated. Subsurface water, on the other hand, is any near-surface water that infiltrates soil and is not absorbed into the ground at a lower level. Ground water and subsurface water are part of the normal hydrologic cycle, which is illustrated in Figure 1.1. Ground water in many locations is used to obtain a supply of drinking water. It also may flow by underground streams or rivers to replenish the supply of drinking water in lakes and other above-ground bodies of water.

Subsurface water is often thought of as more of a problem than a resource, as it often interferes with construction projects or causes damage to existing structures.[1] However, ground water and subsurface water are both important parts of an ecological cycle that helps to maintain the delicate balance of nature of all living things. Ground water storage represents the largest means of fresh water storage. Runoff-producing rains are not by themselves sufficient to reduce continuing outflow of streams and lakes during prolonged drought periods, which often follow periods of precipitation.

The relation between ground water and surface storage can best be described as mutual interdependence. The quantity of usable ground water storage distributed in structures is determined by means of precipitation, evapo-transpiration, geologic structure, and other minor contributing factors. There are two parts that make up the total quantity of ground water supply. The first derives from the hydrologic cycle, and the other consists of water trapped in past ages that is no longer naturally circulated in the cycle.

Introduction

Water seepage to and from surface water sources, its movement to lower basins, and the amount of water transpiration that occurs in an undeveloped ground water basin depend upon the quantity of water in storage and its rate of recharge. Recharge may exceed discharge during periods following abundant rainfall. When recharge exceeds discharge, the excess rainfall increases the amount of water stored in ground water basins. As the pressure of artesian water rises, outflows increase as the gradient to points of discharge become steeper and steeper. Storage decreases from outflow causes water table levels and artesian pressures to decline as recharge no longer occurs. The major fluctuations in storage are seasonal in most undeveloped basins, and there is little change in the mean annual elevation of the water levels. The average annual inflow to storage is thus approximately equal to the average annual outflow, which is defined as the basin yield. An illustration of basic geologic formations is presented in Figure 1.2 to help understand this process.

Ground water distribution is normally divide into zones of aeration and saturation. When all of the voids are filled with water under hydrostatic pressure, it is referred to as the saturation zone. If the spaces are filled partly with air and partly with water, they are referred to as the aeration zone, and may be further subdivided into several subzones as follows:

1. **Soil/water zone:** The soil/water zone commences at the ground surface and travels downward through the major root zone. The total depth of this zone is dependent on vegetation and the type of soil, and is a variable quantity. For the most part, the zone is unsaturated, unless there is heavy infiltration. There are three major types of water that may be encountered in this region: (1) gravitational water, which is excess soil water draining through the soil; (2) capillary water, which is held by surface tension; and (3) hygroscopic water, which is adsorbed from air.

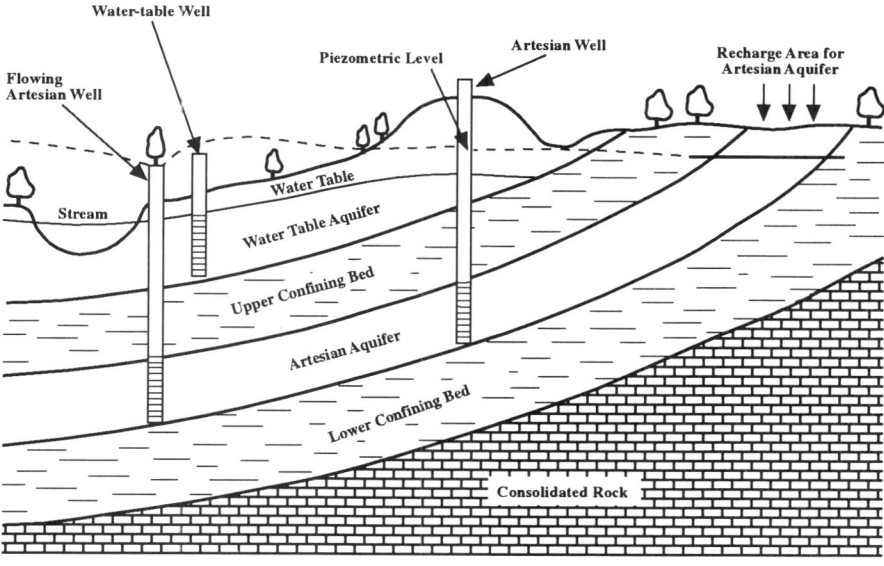

Figure 1.2. Illustration of basic geologic formations. (Source: Michael Frankel, "Sub-Surface Drainage: What is important to the Plumbing Engineer," Plumbing Engineer, 18(2), March 1990, pp. 26-32.)

2. **Intermediate zone:** Starts at the bottom of the soil water zone and extends to the top of the capillary fringe. This zone may be several hundred feet in thickness, or it may be thin or even nonexistent. The zone serves as a means of connecting the near ground-surface region and the near water-table region.
3. **Capillary zone:** Starts from the water table to a height dependent upon the capillary rise that can be generated in the soil. The thickness of the capillary zone is a function of soil texture and may not only vary within a local area, but also by region.

The zone of saturation depends upon water that drains downward below the root zone, which finally reaches a level at which all of the openings or voids in the earth's materials are filled with water. It is this water in the zone of saturation that is referred to as ground water. The upper surface of the zone of saturation is called the water table, unless it is confined by impermeable material. Artesian pressure is caused by water in the zone of saturation that is under an overlapping impermeable formation that causes it to be under a pressure greater than atmospheric pressure. For a well to be artesian, water in the well must occur above the top of the aquifer. An aquifer is an underground layer of permeable rock or soil that permits the passage of water.

Ground water is continually moving within an aquifer even though the movement may be slow. Since irregularities exist in surface topography and underground formations, a water table intersects the surface of the ground or the bed of a stream, lake, or ocean in irregular locations. Thus, ground water moves out of an aquifer or ground water reservoir and to these intersection locations. The water table or artesian pressure surface is sloped from recharge areas to discharge areas. A pressure differential is created by these slopes, causing the flow of ground water within each aquifer. At any point the rate of flow and resistance to movement of water through a saturated formation is a reflection of its slope, and vice versa. Considerable changes typically occur in the elevation and slope of a water table and in artesian pressure level in an underground reservoir due to seasonal variations in the supply of water.

Ground water is no longer the pristine natural resource such as it has historically been perceived. Recent information reveals the fact that ground water is contaminated in widespread geographic areas.[2] Data show that this contamination comes from many different sources and substances, including toxic organic and inorganic chemicals. Because ground water is a primary source of drinking water supply, these toxic chemicals pose unacceptable human health risks. One of the major sources of ground water contamination to this point has been the leaking of underground storage tanks and piping systems.

Ground water that is contaminated by synthetic organic chemicals is particularly disturbing. Contaminant concentrations are often orders of magnitude higher than those found in raw or treated drinking water drawn from even the most contaminated surface supplies, including the Ohio and Mississippi rivers.[3] There are no Federal health standards for most of the synthetic organic chemicals found in drinking water wells. Many of these chemicals are suspected of being carcinogenic or mutagenic. Some create very high health risks with concentrations on the order of 10 parts per billion or below. Levels such as these are typical of surface water contamination of the substances mentioned. Many of these chemicals are tasteless and odorless, even at concentrations that can represent substantial risk.

Ground water found in unconsolidated formations (sand, clay, and gravel) and protected by similar materials from sources of pollution is more likely to be safe than water coming from consolidated formations (limestone, fractured earth, lava, etc.).[4] Water of better sanitary quality can sometimes be obtained by drilling deeper due to limited filtration provided by overlying earth materials. However, due to the geology of certain areas, it is not always possible to find water at greater depths.

1.3 The Concept of Double Containment

A double containment system is by definition an arrangement of a primary storage or transportation vessel (piping system, tank system, or a combined system), contained within a housing that provides a means of secondary containment of a fluid if the primary containment fails. This is illustrated in Figure 1.3. In such an arrangement, primary components must be designed with all of the normal considerations of a single-wall system. Design of the primary system must be according to normal piping design codes and should be treated in the same way that it would have if it were a single-wall piping system.

There is added incentive to design the inside (primary) system in such a manner that failure will not occur. This is due to the fact that a release, or detection thereof, could mean significant repair expense. Secondary containment components (outer piping or tank elements) are then designed to contain a leak, and to withstand any possible effects of a process fluid release (i.e., sudden release due to overpressurization of the inner system, etc.). The two systems together are part of a complex, interacting assembly that is referred to as a double containment system. In a majority of applications, a method of manual or automatic leak detection or sensing of the interstitial space between primary and secondary systems is an important aspect of such a system, as illustrated in Figure 1.3. All told, an effectively designed double containment system, coupled with an appropriate sensing method, will eliminate or minimize to a great extent any possible release of a fluid to a system's external environment. External environment may refer to the surrounding soil in a buried system, the surrounding air in an above-ground system, or the chamber or room in an indoor or enclosed system.

The secondary containment portion of a double containment system may be designed to meet or exceed the capabilities of the inner system, or it may be designed to less stringent requirements, depending on a number of factors. For pressurized systems, a secondary containment system should normally be designed to handle a reasonable amount of pressure; in many applications it is best to rate it to the same specifications as those of the primary system it is encompassing. However, a secondary system can be designed to a lower pressure rating for economy, if the design of the system includes automatic shutoff pumps or system shutoff valves, automatic drainage or atmospheric venting of the annular space, etc. The material selected for a secondary containment system may be the same as its primary system, or a combination may be selected for economy, or for other reasons. In fact, the material requirements for the secondary containment system may dictate that a more expensive material, having greater corrosion resistance, or provided with additional corrosion protection, be added due to external environmental conditions.

1.3.1 Double Containment Piping Systems

To provide secondary containment for piping systems, there are design considerations to be considered that are beyond those that apply to tanks. Whether the system is a pressure transfer pipe or a drain, waste, and vent system will have a significant impact on the

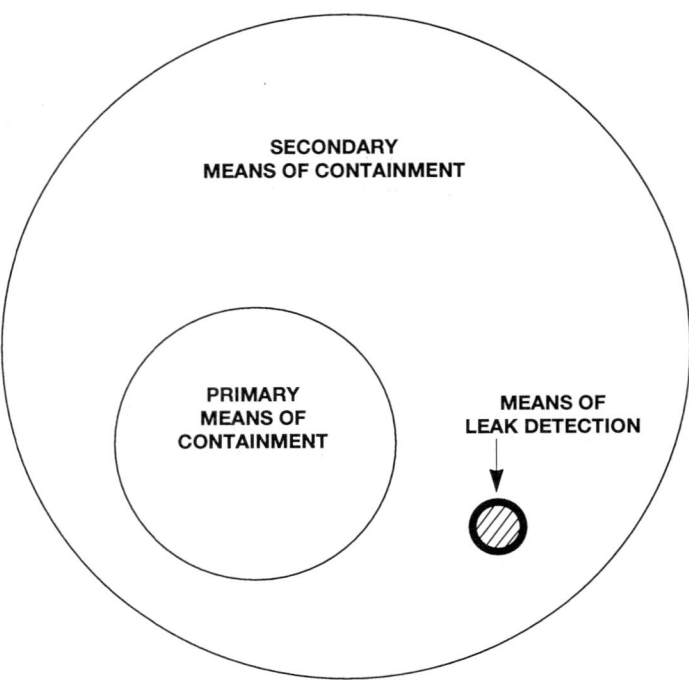

Figure 1.3. Elements of a basic double containment system.

design, layout, material selection, wall thicknesses, leak detection method chosen, and other aspects to be considered. Additionally, a system may be a relatively straight piping run, it may consist of primarily fittings, or it may have parts that are both straight in sections and fitting intensive in others. Also, whether a system is intended for burial or whether it is intended for above-ground use also has a significant impact on its design. Space limitations are an important consideration as well. A piping system that is intended for tight pipe-rack work, or is to be located near other buried structures, will have different installation requirements and different requirements for material selection, wall thicknesses, pressure ratings, etc., in comparison to a system that does not have space constraints.

In terms of material selections for the inner and outer pipes, there are a number of factors that have an effect on what should be selected. For primary piping components, the chief consideration in making its material selection has to due with the ability of the material to withstand the corrosive effects of the inner fluid. The same basic techniques and criteria used for selecting materials for single-wall piping systems must be used for primary piping in a double containment system. However, the interaction of the inner and outer systems, how they are tied together, the preferred joining method, and how joining affects system installation, the compatibility between the secondary system, and its joining methods all factor into the considerations for selecting the "combination of materials" to be used.

For secondary containment piping components, the considerations for choosing the material to be used must include risk analysis. The risk analysis used must take into account the anticipated frequency and duration that it is reasonably expected that the outer material may be in contact with the contained fluid.[5] It also involves weighing the possibility of failure against the sensitivity of the external environment and the relative extent of damage that a "double failure" might cause. Since there may be only infrequent (or no) contact with the corrosive fluid in a properly designed system, and the time of contact once an event occurs may be short in duration, a less expensive material may be an appropriate choice for some applications. However, for many applications the risk created by even the remotest chance of a "double failure" might be too great even to take a chance on using a less corrosion-resistant material.

Leak detection in a double containment piping system usually consists of some means of monitoring the status of the annular space that exists in the system. Detection methods that are used typically measure the presence of a fluid or some change in physical state. Other methods can be used (i.e., flow measurement), but the most effective means is to monitor the status of the system's annular space to detect for a change from a status quo.

1.3.2 Double-Wall Tanks and Vessels

Double containment storage vessels have only recently been introduced. Most double containment vessels have been limited to nonpressure petroleum storage tanks, primarily intended for underground service. A double containment vessel intended for pressure service would be subject to such design criteria as the American Society of Mechanical Engineers (ASME) Boiler and Pressure Vessel Code for its inner and outer shells. Thus, a double-wall pressure-rated vessel would be much more complex in nature, and more expensive, than a double-wall storage tank. Double-wall storage tank design typically consists of a heavier-gauge wall for the primary tank and a lighter-gauge wall thickness for the secondary shell. Tanks of this nature also are normally designed to permit some sort of interstitial monitoring to be facilitated by a monitoring device placed between its two tanks.

Underwriters Laboratories (UL) currently lists two specific double-wall tank designs, types I and II. The type-I design has the exterior shell designed so that it is wrapped in direct contact with its primary vessel. Additionally, the exterior shell need not wrap the full 360 degrees of the primary tank's circumference. A type-II double-wall tank consists of the outer tank being physically separated from its primary vessel. A type-II outer tank is separated from the primary tank by the use of spacers or standoffs, and each shell is a full 360 degrees. A type-II tank lacks the rigidity of a type-I design in buried situations, as backfill provides no support to this type of tank. The inner tank must be treated in much the same manner as a single-wall above-ground tank in the design of the spacer or standoffs. Therefore, the design of a double-wall above-ground tank provides fewer design differences from single-wall tanks, as compared to the underground situation, in this one respect. Inner tanks for petroleum service must be constructed in the United States according to UL 42 specifications, titled "Steel Aboveground Tanks for Flammable and Combustible Liquids."

Detail design considerations for double-wall tanks parallel those of double containment piping. Material selection may be the same or may be different due to economy or to external corrosion considerations. Very often metallic single- and double-wall tanks are provided with a corrosion-resistant outer lamination of fiberglass-reinforced-plastic material, or with an organic coating, for protection against soil corrosion. In terms of pressure ratings, most double-wall tanks are designed with a thinner outer-gauge wall thickness for the secondary tank as described previously. This practice is valid for most nonpressure tanks (0 to 15 psi), but would not be adequate if a secondary containment jacket is required to meet the same pressure rating of the primary tank, as in a jacketed-pressure vessel.

1.4 Leak Detection and Monitoring

While a double containment system can be designed without a leak detection or monitoring system, the effectiveness of a system is considerably lessened without it. There are many methods available to provide sensing of the annulus that exists between primary and secondary systems. These methods are divided into two basic categories: automatic sensing systems and nonautomatic sensing systems.

1.4.1 Continuous Sensing Systems

Continuous sensing systems normally involve a sensing mechanism and an automatic alarm device. The most sophisticated form of sensing consists of continuous line leak detection. Continuous line leak detection is a technology designed to give relatively precise locating abilities for the determination of the location of leaks. For piping systems, this is a very important feature, although it also adds significantly to the initial cost of a system. There are two types of technologies, conductive (resistance) -based systems and those that are based on the measurement of impedance, commonly referred to as TDR-based systems (TDR is a common abbreviation for time-domain reflectometry). Both technologies are effective, although there are subtle differences between the two.

Other forms of continuous sensing are all based on some form of point probes. These include liquid level sensing, pressure sensing, moisture sensing, pH sensing, conductivity (resistivity sensing), vapor detection, ultrasonic sensing, movement sensing, and many others. The various forms of sensing may also be combined for added effectiveness. The advantage of point probe systems is in the ability to customize the system, but an inherent disadvantage is the lack of ability to locate a leak with precision.

1.4.2 Noncontinuous Sensing Systems

Non continuous leak sensing systems consist of visual detection systems, manual detection, inventory monitoring (tanks), and soil and vapor monitoring. Soil and vapor monitoring may also be used on a continuous basis, although it is not very common to do so. Visual detection can be designed easily into above-ground systems, but for underground systems they must be placed in manholes or sumps. Two forms of underground visual monitoring are very inexpensive and can be facilitated readily by having the secondary system flow open at the end of the pipe system, or by designing low point sumps with risers to the ground surface. Manual detection is normally facilitated by installing drip legs with valves in above-ground systems and with either low point sumps with risers to the ground surface in underground systems, or by positioning drip legs with valves in manholes. Manual systems can be a very cost effective and efficient method of leak detection and usually are a good idea to have as a backup or redundant method when other methods are employed. Inventory monitoring is a practical method for tank systems, but for piping systems it becomes more costly and complex, as a material balance must be performed. Noncontinuous systems are for the most part nonspecific in terms of locating leaks unless a system is designed with very frequent locations for visual or manual monitoring.

1.5 Design of Piping Systems

To design a piping system entails many separate tasks, as there are a multitude of internal and external forces, and other factors that may act upon a pipe. There have been many lengthy volumes and dissertations written on each of the various aspects of piping design. This text will highlight each of the various design areas and will additionally relate the design aspects to designing integrated primary and secondary containment pipe systems in a double containment assembly. The major areas of piping design include material selection, fluid dynamics and sizing, heat transfer, stress analysis, and structural analysis. For each of the major areas of design, there are unique aspects that are created when a primary carrier system is placed within a secondary containment housing, since each system will interact with the other. In addition, there are unique characteristics of fluid behavior for fluid flow in an annular space, which has an impact on sizing criteria.

1.5.1 Fluid Dynamics

The study of fluid behavior under flowing conditions is referred to as fluid dynamics. The sizing of piping systems and pumping requirements according to fluid flow is considered a subset of fluid dynamic theory. The flow of fluids under pressure is a common situation that is thoroughly studied by mechanical and chemical engineers as part of their normal curriculum. Although there is no one organization that dictates or governs sizing criteria for piping according to fluid flow, there is a common approach based upon rational methods that have been accepted over the years.

The flow of fluids under gravity conditions is more common to applications that fall within the civil engineering or sanitary engineering disciplines. Therefore, fluid dynamic behavior and sizing criteria for drainage piping is studied primarily by civil engineers. There also is no single set of standards that govern the sizing of pipes operating under gravity flow. However, most piping for these situations is governed by the general rules that have been developed by such professional organizations as the American Society of Civil Engineers (ASCE) and the American Society of Plumbing Engineers (ASPE).

1.5.2 Heat Transfer

Heat transfer theory for piping systems is a well-defined science dating back to the earliest applications for piping systems. Since piping systems are so widely used in heat transfer applications such as chilled water systems, cooling water piping, cogeneration applications, double-pipe heat exchangers, and many others, heat transfer design criteria for piping are widely understood. Many piping systems are insulated to reduce energy costs. When a pipe is insulated, it is often clad with an outer tube or pipe material such as thin-gauge aluminum, steel, or thermoplastic materials for protection of the insulation against weather effects. All of these applications set the basis for heat transfer theory in double containment piping applications.

1.5.3 Mechanical Stress Analysis

Stress analysis of piping system is a function performed by the mechanical engineer. The subject is worthy of devoting entire graduate studies and is thus a highly complex area of piping design. Virtually every piping system is subjected to various forms of stress, although many piping systems are designed without a thorough stress analysis of the system performed. The main thrust of pipe stress analysis is to gain an understanding of the behavior and reactions of piping systems to internal and external forces, temperature effects, and cyclic loads, in order to determine if the pipe will fail due such effects. The most frequent form of stress that is encountered and analyzed in piping systems results from thermal effects to which piping components are subjected. Thermal effects may be the result of a hot or cold fluid being moved through a piping system, from ambient changes in temperature or a combination of the two. Other forms of stress that may be encountered include those imposed by movements of external restraints or anchors, external weights, buoyant forces, internal and external pressure, and many others.

The most sophisticated form of stress analysis available to the mechanical engineer is finite element analysis, whereby individual parts or elements are analyzed to determine the reactions of the various combined stresses imposed upon them. Finite element analysis is the most accurate means to model stresses in a double containment piping system. It is thus a very important tool for this subject matter and is discussed in detail in Chapter 8, Section 8.3, "Simplified and Comprehensive Stress Analysis."

1.5.4 Structural Analysis

Structural analysis involves the consideration of the proper support and reduction of stress at all points of a piping system. The study of this subject matter is normally within the domain of a subdivision of the civil engineering discipline known as structural engineering. Once the proper location, placement, and type of supports are identified, the structural engineer can design and develop a supporting structure to provide support to the piping system, or may identify the appropriate use of existing structures to accomplish the same purpose. In underground piping systems, support is normally provided by proper burial techniques, with the possible exception of the addition of thrust blocks at fitting locations and the design of a support structure for piping/fittings located within underground manways or trenches.

1.5.5 Layout Considerations

In addition to all of the design considerations stated, one must also consider an appropriate layout for a piping system. The concept of piping system layout is a subject that impacts all other areas of design. Each specific change in a single layout detail will create changes in other aspects of a piping system's behavior in terms of stresses, structural loads, heat transfer, fluid flow, etc. Thus layout is interrelated to all other areas of design.

Piping system layout is not a subject covered within the classical academic disciplines of engineering. Rather it is more of an art that is learned through the experiences of every individual designer. Some aspects of system layout are common to all pipe designs; thus specific requirements can be specified in piping design codes such as the ANSI/ASME B31 Code for Pressure Piping. Other requirements will be left to the judgment of the individual designer based on a unique set of project conditions. Once a layout is selected, it must be verified that it presents a workable and safe design by subjecting the system to various means of analysis. If necessary, certain aspects of the initial layout attempt may have to be changed, based on the results of analysis. There is usually more than one safe, workable layout arrangement that may be used for a given project.

1.5.6 Piping Design Codes

Pressure piping design in the United States, Canada, and many parts of the world is governed by the piping code which is written by the ASME, and is titled the B31 Code for pressure piping. The piping code is written by the ASME, but each of its individual sections has been adopted as an American National Standard, and are thus also designated as an ANSI Standard. The B31 Code contains seven separate sections for pressure piping intended for different services. Each of the individual code sections is listed in Table 1.1. In nuclear applications, various subsections of Section III of the ASME BPV Code may apply, the exact subsection of which depends on whether the service is Class I, II or III.

By far, the code that has the most direct applicability to double containment piping systems is the ASME B31.3 Section Code for Chemical Plant and Petroleum Refinery Piping. This code effectively covers pressure piping design for chemical service piping in most industrial applications. It also covers the design of many non pressure piping services that are involved in double containment piping applications. This text emphasizes the relation of the B31.3 code to double containment piping design, and uses its basic philosophy in terms of allowable design stresses, and basis for design, in the design of pressure-retaining components, and for those intended for gravity flow.

In order for the ASME/ANSI B31.3 code to apply to a double containment pipe system, most piping has to operate at a pressure of greater than 15 psi (1 bar), or contain a vacuum. However, the code may still apply to a system having a positive internal pressure less than 15 psi (1 bar), if the fluid being conveyed meets certain requirements in terms of flammability or toxicity, which would apply to many chemicals commonly used in a double containment piping system. The ASME/ANSI B31.3 Code does provide all the information a designer needs to know to design a double containment piping system. However, it is up to the individual designer to interpret the code as to how it instructs one to do so. There are no specific references to a "double containment piping system" or "secondary containment piping component" in the Code. There are specific references to the testing and inspection of outer jacketing, however.

Table 1.1 ASME Code for Pressure Piping, B31

B31.1	Power Piping	1992
B31.2	Fuel Gas Piping	1968
B31.3	Chemical Plant and Petroleum Refinery Piping	1993
B31.4	Liquid Transportation Systems for Hydrocarbons, Liquid Petroleum Gas, Anhydrous Ammonia, and Alcohol	1989
B31.5	Refrigeration Piping	1987
B31.8	Gas Transmission and Distribution Piping Systems	1989
B31.9	Building Services Piping	1988
B31.11	Slurry Transportation Piping Systems	1989
B31G-1984	Manual for Determining the Remaining Strength of Corroded Pipelines: A Supplement to B31, Code for Pressure Piping	

Figure 1.4 illustrates the overall scope of the ANSI/ASME B31.3 Code, as it pertains to chemical and petroleum piping systems, including single-wall and double containment piping that is within a regulated facility. Those lines that are governed by other Codes are likewise indicated in Figure 1.4.

Many nonpressure drainage piping systems are governed by local and/or national plumbing codes. Other nonpressure systems are considered as "process" systems, and are thus out of the jurisdiction of local plumbing codes. In many of these systems, there are no design standards that may govern the design. Thus it is up to the individual designer to use good judgment, and to be sure that they at the very least meet or exceed applicable environmental standards.

1.6 Design of Double Containment Piping Systems

1.6.1 Design Criteria

The proper design of any piping system must include all of the forms of analysis described in Sections 1.5.1 through 1.5.5. However, the design of double containment piping system adds several new twists to the already complex nature of piping design. This is due to the addition of the fluid dynamic requirements for flow in an annulus, and is further due to the added mechanical complexities that result from the interaction of the separate, interacting inner and outer piping systems. In terms of fluid dynamics, the flow of fluids in its annular spaces must be considered for future flushing/drying operations. Flow is more complex in annuli due to frictional losses created by supporting devices, internal anchors, tie-together locations, etc.

As mentioned earlier, heat transfer theory as applied to multiple piping applications applies directly to double containment pipe situations. What is different, yet not unusual, is the use of highly heat-resistant materials such as thermoplastic materials encasing a totally enclosed dead air space, which results in the entrained air behaving as a highly effective form of insulation. While the theory is readily developed, many potential applications exist that are new and unexplored. Stress analysis of these complex situations is perhaps the most perplexing of all the major design areas.

Few individuals have a full understanding of single-wall pipe situations in terms of analysis of their stresses and the reactions thereof. When one considers the interaction of an inner and outer pipe combination, the situation becomes far more complex. In fact, most piping design stress computer codes are not capable of accurately modeling this situation. Structural analysis of double containment piping systems is also more complex, as one has to deal first with the structural supporting requirements of the inner system and then also deal with the support (or burial) of the complex assembly, which tends to be a more rigid assembly than a comparable single-wall pipe. In underground situations, for instance, the structural design for a combined system is governed by the burial situation, yet the structural design of the inner piping is according to above-ground single-wall pipe standards.

The most important individual principle to consider when designing double containment piping systems is that the basic principles of piping design remain unchanged. The systems must first be designed according to the ordinary rules, codes, and considerations of single pipe design. Then all of the complexities of one pipe system interacting within another must be considered as well.

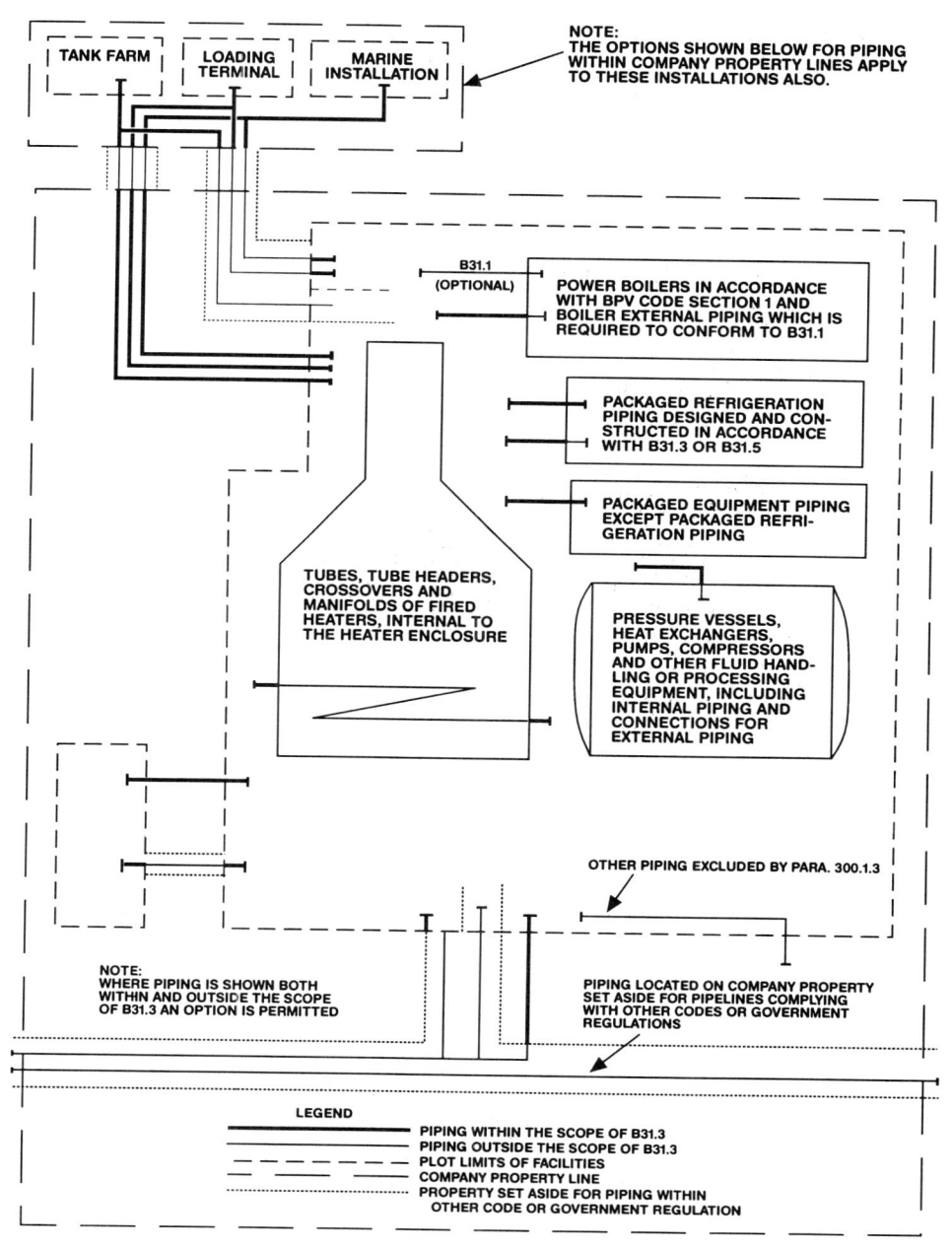

Figure 1.4. Scope of the ANSI/ASME B31.3 Piping Code. (Source: ANSI/ASME B31.3 Chemical Plant and Petroleum Refinery Piping Code, 1993 Edition.)

1.6.2 Layout of Double Containment Piping Systems

In double containment piping layout, a designer is confronted with added challenges, as the layout will have significant impact on the joining techniques chosen and vice versa. A very immediate concern arises from the fact that, by its very nature, double containment piping takes up much more space than single-wall piping. This aspect, when mentioned, seems very obvious, yet designers and installers who have worked with piping systems for years are often surprised by the sheer bulk of the materials. This is a factor that is totally unavoidable if systems are to be designed with a reasonable annulus and an ability to function in the manner in which they were intended.

In addition to the size aspect of the materials, in making layout decisions, a designer is confronted with the challenge that there are two separate piping systems that have to be installed, sometimes with their own unusual techniques. Certain proprietary techniques, specific to double containment pipe systems, can often require more than usual space requirements. Therefore, adequate space must be provided in the layout of the system. Sometimes this can mean a space increase allowance that is well beyond that required of what would have been a relatively small-diameter single-wall piping system.

1.6.3 Use of Double Containment Piping

With all of the added considerations, one has to wonder if it is really worth it to invest the extra effort to design and implement such a system. However, when one considers the potential risks involved in transporting or storing hazardous fluids, and the negative consequences of a release, many of which can be irreversible, it tends to counteract the effort argument. It is the hope of the author that, by consolidating available design information into one comprehensive handbook, users and designers will make the decision that if a fluid handling situation can pose a risk, then the fluid should be prevented from ever getting to the atmosphere or groundwater. The best way to accomplish this is through double containment systems that are designed and installed with consideration given to all that is involved.

References

1. Michael Frankel, "Sub-Surface Drainage: What Is Important to the Plumbing Engineer," Plumbing Engineer, 18 (2), March 1990, pp. 26-32.

2. Robert M. Clark and David L. Tippin, Standard Handbook of Environmental Engineering, Robert A. Corbitt, Editor, Chapter 5, p. 5.10, McGraw-Hill, NY, 1990.

3. U.S. Environmental Protection Agency, Office of Drinking Water, "Planning Workshops to Develop Recommendations for a Groundwater Protection Strategy: Appendices," Washington, D.C., U.S. Government Printing Office, 1980, p. II-5.

4. Robert M. Clark and David L. Tippin, Standard Handbook of Environmental Engineering, Robert A. Corbitt, Editor, Chapter 5, p. 5.11, McGraw-Hill, NY, 1990.

5. W. E. Short, S. J. Brown and C. Sundararajan, "Pipe Risk Analysis: A Case Study," ASME PVP–Vol. 285, Codes and Standards for Quality Engineering, 1994.

Notes

Notes

Chapter Two

Materials of Construction

This chapter covers information on common materials used in double containment piping, double-wall tanks, and other secondary containment applications. The key materials that are covered fall into four major groups: (1) metallic materials; (2) reinforced thermosetting resin/plastic (RTRP) materials; (3) thermoplastic materials; and (4) other nonmetallic materials (i.e., borosilicate glass, ceramic, concrete, masonry, vitrified clay, and others).

The chapter is divided into six parts. Section 2.1 discusses corrosion of metallic materials. Section 2.2 covers the various categories of metallic materials, and discusses their behavior in different media, according to the criteria discussed in Section 2.2. Section 2.3 covers the principles of degradation of plastic materials in corrosive or otherwise aggressive media. Section 2.4 distinguishes the important classes of nonmetallic materials and discusses the properties of certain specific resin types. Sections 2.5 and 2.6 cover corrosion testing and sources of corrosion data, respectively.

2.1 Fundamentals of Metallic Corrosion

Corrosion of metallic substances, in the conventional sense, occurs due to an electrochemical process that causes a gradual alteration or wearing away of a metallic substance. The action is considered as a degeneration of a material, since the process returns a metal to its most stable thermodynamic state. In its broadest sense, corrosion includes the destruction of metals by any method other than that caused by mechanical means.

All metals corrode due to the flow of electricity from one metal to another metal, or other recipient, or between parts of the same metal where the flow of electricity results in a lower energy state. An electrolyte must be present for the process of corrosion to take place. Water, salt water, soil, or any type of ionically charged liquid may serve as an electrolyte. When corrosion occurs, electricity flows from a negatively charged area to a more positively charged area, in the presence of the electrolyte. Therefore, if an electrolyte is present and differences in electrical potential exist on the metallic surface(s), corrosion will occur.

There are many different specific forms of corrosion; however, the different forms may be broadly grouped into five major categories. These are: (1) general or uniform corrosion; (2) galvanic corrosion; (3) stress corrosion cracking; (4) intergranular corrosion; and

(5) microbiologically induced corrosion (MIC). There are also many specific types that fall within these general groupings, some of which are: (1) pitting (galvanic); (2) concentration cell corrosion (galvanic); (3) crevice corrosion (galvanic); (4) corrosion fatigue (general); (5) fretting (general); (6) impingement (general); (7) erosion corrosion (general); (8) embrittlement (general); (9) filiform (general, intergranular); (10) exfoliation (general, intergranular); (11) dezincification (general); (12) graphitization (general); and (13) chemical reaction (including oxidation and reduction, which are also forms of general or uniform corrosion). Some of these forms of corrosion are described in more detail in the following paragraphs. There are other forms of degradation that are unique in some way, and thus are not classified within one of the five major groups.

2.1.1 Conventional (Electrochemical) Corrosion

The five major forms of metallic corrosion occur due to an electrochemical process that is sometimes referred to as *electrochemical corrosion*. Electrochemical corrosion occurs when a difference in potential exists as a result of the presence of a metallic cathode, anode, metal conductor, and an electrolyte (conductive solution such as salt water). Together, these elements make up part of an electrochemical cell, which is illustrated in Figure 2.1. Many times, oxygen is present in the cell, which acts as a depolarizing agent. Electrodes can be different materials or they can consist of the same material if an electrolyte composition is not a constant. In Figure 2.1, electrical current is shown to flow through the metallic conductor and the electrolyte. This causes corrosion to occur at the anode location. In chemical terms, this reaction is defined as an *oxidation-type reaction*. If the reaction were to occur in the opposite direction, then it would be termed *reduction*, resulting in corrosion at the cathode. The metal undergoing electrochemical corrosion does not have to be immersed in a solution, but it does have to be in contact with a wetted medium. Moist soil and moist air are two common examples of a wetted medium.

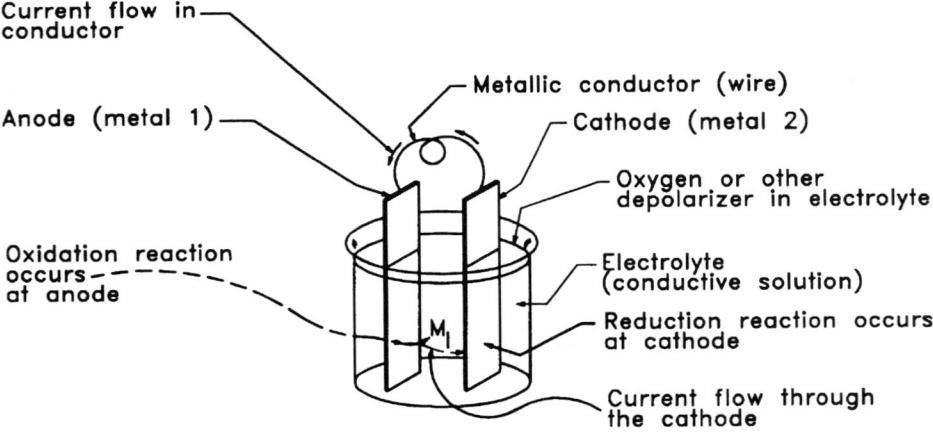

Figure 2.1. Simple electrochemical corrosion cell. (Source: D. M. Berger "Corrosion Principles Can Never Be Forgotten in Organic Finishing," *Met. Finish.* **72**, November 1974.)

Materials of Construction

The presence of oxygen in water greatly enhances its corrosive effects on steel. In fact, if all dissolved oxygen is removed from water by deaeration, water will have a limited effect on steel if the solution pH is maintained at 7 or slightly below. It is for this reason that water is deaerated prior to use as boiler feed water. Also, when an acid serves as an electrolyte to a metal, the rate of corrosion is rapid, and hydrogen gas is evolved. If a metal contains a surface oxide layer, its oxide layer may greatly reduce the rate of corrosion (for instance, in the case of titanium or nickel alloys). These surface films, which are usually inert to some degree, are usually invisible to the naked eye. A very important principle to consider in metallic corrosion is that corrosion is usually nonuniformly distributed over the surface of a metal, unless its oxide film is uniformly affected by a chemical. If an oxide film is uniformly affected, the process is referred to as *general* or *uniform corrosion*, which is discussed in Section 2.1.2.

Electrochemical corrosion can also occur in air. However, the process differs slightly in that the significant presence of oxygen plays an important role. The rate of corrosion in this situation is controlled by the development of surface-insoluble films and the presence of moisture and deposits in and from the atmosphere. The presence of sulfur, nitrogen-based compounds, or salt in air also create different corrosion actions. Corrosion that occurs in air, like most forms of corrosion, is electrochemical in nature. Therefore, it results in a very uniform appearance due to the small size and close arrangements of electrode (anodic and cathodic) locations.

There are a number of factors that contribute to the rate and degree of corrosion of a metallic material. Manufacturing processes that are used, the amount of localized shear

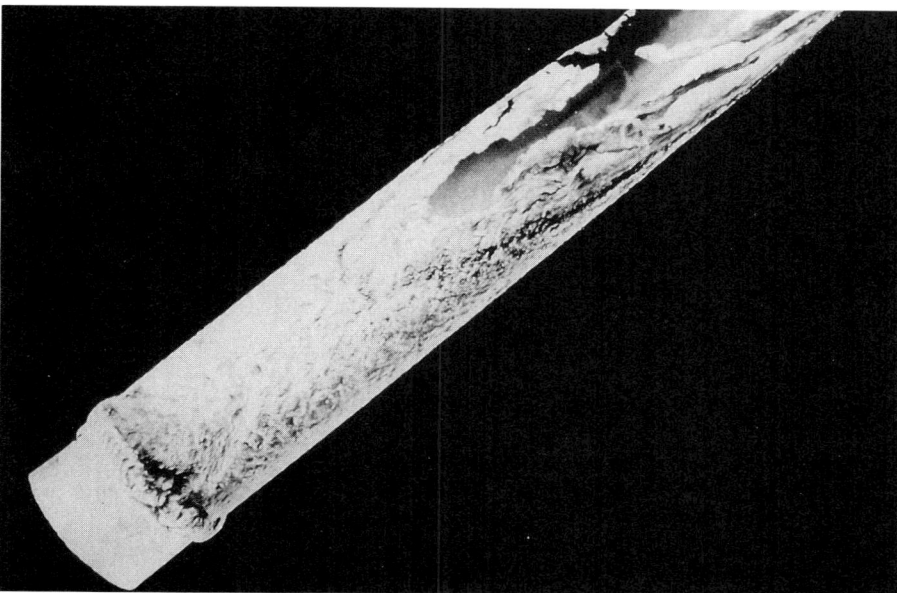

Figure 2.2. Type-316 stainless steel heat exchanger tube subjected to uniform corrosion by sulfuric acid. Note the lower portion, which is Alloy 20Cb-3®, did not corrode. (Source: Trent Tube.)

and torque produced during manufacture, the presence of mill scale resulting from manufacture, and any red iron oxide rust that develops are all contributing factors. Also, the degree of surface smoothness has an effect, as do grain orientation and the relative consistency of the material.

2.1.2 General or Uniform Corrosion

The uniform corrosion of a metal's surface can be expressed as a constant rate of metal loss over a given surface area. This loss is typically expressed as a depth of penetration per year (in/mils per year, or mm per year). Occasionally, it is expressed as a weight loss for a given surface area per year (i.e., mg/dm^2). Corrosion journals that report a rate of corrosion in these terms are conveying information solely about uniform corrosion of a component.

Metals gain their corrosion resistance by the developing a protective oxide film, or other compound, on its surfaces. Films may be passive or active in nature. Most highly corrosion-resistant metals such as stainless steels, nickel alloys, titanium, zirconium, and aluminum have a passive film. Others such as copper and carbon steel have an active film. In order for general corrosion to take place, either uniform breakdown of a passive film or accelerated formation or removal of the active film must occur over the complete surface of a metal.

Figure 2.2 is a photograph of a type-316 stainless steel heat exchanger tube that has been subjected to uniform corrosion on the outside of its tubes, as a result of exposure to sulfuric acid at 30% concentration and 140°F (60°C). In this application, corrosion has taken place from the outside of the tube, working inward. Figure 2.3 illustrates a carbon steel tube that has been subjected to corrosion from the inside out, due to prolonged exposure to plant water having a mildly acidic concentration. As evidenced by the photographs, prolonged exposure under corrosive conditions can lead to eventual failure of the components.

Figure 2.3. A carbon steel pipe that has been subjected to uniform corrosion.

Materials of Construction

Table 2.1 A Partial Listing of the Galvanic Series of Metals and Alloys

Metal	Volts	Material or Alloy
Corroded end (anodic or least noble)		
Magnesium	-1.55	Magnesium
Aluminum	-1.33	Aluminum 2S
Zinc	-.076	Zinc, steel, or iron
Iron	-0.44	Stainless steel (active)
		Alloy 20 Cb-3® (active)
		Hastelloy C (active)
		Nickel (active
Nickel	-0.23	Inconel (active)
		Hastelloy B
Hydrogen	0.00	
Copper	+0.34	Copper
		Monel
		Nickel (passive)
		Inconel (passive)
		Stainless steel (passive)
		Alloy 20 Cb-3® (passive)
Silver	+0.80	Silver
Gold	+1.36	Gold
		Platinum
Protected end (cathodic or most noble)		

2.1.3 Galvanic Corrosion

Galvanic corrosion takes place when dissimilar metals come in contact with each other in the presence of an electrolyte. A galvanic current is set up between two metals, with one metal acting as a cathode and another acting as an anode. There are guides available such as the "Galvanic Series of Metals Guide" to determine which way current will flow whenever dissimilar metals come into contact with each other. Table 2.1 contains a partial list of the galvanic series of metals, which indicates the listed potentials of the metals shown. A dissimilar metal may be a singular metallic piece that has already been corroded in a localized section, as in the case of steel that contains mill scale or rust on its surface. In this instance, the base metal will function as an anode and current will flow from rust (which functions as a cathode), producing further degradation of the steel. Mill scale is cathodic to steel, and an electric current can be easily produced between a steel substrate and its mill scale, as shown in Figure 2.4.

The passive nature of stainless steels is due mainly to the presence of a corrosion-resistant oxide film that forms over its surface. This film tends to protect stainless steel and maintain it in a passive state. The presence of reducing solutions or solutions containing a high chloride content can change an oxide film to an active state and thus produce galvanic corrosion. In atmospheres where the presence of oxygen is eliminated, the same type of process can

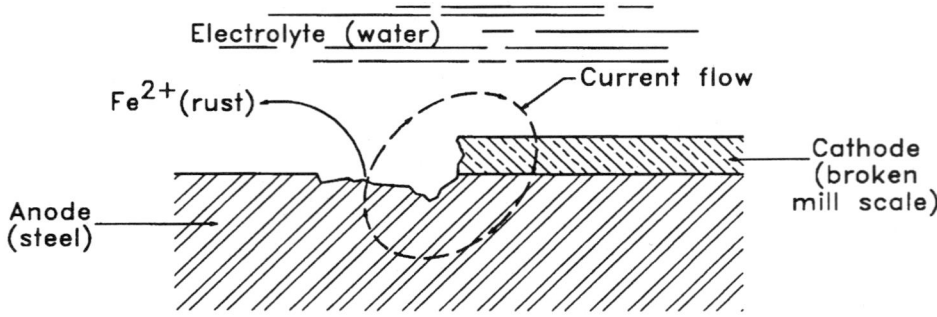

Figure 2.4. Galvanic corrosion between steel and its mill scale. (Source: Jack Kiewit, "The Paint Manual," USDI Bureau of Reclamation, Denver, CO, 1976, Chap. III.)

occur. The presence of direct stray currents in soils may also cause the accelerated failure of stainless steels. A very common occurrence of this is where nearby cathodic protection systems produce stray currents. When this takes place, corrosion can occur very rapidly.

2.1.4 Pitting

Pitting is a specialized form of galvanic corrosion[1] involving a corroding anode area that is small in relation to a much larger protected cathode area.[2] It can occur where there are small pores in a mill scale coating of a metal, and in applied protective coatings. On non-coated pipe, pitting may occur due to small localized imperfections in surface quality. Pitting occurs primarily due to contact between dissimilar materials or due to the presence of areas of a single material where oxygen or ionic concentrations differ. These areas are referred to as a *concentration cell*. Figure 2.5 illustrates how pitting occurs due to a defect in mill scale of a metal substrate. It is important that if a protective coating is to be applied, cathodic and anodic areas must both be covered. If only an anode is coated, and holidays (pinholes) exist in the surface of a coating, intense pitting can take place.

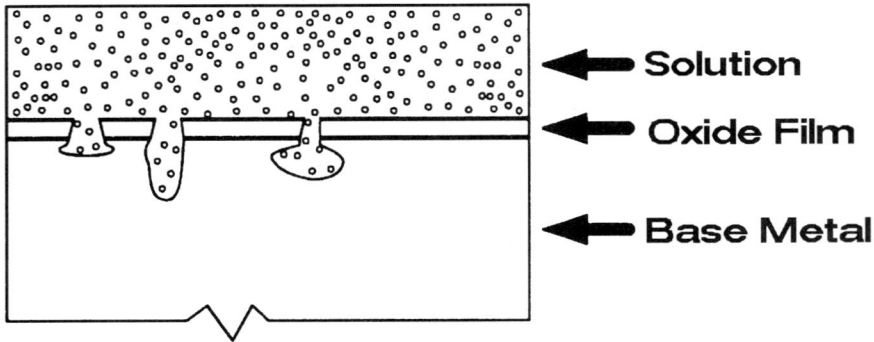

Figure 2.5. Basics of pitting corrosion. (Source: John C. Tverberg, Trent Tube.)

Not all stainless alloys are readily subject to pitting attack. As an example, Figure 2.6 graphically shows the relationship between molybdenum content, pH, and chloride content on pitting corrosion. Below any of the lines shown in Figure 2.6, an austenitic stainless steel alloy having a molybdenum content indicated on the line will not be subjected to a localized attack if the combination of pH and chloride concentration falls below the line. If the combination is above the line, localized attack may occur. Thus, specific additive concentrations of an alloy do make a difference in terms of an alloy's resistance to pitting.

2.1.5 Concentration Cell Corrosion

A concentration cell occurs when areas of a metal are subjected to different concentrations of oxygen or ionic concentrations in solution with water. Figure 2.7 illustrates a specific type of concentration cell corrosion typical in buried pipes. Severe and rapid corrosion, leading to pitting, can be caused. Where differences in dissolved oxygen concentration occur, the part of a metal that is subjected to a low-oxygen atmosphere will behave as an anode.[3] An oxidation-concentration cell of this type is started by depletion of oxygen in a pit (holiday) area. The rate of penetration of a pit is accelerated proportionately. Fabrication operations, as in storage tank construction, can result in cracks produced in the mill scale of a metal, thereby resulting in accelerated corrosion of a tank. Weld areas produce another common area for concentration cell corrosion due to the difference in the concentration of a metal in solution. A situation such as this would be referred to as a metal-ion-concentration cell.

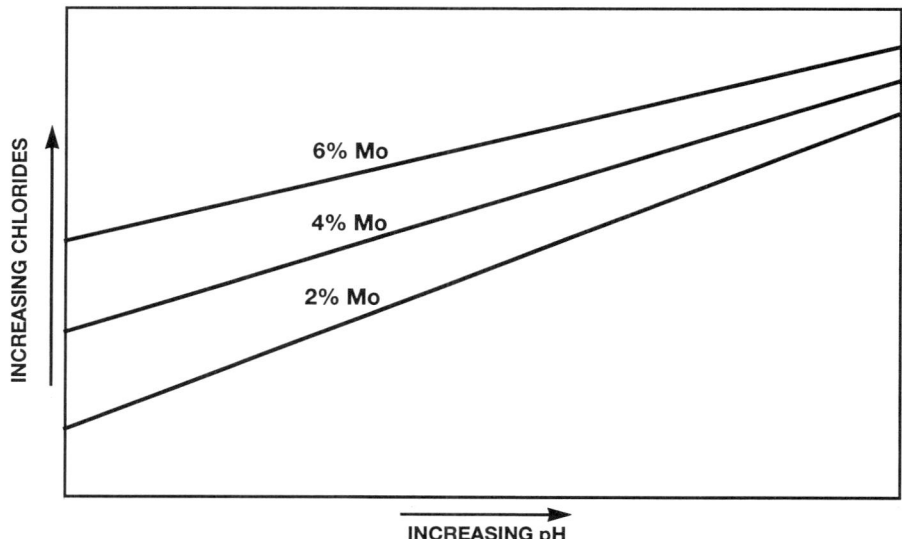

Figure 2.6. The effect of pH, chloride content, and molybdenum content on pitting of austenitic stainless steels. Alloys with Mo level do not experience localized corrosion below the line. (Source: John C. Tverberg, "Choosing Proper Stainless Steel," *JAOCS*, Vol. **65**, no. 11, November 1988, p. 1738.)

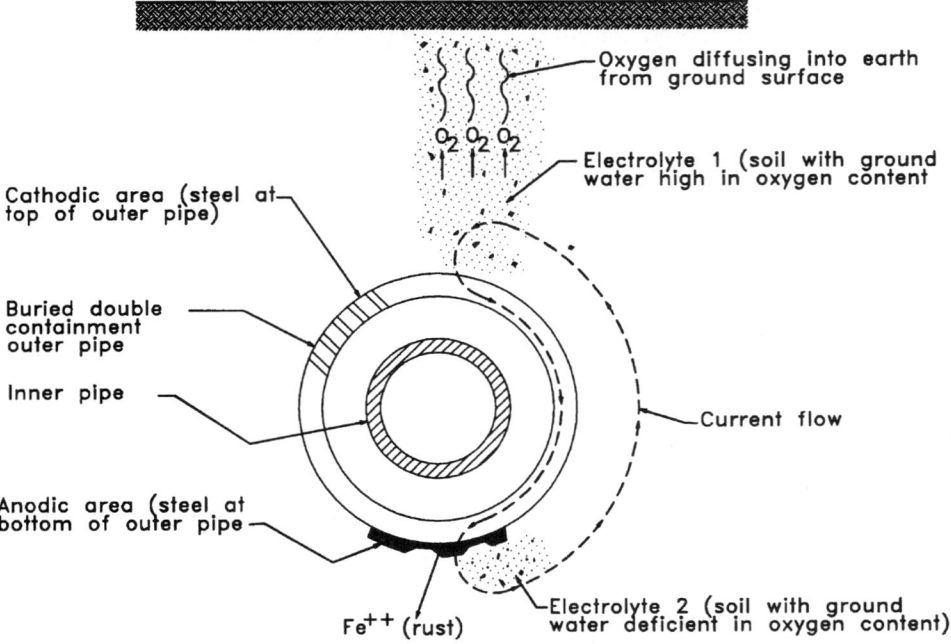

Figure 2.7. Concentration cell corrosion in buried metal pipes due to differences in oxygen concentration. (Source: Jack Kiewit, "The Paint Manual," USDI Bureau of Reclamation, Denver, CO, 1976, Chap. III.)

2.1.6 Crevice Corrosion

Crevice corrosion is a specialized form of galvanic corrosion.[3] Whereas pitting is random, crevice corrosion is highly selective and very predictable. It is necessary only to have a crevice former and an electrolytic solution present. In the crevice corrosion mechanism, it is not necessary for the crevice former to be a metal; essentially any material that tightly presses up against a base metal and creates a crevice will act as a crevice former. Usually, the smaller the crevice, the more readily crevice corrosion is likely to occur due to capillary action. Figure 2.8 illustrates time to initiation of crevice corrosion for a number of stainless and nickel alloys.

Alloys vary greatly in the amount of time necessary for crevice corrosion to occur. For example, in steels that contain chromium (e.g., stainless steels), a higher molybdenum content will result in a longer time to initiation. Higher nickel contents also tend to retard crevice corrosion, as evidenced by Figure 2.8. Crevice corrosion may occur at lower temperatures than pitting, and under ionic concentrations that are substantially lower than occurs as a result of pitting.

At a higher temperature than that shown in Figure 2.8, seawater may cause crevice corrosion to occur in one of the alloys, while it is otherwise shown that it may not occur under the temperature for which the Figure 2.8 on which data are based. The same effect is true with respect to pH content of seawater.

2.1.7 Intergranular Corrosion

Intergranular corrosion is a form of corrosion that takes place along grain boundaries of a metal. An example of a grain boundary is where carbides or other precipitates form due to a metal being exposed for a prolonged period to its sensitizing temperature range. This occurs readily in austenitic stainless steels that contain an appreciable carbon content. Figure 2.9 illustrates the basic terms and process of intergranular corrosion. Figure 2.10 is a photomicrograph showing carbide precipitation at grain boundaries in an austenitic stainless steel. Intergranular corrosion usually manifests itself along grain boundaries in the form of microfissures.

Austenitic stainless steels (i.e., types 304 and 316), when subjected to a temperature range of 1050/1600°F (565/870°C), will have carbides precipitated in the form of chromium carbide. This process is referred to as *sensitization*, the temperature range of which is referred to as the *sensitizing temperature range*. Temperatures of this range are typically reached during welding procedures.[4] Carbon, which exists in relatively low levels, combines with available chromium in its surrounding area. This lowers the chromium content of the alloy in the heat-affected zone, leading to the formation of a passive film on the metal. This creates a galvanic cell, which can lead to rapid grain boundary attack. The process is illustrated in Figure 2.9. Some fluid services that are known to cause corrosion in such grain boundaries include ultrapure water, various acids, and caustic/basic solutions.

To prevent or minimize the effects of intergranular corrosion, there are several measures that may be taken. These include: (1) subjecting a metal to solution annealing after welding (usually in the range of 2000°F; 1093°C), which brings carbides back into solution; (2)

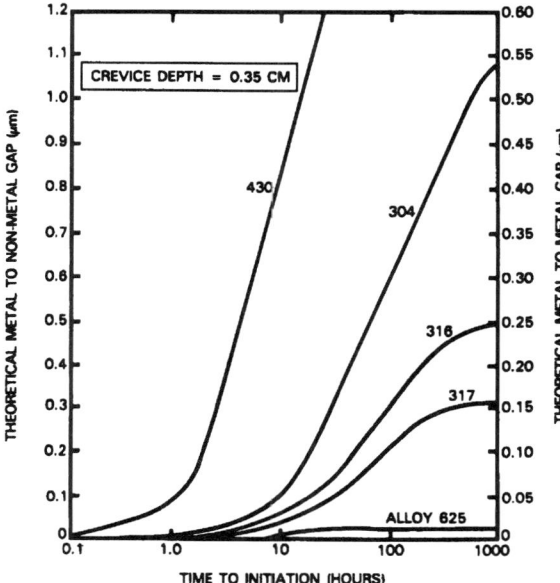

Figure 2.8. The effect of crevice gap on time to initiation for crevice corrosion for various alloys in ambient-temperature seawater. (Source: John C. Tverberg, Trent Tube.)

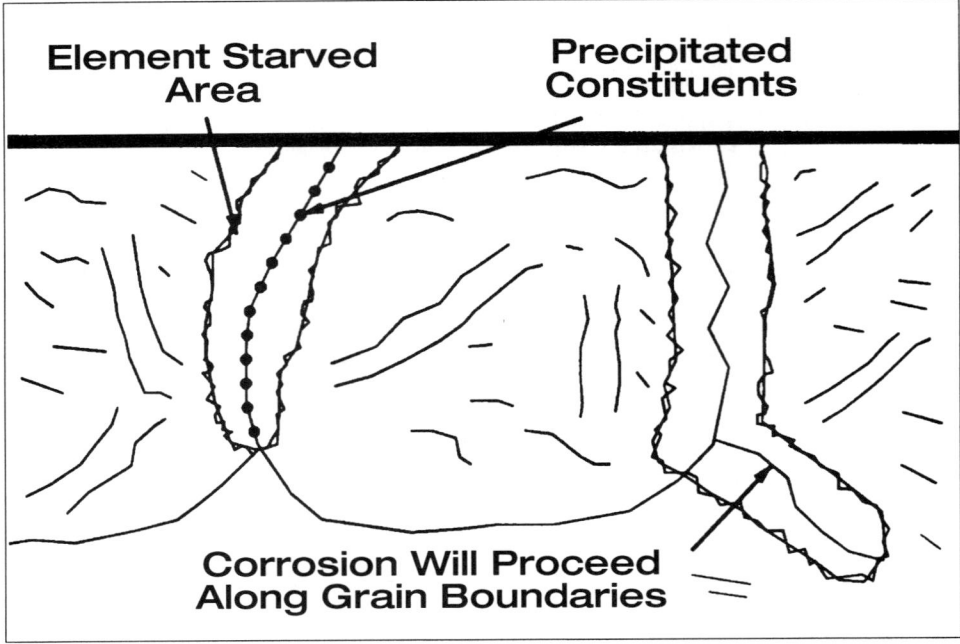

Figure 2.9. Basic terms of intergranular corrosion. (Source: John C. Tverberg.)

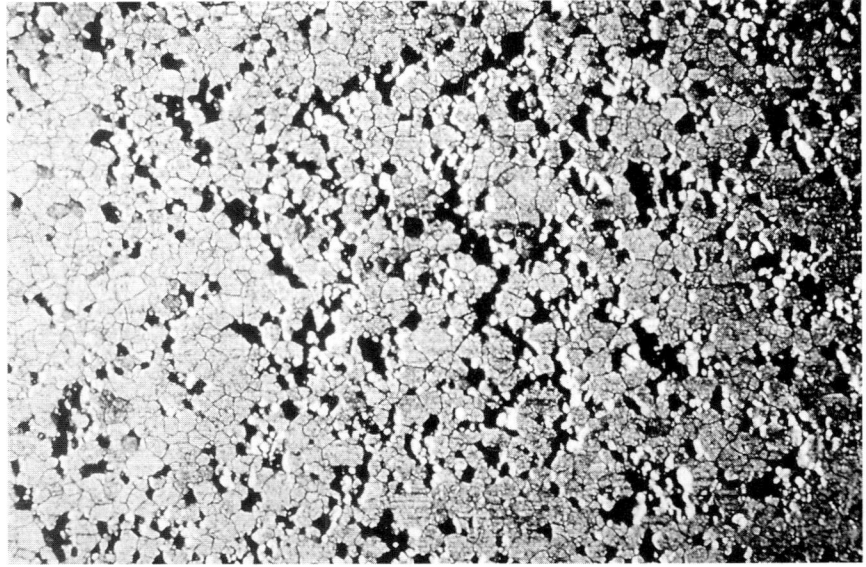

Figure 2.10. Photomicrograph showing carbide precipitation in an austenitic stainless steel. (Source: Carpenter Technology.)

Materials of Construction

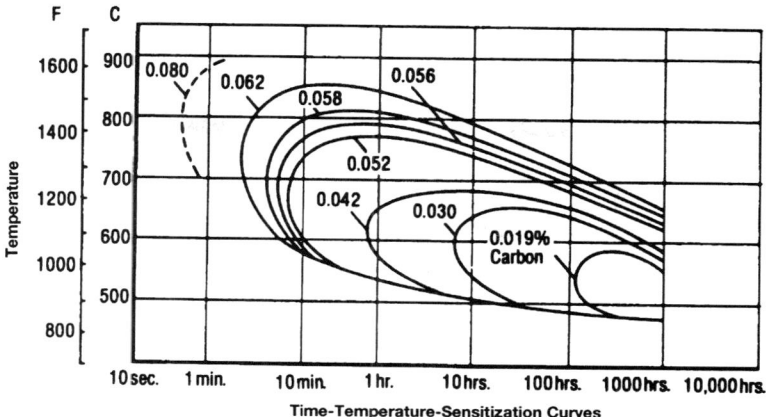

Figure 2.11. The effect of carbon content on carbide precipitation and intergranular corrosion in austenitic stainless steels. (Source: John C. Tverberg, "Choosing Proper Stainless Steel," *JAOCS*, Vol. 65, no. 11, November 1988, p. 1740.)

use of a lower carbon form of an alloy, such as type 304L (0.03 percent carbon), instead of type 304 (0.08 percent carbon); and (3) use of an alloy that contains an element such as niobium, titanium, or columbium that tends to form carbides in preference to chromium, referred to as a *getter*. The best getter tends to be columbium. Any one of the above-mentioned three actions will minimize, and in some cases eliminate, the possibility of intergranular corrosion due to normal field-welding procedures. Figure 2.11 illustrates the effect of carbon content on carbide precipitation in stainless steels. As the carbon content of a stainless steel is lowered, a prolonged exposure time is required to allow harmful levels of chromium carbide to precipitate. In Figure 2.11, chromium carbide precipitation forms in the areas to the right of the various carbon-content curves. Since the time of exposure to temperatures in the range of 1000°F (538°C) during welding is often less than 10 minutes, an austenitic stainless steel with a carbon content of 0.03 percent is substantially less susceptible to grain boundary precipitation and, thus, intergranular corrosion attack.

2.1.8 Stress-Corrosion Cracking

A metal may be subjected to stress-corrosion cracking if it is in contact with a chemical that is known to induce stress cracking in the metal in question, a threshold temperature is exceeded, and the component is under some degree of tensile stress.

Compressive stresses will usually not trigger stress-corrosion cracking of a metal; the metal usually must be subjected to tensile stresses. Additionally, the overall level of stress in a component must approach the material's yield strength in order for stress-corrosion cracking to occur. Stress-corrosion cracking is an important mechanism in double containment piping systems due to the fact that reaction loads (discontinuity stresses) produced at interconnecting components, as a result of differential expansion of inner and outer components, can lead to stresses of the required magnitude to induce stress cracking in the areas where they interconnect (see Section 8.4).

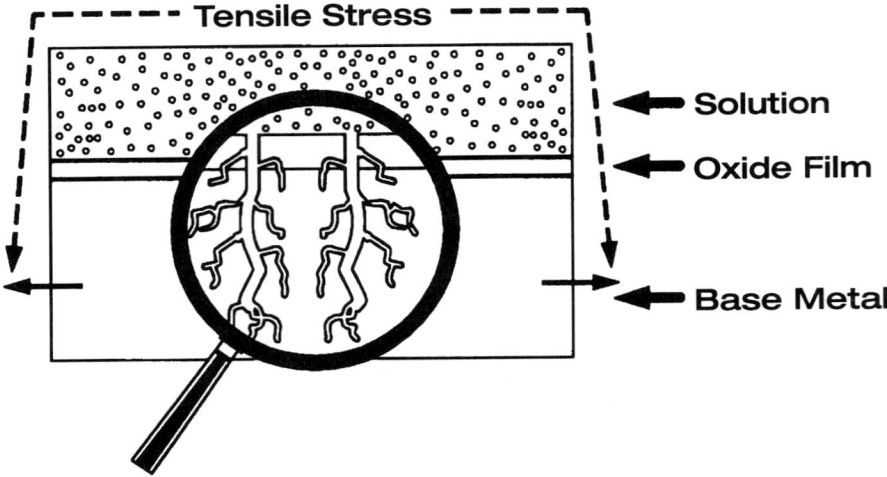

Figure 2.12. The stress-corrosion cracking mechanism. (Source: John C. Tverberg, Trent Tube.)

Figure 2.12 illustrates the stress-corrosion cracking mechanism. Table 2.2 covers a partial list of chemicals that can lead to corrosion-stress cracking in alloys that are listed.

Alloy content plays a very important role in stress-corrosion cracking of metals. Nickel content tends to be a very important element, particularly in austenitic stainless steels. Usually, as nickel content is increased beyond 30 percent, alloys are more resistant to stress-corrosion cracking. Molybdenum also plays a complementary role; the higher the molybdenum content of an austenitic stainless steel alloy, the less the likelihood that stress-corrosion cracking will occur.

When stress-corrosion cracking does occur, the result is a series of trans-granular cracks, which tend to branch out at their ends in a feathery appearance, as illustrated in the example shown in the photomicrograph (Figure 2.13). The structure of the cracks must be studied under a microscope to determine this, as the photomicrograph shows. It usually takes time for a crack to initiate. However, once it does, a crack will act as a site for stress concentration, and will also exposure more surface area, leading to a rapid propagation of stress cracks. Oftentimes, "pits" that develop as a result of a pitting mechanism will act as a stress concentrator, and will lead to subsequent stress-corrosion cracking.

Section 2.3.2.3 "Environmental Stress Cracking," describes corrosion-stress cracking of plastic materials. Plastics are susceptible to corrosion-stress cracking in a manner similar to metals. However, the plastics industry has assigned a unique name, "environmental stress cracking," to describe the exact same effect in plastic materials.

2.1.9 Microbiologically Induced Corrosion

Bacteria and other microorganisms can induce corrosion on metals in a variety of ways. The corrosion effect can either be by indirect means or by direct means, where a microorganism consumes a metal as part of its respiration process. Bacteria and other microorganisms exist that induce a corrosion effect on every metal; not all bacteria affect each metal, however.

Materials of Construction

Table 2.2 Sample List of Alloy Systems Subject to Stress-Corrosion Cracking

Alloy	Environment
Aluminum Base	– Air – Seawater – Salt & Chemical Combinations
Magnesium Base	– Nitric Acid – Caustic – HF Solutions – Salts – Coastal Atmospheres
Copper Base	– Primarily Ammonia & Ammonium Hydroxide – Amines – Mercury
Carbon Steel	– Caustic – Anydrous Ammonia – Nitrate Solutions
Martensitic & Precipitation Hardening Stainless Steels	– Seawater – Chlorides – H_2S Solutions
Austenitic Stainless Steels	– Chlorides-Inorganic & Organic – Caustic Solutions – Sulfurous & Polythionic Acids
Nickle Base	– Caustic Above 600°F (315°C) – Fused Caustic – Hydrofluoric Acid
Titanium	– Seawater – Salt Atmospheres – Fused Salt

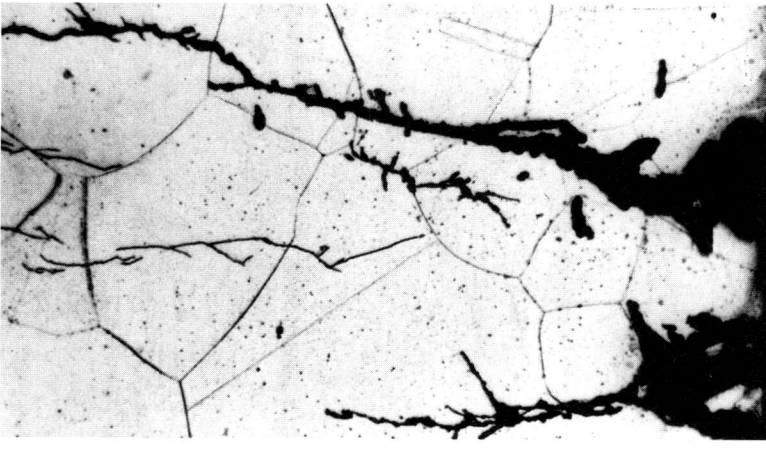

Figure 2.13. Photomicrograph showing trans-granular cracks in a metal that was subject to stress-corrosion cracking. (Source: Carpenter Technology.)

One way that microorganisms may indirectly affect metals is through the excretion of byproducts as a part of their natural respiration process. Common byproducts include: (1) carbon dioxide; (2) hydrogen sulfide; (3) ammonia; (4) organic acids; and (5) inorganic acids. Another indirect mechanism exists, however. This occurs whereby a microorganism produces an organic byproduct that functions as a depolarizer, or as a catalyst for other corrosion mechanisms. In each type of indirect action, it is not the bacteria themselves that corrode a metal, but rather the byproducts that they produce that result in some corrosive action. Bacteria and other microorganisms may also directly affect a metal by using a metal as a metabolic link in its nutritional chain. In this manner, the metal will become directly consumed by the bacteria, which will result in a degradation of it.

Table 2.3 lists a variety of corrosive microorganisms that are known to attack various metals. If a bacteria such as thiobacillus is present in an aqueous solution, it will produce sulfuric acid, which will corrode any metal susceptible to low concentrations of sulfuric acid. An example of such a metal would be ordinary carbon steel, which will be uniformly corroded by low-concentration sulfuric acid solutions at low temperatures.

2.2 Metallic Materials Used in Chemical Services

Sections 2.1 through 2.8 discuss most of the common metallic materials used in piping and tank services for chemical and/or nuclear services. The materials range in both expense and extent of use. However, each of the various materials finds use in various services. Section 2.9 presents some general guidelines as to how to select an appropriate metal for given chemical services.

Table 2.3 Corrosive Microorganisms*

Organism	Action	Problem
- Desulfovibrio - Clostridium - Thiobacillus	- Hydrogen sulfide producers (Sulfate reducers)	- Corrosive to metals - Reduces chromates - Destroys chlorine - Precipitates zinc
- Thiobacillus	- Sulfuric acid producer	- Corrosive to metals
- Nitrobacter	- Nitric acid producers	- Corrosive to metals
- Gallionella - Crenothrix - Spaerotilus	- Converts soluble ferrous ions to insoluble ferric ions	- Produces iron oxide deposits - Increases corrosion

*Source: "Identification & Control of Corrosive Microbiological Organisms Found in Recirculating Cooling Water Systems," by Paul R. Puckorius, Corrosion 78, Paper Number 81.

2.2.1 Copper and Copper Alloys

Copper is classified as a noble metal, and is therefore a corrosion-resistant material by defining it as such. It has properties that include relatively high thermal and electrical conductivity. It is malleable and machinable, but must be alloyed or cold worked to obtain strength due to its inherently low mechanical property values. It is resistant to water and relatively low in cost. Therefore, it has applications in many domestic plumbing applications and industrial service lines.

The American Society for Testing and Materials (ASTM), together with the Copper Development Association and the Society of Automotive Engineers, has developed a five-digit classification system that is part of the Unified Numbering System for metals and alloys. The numbers are five-digit numbers ranging from 10000 to 99999 and are preceded by the letter C (for example, UNS No. C11000 contains a maximum of 99.90% copper by weight). The ASTM lists standards for the various copper grades in ASTM Volume 2.01 of the Annual Book of ASTM Standards.

The important families of copper materials for use in corrosion engineering include the following: copper (minimum 99.3% purity), high copper alloys (minimum 94% copper content), brass (containing zinc as the primary alloying element), bronze (containing tin, aluminum, or silicon as the primary alloying element), and copper nickels (containing nickel as the primary alloying element). The major families of copper for fluid applications are copper, brass, bronze, and copper/nickel alloys. These are described in more detail in the following paragraphs.

2.2.1.1 Coppers

Coppers contain a minimum copper content of 99.3% and may contain one or more of the following elements: (1) antimony; (2) arsenic; (3) bismuth; (4) boron; (5) cadmium; (6) lead; (7) magnesium; (8) nickel; (9) phosphorous; (10) silver; (11) sulfur; (12) tellurium; and (13) zirconium. Copper is used most frequently for electrical purposes, but in terms of fluid handling, it is most commonly used for domestic, commercial, and industrial plumbing purposes. This is due to its resistance to potable water, to its ease of joining by brazing and soldering, and also to its relatively low cost. Other applications for copper include heat transfer tubing for heat exchangers and solar energy applications. The primary reason for its use in these services is due to its high thermal conductivity. Copper is also a standard material for small-bore fuel oil tubing requirements.

When used for plumbing tubing, copper is subject to failure due to pitting and erosion corrosion. Most pitting occurs in either cold water applications or in hot water applications where soft water is used. When hot water has been preconditioned to remove chlorides, sulfates, and nitrate ions by means of deionization or distillation, pitting can occur. It is more prevalent in water that is acidic and contains a low carbon dioxide content (<50 ppm). If these conditions are not present, then pitting in hot water is rare. Pitting of copper in cold water applications can be lessened by preconditioning water prior to its distribution through copper tubes. This pretreatment may include adding a base to neutralize acidic conditions, or aeration to release free carbon dioxide. Velocities in copper piping and tubing must be kept below 5 feet per second or erosion corrosion can occur. Also, avoiding sudden changes in direction and minimizing possible turbulence can also reduce this effect and thus lengthen a copper tube's life.

2.2.1.2 Brasses

Brasses are copper-based alloys that contain zinc as their principal additive. Other minor additives include lead, tin, and aluminum. Alloys that have a zinc content below 15% resist dealloying and are generally more corrosion resistant than those with a greater than 15% content. The low-zinc alloys resist acids, bases, and salts and are resistant to stress-corrosion cracking. However, they are susceptible to attack from oxidizing materials, dissolved oxygen, and compounds that react directly with copper or those that tend to form soluble copper complexes.

2.2.1.3 Bronzes

Tin/bronzes are copper-based alloys that contain tin as their primary alloying element. Phosphorous is always present (usually at less then 0.5%) due to its deoxidizing properties. Other additives include lead, zinc, and nickel. Tin/bronzes are probably the oldest metal alloy known; the longevity of many outdoor structures is a testament to their resistance to atmospheric corrosion.

Aluminum bronzes are copper-based alloys that contain aluminum as their primary additive. The alloys may also contain other additives such as iron, nickel, silicon, manganese, and other elements. Aluminum bronzes are useful where corrosion resistance is important in addition to strength and wear resistance. Adding approximately a 4% nickel content together with a lower aluminum content will tend to prevent dealloying. Some of the major applications for aluminum bronzes include valves and pump parts for saltwater applications and condenser tube sheets for power plants and geothermal applications. Aluminum bronzes are resistant to a wide range of acids and bases.

Both zinc/copper and tin/copper bronzes find their primary use in fluid handling applications as valve bodies, valve parts, pump components, and other mechanical items that have usefulness in plumbing applications. Bronzes are also frequently used in fire protection systems due to their unusually high resistance to water in a stagnant condition, and to microbiologically induced corrosion.

2.2.1.4 Cupro-Nickels

Copper/nickel alloys are solid solution alloys (single phase) that contain nickel as the principal alloying ingredient. Other ingredients may include iron, manganese, silicon, and niobium. The alloys that contain approximately 10% and approximately 30% are the ones that find the most widespread use in corrosion services. The most widespread use for copper/nickel alloys is for piping, tubing, fittings, condenser tubes and plates, and pump castings for saltwater service. Their bio-fouling resistance is excellent, but they may be susceptible to accelerated corrosion if there is a high sulfide content (>0.01 ppm). Also, velocities in copper/nickel alloy tubes should be kept low to minimize erosion corrosion possibilities. The cupro-nickel alloys are highly resistant to stress-corrosion cracking.

2.2.2 Aluminum and Aluminum Alloys

Aluminum is an abundant, low-cost metal with a high strength-to-weight ratio. It is easily processed, fabricated, and can be joined by many common methods. Aluminum and its alloys are chemically resistant to atmospheric corrosion in most atmospheres, and are resistant to many other chemicals. Aluminum also has very high values for electrical and ther-

Materials of Construction

mal conductivities, making it useful for a wide range of applications. Standards for aluminum and aluminum alloy materials, piping, tubing, and related subjects are covered in Volume 2.02 of the ASTM Annual Book of Standards.

Aluminum is classified as an active metal. Its resistance to corrosion strongly depends on the development of a protective oxide film. In aqueous solutions, the thermodynamic conditions under which its oxide film develops are dependent on the relationship between the pH of the solution and its electrical potential. Except below a voltage potential of approximately -1.8 V (as measured on a standard hydrogen electrode scale), hydrogen is passive in the pH range between approximately 4 and 9. The degree of passivity is further dependent upon the temperature, the form of oxide present, and its degree of dissolution.

The thickness of the protective oxide film that forms in water varies according to temperature. The film development results in a very thin, unstructured layer at ambient temperatures, but results in a very thick, highly protective layer as water approaches its boiling point. Beyond a temperature greater than 437°F (225°C), a protective film no longer will develop in either steam or water. At higher temperatures, the reaction progresses rapidly until eventually all the aluminum that is subjected to the condition is converted into oxide.

The oxide layer film of aluminum is almost without exception soluble in both acidic and basic solutions. Some exceptions to this general rule include: (1) acetic acid; (2) sodium disilicate; (3) ammonium hydroxide at greater than 30% concentration by weight; (4) nitric acid at greater than 80% concentration by weight; and (5) sulfuric acid, at 98 to 100% concentration. In contact with solutions 3, 4, and 5 the oxidizing nature of the solution contributes to the maintenance of its oxide layer. Other than the exceptions noted, acids and bases tend to corrode aluminum.

Most corrosion of aluminum occurs electrochemically. There is an oxidation (anodic reaction), which leads to a dissolution of the metal, coupled with a reduction (cathodic reaction) of an environmental species. Aluminum commonly corrodes due to defects in its protective oxide film. As purity is increased, there are fewer defects that occur in its protective film, and thus an improvement in its corrosion resistance. There is never a complete elimination of defects. Therefore, even the purest aluminum is subject to corrosion. Some of the more common types of corrosion that aluminum experiences are pitting, galvanic corrosion, exfoliation corrosion, and stress-corrosion cracking. These various forms of corrosion as they affect aluminum are described in more detail in Sections 2.2.2.1 through 2.2.2.6.

2.2.2.1 Overall Chemical Resistance of Aluminum

Aluminum alloys are compatible with most anhydrous inorganic salts, and with aqueous solutions of these salts, when the concentrations are within its passive range. Halide salt aqueous solutions may cause pitting to occur if an aluminum alloy is polarized to its pitting potential (see Section 2.2.2.4). Aluminum alloys are not suitable for most inorganic acids and bases, but are resistant to many organic acids. Resistance in general to most organic chemicals is quite good. Applications involving halogenated organic compounds should be given special consideration prior to use. Reactions may occur, producing an aluminum alkyl in some instances. Reactions may even be autocatalytic, and can even proceed in a highly exothermic fashion. Also, consideration should be given to mixtures of organic compounds and additions of inorganic compounds or elements in contact with aluminum. Effects of combined compounds on aluminum may be entirely different from their individual effects.

2.2.2.2 Stress-Corrosion Cracking of Aluminum

Aluminum alloys that contain levels of elements such as copper, magnesium, silicon, and zinc are susceptible to stress-corrosion cracking. Whenever soluble alloying agents precipitate along grain boundaries, the area is anodic to the surrounding microstructure. Corrosion can propagate selectively along a grain boundary. Therefore, stress-corrosion cracking can be minimized in these alloys through metallurgical treatments, or by producing a more homogeneous microstructure, having a minimum precipitation of soluble alloying elements. Stress-corrosion cracking in aluminum can normally be retarded greatly if cathodic protection is applied.

Stress-corrosion cracking of an aluminum alloy is related to the magnitude and duration of stresses acting on its surface. In most alloys, it is also determined by the direction of stresses that are applied. The direction of stresses acting on a cast alloy usually have little effect due to the equiaxial nature of the grain structure. Stress-cracking resistance of aluminum alloys is a factor of the specific environment to which it is subjected. A chemical stress-cracking agent, its concentration, the amount, duration, and direction of stresses that are applied, and the temperature, types, and amounts of impurities or other additives are all factors that make up the specific environment. Testing and stress-corrosion data for aluminum and other metals must always be interpreted very carefully.

2.2.2.3 Galvanic Corrosion of Aluminum

The galvanic corrosion of aluminum and aluminum alloys is dependent on its degree of polarization, within a given galvanic cell. Where contact is made with a metal that is more cathodic than aluminum, and aluminum is polarized to its pitting potential, a small increase in potential results in a large increase in the resultant corrosion current. Aluminum may be used in contact with stainless steel and chromium where it is exposed in atmospheric environments or other mild conditions. A method of minimizing potential galvanic corrosion of aluminum where it is necessary to contact another metal is to maximize the exposed area that is in contact with a more cathodic metal. By maintaining a high ratio of contact, current density is reduced on aluminum. When protective coatings are considered as a means to prevent the occurrence of galvanic corrosion, coatings should either be applied to the more cathodic metal or to both metals. They should never be applied to aluminum alone, due to the occurrence of defects in a coating, which can never be completely eliminated.

2.2.2.4 Pitting of Aluminum

When aluminum is in contact with a solution that is within its pH range, it may still be subject to corrosion of the pitting type. Pitting in aluminum is dependent on its relation to a value known as its pitting potential. Pitting potential corresponds to a steady level or plateau that occurs on a plot of potential versus percent concentration of added element. Pitting will not occur in aluminum that is polarized to a level below its pitting potential. Pitting can be substantially reduced, or prevented, by removal of oxygen or other reducing species that may be present. When reducing species are removed, insufficient reactions occur on its surface to cause polarization to its pitting potential. An increase in acidity or alkalinity beyond its passive range usually has the result of decreasing potential pitting. This is due to an increase in the uniformity of corrosion attack that results. An example of pit-type corrosion of an aluminum sheet is shown in the photograph in Figure 2.14.

Figure 2.14. Pit-type corrosion in a block of aluminum material. (Source: Trent Tube.)

2.2.2.5 Exfoliation Corrosion of Aluminum

Aluminum products that have a pronounced directional structure in which grains are highly elongated and a high area-to-thickness ratio is present are susceptible to exfoliation corrosion. When these conditions are present in aluminum, it may decompose selectively along its boundary layers. The effect is to cause a leafing, or delamination, of the material. Its resistance to exfoliation corrosion may not be predicted by its resistance to intergranular attack. Exfoliation is more prevalent and more severe in heat-treatable wrought alloys. This form of corrosion may be prevented by the same manufacturing and metallurgical techniques that are used to prevent stress-corrosion cracking.

2.2.2.6 Corrosion Considerations for Welding of Aluminum

The welding process, the alloy being welded, and the alloy of filler materials selected must all be considered in terms of their effect on the corrosion resistance of the finished product. The occurrence of galvanic cells should be avoided by minimizing differences in potential among the parent alloy, in regions where microstructural changes are present (such as in heat-affected regions) and in filler materials. Heat-affected regions may be annealed by postweld heat treatment to minimize any potential corrosive effects. In most non-heat-treatable alloys, weldments normally have good resistance to corrosion without undergoing any secondary treatment.

Wrought alloys of aluminum can be categorized into two types: (1) heat treatable and (2) non-heat-treatable types. Heat-treatable types are strengthened by either: (1) a precipitation or ageing process, at natural ambient temperatures, or at temperatures of 240 to 380°F (115 to 195°C); (2) a solution heat treatment at 860 to 1050°F (460 to 565°C), or (3) quenching to retain them in a solid solution. Non-heat-treatable types are strengthened through a process that results in strain hardening. The results of strain hardening may be supplemented through solid solution and dispersion hardening.

2.2.3 Carbon Steel and Low-Alloy Steel

Carbon steel is the most common, versatile, and least expensive of all metals that are used in the chemical process industry. It possesses high ductility, and is capable of being processed with cold-forming techniques. Perhaps the most important quality of steel is that it is the most readily weldable of all metals that are used for industrial applications. Steel is strong and relatively heavy (three times that of aluminum and two-thirds that of lead). Its tensile strength can vary from as low as 55,000 psi (3,900 bar) for low-carbon steel to 100,000 psi (7,092 bar) for high-carbon steel that has been annealed.

Standards and specifications concerning steel, steel alloys, and its products are covered in Volumes 1.01 through 1.06 of the ASTM Annual Book of Standards. The Iron and Steel Institute, the American Society of Mechanical Engineers, and the American National Standards Institute also issue standards for carbon steel and its various alloys. Certain elements are always present in steel due to processing and production techniques used. These include: (1) carbon; (2) manganese; (3) phosphorous; (4) sulfur; (5) silicon; and (6) traces of aluminum, nitrogen, and oxygen. Various other elements may be added to form alloys. These alloying elements include chromium, copper, molybdenum, nickel, and vanadium.

2.2.3.1 Overall Corrosion Resistance of Steel

As with most metallic substances, the corrosion resistance of steel is dependent on the formation of a corrosion-resistant oxide film on its surface. The resistance of the oxide layer that forms on steel offers it only limited amounts of protection. Perhaps the most notorious corrosive agent with respect to steel is any type of dilute acid. For instance, carbon steel should never be used with sulfuric acid below 90% weight concentration. However, it may be used with concentrations between 90 and 98%, at temperatures up to the boiling point. Steel should never be used with highly oxidizing acids such as nitric, hydrochloric, or phosphoric acid at any concentrations.

In terms of mineral bases, steel may be used up to 75% concentration by weight, and up to 212°F (100°C), if iron contamination is permissible for the caustic solution. It is sometimes necessary to use stress relieving in order to minimize effects from caustic embrittlement. Also, steel corrodes at a very slow rate in the presence of salt water and brines. Again, it may be used if iron contamination is permissible. Steel is relatively unaffected though by fresh water, and by many organic chemicals. Cathodic protection will significantly increase the service life of steel in fresh and salt water applications, and in direct buried applications.

The addition of small amounts of alloying materials has a significant effect on the corrosion resistance of the steel to atmospheric corrosion. Low-alloy steels contain small amounts of one or more alloying elements. While its corrosion resistance to atmospheric corrosion is greatly enhanced, its resistance to liquid electrolytes is not significant. The addition of alloying elements results in the formation of a dense rust film, which is affected by acid and alkaline solutions. There is, however, a significant increase in strength, allowing piping and vessels to be made with thinner walls than if ordinary carbon steel were used. A typical low-alloy steel is AISI 4340, which contains the following alloying element content: (1) carbon 0.4%; (2) manganese 0.7%; (3) nickel 1.85%; (4) chromium 0.8%; and (5) molybdenum 0.25%. There are many types of other combinations, both standard AISI grades and proprietary formulations as well. A steel is classified as a "low alloy" if its chromium content is below 11%. A chromium content of 12% or above classifies a steel as a "stainless steel."

2.2.4 Stainless Steel and Intermediate-Alloy Steel

Among the most widely used materials for corrosion resistance is a group of steel alloys referred to as the *stainless steels*. Stainless steels contain a minimum content of 12% chromium, as noted previously. There are more than 70 different types of standard stainless steels, in addition to many special alloys. These alloys of steel are produced both in the wrought form (American Institute of Steel and Iron types) and through casting techniques (American Casting Institute, ACI, types). The characteristics of stainless steels include a base ferrous content, 12 to 30% chromium content, nickel content ranging from 0 to 22% content, and minor amounts of carbon, columbium, copper, molybdenum, selenium, tantalum, and titanium. There are three major types of stainless steels and two relatively new types. The three major types are: (1) martensitic alloys; (2) ferritic alloys; and (3) austenitic stainless steels. Two relatively new groups are: (4) duplex alloys; and (5) precipitation hardening alloys. Stainless steels in general are heat resistant, corrosion resistant, and relatively resistant to leaching or other contaminating effects.

2.2.4.1 Martensitic Stainless Steels

Martensitic alloys typically have a chromium content between 12 and 20% and contain minor amounts of carbon and other additives. The corrosion resistance of martensitic alloys is in general the least of the three major types, particularly in comparison to austenitic stainless steels. The corrosion resistance of martensitic stainless steels is generally good, however, in atmospheric conditions, fresh water, and organic chemicals.

2.2.4.2 Ferritic Stainless Steels

Ferritic stainless steels typically have a chromium content between 15 to 30%, a nickel content of less than 3%, and a low carbon content (0.1%). The higher the chromium content, the higher the chemical resistance of a ferritic alloy. Ferritic alloys are versatile, and can be fabricated by all standard methods due to their ductility. The corrosion resistance of ferritic alloys is good, although reducing acids do readily attack them. Alloys such as type-430 perform very well in nitric acid service. Ferritic alloys also have natural resistance to scaling, and are highly resistant to high-temperature oxidation.

2.2.4.3 Austenitic Stainless Steels

Austenitic stainless steel alloys exhibit the best corrosion resistance in general among stainless steels. The typical makeup of an austenitic stainless steel is: (1) low carbon content (0.08%); (2) chromium content of 16 to 26%; and (3) nickel content of 7 to 22%. Typical examples of austenitic stainless steels include types-304 and -304L, and types-316 and -316L stainless steels. Welding of austenitic steels may readily be performed, although there are some special considerations. The welding heat must be controlled, or chromium carbide precipitation may occur. This effect will deplete some chromium in localized regions, and thus will lower its corrosion resistance to process fluids along grain boundaries. Carbides can be put back into solution through heat treatment. In severe applications, this practice is highly encouraged to reduce the possibility of intergranular corrosion, although it is not always practical for field welds. Low-carbon types such as 304L and 316L have 0.3% maximum carbon and tend to minimize the possibility of intergranular corrosion.

Austenitic stainless steels demonstrate wide-scale resistance to nitric acid at most concentrations and temperatures. In general, corrosion resistance and resistance to stress-corrosion cracking goes up with nickel content for the austenitic stainless steels. Caution is advised when using austenitic alloys such as type-316 in sulfuric acid service. Type-316 should only be used at less than 5% and greater than 85% concentration, and at temperatures below the boiling point. Reducing media and chloride ions destroys its surface oxide film, which otherwise protects an austenitic stainless steel; thus, reducing media and chloride ions attack these materials readily. High tensile stresses, in conjunction with chloride ions, will cause stress-corrosion cracking, sometimes in a severe and rapid manner, depending on the overall conditions.

2.2.4.4 Duplex Stainless Steel Alloys

Duplex alloys are a relatively new type of stainless alloy that are gaining widespread use. A duplex alloy has some of the characteristics of both ferritic and austenitic alloys, due to the presence of both ferrite and austenites in their crystalline makeup. A duplex alloy has a nickel content between 4 and 7%, thus placing it between the two levels that define the ferritic and austenitic groups. Duplex alloys typically have high strength and possess good chemical resistance. An example of a duplex alloy is alloy 2205, which has the following makeup: (1) 21.0/23.0% Cr; (2) 4.5/6.5% Ni; (3) 2.5/3.5% Mo; (4) 0.030% maximum C; (5) 0.08/0.20% N; and (6) the balance is Fe. The listed tensile strength and yield strength for alloy 2205 are 90 ksi (6.4 bar) and 65 ksi (4.6 bar), according to ASTM A-789.

2.2.4.5 Precipitation Hardening Stainless Steels

Precipitation hardening stainless steels are a special form of stainless steels that may be hardened through the use of heat. Examples of precipitation hardening steels include alloy 350, alloy 355, alloy 718, and alloy 17-7 PH. Precipitation hardening steels are characterized by unusually high strength and find applications mainly in tubing.

2.2.4.6 Intermediate Alloys

A special group of alloys has been created to improve upon the chemical resistance of austenitic stainless steel alloys. These alloys are referred to as medium or intermediate alloys. Such alloys normally contain higher concentrations of nickel and chromium than an austenitic alloy. Intermediate alloys show improved resistance to sulfuric acid at a wide range of concentrations and temperatures. The alloy that has shown increasing popularity for sulfuric acid handling is alloy 20 (a group of alloys containing 20% chromium, 24 to 33% nickel, and minor amounts of molybdenum and copper). Figure 2.15 illustrates two bolts that were subjected to hot sulfuric acid for identical periods. The one on the left is type-316 stainless steel; the one on the right is alloy 20Cb-3® stainless steel. As can be

Materials of Construction

Figure 2.15. Example of bolts subjected to hot sulfuric acid. (Right: Type-316 stainless steel; left: alloy 20Cb-3®.) (Source: Carpenter Technology.)

seen, alloy 20Cb-3® is much more suitable for the stated service. (20Cb-3 is a registered trademark of Carpenter Technology Corporation.)

In addition to good sulfuric acid performance, alloy 20 is useful for nitric acid below 205°F (95°C) and 60% concentration, and hydrochloric acid at cold temperatures and concentrations below 20%. Some alloys that are considered as medium alloys are often classified instead as nickel alloys (i.e., Inconel and Hastelloy), since they contain appreciable amounts of iron but have nickel as their major constituent (except for Inconel alloy 800).

2.2.5 Nickel and Nickel Alloys

The element nickel has some unique electrochemical properties that explain why it serves as the base element for an important group of corrosion-resistant alloys. Several of these alloys are used by the chemical industry for aqueous corrosion service involving difficult-to-handle chemicals. Nickel is metallurgically compatible with a number of other metals, including copper, chromium, molybdenum, and iron. For this reason, nickel can be alloyed with one or more or these elements to form alloys with wide ranges in composition. Pure nickel's unique electrochemical properties allow it in reducing acid systems to assume an open-circuit potential equal in magnitude to platinum's potential. It does not readily liberate hydrogen, and as such requires an oxidizing ion species to be present in order for corrosion to proceed.

Pure nickel is either designated as alloy 200 or 201, with the latter having a lower carbon content. A lower carbon content prevents graphitization at high temperatures, allowing it to maintain ductility. A pure nickel with a lower carbon content is therefore preferred at temperatures over 600°F (316°C).

Nickel does possess the ability to form a passive film, although its film is not very stable. While the general corrosion resistance of nickel-based alloys is excellent, they also possess an unusually high resistance to stress-corrosion cracking, pitting, and crevice corrosion. As the molybdenum content is increased, there is a subsequent increase in resistance to localized corrosion effects. The chromium content also has an effect on these situations, particularly where there is an oxidizing environment. Nickel has excellent resistance to alkalis over a wide range of temperatures. As the concentrations and temperatures in alkaline environments increase, the corrosion resistance of nickel decreases slowly. Its resistance in those environments approach that of zirconium. Under high amounts of stress, nickel (and nickel alloys) is susceptible to stress-corrosion cracking in caustic environments. It is resistant to stress-corrosion cracking in typical chloride environments that give other materials problems. Nickel is resistant to dilute acids, but the addition of oxidizers such as oxygen or salts has a big effect on its rate of corrosion. Nickel is resistant to anhydrous ammonia or weak ammonium hydroxide environments, but strong concentrations can cause rapid attack.

In addition to the two pure nickel alloys, there are 12 major alloys that are commercially available, and 5 major cast alloys. The twelve wrought alloys, in order of increasing chromium content include: (1) Monel alloy 400; (2) Monel alloy K-500; (3) Hastelloy alloy B-2; (4) Inconel alloy 600; (5) Inconel alloy 800; (6) Inconel alloy 825; (7) Hastelloy alloy G; (8) Hastelloy alloy X; (9) Inconel alloy 625; (10) Hastelloy alloy C-276; (11) Hastelloy alloy C-4; and (12) Hastelloy alloy S. The five major cast alloys include: (1) Alloy B; (2) Chlorimet 2; (3) Alloy C; (4) Chlorimet 3; and (5) Hastelloy C-4C.

2.2.6 Titanium and Titanium Alloys

Titanium is the ninth-most-abundant element in the earth. Thus it is readily available as a material of construction. Figure 2.16 shows titanium tube hollows being processed on a

Figure 2.16. Titanium tube hollows being processed on a Pilger mill. (Source: Teledyne Wah Chang, Albany.)

Materials of Construction 43

pilger mill to seamless tube and pipe. Figures 2.17 and 2.18 show titanium tubes being deburred after cutting to length and going through a finishing step, respectively.

The American Society of Testing and Materials publishes standards for titanium in Volume 2.04 of its Annual Book of Standards. These specifications are covered in Table 2.4. Of all the titanium grades specified for corrosion-resistance service, ASTM grade 2 is most frequently used. Of all the grades, types 1, 2, 3, and 7 are used for chemical plant and oil refinery use. ASTM type 7 is a titanium/palladium alloy that offers improved corrosion resistance compared to the unalloyed grades. Titanium is a light metal that is slightly more than half as dense as ferrous or copper-based alloys. Titanium has a substantially lower modulus of elasticity compared to steel, as well as a much lower coefficient of thermal expansion. Therefore, special attention must be given to its unique properties when designing and analyzing piping systems made of titanium.

Titanium performs well in resisting the corrosive effects of many chemicals due to a very stable oxide film that develops on, and adheres to, its surface. In titanium, corrosion resistance is highly controlled by this oxide film, which develops when it is exposed to air. Its relatively thin film is attacked by only a few substances such as hydrofluoric acid. Titanium has a strong affinity for oxygen and moisture. Therefore, it is capable of re-forming the film over pores or ruptures when moisture or oxygen is present. For this reason, titanium is excellent in wet chlorine service, when it contains above 0.5% moisture content. Titanium is extremely resistant to erosion corrosion due to the strength of its film. It is capable of withstanding salt water velocities (as high as 90 ft/s).[5] Titanium is resistant to water of all types, including steam (and corrosive steam) and is very resistant to fouling and buildup of deposits of any type.

Figure 2.17. Deburring of titanium piping. (Source: Teledyne Wah Chang, Albany.)

Figure 2.18. Finishing of titanium piping. (Source: Teledyne Wah Chang, Albany.)

Titanium is resistant to galvanic corrosion, unless it is in the presence of a reducing medium and it is in contact with a more noble metal. Titanium has widespread resistance to salt water and chloride solutions. In some cases, pitting may occur in chloride solutions where the right combination of conditions are present. Titanium has excellent resistance to chlorine, wet chlorine gas, chlorinated brines, and hypochlorites. Titanium shows good general resistance to organic chemicals and organic acids. Titanium also has excellent resistance to oxidizing acids such as nitric and chromic acids. On the other hand, reducing acids such as hydrochloric, sulfuric, and phosphoric acids attack titanium.

Dry chlorine gas will also attack titanium. It is susceptible to hydrogen embrittlement in alkaline solutions above their boiling point and in acidic conditions where hydrogen is generated on its oxide surface.

2.2.7 Tantalum and Tantalum Alloys

Tantalum is an element that has found industrial usage since the beginning of the twentieth century. It is classified as a group-five element, and has an atomic number of 73. Tantalum's physical properties are similar to those of mild steel, except for a distinctly

Materials of Construction

Table 2.4 ASTM Specifications for Titanium and Titanium Alloys

ASTM B 265-76	Titanium and Titanium Alloy Strip, Sheet and Plate
ASTM B 337-76	Seamless and Welded Titanium and Titanium Alloy Pipe
ASTM B 338-76	Seamless and Welded Titanium and Titanium Alloy Tubes for Condensers and Heat Exchangers
ASTM B348-76	Titanium and Titanium Alloy Bars and Billets
ASTM B 363-76	Seamless and Welded Unalloyed Titanium Welding Fittings
ASTM B 367-69	(1974) Titanium and Titanium Alloy Castings
ASTM B 381-76	Titanium and Titanium Alloy Forgings

higher melting point. It can be welded and formed by a number of different techniques. Since tantalum is a reactive metal, special attention must be paid to any area that comes in contact with heat. Four special techniques have been developed for welding: (1) electron beam welding; (2) flow-purged chamber welding; (3) dry box or vacuum purge chamber welding; and (4) open air welding.

Tantalum exhibits inertness to a very wide range of organic and inorganic chemicals at temperatures below 300°F (150°C). Two notable exceptions to this are hydrofluoric acid and fuming sulfuric acid. Table 2.5 displays a listing of various chemicals to which tantalum is completely inert up to at least 300°F. Tantalum is attacked by concentrated alkaline solutions at ambient temperatures. It may be oxidized at high temperatures by various gases, but it is very resistant to oxidation by gases at ambient temperatures. Fused sodium hydroxide and potassium hydroxide can dissolve tantalum.

2.2.8 Zirconium and Zirconium Alloys

Zirconium is the fortieth element in the Periodic Table and is relatively abundant in the earth's crust. It is sixth in order of abundance of the engineering metals. Figure 2.19 shows zirconium round stock being reduced on a GFM rotary forge. The forge is capable of handling round or square forgings up to 15 3/4 inches in diameter, 33 feet in length, and 4400 pounds in weight. Zirconium originally found its usefulness in nuclear reactor service cooling bundles. However, due to its relative abundance and excellent chemical resistance, it is becoming increasingly popular in chemical piping service. Zirconium resembles titanium in corrosion resistance, except that it is more resistant to hydrochloric acid. There are a number of alloys of zirconium that exist, referred to as Zircaloys. Zircaloys are similar to pure zirconium, but exhibit improved mechanical properties.

Standards for zirconium materials are covered by ASTM in Volume 2.04 of the ASTM Annual Book of Standards, and also through the American Society of Mechanical Engineers. Industrial grades that are available include Grade UNS R60702 (Grade 702, Grade 11, commercial grade) for ordinary corrosive chemical service, Grade UNS R60704 (Grade 704), an alloy with 1.5% tin and 0.2% iron plus chromium for chemical service where added strength is needed, and Grades R60705 (Grade 705) and R60706 (Grade 706),

Table 2.5 Chemical Resistance of Tantalum [3]

Partial List of Compatible Chemicals			
Acetic acid	Carbonic acid	Hydrogen chloride	Phosphoric acid, < 4 ppm F
Acetic anhydride	Carbon dioxide	Hydrogen iodide	Phosphorus
Acetone	Chloric acid	Hydrogen peroxide	Phosphorus chlorides
Acids, Mineral (except HF)	Chlorinated hydrocarbons	Hydrogen sulfide	Phosphorus oxychloride
Acid salts	Chlorine oxides	Iodine	Phtalic anhydride
Air	Chlorine water and brine	Hypochlorous acid	Potassium chloride
Alcohols	Chlorine wet or dry	Lactic acid	Potassium dichromate
Aluminum chloride	Chloroacetic acid	Magnesium chloride	Potassium iodide, iodine
Aluminum sulfate	Chrome plating solutions	Magnesium sulfate	Potassium nitrate
Amines	Chromic acid	Mercury salts	Refrigerants
Ammonium chloride	Citric acid	Methyl sulfuric acid	Silver nitrate
Ammonium hydroxide	Cleaning solutions	Milk	Sodium bisulfate, aqueous
Ammonium phosphate	Copper salts	Mineral oils	Sodium bromide
Ammonium sulfate	Ethyl sulfate	Motor fuels	Sodium chlorate
Amyl acetate	Ethylene dibromide	Nitric acid, industrial fuming	Sodium chloride
Amyl chloride	Fatty acids		Sodium hypochlorite
Aqua regia	Ferric chloride	Nitric oxides	Sodium nitrate
Barium hydroxide	Ferrous sulfate	Nitrogen	Sodium sulfate
Body fluids	Foodstuffs	Nitrosyl chloride	Sodium sulfite
Bromine, wet or dry	Formaldehyde	Nitrous oxides	Sugar
Butyric acid	Formic acid	Organic chlorides	Sulfamic acid
Calcium bisulfate	Fruit products	Oxalic acid	Sulfur
Calcium chloride	Hydriodic acid	Oxygen	Sulfur dioxide
Calcium hydroxide	Hydrobromic acid	Perchloric acid	Sulfuric acid, under 98%
Calcium hypochlorite	Hydrochloric acid	Petroleum products	Water
Carbon tetrachloride	Hydrogen	Phenols	

Notes:
[1] Tantalum is totally inert to the materials up to a minimum of 302°F (150°C).
[2] Tantalum is affected by hydrofluoric acid and fuming sulfuric acid under 302°F (150°C).
[3] From "Corrosion and Corrosion Protection Handbook," by Philip A. Schweitzer, Marcel Dekker, New York, 1983.

both alloys with niobium. Special attention must be paid to the welding of zirconium, as an inert atmosphere is a prerequisite.

Zirconium has excellent corrosion resistance to many organic and inorganic solutions. These include mineral acids, molten alkalis, strong bases, and most organic and salt solutions. It also has excellent resistance to oxidation up to 750°F (399°C) in environments of steam, air, oxygen, nitrogen, and carbon or sulfur dioxide. Its excellent chemical resistance is due to a very inert passive oxide film that forms on its surface upon contacting air. Ingredients that dissolve this film, thereby attacking zirconium, include: (1) hydrofluoric acid; (2) concentrated sulfuric acid; (3) aqua regia; (4) wet chlorine; and (5) ferric and cupric chloride solutions.

2.2.9 Cast Alloys

There is a broad range of cast alloys that are available for use in corrosive chemical services. Iron/carbon and stainless steel cast materials account for the majority of corrosion applications. This is true despite the fact that there are also carbon and low-alloy steels, cast iron, nickel-based alloys, aluminum alloys, copper-based alloys, titanium, zirconium, and magne-

Figure 2.19. Zirconium round stock being reduced on a GFM rotary forge. (Source: Teledyne Wah Chang, Albany.)

sium alloys available for industrial uses. For virtually every wrought alloy composition, there is a cast equivalent available. There are many alloys available only as cast materials.

Cast irons are one of the most economical materials available, although they do not contain much inherent mechanical strength or corrosion resistance. High-silicon cast irons contain a 13 to 16% silicon composition and have excellent corrosion resistance. These materials, known as "Duriron" and "Corrosiron," while having improved chemical resistance, are weakened mechanically as compared to cast iron. High-silicon irons also do not stand up well to shock and impact since they are very brittle materials. Their hardness does make them good choices for erosion-corrosion services, however. Duriron and Corrosiron are very resistant to oxidizing and reducing environments due to the formation of a strong, passive oxide film on their surface.

A cast alloy that shows improved mechanical strength over high-silicon alloy irons is Hastelloy D, with 82% nickel, 3% copper, 9% silicon, and 2% iron. Another group of cast iron alloys are called Ni-Resist, with nickel contents ranging from 13.5 to 36% and some with 6.5% copper. Nickel alloy cast materials have generally better toughness and superior impact resistance as compared to cast irons. Nickel alloy castings can be welded, as well as machined, making them versatile materials. A ductile variety and a hard variety known as Ni-Hard are available. These alloys have excellent chemical resistance except for service in oxidizing acids such as nitric acid. Some grades are capable of withstanding temperatures of up to 1500°F (816°C).

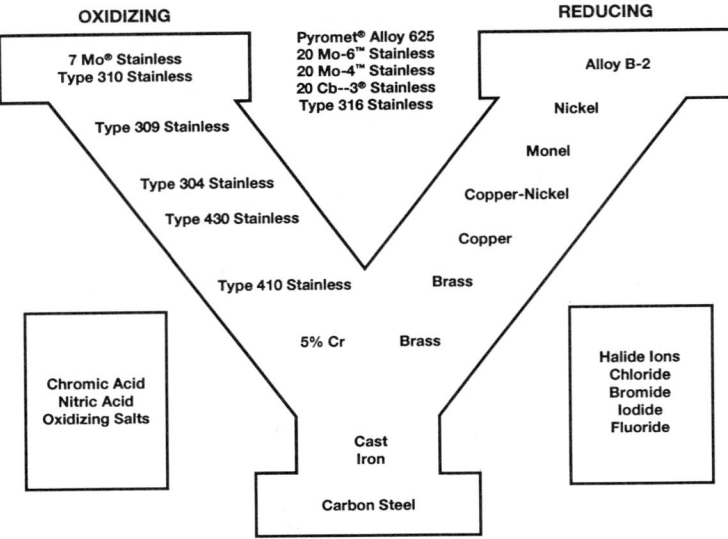

Figure 2.20. The "Y" of corrosion. (Source: John C. Tverberg, Trent Tube.)

2.2.10 Selection of Alloys for Corrosive Services

To select a metallic material for corrosive services, many factors need to be considered. These include applicable code requirements, or limitations, limitations of the authority having local jurisdiction, expected performance based on fluid design factors, availability, installation criteria, and cost. Cost is often viewed as the primary criterion by many. However, in weighing the cost of components, the true cost must be taken into account, including maintenance and replacement costs and liability costs, should a less-than-adequate material be selected.

Figure 2.20, which is referred to as the "Y" of corrosion, is a useful general reference for comparing several common metal alloys in terms of their performance in various media. The right-hand branch of the "Y" lists materials that can handle reducing media, in order of increasing capability as one moves from the bottom to the top of the branch. The left-hand branch of the "Y" lists materials that can handle oxidizing media. Specific alloys tend not to work with the media listed on the opposite branch side of the "Y." The alloys listed between the two branches at the top are suitable for both oxidizing and reducing media.

Table 2.6 also acts as a general selection guide in determining a suitable alloy based on various types of fluid design conditions. Both in Figure 2.20 and Table 2.6, the choices listed are not the only appropriate metal choices. Also, all factors must be considered, including stress levels, etc. To determine an appropriate metallic material choice, a corrosion expert must always be consulted and/or testing conducted.

Table 2.6 Stainless Steel Selection Guide

Corrodent Severity	General Corrosion			Localized Corrosion	
	Sulfuric Acid	Phosphoric Acid	Nitric Acid	Pitting or Crevice	Chloride SCC
Very severe to severe	20 Cb-3[a]	20 Mo-6[b] 20 Mo-4	7 Mo PLUS 20 Mo-4 20 Mo-6 SEA-CURE	SEA-CURE 20 Mo-6	20 Mo-6 20 Mo-4 20 Cb-3 SEA-CURE
Moderate to severe	20 Mo-4 20 Mo-6	20 Cb-3	20 Cb-3	20 Mo-4	
Moderate		SEA-CURE 7 Mo PLUS	Type 316 Type 304	7 Mo PLUS 20 Cb-3	
Mild to moderate	SEA-CURE 7 Mo PLUS	Type 316		Type 316	
Mild		Type 316		Type 304	

[a] 20 Cb-3 is a registered trademark of Carpenter Technology Corp.
[b] 20 Mo-4 and 20 Mo-6 are trademarks of Carpenter Technology Corp.

2.3 Degradation of Nonmetallic Materials in Corrosive Media

The manner in which nonmetallic materials are adversely affected by corrosive media differs immensely from that of metallic compounds. In metallic material behavior, the resistance of a metallic substance depends almost entirely on the formation of an oxide layer and the relative passivity of that layer to the chemical solution in question. Nonmetallic materials, on the other hand, are normally atomically stable and do not form an oxide layer when they are in contact with air. Therefore, the analysis of a nonmetallic material's resistance to corrosion is dependent solely on the resistance of the base material itself, and not an oxide layer.

2.3.1 Degradation of Plastics in Corrosive Environments

2.3.1.1 Behavior of Plastic Materials in Corrosive Environments

One of the few similarities that plastic materials have with metallic materials is that the exact corrosion resistance of a given grade of material varies according to its exact chemical makeup and method of manufacture. There are many different families of thermoplastics and thermosetting resins. The chemical resistance between any two families varies greatly, as do comparative physical properties. Further, within any given family of plastic, each resin type has its own unique characteristics as compared to another type within a same family. Even within a given resin type, behavior will vary according to each type of specific resin formulation. For instance, each given thermoplastic will vary according to whether it is a homopolymer or a copolymer (of two or more monomers), the type and concentration of additives it is blended with, the process of polymerization by which it is made from, and other factors. These different makeup factors will manifest themselves in a variance of such physical properties as density, molecular weight, degree and amount of side branching, degree and amount of crystallization, molecule size, impurity level, thermal stability, and many others. Many of these characteristics in turn have a bearing on the ability of a specific resin to resist a given corrosive chemical condition. Thermosetting resins vary in a similar manner according to the type of curing process, base resin makeup, method of manufacture, reinforcement material used, etc.

Therefore, it is imperative that when the chemical resistance of a plastic is being reviewed, specific resin makes or grades are investigated. For example, reviewing the chemical resistance of a generic type of resin such as "polypropylene" must include a review of the specific type or grade of "polypropylene" to be used. Resistance to a certain chemical may differ, depending on whether its grade is a homopolymer or copolymer (including what co-monomer it is alloyed with, and the percentage of co-monomer it contains), the density of the polymer, degree of side branching, crystallinity, molecular weight, etc.

There is a variety of different ways that chemicals degrade plastic materials. These include: (1) softening; (2) swelling; (3) delamination (applies to reinforced thermoset plastics only); (4) weeping (reinforced thermosetting plastics only); (5) discoloration; (6) dissolving; (7) embrittlement; (8) stress cracking; (9) crazing; (10) charring; (11) blistering; (12) weight gain or loss, and others. These various effects may occur as a result of one or more of the following general mechanisms: (1) absorption; (2) environmental-stress cracking (ESC); (3) chemical reaction; (4) solvation; and (5) plasticization. The behavior of an aggressive chemical, which manifests itself in one of these mechanisms, results in a specific plastic being affected by one of the aforementioned effects. These mechanisms can act in combinations, and when they do, it becomes very difficult to detect which ones are occurring.

Environmental-stress cracking is virtually the same process that occurs in stress-corrosion cracking of metals. The plastics industry has developed the use of the term environmental-stress cracking to refer to the process as it occurs in a plastic material. The plastics industry, by developing a unique term, thereby avoids the use of the term *corrosion* as being associated with plastic materials in any way. The resistance of a given plastic material to environmental-stress cracking is commonly referred to as its ESCR, with the R standing for resistance.

Some of the specific degradation mechanisms that fall under the general heading of a chemical reaction mechanism include: (1) alkylation; (2) oxidation; (3) reduction; (4) hydrolysis; (5) halogenation (chlorination); (6) dehydro-halogenation; and (7) radiation. There are cases where a potential reaction mechanism can be predicted due to the structure of a plastic material under consideration and the chemical involved. In most situations where a prediction cannot be made, testing must be performed to determine the suitability of a plastic material in question.

2.3.1.2 Plastic Material Selection Criteria

There are six major factors that have an effect as to whether a chemical will affect a plastic and the rate of its possible chemical attack. These are: (1) type of chemical or corrosive agent; (2) concentration of the chemical; (3) temperature profile at which the chemical contacts a plastic material; (4) amount of time that a chemical agent comes in contact with a plastic; (5) overall level of stress that a plastic material is under while exposed to the chemical; and (6) the surface quality of the finished plastic component. The combined effect of each of these six factors determines whether a chemical will have an effect on a component made from a specific grade of plastic resin. Therefore, when making a material selection involving a nonmetallic material, all six of these factors should be considered to determine an appropriate choice.

Most corrosion information that is available for nonmetallic materials only take into account the first three factors (or in some cases only #1 and #3). Also, information in many common sources is based on general resin types, as opposed to specific grades. Therefore, testing under proposed conditions (using accelerated stress-cracking tests) is the best way to make a material determination. The six major factors and their effect on the chemical resistance of plastic material performance is described in more detail in the following.

2.3.1.3 Temperature Effects and Degradation of Plastics

Almost without exception, an increase in the temperature of a system will have an accelerating effect on all of the different attack mechanisms described previously. In those situations where a chemical attacks a material at ambient or low temperatures, an increase in system temperature will increase the magnitude and rate at which an attack occurs. A relatively small increase in temperature will cause a polymer to expand significantly. The resulting expansion will cause an increase in available surface area and spacing of individual molecules. This results in an increase in permeability and solubility of the molecules, resulting in faster attack. Also, an increase in temperature means that energy is added to the system, lowering the activation energy required for a reaction to proceed. The level of energy that is added to the system in some cases exceeds the required activation energy needed for a chemical reaction to proceed in the first place. Therefore, a material may be attacked by a chemical at a higher temperature but not at a lower temperature. Though this description refers specifically to chemical reaction mechanisms, temperature has a similar effect with respect to solvation, absorption, plasticization, and environmental-stress cracking.

As a temperature is increased a level may be reached where stress cracking can occur. When this happens, there will be a sudden and distinctive change in the stress-cracking resistance behavior of a plastic material. At the transition temperature (and at temperatures higher than the transition temperature), a material is subject to environmental-stress cracking. Below the transition temperature, a material may be completely unaffected by the chemical, if the total stress in the system remains at a constant value.

The importance of this behavioral characteristic of plastic materials is that information on stress-cracking resistance at room temperature should not be used to make a judgement of stress-cracking resistance at higher temperatures. Extrapolation of stress-cracking information, as well as other chemical effects, must be strictly avoided when making material selections.

2.3.1.4 Concentration Effects and Degradation of Plastics

The concentration of a particular reagent usually has a direct effect on the amount and rate of chemical attack on a plastic material. There are a few exceptions to this general rule, such as with some concentrations of mineral acids. Also, with respect to certain organic chemicals, the level of concentration seems to have little or no effect. However, the general rule is that if a chemical tends to have some sort of effect on a plastic material, then at greater concentrations there is a greater driving force and thus an increase in effect.

In a similar fashion to the effect of temperature, there can be a certain threshold level in the concentration profile where there is a significant increase in effect. This is particularly true with respect to the environmental-stress-cracking mechanism. As with temperature, the concentration of a certain stress-cracking agent can be increased with little or no effect until a "threshold" level is reached. At this threshold level of concentration, or above it, environmental-stress cracking can manifest itself suddenly and rapidly. Therefore, it is not recommended to predict stress-cracking resistance or chemical resistance at a level of concentration substantially different from a known data point. Extrapolation must not be performed from a lower concentration to a higher one, in any case. Extrapolation between two data points is more accurate, but accuracy does increase with additional data points. Testing under project design conditions is always the best way to make a determination. Interpolation can be valid between known data points, particularly where a large number of data points exist.

2.3.1.5 Surface Quality, Manufacturing Effects, and Degradation of Plastics

The method used to manufacture a product, the quality of tooling used, product geometry and thickness, type of material, temperature of manufacture, and cooling process used all have an effect on a final plastic product. Plastic components can vary in surface quality, and can vary in the amount of residual internal stress inherent in each part. Manufactured-in stresses are a form of energy, which equates to at least a portion of the activation energy needed for a given chemical reaction to proceed. All reaction processes result in some form of residual stress, including injection molding, extrusion, compression molding, centrifugal casting, spiral (filament) winding, etc. This means that less energy is needed to cause a reaction to occur than if manufactured-in stresses are kept to a minimum.

Surface smoothness of a manufactured product also plays a significant role in terms of chemical resistance of finished plastic products, an often overlooked fact. Surface irregularities in a product translate into increased surface area, and can also act as stress concentration areas for chemical attack to be initiated. Therefore, the greater the amount of surface irregularities, the greater the likelihood of chemical attack to occur. In addition, surface irregularities in piping systems mean additional frictional losses compared to a perfectly smooth pipe. Greater frictional losses mean greater pumping pressures required, adding to internal pressure stresses of the system, resulting in higher resulting residual stress levels, and an increased likelihood for chemical attack to occur.

2.3.2 Chemical Reaction Mechanisms

Chemical attack can occur through a variety of attack mechanisms including: (1) alkylation; (2) oxidation; (3) reduction; (4) hydrolysis; (5) halogenation; (6) dehydrohalogenation, and others. The form of mechanism that will occur in any given situation depends on the chemistry between each specific material and chemical involved. If a reaction is initiated at the end of a polymer chain, a chain reaction may be initiated, leading to a complete depolymerization or "unzipping" of a polymer. If the active sites where attack is initiated occur randomly along its chain, then a polymer will be subjected to separation or "scissioning" at such sites.

If chemical reaction rates proceed rapidly for a certain chemical reaction, then the reaction can be readily detected through ordinary means. If a chemical reaction rate between a potential chemical reactant and a polymer is very slow, then detection through testing can be very difficult. Determination of the presence of chemical reactions is normally facilitated by testing or measuring physical properties. Physical properties for which tests are made may include one or more of the following: (1) molecular weight; (2) density; (3)

tensile strength; (4) flexural strength; (5) overall dimensions; and (6) general appearance, including color and other factors.

A slower reaction can be better detected by measuring a plastic material's short-term properties, such as tensile and flexural strength. There is no preset level of property change that determines when a material has been affected adversely to the point where it may no longer be used. Rather, it is normally up to each individual designer to make such a determination. However, a significant quantitative change in tensile creep rupture will normally preclude the use of such a polymer in the engineering design of a piping or tank system.

2.3.2.1 Solvation

Polymers are normally susceptible to solvation by certain organic solvents. When solvation does occur, a polymer is subject to weight gain, swelling, dimensional changes, or possibly all three. This is due to a polymer acting as a solvent and the chemical acting as a solute. The effect is normally readily detected by measuring a product for dimensional or weight change. Solvation can also cause a dissolution of a polymer if it acts as a solute, and the chemical acts as the solvent. In this instance, the change will be a loss in dimension or weight. A typical example of solvation occurs when a polyolefin such as polypropylene contacts a light hydrocarbon such as jet fuel. The polyolefin will swell and gain weight due to some absorption of the hydrocarbon.

2.3.2.2 Plasticization

Plasticization occurs when a solvent chemically attaches itself to a molecule through the formation of secondary chemical bonds. A polymer is normally less than completely soluble under such conditions in the solvent. The solvent becomes selectively absorbed into the surface of the polymer in a weak location. The result of plasticization is the formation of a modified polymer that exhibits significantly lower mechanical properties, lower glass-transition temperature, and other properties. To a much lesser extent than occurs in solvation, a polymer might also experience some gain in weight and dimensions. A measurement of the glass transition temperature, coupled with measurements of physical properties, provide the best measurement and means of determination for the presence of this mechanism.

2.3.2.3 Environmental-Stress-Cracking Mechanism

Environmental-stress cracking is a difficult mechanism to predict, and is not readily detectible by performing simple tests, primarily due to the infinite variety of conditions that can exist. A polymer must be subjected to stress in addition to a known stress-cracking agent, at specific concentrations and temperatures, in order for the mechanism to manifest itself. A variety of tests have been designed, however, that allow detection to be predicted.

When it does occur, the result is the initiation of cracks or crazes in the surface of a polymer. This reaction occurs when a polymer reaches a level of stress (strain) referred to as the *critical stress* (critical strain). The significance of the values for polymers as they affect design of plastic piping systems has not yet been fully established. However, the creep rupture test can be performed, with the specimen subjected to the media and stress to produce design data. Therefore, the creep rupture test, when applied with media under stress, is the preferred method of testing. (See Section 2.5.2.)

There are several factors that collectively determine the rate at which a reagent may stress crack a plastic material. These factors include: (1) the reagent; (2) its concentration; (3) the polymer; (4) temperature of exposure to the reagent; (5) the total amount of stress in the polymer specimen; (6) surface roughness; (7) time of exposure; (8) morphology of the polymer; (9) type of reaction (if any); and possibly others. The rate at which a reagent will stress crack a given polymer varies considerably, and it is not yet completely understood as to why.

Environmental-stress cracking generally occurs as a reagent, due to its chemical action on a polymer, creates a localized weakening on its surface. Tensile or compression, rotational, and other stresses lead to the localized failure, opening a crack and creating a concentration area for further penetration. The crack then behaves as an apex, where stress is concentrated, causing propagation to proceed in a rapid manner. An initiated crack continues to propagate until catastrophic failure of a material is achieved, and the level of stress reaches a much reduced level.

Crack initiation may occur due to selective absorption along local regions of a polymer, or as a result of selective solvation in localized areas of a polymer, or by complexing along active sites of a polymer. Selective absorption causes plasticization to occur in amorphous regions of a polymer. When a reaction selectively solvates a polymer along regions, it leaves a weak point in its surface, resulting in a crack. Complexing a reagent along active sites of a polymer chain leads to an alteration of local crystallinity, coupled with a reduction in strength at those sites. Whatever the reason for the first crack to appear, propagation of cracks always proceeds in the same fashion. This reason gives validity to the theory of a crack acting as a stress concentration area that tends to amplify the magnitude of stress.

Another mechanism for stress cracking occurs when a polymer becomes chemically altered through selective reaction. When this occurs, a material experiences a change in local crystal size, and a subsequent irregular density pattern in its morphology. This results in tensile and compressive stresses, with shear occurring, thereby opening a crack. Once a crack occurs, propagation occurs in the same manner as described earlier. A crack acts as a stress concentrator, lowering the activation energy needed for reaction to take place. Catastrophic failure is again the end result.

A typical example of this mechanism occurs when free-radical (nascent) chlorine is exposed to non-fully-fluorinated fluoropolymers such as polyvinylidene fluoride (PVDF), when it (a PVDF part) is under stress. The result is chlorination of PVDF, resulting in a more dense chloro-fluoro-polymer in random locations. Stress cracking can result in premature failure in some cases.

Figure 2.21 shows an example of a copolymer polypropylene (propylene/butylene) component that has failed as a result of environmental-stress cracking in a plating solution containing copper ions at 130°F (55°C) maximum temperature. Copper ions are a notorious stress-cracking agent for polypropylene materials, except for polypropylene materials of the copper-stabilized variety. The figure is a view inside a reducer-tee section. Notice that the stress cracks only appear in internal heat-element butt-weld beads. This is due to the fact that heat-element butt welding produces large residual stresses in thermoplastics.[6,7] In this instance, the material in the weld bead was under a level of stress above the threshold level

Materials of Construction 55

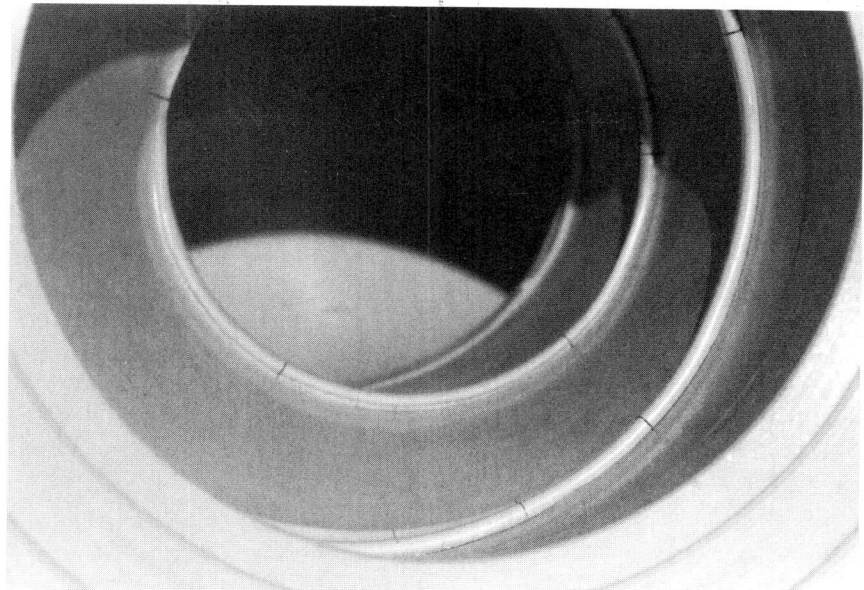

Figure 2.21. A copolymer polypropylene component (propylene-butylene copolymer) that has failed as the result of environmental stress cracking. The cracks are visible along the butt-welded joints.

of stress necessary for stress cracking to occur; the remainder of the material was under this threshold level. When the part failed, the failure manifested itself in the form of a fine spraying of fluid through a crack that propagated directly to the outside of the weld; catastrophic failure did not result, since copolymer polypropylene is a highly ductile material at this temperature, and the spraying was noticed early.

2.4 Nonmetallic Materials Used in Piping and Tanks
2.4.1 Thermoplastics
2.4.1.1 Vinyl-Based Resins

There are three basic vinyl-based chlorinated polymers that are used in the fluid-handling industry. These are polyvinyl chloride (made from the suspension process, referred to as PVC), postchlorinated PVC (CPVC), and vinylidene chloride polymer. Of these three, polyvinyl chloride has found widespread acceptance in the building and construction market. Over three billion pounds per year of suspension-grade resin produced in the United States each year is used for PVC pipe production.[8]

Polyvinyl chloride, like most thermoplastic resins, is completely inert to normal atmospheric conditions, with the exception of ultraviolet light exposure, by which it may be degraded. It can be used from 0 to 140°F (-18 to 60°C), although below 40°F (4°C) its impact properties are lowered significantly. PVC, like all polymers, is completely immune to soil corrosion and atmospheric corrosion of all types, in the conventional electrochemical sense. However, it is potentially affected by soils that contain a trace, or high amount, of certain organic chemicals. PVC shows good general chemical resistance, but it can be attacked by strong acids, certain chlorinated hydrocarbons, aliphatic hydrocarbons, aromatic hydrocarbons, alcohols, and other compounds.

Postchlorinated polyvinyl chloride (CPVC) is produced by a postchlorination process, where more chlorine is added to the base PVC molecule. The effect of this process is to raise the glass-transition temperature of the polymer significantly. As a result, CPVC has a maximum working temperature that is 80°F (45°C) higher than PVC, making its maximum rating 210°F (99°C). The chemical resistance of CPVC is almost identical to that of ordinary PVC. However, its flammability and resistance to combustion are greatly improved.

Vinylidene chloride is used solely as a lining material, and is marketed under the trade name "Saran" (Dow Chemical Company). It can be used to a temperature of 175°F (80°C) and has very good general chemical resistance. It has excellent resistance to halogenated acids at a variety of concentrations and temperatures. It is also resistant to sulfuric acid at low concentrations, but is not resistant to concentrated sulfuric acid.

2.4.1.2 Polyolefins

Polyolefins are a group of polymers that include polyethylene, polypropylene, and polybutylene. All three are polymers of basic olefin monomers: (1) ethylene ($CH_2=CH_2$); (2) propylene ($CH_3CH_2=CH_2$); and (3) butylene ($CH_3CH_2CH_2=CH_2$), respectively. All three are highly crystalline, lightweight thermoplastics that have good chemical resistance, toughness, dielectric properties, water vapor impermeability, and relatively high glass-transition temperatures.

2.4.1.2.1 Polyethylene

High-density polyethylene (HDPE) and high-molecular-weight high-density polyethylene (HMW-HDPE) have found widespread acceptance in industrial, gas, and petroleum handling services on a worldwide basis. Approximately 80 percent of all natural gas distribution piping is now installed using high-density polyethylene piping. Pressure pipe applications for HMW-HDPE include a variety of industrial chemical and mining uses. Additional uses include sewer lines, oil patch piping, domestic water uses, chemical sewer piping, landfill piping, and others. Polyethylene has excellent resistance to a wide range of chemical reagents, and, depending on its grade, can be used up to 180°F (82°C). It has excellent property retention at low temperatures and is rated to a minimum of -30°F (-34°C) according to the 1993 Edition of the ANSI/ASME B31.3 Code for Chemical Plant and Oil Refinery Piping.

High-density and HMW high-density both have a greater resistance to environmental-stress cracking than other grades of polyethylene. Polyethylene is susceptible to attack by ultraviolet light, but there are many resins available with additives to protect against uv attack. Polyethylene performs very well in the handling of abrasive slurries, due to its ability to resist erosion corrosion.

Ultra-high-molecular-weight polyethylene (UHMWPE) is available with chains 10 to 20 times longer than high-density grades. While this adds to the product's impact and abrasion resistance, it also makes the product more difficult to process. UHMWPE has 10 times the abrasion resistance of carbon steel.

2.4.1.2.2 Polypropylene

Polypropylene has a low specific gravity and good resistance to chemicals and fatigue. Polypropylene has one of the widest ranges of pH capability of nonfluorinated polymers. Therefore, it has become recognized as a material of choice in acid-waste piping systems for laboratory and research facilities. Polypropylene is normally compounded with fillers, additives, or copolymerized with ethylene or butylene to form alloys with increased impact resistance. There are homopolymer grades available without any additives for service in high-purity applications.

Polypropylene is suitable for service from 30 to 210°F (-1 to 99°C). Random copolymers with butylene can withstand occasional temperatures of up to 280°F (138°C). As mentioned, polypropylene has widespread resistance to many acids and bases, although strong acids and oxidizing acids at elevated temperatures can attack polypropylene. Polypropylene is also not suitable in most services with chlorinated hydrocarbons and oxidizing compounds and is susceptible to stress cracking by copper and other metallic ions. Polypropylene has been known to be catastrophically affected by contact with pure xylene. It is, however, one of the best materials for handling sodium hydroxide at all concentrations up to its maximum working temperature.

2.4.1.2.3 Polybutylene

Polybutylene is a relatively semicrystalline thermoplastic resin compared to the other polyolefins. Polybutylene is rated from 0°F (-18°C), up to a maximum temperature of 210°F (99°C). Polybutylene shows a good retention of properties throughout its range. It is generally resistant to acids, bases, organic chemicals, paraffinic and naphthenic oils, detergents, and various inorganic chemicals. There is a noticeable decrease in its chemical resistance at elevated temperatures. Polybutylene has high resistance to environmental-stress cracking, tensile creep, and abrasion. The largest present market for polybutylene is in plumbing and fire-protection sprinkler distribution piping markets. In plumbing applications, it has found widespread use in residential and commercial plumbing and hydronic heating applications. The butylene monomer is readily copolymerized with propylene, and when added to propylene it improves polypropylene's heat-deflection temperature, elongation characteristics, and its resistance to environmental-stress cracking.

2.4.1.3 Fluoropolymers

Fluoropolymers are a general group of polymers that include both fully fluorinated polymers and partially fluorinated resins. Of the fully fluorinated polymers, there are three basic types: (1) polytetrafluoroethylene (PTFE); (2) fluorinated ethylene/propylene polymer (FEP); and (3) perfluoroalkoxy alkane polymer (PFA). The partially fluorinated polymers also include three choices for industrial usage: (1) polyvinylidene fluoride (PVDF); (2) polyethylene/chlorotrifluoroethylene copolymer (ECTFE); and (3) polyethylene/tetrafluoroethylene copolymer (ETFE). All six exhibit similar properties in that they are crystalline thermoplastics with a high degree of widespread chemical resistance. In general, the chemical resistance of fluoropolymers tends to increase with the relative percentage of H-F bonds that make up the molecule. Thus, the fully fluorinated polymers have the better overall chemical resistance. Mechanical strength tends to be inversely related to the same ratio.

2.4.1.3.1 Polyvinylidene Fluoride

Polyvinylidene fluoride has been used for industrial purposes since the early 1960s. It is a high-molecular-weight, crystalline material that is capable of being used from 0 to 280°F (-18 to 138°C) continuous operating temperature. It has the best mechanical properties of all the fluoropolymers, including rigidity, hoop strength, creep resistance, and wear resistance. It has excellent resistance to most organic and inorganic chemicals and solvents, and is particularly resistant to strong oxidizers and halogens such as liquid bromine and bromine salt solutions. It is very resistant to strong acids, but it is affected by most solutions with a pH greater than 11, and is readily subject to stress cracking with concentrated sodium hydroxide and ammonium hydroxide solutions. It is also subject to chlorination by free-radical chlorine with resultant stress cracking. It is not recommended for sodium hypochlorite if natural, unpigmented resin is used. This is due to ultraviolet light activation of nascent chlorine molecules, freed from sodium hypochlorite molecules in solution. Other forms of chlorinated solutions also can create problems if free-radical (nascent) chlorine exists in trace amounts, or if sunlight passing through the walls of natural, unpigmented PVDF grade resins dissociates chlorine molecules. It has received widespread acceptance in ultrapure water handling due to its capability of being produced and processed with a high degree of purity and inertness.

2.4.1.3.2 Ethylene-Chlorotrifluoroethylene Copolymer

Ethylene/chlorotrifluoroethylene copolymer is a predominantly 1:1 copolymer of ethylene and chlorotrifluoroethylene monomers, resulting in high strength and temperature ratings. Its mechanical strength is high for a fluoropolymer, although it is somewhat lacking in hoop strength and stiffness compared to PVDF. It does exhibit much of the same resistance to chemicals that PVDF has, and fares even better then PVDF with respect to strong bases and chlorinated compounds. It is an excellent choice of material for hot, concentrated bases. It may be subject to further chlorination and stress cracking from free-radical chlorine if it is under a high degree of stress, although the level of stress required is much greater than that required for PVDF. Unlike PVDF, ECTFE does retain much of its impact strength at very low temperatures, with useful properties down to cryogenic temperatures, although it is not intended for piping service at temperatures below -40°F (-40°C). It has an upper working temperature of 300°F (149°C). ECTFE has become increasingly popular as a tank and vessel lining material for services involving highly corrosive fluids.

2.4.1.3.3 Ethylene-Tetrafluoroethylene Copolymer

Of all the non-fully-fluorinated fluoropolymers, ethylene/tetrafluoroethylene copolymer exhibits the best overall chemical resistance, approaching that of a fully fluorinated polymer. ETFE is a predominantly 1:1 alternating copolymer of ethylene and tetrafluoroethylene monomers. ETFE has a temperature rating that is higher than both PVDF and ECTFE, but it lacks some of the mechanical strength that PVDF and ECTFE has. ETFE has very high impact resistance and has useful properties down to the cryogenic region. ETFE is commonly used in molded parts and as a lining material for pumps, valves, and other chemical process equipment.

2.4.1.3.4 Polytetrafluoroethylene

PTFE is a fully fluorinated crystalline polymer that has excellent chemical resistance properties. It is useful from cryogenic temperatures to 500°F (260°C), and is virtually inert to all chemicals throughout its working temperature range. Its impact resistance is high, but it has relatively low tensile strength, creep resistance, and resistance to wear. It is used in the chemical process industry primarily as a lining material for carbon steel piping, pipe fittings, valves, pumps, and for other vessels and fluid handling equipment. It has a coefficient of friction lower than most materials, and it has a very high limiting oxygen index.

2.4.1.3.5 Fluorinated Ethylene/Propylene Polymer

FEP is a copolymer of tetrafluoroethylene and hexafluoropropylene. It is also a fully fluorinated polymer, and thus shares the excellent resistance to chemical attack that PTFE does. FEP has low frictional properties and is useful from cryogenic temperatures up to 392°F (200°C). As a lining material for piping, it is rated to a maximum of between 300 and 400°F (149 to 204°C), depending on the source consulted. It is a soft plastic, with the lowest tensile strength, wear resistance, and creep resistance of all the fluoropolymers. Its application in fluid handling situations is strictly limited to use as a lining material for piping and other types of chemical process equipment.

2.4.1.3.6 Perfluoroalkoxy Alkane Polymer

PFA is a fully fluorinated polymer that has the highest mechanical strength of the three basic types. It has an upper working temperature of 450 to 500°F (232 to 260°C), and shares all of the chemical inertness that PTFE possesses. PFA can be produced as a highly pure resin, making it ideal for use in high-purity water at temperatures approaching the boiling point. PFA is used as a lining material, but it is the only fully fluorinated polymer that is produced as rigid standalone piping and tubing material.

2.4.1.4 Ketone-Based Polymers

Ketone-based resins are a relatively new class of polymers. They are thermally stable products that maintain their mechanical properties at their upper temperature limits. The most commercially useful product thus far is poly(aryl)etheretherketone, otherwise known as PEEK. Other types of polyketones include polyetherketone (PEK) and the isomer of PEEK, polyetherketoneketone (PEKK). PEEK is a highly crystalline polymer with excellent mechanical properties, including flexural, impact, tensile, and fatigue properties. It has an upper working temperature of approximately 480°F (249°C). It is chemically resistant to most organic solvents, and has very low water absorption properties. It is also very resistant to nuclear radiation due to its molecular stability. PEEK also is made to exceptionally high-purity standards and tends not to leach any products into ultrapure water, up to the boiling point of water. Since there is no fluorine content in the structure of PEEK, there is negligible fluorine leached in contact with hot deionized water, unlike that experienced with most fluoropolymers. Polyetheretherketone may be joined by heat-element thermal butt welding, but it must be done in a moisture-free atmosphere, making fabrication of the material very difficult.

2.4.1.5 Other Thermoplastics

There are other thermoplastic materials used for fluid handling, but not many that are used in significant quantities for industrial services. Perhaps the most significant product is acrylonitrile/butadiene/styrene copolymer (ABS), which is used extensively in plumbing service. There is a high-ductile grade of ABS (ASTM 55535) that is available for industrial compressed air services and other industrial chemical services. Two possible piping materials of the future include such phenyl-based resins as polyphenylene sulfide (PPS) and polyphenylene oxide (PPO). Both of these products have excellent chemical resistance, high temperature ratings, and unusually high resistance to combustion for a plastic.

2.4.2 Reinforced Thermosetting Resin Plastics (RTRP)

A thermosetting plastic material is one that is polymerized (cured) by application of heat or chemical means and is thus changed into a substantially infusible or insoluble final product. Unlike thermoplastic products, they cannot be readily reshaped by heating them to their melting range once they are cured. The products will normally degrade once they are heated to such temperatures. Among the thermoset plastic materials are epoxies, polyesters, furans, and phenolic resins. The products manufactured from these resins for fluid-handling purposes are always provided with additional reinforcement of some type to form a composite matrix. The most common type of reinforcement material is E glass, but occasionally other materials such as graphite, carbon fibers, and ceramic fibers have been used. For applications involving a highly corrosive reagent, the inner wall is often a lining of 100% resin to act as a corrosion-resistant barrier. In some cases the lining may even be a thermoplastic material; when an inner corrosion liner is a thick thermoplastic substrate (e.g., schedule 40 pipe, 1/4 inch or 6 mm sheet), the fabricated product is referred to as a dual-laminate material.

2.4.2.1 Reinforced Epoxy Resins

The term *epoxy* usually refers to a variety of cross-linking materials that contain the oxirane (epoxy) group. The oxirane group may be cured by a wide range of curing agents or hardeners. Standard epoxy resins are based on epichlorohydrin and bisphenol-A as starting materials. There are also some more advanced types based on aromatic amines, aminophenols, phenols, and formaldehyde. To cure an epoxy at room temperature, an aliphatic polyamine or polyamide is normally used as the curing agent. Curing at elevated temperatures can be accomplished with one of the following: (1) anhydrides; (2) carboxylic acids; (3) phenol novolac resins; (4) aromatic amines or melamine; (5) urea; and (6) phenol/formaldehyde condensates. Cured epoxy resins are amorphous and have properties that vary with temperature. Due to epoxy resin systems being of the thermosetting type, they are creep resistant. Piping made from epoxy materials have an upper working temperature of 250°F (121°C). Epoxies are widely used for oil and gasoline service. However, except for specially formulated grades, they are susceptible to attack from gasohol and other synthetic fuels. In general, epoxies have good chemical resistance throughout their temperature range.

2.4.2.2 Reinforced Polyester and Vinyl Ester Resins

Polyesters are a general class that encompass a number of subgroups including: (1) general-purpose polyester resins (not used in chemical process equipment); (2) isophthalic polyester resins; (3) high-performance polyester resins; and (4) vinyl esters. Polyesters are very versatile in their range of properties and in their potential applications. An unsaturated polyester resin normally consists of an unsaturated polyester dissolved in a cross-linking monomer that contains an inhibitor, which prevents cross-linking until it reaches a desired level. Unsaturated polyester is derived as a condensation product of an unsaturated dibasic acid and a glycol. The extent of unsaturation can be controlled by including a certain amount of a saturated dibasic acid as well. This can be phthalic anhydride, isophthalic acid, or adipic acid. The glycol can be propylene, ethylene, diethylene, dipropylene, neopentyl glycol, or any combination thereof.

Styrene, vinyl toluene, methyl methacrylate, di-methyl styrene, and diallyl phthalate are all monomers that may be used for cross-linking purposes. Some of the typical inhibitors include hydroquinone, parabenzoquinine, and tertiary butyl catechol. High-temperature catalysts usually consist of an organic peroxide such as peroxyesters or benzoyl peroxide. If the resin is to be fabricated at room temperature, the addition of a promoter to the catalyst is also required. Some typical combinations for room-temperature fabrication include: (1) methyl ethyl ketone peroxide and cobalt octoate (promoter) and (2) benzoyl peroxide and diethylaniline or dimethylaniline (promoter).

2.4.2.2.1 Isophthalic Polyesters

Isophthalic polyester resins have good resistance to organic solvents, particularly hydrocarbons. For this reason they are used extensively as a material for underground storage tanks and underground transfer piping. They are resistant to a variety of weak acids, are highly resistant to salts of many bases, and show resistance to solvents, including: (1) amyl alcohol; (2) ethylene glycol; (3) formaldehyde; (4) gasoline; (5) kerosene; and (6) naphtha. Isophthalic polyester resins are not resistant to: (1) acetone; (2) amyl acetate; (3) benzene; (4) carbon disulfide; (5) alkaline salt solutions based on potassium and sodium; (6) hot deionized water or distilled water; and (7) strong oxidizing acids. They generally have an upper temperature limitation of approximately 150°F (66°C).

2.4.2.2.2 High-Performance Polyester Resins

High-performance polyesters are either of the bisphenol type or they are chlorinated polyesters. They can be used to a maximum working temperature of 250°F (121°C). In addition, they have excellent resistance to acids, including strong oxidizing acids. The chlorinated polyesters are resistant to saturated hydrocarbons, but not to all unsaturated hydrocarbons. They also are readily attacked by solvents such as carbon disulfide. When used in a dry condition, they can be used up to 300°F (149°C). The high-chemical-resistance polyesters are very resistant to wet and dry chlorine gas, brine solutions with chlorine, sodium hypochlorite, trichloracetic acid, hypochlorous acid, chromic acid plating solution, and certain nitrate solutions including lead, nickel, and zinc.

2.4.2.2.3 Vinyl Ester Resins

Vinyl esters are a special type of polyester that offer high temperature and chemical resistance in comparison to the other types. Bisphenols actually have a better resistance overall except for chlorine-related compounds. Vinyl ester resins have good resistance to acids and bases up to their temperature limit. They are highly resistant to strong oxidizing solutions and surpass other polyesters and epoxy materials for chlorine, chlorine dioxide, chlorine-caustic solutions, calcium and sodium hypochlorite, and bleach solutions. For applications that are for bleach service, a benzoyl peroxide dimethylaniline catalyst promoter is usually used. The physical properties of vinyl esters are very similar to polyesters. Another similarity lies in the carbon double bond that occurs in both. These unreactive double bonds may be attacked by oxidizers, or may be subject to halogenation. In a typical vinyl ester molecule, double bonds are located away from carbon atoms, thereby making its structure more inert.

2.4.2.3 Reinforced Furan Resins

Furan is a product that has excellent solvent resistance and a temperature rating to 400°F (204°C). Their resistance to solvents include the following: (1) acetone; (2) benzene; (3) carbon tetrachloride; (4) chloroform; (5) carbon disulfide; (6) ethyl alcohol; (7) fatty acids; (8) methyl ethyl ketone; (9) toluene; and (10) xylene. Furan is fabricated with highly specialized methods. This is because setting up depends on a condensation reaction rather than the rapid type of polymerization that occurs with polyesters and epoxies. If fabricated using inadequate techniques, fabrication can result in having severe internal stress in finished structures. The ultimate result of excess stress buildup is a catastrophic failure of a fabrication. Furan has very good fire-retardant qualities, with decent numbers for smoke evolution as well. Furan is not as resistant to chlorine and chlorine-related compounds as polyesters are. A very good alternative use of furan resins involves using a dual-laminate structure with furan as an inner liner and a polyester overlay. This allows for the chemical resistance of furan to be used for contacting the inner fluid, while at the same time having the mechanical strength of a glass-reinforced polyester.

2.4.2.4 Reinforced-Phenolic Resins

Phenolic resins are heat-cured thermosetting plastics produced through the reaction of phenol and formaldehyde. A phenolic's cross-linked structure provides it with heat resistance, dimensional stability, creep resistance, and other good properties. They have good resistance to flame and smoke spread. Phenolic resins do not have good impact resistance, and as a result do not find much usage in the chemical industry.

2.4.3 Borosilicate Glass

Borosilicate glass has excellent chemical resistance overall, and is recommended for service up to 450°F (232°C). It is subject to attack by hydrofluoric acid and strong bases of all types. Glass is very brittle and can be damaged by thermal shock. For these reasons, it is usually limited to those applications where it is desirable to have transparency. Glass linings are resistant to all concentrations of hydrochloric acid up to 300°F (149°C), all concentrations of nitric acid up to their boiling point, and concentrated sulfuric acid up to

450°F (232°C). Acid-resistant glass with improved alkali resistance (up to a pH of 12) is available, as well as thermal-shock-resistant glass steel.

2.4.4 Concrete

Concrete is a very important construction material for use in applications requiring secondary containment. In underground secondary containment, concrete may be used to construct secondary containment trenches, manways, manholes, chambers, sumps, pits, and a variety of other specialized housings. In above-ground applications, it may be used as a means of constructing secondary containment dikes, berms, chambers, channels, open trenches, and other specialized housings. Chapters 11, 14, and 16 discuss the application of concrete structures in secondary containment in more detail.

The makeup of concrete involves an aggregate mixture of inert reinforcing particles in an amorphous mix of hardened cement paste. One of the most common forms of concrete, that which is made of portland cement, has only limited chemical resistance to acids and bases and organic chemicals. It may also fail in a mechanical manner due to absorption of crystal-forming solutions such as brines. Concrete is typically highly porous, although there is a wide variety of porosities available, including recently introduced "zero-porosity" concretes. It typically becomes alkylated by direct ground-water exposure, resulting in degradation over time. For these reasons and others, concrete is required to have some means of coating or lining material applied to it when used in immersion applications, and in applications where it serves as a means of secondary containment, according to most authorities having local jurisdiction. Chapter 16 describes various lining and coating systems that are available to protect concrete.

Some concretes possess enhanced characteristics in terms of corrosion-resistant cements. For instance, concrete that is based on calcium aluminate cement can withstand a wide variety of chemicals. Also available is a polymer concrete that uses an epoxy, polyester, or vinyl ester thermosetting resin in approximately 90% aggregate mix to form a corrosion-resistant structural material. Even when concrete is constructed of corrosion-resistant materials, it may still be subjected to cracking and splitting due to ground movements and settling. It is typically best to add a further means of coating or lining to add protection.

2.4.5 Masonry

Brick construction can be used for many severely corrosive conditions under which high alloys would fail. Common brick materials are made from carbon, red shale, or acid-proof refractory materials. Red-brick shale is not used above 350°F (177°C) because of spalling. Acid-proof refractories can be used up to 1600°F (870°C).

A number of cement materials are used with brick. Standard are phenolic and furan resins, polyesters, sulfur, silicate, and epoxy-based materials. Carbon-filled polyesters and furans are good against nonoxidizing acids, salts, and solvents. Silica-filled materials should not be used against hydrofluoric or fluorosilic acids. Sulfur-based cements are limited to 200°F (93°C), while plastic resins can be used up to about 350°F (177°C). The sodium silicate-based cements are good against acids to 750°F (399°C).

Differential thermal expansion of the brick, its joints, and a vessel substrate necessitates an intermediate lining of lead, asphalt, rubber, or plastic. This membrane functions as a barrier to protect the substrate from corrosion damage. A special prestressed brick design that maintains the brick in compression by using a controlled-expansion resinous mortar and brick bedding material precludes the use of an elastomeric membrane.

2.5. Corrosion Testing Methods

2.5.1 Metals

There are a number of techniques that provide reliable corrosion data for the measurement of metallic corrosion. Many are simple tests of weight change, or are observational in nature, and can be used for nonmetallic materials as well. There are techniques that are specific to metal. These include hydrogen diffusion, measurements of electrical resistance, or other electrical properties. Whenever it is possible to install a small section of material into an existing system, it should be done instead of laboratory testing. If this type of test can be designed and carried out correctly, then the results will usually give a fairly accurate idea of what is to be expected of a material. In certain situations, this is not always possible due to economic or time constraints. If a test is planned far enough in advance, it can save on maintenance expenses, more than recovering its cost over a project's lifetime.

2.5.1.1 Preliminary Testing

Prior to the design of any new chemical plant, laboratory corrosion tests must be performed for those services where no known corrosion data exist. By testing various materials under laboratory conditions, those products that will definitely not work can be ruled out immediately. Field conditions should be duplicated as nearly as possible to the actual conditions in order to screen possible choices thoroughly. Tests should include stressed samples if there will be stresses present in actual conditions. Many times, the corrosive effects of a condensate of a solution's vapors will be more corrosive than actual fluid conditions. Therefore, provisions should be made to expose the samples to air and care should be taken to regulate the temperature of the samples. The National Association of Corrosion Engineers (NACE) has established standards for the testing of metals for corrosion.

In the performance of a laboratory test, time of exposure has a great effect on results. There are instances where a material is attacked rapidly at first, followed by a period of slower attack. The reverse can also occur, involving a slow attack until a metal's protective oxide film is worn away, followed by rapid attack. While there is no definitive way to determine the length of time that a product should be exposed to a reagent, the possibilities of a changing effect should be kept in mind. Some of the important factors to be recorded during the course of the test include: (1) reagent concentration; (2) test solution volume; (3) temperature; (4) time of test; (5) type of apparatus; (6) types of samples; (7) cleaning technique used; (8) weight; (9) type of corrosive effect; and (10) rate of corrosion per sample.

Some of the measuring techniques that can be used to determine changes in dimension include: (1) ultrasonic measurement; (2) measurement by eddy current techniques; and (3) microscopic examination. The most common dimensional change, change in specimen weight, can be measured readily by weighing samples or by measuring fluid level after immersion in a bath. However, a change in the physical size of a sample can also be detected by the other means stated.

There are other specialized tests that can be performed that include measurement of hydrogen diffusion, electrical resistance, and other electrical techniques. Hydrogen diffusion can be detected by using pressure measurements, or by monitoring hydrogen using electrochemical techniques. This is normally performed by attaching a probe that measures the amount of diffusion across a palladium foil and the subsequent oxidation of hydrogen by measuring the current needed to oxidize hydrogen. Other electrochemical techniques that are useful include the use of polarization curves, linear polarization, and the use of a zero-resistance ammeter.

2.5.2 Corrosion Testing for Plastic Materials

There are a variety of laboratory tests that can be performed for thermoplastic and reinforced thermosetting resin-plastic materials. The same tests that are used for preliminary testing of metals can be used for plastics as well. These tests involve simple exposure and measurement of changes in one or more physical properties. Samples must always be inspected visually as to the type and extent of effect that a chemical has had. These tests are useful primarily as a means of prescreening materials to eliminate those that definitely will not work. Simple immersion is generally not useful for detecting environmental-stress cracking. For stress cracking to occur, specimens must be subjected to a reagent while under load.

The most commonly used chemical resistance tests that are used for plastics include the procedure outlined in ASTM D-543, "Resistance of Plastics to Chemical Reagents." Another common test for reinforced thermosetting resins is ASTM C-581, "Chemical Resistance of Thermosetting Resins Used in Glass Fiber Reinforced Structures."

There are several stress-cracking tests for polymers that have been developed, including: ASTM D-1693 (Environmental-Stress Cracking of Ethylene Plastics), and adaptations of the tensile creep test procedure, ASTM D-2990, to environmental-stress cracking.

The oldest stress-cracking test developed for a polymer material is ASTM D-1693. In this test, a specimen with a thickness in the range of 0.070 to 0.120 inches is bent into a U shape. At the point of highest stress concentration at the bend, a slit of undefined length and width is made using a razor blade. The bent assembly is then immersed in a reagent

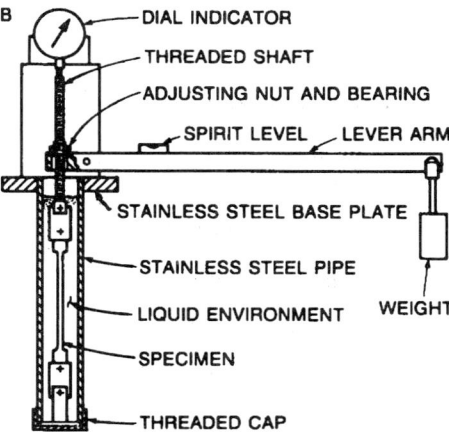

Figure 2.22. Method for measuring environmental stress-cracking resistance of plastics at room temperature. (Source: ASTM D 2990.)

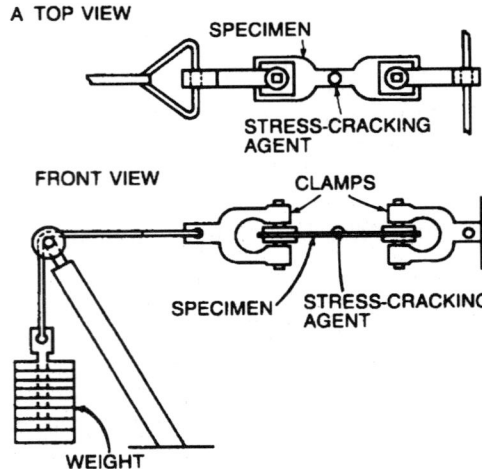

Figure 2.23. Method for measuring environmental stress-cracking resistance of plastic materials at elevated temperatures. (Source: ASTM D 2990.)

bath. If any further cracks appear, then the time for them to appear is taken as an indication of the rate of stress cracking. The slit depth, temperature, and thickness of the specimen can be varied. This test can be used for a material such as polyethylene with a low enough modulus of elasticity to allow a material to be bent in half. This test tends to produce a widespread distribution of failure times. Therefore, it is not used very often. Instead, tests based on tensile creep rupture are much more common.

Figure 2.22 illustrates a method for measuring environmental-stress-crack resistance at room temperature by measuring tensile stress while in contact with a reagent. Figure 2.23 illustrates a more versatile test setup that is more useful for measuring stress cracking at elevated temperatures. It also permits measurement of strain and modulus during

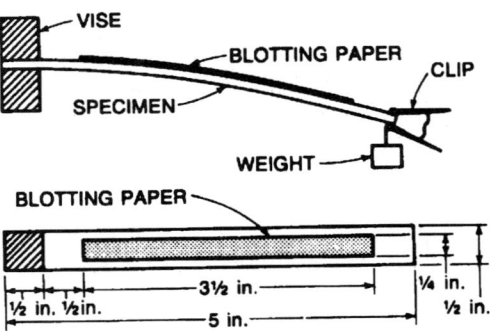

Figure 2.24 Cantilever-type test for measuring ESCR of plastic materials. (Source: "Selecting Plastics for Chemical Resistance," Modern Plastics Encyclopedia, McGraw-Hill, 1990 Edition, p. 432.)

the test. A simple dial gauge or electronic transducer indicates rupture and measures creep strain.

A cantilever-type test can be used, as illustrated in Figure 2.24. A blotting paper strip is used that runs the entire length of the sample and is saturated with a test reagent. A beam is then bent by applying a weighted device to an end. Stress in the bar varies from zero at the free end to its maximum level at the clamped end. A stress-cracking agent will produce cracks up to the point where stress is below the critical level. Each plastic material has its own characteristics, which can be calculated according to:

$$S_c = \frac{6Fl}{bt^2} \qquad (2.1)$$

where:
S_c = critical stress, psi
F = weight, lb
l = distance along the bar between the free end and stress-crack closest to it, in
b = width of bar, in
d = thickness of bar, in

A weight must be used that produces cracking near the midpoint of the bar. Determination of this weight may require some trial and error. Another alternative for screening large numbers of reagents is to hang a weight that simulates the maximum stress level of the system at approximately the midpoint of the bar. Materials that show any amount of stress cracking can be ruled out for service, or appropriate safeguards taken to reduce stress in the proposed system.

Another way to accomplish the same effect is to place the specimen under stress by bending it. A common approach to this procedure is to use a specimen in the shape of an ellipse, as shown in Figure 2.25. Specimens can be clamped at each end of an ellipsoidal surface, forcing it to conform and bend to the surface. The specimen is then wetted by the reagent by laying down a wetted strip as described in the cantilever test. Strain in the speci-

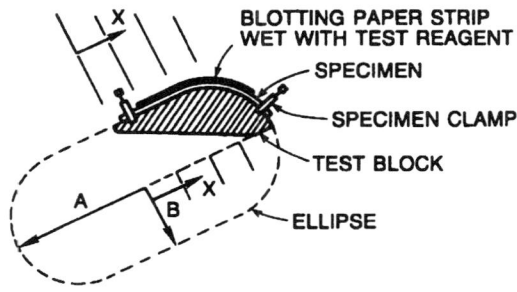

Figure 2.25. Ellipsoidal specimen bending test for plastic materials. (Source: "Selecting Plastics for Chemical Resistance," Modern Plastics Encyclopedia, McGraw-Hill, 1990 Edition, p. 432.)

men varies with the curvature, and the critical strain at which stress cracking will initiate can be calculated by:

$$e_c = \frac{6Fl}{Ebt^2} \qquad (2.2)$$

where:
e_c = critical stress, psi
F = weight, lb
l = distance along the bar between the free end and the stress-crack closest to it, in
E = modulus of the material (usually the short-time flexural modulus, psi
b = width of bar, in
d = thickness of bar, in

Some of the limitations of this test include the fact that any one design will produce a relatively narrow range of strains. Also, it may be difficult to test very rigid plastic materials by this method. For very rigid materials, the cantilever test appears to be the most practical. To determine the critical strain at which stress cracking will result from the ellipsoidal bent specimen shown in Figure 2.25, the following formula can be used:

$$e_c = \frac{bt}{2a^2}\left[1 - x^2\left(\frac{1}{a^2} - \frac{b}{a^4}\right)\right]^{-\frac{3}{2}} \qquad (2.3)$$

where:
e_c = critical strain, in/in
a = semimajor axis of ellipse, in
b = semiminor axis of ellipse, in
x = distance along the major axis, in
t = sample thickness, in

2.6 Sources of Corrosion Data

The are a number of sources of corrosion data that can be consulted, many of which can be used for preliminary screening purposes. The first place to which to refer is field experience from the same location or process site if it exists, or at a similar facility in another location. When this is not possible, then information may be available from manufacturers of the raw materials under consideration. This information should be sought first, prior to consulting data on generic material grades. When this is not possible, information can then be sought from a number of different sources that provide data on generic material grades.

Of the information that is available on generic resin grades, some information is in quantitative form, and some is in qualitative form. Information tends to exist in quantitative form more so with respect to metallic substances (mostly limited to general or uniform corrosion), while information for nonmetallic materials tends to be mostly qualitative in nature. The largest collection of independent data is available through the National Association of Corrosion Engineers' "Corrosion Data Surveys."[9,10] The approach for metals is based more on a quantitative approach, as compared to the approach used in the nonmetallic section. *Modern Plastics Magazine* (McGraw-Hill) provides quantitative data for plastic materials, by manufactured resin brand, in their "Chemical Resistance Tables."[11] Other useful references exist.[12-14]

References

1. John C. Tverberg, "Choosing Proper Stainless Steel," *JAOCS*, Vol. 65, no. 11 (November 1988), p. 1736.

2. Philip A. Schweitzer, "Corrosion and Corrosion Protection Handbook," Marcel Dekker, Inc., New York, 1983, p. 9.

3. Philip A. Schweitzer, "Corrosion and Corrosion Protection Handbook," Marcel Dekker, Inc., New York, 1983, p. 10.

4. Tverberg, op. cit.

5. G. J. Danek, Jr., "The Effect of Sea-Water Velocity on the Corrosion Behavior of Metals," *Nav. Eng. J.* 78(5): 763, 1966.

6. J. R. F. Andrews and M. Bevis, "The Butt Fusion Welding of PVDF Pipe Systems; Part 1 The Butt-Fusion of PVDF Pipe Systems," *J. Mat. Sci.* **19** (1984) 645-652.

7. P. Barber and J. R. Atkinson, *J. Mater. Sci.*, 1131 (1972).

8. *Modern Plastics Encyclopedia*, 1990 Edition, McGraw-Hill, New York, NY.

9. Corrosion Data Survey, Nonmetals Section, 5th Edition, NACE, Houston, TX, (1981).

10. Corrosion Data Survey, Metals Section, 6th Edition, NACE, Houston, TX, (1985).

11. Chemical Resistance Tables, *Modern Plastics Encyclopedia,* McGraw-Hill, New York, NY, (1989).

12. Compass Corrosion Guide, La Mesa, CA, (1983).

13. Handbook of PVC Pipe, 3rd Edition, Uni-Bell Plastic Pipe Association, Dallas, TX, (1979).

14. Plastic Pipe Institute, Technical Report TR-19, Washington, D.C.

Notes

Chapter Three

System Selection

3.1 General Statements

For every application involving double containment piping, a designer is presented with numerous combinations of materials from which to select and a wide variety of layout alternatives. There may be several metallic and nonmetallic piping materials that are equally capable of handling the chemicals involved. For this reason, it is usually difficult to narrow the choice of inner and outer materials to a single combination. The decision often is made subjectively, based upon each designer, facility owner, or installer having had more experience with a certain material or joining method. Such decision makers tend to have personal preferences, developed over the course of time by working with various materials. There are a vast number of material and system configurations to consider on any given project. However, this aspect of double containment piping may be viewed in a positive way. The vast number of choices presents a designer with an opportunity to provide a solution for every application. A suitable design can be achieved in every case. The final choice of material and system configuration usually is determined by price and code criteria, and to a certain extent, by personal preference.

A complete double containment piping system is one that is engineered, designed, installed, started up, and operated successfully. To engineer a double containment piping system fully involves much more than simply procuring "pre-engineered" components. The term *pre-engineered* is a widely used marketing term that usually refers to products that are sold as prefabricated components, which are in some cases available in standard sizes and materials. This term is often misused, as a surprising number of "pre-engineered" systems available from even the largest sources have been conceptualized and sold with little to no engineering having been performed on the system components. However, double containment piping systems by their nature usually have at least some unique requirements, necessitating a custom design to some extent. Thus designers must use good judgement when procuring "pre-engineered" components as part of their system. They must always make sure that such systems be fully designed and analyzed, as described in Chapters 4 through 10 of this text.

Double containment piping systems must be designed, fabricated, inspected, and tested to applicable codes and regulations. In the United States, most chemical and petroleum piping systems are subject to the rules of the ANSI/ASME B31.3 Chemical Plant and Petroleum Refinery Piping Code, including double containment piping systems. The ANSI/ASME B31.3 Code is a subset of the ANSI/ASME B31 Pressure Piping Code, which is described in more detail in Section 1.5.6, "Piping Design Codes." The ANSI/ASME B31.3 Code references various applicable product manufacturing standards, performance standards, and material specifications that are to be used as a part of a double containment piping system. The Code covers all of the rules necessary to complete a double containment piping system. However, the ultimate responsibility of interpreting and applying the ANSI/ASME B31.3 Code, in order to arrive at a completed double containment piping system that is safe and reliable, rests with each individual owner or owner-appointed representative.

Any two combinations of materials can be effectively combined for a given double containment piping application. However, there are many aspects involved to make this possible. Materials must be properly selected for the given application. System components must be procured; it must be known early on in a project whether they are commercially available in the sizes and pressure ratings required. They must also be available in the time frame required by the project. A system must always be engineered with sufficient detail, giving consideration to all applicable factors. All regulations governed by the local jurisdiction and appropriate design code rules must be satisfied. Ultimately, whoever is to take responsibility for the design and operation of the facility must be satisfied that the design is a safe and workable solution.

3.1.1 Considerations for Material and System Selection

There are many variables that must be taken into consideration in the selection of a system. These include:

1. Authority having local jurisdiction.
2. Applicable design codes, product standards, and regulations.
3. Amount, type, and concentration of chemical or chemicals to be transported.
4. Base temperature profile of the fluid.
5. Distance to be moved.
6. Required layout of the system, including branch connections, etc.
7. Initial pressure of the system.
8. Required termination pressure of the system.
9. Gravity versus pressure flow.
10. Ambient conditions.
11. Required lifetime of the primary system.
12. Expected duration of contact of the secondary containment with the primary fluid during system upset.
13. Leak detection method chosen.
14. Required lifetime of the secondary containment piping system.
15. Future requirements of the system.
16. Others.

From a determination of these variables, a system's basic parameters can be established, choices can be identified, and an economic analysis performed to determine a single best

System Selection

system selection. Alternative means of secondary containment, beyond the use of double containment piping, should be considered in any economic analysis.

Once a basic conceptual outline for a system has been established by the parameters listed and confirmed as a viable choice by an economic analysis, specific design factors need to be identified. Some of the variables that need to be defined in order to establish the system's specific design factors, and therefore its design basis, are:

1. Material of the primary piping system.
2. Design stress (and physical properties) of the primary piping material.
3. Required wall thickness of the primary piping system.
4. Required joining method for primary piping and components.
5. Material of the secondary containment piping system.
6. Design stress (and physical properties) of the secondary piping material.
7. Required wall thickness of the secondary containment piping system.
8. Required joining method for secondary containment piping and components.
9. Required fitting types (drainage versus pressure).
10. Sequence of joining.
11. Types and locations of secondary closures.
12. Mating requirements at termination areas.
13. Venting and drainage requirements.
14. Required type of leak detection.
15. Compartment, access, and cleanout requirements.
16. Required leak detection connections.
17. Support spacing requirements for the primary piping system.
18. Thermal expansion compensation and flexibility requirements.
19. Amount, type, and location of interconnecting points.
20. Heating and cooling requirements.
21. Commercial availability of the primary and secondary containment piping and components.

If one were to carry out all of the above work, the system would still not be designed. An approximate layout drawing could be prepared; one would also have a good engineering basis, but one would not have designed the system. One would still need to engineer the system, and would further have to fabricate correctly and successfully install the system.

In double containment piping, determination of a reasonable cost estimate can only be made once there is a firm idea of the basic system requirements. Once these are known, the system selection and detailed engineering can proceed. Having decided on the type of system (gravity drainage versus pressure transfer), materials of construction can then be selected. Diameters can be initially sized using known fluid-dynamic techniques (see Sections 4.2 and 4.3), and the optimum diameters then selected based on economics decisions involving energy costs (in the case of pressure transfer, see Section 4.2.3). The types of joining methods to be used should be resolved early on so that the actual product types to be used can be better defined. Then the steps of the detailed design covered in Chapters 5, 6, and 7 can be followed to determine wall thickness requirements of components, etc.

3.2 Classification of Material Combinations

This section presents an overview of combinations of systems that may be used in typical applications involving double containment piping. Many of the combinations that are discussed have yet to be used in commercial practice. However, all the combinations discussed can be effectively combined if they are engineered with sufficient detail, and consideration is given to all the factors that apply. Commercial raw materials, fabrication technology, and joining methods exist to allow any of them to be used. There are potential applications that make the use of any of these material combinations a viable option.

There are seventeen major groupings of inner-outer combinations of materials that can be considered, including four groups that have identical (homogeneous) combinations of: (1) metallic materials (Group 1, Type 1 Material Systems); (2) reinforced thermosetting resin plastic (fiberglass) materials (Group 1, Type 2 Material Systems); (3) thermoplastic materials (Group 1, Type 3 Material Systems); and (4) other nonmetallic materials (Group 1, Type 4 Material Systems). These four categories where the inner and outer materials are composed of the same materials are classified as homogeneous material systems. For a material combination to be classified as homogeneous, the inner and outer materials must be substantially of identical specification, not just from the same material category. In other words, a system that consists of type-316L stainless steel inside type-316L stainless steel is a homogeneous combination, whereas one that is type-316L stainless inside ASTM A-53 carbon steel is not. Homogeneous material systems are discussed further in Section 3.2.1.

The second example in the previous paragraph points out there are also hybrid material systems in which both inner and outer materials are within the same class. It is typical of a hybrid-material system within the same basic family of materials, metallic materials (see Section 3.2.2.1, "Group 2, Type 1 Material Systems"). This group of hybrid materials is classified as Group 2 materials and is discussed further in Section 3.2.2. Group 2 combinations include: (5) dissimilar inner and outer metallic materials (Group 2, Type 1); (6) dissimilar inner and outer reinforced-thermosetting-plastic materials (Group 2, Type 2); (7) dissimilar inner and outer thermoplastic materials (Group 2, Type 3); and (8) dissimilar inner and outer other nonmetallic materials (Group 2, Type 4).

Different materials and different classes of materials can be further differentiated into nine separate inner and outer material hybrid material combinations within separate families, as follows: (9) metallic inside thermoset materials (Group 3, Type 1); (10) metallic inside thermoplastic materials (Group 3, Type 2); (11) metallic inside other nonmetallic materials (Group 3, Type 3); (12) thermoset inside metallic materials (Group 3, Type 4); (13) thermoset inside thermoplastic materials (Group 3, Type 5); (14) thermoset inside other nonmetallic materials (Group 3, Type 6); (15) thermoplastic inside metallic materials (Group 3, Type 7); (16) thermoplastic inside thermosetting materials (Group 3, Type 8); and (17) thermoplastic inside other nonmetallic materials (Group 3, Type 9).

Because each grouping can be further differentiated, there is an almost limitless number of possible combinations of double containment pipe systems. For this reason, the discussion in this chapter is limited to a general discussion of these 17 major groups and subgroupings, and the services for which they should be used. Size ranges that are available within any specific combination vary depending upon the commercial availability of the individual single-wall piping and components. Also, joining technologies for certain types of materials may further limit potential size availability for a given material system.

System Selection

The suitability for a specific combination must be established for every application during a system's selection process. The choice must be established based on many factors involved in the process. A final decision on materials to be used must be reviewed by a competent professional, having knowledge or experience in the science of corrosion of materials. Once a combination is selected, a system must still be designed, analyzed, procured, fabricated, installed, inspected, tested, and started up in accordance with the recommendations presented in Chapters 4 through 13 of this text. It is highly recommended that work to be performed on any double containment piping project be under the guidance of, or be reviewed by, a competent Professional Engineer who is registered in the country or jurisdiction where the work is being performed.

3.2.1 Group 1 Homogeneous Material Systems (Identical Inner and Outer Material Combination)

These systems may be used within the limitations of their size ranges and available joining technologies. There are many reasons that the same type, grade, and class (ASTM or ANSI designation) of material may be selected for both a primary piping system and its associated secondary containment piping system. This may be as a result of certain measurable physical characteristics of the materials and their reactions to specific fluid services. However, they may also be based upon factors that may not be quantified such as the ability to connect inner and outer system components (i.e., at termination/transition areas), relative ease of fabrication, maintenance and repair ability, overall simplicity, or sometimes due to restrictions by codes or standards. A homogeneous combination of materials is also frequently the combination that results in the least cost. Attention should be given in such systems to the issue of "separation" of components (see Chapter 7). In some cases, a system may be restricted to a homogeneous construction by local building codes, standards, or regulations of the local authority having jurisdiction.

3.2.1.1 Group 1, Type 1 Material Systems (Metallic Materials)

This material group is best suited for above-ground services where nonmetallic materials are restricted from use by design, building, or fire codes, or where elevated pressures and/or temperatures are involved. Metallic-metallic systems are also used for underground systems; however, they typically must be provided with coating and/or cathodic protection in applications where there is an active soil, except for applications where a corrosion-resistant alloy is used for the outer pipe. All-carbon-steel systems are common in underground services for applications such as oily water sewers, since carbon steel is a relatively low-cost material.

Many times, both metallic and nonmetallic materials may be suitable for the fluid services. In this case, an all-metallic system may be an appropriate choice if a majority of a facility's workers have had more fabrication expertise involving welding and assembly of metallic piping components, as opposed to plastic systems. However, there are many other technical reasons why an all-metallic system may be chosen. Metals typically provide a greater degree of safety, are structurally stronger in most instances, and can better withstand impact in many applications. Metallic materials are typically better suited to extremes of temperatures as well.

Table 3.1 lists some common metal materials for use in Group 1, Type 1 double containment piping systems. These include carbon steels, stainless steels, intermediate alloys,

Table 3.1 Common Metallic Materials Used in Group 1, Type 1 Systems And Group 1, Type 2 Systems

Carbon steel	Stainless steel	Ni and Ni alloys	Cu and Cu alloys
Type A53	304	200	C10200
Type A106	304L	201	C12000
	316	400	C12200
	316L	600	C14200
	317	800	90Cu-10Ni
	317L	G1	70Cu-10Ni
	347	C-4	80Cu-20Ni
	405	X	
	410	C-276	
	430	B-2	
	446	625	

nickel and nickel-based alloys, but could also include copper alloys, titanium, zirconium, tantalum, and aluminum alloys. These are general groups that contain many specific products. It must be remembered that for a system to be considered as a Group 1 material system, it must consist of substantially the same material for both inner and outer components (and interstitial components for that matter). Otherwise it may be classified as a Group 2, Type 1 all-metallic system.

The use of Group 1, Type 1 systems in underground services is occasionally avoided, due to the requirement for coating and cathodically protecting its secondary containment piping (where an active soil is present). To eliminate the need for cathodic protection, a metallic primary piping system requiring secondary containment in underground services is often fitted with a nonmetallic secondary containment jacket. The outer plastic jacket acts as a substitute for a metallic outer jacket in order to eliminate the need for coating and cathodic protection. It is for this reason that Group 3, Types 1 and 2 material combination systems are often used in underground double containment piping services (refer to Sections 3.2.3.1 and 3.2.3.2.)

The underground storage tank (UST) laws of many individual U.S. states (under Title 40, Code of Federal Regulations, Part 280; RCRA requirements) provide that single-wall metallic piping in petroleum product services, which are coated and equipped with cathodic protection and are further subjected to frequent tightness tests, are acceptable in place of a double containment piping system having interstitial monitoring. However, piping in hazardous chemical service must be provided with an acceptable means of secondary containment and be monitored for leaks. According to the U.S. Resource Conservation and Recovery Act, hazardous chemicals are referred to in Title 40, CFR, Part 281, subpart D.

3.2.1.2 Group 1, Type 2 Material Systems

Applications of double containment systems that use the same reinforced thermosetting resin plastic (RTRP) material for both primary and secondary containment piping are very common in underground and above-ground applications involving gasoline, fuel oil, jet

System Selection

fuel, organic and inorganic chemicals, acid/chemical waste, and emergency spill containment. RTRP materials are a very versatile group of materials, having characteristics that include the good, general chemical resistance of plastics and the relatively high strength of metallic materials. They are also available in a wide range of sizes and are conducive to fabrication, which is a plus for double containment piping systems.

While many of the applications feature the same primary and secondary RTRP material, they often use different joining techniques for the primary and secondary containment components. A typical example includes adhesive bonding for the inner components, and butt and wrap for outer components. This is due to the difficulty in adhesive bonding techniques for larger pipe sizes, particularly those above 8 in. in diameter, although the technique is offered up to 16 in. diameter by some manufacturers. As in large-diameter single-wall RTRP piping, butt-and-wrap techniques can be used to join large-diameter secondary containment piping instead of using adhesive bonding. Some manufacturers offer split-and-mechanically-rejoined components for applications where low cost is essential, and the requirements for a secondary containment system are minimal (< 32 psi rating; < 2.3 bar rating).

Table 3.2 lists the possible choices for major RTRP resin categories used in homogeneous combination systems. However, one must be aware that there are many variables involved in determining specific characteristics of RTRP components, including resin type, specific type of curing agent used, type of reinforcement material, and method of reinforcement, including angle of winding if a component is filament wound, among others. One must thoroughly investigate the capability of procuring commercially available RTRP components having desired performance characteristics that are specified for each project.

3.2.1.3 Group 1, Type 3 Material Systems (Thermoplastic Materials)

Thermoplastic materials offer a variety of easy-joining options for systems that consist of identical materials for primary and secondary containment systems and interstitial components. Since they are weldable, fusible, and bondable by a variety of methods, there are a wide number of choices available for installation and closure of a system for certain materials. Methods have been developed to enable simultaneous heat fusion and simultaneous solvent cementing of primary and secondary components to each other.

The various bonding (joining) techniques that are most commonly applied to thermoplastic piping components include: heat element thermal butt fusion, heat element thermal socket fusion, heat element thermal saddle fusion, heat element thermal sidewall fusion, electrofusion, hot-gas welding, extrusion welding, solvent cementing, and adhesive bonding. Mechanical joining techniques are also an option for thermoplastic components.

Table 3.2 Common Fiberglass-Reinforced Resin Types for Use in Group 1, Type 2 Systems

Epoxy
Polyester
Vinyl ester
Furan

Table 3.3 Common Thermoplastic Resins for Use in Group 1, Type 3 Systems

Type	Generic name	Abbreviation
Vinyl-based polymers	-Poly vinyl chloride	PVC
	-Post chlorinated polyvinyl chloride	CPVC
Polyolefins	-High density polyethylene	HDPE
	-Homopolymer polypropylene	PP-h
	-Copolymer polypropylene	PP-c
Fluoropolymers	-Polyvinylidene fluoride	PVDF
	-Ethylene-chlorotrifluoroethylene copolymer	ECTFE
	-Ethylene-tetrafluoroethylene copolymer[1]	ETFE
	-Perfluoro-alkoxy-alkane polymer[2]	PFA

Notes:
[1] Available as custom-produced piping, tubing, and fittings only.
[2] Commercially available up to 3 in. nominal diameters only (European equiv.=90 mm); custom produced in larger sizes.
[3] Other miscellaneous materials for theoretical use include; acrylonitrile-butadiene-styrene copolymer (ABS), polybutylene (PB), polyetherether ketone (PEEK).

Common thermoplastic resins that are used for service in Group 1, Type 3 double containment piping systems are shown in Table 3.3.

Thermoplastics are useful in underground systems due to a high degree of resistance to soil corrosion and are resistant to a wide range of chemicals. Thermoplastic piping is limited in many above-ground services due to code restrictions, flammability and smoke-related building code restrictions, temperature limitations, and lack of mechanical and structural strength. In above-ground work, a need for additional safeguarding in many fluid services translates into added costs. This can offset cost advantages that thermoplastic materials otherwise have. However, the difference is less than for single-wall piping, due to the increased rigidity of a pipe within a pipe, and thus the ability for Group 1, Type 3 systems to allow comparatively wide support spans.

3.2.1.4 Group 1, Type 4 Material Systems (Other Nonmetallic Material Systems, Including: Borosilicate Glass, Dual-Laminate, Concrete, Lined or Coated Concrete, Ceramic, and Vitrified-Clay Materials, and the Like)

Construction of primary and secondary containment pipes having identical materials to one of those listed is not common or practical for most services. The most likely system that may be encountered is a borosilicate glass primary piping system within a borosilicate glass secondary containment system. Applications for this system are limited to situations where visual observation of the inner fluid is beneficial, yet safety or heat-transfer considerations dictate that such a system should be used. (An example is in research applications involving critical fluids.) Otherwise, there is little need to use the other possibilities such as dual laminates within dual laminates or concrete within concrete. Dual laminates for inner and outer piping would usually be replaced by either a straight RTRP or thermoplastic material for either a primary or secondary containment portion of the piping system. Use of a concrete within a concrete system is impractical because of structural, construction, and economic reasons.

3.2.2 Group 2 Hybrid Material Systems (Inner and Outer Materials Consist of Different Materials from within the Same General Family of Materials)

These systems may be used within the limitations of their size ranges and available joining technologies. The most common reason that one type, grade, or class (ASTM or ANSI designation) of material is selected for a primary piping system and a different material from within the same class is selected for its associated secondary containment piping system is for reasons of economy. However, such a combination of materials may also be selected due to certain technical requirements of the secondary containment pipe that are not imposed on the inner pipe system (i.e., impact resistance during extremely cold temperatures). These types of systems are not as simple and straightforward as Type 1 systems; this tends to offset a portion of the overall cost savings provided by such a combination. Factors such as the ability to connect inner and outer system components (i.e., at termination/transition areas), relative ease of fabrication, the ease of maintenance, and the ability to repair such a system readily are affected. Designers must be aware of restrictions against a substitution of secondary containment materials by building codes, standards, or regulations of the local authority having jurisdiction.

3.2.2.1 Group 2, Type 1 Material Systems (All-Metallic Materials)

Use of combinations of metallic materials are ideally suited to situations where a metallic material is required for both the primary service and secondary containment piping due to code restrictions, corrosion properties, or mechanical properties. An all-metallic combination system can be used both above ground and underground. However, coating and/or cathodic protection may be required for directly buried secondary containment metallic pipes. Special attention must be paid to interconnecting points and crevices where galvanic corrosion effects are likely to occur (in the event of a leak into the annulus). Some combinations of metallic materials may be directly welded together, whereas others may not. Note that all-metallic hybrid materials are usually combined in terms of a more expensive material being contained within a less expensive material of construction.

In most applications, it makes sense to have a material of greater expense housed within a material of lesser expense. There are some unusual circumstances, such as applications involving external corrosion, where an expensive material for use in secondary containment piping components may result in a logical choice. An example is a double containment piping system whose exterior is subjected to a salt water atmosphere, as is often the case for piping on an offshore oil platform. A Cu-Ni material could be an appropriate choice for a secondary containment material, depending on whether it is also compatible for service with the service chemical. This may be a more expensive choice than the material of the primary system, if for instance A-53 carbon steel is the material used for the primary components.

3.2.2.2 Group 2, Type 2 Material Systems (All-RTRP Materials)

This group is limited to a few basic combinations of major resin types, most of which feature an epoxy or polyester resin as the material for secondary containment components.

Table 3.4 Typical Combinations of Thermosetting Resin Groups for Use in Group 2, Type 2 Systems (in Decreasing Order of Likely Use)[1]

Primary material	Secondary containment material	Commercially available
Polyester	Epoxy	Y
Vinyl ester	Epoxy	Y
Vinyl ester	Polyester	Y
Furan	Epoxy	N
Furan	Polyester	N
Furan	Vinyl ester	N

Notes:
[1] Some combinations listed here represent theoretical combinations only. They have yet to be tried in practice as of this writing.

Epoxy and polyester resins are typically the lowest-cost reinforced thermosetting resin materials available. Therefore, they represent economical choices when selected as materials of construction for secondary containment components, when using a dissimilar reinforced thermosetting resin for primary piping components. There are numerous proprietary resins available to be used as RTRP materials. Therefore, it is possible to obtain inner and outer piping components of varying resin formulation commercially. If this occurs, such a system could theoretically be grouped under this heading. However, they should only be grouped as a Group 2, Type 2 combination if the resins are substantially different, and typical bonding techniques could not produce a suitable bond. It is not generally advisable to combine resin grades, if it can be avoided. Table 3.4 lists combinations of major reinforced thermosetting resin groups for double containment piping applications for use in Group 2, Type 2 applications.

3.2.2.3 Group 2, Type 3, Material Systems (All Thermoplastic Materials)

Material systems (all-thermoplastic materials) arrangements of dissimilar thermoplastic resins are common for underground acid waste and emergency spill containment systems. Hybrid material thermoplastic systems are also frequently used for above-ground acid and caustic transfer lines. The main incentive for combining materials is economics, although some polyolefins such as polypropylene and polyethylene offer joining techniques for secondary containment components that simplify the ability to install and repair a system. Several unique, proprietary products and techniques have been developed for material products under this heading; most consist of polyolefin and/or vinyl-based polymers, and even some fluoropolymers. Some joining techniques have been developed for purposes of simplifying installation to some extent (e.g., simultaneous fusion). Thermoplastics are normally combined in terms of a more expensive material being contained within a less expensive one. Only where conditions of severe external corrosion exist would a more expensive material be required for a secondary containment system.

3.2.2.4 Group 2, Type 4 Material Systems (All Other Nonmetallic Materials, Including: Borosilicate Glass, Dual-Laminate, Concrete Materials, and the Like)

Glass, ceramic, concrete, lined concrete, and vitrified clay are useful secondary containment materials in a double containment piping system. Dual-laminate materials, on the other hand, are most commonly useful as a primary piping material. It would not be practical to place one of these materials inside a dissimilar material group from this category. Most of these materials are best suited for use in a double containment piping system in conjunction with an inner metallic, RTRP, or thermoplastic piping system. Glass can be used for above-ground portions of the secondary containment with almost any primary material selected (except concrete, ceramic, clay, etc.), for purposes of visual inspection of an annulus and associated primary pipe system. However, the limitations and special precautions needed for glass piping must be taken into account in designing systems having this feature.

3.2.3 Group 3, Hybrid Material Systems (Inner and Outer Materials Consist of Different Materials from within Different Families)

These systems are limited by size ranges that are available and limitations imposed by the use of substantially different joining techniques. The common reasons for combining primary and secondary containment materials of different classes is for reasons of economy, or for structural or external corrosion concerns. These types of systems are even more complex than a Group 2 hybrid; the reason has to do with the complexity of interconnections and the use of joining concepts that are totally different and highly incompatible. This tends to offset a portion of the overall cost savings provided by such a combination. However, such systems do tend to be among the most economical choices and are very common, especially Group 3, Types 1, 2, 4, and 7. Group 3, Types 1 and 2 are usually selected to eliminate any need to apply coating and cathodic protection to metallic pipes in underground service; Group 3, Types 4 and 7 are usually selected for reasons of providing a primary pipe with adequate structural safeguards. Designers must be aware of restrictions against using combinations of material class for inner and outer systems, by building codes, standards, or regulations of the local authority having jurisdiction.

3.2.3.1 Group 3, Type 1 Material Systems (Metallic inside RTRP Materials)

This combination is an effective way to lower costs where metallic piping has been selected for a primary system and an RTP material is sufficient for fluid service based upon expected duration of contact for secondary containment piping components. This combination is particularly useful for underground systems where a metallic material is suitable for primary service, and the economics of providing coating and cathodic protection for metallic secondary containment piping makes it attractive to use nonmetallic materials instead. This combination can also be used above ground, where code restrictions do not prevent the use of a nonmetal. Potential applications are in fact limited only by applicable codes or the restrictions of the authority having jurisdiction. Typical combinations involving major metal/RTRP general classifications are shown in Table 3.5.

Table 3.5 Typical Combinations of Metallic inside Reinforced Thermosetting Resins for Use in Group 3, Type 1 Systems

Primary material	Secondary Containment Materials		
	Epoxy	Polyester	Vinyl ester
Carbon steel	X	X	XX
Low & intermediate alloy steel	X	X	XX
Stainless steels	X	X	XX
Nickel & nickel alloys	X	X	X
Copper & copper alloys	X	X	XX
Titanium	X	X	X
Zirconium	XX	XX	XX
Tantallum	XX	XX	XX

Legend: X – Represents a likely combination.
XX – Represents an unlikely, unusual, or rare combination.

3.2.3.2 Group 3, Type 2 Material Systems (Metallic inside Thermoplastic Materials)

Metallic materials inside thermoplastics find their widest use in underground systems, to eliminate any need for coating and cathodic protection otherwise required for a metallic secondary containment pipe. There are many mechanical, structural, and code reasons for desiring a metallic material (i.e., pressure rating or burial loads) for a primary piping system. Use of a thermoplastic material not only eliminates a need to coat and cathodically protect metallic secondary containment piping components, it oftentimes yields easier secondary containment installations and occasionally lowers initial material costs as well. Use of this combination in above-ground systems is often limited by building and design code restrictions, temperature limitations, and flammability and combustion concerns. In above-ground installations, use of a metallic inner material enables designers to take advantage of the relative rigidity of such a combination. Thus, support spans for a combined system of this type may be at least as generous as that required for its metallic primary piping system; often, allowable support spans may be even longer. This same support span determination would otherwise yield an unusually long distance for a thermoplastic pipe material, offsetting one of thermoplastic piping's most common cost disadvantages.

A special application of this type of arrangement is where old carbon steel natural gas piping is retrofitted with high-density polyethylene (HDPE) pipe, external to the existing system, while the existing system is still operational. HDPE is fitted over such a pipe, with welds made by using split heating elements having a hole large enough for the interior piping, so that the interior piping does not come in contact with the heating element. In both the United States and Europe, this method, referred to as annular welding by some, has been used extensively in the Southern United States for low-pressure natural gas distribution pipe retrofitting (<150 psi; <10 bar). The same procedure has also been used for making secondary containment piping butt welds in certain all-thermoplastic double containment pip-

System Selection

ing systems (i.e., PVDF/PVDF, PVDF/polypropylene, PVDF/HDPE, polypropylene/polypropylene, polypropylene/HDPE, and HDPE/HDPE). Figures 3.1 and 3.2 illustrate this welding method as it is applied to thermoplastic secondary containment piping.

3.2.3.3 Group 3, Type 3 Material Systems (Metallic inside Other Nonmetallic Materials)

Metallic piping materials may be placed inside borosilicate glass materials for aboveground services to provide visual leak detection capabilities. All of the precautions needed for glass, due to its lack of ductility, need to be taken into account in such a design. In some applications, glass may be used in selective locations in order to provide limited visual leak detection.

Use of dual-laminate materials as an external piping material for secondary containment of metallic primary materials is an overkill for most applications. If a dual laminate is an acceptable material for secondary containment components, then the base material otherwise used for the dual laminate would normally be acceptable as a standalone material.

Metallic piping can be run inside concrete secondary containment piping. However, placing a metallic pipe in a concrete trench is a better arrangement. Concrete is required to be lined, coated, or protected by other means when used as a structural material for secondary containment purposes, under most local jurisdictions.

Concrete may be poured directly around a primary metallic piping system to encase a pipe and act as a means of secondary containment. However, concrete must form a shell around the external circumference of the primary pipe and have true separation between the materials. If the concrete completely encases a primary piping system, then the metallic primary piping system in effect has been provided with a means of secondary containment. However, there are both technical and practical reasons why this approach is not recom-

Figure 3.1. A polyolefin outer pipe is being heated by means of a split-hinged heater plate that contains a center hole surrounding an inner metallic pipe.

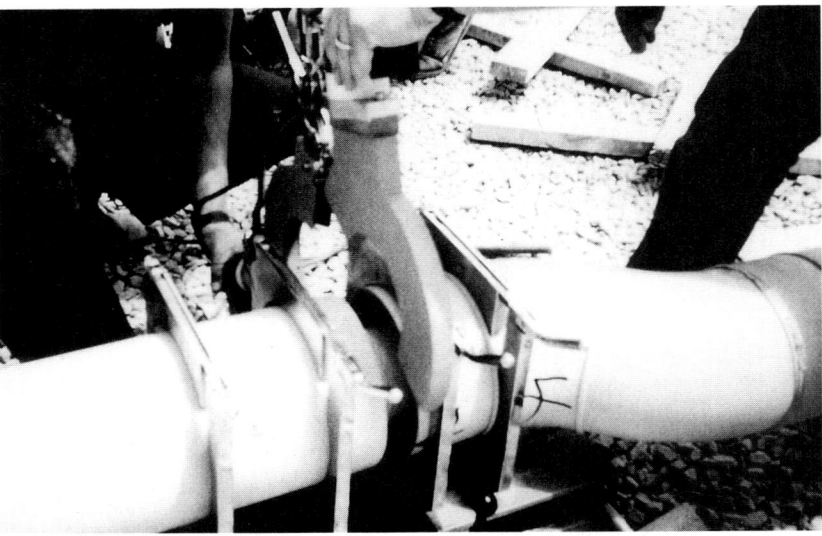

Figure 3.2. The split and hinged heater plate is shown as being removed after the outer pipe ends have been heated, and are ready to be immediately joined together to form the outer butt-fused joint.

mended. Testing the seal of the enclosed chamber that is created is very difficult. Moreover, it is very difficult to pour concrete in such a fashion that there is a guarantee of a seal around the entire pipe. There is no means to incorporate leak detection in most applications. Repair and maintenance of such an arrangement is also highly complex and difficult. It would not be considered satisfactory to meet UST laws due to a lack of interstitial leak sensing capability and due to a lack of further coating and/or lining for the bare concrete.

3.2.3.4 Group 3, Type 4 Material Systems (RTRP inside Metallic Materials)

RTRP primary piping materials contained within a surrounding metallic secondary containment system can be used for above-ground services where an RTRP material is best suited to handle the fluid service. A metallic material is provided to safeguard its associated RTRP primary system, and it allows for long horizontal spans, which helps reduce overall system cost. This arrangement can be used as a means of safeguarding a reinforced thermosetting resin material for hazardous, flammable or lethal services (i.e., category "M" services, according to the ASME B31.3 Code). In a design of this type, metallic secondary containment components must be completely capable of containing a service fluid under pressure during the event of a failure of the primary piping, for a safe period of time; it is best to select a time period conservatively to add an adequate degree of safety factor.

These combinations may also be used for underground systems where added structural integrity of metallic outer pipe is either desired or necessary. In most instances, a coated and cathodically protected carbon steel material could be used as the material of construction of the secondary containment piping, although stainless steels might be appropriate for some fluid services. Potential combinations for Group 3, Type 4 systems are only limited by available single-wall RTRP and metallic components.

3.2.3.5 Group 3, Type 5 Material Systems (RTRP inside Thermoplastic Materials)

A combination involving an RTRP material housed inside a thermoplastic can be used in some instances to reduce the cost of a system. It is particularly useful for underground services, particularly where PVC, polyethylene, or polypropylene are selected as a material for secondary containment components. These combinations can also be used for above-ground and underground large-diameter services. They represent an economical choice in these situations. Most of the cost savings result from the relative ease of joining thermoplastic secondary containment materials, particularly in large diameters, as opposed to RTRP materials. When this arrangement is used in above-ground services, significantly increased spans are allowed compared to regular single-wall thermoplastic pipe. In fact, spans will typically exceed values required for RTRP piping, if the primary piping is affixed with adequate interstitial supports.

3.2.3.6 Group 3, Type 6 Material Systems (RTRP inside Other Nonmetallic Materials)

The same discussion in Section 3.2.3.3, "Metallic inside Other Nonmetallic Materials," applies to RTRP materials that are to be used inside such nonmetallic materials as glass, concrete, dual-laminate materials, or others.

Reinforced thermosetting-resin coatings applied to the inside diameter of a previously existing concrete piping can be considered as a form of double containment if the existing concrete pipe is still intact and without leaks. This process is used extensively for upgrading old sewer and water pipes and is of limited use in containment applications. Such an arrangement lacks an inherent ability to monitor interstitial spaces for leaks. Therefore, there is no way to provide leak detection in such systems.

3.2.3.7 Group 3, Type 7 Material Systems (Thermoplastic inside Metallic Materials)

The same basic discussion in Section 3.2.3.4, "RTRP inside Metallic Materials," applies in principle to thermoplastic materials to be used inside a metallic material. However, there are special cases that deserve further discussion. One such case involves lined-steel systems that contain sophisticated means for safeguarding flanges (see Section 11.8 and Figures 11.126-127). In such arrangements, an external metal housing might be considered as secondary containment if there is a distinguishable annular space that exists between primary and secondary systems, which further allows monitoring. However, this arrangement lacks some of the design and operational flexibility of a "typical" double containment pipe assembly which has a "true" annulus. Also, for a lined metallic piping to serve as a true double containment piping assembly, any vent holes provided on the secondary containment jacket must be provided with couplings or other fluid recovery means.

The design of lined metallic piping is normally based on outer metallic materials only, as covered in the ASME B 31.3 (Chemical Plant and Petroleum Refinery Piping Code). Its thermoplastic liner normally is not designed for pressure. Also, most lined piping is designed with weep holes that would allow some amount of fluid to escape or vent to the atmosphere. Therefore, in a technical sense, this arrangement cannot be classified as double containment since it only possesses one "primary" means of containment and lacks a means of secondary containment, unless provided with the means described in the previous paragraph. Lined metallic piping systems do represent an excellent means of safeguarding metallic piping against corrosion. Their use in direct burial requires that external coatings and/or cathodic protection be provided to the pipe/fittings, flanges, bolts. A good compromise in many below-ground systems involves placing lined steel piping within lined concrete or polymer concrete trenches. For this type of arrangement, the user should consult the local authorities as to the acceptability of such practice as a means for complying with local building codes and regulations.

3.2.3.8 Group 3, Type 8 Material Systems (Thermoplastic inside RTRP Materials)

This arrangement is normally selected where support of a system is a concern, burial loads of a system are a concern, the joining methods of the RTRP secondary containment piping systems are preferred, or codes require the use of an RTRP secondary containment system. Where codes require a UL system, the local authority having jurisdiction and the insurance technical representative should be consulted as to the appropriateness and acceptability of such an arrangement.

In some dual-laminate systems having flanges that are safeguarded, they may be classified by some as a double containment piping system. In most cases, since a secondary containment material is permanently fused to a primary material that has been chemically etched, an annulus is not readily present. Also, a true annulus does not exist, making it very difficult to provide a means of leak detection. Therefore, this arrangement usually is not classified as a double containment piping system by an authority having local jurisdiction.

3.2.3.9 Group 3, Type 9 Material Systems (Thermoplastic inside Other Nonmetallic Materials)

The same discussion in Section 3.2.3.3, "Metallic inside Other Nonmetallic Materials," applies to thermoplastic materials that are to be used inside such nonmetallic materials as glass, concrete, dual-laminate materials, and the like.

Thermoplastic piping slip lined within a previously existing concrete piping (slip lining) can be considered as a form of double containment if the existing concrete pipe is still intact and without leaks. This process is used extensively for upgrading old sewer and water pipes. Examples of this slip lining procedure are shown in Chapter 16. A similar comparison may be made with methods for retrofitting old sewers using a PVC liner that is inflated inside the concrete sewer.

3.2.4 Dimensional Tables for Double Containment Pipe Combinations

Dimensional data for combinations of primary and secondary containment piping based on minimum annuli are shown in Appendix A, Tables A-1 through A-20. If leak detection cable (also referred to as continuous sensing and locating systems) is used, a relatively large annulus is typically required; the value required for an annulus that will house conductive-based (resistance-based) leak detection cable is: $AS > 0.75$ in. (19 mm); that required for time-domain-reflectometry-based systems (also referred to as TDR-based or impedance-based systems) may be as great as $AS > 1.5$ to 2.00 in. (50.80 mm), where AS is the cross-sectional annular area. (AS is an abbreviation for "Annular Space").

There are technical reasons that contribute to determining a desired annulus size for a double containment pipe assembly/system. These include issues involving annular flow,

System Selection

thermal expansion, and/or heat-transfer concerns. Installation and fabrication issues also typically impose minimum size constraints on system components/annular space requirements. Tables A-1 through A-20 of Appendix A should be used only as a basic starting point for size selection, or for reference purposes when values are needed in calculations. For each new project, a detailed analysis of the requirements for outer component size must be made. Included in the analysis should be design considerations stated in Chapters 4 through 10 and layout, installation, fabrication, and testing issues discussed in Chapters 11 through 13. Only then can an optimum size combination be selected. Optimum size will allow service fluids to be safely transported but will simultaneously have the sum of all associated costs minimized [including energy costs (for pumping), material costs, and operating expenses]. The final responsibility for selecting appropriate sizes is always that of the designing engineer of record.

3.2.5 Commercial Availability of Pipe and Components

There are many possible combinations of sizes and wall thicknesses that may be combined, based on geometry, without respect to material of construction. However, the actual commercial availability of sizes, thicknesses, and pressure ratings for each material combination varies. Moreover, the limitations of certain components and joining methods available in any given material add further restrictions to the possible number of combinations that may be used on any one project. The commercial availability of all components must be researched prior to the start of every double containment piping design, early on in the selection process. This will avoid unnecessary redesign and will minimize wasted engineering effort.

Notes

Chapter Four

Fluid Dynamics and Sizing Analysis

4.1 General Fluid Considerations

4.1.1 Fluid Types

Real fluids consist of two types, Newtonian and non-Newtonian fluids. Newtonian fluids have viscosities that are independent of rate of change of shear stress. The relative viscosity of a Newtonian fluid does not change with flow. Non-Newtonian fluids have viscosities that are dependent on the rate of change of shear stress.

Most gases and thin liquids are typical Newtonian liquids. Non-Newtonian fluids include gels, emulsions, suspensions, and gases near the critical points. Most of the fluids encountered in the design and operation of double containment piping systems are Newtonian fluids.

4.1.2 Fluid Properties

4.1.2.1 Density and Specific Gravity

Density defines the amount of fluid that occupies a certain volume. The mass density is defined as the amount of fluid mass per given unit volume. The weight density (also called specific weight) is referred to simply as the "density" in engineering practice. It is defined as the weight of fluid that occupies a certain volume. By measuring the amount of volume that a certain weight of fluid occupies, the reciprocal of the weight density can be determined, which is termed the specific volume. The weight density divided by the gravitational constant is the mass density.

The ratio of the density of a fluid to the density of pure water is called the specific gravity of a fluid. For a gas, specific gravity can be found by the ratio of the density of the gas to the density of air at standard temperature and pressure (STP). The specific gravity of a liquid or solid can be calculated, if the weight density of the fluid is known, by:

$$\text{specific gravity}_{(liquid\ or\ solid)} = \frac{\rho}{62.45} \qquad (4.1)$$

where:
ρ = weight density, lbs/ft³
62.45 lbs/ft³ = density of water at STP

The specific gravity of a gas is calculated in the same manner as that of liquids and solids, once the weight density of the gas is known. The weight density is normally determined by the ideal gas law, at pressures below the critical pressure of a gas, knowing the conditions (volume, temperature, and pressure) at which a gas exists. The specific gravity of a gas is therefore determined by:

$$specific\ gravity_{(gas)} = \frac{\rho}{0.0749} \qquad (4.2)$$

$$= \frac{P}{0.0749\,RT}$$

where:
ρ = weight density, lbs/ft³
P = absolute pressure, psia
T = absolute temperature, °Rankine
R = ideal gas law constant = 10.73
0.0749 lbs/ft³ = density of air at STP

4.1.2.2 Viscosity

The viscosity of a fluid is a coefficient that measures the resistance of a fluid to flow. It is defined as the ratio of the shear stress of a fluid to its velocity distribution. The viscosity coefficient is a highly temperature-dependent property.

Fluid viscosities are entirely independent of a system's pressure. Liquids and gases tend to behave in opposite manners with respect to their viscosities, when a change in fluid temperature occurs. The viscosity of a liquid decreases with an increase in fluid temperature, while the viscosity of a gas increases with an increase in fluid temperature.

There are two types of viscosities that are used in practice, relative viscosity and kinematic viscosity. Relative viscosity, sometimes referred to as dynamic viscosity or absolute viscosity, is equal to the coefficient of viscosity.

When the relative viscosity of a fluid is divided by its mass density, the kinematic viscosity is determined, as in:

$$v = \mu \frac{g}{\rho} \qquad (4.3)$$

where:
v = kinematic viscosity, ft²/s
μ = dynamic viscosity, centipoise
ρ = fluid density, lb/ft³

4.1.2.3 Vapor Pressure

The pressure created by free molecules at a liquid's surface, or by a fluid on its surface molecules, is referred to as the vapor pressure of the fluid. Every fluid has a distinct vapor pressure that varies according to its temperature. When the vapor pressure of a fluid is equal to the local ambient pressure, boiling will occur. Therefore, the boiling point of a fluid depends upon its temperature (vapor pressure at a given temperature) and the external pressure to which it is subjected. In many calculations involving fluid flow and piping, it is essential to know the vapor pressures and boiling points of the fluids involved. Some examples are the determination of the type of flow (liquid, gaseous, or multiphase) and in the determination of the available net positive suction head (NPSH) (see Section 4.2.1.8) for pump suction pipe sizing.

4.1.3 Fluid Statics

4.1.3.1 Pressure

The pressure of a piping system is measured by using some reference point. There are two reference points normally used in engineering calculations, zero absolute pressure (complete vacuum) and standard atmospheric pressure (the measurement of the pressure of the earth at sea level, or approximately 14.7 psia; approximately 1 bar).

When zero absolute pressure is used as the reference point, the pressures are termed absolute pressures. When measured in units of pounds per square inch, they are given the designation psia (where the letter "a" stands for absolute).

If pressure is measured above atmospheric pressure, it is called a gauge pressure. When measured in units of pounds per square inch, they are given the designation psig (where the letter "g" stands for gauge). Mechanical gauges used in installations may read gauge or absolute pressures. It should not be assumed that the pressure read on a gauge is gauge pressure. If there is confusion as to what pressure is being read, the manufacturer of the gauge should be consulted.

When pressures below atmospheric exist (vacuum conditions), there can be confusion as to the units involved. When vacuum pressures are measured in terms of gauge pressures, they will be assigned a negative value, up to a maximum value of -14.7 psig (-1.0 atmospheres gauge pressure). When vacuum pressures are measured using zero absolute pressure as the reference point, the units will be positive values with 0 psia (0 atmospheres absolute) being absolute vacuum, and just below +14.7 psia representing the other extreme (very little vacuum). However, the most commonly used units for vacuum pressure conditions are not psi or atmospheres, but rather millimeters of mercury, inches of water, and feet of water. These units are normally referenced to a zero absolute pressure reference point and thus have positive values (i.e., 7.35 psia = 14.95 mm Hg = 408 in. of H_2O = 34.0 ft of H_2O).

4.1.3.2 Static Pressure

The pressure of a body of fluid in static equilibrium is the static pressure of the fluid. At any point in a fluid at rest, pressure acts with equal intensity in all directions.

4.1.3.3 Pressure Measurement

There are many different devices used to make pressure measurements. These include: liquid-column manometers, multiplying gauges and mechanical pressure gauges.

Discussion of the actual operation of each individual device is not presented in this text. For a detailed presentation on pressure measurement, many excellent references exist.[1-8]

Types of liquid-column manometers used include: open or differential U tubes, inverted differential U tubes, closed U tubes, and mercury barometers. There are four basic methods of magnifying the readings of pressure measuring gauges. These include: (1) change of manometric fluid; (2) inclined U tubes; (3) the use of draft gauges where one leg of a U tube reservoir has a much larger bore than the tubing that forms the inclined leg; and (4) use of a two-fluid U tube.

The most common type of gauge that is used commercially is a mechanical type called a Bourdon-tube gauge. In a Bourdon-tube gauge, the pressure is measured by the movement of an oval bent tube in an arc of a circle, and closed at one end. Bourdon-type gauges are useful for pressures ranging from zero absolute pressure to 100,000 psia (700 MPa absolute).

4.1.4 Fluid Dynamics

4.1.4.1 Measurement of Fluid Flow

There are many types of devices and meters used to measure the flow of fluids in pipes. These include: (1) pitot tubes; (2) aenometers; (3) venturi meters; (4) orifice meters; (5) rotameters; (6) mass flowmeters; (7) weirs; and (8) externally operated (non contact) vibration-based flowmeters.

Detailed discussion of the actual operation of each individual device is not presented in this text. For a detailed presentation on flow measurement, the following references are suggested:

1. Perry and Green, *Perry's Chemical Engineering Handbook*, 6th ed., McGraw-Hill, New York, 1984.

2. Considine, *Process Instruments and Controls Handbook*, 2nd ed., McGraw-Hill, New York, 1974.

3. Addison, *Hydraulic Measurements*, 2nd ed., Wiley, New York, 1949.

4. ASME Research Committee on Fluid Meters Report, Fluid Meters Report, Fluid Meters/Their Theory and Application, 6th ed., 1971.

5. ASME Power Test Code, Part 5: "Measurement of Quantity of Materials," 1959.

6. Dean, *Aerodynamic Measurements*, M.I.T., Cambridge, MA, 1953.

7. Landesburg, Pease, and Taylor, *Physical Measurements in Gas Dynamics and Combustion*, Princeton, Princeton, NJ, 1954.

8. Ower and Pankhurst, *The Measurement of Air Flow*, 5th ed., Pergamon, Oxford, 1977.

9. Spink, *Principles and Practice of Flowmeter Engineering*, 9th ed., Foxboro, MA 1967.

10. Marks, *Mechanical Engineering Handbook*, 8th ed., McGraw-Hill, New York, 1978.

4.1.4.2 The Overall Energy Balance (Bernoulli's Theorem)

The first law of thermodynamics states that the total energy of any system and its surroundings is conserved, and can therefore be accounted for. From this principle, a mechanical energy balance can be made around a system to take into account all of the energy factors involved in the system.

Fluid Dynamics and Sizing Analysis

The mechanical energy form of Bernoulli's theorem for a flowing system can be written as:

$$\frac{z_1 g}{g_c} + \frac{v_1^2}{2g_c} - \int_1^2 V\, dP - J + W_c = \frac{z_2 g}{g_c} + \frac{v_2^2}{2g_c} \tag{4.4}$$

where:
z = head, ft
g = gravitational constant = 32.2 ft/s^2
g_c = dimensional constant
v = fluid velocity, ft/s
V = volume, ft^3
P = pressure, psi
J = specific energy, ft lbs/lb
W_c = external work, ft - lb$_f$

For liquid systems, since the specific volume is substantially constant, Eq. (4.4) reduces to:

$$\frac{z_1 g}{g_c} + \frac{v_1^2}{2g_c} - V(P_2 - P_1) - J + W_c = \frac{z_2 g}{g_c} + \frac{v_2^2}{2g_c} \tag{4.5}$$

For gaseous systems, the exact value of the term $\int(V\, dP)$ depends upon whether the flow is isothermal, adiabatic, or some other flow condition.

4.1.4.3 Momentum Transfer

To analyze the forces involved in fluid systems, an overall momentum balance can be performed on a given control volume of fluid within the system. When two masses exchange momentum, the total amount of momentum in the system remains the same. This principle, known as the principle of conservation of momentum, can be used to determine the forces on branches, bends, reducers, and other components in the system. For fluids with steady flow, and a uniform velocity, the equation of momentum is:

$$F = w \frac{Dv}{g_c} = \rho g \frac{Dv}{g_c} \tag{4.6}$$

where:
F =
w = weight, lbs
v = fluid velocity, ft/s
ρ = fluid weight density, lbs/ft^3
g = gravitational constant = 32.2 ft/s^2
g_c = dimensional constant

The forces exerted by a fluid on a bend are shown in Figure 4.1(A). They can be calculated from:

$$F_x = p_1 A_1 - p_2 A_2 \cos q + \left(\frac{pg}{g_c}\right)(v_1 - v_2 \cos q) \tag{4.7}$$

where:
F_x = force in the x direction, lbs
A = area of concentration, ft^2
θ = angle of the change in direction of fluid flow, degrees
p = pressure, lbs/ft^2

$$F_y = \frac{wg}{g_c} + \left(p_2 A_2 + \frac{pgv_2}{g_c}\right) \sin q \tag{4.8}$$

where:
F_y = force in the y direction, lbs

Forces exerted by a fluid on oblique flat plates (similar to that occurring in a lateral branch) are shown in Figure 4.1(B). They can be calculated from the following equations:

$$F = \frac{pgv}{g_c} \sin q \tag{4.9}$$

$$q_1 = \left(\frac{q}{2}\right)(1 + \cos q) \tag{4.10}$$

where:
q_1 = flow rate in first run direction, ft^3/hr
q = flow rate in branch, ft^3/hr
F = force exerted on the flat plate, lbs

$$q_2 = \left(\frac{q}{2}\right)(1 - \cos q) \tag{4.11}$$

where:
q_2 = flow rate in second run direction, ft^3/hr

Forces on stationary normal flat plates (as in a tee branch) can be calculated from:

$$F = \frac{pgv}{g_c} \tag{4.12}$$

$$q_1 = q_2 = \frac{q}{2} \tag{4.13}$$

Fluid Dynamics and Sizing Analysis

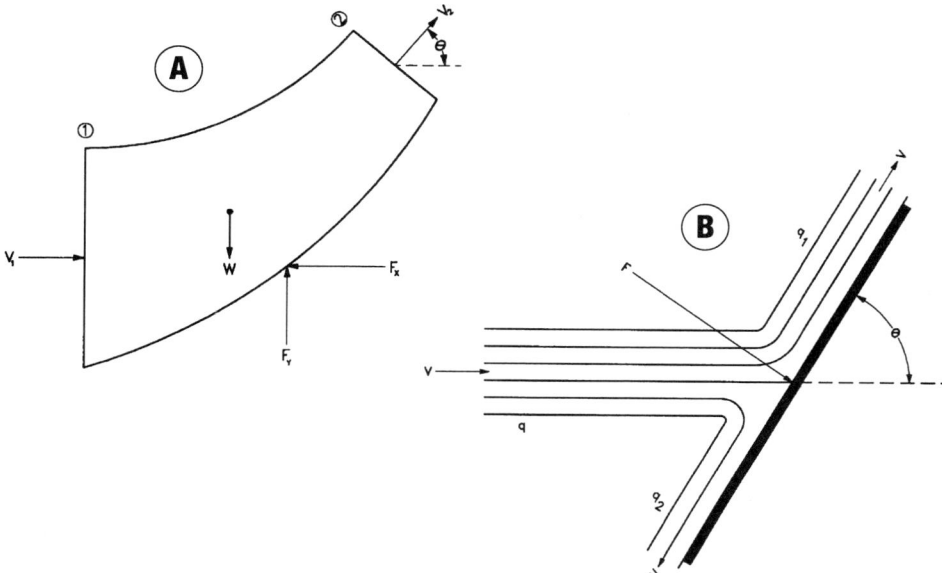

Figure 4.1. A. Components of forces exerted by a fluid on a bend. B. Components of forces exerted by a fluid on oblique flat plates.

Equations (4.7) through (4.13) apply to primary (inner) pipe components only. For flow that occurs in an annulus, the determination of forces that are developed at bends and other components require numerical techniques (i.e., computational fluid dynamics). Due to the complexity of the calculations, it is not presented here. They also will not be as significant, and may be expected only during flushing operations and the like.

Momentum transfer due to annular flow is usually not significant. However, a part of the forces calculated in Eqs. (4.7) through (4.13) due to momentum transfer in primary components will also act upon outer pipe components, if they are permanently connected to corresponding inner pipe components via interstitial supports. These effects must be considered when analyzing the strength of outer pipe elbows, tees, laterals, etc.

4.2 Flow in Pressure Pipes

4.2.1 Primary (Inner) Piping Considerations

4.2.1.1 Types of Flow

The manner in which a fluid flows in a circular pipe that is completely full is dependent upon the inside diameter of the pipe, its relative smoothness, the fluid viscosity and density, and the velocity at which the fluid is flowing. The manner of flow affects the pressure of the fluid at any given point in the piping system due to frictional losses that occur as a result of fluid flow. A flowing fluid will lose pressure due to the friction between the flowing fluid and component walls. Different types of flow (i.e., turbulent versus laminar) will produce different amounts of frictional losses.

Table 4.1 Classification of Fluid Flow According to Reynolds Number

Type of flow	Reynolds number range
Laminar flow	$N_{RE} < 2100$
Transition region	$2100 < N_{RE} < 3000$
Turbulent flow	$N_{RE} > 3000$

The variables described here can be grouped into a dimensionless ratio referred to as the Reynolds number. This dimensionless grouping compares the dynamic forces of a fluid's mass flow to its shear stress due to viscosity.

Some of the more common ways of defining the Reynolds number are shown:

$$N_{RE} = \frac{D_e v \rho}{\mu g} = \frac{D_e G}{\mu} = \frac{D_e v}{\mu} \tag{4.14}$$

where:
N_{RE} = Reynolds number, dimensionless
D_e = equivalent diameter, ft
ρ = density, lbs/ft^3
υ = kinematic viscosity, sq. ft/s
μ = dynamic viscosity, centipoise
v = velocity, ft/s
g = gravitational constant = 32.2 lbs/ft^3
G = mass flow rate = lb/h

There are four different types of flow to be considered in piping system analysis: (1) laminar or streamline flow; (2) flow in the transition region; (3) turbulent flow, and (4) uniform flow. The first three are actually encountered in piping systems. Uniform flow is an idealized form of flow that is assumed in many simplified engineering analysis. Figures 4.2 through 4.4 illustrate the velocity distribution profile for uniform flow, laminar flow, and turbulent flow, respectively. The nature of the flow is determined by the magnitude of the Reynolds number, as shown in Table 4.1.

A fluid flowing in the transition region is assumed to be turbulent for purposes of calculation. The range for transition region varies from 2100 to 10,000, depending on many circumstances. A pipe designed for turbulent flow will have excess capacity for fluids in laminar flow.

4.2.1.1.1 Laminar Flow

The kinetic energy of a flowing fluid can be found from:

$$E_K = \frac{1}{2} \int_{y=0}^{d} v^2 M \tag{4.15}$$

Fluid Dynamics and Sizing Analysis

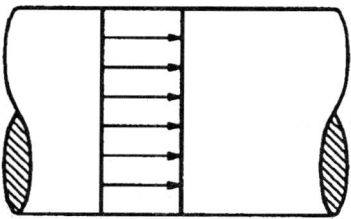

Figure 4.2. Velocity distribution for a fluid flowing in uniform flow.

where:
y = distance, ft
M = a function of d_y and the velocity at the point of flow selected

The actual kinetic energy can be calculated from by using a value for α found by assuming a constant density:

$$E_K = \alpha \frac{1}{2} m \bar{v}^2 \qquad (4.16)$$

where:
E_K = kinetic energy
α = ratio of observed kinetic energy to the energy of a uniformly distributed flow
m = mass, lbs
$\bar{v}$ = average velocity, ft/s

$$\alpha = \frac{1}{A} \int_0^d \left(\frac{v}{\bar{v}}\right)^3 dA \qquad (4.17)$$

where:
A = cross sectional area, ft^2

4.2.1.1.2 Uniform Flow

Figure 4.2 illustrates the profile for a fluid flowing under uniform flow. This form of flow is for the ideal case where the ratio of the observed kinetic energy of a fluid to the kinetic

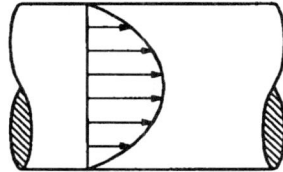

Figure 4.3. Velocity distribution for a fluid flowing in laminar flow.

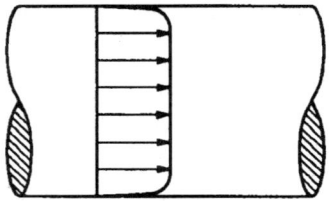

Figure 4.4. Velocity distribution for a fluid flowing turbulent flow.

energy of a uniformly distributed flow (α) is equal to one. This distribution of fluid flow represents an ideal case for simplified engineering analysis only, and is never encountered in actual practice. It is shown here for means of comparison only.

4.2.1.1.3 Laminar Flow

Figure 4.3 illustrates the velocity distribution profile for a fluid flowing in laminar flow. The distributions are parabolic in shape, because of the predominance of viscous effects over inertial forces; they will vary depending on the value of the Reynolds number. However, the ratio of the observed kinetic energy of the fluid to the kinetic energy of a uniformly distributed flow (α) is always equal to two.

4.2.1.1.4 Turbulent Flow

Figure 4.4 illustrates the profile for a fluid in turbulent flow. Velocity distributions in turbulent flow are ellipsoidal in shape, due to the predominance of inertial over viscous forces. The ratio of the observed kinetic energy of the fluid to the kinetic energy of a uniformly distributed flow (α) typically varies between 1.02 and 1.15. For smooth tubes, the velocity profile can be determined by Nikuradse's equation:

$$v_y = v_c \left(\frac{y}{r_o} \right)^n \tag{4.18}$$

where:
v_y = velocity at location y, ft/s
y = distance from the closest outside diameter to the point in the fluid profile, ft
v_c = the center velocity, ft/s
r_o = the outside radius, ft
$n = 1/7$, $N_{RE} < 100{,}000$
$ 1/8$, $N_{RE} < 400{,}000$

There are many other empirical equations to determine the velocity profile of a fluid flowing under turbulent flow in a smooth, circular pipe. The velocity profile is not a significant factor in most piping applications. While it is important to know the type of flow (i.e., turbulent versus laminar), the details of the velocity profile itself are not used in most piping calculations. To understand this subject further, the reader is encouraged to consult other references.[9,10]

4.2.1.2 Velocity Equivalents

Many times in engineering calculations the velocity term is needed but is not readily known. The velocity of the flowing fluid can be calculated using other flow characteristics. Many of the common equivalences are shown in:

$$v = \frac{q}{A} = 183.3\frac{q}{d^2} = 0.408\frac{q}{d^2} \tag{4.19}$$

$$= 0.286\frac{b}{d^2} = 183.3\frac{w\overline{V}}{d^2} = 0.0509\frac{W\overline{V}}{d^2}$$

$$= 0.00389\, q\frac{S_g}{\rho d^2}$$

where:
W = mass rate of flow, lb/h
S_g = specific gravity of a gas relative to that of air
ρ = weight density, lb/ft^3
q = volumetric flow rate, ft^3/h
b = volumetric flow rate, barrels/h (= 42 gallons/h)
d = inside diameter, in.
A = cross-sectional area of pipe, ft^2

4.2.1.3 Friction Loss Calculations

When fluid flows in a pipe under pressure, frictional losses always occur due to the friction of fluid particles rubbing against each other, and along the inner walls of the pipe. Friction loss in a pipe represents a net loss in the total available work energy of a fluid.[11] When pressure energy is converted into another form of energy such as heat, pressure drop occurs. Friction loss and pressure drop are related by the Bernoulli equation, Eq. (4.4). One can predict the pressure drop in a piping system by systematically calculating the pressure drops for individual components, and then summing them. When there are branch lines in the system being analyzed, this method is used. Alternatively, one can sum the friction loss resulting from each individual resistance, and then calculate the overall pressure drop using Eq. (4.4).

With both approaches, it is common to convert individual components into an equivalent length of pipe, and then calculate the resistance or pressure drop for straight lengths of pipe only. Converting individual components of a piping system into equivalent straight lengths is discussed in more detail in the subsection entitled, "Equivalent Lengths of Pipe for Components."

The Darcy formula is commonly used for calculation of pressure drop. The equation can be used for all flow conditions (laminar or turbulent flow), except where conditions of liquid cavitation exist. Equation (4.20) states the Darcy equation with pressure drop in terms of feet of fluid flowing: Eq. (4.21) states it with the pressure term being expressed in terms of pounds per square inch:

$$h_f = f \frac{L}{d} \frac{v^2}{2g} \qquad (4.20)$$

$$\Delta P = f \frac{L}{d} \frac{v^2}{2g} \frac{\rho}{144} \qquad (4.21)$$

where:
h_f = head loss due to friction, ft
L = length of pipe, ft
ΔP = pressure loss due to friction, psi
f = friction factor
v = velocity, ft/s

Equations (4.20) and (4.21) show pressure drop as being proportional to the square of the velocity, the length of pipe being analyzed, and is inversely proportional to the diameter of the pipe and the gravitational constant. The Darcy formula is useful for straight horizontal, vertical, or sloping pipe with a constant diameter, and for transporting fluids that have relatively constant properties. The friction factor terms used in Eqs. (4.20) and (4.21) are the Darcy or Moody friction factors. The Fanning friction factor, also used in engineering practice, is 25% of the value of the Darcy friction factor. For vertical or sloping pipes, the elevation term of the Bernoulli equation must not be neglected in calculating the overall pressure drop of the system.

For laminar flow situations ($N_{RE} \leq 2100$), the friction factor can be calculated by the following simplified empirical relationship:

$$f = \frac{64}{N_{RE}} \qquad (4.22)$$

When Eq. (4.22) is substituted into the Darcy equation with pressure drop in psi, Eq. (4.21) reduces to the Poiseuille equation for pressure drop in laminar flow:

$$\Delta P = 0.000668 \frac{\mu L v}{d^2} \qquad (4.23)$$

where:
μ = relative viscosity, centipoise

Equation (4.23) relates pressure drop to the diameter, length, fluid velocity, and viscosity of the fluid (Reynolds number variables), and not to the roughness of the pipe walls. In turbulent flow conditions, the friction factor and the pressure drop are dependent upon the ratio of the roughness of the pipe interior surface to the length of the exposed surface. This ratio is referred to as the relative roughness of the pipe. Since the smoothness of a pipe's interior bore is normally a constant, regardless of size, a smaller-diameter pipe has a larger relative roughness term. Therefore, smaller pipes will produce a larger pressure drop than larger-diameter piping manufactured to a similar finish.

Fluid Dynamics and Sizing Analysis

To calculate friction factor for pipes flowing under turbulent flow conditions, the Colebrook and White equation is used:

$$\frac{1}{\sqrt{f}} = -2 \log \left(\frac{e/D}{3.7} + \frac{2.51}{N_{RE}\sqrt{f}} \right) \tag{4.24}$$

where:
f = the Darcy friction factor
e = roughness characteristic
D = outside diameter, ft

The Colebrook and White equation is valid over the complete range of ε/D encountered in practice. Since Eq. (4.24) is implicit in f, trial and error calculations, or computer modeling, are required for its solution. The friction factor is normally found from a graphical presentation of the solution of this equation, known as the Moody friction factor chart (sometimes called the Darcy friction factor chart).

The Moody chart, originally developed by L. F. Moody,[12] is shown in Figures 4.5 and 4.6. Note that Figures 4.5 and 4.6 show values of Fanning friction factors. To obtain the Darcy or Moody factors, multiply the values from the chart by 0.25. Relative roughness of commercial piping over a wide range of diameters is shown in Figure 4.7. Most commercial plastic piping is roughly equivalent to drawn tubing in terms of smoothness of the pipe internal bore. However, many recently developed materials, including fluoropolymer piping and polyketone piping, are produced with an internal finish superior to drawn tubing. The friction loss calculations of the specific material should be obtained from the manufacturer of the specific product.

For smooth pipes, the relative roughness term becomes negligible, and the equation reduces to Prandtl's equation:

$$\frac{1}{\sqrt{f}} = -2 \log \left(\frac{2.51}{N_{RE}\sqrt{f}} \right) \tag{4.25}$$

For rough pipes, the relative roughness term becomes the controlling factor, and the equation reduces to the von Karman equation:

$$\frac{1}{\sqrt{f}} = -2 \log \left(\frac{e/D}{3.7} \right) \tag{4.26}$$

The Sachem Equation, Eq. (4.27), can be used to determine the friction factor; it is explicit in f, and accurate enough for design purposes. In many instances, the friction factor can be determined more accurately by calculation than by reading from Figures 4.6 and 4.7.

$$f = \left\{ -2 \log \left[\frac{e/D}{3.7} - \frac{5.02}{N_{RE}} \log \left(\frac{e/D}{3.7} + \frac{14.5}{N_{RE}} \right) \right] \right\}^{-2} \tag{4.27}$$

For rough calculations of pipe diameter, Eq. (4.28) can be used. This equation is represented by the dotted line (between transition zone and complete turbulence) in Figure 4.5.

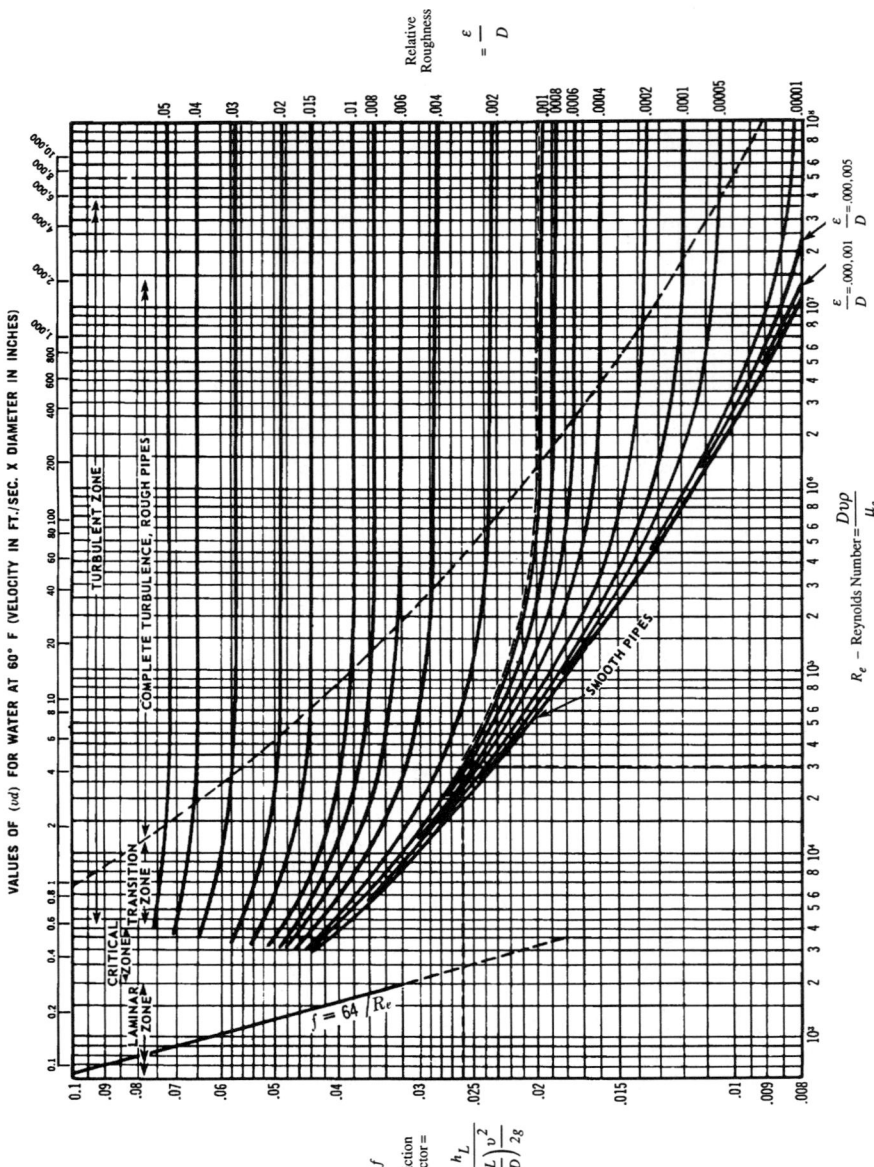

Figure 4.5. The Moody chart for determining friction factor for any type of commercial pipe.[12] (Source: Crane Co.)

Fluid Dynamics and Sizing Analysis

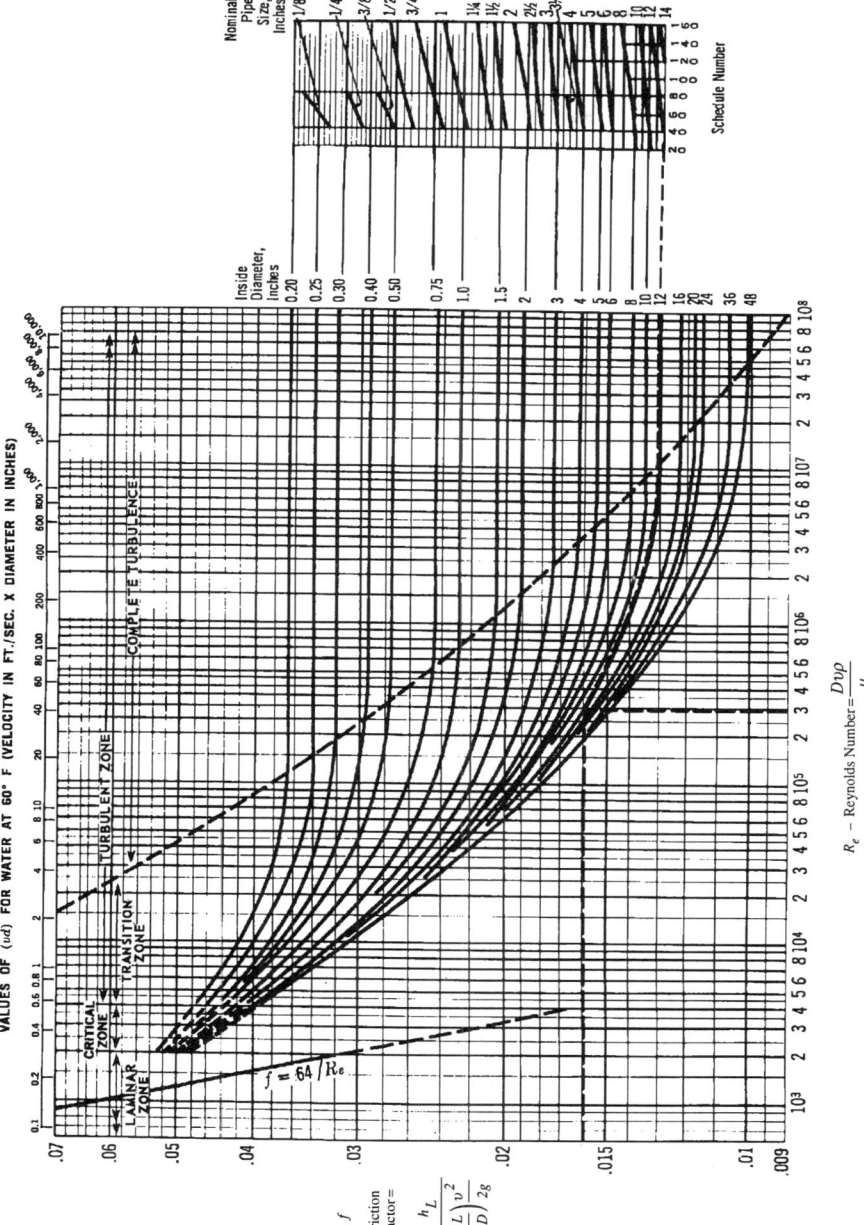

Figure 4.6. The Moody chart for determining friction factor for clean commercial steel pipe.[12] (Source: Crane Co.)

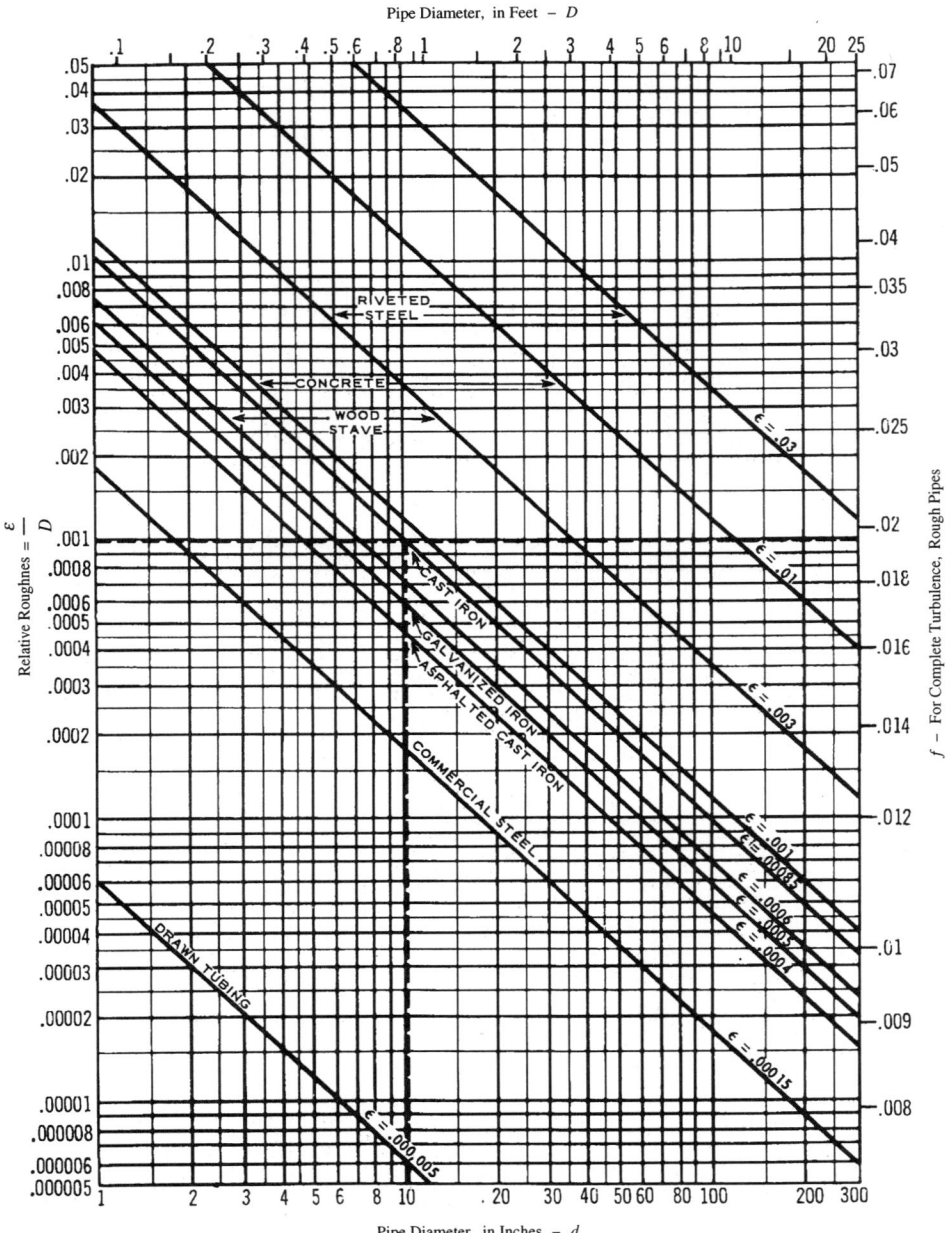

Figure 4.7. Relative roughness and friction factors of commercial pipe for complete turbulence.[12] (Source: Crane Co.)

Fluid Dynamics and Sizing Analysis

$$f = \frac{0.01}{N_{RE}^{0.16}} \tag{4.28}$$

4.2.1.3.1 Hazen and Williams Method

A totally different method for calculating pressure drop has been used extensively in applications involving nonmetallic piping and civil-engineering-type applications (storm sewer design, for example). The Hazen and Williams method involves the use of a general simplified friction loss coefficient, and is valid for turbulent flow. The coefficient is typically determined by experience or simple experimentation.

Most manufacturers of nonmetallic piping promote the use of a Hazen and Williams friction loss coefficient of 150 for their materials. This value, in many situations, oversizes the piping diameter. This is particularly true of pipes with an extremely smooth bore, such as fluoropolymers, polyketones, and other materials with a glasslike finish or better.

For plumbing or drainage-type applications, the C factors for materials can be found in building design codes and plumbing standards or from drainage pipe manufacturer's data or other standards. The Hazen and Williams formula is shown the following in equation for pressure drop described in terms of feet of fluid flowing:

$$h_f = 0.2083 \left(\frac{100}{c}\right)^{1.85} \left(\frac{q^{1.85}}{d^{4.87}}\right) \tag{4.29}$$

where:
h_f = head loss due to friction, ft
c = Hazen and Williams constant
q = volumetric flow rate, ft^3/s
d = inside diameter, ft

and in the following equation for pressure drop in terms of psi:

$$\Delta P = 0.09 \left(\frac{100}{c}\right)^{1.85} \left(\frac{q^{1.85}}{d^{4.87}}\right) \tag{4.30}$$

where:
ΔP = pressure drop, psi

Proper modification of the Darcy or Hazen and Williams equations reveal that the pressure loss of a flowing fluid is proportional to the fifth power of the internal diameter for a given type of pipe. Therefore, when the pressure drop is known for a given diameter of piping, the pressure drop in another diameter pipe, with similar internal bore surface finish, can be quickly estimated by applying the ratio of the fifth power of the diameters in question. This relationship is demonstrated in:

$$\Delta P_2 = \Delta P_1 \frac{d_1^5}{d_2^5} \tag{4.31}$$

4.2.1.4 Friction Loss Determination for Compressible Flow

Gas flowing through a pipe system is not always in compressible flow. For compressible flow to occur, the pressure drop of the fluid must be high enough to result in a 10% reduction or more in the density of the flowing gas. The flow of gases, like liquids, may be laminar or turbulent according to the same criteria. There is a third type of compressible flow that occurs at very low quantities of flow called molecular flow. When the mean free path of the gas is between 1 and 65% of the channel diameter, then slip flow occurs. Slip flow is a type of flow that is intermediate between pure molecular flow and pure laminar flow. Both slip flow and molecular flow are normally encountered only in vacuum situations (e.g., vacuum drying of an annular space in a double containment piping system), and thus are of limited importance in double containment piping design.

Since the density of gas changes in turbulent flow, the Darcy equation must be used in its differential form when applied to a gas:

$$\frac{dF}{dx} = \frac{fv^2}{2g_c D} = \frac{fv^2}{8g_c R_H} = \frac{fG^2}{8g_c r^2 R_H} \tag{4.32}$$

where:
R_H = hydraulic radius, ft

For isothermal flow in horizontal round pipes of sufficiently long lengths the following can be used to determine the pressure drop:

$$w = \frac{p}{8}\sqrt{\frac{(p_1^2 - p_2^2)g_c D^5 m}{fLRT}} \tag{4.33}$$

For adiabatic flow in horizontal pipes, the analysis is more complex. Levenspiel[13] presents nomographs for gases flowing under conditions of adiabatic turbulent flow.

4.2.1.5 Pressure Drop through Fittings and Valves

There are two basic approaches to the calculation of pressure drop through a fitting or valve. One can calculate the individual pressure drops of every fitting and valve component, and other points of pressure drop (i.e., entrance and exit locations), and then sum the individual drops. Alternatively, the resistance of each individual component can be converted into an equivalent length of pipe, and their values summed. The overall pressure drop can then be calculated using this sum. To calculate the pressure drop through an individual valve or fitting, the following equation is used:

$$K = \frac{\Delta P}{v^2/2g_c} \tag{4.34}$$

where:
ΔP = total frictional losses less frictional loss for centerline length of straight pipe, psi

Values of K for many common fittings are shown in Table 4.2. Many valve manufacturers express the flow capacity of valves in terms of a flow coefficient, C_V. The relationship of flow coefficient C_V to K is shown in:

Fluid Dynamics and Sizing Analysis

Table 4.2 Values of K for Various Common Fittings and Values for Turbulent Flow

Type of fitting or valve	Values of K
45° ell, standard	0.35
45° ell, long radius	0.2
90° ell, standard	0.75
Long radius	0.45
Square or miter	1.3
180° bend, close return	1.5
Tee, standard, along run, branch blanked off	0.4
Used as ell, entering run	1.0
Used as ell, entering branch	1.0
Branching flow	1*
Coupling	0.04
Union	0.04
Gate valve, open	0.17
3/4 open	0.9
1/2 open	4.5
1/4 open	24.0
Diaphragm valve, open	2.3
3/4 open	2.6
1/2 open	4.3
1/4 open	21.0
Globe valve, bevel seat, open	6.0
1/2 open	9.5
Composition seat, open	6.0
1/2 open	8.5
Plug disk, open	9.0
3/4 open	13.0
1/2 open	36.0
1/4 open	112.0
Angle valve, open	2.0
Y or blowoff valve, open	3.0
Plug cock, θ = 5°	0.05
10°	0.29
20°	1.56
40°	17.3
60°	206.0
Butterfly valve, , θ = 5°	0.24
10°	0.52
20°	1.54
40°	10.8
60°	118.0
Check valve, swing	2.0
Disk	10.0
Ball	70.0
Foot valve	15.0
Water meter, disk	7.0
Piston	15.0
Rotary (star-shaped disk)	10.0
Turbine-wheel	6.0

*This is pressure drop (including friction loss) between run and branch, based on velocity in the mainstream before branching. Actual value depends on the flow split, ranging from 0.5 to 1.3 if mainstream enters run and from 0.7 to 1.5 if mainstream enters branch.

$$C_v = \frac{C_1 d^2}{\sqrt{K}} \tag{4.35}$$

Equivalent lengths of pipe for fittings may be determined by using Figures 4.8 and 4.9, once a value of K is known. The relationship between "equivalents of resistance coefficient K" and "flow coefficient C_v" is shown in Figure 4.10. These equivalent lengths are usually determined experimentally or by the experience of component manufacturers. To determine the equivalent length of pipe for a component, information on the frictional loss characteristics of the respective piping must also be known. Equivalent length is a very literal term; the value assigned to a component means that the designated length of straight pipe would under comparable conditions produce the same amount of frictional loss.

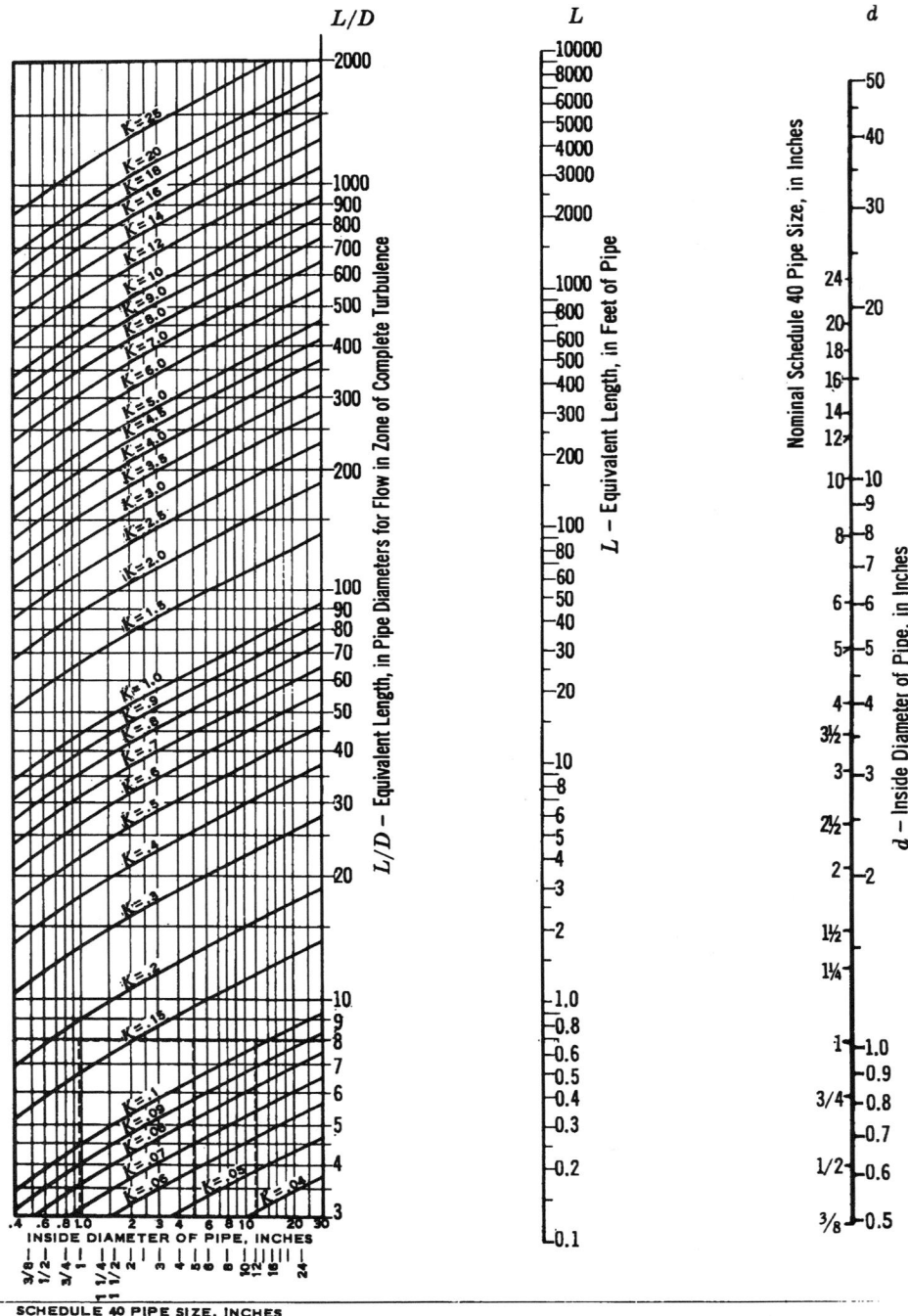

Figure 4.8. Equivalent lengths L and L/D (in feet of pipe) and resistance coefficient K. (Source: Crane Co.)

Fluid Dynamics and Sizing Analysis

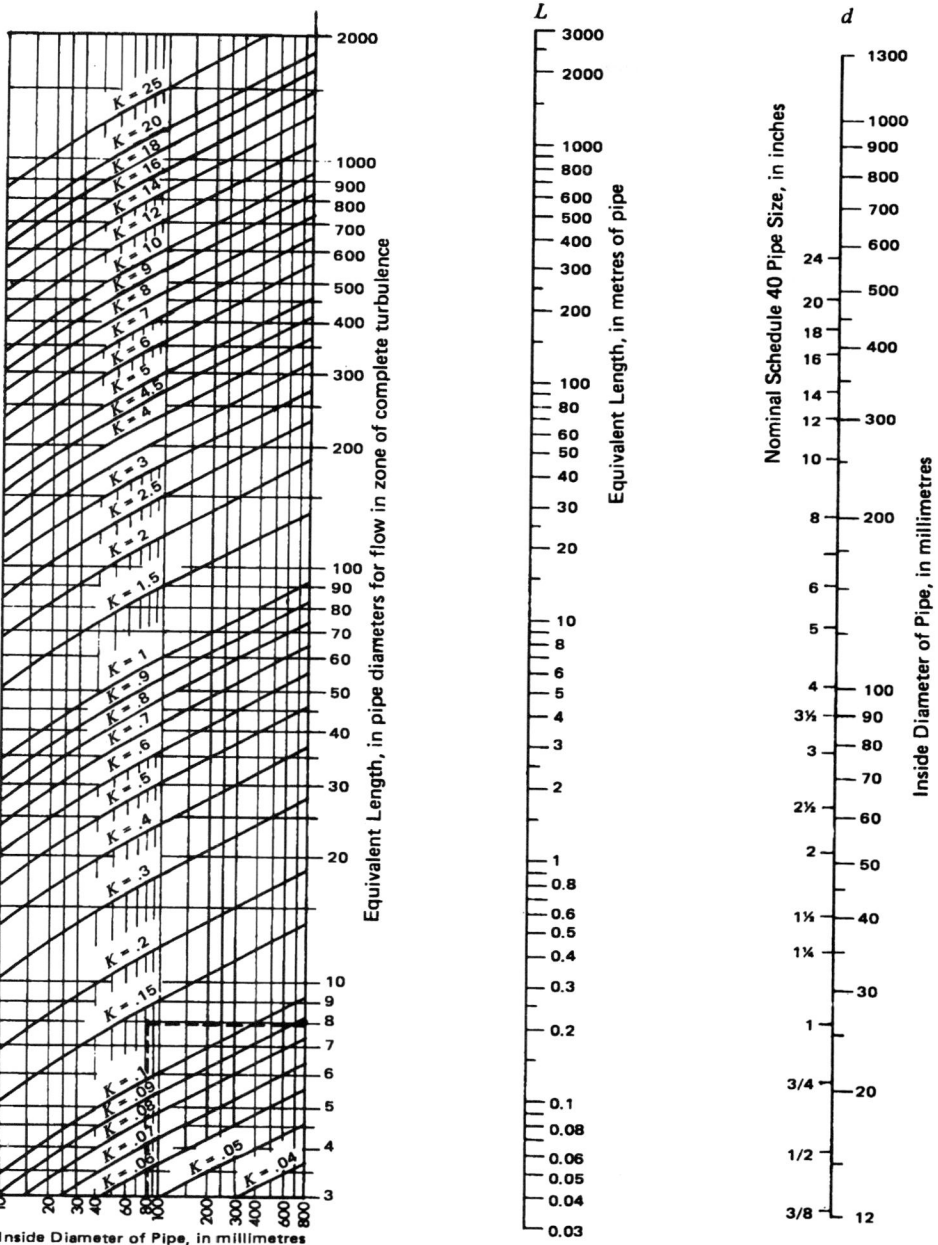

Figure 4.9. Equivalent lengths L and L/D and resistance coefficient k, in meters of pipe. (Source: Crane Co.)

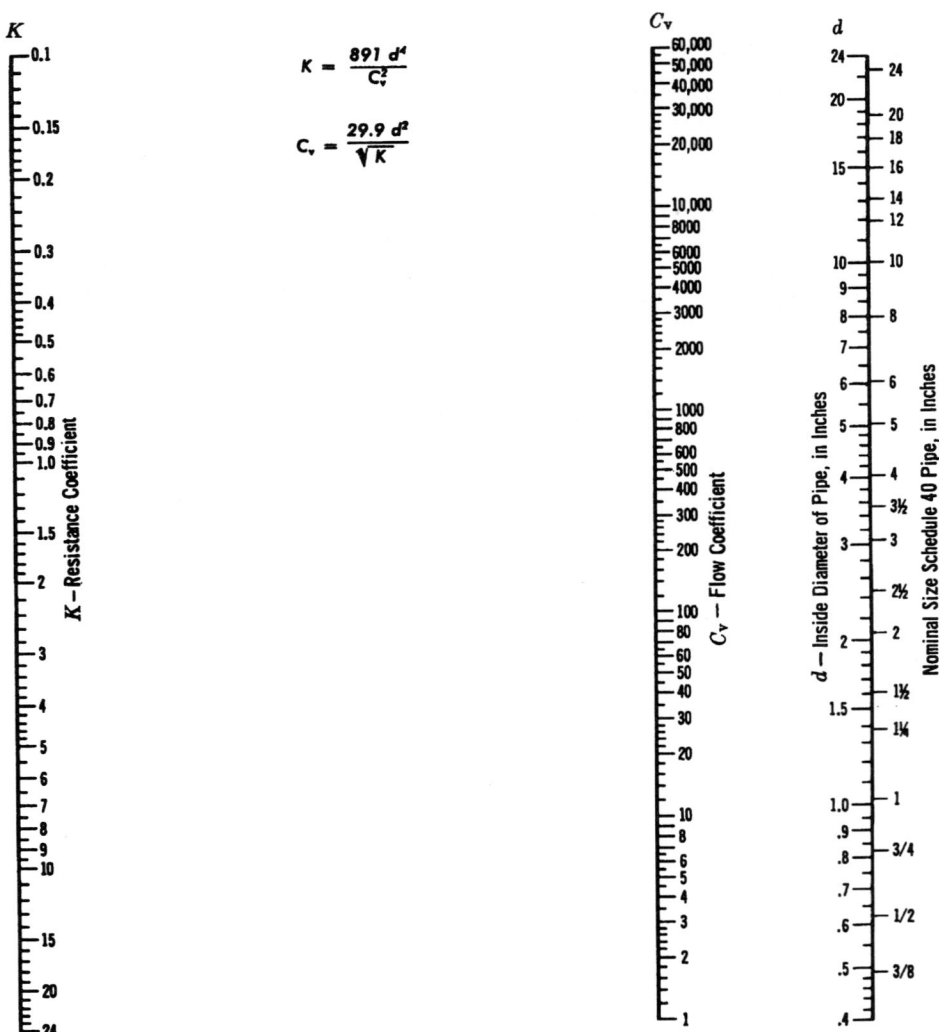

Figure 4.10 Equivalents of resistance coefficient K and flow coefficient C_v. (Source: Crane Co.)

Fluid Dynamics and Sizing Analysis

Equivalent lengths are available from most manufacturers and other sources, based on turbulent flow conditions. These data are valid for flow through primary components only; information does not exist for equivalent "secondary containment pipe lengths" for flow through the annulus of secondary containment components. For laminar flow, the data may also be invalid, particularly at Reynolds numbers below 500.[14] Many other good sources of laminar flow friction loss data can be found in various reports.[15-17]

4.2.1.6 Pressure Drop Due to Entrance and Exit Losses

Pressure loss through an entrance or exit depends on the shape of the opening. For turbulent flow in a sharp-edged entrance, the pressure loss can be calculated by

$$\Delta P = K_c \frac{V_2^2}{2g_c} \tag{4.36}$$

For a rounded or trumpet-shaped entrance, having a radius greater than 15% of the pipe diameter, a contraction coefficient of $K_c = 0.1$ can be used, since frictional loss will be smaller. Various values of coefficients for sudden contractions in turbulent flow conditions are shown in Table 4.3.

For laminar flow involving a sudden contraction, the friction loss can be expressed in terms of an equivalent length of straight pipe:

$$\frac{L_e}{d} = a + bN_{RE} \tag{4.37}$$

For a sudden enlargement and exit loss, the following equation is used for turbulent flow ($7.5° < \alpha < 35°$). For trumpet-shaped enlargements in turbulent flow, the friction loss will be less than in straight pipes of equal length (as much as 20 to 60% less).[18] For laminar flow, the pressure drop can be estimated by:

$$\Delta h = K \frac{(V_1 - V_2)^2}{2g_c} \tag{4.38}$$

$$\left(\frac{L}{D}\right)_e = \frac{1}{6n \tan(\alpha/2)} \left[1 - \left(\frac{D_2}{D_1}\right)^{3n}\right] \tag{4.39}$$

where:
K, n = material constants (see p. 5-26 and 5-36 of Perry's Chemical Engineering Handbook, 6th Edition)
α = total angle between converging walls

Table 4.3 Sudden Contraction-Loss Coefficients for Turbulent Flow

A_2/A_1	0	0.2	0.4	0.6	0.8	1.0
K_c	0.5	0.45	0.36	0.21	0.07	0

Note: A_2 is the smaller diameter, A_1 is the larger diameter.

4.2.1.7 Pressure Drop across Internal Restrictions

Fluid flowing around internal weld protrusions or beads (in nonmetallic heat-element butt-fused pipe, for instance) and in socket weld gaps (as in solvent-cemented joints in nonmetallic piping, for instance) will experience a pressure drop as a result of these internal restrictions. The pressure loss will depend on the size of the internal gap/bead and the type of flow. For internal restrictions, the pressure drop can be calculated by using the equations developed for calculating pressure drops across an orifice. For a gap in a pipe (i.e., at socket joints), the area can be treated the same as a sudden expansion followed by a sudden contraction. In small-diameter pipes, both of these occurrences should be considered, as they can result in a significant pressure drop.

4.2.1.8 Pump Suction Pipe Sizing

Determination of the inlet diameter for a pump is limited by the requirements of the pump. Most pumps need a minimum net positive suction head in order to prevent a fluid from vaporizing (particularly centrifugal types). When a fluid vaporizes, it forms a bubble or void, which collapses as the pump pressurizes the liquid back above its vapor pressure. The resulting implosion is potentially damaging to an impeller face of a centrifugal pump, and can have other damaging effects as well.

To determine the available NPSH of a fluid, Bernoulli's equation can be used to perform an energy balance of the inlet (upstream) portion of the system; the resulting equation is shown in terms of feet of fluid:

$$NPSH = h_a \pm Z + h_f + h_e - h_v \qquad (4.40)$$

where:
h_a = absolute pressure at surface of liquid, ft
Z = static elevation of fluid above centerline of pump, ft
h_f = friction losses in the suction piping, ft
h_e = entrance losses in the suction piping, ft
h_v = absolute vapor pressure of fluid at pump temperature, ft

4.2.1.9 Parallel Flow (Complex Pipes)

When fluid flows in a parallel network of two or more pipes, the individual flow rates can be easily found because the pressure drops through each leg are equal. Figure 4.11 illustrates a parallel circuit of pipes. Even if different diameter sizes are used, the pressure drop through each will always be the same. This is a basic principle of fluid flow. It is similar to the principles of electricity in that there are equal voltage drops across parallel legs of a parallel circuit. Voltage and pressure are both terms for measurement of the potential energy in a circuit (or system). Equation (4.42) (based on the Darcy equation) and Eq. 4.43 (based on the Hazen and Williams equation) are both derived by equating the pressure drop variables of the individual legs, and then simplifying each equation. They both provide similar results.

If more than two parallel legs exist, a similar approach can be used. To determine the total pressure drop of a parallel system, Eq. (4.46) can be used.

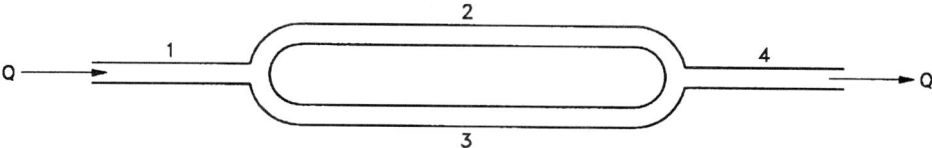

Figure 4.11. A simple parallel pipe circuit.

$$R = \frac{q_3}{q_2} \tag{4.41}$$

$$= \left[\left(\frac{l_2}{l_3}\right)\left(\frac{d_3}{d_2}\right)^5\right]^{\frac{1}{2}} \tag{4.42}$$

$$= \left[\left(\frac{l_2}{l_3}\right)^{1.08}\left(\frac{d_3}{d_2}\right)^{5.26}\right]^{\frac{1}{2}} \tag{4.43}$$

$$v_2 = \frac{q_2}{448.8 A_2} \tag{4.44}$$

$$v_3 = \frac{q_3}{448.8 A_3} \tag{4.45}$$

$$h_f = h_1 + h_2 + h_4 = h_1 + h_3 + h_4 \tag{4.46}$$

4.2.2 Two-Phase Flow

For two-phase flow (liquid and gas flowing at the same time), the pressure drop differs from that of each phase flowing individually. While there has been much research on this subject, no simplified general correlation has been developed to date. The analysis for two-phase systems depend on whether a pipe is horizontal or vertical, and on the nature of flow occurring.

For horizontal pipes, the following general flow pattern types can be encountered: (1) froth or bubble flow; (2) stratified flow; (3) plug flow; (4) slug flow; (5) wavy flow; (6) spray or dispersed flow; and (7) annular flow (refers to annular flow of two phases in a single-wall pipe, not to be confused with the annular space of a double containment piping system).

For vertical pipes, the various general flow pattern types are separated into up-flow and down-flow. The up-flow patterns include: (1) ripple or wave flow; (2) piston, plug, or slug flow; (3) froth flow; (4) bubble or aerated flow; (5) mist flow; and (6) film (annular) flow. Down-flow refers to one specific pattern of flow. However, the type of pipe makes a difference when conditions of down-flow exist (for example: straight pipe, helically coiled tubes, drain and overflow pipes, etc.).

Two-phase flow is a highly complex subject requiring an analysis of each of the different patterns of flow and the fluids involved. For a comprehensive discussion of this subject, an excellent reference is: Perry and Green, Chemical Engineering Handbook, 5th ed., McGraw-Hill, Chapter 5, pp. 5-40/5-48.

4.2.2.1 Preliminary Pipe Sizing

Many times the design of a piping system begins with a knowledge of a volume and rate of fluid, which needs to be conveyed a certain distance. Initially, the diameters required to facilitate this flow may not be known. If a project is a duplicate of, or similar to, a previously operating system, selecting the initial diameter size for each service line and branch is easy.

However, a completely new system requires a trial and error procedure. There are empirical methods for both liquids and gases that determine a maximum obtainable velocity for a fluid, and subsequently, a minimum corresponding diameter. They serve as a good starting point. There are also methods that allow a designer to select a preliminary pipe size.[19] These estimate the diameter sizes as a function of their flow rates and densities.

For liquid services involving clear fluids, the maximum velocity can be determined by Eq. (4.47). If the fluid is a corrosive or erosive fluid, however, the value calculated from Eq. (4.47) should be divided by a safety factor of two.

$$v = \frac{48}{(\rho)^{1/3}} \qquad (4.47)$$

where:
v = velocity, ft/s
ρ = density, lb/ft^3

Once maximum velocities are known, the minimum diameters for liquids can be calculated by Eq. (4.48) for clear liquids, and by Eq. (4.49) for corrosive/erosive liquid service. A safety factor has been built into Eq. (4.49) to account for corrosive or erosive actions of such a fluid.

$$d = 1.03 \frac{W^{1/2}}{\rho^{1/3}} \qquad (4.48)$$

$$d = 1.457 \frac{W^{1/2}}{\rho^{1/3}} \qquad (4.49)$$

where:
d = inside diameter, ft
W = flow rate, ft^3/h

Fluid Dynamics and Sizing Analysis

The maximum velocity for a fluid in a pipe will result when the minimum required diameter is used. What is much more important to a system design is the typical diameter used for a designated amount of flow. To select a typical diameter based upon a typical, required, or target velocity, Eq. (4.50) can be used. If typical, required, or target diameters are not known, Eqs. (4.51) for pressure systems and (4.52) for drainage or pump suction piping may be used to estimate typical diameters. It is probably a wise idea to use Eq. (4.51) or (4.52) to estimate minimum required diameters, and contrast the values with those found by the use of Eq. (4.50). This will serve as a quick way to check if an error is made in using the equations, if substantially different values are found.

$$d = ^{0.034}\sqrt{\frac{v}{5.6}} \tag{4.50}$$

$$d = 2.607\left(\frac{W}{\rho}\right)^{0.434} \tag{4.51}$$

$$d = 3.522\left(\frac{W}{\rho}\right)^{0.434} \tag{4.52}$$

Once preliminary sizes are selected, pump requirements can be determined. Further analysis is needed to determine actual sizes, including analysis of the mechanical integrity of the system and economic analysis. The best diameter is found by using a pipe that minimizes the total of all costs (including energy) and conveys a fluid in a safe and reliable manner.

For gaseous or vapor flow, maximum velocity is limited by the sonic velocity of the gas (vapor). The determination of sonic velocity can be found from:

$$v_s = (WgZRT)^{1/2} \tag{4.53}$$

where:
v_s = sonic velocity, ft/s
Ω = molecular weight, lbs/lb mols
R = gas law constant
Z = compressibility factor
T = absolute temperature, °R
g = gravitational constant = 32.2 ft/s^2

In the following equation substituted actual values for the ideal gas law and the gravitational constant have been used:

$$v_s = 223\left(\frac{WZT}{m}\right)^{1/2} \tag{4.54}$$

where:
m = mass, lbs

Equation (4.55) can be used to calculate the minimum diameter for services involving a clear gas. For services involving a corrosive gas, the value determined by the following equation should be increased by about 50% to provide a safety factor:

$$d = 0.585 \left(\frac{W}{\rho}\right)^{1/2} \left(\frac{m}{WZT}\right)^{1/4} \tag{4.55}$$

Typical velocity can be found by using Eq. (4.56), which is a modification of Eq. (4.53). Equation (4.57) can be used as a starting point to estimate the typical diameter for pressurized gas flow. Equation (4.58) should be used for suction or vent piping.

$$v = 148.7 \left(\frac{WZT}{m}\right)^{1/2} \tag{4.56}$$

$$d = 1.065 \frac{W^{0.408}}{\rho^{0.343}} \tag{4.57}$$

$$d = 1.414 \frac{W^{0.408}}{\rho^{0.343}} \tag{4.58}$$

4.2.2.2 Water Hammer

A column of fluid that undergoes a sudden change in fluid velocity will be subject to a phenomenon called water hammer; this phenomenon manifests itself as a pounding of the line. The effect of a water hammer is directly related to the dynamics of the flow condition.

Fluid velocities change due to a number of possible occurrences, including: (1) starting and stopping of pumps; (2) the closing of valves; (3) change in turbine or motor speeds; (4) change in liquid feed tank elevation; (5) wave action in a feed tank; (6) liquid column separation; and (7) entrapped air.

To calculate the maximum pressure rise due to a sudden change in fluid velocity, Eq. (4.59), the Joukowsky, or water hammer, equation, should be used:

$$h_{wh} = a \frac{(Dv)}{g_c} \tag{4.59}$$

The velocity of the wave propagation term can be calculated from:

$$a = \sqrt{\frac{1}{\frac{r}{g_c}\left(\frac{1}{K} + \frac{D}{bE}\right)}} \tag{4.60}$$

The maximum pressure rise will develop if the change of velocity occurs within the time frame that it takes for a wave to travel from the point of initiation to the point of reflection and back. This value can be determined from:

$$\tau = \frac{2L}{v} \tag{4.61}$$

where:
τ = time, s
L = length, ft
v = velocity, ft/s

If the time it takes the velocity to change exceeds τ, then the amount of pressure rise will be less than the maximum that will occur. The value of τ therefore sets the time of opening of a valve that will result in the maximum possible pressure rise. A time of less than τ for valve opening or closing will also result in the maximum possible pressure rise occurring in a system. One of the objectives in determining the opening (closing) time for a valve is to select a time that exceeds the value of τ by as much the process allows, while considering the economics. The value of τ either refers the time for an actuator to work, or it could also mean the time for a plant operator to manually open or close a valve.

For standard steel pipe, the value of the wave velocity is 1220 m/s (4,000 ft/s). There are many references that present charts and methods for calculating actual pressure rises for times other than θ. The charts and equations for calculating water hammer pressures other than the maximum are known as the Allievi Equations (and charts). In 1933, there was a joint symposium between the American Society of Mechanical Engineers and the American Society of Civil Engineers, termed the Symposium on Water Hammer, whereby much of the information in use today was presented. Other good basic references exist.[20]

Surges can be of a transient or oscillating nature. Transient-type surges result from the opening or closing of valves, centrifugal pump starting or stopping actions, and many other factors. If oscillating conditions are caused by the actions of a control valve, or reciprocating pump, a line could potentially reach its harmonic vibration frequency, and be subject to severe catastrophic failure. Precautions must be taken to protect against this type of occurrence. In all systems, there is the need to keep total surges below levels where damage can occur to a piping system.

Negative surges can be equally damaging.[21] Since they can result in cavitation as the fluid momentarily falls below its vapor pressure, a piping system and its components can experience catastrophic failure. Many times in a piping system, there is noise in the area of elbows that sounds like gravel. Often this is the result of a fluid flowing at too great a velocity in an elbow, with resulting cavitation.

4.2.3 Pressure Pipe Sizing Economics

Selection of the actual inside diameter of piping to be used for a particular service is an economic decision that balances the cost of installing and maintaining a certain size of pipe against energy costs for pumping. The larger a pipe, the higher the cost for materials, installation, and maintenance. However, this is offset by a corresponding decrease in energy needs to move the fluid through the system.

The primary variable affecting energy cost is the velocity of the fluid.[21] To determine the velocity of a fluid flowing through a pipe, the flow rate, pipe outside diameter, and wall thickness must be known. The flow rate is generally fixed, and is the starting point in any piping design. The rules for determining required wall thickness rules for piping are presented in Chapters 5 and 6 for primary pipes, and in Chapter 7 for secondary containment pipes. Selecting a nominal pipe diameter size will set its inside diameter requirements, since the required wall thickness to maintain pressure are defined by the rules of Chapters 5 through 7. Since flow rate is a known quantity, the fluid velocity is then defined.

For a variety of diameter sizes, the total costs can be determined and plotted as a series of curves. The curves can then be translated into total cost equations and the equations differentiated to determine the optimum diameter.[22] Smith also suggests a straightforward procedure for selecting optimum diameter, once a construction material has been selected. A number of economical pipe diameter charts have been developed based on the same principles.[23,24] For a more comprehensive discussion on this subject, it is suggested that the reader refer to: Peters and Timmerhaus, Plant Design and Economics for Chemical Engineers, 3rd ed., McGraw-Hill, New York, 1980, Chapter 10.

4.3 Pressure Flow in Annuli (Secondary Containment Piping Considerations)

4.3.1 General Comments

In the design of double containment piping systems, there are one or more inner primary pipes positioned within a secondary containment pipe, creating an annular space. The ability of a primary fluid, or a flushing fluid, to flow effectively in the annular space in the event of a leak must be considered. Indeed, the flow capability of a fluid in the annular space is one of the major criteria in establishing the proper inside diameter of a secondary containment pipe. Additionally, there is the need once a leak occurs for the annular compartment to be flushed with water or other fluid to decontaminate the annulus. Once an annulus has been decontaminated and drained, there may be the need to blow clean, dry gas or air through an annulus to dry it. In systems where gases are handled, there is the need to remove the gaseous content by pulling a vacuum, or by blowing a compressed clean, dry gas through it. All of these considerations must be analyzed when designing a double containment piping system. One must consider friction losses and flow disturbances created by supporting devices, baffles, and the outside diameters of the primary pipe or pipes.

4.3.2 Equivalent Diameter and Hydraulic Radius

The same basic equations to determine frictional losses and pressure drops that are used in normal piping are used for annuli. However, the diameter term must be replaced by an appropriate factor to model the flow properly. The Darcy equation can be used to calculate friction factor and pressure drop in a circular annulus, if the term for diameter is replaced by a hydraulic diameter. Hydraulic diameter is calculated by multiplying the hydraulic radius by four, given by:

$$R_H = \frac{(d_2 - D_1)}{4} \tag{4.62}$$

where:
R_H = hydraulic radius, ft
d_2 = inside diameter of the secondary containment pipe, ft
D_1 = outside diameter of the primary pipe, ft

There are other annular configurations beside circular to consider. Annular configurations involving longitudinal fins and eccentric annuli may also be analyzed.[25] For multiple inner pipe systems within a circular common outer pipe, the hydraulic radius is calculated as the free volume in the tube bank divided by the exposed surface area of the tubes.

4.3.3 Annulus Reynolds Number

Calculating the Reynolds number in an annulus depends on whether there are annular orifices or baffles present. Without any annular orifices, the Reynolds number is calculated by using the hydraulic diameter. However, if there is a baffle containing an orifice in the annular compartment, the Reynolds number is calculated by:

$$N_{RE} = \frac{(d_2 - D_1)G}{v} \tag{4.63}$$

where:
G = the mass velocity through the orifice opening, lbs/s
v = kinematic viscosity, centipoise

4.3.4 Friction Factors and Pressure Drop Calculations

The drag force for two-dimensional flow past a continuous cylinder of a given unit area is proportional to the density and the square of the velocity. The force is calculated from a momentum balance, assuming the momentum for the x vector is reduced proportionally at all Reynolds numbers. To estimate the drag force use:

$$F'_d \propto \frac{\rho u_o^2}{2g_c} \frac{A}{L} \tag{4.64}$$

where:
F'_d = the drag force per unit length
A/L = the frontal area per unit length
u_o = center velocity, ft/s

The ratio of the area of a cylinder to its length is proportional to the cylinder's diameter. Therefore, the equation can be reduced to:

$$F'_d \propto \frac{\rho u_o^2 D}{2g_c} \tag{4.65}$$

As the Reynolds number changes, the incremental decrease in the x vector momentum is affected by the drag coefficient, and Eq. (4.65) becomes:

$$F_d' = C_D \frac{\rho u_o^2 D}{2g_c} \tag{4.66}$$

where:
C_D = drag coefficient

The critical length Reynolds number at which the boundary layer becomes turbulent is defined by:

$$(N_{RE})_{x,crit} = \frac{V\rho x}{\mu} = 200,000 \tag{4.67}$$

For a laminar boundary layer, the drag force can be calculated from:

$$D = \left(\frac{\rho}{g_c}\right)V^2\theta \tag{4.68}$$

where:
θ = momentum area [provided in Fig. 5-72 of the 6th Edition of Perry's Chemical Eng. Handbook, and from Sakiadia, Am. Inst. Chem. J., 7, 467 (1961)]

Experience with design of double pipe heat exchangers in turbulent annular flow suggests that actual friction factors may be 5 to 10% greater than those calculated using the hydraulic diameter.[26] For additional information on this subject, other references are suggested.[27-29]

For laminar flow in annular channels, pressure loss can be calculated using Eq. (4.69). This will be the likely flow condition in a design involving pipes contained in a secondary containment trench. For significant flow to occur, a leak in one of the pipes must produce a relatively large amount of flow, or a large volume of flushing fluid would have to be used. Flow must completely surround the pipes for this equation to apply, and the flow must be laminar as described. Laminar flow in an annular channel is covered in detail in suggested readings.[30]

$$\frac{D_o \Delta p}{4L} = K\left(\frac{2n+1}{2n}\right)^n \frac{1}{(1+k)^n(1-k)^{1+2n}} \left(\frac{32q}{\pi D_o^3}\right)^n \tag{4.69}$$
$$\text{for } n > 0.25 \text{ and } k > 0.5$$

where:
$k = D_i/D_o$
D_o = the outer diameter of the annulus
D_i = the inner diameter of the annulus
K, n = material constants, (see p. 5-26 of Perry's Chemical Engineering Handbook, 6th Edition)

The design of multiple inner pipe systems contained within a single common secondary containment pipe is very similar to the design of shell and tube one-pass heat exchangers. Many references present methods for calculating the pressure drop in shell and tube heat exchangers.[31-33]

Fluid Dynamics and Sizing Analysis

Equation (4.70) can be used to calculate the friction loss in a fluid flowing under turbulent flow past a bank of N rows of tubes. In this type of system, turbulent flow exists for Re > 500; laminar flow exists when $N_{RE} < 100$, and the transition region is from $100 < N_{RE} < 500$.

$$f' = 0.99\left(\frac{\rho u_b D}{\mu}\right)^{-0.2} \quad (4.70)$$

where:
f' = friction loss, ft
u_b = average velocity, ft/s^2

There are many other equations that exist to calculate frictional losses in shell and tube heat exchangers. A comprehensive survey of many of the equations that have been developed is given in Chapter 11 of Knudsen and Katz et al.[31] For six pipes positioned within a hexagonal secondary containment channel, friction factors and hydraulic diameters are reported in other readings.[34]

4.3.5 Effect of Annular Baffles and Orifices

Oftentimes, interconnecting parts and/or interstitial supports are used that partially restrict flow across the annulus cross-sectional area. Depending on the geometry of the annular obstruction, one can calculate the pressure loss for an annular orifice, by treating openings as annular orifices. Frictional losses for annular flow through annular orifices may be determined by using Eq. (4.71). A cross-sectional view of an annulus having one annular orifice that is otherwise solid in the cross section is shown in Figure 4.12.

$$w = q_1 r_1 = CYA_2 \sqrt{\frac{2g_c(p_1 - p_2)\rho_1}{1 - \beta^4}} \quad (4.71)$$

$$= KYA_2 \sqrt{2g_c(p_1 - p_2)\rho_1}$$

where:
C = conductance, ft^3/s
K = fluid bulk modulus of elasticity, lb$_f$/ft^2
Y = expansion factor, dimensionless
A_2 = cross-sectional area of the annulus, ft
β = ratio of diameters ($= D_1/d_2$), dimensionless

For an annular Reynolds number found from Eq. (4.63) in the range of 100 to 20,000, a K value of 0.63 to 0.67 can be used.[35] These numbers were developed for β ($= D_1/d_2$) in the range of 0.95 to 0.966. In systems where the β value is smaller, a different value for K may apply.

For annular baffles containing more than one hole, pressure drop can be modeled by using a combination of the data that exist for perforated plates and annular orifices, and applying good engineering judgment. For examples of common interstitial supporting devices, refer to Section 9.3.1.2.

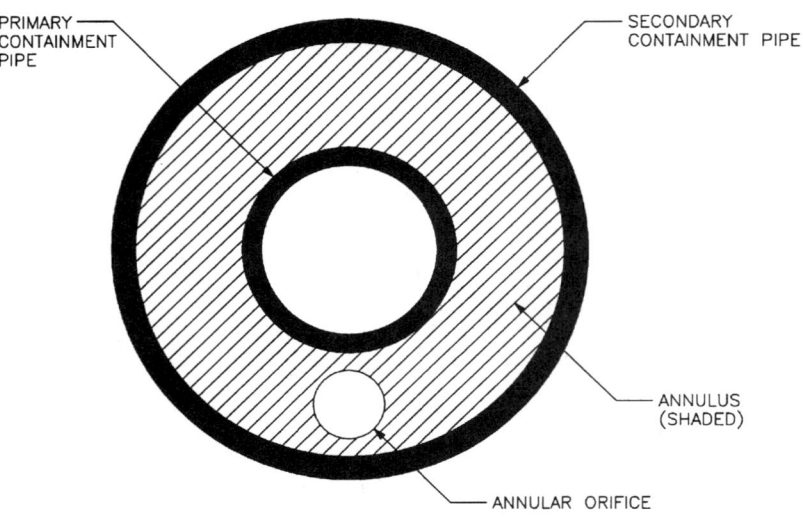

Figure 4.12. Illustration showing an annular opening (orifice) in an annulus (shaded portion).

4.3.6 Flow around Annular Objects and Vortex Shedding in an Annular Space

One of the most common objects around which a liquid in an annular space must flow is a thin, flat plate or fin. This is due to the use of a three-, four-, five- or six-legged spider, or having short longitudinal fins welded directly to the outer wall of a primary pipe as a means of centering and support (as shown in Figure 9.8). When a double containment pipe is designed in such a fashion, potential flow in the annular space under upset conditions, or during flushing procedures, may be subject to vortex shedding.

Vortex shedding occurs when fluids flow past an object, causing vortices to be shed periodically. Pipes can be subject to damaging vibrations under such conditions, particularly if the frequency of the vortex shedding is close to the natural harmonic frequency of the pipe. Vortex shedding is one of the most interesting dilemmas inherent in double containment piping. While there have been few reported troubles of vortex-shedding-induced vibrational damage, the annular spaces of installed systems will only be used in the event of failure and subsequent flushing procedures. Most applications involving double containment piping have not been installed long enough to reach a point of failure; these concepts may not have had a chance to have been experienced.

Vortex shedding at the very least presents a potential problem, and must be considered in the design of a system. If vortex shedding is found to be a potential problem, the velocity of the annular flow must be lowered. This can be accomplished either by increasing the diameter of a secondary containment piping, or by using annular structures of different geometrical configuration.

The velocity in a vortex that is shed will typically be 86% of the free-stream velocity of the fluid. The frequency of the vortices that are shed can be calculated from the Strouhal number, N_S, which is a function of the Reynolds number, and can be calculated by Eq. (4.72). Below a Reynolds number of 40, vortices are difficult to detect. Between a

Fluid Dynamics and Sizing Analysis

Reynolds number of 40 and 500, the Strouhal number is directly related to the Reynolds number. Above $N_{RE} = 10^5$, the Strouhal number increases at a very fast rate.

$$N_s = \frac{fD}{V} = 0.19, \quad 500 < N_{RE} < 10^5 \tag{4.72}$$

The frequency of the vibration of an object is equal to the frequency of vibration of vortex shedding, unless the frequency of vortex shedding is within 80 to 120% of the object's natural resonance. It then takes place at the object's natural harmonic frequency.[36]

When vortex shedding occurs, an alternating lateral force acts normal to the direction of flow. The maximum lateral force acting in a direction away from the terminal vortex can be calculated by the following equation. Its magnitude is roughly twice the anticipated force due to fluid drag (note: C_K is von Kármán coefficient, F_K is von Kármán or lateral force).

$$F_K = C_K A_r \frac{V^2}{2g_c} \tag{4.73}$$

For flow past a finite flat plate, such as that of a spider or vane-type interstitial support, the critical Reynolds number at which annular flow becomes turbulent is found from Eq. (4.74). The total drag for laminar flow along this type of object is given in Eq. (4.75).[37,38] For a turbulent boundary layer where $(N_{RE})L$ is less than 10^7, the total drag is found from Eq. (4.76).[39,40] In Eqs. (4.75) and (4.76), "b" is a coefficient.

$$(N_{RE})_{x,crit} = \frac{V\rho x}{\mu} = 500,000 \tag{4.74}$$

$$D = 1.328 bLV^2 \left(\frac{\rho}{g_c}\right)(N_{RE})_L^{-0.5} \tag{4.75}$$

$$D = 0.072 bLV^2 \left(\frac{\rho}{g_c}\right)(N_{RE})_L^{-0.2} \tag{4.76}$$

A fluid may also have to flow past a thin flat continuous plate, such as found in a thermoplastic combination inner and outer pipe having continuously extruded longitudinal vanes. The critical Reynolds number at which the flow becomes turbulent for a continuous flat surface may not be the same as that determined by Eq. (4.74), but may instead be higher.[41] For laminar flow under this condition, the total drag is calculated by Eq. (4.77); for turbulent flow it can be determined by Eq. (4.78). The numbers determined by Eq. (4.79) may be as much as 15% lower.[42] Many other excellent references on this subject are recommended readings.[43,44]

$$D = 1.776 bLV^2 \left(\frac{\rho}{g_c}\right)(N_{RE})_L^{-0.5} \tag{4.77}$$

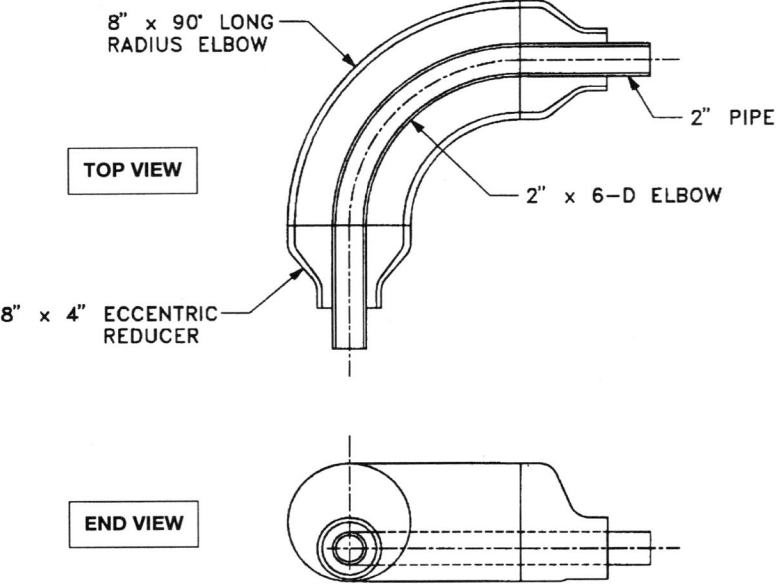

Figure 4.13. Localized low points in an annulus can be minimized by using eccentric increasers where a localized increase in diameter of the outer jacket is needed.

$$D = 0.056 b L V^2 \left(\frac{\rho}{g_c}\right)(N_{RE})_L^{-0.2} \qquad (4.78)$$

Primary piping may be positioned in an eccentric fashion within a much larger outer pipe. Use of such an arrangement can provide added room to alleviate expansion, and it may be found in expansion loops, expansion offsets, expansion swing joints, and changes in direction. Such an arrangement is particularly useful in underground piping. Localized low points in an annulus may be minimized by positioning eccentric reducers accordingly, as shown in Figure 4.13. To determine the friction factors and pressure losses in an eccentric annulus, consult Knudsen and Katz, Fluid Dynamics and Heat Transfer, pp. 193-205, McGraw-Hill, New York, 1958.

An eccentric annulus will rarely become totally filled with fluid. Systems having an eccentric annulus are usually designed for gravity-drainage flow, as in a chemical sewer-type system (i.e., oily-water collection systems). However, such arrangements can also be used for systems requiring a pressure-rated annulus.

4.3.8 Gas Flow Through Annuli

Analysis of the flow of compressible fluids in the annular space of a double containment piping system follows the same basic procedures as in round circular pipes. The hydraulic radius and hydraulic diameter used are the equivalent diameter for the calculation of frictional losses. A compressible fluid flowing across objects may also result in vortex shedding, with potentially damaging results, as described in Section 4.

4.3.9 Vacuum Flow in a Circular Annulus

Vacuum may be used in annuli for insulating purposes, or to remove contaminants and allow an annular space to dry after a failure of a primary system. The capacity of an annular space to allow for flow under vacuum conditions is the conductance of the pipe, which is the reverse of its resistance. Conductance is normally measured in terms of cubic meters per second (cubic feet per second), as is defined by:

$$C = \frac{q'}{\Delta P} \qquad (4.79)$$

where:
C = conductance
q' = flow rate, ft^3/h
ΔP = pressure drop, psi

Using vacuum is a good option for the evacuation of an annular space following the release of a fluid with a high vapor pressure (or low boiling point). To calculate the conductance of an annular space, one can use[45]:

$$C = C_1 K \frac{(D_1 - D_2)^2 (D_1 + D_2)}{L} \sqrt{\frac{T}{M}} \qquad (4.80)$$

where:
K = fluid bulk modulus of elasticity, lb_f/ft^2
T = temperature, °R
M = molecular weight, lb/lb mol
L = length, ft

4.4 Flow of Fluids in Chemical Process Sewer and DWV Piping

The design of piping for gravity-flow conditions is inherently different from that of pressure piping. In pressure piping design, one selects a diameter that allows the fluid to be delivered at a safe velocity while optimizing the total of materials, installation, maintenance, and pump energy costs. For drainage piping, a pipe diameter and layout that allows the fluid to drain as smoothly and rapidly as possible, without noise, and with as little chance of backup as possible is chosen. Where dangerous sewer gasses pose a risk, the pressure of drainage piping must be maintained within a very tight range of fluctuation (±1 in. of water) to ensure that traps remain sealed. This tight tolerance over internal pressure is maintained by proper venting and effective layout techniques. In pressure piping, there is no such requirement, as the system is always pressure tight.

In the design of gravity-drain double containment piping systems, both the primary piping system and the annular space between the two pipes must be designed for effective drainage capability. Any design of a double containment drainage system must ensure that its primary system will allow ready drainage of fluids, and, in the event of a leak, so will its secondary containment system.

4.4.1 Applicable Building Codes and Design Standards

Design of chemical waste, drain, and vent systems are often governed by local building codes. One must consult the authority having local jurisdiction to determine what local building codes cover such systems. These rules or those of the insurance underwriter usually take precedence over other design rules, including those presented in this text.

Some municipalities do not cover chemical sewer systems, and instead classify them as process systems, the design of which is left up to the owner. In the United States and Canada, depending on the fluid services, system pressure, and location of the system, ANSI/ASME B31.3 Code rules may apply.

4.4.2 Primary (Inner) Piping Considerations

4.4.2.1 Gravity Flow in Vertical Risers

Liquid waste is normally introduced into a vertical stack by use of a long-turn-tee-wye (45° lateral with 1/8th bend), or sanitary tee. The liquid may or may not fill the internal cross-sectional area of the stack piping. As the liquid enters the stack, gravity immediately propels it downward. Within a very short time, it swirls around the inside diameter of the vertical stack pipe, assuming the form of a sheet. The liquid continues to accelerate downward, slowed only by the action of frictional forces of the liquid waste along the inside walls of the vertical stack piping.

The liquid waste reaches a maximum rate of flow when the frictional forces acting on the liquid are equal to the force of gravity. From that point until it reaches the bottom of the stack, the liquid remains relatively unchanged both in shape and velocity. The maximum velocity, otherwise known as terminal velocity, is achieved within a relatively short period of time and at a relatively short distance, known as terminal length.

Terminal velocity can be calculated by Eq. (4.81) and terminal length by Eq. (4.82). A typical terminal velocity for water is 10 to 15 ft/s; the corresponding terminal length is approximately one building story height (10 to 15 ft).[46]

$$V_T = \frac{q^{2/5}}{3.0d} \qquad (4.81)$$

where:
V_T = terminal velocity, ft/s
q = flow rate, ft^3/s
d = inside diameter, ft

$$L_T = 0.052(V_T)^2 \qquad (4.82)$$

where:
L_T = terminal length, ft

Liquid that falls along the inside diameter of the stack assumes the shape of a ring, and flows along with air in the core of the ring. As this core of air is dragged down the stack along with the liquid waste, a pressure reduction occurs in the stack due to air being dragged along by the fluid. If the air is not replaced, a vacuum condition will be created in the stack.

Fluid Dynamics and Sizing Analysis

Table 4.4 Maximum Capacities of Vertical Stacks

Pipe size, in.	Flow in GPM		
	r = 1/4	r = 7/24	r = 1/3
2	18.5	23.5	--
3	54	70	85
4	112	145	180
5	205	270	324
6	330	435	530
8	710	920	1145
10	1300	1650	2055
12	2050	2650	3365

Air can be replaced by venting the stack through the roof through a "vent extension" (see Section 4.4.2.5.2), or through an "accompanying" vent stack (see Section 4.4.2.5.1). If a branch discharging fluid is encountered on the way down the stack, fluid will disrupt the falling liquid waste and air. Back pressure is then created in the discharging branch, with a resulting increase in the rate and velocity of flow out of the stack and drain.

The capacity of a stack may be determined by the ratio of the area occupied by the fluid to the total cross-sectional area of the pipe. The capacity of the stack as a function of this ratio and the stack diameter, is given by:

$$q = 27.8 \, r^{5/3} d^{8/3} \tag{4.83}$$

where:
r = ratio of the cross-sectional area of the sheet of water to cross-sectional area of the stack.

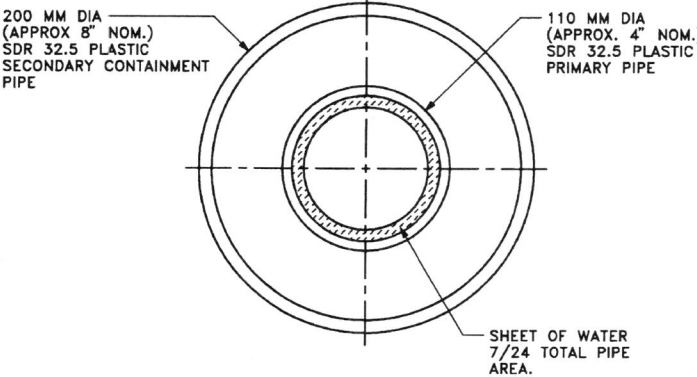

Figure 4.14. Cross-sectional view of a vertical stack that is occupied by a fluid on 7/24th's of its diameter under typical flow conditions once the terminal length is reached.

Values for flow rates based on flow ratios of 1/4, 7/24, and 1/3 are shown in Table 4.4. A stack that has a cross-sectional area that flows 7/24 full is illustrated in Figure 4.14.

As liquid waste continues to be discharged into a stack, the ratio of cross-sectional area of the sheet of waste to the cross-sectional area of the stack increases. If the stack allows a ratio greater than 0.33 to occur, the sheet of water will overcome the frictional resistance of the air mass. At some higher ratio, the liquid will diaphragm across the stack's entire cross-sectional area, creating back-flow and pressure differentials that could exceed ±1 in. water.

Drainage stacks should be based upon a ratio that never exceeds 1/3 full for water. Most plumbing codes base stack requirements on a ratio of 7/24. When handling acid and other chemical wastes, the design should be conservative, since the pipes are carrying fluids that are highly corrosive. Therefore, the design of a stack to handle chemical and other acid wastes according to the above criteria will be based on a conservative design, since the greater densities of the fluids help them to overcome the frictional resistance of air. A cross-sectional ratio of 1/4 to 7/24 ensures that a falling liquid will not diaphragm across a pipe's internal diameter and form slugs (unless it is interrupted by flow from another branch).

At the bottom of a vertical stack, a fluid normally enters a horizontally sloped building drain pipe through a 90° elbow located at the base of the stack. A fluid enters a horizontal building drain pipe at a relatively high velocity, but its velocity decreases rapidly, though all gravity-fed drain pipes usually slope vertically to some extent (common values are 1/8 to 1/4 in./ft; approximately 1% to 2%).

A fluid will continue to slow and fill a cross section until it suddenly completely fills a pipe's cross-sectional area. This is called hydraulic jump, occurring anywhere from immediately after the elbow at the base of the annular stack, and up to ten diameters from the base elbow, as shown in Figure 4.15. The magnitude of jump varies with pipe diameter and quantity of flow.

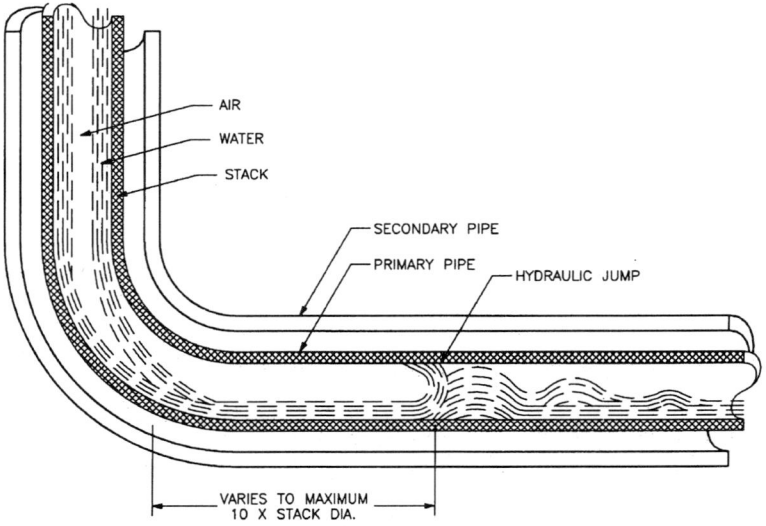

Figure 4.15. Illustration of hydraulic jump in the primary pipe of a double containment piping system.

Table 4.5 Minimum Recommended Trap Size for Fixtures

Fixture	Trap Size, in.
Cup sink, laboratory	1.5
Funnel drain	1.5
Laboratory sink	2
Sinks:	
Surgeon	1.5
Flushing rim (with flush valve)	3
Service (trap standard)	3
Service (P trap)	2
Pot, scullery, etc.	1.5
Sterilizers	2

Source: A. Steele, Engineered Plumbing Design, Construction Industry Press, Elmhurst, IL, 1982.

Immediately after the point where hydraulic jump occurs, the pipe flows full until frictional resistance of the inner walls of the drain pipe slow the fluid to the point where uniform flow occurs. Any offset of greater than 45° can slow a fluid from its terminal velocity condition, resulting in hydraulic jump. Consecutive offsets may produce similar results.

4.4.2.2 Fixture Outlets and Fixture Loads

An outlet drain pipe from a fixture that feeds into a chemical waste piping system should be designed so that the outlet does not flow completely full. An outlet designed to accommodate approximately half-full flow will prevent self-siphonage and the loss of trap seals. The minimum size of a fixture drain is also set by the size of the fixture trap, if there is one associated with the drain. Minimum recommended sizes of traps for commonly encountered fixtures in a chemical sewer-type system are shown in Table 4.5.

The pressure at a fixture drain is equal to the height of fluid above it. The pressure will be atmospheric for a drain discharging at its design capacity of approximately half-full. The average rate of flow from a fixture can be determined by:

$$q = 13.17 \, d^2 h^{1/2} \qquad (4.84)$$

where:
h = height, ft
d = inside diameter of the drain, ft

In chemical-waste/sewer piping systems, several types of fixtures may be encountered (i.e., laboratory sinks, fume hood work station sinks, emergency showers, floor drains, process equipment outlets, and others). The discharge from such fixtures is normally stated by plumbing engineers in terms of fixture units.

A fixture unit is a measure of the flow from fixtures to chemical waste piping (or plumbing drain waste and vent piping), expressed in units of 7.5 gpm. The total discharge (in gallons per minute) of the fixture at maximum expected frequency of discharge, may by divided by 7.5 gpm to determine "fixture units." The value of 7.5 gpm is

Table 4.6 Fixture Units for Various Fixtures

Fixture type	Fixture-unit value as load factors
Combination sink and tray	3
Floor drains	1
Sinks:	
Surgeon	3
Flushing rim (with flush valve)	8
Service (trap standard)	3
Service (P-trap)	2
Pot, scullery, etc.	4
Wash sink (circular or multiple), ea. set of faucets	2

Source: A. Steele, Engineered Plumbing Design, Construction Industry Press, Elmhurst, IL, 1982.

a commonly accepted value based on experience, and is the value used in most plumbing codes. Some municipalities require different values; it is up to the project manager of record who has responsibility for a given project to verify that the appropriate building codes are used by designers.

The fixture unit is a dimensionless value that defines a unit of flow for fixtures, and is an important value in many plumbing engineering calculations. The reason that fixture units are used in plumbing design is that a variety of fixtures may be used on each design, each having unique discharge-rate characteristics. Since chemical waste piping systems function in the same manner as plumbing (drain, waste, and vent) piping, they perform best when they are engineered using techniques that are applicable to sanitary drain, waste, and vent piping. Therefore, the fixture unit is important for chemical sewer system design.

Values of fixtures units for a variety of fixtures involved in chemical-waste/sewer piping systems are listed in Table 4.6.

The maximum permissible fixture unit loads for branches, vertical stacks, and horizontal headers (building drains) are shown in Appendix B, Tables B-1 through B-3.

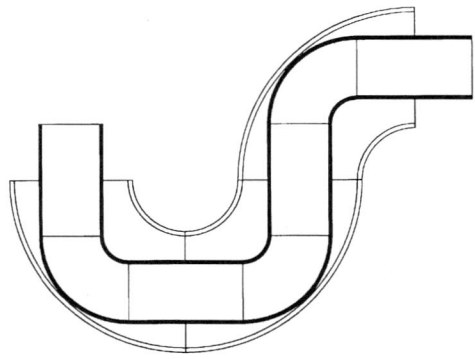

Figure 4.16. Typical double containment P-trap fabricated from butt-welding fittings (refer to Figure 7.7).

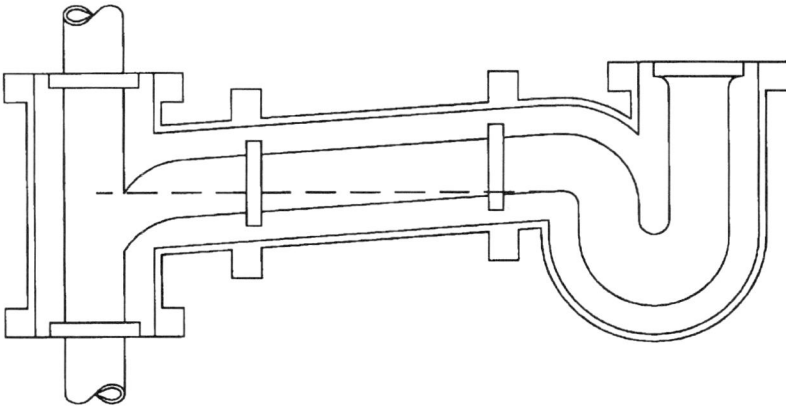

Figure 4.17. Example of a double containment P-trap, whereby the primary trap is vented.

4.4.2.3 Fixture Traps

Traps are provided in piping systems to prevent the back flow of dangerous, corrosive, or regulated sewer gases out of fixtures. A minimum seal of 2 in. (50 mm) of water must be maintained in a trap. Since the requirement for the design of both chemical waste piping and sanitary drain waste and vent piping system is to maintain a pressure fluctuation of ±1 in. of water (25 mm of water), a trap with a 2 in. (50 mm) seal will function with a 100% safety factor at all times. A minimum of a two-times safety factor is generally considered to be adequate if a system is adequately vented (refer to Section 4.4.2.5). A typical double containment P-trap is illustrated in Figure 4.16. If a trap is vented as illustrated in Figure 4.17, the vent pipe opening should never be placed below the level of the weir of the trap's crown, or there will be an increased possibility of self-siphonage of the trap.

4.4.2.4 Gravity Flow in Horizontal Branches

The rate of flow in a branch is the sum of all the individual flows that are discharged into it, including other sub-branches. The design of a branch should be based upon uniform flow, with the same criteria as for a horizontal building drain. Design of horizontal drains is discussed in Section 4.4.2.6.

Branches that are less than five feet in length are an exception. Branches that are this short will not permit a draining fluid to achieve uniform flow; surging conditions will exist.

The flow in a branch should never exceed that of a stack or drain into which it discharges. Also, a branch must never be designed so that it will disrupt the sheet of fluid in a stack to an extent that back pressure will result in the stack. Proper selection of the size of a stack will ensure that the total flow from branches discharging simultaneously will not result in significant pressure fluctuations due to back flow.

4.4.2.5 Vent Design

The characteristics of liquid waste as it falls down a stack are described in Section 4.4.2.1, "Gravity Flow in Vertical Risers." Air that is dragged down the center core of stacks must be vented, or pressure fluctuations in the system of greater than ±1 in. of water (±25 mm of water) may result (also see Section 4.3.9). At every point in a vertical stack where liquid is flowing uniformly at its terminal velocity and the flow is suddenly disrupted, the accompanying air flow is also brought to an abrupt halt, creating a pressure fluctuation in the system. A pressure fluctuation of more than 2 in. (50 mm) of water column will cause standard-dimension traps at fixtures to lose their seals, creating the possibility that dangerous acid waste or solvent fumes can escape back through the fixture.

The proper venting of a chemical sewer waste system will be such that pressure fluctuations are kept to a level of ±1 in. of water column (25 mm of water column), under the worst possible conditions. This will ensure that the seals of all fixture traps are maintained, and that the possibility of dangerous, corrosive, or regulated gases escaping in the building is kept to a minimum.

4.4.2.5.1 Attendant Vent Stacks

For a vertical drain stack that is designed based upon the condition of a cross-sectional ratio of 7/24, the remaining 17/24 of the diameter is filled with air. Air flow and fluid flow versus vertical stack diameter is shown in Table 4.7. This air is dragged down the column along with the fluid and must have a place through which to flow at the base of the stack. A vent stack is normally provided at the base of all tall stacks in order to provide an avenue of escape for the flowing air. Such a vent stack must have a large enough diameter such that air is relieved and there is less than ±1 in. (±25 mm) water column pressure fluctuation in the system. In addition, the air must be replaced or a vacuum will be created and siphonage of the traps will occur. This is accomplished by extending the drainage stack all the way through the roof (this is referred to as a vent extension, described in Section 4.4.2.5.2).

To calculate the maximum length of vent piping based upon the pressure fluctuation requirement of ±1 in. (±25 mm) of water column, the Darcy equation can be used.

$$L = \frac{2226 d^5}{fq^2} \qquad (4.85)$$

where:
L = length, ft
d = inside diameter, ft
f = Darcy friction factor (see Section 4.2.1.3)
q = flow rate, ft^3/s

An attendant vent stack should be provided for every primary drainage stack in a chemical sewer system that drains fluid more than two building stories in height. It should extend all the way to a building's roof, unchanged in diameter size. An attendant vent stack can be connected to a vent extension of the primary stack (see Section 4.4.2.5.2) so that there is only a single penetration through the roof. Attendant vent stacks and vent extensions of a primary pipe in a double containment piping system may be combined into a single vent header.

The minimum diameter of a vent stack is one-half the diameter of the stack to which it is attending. An attendant vent stack serves multiple purposes, by also providing air to a stack at the various branch and connections and fixtures, through accompanying vent

Table 4.7 Air Flow and Water Flow in Vertical Stacks

Diameter of drainage stack, in.	Water flow, gpm	Air flow, gpm
2	23.5	57.1
3	70	170.1
4	145	352.4
5	270	656.1
6	435	1057.1
8	920	2235.6
10	1650	4009.5
12	2650	6439.5

Source: A. Steele, Review Manual for the CIPE Exam, ASPE, Sherman Oaks, CA, 1984.

branch cross-connections (relief vents). In buildings with more than ten stories, a vertical drain stack should at a minimum be provided with a relief vent at each tenth branch interval (counting from the top downward).

In some cases, individual fixtures may have to be vented because of local rules of the authority having local jurisdiction. However, there are other alternative methods to vent fixtures, including: (1) wet venting; (2) stack venting; (3) circuit and loop venting,; and (4) combination waste and vent venting. For a comprehensive discussion of these methods, consult: Steele, Engineered Plumbing Design, Construction Industry Press, Elmhurst, IL, 1982. When each primary pipe portion of a double containment pipe fixture trap is vented, the design is referred to in common plumbing engineering terminology as a "continuous" vent design. The recommended maximum distance of a vent from a given fixture trap, regardless of how it is vented, is shown in Table 4.8.

For flow in horizontal branches, branches are normally designed to flow approximately one-third to one-half full. Therefore, the flowing waste in a branch will drag most of the air in the upper half of the pipe along with it. The rate of air flow for horizontal branches of various diameters at a variety of slopes based upon this assumption are shown in Table 4.9.

Accompanying vent stacks and the vent extensions of chemical waste stacks can be combined into a single vent header extending through a single penetration in the roof. A vent header sizing table is included in Appendix B, Table B-4. The sizing of vent headers involves knowing the total number of fixture units feeding the vent stacks and the number of vent extensions that connect into the header. For a given number of units, the diameter

Table 4.8 Vent Distance from Traps

Size of fixture drain, in.	Maximum distance of vent to trap, in.
1.25	30
1.5	42
2	60
3	72
4	120

Source: A. Steele, Review Manual for the CIPE Exam, ASPE, Sherman Oaks, CA, 1984.

Table 4.9 Rate of Air Flow in Horizontal Branches

Diameter of drain, in.	Slope, in. per ft	Rate of flow, gpm
1.5	1/4	6.0
2	1/4	8.8
2.5	1/4	15.5
3	1/4	25.5
4	1/8	38.0
5	1/8	69.0
6	1/8	112.0
8	1/8	240.0

Source: A. Steele, Review Manual for the CIPE Exam, ASPE, Sherman Oaks, CA, 1984.

and maximum permissible length of a vent can be determined. For a horizontal header, twenty percent of the total lengths shown in Table B-4 of Appendix B are used.

4.4.2.5.2 Vent Extensions

The air that is dragged down a vertical waste stack along with a fluid must be replenished or a vacuum will be pulled. A vertical stack may provide replenishment for the "dragged" air by having its top extended through a building roof. This is termed a vent extension, which is usually sized to the same diameter as the rest of the stack, unless vent extensions of other stacks or accompanying vent stacks tie into it. Vent extensions and combined vent extension headers must be sized properly, or a vacuum will be created in the stack, and siphonage of fixture traps will occur.

4.4.2.5.3 Venting of Closed Sumps

In many chemical sewer waste systems, the waste discharges from the bottom of a vertical stack into a collection sump, collection storage tank, or neutralization basin. If units such as these are closed air-tight, they also must be vented. The permissible length of such a vent can be determined by Eq. (4.85). In practice, a diameter size of 3 in. has been found sufficient for sanitary ejectors and sumps.[47]

4.4.2.5.4 Natural Draft Circulation in Vent Systems

In a properly designed chemical waste sewer vent system, air will flow naturally and prevent the buildup of foul odors. The circulation is created because the air pressure of the ambient air outside the system is normally slightly different from the stagnant air inside the vent system. This difference is due to differences in temperature between the two air masses. Cool air, being more dense, tends to displace warm air that is less dense, inducing natural draft circulation.

The formula for natural circulation is given by Eq. (4.86). Under conditions of natural draft circulation, the frictional loss of the air on the walls of the vent piping will typically be just small enough to allow the natural flow to occur.

$$H = 0.1925(w_0 - w_1)H_s \qquad (4.86)$$

Fluid Dynamics and Sizing Analysis

Table 4.10 Values of S and $S^{1/2}$

Slope, in. per ft	S, ft/ft	$S^{1/2}$
1/8	0.0104	0.102
1/4	0.0208	0.144
1/2	0.0416	0.204

4.4.2.6 Gravity Flow in Horizontal Drain Pipes

As fluid flows in a sloping horizontal pipe under gravity flow, it will normally fill only a portion of its diameter. The flow is a result of the gravitational force generated by the slope of the pipe and the height of fluid. The same is true for fluid flowing within an open trench or open channel of any shape. For a constant amount of flow, the fluid will always reach an equilibrium condition that is referred to as uniform flow.

Uniform flow is achieved in a pipe or open channel having a constant geometry and a fixed slope. When the slope of a fluid surface matches that of its pipe (or open channel), it has achieved uniform flow. The most commonly used equation for determining the velocity in an open conduit of constant shape and size under uniform flow conditions is the Manning equation:

$$V = \frac{1.486}{n} R^{2/3} S^{1/2} \tag{4.87}$$

To determine the quantity of flow, one may use:

$$Q = AV \tag{4.88}$$

By replacing the velocity term in Eq. (4.87) with $V = Q/A$ [rearrangement of Eq. (4.88)], the flow rate form of the Manning equation is derived:

$$Q = A \frac{1.486}{n} R^{2/3} S^{1/2} \tag{4.89}$$

The hydraulic radius term (R) in Eqs. (4.87) and (4.89) is the ratio of the cross-sectional area of flow to the wetted perimeter of a pipe's exposed surface. Tables B-5 and B-6 of Appendix B give values for circular pipes of full and half-full flow for R, $R^{2/3}$, and their cross-sectional areas. The values of S and $S^{1/2}$ are shown in Table 4.10.

The flow rates for a variety of nominal diameter sizes, determined using Eq. (4.88), at half- and full-flow conditions are given in Table B-7 of Appendix B for a slope of 1/8 in./ft (1%) for sanitary drains. Values for storm drains at a slope of 1/4 in./ft (2%) are given in Table B-8 of Appendix B. The minimum velocity needed to achieve scouring action in a pipe is 2 ft/s. When the velocity falls below this level, suspended solids such as sand, grit, and metallic particles will begin to deposit in the pipe. If solids pose a problem, the minimum diameter of a horizontal drain pipe should be such that the velocity of a draining fluid does not fall below 2 ft/s.

4.4.3 Shapes of Horizontal Drainage Pipes

A flowing fluid, as previously mentioned, must be maintained at a minimum of 2 ft/s in order for scouring action to take place. For horizontal drainage piping handling wide fluctuations in flow, there can be periods of relatively little flow. The same is true for pipes that are infrequently used. In round pipes or rectangular-shaped trenches, the fluid velocities can drop below 2 ft/s under low flow. One solution is the use of lined-concrete structures of alternative shape, such as oval, egg, elliptical, triangular, etc. Some of these alternative geometries are pictured in Figure 4.18. Each has a shape such that its cross-sectional area becomes smaller as the fluid level becomes less.

4.4.4 Surcharging of Horizontal Drainage Pipes

The capacity of a drainage pipe may be higher than that possible under conditions of gravity uniform flow through the use of surcharging. Surcharging is a method of accomplishing this feat, by using risers or manholes, as illustrated in Figure 4.19. When flow in the drainage pipe is increased beyond its design capacity, the fluid level rises in the manhole. This imposes a pressure head on the system in excess of atmospheric pressure, which will increase the velocity and flow capacity beyond that of ordinary full-flow, gravity conditions.

The magnitude of surcharging is expressed as feet or meters of maximum anticipated surcharge (measured from the highest level of liquid, under extreme conditions, to the crown elevation of the drain pipe). Surcharging is useful where increased future demand is a possibility. There is also economic incentive, since one can decrease the size of a conduit and still be able to achieve a specified flow capacity.

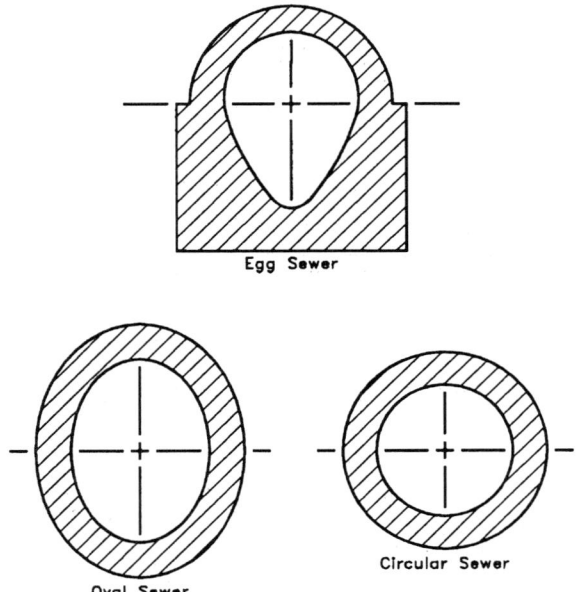

Figure 4.18. Alternative shapes for horizontal drainage pipes. (Source: ASPE Databook.)

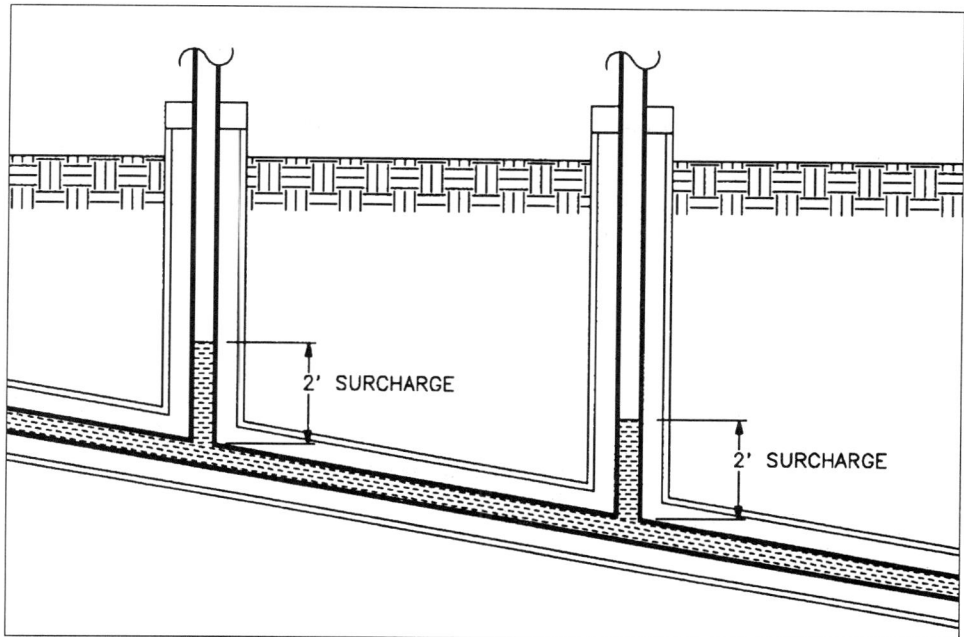

Figure 4.19. Illustration of primary piping with a 2 ft. surcharge.

4.4.5 Spill-Containment Systems

Spill containment systems are designed similar to storm water systems since they both are intended to drain an open area where a fluid rapidly accumulates. If waste from process equipment or lab fixtures are also to be drained into the same system, they are referred to as a combined system. They should be kept separate from storm water systems whenever possible. However, in outdoor systems they must be designed to accept rain water in containment areas that are typically separated from the rest of the facility by dikes or berms.

Spills must be drained from an area at the same rate at which they will accumulate in the event of a system upset, plus the amount from a rain storm of some return period (e.g., a 10 to 25 year storm) for outdoor installations. The horizontal headers underground must also be sized to handle combined flow from other branches or services, if applicable. The rate of drainage from a designated spill area must be determined for each specific application. Both the maximum expected rate and frequency of occurrence must be determined. To determine the quantity of surface runoff, the rational method can be used:

$$Q = CIA \tag{4.90}$$

where:
Q = flow rate, ft^3/s
I = intensity, ft^3/s^2
A = area of collection, ft^2
C = coefficient, dimensionless

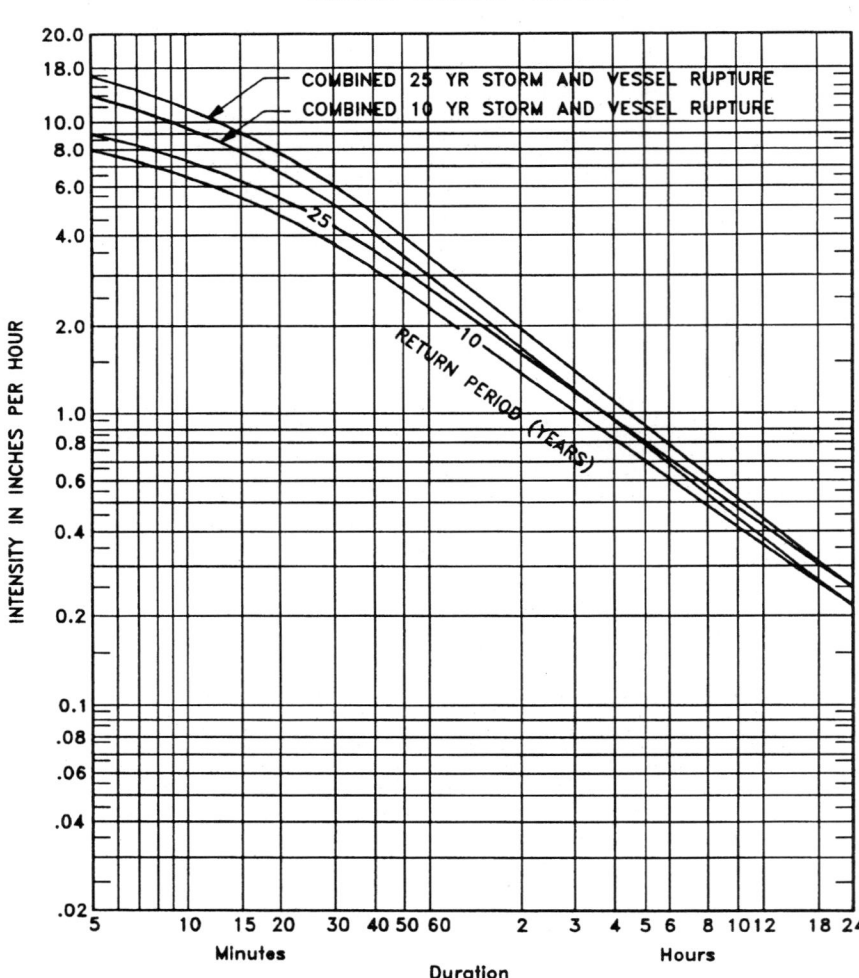

Figure 4.20. An example of the relationship of inches of rain water accumulation per hour to area of flow, along with example combined data due to a theoretical vessel rupture.

In this equation, the intensity term represents the intensity of accumulation of a spill, and replaces the intensity (of rainfall) used in storm water sizing. A plot of intensity of rainstorms to return period and time of concentration is shown in Figure 4.20. For a specific outdoor application, a combined plot of rainstorms and spill intensity can be drawn up to determine the value of I to be used in Eq. (4.90). The time is defined as the time it takes for the area to start contributing to the flow at a given point. Runoff coefficients are shown in Table 4.11 for a variety of surfaces. In determining the flow area, it is recommended that 50% of vertical walls be included in the area if the spill involves spraying or drainage against a wall, or against the walls of a dike or berm.

Fluid Dynamics and Sizing Analysis

Table 4.11 Runoff Coefficients

Type of Surface	Description	C
Grass	Rough grass or undeveloped ground	0.10
Grass	Average developed lawn	0.20
Grass	Sparse lawn on hard ground	0.25
Paved	Rough surface on flat slopes	0.80
Paved	Average pavement	0.85
Roof	Flat roof with provisions for standing water	0.85
Roof	All other roof surfaces	0.95
Composite	Average values for entire site area (value depends upon relative amount of lawn area)	0.55 to 0.75

Source: A. Steele, Review Manual for the CIPE Exam, ASPE, Sherman Oaks, CA, 1984.

In many applications where the system is designed to handle failure of a tank or vessel, a large amount of fluid will accumulate over a very short period of time. In an outdoor application, the drain piping can be sized to drain 110% of the amount of fluid that the tank or vessel can contain at any one time, plus the rainfall based on a 10-25 year storm.

The maximum recommended flow for spill-containment collection piping is shown in Table 4.12, based upon the area to be drained and the slope of the piping. These values are the same as those recommended for storm water piping systems.[48-50]

4.4.6 Combined Chemical Waste (DWV) and Spill-Containment Systems

In many designs, a single system will be used to drain chemical waste and collect spills from a series of floor drains. If the spills are relatively minor in nature, then the system can be designed as a typical chemical waste system. However, if emergency spill floor drains must accommodate a major spill, such as can occur when an above-ground tank or vessel rup-

Table 4.12 Maximum Suggested Loads for Spill Containment Piping

Pipe diameter, in.	Horizontal piping Drained area, sq. ft. for various slopes		
	1/8	1/4	1/2
2	---	---	---
2.25	---	---	---
3	822	1160	1644
4	1880	2650	3760
5	3340	4720	6680
6	5350	7550	10700
8	11500	16300	23000
10	20700	29200	41400
12	33300	47000	66600
15	59500	84000	119000

Note: Values equivalent to those shown for storm water piping in Table 7-1 of A. Steele, Engineered Plumbing Design Construction Industry Press, 2nd Ed., Elmhurst, IL, 1982.

tures, the system should be designed as a spill-containment-type system. One must convert fixture unit loads of the chemical waste system portion into equivalent square feet of drained area. A procedure based upon sanitary system/storm water systems is as follows[51-53]:

1. When the total fixture unit load is 256 FU or less, use a minimum of 1,000 square feet of area (93 square meters of area).
2. If the fixture unit load is greater than 256, multiply the total fixture unit load by a conversion factor of 3.9 square feet per fixture unit (0.36 square meters per fixture unit).
3. If the flow is continuous or intermittent full flow, as in the case of discharge from process equipment, use a conversion factor of 24 square feet per gpm (2.23 square meters per gpm) to convert to equivalent drained area.

Spill-containment systems, combined or otherwise, may be governed by local building codes in many municipalities. For a discussion on the relationship of building codes to the design, refer to Section 4.4.1, "Applicable Building Codes and Design Standards." In the United States, storm water rules under the clean water act may apply.

4.5 Gravity Flow in Annuli and Venting of Annuli

4.5.1 Principles of Gravity Flow in an Annulus

Fluids flow in an annulus in the same manner as they do in a circular pipe or open channel. Differences arise from the geometrical configuration and flow disruption by supports, baffles, and other devices. The difference in geometry can be taken into account by using Eq. (4.83) for vertical flow; Eq. (4.89) may be used for horizontal flow. The hydraulic diameter and hydraulic radius terms may be substituted for diameter [Eq. (4.83)] and radius [Eq. (4.89)] in the respective equations. The flow will also be affected by the drag that occurs of the fluid flowing over the inner piping. In modeling and analyzing the overall velocity and flow profile, one must remember that the additional pressure drop resulting from interstitial supports and baffles should be taken into account. Enough information exists to approximate the impact of the additional restrictions to flow.

4.5.1.1 Vertical-Flow Considerations

Fluid flow in the secondary containment portion (annulus) of a double containment vertical stack is similar to that of its associated primary stack. As a fluid enters the annular space during a system upset, it will swirl around the inner wall of the outer pipe and form a donut-shaped sheet. If the annular space is large enough, the flow will achieve terminal velocity if there are no baffles or restrictions encountered over the terminal length. The terminal velocity and terminal length can be calculated using Eqs. (4.81) and (4.82), respectively. The diameter in Eq. (4.81) should be the inside diameter of the outer pipe.

Equations (4.81) and (4.82) may be used if sufficient annular space and sufficient uninterrupted annular length are allowed, such that diaphragming will not occur. To determine if there is sufficient annular space to prevent diaphragming, the cross-sectional ratio should first be calculated. In a typical system, one can calculate the cross-sectional ratio (which the draining fluid occupies) in the annulus based upon the maximum anticipated flow. Maximum anticipated flow will occur when a total failure of the primary system occurs, and the maximum flow of the primary system is discharged into the annulus.

For a system that has been designed based upon a specified cross-sectional ratio for the primary stack, the cross-sectional ratio of the secondary can be readily determined. To do so, it must be assumed that there is sufficient annular space to allow for a donut-shaped sheet of fluid to form. Moreover, the space occupied by the inner primary stack must be assumed not to cause a significant drag on the accompanying air flow in the center core, which would otherwise interrupt the profile of the donut-shaped sheet of fluid. If these conditions are met, and the annulus is vented to the atmosphere, the inner primary stack can be ignored and the calculation becomes simple. If these conditions are not met, slug flow will occur.

For example, a 4.33 in. (110 mm) actual OD SDR 32 polypropylene pipe (actual inside diameter = 4.05 in.) within a 7.87 (200 mm) SDR 32 polyethylene pipe (actual inside diameter = 7.38 in.) is an arrangement that may be used as a vertical double containment drainage stack. (Note: these are common European HDPE pipe sizes for chemical drains.) Figures 4.14 and 4.21 present a cross-sectional illustration of this example. If the inside diameter is designed for a maximum of 7/24 fluid-to-air ratio, the cross-sectional area of the donut-shaped sheet of fluid is 0.026 ft^2 (0.0024 m^2). The 200 mm secondary containment pipe has a total "free" cross-sectional area of 0.30 ft^3 (0.0279 m^2), taking into account the area occupied by the associated primary pipe stack. If the maximum amount of fluid that the primary (110 mm) pipe was designed to handle were to be discharged into the annulus, the cross-sectional ratio would be equal to 0.026 ft^2/0.30 ft^3 = 0.087. In the event of a leak, the donut-shaped ring of fluid falling down the secondary containment pipe would occupy a maximum of about 1/12th of the cross-sectional area of the 200 mm pipe, well within the 1/4 to 1/3 recommended for a typical single-wall pipe. Since the space occupied by the primary pipe is only 0.10 ft^3, there is still 0.3ft^3 -

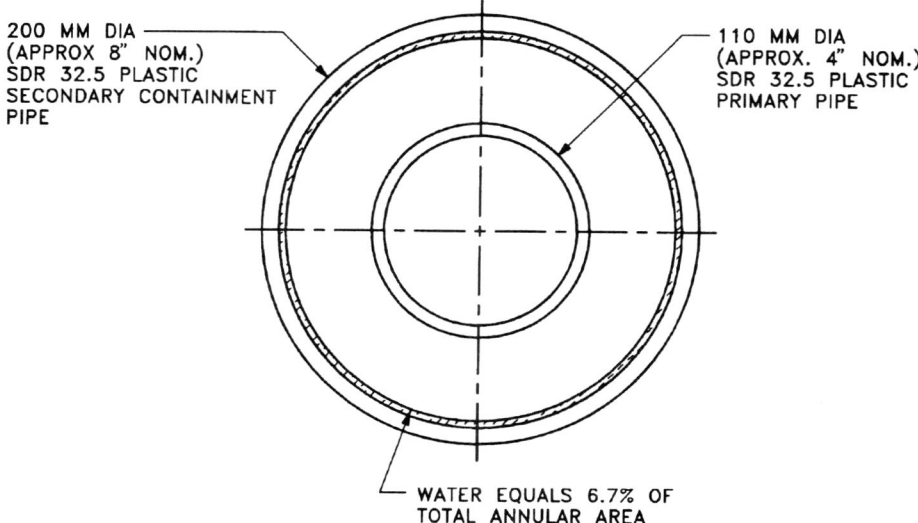

Figure 4.21. Example of a cross-sectional view of a vertical stack where the annulus is occupied by an amount of fluid which is equivalent to 7/24 of the inside diameter.

0.1ft³ - 0.026ft³ = 0.174 ft³ (0.0162 m²) of free air space occupied by the donut-shaped ring of air space between the sheet of water and the 110 mm pipe.

With the free air space that exists in the example, the assumption that fluid will drain with the profile shown in Figure 4.21 is reasonable, unless there are other annular restrictions. However, other arrangements without such generous annular spaces may not result in a flow profile under which terminal velocity is reached. When there is inadequate space, the drag of air along the inner pipe outer walls will cause the fluid to diaphragm across the entire open area, causing slug flow to occur with accompanying back pressure.

For typical stack-type drainage flow to occur in an annulus, the system must have both sufficient space and uninterrupted length. The vertical stack must have an uninterrupted length equal to the terminal length, which is the same value as that of a normal single-wall pipe stack described earlier (see Section 4.4.1). Supports, baffles, and other internal restrictions will impact the flow profile and will tend to cause slugs to form. As soon as a device of this type is encountered, the flow will become interrupted. Whenever a double containment chemical waste vertical stack is designed, the use of interstitial supports and centralizing devices should be minimized. The number of these devices should be the minimum required to provide lateral support and prevent damaging vibrations from occurring. They should also contain as much open area as possible.

4.5.1.2 Venting of Annuli

For drainage of a drain pipe system's annular space to proceed smoothly and rapidly, air must be able to escape. There must also be a replacement of the air that is dragged down the stacks and horizontal drains, or a slight vacuum can be created.

4.5.1.2.1 Continuous Venting for Annuli

Continuous open venting and circulation of air in a drain pipe system's annulus is difficult if leak detection is to be provided. Many leak detection systems function best when they are kept airtight and free of moisture. Allowing an annular space to be continuously vented to the atmosphere will allow moist air to enter the system. Instead, separate vent valves can be provided that can be opened in the event of a detected leak. Although less than ideal, it is a good compromise.

The proportion of air that must be vented from an annulus, based on the percentage of available total cross-sectional area, is less than for a primary stack. This is a result of a secondary containment stack having much of its internal area occupied by an internally positioned primary pipe. Though the area percentage is less, the total quantity of air to be vented is often greater than that of a primary pipe. For the previous example, the quantity of free air space in an annulus is approximately three times as great as the free air space of a primary pipe (0.174ft³ as contrasted to 0.063ft³ for the primary pipe). Since air needs to be replaced for drainage to occur smoothly, the secondary containment stack should have a vent extension like the primary pipe. If it is not considered necessary to have smooth and rapid drainage occur in the annulus, a vent extension or connection for the annulus is not required.

A secondary containment annulus vent system may be tied into, or combined with, that of the primary piping, if continuous venting is desired, and the authority having local jurisdiction allows this to be done. The vent headers can be sized by using sound analysis

Fluid Dynamics and Sizing Analysis

techniques and good engineering judgment, using the data shown in Table B-4 of Appendix B for sizing vent headers. The user must be aware that this layout approach may result in condensation of atmospheric moisture in the annular spaces, and that false leaks may be reported. In general, it is not a recommended practice to design an annulus to be continuously vented, and the reader should avoid doing so whenever possible.

4.5.1.2.2 Static Vents

In a pressure or drainage double containment pipe system, annular spaces (including each separate annular compartment) must be equipped with vents at all high points to allow the system to drain smoothly and rapidly in the event of a leak or failure of the primary system. Static vents are typically controlled by using a quarter-turn ball valve. The vents also serve a dual purpose as a location for a hose connection to allow flushing and decontamination of the system once a leak occurs. It will also allow air to be vented out of an annular space in order to perform a hydrostatic test. When conducting a hydrostatic test, it is important that all air be thoroughly vented out of the system for safety reasons.

4.5.1.2.3 Static Drains

In all double containment pipe systems, the annular space (including each separate annular compartment) should be equipped with low-point drains to allow drainage of the system in the event of a leak or failure of the primary system. Drainage of fluid from an annulus to an atmospheric drain line is affected by the area and shape of the opening. As fluid drains from an annular space, the total head of the fluid is converted to kinetic energy at the tank opening according to:

$$V_o = C_o \sqrt{2gh} \quad (4.91)$$

where :
V_o = velocity of fluid through the opening = ft/s
C_o = constant that depends on the shape of the opening
g = gravitational constant = 32.2 ft/s^2
h = height of fluid, ft

The quantity of fluid that will flow from an opening that has an area A_o is given by:

$$Q_o = (C_o A_o) V_o \quad (4.92)$$

where :
Q_o = flow rate through the opening, ft^3/s
A_o = cross-sectional area of the opening, ft^2

Substituting the velocity term in Eq. (4.91) into Eq. (4.92) results in:

$$Q_o = C_o A_o \sqrt{2gh} \quad (4.93)$$

When a fluid flows through a drain, there will be an exit loss that can be determined by Eq. (4.94). If the fluid flows through a drainage hose, one must consider the additional frictional losses occurring in the downstream piping of the drain.

$$h_f = \left(\frac{1}{C_o^2} - 1\right)\frac{V_o^2}{2g} \tag{4.94}$$

As the liquid is being drained, the static head will diminish over time, and the discharge rate through the orifice will decrease. The relationship for the annular space of a horizontally positioned pipe is shown in:

$$Q\,dt = -A_a\,dh \tag{4.95}$$

where:
A_a = cross-sectional area of the annulus, ft^2

The time to empty a horizontally positioned annular space is determined by Eq. (4.96). Integrating Eq. (4.96) allows for easy calculation of the time needed to drain a horizontally positioned annular space, resulting in Eq. (4.97):

$$t = \int_{h_1}^{h_2} \frac{A_a\,dh}{C_o A_o \sqrt{2gh} - V_{in}} \tag{4.96}$$

$$t = \frac{A_a(h_2 - h_1)}{C_o A_o \sqrt{2gh} - V_{in}} \tag{4.97}$$

Where an annular space containing a pressurized fluid is drained without venting to the atmosphere, the total velocity of the fluid will be determined by Eq. (4.91), with the term for h being replaced by the pressure of the fluid in feet of head. The rate of discharge and time to discharge can be calculated from Eqs. (4.93) and (4.97), respectively, by substituting the total pressure in feet of head for the term for h in the equations.

When flushing fluid is being discharged and continuously replenished, the time to fill the annular space can be determined by using Eq. (4.98), assuming that the flushing rate is greater than the rate of discharge. When the flushing rate is not greater, the annular space will not fill up.

$$t = \int_{h_1}^{h_2} \frac{A_a\,dh}{C_o A_o \sqrt{2gh} - V_{in}} \tag{4.98}$$

4.5.1.3 Horizontal-Flow Considerations

Under uniform conditions in a horizontally positioned sloping circular annulus, where the primary pipe is concentrically positioned within the secondary, drainage flow can be determined by the Manning equation [Eq. (4.87)], using the hydraulic diameter of the annulus.

Where the fluid height reaches the bottom of the primary pipe, one must also account for the drag that occurs from the fluid flowing against the outside diameter of the inner pipe. The effect of interstitial supports, baffles, and other annular devices must also be taken into account. The adjusted Manning equation, which is valid only for situations where the fluid height does not reach the primary pipe, is shown in:

$$V = \frac{1.486}{n} R_H^{2/3} S^{1/2} \qquad (4.99)$$

4.5.2 Open-Channel Flow (with Flow around a Continuous Circular Pipe) in Secondary Containment Trenches

Quite frequently, a primary pipe is placed within a containment trench or channel, in lieu of placing it within a secondary containment pipe. Secondary containment trenches are discussed further in Section 16.4. Trenches may be rectangular, oval, egg shaped, semicircular, elliptical, trapezoidal, and triangular in shape. For laminar flow, analytical equations for the determination of velocity profiles and distributions may be found in: Straub, Silberman, and Nelson, Trans. Am. Soc. Civ. Eng. 123, 685-714 (1958). For turbulent flow, several excellent sources exist.[54,55] It has been reported for turbulent flow in elliptical channels that the friction factors are greater than those for pipe.[56] References for the prediction of friction factors for turbulent flow in these channels include: Rehme, Int. J. Heat Mass Transfer 16, 933-950 (1973); Malak, Hejna, and Schmid, Int. J. Heat Mass Transfer 18, 139-149 (1975). The calculation of velocity through open channels was presented earlier in this chapter in Section 4.4.2.6.

When fluid flows in a secondary containment trench containing one or more primary pipes, the fluid will be subject to boundary layer flow over and around the pipes and supports. The drag and vortex shedding that occurs should be included in modeling the velocity and fluid profile.

4.5.3 Surcharging of Primary Pipes and Their Annuli

Surcharging can occur in an annulus as it does in primary piping. There are two possible situations with surcharging in an annular space. Both the primary and secondary containment pipes may be surcharged as shown in Figure 4.22. However, secondary containment pipes may also be surcharged without any surcharging occurring in its primary pipe, as shown in Figure 4.23.

The presence of an inner primary riser or manhole does not affect the extent to which a secondary containment pipe is surcharged. What matters is the height of the fluid in the secondary containment riser or manhole, since the fluid pressure is dependent only on height, not on the width or cross-sectional area it occupies.

In the design of many double containment systems, surcharging is often designed into the system by use of access port risers for leak detection access, secondary containment risers, and frequently through the use of manholes. In the event of a leak, although the ability of the secondary pipe to drain the fluid is less than the primary, surcharging aids the flow, and therefore aids drainage of an annular space.

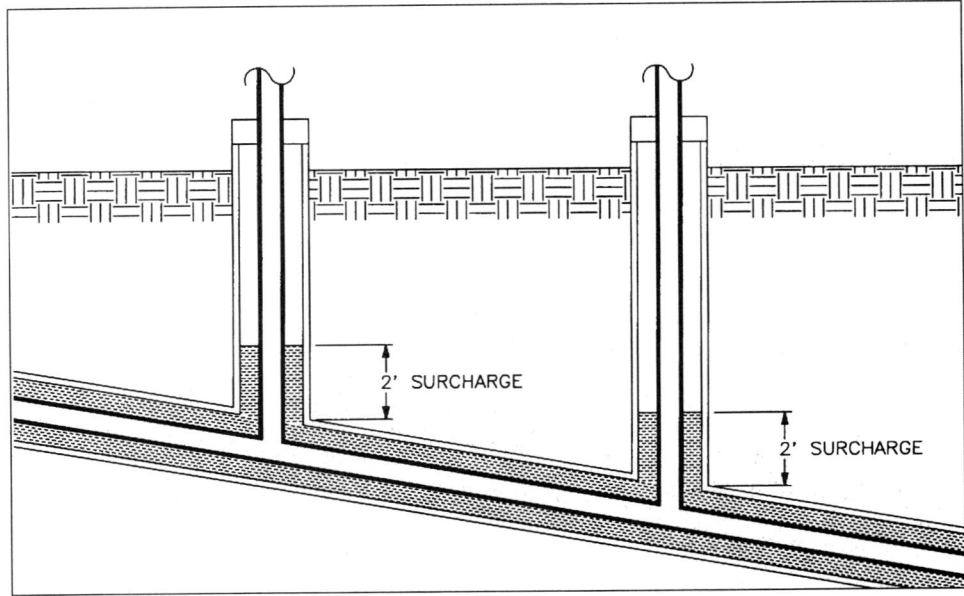

Figure 4.22. Illustration of secondary containment piping with a 2 ft surcharge where brances exist.

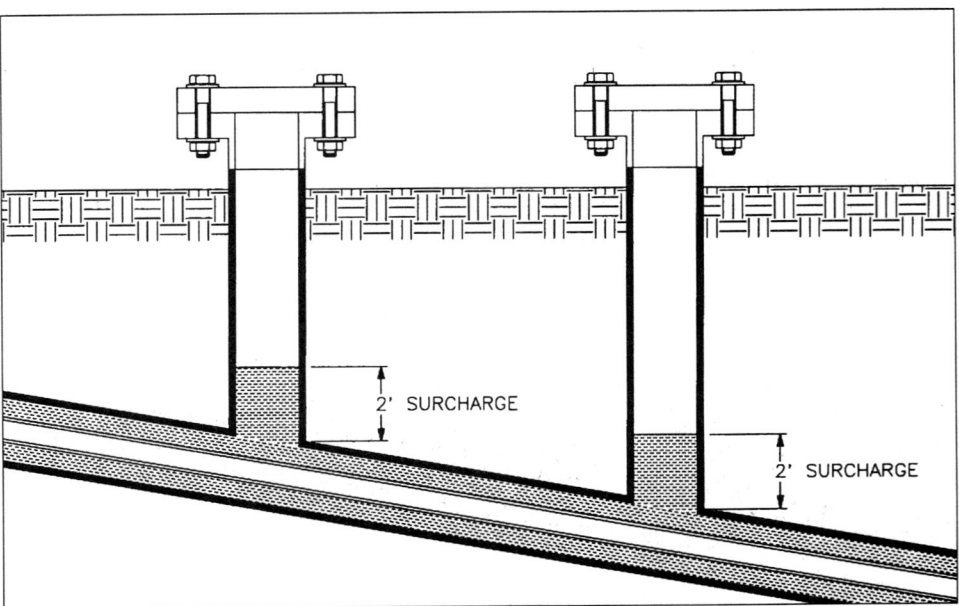

Figure 4.23. Illustration of a 2 ft surcharge in secondary containment piping by means of access ports.

References

1. Perry and Green, Perry's Chemical Engineering Handbook, 6th ed., McGraw-Hill, New York, 1984.
2. ASME Power Test Code, Part 2, PTC 19.2-1964.
3. Considine, Process Instruments and Controls Handbook, 2d ed., McGraw-Hill, New York, 1974.
4. Doolittle, Mechanical Engineering Laboratory, McGraw-Hill, New York, 1957.
5. Jones, Instrument Technology, Vol. 1, 3d ed., Butterworths, London, 1974.
6. Sweeney, Measurement Techniques in Mechanical Engineering, Wiley, New York, 1953.
7. Baumeister, Avallone, and Baumeister, Marks' Mechanical Engineering Handbook, 8th ed., McGraw-Hill, New York, 1978.
8. Harland, Mach. Des. 40(22), 69-74 (Sept. 19, 1968).
9. H. Schlichting and Kestin, Boundary Layer Theory, 7th ed., McGraw-Hill, New York, 1979.
10. J. G. Knudsen and D. L. Katz, Fluid Dynamics and Heat Transfer, McGraw-Hill, New York, 1958.
11. Perry et. al., op. cit., pp. 5-21.
12. L. F. Moody, "Friction Factors for Pipe Flow," Transactions of the American Society of Mechanical Engineers, Volume 66, November 1944, pp. 671-678.
13. Levenspiel, Am. Inst. Chem. Eng. J. 23, 402 (1977).
14. Kittredge and Rowley, Trans. Am. Soc. Mech. Eng. 79, 1759-66 (1957).
15. Beck and Miller, J. Am. Soc. Nav. Eng. 56, 62-83 (1944).
16. Beck, ibid., 235-271, 366-388, 389-95 (1944).
17. Karr and Schutz, J. Am. Soc. Nav. Eng. 52, 239-56 (1940)
18. Gibson, Hydraulics and its Applications, 5th ed., Constable, London, 1952, p. 95.
19. G. R. Kent, "Preliminary Pipeline Sizing," Chemical Engineering Magazine, Sept. 25, 1978, p. 19.
20. Angus, Hydraulics for Engineers, 3d ed., Pitman, Toronto, 1943, pp. 283/284, 29.
21. Perry's Chemical Engineering Handbook, 6th Edition, p. 5-60.
22. Byron D. Smith, "Sizing the Economical Pipe," Chemical Engineering, Sept. 1989, pp. 201-4.
23. Generaux, Chem. Metall. Eng. 44(5), 241-48 (1937).
24. Johnson and Maker, Proc. Tenth Mid-Year Meet. Am. Pet. Inst., Sec. III, pp. 7-23 (1940).
25. Knudsen and Katz, op. cit., pp. 193-205.
26. Perry and Green, Chemical Engineering Handbook, 6th Edition, Chapter 5, pp. 5-25, McGraw-Hill, New York, 1984.
27. Brighton and Jones, J. Basic Eng. 86, 835-42 (1964).
28. Okiishi and Serovy, J. Basic Eng. 89, 823-36 (1967).
29. Lawn and Elliot, J. Mech. Eng. Sci. 14, 195-204 (1972).
30. Bird, Armstrong, and Hassager, Dynamics of Polymeric Fluids, Vol. 1: Fluid Mechanics, Wiley, New York, 1977.

31. Knudsen and Katz, op. cit., Chapter 11, McGraw-Hill, New York, 1961.

32. D. A. Donohue, "Heat Transfer and Pressure Drop in Heat Exchangers," Ind. Eng. Chem. 41, 2499 (1949).

33. Bennett and Myers, Momentum, Heat and Mass Transfer, McGraw-Hill, 2nd ed., New York, 1974.

34. Rheme, Int. J. Heat Mass Transfer 15, 2499-517 (1972).

35. Bell and Bergelin, Trans. Am. Soc. Mech. Eng. 79, 593-601 (1957).

36. Perry and Green, Chemical Engineering Handbook, 5th ed., McGraw-Hill, New York, 1984, Chapter 5, pp. 5-56.

37. Prandtl and Tietjens, Applied Hydro- and Aero-Mechanics, McGraw-Hill, New York, 1934.

38. Schlichting, op. cit.

39. Prandtl and Tietjens, op. cit.

40. Schlichting, op. cit.

41. Tsou, Sparrow, and Kurtz, J. Fluid Mech. 26, 145-61 (1966).

42. Perry, Chemical Engineering Handbook, Chapter 5, pp. 5-58.

43. Den Hartog, Proc. Nat. Acad. Sci. 40, 155-57 (1954).

44. Goldwag and Berry, J. Eng. Power, 90, 213-17.

45. Guthrie and Wakerling, Vacuum Equipment and Techniques, McGraw-Hill, New York, 1949, pp. 37-38, 52-53.

46. ASPE Data Book, American Society of Plumbing Engineers, Vol. 1, pp. 11, Thousand Oaks, CA (1988).

47. Steele, Review Manual for the CIPE Examination, ASPE, Sherman Oaks, 1984, p. 54..

48. Steele, Review Manual for the CIPE Review Manual, p. 33, ASPE, Sherman Oaks, 1984.

49. Steele, Engineered Plumbing Design, p. 65, Construction Industry Press, Elmhurst, IL, 1982.

50. ASPE Databook, Chapter 2, American Society of Plumbing Engineers, Thousand Oaks, CA, 1990.

51. Steele, Review Manual for the CIPE Exam, p. 35, ASPE, Thousand Oaks, CA, 1984.

52. Steele, Engineered Plumbing Design, p. 70, Construction Industry Press, Elmhurst, IL, 1982.

53. ASPE Databook, Chapter 2, American Society of Plumbing Engineers, Thousand Oaks, CA, 1990.

54. O'Brien and Hickox, Applied Fluid Mechanics, McGraw-Hill, New York, 1937.

55. Chow, Open-Channel Hydraulics, McGraw-Hill, New York, 1959.

56. Cain and Duffy, Int. J. Mech. Sci. 13, 451-59 (1971).

Notes

Notes

Chapter Five

Design of Metallic Primary Components

This chapter presents information concerning the mechanical design of primary (also referred to as "core") piping and components that are constructed of metallic materials. Design issues for components that serve as a primary means of containing process fluids under pressure are presented here. This includes components that serve as an integral part of both the primary and secondary containment systems. Components of this type are referred to in this chapter, and elsewhere in this book, as "interconnecting components."

The approach to component design in this chapter is based on techniques and methods generally used for single-wall piping components in chemical services (i.e., ANSI/ASME B 31.3 Code services). The philosophy of this chapter is that the basic rules and principles of ordinary piping should not differ or be changed for primary pressure retaining components in a double containment piping system. Primary piping system components in a double containment piping system represent the "primary" means of containing a service fluid. Therefore, its design must be based on well-proven engineering principles that are applied to single-wall systems. The material covered here is highly technical in nature, and should be applied by a competent professional. However, every effort has been made to present the information in a manner that assists in understanding important design aspects to be considered for every double containment piping project involving metallic primary components.

5.1 Design Criteria and Conditions

5.1.1 Design Conditions

In the design of a chemical piping system, regardless of the construction materials, the conditions selected as the basis for design should include all applicable forces and effects. This is true for both pressure and gravity drainage systems alike. These effects include ambient effects, dynamic effects, and thermal expansion effects, among others. Ambient effects should include consideration of all heat-transfer effects, pressure effects, and corrosive and erosive actions. Certain dynamic effects that can be achieved under extreme conditions should also be given consideration. Geysering, water hammer, and cavitation are examples of these. Unusual thermal effects also may occur when piping is subjected to differential temperatures that vary greatly between internal and external conditions (or

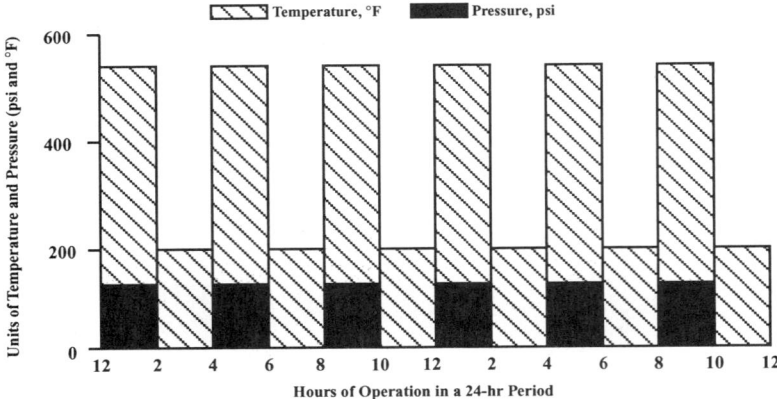

Figure 5.1. Example of a typical daily temperature/pressure cyclic profile for a jacketed pipe.

between primary piping and secondary containment piping), or to internal conditions resulting from two-phase flow. These effects must be considered, or primary piping may be subject to damage at points of restraint, such as unidirectional restraints, anchors, branch connections, as well as other interconnecting locations.

5.1.2 Design Pressure

To determine the design pressure of each component in a system, an analysis must first be performed to establish the temperature/pressure profile of the system. The temperature/pressure profile for any system may be determined from knowledge of the process operation for which it is intended, and from knowledge of typical weather patterns in the climate in which the system is installed. It is determined based on knowledge of how and where the pipe is to operate. Examples of temperature/pressure profiles are presented for a system, on a daily and weekly basis, respectively, in Figures 5.1 and 5.2.

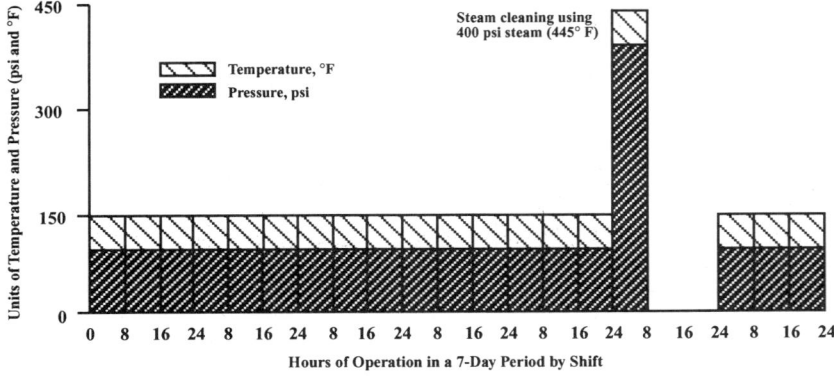

Figure 5.2. Example of a weekly temperature/pressure profile for a system. The most severe temperature/pressure condition occurs during steam cleaning, thereby setting the system design basis.

Design of Metallic Primary Components 153

The pressure/temperature condition of a piping system may vary significantly over time, as indicated by the example. An analysis of this profile can determine the condition of coincident pressure and temperature that results in the greatest required wall thickness for each component or the highest component rating, whatever is applicable. The pressure (of the coincident pressure/temperature condition) that results in the highest value is that selected as the basis for the design pressure. A suitable safety margin is typically added to this value by a designer to result in the design pressure of the primary (core) system.

When more than one set of pressure/temperature conditions exist for different components in a piping system, there may be different design pressures established for the different components in the system. This will occur only if the governing pressure/temperature conditions that apply to individual components differ.

Provision should be made to contain safely or relieve any excess pressures to which a primary pipe may be subjected. Secondary containment piping in double containment piping systems can be used for this purpose, even though it may not be intended as its primary function. Where it is used for this purpose, a secondary containment pipe must be designed to provide relieving capacity for the entire carrier pipe that it protects. Additionally, each individual compartment in an annulus between primary and secondary containment pipe sections must be designed to relieve the capacity of the entire carrier pipe system that is contained within. Typical sources of internal overpressurization may include failure of control devices, process upset, dynamic effects, and ambient effects. Relief protection requirements for a secondary containment portion of a double containment piping system is discussed in Chapter 7 (see Section 7.7).

5.1.3 Design Temperature

The design temperature for each component in the primary (core) piping system is the same temperature that establishes the design condition discussed in Section 5.1.2, under the heading "Design Pressure." It is the temperature of the coincident pressure/temperature condition that results in the greatest wall thickness and highest component rating.

For piping that is externally insulated, the component design temperature should be selected based on the fluid temperature unless measurements from tests, experience, or analysis support evidence otherwise. In many applications involving double containment piping, primary piping is of the externally insulated type by virtue of an insulating dead-air space or vacuum between the two pipe systems. When a component is uninsulated, the following general rules apply for pipe where the outside surface is subject to ambient cooling: Component temperature should be taken as the same as the fluid temperature for fluid temperatures less than 150°F (65°C). Where fluid temperature is equal to or exceeds 150°F (65°C), average wall temperature can be determined by calculation, or by field test, in order to establish a lower design temperature than that of the fluid temperature. Otherwise, design temperature may be taken as 95% of the fluid temperature for pipe and components with similar wall thicknesses.

Components such as flanges, lap-joint flanges (refer to Section 5.2.1.6 for additional requirements for flanges), and bolting may be assigned lower design temperature ratings (respectively, 90%, 85%, and 80% of the fluid temperature). For piping that is internally insulated, its design temperature should be based upon heat-transfer data that are the result of field experience or tests, or it can be determined by heat transfer calculations.

The minimum design temperature of primary components should be the lowest possible temperature condition that a component may experience. This may be either the core fluid temperature, ambient temperature, or temperature of a system's annular space or heat-transfer fluid. The temperatures of all three should be considered and the minimum selected as the minimum design temperature. Consideration also should be given to special ambient effects that are described in more detail in Section 5.1.4.

5.1.4 Ambient Effects

The effects of all internal and external conditions of a piping system should be considered, particularly those that produce a heating or cooling effect. A common problem that creates an internal effect occurs when a vapor condenses on the inside of a pipe, for example, after steam cleaning takes place. When this occurs, a vacuum can be created. For piping with this condition, the design of components should be such that they can withstand the vacuum. Otherwise, provision for vacuum relief must be designed into the system. An example of an external ambient effect involves solar heating of a stagnant fluid within a component, which may overpressurize a pipe, unless pressure relief is provided. Another external ambient effect involves low ambient temperatures, where displacement stresses are caused or where icing may result. Both of these effects may result in operational problems for a pipe system. Other examples of operational problems that may result from these effects include the failure of safety-related devices such as shut-off valves, control valves, and pressure or safety relief devices.

A unique ambient effect exists in double containment piping systems whereby a primary pipe can be subjected to a differential pressure if a surrounding annulus contains a pressure that is greater than the pressure of the core fluid. This occurs in double pipe heat exchangers and in jacketed process piping, and can occur while pressure testing secondary containment pipe system (annulus) of a double containment piping system. Hydrostatic testing procedures should be designed for primary (core) and secondary containment (jacket) pipe systems in a such a manner that problems due to overpressurization from external sources may be avoided. Chapter 13, "Inspection, Examination, and Testing," Section 13.3 covers testing methods and testing sequence pertaining to double containment piping systems. In a design, external pressure from all potential sources should be considered for each component. The same condition occurs in single-wall pipe under submerged conditions. In submerged piping, external pressures must be considered in addition to buoyant forces.

5.1.5 Weight Effects

In addition to dynamic forces, there are permanent forces that arise from dead weight and loads that vary (live loads). Static weight load, or dead-weight load, consists of the weight of piping components, insulation, and any superimposed objects in above-ground piping. In underground piping, it consists of the burial load as calculated for static burial conditions according to Marston's equations (Eq. 9.34 for rigid pipes, and Eq. 9.35 for flexible pipes). For underground systems, live loads include concentrated and distributed loads that may be imposed by vehicular traffic over or in the area of a buried pipe. Live loads imposed by vehicular traffic must be considered in the design of an underground piping system, whenever pipes are buried to a shallow depth, and such traffic is a possibility. Chapter 9 provides further discussion concerning the subject of burial load calculations and burial analysis techniques for buried double containment pipes.

In double containment piping systems, the outside (secondary containment) pipe will be directly subjected to the static-soil and live loads; however, certain primary piping components will see a portion of the loads due to interstitial supports and other interconnecting components, which transmit a portion of the loads to the primary system. These loads must be considered for the primary piping system.

5.1.6 Dynamic Effects

There are many dynamic effects to be considered in the design of piping. These include, but are not limited to: (1) impact on a pipe component wall as the result of internal or external forces, (2) vibrational effects, (3) seismic loading, (4) discharge effects, and (5) wind actions. The effects of impact can be created from a number of sources, including sudden changes in fluid velocity, slug flow, liquid flashing (cavitation), and geysering. Vibration of a piping component can result from pressure pulsations, acoustic resonance, mechanical vibration of attached equipment, wind loads, and the same sources that also cause impact. Seismic forces will vary according to the geographic location of a piping system. It is recommended to use analysis procedures outlined in ANSI A58.1, or as defined by the Uniform Building Code, or the local authority having jurisdiction.

When fluids are discharged from a pipe, there are reaction forces created by the sudden change in energy. These effects, which are likely in a double containment system in which a primary pipe under high pressure experiences failure, must be taken into consideration in the design. Wind also imposes loads upon a pipe. An analysis of wind effects for aboveground pipe can be made using the procedure outlined in ANSI A58.1, or by following the rules of the Uniform Building Code or the authority having jurisdiction.

5.1.7 Thermal Expansion and Contraction Effects

Many types of loads result from thermal effects. All of these need to be considered as acting concurrent with other forces that are imposed on each component in a piping system. When free expansion and contraction of the piping is restricted by use of restraints, anchors, soil, or interconnecting components in a double containment piping system, thrusts and moments that result must be taken into account in an analysis. In double containment piping, loads and discontinuity stresses will occur due to differential thermal expansion and must be considered in the design of components. In many cases involving double containment (or jacketed) piping, differential thermal expansion will occur even though the material used for the secondary containment pipe has the same coefficient of thermal expansion as its primary components. Where an annulus is provided with insulation, or where an annulus contains air that is contained within an uninsulated secondary containment pipe, the temperature difference of the two pipes may be substantially different.

Temperature gradients may also exist in a relatively thick single-wall pipe, or across a pipe wall that consists of a material that has low thermal conductivity. Temperature gradients may also arise from internal conditions such as two-phase streamline flow in drainage piping. Regardless of the cause, the temperature profile of each component and any interaction between components must be taken into account in an analysis. Chapter 8 is devoted to the discussion of thermal expansion issues pertaining to double containment piping, including methods of analysis. Simplified equations for analysis of pipes when elastic behavior is present are presented in Sections 5.1.13 and 5.5.1.6 of this chapter.

5.1.8 Other Considerations

In addition to the effects already stated, other considerations include cyclic effects, forces imposed by movements of supports, anchor and terminal movements, reduced ductility effects, and in low-temperature service, condensation and oxygen enrichment. Condensation is an issue if the annular spaces of a double containment piping system contains moist air. If an inert gas, dry air, or vacuum is instead placed in an annular space, the effects of condensation on primary piping no longer need to be considered. It is, however, a rare application of double containment piping for oxygen enrichment to present a concern.

Cyclic effects include those involving pressure, temperature, or dynamic loadings. Movement of piping supports and the like can result from thermal expansion of equipment to which the piping is attached, settling, wind forces, or, in the case of submerged or marine piping, as the result of tidal actions. Welding, heat treating, and other fabrication procedures, as well as temperature shock resulting from internal fluid behavior, can reduce a material's ductility. In double containment piping systems, adequate shielding of the primary piping is normally provided by its associated secondary containment pipe components. Extra special precautions do need to be taken, however, where a "double failure" can result from a severe blow or other occurrence, when its materials are no longer in a ductile state. In any metal piping service below -312°F, special precautions should be taken to provide additional shielding or insulation of its components. This condition is rare for a double containment piping system. However, an example is in the use of a double-walled aluminum system, with a vacuum applied in the annulus (as a means of insulation), for use in cryogenic gas services in aeronautic and space applications (i.e., rocket-launch applications where liquefied hydrogen is used as a fuel). The use of materials possessing adequate low temperature ductility is absolutely critical when such extremely low temperatures are involved.

5.1.9 Design Criteria

In establishing the basis for the design of a piping system, several design factors need to be established. These include the pressure and temperature ratings described in Sections 5.1.2 and 5.1.3, but also include allowable stress criteria, and any design allowances. These factors should include minimum design values, as well as any permissible variations as they apply to the design of a piping system.

5.1.10 Pressure/Temperature Design Criteria

There are many established, recognized standards that establish the pressure/temperature ratings of common elements for metal piping components. Most of these standards are established by the American Society of Mechanical Engineers (ASME), the American National Standards Institute (ANSI), the American Petroleum Institute (API), the American Water Works Association (AWWA), the U.S. Navy Engineering Command, under the Defense Department (DOD), the Manufacturer's Standardization Society (MSS), and the Society of Air Conditioning Engineers. A summary of applicable component standards that are "listed" in the ANSI/ASME B31.3 Chemical Plant and Petroleum Refinery Piping Code (1993 Edition) is shown in Table C-1 in Appendix C of this text. For components that are "listed," but are not assigned a pressure/temperature rating, the ANSI/ASME B31.3 Chemical Plant and Petroleum Refinery Piping Code requires that their pressure rating be based on 87.5% of the nominal thickness of seamless pipe that corresponds to the pressure class, schedule, or weight of the fitting, less any allowances. (Examples of allowances would be threading allowance, corrosion allowance, etc.)

Most metallic double containment piping systems involve the use of at least some components that are not "listed" in Table C-1 of Appendix C. Components of this type are referred to throughout this text, and by the ANSI/ASME B31 Codes, as "unlisted components." The design of these elements should be in accordance with known design techniques (such as those suggested for the design of other standard "listed" components by ASME, etc.) and verified by detailed numerical stress analysis. The results of the design and analysis should be supplemented by experimentation, proof test, or demonstrated to have performed successfully in extensive service, under comparable conditions with similar materials and dimensions, and should also be given a detailed analysis. The ASME Boiler and Pressure Vessel Code outlines methods for proof testing and experimental stress analysis in Section VIII of the BPV Code (Division I, Part UG-101 for proof testing, Division 2, Appendix 6 for experimental stress analysis). These methods outline the types of tests to be performed, analytical methods that must be used, and documentation that must be recorded.

Certain services will have occasional variations of temperature and/or pressure that exceed normal operating levels. Normally, the most severe coincident pressure and temperature that result in the greatest component thickness are selected as the design basis. However, there are conditions where allowances for pressure and temperature variation will result in an adequate design. These allowances permit the design of a system for higher allowable pressures for short-term events. These allowances are limited to ductile materials where there are fewer than 7000 cycles, the nominal pressure stress does not exceed the yield strength at that temperature, and the combined longitudinal stresses do not exceed 1.33 times the allowable stress. If the allowable stress exceeds two-thirds of the yield strength of a material, the allowable stress value should be reduced as stated in the ASME B 31.3 code. The intent of the reduction is to prevent stresses during the short-term event from exceeding the yield stress of the material.

If these conditions are met, and there are not more than 7000 cycles in the life of the piping system, then its design pressure rating can be exceeded by: (1) 33% for each 10 hour or less occurrence up to 100 hours per year; or (2) 20% for each 50 hour or less occurrence up to 500 hours per year (including self-limiting events). The maximum operating pressure must not exceed the pressure to which a component has been tested. The effects of these variations should be verified by analysis and are subject to the approval of the owner, prior to acceptance. These effects are considered short-term variations per the ANSI/ASME Chemical Plant & Petroleum Refinery Piping Code.

Combined effects of all sustained and cyclic effects must be considered when making a determination as to allowances. Temperature variations to lower temperatures may require impact testing to determine if a material is suitable for the lowest temperature it will experience in service. (Note: The minimum temperature should be the lowest temperature expected in service.)

5.1.11 Allowable Stresses and Other Limits

The allowable stresses for some metals are listed in Table A-1 of Appendix A of the ANSI/ASME B31.3 Chemical Plant and Petroleum Refinery Piping Code, in the current edition of the Code. Similar values are provided in Appendix A of the ANSI/ASME B31.1 Power Piping Code. The values given in Appendix A are for stresses in tension for commonly used metallic materials in chemical fluid services. To determine values of stresses in shear, and stresses in bearing for materials listed in Appendix A of the ANSI/ASME B31.3 Code, multiply the values for S by 0.80 and 1.60, respectively. It is highly recommended to reference the current edition and addenda of these Codes, since values are often updated.

The data in Appendix A of the ANSI/ASME B31.3 Code are grouped according to materials and product form, with temperatures up to 600°F (316°C). If stress values for bolting materials are needed, the values reported in the applicable edition of the ASME B31.3 Code, or other applicable code/reference should be consulted. For calculations that require allowable stresses in compression, values should not exceed those in Appendix A for S, or the allowable stress based on buckling per the rules of ASME BPV Code Sec. VIII, Div. 1.

For most metallic materials used for piping components, the basic allowable stress is selected from the lowest of the following values: (1) the lower of one-third the specified minimum tensile strength or one-third the tensile strength at temperature; (2) the lower of two-thirds the specified minimum yield strength or two-thirds the yield strength at temperature; (3) average stress for a creep rate of 0.01% per 1000 hours; (4) two-thirds the average stress for creep rupture at the end of 100,000 hours; (5) the minimum stress for creep rupture at the end of 100,000 hours times 0.8.

For austenitic stainless steels and nickel alloys with a similar stress/strain behavior (other than flanged joints and other deformation-sensitive components), the lower value of two-thirds times the specified minimum yield strength or 90% of the yield strength at design temperature is used rather than (2), above. This higher percentage of yield strength is basically permitted for materials that do not have well-defined yield points.

Structural-grade materials require an additional safety factor and may be determined by multiplying the lowest value determined by the criteria stated above by 0.92. Cast iron and malleable iron require that their values are determined in a different manner due to their comparative brittleness. For malleable iron, the allowable stress value can be determined from the lower of the specified minimum tensile strength at room temperature or the tensile strength at design temperature times 0.2. The allowable stress for cast iron is selected from the lower of the same two values, multiplied instead by a factor of 0.1.

In hoop stress calculations for internal pressure involving components that are made from cast materials or that have longitudinal welds, the product SE appears in some equations. This product means that the value for S must be coupled with a quality factor E in order to correct for potential defects. The casting quality factor, E_c, can be determined from Table C-2 of Appendix C (of this text) according to the type and extent of supplementary examination. The weld joint quality factor can be determined from Table C-3 of Appendix C according to the type of joint and the type and extent of subsequent examination.

5.1.12 Limits on Sustained Longitudinal Stresses in Primary Piping

In general, primary piping is first sized so that the selected wall thickness is adequate to withstand loads safely due to internal and external pressure. This sizing is generally based on hoop stress considerations and is discussed later (refer to Section 5.2.1.1). The primary system must also be designed such that the sustained longitudinal stresses in the piping is less than its allowable stress, S_H, at the temperature of the condition being evaluated.

Examples of loads that cause these sustained stresses are weight, wind, thermal expansion, and pressure. Earthquake forces are also generally treated as sustained loads, although they have similar characteristics to displacement-controlled thermal expansion loads.

In calculating the sustained stress, mechanical (e.g., threads), erosion, and corrosion allowances must be subtracted from the primary pipe wall thickness. However, in ASME B 31.3, unlike the pressure design equations, the nominal pipe mill wall thickness (without subtracting mill tolerance) is used in the calculation of stress due to sustained loads. However, erosion, corrosion, thread depth, groove depth, and other allowances need to be considered. No weld joint quality factors are used in the longitudinal stress calculation.

In calculating the sustained stress, the B 31.3 Code does not address the issue of use of stress intensification factors used in flexibility analysis, to be discussed later (refer to Section 5.5.1.6), in the calculation of stress due to sustained loads. This decision is left up to the designer. It is conservative to use them and may be nonconservative to ignore them. Some designers, using the B 31.1 Code for guidance, use 75% of the stress intensification factors in the sustained stress calculation.

The total of longitudinal stresses, S_L, when considered concurrently with occasional loads from seismic or wind effects, can be 33% more than the basic allowable stresses stated in Appendix A of the ANSI/ASME B31.3 Code. For castings you need to include the casting quality factor from Appendix C-3, of Appendix C. However, you are not allowed to exceed the yield stress. This is to prevent stresses due to occasional loads from exceeding the yield strength of a material.

5.1.13 Limits on Longitudinal Displacement Stresses in Primary Piping

Stresses due to displacements such as the thermal expansion of piping are permitted to exceed the yield stress of the piping material because they are strain controlled. Although the piping may yield on initial heating, residual stresses are established on cooling that lead to substantially elastic action. The total stress in the piping, including thermal stress, is limited such that this shakedown to elastic cycling occurs, as well as to prevent fatigue failure. The equation that ASME B31 Code calculations are based on is:

$$S_E + S_L < f[1.25(S_c + S_h)] \qquad (5.1)$$

where:
S_E = the computed displacement stress range, psi
S_L = the sum of sustained longitudinal stresses, psi
f = stress range reduction factor for the displacement cycle conditions for noncorroded pipe for the total number of years the system is expected to be in active operation, found from Table 5.1
S_c = the basic allowable stress at the minimum temperature of the metal for the displacement cycle under analysis, psi
S_h = the basic allowable stress at the maximum temperature of the metal for the displacement cycle under analysis, psi

The sum of the sustained and displacement stresses is limited to $1.25(S_c + S_h)$ [Eq. (5.1)] for shakedown. Further, this allowable stress is further reduced by the factor f (from Table 5.1) for cycles greater than 7,000 due to fatigue considerations. This basic equation is converted into two code equations. The first assumes that $S_L = S_h$, and is as follows:

$$S_A = f(1.25S_C + 0.25S_h) \qquad (5.2)$$

where:
S_A = allowable displacement stress range, psi
S_C = the basic allowable stress at the minimum temperature of the metal for the displacement cycle being analyzed, psi
S_h = the basic allowable stress at the maximum temperature of the metal for the displacement cycle being analyzed, psi
f = the stress range reduction factor for the displacement cycle conditions for noncorroded pipe for the total number of years the system is expected to be in active operation, found from Table 5.1

The second equation, considered the liberal stress equation, considers that sustained longitudinal stresses may be less than their allowable S_h, as follows:

$$S_A = f\left[1.25(S_C + S_h) - S_L\right] \qquad (5.3)$$

Once the value of S_A is calculated, it can be compared with the computed displacement stress range, S_E, which should never exceed the allowable stress range of a material under consideration. If it is, then a material with a larger S_A should be selected, or the design of the system must be altered to result in a lower S_E.

These equations are based on thermal displacement cycles and are not applicable to high cycle applications (> 2,000,000 cycles) such as due to vibration. Evaluation of vibration generally requires consideration of significantly higher cycles, based on endurance limits.

If the computed stress range varies, the greatest computed stress range must be used for S_E. The equivalent number of full displacement cycles that must be used in such cases, N, can be calculated from:

Table 5.1 Stress-Range Reduction Factors f*

Cycles N	Factor F
7,000 and less	1.0
Over 7,000 to 14,000	0.9
Over 14,000 to 22,000	0.8
Over 22,000 to 45,000	0.7
Over 45,000 to 100,000	0.6
Over 100,000 to 200,000	0.5
Over 200,000 to 700,000	0.4
Over 700,000 to 2,000,000	0.3

* From the ASME B31.3 Code-1993 Edition.

$$N = N_E + \sum \left[r_i^5 N_i \right] \text{ for } i = 1, 2, \dots, n \tag{5.4}$$

where:
N_E = number of cycles of maximum computed displacement stress range, S_E
$r_i = S_i / S_E$
S_i = any computed displacement stress range smaller than S_E
N_i = number of cycles associated with displacement stress range S_i

5.2 Pressure Design of Primary (Core) Piping Components

The design of pressure-containing components, other than components that are "listed" (i.e., those in Table C-1 of Appendix C), should take into account the properties of the material, and all applicable effects listed earlier in this chapter. The major objective in designing pressure-containing components is that they should be constructed with a sufficient thickness and with adequate reinforcements in required locations, such that they can withstand pressure/temperature and fatigue requirements, and all other applicable effects, considering the properties of the material. For those elements and components that are "listed," the pressure/temperature ratings of such components are typically established in the referenced specification.

5.2.1 Pressure Design of Components

5.2.1.1 Straight Pipe

For straight piping under an applied internal pressure, the minimum required thickness, excluding required mechanical allowances, can be determined from any of Eqs. (5.5) through (5.8), for situations where the resulting thickness is less then 1/6th of the actual outside diameter of the piping and the ratio of P/SE is less than 0.385. For situations that do not meet the criterion t < D/6 or P/SE < 0.385, additional considerations involving fatigue and thermal stress effects and "theory of failure" have an effect on the required thickness, and must be considered.

$$t = \frac{PD}{2(SE + PY)} \tag{5.5}$$

$$t = \frac{PD}{2SE} \tag{5.6}$$

$$t = \frac{D}{2}\left(1 - \sqrt{\frac{SE - P}{SE + P}}\right) \tag{5.7}$$

$$t = \frac{P(d + 2c)}{2[SE - P(1 - Y)]} \tag{5.8}$$

where:
P = internal design gauge pressure, psi
D = outside diameter of pipe, in.
E = quality factor from Table A-1A or A-1B of the ANSI/ASME B31.3 Code (Tables C-2 and C-3 of Appendix C of this text)
S = stress value for material from Table A-1 of Appendix A of the ANSI/ASME B31.3 Code, psi
t = pressure design thickness determined for internal or external pressures, or the greater of the two if they both apply
c = sum of mechanical allowances (thread or groove depth) plus corrosion, erosion, and other allowances. Where the tolerance is not specified for machine or grooved surfaces, it is safe to assume the tolerance to be 0.02 in.(0.5 mm) in addition to the specified cut depth
Y = coefficient for effective stressed diameter

For a straight pipe that is subjected to external pressure, its minimum thickness can be determined by the external pressure design procedures covered in the ASME Boiler and Pressure Code, Section VIII, Division 1, UG-28 through UG-30. For thick pipes, when D_o/t is less than 10, the following modification is made to these procedures. The values for tensile stress from Appendix 1 of the ANSI/ASME B31.3 Code (S) that should be used in determining the value of P_{a2} shall be the lesser at design temperature of 1.5 times the stress value from Appendix 1 of the ANSI/ASME B31.3 Code, or 0.9 times the yield strength (which is tabulated in Section VIII, Division 2 of the ASME Boiler and Pressure Vessel Code). As an alternate to providing a thicker pipe wall, ring stiffeners may be added. Their design requirements are provided in Section VIII, Division I of the ASME Boiler and Pressure Vessel Code.

Once a pipe's design thickness for withstanding pressure has been calculated, its minimum required thicknesses can be determined from Eq. (5.9). (If a pipe can experience both internal and external pressures, thicknesses should be calculated for both conditions and the larger value chosen.) Minimum thicknesses required for pipe is usually larger due to mechanical and corrosion allowances. The nominal wall thickness of a pipe selected must be greater than or equal to the minimum required thickness, plus manufacturer's minus tolerance on wall thickness, which is generally 12.5% (i.e., as required by ASTM A512).

$$t_m = t + c \qquad (5.9)$$

where:
t_m = minimum required thickness, including all allowances, in.
t = pressure design thickness determined for internal or external pressures, or the greater of the two if they both apply, in.
c = the sum of mechanical allowances (thread or groove depth) plus corrosion, erosion, and other allowances – where the tolerance is not specified for machined or grooved surfaces, it is safe to assume the tolerance to be 0.02 in. (0.5 mm) in addition to specified cut depth, in.

The values of Y for various metal classes where t is less than D/6 are shown in Table 5.5. For temperatures within the limits of Table 5.5, interpolation may be used to determine the value for Y. For thicknesses $\geq$ D/6, the value for Y may be calculated by:

Design of Metallic Primary Components

$$Y = \frac{d+2c}{D+d+2c} \tag{5.10}$$

where:
D = actual outside diameter of the pipe, in.
d = actual inside diameter of the pipe equal to the maximum allowable value per specifications, in.

5.2.1.2 Curved and Mitered Segments of Pipe

When piping is fabricated into a curved segment by means of bending, the required minimum thickness of pipe after bending is the same as for the base pipe. Elbows are almost always manufactured in accordance with accepted, listed standards (e.g., ANSI B16.9 for butt welding short- and long-radius elbows). ANSI B16.5 flanged elbows have the same pressure rating as the flange, and ANSI B16.9 butt-welding elbows have the same pressure rating as pipe of matching nominal diameter and thickness.

Elbows that are produced from mitered sections of straight pipe (having an angular offset ≥ 3°) can be calculated by the following procedures for multiple- and single-miter bends.

For multiple-miter bends, where the angle of miter cut does not exceed 22.5°, use Eqs. (5.11) and (5.12) to calculate the maximum allowable internal pressure (P_m) and select the lesser of the two values. For single-miter bends, the value of P_m can be determined from Eq. (5.11) or (5.13) for miter cut angles less than or equal to 22.5° or exceeding 22.5°, respectively. In Eqs. (5.11) through (5.13), the thickness T should not extend beyond the inside crotch of the end miter welds a distance less than the value calculated in Eq. (5.14) for M. Miters with angular offsets of 3° or less do not require any special consideration and are designed the same as straight pipe.

Figure 5.3 illustrates terms relating to mitered bends.

$$P_m = \frac{SE(T-c)}{r_2}\left(\frac{T-c}{(T-c)+0.643\tan q\sqrt{r_2(T-c)}}\right) \tag{5.11}$$

$$= \frac{SE(T-c)}{r_2}\left(\frac{R_1 - r_2}{R_1 - 0.5r_2}\right) \tag{5.12}$$

Table 5.5 Values of Coefficient Y for $t < D/6$

Materials	Temperature, °F (°C)					
	≤900 (482)	950 (510)	1,000 (538)	1,050 (566)	1,100 (593)	≥1,150 (621)
Ferritic steels	0.4	0.5	0.7	0.7	0.7	0.7
Austenitic steels	0.4	0.4	0.4	0.4	0.5	0.7
Other ductile metals	0.4	0.4	0.4	0.4	0.4	0.4
Cast iron	0.0	---	---	---	---	---

* From the ASME B31.3 Code-1993 Edition.

$$P_m = \frac{SE(T-c)}{r_2}\left(\frac{T-c}{(T-c)+1.25\tan q\sqrt{r_2(T-c)}}\right) \quad (5.13)$$

where:

M = larger of $2.5(r_2T)^{0.5}$ or $\tan\theta(R_1 - r_2)$

c = sum of mechanical allowances (thread or groove depth) plus corrosion, erosion, and other allowances where the tolerance is not specified for or grooved surfaces, it is safe to assume the tolerance to be 0.02 in. (0.5 mm) in addition to the specified cut depth, in.

E = quality factor from Table 5.3, or Table 302.3.4 of the ANSI/ASME B31.3 Code (designer should reference current edition and addenda of the Code)

P_m = maximum allowable internal pressure for miter bends, psi

r_2 = mean radius of pipe using nominal wall thickness $\overline{T}$, in.

R_1 = shortest distance from the pipe centerline to the intersection of the planes of miter joints [the minimum value of R_1 should be used, determined from Eq. (5.14)], in.

S = stress value for material from Appendix 1 of the ANSI/ASME B31.3 Code, psi

T = pipe wall thickness (measured or minimum allowable per specifications);

q = angle of miter cut, deg.

α = angle of change in direction at miter joint, which is 2θ, deg.

$$R_1 = \frac{A}{\tan q} + \frac{D}{2} \quad (5.14)$$

where:

$A = 1.0$ for $(T\text{-}c) \leq 0.5$, $A = 2(T\text{-}c)$ for $0.5 < (T\text{-}c) < 0.88$, and $A = [2(T\text{-}c)/3] + 1.17$, for $(T\text{-}c) \geq 0.88$

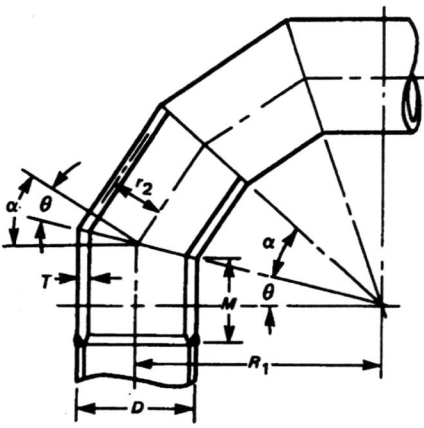

Figure 5.3. Nomenclature for miter bends. (Source: ANSI/ASME B31.3 – 1993 Edition.)

5.2.1.3 Branch Connections

Whenever a pipe is provided with an opening to facilitate a branch connection, it is weakened by the opening. There are many possible combinations of sizes and configurations of branch connections; these are a common component of most piping systems. Therefore, the design of branch connections deserves special attention to ensure a properly functioning primary piping system.

Branch connections include: (1) fittings such as tees, laterals, crosses, and extruded outlets; (2) welding outlet fittings, including cast or forged nozzles, couplings, and integrally reinforced weld-on and weld-in fittings such as weld-o-lets and sweep-o-lets (each having butt-welding, socket-welding, threaded, or flanged ends for pipe joining); and (3) branches made by welding a branch pipe directly into a run pipe. For branches having a branch angle greater than 45°, there are established procedures for determining the pressure design of such components. For branches with a value of β (See Figure 5.4) less than 45° and for configurations not described here, special considerations need to be met. These special branch connections must be designed in a consistent manner to the standard branches described here.

Consideration should be given to all dynamic, thermal, and cyclic effects in all branch designs. However, for fabricated branches which do not comply with area replacement rules, additional verification should be provided. The design of fabricated branch connections may be based upon calculations provided in this section, but they can also be substantiated by either extensive and successful field experience under similar conditions, experimental stress analysis, or proof test. These are outlined in the ASME Boiler and Pressure Vessel Code, Section VIII, Division 2, Appendix 6 (experimental analysis), and Section VIII, Division 1, UG-101 (proof testing), respectively.

In addition to pressure loadings, the design of a branch should take into account all forces and moments that may be applied to it. These forces and moments may be the result of thermal expansion and contraction, dead weight and live loads, movement of piping supports, and other external and internal effects. For situations where branch piping represents a significant reduction in diameter (greater than 3:1), it must be provided with adequate flexibility. Also, a branch may be stiffened by the use of ribs, gussets, or clamps if the design takes into account all applicable forces and effects. These stiffeners may be added for external loads and/or stiffening to minimize vibration of branch connections to a vibrating line (reducing the cyclic loads on the branch connection to the pipe). They are not to be counted as contributing to the required reinforcement area (for withstanding pressure) of the branch.

Unless the wall thickness of a pipe is sufficiently thick to reinforce a branch opening, as well as to maintain pressure within a pipe, it is necessary to provide reinforcement to the area of such a branch. For pipe sizes NPS 2 in. or less, couplings or half-couplings, welded directly to the run pipe, are often used for reinforcement. These are considered as having sufficient thickness if the branch diameter does not exceed 25% of the run piping diameter, and the coupling has a rating of at least Class 2000. Minimum coupling thickness should also be at least as great as the run pipe. Also, it is not necessary to provide reinforcement to a branch connection when using a fitting listed in Table C-1 of Appendix C to facilitate the branch. A third application where reinforcement is provided on the fitting occurs when an integrally reinforced branch connection is welded to a run pipe; if it is an unlisted component, this can be done only if the metallic material is listed in Table A-1 of Appendix A of the ANSI/ASME B31.3 Code, and the design takes into account all applicable forces and effects, and its design is substantiated by either extensive and successful field experience under similar conditions, or experimental stress analysis, proof test, or detailed numerical analysis. Qualification of fittings in this manner is typically done by the manufacturer.

For those situations where reinforcement is necessary, the required reinforcement area is determined by Eq. (5.15). Refer to Figure 5.4 for an illustration of branch connection terms concerning reinforcement.

$$A_1 = t_h d_1 (2 - \sin \beta) \tag{5.15}$$

where:
A_1 = the required reinforcement area for internal pressure design (for external pressure design, the required area is 0.5 times A_1), in.^{2}
t_h = pressure design thickness of header, in.
d_1 = effective length removed from the pipe at a branch, in.
β = smaller angle between axes of branch and run, deg.

The equation that must be satisfied to determine if there is sufficient reinforcement is Eq. (5.16). The basic philosophy is that the area that is removed from the pipe by the penetration of the branch connection is replaced by excess metal available within a reinforcement zone. This excess metal can be in the run or branch pipe, or in the added reinforcement.

$$A_2 + A_3 + A_4 \geq A_1 \tag{5.16}$$

where:
A_2 = the area available for branch reinforcement in the run pipe, in.^{2}
A_3 = the area available for branch reinforcement in the branch pipe, in.^{2}
A_4 = the area available for branch reinforcement in pads or other attached reinforcement (including welds), in.^{2}

A_2 is the area resulting from excess thickness in the run pipe wall, and can be calculated from Eq. (5.17). A_3 represents the same situation in the branch wall, and can be determined from Eq. (5.18). For longitudinally welded pipe, the weld joint quality factor E should be used in calculating the pressure design thickness t, except in the case of the run pipe, where it need not be included if the branch does not intersect the weld. A_4 is the area provided by welds and the added reinforcement. Reinforcement may be added in the form of a ring of reasonably constant width and may be made from a different material provided it is compatible with the pipe and branch in terms of weldability, heat treatment, resistance to galvanic corrosion, thermal expansion, and other pertinent factors. Its allowable stress may be greater, but if it is less, its calculated available area for reinforcement must be reduced by the ratio of allowable stresses in determining A_4.

Only metal added within the reinforcement zone is considered to contribute to reinforcement of the branch connection for pressure. The reinforcement zone is defined by a

Design of Metallic Primary Components

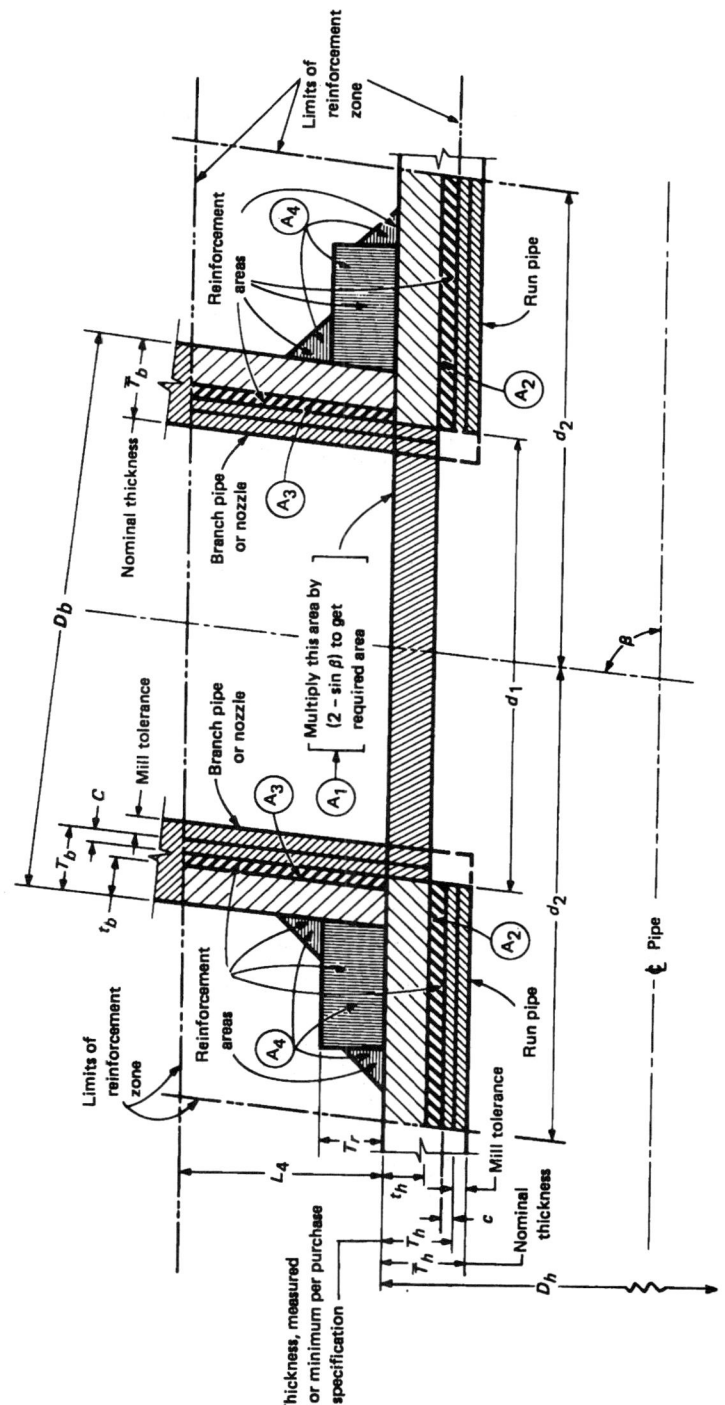

Figure 5.4. Branch connection reinforcement criteria and nomenclature. (Source: ANSI/ASME B31.3 – 1993 Edition.)

parallelogram with length d_2 on either side of the centerline of the branch pipe and with a height L_4 from the minimum expected inside diameter.

$$A_2 = (2d_2 - d_1)(T_h - t_h - c) \tag{5.17}$$

$$A_3 = \frac{2L_4(T_b - t_b - c)}{\sin \beta} \tag{5.18}$$

Where multiple branches exist, special consideration should be given to those cases where the reinforcement zones of those branches overlap. The distance between the centerline of each of the branches should be at least 1.5 times their average diameter for any two branches. Also, the area of reinforcement between any two openings should be at least half of the total reinforcement required by both openings. Care should be taken to ensure that the areas of the metal cross sections are not double counted in evaluating required reinforcement areas and available areas.

Branch connections where one or more outlets are formed by extrusion, using a die or dies to control the radii of the extrusion, are referred to as extruded outlet headers. The height of an extruded outlet, h_x, should be greater than or equal to the external radius of such an outlet, r_x. The radius, r_x, has a minimum and maximum required value. Minimum r_x is normally the lesser of $0.05D_b$ or 1.50 in. (38 mm). Maximum D_b depends on the outside diameter size, D_b, of the branch pipe. For D_b less than 8 in. NPS (approx. 203 mm OD), r_x should be a maximum of 1.25 in. (32 mm), and for D_b greater than or equal to 8 in. NPS (approx. 203 mm OD), its value can be determined by multiplying D_b by 0.1 and adding 0.5 in. (13 mm). For complex situations involving multiple radii, use the best fit radius over a 45° arc as the maximum allowable radius.

The required area of reinforcement for an extruded header can be calculated by Eq. (5.19). The available area for reinforcement of such headers can be determined from Eq. (5.20). Figures 5.5 through 5.8 illustrate the various terms for extruded outlets.

$$A_1 = Kt_h d_x \tag{5.19}$$

where:

A_1 = area required for branch reinforcement, in.^{2}
K = factor determined by ratio of branch diameter to run diameter:

$$K = \begin{cases} 1.00, & D_b/D_h > 0.60 \\ 0.6 + 2/3(D_b/D_h), & 0.60 \geq D_b/D_h > 0.15 \\ 0.70, & D_b/D_h \leq 0.15 \end{cases}$$

t_h = pressure design thickness of header, in.
d_x = the design inside diameter of the extruded outlet, measured at the level of the OD of the header, considering allowances and tolerances, in.

Design of Metallic Primary Components

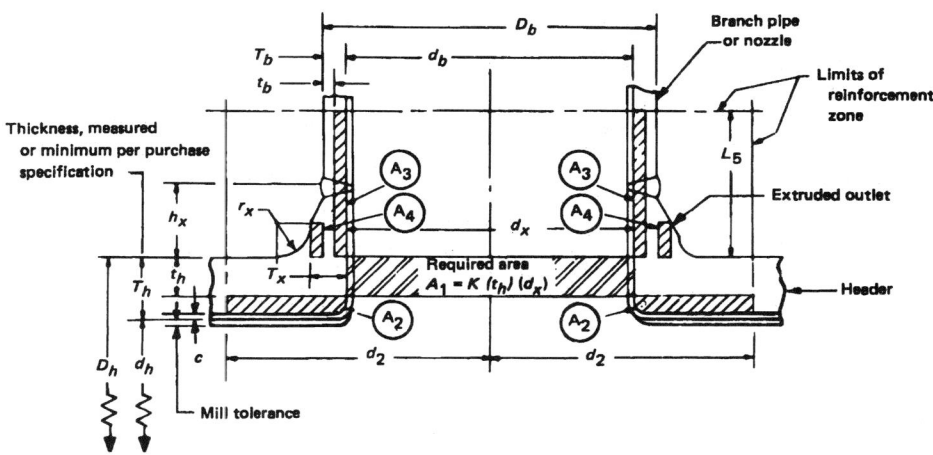

Figure 5.5. Extruded outlet header nomenclature for condition where K=1.00. (Source: ANSI/ASME B31.3 – 1993 Edition.)

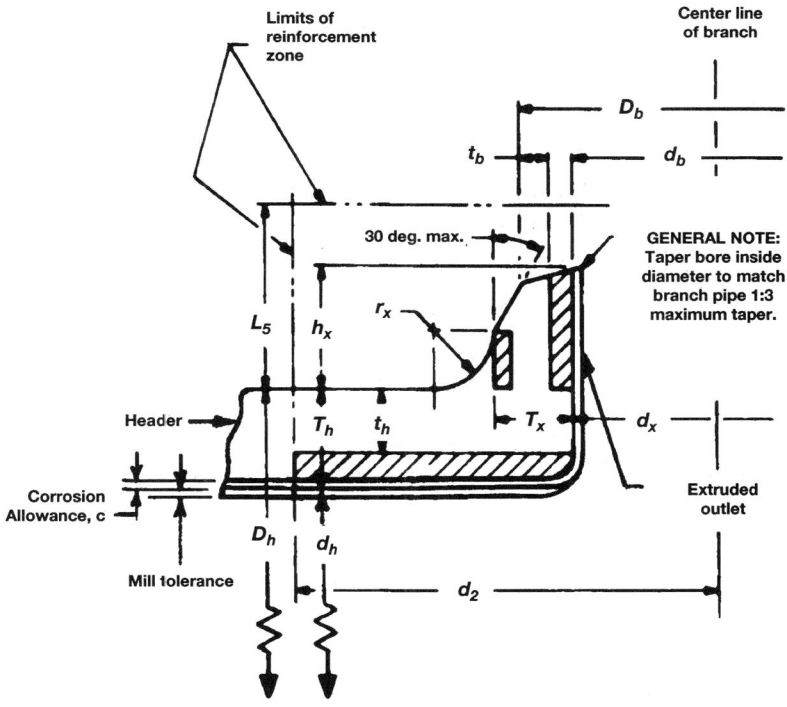

Figure 5.6. Detail of extruded outlet header crotch. (Source: ANSI/ASME B31.3 – 1993 Edition.)

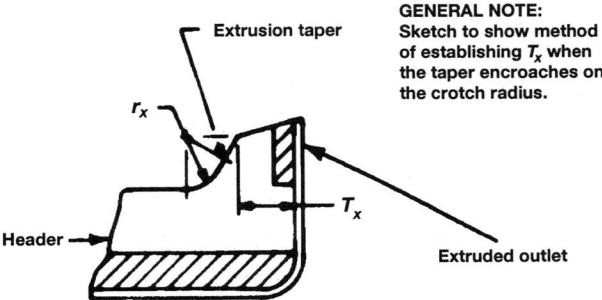

Figure 5.7. Detail of extruded outlet header crotch. (Source: ANSI/ASME B31.3 – 1993 Edition.)

$$A_2 + A_3 + A_4 \geq A_1 \tag{5.20}$$

where:
A_2 = the area available for branch reinforcement in the run pipe
A_3 = the area available for branch reinforcement in the branch pipe
A_4 = the area available for branch in excess thickness of the extruded outlet lip

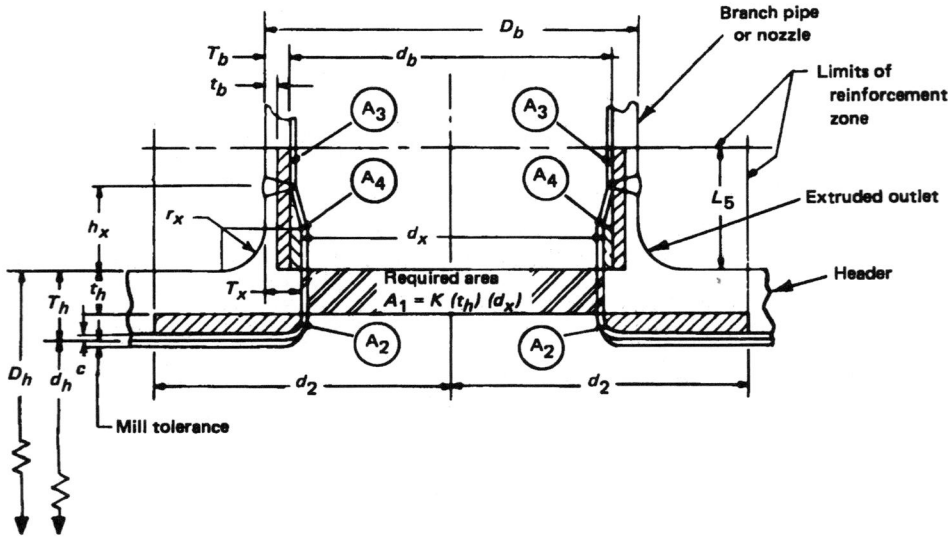

Figure 5.8. Extruded outlet header nomenclature for condition where K=1.00 and $d_x < d_b$. (Source: ANSI/ASME B31.3 – 1993 Edition.)

The nomenclature used for extruded outlet headers is as follows:

d_x = the design inside diameter of the extruded outlet, measured at the level of the outside surface of the header. This dimension is taken after removal of all mechanical and corrosion allowances, and all thickness tolerances, in.

h_x = height of the extruded outlet. This must be greater than or equal to r_x, except as shown in Figure 5.6., in.

L_5 = height of the reinforcement zone = $0.7(D_b T_x)0.5$, in.

T_x = corroded finished thickness of extruded outlet, measured at a height equal to r_x above the outside surface of the header, in.

d_2 = half-width of the reinforcement zone (equal to d_x), in.

r_x = radius of curvature of the external contoured portion of the outlet, measured in the plane containing the axes of the header and branch, in.

The area A_2 represents excess metal available for reinforcement in a header wall, and can be calculated from Eq. (5.21). A_3 represents the same in a branch pipe wall, and can be determined from Eq. (5.22). A_4 is the area provided by the excess thickness in the extruded outlet lip and can be determined from Eq. (5.23).

$$A_2 = d_x(T_h - t_h) \tag{5.21}$$

$$A_3 = 2L_5(T_b - t_b) \tag{5.22}$$

$$A_4 = 2r_x(T_x - T_b) \tag{5.23}$$

Where multiple openings exist, special consideration should be given to those cases where reinforcement zones of those branches overlap. The distance between the centerline of each of the branches should be at least 1.5 times their average diameter for any two branches. Also, the area of reinforcement between any two openings should be at least half of the total required by both openings. Caution should be taken so that the areas of the metal cross sections are not double counted in evaluating required reinforcement areas and available areas.

5.2.1.4 Closures

Standard closure designs, manufactured according to listed standards, are typically used where closures are required. These have the same pressure ratings as the matching pipe. If a custom designed closure is necessary, detailed procedures for the design of closures are provided in the ASME Boiler and Pressure Vessel Code, Section VIII, Division 1. The procedures described therein depend on the geometry of the closure and whether it is concave or convex to the pressure side of the closure. The various geometries include ellipsoidal, torispherical, hemispherical, conical (with no transition to knuckle), toriconical, or flat. The

Table 5.2 ASME BPV Code References[1] for Closures

Type of closure	Concave to pressure	Convex to pressure
Ellipsoidal	UG-32(d)	UG-33(d)
Torispherical	UG-32(e)	UG-33(e)
Hemispherical	UG-32(f)	UG-33(c)
Conical (no transition to knuckle)	UG-32(g)	UG-33(f)
Toriconical	UG-32(h)	UG-33(f)
Flat (pressure on either side)	--- UG-34	---

1. Paragraph numbers are from the ASME BPV Code, Section VIII, Division 1.

thickness of the closure can be determined from Eq. (5.9). References to ASME Boiler and Pressure Vessel Code Sections that pertain to closure design are provided in Table 5.2.

Some closures may contain openings, and when they do, they need to be designed with sufficient reinforcement to counteract the weakening created by the opening. Openings in closures that are less than one-half of the inside diameter of the closure should be designed as a reducer or as a flange if the intended closure is flat. The reinforcement for a closure containing an opening can be designed by the techniques described in the ASME Boiler and Pressure Vessel Code, or by the same methods described for branch connections. When rules for branches are followed, references to the run and header pipe, including the subscript h, one should apply to the closure. If a closure contains curves, the contour of the closure will serve as a reference point for the boundaries of the reinforcement zone. In this case, the dimensions of the reinforcement zone should be measured parallel and perpendicular to the contours of the closure surface.

When multiple openings are designed into a closure, the same principles of reinforcement for multiple branches can be applied to achieve a satisfactory design.

5.2.1.5 Reducers

A reduction in pipe diameter may be accomplished with the use of a concentric or eccentric reducer (as opposed to a reducing branch or reducing tee). Some concentric and eccentric fittings are contained in the standards listed in Table C-1 of Appendix C, and thus can be used as pressure fittings as described in those standards. Flanged end reducers (ANSI B16.5) have the pressure rating of the flange, and butt-welding reducers have the same pressure rating of the matching pipe. Concentric reducers that contain conical sections, reversed curve sections, or a combination of the two can be designed according to the same methods described in the ASME BPV Code, Section VIII, Division 1, for conical and toriconical closures. For those fittings that are not listed in Table C-1 of Appendix C, or

Design of Metallic Primary Components

designed per the previous sentence, the design should be in accordance with known techniques and verified by calculation or analysis. The design should be supplemented by experimental stress analysis, proof test, numerical stress test, or reference to extensive successful service under comparable conditions using similar materials and dimensions.

5.2.1.6 Flanges and Blanks

Standard flange designs per ANSI B16.5 are often used in piping systems. Other flange standards include ANSI B16.1 for cast iron flanges, ANSI B16.42, MSS SP-44, and API 605. The flange geometry is standardized for each pressure class; pressure ratings for each pressure class depend on the material and the temperature. Table 5.8 provides pressure ratings per ANSI B16.5 for 150 and 300 psi flanges of selected materials and temperatures.

Flange design for metallic piping systems is a well-defined practice that is covered thoroughly in the ASME Boiler and Pressure Vessel Design Code. Flange design is covered in Section VIII, Division I, Appendix 2 of the ASME BPV Code. The stress values for the flange material used in the procedures should be taken from Appendix 1 of the current edition and addenda of the ANSI/ASME B31.3 Code. For the bolt stresses used, the values also listed in Appendix A-2 of the ANSI/ASME B31.3 Code should be used. For flanged joints where the flange faces are in contact outside the bolt circle, with or without a gasket sandwiched in between, the procedures outlined in Appendix Y of the ASME BPV Code should be used.

The thickness of a blind flange, including manufacturer's minus tolerance, can be determined by the rules of the ASME BPV Code, Section VIII, Division 1, UG-34, followed by the application of:

$$t_m = t + c \tag{5.24}$$

where:
t = pressure design thickness, in.
c = the sum of mechanical allowances plus corrosion, erosion, and other allowances, in.

The minimum required thickness of a permanent blank can be determined by using Eq. (5.25). Various types of permanent blanks are illustrated in Figures 5.9 through 5.11.

$$t_m = d_g \sqrt{\frac{3P}{16SE}} + c \tag{5.25}$$

where:
d_g = the inside diameter of gasket for raised or flat face flanges, or the gasket pitch diameter for ring joint and fully retained gasketed flanges, in.
E = quality factor from Table C-2 or C-3 of Appendix C, where applicable
P = design pressure
S = the stress value for materials from Table A-1 of Appendix A of the ANSI/ASME B31.3 Code, psi
c = the sum of corrosion, erosion, and other allowances, in.

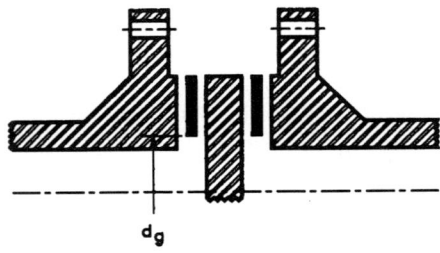

Figure 5.9. Example of a permanent blank. (Source: ANSI/ASME B31.3 – 1993 Edition.)

5.3 Selecting Primary (Core) Piping Components for Fluid Services

In the design of the primary (core) piping system, it is essential that designers be aware of material limitations in specific fluid services, with respect to each component. The suitability of any material in a given service will depend upon many factors. These include the relative corrosivity of the fluid on the material, operating conditions that exist that may enhance corrosive effects of a fluid on a material, physical properties of a material over the full range of anticipated operating temperatures, and the type of component to be used. This section presents the designer with some specific guidelines and limitations for selecting metallic materials for component design, in order for a primary piping system to transport a chemical safely over the design life of the double containment piping system.

The ANSI/ASME B31.3 piping code provides the designer of chemical and petroleum piping systems with many specific restrictions and limitations. For example, components may be permissible in cooling water service that are not otherwise permissible in flammable liquid service, or not permissible in other services with a significant potential for fatigue failure. Fabrication and inspection requirements also vary for components, based on the severity of the service. The ANSI/ASME B31.3 base code provides for design and selection of components based on "normal" service; it also provides specific modifications for other fluid service classifications and for "severe-cyclic" service (i.e., > 7,000 cycles, and the computed stress range exceeds 80% of the allowable displacement stress range).

All fluid services covered by the piping Code are considered "normal" other than those classified in Category D, which is less restrictive, and category M, which is more restric-

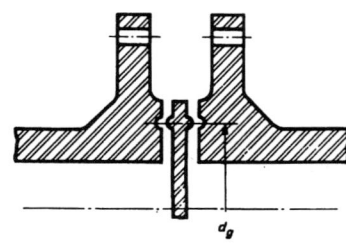

Figure 5.10. Example of a permanent blank. (Source: ANSI/ASME B31.3 – 1993 Edition.)

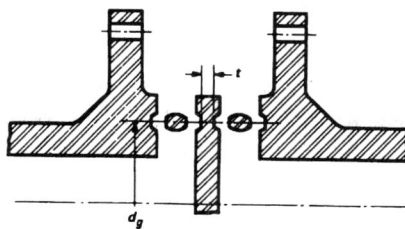

Figure 5.11. Example of a permanent blank. (Source: ANSI/ASME B31.3 – 1993 Edition.)

tive. Category D and Category M fluid services, and normal fluid service, are all defined in paragraph 300.2 of the ANSI/ASME B31.3 Code. Designers are encouraged to refer to these definitions, as these fluid services are referred to throughout this text. The ANSI/ASME B31.3 Code requires each end user to define fluid services for piping systems at their facilities that are to be in compliance with the ASME B31.3 Code. Fluid services are not selected for the end user by ANSI or ASME, or by any other consensus organization or standard in North America; it is up to the end user to do so. There are various local and federal government regulations, which vary by regulation and jurisdiction, that do define some hazardous or toxic chemicals. However, these definitions or classifications are not directly correlated to the Piping Code.

Severe cyclic service describes a piping system, irrespective of fluid service, that is subject to a high number of cycles, and has a high displacement stress range. A piping system with these two conditions is subject to potential fatigue failure, and is thus classified as having "severe cyclic service." For a specific definition of severe cyclic service, refer to paragraph 300.2 of the ANSI/ASME B31.3 Code.

5.3.1 Primary (Core) Pipe

Chapter 2, "Materials of Construction," gives a comprehensive discussion of the chemical performance and limitations of all metallic materials typically used in chemical plant and petroleum refinery applications. Piping materials must be selected based on their ability to handle a fluid service safely, taking into account all aspects of the engineering design. Specific limitations and recommendations for primary (core) piping materials are as follows.

Table 5.3 lists grades of carbon steel pipe that contain limitations as to the type of services in which they may be used. Table 5.4 lists grades of pipe that may be used in pipe for "severe-cyclic" conditions. These include certain grades of carbon steel pipe, low- and intermediate-alloy steel pipe, stainless steel pipe, copper and copper alloy pipe, nickel and nickel alloy pipe, and aluminum alloy pipe.

5.3.2 Primary (Core) Fittings

There are many rules that should be followed when selecting fitting types for pressure service. The physical properties of a fitting material, its method and quality of manufacture, the fluid properties, operating conditions, and other factors all have a bearing on selection. In general, only fittings that are of the forged type, wrought (factor $E_j \geq 0.90$), or cast (factor $E_c \geq 0.90$) may be used in severe-cyclic service.

Table 5.3 Grades of Carbon Steel Pipe [1] That May Be Used under Severe-Cyclic Conditions, According to ANSI/ASME B31.3

API 5L	Grade A or B, seamless
API 5L	Grade A or B, SAW, str. seam, $E_j \geq 0.95$
API 5L	Grade X42, seamless
API 5L	Grade X46, seamless
API 5L	Grade X52, seamless
API 5L	Grade X56, seamless
API 5L	Grade X60, seamless
ASTM A 53	seamless
ASTM A 106	
ASTM A 333	seamless
ASTM A 369	
ASTM A 381	$E_j \geq .090$
ASTM A 524	
ASTM A 671	$E_j \geq 0.90$
ASTM A 672	$E_j \geq 0.90$
ASTM A 691	$E_j \geq 0.90$

1. Casting or joint factors E_c or E_j specified for cast or welded pipe which do not correspond with E factors in Table A-1A or A-1B (B31.3 Code) are established in accordance with paras. 302.3.3 and 302.3.4 of the ASME B31.3 Code-1993 Edition.

Table 5.4 Grades of Alloy and Nonferrous Pipe [1] That May Be Used under Severe-

ASTM A	333, seamless
ASTM A	335
ASTM A	369
ASTM A	426, $E_c \geq 0.90$
ASTM A	671, $E_j \geq 0.90$
ASTM A	672, $E_j \geq 0.90$
ASTM A	691, $E_j \geq 0.90$
ASTM A	268, seamless
ASTM A	312, seamless
ASTM A	358, $E_j \geq 0.90$
ASTM A	376
ASTM A	430
ASTM A	451, $E_c \geq 0.90$
ASTM B	42
ASTM B	466
ASTM B	161
ASTM B	165
ASTM B	167
ASTM B	407
ASTM B	210, Tempers 0 and H112
ASTM B	241, Tempers 0 and H112

1. Casting or joint factors E_c or E_j specified for cast or welded pipe which do not correspond with E factors in Table A-1A or A-1B (B31.3 Code) are established in accordance with paras. 302.3.3 and 302.3.4 of the ASME B31.3 Code-1993 Edition.

The following fitting types should not be used under any circumstances for severe-cyclic conditions. Pipe bends prepared from corrugated or creased materials and miter bends that contain a change of direction greater than 22.5° at a single joint should also be restricted from use in severe-cyclic conditions. A miter bend containing a single joint that makes a change in direction greater than 45° should only be used in Category D service. Pipe bends, made with appropriate procedures covered in Chapter 12, "Fabrication, Erection, and Assembly," are suitable in the same service as the pipe from which it is made. Fittings conforming to MSS SP-43 proprietary "Type C" lap-joint stub end welding fittings and flared laps should not be selected for severe-cyclic services under any circumstances. Also, low-yield-strength bolting should not be used on flanges that are intended for severe-cyclic conditions.

Fabricated laps are suitable for use in normal fluid services where the outside diameter is in accordance with ASME B16.9 lap-joint stub end dimensions, its lap thickness is equal to at least the thickness of the pipe, and the allowable stress of the lap material meets or exceeds that of the pipe, and weld procedures outlined in Chapter 12 are followed. Flared laps may also be used for normal fluid services provided they are adequately constructed. The conditions that determine adequacy of flared laps include ASME B16.9 lap-joint stub end dimension equivalence, its fillet radius must be less than or equal to 1/8th in., and its thickness at any point being greater than or equal to 0.95 times the minimum required pipe wall thickness times the ratio of pipe outside radius to lap thickness. Its design should use known calculation techniques and verified by calculation or analysis. The pressure design of flared lap joints must be supplemented by experimental stress analysis, proof test in accordance with ASME BPV Section VIII, Div. 2, or by reference to extensive successful service under comparable conditions using similar materials having similar dimensions.

All listed flanges, blanks, and gaskets are suitable for normal fluid services. Slip-on flanges should be avoided in services where there are many large-temperature cycles because of differential expansion between the pipe and the flange. However, they may be used if they are provided with adequate insulation. Double-welded slip-on flanges are always preferable to single-welded slip-on flanges. However, single-welded slip-on flanges are generally permissible except in the following services, where they must be double welded. These services include: (1) services subject to erosion, crevice corrosion, and cyclic loading; (2) services where the fluids involved are flammable, toxic, or damaging to human tissue; (3) "severe-cyclic" services; and (4) where they may be subject to temperatures below -150°F (-101°C). In many countries, such as Great Britain, their Codes do not permit the use of single-welded slip-on flanges in any chemical fluid services.

Without additional safeguards, any flange that is subject to severe-cyclic service should be a welding neck flange (i.e., ASME B16.5, API 605 type, or equivalent). Flanges designed for use with a flared metallic lap should have a beveled or rounded face and bore intersection (approximately a 1/8th in. bevel).

In any piping system, there are always valves and other specialty components such as traps, strainers, separators, measuring devices, and other pressure-containing components. Some of these components are listed in Table C-1 of Appendix C; most of them may be used in normal fluid services. Bolted bonnet valves, where bonnets use fewer than four bolts or a U bolt to secure the bonnet to the body, should be restricted from use in Category M fluid services. Valves and other components that are not listed in Table C-1 of Appendix C can have their pressure ratings established by the procedures described in Appendix F of ASME B16.34. Otherwise, they may also be qualified for use by establishing their design in accordance with known techniques and verified by calculation or analysis. The design should be supplemented by experimental stress analysis in accordance with ASME BPV Section VIII, Div. 2, proof testing in accordance with ASME BPV Section VIII, Division 1, Part UG-101, or by reference to extensive successful service under comparable conditions, using similar materials and dimensions.

5.4 Selecting Joining Methods for Fluid Services

5.4.1 General

In the selection of each connection for every component in a piping system, consideration must be given to the material, fluid service, resulting joint tightness and pressure rating of the joint, operating conditions, and all other internal and external effects and loadings. The types of connections normally considered for metallic components include: (1) welded joints; (2) flanged joints; (3) expanded joints; (4) threaded joints; (5) tubing joints; (6) caulked joints; (7) soldered and brazed joints; and (8) others.

5.4.2 Welded Joints

The procedures for welding, including preheating and heat treatment, are described in Chapter 12, "Fabrication, Erection, and Assembly." Chapter 13, "Inspection, Examination, and Testing," covers requirements for the examination of welds and presents detailed acceptance criteria. Welding is normally differentiated for services by the level and extent of examination and inspection that is provided to a weld during and after a weld is made. These requirements are described in more detail in their respective chapters.

Socket welds must be excluded in any service where crevice corrosion is judged to be a potential problem, or where the service involves erosive fluids. Where socket-welded joints are used, they should conform to ASME B16.11 or ASME B16.5 (flanges). If a service involves "severe-cyclic" conditions, socket welds should be restricted to sizes 2 in. or less. A drain or bypass in a component may be attached by the use of a "socket weld," if the socket dimensions are equal those of Figure 5.12. Socket-weld components are joined by using fillet-welding techniques as described in Chapter 12. Fillet welding may also be used to attach slip-on flanges, to attach reinforcement pieces, to supplement primary welds and

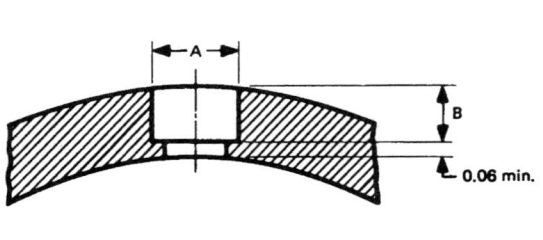

Connection size, NPS	Minimum diameter of socket A, in.	Minimum depth B, in.
3/8	0.69	0.19
1/2	0.86	0.19
3/4	1.06	0.25
1	1.33	0.25
1-1/4	1.68	0.25
1-1/2	1.92	0.25
2	2.41	0.31

Figure 5.12. Socket connections for welding. (Source: ANSI/ASME B16.5 1988 Edition.)

in any manner that prevents the failure of joints. Seal welds are used to prevent leakage of threaded joints. If a threaded joint is to be seal welded, thread compound or tape should not be used, as it can result in poor welds.

5.4.3 Flanged Joints

Flanges are generally used in piping systems to provide joints that can be readily disassembled, and to provide a means of connection to other flanged components. For example, they are frequently used in conjunction with valves that may need to be removed in the future for maintenance or replacement. They are also used at locations that may require blinds. Flanged joints may also be used to join piping, fittings, valves, or accessories that have different pressure ratings. They may also be used to join dissimilar materials, including metallic and nonmetallic materials. When metallic flanges of different ratings are joined, the bolting torque should be limited to prevent overstressing the lower-rated flange. The pressure rating of the assembled joint will be limited to that of the flange having the lower pressure rating. When metallic flanges are joined to nonmetallic flanges, gaskets used should be full faced and the flange flat faced. Care should be taken so that a nonmetallic flange is not overstressed.

5.4.4 Expanded Joints

Expanded joints are generally used in noncritical applications (such as food and beverage production). However, they have been used in applications involving termination flanges for jacketed-process piping systems. Expanded joints usually involve attachment of slip-on-type flanges to pipes (i.e., as in attachment of a termination flange to a core pipe), which allows for rapid assembly of the joint as compared to a welded flange. The use of expanded joints should be restricted to services where "severe-cyclic" conditions are not present and under conditions where separation of the joint is not very likely. In applications involving Category M fluids, safeguarding of expanded joints is required according to the ASME B31.3 Code. Prior to using such a joint, consideration should be given to the effects of all internal and external effects as they might affect the integrity of the joint, including vibration, differential thermal expansion, and the like.

5.4.5 Threaded Joints

Threaded joints are only used in small-diameter piping. There are many limitations and restrictions regarding the use of threaded joints in pressure piping services involving chemicals. Therefore, they usually represent a poor selection for joints in primary (core) piping components. Threaded joints should not be used in "severe-cyclic" conditions, services with the likelihood of crevice corrosion, erosive services, or where thermal expansion or mechanical forces (e.g., valve tightening) are deemed to present a potential problem.

For threaded joints that are to be seal welded after assembly, they should be assembled without the use of any sealing compound. When threading is required, the use of taper-threaded joints is always preferred over straight threads. If a coupling has a straight female thread, it should be used with a male tapered thread and restricted to Category D service. Straight-threaded joints may be used in joints that do not require the threads to seal, such as in union arrangements, which depend on the compression of a gasket or "O" ring to facilitate the seal. Examples of straight-threaded joints of this type are shown in Figure 5.13. This

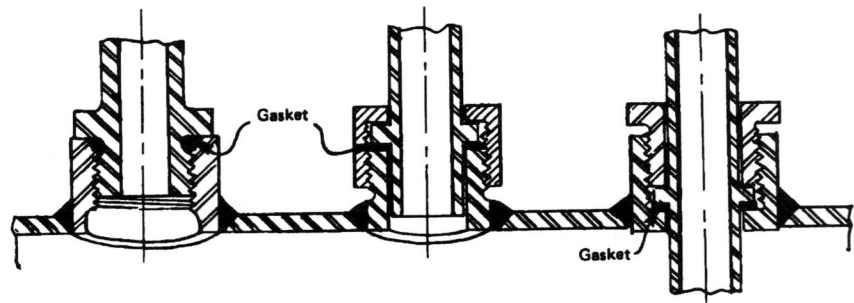

Figure 5.13 Typical threaded joints using straight threads. (Source: ANSI/ASME B31.3 – 1993 Edition.)

type of joint should be safeguarded if "severe-cyclic" conditions exist, and where they are subject to external loadings. Threaded joints of dissimilar materials (e.g., a carbon steel and an austenitic stainless steel) should be avoided because of galvanic and mechanical concerns.

5.4.6 Tubing Joints

When tubing is used in piping systems, consideration should be given to the fittings used and their various types of joining methods. The various fittings used with tubing include flared, flareless, and compression-style tubing fittings. Tubing fittings usually are subject to frequent assembly and disassembly, and the effect of this upon the integrity of the joint should be considered. Consideration must also be given to cyclic conditions, vibration, shock, thermal expansion and contraction, thrust, and other factors as to how they may affect the reliability of the tubing joints. Tubing can be used in normal fluid service, provided the tubing is suitable for the service conditions, and they are safeguarded when "severe-cyclic" conditions are present.

5.4.7 Caulked Joints

Caulked joints of the bell-and-spigot type should be limited to Category D service. Caulked joints are typically used in underground sewer piping and underground water piping applications and as such are not readily applicable to primary piping services in a double containment piping system. Additional precautions and safeguards, such as adequate anchoring to prevent joints from being pulled apart by internal pressure, must be provided to prevent failure of these types of joints due to internal pressure thrust under normal operating conditions.

5.4.8 Soldered and Brazed Joints

Soldering and brazing procedures are covered in Chapter 12, "Fabrication, Erection, and Assembly." Fillet-brazed and soldered joints are not permitted for chemical services in piping systems that must be designed according to ASME B31.3, and should not be used in general for chemical piping services. Where brazing alloys and solders can be exposed to fire or high temperatures, safeguarding methods, including external insulation or other means to protect against fire exposure should be provided. Soldered joints should be limited to Category D services, such as water piping. Brazed and braze-welded joints are considered suitable for normal fluid service.

5.4.9 Other Joints

There are other joining methods that are used to join metallic systems. Gasketed bell-and-spigot-type, packed-gland-type, and grooved-compression joints are three examples of these. Whenever using a joining method such as the first two, one should pay attention to ensure joint integrity through the use of mechanical or welded interlocks, or other methods. This is particularly important where a fluid service is flammable, toxic, or damaging to human tissue, or where "severe-cyclic" conditions are present. Safeguarding is required for bell-and-spigot and gland-type joints in "severe-cyclic" service, or where cyclic conditions are present. In general, it is not a wise idea to use these types of joints for primary (core) service in a double containment piping system.

5.5 Flexibility and Support

5.5.1 Piping Flexibility

5.5.1.1 General

Piping systems that are not provided with adequate flexibility are subject to premature failure due to problems that can develop during normal operating conditions. Thermal expansion or movements of piping supports and terminals can cause failure of piping or its supports due to overstress or fatigue. These effects may also cause excessive thrusts and moments in piping, resulting in failure and subsequent leakage at joints or harmful stresses or distortions, including overload of equipment connected to it. Therefore, piping should be designed with sufficient flexibility in order to reduce or minimize the possibility of these occurrences. A proper design and layout will result in a system whose computed stress range is within the allowable stress range for the material under consideration, as determined by Eq. (5.2) or (5.3). A successful design also will not cause detrimental reaction forces to occur at supports or connected equipment. Any possible movements will also be restricted to be within prescribed limits of the materials involved.

Terminal load limits may be based on equipment-vendor-set limits, reference standards such as API 610 for centrifugal pumps, or determined by analysis, such as stress analysis of vessel nozzles. Flexibility calculations performed in computing stress range should determine these reaction forces as well. The extent of thermal expansion must be considered in avoiding potential interferences, including interferences between primary (core) and secondary containment components.

This section presents various procedures and data for determining the requirements for flexibility in a primary (core) piping system, to assure that the primary portion of a double containment piping system meets all of these requirements.

In double containment piping systems, unique problems arise since primary and secondary containment pipes may be interlocked in locations. They may also be contacted at various points in the radial direction due to the use of interstitial supports or standoffs. Therefore, the two systems must be considered as a whole, considering that a portion of the loads imposed on each will be acting concurrently upon the other. Chapters 7 and 8 provide a more detailed discussion of this topic. In Chapter 8, details are presented as to methods for providing flexibility and compensation for displacements that occur, in order to result in the least computed stress range for a completed double containment piping assembly (see Sections 8.4 and 8.8).

5.5.1.2 Displacement Strains

There are a variety of displacements (strains) that can occur that may result in bending, torsional, and axial stresses. Each of these need to be considered in performing a flexibility analysis. Thermal displacements result from dimensional changes any time a change in temperature occurs. At any given temperature, piping will expand or contract to a predicted length, if left totally unrestrained, in order to assume its lowest-energy state. When it is restrained from assuming this position, the difference in its actual position and its stress-free position results in a displacement strain with a resulting stress. The restraints in any given instance may or may not be assumed to be rigid. If they are not rigid restraints, the displacement of the restraint may reduce this imposed displacement and, thus, the stress in the piping system.

In some cases an external effect, such as a thermal expansion of the equipment to which the pipe is attached, will cause movements of the pipe that must be considered. Movement originating from the settling of backfill materials above or below a buried piping system can be considered as a one-time occurrence, and if localized strain and end reactions are controlled, displacement does not necessarily have to be considered as part of the total displacement strain. All loading conditions must be considered. Conditions that act concurrently must also be combined in the flexibility analysis. Conditions that may or may not act concurrently should be evaluated each way to determine which combination is critical. An example of this is multiple-pump (in parallel) systems with common headers where individual pumps may or may not be on, and the associated suction and discharge piping for each pump may or may not be at the process temperature (depending on if the pump is on). In this example, the critical operating condition must be determined. In some cases, a piping system may have multiple operating temperatures. For these cases, an equivalent number of full temperature cycles is calculated, as shown in the definition of N_E from Eq. (5.4). Cases involving conditions other than the simply operating temperature can also be evaluated using the same equation, simply by substituting the calculated stress for temperature in the computation of equivalent number of cycles. Possible support movements must be carefully included. For example, thermal expansion of the vessel to which a pipe is attached may reduce stress in that pipe. However, it may be possible for the same pipe to heat up before the vessel. In this case, the governing thermal expansion case could be with the pipe hot and the vessel cold. It is the maximum stress range that is of concern, not the stress at the operating condition.

The total of all displacement strains should be determined for the various parts of a system, since they all have equivalent effects and cause proportional deformations. It should be remembered that for double containment piping systems, loads imposed on the outer pipe will be partially transferred to the inner piping if there are rigid interconnecting components and supports in the interstitial space. Further, differential expansion between the inner and outer piping can be a critical consideration.

There are instances where displacement stresses are allowed to accumulate to a sufficient level where you can exceed yield stress of portions of a system, without deleterious effect on the system. The reason has to do with a phenomenon that is similar to cold springing. As a system reaches its highest level of displacement at the extreme design temperature (high or low temperature), some amount of creep or yielding will occur with a subsequent reduction of stress (relaxation effect). As the system returns to the installed temperature at the end of the cycle, the system will reverse and redistribute its stresses. This situation is referred to as self-springing.

5.5.1.3 Displacement Stresses

The objective in determining the layout of any system should be to minimize the total displacement strains and distribute them in such a manner that there are no excessive strains at any one point. This even distribution of displacement strains will result in a structurally balanced system and allow strains to be considered as proportional to the elastically calculated stress range due to displacements. If the system is not balanced, an excessive amount of displacement strain may occur at one or more points in the system. In this case, the stresses can not be considered as proportional to the total displacement strains in the system. When the piping layout results in this condition, there can be resulting problems at the most susceptible locations in the system. The situation will be even worse if the system is operated in the creep range, resulting in a concentration of creep strain accumulation at these locations. This condition is typically called elastic followup or strain concentration.

The elastic followup generally has the following components. There is a flexible but relatively low-stressed system tied in with a stiff, but relatively high-stressed, local part of the system or component. Examples of local regions could be a highly stressed elbow or a hot-wall section in a cold-wall piping system. The local section could be relatively highly stressed because stresses are significantly higher than in the rest of the system, or it could be hotter or constructed of a lower-strength material than the rest of the system. In performing the flexibility analysis of the piping system, all of the components are assumed to behave elastically, stresses are calculated, and strains are assumed to be proportional to the elastically calculated stresses. The local section of concern may be calculated to absorb only a small proportion of the total system displacement because it is stiff relative to the remainder of the system. Note that it could be stiffer even if it is smaller in diameter and/or wall thickness simply because it is shorter.

When elastic followup occurs, the local section yields significantly before the rest of the system as it is heating up, or creeps as a significantly higher proportion of the displacement-induced strain in the system is absorbed by the local section than was elastically calculated. Yielding substantially reduces the stiffness. The stiffer the local region was (on an elastic basis) when compared to the remainder of the system, the higher the strain concentration may be. This is because a larger proportion of the displacement strain would be calculated on an elastic basis to be in the remainder of the system, but could end up in the local section due to elastic followup. The problem that can result is early fatigue because higher cyclic strains than elastically predicted occur and, in the creep regime for the material, creep-rupture failures occur because displacement stresses do not relax in the local region in the same manner they would in a balanced system. While local yielding is permitted in piping systems (as described in the next paragraph) because residual stresses develop in the initial operating cycles that lead to subsequent elastic behavior, this may not occur in unbalanced systems.

An unbalanced system is possible for many reasons. Some of the common reasons in single-wall piping include: (1) systems where small pipe runs that are highly stressed run in series with restrained or relatively large piping; (2) piping components contain a local reduction in wall thickness; (3) local use of a material having a reduced yield strength; and

(4) a system containing short offsets or changes in direction intended to absorb a large displacement from a longer adjacent run. In addition to these possibilities, in a double containment piping system, anchor-type connections between the primary and secondary pipe are subject to elastic follow-up caused by differential thermal expansion between the pipes. Unbalance can be minimized to the greatest extent possible in a double containment piping system through effective layout of the system and through the proper applications of the techniques discussed in Chapter 8.

The use of cold springing is not a realistic option for achieving a balanced system in a double containment piping system. It is very difficult to implement in practice, and is discussed in more detail in the subsequent paragraph.

5.5.1.4 Cold Spring

Cold spring is generally used to reduce terminal loads on equipment. It is a method whereby the piping is intentionally deformed in the installation process so that the final end reactions that occur in the extreme condition are reduced. This method does not affect the stress-strain range during heat up and cool down and thus has no benefit with respect to the allowable cyclic displacement in the system. Therefore, the application of cold spring will not affect the values determined in stress range calculations. However, cold spring can reduce the magnitude of the stress in piping and terminal loads in the hot condition. Part of the load is essentially shifted to the cold condition by cold springing. Cold springing, for example, may be used on pump-associated piping to reduce the maximum loads on the pump to within the vendor-specified allowable loads.

Use of cold spring is generally not recommended because it is difficult to achieve properly in practice. Cold springing requires considerable force to adjust in the field properly. Therefore, the field capabilities of cold springing must be thoroughly investigated prior to specifying cold spring. Because of the difficulty in properly achieving desired cold spring levels, the ASME B31.3 Code only permits taking credit for two-thirds of the intended cold spring in calculating terminal reaction loads, as discussed later in this chapter. Cold spring is difficult to implement in single-wall piping, is even more difficult to achieve in double containment piping, and is therefore not generally recommended for such systems.

5.5.1.5 Displacement Stress Range

Although the magnitude of stress and reactions between the extreme displacement condition and the as-installed condition is usually constant during each cycle of operation, for a given pair of displacement conditions, the magnitude at each condition may change due to self-springing resulting from creep or yielding. The difference that is calculated at each condition remains the same. The difference that is calculated between the two extreme displacement conditions is called the displacement stress range, S_E. This, coupled with the allowable stress range, S_A, to which it is compared, are perhaps the most important factors in performing a flexibility analysis on a piping system.

5.5.1.6 Flexibility Analysis of Primary Piping

A formal stress analysis need not be performed if there is a successfully operating system that is similar to the system being analyzed. Also, if there is a previously analyzed system that allows the designer to make a comparison to the system in question and judge it adequate, then a formal analysis is also not necessary.

Where a formal analysis is required, a simplified, approximate, or comprehensive analysis should be performed, using simplified equations of the type described in this section. If a simplified or approximate method is selected, such as the methods outlined in this section, it should be used only within the limits for which it is known to be reliable. There are a variety of comprehensive methods that include analytical techniques (performed manually or via computer), or using charts that have been developed by these same techniques. With the advent of the personal computer, most analysis performed today is computer generated. However, the method selected must be capable of providing an evaluation of all forces, moments, and stresses caused by displacement strains, taking into account stress-intensification factors for all elements of the system other than straight pipe. Other components and elements generally contain flexibility relative to straight pipe, and this flexibility may be accounted for in the analysis.

In the calculation of its flexibility, a piping system should be treated as an entire system between anchor points. (An anchor point refers to positions along a piping system, including component locations, that are clamped to rigid structural parts or are secured by other means, to prevent movement of the piping at that point under all applicable loads.) In a double containment piping system, points where the system are interconnected may function as anchor points on the primary piping and should be treated as such, when appropriate. These interconnection points may approach the behavior of an anchor for analysis purposes. The degree of restraint on a primary piping system imposed by the friction of interstitial supports acting on the secondary containment pipe inside wall must be taken into account in flexibility calculations, only if it is deemed to be significant. All displacements must be considered over the full temperature range. It is critically important in performing a flexibility analysis of double containment pipes to include any potential interaction between the secondary containment (jacket) and primary (core) pipe in the analysis.

5.5.1.6.1 Calculation of Flexibility Stresses

The computed stress range, S_E, can be calculated by Eq. (5.26) and then compared to the allowable stress range, S_A, to determine whether the flexibility of a piping layout is adequate. In a piping flexibility analysis performed in accordance with the ASME B31.3 Code, stresses due to bending and torsional moments are calculated and compared to the allowable displacement stress range, S_A. Calculations are based on nominal pipe wall thickness (without subtracting erosion, corrosion, or other allowances), and the stress calculations are based on using the modulus of elasticity at the installation temperature. This approach is based on the fact that displacement stresses due to axial loads are not generally significant in typical petroleum refinery and chemical plant piping systems. However, the designer must be aware of situations where they become critical. For example, if a straight pipe was run between two anchors and subjected to a temperature change, a high axial load and resulting stress can easily be developed. However, the equation in the code would calculate zero stress because only bending and torsional moments are considered. This situation is readily developed in double wall piping systems if the primary and secondary pipes are anchored together in two consecutive locations in a straight run and are at two different temperatures. The designer must consider the axial load contribution to the displacement stress range if such conditions are present. Where conditions are such that stresses due to axial loads are significant, that stress should be added to the bending stress as shown in Equation 8.7. However, the designer must also determine what stress intensification factor to use for the axial loads, since none are provided for in the ANSI/ASME B31 Codes.

The simplified equation provided by the ANSI/ASME B31.3 Code and applicable to most piping systems is presented as:

$$S_E = \sqrt{S_b^2 + 4S_t^2} \qquad (5.26)$$

where:
S_b = the resultant bending stress, computed using the as-installed modulus of elasticity, E_a (note: some components have different stress intensification factors for in-plane and out-of-plane bending; calculation procedures for these components are outlined in the subsequent paragraphs), psi
S_t = torsional stress, computed using the as-installed modulus of elasticity (E_a), $S_t = M_t/2Z$, psi
M_t = torsional moment, in.-lbs
Z = the section of modulus of pipe, in.^{3}

5.5.1.6.2 Calculation of Flexibility Stresses in Elbows, Bends, Miter Bends, and Full-Size Outlet Branch Connections

The resultant bending stresses, S_b, for elbows, bends, miter bends, and full-size outlet branch connections can be calculated from Eq. (5.27) and then substituted in Eq. (5.26). Figure 5.14 illustrates the various moments related to elbows, bends, and full-size outlet branch connections.

$$S_b = \frac{\sqrt{(i_i M_i)^2 + (i_o M_o)^2}}{Z} \qquad (5.27)$$

where:
S_b = resultant bending stress, psi
i_i = in-plane stress intensification factor from Appendix D of the ANSI/ASME B31.3 Code
i_o = out-of-plane stress intensification factor from Appendix D of the ANSI/ASME B31.3 Code
M_i = in-plane bending moment, in.-lbs
M_o = out-of-plane bending moment, in.-lbs
Z = section modulus of matching pipe, (not of the fitting, itself), in.^{3}

5.5.1.6.3 Calculation of Flexibility Stresses in Reducing-Outlet Branch Connections

The resultant bending stress, S_b, for reducing outlet branch connections can be calculated by using Eqs. (5.28) and (5.29), with the following modifications. Figure 5.15 illustrates the various moments related to reducing-outlet branch connections. The header portion of the reducing branch connection (legs 1 and 2) simply uses Eq. (5.28). The reducing branch uses Eq. (5.29), which substitutes an effective section modulus Z_e for the section modulus Z.

$$S_b = \frac{\sqrt{(i_i M_i)^2 + (i_o M_o)^2}}{Z} \tag{5.28}$$

$$S_b = \frac{\sqrt{(i_i M_i)^2 + (i_o M_o)^2}}{Z_e} \tag{5.29}$$

where:
S_b = resultant bending stress, psi
Z_e = effective section modulus for branch, $= \pi r_2^2 T_s$, in.^{3}
r_2 = mean branch cross-sectional radius, in.
T_s = effective branch wall thickness, lesser of T_h and $(i_i)(T_b)$, in.
T_h = thickness of pipe matching run of tee or header exclusive of reinforcing elements, in.
T_b = thickness of pipe matching branch, in.
i_i = in-plane stress intensification factor from Appendix D of the ANSI/ASME B31.3 Code
i_o = out-of-plane intensification factor from Appendix D of the ANSI/ASME B31.3 Code

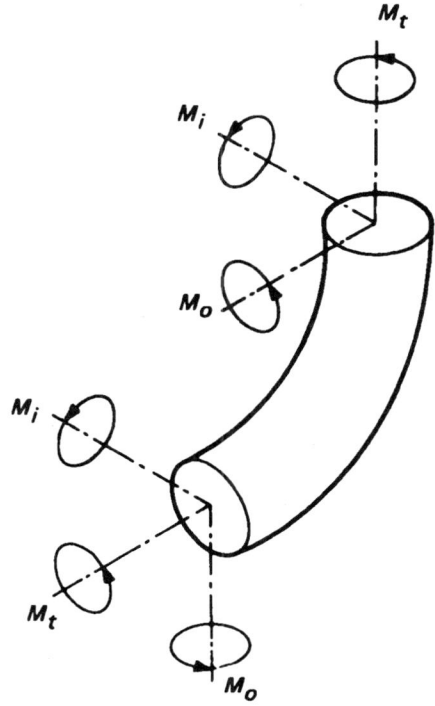

Figure 5.14. Moments in primary pipe bends. (Source: ANSI/ASME B31.3 – 1993 Edition.)

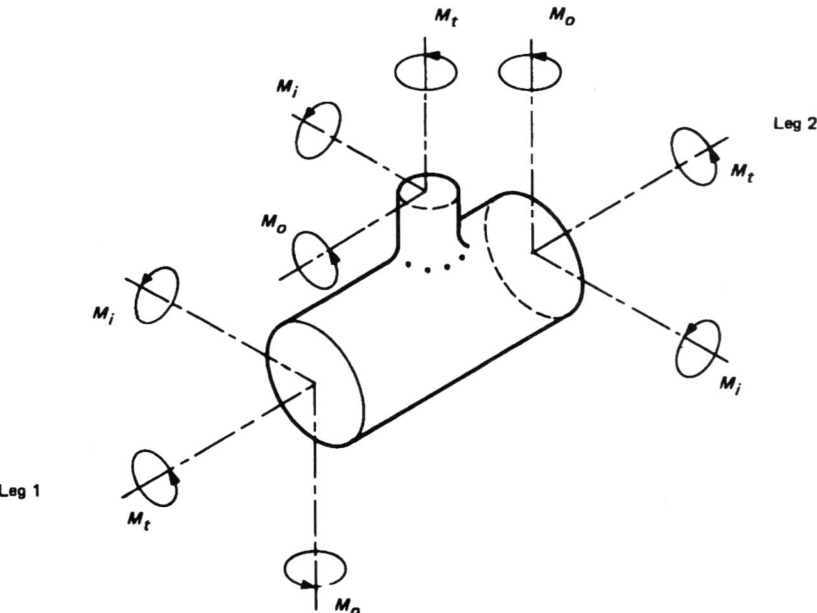

Figure 5.15. Moments in primary pipe branch connections. (Source: ANSI/ASME B31.3 – 1993 Edition.)

The allowable stress range S_A is computed in accordance with Eqs. (5.2) and (5.3). The analyst should be aware that additional precautions, including more comprehensive weld inspections, will be required if there are more than 7,000 cycles and the computed stress range exceeds 80% of the allowable displacement stress range (e.g., "severe cyclic" service).

5.5.1.7 Reactions Involving Primary Components

The proper design of restraints and supports for piping systems involves the calculation of reaction forces and reaction moments. These values are also useful for evaluating the effects of piping displacements on connected equipment. For piping systems without cold spring, these reaction loads may be computed for the extreme temperature condition, using the modulus of elasticity at that temperature. Evaluation of reaction loads for piping systems with cold spring is more complex. Further, it is necessary to determine possible reaction loads at the installed temperature condition, even if there is no cold spring. The effect of self-springing must be considered. To evaluate these situations, the maximum reaction range, R, is calculated based on the extreme displacement conditions and the installed modulus of elasticity, E_a. A simplified approach for determining the reaction loads with cold spring or self-springing follows.

For a two-anchor system without any intermediate restraints (as in a simplified short-run double containment piping system), the maximum instantaneous values of reaction forces and moments can be calculated from Eqs. (5.30) and (5.31). Equation (5.30) is based on the maximum or minimum metal temperature (whichever produces the larger value) for extreme displacement conditions R_m.

$$R_m = R\left(1 - \frac{2C}{3}\right)\frac{E_m}{E_a} \tag{5.30}$$

where:
C = cold spring factor ranging from zero for no cold spring to 1.0 for complete cold spring
E_a = modulus of elasticity at installed temperature, psi
E_m = modulus of elasticity at maximum or minimum metal temperature, psi
R = range of reaction forces or moments (from flexibility analysis) based on the full displacement stress range and based on E_a, lbs
R_m = estimated instantaneous maximum reaction force or moment at maximum or minimum metal temperature, lbs

For the original condition, R_a, the calculation is based on the expected temperature at which the piping is to be assembled. R_a is equal to CR or C_1R, whichever is greater, and C_1 is determined by:

$$C_1 = 1 - \frac{S_h E_a}{S_E E_m} \tag{5.31}$$

where:
C_1 = estimated self-spring or relaxation factor, zero if the value of C_1 is negative
R_a = estimated instantaneous reaction force or moment at the installed temperature due to self-springing of the piping, lbs
S_E = computed displacement stress range, psi
S_h = basic allowable stress at maximum or minimum metal temperature during the displacement cycle under analysis, psi

Complex systems may not be modeled by Eqs. (5.30) and (5.31), and must instead be modeled with a more comprehensive analysis. The analysis should take into account the location, type, and extent of overstrain and the resulting effect it has on the distribution of stress and subsequent reactions.

5.5.2 Support of Primary Piping

The design of support structures for piping systems are covered in Chapter 9, "Structural Considerations." The design of supports and supporting structures should take into account all internal and external loads that act concurrently on a pipe and are transferred to the supports. The weight of the piping should include the basic material plus the internal fluid. Where the fluid is a gas, the weight of the test fluid (water) should be used in the support calculations if the test performed is a hydrostatic-type test.

For piping that is provided with inadequate support, there are many things that can go wrong. Excess stresses, leakage at the joints, and excessive thrusts and moments on connected equipment may be caused, just the same as if inadequate flexibility is provided.

However, there are many other possible causes of such effects, including: (1) detrimental stress in supporting or anchoring elements; (2) resonance coupled with fluid-induced vibration or pressure pulsations in a line; (3) restraints counteracting a proper flexibility design that would otherwise alleviate thermal expansion; (4) disengagement from piping supports; (5) excessive sag or distortion of piping; and (6) excessive heat flow. Generally, simple methods and good engineering judgment may be used to locate and design pipe supporting elements. However, a more detailed analysis is occasionally required. When it is, the stiffness of supports should be accounted for, as well as the stresses, moments, and reactions, in order to arrive at the design of the supporting elements. Details of supporting devices and other related matter are further discussed in Chapter 9, "Structural Considerations."

5.6 Auxiliary Piping Systems

Two common forms of piping that are normally used in conjunction with primary pressure piping systems include instrument piping and pressure-relief piping systems. Instrument piping is used to connect instruments to other piping or equipment, or it consists of control piping that is used to connect air or hydraulically operated control apparatus. Normally, instruments are treated separately from other primary and secondary containment piping, including permanently sealed fluid tubing systems that serve as temperature or pressure response systems. Pressure relief piping is used either directly as overpressure protection without an auxiliary pressure relief device, or it is used as the discharge piping for those types of devices.

Instrument piping should be designed in accordance with the same requirements and considerations as ordinary pressure piping. Additionally, instrumentation that is attached directly to such piping should be designed to be able to maintain the integrity of the overall system. Since most instrument piping contains static fluids, they are readily subject to freezing and should thus be provided with heat tracing where this is a possibility. If it is required to drain instrument piping that contains hazardous or toxic fluids, care should be taken to dispose of those fluids properly. Since most double containment systems use fluids of this type, instrument piping for these systems should be designed with this feature either built in, or blow-down procedures should be given emphasis in the maintenance and operating procedures and manuals.

Pressure-relief piping for primary carrier piping in double containment piping systems is unique in that the secondary containment piping is usually also designed for this purpose in case of catastrophic failure. In that sense, secondary containment piping gives an added sense of protection in case of a malfunctioning relief valve or device. Secondary containment piping systems also provide designers with an option of allowing discharge piping from safety relief valves and other relief devices to be safely discharged into it. Since most double containment piping is for hazardous or toxic fluid transport, this is a realistic option for a designer. In some instances, a relief valve may be discharged directly into a chamber constructed as an integral part of a secondary containment system, eliminating the need for discharge piping altogether. It should be noted that in above-ground double containment piping where fire or some other event can cause a fluid in a primary piping to overpressurize, it is a must to have pressure-relief valves installed on the primary carrier pipe, in addition to having a secondary containment piping outer wall, depending on the rules of the authority having jurisdiction.

The sizing of safety-relief valves and other relief devices, according to the ANSI/ASME B 31.3 Code is covered in the ASME Boiler and Pressure Vessel Code, Section VIII, Division I, UG-125(c), UG-126 through UG-128, and UG-132 through UG-136, excluding UG-135(e) and UG-136(c). The reader should also be aware of the requirements stated in ANSI/ASME B31.3 para. 322.6.3 "Pressure Relieving Devices", where the B31.3 Code applies. In the requirements referenced in the ASME BPV Code, Section VIII, Div. I, the phrases "maximum allowable working pressure" and "vessel" are the equivalent, respectively, of the "design pressure" and "piping system" nomenclature used in this chapter. For a liquid thermal expansion relief device which protects only a blocked-in portion of a piping system, the set pressure is not allowed to exceed either the system test pressure or 120% of design pressure, according to the ANSI/ASME B31.3 Code 1993 Edition. The maximum system pressure during a pressure relieving event should be based on the criteria stated in the ASME BPV Code, Section VIII, Div. I, except that the allowances discussed in Section 5.1.10 in this chapter can be met, so long as the conditions discussed in Section 5.1.10, and para. 302.2.4 of the ANSI/ASME B31.3 Code are met. Reactions on piping when safety-relief devices discharge should be such that they do not cause harm to the overall system and should be taken into consideration during the design of both primary and secondary containment portions of a double containment piping system. The use of block valves between piping being protected and its relief device should be limited to those situations where such a valve cannot prevent the piping system from being properly relieved.

Notes

Chapter Six

Design of Nonmetallic Primary Components

This chapter presents information concerning the design of primary (also referred to as "core") piping and components that are constructed of nonmetallic materials. Design issues for components that serve as a primary means of containing process fluids under pressure are presented here. This includes components that serve as an integral part of both the primary and secondary containment systems. Components of this type are referred to in this chapter, and elsewhere in this book, as "interconnecting components."

The approach to component design in this chapter is based on techniques and methods generally used for single-wall nonmetallic piping components for use in chemical piping services. The philosophy of this chapter is that the basic rules and principles of ordinary single-wall piping should not differ or be changed for primary pressure-retaining components in a double containment piping system. Primary piping system components in a double containment piping system represent the "primary" means of containing a service fluid. Therefore, the design of its components must be based on well-proven engineering techniques and principles that are applied to single-wall nonmetallic components. While the material covered here is highly technical in nature, and should be applied by a competent professional, every effort has been made to assist readers in understanding important design criteria to be considered on any project involving nonmetallic primary components. Design and analysis procedures, data, and concepts are provided for components made from several categories of materials. These material groups include thermoplastic, plastic-lined metallic, reinforced-thermosetting-resin plastic (RTRP), dual laminates, borosilicate glass, and others. Emphasis is given to thermoplastic and RTRP materials, which account for the vast majority of nonmetallic piping used in double containment piping systems.

Two very important points need to be made concerning the use of nonmetallic piping systems for double containment chemical service. First, they should not be used in "severe-cyclic" services, unless it can be demonstrated that an identical previously existing system has functioned successfully. Second, nonmetallic piping that is contained within a metallic secondary containment piping system, where both systems are properly designed, represents a form of safeguarding a nonmetallic system. Therefore, a system as just described might allow nonmetallic materials to be used in a fluid service for which they would be otherwise excluded. It is completely the responsibility of each designer to determine the acceptability of such an approach, based upon the fluid services, applicable design codes, regulations that apply, and the rules of the authority having local jurisdiction.

6.1 Design Issues for Nonmetallic Primary Piping Components

6.1.1 Design Conditions

In the design of any nonmetallic piping system, the conditions selected as the basis for design should include all applicable forces and effects. This is true for both pressure and gravity drainage systems alike. These effects include ambient effects, dynamic effects, and thermal expansion effects, among others. Ambient effects should include consideration of all heat-transfer effects, pressure effects, and corrosive and erosive actions. Certain dynamic effects that can be achieved under extreme conditions must also be given consideration. Geysering, water hammer, and cavitation are examples of these. Unusual thermal effects may occur when the piping is subjected to differential temperatures that vary greatly between internal and external conditions (i.e., between primary components and secondary containment components), or are due to internal conditions resulting from two-phase flow. These effects should be considered, or primary piping components may be subject to damage at points of restraint, such as unidirectional restraints, internal anchors, branch connections, as well as other "interconnecting component" locations.

6.1.2 Design Pressure

To determine the design pressure of each component in a system, an analysis must first be performed to establish the temperature/pressure profile of the system being designed. The temperature/pressure profile for any system may be determined from knowledge of the process operation for which it is intended, and from knowledge of typical weather patterns in the climate in which the system is installed. It is not a calculated amount, and is strictly determined based on knowledge of how and where the pipe is to operate. Examples of temperature versus pressure profiles are presented for a system, on a daily and weekly basis, respectively, in Figures 5.1 and 5.2.

The pressure/temperature condition of each component may vary significantly over time, as shown in the example. An analysis of this profile can determine the condition of coincident pressure and temperature that results in the greatest required wall thickness and, hence, the highest component rating. The pressure of the pressure/temperature condition that results in the highest value should be selected as the basis for the design pressure. A suitable safety factor is typically added to this value by a designer to result in the design pressure of the primary (core) system. When more than one set of pressure/temperature conditions exist in a piping system, there may be different design conditions for different components in the overall system. This will occur whenever governing pressure/temperature conditions that apply to individual components differ.

Provision must be made to contain or relieve safely any excess pressures to which primary piping components may be subjected. Secondary containment piping in double containment piping systems can be used for this purpose, even though it may not be intended as its primary function. Where it is used for this purpose, a secondary containment pipe should be designed to provide relieving capacity for the entire carrier pipe it protects. Also, each individual compartment in an annulus between primary and secondary containment pipes should be designed to relieve the capacity of the entire carrier pipe system that is contained within. Typical sources of internal overpressurization may include failure of control devices, process upset, dynamic effects, and ambient effects. Relief-protection requirements for the secondary containment portion of double containment piping systems are discussed in Chapter 7.

6.1.3 Design Temperature

The design temperature for each component in a primary (core) piping system is the temperature of the design condition discussed in Section 6.1.2, under the heading "Design Pressure." Namely, it is the temperature of the coincident pressure/temperature condition that results in the greatest wall thickness and highest component rating.

The minimum design temperature of nonmetallic primary components should be the lowest possible temperature condition that they may experience in service. This temperature may be the core fluid temperature, ambient temperature, or the temperature of the annular space (or heat-transfer fluid if the system is used as a jacketed process piping system). The temperatures of all three should be considered and the minimum selected as the design temperature. Consideration also should be given to special ambient effects, which are described in more detail in Section 6.1.4. Caution should be used in selecting nonmetallic piping for low temperatures due to the relatively brittle nature of most nonmetals at low temperatures. In no case should a material be used below its recommended low service design temperature rating, as stated in the ANSI/ASME B31.3 Chemical Plant and Petroleum Refinery Piping Code, or according to the manufacturer, whichever is greater.

6.1.4 Ambient Effects

The effects of all internal and external conditions that may be imposed upon a nonmetallic piping system should be considered, particularly those that produce a heating or cooling effect upon it. Vapor condensation is a rare occurrence in nonmetallic systems due to relatively low-temperature fluids that are conveyed. However, there is occasion where low-pressure steam for transport or used for steam-cleaning purposes can condense, creating a vacuum, with resulting collapse potential of the nonmetallic pipe wall. An external effect that can cause internal problems occurs when low external temperatures cause a stagnant fluid to freeze and expand, resulting in failure of the pipe wall or other components. These internal and external effects, as well as all others that apply, should be taken into account in the design of the system.

A unique ambient effect exists in double containment piping systems where the primary pipe is under a differential pressure created by a greater external pressure. This occurs in double pipe heat exchangers and in jacketed process piping, and can occur while pressure testing secondary containment portions of a double containment piping system, or when an annular space in a double containment piping system is being flushed during repair situations. Chapter 13, "Inspection, Examination, and Testing," discusses testing issues pertaining to double containment piping systems, including testing methods and testing sequence. In the pressure design, external pressures should be considered for each component, if there is any potential for an external pressure to be imposed on the component. The same effect occurs in single-wall pipe under submerged conditions. In submerged piping, external pressures must be considered in addition to buoyant forces.

6.1.5 Weight Effects

In addition to dynamic forces, there are static forces that arise from external loads that must be considered. The static weight load, or dead-weight load, consists of the weight of piping components, insulation, and any superimposed objects in above-ground piping. In underground piping it consists of the burial load as calculated according to the Marston equations [Eqs. (9.34) and (9.35)], for static soil loads. For underground systems loads may also include concentrated and distributed loads that may be imposed by vehicular travel over the area of the buried pipe. Burial analysis is further discussed in Chapter 9, Section 9.4.1.

In double containment piping systems, the outside (secondary containment) pipe will be directly subjected to the static-soil and live loads; however, certain primary piping components will see a portion of the loads due to interstitial supports and other interconnecting components, which transmit a portion of the loads to the primary system. These loads must be considered for the primary piping system.

6.1.6 Dynamic Effects

There are many dynamic effects that must be considered in the design of nonmetallic piping, if they are present. For single-wall pipes these include, but are not limited to: (1) impact of a pipe wall due to internal or external forces; (2) vibration; (3) seismic loading; (4) discharge effects; and (5) wind actions. The effects of impact can be created from a number of sources, including sudden changes in fluid velocity, slug flow, cavitation or liquid flashing, and geysering. Liquid flashing and geysering are rare for nonmetallic piping applications. Vibration of a piping component can result from pressure pulsations, pump motor resonance, wind loads, and the same sources that can create impact. Seismic forces will vary according to the geographic location of the piping system. It is recommended to use the analysis procedures outlined in ANSI A58.1, or as defined by the Uniform Building Code or the local authority having jurisdiction. When fluids are discharged from a pipe, there are reaction forces created by the sudden change in energy. These effects must be taken into consideration in the design. Wind also imposes loads upon a pipe. An analysis of wind effects for above-ground pipe can be made using the procedure outlined in ANSI A58.1, or by following the rules of the Uniform Building Code or the authority having local jurisdiction. In double containment piping system design, consideration should be given to loads imposed upon the secondary containment piping that are transferred to the primary piping due to interconnecting components and interstitial supports and standoffs.

6.1.7 Thermal Expansion and Contraction Effects

Many types of loads result from thermal effects; for nonmetallic systems, the relative effects of such loads are significant over small temperature variations compared to metallic materials. All of these loads need to be considered as acting concurrent with other forces imposed on each component in a piping system. If a nonmetallic piping systems is to be constrained from free expansion and contraction, the ANSI/ASME B31.3 Chemical Plant and Petroleum Refinery Piping Code requires a formal analysis, unless a previously duplicate system can be shown to operate successfully under identical conditions, or can be judged to be adequate by comparison with previously analyzed systems. However, when free expansion and contraction of the piping is restricted by the use of interstitial supports, or internal anchors, or other means in a double containment piping system, axial loads and moments will be significant, and must be taken into account in any analysis that is performed. In double containment piping and dual-laminate materials, loads and resulting discontinuity stresses will occur due to differential thermal expansion; these must also be considered in the design. In double containment piping, this effect will occur even if the material used for the secondary containment piping has the same coefficient of thermal expansion as that of the inner components. This is a certainty in systems that are properly buried. Where an annulus is provided with insulation, and particularly where an annulus contains entrapped air, or the outer pipe is externally insulated or buried, the temperature difference of the two buried pipes can still be substantially different. Temperature gradients will also exist in a relatively thick-wall pipe, or across a pipe wall with materials that have relatively low thermal conductivity. Temperature gradients may also arise from internal conditions such as two-phase flow, although it is a rare condition for applications involving nonmetallic piping. Temperature gradients are normal in chemical process sewers where the main is designed to flow less than full. Regardless of the cause, the temperature profile of each component should be taken into account in in analyzing for discontinuity stresses.

6.1.8 Other Considerations

In addition to the effects already stated, other considerations include cyclic effects and forces imposed by movements of supports, anchors, terminal movements, interconnecting components, and also reduced ductility effects in low-temperature service, among others. Cyclic effects include those involving pressure, temperature, or dynamic loadings. Movement of piping supports and the like can result from flexibility or thermal expansion, settling, wind forces, or, in the case of submerged or marine piping, as the result of tidal actions. Fusion and bonding procedures can reduce a nonmetallic material's ductility and should be taken into consideration. In any service where the piping components are used in temperatures that involve component materials having reduced ductility, and therefore low impact resistance, special precautions must be taken, including providing adequate shielding or insulation of the piping components, or other precautions. By the nature of double containment piping systems, this feature is built in to most systems by the virtue of the secondary containment feature. Extra precautions must be taken, however, where a double failure can result from a severe blow or other occurrence.

6.2 Design Criteria for Nonmetallic Primary Piping Components

6.2.1 General

In establishing the basis for the design of nonmetallic primary piping components in a double containment piping system, several design factors need to be established. These include the pressure and temperature ratings, stress criteria, design allowances, and minimum design values, as well as any permissible variations as they apply to the design of these components. In addition, the limitations of nonmetallic material properties typically are of a much lower magnitude than metallic materials; therefore, the designer needs to be aware of these limitations. In selecting nonmetallic components, the designer should be satisfied as to the adequacy of the raw material and as to the integrity of the manufactured component. In making these judgments, the designer should carefully judge all of the following physical properties of the raw material:

1. Tensile strength;
2. Compressive strength;
3. Flexural strength;
4. Shear strength;
5. Long- and short-term modulus of elasticity (at design temperature);
6. Creep rate at design conditions;
7. Design stress and its basis;
8. Poisson's ratio (longitudinal and circumferential directions);
9. Ductility;
10. Plasticity;
11. Elongation characteristics;
12. Impact properties, including notch sensitivity;
13. Thermal shock properties;
14. Temperature limits;
15. Transition temperatures (brittle, glass, melting, vaporization);
16. Porosity;
17. Permeability; and
18. Purity.

The designer should also give careful consideration to characteristics of the final product, including:

1. Method of manufacture;
2. Testing methods;
3. Methods of making joints and the efficiency thereof; and
4. Possibility of deterioration in service.

6.2.2 Pressure-Temperature Design Criteria

There are many established, recognized standards that set the pressure-temperature ratings of common nonmetallic piping components. In the United States and Canada, most of these standards are established by the American Society for Testing and Materials (ASTM), the American Petroleum Institute (API), the American Water Works Association (AWWA), and the National Institute of Standards and Technology (NIST). A listing of component standards that are applicable to double containment piping systems, and referenced in the 1993 edition of the ANSI/ASME B31.3 Code, is shown in Table D-1 of Appendix D. For components that are listed in Table A326.1 of the ANSI/ASME B31.3 Code (Table D-1 of Appendix D of this text), and do not have a specific rating stated within the referenced standard, the rating of the component may be found by applying the methods covered in this chapter (refer to Section 6.3, "Pressure Design of Nonmetallic Primary Components"). Table B-1 of Appendix B of the ANSI/ASME B31.3 Code states design stress ratings of various thermoplastic materials, and Tables B-2 and B-3 of the ANSI/ASME B31.3 Code state listed specifications for RTRP materials. Figures 6.1 and 6.2 indicate the acceptable working temperature ranges for common RTRP and thermoplastic materials, respectively.

Oftentimes, components that are to be used in a nonmetallic primary piping system will not be listed in Table D-1 of Appendix D. Nonmetallic piping system components of this type are referred to in the ANSI/ASME Code, and throughout this text as "unlisted compo-

Design of Nonmetallic Primary Components

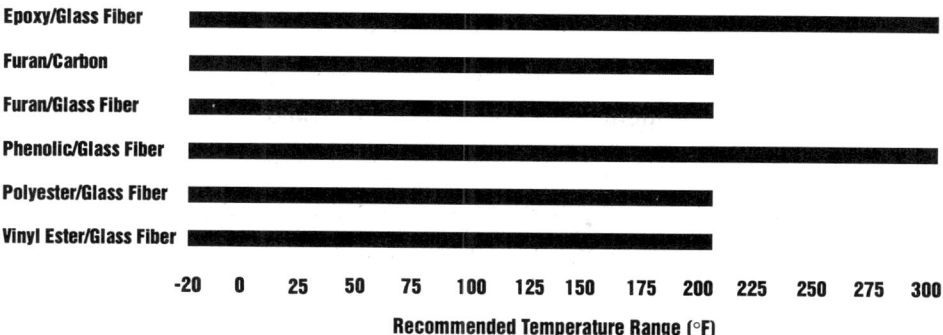

Figure 6.1. Recommended working temperature range for various RTRP materials according to the 1993 edition of the ANSI/ASME B31.3 Code.

nents." The design of these components should be in accordance with known techniques. In addition, they should be substantiated through demonstrated extensive successful service experience under comparable design conditions having similarly proportioned components of like material, or by performance testing of the component. Any performance test should simulate design conditions, and be performed for a comparable number of cycles. Performance tests should be performed for a period of time that is sufficient to determine the acceptability of the component or joint that is under analysis, for its design life. All applicable ambient and dynamic effects described in the beginning of this chapter should be taken into account in the design of an "unlisted" component.

Many services involve occasional temperature variations of temperature and/or pressure that exceed normal operating levels in a piping system. Nonmetallic piping systems should not be designed to allow variations that exceed the design ratings of any components in the system. The only exception to this rule is for metallic piping components that are lined

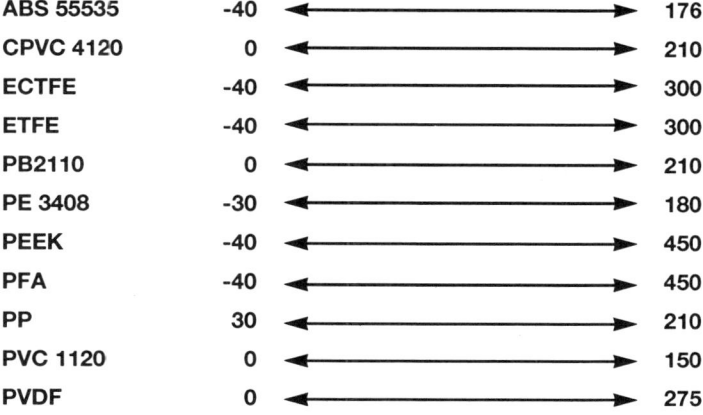

Figure 6.2. Recommended working temperature range (°F) for various thermoplastic materials according to the 1993 edition of the ANSI/ASME B31.3 Code.

with nonmetals. Metallic piping components lined with nonmetallic materials are typically designed according to the same design criteria outlined for metallic piping systems described in Chapter 5 (Section 5.1.10). However, the suitability of lining materials for increased design conditions should be established through prior successful operating experience, or through performance testing under like conditions, in order for the hybrid component as a whole to receive such design allowances.

6.2.3 Allowable Stresses and Other Limits

The hydrostatic design stresses (HDS) for thermoplastic materials are listed in Table B-1 of Appendix B of the ANSI/ASME B31.3 Code. Tables B-2 and B-3 of Appendix B of the ANSI/ASME B31.3 Code list specifications for RTRP materials. For those materials that do not have an associated ASTM specification for piping, there are no design stresses listed. Table 6.1 lists various ASTM specifications that establish ratings for laminated-reinforced, filament-wound, and centrifugally cast reinforced-thermosetting-resin pipe and reinforced-plastic-mortar pipe. Tables B-4 and B-5 of Appendix B of the ANSI/ASME B31.3 Code lists established pressure ratings for concrete pipe and for borosilicate glass piping. These HDS values, allowable stress criteria, and pressures are based on applications with water or fluids that do not have a significant effect on the properties of the material. The upper temperature limits of each of these materials are typically reduced at higher pressures, depending on the fluid and the expected service life. Lower temperature limits are not as materially affected by strength (the HDS values may be used at lower temperatures), but may be appreciably affected by the environment. The recommended limits apply only to the general materials listed. It is recommended to consult with a manufacturer of specific grades of materials to be used. The final properties of each proprietary brand of resin will vary. The values indicated in all of the referenced tables may be used in certain design calculations, but the use of hydrostatic design stresses for calculations other than to determine pressure has not yet been verified by the ASME or other standards organization, although this is often the common practice.

6.2.4 Bases for Allowable Stresses and Pressures

6.2.4.1 Thermoplastic Materials (Including Liners for Dual Laminates)

The accepted basis in the United States for establishing the hydrostatic design stress value for piping made from thermoplastic resins is in accordance with ASTM D 2837. Hydrostatic design values for some thermoplastic materials, at a variety of temperatures, are stated in Table B-1 of Appendix B of the ANSI/ASME B31.3 Code. Designers should always consult with manufacturers of a raw material under consideration to determine whether the values of the specific grade of resin to be used varies from Table B-1 of the ANSI/ASME B31.3 Code.

Table 6.1 Listed Specifications for Filament Wound and Centrifugally Cast Reinforced Thermosetting Resin and Reinforced Plastic Mortar Pipe

Spec. nos. (ASTM except as noted)		
D 2517	D 2997	D 3754
D 2996	D 3517	AWWA C950
		C582

6.2.4.2 Reinforced Thermosetting-Resin Plastic Materials (Laminated, Filament-Wound, Centrifugally Cast)

The basis for establishing design stresses for a specific material varies, depending on the method of manufacture of the material. For materials whereby the lamination is produced by hand layup (often referred to as "custom-made" pipe), the value should be 10% of the minimum tensile strength specified, according to ASTM C 582. These values should be used only within the temperature range of -20 through 180°F.

For machine-made components, including filament-wound and centrifugally cast materials, the hydrostatic design basis stress (HDBS) values are determined according to the procedures set forth in ASTM D 2992 for a temperature of 73.4°F (23°C). To determine the hydrostatic design strength (HDS) from the hydrostatic design basis strength, the HDBS should be multiplied by a service design factor that is selected for the application. The procedure for selecting the service design factor is also described in ASTM D 2992. The service design factor, F, should not exceed 1.0 when using the cyclic HDBS, and should be a maximum of 0.5 when using the static HDBS.

The design stresses of dual-laminate materials should be based solely on the base material, without any additional increase in design stress resulting from the laminated reinforcement layers. While the additional reinforcing laminate layers provide added structural strength, they cannot be considered to add to the design stress of the material due to the inconsistencies inherent in fabricating such products. For example, a thermoplastic liner such as polyvinylidene fluoride that is reinforced with a vinyl ester spiral-wound lamination should have its design stress determined solely on the basis of its thermoplastic liner material.

6.2.4.3 Borosilicate Glass and Concrete Materials (and Other Materials)

The pressure ratings of materials made according to Table B-4 (concrete) and B-5(borsilicate glass) of Appendix B the ANSI/ASME B31.3 Code are conservatively based on physical properties of such materials conforming to the referenced specifications only. The pressure ratings of other materials not described in this chapter should be designed in accordance with known techniques. They should be substantiated with either extensive successful service experience under comparable design conditions with similarly proportioned components of like material, or they should be subjected to a performance test. The performance test should simulate design conditions, and should be performed for a period of time sufficient to determine the acceptability of the component (or joint). All applicable ambient and dynamic effects described in the beginning of this chapter should be taken into account in order to establish the pressure ratings of these unlisted components.

6.2.5 Stresses Due to Sustained Loads

In general, the limits of stresses of nonmetallic piping as a result of internal pressure are based on the design stress of the material involved, its wall thickness, and the outside diameter of the pipe. The exact relationship depends on the type of nonmetallic material, as is indicated in Eqs. (6.2) through (6.4), shown in Section 6.3.1.1. The determination of wall thicknesses and the equations are presented in more detail in Section 6.3, "Pressure Design of Nonmetallic Primary Piping Components."

The limits of stresses due to external pressure also depend on the material involved. For thermoplastic piping, the limits are based upon the criteria established in ASTM D 2321 (or

AWWA C900). To determine the maximum allowable buckling of thermoplastic pipe, the amount of strain and possible buckling of the piping should be considered. In no case should the allowable diametrical deflection exceed 5% of the inside diameter of the pipe, even though some materials can withstand as much as 30% deflection on a short-term basis. Some thermoplastics are very rigid and can withstand less than 5% deflection. Their ratings for allowable diametric deflection are less than 5% in those cases. For reinforced thermosetting-resin and reinforced-plastic-mortar piping, the limits are based on the criteria established in ASTM D 3839 and also by Appendix A of ANSI/AWWA C950. RTP and RPM also should never be assigned allowable diametrical deflections that exceed 5% of the inside diameter of the pipe.

Nonmetallic piping constructed of other materials should be subjected to a crushing test (ASTM C 14) or a three-edge-bearing test (C 301). To determine the allowable load, first establish the minimum value through one of the tests, and then multiply the minimum value by 25%.

6.2.6 Stresses Due to Occasional Loads

Occasional loads consist of such items as wind, earthquake, and other temporary loads. These loads should be considered as acting concurrently with other loads (i.e., from internal pressures, weight, etc.) if they are part of the design conditions. The sum of the combined loads should be less than the applicable limits for the piping material under analysis. Loads that would not act together under any normal circumstances do not have to be analyzed as acting concurrently. For example, wind and earthquake loads need not be considered as acting concurrently. Also, stresses due to pressure testing do not need to be considered to be acting concurrently with occasional loads such as earthquakes and wind.

6.2.7 Allowances

Once the theoretical required wall thickness is determined, based on the design stress rating of the material and the diameter size of the component, the minimum required thickness may be determined. The minimum required thickness must take into account certain design allowances. These include corrosion, erosion, and thread or groove depth, where applicable. Also, there are a variety of loads that may require the wall thickness of a component to be increased in order to maintain its mechanical integrity. The wall thickness might have to be increased to prevent overstress, collapse damage, or buckling from loads due to supports, ice formation, burial load, or a variety of other causes. In some cases, the increase in a component's wall thickness may not necessarily increase the strength of the component. In fact, an increased wall thickness in some cases may lead to an increase in brittleness or local stress. Where this is the case, a component's strength may be increased with the use of stiffeners, braces, or other additional supports.

6.3 Pressure Design of Nonmetallic Primary Piping Components

6.3.1 Straight Pipe

To determine the required thickness of straight sections of primary pipe, it is necessary to determine the pressure design wall thickness of the pipe. First, it is necessary to calculate the internal design pressure wall thickness [Eqs. (6.2) through (6.4)] and also the external

Design of Nonmetallic Primary Components

pressure design wall thickness (see Section 6.3.1.3). The greater wall thickness value determined by the two different procedures is then selected as the design wall thickness. Then, the minimum required wall thickness can be determined by:

$$t_m = t + c \tag{6.1}$$

where:
t_m = minimum required thickness including mechanical, corrosion, and erosion allowances, in.
t = pressure design thickness, determined from internal pressure design equations; Eqs. (6.2) through (6.4), or by the procedures in Section (6.3.1.3) for external pressure design wall thickness, whichever is greater, in.
c = sum of mechanical allowances (thread or groove depth) plus corrosion and erosion allowance. For threaded components, nominal groove depth (h dimension as stated in ASME B1.20.1 or equivalent) should be used. For machined or grooved surfaces that do no have a specified tolerance stated, a tolerance of 0.02 in. (0.5 mm) can be used in addition to the specified depth of the cut, in.

6.3.1.1 Straight Pipe under Internal Pressure

The minimum required wall thickness due to internal pressure can be calculated by Eqs. (6.2) and (6.3), depending on the type of material to be used. The values of allowable design stress listed in Tables B-1, and those determined by the referenced ASTM standards listed in Tables B-2, and B-3 of Appendix B of the ANSI/ASME B31.3 Chemical Plant and Petroleum Refinery Piping Code should be used as a starting point. However, the manufacturer's recommended values of the specific materials to be used should be used for the design basis, if the value results in a greater wall thickness. Equations (6.2) and (6.3) are various forms of ISO Equation 161/1 for determining plastic piping required wall thicknesses based on internal pressures. Equation (6.2) is used for thermoplastic piping, whereby the value for S is derived from a hydrostatic design basis, using a service design factor. The hydrostatic design basis is usually determined from regression data based on a minimum of 100,000 hrs. (11.4 yrs.) of applied hoop stress. Equation (6.2) is also used for laminated RTR pipe ("custom-made" pipe). Equation (6.3) is used for filament wound and RPM pipe.

$$t = \frac{PD}{2S + P} \tag{6.2}$$

$$t = \frac{PD}{2SF + P} \tag{6.3}$$

where:
F = service (design) factor, as described in Section 6.2.4.2, dimensionless
P = internal design gauge pressure, psi
D = outside diameter of pipe, in.
S = allowable design stress from applicable Table in Appendix B of the ANSI/ASME B31.3 Code, or as determined from Section 6.3.1.3, psi

Table 6.2 Recommended Temperature Limits[1] – Thermoplastics Used as Linings

Materials[2]	Minimum		Maximum	
	°F	°C	°F	°C
ECTFE	-325	-198	340	171
ETFE	-325	-198	300	149
FEP	-325	-198	400	204
PFA	-325	-198	500	260
PP	0	-18	225	107
PTFE	-325	-198	500	260
PVDC	0	-18	175	79
PVDF	0	-18	275	135

Notes:
1. These temperature limits are based on material tests and do not necessarily reflect evidence of successful use as piping component linings in specific fluid services at these temperatures. The designer should consult the manufacturer for specific applications, particularly as temperature limits are approached.
2. See paragraph A326.3 of the ANSI/ASME B31.3 Chemical Plant and Petroleum Refinery Piping Code, for a complete description of material abbreviations.

6.3.1.2 Metallic Piping Lined with Nonmetals (for Primary Piping Applications)

For metallic piping that is lined with nonmetals in applications subject to internal pressures, the design minimum required thickness should be calculated only for the metallic outer piping (up to the temperature limits of the liner). The techniques for determining the minimum required thicknesses for metallic primary piping components under internal and external pressures are thoroughly outlined in Section 5.2 in Chapter 5. The thickness of a liner material is independent of the pressure requirements, but such a liner should only be used up to its maximum design temperature. Recommended temperature limits for materials used as lining materials are shown in Table 6.2. Designers should consult with each manufacturer of lined pipe material for their specific recommendations based on each specific application.

Since all lined metallic piping requires venting of the liner, this type of piping is generally not recommended as primary piping in a double containment piping system. However, it is readily suitable as a primary pipe where a means of secondary containment other than straight pipe, such as a secondary containment trench, or dike, is used.

6.3.1.3 Straight Primary Pipe under External Pressure

For thermoplastic pipe, the various types of reinforced thermosetting-resin pipes, glass, and concrete pipe, the external pressure design should be in accordance with known techniques. The thicknesses required should be substantiated either by extensive successful service experience under comparable design conditions and similarly proportioned components of like material, or they should be subjected to a performance test. The performance test should simulate design conditions, and should be performed for a period of time sufficient to determine the acceptability of the material in the component form under analysis. All

Design of Nonmetallic Primary Components

applicable ambient and dynamic effects described in the beginning of this chapter should be taken into account in determining external design thicknesses.

The minimum required wall thicknesses due to internal pressure of dual-laminate materials should be determined for the primary liner, without any additional contribution of design strength credited to the external reinforcement material. While additional reinforcing laminate layers provide added structural strength, they cannot be considered to add to the design stress of the material due to the inconsistencies inherent in fabricating such products. For example, a polyvinylidene fluoride thermoplastic liner reinforced with a polyester hand-layup lamination should have its minimum required thickness determined solely on the basis of its thermoplastic liner material.

6.3.1.3.1 External-Pressure Calculations for Minimum Thickness of Nonmetallic Piping

For thermoplastic piping and reinforced thermosetting-piping materials, there is an analytical approach that has received general acceptance worldwide. The theory is based on the assumption that there are two modes of failure that may occur from external pressure loads. One is for thick-wall pipe (generally D/t < 50), where collapse is caused by a compression mode of failure of the pipe material. The maximum external pressure for this type of condition can be determined by:

$$\Delta P = \frac{\sigma}{2D^2}\left(D^2 - d^2\right) \qquad (6.4)$$

where:
ΔP = collapse pressure, or the differential between external and internal pressure, psi
σ = compressive strength, psi
D = outside diameter of pipe, in
d = inside diameter of pipe, in

The other mode of failure is based on thin-walled pipe (generally D/T > 50), where elastic instability of the pipe wall causes failure to occur. This pressure can be determined as follows:

$$\Delta P = \frac{2cE}{(1-\mu^2)}\left(\frac{t}{D_m}\right)^3 \qquad (6.5)$$

where:
μ = Poisson's ratio, dimensionless
D_m = mean pipe diameter, in.
E = modulus of elasticity, psi
c = safety factor (usually 0.5 for plastics)
t = wall thickness, in.

Both equations have been used historically by the plastic piping industry to determine external pressure and vacuum-resistance capabilities, although the equations have never been verified for use in vacuum applications by an independent consensus code body. The usual approach is to determine values from both equations, with a known value for external pressure differential (ΔP); the required minimum wall thickness of the pipe is determined from the equations, and the minimum of the two values selected as the basis for required minimum design thickness. The result should always be substantiated by one of the means described earlier in this chapter.

The minimum required thicknesses as a result of external pressures of dual-laminate materials should be determined for the liner, without any additional contribution of design stress resulting from the reinforcing material. While the additional reinforcing laminate layers provide added structural strength, they cannot be considered to add to the design stress of the material due to the inconsistencies inherent in fabricating such products. For example, a PVC thermoplastic liner reinforced with an aromatic amine cured-epoxy spiral-wound lamination should have its minimum required thickness determined solely on the basis of its thermoplastic liner material. For RTRP materials, the procedure described in Appendix A of AWWA C950 is used.

6.3.1.3.2 Metallic Piping Lined with Nonmetals under Vacuum Conditions

For lined metallic piping in applications involving any potential for exposure to external pressures, the minimum required thickness of both the liner material and the base material must be calculated. The techniques for determining the minimum required thicknesses for metallic piping under external pressures are presented in Chapter 5 (see Section 5.2.1.1). The techniques applicable for thermoplastic materials are described in Section 6.3.1.3.

6.3.2 Curved Sections of Nonmetallic Pipe

Piping that is fabricated into long-radius bends by bending straight pipe (with or without the application of heat) should have a minimum required thickness, t_m, equal to the requirement for the piping, measured after the bending procedure has been finished. This includes all applicable corrosion, erosion, and other allowances inherent in Eq. (6.1). Fabricated bends from curved sections of pipe should be substantiated either by extensive successful service experience under comparable design conditions with similarly proportioned components of like material, or they should be subjected to a performance test. The performance test should simulate design conditions, including cyclic conditions, for a period of time sufficient to determine the acceptability of the component (or joint) for its required design life. All applicable ambient and dynamic effects described in Section 6.1 should be taken into account in the design of such a component. Bends of this type should have 8% or less flattening in the bent section for internal pressure applications and a maximum of 3% flattening in the bent section for applications involving external pressures. Flattening is measured as a change in the outside diameter of the pipe that is bent to form the final piece.

6.3.3 Nonmetallic Miter Bends and Nonmetallic Corrugated Bends

Miter bends include bends that are fabricated from mitered sections of straight pipe, joined together by heat-element thermal-butt-fusion, extrusion-welding or hot-gas-welding techniques (thermoplastic materials), or by butt-and-wrap techniques (RTRP materials). Fabricated miter bends from mitered sections of pipe should be substantiated either by extensive successful service experience under comparable design conditions with similarly proportioned components of like material, or they should be subjected to a performance test. The performance test should simulate design conditions, including cyclic conditions, and should be conducted for a period of time sufficient to determine the

Design of Nonmetallic Primary Components

acceptability of the component (or joint) for its required design life. All applicable ambient and dynamic effects described in Section 6.1 should be taken into account in the design of such a component.

It is generally accepted practice that thermoplastic miter bends produced using heat-element thermal-butt-fusion techniques that are not substantiated by one of the means discussed may be considered to have a pressure rating equal to that of the matching pipe, if it is produced from a pipe having a minimum thickness in the miter section that is more than 25% greater in thickness compared to the minimum thickness of the piping to which it is to be joined. The ASTM F-17 Committee is currently studying the possibility of a new standard for mitered thermoplastic elbows produced on this basis. Bends that are manufactured using other designs or techniques (corrugated or creased piping) should be qualified in the same manner as for fabricated miter bends.

6.3.4 Nonmetallic Branch Connections

Whenever a pipe is provided with an opening to facilitate a branch connection, it is weakened by the opening. There are many possible combinations of branch sizes and configurations that can be used for branch connections. Therefore, the design of branch connection deserves special attention to ensure a properly functioning piping system. Branch connections include: (1) fittings such as tees, laterals, and crosses; (2) outlet fittings such as nozzles and adapters (each having butt welding, socket welding, threaded or flanged ends for joining); and (3) branches made by welding (bonding) a branch pipe directly into a run pipe. Where cyclic conditions exist, the use of directly welding (bonding) branch piping should be avoided for nonreducing branches. For nonreducing branches, the use of nonfabricated fittings such as tees or laterals is highly recommended.

Where it is determined to be necessary, branch connections in nonmetallic piping systems should be provided with reinforcement in the area of the opening. The amount of reinforcement added to the "reinforcement area" should be determined with the same basic techniques described in Section 5.2.1.3, taking into account the differences due to nonmetallic material behavior. The amount of reinforcement should be substantiated either by extensive successful service experience under comparable design conditions with similarly proportioned components of like material or by a performance test. Such a performance test should be conducted under design conditions including simulated dynamic and creep-causing effects, for a period of time sufficient to determine the acceptability of the component or the joint for its design life. All applicable ambient and dynamic effects described in Section 6.1 should be taken into account in the design of such components of the primary piping system.

If a branch connection consists of a "listed" tee from among the component standards in Table D-1 of Appendix D, then it can be assumed that the branch has adequate strength to withstand the internal pressures and external pressures to which it will be subjected, as long as the loads are within the specified limits of the components. When these listed fittings are used, it is not necessary to perform additional calculations. It is also not necessary to substantiate the design of the fitting through performance testing as described in the previous paragraph.

There are additional loadings to consider, besides internal and external pressures, when designing a branch connection. These include external forces and movements that are the result of thermal expansion and contraction, dead and live loads, and movement of piping supports and terminals. The effect of these additional forces should be considered in the design of branches. Branch connections made by attaching the branch piping directly into the wall of the run pipe should be avoided, if a "listed" component can be used. However, where the branch is of the reducing type, particularly where large reductions are accomplished, and the piping is not subjected to repetitive stresses, this type of connection is commonly used. Where the branch does represent a significant reduction, the branch piping should be provided with adequate flexibility to protect against movements of the larger line. Also, ribs, gussets, and clamps are often used as a means of stiffening a branch connection. When they are, their areas should not be counted as contributing to the reinforcement area in calculations. Ribs, gussets, and clamps may be used to increase the pressure rating, but the rating of the component must be substantiated through successful experience or via proof testing.

In some cases, supporting plates and other devices in certain double containment branch connection designs may function as a rib or gusset; thus, any potential gain in strength from such a member should not be counted in branch reinforcement calculations, and must instead be substantiated by the means stated in the previous paragraph.

6.3.4.1 Fabricated Nonmetallic Primary Branch Connection Fittings

Fabricated branches include branches that are fabricated from mitered sections of straight pipe, joined together by heat-element thermal-butt-fusion, extrusion-welding, or hot-gas-welding techniques (thermoplastics), or by contact molding, known as hand lay-up (RTRP materials). Fabricated miter bends from mitered sections of pipe should be substantiated either by extensive successful service experience under comparable design conditions with similarly proportioned components of like material, or by a performance test. Such a performance test should be conducted under design conditions including simulated dynamic, cyclic and creep-causing effects, for a period of time sufficient to determine the acceptability of the component or the joint for its design life. All applicable ambient and dynamic effects described in Section 6.1 should be taken into account in the design of such components. Fabricated branch connections produced using heat-element thermal-butt-fusion techniques may be considered as being provided with reinforcement by using a pipe having a minimum thickness that is substantially greater than the minimum thickness of the piping to which it is to be joined. A branch connection fabricated in this manner still has to be substantiated by one of the means stated here.

6.3.5 Closures

Nonmetallic closures that are not listed as in Table D-1 of Appendix D should be qualified in the normal manner of an "unlisted component," as stated in Section 6.2.2. The design of these elements should be in accordance with known techniques. They should be substantiated either by extensive successful service experience under comparable design conditions with similarly proportioned components of like material, or by a performance test. Such a performance test should be conducted under design conditions including simulated dynamic, cyclic, and creep-causing effects, for a period of time sufficient to determine the acceptability of the component or the joint for its design life. All applicable ambient and dynamic effects described in Section 6.1 should be taken into account in the design of such an element.

6.3.6 Pressure Design of Flanges

Procedures for designing nonmetallic flanges for use with flat-ring gaskets are generally the same as those covered in the ASME Boiler and Pressure Vessel (BPV) Code, Section VIII, Division 1, Appendix 2, which are rules based on metallic materials. However, the use of this procedure has never been substantiated for nonmetallic flange design by the ASME or any other consensus standard code body. The allowable stresses and temperature limits stated in this chapter should be substituted for those values stated in the ASME BPV Code when using the methods of analysis presented there.

For flanges that are used with full-face gaskets that extend beyond the bolts, and for flanges that have solid contact beyond the bolts, the techniques in the ASME BPV Code, Section VIII, Division 1, Appendix Y should be used. The reason has to do with the fact that the forces and reactions in those joints differ from those using flat-ring gaskets. Flanges that are not listed in Table D-1 of Appendix D of this text should be qualified in the same manner as an "unlisted component." The design of these flanges should be in accordance with known techniques. They should be substantiated either by extensive successful service experience under comparable design conditions with similarly proportioned components of like material, or by a performance test. Such a performance test should simulate design conditions, including cyclic conditions, for a period of time sufficient to determine the acceptability of the component for its design life. All applicable ambient and dynamic effects described in the beginning of this chapter should be taken into account in the design of flanges, since they are deformation-sensitive components.

6.3.6.1 Minimum Required Thickness of Thermoplastic Blind Flanges

To calculate the minimum required design thickness of a blind flange, the rules of the ASME BPV Code, Section VIII, Division 1, UG-34 may be used. Once the minimum required pressure design thickness, t, is determined, the actual minimum required thickness, t_m, can be determined by:

$$t_m = d\sqrt{CP/SE + 1.9Wh_G/SEd^3} + c \qquad (6.6)$$

where:
C = a factor depending upon the method of attachment of head, shell dimensions, and other items, dimensionless (the factors for welded covers also include a factor of 0.667 which effectively increases the allowable stress for such constructions to $1.5S$)
d = diameter, or short span, in.
E = joint efficiency
h_G = gasket moment arm, equal to the radial distance from the center line of the bolts to the line of the gasket reaction
P = internal design pressure, psi
S = maximum allowable stress value, psi
t = minimum required thickness of flat head or cover, in.
W = total bolt load, provided from ASME BPV, Sec. VIII, Div. 1, part UG-34, given for circular heads in formulas (3) and (4), 2-5 (e), lbs

The minimum required thickness of a permanent blank can be calculated by:

$$t_m = d_g \sqrt{\frac{3P}{16SE}} + c \tag{6.7}$$

where:
d_g = inside diameter of gasket for raised or flat face flanges, or the gasket pitch diameter for ring joint and fully retained gasketed flanges, in.
E = the quality factor (from the manufacturer), dimensionless
P = design gauge pressure, psi
S = bolt design stress, psi
c = sum of mechanical allowances (thread or groove depth) plus corrosion and erosion allowance. For threaded components, nominal groove depth (h dimension as stated in ASME B1.20.1 or equivalent) should be used. For machined or grooved surfaces that do no have a specified tolerance stated, a tolerance of 0.02 in. (0.5 mm) can be used in addition to the specified depth of the cut, in.

6.3.7 Reducers

A reduction in pipe diameter size may be accomplished with the use of a concentric or eccentric reducer fitting. Concentric and eccentric reducers that are not listed in one of the standards referenced in Table D-1 of Appendix D should be qualified in the same manner as an unlisted component. The design of unlisted reducers should be in accordance with known techniques. They should be substantiated either by extensive successful service experience under comparable design conditions with similarly proportioned components of like material, or by a performance test. Such a performance test should be conducted under design conditions including simulated dynamic and creep-causing effects, for a period of time sufficient to determine the acceptability of the component for its design life. All applicable ambient and dynamic effects described in the beginning of this chapter should be taken into account in the design of such reducers.

6.3.8 Pressure Design of Other Components

Other components listed in one of the referenced standards in Table D-1 of Appendix D can be assumed to be capable of being used for pressure purposes up to their stated limits in those specifications. Their designs are not required to be substantiated through analysis, experience, or proof testing. Other components to be used as pressure-containing elements, not listed in a standard referenced in Table D-1 of Appendix D, should be qualified in the same manner as "unlisted components." The design of these other unlisted elements should be in accordance with known techniques. They should be substantiated either by extensive successful service experience under comparable design conditions and similarly proportioned components of like material, or by a performance test. The performance test should be conducted under design conditions including simulated dynamic and creep-causing effects, for a period of time sufficient to determine the acceptability of the component, and/or joint, over its design life. All applicable ambient and dynamic effects described in the beginning of this chapter should be taken into account in the design of such elements.

6.4 Selecting Nonmetallic Piping Components for Fluid Services

In general, piping, fittings, valves, and other components that are listed in a component standard referenced in Table D-1 of Appendix D should be used wherever possible in the design of a nonmetallic primary piping system. The material of the pipe and other components should be determined to be suitable for the manufacturing process and the fluid service, considering all applicable effects. Chapter 2, "Materials of Construction," prescribes a comprehensive discussion of the chemical performance and limitations of nonmetallic materials. This section presents the designer with some specific guidelines and limitations for selection of nonmetallic components, based on the fluid services involved. Several specific requirements and limitations for fluid services of nonmetallic materials and metallic materials lined with nonmetals are described in the following paragraphs.

Unlisted bends (curved pipe, mitered bends, corrugated bends, etc.) that are made in accordance with the techniques described in Sections 6.3.2 and 6.3.3 and verified for pressure design as described in those sections can be used for the same service as the piping for which it is intended.

Fabricated laps are suitable for use in nonmetallic piping for normal fluid service if the lap material is compatible with the fluid being transported. Also, the outside diameter of the lap should be based on ASME B16.9 lap-joint stub end dimensions, the lap thickness should be at least that of the pipe, and the lap material should be listed in Table D-1 of Appendix D and should have an allowable stress equal to, or greater than, the pipe. Flared laps should never be used in a nonmetallic piping system involving chemical services.

Special attention should be paid to pressure design involving fabricated branch connections, particularly where the branch is welded (bonded) directly to the run pipe. Be sure to verify that such a design is feasible with the method described for unlisted components (defined in Section 6.2.2) or Section 6.3.4 for the design of branch connections.

Listed nonmetallic valves are suitable for normal fluid services except for valves that contain bolted bonnets, with bonnets secured to the bodies by less than four bolts (or U bolts). These valves should only be used for Category D fluids. The pressure design of some valves may be established by the method described in ASME B16.34, Appendix F. If not, the valve pressure design should be verified by the same methods as described for "unlisted components" in Section 6.2.2.

All nonmetallic flanges should contain a suitable gasket, facing, and bolts and should be capable of developing the full rating of the required joint (usually 150 or 300 psi, for nonmetallic materials). All applicable forces and effects should be taken into consideration in selecting the appropriate type of flange connection based on the fluid service involved. Most nonmetallic flanges vary greatly in design from manufacturer to manufacturer. Therefore, it is highly recommended to check with the manufacturer of the flange to be used to determine specific requirements prior to final selection and application. Threaded flanges are allowed for nonmetallic flanges for chemical service and are described in more detail in Section 6.5. Metallic piping with nonmetallic liners should be checked to be sure that its required seating load is compatible with the flange rating and facing. The strength of such a flange and its bolting are also factors when using the plastic as the gasket in lined-metallic piping services.

In terms of bolting requirements for nonmetallic flanges, the same requirements for metallic flanges apply. It is strongly recommended to avoid the use of nonmetallic bolting materials for use with nonmetallic flanges (or valve bonnets, instrumentation housing, etc.), even though they are available. Although nonmetallic bolting is corrosion resistant, it lacks strength compared to metallic bolts, and, as such, should be strictly prohibited from use in flanged joints. Tapped holes for bolts used for pressure-retaining purposes in nonmetallic components should be qualified by the methods described for "unlisted components" in the beginning of this chapter.

6.5 Selecting Nonmetallic Piping Joining Methods for Primary Pipe Fluid Services

In general, the type of joints selected for a particular service should suit the material of the piping and the fluid service involved. Due consideration should be given to joint tightness and mechanical strength under expected service and test conditions of pressure, temperature, and external loading.

6.5.1 Bonded Plastic Pipe Joints for Primary Piping Service

The various types of bonding methods for plastic materials and the qualification of procedures and bonders are described in Chapter 12, "Fabrication, Assembly, and Erection." Detailed inspection procedures and requirements for plastic materials are described in Chapter 13, "Inspection, Examination, and Testing." Specific requirements and limitations for the various types of joints as they apply to different fluid services are described in the following.

Joints that are subjected to a visual examination, as described in Chapter 13, as the only means of examination of the joint should be limited to use in Category D fluid service. Various welds that are used for repair services such as fillet welds or extrusion welds may be used, but the requirements for bonding qualifications described in Chapter 12 are applicable to these types of welds (bonds). Welding personnel should be qualified to the specific bonding procedure used, prior to undertaking any repairs. Seal welds (bonds) should be limited to threaded joints, for situations where such welds (bonds) are necessary to prevent leakage of a joint. It should be demonstrated that there will be no adverse effects in implementing either repair procedures or seal welds (bonds) prior to their use.

Flanged joints and their applicability, considerations, and limitations for fluid services are described in Chapter 5. Since nonmetallic materials are restricted from use in "severe-cyclic" services in accordance with the ANSI/ASME B31.3 Code, the use of expanded joints are limited to applications where such conditions are not possible. When expanded joints are used, consideration should be given to their tightness when vibration or differential expansion and contraction or external mechanical loads exist.

Threaded joints for thermoplastic piping should only be used for piping that is at least as thick as Schedule 80, as defined in ASTM D-1785. For piping that is directly threaded, it should be limited to pipes that have at least a 2 in. diameter, Schedule 80 (approximately the thickness of 63 mm OD, SDR 11). All male threads used in threaded joints for thermoplastic pipes should have dimensions as required by ASME B1.20.1, should be of the NPT type, and should also conform to applicable standards listed in Table D-1 of Appendix D. Consideration should be given in the selection of thread lubricant or sealant, in that it is compatible with the material and unaffected by the fluid that it is to contact. Threaded joints should be restricted from use with reinforced plastic mortar (RPM) piping.

Design of Nonmetallic Primary Components

There are many types of plastic tubing fittings that are used to connect plastic tubing, including flared, flareless, and compression types. For plastic tubing fittings, compression types are the most common. In selecting the appropriate type of fitting and joining method for a specific chemical service, designers should consider assembly and disassembly of the tubing, cyclic loading, vibration, shock, and the effects of thermal expansion and contraction on the integrity of the connection.

Special joints such as caulked-type joints and joints of the bell-and-spigot-type or packed-gland joints should be restricted from use with Category M fluids, and instead should be limited to use in Category D services. Flexible elastomeric sealed joints should conform to the requirements of ASTM D-3139, except for RTR and RPM types, which should conform to ASTM D-4161. Safeguarding is required for bell-and-spigot-type and gland-type joints when used in "severe-cyclic" service, or whenever cyclic conditions of any type are present. In general, it is not a wise idea to use these types of joints for primary (core) service in a double containment piping system.

6.6 Flexibility and Support of Nonmetallic Primary Piping

6.6.1 Flexibility of Nonmetallic Piping

6.6.1.1 General

The stress-strain behavior of most nonmetallic materials differs considerably from that of metallic materials. The assumption that displacement strains are proportional to stress over a wide range, and, therefore, the presence of substantially elastic behavior, is not a valid assumption for most nonmetals. Thermoplastic materials (and to a lesser extent thermosetting materials) experience substantial relaxation or creep when subjected to stress. They are technically classified as viscoelastic materials, as they are neither purely elastic or purely inelastic. In other words, they have what can be described as a fading memory in the sense that they have a decreasing ability to recover their original dimensions over time, when a stress is applied to such a material and then subsequently released. Also, some of the physical properties of the materials vary according to the orientation of the molecules; Poisson's ratio is an example of one type of property that varies in this respect, for many materials. For these reasons, the material behavior of nonmetallic materials is less well defined for mathematical analysis, and the simplified forms of analysis provided for metallic materials in Chapter 5 are generally not valid for these materials.

The object in a flexibility design of a nonmetallic piping system used as a primary pipe is to design the system with a layout such that flexural stresses resulting from displacements due to thermal expansion, contraction, and other movements and reactions are minimized. As in the process of increasing flexibility in metallic piping systems, attention must be paid to supports, terminals, restraints, and particularly to interconnecting parts and intermediate restraints in double containment piping systems. However, a factor that complicates matters for nonmetallic materials such as thermoplastic (and to a lesser extent RTRP) materials is that they experience unusually large magnitudes of thermal expansion and contraction over relatively small increases and decreases in temperature. This is due to the relatively high thermal expansion coefficients typical of nonmetallic materials, including certain materials such as the polyolefin thermoplastics (polyethylene, polypropylene, and polybutylene). They can be as much as twenty times, or more, than those of certain metallic materials. The Plastic Piping Institute (PPI) has a detailed report concerning thermal expansion issues pertaining to plastic piping materials in their PPI Technical Report TR-21.

6.6.1.2 Displacement Strains and Stresses

When nonmetallic piping is restrained from movement due to external movement or thermal expansion, a displacement strain is imposed on the piping components in much the same way that such strains are imposed on metallic piping system components. However, the values of these strains cannot be accurately predicted as they are in metals, because the amount of strain in relation to stress will vary with time. Therefore, predicting stresses throughout a nonmetallic piping system, based on the level of strain, may result in inaccuracies and therefore is not generally recognized as a valid practice. The ANSI/ASME B31 Codes state that simplified equations for elastic materials do not apply in this condition. When elastic behavior cannot be assumed, or when unbalanced stresses occur in a nonmetallic piping system, stresses cannot be considered proportional to displacement strains.

There are some unique characteristics of the behavior of nonmetallic materials to consider. Thermoplastic piping and some reinforced thermosetting piping are not subject to immediate failure from displacement strains, but they are subject to possible distortions or deformations. These distortions may continue upon repeated thermal cycling or prolonged exposures. In some cases, distortions may be progressive in nature, resulting in failure of the material due to the thinning of the wall or localized weakening of the material. In other nonmetallics, such as borosilicate glass, ceramic materials, and others, materials can develop high displacement stresses, and can subsequently fail due to their highly rigid behavior.

The assumption that nonmetallic materials exhibit substantially elastic behavior is valid in some limited cases. However, in the majority of instances it is not a valid assumption. When elastic behavior cannot be assumed or when unbalanced stresses occur in a piping system, stresses cannot be considered proportional to displacement strains. There are generally two types of reactions that occur. In thermoplastic and RTRP materials, relatively low values of displacement strains generally produce stresses of the plastic type. High values of displacement strains also tend to produce stresses of this type, but to a much greater magnitude. The other type of reaction occurs for the rigid nonmetallics, where strains initially produce relatively large stresses and overstrain generally results in failure, as opposed to plastic deformation. Therefore, rigid nonmetallics such as cement, borosilicate glass, and ceramics should be designed to keep displacement-strain values to a minimum. The possibility of overstrain developing in the piping system can be minimized to a great extent through effective layout of the system and through the use of expansion joints or devices. These concepts are discussed in more detail later in this section and in Chapter 8.

Another method that is used to balance the magnitude of stress under an extreme displacement condition is the selective use of cold spring. This is a method that is used in some single-wall pipe applications. Cold spring is a method whereby piping is intentionally deformed or preloaded during assembly. When the extreme design temperature is reached during operation, the total amount of displacement is limited. It can result in a more even distribution of stresses throughout a piping system, if selectively applied in the right locations. One of the more important benefits of this technique occurs during startup, as there is less deviation of hangers from their original settings and less deviation from as-installed dimensions during initial operation. There should be no credit taken for cold spring in stress range calculations, as the overall magnitude of displacement remains unchanged. Unlike metallic piping, credit is not permitted in the ANSI/ASME B31.3 Code for the calculation of thrusts and moments in nonmetallic pipe applications.

Cold spring is very difficult to implement properly in single-wall piping applications; it is generally not recommended for primary pipes in double containment piping assemblies.

6.6.1.3 Flexibility-Analysis-Related Properties for Thermoplastics

Table C-5 of Appendix C of the ANSI/ASME B31.3 Code for Chemical Plant and Petroleum Refinery Piping lists average thermal expansion coefficients for a variety of nonmetallic materials. Supplementary data can be obtained from the National Institute of Standards and Technology (NIST) and the annual edition of the Modern Plastics Encyclopedia published by Modern Plastics Magazine. It is always recommended for thermoplastic and thermosetting-resin materials to obtain precise data if they are available from the manufacturer of each specific resin to be used. Designers should also be aware that the thermal expansion coefficients of many plastics are nonlinear over relatively small temperature ranges. The values for the tensile modulus of elasticity of various nonmetals are listed in Table C-8 of Appendix C of the ANSI/ASME B31.3 Code, for laboratory rate-of-strain conditions. The actual values of modulus of elasticity of plastics can vary greatly by specific grade of resin used, or by the specific course of the strain imposed. The manufacturer of the specific resin to be used should be consulted for more accurate values of E to be used under the actual conditions. Poisson's ratio for nonmetals can also vary widely, including with the orientation in which it is measured. Therefore, the value of 0.3 normally used for calculations involving metals should not be assumed for calculations involving nonmetals. In fact, most plastics have a Poisson's ratio that is not constant (i.e., common values for thermoplastics vary between 0.3 and 0.4). The simplified formulas used in stress analysis of metallic components are not recommended for use with nonmetallic materials for these reasons. A more comprehensive and nonlinear form of analysis is usually more appropriate. In all flexibility calculations, the outside diameters and nominal thicknesses are to be used, depending on the method chosen.

A formal analysis should be performed on a nonmetallic primary piping system, with few exceptions. If there is a duplicate system that exists or is closely related to a successfully operated system, or where a system can be judged adequately by an analysis that has already been performed on a similar system, or if the system in question has adequate inherent flexibility, there does not have to be a formal analysis performed. If a formal analysis is required, a designer can perform a simplified, approximate, or comprehensive stress analysis, using a method shown to be valid for each specific case. There are no simplified analysis techniques presented here for nonmetallic piping systems. It is suggested that the designer consult an authority who is familiar with methods such as finite element analysis of nonlinear-behaving materials, to satisfy the stress analysis requirements. Alternatively, proof testing may be performed under conditions that simulate the intended operating conditions, for a long enough period of time to yield accurate results.

In the instance that substantial elastic behavior can be demonstrated, due to an adequate internally flexible layout, then the simplified methods presented in Chapter 5 for metallic materials can be used (see Section 5.5.1.6). In those cases where traditional simplified stress analysis methods can be demonstrated to be valid, reaction moments can also be calculated in the same manner as Chapter 5 for metallic piping components. Special attention must always be paid to the displacement or rotation of piping with respect to internal supports and close-tolerance locations (e.g., at elbows within elbows, tees within tees, etc.). This is particularly important where large reductions occur; the small branches involved are subjected to large displacement strains imposed by the larger pipe run in such instances.

One method of designing nonmetallic piping in situations involving thermal expansion and contraction or external movements is to provide adequate flexibility in the layout of the system. Oftentimes, the best means of accomplishing this is through the use of natural changes of direction that occur in a piping system. In many cases, the changes of direction provide inherent flexibility without additional considerations, due to relatively small bending and torsional strains that occur through displacement. The use of changes in direction, as well as the selective use of expansion loops, offsets, swivel joints, expansion joints, and restraining techniques is described in Chapter 8 (see Sections 8.5 through 8.8).

In addition to the layout techniques involved, methods for calculating required lengths, required activation forces, and other considerations are all presented in Chapter 8. The selective use of internal and external anchors, as well as the determination of points of restraint, are all discussed in Chapter 8 as they specifically apply to double containment piping systems.

6.6.1.4 Support of Primary Piping Systems

The details of support requirements for nonmetallic piping systems are thoroughly presented in Chapter 9, "Structural Considerations." There the details of calculations, as well as details for guides, supports, anchors, and terminals, are presented. In general, piping should be supported, guided, and anchored so as to prevent damage to the piping. Point loads, narrow areas of contact, and possibilities of gouging or notching of pipes should be avoided. This can be accomplished by provided supporting devices, such as saddles, having wide bearing loads, and using an elastomeric padding material between the support and piping component being supported. Valves and other heavy components or assemblies should be supported to prevent the transmission of excess loads to associated piping components. Always consult with the manufacturer of specific components such as valves to determine and provide for any unique considerations.

For thermoplastic and thermosetting-resin primary piping systems, interstitial supports should be spaced to avoid excessive sag in the pipe at the design temperature. Consideration should be given to movements created by thermal expansion and creep effects of the material. Plastic piping support spacing requirements are typically much closer than metallic support spacing requirements, particularly at elevated temperatures. Thus, continuous supports are often used in single-wall piping systems; continuous support is not typically needed due to the added stability provided by means of the double containment piping assembly. Rigid or brittle nonmetallic piping materials must be capable of expanding freely, as relatively small displacements can cause initially high levels of stress and subsequent failure. A good general rule for borosilicate glass is to use no more than one anchor per straight run without adding an expansion alleviation device between anchors.

6.6.1.5 Auxiliary Piping Systems

Two common forms of piping that are normally used in conjunction with primary piping systems include instrument piping and pressure-relief piping systems. Instrument piping is used to connect instruments to other piping or equipment, or it consists of control piping that is used to connect air or hydraulically operated control apparatus. Instruments are normally treated separately, including permanently sealed fluid tubing systems that serve as temperature or pressure-response systems. Pressure-relief piping is used either directly as overpressure protection without an auxiliary pressure relief device, or it is used as discharge piping for those types of devices.

Instrument piping should be designed in accordance with the same requirements and considerations as ordinary primary pressure piping. Additionally, instrumentation that is attached directly to the piping should be designed to be able to maintain the integrity of the system. Since most instrument piping contains static fluids, they are readily subject to freezing and should thus be provided with heat tracing where this is a possibility. If it is required to drain instrument piping that contains hazardous or toxic fluids, care should be taken to dispose of those fluids properly. Since most double containment systems are used to handle chemicals of this type, instrument piping should be designed with such a feature built in, or blow-down procedures should be given emphasis in the maintenance and operating procedures and manuals.

Pressure-relief piping for primary carrier piping in double containment piping systems is unique in that the secondary containment piping is also designed for that purpose, in case of catastrophic failure. In that sense, the secondary containment piping gives an added sense of protection in case of a malfunctioning relief valve or device. The secondary containment piping also provides a designer the option of allowing discharge piping from safety relief valves and other relief valves to be safely discharged into it. Since most double containment piping is for chemical transport, this is an option for a designer. In some instances a relief valve may be discharged directly into a chamber constructed as an integral part of the secondary containment system, eliminating the need for discharge piping altogether. In above-ground double containment piping where fire or other events can cause a fluid in the primary piping to overpressurize, it is necessary to have pressure-relief valves installed on the primary carrier pipe, in addition to having a secondary containment piping outer wall.

The design of pressure-relief valves and other relief devices is covered in the ASME Boiler and Pressure Vessel Code, Section VIII, Division 1, UG-126 through UG-128 and UG-132 through UG-136, excluding UG-135(e) and UG-136(c). In those requirements, the phrases "maximum allowable working pressure" and "vessel" are the equivalent, respectively, of the "design pressure" and "primary piping system" nomenclature used in this chapter. UG-125(c) sets the basis for establishing the relief set pressure and the maximum relieving pressure, except where overpressurization is permitted, as described in Section 6.2.2. Reactions that occur in piping when safety relief devices discharge should be such that they do not cause harm to the overall system and should be taken into consideration in the engineering design. The use of stop valves between piping being protected and relief devices should be limited to those situations where the design of such valves cannot prevent the piping system from being properly relieved.

Notes

Chapter Seven

Design of Secondary Containment Components (Metallic and Nonmetallic Materials)

7.1 Introduction

Secondary containment piping components must be capable of containing a process fluid for a sufficient period of time, whenever a system's primary means of containment experiences a failure. In many applications involving a pressurized primary system, secondary containment systems have the potential of developing significant pressures within a short period of time. In nonpressure applications and in certain pressure piping systems with open-ended annuli, secondary containment systems will not be capable of achieving significant pressures; however, a secondary containment system must always be capable of preventing leaks of corrosive or toxic chemicals to the atmosphere for a reasonable length of time, regardless of the pressure involved. In systems that are designed to be in accordance with the requirements of the Resource Conservation and Recovery Act (RCRA) in the United States, secondary containment piping systems have defined minimum performance requirements to meet.

In order to provide a high degree of protection, secondary containment components should be designed using a similar philosophy to that inherent in Chapters 5 and 6 for primary piping components. Each secondary containment component must take into account pressure, other stresses, and forces that may be imposed upon them, both during normal operation and during upset conditions. Where applicable, they must be analyzed and qualified in the same manner as primary components, including calculations and experimental analysis performed on primary piping system components. This gains added importance for "unlisted" components, "listed" components that are modified, and those that are subjected to unusual forces due to interactions between primary and secondary containment systems.

Three differences between the design of ordinary single-wall piping components and secondary piping components in a double containment system are readily apparent.

First, the secondary containment portion of a double containment system is designed to be used only in the event of a failure in the primary piping system. When it is required to perform this function, the time required to contain the fluid is typically much shorter than the overall service life of the primary piping system. Additional safeguards, including automatic pump shutoff switches and other safety features, can prevent the secondary containment piping from reaching the same pressure condition as that of the primary piping, which can potentially lower required design conditions of the secondary components. What this means is that, occasionally, a secondary containment piping system may potentially be designed with materials having lower corrosion resistance, less corrosion allowance, lesser temperature and pressure capabilities, and an overall service design life less than that of the primary piping system. While in theory this may be true, the authority having local jurisdiction may not allow a secondary containment system to be designed based on lower design criteria, particularly in systems to be designed to RCRA requirements. It is the designer's responsibility to determine whether it is appropriate or permitted to do so.

The second main apparent difference is that the design of secondary piping components must take into account any and all interactions between the secondary piping components and the inner primary piping components. This includes forces that the primary components may exert on the secondary containment components and forces that the secondary components may exert on the primary components. Normally, these interactions are concentrated where interconnecting components and interstitial supporting devices are located, and predominate in systems that undergo differential thermal expansion. Thus, the design of the outer components must take into account these interactions, in addition to ordinary design criteria.

A third major difference in the design of secondary containment piping systems is not immediately obvious, since the design component must take into account all of the features necessary to test, operate, maintain, and repair the completed double containment assembly. This includes the ability to vent and drain the annulus in the event of a leak. Many secondary containment systems have been designed only to resist the corrosive effects, temperatures, and pressures of the fluid transported in the primary piping system. However, without additional provisions to drain, vent, flush, dry, purge, and test the annular space, there will be a great deal of control lost over the safe operation of the system. This includes the safe removal of fluid that has collected in the annulus. Without the ability to drain, flush, or purge the annulus readily, a secondary containment pipe system containing corrosive or hazardous fluids must be cut into, thereby resulting in a spill; such a spill can often be of a reportable magnitude. To allow a spill to occur is contradictory to the very concept of providing secondary containment. Consideration should be given to allowing an annular space to be flushed (decontaminated), prior to undertaking maintenance and repair of a system for the protection of the workers involved, and for the fluid to be removed in a controlled manner.

The design of secondary containment components many times will affect the design of primary components of the system. For instance, pressure-testing requirements of the outer secondary containment system will impose external pressures on the inner primary containment system. This will mean in some cases that primary components may have to be increased in thickness and have reinforcement added in certain components to withstand the application of the external pressures imposed upon them. Another example involves systems in which differential thermal expansion conditions exist; discontinuity stresses will result at interconnecting component locations, requiring primary components to be designed with the ability to withstand these effects. Designers must understand that since two piping systems are involved, double containment piping system design is by its nature an interactive process.

7.1.1 Scope

This chapter covers all of the design and fluid service considerations for secondary containment components and considerations that are unique to designing double containment piping assemblies. All of the ordinary rules described in Chapters 5 and 6 apply in principle to the design of secondary containment components. Therefore, the discussion of the rules stated in those chapters are not duplicated here. For such items as pressure design of elbows, bends and mitered bends, and reinforcement requirements for branches, one should also consult the chapter dealing with the type of material from which the component in question is constructed (Chapter 5 for metals, Chapter 6 for nonmetals). What is covered in this chapter are those items pertaining to the design of secondary containment piping that are not covered in the previous two chapters. These include the pressure-retaining characteristics of specialized secondary containment components, and issues that arise due to interaction between inner and outer systems.

The discussion in this chapter covers components manufactured from both metallic and nonmetallic materials. The discussion applies to both homogeneous and hybrid material systems (refer to Chapter 3 for a definition of these).

7.2 Conditions and Criteria for Secondary Containment Components

7.2.1 Design Conditions

7.2.1.1 General

The determination of the design temperature, design pressure, and the forces applicable to the design of secondary containment piping components follows the same basic principles described in Chapter 5 ("Design of Metallic Primary Piping Components") and Chapter 6 ("Design of Nonmetallic Primary Piping Components"). The same is true of the considerations given to various effects of the secondary components and the loads that are imposed on them.

The philosophy inherent in Chapters 5 and 6 may mislead designers to conclude for some applications that design temperature and pressure of secondary containment system components may be lower than those of the associated primary system components. However, designers are advised also to weigh the relative risk of experiencing "double failures," which increase as a result of using secondary containment components that have reduced pressure and temperature ratings. A risk analysis must be performed that not only takes into account design factors, but also the overall economics associated with such a risk.

The chance of a "double failure" occurring will increase with any lowering of design conditions, since the component safety factor will be lessened. The potential cost associated with a double failure can be estimated, as can the likelihood of such a failure. Even though the chance for a double failure will be relatively remote if good design and fabrication techniques are followed, the resulting cost must be included in a risk analysis. The cost can be determined by multiplying the chance of occurring times the estimated cost incurred as a result of a "double failure," including such things as government-imposed fines, cleanup costs, repair costs, and third-party litigation. In systems that are designed with high integrity, the risk will be low, as will the estimated cost due to the risk.

7.2.2 Design Pressure

To select the design pressure of each secondary piping component, the same rules of Chapters 5 (see Section 5.1.2) and 6 (see Section 6.1.2) apply. However, a designer must also take into account the pressure required to flush the annulus during repair procedures, if this is a part of planned corrective action procedures. In some cases, this pressure can be the pressure at the coincident temperature that results in the greatest required thickness of secondary containment components. When this occurs, it is the flushing fluid pressure that must be selected as the design pressure of the applicable components.

A secondary containment casing is designed to encase a primary piping system completely. Therefore, it will be the recipient of any escaped fluids due to excess pressures that develop in the internal piping. If a primary piping system is designed to have its source of pressure immediately cut off when a leak into an annular space occurs, a secondary containment piping system may not have to be designed to the same pressure as the primary piping. The design pressure for the secondary containment may be lower than the primary piping if it will never be able to become pressurized to the same extent as the primary piping, and the flushing pressure does not exceed primary piping design pressure. However, where the source of pressure is designed to be cut off automatically in the event of a leak, a designer may establish the pressure rating of the secondary containment casing to be the same as that of the primary system, as a conservative measure. Instrumentation designed to cut off pressure automatically may fail; therefore, consideration must be given to this possibility.

The provision of a secondary containment casing does not eliminate the need to have safety relief devices installed on a primary piping system, where they are required by code. However, a secondary containment piping (annular space compartment) does provide a possible place to discharge released fluids safely, as an optional design.

Where an annulus is intended to receive fluids from overpressurization of primary pipe components, and a system contains separate annular compartments (zones of secondary containment), each compartment must be provided with the capability of safely relieving or draining the contents of the entire primary piping system. This is true unless there is some provision for the shutoff of fluids at their source (at the outlet of the pump or at the discharge of the tank or process vessel). If the flow of the fluid can be automatically shut off, then it may only be necessary to accommodate the total amount of fluid that is upstream of the compartment location, including the entire compartment area, prior to its source of pressure being shut off. Each annular compartment in the entire secondary piping system must be designed for this capability.

In some systems it is not required to accommodate the fluid downstream of a secondary containment compartment area that is present at the time of upset. This is true if all of the following apply: (1) The system is pitched (sloped) to drain to the end; (2) the source of internal pressure of the primary piping is automatically shut off; (3) the fluid will flow by gravity, unrestricted, to the end of the line; and (4) the primary piping is vented. In this case it is still good practice to design each compartment to accommodate all of the fluid of the primary system that is upstream of the compartment location, including the entire compartment area.

It should be remembered that vacuum or external pressure design criteria must also be considered for secondary containment components, where it is applicable. Vacuum can be created in the annular space where drainage occurs in a system that does not have adequate venting. It can also occur where a leak of the fluid being transported vaporizes in the annular (interstitial) space and it is possible for the vapor to condense suddenly. Vacuum may also be intentionally selected as a means of evacuating the annular space compartment, particularly where hazardous, flammable, or toxic gases are being transported in the primary piping. It may also be maintained on a continuous basis as a means of insulating the primary service, which is common in many cryogenic gas services.

7.2.3 Design Temperature

To select the design temperature of a secondary containment system, the same rules of Chapters 5 (see Section 5.1.3) and 6 (see Section 6.1.3) apply. However, the designer must also take into account possible higher temperatures required for drying of the annulus (interstitial space) during repair procedures, and after initial hydrostatic testing, if warm air or gas is used to accelerate the drying process. The temperature of a hot gas that is introduced for this purpose can in some cases be the temperature at the coincident pressure that results in the greatest required thickness or component rating of secondary containment piping components. In this case, the temperature of a hot gas used for drying should be selected as the design temperature of the secondary containment piping components. As in primary piping, the minimum design temperature should be the lowest temperature expected of the components in service. Many times for secondary containment systems, the minimum design temperature will be lower than the primary piping for which it is being designed. This is due to the temperature gradient created by the use of a material having a low thermal conductivity for the secondary containment such as a thermoplastic or RTRP material) and the insulating dead air entrapped within the annular space. For this reason, it is always recommended to perform a heat-transfer calculation or test to determine the temperature of the outer wall.

7.2.4 Ambient Effects

The same ambient effects discussed in Chapters 5 (see Section 5.1.4) and 6 (see Section 6.1.4) need to be considered when designing secondary containment piping and secondary piping components. Internal ambient effects can be minimized through the introduction of an inert gas or vacuum in the annular space. If a leak detection system is used, the introduction of an inert gas will greatly enhance its proper functioning, and is usually recommended anyway. The type of gas should be carefully selected so that the piping materials (primary and secondary) are unaffected. Normally for all nonmetallic systems, dry air or nitrogen can be used. However, if a metallic material is used, argon is another potential choice. Many metals are subject to embrittlement from hydrogen, oxygen, or nitrogen, particularly when the atmosphere is moisture free.

Internal effects that can occur during the limited period of time when a process fluid enters an annular space during upset must also be taken into consideration. The same is true of the possible effects of external ambient conditions on flushing and testing fluids, and on gas used for drying. All external ambient effects that are normally considered for single-wall piping systems, and primary piping portions of double containment systems, should also be considered for secondary containment piping components. Any reactions transmitted from secondary containment components to primary components as a result of these ambient effects must also be taken into account.

7.2.5 Dynamic Effects

There are dynamic effects that need to be considered in the design of secondary containment components, which are different from those that must ordinarily be considered for primary piping (or single-wall piping). These include: (1) the effect of a catastrophic release of fluid from the primary piping that is under pressure, impacting the outer wall upon its release; (2) the sudden introduction of flushing water into an otherwise empty annulus; (3) the discharge of flushing fluid through low-point drains; (4) the introduction of drying air at a high velocity and high temperature; (5) vortex shedding of flushing fluid moving at a high velocity through the annulus and around vanes or other annular restrictions (see Section 4.3.6), and (6) a sudden rise in annulus pressure in certain jacket applications. In addition, all of the other dynamic forces described in Chapters 5 and 6 require consideration. Be sure to consider the effect that the transfer of dynamic loads via interconnecting parts and interstitial supporting devices will have on the associated primary piping system.

7.2.6 Weight Effects

The effects of all static and dynamic weight loads must be considered in the design of secondary containment components. These loads are discussed in Chapters 5 (see Section 5.1.5) and 6 (see Section 6.1.5). Since many double containment piping systems are intended for buried service, these effects must be carefully analyzed. For nonmetallic systems, the effect of the added rigidity of interstitial supports must be included in the analysis to be performed. Since the resistance of flexible pipes depends on the ability of the wall to deform, increased rigidity due to rigid interstitial supports becomes important, as it actually lessens the ability of a flexible pipe to withstand an external load. This subject is discussed in more detail in Chapter 9, "Structural Design" (see Section 9.4.1.4).

7.2.7 Thermal Expansion and Contraction Effects

As discussed in Chapters 5 and 6 (see Sections 5.1.7 and 6.1.7), there are many types of loads that result from thermal effects. All of these loads need to be considered as acting concurrent with other forces imposed on each component in the piping system. In double containment piping, there are additional considerations that must be taken into account in some components. First, interconnecting components (e.g. internal anchors) are most accurately modeled by considering the system as a whole, taking into account the interrelationship of their primary and secondary containment portions. The effect of primary components imposing loads upon secondary piping materials and the effect of secondary components imposing loads upon primary components can be taken into account as acting concurrently in a combined, complex model using FEM Techniques. When this type of modeling approach is performed, each of the primary and secondary components can have all pertinent stresses identified and determined, including the resultant localized discontinuity stresses. Thus, all components can be correctly studied to determine if they will function within their acceptable stress limits. This is discussed further in Chapter 8 (see 8.4).

Perhaps one of the most widely held misconceptions involving double containment piping systems is that there is an inherent disadvantage if a system contains materials having different values of the coefficient of thermal expansion (i.e., as in a hybrid system). The fact is that problems can arise even if the primary and secondary containment materials are constructed of the exact same materials. Very rarely are primary and secondary containment portions of a double containment assembly subjected to the exact same temperature increases or decreases, due to large temperature gradients that typically exist between inner and outer pipes. Even when they are, in a properly buried system, the soil will totally restrain secondary containment piping. Dead air or vacuum contained in annular spaces of a "closed-annulus" system functions as an excellent means of insulation at close to room temperatures, exceeding that of most commercial insulation materials (i.e., urethane foam, mineral wool, fiberglass, foam rubber, etc.) at close to ambient temperatures. Unless unusual circumstances exist, both inner and outer piping will be subjected to different thermal stress levels (i.e., differential thermal expansion) and discontinuity stresses at points of interconnection. If unusual circumstances do exist that cause the two piping systems to be subject to an equal amount of expansion or contraction, both systems still need to be analyzed to determine if there is adequate flexibility, and whether they are within their allowable stress range. For further discussion, refer to Chapter 8 (see Section 8.3.1).

Double containment piping systems present unique challenges to designers since primary and secondary systems are normally subject to different magnitudes of potential expansion or contraction. This condition is referred to as differential thermal expansion, and can cause failure in either primary or secondary system components due to overstress or fatigue, if the condition is not provided with proper compensation.

In extreme instances, differential thermal expansion can cause failures in both primary and secondary systems, defeating the overall purpose of providing secondary containment. Termed a "double failure," a typical sequence involves primary components that become restrained against secondary containment components, resulting in failure of secondary containment elements (i.e., due to the failure of an outer pipe weld or bond under stress, mechanical failure of a secondary component, etc.) that go undetected. Displacement strains that result from the prevention of natural movements of primary piping components and discontinuity stresses that occur cause failure and leakage at a primary system joint(s), or mechanical failure in the wall of primary piping or one or more of its components. In some cases, the excessive stresses created by the displacement strains imposed upon the primary piping system cause premature failure due to accelerated corrosion attack on the primary piping materials. A failure of the primary for one of these reasons, combined with a previously undetected failure of the secondary system, will result in a "double failure" and subsequent release of fluid to the external surroundings. These are typical examples of a "double failure"; there are many other possible ways that a "double failure" can occur.

A proper design and layout will result in a system whose computed stress range for primary and secondary components, based on a model of a complete, interacting assembly, results in a situation whereby the allowable stress ranges for all components are not exceeded. In order for this to occur, a system must be laid out with an adequate degree of flexibility so that failure due to detrimental stresses or distortion, or from leakage at joints, can be avoided. Also, expansion or movement of primary and/or secondary containment piping should not be allowed to cause failure of interstitial supports, since these parts function as structural supports of the primary piping system, and prevent buckling failure of the primary system due to elastic instability.

Certain interconnecting components constructed from homogeneous materials lack "separation" between primary and secondary containment systems. In effect, they are not surrounded with a means of secondary containment, and function as a pressure-retaining component of the primary system. Thus, components of this type are technically classified

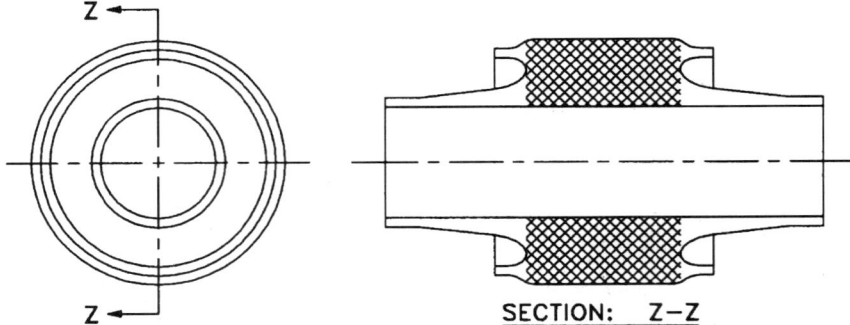

Figure 7.1. A homogeneous-construction internal anchor fitting lacking secondary containment in its center portion (cross-hatched area). (U.S. Patent #'s 4,930,544 and 5,141,261.)

as a primary piping system component. These types of components are discussed in Chapter 11 (see Section 11.1.1.4). Examples of these components are also illustrated in Figures 7.1 through 7.6, 7.19, 11.4 through 11.7, 11.21, 11.26, 11.27, 11.49 and 11.50.

Figure 7.1 indicates the portion of a simple, homogeneous-construction fitting used for internal anchoring or termination transitions that lacks secondary containment in the portion indicated by cross-hatching. This feature is integral to all homogeneous components, not just to the simple geometry shown here. Such a component must be designed as a component of the primary system, and the entire component must be designed according to the criteria of Chapters 5 or 6, depending on the material.

7.2.8 Displacement Stresses in Double Containment Assemblies

The objective in determining an appropriate layout of a double containment piping system is to minimize the total displacement strains of both the primary and secondary piping systems. Also, the displacement strains of both systems should be distributed in such a manner that there are no excessive strains at any one point. This even distribution of displacement strains will result in a "balanced" system, allowing stresses to be somewhat proportional to total displacement strains in the system. Conversely, if a layout results in an "unbalanced" system, there may be resulting problems at areas of concentrated localized discontinuity stress (i.e., internal anchors, termination arrangements, etc.).

An "unbalanced" double containment piping system is possible for many reasons. The common ones include: (1) systems where the primary and secondary containment materials are fixed together at pipe ends by the use of bonded collars, support discs, plates, or laps; (2) systems that are permanently restrained by the use of consecutive internal anchors; (3) systems that contain improper spacing of interstitial supports or completely lack support; and (4) where inner and outer components lack adequate space for differential movements and become restrained against each other. "Unbalance" can be minimized to the greatest extent possible through effective layout, and through the proper application of the techniques discussed in Chapter 8. The use of cold spring to assist in balanced operation of either primary or secondary systems must be analyzed in terms of its effect on the completed assembly before it can be determined to be feasible.

7.2.8.1 Differential versus Nondifferential Thermal Expansion

There are two types of thermal expansion situations that may arise in a double containment piping system. The first type occurs in above-ground systems only, whereby the primary piping system is subjected to a nearly identical magnitude of thermal expansion or contraction (termed nondifferential thermal expansion/contraction). There are three ways that this type of occurrence can result: (1) where primary and secondary piping is constructed of the same material and the systems are totally restrained to each other, or (2) there is sufficient external insulation surrounding the secondary pipe in homogeneous material systems, or (3) where the materials are different, and a temperature gradient results in an even amount of expansion. These situations may only occur where both inner and outer pipes are free to experience such expansion and contraction. They cannot occur in underground piping.

The most likely way that nondifferential thermal expansion occurs in a nonrestrained system is when dissimilar materials are used (which is the third means described in the previous paragraph). An example is where a primary piping material that undergoes a relatively small amount of thermal expansion conveys a hot fluid, and is contained within a secondary containment material that has both a low thermal conductivity and a comparatively high degree of thermal expansion. In this situation, the secondary containment piping will see a much lower temperature than the primary, yet when it does, it grows substantially more. In this case, the total amount of thermal expansion that occurs tends to be more even. In above-ground situations, this type of occurrence is beneficial to a designer. A layout that allows an adequate degree of flexibility in both systems may be accomplished without using expensive and complex techniques. This can be achieved by compensating for the outside piping and allowing the primary pipe system to flex along with its associated outer system. In fact, above-ground systems can be purposely designed to be coupled together and allowed to flex together by installing flex connectors or flexible grooved mechanical couplings on the outer jacket, causing the inner and outer jackets to "grow" and "shrink" together. In underground systems, outer pipes are typically totally restrained from moving; actual system conditions will be of the differential type.

Differential thermal expansion/contraction occurs when primary and secondary piping systems are each subjected to temperature conditions that suggest that different magnitudes of thermal expansion and contraction will occur. As described in Section 7.2.7, this situation can arise regardless of whether the materials of construction are the same. In underground piping, secondary containment piping is restrained from expanding or contracting. Thus, an underground system must be treated as if differential expansion conditions exist, even when conditions conducive to nondifferential thermal expansion exist.

For a further discussion of the effects of differential and nondifferential thermal expansion, the reader should refer to Chapter 8, "Thermal Expansion Considerations"

7.2.9 Effects of Support, Anchor, and Terminal Movements

The effects of movements of external supports, anchors, and terminals must be analyzed according to the same principles described for primary piping in Chapters 5 and 6. Whenever a secondary containment pipe component experiences a movement of one of these external elements, part of the force imparted to the pipe will be transmitted to the primary pipe via points of interconnection. A designer must also be concerned with these effects, and that of movements of the primary and secondary containment pipes as they affect interstitial supports and interconnecting points (i.e., terminations and internal anchors). In most systems, movements of external supports, anchors and connected equipment will cause only a minor load to be transmitted to the primary pipe, except for systems which are designed in a fully restrained manner.

When primary and secondary containment pipes are tied together via interconnecting components, the system becomes a complex entity with loads being transmitted from the primary to the secondary systems and the secondary to the primary systems. Interstitial supports in such systems behave as points that partially restrict torsional and bending movements, limiting movements of the primary piping chiefly to the longitudinal direction. Such a system must be analyzed to determine that the structural integrity of interstitial supports, internal anchors, and termination fitting locations are not damaged. Failure of these parts can result in subsequent failure of the primary system.

When some means of internal anchoring method is used to anchor inner and outer pipe systems together, movements of external supports, anchors, and terminals will cause movements or loads of internal anchors to occur, which will in turn have an effect on the inner pipe. An internal anchoring arrangement may or may not become damaged through such "external" movements imposed on a secondary containment piping system. The effect on an internal-anchoring arrangement must be studied by analyzing localized discontinuity stresses at the location in question. However, the effect of movements of internal-anchoring arrangements on a primary piping system must also be studied for its effect on other primary components. This is discussed further in Sections 7.5 and 7.6.

In homogeneous components that do not contain "separation" between inner and outer systems, such components can theoretically achieve failure through to the outside of the secondary containment system (see Section 7.3.3), which is one type of system "double failure." Components that contain true and complete "separation" between inner and outer systems, and maintain a zone of secondary containment, are not readily subject to this type of catastrophic double failure. However, they still require equivalent analytical attention to that required of homogeneous, non-secondarily-contained components.

7.2.10 Other Effects

Other effects described in Chapters 5 and 6 such as reduced ductility effects, cyclic effects, and air condensation apply to secondary containment systems, and to a double containment system's annulus. Secondary containment materials must be analyzed by the same criteria described in those chapters. When materials that lack ductility are used for secondary containment piping components, the system must be laid out with an adequate degree of flexibility. Movements of inner primary materials, or differential thermal stresses, should not be allowed to cause detrimental strains to be imposed on the outer secondary containment materials. Borosilicate glass is such a material, and must be designed in such a manner when used for secondary containment piping components. Secondary containment piping is rarely subjected to "severe-cyclic" conditions. However, in circumstances where such conditions are present, all precautions described in Chapters 5 and 6 must be taken into consideration, and the same limitations or exclusions will also apply to the secondary containment piping system.

7.3 Design Criteria for Secondary Containment Components

7.3.1 General

The design factors used in establishing the design basis for secondary containment piping components are the same as those described in either Chapter 5 (metallic materials) or in Chapter 6 (nonmetallic materials), depending on the material used. These include the pressure and temperature ratings, stress criteria, and any design allowances that apply.

7.3.2 Pressure/Temperature Design Criteria

Tables B-1 and C-1 of Appendices B and C reference component standards that are listed in the 1993 edition of the ANSI/ASME B 31.3 Chemical Plant & Petroleum Refinery Code. The pressure ratings of the components listed in these tables apply if they are used as a component of the secondary containment system, unless they are altered in some way (e.g., by splitting and rewelding or rebonding the halves or segments together). For metallic piping components that do not have specific ratings in Table B-1 of Appendix B, the rating of the component should be based on 87.5% of the nominal thickness of seamless pipe that corresponds to the pressure class, schedule, or weight of the fitting, less any allowances.

Many components used in secondary containment piping systems are "unlisted components"; many more than are typically found in a primary system or in a standard single-wall piping system. "Unlisted components" of a secondary containment piping system should be designed in accordance with known techniques and verified by calculation or analysis. The design and analysis of such components should be supplemented by experimentation, proof test, or extensive service under comparable conditions using similar materials with similar dimensions. The ASME Boiler and Pressure Vessel Code outlines proof testing and experimental stress analysis in Section VIII, Division 1, UG-101 for proof tests, and Section VIII, Division 2, Appendix 6 for experimental stress analysis.

In many double containment piping systems, standard piping components are altered to have primary and secondary containment components permanently attached to each other via welded or bonded supports or attachments. Even though individual primary and secondary piping components of such assemblies may each be "listed" components in and by themselves, they are "unlisted" components once they are welded or bonded together as an assembly. These components may become derated in terms of pressure rating due to localized weakening resulting from welding and bonding; they almost always experience a reduction in flexibility as well. The pressure rating of these types of components should be verified by proof testing or experimental stress analysis prior to establishing their pressure ratings.

7.3.3 Containment and "Separation" Issues

Three important questions must be asked for every secondary containment component of a double containment system: (1) Is the component fabricated starting from separate components that wind up as part of a weldment or assembly? (2) Does any portion of the secondary containment component serve as a means of primary containment of the service fluid? and (3) Does the component possess a zone or area of containment for the entire 360° around the circumference of the primary containment, at all points along its length? If the answers to these questions are: (1) Yes, (2) No, and (3) Yes, then the component is a true secondary containment component, with complete "separation" between inner and outer systems. If the answers to any of the three questions are: (1) No, (2) Yes, and (3) No, then the component is not a secondary containment component with complete "separation" between the inner and outer systems. It is instead classified as a primary component of the double containment piping system, and is not a secondary containment component. It is subject to design criteria of primary components; thus it must be designed according to the considerations stated in Chapter 5 or 6 of this text, depending on the material is a metallic or nonmetallic material, respectively.

The importance of determining whether "separation" exists in a double containment piping system component is that a component that does not possess "separation" between primary and secondary containment systems can experience failure through the continuous wall (i.e., a "double failure") if the right combination of design conditions is present. In fact, if failure occurs at this point, fluid may never even enter the annulus, and the leak may therefore never be detected. If the component does not possess a "separation" feature, then the component must be considered an "unlisted" component of the inner primary piping system and subject to the same criteria for design as all other "unlisted" primary components (see Sections 5.1.10 and 6.2.2). Certain homogeneous interconnecting components (i.e., internal anchors and termination fittings of homogeneous construction) fall into this category (see Figures 7.1 through 7.6, 7.19, 11.4 through 11.7, 11.21, 11.26, 11.27, 11.49 and 11.50). Thermoplastic piping and other components that are extruded with a double wall having longitudinal vanes or ribs also fall into this category.

If a component may be described as possessing "separation," the secondary containment portion of such a component can be designed in accordance with the considerations stated in this chapter. The portion of the component that serves as a pressure-retaining part of the primary system must be designed according to the criteria in Chapter 5 or 6, depending on the material used. Examples of interconnecting components possessing "separation" are shown in Figures 7.18, 11.11, 11.44, 11.45 and 11.51 through 11.54.

For homogeneous-material-interconnecting components that lack "separation," interaction occurring between inner and outer systems produces discontinuity stresses that are concentrated at the component (in the form of localized stresses) and must be considered in their design. Of primary importance in analyzing the validity of such components is the possibility that stresses may accelerate a corrosion attack on their inner wall (the inside diameter of the "inner" or "primary" portion of the component), leading to a "double failure" from the extreme inside diameter through to the extreme outside diameter. There are many corrosion mechanisms that are dependent upon the level of stress present, the most notable of which is stress-corrosion cracking (metallic materials, described in Section 2.1.8) and environmental-stress cracking (plastic materials, described in Section 2.3.2.3). Since interconnecting parts are components that by their design will experience high relative stress values (see Section 8.4.3), this issue is critical for components that lack "separation."

Figures 7.2 through 7.4 illustrate an example three-step scenario illustrating how a corrosion-stress cracking mechanism may lead to a catastrophic "double failure" of a homogeneous-material internal-anchor fitting that lacks "separation." A "double failure" can result since such a component does not possess a zone of secondary containment at the point at which "separation" does not exist, which also tends to serve as the point of highest stress, and, thus, the point of initial corrosion attack (see Section 8.4.3). It should be noted that a leak of this type may never be detected by the leak detection system.

Design of Secondary Containment Components (Metallic and Nonmetallic Materials) 231

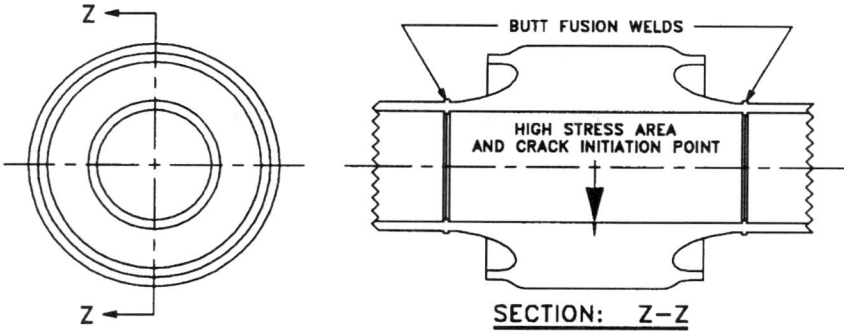

Figure 7.2. Step one: A crack initiates after a period of exposure at the area of highest residual stress in the component, if the stress is over a threshold level. This area will either be at the butt fusion welds, or in the center of the component as shown here.

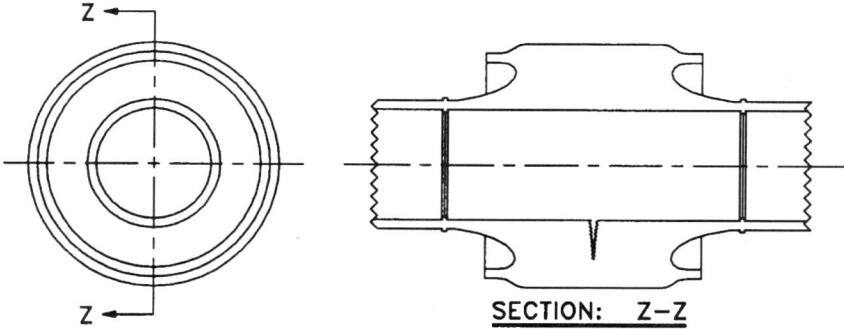

Figure 7.3. Step two: The crack continues to propagate at the point where it initiates. A crack that develops in the center of the component would be worst than at the welds, as continued propagation can lead to an undetected leak, as shown in step three.

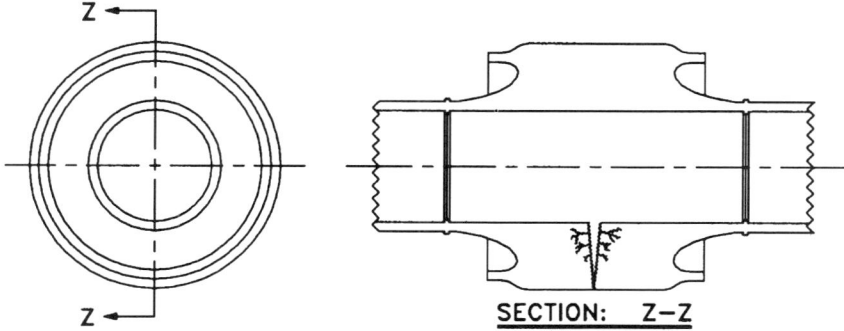

Figure 7.4. Step three: The crack propagation continues until catastrophic failure occurs. All geometries of homogeneous interconnecting components are susceptible to such an attack. It is not limited to the specific geometry shown here.

7.4 Pressure Design of Secondary Containment Piping Components

7.4.1 General

The design of pressure-containing secondary containment components, other than unmodified "listed" components (listed among the consensus standards in Tables B-1 and C-1 of Appendices B and C, respectively), should take into account the material of construction's properties and all applicable effects described in this chapter, and those that are applicable from Chapter 5 or 6. The main design objective for designing secondary containment components that are intended to be pressure rated, or designed to withstand thermal expansion, is that they be constructed with a thickness and reinforcement in locations such that they can withstand pressure and temperature requirements concurrent with all other applicable effects imposed on the secondary containment system of which they are a component.

One major difference between the design of these components and those selected for the design of primary components is that they may only be required to contain a pressure, and the corrosive effects of a fluid, for a brief duration. Additionally, this brief duration is infrequently required in a well-designed system. Also, many secondary containment systems may be intended to see only low pressure (< 15 psi; 1 bar), due to safeguards provided as a part of the primary piping. Under certain conditions, a secondary containment system may be permitted to be designed with an "open-ended" annulus, thereby rendering such a system substantially incapable of developing any significant pressure. For these reasons and others, a designer may be permitted to design such secondary containment components to have a lower pressure rating and/or corrosion allowance than associated primary system components, unless special circumstances suggest otherwise, or the authority having local jurisdiction does not allow it.

If a lower rating is selected for a secondary containment system, the possibility of overpressurization caused from a release of fluid from a primary system rated to higher pressures must be prevented by adding safeguards. An overpressurization of primary pipes can occur either during operation or during a hydrostatic test prior to startup. If a primary system is tested to a higher pressure than that for which its secondary containment system is rated, overpressurization of its secondary containment portion can result. If a secondary containment system is filled with an incompressible fluid during the test of its associated primary piping system, and a leak develops in the primary pipe system, the secondary piping system will become pressurized to the same extent as its primary system. The effect will be virtually instantaneous, unless the annulus is filled with a compressible fluid, but in any event it may result in rapid failure (see also Section 13.3.2). This situation must be prevented from occurring whenever lower-pressure-rated secondary containment piping components are selected.

For components described in Chapters 5 and 6 that are used as unaltered components, such as straight pipe, curved and mitered segments of pipe, bends, branch connections, reducers, flanges, etc., the design techniques, calculations, references, and ratings presented in Chapter 5 (metallic materials) and Chapter 6 (nonmetallic materials) can be used. If these components are changed or welded or bonded to their associated primary piping via supports, or changed by some other fabrication method (e.g., splitting and rewelding), they must be treated as "unlisted" components. Sections 5.1.10 and 6.2.2 define the term "unlisted" components for metallic and nonmetallic material construction, respectively.

The design of each of the major aspects of a secondary containment system is presented in the following sections. Many components that are unique to secondary containment systems are covered.

7.4.2 Containment of Straight Pipe

In sections of straight pipe where standard pipe is used in an unaltered manner, piping may be treated in the same manner as straight primary pipe, which is described in Chapters 5 and 6 (see Section 5.2.2 and 6.3.1.1). If ends of secondary containment pipe are welded (bonded) to their associated primary pipes via supports, or are modified in any other way, then the straight pipe lengths must be qualified as "unlisted" components. Pipe that is altered in some fashion may become derated in terms of pressure-retaining capability.

Figure 7.5 illustrates a cross-sectional view of straight thermoplastic pipe that comprises an inner and outer pipe assembly that is extruded in one piece. This specialized pipe extrusion contains annular pockets that encompass a large portion of its annular volume. However, "separation" is not provided along each of its thin longitudinal vanes, and a zone of secondary containment for the primary system along these thin portions. To be consistent with the philosophy of this text, a one-piece extrusion of this type is classified as an "unlisted" component of the primary system, and its design based on the criteria for a thermoplastic primary component. In such an extrusion, there are annular pockets between adjacent longitudinal vanes that provide secondary containment for the primary pipe over most of its surface. Figure 7.6 is a photograph of a cross section of a typical joint (bond) of consecutive Figure 7.5 sections that have been joined by simultaneous thermal-butt fusion.

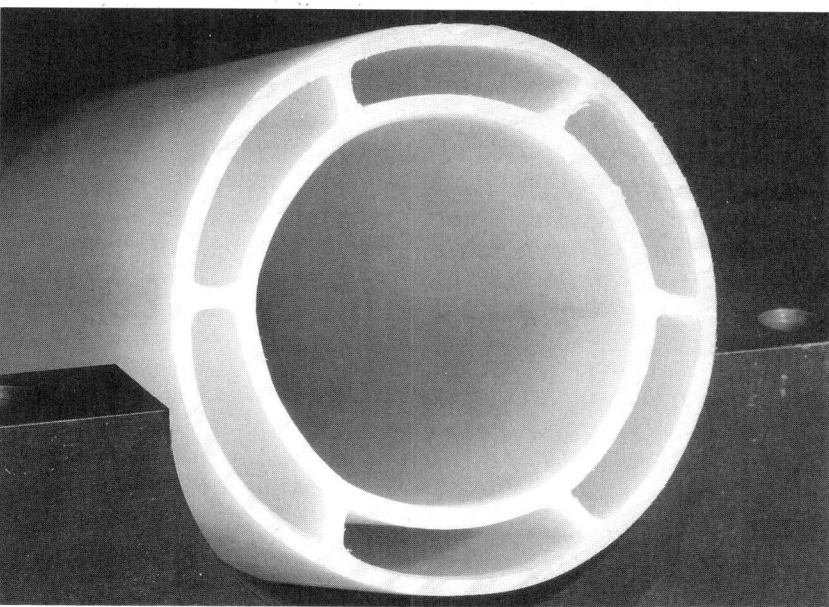

Figure 7.5. Cross-sectional view of a unitarily extruded thermoplastic pipe having integral secondary containment. (Source: PolyFlowLines, Inc., U.S. Patent #4,779,652.)

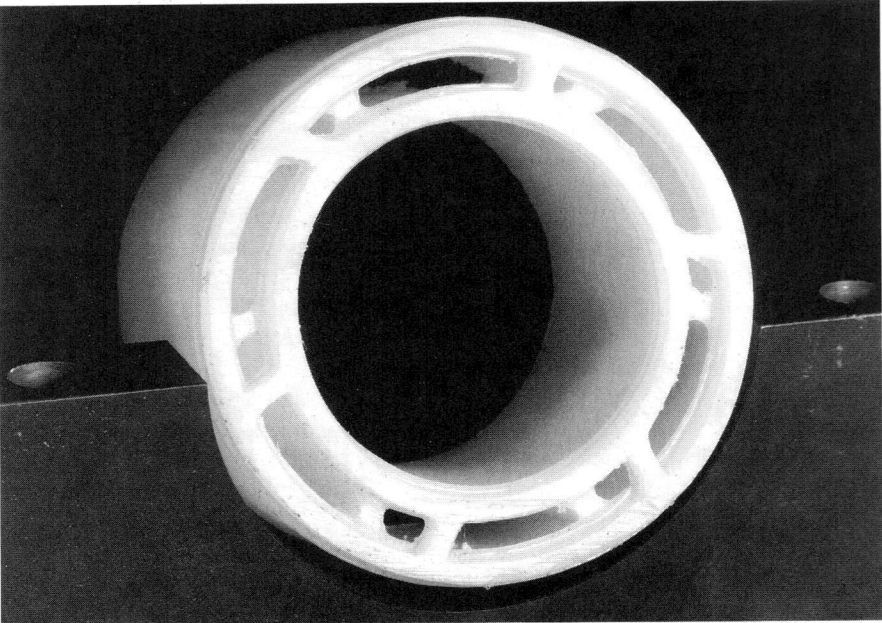

Figure 7.6. Cross-sectional view of simultaneously butt-welded pipes of the type shown in Fig. 7.5. (Source: PolyFlowLines, Inc., U.S. Patent #4,779,652.)

7.4.3 Containment of Elbows

Perhaps the single most important issue in internally flexible systems with respect to containment of elbows has to do with the relative radius of the elbows after assembly is complete. This issue directly affects the ability of a system to withstand the rigors of differential thermal expansion. It is addressed in further detail in Sections 8.6.3 and 8.6.4. The three possible relative orientations of elbow radii are illustrated in Figures 7.7, 7.8, and 7.9. In Figure 7.7 the common arrangement is illustrated whereby the primary elbow has a shorter radius than that of a corresponding secondary containment elbow. This arrangement is beneficial in cryogenic applications of jacketed piping, and in applications where expansion of the secondary pipe relative to the primary pipe in above-ground applications is the controlling design factor. Figure 7.8 illustrates an elbow relationship, whereby the radii are equal throughout the curved portion of the elbow. This arrangement is beneficial in systems where both differential expansion and contraction of the inner and outer systems relative to each other must be tolerated. Figure 7.9 illustrates an arrangement whereby the primary elbow has a longer radius than that of the corresponding secondary containment elbow. This arrangement is beneficial in applications where a greater amount of differential expansion is to be expected on the primary piping. This is the most common design condition in double containment pipes. An example would be a buried system whereby hot chemicals are being transported in the primary pipe. However, this arrangement is also the preferred arrangement whereby the design involves differential contraction of the secondary containment system relative to the primary pipe. This is the second most common-

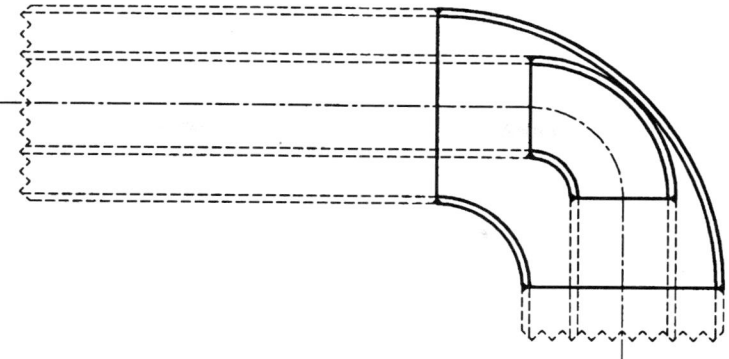

Figure 7.7. Example of a double containment elbow arrangement whereby the primary elbow has a shorter radius than that of the corresponding secondary containment elbow.

ly encountered design condition in double containment piping systems. An example involves above-ground systems that are installed in the summer months, whereby the secondary containment jacket is subject to contraction due to winter temperatures, yet the primary pipe conveys a constant temperature fluid. Therefore, the type of elbow arrangement shown in Figure 7.9 is a highly important arrangement in double containment piping, which is beneficial to the two most common design conditions.

In addition to concerns of relative radii, one must also consider the method of manufacture and assembly of such elbows. The choices for providing a means of secondary containment for primary pipe elbows include the use of the following secondary containment components: (1) "listed," "off-the-shelf" elbow fittings, whereby the primary fitting is "freely floating" in space; (2) fabricated elbows by the use of mitering or bending techniques applied to modify straight pipe for such purposes (see discussion in Section 11); (3) proprietary elbows that are manufactured using conventional techniques (see Section 11.1.2.1 for discussion of these elbow fabrication concepts; (4) use of split-and-rewelded

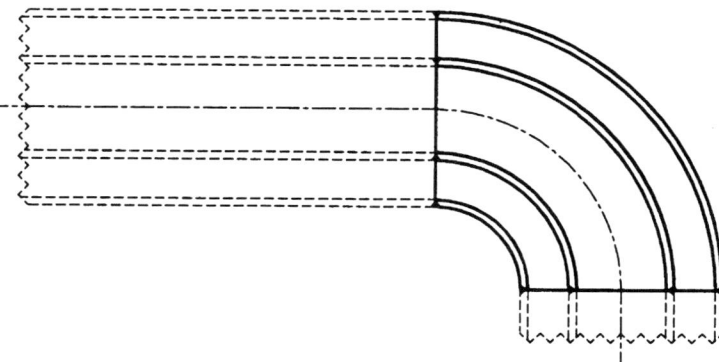

Figure 7.8. Example of an elbow where the inner and outer radii are equal throughout their curved portion.

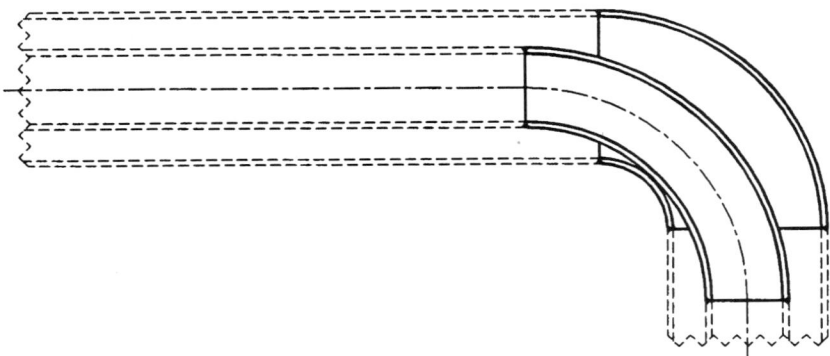

Figure 7.9. Example of an elbow whereby the inner elbow has a radius which is longer than that of the corresponding secondary containment elbow. (patent pending)

(rebonded) standard elbows; (5) use of split-and-rejoined proprietary elbows; (6) use of secondary containment tees, laterals, or crosses that house a primary piping elbow (usually limited to areas that require access); and (7) use of manholes, sumps, or other containment vessels or structures, which are typical of underground systems (refer to Figures 11.99 for an example of this type of arrangement.

For components described in items 1, 2, and 3 the considerations described in Section 5.2 or 6.3 (item 1); Section 5.2.1.2, 6.3.2 (item 2), or 6.3.3, and Section 5.1.10 or 6.2.2 (item 3) may be used for qualification of these components. The remainder of the items must be qualified as "unlisted" components, as described in Section 5.1.10 or 6.2.2 depending on the material involved.

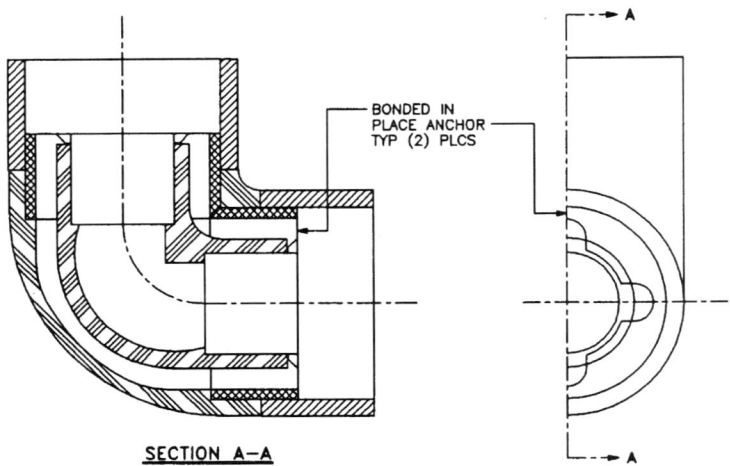

Figure 7.10. An example of a double containment elbow assembly, involving a secondary containment elbow that is secured (restrained) to its associated primary elbow by means of a bonded-in-place support plate. (Source: Fibercast Corp., U.S. Patent #4,886,305.)

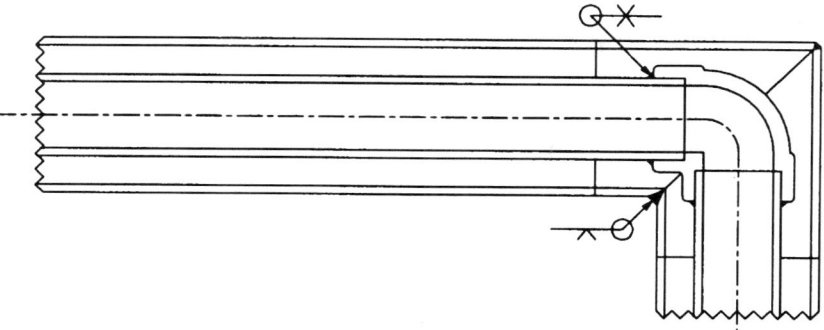

Figure 7.11. Example showing the use of a single-mitered secondary containment elbow to add space to a respective primary elbow.

If secondary containment elbows are welded (bonded) to their associated primary elbows via interstitial members, or are modified in any other way, then they must also be qualified as "unlisted" components. Elbows that are connected in this fashion may become derated in terms of pressure-retaining capability. An example of proprietary fabricated assembly with inherent mechanical integrity is illustrated in Figure 7.10. Elbow arrangements of this type are beneficial in certain RTRP systems where the design of compression molded elbows and straight socket joints do not allow for a large amount of bending, without risking joint failure.

In order to add room for expansion of a primary elbow and its adjacent piping, the designer also has the option of increasing the localized diameter of the secondary containment components in the area of the elbows. The relative radius issues still arise, and simply adding a larger elbow does not automatically increase space in a desired direction. The designer can also modify the primary piping elbows using one or more means in addition to adding space. Examples of several design choices for increasing the outer size of elbows around the primary, using mitering techniques and/or concentric reducers with a larger elbow, are illustrated in Figures 7.11 through 7.15.

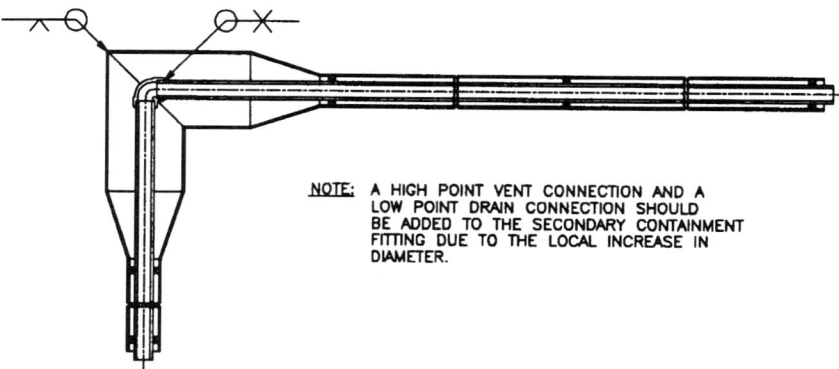

Figure 7.12. Example of the use of a locally increased outer elbow section produced using a single-mitered center joint to add space to a respective primary elbow.

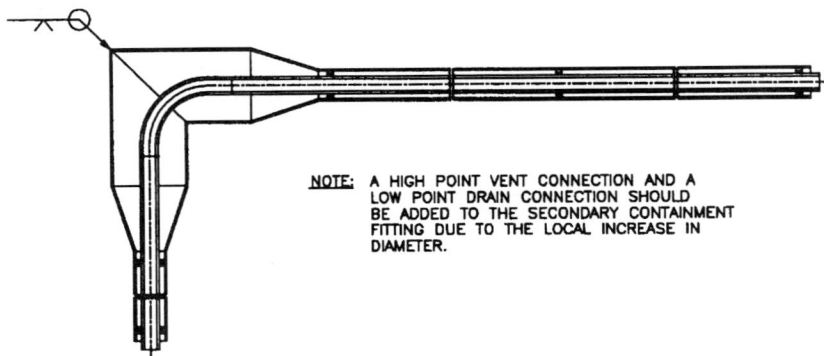

Figure 7.13. Example of the use of a locally increased outer elbow section produced using a single-miter design with an extra long radius primary bend to add more space.

7.4.4 Containment of Branch Connections

The choices for providing a means of secondary containment for primary pipe branch fittings include the use of the following secondary containment components: (1) "listed," "off-the-shelf" branch fittings that consist of primary branch fittings installed inside associated secondary containment branch connection fittings, where such fittings are unaltered in any way, the primary fittings are "freely floating" in space, and sizes and fabrication methods permit this to be the case; (2) secondary containment branch connections fabricated around a standard primary fitting, using straight secondary containment pipe; (3) secondary containment branch connections fabricated in conjunction with a fabricated primary fitting, where both are fabricated from their respective straight pipes; (4) standard "listed," "off-the-shelf" components that are split and rejoined around their associated primary fitting; and (5) manholes, sumps, or other containment vessels or structures, typical of underground systems (refer to Figures 11.95 through 11.98 for examples of these arrangements).

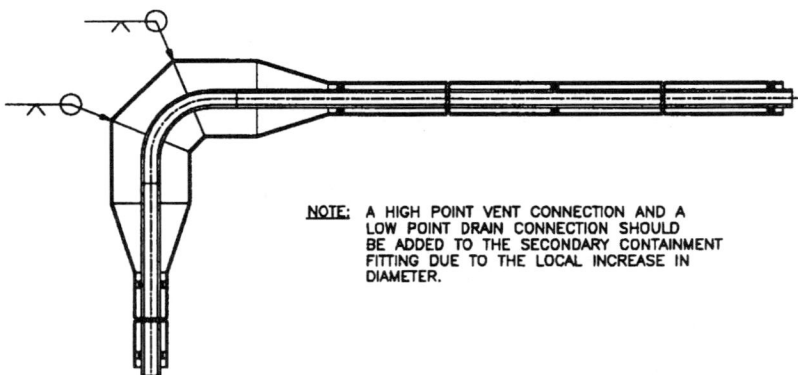

Figure 7.14. Example of the use of a locally increased outer elbow section produced using a multiple-miter design and incorporating an extra long radius bend on the primary.

Design of Secondary Containment Components (Metallic and Nonmetallic Materials) 239

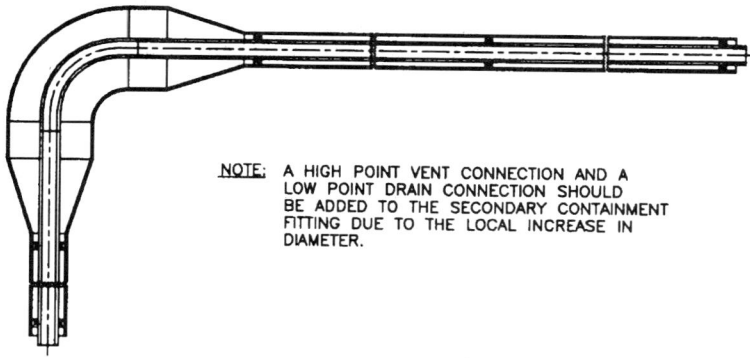

Figure 7.15. Example of the use of a locally increased outer elbow section using a standard short radius 1-D bend, and a matching radius primary elbow, to result in a True Radius™ combination.

For secondary containment fittings of this type, the considerations described for qualification in Section 5.2.1.3 or 6.3.4 apply, except for item 5 (refer to Section 11.4.2).

If secondary containment branch fittings are welded (bonded) to their associated primary branch fittings via interstitial members, or are modified in any other way, then they must also be qualified as "unlisted" components. Branch fittings that are connected in this fashion may sometimes be derated in terms of pressure-retaining capability. An example of a fabricated assembly made from "listed" components is illustrated in Figure 7.16.

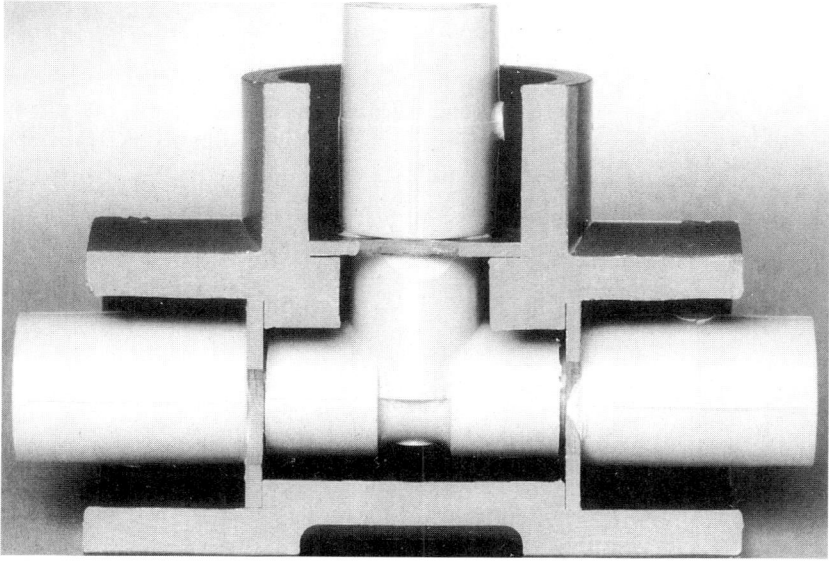

Figure 7.16. An example of a double containment tee assembly involving a secondary containment tee that is secured (restrained) to its associated primary tee by means of a welded-in-place support plate. (Source: Guardian division of Nibco, Inc.)

7.4.5 Containment of Reducers

A primary pipe concentric or eccentric reducer may be contained inside a secondary containment reducer having a transition equivalent to the secondary containment pipes on its adjoining sides. A primary pipe reducer may also be housed inside a continuous straight pipe section; that is, a 4 x 2 in. reducer (approx. 100 mm x approx. 63 mm) may be housed inside a continuous 8 in. (approx. 200 mm) pipe section, instead of being housed within a secondary containment 8 x 4 in. (approx. 200 mm x approx. 100 mm) concentric reducer. If primary fittings are housed within a standard secondary containment reducer, then the considerations described in Section 5.2.1.5 or 6.3.7 apply. If the primary reducer is welded (bonded) to the associated means of secondary containment, then the assembly must be qualified as an "unlisted" component.

Eccentric reducers present a greater challenge to designers, if the requirement is to contain an eccentric reducer inside an eccentric reducer. Virtually any dual-eccentric reduction of this type requires some modification to standard off-the-shelf components. Containment of straight primary-pipe sections inside eccentric secondary containment reducer fittings is often done where a localized increase in diameter is required for purposes of providing additional layout space for expansion flexibility (e.g., see Section 8.7). This type of arrangement will result in the primary pipe's being eccentrically positioned inside the enlarged secondary containment pipe section; this is the usual intention for this type of arrangement.

Eccentric reductions of primary pipes inside straight nonreduced sections of secondary containment pipe sections are also possible, although reduction in this fashion will result in the reduced-diameter section being eccentrically positioned within the unchanged-diameter secondary containment pipe section.

7.4.6 Containment of Flanges and Flanged Components

It is not common to have primary flanges secondarily contained inside a secondary containment flanged section. This type of approach requires a secondary containment housing that is of a comparatively large diameter. An example of how this would appear is

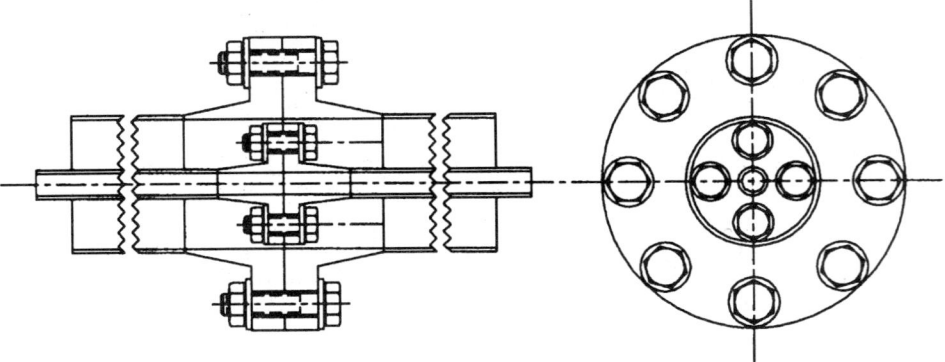

Figure 7.17. Illustration of a primary flange contained inside an outer pipe, also joined by means of a flange. Note that the outer pipe must be relatively large in diameter to accommodate the primary flange.

illustrated in Figure 7.17. This type of arrangement is simply two standard flanges, one housed within the other. If a standard primary flange is housed within a standard secondary containment flange, the flanges are each standard, and may be designed in accordance with Section 5.2.1.6 or 6.3.6. If the primary flange is welded (bonded) to the means of secondary containment, then the assembly must be qualified as an "unlisted" component. Termination flanges, which involve both pipes being joined to the same flange (including insert flanges), are examples of modified, and therefore "unlisted," flanges (components).

Three more likely arrangements where a flange is to be incorporated on the primary pipe are illustrated in Figures 11.43, 11.61 and 11.96. The outer flange housing in Figure 11.61 is an "unlisted" component of the secondary containment, which in some cases has to be designed to the Boiler and Pressure Vessel Code, since it is in effect equivalent to a small pressure vessel. Figure 11.43 sometimes uses a termination fitting of the type described in Section 7.4.8, although it can be used without adjacent termination fittings.

7.4.7 Internal Anchoring Components and Other Anchoring Methods

A variety of components to anchor the primary pipe components internally in a double containment piping system are presented in Chapter 11, Section 11.1.1.4.2. There are three basic types of internal anchoring methods that are described in Section 11.1.1.4.2: Type 1 (those that are constructed of a homogeneous material, lacking "separation" and a "secondary containment feature"); and Type 2 (those that do contain separation and a secondary containment feature). Type 3 internal anchors include fabricated assemblies where the part can not be classified as a pure Type 1 or Type 2 internal anchor (see Figures 1.44 and 11.45 for examples). All types of internal anchor methods are considered as "unlisted" components.

Type 1 internal anchoring components are "unlisted" components of the primary piping system, examples of which are shown in Figures 11.49 and 11.50. Type 2 internal anchoring components contain portions that are "unlisted" components of both the inner and outer systems. Figure 7.18 illustrates an example of a Type 2 internal anchor component (see also Figures 11.11, and 11.51 through 11.54). The component illustrated in Figure 7.18 is provided with added reinforcement for both its primary and secondary containment portions, which are indicated in the figure.

Figure 7.19 illustrates a Type 1 internal anchor component that is provided with transitional reinforcements and smooth fillets to minimize the buildup of localized discontinuity stresses. A reinforcement area is also shown on the external portion of the outer wall; this must be added if the secondary containment system is required to be designed to handle significant annular pressures, and it tends to add resistance to bending loads of its inner portion.

Other anchoring methods exist to anchor primary and secondary containment systems together using "external" methods. These are described and illustrated in Section 11.1.1.4.1. Assemblies of this type must always be qualified as "unlisted" components, according to Section 5.1.10 or 6.2.2.

7.4.8 Termination and Initiation Transitional Components

A variety of termination and initiation transitional methods are presented in Chapter 11, Section 11.1.1.2.1, which are used to make a transition from a double containment pipe section to a single-wall pipe section or connection. There are three basic types of rigid

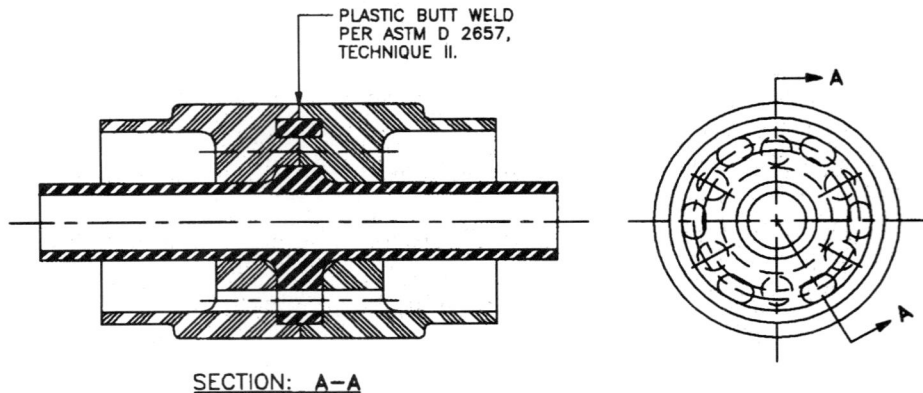

Figure 7.18. Illustration of a proprietary Type 2 internal anchor. (U.S. Patent #5,085,471.)

termination/initiation methods that are defined in Section 11.1.1.2.1: (1) Type 1 termination/transition components are constructed of a homogeneous material in one piece, lacking "separation"; (2) Type 2 components do contain separation and a secondary containment feature; and (3) Type 3 fabricated assemblies that comprise a termination or initiation transition. There are also terminational transition fittings that function as an expansion joint. All three types of termination/initiation transitional methods are considered as "unlisted" components, since all are interconnected in some manner. The issue of separation in these types of components is not as important as it is in other components that occur midline (i.e., internal anchoring arrangements), since immediately adjacent to the transition will be positioned single-wall components; thus, there is usually some other means of secondary containment in the position of the transition (i.e., sump, building interior, etc.).

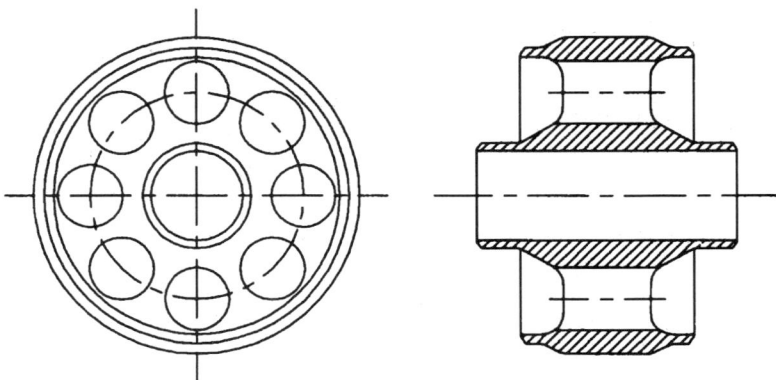

Figure 7.19. Illustration showing a Type 1 internal anchor with transitions and reinforcements that minimize stresses on the part. (U.S. Patent #5,141,261 and #4,930,544.)

Design of Secondary Containment Components (Metallic and Nonmetallic Materials)

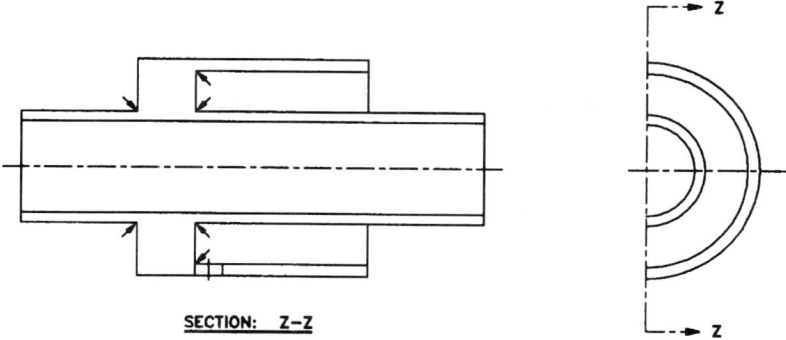

Figure 7.20. A simple Type 1 termination/initiation component having stress risers at critical corners.

Type 1 termination/initiation components are "unlisted" components of the primary piping system. Type 2 and Type 3 termination/initiation components contain portions that are "unlisted" components of both the inner and outer systems. A typical Type 1 termination/initiation component is illustrated in Figure 7.20. This component contains inherent stress risers, as indicated by arrows in the illustration. This is not a desirable feature, as high discontinuity stresses will become concentrated at these points whenever substantial temperature changes occur, which may lead to a catastrophic "double failure" of the component. A Type 1 component that results in a lower accumulation of localized stresses under comparative design conditions, and thus less likelihood of failure as compared to Figure 7.20, is illustrated in Figure 7.21, as a result of its structural design with strategically positioned fillets, contours, and reinforcements.

Several examples of Type 2 termination components are illustrated in Chapter 11 (e.g. Figures 11.11. Typical Type 3 components include such components as the termination flanges and closure rings. Many variations of these components exist (see Section 11.1.1.2.1).

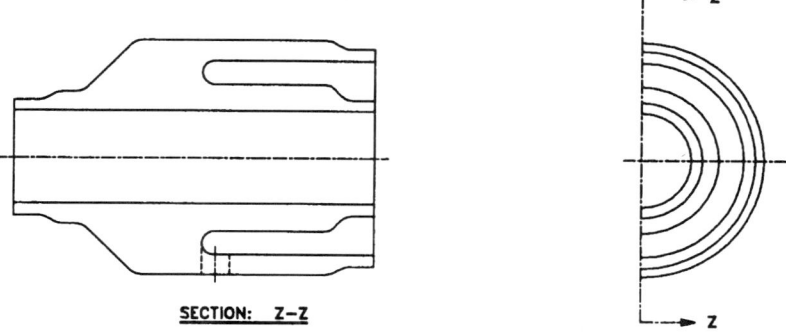

Figure 7.21. A Type 1 termination/initiation component that is designed to have reduced stresses due to differential thermal expansion as compared to the part in Fig. 7.20. (U.S. Patent #5,141,261.)

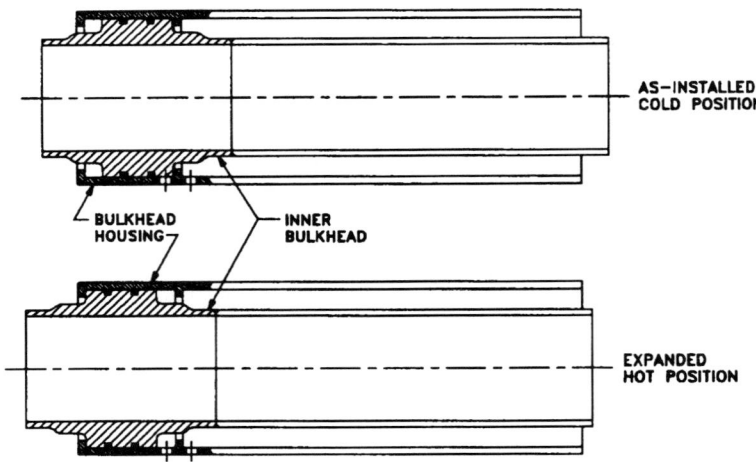

Figure 7.22. A flexible termination fitting designed to act as a combination termination/transition, and piston-style expansion joint. (U.S. Patent #5,141,261.)

An example of an expansion-joint-type component is a fabricated flexible termination fitting, shown in Figure 7.22, 11.12, and 11.14 through 11.16. These arrangements are unique because it is the only Type 3 termination/initiation arrangement that contains only a secondary containment portion; it is applied to a unaltered section of straight, "listed" primary pipe to complete the termination/initiation transitional arrangement. Thus, only the secondary containment portion of such an arrangement needs to be qualified, although to do so often requires that it be modeled and tested with its associated primary pipe.

7.5 Selecting Secondary Containment Piping Components and Joining Methods for Fluid Services

All of the considerations, techniques, and limitations that are described in Chapter 5 (metallic materials) and Chapter 6 (nonmetallic materials) apply equally to the selection of components and types of joints to be used as part of a secondary containment system. The expected time of contact of a given secondary containment system with its primary service fluid is typically of a much shorter duration than that of its primary piping system. Additionally, contact with a service fluid is likely to be expected very infrequently in a well-designed system. Therefore, the material selected for secondary containment components may theoretically be somewhat less chemically resistant to a service fluid than an associated primary pipe material in the same application. However, designers must always be aware that the authority having local jurisdiction may not allow a material of this type to be substituted. In the United States, the requirements of the Resource, Conservation and Recovery Act must be interpreted for every given application, where it is appropriate.

When a primary piping is subjected to "severe-cyclic" conditions, but its secondary containment piping is not, a material otherwise not allowable for "severe-cyclic" services for the primary pipe by the Code (i.e., ANSI/ASME B 31.3) may be allowable for the secondary containment system, if such material is chemically resistant to the primary fluid for the maximum anticipated duration of contact. However, an analysis must be used that models both inner and outer systems in simultaneous fashion, to be sure that the cyclic effect of the primary system does not cause the outer to fail prematurely (i.e., due to fatigue failure).

7.6 Secondary Containment Piping Support

The design of external support structures for double containment piping assemblies is discussed in Chapter 9, "Structural Considerations" (see Section 9.3). The design of external supports should take into account the increased rigidity of the assembly. The external support structure should also take into account both external and internal loads acting concurrently on the assembly as they are transferred to the supports and supporting structure.

The weight of the assembly should be based on the heaviest expected condition. This typically includes the primary and secondary containment piping components, the internal fluid, and the weight of the heaviest fluid or gas introduced in the annular space. In applications where the process fluid in the primary piping and/or annular space is a gas, the weight of the test fluid should be the value used in support space calculations, if a hydrostatic test is performed. Flushing fluid or leaked process fluid may represent the heaviest fluid condition to be expected in the annulus of some systems that use only a pneumatic test of the secondary containment system. Other details of support are stated in Chapter 9.

Chapter 9 also presents a discussion of interstitial support design (see Section 9.3.1.2) for supporting primary piping components that are suspended within associated secondary containment piping in a double containment assembly. Such supports provide structural stability of the primary pipe system due to the effects of dead-weight bending. However, they also guide the primary pipe under conditions of expansion, and allow the outer pipe to prevent buckling failure due to elastic (or inelastic) instability of the primary pipe due to expansion, via these supports.

7.7 Auxiliary Piping Systems

Auxiliary piping systems that are associated with a secondary containment piping system include such tertiary services as instrument piping, flushing piping, drain piping, vent piping, drying gas supply piping, and evacuation piping for vacuum removal of fluid. Instrument piping may be used in conjunction with either instruments used to measure pressure, temperature, pH, or other physical measurement, or it is used in conjunction with probe leak-detection systems. The purpose of instrument piping is either to connect instruments to other piping or pieces of equipment or it may be used to connect air or hydraulically operated control apparati (referred to as control piping when used in this capacity).

Instrument piping designed for a secondary containment piping system must be designed in accordance with the same requirements and considerations described in the rest of this chapter. It should be designed to maintain the integrity of its associated secondary containment system, particularly where it is connected directly to secondary containment piping. Since most instrument piping is intended to contain static fluids, it is readily subject to freezing, and thus should be provided with heat tracing where possible. Since most double containment piping systems that have associated instrument piping require it to contain hazardous fluids, they must be designed so that the fluids may be properly disposed of when the instrument piping is drained. Where it is possible and feasible, the system should be designed to have this automatic capability.

Drainage piping that is at the low end of a double containment piping system's annular space, or at the low end of any annular compartment, must be designed in a manner such that it maintains the integrity of the secondary containment piping. Low-point annular drains are typically provided with a valve that is situated in a valve box or concrete pit/sump (in the case of underground systems), which can be later accessed for safe removal of the fluid. This avoids the need to install a tertiary drainage pipe to drain the annulus, which itself may have to be provided with secondary containment in some jurisdictions.

Flushing piping, vent piping, and purge gas piping (or evacuation piping) must all be designed so that the integrity of its associated secondary containment piping system remains intact. Many times these associated service piping systems originate from a secondary containment pipe system as a valve; often there is no permanent piping connected to such valves, and flexible hoses or pipes are attached in the event that they are required to be used. In any event, a design should take into account large reactions that are often produced by the sudden introduction of flushing or purge fluids into the drain pipe or hose.

Notes

Notes

Chapter Eight

Thermal Expansion Considerations

8.1 Thermal Expansion

Materials, when subjected to a change in temperature, experience a change in volume with proportional changes in dimension. A material subjected to a temperature increase will generally undergo thermal expansion. Conversely, a material that sees a decrease in temperature will undergo thermal contraction. The average change in length of a material per unit length based on a 1°F or 1°C rise in temperature is known as its linear coefficient of thermal expansion. Over a wide temperature range, the coefficient of linear thermal expansion for most materials is nonlinear and generally increases with increasing temperature. Actual values should be used for calculation whenever they are available. Average thermal expansion coefficient data are available for most common materials in ASME Pressure Piping and Boiler/Pressure Vessel Codes and other sources. Note, however, that values for plastic materials vary greatly by specific resin made by each manufacturer. The manufacturer of the plastic resin must be asked for appropriate values for their material.

Equation (8.1) is a general equation that can be used for calculating the thermal expansion of a material, when its coefficient is nearly linear over the temperature range being considered. For those situations where the coefficient is not linear and empirical data are available (i.e., values for the material under consideration over the temperature range), Eq. (8.2) can be used to predict more accurately the growth (or shrinkage) of a material.

$$\Delta L = 12 \alpha L \, \Delta T \tag{8.1}$$

where:
L = length of piping, ft
ΔL = change in length due to thermal expansion, in
ΔT = temperature change, °F
α = coefficient of thermal expansion, in./in./°F

$$dL = 12 \int \alpha(T) dT \tag{8.2}$$

8.1.1 Overview of Differential versus Nondifferential Expansion

There are two basic thermal expansion relationships that may be present in double containment piping applications. When both primary and secondary containment pipe components experience different magnitudes of expansion or contraction, it is referred to as differential thermal expansion. By contrast, in designs where equal magnitudes of movement in the pipes occur, the condition is referred to as nondifferential thermal expansion or contraction. In designs where the primary and secondary containment systems are restrained to one another, inner and outer pipes may be forced to maintain nearly equal lengths by virtue of the restrained design. However, this situation does not represent nondifferential thermal expansion. Instead, the conditions created by such a design can be referred to as "restrained differential thermal expansion."

Nondifferential thermal expansion can result if inner and outer systems are identical materials and are subjected to nearly equivalent temperatures. One example is an above-ground application involving a tank that supplies liquid to a double containment piping system, which changes temperature with its surrounding environment. When the outside environment changes, the fluid temperature in the tank will also change over time. Inner and outer pipes will, over time, tend to be exposed to approximately the same temperatures, since the outer pipe will change with the outside environment, and the inside pipe will change with the fluid it is conveying. If they are constructed of the same materials, they will be subjected to a nearly equal amount of thermal expansion, given enough time.

However, under normal conditions, a large volume of liquid in the tank will not change temperature as rapidly as the secondary containment piping does, particularly when the ambient temperature change is large and sudden. During the time when the temperature of the fluid in the tank is catching up with the ambient temperature, the inner and outer pipes will tend to grow to different dimensions, if they are constructed of the same material, and they are not restrained to each other. Thus, during the transient time, pipes will undergo differential thermal expansion/contraction.

In open air, ambient heating or cooling of the outer pipe may result in a greater temperature increase (or decrease) than that of the inner pipe. This condition is also referred to as differential thermal expansion/contraction, and is defined further in subsequent paragraphs of this section and discussed throughout this chapter. There are several other ways in which differential thermal expansion can be caused.

Nondifferential thermal expansion is also possible if the primary and secondary containment pipes are constructed of different materials (each having different thermal expansion coefficients), provided a unique combination of factors exist. However, the right combination of temperatures for the inner and outer pipes must exist in order for their respective thermal expansion coefficients to produce a nearly equal amount of thermal expansion in each of the pipes. Regardless of the magnitude of their growth/shrinkage, the rate at which they grow/shrink will usually be different, due to the dissimilar thermal conductivities, specific heats, and rates of heat loss for the different pipes. During the heat up or cool down transients, the pipes will be subjected to differential thermal expansion.

Nondifferential thermal expansion in above-ground systems can occur in systems where primary and secondary containment pipes are not fully restrained. Outer pipes tend to be substantially restrained from expansion and contraction movements in most underground double containment piping systems, and therefore, differential thermal expansion is the

Thermal Expansion Considerations

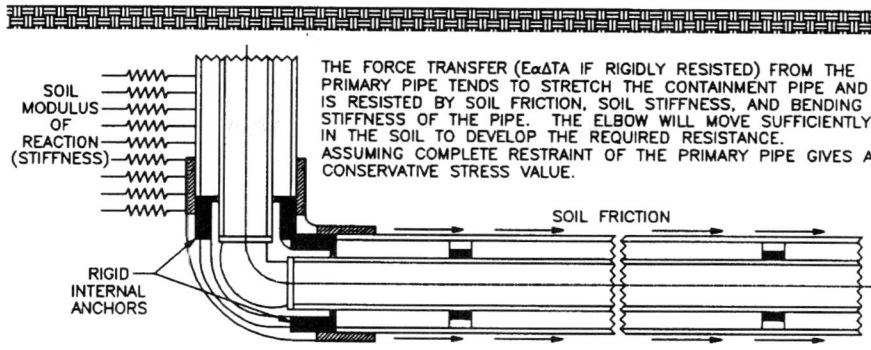

Figure 8.1. Illustration of the factors involved in a buried double containment pipe that is subject to changes in temperature. A restrained RTRP system is shown in this example. (U.S. Patent #4,886,305.)

more common concern in underground systems where the primary piping design allows for freedom of movement. Even if the below-ground system permits movement of the containment pipe, differential thermal expansion is still significant, unless the primary and secondary containment systems are restrained to each other. However, the design case for any systems that have primary and secondary containment systems anchored (restrained) to each other is referred to as restrained differential thermal expansion/contraction, as opposed to true nondifferential thermal expansion/contraction. In this case, the stresses that result from restraining the differential movement become significant. Standard burial practice for pipes requires that consistent and thorough backfilling procedures be carried out on soil that covers a piping system. The friction of compacted soil acting on the outside surface of a secondary containment pipe will act to either restrain or arc restrain the outer pipe, if correct burial procedures have been followed (except for small pipes, < 2 1/2 in. nominal diameter; approximately 75 mm outside diameter).

Actually, restraint of buried pipe varies from total restraint at locations remote from the ends of the straight runs to lesser restraint at the end of straight runs (unless concrete thrust blocking is applied). The degree of restraint depends on soil/pipe friction and the modulus of soil reaction. However, it is generally conservative for buried double containment piping. to assume that it is completely rigid. Figure 8.1 illustrates factors to be considered in the analysis of buried double-wall pipes.

Procedures for more exact evaluations of buried piping are provided in Appendix VII of the ANSI/ASME B31.1 Piping Code. Since the outer pipe is restrained from movement in this situation, any time its primary system undergoes any temperature change, such a system will experience differential thermal expansion/contraction, unless primary and secondary containment pipes are fully restrained to each other, in which case restrained differential thermal expansion/contraction is the design case.

Thus, in underground systems, nondifferential thermal expansion is not possible. Either differential thermal expansion, or restrained differential thermal expansion conditions are the norm. Even in above-ground systems, true, complete nondifferential thermal expansion tends to be the exception as opposed to the rule. It is rare to exist as a naturally occurring condition for double-wall pipe systems.

In double containment piping systems, differential expansion and contraction is a much more common design condition. Thermal stresses become highly significant in such systems when one or more of the pipes are restrained, or when sufficient flexibility is not provided in the layout. If sufficient flexibility exists in a pipe system, it will not fail due to strains that are imposed on components as a result of thermal stresses. A system that is inadequately designed may be subject to catastrophic failure under such conditions. However, this does not mean that a restrained design will not work. If a restrained design will work, then such a system can be described to have adequate flexibility. In fact, in some RTRP systems it can be a more appropriate choice based on the style of components and their method of manufacture and joining method used.

Differential thermal expansion can occur even when all components of a primary and secondary containment piping system are made from the same material. The main reason is that a closed annulus will typically contain entrapped, stagnant air or vacuum, or in some designs, commercial insulation. (Entrapped air space is a better insulator than commercial insulation at close to room temperatures; vacuum remains an effective insulator even at elevated temperatures.) Discontinuity stresses and large displacement strains can result if primary and secondary containment piping systems are fixed together, potentially causing problems at fittings, joints, points of interconnection, and in the walls of the piping. It is best dealt with by designing adequate flexibility into the system, in order to alleviate potential buildup of concentrated localized discontinuity stresses. A system with adequate flexibility will have stresses that are balanced to the greatest extent possible, and therefore operates within acceptable limits. Double containment piping systems thus fall into four basic categories with respect to how differential thermal expansion is accommodated.

Primary Fixed and Secondary Containment Fixed - The secondary containment pipe is restrained from thermal expansion or contraction by the ground (for buried systems) or other restraint, and the primary pipe is restrained from thermal expansion by anchors to the secondary containment pipe. The internal anchors between the primary and secondary containment pipes and the anchors on the secondary containment pipe may be subject to high loads due to differential thermal expansion.

Primary Free and Secondary Containment Fixed - The secondary containment pipe is restrained from thermal expansion by the ground (for buried systems) or other restraints and the primary pipe is free to expand within the secondary containment pipe. The primary pipe can usually be evaluated independent of the secondary containment pipe. However, sufficient clearances must exist between the primary and secondary containment components to provide for sufficient flexibility to accommodate this differential thermal expansion (contraction).

Primary Fixed and Secondary Containment Free - The secondary containment pipe system is free to expand/contract; however, the primary pipe is anchored to the secondary containment pipe, so they must move together. Overall deflection depends on relative stiffness and temperature. Primary and secondary containment pipes must be evaluated together.

Primary Free and Secondary Containment Free - Both the primary and secondary containment pipes are free to expand. The primary pipe may be evaluated independent of the secondary containment pipe; however, the weight of the primary pipe must be evaluated in the evaluation of the secondary containment pipe. Sufficient clearance between the primary and secondary containment components must be provided to accommodate differential thermal expansion/contraction.

These four basic designs may be further differentiated by the magnitude of the temperature changes of the primary and secondary containment pipes, and whether the system is below ground or above ground. Further, different parts of a system may be designed to fall within different groupings (e.g., a system that is designed as partially restrained, and that is to be partly buried and partly installed above ground). See Section 8.8, "Classification of Systems and Application of Compensation Techniques."

8.1.2 Nondifferential Thermal Expansion

Nondifferential thermal expansion is primarily limited to above-ground installations where core fluid temperatures remain close to external ambient conditions, the system is externally insulated, and both pipes are free to experience movements. It may also occur in systems where a marginal temperature gradient exists between inner and outer pipe systems (i.e., a metallic system handling high-temperature core fluids and further provided with external insulation). In either case, differences in expansion between inner and outer systems are minimal.

When nondifferential thermal expansion of an above-ground system occurs, the entire system may be compensated by treating its secondary containment portion as if it were a single-wall above-ground pipe system. Stresses in its inner system will tend to become balanced as a result. Common thermal expansion compensation methods (loops, offsets, etc.) can be used; however, expansion joints may also be beneficial if they are applied to inner and outer systems to compensate equally the growth/shrinkage in both pipes.

Above-ground double containment piping systems can be designed using a compensation method that forces the system to experience substantial nondifferential thermal expansion in applications where conditions otherwise suggest that differential thermal expansion should occur. This is accomplished by selectively anchoring the primary and secondary containment pipes to each other, and incorporating expansion-alleviating devices into the secondary containment jacket. This is not actually nondifferential thermal expansion; it is actually differential thermal expansion with a means provided to accommodate it. In doing so, a situation is created whereby secondary containment jacket will flex or constrict with corresponding changes in the primary pipe. The outer jacket effectively acts in an accordionlike manner. The two systems can be laid out in this manner to compensate for the overall flexibility of the combined systems. If the outer jacket is laid out and supported to allow for adequate flexibility, the expansion/contraction of the primary piping system will tend to also be compensated for as well.

The location of the internal anchors for the type of system described in the previous paragraph is highly important, as improper placement can lead to a reduction in flexibility of the primary piping system at elbows. Devices that are appropriate for use in the secondary containment system to allow the pipe to grow and shrink in an accordionlike manner include bellows-type expansion joints, piston-style expansion joints, and the use of flexible-style grooved mechanical couplings. An example use of grooved mechanical couplings in secondary containment systems as a means of compensating differential thermal expansion is illustrated in Figure 8.16, in Section 8.6.1 of this chapter. There is one further important point to be made with respect to systems designed with flexible secondary containment piping systems having accordion-acting behavior. This type of layout assists in minimizing the size of the outer jacket by preventing differential movements of primary and secondary containment fittings relative to each other. This benefit helps to greatly reduce the overall cost of above-ground systems. This type of system layout is not an option in underground system design, unless the double containment piping system is installed in a tertiary trench that permits axial movement of the containment piping.

Temperature conditions suitable for nondifferential thermal expansion effects may be present in underground pipes. Where such conditions are present in an underground system, thermal expansion will be of the differential type, due to the restraint imposed by the soil friction acting on the secondary containment piping portion. In an underground system where inner and outer systems are permanently anchored to each other at ends of straight pipe sections (i.e., a totally restrained system), and if temperature conditions are right for nondifferential expansion to be present, discontinuity stresses and high displacement strains will result throughout the system, achieving high localized values at points of interconnection. The effect can be catastrophic for metallic or plastic systems if temperature changes are relatively large; at temperatures common to double containment piping systems, such conditions are more severe for plastic piping.

8.1.3 Differential Thermal Expansion

Systems that are subject to differential thermal expansion require that their inner and outer pipe systems be analyzed both as individual pipe systems and as a complex, interacting assembly. Each pipe system, considering all applicable loads acting concurrently on them, must be within the required stress and or load limits of its components, based on the materials from which they are manufactured. Typical calculations are described in Sections 5.1.13 and 5.5.1.6.1 of Chapter 5 for metallic piping components. For nonmetallic components, and for components of the secondary containment system, the calculations are often inherently complex (see Sections 6.6 and 7.2.8).

Under this condition, an inner pipe system always induces stresses on its associated outer system. However, outer pipe components may also induce stresses on the associated inner components. These stresses are transmitted through points of interconnection and centralizing devices (interstitial supports). Close attention must be paid to points of interconnection and other areas of contact where concentrated localized discontinuity stresses may develop. Double containment elbows, branch fittings, and other pieces whose inner and outer components come in contact with each other due to differential thermal expansion movements are a typical area where concentrated localized discontinuity stresses develop. Figures 8.2 and 8.3 illustrate an example involving elbows. Concentrated localized stresses create system imbalance and can lead to a total failure of a system (i.e., a "double failure").

System imbalance and discontinuity stresses may occur under differential thermal expansion conditions when inner components contact their corresponding secondary containment components, as Figure 8.3 illustrates. Interconnecting points are also typical areas where a concentration of localized discontinuity stresses may develop. Typical points of interconnection include: (1) termination arrangements; (2) internal anchors; and (3) baffles and welded (bonded) centralizing devices, where such components are welded (bonded) to inner and outer components, in both restrained and nonrestrained systems.

Interconnecting components, areas, or assemblies such as these must be treated as components that serve a dual role as part of both primary and secondary containment pipe systems. In interconnecting components that are constructed in one piece from a

Thermal Expansion Considerations

Figure 8.2. Relative position of inner and outer elbow sections involving a standard radius combination, prior to thermal expansion.

homogeneous material, a zone of containment does not exist in the component itself, and also can be described as lacking "separation" (see Section 7.3.3). Thus, components of this type lack a means of secondary containment and act as a primary means of containment only. These types of components must be considered as a component of the primary piping system only, since failure can occur directly to the outside environment, and without leak-detection (see Figures 7.2 through 7.4, and the accompanying discussion in Section 7.3.3).

If a homogeneous interconnecting component is inadequately designed, failure to its surrounding external environment may occur prematurely. A double pipe system having such a component may experience failure sooner than a comparable single-wall piping system that is adequately designed for the same set of service conditions. Examples of internal-anchor components, including both one-piece homogeneous parts and parts that feature a zone of secondary containment inherent in the component, are illustrated in Figures 11.44 through 11.54 of Section 11.1.1.4.2.

Figure 8.3. The arrows in the illustration indicate the relative movement of the primary elbow section to the corresponding secondary containment elbow section, as the result of differential thermal expansion.

Compensation alternatives exist to add inherent flexibility for each situation; some compensation techniques are an alternative for all of these conditions. These compensation alternatives are discussed in subsequent sections of this chapter (refer to Section 8.8, "Classification of Systems and Application of Compensation Techniques," and Section 8.6, "Methods for Alleviation"). To understand where they are appropriately used, the reader should consult Section 8.8, "Classification of Systems and Application of Compensation Techniques." The subject is further addressed in other sections of this text. In particular, Chapter 11, "Layout Concepts for Double Containment Piping" also addresses this subject matter in Section 11.1.1.5, "Systems Requiring Flexibility for Thermal Expansion". The layout selected for any system will have a direct bearing on the performance of the system. Chapters 5, 6 and 7 also address issues of thermal expansion and contraction as it relates to primary components (Chapters 5, 6) and secondary containment components (Chap. 7.)

8.2 Axial Stress Calculations

In traditional above-ground single-wall chemical plant piping systems, axial stresses are typically not significant if a system is designed with an adequate amount of inherent flexibility. Bending and torsional stresses tend to be significant, and as a result, they are included in the simplified equations presented in the ANSI/ASME B31 Codes [see Eqs. (5.26) through (5.29)], in which axial stresses are not included. However, the opposite is true in single-walled buried systems, and in double containment piping systems that are designed in a fully or partially internally restrained manner. Axial stresses in these systems must be accounted for in simplified stress analysis, in addition to bending and torsional stresses, where bending and torsion are judged to be significant. In nonmetallic double containment piping systems, internally restrained systems are common, particularly in buried systems, whereby the secondary containment piping system is also at least partially restrained. Since axial stresses can play an important role in double containment piping systems, this section presents simplified equations to determine average axial stresses, and to determine combined stresses for a number of different conditions.

The axial stress, when significant, must be added to the bending stress in Eq. (5.26), resulting in Eq. (8.7). Unfortunately, no stress intensification factor for axial loads is provided in the ANSI/ASME B31 Codes. In the absence of better information, one approach is to use the bending stress intensification factors for axial loads.

When the secondary containment pipe is completely restrained in the axial direction and the primary pipe is restrained from thermal expansion/contraction by internal anchors, the thermal expansion of the primary and secondary containment pipes are completely restrained. Elastically calculated thermal expansion induced axial load in the primary pipe is:

$$\sigma_A^T = E\alpha \Delta T \tag{8.3}$$

where:

σ_A^T = stress in pipe due to axial loads caused by thermal expansion, psi
E = elastic modulus of primary pipe, psi
α = coefficient of thermal expansion of primary pipe, in./in./°F
ΔT = primary pipe max. operating temperature minus installation temp., °F
A, T = subscripts denoting "due to axial loads" and "as the result of thermal expansion," respectively

Thermal Expansion Considerations

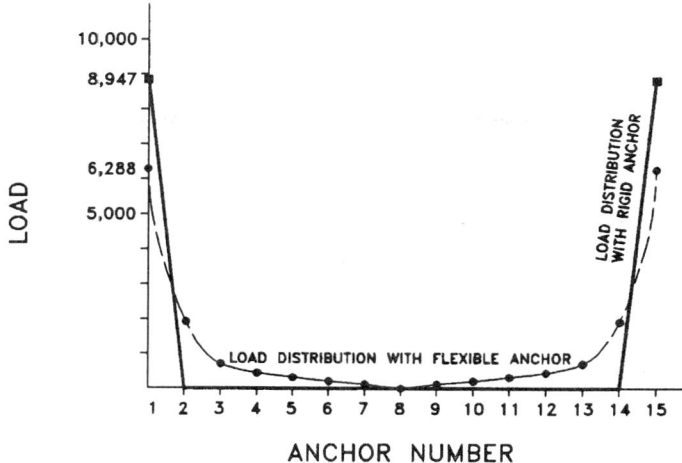

Figure 8.4. Example of a fully restrained system using multiple internal anchors in sequence. The load σ_A^T is transferred over a few internal anchors on either end. This figure is based upon a 10" Sch. 80 PP/12" Sch. 40 PP buried chemical sewer subjected to a temperature range of 80°F. (Source: Becht Engineering.)[1]

Stress in the secondary containment pipe depends on the location of the internal anchors relative to the external anchors. Neglecting the load transfer from the internal to external anchors, stress in the secondary containment pipe will also be per Eq. (8.3); however, the primary elements would be replaced with secondary containment elements.

A load, equal to σ_A^T times the area of the primary pipe wall, must be transferred through the internal anchors from the primary pipe to the secondary containment pipe. If the anchors are rigid, the transfer occurs at the last anchors at the end of each straight run. If the anchors are partially flexible, the load may be transferred over a few anchors at each end, as illustrated in Figure 8.4. The reduction in load is caused by compliance of the internal anchors permitting some expansion/contraction, partly relieving the thermal load.

When the secondary containment pipe is permitted to extend because of system flexibility and the primary pipe is internally anchored to the secondary containment pipe, the change in length of both must be the same.[2] From compatibility follows:

$$\alpha_P \, \Delta T_P \, L + \frac{FL}{A_p E_p} = \alpha_S \, \Delta T_S \, L - \frac{FL}{A_S E_S} \tag{8.4}$$

where:
F = tensile force in primary pipe, lbs. From equilibrium, compressive force in secondary pipe has the same magnitude
α = coefficient of thermal expansion, in./in.°F
ΔT = operating temperature minus installation temperature, °F
L = pipe length, in
A = pipe area, in.^{2}
E = elastic modulus, psi
P, S = subscripts denoting primary and secondary system

The axial load transfer between the primary and secondary containment pipe can be determined from:

$$F = (\alpha_S \Delta T_S - \alpha_P \Delta T_P)\left(\frac{A_P E_P A_S E_S}{A_P E_P + A_S E_S}\right) \quad (8.5)$$

From which the axial compressive stress in the primary is simply:

$$\sigma_p = (\alpha_p \Delta T_p - \alpha_s \Delta T_s)\frac{A_s E_s}{A_p E_p + A_s E_s} E_p \quad (8.6)$$

Other factors that will change these loads would be found in a detailed flexibility analysis. This includes the effect of partial restraint of the pipe that expansion loops impose, and additional change in length caused by internal pressure.

In a double containment piping system, where the primary pipe is perfectly guided by means of a frictionless support, both primary and secondary containment systems are anchored to each other, and the outer pipe system is well guided and free to move, the axial force acting on the inner pipe will be equal and opposite to the axial force acting on the outer pipe, if there is a negligible amount of bending and torsion of the primary pipe. If the inner pipe is in axial tension, the outer pipe will be in axial compression and vice versa, depending on the combination of conditions. However, these effects assume conditions that are unrealistic to expect in actual practice. Primary pipes are always much less than perfectly guided, and supports used are by no means frictionless.

In actual practice, since some tolerance may exist between the outside dimension of internal centralizing devices, there is some limited bending and torsion of the primary pipes. This is illustrated in Figure 8.5.

To determine combined stress due to axial, bending and torsional effects acting concurrently on the pipe, the Becht equation may be used:

$$S_E = \sqrt{(S_b + S_{axial})^2 + 4S_t^2} \quad (8.7)$$

where:

$$S_{axial} = \frac{F}{A}(i_a)$$

F = axial force, which is dependent on the degree of restraint, lbs
A = cross sectional pipe area, in.^{2}
i_a = axial stress intensification factor (note: this factor is not currently provided in ANSI/ASME B31 Codes)
S_b = bending stress, psi
S_t = torsional stress, psi
S_E = computed stress range, psi

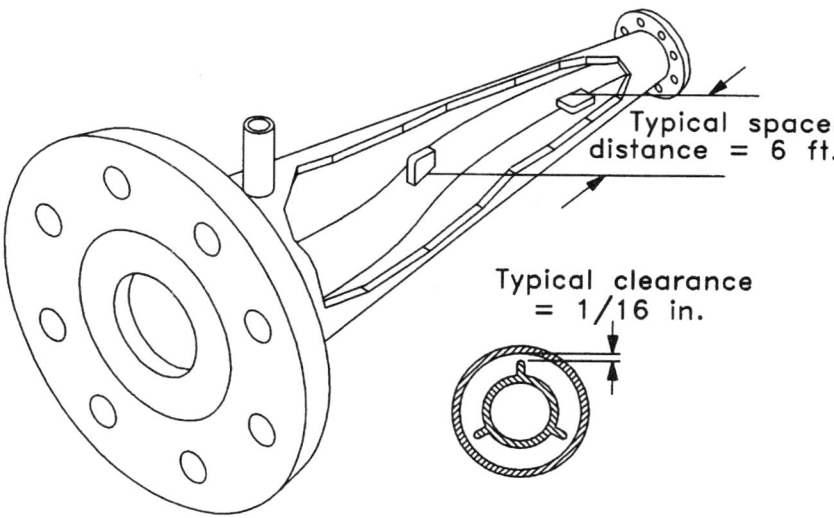

Figure 8.5. In double-walled pipes, there is typically some tolerance between the O.D. of interstitial supports and the I.D. of the secondary containment pipe. This results in some limited bending and torsion of the primary pipes when a hot fluid is transported through it. (Source: Reprinted with permission from **CHEMICAL ENGINEERING** June 1993, Copyright 1993 by McGraw-Hill, Inc., with all rights reserved.)

8.2.1 Axial Stresses in Directly Buried Thick-Walled Secondary Containment Pipes

In certain thick-walled thermoplastic pipes, the temperature across the wall of the pipe will not be linear. Examples include thick-walled thermoplastic piping (e.g., SDR 11 HDPE and PP) and RTRP pipes (e.g., vinyl ester piping with a 110 mil reinforced corrosion barrier, plus a filament wound structural layer). Equations (8.8) and (8.9) provide equations to calculate axial stresses for nonmetallic piping that does not have a linear temperature profile across its wall. If this type of piping is used as a secondary containment pipe in a double containment piping system, the system is highly complex, and stresses are best analyzed by using a comprehensive analysis or by performing experimental analysis.

$$S_c \text{ at outer surface} = \frac{E\alpha(T_2 - T_1)}{2(1-\mu)ln\left(\frac{D}{d}\right)}\left[1 - \frac{2d^2}{D^2 - d^2}ln\left(\frac{D}{d}\right)\right] \quad (8.8)$$

where:
T_2 = temperature of hot fluid, °F
T_1 = soil temperature, °F
S_c = circumferential stress (compressive on inner surface, tensile on outer), psi
E = modulus of elasticity, psi
α = coefficient of thermal expansion, °F^{-1}
μ = Poisson's Ratio, dimensionless

$$S_c \text{ at inner surface} = \frac{E\alpha(T_2 - T_1)}{2(1-\mu)ln\left(\frac{D}{d}\right)}\left[1 - \frac{2d^2}{D^2 - d^2}ln\left(\frac{D}{d}\right)\right] \qquad (8.9)$$

8.3 Simplified and Comprehensive Stress Analysis

Piping systems that are not provided with adequate flexibility are subject to failure due to problems that can develop during operating conditions. These problems are highlighted in the design chapters for metallic components (Chapter 5), nonmetallic components (Chapter 6), and for secondary containment components (Chapter 7). The subject of displacement stresses and strains is also discussed in these three chapters.

A proper design and layout for a double containment piping system will result in a system whose computed stress range for both primary and secondary containment systems, based on a combined, interactive model, are within the allowable stress ranges for all components in each system., and will prevent detrimental chemical effects from occurring. A successful design will also not cause detrimental reactions to occur at supports or connected equipment; movements of this type must be within prescribed limits of each system in a successfully designed system. For a double containment piping system, a designer needs to be concerned with the type, location, and relative position of both internal and external supports, and with internal and external connections to equipment.

Double containment piping presents unique problems to the designer in that primary and secondary containment systems are usually interlocked in locations. They are also contacted at various points between interlocking locations by means of interstitial supports or standoffs. Therefore, the two systems must be considered as a whole when modeling and analyzing inner and outer systems for flexibility, since such a system is essentially a mechanical assembly. To model an "assembly" of this type properly, any interaction between pipes at interconnecting points, at centralizing devices (interstitial supports), and at other points of interconnection (i.e., where inner and outer elbows or branch fittings contact each other) must generally be included the analytical model.

Where "separation" does not exist in interconnecting components (e.g., see Figures 11.49 and 11.50), such components must be considered as part of the primary system in the analysis to be performed. The allowable stress range of the primary component materials should be used as the design basis to judge the adequacy of such components. However, their effect on the associated secondary containment system must also be evaluated.

In contrast, interconnecting components that possess an inherent secondary containment capability (e.g., see Figures 11.51 through 11.54) require only that the primary portion of the component assembly be considered as part of the primary system. The effect of the outer portion acting on the component's primary portion needs to be considered in the analysis of such a component, however. The analysis of such components should compare computed stress ranges to the allowable stresses for both of its "separated" inner and outer portions.

Figure 11.52 typifies a component of the type described in the previous paragraph. In this example, its inner and outer portions are constructed of dissimilar materials (Type 304 stainless steel inside an aromatic amine cured epoxy RTRP material). The stresses calculated for the primary portion must be compared to the allowable stress range for Type 304 stainless steel components of the primary system, determined by Eq. (5.2) or (5.3); the stresses calculated for the secondary containment component must be compared to the allowable stresses of the aromatic amine cured epoxy material obtained from the resin manufacturer or other reliable source. An analysis should always include a thorough study of localized discontinuity stresses for interconnecting components of this type (see Section 8.4.3), and should consider any loads imposed on the inner portion by its outer portion.

8.3.1 Flexibility Analysis

The main objective of a flexibility analysis is to verify that sufficient flexibility of the piping system has been included in the design. Sufficient flexibility is achieved when the piping or its supports are not subjected to damaging or severe displacement strains due to thermal expansion. While a minimum level of acceptable flexibility can be provided, it is often best to design as much flexibility as possible into the layout of a system, as is required to achieve a balanced design with resulting low accumulation of stresses. A material that is under high stress will be more susceptible to mechanical failure or to corrosion attack by a variety of mechanisms (particularly from stress-cracking mechanisms; see Sections 2.1.8 and 2.3.2.3). Accordingly, the lower the level of stress in the system, the less chance there is of a system failure, and the longer the resulting life expectancy.

The design chapters for metallic materials (Chapter 5) and nonmetallic materials (Chapter 6) cover detailed considerations for flexibility requirements (Sections 5.5.1.6 and 6.6, respectively). Thus, these considerations are not repeated here. Chapter 5 presents specific equations for the simplified analysis of pipes, branches, bends, etc. for metallic components. Requirements of nonmetallic materials in Chapter 6 should be reviewed in detail, as the use of simplified equations do not readily apply to these materials. Chapter 11 and sections of this chapter (see Sections 8.6 and 8.8) present many methods for increasing flexibility in a double containment piping system, and for dealing with differential thermal expansion. It should be noted that totally restrained, partially restrained, and totally flexible systems all represent appropriate designs, depending on the application.

In general, a flexibility analysis should include a detailed check of all typical components (pipes, bends, elbows, etc.) subject to a given set of design conditions. If the components of like material and geometry are subject to different boundary conditions, then the parts should be analyzed under the conditions that produce the requirement for greatest component thickness and amount of reinforcement from Chapters 5 and 6 (see Sections 5.1 and 6.2).

8.3.2 Detailed Piping Flexibility Analysis of Double Containment Piping Systems

Detailed piping flexibility analysis can be performed on any number of commercially available computer programs. Modeling techniques are similar to conventional flexibility analysis of single-wall pipes, except that two pipes are modeled.

The primary pipe is connected to the secondary containment pipe at points of restraint and external restraints imposed on the secondary containment pipe. This is typically accomplished by placing nodes in both primary pipes at any point where they are tied together via guides, supports, or internal anchors and coupling the appropriate degrees of freedom at the nodes. For example, movement in all three global directions may be coupled at a location of an internal anchor, and possible rotations also. Only ver-

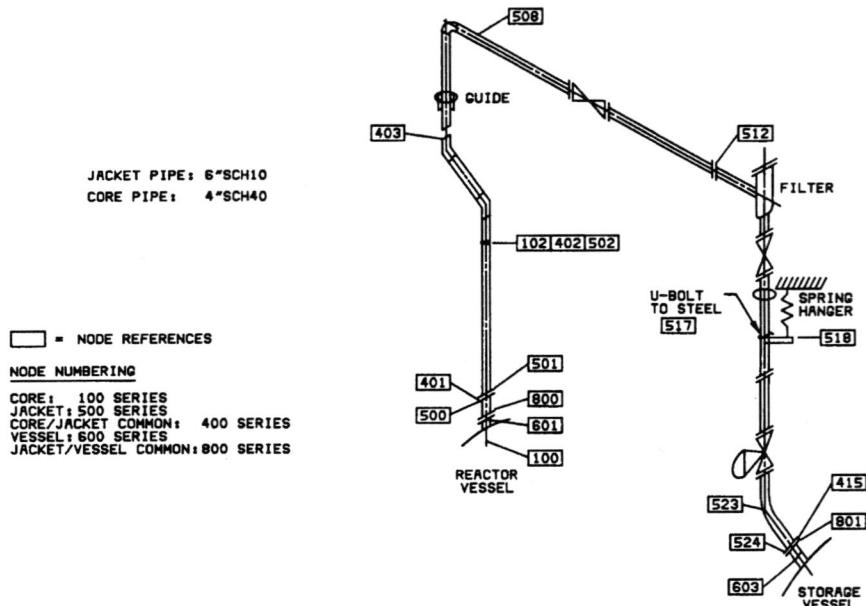

Figure 8.6. Isometric drawing of a representative jacketed process pipe for which a flexibility analysis has been performed, with node points indicated. (Source: Short, W. E. III, "Failure Assessment and Redesign of a Jacketed Piping System, a Practical Approach," 1988 ASME PVP Conference.)[3]

tical movements would be coupled at points of support of the primary pipe. Due to the large number of guides and supports, flexibility analysis models of double-wall piping have quite a few nodes. An example flexibility analysis has been carried out on a typical stainless steel jacketed process piping system, which is illustrated in Figure 8.6.[3] Since jacketed process piping systems are similar in many respects to double containment piping systems, the example jacketed systems is suitable as a means of comparison to a double containment piping system. The example is interesting in that it involved a failure of inadequately designed termination components and their subsequent redesign. (Refer to Figures 11.17, 11.18, 11.22 and 11.23, which are the before and redesigned components.)

Table 8.1 Original Design vs. Actual Operating Conditions for Example Jacketed Piping System of Figure 8.6

	Core pipe			Jacket pipe		
	Pres. psig	Temp. °F	Duration	Pres. psig	Temp. °F	Duration
Original design (steady-state)	150	600	S-S	150	600	S-s
Actual operating (cyclic)	0-35-0 *2 h	270-520-270 *4	*	0-120-0	AMB-570-AMB	S/U &S/D
	0-60-0	270-520-270	8 min.			

Thermal Expansion Considerations

Table 8.2 Selected ASME Code Material Stress Values

Material (304SS)	Stress values, KSI				
	Ultimate [1] 100°F S_{uts}	Yield [1] @ 100°F S_{ymin}	Yield [2] @ 600°F S_{y600}	Tensile [1] S_a 100°F	S_a 600°F
Flange SA-182-F304	75.0	30.0	18.3	20.0	16.4
Pipe SA-312-TP304	75.0	30.0	18.3	20.0	16.4
Slate SA-240-TP304	75.0	30.0	18.3	20.0	16.4

Notes: [1] from the ANSI/ASME B31.3 Code, 1984 Edition; [2] from the ASME BPV Code, Sec. VIII, Div. 2, 1986 Edition.

Based on data in Table 8.1, approximate stress levels have been calculated at certain selected node points from those shown in Figure 8.6 for four different design cases, as listed in Table 8.3. The stress levels have been calculated using the ANSI/ASME B31.3 Piping Code, equations of which are included in Chapter 5 of this text. Computed stress ranges were determined by use of the DYNAFLEX[4] mainframe version pipe stress software.

Selected ANSI/ASME B31.3 Code material stress values for materials involved are listed in Table 8.2. Based on these values, allowable stresses for piping and jacket are computed due to thermal expansion to be 30,590 and 30,028 psi, respectively, by application of the least conservative B31.3 Code formula, which is Eq. (5.2).

In many jacketed process and double containment piping systems, applications involve cyclic conditions (including both pressure and thermal cycles); thus, cyclic effects must be taken into consideration. In the example presented, though conservative to use for piping flexibility analysis, the fatigue cycles in the ASME Boiler and Pressure Vessel Code, Section VIII, Division 2, Appendix 5 were used as a basis.

Table 8.3 Approximate Stress Levels at Points of Interest for Various Conditions Modeled

Model	Description	Stress (PSI) At			
		Closure failure (Node 403)	U-Bolt restr. (Node 517)	Spring hgr. (Node 518)	45'El above RX (Node 523)
A	With spring hanger & U-bolt restr. 520°F core, 570°F jkt, 850°F Stg. Vessel inl.	64,000 to >S_{uts}	36,800	33,300	>>S_{uts}
B	With spring hanger but remove U-bolt, 520°F core, 570°F jkt, 650°F Stg. Vessel inl.	64,000 to >S_{uts}	—	4,900	21,000
C	As Model A, except 520°F core & RX out., 570°F jkt. & Stg. Vessel inl.	25,000 to 33,8000	>S_{uts}	>S_{uts}	>>S_{uts}
D	As Model B, except 520°F core & RX out., 570°F jkt. & Stg. Vessel inl.	24,800 to 33,8000	—	4,400	18,000

Double containment piping systems typically have a larger number of unique and complex "unlisted" components as compared to single-wall piping systems. Assembled components in some cases must be analyzed separately (e.g., as in an elbow contained within an outer elbow that is "freely floating"and designed with sufficient clearance to prevent contact). However, it is common for some double containment components to consist of permanently interconnected primary and secondary containment components, fixed via welded (bonded) interstitial members. Permanently interconnected component assemblies are unique as compared to the single-wall components that make up the parts; thus, they should be modeled as one unique assembly.

Any components that have interconnecting aspects, or that come in contact with other components, must be modeled differently than single-wall components. The most accurate way to model such parts is to model them as acting concurrently upon each other, and to determine the magnitudes of localized discontinuity stress by the use of the finite element method. An overview of finite element methods as applied to double containment piping system analysis is presented in Section 8.3.3.

8.3.3 Finite Element Analysis (Comprehensive Analysis) of Double Containment Piping Components

Conducting a finite element analysis of piping systems involves the use of the finite element method to study the stresses in individual components or complete pipe spools. The finite element method is a numerical analysis technique that is used to obtain approximate solutions to complex engineering problems such as the local stresses in a piping component. The finite element method models the solution region of a component as being built up of many small interconnected subregions or elements. By modeling a piping component in this fashion, an approximation to the governing equations can be found. The solution region can therefore be analytically modeled or approximated by replacing it with an arrangement of discrete elements. Unknown field variables may then expressed in terms of assumed approximating functions within each element, effectively reducing the problem to one of a finite number of unknowns.

Approximating functions are defined in terms of values of the field variables at specified points called nodes or nodal points. Nodal points most often lie on the element boundaries where adjacent elements connect. Each element may have interior nodes as well. The nodal values of the field variables become the new unknowns of the finite element model.

Once the nodal values are found, interpolation functions can define field variables throughout the assemblage of elements. The nature of the solution and the degree of approximation depend on the number of elements and nodes and their size. However, they also depend on the interpolation functions selected, after which matrix equations can be selected that express the entire solution region or system. The resulting simultaneous equations may then be solved rapidly via computer program. If the equations are nonlinear, the solution is more difficult, but can still be solved readily by computer calculation. From the computations of displacements of piping components, oftentimes other values are computed, such as the amount of strain, shear stress, and other variables.

8.4 Stresses in Interconnecting Components

8.4.1 General Nature of Interconnecting Components

An interconnecting component in a double containment piping system includes any component where the primary and secondary containment component comes in contact with each other, causing the inner and outer systems to become restrained at one or more points. Systems may either be physically interlocked at that point (e.g., by an internal-anchoring arrangement), or each may be simply restrained from movement due to the layout of the system and resulting contact that occurs between inner and outer components. In any case, an interconnecting component will become an area where stresses may become concentrated. The concentrated strains that result may be in excess of what a component may be able to withstand. Thus, interconnecting components may be subject to premature failure in a system that is not laid out with adequate flexibility.

Internal anchors may be subject to strain concentration or elastic follow-up. Therefore, some critical attention to the nature of the loads is appropriate. Thermal expansion induced stresses are often permitted to exceed yield stress based on the assumption that deformation, either elastic or inelastic, will relieve the load. Consider, however, a long straight run of pipe with anchors on each end. When the pipe is heated, compressive stresses develop so the elastic compressive strain offsets the thermal expansion strain. The pipe is like a long, compressive spring. If an anchor yields, the pipe can continue to push it until it fails the component by overstrain. The axial deformation can be extreme relative to the anchor while being insignificant relative to the long pipe.

Therefore, it is important to treat thermal expansion loads on rigid internal anchors as primary loads. General membrane and bending stresses in the anchors should be limited to primary stress limits, generally less than the yield stress. In an FRP anchor that is bonded between primary and secondary containment pipes, the strength of the anchor depends to a large extent on the bond strength. This type of bond should never be allowed to experience a stress imposed by the thermal end load of the pipe that exceeds the shear strength of the bond.

8.4.2 Types of Interconnecting Components

There are many types of interconnecting components to be considered in a double containment piping system. These include such components as: (1) pipes that are extruded in one piece with longitudinal ribs; (2) fittings that are molded, cast, forged, formed, or machined from a homogeneous material; (3) primary and secondary containment components that are permanently affixed together via welded (bonded) supports, collars (e.g., see Figures 7.10, 7.16 and 8.12), branch connections or instrument taps (e.g., see Figures 8.7, 8.8 and 11.39) or other members, (e.g., fittings and pipes prepared for simultaneous fusion (see Section 12.3.3)), also, RTRP systems that use tightly fitting adhesively bonded collars at the end of straight sections (see Figures 8.1 and 8.12); (4) the contact points created where primary and secondary components come in contact with each other due to differential thermal movements; (5) internal-anchor arrangements; (6) external-anchor arrangements, and; (7) termination and initiation arrangements.

Guided supports of the collar or spider style may also be broadly classified as an interconnecting component of the #4 type described in the previous paragraph, due to the contact points they create. Forces may be transmitted from each system to each other through interstitial supports of these types. Forces due to thermal stresses at the contact points of internal supports tend to be low, since collar or spider style supports that are attached to primary pipes allow the primary pipe to move longitudinally, radially, and torsionally, to a limited extent resulting in some compliance and lowering of stresses. However, forces due to earth loads in buried pipes may cause high discontinuity stresses at the locations where these types of supports are installed by ovalizing the outer pipe sufficiently to bring it in contact with the internal support.

8.4.3 Discontinuity Stresses in Interconnecting Components

It is emphasized throughout this text that it is important to understand the nature and magnitude of concentrated localized (discontinuity) stresses that may be present in interconnecting double containment components. To understand the effects of these stresses fully, it is recommended to model these types of parts using finite element analysis techniques (see Section 8.3.3), or by conducting experimental analysis, as outlined in the ASME BPV Code, Section VIII, Div. 1, Appendix 2. However, one can determine a rough estimate of the discontinuity stresses at jacket/ring intersections of interconnecting components such as termination flanges, closure rings and internal anchors (see Figure 8.9) by using the following simplified equations[5]:

$$M = \frac{1.7 P_S R_S t_S + 2 E_S (\alpha_p - \alpha_r)(\Delta T_r) t^2}{6.61 + \dfrac{10.288}{(R_p/t^3)^{1/2}} \left[\dfrac{1.3 K^2 + 0.7}{K^2 - 1} \right]} \frac{1}{w} \tag{8.10}$$

where:
M = moment, in-lbs.
P_S = pressure of the secondary pipe, psi
R_S = radius of the secondary pipe, in
t = thickness of the secondary pipe, in
E_S = modulus of elasticity of the secondary pipe, psi
α_p = coefficient of expansion of the primary pipe, in/in °F^{-1}
α_r = coefficient of expansion of the material of the interconnecting ring
   section of the interconnecting component, in/in °F^{-1}
ΔT_r = temperature difference between the average interconnecting ring
   temperature and the outside, °F
R_p = radius of the primary pipe, in
K = ratio of the interconnecting ring radius to the radius of the secondary pipe
W = width of the interconnecting ring, in, (Note: where a termination
   flange is used in a jacketed pipe, the width should be taken as the
   combined width of the two flanges, see Figure 8.9)

$$V = 2 \lambda M \tag{8.11}$$

where:
V = shear, lbs
$\lambda = 1.285/\sqrt{R_S t}$

Thermal Expansion Considerations

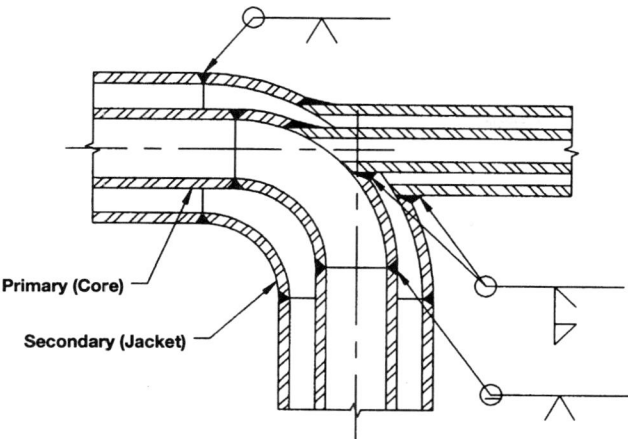

Figure 8.7. Illustration of a piping stab-in on the heel of the primary and secondary elbow as typically used in jacketed piping systems. (Source: Controls Southeast.)

$$\sigma_{1C}(longitudinal) = \pm \frac{6M}{t^2} \tag{8.12}$$

$$\sigma_{2C}(circumferential) = \frac{PR_S}{t} - \frac{2V}{t}\lambda R_S \pm \frac{2M}{t}\lambda^2 R \tag{8.13}$$

Using the + 2M/t value in 8.13 provides the stress for the outside circumference, and using the - 2M/t value provides the stress for the inside circumference of the interconnecting ring. An equation for radial stress is not provided, however, it must be determined if

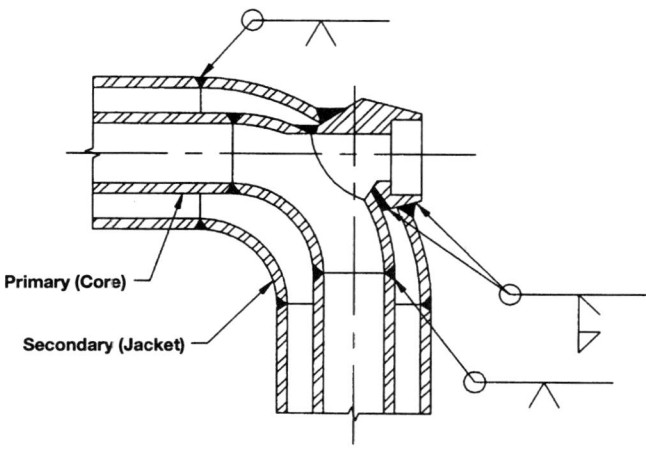

Figure 8.8. Illustration of a socket weld Elbolett® on a primary elbow. (Source: Controls Southeast.)

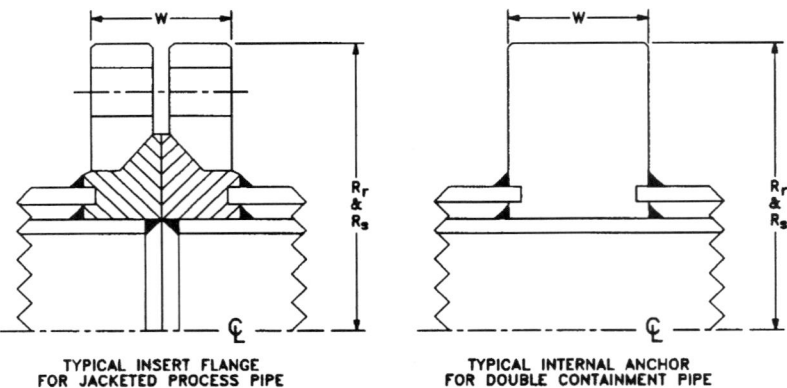

Figure 8.9. Examples of interconnecting components illustrating the terms used in Equation 8.10, shown on left is mating termination insert flanges and on the right is a welded internal anchor. (Source: Adapted from "How to Stress Analyze Jacketed Piping", Getz, Richard, Hydrocarbon Processing, 1978.)

there is a significant temperature gradient across the ring portion, as would be the case in a thermoplastic material. The stresses determined by using Eqs. (8.12) and (8.13) can then be evaluated along with other concurrent stresses to determine an estimate of the combined stresses at the interconnecting juncture. Such stresses include the discontinuity stresses from Eqs. (8.12) and (8.13), radial stresses due to a non-uniform temperature, axial column loading due to differential expansion and/or contraction, pressure stresses, dead weight bending and stresses imposed from thermal external bending loads.

It should be kept in mind that these equations represent a simplified approach. To be accurate, the designer must either model the component using the finite element method, or perform experimental stress analysis.

An example comprehensive finite element analysis was performed on the thermoplastic multi-part internal anchor shown in Figure 11.50. This fitting is classified as an interconnecting component described by #5 of Section 8.4.2, and is further classified as a Type 2 internal anchor according to Section 7.4.7. This form of internal anchor, using the relative geometries of the internal bulkhead and bulkhead housings have been proven by finite element analysis to be a highly efficient design as an anchor. This can be concluded due to the low accumulation of stresses in the internal bulkhead, as compared to many other alternative Type 1 designs. In fact, the stresses which result from an imposed longitudinal (axial) stress of 1,000 psi tend to be lower than that of the corresponding adjacent pipe. What is of further interest in the conclusion of the study is the relation of the calculated stress to the relative modulus of elasticity (Young's Modulus) of the differing primary and secondary containment materials, this is illustrated in Figure 8.10. The graph shows that the calculated stress depends greatly upon whether the as-installed modulus is used, or the modulus at system temperature is used. From the graph is is shown that the lower the value of the modulus of the primary pipe, the higher the calculated stress will be. The same also holds true for the containment pipe. This information can be used for this particular fitting design for a wide range of combinations of thermoplastic materials.

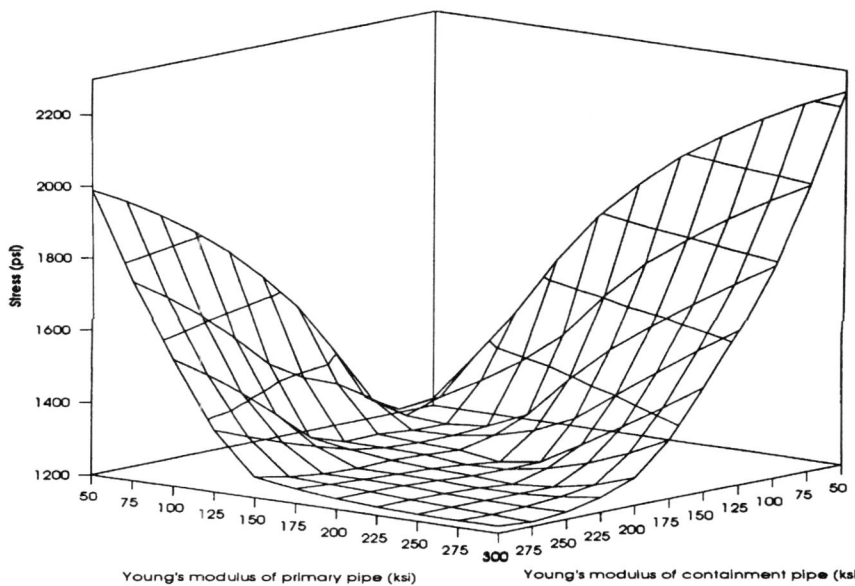

Figure 8.10. Maximum Von Mises stress in a 6" PVDF by 10" HDPE anchor fitting (ref. Figure 11.51), as a function of variations of Young's Modulus in the internal bulkhead and the anchor housing. Loading is based on 1,000 psi axial stress on the primary piping in all cases. (Source: Becht Engineering.)[6]

8.5 Totally Restrained Double Containment Piping Systems

A system designed in a totally restrained manner must have straight sections of piping provided with sufficient lateral guiding such that the pipe will not excessively buckle at any one point. In double containment piping systems, primary pipes may be prevented from experiencing buckling-type failure by use of interstitial supports and/or internal and external anchors. If the piping materials possess substantial elasticity, the frequency of lateral guides or anchors required to prevent failure can be determined using simplified methods involving structural beam-theory methods.

For materials that do not demonstrate substantial elastic behavior, a more comprehensive approach must be used, taking into account the nonlinear behavior of the material under prolonged elevated temperatures and/or cyclic loading, and the complex, nonlinear, time, and temperature-dependent behavior of the material under consideration. Some inelastic materials (especially RTRPs) also possess physical properties that differ depending on the direction (orientation) being considered. Thus, they can be described as having anisotropic properties. These factors make it difficult to determine accurately whether or not RTRP materials, and other nonmetallic materials, will fail when totally restrained under a given set of conditions. If elastic behavior is met, the ability of the pipes to withstand failure due to the combined effects of bending, torsion, and axial loads can be determined using the simplified methods described in Chapter 5, and those for determining axial loads in Section 8.2. The calculation for estimation of the total displacement stress SE is given in Eq. (5.27) (Section 5.5.1.6.1). This value does not take into account axial stresses, which should be added to the bending stress, as shown in Eq. 8.7. The resultant computed value, including the axial stress, may be compared to the allowable stress of the material being considered, as determined by Eqs. (5.2) or (5.3). For nonmetallic materials, a comprehensive analysis may be required as described in Section 8.3.3. However, it is the responsibility of the designer to prove the validity of the analytical method used. The effects of buckling, and also localized discontinuity stresses, must be taken into account. The maximum distance between interstitial guides (centralizing devices) sets a limiting condition in order to prevent primary pipes from failure due to buckling between the guides. The maximum distance between interstitial guides can be determined by use of the equation for buckling of a column with conditions of pinned ends (both ends). Equation (8.14), which is the pipe form of the equation, can be used to calculate this maximum distance, L, between lateral guides (centralizing devices or interstitial supports). The value for L that is found from this equation will establish the maximum distance that centralizing devices (interstitial supports) can be spaced, in order to prevent a buckling-type failure. This value should always be compared to the calculated spacing for vertical support of the pipe due to dead weight based on an acceptable amount of sag [refer to Eqs. (9.29) through (9.32)] and the minimum of the two values used. Figure 8.11 illustrates some of the terms of the equation.

$$L = 0.2617 \left(\frac{I}{\alpha A \, \Delta T} \right)^{1/2} \tag{8.14}$$

where:
L = maximum distance between consecutive supports, beyond which buckling failure may be risked, in.
I = moment of inertia, in.^{4}
A = cross sectional area, in.^{2}

The less the distance between supports, the greater the amount of support (and guiding) provided to the system. However, there are limitations as to how frequently centralizing devices (interstitial supports) may be spaced; the reasons have to do with both economics (the cost of adding supports) and thermal expansion considerations. Too many centralizing devices will provide a greater surface area of contact, and due to the weight of the piping system will have some effect on reducing free allowable movement. Providing additional supports will also result in a higher cost for the system. Thus, the optimum number of supports to be used in an unrestrained system is the minimum number that will both provide adequate support in the primary pipe and prevent its buckling.

In double containment piping design, interstitial supports will function as lateral guides if they are attached to the primary pipe only, and allow it to move freely in the longitudinal direction. Supports that are permanently welded (bonded) to both pipes function as a point of restraint (anchor), instead of as a guide. If interstitial supports are designed to act as axial guides, and allow longitudinal movements of a primary pipe, Eq. (8.14) can be used to calculate the maximum span between them to prevent excessive buckling.

It should be remembered that interstitial supports must always allow for some tolerance in their radial direction. This is to allow outer piping to be inserted over the supports, or to allow primary piping with attached supports to be inserted into its associated outer pipe.

Thermal Expansion Considerations

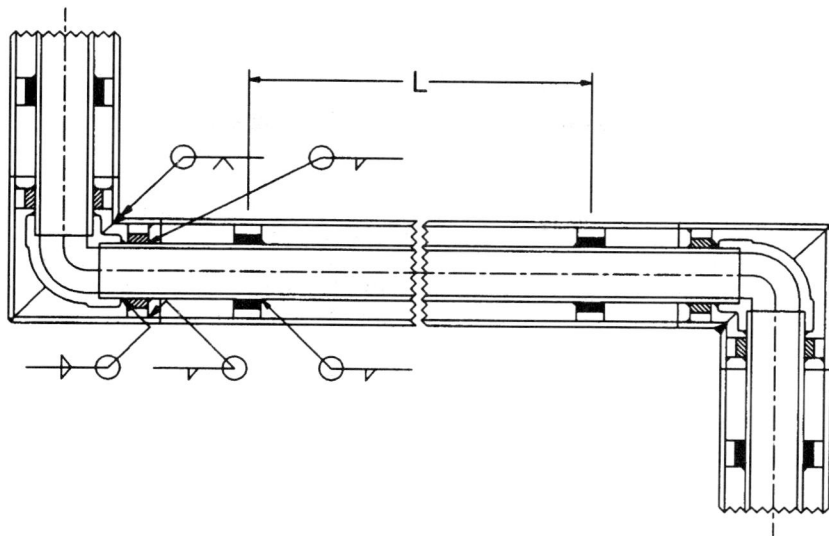

Figure 8.11. Illustration of a totally restrained section of straight double containment metallic piping. [L indicates the distance between internal axial guide supports used in Eq. (8.14).]

This tolerance usually would be considered insignificant in any analysis, as it will typically allow only a limited amount of bending and torsion to take place which is insignificant.

8.5.1 Elastic-Plastic Follow-up in Interconnecting Components

Thermal expansion stresses in piping are permitted to exceed yield stress because they are assumed to be deformation controlled. That is, it is assumed that the maximum strain that will occur at any point in the system will not exceed the elasticity calculated strain due to thermal expansion. Plastic strain is assumed to relieve the thermal expansion. However, in cases with elastic follow-up, local elastic-plastic-creep strain caused by thermal expansion can substantially exceed the elastically calculated strain. Interconnecting anchors in a fully restrained system is a case of elastic follow-up. In this case, yielding and deformation of the anchor will not result in proportionate reduction in the thermal expansion load and stress, since the piping acts as a long spring with stored thermal expansion, continuing to push the anchor. The piping may continue to drive the anchor as it yields, until the thermal expansion of the piping occurs (i.e., is no longer restrained) and deforms the anchor. In cases such as these, the thermal expansion loads must be treated as primary loads. Stresses are then generally limited to less than yield stress.

8.5.2 Fully Restrained Nonmetallic Systems

Primary components in a double containment piping system may be considered as being completely restrained if they are fixed at the ends of straight sections. Two common examples of this include simultaneously welded thermoplastic systems and RTRP systems that include adhesively bonded annular collars at the end of straight sections of primary pipes. Both of these methods are described in Chapter 12. They have not been in use long enough or received sufficient study to understand fully the long-term actual performance of completed assemblies that use these methods. Both methods of joining have been used as a means of reducing overall installation costs in double containment piping systems.

The two types of systems described are examples of fully restrained plastic systems. There are many potential drawbacks to these types of systems. Plastic systems that are restrained from allowing free expansion/contraction to take place are subject to creep-strain accumulation and plastic-elastic followup, which may result in a premature failure of such a system. Also, secondary containment piping in some of these systems are restricted to some extent from free diametrical deflection; this is an important issue with respect to burial capability. Diametric deflection of flexible pipes are discussed in Section 9.4.1.4. Totally restrained systems that make use of adequately designed internal anchors are better overall in terms of assuring a safe design that can be analyzed.

Simultaneously welded systems do not meet the inspection, examination, and testing criteria of codes such as the ASME B31.3 or B31.1 Codes, where these codes apply. It covers most double containment piping system applications other than certain underground chemical sewers and fuel applications in the United States and Canada (see Section 13.3.3).

To use a plastic double containment piping system that features a fully restrained primary pipe system, a comprehensive stress analysis should be performed. Sometimes, a case history that involves comparable conditions can act as a qualifying means to allow a system to be used, without any comprehensive analysis. When this is true, the case history should be further supplemented by some form of analysis. Whether analysis alone is used, or analysis in conjunction with case histories, the qualification means must be shown to be valid for the given material under the specified design conditions. A system can also be judged to be adequate by comparison to a previously analyzed system.

If pressures and temperatures are low enough, a restrained-plastic system can be acceptable. In such a system, the combined stresses that would accumulate in the pipe must be less than the material's allowable stress rating, based on long-term expectations. For this to be true the internal pressure and temperatures must not be excessive, temperature variations have to be within safe limits, and cyclic effects must additionally be considered. The magnitude of internal pressure stress in the system is highly important, as it is considered a primary stress, although it is the thermal stress that tends to be of a much larger magnitude. The thermal stress is often considered a secondary stress where materials that can yield are being used. It is necessary in all cases to consider the combined effects of pressure and temperature, in addition to all other concurrent forces. Since there has been insufficient long-term service histories for restrained-type plastic double containment piping systems, a comprehensive analysis must be performed whenever such a design is to be considered. One should also not forget to consider the effects that stress may have in terms of enhanced chemical attack. A typical fully restrained section between consecutive elbow components using RTRP materials is illustrated in Figure 8.12.

Fully restrained systems are perhaps most common in RTRP systems that are based on machine-made (commodity) pipe and fittings. The major reason for the tendency to use this type of design is that compression-molded socket fittings, which are made to allow for joining to adjacent pipe sections by adhesive bonding techniques, are highly rigid structures, whose physical properties often are substantially below that of the exceptionally strong centrifugally cast or filament-wound pipe. The fittings themselves often can withstand some bending loads that may be imposed on them by the accompanying pipe.

Thermal Expansion Considerations

The fittings in effect are rigid structures and experience little to no flexure when subject to bending loads. However, in systems designed using traditional flexible layout methods, failure is actually most likely to occur in the joints (adhesive bonds) between the fittings and the pipes and not in the fittings themselves. In systems with a large amount of ΔT (e.g., 100°F and higher), the fittings themselves can also fail if they are inadequately designed to handle the resultant loads, or if the system is designed with loads in excess of the fitting's capability. Therefore, restrained RTRP systems can provide a viable alternative, if designed and installed correctly. Attention in restrained RTRP double containment piping systems should be given to the strength of adhesive bonds, internal liner strength, and the fitting strength, based on the material selected and the laminate construction.

Therefore, in both single-walled RTRP systems, and in double containment RTRP systems, the better way to design the system is often based on anchoring the system at the ends of every straight run (as shown in Figure 8.12), thereby isolating the fittings from expanding or contacting pipes, and by further making sure that an adequate supporting scheme is used to prevent buckling and point loading. This is particularly true for systems which use socket based "machine-made" components.

By contrast, some example combinations of restrained thermoplastic provide some interesting insight as to the behavior of such systems. In the first example, a flexible material, HDPE, is contained within HDPE using SDR 11 pipes. Inner and outer pipes are fully restrained to each other at the end of every straight run using a Type 2 internal anchor (shown in Chapter 11, Figure 11.51). The system is fully buried, and in such a system the containment pipe may also be anchored by the surrounding soil. The primary pipe is restrained by means of the internal anchors at the ends the straight runs.

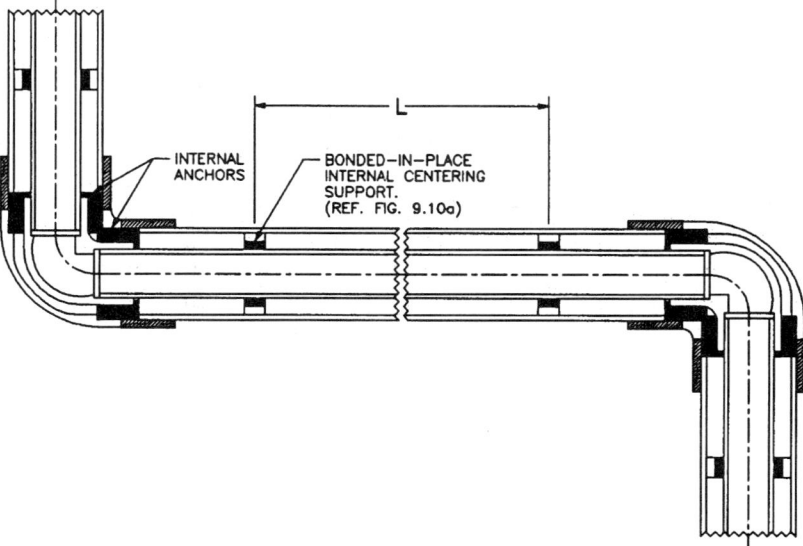

Figure 8.12. Illustration of a totally restrained section of RTRP pipe that is fully restrained at the end of each straight length of pipe. (Source: Fibercast Co., U.S. Patent #4,886,305.)

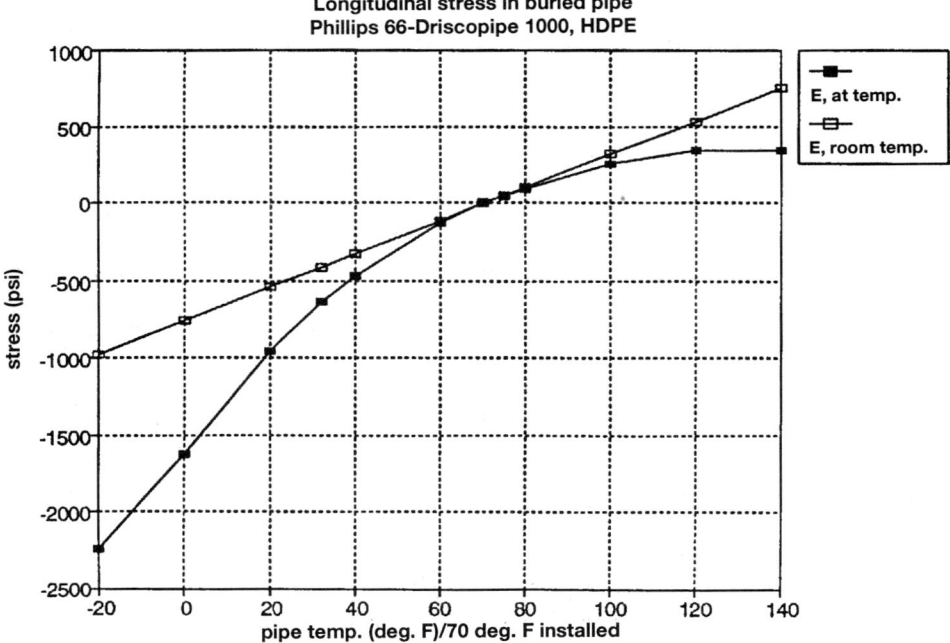

Figure 8.13. Longitudinal stress due to restrained thermal expansion in the primary pipe of a buried restrained double containment piping system which uses HDPE materials. (Source: Becht Engineering.)[7]

Restrained thermal expansion creates compressive longitudinal stress in the primary pipe. The amount of accumulated longitudinal stress in such a restrained straight pipe section is in direct proportion to the coefficient of thermal expansion, the temperature differential, and the modulus of elasticity. The results of this example are shown in Figure 8.13, which also shows the effect of room temperature versus elastic modulus.

The second example illustrates the effect of using a restrained design in an aboveground system. The example chosen is based on the use of a 6" SDR 11 PVDF material inside of a more flexible SDR 11 HDPE (using a 10" diameter). In aboveground double containment piping systems, the containment pipe is usually not fully restrained, as was the case in the example considered. In such a system thermal expansion of the primary pipe stretches the containment pipe in tension and causes itself to be under compression. Coefficient of thermal expansion, temperature differential and pipe stiffness of both primary and containment pipes influence the amount of longitudinal stress in either pipe. Figure 8.14 illustrates the variations of the primary pipe longitudinal stress for the PVDF/HDPE combination in this aboveground example.

8.6 Methods for Alleviation of Expansion and Contraction

All of the methods that are applied to single-wall pipes to alleviate thermal expansion/contraction and distribute stresses in a uniform manner, also apply in principle to double containment pipe assemblies. However, the manner in which they are physically applied, and

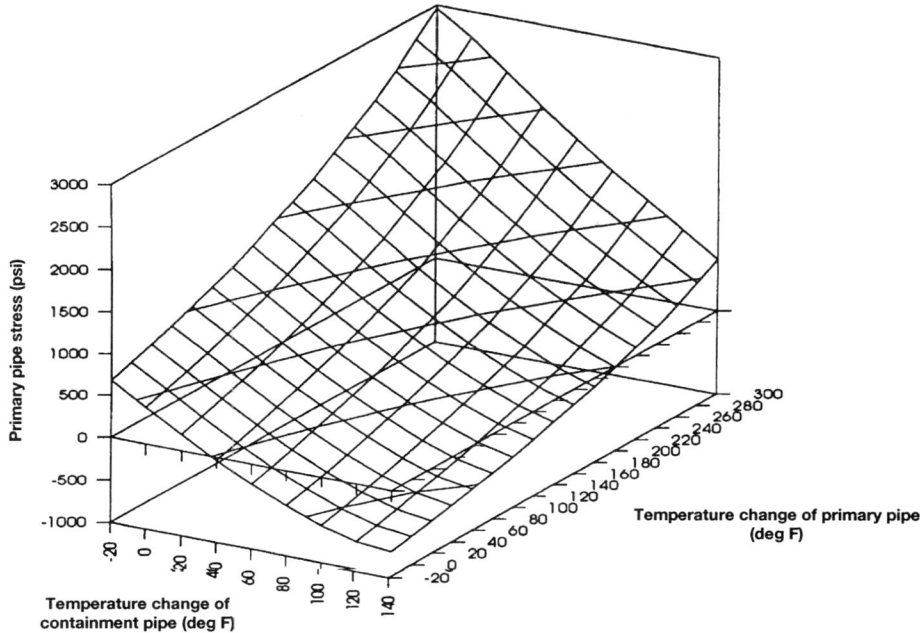

Figure 8.14. Primary pipe longitudinal stress due to restrained thermal expansion in an aboveground double containment piping system using a 6" φ, 0.602" thick wall PVDF (SDR 11) primary pipe and a 10"φ, 1" thick wall HDPE (SDR 11) secondary containment pipe. (Source: Becht Engineering.)[8]

the fabrication/layout techniques used, differ considerably. Also, not all techniques may be used in comparable situations.

This section describes how ordinary methods may be physically applied to double containment pipe assemblies. Section 8.8, "Classification of Systems and Application of Compensation Techniques," discusses where these different compensation alternatives may be used to compensate expansion/contraction of inner and/or outer pipe systems, based on the nature of the design condition.

8.6.1 Expansion Joints in Double Containment Systems

Expansion joints may be used in principle on both primary and secondary containment piping systems. They may be used in both primary and secondary containment portions of nonrestrained systems. However, there are many limitations to their use. If expansion joints are to be used as part of a primary piping system, ready access to the expansion joint must also be included in the design. This means provision of a tank or access device as a component of the secondary containment system. In underground systems, the use of expansion joints for compensation of primary pipes is limited to manholes or where a trench is used as a means of secondary containment. On the other hand, expansion joints may be readily applied to secondary containment pipes in above-ground, nonrestrained systems, to compensate for differential thermal expansion. This is true regardless of which pipe is experiencing the larger temperature change.

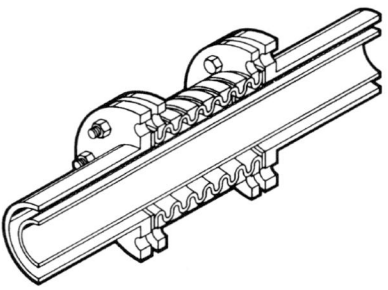

Figure 8.15. Illustration of a standard expansion joint applied to a secondary containment piping system as a means to accommodate expansion/contraction differential between the primary and secondary containment pipes. (Source: Asahi/America.)

An example of a standard expansion joint applied to a secondary containment pipe is illustrated in Figure 8.15. A more efficient method, which eliminates the need for expensive secondary containment flanges and takes up less space, is the internally guided, "simple" expansion joint using sleeve couplings, shown in Figures 11.55 and 11.56.

Expansion joints are not practical where nondifferential thermal expansion/contraction exist. In such conditions, they would be required on both inner and outer pipes, along with internal anchors. Also, where primary and secondary containment pipes are permanently welded together to result in a permanently restrained system, expansion joints will not compensate expansion/contraction, nor will they prevent buildup of concentrated localized stresses in components, unless they are provided between each pair of anchors.

Flexible-style grooved mechanical joints in the secondary containment system can also serve as the functional equivalent of expansion joints. Each joint will allow for a certain limited amount of expansion and contraction movement, with the exact amount depending on the specific style and size of the coupling being considered. An example of the use of typical grooved mechanical couplings is illustrated in Figure 8.16. It should be pointed out that while grooved mechanical couplings will serve effectively in this manner, they do have to be supported to avoid overbending the joint area with resulting failure of the joint. Some credit due to the added rigidity provided through the use of tightly fitting internal centering supports may reduce the overall need for support, although this is not advised, except when using internal centering supports specifically designed for this purpose. It is recommended to contact the manufacturer for specific recommendations regarding supporting of each proprietary brand and type of grooved mechanical coupling for use in double containment piping applications.

When designing a system having expansion joints, special attention needs to be paid to the activation forces required of the joint. These must be compared to the activation forces that the piping will produce, and the two numbers then compared to determine if the design is a feasible one. The allowable activation force for expansion joints depends on the thermal forces (end loads) developed by piping to be compensated, as well as the support spacing that is provided to the pipe. When using expansion joints, internal and external supports must be selected based on their ability to guide movements of the pipe in the axial direction, to the greatest extent possible. In double containment pipes, interstitial supports can serve as internal guides for primary piping.

Thermal Expansion Considerations

To calculate the maximum allowable activation forces required for expansion joints, Eq. (8.15) may be used. This equation is based upon that used in structural beam theory for the buckling of a column with an end condition of pinned and pinned. To determine the required activation force for an expansion joint, the manufacturer of each specific component must be consulted.

$$F_b = \frac{\pi^2 EI}{L^2} \tag{8.15}$$

where:
F_b = critical buckling force, lb
I = moment of inertia, in.^{4}
L = length of piping segment, in.
E = modulus of elasticity, psi

An expansion joint should be selected and installed such that it can accommodate all movements of the piping, no matter in which direction they occur. A certain amount of preset is often required of the expansion joint in order to accommodate multidirectional movements.

An expansion joint must be anchored properly to protect against failure due to excessive pressure thrusts upon startup. Intermediate anchors must be provided at a point between consecutive expansion joints to direct expansion and contraction in the direction of the expansion joint. Intermediate anchoring may be achieved by the use of one of the methods shown in Sections 11.1.1.4.2 (internal anchoring) and 11.1.1.4.1 (external anchoring).

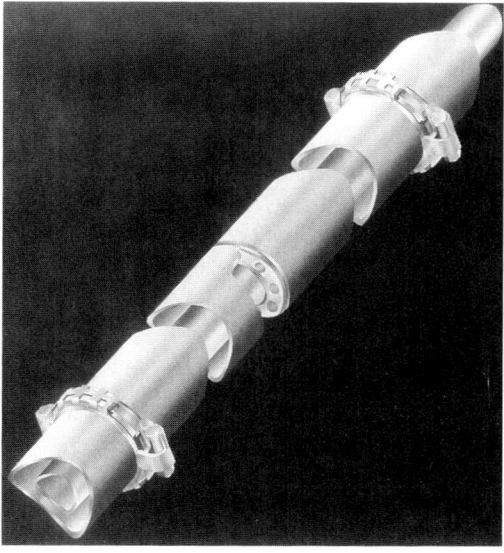

Figure 8.16. Example of the use of flexible-style grooved mechanical couplings as a means of accommodating differential thermal expansion and contraction in double containment piping. In this type of design, the inner and outer piping sections are anchored together in selected locations, thereby enabling the outer jacket to grow and shrink with the primary pipe in an accordionlike manner. (Source: DCS Corp.)

As mentioned previously, appropriate guides should be installed to direct piping movement into an expansion joint (or to assist an expansion joint to expand when the pipe is contracting). The industry minimum standard for guide spacing at the entry to expansion joints involves the first guide being placed a distance that is a multiple of four times the nominal diameter of the pipe and the second guide placed a distance that is a multiple of fourteen times the nominal diameter of the pipe, from the entry point to the expansion joint. This minimum standard was originally stated by Crocker almost 60 years ago. It is still used today, regardless of the piping material used.

Whenever expansion joints are to be used, the standards of the Expansion Joint Manufacturer's Association (EJMA Standards) and the rules stated in Appendix X of the ASME B 31.3 Code for Chemical Plant and Petroleum Refinery Piping (metallic-bellows expansion joints) should be followed. These two sources present useful and important guidelines that help assure a safe working design.

8.6.2 Use of Directional Changes (Elbows) to Accommodate Expansion of Double Containment Pipe Straight Sections

The natural changes of direction (elbows) that exist in any system can be used to accommodate thermal expansion/contraction. Additional changes of direction may also be added to the layout to increase the amount of flexibility inherent in a system. By doing so, stresses will be minimized and the distribution of stresses will be better balanced. Directional changes, like loops and offsets, can be positioned in a horizontal or vertical configuration. Like the two-elbow offset, they also can be rotated at any angle. An elbow used in this fashion is sometimes referred to as "freely floating."

Figure 8.17 illustrates the important aspects of a directional change with respect to expansion/contraction in a double containment piping system.

Two major design considerations are involved when using an elbow to accommodate thermal expansion/contraction. The first involves whether or not an inner elbow will come in contact with its associated outer elbow due to differential movements of the two systems. If they do contact each other, substantial stresses may develop in both primary and secondary containment components, which may lead to a premature "double failure." To determine if this is possible, and thus if sufficient space exists, a dimensional analysis must be conducted. This issue is discussed in more detail in Section 8.6.3.

A second major design issue involves the closest point of lateral restraint from the inner and/or outer elbow. Each elbow that requires movement must have its closest points of lateral restraint, on each side, far enough away so that adequate flexibility is provided to the elbow. The closest point that will provide lateral restraint to a primary pipe elbow is where inner and outer components come in contact with each other, whenever they are allowed to do so. If a system is designed such that inner and outer component contact is prevented through adequate annular space, the interstitial support that is closest to the elbow on each side will become the point of closest lateral restraint in most applications. There are interstitial supports that do not function as point of lateral restraint in horizontally positioned systems. These are referred to as "flexibility" supports; examples of which are provided in Figures 9.14 through 9.16, and in Figure 8.37.

The analysis of above-ground secondary containment elbows may be treated similarly to the way that above-ground single-wall-piping elbows are. However, the added rigidity that

Thermal Expansion Considerations

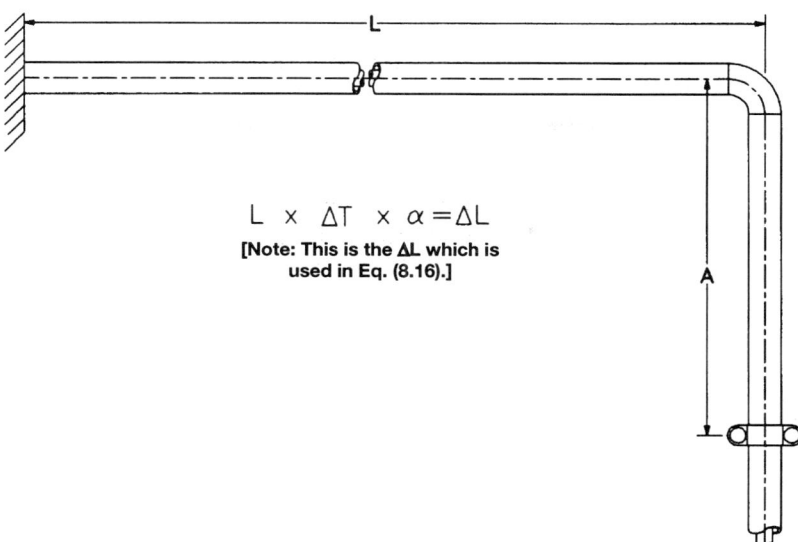

$$L \times \Delta T \times \alpha = \Delta L$$
[Note: This is the ΔL which is used in Eq. (8.16).]

Figure 8.17. Important aspects of a directional change assembly with respect to expansion/contraction in a double containment piping system.

interstitial supports provide, in terms of adding resistance to bending, must be included in any design. If a system is permanently and rigidly interconnected with frequency (i.e., a simultaneously fused system or a restrained-fitting RTRP system), there will be greatly added resistance to bending for the secondary containment elbows. Underground, direct-buried piping is permanently restrained from moving in most sizes (see Section 8.6). Therefore, layout options for the use of secondary containment elbows where they are directly buried do not readily exist.

Figures 8.18 and 8.19 illustrate a horizontally positioned double containment directional change assembly designed to accommodate a specified amount of differential thermal expansion, the amount of which will vary by material type and joining method, and without having to increase the diameter.

The directional change subassemblies in Figures 8.18 and 8.19 combine certain elements (e.g., internal flexibility supports, elbows with the primary elbow having the greater radius, internal anchors, and others) into the design of a patent pending horizontally positioned directional change subassemblies that allows a system to achieve the lowest stressed state (LSS) in systems where the primary pipe experiences the greater magnitude of expansion or contraction. These directional change assemblies, along with expansion loops and offsets designed in similar fashion can be essential parts of a completed internally flexible system designed to achieve minimal stress in a system. Figures 8.18 and 8.19 illustrates the basic combination of elements in a directional change assembly which allow an elbow subassembly to be designed in the "least-stressed-state," or "LSS," when the primary pipe is subjected to thermal expansion and/or contraction.

Such directional change assemblies are designed by implementing the following procedures. In an internally-flexible system, the minimum cost for the system will always be

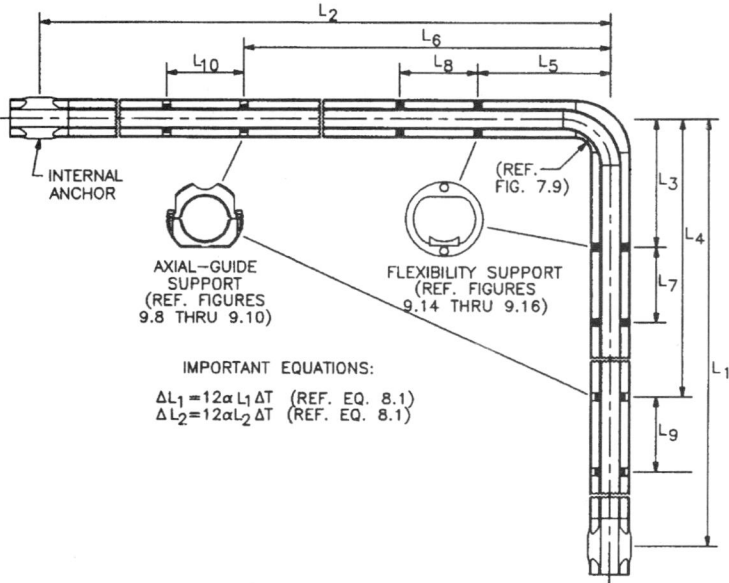

Figure 8.18. A section of a double containment system incorporating "internal flexibility" by means of specialized elements that maximize flexibility of the primary piping and minimize thermal stresses in the system (patent pending).

achieved by making use of the naturally existing change of directions that are provided for in the desired layout, without having locally or globally to increase the outside diameter of the secondary containment system beyond the minimum jacket size required for the system as described in Chapter 3, or add any expansion loops. Therefore, this is a starting goal for any system whereby the goal is to make use of the flexibility that changes of direction may provide.

In the first step of the patent-pending method for designing Figure 8.18-type assemblies, the locations of internal anchors are selected; an intermediate location between consecutive changes in direction is selected. However, intermediate internal anchors need not be provided if the straight lengths between consecutive elbows is relatively short in comparison to other straight lengths in the system. In fact, intermediate anchors should be avoided in certain locations where it may create an overly stresses short offset off of a long straight length.

Next, calculate the change of dimension due to linear thermal expansion/contraction from each intermediate anchor to an elbow, from each direction, L_1 and L_2. The amounts calculated are indicated as ΔL_1 and ΔL_2, based on using Eq. (8.1). Next select an appropriate elbow style (refer to Section 7.4.3) that provides room for growth in the desired direction. The combinations to consider include designs where the inner and outer elbow radii are equal (Figure 7.8), and combinations where the inner radius is less than that of (Figure 7.7) or greater than that of (Figure 7.9) the corresponding outer elbow. A dimensional analysis must then be conducted based on the elbow selected, and the determined values of ΔL_1 and ΔL_2. Procedures for conducting a dimensional analysis for elbows is covered in Section 8.6.3. Examples of elbow dimensional analysis are provided in Figures 8.23 through 8.25.

Thermal Expansion Considerations

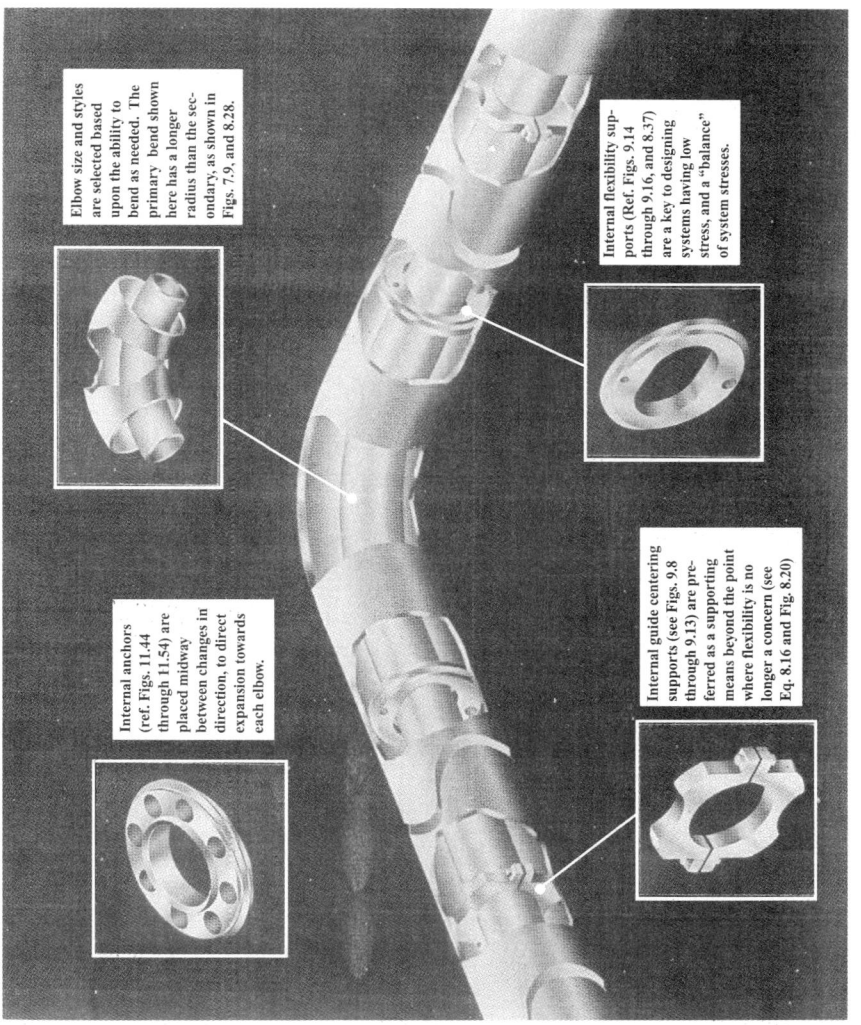

Figure 8.19. A section of a double containment system incorporating "internal flexibility" by means of specialized elements that maximize flexibility of the primary piping and minimize thermal stresses in the system (patent pending). (Source for illustration: DCS Corp.)

STEP 1 — Locate intermediate internal anchor locations.

STEP 2 — Select an elbow type and assembly configuration (e.g., a longer radius primary with short radius secondary), prior to designing an increase in the diameter of the secondary containment elbow.

STEP 3 — Determine ΔL by Eq. (8.1) for each side. The ΔL calculated on one side will be used as the basis to locate supports on the opposite side.

STEP 4 — Determine if the elbow has sufficient space by using the dimensional analysis outlined in Figure 8.22

If insufficient space exists. (loops back to Step 2)

If sufficient space exists.

STEP 5 — Determine the closest point for which lateral restraint can be tolerated on one side (L_6), based on: (1) A specified amount of stress that is acceptable for the elbow, (2) the ΔL determined for the opposite side which is ΔL_1; and (3) using Eq. 8.16.

STEP 6 — Repeat Step 5 to determine L_4 for the opposite side. Remember when using Eq. (8.16), to use the ΔL determined for the opposite side than that under consideration (which is ΔL_2).

Go to Step 7

Thermal Expansion Considerations

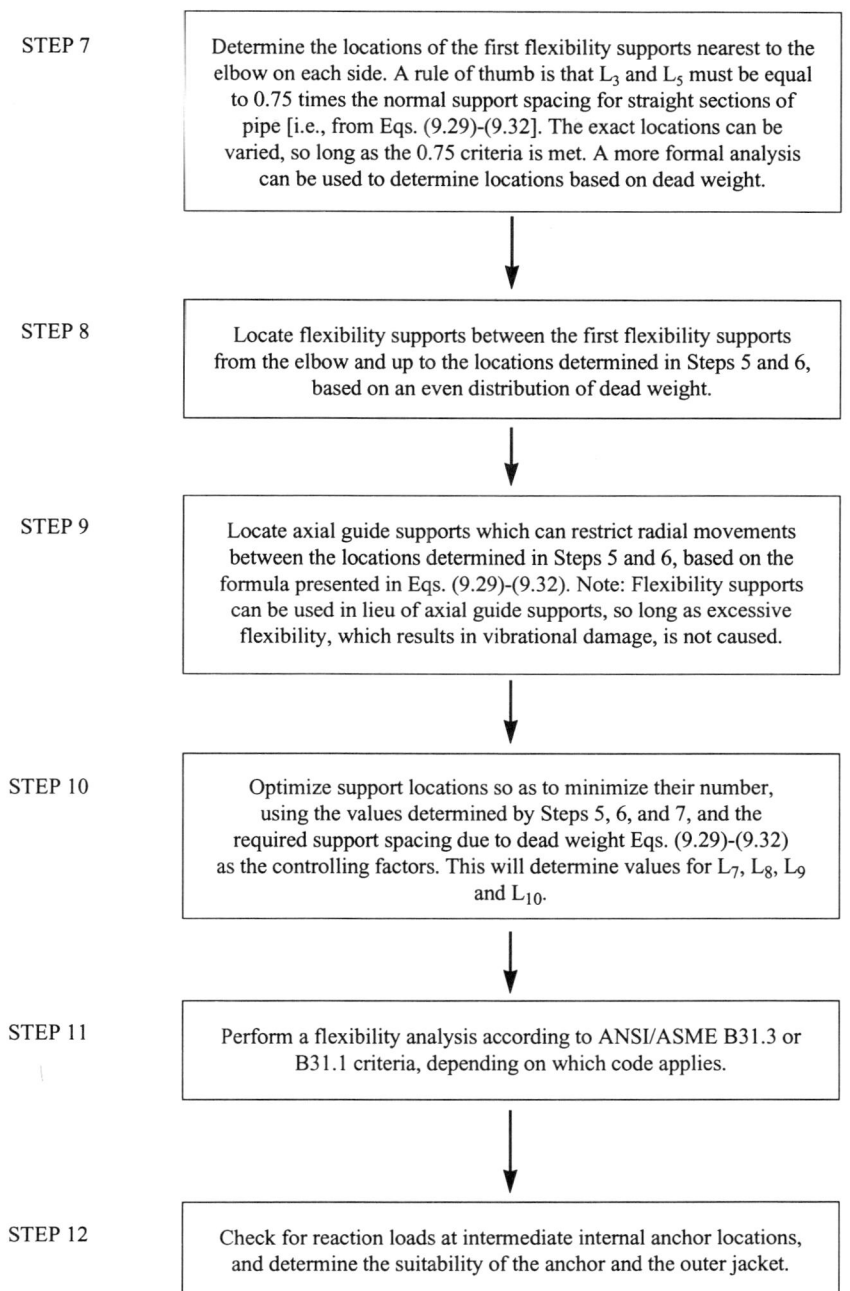

Figure 8.20. Flow chart for designing changes of direction so as to minimize stresses of the primary system in nonrestrained systems undergoing differential thermal expansion (patent pending).

The elbow types represented by Figures 7.8 and 7.9, 8.18, and 8.19 allow for additional room toward the outside portion of an elbow assembly. These types are best for systems that have design conditions whereby the primary pipe is subject to a greater amount of thermal expansion than that of the corresponding outer, as in an internally flexible underground system designed to handle hot fluids (designated class U21XY, where X = 4th digit value and Y = 5th digit value from Table 8.7)), and aboveground systems whereby hot fluids are transported or the secondary containment pipe is subject to an anticipated greater amount of thermal contraction than that of the primary during cold seasons (classes A21XY or A22XY). These applications are among the most commonly encountered design conditions, thereby making these elbows types important choices in double containment piping design.

If the dimensional analysis concludes that based on the elbow combinations there is insufficient room to provide for the required range of movements so as to prevent contact, then options must be decided on. The designer has the choice of selecting another elbow combination (e.g., a longer-radius primary in place of a concentric primary for a class U21XY system), locally expanding the secondary component dimensions in the area of the elbows (or globally), adding additional changes of direction to accommodate the resulting smaller amounts of dimensional change between consecutive elbows, or altering the design altogether (e.g., using a restrained system, or conversely, providing additional flexibility by means of expansionary devices on the inner and/or outer pipe system, depending on the class of system that the section under consideration falls within; see Section 8.8).

Assuming there is sufficient space in the analysis conducted, the designer must next determine appropriate types and locations of internal centering devices. This includes determining the positions of supports that function as axial-guide-only devices, and those that function as flexibility supports (i.e., internal centering devices that provide vertical support in horizontal assemblies, but allow both axial and radial movements within prescribed limits. Examples are provided in Figures 9.14 through 9.16, and 8.37).

Starting from each elbow, and working toward internal anchors, determine the location of the minimum length from the elbow whereby a first axial-guide-only support can be located based on a specified minimum stress, and inside of which rigid supports (i.e., those that do not allow radial movements of the primary pipes) cannot be located, and internal flexibility supports (i.e., those that do permit radial movements) must instead be used. These values are shown as L_4 and L_6 in Figure 8.18, and are estimated by using Eq. (8.16) which is a cantilever beam equation. The value of L in the equation is taken as the value from the opposite side under consideration (the value of L_2 is used to estimate L_4; the value of L_1 is used to estimate L_6, etc.). Once this absolute dividing line is approximated by the following cantilever beam equation, both flexibility supports and axial-guide-only supports can be located.

$$L = \sqrt{\frac{1.5 \Delta LED}{S}} \qquad (8.16)$$

where:
L = minimum distance of support that restricts lateral movement, in.
DL = amount of pipe growth from the side opposite that being considered, in.
E = modulus of elasticity at the temperature, psi
D = outside diameter, in.
S = design stress level, psi

Thermal Expansion Considerations

The designer can proceed by determining initial locations for the first flexibility support to be located on both sides of the primary elbow. A conservative approach is one where the primary elbows are vertically supported in horizontal assemblies by supports located no more than 0.75 times that of the support spacing that is determined to prevent sagging of straight pipe sections due to dead weight bending (Eqs. 9.29 through 9.32). In other words, $L_3 + L_5$ must be less than or equal to 0.75 [value from Eqs. (9.29)-(9.32)]. Once these supports are located, the locations and types of the remainder of the supports can be determined by using a trial and error method, assuming that rigid supports can be placed no closer than L_4 and L_6, and internal flexibility supports be located within 0.75 (times the normal support spacing for straight pipe). Other than these requirements, the final selection and placement of the types and locations of internal centering supports be located by minimizing the number required, the stresses that result, and while providing sufficient structural support to the primary pipe so as to prevent sagging or buckling, and to prevent damage due to vibration.

The procedure described above is summarized in the flowchart presented in Figure 8.20. The final result of the procedure must always be verified by performing a flexibility analysis on the assembly. If the resulting stresses are not within acceptable limits, the design must again be altered (e.g., selecting different radii, adding more bends, going to a restrained design, etc.). All intermediate internal anchors should be studied by means of a comprehensive analysis to determine the effects of localized stresses or otherwise qualified (see Section 5.1.10 and 6.2.2). The effect of the reaction forces on the secondary jacket at intermediate internal anchor locations should also be checked.

Equation (8.16) is used to calculate the minimum allowable distance of lateral restraint from the elbow, which is based on the guided-cantilever-beam theory. (This is the same calculation used for expansion loops; see Section 8.6.6.) The distance should be calculated for both sides of an elbow, as it will see expansion from two directions. If the lengths of piping on either side of an elbow are different, then the distance L (L_4 and L_6 in Figure 8.18) will be different on each side. L from Eq. (8.16) is based on the opposite run being considered and determines the minimum distance for which lateral movements must not be restricted.

When using directional changes for this purpose, any additional flexibility inherent in the elbows is not included in this calculation. Therefore, such calculations tend to be conservative. In piping, flexibility inherent in elbows can be significant.

In Europe, a common approach to determine the closest allowable point of lateral restraint for thermoplastic piping involves a simplified calculation. The basic formula is also used in calculations for expansion loop sizing and in determining offset length. An all encompassing constant C is used to account for the expansion characteristics of various

Table 8.6 Constants for Common Thermoplastic Piping Materials for Use in DVS Loop Sizing Equations (Values from DVS 2210 Teil 1)

Material	Constant
PVC-U	33.5
PP	30
HDPE	26
PVDF	21.7

thermoplastic materials, as shown in Equation 8.17. Values for X are shown in Table 8.6. This equation tends to yield different results than that determined by Equation 8.16. The results from either equation should be verified by performing a flexibility analysis, using the stress intensification factors described in Section 5.5.1.6 (for metallic components).

$$L = X\sqrt{D\Delta L} \tag{8.17}$$

where :
L = maximum distance between consecutive supports, beyond which buckling failure may be risked, mm
X = all encompassing constant; from Table 8.6
D = outside diameter, mm
ΔL = change in length, mm

8.6.3 Allowable Space in Double Containment Elbows; Conducting a Dimensional Analysis

To use a directional change in a double containment system to accommodate expansion, the diameter of the secondary containment section of the elbow might require enlargement to allow its associated primary elbow to be deflected properly. A top-section view of a horizontally positioned arrangement which uses a local enlarged diameter is shown in Figure 8.21. For double containment elbows, the allowable space required must be considered based on both of the straight sections of pipe leading into the elbow. Not all situations require the use of expanded secondary containment sections. To determine if sufficient space exists in the elbows, one may follow the procedure illustrated by Figure 8.22, described as follows.

First, determine the amount of linear expansion, using Eq. (8.1), of the straight pipe section on one side of the entry to the elbow for both inner and outer pipes. (It should be performed for the primary only if the secondary containment piping is restricted from

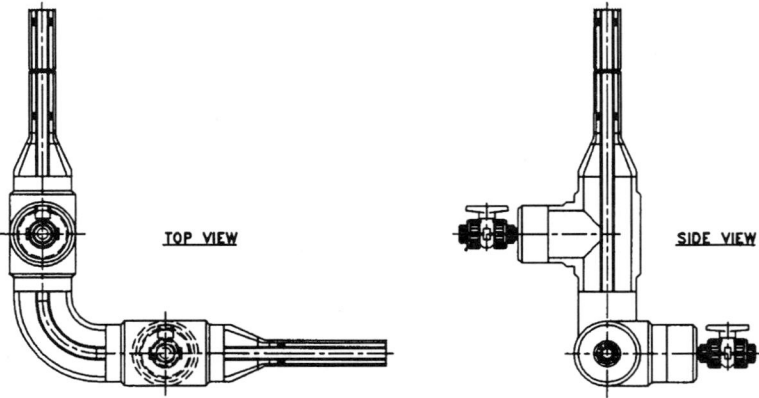

Figure 8.21. Top and side views of a horizontally positioned double containment elbow section that has a locally enlarged secondary containment elbow.

Thermal Expansion Considerations

growing/shrinking.) The values should be determined based on the distance from the closest fixed point (anchor location) to the entry of the elbow. The difference in growth between the primary and secondary piping may then be found by subtracting the smaller value from the larger value; this value is known as the amount of differential thermal expansion (there is a unique value for each side). Add the total of twice the value of differential expansion to the outside diameter of the primary pipe components at the anticipated contact point(s). It is wise that this amount be increased by an appropriate safety factor (10 to 25%). The total value must be compared to the actual inside diameter of the secondary containment components at the contact point(s) (Figures 8.23 through 8.25 provide examples; further examples are provided in Appendix F of this text). The problem can be tackled in a simple manner by drafting the components to be used, and determining by physical measurement or visual observation, if there is sufficient space. (Remember to include safety factor to account for some misalignment during fabrication and assembly.) The arc of the elbows and all likely contact points should be considered in the dimensional analysis. If adequate clearance is inherent in the component arrangement, then secondary containment component diameters do not need to be increased. If the amount of clearance is inadequate, then the inside diameter of secondary containment components needs to be increased until proper clearance is provided. Since the required pressure rating of secondary containment components is usually a known quantity, the actual diameter of the increased section can be determined by the minimum inside diameter required.

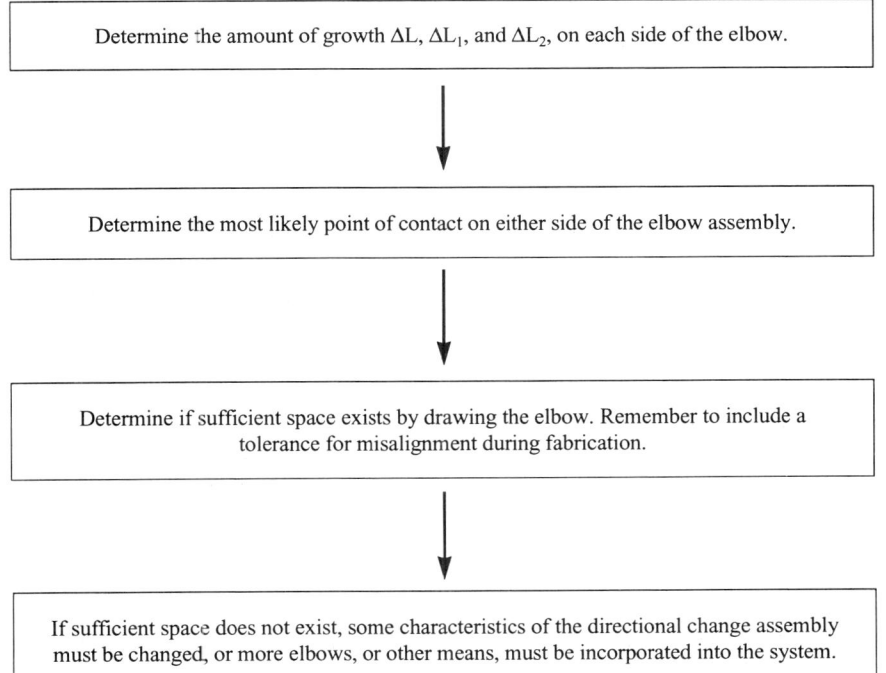

Figure 8.22. Flow chart of conducting a dimensional analysis for elbow sections.

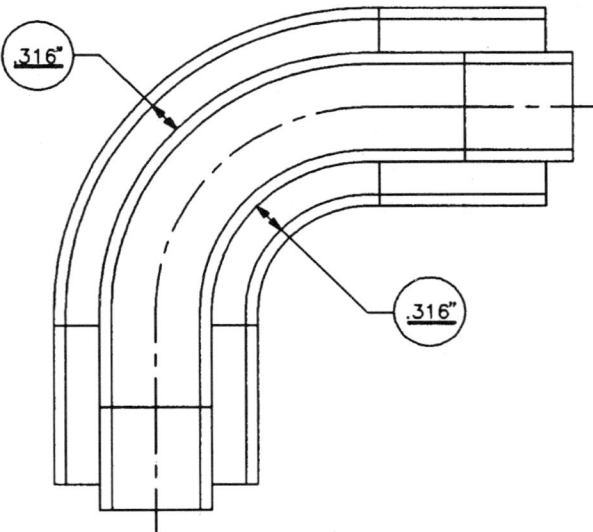

Figure 8.23. Dimensional analysis of a 3/4 in. diameter steel long-tangent, butt-welding primary elbow with a 4-D radius inside of an 1 1/2 in. long radius butt-welding elbow. Taking into account a ±0.125 in. misalignment during fabrication, this elbow is designed to accommodate a ΔL of 0.191 in., in any direction, due to the equal spacing in the elbow.

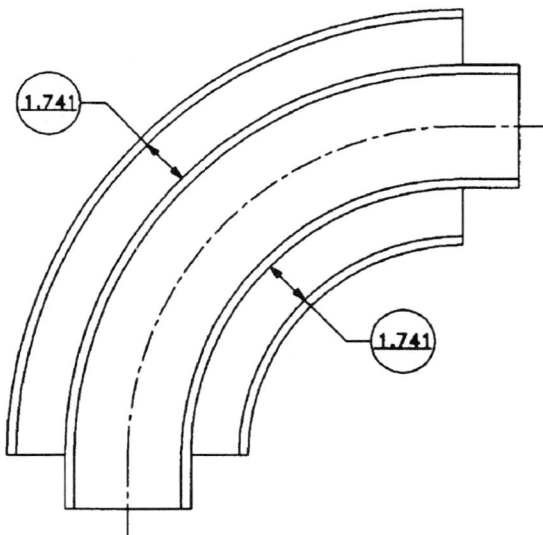

Figure 8.24. Dimensional analysis of a 4 in. diameter steel long-tangent butt-welding primary elbow with a 3-D radius inside of an 8 in. Sch. 40 long radius butt-welding elbow. Taking into account a ±0.25 in. misalignment during fabrication, this elbow is designed to accommodate a maximum ΔL of 1.49 in. in any direction, due to the equal spacing in the elbow.

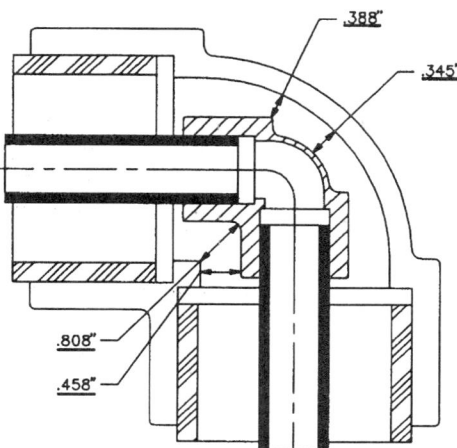

Figure 8.25. Dimensional analysis of a 3/4 in. diameter PVC Sch. 80 socket solvent cementing elbow inside of a 2 1/2 in. diameter PVC Sch. 80 socket solvent cementing elbow. Due to the irregular geometry and socket dimensions, the closest point of contact and allowable room for movement depends on the relative movements.

8.6.4 Relative Orientation of Components in Flexible Double Containment Elbow Arrangements

The orientation of the primary elbow within a respective secondary containment elbow can vary depending on the relative radii of the elbows and the relative diameters (including the comparative sizes of the piping adjacent to the elbows). In certain elbow arrangements, the relative cross-sectional orientation of the respective components can change from that of being concentrically positioned to more of an eccentric orientation. In systems whereby the secondary containment diameter is enlarged in the area of the elbow, there are additional considerations involving the orientation of the primary pipe within the secondary containment, such as whether localized low points or high points are created, and what unique problems these may pose to the application. These considerations depend on the positioning of the eccentric or concentric reducers that are used in the installation.

Elbows that maintain their concentricity during system operation are found in designs that are fully restrained. An example is illustrated in Figure 8.26 for a metallic system that is anchored at both ends by the use of a welded internal anchor system. The relative cross-sectional orientation of the primary and secondary containment portions are shown in the cross-sectional view portion of Figure 8.26. The concentric cross-sectional orientation is maintained throughout the curved portion of the fitting assembly.

Other elbows for all practical purposes may be substantially concentric, but in reality most are eccentric throughout. For example, in an elbow section that involves an elbow such as that shown in Figure 7.8 whereby the inner and outer elbows have a radius and a centerline-to-end radius that is equal throughout the curved portions of the elbows, and whereby the secondary containment elbow area is not locally increased at any one point, the relative orientation of the primary component will always be substantially concentric to that of the secondary containment throughout. In practice, the relative orientation is typi-

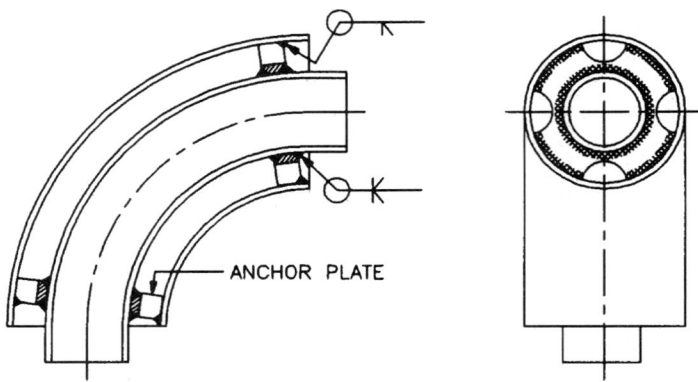

Figure 8.26. An example of an elbow assembly in which the elbows have a matching radius in the curved portion of the fittings, and the elbows are anchored at either end by means of welded partition plates.

cally somewhat eccentric at all points, and with the most eccentric point being the furthest point from the nearest internal support. This is due to two reasons. Internal centering supports by design do not perfectly center primary pipes within secondary pipes in most common designs so that the primary components can be inserted into the secondary containment casing during assembly. Also, primary pipes, like all pipes, will sag between supports by an amount determined by Eqs. (9.29)-(9.32), although this consideration is negligible in systems that are adequately supported.

Eccentric orientation at a given point in an elbow may be created by having a primary elbow that has a longer or shorter radius that that of the corresponding elbow. Examples of these are shown in Figures 7.8 and 7.9. The cross-sectional orientations, taken at the end of the curved portion of the fittings and at the centerlines are shown in Figures 8.27 and 8.28. Note that the cross section taken at the centerline of the fitting assemblies shows that the components are eccentric to one another at that point, while at the end of the curved portions, the fittings are substantially concentric.

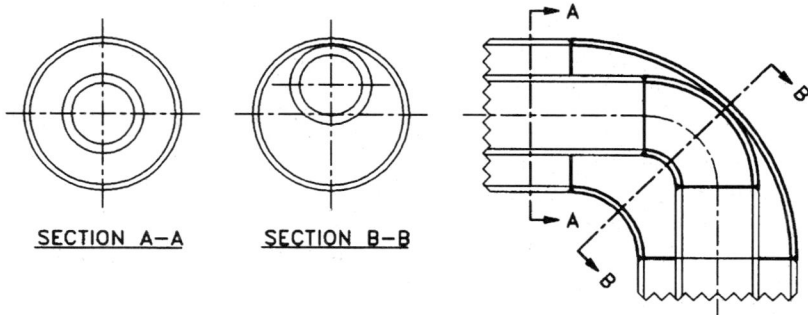

Figure 8.27. Cross-sectional views taken at opposite reference points for an elbow assembly in which the primary elbow has a shorter radius than that of the corresponding secondary containment elbow.

Thermal Expansion Considerations

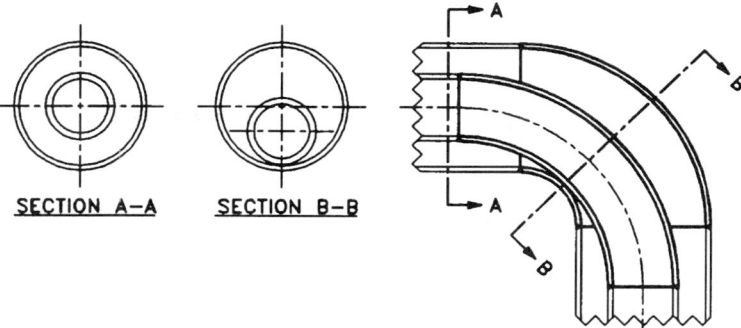

Figure 8.28. Cross-sectional views taken at opposite reference points for an elbow assembly in which the primary elbow has a longer radius than that of the corresponding secondary containment elbow. (Patent Pending.)

In certain applications, there may not be sufficient room in the elbow assemblies to accommodate required movements of the primary pipe, in a flexible system design. In this case, the secondary containment system in the area of the elbow must be enlarged, and to a distance on both sides of the elbow to prevent a predetermined contact based on a calculated amount of dimensional change, taking into account the geometry of the components. In this event, the secondary containment piping system can be locally enlarged either by using concentric or eccentric reducers to result in the desired enlargement.

An example of a section that is locally enlarged by using a concentric increase is provided in Figure 7.15. Note that in the example, the radius of the primary elbow is such that the primary component maintains a substantially concentric orientation throughout the section. Consideration should be given to the localized high points and low points that are created in the annulus, and suitable drains and vents should be provided. Such an arrangement due to these concerns is better suited to above-ground applications where ready use can be made of the low-point monitoring points, although the arrangement can be used underground, if suitable drain and vent provisions are made.

Examples of the use of eccentrically enlarged sections of elbows are provided in Figures 8.29 and 8.30. In Figure 8.29, the orientation of the eccentric reducers directs the added space toward the outside, where the added space is typically needed. Figure 8.30 illustrates that by rotating the eccentric reducer slightly, low spots in the annulus may be avoided (the specified slope of the system should be considered), while still providing a certain amount of added room for required movements. In an arrangement such as Figure 8.29, if continual annular drainage can be achieved, there is no need to provide for low-point drains in the elbow section. Such an arrangement is ideal for underground systems.

In elbow sections where a localized increase is used, the amount of enlarged secondary containment piping on either side of the elbow should be limited to the amount at least 2" in. short of that which would otherwise necessitate an internal flexibility support to be used (see Figures 8.37, and 9.14 through 9.16). The reason has to do with avoiding the need to incorporate internal flexibility supports that need to be specially machined, and which would add expense.

8.6.5 Support Requirements in Double Containment Elbows

Direct support should normally be avoided for primary piping elbows in double containment expansion elbow sections, where possible (see 7.4.3), except in machine-made RTRP systems. If direct support is required, interstitial supports designed for this purpose should provide vertical support, without restricting required bending and torsional movements. An example of a support designed to support primary components in double containment expansion loop/offset assemblies are shown in Figures 8.37, and 9.14 through 9.16.

As in a single-wall pipe change of direction (elbow), it is recommended to provide anchors at the midpoint between consecutive changes in direction. In a double containment piping system, anchoring is provided by an interconnecting component such as an internal anchor. It can also be facilitated by terminating the zone of secondary containment, thus providing a means of external anchoring. The use of the first method is advisable wherever possible, particularly in underground systems, in order to maintain a continuous zone of secondary containment.

8.6.6 Expansion Loop Design for Double Containment Pipes

Where other methods are not suitable, expansion loops offer an alternative method to compensate for thermal expansion/contraction. Loops may be positioned vertically or horizontally, and may vary in configuration. They may be used in horizontal or vertical piping sections. One conventional approach for calculating the size of an expansion loop in a single-wall pipe is by the use of guided-cantilever-beam theory. The approach assumes that piping demonstrates substantial elasticity and assumes limited piping rotation of the ends of each straight run. It also assumes that the loop consists of two cantilever beams, each with one end fixed and concentrated loads applied at their free ends. Figure 8.31 shows a stan-

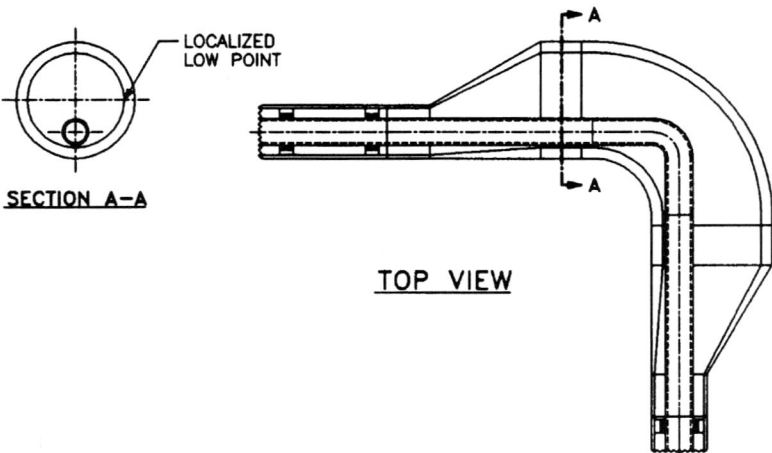

Figure 8.29. Example of a double containment elbow assembly whereby the secondary containment portion is locally enlarged by means of an eccentric reducer (increaser), with all of the added room directed outward.

Thermal Expansion Considerations

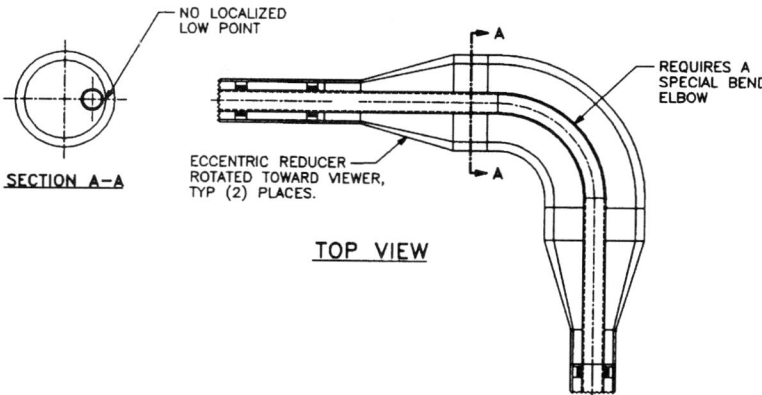

Figure 8.30. Example of a double containment elbow assembly whereby the secondary containment portion is locally enlarged by means of an eccentric reducer rotated to a degree, so as to allow added room for the primary pipe in the outward direction, but at the same time preventing a localized low point.

dard four-elbow 90° expansion loop in a double containment piping system. Figure 8.32 illustrates an expansion loop assembly involving four 45° elbows.

To calculate the loop size using the guided cantilever method, the maximum temperature change between the installation temperature and the minimum and maximum operating temperatures needs to be established. Oftentimes this means having to make a conservative prediction as to what expected installation temperature will produce the greatest ΔT, based on the time of year and location. The guided-cantilever-beam equation to calculate the length of the two parallel legs (shown as leg "L_A" of the loop shown in Figure 8.31) is:

$$L = \sqrt{\frac{0.75(\Delta L)(E)(D)}{S_b}} \qquad (8.18)$$

where:
L = length of parrallel legs of the expansion loop, in.
ΔL = maximum deflection of the parrallel legs, based on a loop having equal adjacent tangents, in.
E = modulus of elasticity, psi
D = pipe O.D., in.
S_b = maximum bending stress, psi

In the guided-cantilever beam-theory approach, the normal length that is used for the middle leg, "L_B," is usually one-half of the value calculated for leg "L_A." A longer normal length adds flexibility to the loop; a shorter normal length decreases loop flexibility. In the design of expansion loops with this approach, the additional flexibility inherent in elbows is not included. Therefore, such calculations tend to produce conservatively sized loops. In metallic and nonmetallic piping, the additional flexibility inherent in elbows can be significant.

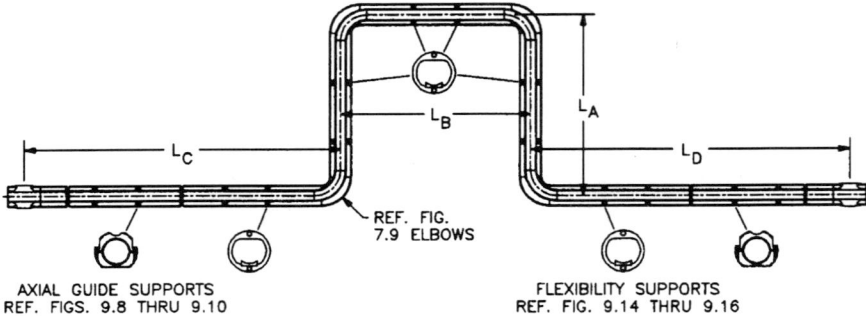

Figure 8.31. A double containment expansion loop assembly in which a local increase in secondary containment diameter is not used. In this patent pending arrangement, maximum use of internal flexibility is achieved by the style of components selected (e.g., Figure 8.28-type elbows), and due to their relative placement (patent pending).

Many other methods exist to determine the sizes of expansion loops, and to calculate the resulting reaction forces and bending stresses that are produced. Some of the more common methods, many of which historically were proprietary methods, include: (1) the Grinnell method; (2) the Kellogg method; (3) the Blaw-Knox method; (4) the Tube Turns method; and (5) the Truong Method. The first four methods described were developed by the companies having the same name. The Truong method was developed by Dr. Quy Truong of Houston, Texas. It is a relatively new method, but one that is gaining popularity due to the comparative degree of accuracy for estimating bending stresses that it allows.

The Grinnell method gained popularity since it uses tabular values published for various constants. Tables E-2 through E-5 of Appendix E presents Grinnell data, which may be related to Figure 8.31. From these values, the reaction force may be calculated by Eq. (8.19), and the resulting maximum bending stress from Eq. (8.20). For loops where the tangents are not equal on either side, reaction forces for F_x and F_y must be determined. The value of the constant C in the two equations represents a combination of the modulus of elasticity and the coefficient of expansion of the material of the pipe (see Table E-1 of Appendix E).

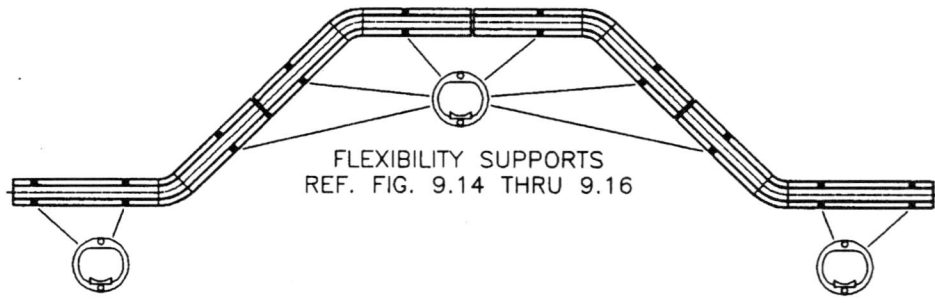

Figure 8.32. Example of an expansion loop assembly involving four 45° elbows (patent pending).

Thermal Expansion Considerations

$$F_x = k_x C \left(\frac{I_p}{L^2} \right) \tag{8.19}$$

where:
F_x = reacting force, lbs
C = constant, based on the modulus of elasticity and the coefficient of expansion of the material, see Table E-1 of Appendix E
k_x = value from Table E-2 through Table E-5 of Appendix E
L = length of the tangent, ft
I_p = moment of inertia, in.^{4}

$$S_b = K_b C \left(\frac{D}{L} \right) \tag{8.20}$$

where:
S_b = maximum bending stress, psi
K_b = value from Table E-2 through E-5 of Appendix E
L = length of tangent (based on a loop having equal tangents, ft
D = diameter, in.
C = constant (based on the modulus of elasticity and the coefficient of expansion of the material), see Table E-1 of Appendix E

For the calculation of thermoplastic piping expansion loops, a simplified approach is used in some European countries. Based on the West German DIN standards, a simplified calculation is presented in Eq. (8.17). In this equation an all-encompassing constant X is used to account for the expansion characteristics of certain thermoplastic materials. Values for X are shown in Table 8.6. This equation tends to result in longer loops than those calculated using the guided-cantilever-beam approach. The suitability of the results from either equation should be verified by performing a flexibility analysis on the system (see Sections 8.3.1 and 5.5.1.6).

8.6.7 Allowable Space and Bending in Double Containment Piping Expansion Loop Entry Elbows

To provide an expansion loop in a double containment piping assembly, the secondary containment section of the loop may require a diameter larger than that used in straight pipe sections, if sufficient space does not already exist. The reason for the allowable space is to allow the primary piping "A" legs to cantilever properly, thereby allowing bending in a desired manner. A section view of a horizontally positioned double containment expansion loop assembly, where the outside pipe diameter has been locally enlarged, is presented in Figures 8.33 through 8.36. If it is the secondary containment piping that requires the expansion alleviation, the same type of arrangement can be used. For secondary containment piping that is contracting/expanding, primary elbows/pipe in the loop must be free to flex along with its associated outer legs. Otherwise, the primary and secondary containment piping components may impact each other, causing unintended restraint and possibly high load losses. Designers must be aware that by increasing secondary containment diameter in a localized

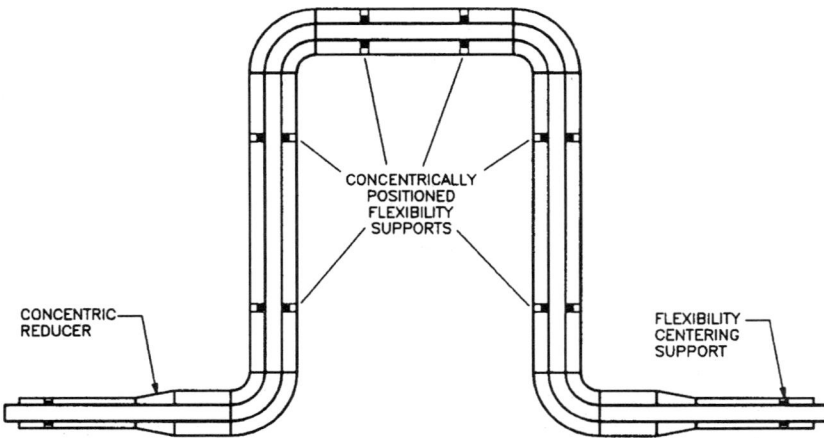

Figure 8.33. A double containment expansion loop assembly in which a local increase in secondary containment diameter is achieved by using concentric reducers, and elbows with matching radii (patent pending).

fashion, wall thickness may also be increased in the expanded section. This may add resistance to bending in the loop itself; therefore, a loop designed in this manner to accommodate expansion/contraction of the secondary systems will have to be to account for this.

Not all situations require the use of expanded secondary containment sections. The diameter of secondary containment components in the loop assembly only need be expanded if the amount of thermal expansion of the primary piping straight sections between the nearest fixed point and the loop dictates that interference will occur between inner and outer entry elbows (due to the expected differential growth). To determine the size of additional space required and thus the amount that secondary containment must be expanded, the same type of dimensional analysis stated in Section 8.6.3 can be used.

The main purpose of expansion assemblies is to provide a loop where perpendicular "A" legs can deflect and absorb the expansion freely, without any restrictions. Special attention must therefore go towards the design of interstitial supports that are positioned in the assembly. If ordinary interstitial supports are used, the desired lateral bending of the loop legs will be restricted, reducing flexibility and increasing stress. If the interstitial support does not permit lateral movement of the pipe, it cuts the length of the pipe that can flex to the distance between the elbow and the first support. The impact of reducing the length of the pipe that can flex can be readily observed in Eq. (8.20). Interstitial supports that are specially designed to allow for proper lateral bending of the loop legs should be used (see Section 8.6.9) where required.

8.6.8 Orientation of Pipes in Double Containment Expansion Loops

There are several design options for laying out expansion loops in double containment piping systems. The loops may have a nonincreased diameter of the secondary containment in the loop section, which is the preferable arrangement, and is illustrated in Figures 8.31 (using 90° elbows) and 8.32 (using 45° elbows). This is due to the avoidance of localized

Thermal Expansion Considerations

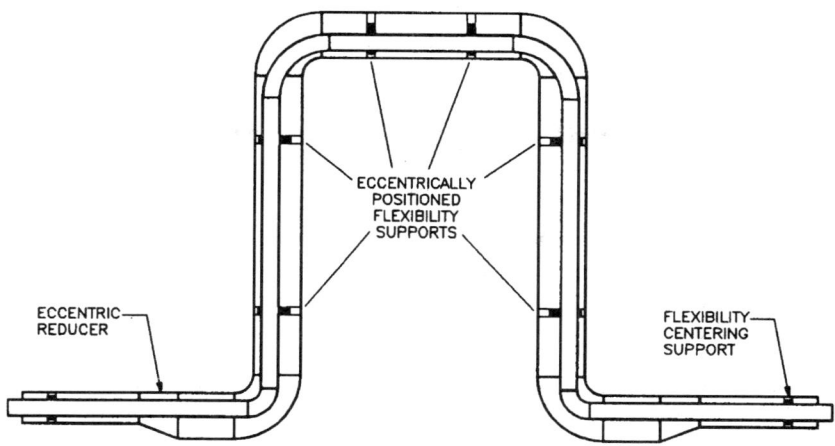

Figure 8.34. A double containment expansion loop assembly in which a local increase in secondary containment diameter is achieved by using eccentric reducers, and elbows with matching radii.

low and high points in the annulus, and to the fact that it is a less expensive arrangement to fabricate and install than the alternative. In such an arrangement, the primary pipe is concentrically positioned through most of the straight sections, and throughout the assembly if concentric elbows are used.

Alternatively, the secondary containment jacket may be locally increased in the area of the loop to increase the allowable room for movement of the primary system. The secondary containment section may be locally increased and decreased at each elbow, or by increasing at the entry and exit to the loops. Also, the increase is either by means of the use of concentric reducers, which may result in a substantially concentric orientation, or by means of eccentric reducers, which results in a substantially eccentric orientation in both straight pipes and elbow components. The relative positioning of the eccentric reducers determine where the added room is provided, and to what extent, if any, low points are created. Examples are provided in Figures 8.33 through 8.36.

8.6.9 Expansion Loop Interstitial Support Design

The types of supports recommended in a double containment expansion loop should provide vertical support, without lateral restriction. In loops where the secondary section is not enlarged, a standard flexibility support of the types shown in Figures 9.14 through 9.16 is used. Figure 8.37 is an illustration of a fabricated internal flexibility support specially designed for use in double containment expansion loop assemblies, whereby the secondary containment pipe is locally increased in diameter. These supports provide vertical support without vertical restriction, and further prevent buoyant effects on the primary piping when an annulus is filled with fluid.

8.6.10 Anchoring Methods

As in single-wall pipe expansion loop design, it is recommended either to provide intermediate anchors at the midpoint between loops, or main anchors at the middle of leg "L_B" of

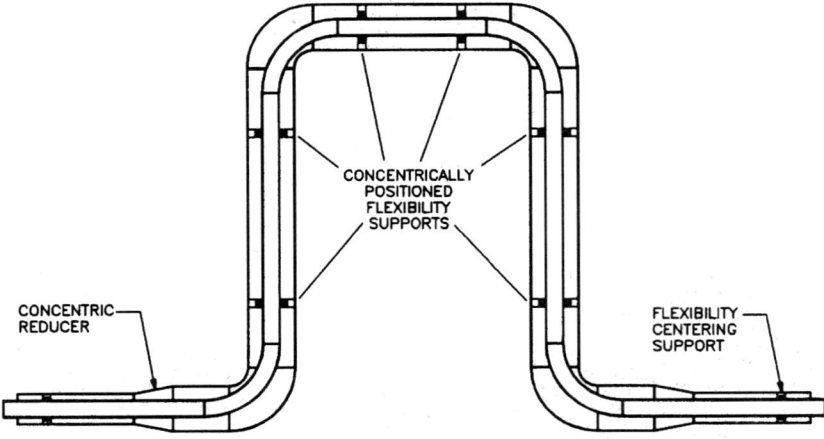

Figure 8.35. An expansion loop assembly in which a local increase in secondary containment diameter is achieved by using concentric reducers. What is different here as compared to Figure 8.32 is that the radii of the primary elbows are longer than the corresponding secondary containment elbows (patent pending).

the loop itself. In the most conservative design, anchorage can be provided at both main and intermediate locations. In a double containment piping system, anchoring may be provided by several means, including the use of internal-anchor fittings (shown in Sections 11.1.1.4 and 7.4.7). It can also be facilitated by terminating the zone of secondary containment and providing an anchor on the primary pipe at the termination location (referred to as external anchoring in Sections 11.1.1.4.1 and 7.4.7). The use of the first method is advisable wherever possible, particularly in underground systems, due to the desirability of maintaining a continuous zone of secondary containment, without having to use flanges.

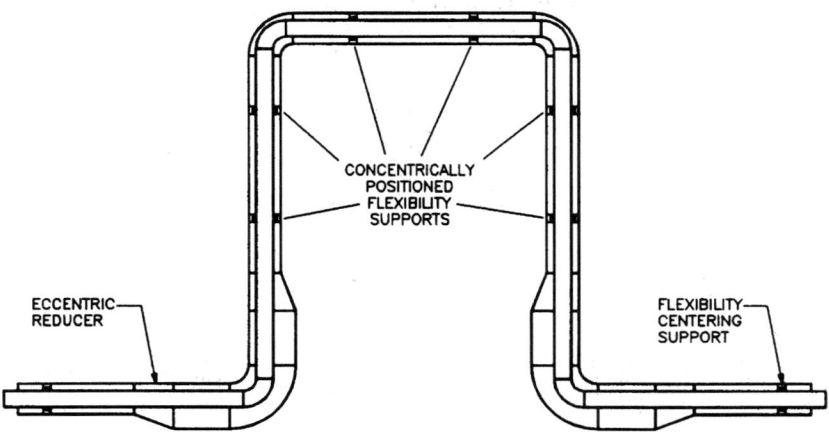

Figure 8.36. Example of an expansion loop where only the localized areas of the entry and exit elbows are enlarged in diameter, in this example using eccentric reducers (patent pending).

Thermal Expansion Considerations

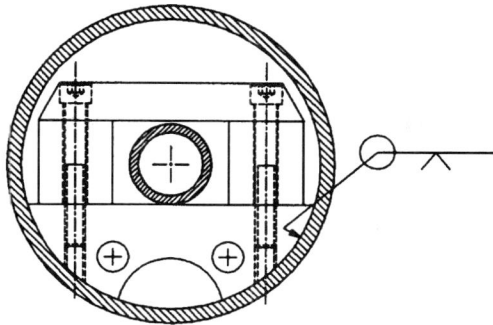

Figure 8.37. Illustration of a fabricated internal flexibility support for use in expansionary elbow and expansion loop sections where a local increase in diameter is used. (U.S. Patent #5,197,518.)

8.6.11 Alternate Expansion Loop Arrangements

The configurations shown in Figures 8.31 through 8.36 are based on the four-elbow expansion loop. A variation of the four-elbow loop is the six-elbow type of expansion loop, illustrated in Figure 8.38. To calculate a loop of this design, the extra vertically positioned parallel legs are not counted in the calculations; their lengths are usually insignificant. Therefore, the design technique used for sizing the length remains the same as for the four-elbow loop. A six-elbow loop arrangement is useful where there is a lack of space to provide a normal four-elbow loop. This type of loop is often used in above-ground pipe racks where lack of available horizontal space exists in the same plane.

A variation of the four-90°-elbow expansion loop is where there are four 45° elbows used in place of the 90° elbows. This type of loop is useful for vertical loops in drain piping, so that drainage will continue in an undisturbed manner, and is illustrated in Figure 8.32. It is also useful for underground pipes, in order to save on the total width of a loop. When calculating a loop of this design, the length of the "A" legs needs to be multiplied by a factor of 1.414 (the value of the square root of two) to provide the same amount of compensation as that required of the four-90°-elbow loop.

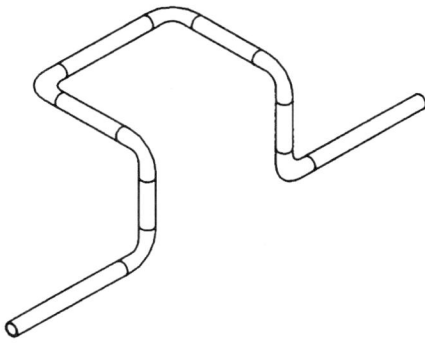

Figure 8.38. Illustration of a six-elbow expansion loop.

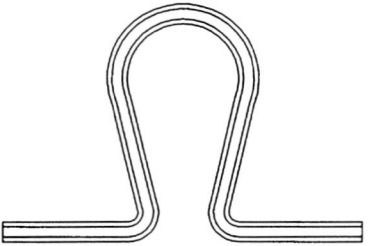

Figure 8.39. Illustration of an expansion loop made from fabricated bends.

A third variation of expansion loop exists where pipe is bent into a semi-circle, without any elbows used at all, as illustrated in Figure 8.39. While a loop of this type is useful for small-diameter pipes or flexible piping, it is not feasible for most double containment pipes. This is due to the added rigidity that the double containment piping physical structure provides. The structural aspects of bending double containment pipes is covered in Chapter 9, "Structural Considerations," and bending issues are discussed in Chapters 5, 6 and 7.

8.6.12 Expansion Offset Design for Double Containment Piping

An expansion offset assembly is very similar to an expansion loop. Their design requires the same general rules applied to expansion loop design/sizing. Expansion offsets can also be positioned in either a horizontal or vertical configuration. The conventional method for calculating the size of expansion offsets in a single-wall pipe is by the use of guided-cantilever-beam theory. The approach assumes that the piping demonstrates substantial elasticity and assumes the ends of the pipe are restrained against rotation. It also assumes that a pipe offset behaves as a cantilever beam, with one end fixed and a concentrated load applied at its free end. Figure 8.40 illustrates an example of a double containment expansion offset which does not require a localized enlargement of the secondary pipe.

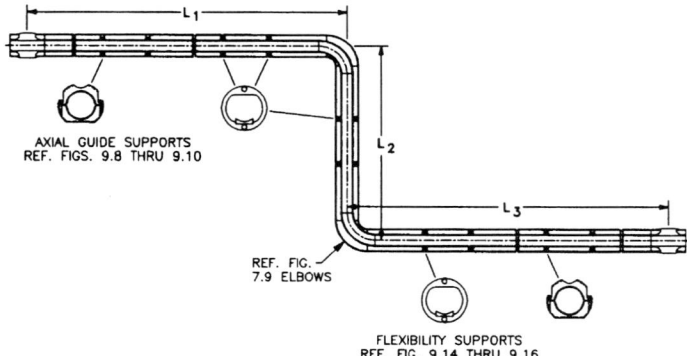

Figure 8.40. Example of an expansion offset section for a double containment pipe system. This section contains the necessary features for complete internal flexibility, and if properly designed, will maintain low stresses in the operating pipe (patent pending).

Thermal Expansion Considerations

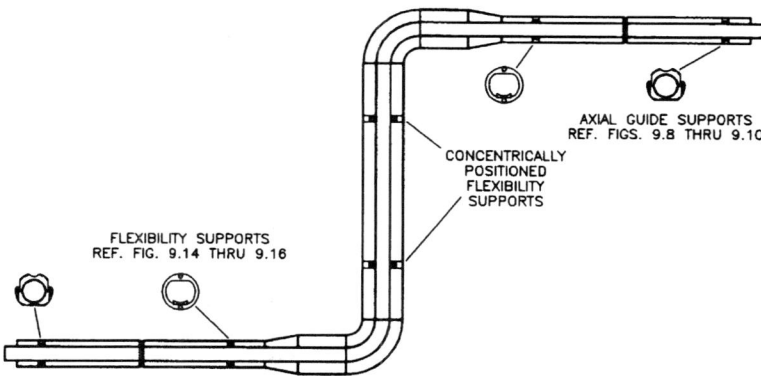

Figure 8.41. Example of the use of an expansion offset assembly that incorporates a local increase in diameter of the secondary containment pipe, using concentric reducers to allow room for a desired amount of movement in the primary pipe system (patent pending).

To estimate reacting forces and the maximum bending stresses of Z-bend offsets (shown in Figure 8.40) using the Grinnell method, Equations 8.21, 8.22 and 8.23 may be used. Table E-6 of Appendix E presents the Grinnell data for k_x, k_y, and k_b, which may be related to Figure 8.40. The value of the constant C in the two equations represents a combination of the modulus of elasticity and the coefficient of expansion of the material of the pipe (refer to Table E-1 in Appendix E). It should be remembered that internal supports of the flexibility type (see Figure 8.37, and 9.14 through 9.16) allow piping to flex where needed.

$$F_x = k_x C \left(\frac{I_p}{L^2} \right) \tag{8.21}$$

$$F_y = k_y C \left(\frac{I_p}{L^2} \right) \tag{8.22}$$

$$S_B = k_B C \left(\frac{D}{L} \right) \tag{8.23}$$

where:
F_x = reacting force in the "x" direction, lbs
F_y = reacting force in the "y" direction, lbs
k_x = value from Table E-6 of Appendix E
k_y = value from Table E-6 of Appendix E
k_B = value from Table E-6 of Appendix E
C = coefficient from Table E-1 of Appendix E
I_p = moment of inertia, in.^{4}
$L = L_1 + L_3$ from Fig. 8.40, ft
D = outside diameter, in.
S_B = maximum bending stress, psi

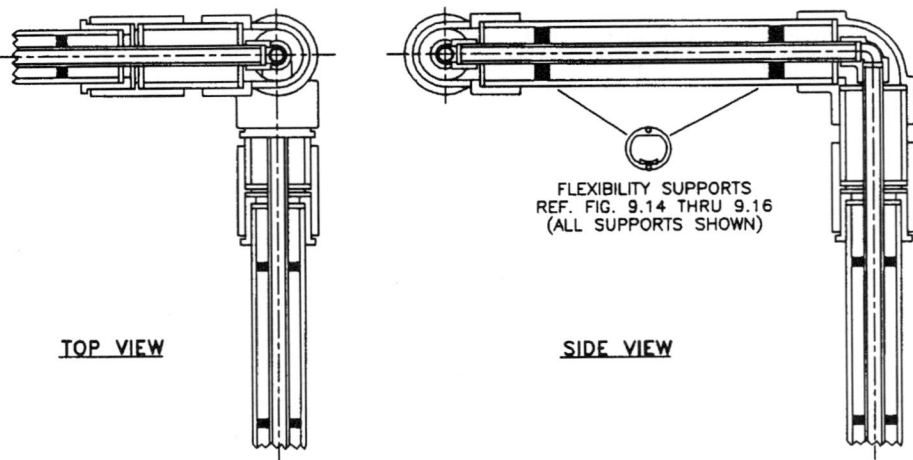

Figure 8.42. Example of a multi-plane two elbow offset.

In European countries, a simplified equation like the one for expansion loops is used for thermoplastic piping offsets, stated in Eq. (8.17), where an all-encompassing constant X is used to account for the expansion characteristics of certain thermoplastic materials. Values for X are shown in Table 8.6. This equation tends to result in longer offsets than that calculated in the guided-cantilever-beam approach. The suitability of the results from either equation should be verified by performing a stress analysis on the system.

The same allowable-space issues described for expansion loops in Section 8.6.6 apply to double containment pipe offsets, an example of which, designed with an increased outer diameter section, is shown in Figure 8.41. The issues regarding interstitial support design (i.e., allowing lateral restriction while providing vertical support), described in Section 8.6.9, apply equally to double containment offset assemblies. The same is true of anchoring needs (see Section 8.6.10).

Not all situations require the use of expanded secondary containment sections. To determine if sufficient space exists in the entry elbows in order to avoid contact between them, use the procedure described in Section 8.6.3. The same considerations for eccentric versus concentric positioning of the double containment expansion sections described in the previous section apply to offsets as well (refer to Section 8.6.8).

The two-elbow expansion offset shown in Figures 8.40 and 8.41 can be positioned vertically or horizontally, and can also be rotated at a variety of angles. No matter the angle to which it is positioned, piping in the offset is always in one plane. An example of a two-elbow offset with a change in planes is illustrated in Figure 8.42. A variation of the two-elbow offset is the three-elbow offset, which can be used to position the runs of pipe at different elevations. A three-elbow offset that has very short lengths of pipe is sometimes referred to as a swing joint.

8.6.14 Allowable Space in Branch Fittings

Branch fittings (i.e., tees, laterals, crosses) present many of the same concerns in terms of allowable space as elbows in systems undergoing thermal expansion (also refer to Section

Thermal Expansion Considerations

7.4.4). The only difference is that there are additional adjacent lines to consider. In a tee or lateral there are three thermally expanding/contracting straight pipes leading in; in a cross or double lateral there are four. The same type of dimensional analysis must be used. If contact is anticipated at any point, the diameter of the secondary containment branch must be increased to allow sufficient room in order to avoid component contact. This is so unless it is determined through analysis that interference will not result in either premature or long-term failure of the components.

8.6.15 Use of Cold Spring

In Chapters 5 and 6, the concept of cold spring was introduced. Cold spring involves the intentional deformation of piping during installation in order to produce a desired initial displacement stress. This intentional "preloading" of piping systems can help reduce the overall magnitude of displacement stresses and strains in any one direction during the operation of a piping system. The method is particularly effective in reducing initial strain during startup of a system, which is the time when many systems are subject to failure. The method also produces less likelihood of deviation from as-installed dimensions during initial operation. Hangers, internal or external supports, and interconnecting devices in a double containment piping system will not be displaced from their original settings to a great extent, depending on their design.

Cold spring is a valid method that in theory may be used for controlling expansion in primary pipes, secondary containment pipes, or for a double containment pipe assembly. It is valid for conditions where nondifferential thermal expansion conditions exist, or where both pipes are permanently affixed together. No credit should be given for cold spring in stress calculations (refer also to the discussion on cold spring in Sections 5.5.1.4 and 6.6).

While in theory it may be used, it is very difficult to apply a specified amount of cold spring in primary piping for differential thermal expansion conditions, during the fabrication of a double containment pipe assembly. For this reason, cold spring is of limited application in double containment piping systems.

8.6.16 Snaking of Buried Pipes

Snaking of underground pipes is a method of providing cold spring to a buried pipe, providing additional pipe in the trench to compensate for thermal expansion of the line. It is used frequently in small-diameter, nonmetallic, buried, single-wall piping, where soil friction is unlikely to restrain such a piping system from moving as it changes dimension. Normally, changes in dimension result because piping has been in the sun during warm periods of the day and then covered with a cooler soil at the end of a day. While it is generally recommended to avoid burying pipes until they are approximately at the same temperature as the soil, backfilling is commonly performed on hot, sunny afternoons on many projects.

Pipes may be purposely deflected into a snakelike, sine wave pattern to compensate for this condition, a method called "snaking." This method is not feasible for pipes larger than 3 in. (90 mm) due to its having a relatively large surface area, since there is sufficient soil friction to restrain it from moving (if proper burial techniques are followed). Since most secondary containment piping systems have a diameter equal to, or larger than, 3 in. (90 mm), the practice of snaking has limited application for double containment pipe systems. Interstitial supports and interconnecting parts add rigidity to double containment piping, thereby adding further resistance to its being deflected into a sine wave pattern. Longitudinal bending analysis of pipes is covered in Chapter 9, "Structural Considerations," Section 9.2.

8.7 Use of Heat Transfer as a Controlling Medium

8.7.1 Use of Heat Tracing to Limit Negative Temperature Changes of the Secondary Containment Piping

In many applications involving double containment piping systems, temperature changes will be largest in the negative direction from the installation temperature. This will mean that the controlling design conditions will be the contraction magnitude of the system that may be expected. It frequently is the condition when the system is installed during a warm period of the year in a region subject to seasonal changes. The effect of these conditions is often a substantial contraction of the installed components once cold temperatures arrive. The conditions may be insignificant for metals; however, the stresses that are produced could be in excess of the allowable design stresses for nonmetallic components. When this occurs, damaging tensile stresses can be imposed on welds, fittings, and other piping components. Since tensile stress ratings for most materials are less than their compressive strengths, the effects can be damaging to the system. In addition, many materials become increasingly brittle at cold temperatures and under added tension; increased brittleness usually means lower impact strength for components. Many commonly used thermoplastics (e.g., PVC and polypropylene) become brittle as freezing temperatures are approached.

One valid way to compensate for thermal contraction is to limit the amount of temperature change. This can be achieved by using heat tracing (either electrical or steam tracing) to limit contractions such that they are insignificant. Heat tracing of outer piping will help to maintain its temperature at safe levels, minimizing its contraction. In some cases, the use of heat tracing with a high enough set temperature will change the design conditions so that positive temperature changes from the installation temperature become the controlling design condition. (It is always a wise idea to plan for the likelihood that the heat tracing system may be inoperable during the time it is needed most. In other words, while it will help with the operation of the system, it is wise to design the piping system to accommodate the full amount of contraction in the event of a heat tracing system failure.)

Heat tracing is most effectively applied to the outside diameter of a secondary containment piping system and used along with external insulation and a covering that surrounds and protects it. In many cases, this will result in the most economical approach. It is possible to provide heat tracing in an annular space, which may eliminate the need for external heat tracing if a nonmetallic material having a low thermal conductivity is the material of the secondary containment system. However, providing heat tracing in the annular space tends to complicate systems, creating operational problems in the event maintenance of the heat tracing system or its components is required, or if a primary system leak occurs.

Detailed design requirements for heat tracing are normally provided by the manufacturers of commercially available heat tracers. Other design information for calculations, etc. can be obtained from the standards of the Institute of Electronics and Electrical Engineers (IEEE). When nonmetallic piping is involved, it is highly recommended to use electrical heat tracing that is of the self-regulating type. In all nonmetallic systems, it is also wise to use a thermostatically controlled heater to prevent overheating of nonmetallic pipe, regardless of the type of heat tracer used.

8.7.2 Use of Upstream Heat Exchangers to Limit Positive Temperature Changes of the Primary Piping

In many applications involving double containment piping systems, temperature changes will be largest in the primary system in the positive direction from the set installation temperature. This is typically caused by having a hot fluid discharged through the primary piping, meaning that the controlling design conditions will be potential thermal expansions of the primary system to be expected. This condition may be addressed in some designs by adding a heat exchanger upstream of the initial entry point of the pipe (i.e., upstream of the pump, inside a storage tank, etc.). By cooling the fluid prior to discharge through the piping, less thermal expansion will occur, and a simpler layout can be implemented. The decision on when it is appropriate to involve an upstream heat exchanger involves an economic comparison of using other compensation techniques versus the capital and operating costs of adding the heat exchanger. Heat exchangers of the double-walled piping type are discussed further in Section 10.2.

8.8 Classification of Systems and Application of Compensation Techniques

As described in the beginning of this chapter, and in Chapter 7, thermal expansion and contraction in double containment piping systems can occur in two different ways. These are: (1) differential thermal expansion, where primary and secondary containment pipes are subject to different magnitudes and rates of growth, and (2) nondifferential thermal expansion, where inner and outer pipes grow at roughly the same rate and magnitude. Differential thermal expansion is more common than nondifferential thermal expansion. There are many more classifications of design conditions that are subjected to differential thermal expansion conditions, some of which involve above-ground design conditions and some of which cover underground design conditions. Thus, for systems subject to nondifferential thermal expansion, there are numerous possibilities, depending on whether the system is above ground or underground.

In the following section, six broad classifications of expansion conditions for systems are discussed. To be specific, these six broad classifications can be further delineated by assigning a five digit alphanumeric character, from Table 8.7. (In Table 8.7, the designation "A" refers to above-ground systems, and "U" refers to underground applications.) The third character refers to the condition of the primary pipe as to whether it is free to flex (designated by a "2") or as to whether it is fixed (designated by a "1"). Similarly, the fourth and fifth characters refer to the magnitude of temperature changes of the primary and secondary containment pipes, respectively.

Various methods for thermal expansion alleviation that are appropriate in many of these design classes are discussed in Sections 8.8.1 through 8.8.6. Only systems that involve thermal expansion or contraction in either the inner or outer system (or both) are considered; those that experience negligible changes in dimension/temperature are simpler in nature and do not require compensation.

Table 8.7 Five Digit Cell Classification System for Classifying Double Containment Piping System Sections According to Their Design Conditions

1st Digit Location of system		2nd Digit Condition of primary		3rd Digit Condition of secondary		4th Digit Temp. change basis for stress range of primary (°F)		5th Digit Temp. change basis for stress range of secondary (°F)
A	Designates an above-ground portion of an installation	1	Restrained primary pipe	1	Restrained secondary containment pipe	0	0-10	0	0-10
						1	$10 < \Delta T \leq 25$	1	$10 < \Delta T \leq 25$
		2	Primary pipe free			2	$25 < \Delta T \leq 50$	2	$25 < \Delta T \leq 50$
U	Designates an underground portion of an installation			2	Secondary containment pipe free	3	$50 < \Delta T \leq 75$	3	$50 < \Delta T \leq 75$
						4	$75 < \Delta T \leq 100$	4	$75 < \Delta T \leq 100$
						5	> 100	5	> 100

Note: Any double containment piping system can involve several different sections, each of which can have different design conditions, and thus, the different sections would be assigned a different cell class. For instance, an aboveground portion of a system could be a type A2234, while an underground section of the same pipe system could be a type U2130. In other cases, there might be only a single cell class that applies to the whole system. Classifying system design criteria by cell class is helpful to identify design possibilities readily.

8.8.1 Type "A" Conditions - Considerations for Primary Piping, Above-Ground Systems

Hot fluids are often transported in the primary piping of an above-ground double containment piping system. Even where a fluid is not initially hot, heat gained by the fluid from a centrifugal pump may be appreciable, causing conditions conducive to differential thermal expansion to take place in the system. A full range of possible compensation alternatives exist for either situation. The various alternatives include: expansion loops, offsets, use of changes of direction, swing joints, use of expansion joints applied to the secondary containment piping, expansion joints or flexible-style grooved mechanical couplings applied to the primary piping that are housed in an accessible containment vessel, use of a restrained design, use of cold spring, use of heat exchangers to cool the media prior to transporting, and combinations of these techniques.

Wherever an anchor might be provided in a single-wall piping system, an internal anchor or other anchoring method must be provided to anchor a primary pipe in the comparable situation in a double containment piping system. If inner and outer pipe systems are interconnected, the point of interconnection will act as an anchor on the primary piping. This is true whether or not the secondary containment pipe is anchored externally; it is advisable under most conditions at least to guide secondary containment piping frequently, to prevent excessive bending of both pipes. However, it is usually best to anchor the secondary containment pipe externally in the immediate area of the internal anchor, which will produce the least stresses on the outer jacket. Anchoring may be accomplished either internally or by terminating a zone of secondary containment, since any termination point is a point of rigid interconnection between inner and outer systems. An external anchor may be added to single-wall components (flanges) in the terminated area (outside the secondary containment zone); this arrangement is then termed a double-termination/anchor arrangement (e.g., see Figure 11.41 through 11.43, 11.47 and 11.48). For more discussion concerning anchoring, refer to Sections 8.6.10 and 11.1.1.4.

In Type "A" conditions where large temperature changes occur, the use of a totally restrained system is not advisable for most nonmetallic systems. If a nonmetallic system is to be subjected to Type "A" conditions, it is not advisable to have inner and outer pipes restrained together, or high discontinuity stresses may develop in the primary components. In order to keep inner and outer thermoplastic systems from being totally connected, staggered welding (bonding) techniques must be used (staggered welding is the process of joining inner pipes and outer pipes separately, as discussed in Section 12.3.2). In a system with Type "A" conditions to be designed in an unrestrained fashion, interstitial supporting devices that allow freedom of movement to occur without producing excessive friction must be used. Thus, it is not advisable to use simultaneous fusion-joining techniques in nonmetallic systems, or consecutively bonded annular collars at the end of straight sections, subjected to these types of conditions.

8.8.2 Type "U" Conditions - Considerations for Primary Piping, Underground Systems

A condition in which larger dimensional changes are anticipated in primary piping is a common design situation involving double containment piping systems. This usually occurs whenever hot, spent process fluids are to be transported to a distant waste treatment site (e.g., spent semiconductor etching solutions). All of the compensation techniques that are used to compensate systems having Type "A" conditions (Section 8.8.1) also may be used for underground systems, with the exception of expansion-type devices or joints placed on the secondary containment pipe. If the piping is properly buried, the outer piping of such systems will be restrained from movement.

Internal-anchor fittings may be used as a means of anchoring primary piping in systems of this type, without disrupting the secondary containment zone. When expansion loops, offsets, or changes of direction are used as a means of compensating for thermal expansion/contraction, expanded secondary containment sections (if they need to be expanded) should be oriented in an eccentric fashion. In other words, eccentric reducers should be used to expand the secondary containment areas, as opposed to using concentric reducers, to avoid low points in the completed assembly. Refer to Section 8.6.8 for further discussion on eccentric versus concentric orientation. Venting requirements of the annulus in any eccentrically oriented expanded section must be addressed in the design and layout.

In Type "U" conditions where large temperature changes occur, the use of a totally restrained system is not advisable for most nonmetallic systems. If a nonmetallic system is to be subjected to Type "U" conditions, it is usually not advisable to have inner and outer pipes totally restrained together, or high discontinuity stresses may develop in the primary components, leading to failure due to creep-strain accumulation upon repeated cycles. In order to keep inner and outer pipes from being totally connected in thermoplastic systems, staggered welding (bonding) techniques must be used. For a system having Type "U" conditions to be designed in an unrestrained fashion, interstitial supporting devices must be used that allow freedom of movement to occur without producing excessive friction and that also allow proper diametrical deflection of the outer pipe. Thus, it is not advisable to use simultaneous fusion joining techniques, or consecutively bonded annular collars at the end of straight sections, in nonmetallic systems subjected to these types of conditions.

Other options include the use of manholes to house double containment or single containment fittings. Refer to Sections 16.4 and 11.4.2 for a discussion on alternative secondary containment structure and secondary containment manhole design, respectively. It is common practice to terminate secondary containment piping in systems at the entry to secondary containment manholes in order to house single fittings within a manhole. (The manhole itself provides secondary containment for the primary fittings, if the requirements of the authority having local jurisdiction are met.) Manholes provide an ideal arrangement to incorporate expansion joints in order to accommodate movements of a primary system.

Manholes should be provided with a leakproof lining or coating where applicable or required, or constructed of a corrosion-resistant structural material. If this is done, and all penetrations of the manhole are sealed, a continuous zone of secondary containment can be maintained, even if the secondary containment piping has been terminated upon manhole entry.

In cases involving a large temperature increase, double containment piping can be replaced with single-wall piping in a lined concrete trench as a better overall alternative. Unlined concrete trenches have been used in some localities, although it is not recommended by this text, nor in most jurisdictions in the United States. The issues involving the use of trenches as part of a containment system are discussed in more detail in Chapter 16.

8.8.3 Type "A" Conditions - Considerations for Secondary Containment Piping, Above-Ground Systems

In a system that experiences this type of condition, thermal expansion is easily controlled by providing expansion joints, attached to the secondary containment piping by means of flanges, as shown in Figure 8.15. The drawback to providing ordinary expansion joints on this type of system is that their sizes tend to be very large and, by the nature of the chemicals being handled, may require the use of a convoluted PTFE bellows, or other highly corrosion-resistant material. These types of expansion joints can be very expensive and in large diameters would add significant expense into the system. A less expensive and efficient approach is to use "internally guided," simple expansion sleeve couplings of the type shown in Figures 11.55, and 11.56. A more practical solution may be to use flexible-style grooved mechanical couplings, which function as an expansion joint on the outer jacket.

In many applications, a satisfactory approach is to anchor secondary containment pipe externally at selective locations and place intermediate guides between fixed points, so as to prevent buckling failure. This results in a restrained design of the secondary containment system; in the case of metallic secondary containment piping, it is usually an acceptable choice under the levels of temperature change to be expected in Type "A" conditions. In the case of nonmetallics, the acceptability of this practice depends on the specific material being used and the overall levels of stress involved. It is up to the user to determine the acceptability of using this technique for metallic and nonmetallic secondary containment piping, through analysis.

In Type "A" conditions where large temperature changes occur, the use of a totally restrained system is not advisable for most nonmetallic systems. If a nonmetallic system is to be subjected to Type "A" conditions, it is not advisable to have inner and outer pipes restrained together, except for FRP systems, or high discontinuity stresses may develop in the primary components. In order to keep inner and outer thermoplastic systems from being totally connected, staggered welding (bonding) techniques must be used (staggered welding is the process of joining inner pipes and outer pipes separately, as discussed in Section 12.3.2). In a system with Type "A" conditions to be designed in an unrestrained fashion, interstitial supporting devices that allow freedom of movement to occur without producing excessive friction must be used. Thus it is not advisable to use simultaneous fusion joining techniques, and in some cases, consecutively bonded annular collars at the end of straight sections in nonmetallic systems subjected to these types of conditions.

If the more severe temperature change is due to a decrease in temperature from the installation temperature, the components may be placed under a tensile stress that may be excessive for the materials involved. A suitable method for controlling thermal contraction is to use heat tracing as a means of limiting the overall decrease in temperature. The application of heat tracing for this purpose is described in Section 8.7.1 and in Chapter 10.

8.8.4 Type "U" Conditions - Considerations for Secondary Containment Piping, Underground Systems

Cases that involve this potential combination of factors usually arise due to negative swings resulting from burying a pipe that has developed a high surface temperature due to direct sun exposure. When a cool covering is later placed over the hot pipe during backfilling procedures, there is a resulting initial contraction of the material. The piping can then be subject to additional contraction if the soil further decreases in temperature during the colder months, and the pipe is buried at shallow depths.

In these types of systems, proper burial techniques (always important for double containment pipes anyway) will prevent the pipe from moving due to the action of soil friction on the pipe walls. Also, by simply waiting until early morning to bury the pipe, potential temperature changes can be minimized to a great extent. Thus, it is usually advisable to follow this practice. For direct-buried secondary containment pipes, the use of expansion joints, offsets, loops, etc. will not assist in the alleviation of thermal expansion/contraction in Type "U" conditions.

Because the temperature change is usually relatively small, buried metallic pipe can usually be classified as fully restrained for this condition. A temperature change of more than 200°F (392°C) is required to reach yield stress in most metallic materials.

For extreme conditions, the use of a concrete trench to house the entire system may be warranted. This will avoid a secondary containment pipe system being subject to severe and damaging displacement strains. A trench may house a double containment pipe system in a tertiary-containment fashion. It may also house a single-wall pipe and act as a means of secondary containment, in lieu of having a secondary containment pipe. Chapter 16 discusses the use of trenches as a design alternative for double containment piping systems.

8.8.5 Nondifferential Thermal Expansion/Contraction, Above-Ground Systems

When nondifferential thermal expansion of above-ground systems is expected, an entire double containment system can be easily compensated for by allowing for sufficient flexibility of the completed assembly. In most systems of this type, thermal stresses will be well balanced throughout the assembly if one designs for flexibility of the secondary containment system. For this to occur, primary and secondary piping must be equally compensated; stresses in both systems will then be alleviated and the stresses and displacement strains will be well balanced throughout. The only method that must be avoided is the use of expansion joints applied to only the inner or outer pipe systems, as this will create conditions of differential thermal expansion. In systems of this type, which are also are rigidly welded (bonded) together (totally restrained), the combined system will be very rigid and will tend to resist bending, making expansion alleviation a more difficult task.

8.8.6 Nondifferential Thermal Expansion/Contraction, Underground Systems

Nondifferential thermal expansion of underground piping can be treated the same as that of the differential thermal expansion case, unless the pipes are totally restrained (i.e., simultaneously welded systems). Except for totally restrained systems, all of the same methods of compensation suggested for the underground case involving differential thermal expansion (see Section 8.8.2) can be effectively implemented for systems having non-differential conditions. If the total amount of thermal expansion is small enough, then having both systems permanently attached to one another, and thus both restrained by the soil friction, will not represent a problem.

In non-differential expansion/contraction conditions where large temperature changes occur, the use of a totally restrained system is not advisable for most nonmetallic systems. If a nonmetallic system is to be subjected to such conditions, it is usually not advisable to have inner and outer thermoplastic systems totally restrained together, or high stresses may develop in the primary components. In order to keep inner and outer pipes from being totally connected, staggered welding techniques must be used. For a system having these conditions to be designed in an unrestrained fashion, interstitial supporting devices that allow freedom of movement to occur, without producing excessive friction, and that also allow proper diametrical deflection of the outer pipe must be used. Thus it is not advisable to use simultaneous-fusion joining techniques or consecutively bonded annular collars at the end of straight sections in nonmetallic systems subjected to these types of conditions, when large magnitude temperature changes occur.

References

1. C. Becht IV, C. Benteftifa, Analysis of Polypropylene Termination Fittings," Becht Engineering, Liberty Corner, NJ, 1992.

2. Carl E. Martin, "Thermal Expansion Loads in Double Containment FRP Systems," Fibercast Company, Sand Springs, OK, 1990.

3. W. E. Short II, "Failure Assessment and Redesign of a Jacketed Piping System, a Practical Approach," ASME PVP - Vol. 139, Book No. 400423, 1988.

4. Dynaflex Users Manual, Mod. 205E, 1984, Scientific Software - Intercomp.

5. Richard Getz, "How to Stress Analyze Jacketed Piping," Hydrocarbon Processing, February 1978.

6. C. Becht IV, Y. Chen, "Double Containment Anchor Fitting Stress Evaluation," Becht Engineering, Liberty Corner, NJ, 1994.

7. C. Becht IV, et. al., op. cit.

8. C. Becht IV, et. al., op. cit.

Notes

Notes

Chapter Nine

Structural Considerations

9.1 General

There are many structural aspects of double containment piping systems that must be addressed in an engineering design. This includes considerations involving external support of above-ground systems, longitudinal bending of above- and underground systems, and burial considerations for underground piping. The considerations presented in this chapter are applicable to single-wall piping systems, primary portions of double containment piping systems, and to double containment piping assemblies.

Many structural concerns for double containment piping have yet to be studied or considered on a wide scale. This is particularly true of burial considerations, where analysis that have been developed for flexible single-wall pipes may no longer apply once a pipe is housed within another and separated by interstitial supports. The added rigidity and stiffening effect that the interstitial supports and interconnecting components provide to the complex structure can change the nature of the problem, namely, the analysis of burial performance of the direct-buried pipe.

The moment of inertia and centroid of a double containment assembly may sometimes be markedly different from a comparable secondary containment pipe considered alone. Since there are many structural members present in assembly (i.e., interstitial supports), inner and outer pipes tend to add further stability via loads transmitted to each other. In comparison to a single-wall pipe, double containment assemblies can be more rigid, easier to support, harder to bend, less resistant to soil loads, and may have other characteristics that are different.

Basic structural concepts, techniques, and methods for analysis are presented in this chapter. However, designers are to be cautioned that every system is unique, and that an appropriate expert must be consulted in order to model correctly the structural aspects of a double containment piping system.

9.2 Longitudinal Bending Analysis

In some circumstances, there is a need to deform pipe intentionally by bending, and in other circumstances, pipe bending cannot be avoided. Such purposes may include snaking of a pipe in direct-burial situations, bending of a pipe in underground situations to go

Figure 9.1. Example of a double containment piping system that has been deflected into position in order to conform to a building structure and tree line.

around structures, and the situation when a pipe is laid in a trench after being welded outside of the trench. An example of a double containment piping system that has been cold bent for placement in a trench to go around a building structure is shown in the photograph of Figure 9.1. The piping system is bent to follow the trench line, thereby avoiding the need to install an elbow, and instead use straight pipe. So long as the piping is not bent to a more severe angle than its minimum bending radius, and the stress due to bending does not exceed the allowable stress of the material, the pipe can be bent into position. This is a common procedure for long buried pipes.

Pipes that are subjected to intentional bending can be analyzed by calculation of a pipe's bending stress, and comparing that stress to the allowable stress of the material. The determination of allowable stress range for metallic components are covered in Chapter 5, and therefore are not repeated here [see Eqs. (5.2) and (5.3)]. Values of allowable design stresses for most common metallic piping materials, at a variety of temperatures, are stated in Appendix A of the ANSI/ASME B31.3 Code for Chemical Plant and Petroleum Refinery Piping, and Appendix A of the ANSI/ASME B31.1 Code for Power Piping (refer to the current editions of these codes). The basis for allowable stresses for nonmetallic materials are discussed in Chapter 6; values for design stresses for a few nonmetallic materials are shown in Appendices B of ANSI/ASME B31.3 and III of ANSI/ASME B31.1.

For purposes of bending analysis of nonmetallic materials, allowable stresses should be reduced to one half the value usually used in Equations 6.2 and 6.3. This is due to nonmetallic materials having welded (bonded) joints that can experience end thrusts (from applied internal pressure), which impose a tensile stress of up to one-half of the hoop stress. Based on this approach, the allowable bending stress for a nonmetallic material for purposes of longitudinal bending analysis can be determined by Eqs. (9.2)-(9.4). For non-pressure-rated, nonmetallic piping that uses gasketed connections (with proper thrust blocking), the allowable stresses shown in Appendix B of the ANSI/ASME B31.3 and Appendix III of ANSI/ASME B31.1 can be used without any additional reduction factor applied.

Structural Considerations

$$S_B = \frac{HDS}{2} \tag{9.1}$$

where:
S_B = suggested allowable stress in bending for a nonmetallic pipe, psi
$HDS = S$ = hydrostatic design stress, found from Eqs. (6.2) or (6.3), psi

Finite bending of pressurized tubes has been discussed by Reissner.[1] Much of the analysis presented here is based on the Reissner report, along with methods of classical structural theory.[2-5]

Stresses due to thermal expansion and contraction, internal and external pressure, and weight effects should be considered as acting concurrent along with bending of the pipes, if these effects are part of the design. As discussed in Chapters 5, 6, and 7, the effects of all forces must be considered in the analysis of a piping system. The discussion here treats solely the bending analysis of the piping, as if bending were acting alone. However, these other forces must be taken into account where they apply, to conduct a proper model and analysis of an actual double containment piping system.

The equations presented in this section are for bending analysis of single-wall pipes. The bending analysis of double containment assemblies is more complex, as the effect of interstitial supports and other interconnecting components act to stiffen both pipes and transfer loads between each pipe. As a result, it may take a greater force to deflect a combination of pipes, since the added rigidity provides the complex assembly with increased resistance. The situation becomes more complex when different materials, each having its own allowable bending stress, modulus of elasticity, and Poisson's ratio are combined. Also, a primary pipe may be subject to a continuous internal pressure, whereas its associated secondary containment housing may coexist in a state where it is not subjected to an internal pressure. This means that inner and outer pipe systems may require different forms of analysis, though they constitute a combined system. A structural engineer, with expertise in modeling complex double containment assemblies, should be consulted to study this situation properly.

9.2.1 General Methods, Single-Wall Piping Considerations

The relationship between the moment that is induced by the longitudinal bending of a pipe and its corresponding bending stress is found from Eq. (9.2). To determine the moment of inertia of a single-wall pipe, Eq. (9.3) can be used.

$$M = \frac{S_b I}{C} \tag{9.2}$$

where:
M = moment in bending, in.-lbs
I = moment of inertia, in.^{4}
C = ratio of D/2, in.
D = outside diameter, in.
d = inside diameter, in.

$$I = \frac{\pi}{64}\left(D^4 - d^4\right) \tag{9.3}$$

For very thick single-wall pipes, a bent length of pipe will conform approximately to a circular arc after bending. The minimum radius of the bending circle can be found by Timoshenko's equation:

$$R_b = \frac{EI}{M} \quad (9.4)$$

where:
R_b = minimum radius of bending, in.
E = modulus of elasticity, psi

Substituting for the moment of inertia in Eq. (9.4), and rearranging Eq. (9.2) to solve for I, results in:

$$R_b = \frac{ED}{2S_b} \quad (9.5)$$

This equation suggests that the minimum radius of bending has no relationship to the wall thickness of a pipe. In thick-walled pipes this is approximately true. However, in thin-walled pipes, the actual bending that can be achieved is greater than would be expected by application of the equations above. The reason is that both the tensile forces at the convex side of the pipe and the compressive forces of the concave side of the pipe have resultants towards the neutral axis. Therefore, circular cross sections of the pipe tend to become flattened and elliptical in shape. This distortion affects the strain of the longitudinal fibers of thin-walled pipe. The fibers of a thin-walled pipe farthest from its neutral axis do not share in the stresses indicated by ordinary bending theory. The effect on bending stress is therefore the same as if the moment of inertia were decreased. Therefore, for thin tubes, the following revised equation can be used[6]:

$$R_b = \frac{kEI}{M} \quad (9.6)$$

Where k is determined by the following formula:

$$k = 1 - \frac{9}{10 + 12\left(\frac{tr}{a^2}\right)^2} \quad (9.7)$$

where:
t = pipe wall thickness, in.
r = radius, in.
a = constant

The effect of the flattening depends on the magnitude of the ratio tr/a^2. The flattening on the stress distribution has been shown by Karman to be determined by the following two equations:

Structural Considerations

$$\sigma = \frac{My}{kI}\left(1 - \beta\frac{y^2}{a^2}\right) \tag{9.8}$$

where:
σ = stress, psi

$$\beta = \frac{6}{5 + 6\left(\dfrac{tr}{a^2}\right)^2} \tag{9.9}$$

The maximum stress is obtained from:

$$\sigma_{max} = k_1 \frac{Md}{2I} \tag{9.10}$$

k_1 can be determined by Eq. (9.11). Values of k_1 are shown in Table 9.1.

$$k_1 = \frac{2}{3k\sqrt{3\beta}} \tag{9.11}$$

To determine the central angle subtended by the length of pipe, Eq. (9.12) can be used. The angle of lateral deflection is determined by Eq. (9.13). The offset at the end of the pipe, determined from the tangent to the circle, is determined by Eq. (9.14). To calculate the lateral offset force necessary to bend a pipe, Eq. (9.15), which is used for cantilever beams, is also used for the single-wall pipe. All of the various terms included in Eqs. (9.12)-(9.15) are illustrated in Figure 9.2.

$$\beta = \frac{360L}{2\pi R_b} = \frac{57.30L}{R_b} \tag{9.12}$$

$$\alpha = \frac{\beta}{2} \tag{9.13}$$

$$A = 2R_b\left(\sin\frac{\beta}{2}\right)^2 = 2R_b(\sin\alpha)^2 \tag{9.14}$$

$$P = \frac{3EIA}{L^3} \tag{9.15}$$

Table 9.1 Values of k_1 for Various Values of tr/a^2

tr/a^2	0.3	0.5	1.0
k_1	1.98	1.30	0.88

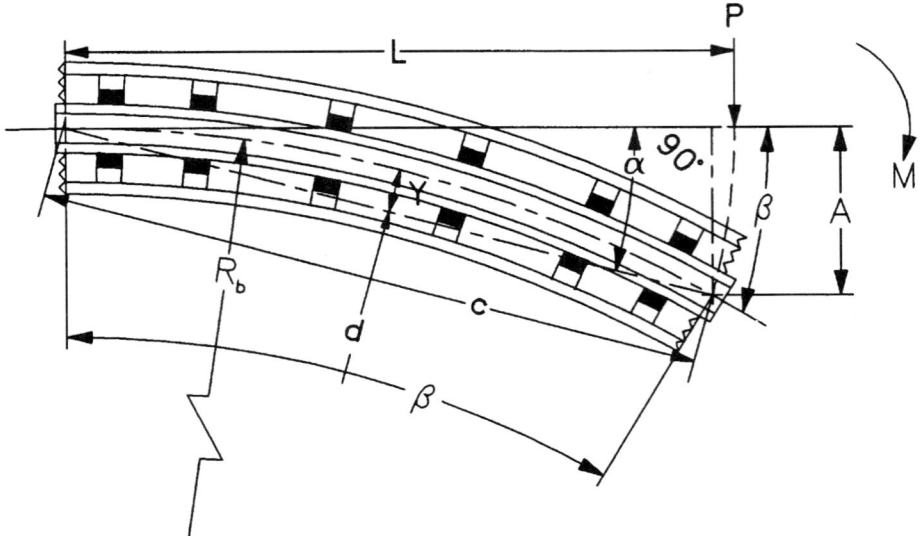

Figure 9.2. Illustration of a single-wall pipe bend showing important terms.

where (refer also to Fig. 9.2):
L = length of pipe being bent, in.
R_b = minimum radius of bending, in.
β = central angle subtended by the length of pipe, degrees
α = angle of lateral deflection, deg.
A = offset at end of pipe, determined from the tangent to the circle, in.
P = force necessary to bend the pipe to achieve the offset A, lbs
I = moment of inertia, in^4

For thin-walled pipes, the methods of Reissner can be used to determine the allowable degree of bending based on its diametrical deflection.[7] These methods are based on the fact that a thin-walled pipe will ovalize into an approximately elliptical shape, as described earlier. To determine the amount of diametric deflection achieved, the percentage of deflection found from Eq. (9.16) or (9.17) can be used.

$$\delta = \frac{\Delta}{D} \tag{9.16}$$

$$\% \text{ deflection} = 100\delta = 100\frac{\Delta}{D} \tag{9.17}$$

where:
D = outside diameter, in.
Δ = change in diameter, in.
δ = ratio

Structural Considerations

Reissner's correlation is expressed in Eq. (9.18) with the value for λ determined by Eq. (9.19), and $A_1 a^2$ defined by Eq. (9.20). The usual amount of acceptable deflection for a nonmetallic pipe is 5%, although for some types the acceptable amount may vary. For metallic materials, the amount of acceptable deflection can be determined by stress-versus-strain formulas.

$$\delta = \frac{\Delta}{D} = -(A_1 a^2)\left(\frac{2}{3} + \frac{71+4\lambda}{135+9\lambda}(A_1 a^2)\right) \tag{9.18}$$

$$\lambda = \frac{12(1-\mu^2)PD_M^3}{8Et^3} \tag{9.19}$$

$$(A_1 a^2) = \frac{1}{16}\left(\frac{18(1-\mu^2)}{12+4\lambda}\right)\frac{D_M^4}{R^2 t^2} \tag{9.20}$$

where:
μ = Poisson's ratio, dimensionless
P = internal pressure, psi
D_M = mean diameter, in.
E = modulus of elasticity, psi
t = pipe wall thickness, in.
R = radius of bending, in.

When a specified change in direction in a pipe exceeds the permissible deflection angle for a given pipe length, the bend can be accomplished by the use of a multiple bend. A multiple bend of a single-wall pipe is illustrated in Figure 9.3. Equations (9.21)-(9.25) show calculations for determining the various offsets for a multiple-bend single-wall pipe.

$$A_1 = C(\sin \alpha) \tag{9.21}$$

$$A_2 = C(\sin \alpha + \sin 2\alpha) \tag{9.22}$$

$$A_3 = C(\sin \alpha + \sin 2\alpha + \sin 3\alpha) \tag{9.23}$$

$$A_4 = C(\sin \alpha + \sin 2\alpha + \sin 3\alpha + \sin 4\alpha) \tag{9.24}$$

$$A_n = C(\sin \alpha + \sin 2\alpha + \sin 3\alpha + ... + \sin n\alpha) \tag{9.25}$$

where:
A_1, A_2, A_3, A_4, A_n, C, and α relate to Figure 9.3

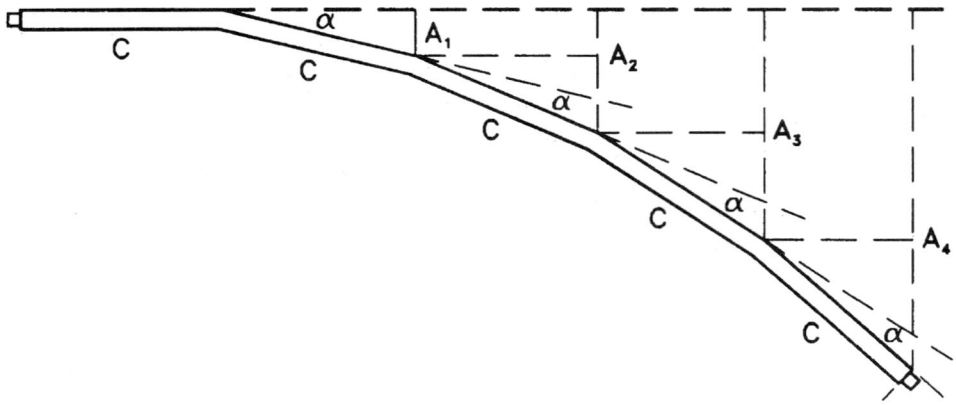

Figure 9.3. Illustration of a single-wall multiple pipe bend showing important terms.

To calculate the bending strain of piping, the following relationship may be used:

$$\varepsilon = \frac{S_b}{E} = \frac{D}{2R_b} \tag{9.26}$$

where:
ε = strain, in./in.
S_b = bending stress, psi
E = modulus of elasticity, psi
D = outside diameter, in.
R_b = radius of bending, in.

9.1.2 General Methods; Double Containment Piping Assemblies

For a double containment pipe assembly that has inner and outer pipes constructed of the same materials, its moment of inertia can be found by adding the individual moments of inertia to obtain a corrected moment. This is stated as:

$$I = \frac{\pi}{64}(D_2^4 - d_2^4) + \frac{\pi}{64}(D_1^4 - d_1^4) \tag{9.27}$$

where:
I = moment of inertia, in.^{4}
D = outside diameter, in.
d = inside diameter, in.

note: the subscripts 1 and 2 refer to the primary and secondary containment pipes, respectively

Equation (9.27) assumes a void area comprised by an annulus that surrounds 360° of the circumference of the primary pipe. At locations where interstitial supports exist, the cross-sectional moment of inertia can be determined by treating the cross section as a solid section, and then subtracting any open areas of the annular space. For a primary pipe that has n vanes of straight configuration welded to its exterior that extend approximately to the inside diameter of the secondary containment piping, the cross-sectional moment of inertia can be determined by adding the individual moments of the primary pipe, the vanes, and the secondary containment pipe using the center of the primary pipe as the centroid.

The same approach can also be used for a double containment pipe that has continuous vanes along its annulus. At locations where a collar-type interstitial support is used (see Figures 9.10 and 9.11), or where an internal anchor component is used in the design (see Figures 11.44 through 11.54), the moment of inertia can be estimated by treating a pipe as a solid component at that point, and subtracting the void areas. This can be found from:

$$I = \frac{\pi}{64}(D_2^4 - d_1^4) \tag{9.28}$$

For most applications of double containment piping systems where the same inner and outer materials are used for primary and secondary containment portions, and interstitial supports and interconnecting components are spaced a reasonable distance apart, Eq. (9.27) may be applied to the system as a whole to determine its corrected moment of inertia. For double containment piping systems involving combinations of materials, the simplified equations presented here would not apply.

For most situations, the minimum bending radius of the assembly can be calculated from Eq. (9.5) using the corrected moment of inertia from Eq. (9.27). If dissimilar materials are used for primary and secondary containment pipes, then the allowable bending stress of each material must be used in Eq. (9.5) and results checked to determine if the amount of bending is too severe.

9.3 Support of Primary Piping and Nonburied Assemblies

9.3.1 Primary Piping Considerations

9.3.1.1 Determination of Support Spacing - General Methods

All piping must be properly supported in order to prevent it from sagging and from possible deformation due to the dead weight of the components and the fluids that the piping system conveys. The determination of support spacing for piping of all materials historically has been determined on a trial basis, or has been based on the theory of cantilever beams.

When cantilever beam theory is used, the calculations vary according to the total number of supports in the system. For a system with more than five supports, the support spacing reaches a limit beyond which the requirement remains a constant. This condition (where five or more supports on a single run of pipe exist) is referred to as an "n-span" condition.

For a short section of pipe that has two supports over its length, the pipe is referred to as a one-span pipe. The cantilever-beam equation to calculate the maximum distance between supports for a one-span pipe, based upon a specified amount of sag between supports, y, is determined by:

$$L = \left(\frac{76.92\,yEI}{W}\right)^{0.25} \tag{9.29}$$

where:
L = distance between supports, in.
y = sag between supports, in.
E = modulus of elasticity, psi
I = moment of inertia, in.^{4}
W = weight of pipe per unit length, lb/linear in.

For a section of pipe that has three supports over its length, the pipe is referred to as a two-span pipe. The equation to calculate the maximum distance between supports for a two-span pipe, based upon a specified amount of sag between supports, is determined by:

$$L = \left(\frac{185.2\,yEI}{W}\right)^{0.25} \tag{9.30}$$

For a section of pipe that has four supports over its length, the pipe is referred to as a three-span pipe. The equation to calculate the maximum distance between supports for a three-span pipe, based upon a specified amount of sag between supports, is determined by:

$$L = \left(\frac{144.9\,yEI}{W}\right)^{0.25} \tag{9.31}$$

As described above, a pipe that has five or more supports along its continuous length is referred to as an "n"-span pipe. The equation to calculate the maximum distance between supports for an "n"-span pipe, based upon a specified amount of sag between supports, is determined by:

$$L = \left(\frac{153.8\,yEI}{W}\right)^{0.25} \tag{9.32}$$

The amount of acceptable sag between supports is typically greater for a plastic pipe than it is for a metallic pipe (or other rigid nonmetallic pipe). The amount of acceptable sag for most piping materials are based upon an acceptable strain. Even though the allowable sag between supports is less for a metallic material than for many nonmetallic materials (especially thermoplastic and reinforced-thermosetting-plastic materials), support spans that can be achieved are much greater due to the larger moduli of elasticities of metallic materials.

In the calculation of support spacing, it is very important to include the weight of the fluid to be conveyed in determining the value for weight per unit length in the equations. The weight of the heaviest fluid to be encountered in a pipe is the value on which that unit weight must be based. This is true even if the fluid used to pressure-test the line is the heaviest fluid to be supported. Also, be sure to use the lowest value of the modulus of elasticity that the system will experience in service. For a thermoplastic material, this value can drop substantially, even for a small rise in temperature.

Structural Considerations

Examples of various external supports and hangers, and external anchor and guide arrangements for all types of piping are shown in Figures 9.4 through 9.7. Piping supports that are selected for use should not cause excessive bearing loads to occur on the pipe being supported, and should be designed to allow adequate flexibility where required. Conversely, anchors and other supports that are designed to restrain the pipe must be selected based upon the ability of the support and its associated supporting structure to withstand the maximum loads that supported piping might impose on them. These effects must be considered as acting concurrent with all other effects in performing a flexibility analysis on the system. This subject is covered in more detail in Chapters 5 through 8 (see Sections 5.1, 6.1, 7.2, and 8.3.2).

Various types of spring hangers can be used to support piping where vertical expansion of a piping run is a concern. Spring hangers can be of the variable or constant support type.

9.3.1.2 Interstitial Supporting Devices

There are many names given to devices used to center a primary pipe system, and components thereof, within its associated secondary containment housing, thereby lending support to a primary system. Some of the common names include: (1) interstitial supports; (2) internal supports; (3) annular supports; (4) spacers; (5) standoffs; (6) centralizing devices; (7) centralizers; (8) spidering devices; (9) spiders; and (10) many others. (Note: Spider® Clip is a registered trademark of Asahi/America, Inc.). Interstitial or internal supports are the preferred term in this text, since these devices are positioned in the interstice, and have a primary purpose of lending structural support to the primary system.

There are several purposes that these components are intended to provide. Aside from providing support, these devices must also properly center a primary pipe system, where it is essential. These components must also properly distribute stresses between the two systems such that an even balance of stresses is maintained, and a concentration of localized discontinuity stresses is avoided or minimized, namely, point loads that are concentrated at a support. Oftentimes when thermal expansion is involved, these supports are intended to serve as guides, which allow movements of primary piping, and direct or limit such movements to occur in the longitudinal direction (axial direction).

Interstitial supports must be designed to provide these functions, while also allowing for free flow to occur in a system's annulus. In systems that use leak detection cable, they must also allow cable to be continuously run through them, along the 6 o'clock position of the annulus. In underground applications, attention must be paid to the amount of diametrical deflection that is allowed to the primary piping, which is a factor in the engineering design of an interstitial supporting device.

Thus, the design of these supports is not completely arbitrary. They are a central element in the successful design and operation of a double containment piping system. The design of these components affect many other factors of the system, including calculated stress ranges, annular-flow capacity, and other criteria.

9.3.1.2.1 Types of Interstitial Supporting Devices

There are many specific and proprietary designs of interstitial supports that are used for double-containment piping systems. However, these types may be grouped into four general categories: (1) vane or fin-type; (2) support pads; (3) welded or bonded collar attachments; (4) mechanically joined attachments, and (5) flexibility supports.

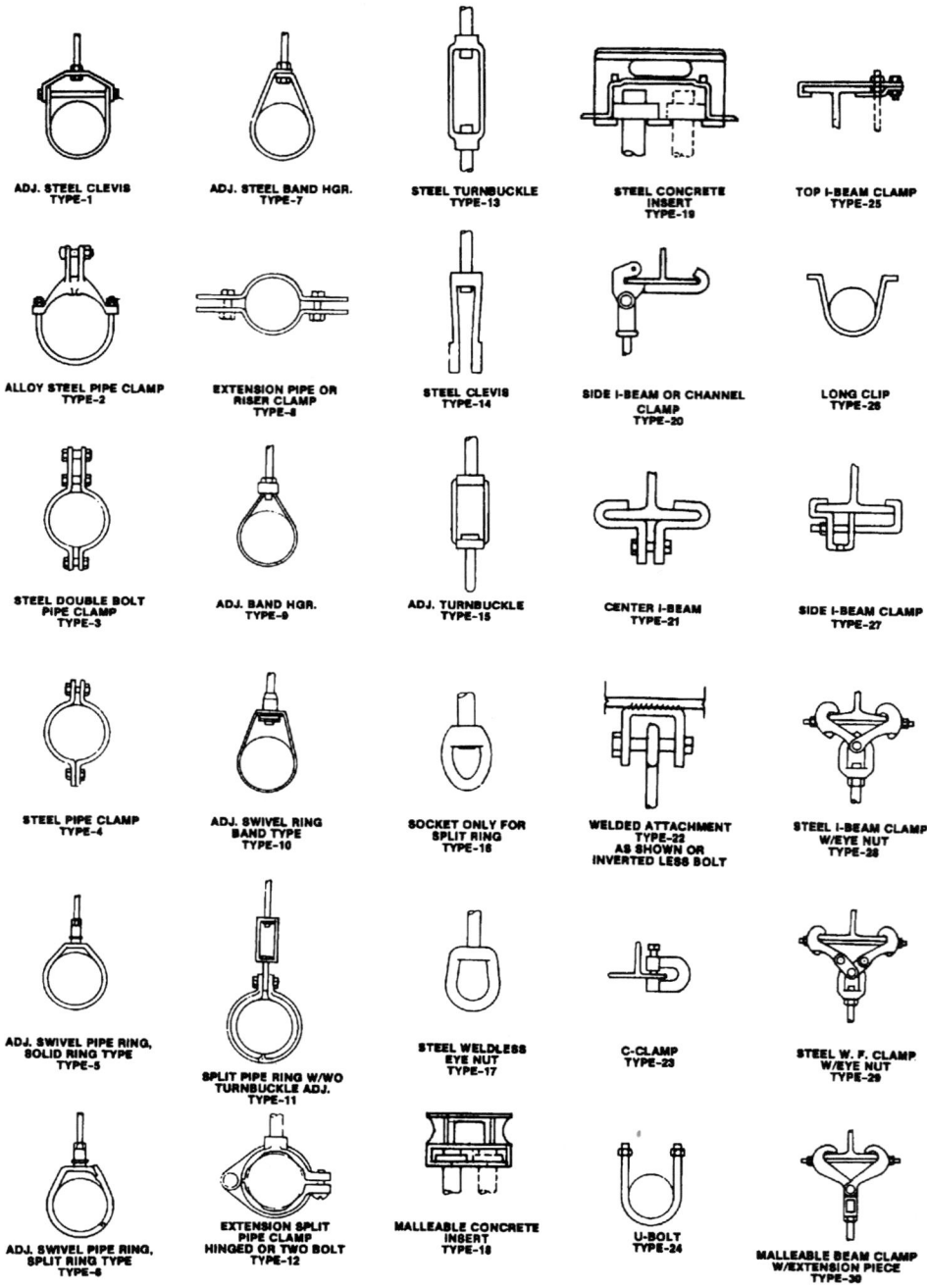

Figure 9.4. Examples of external hangers for external support of piping. (Sources: MSS SP-58-1988; ASPE Data Book, Chapter 13; Grinnell.)

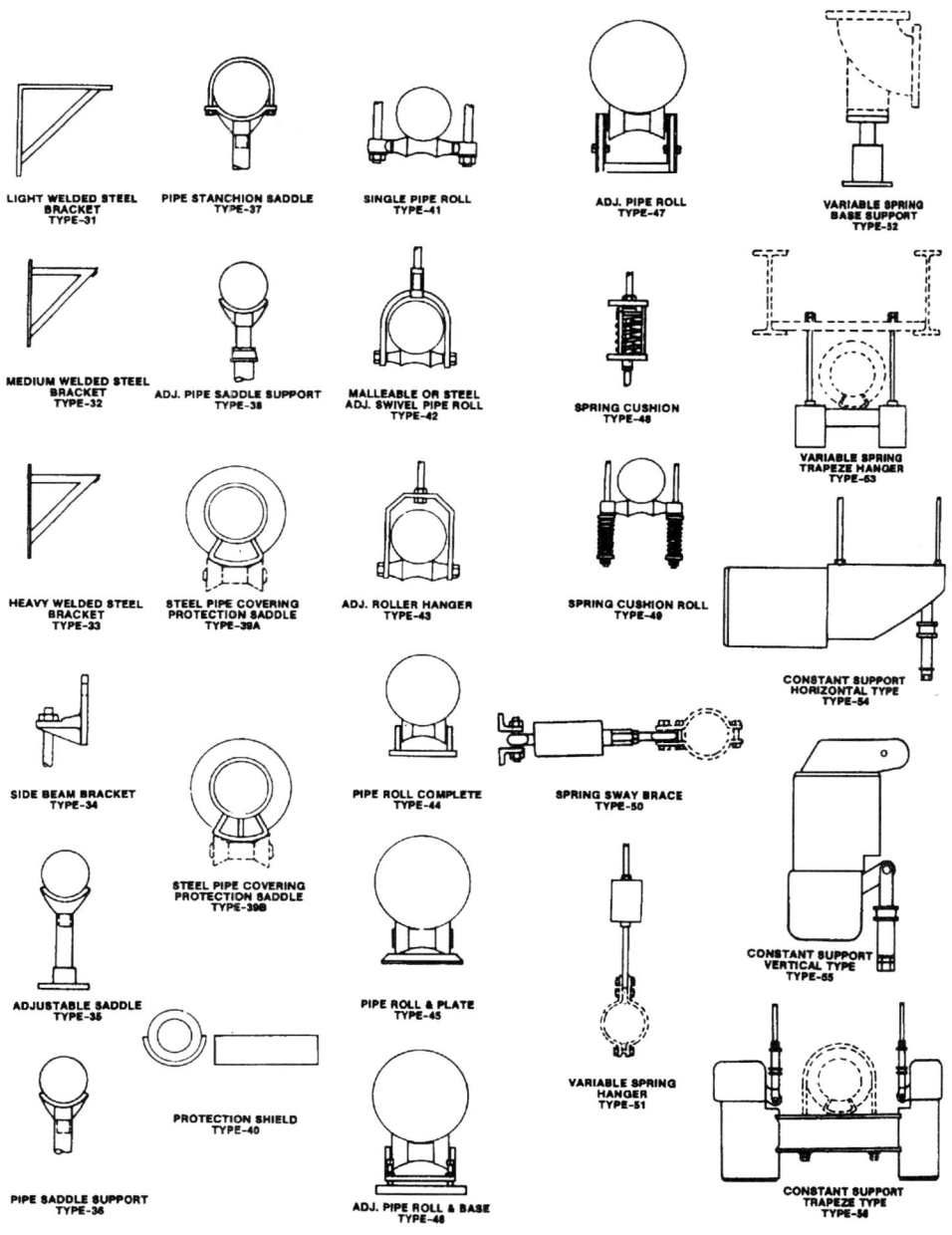

Figure 9.4. continued Examples of external hangers for external support of piping. (Sources: MSS SP-58-1988; ASPE Data Book, Chapter 13; Grinnell.)

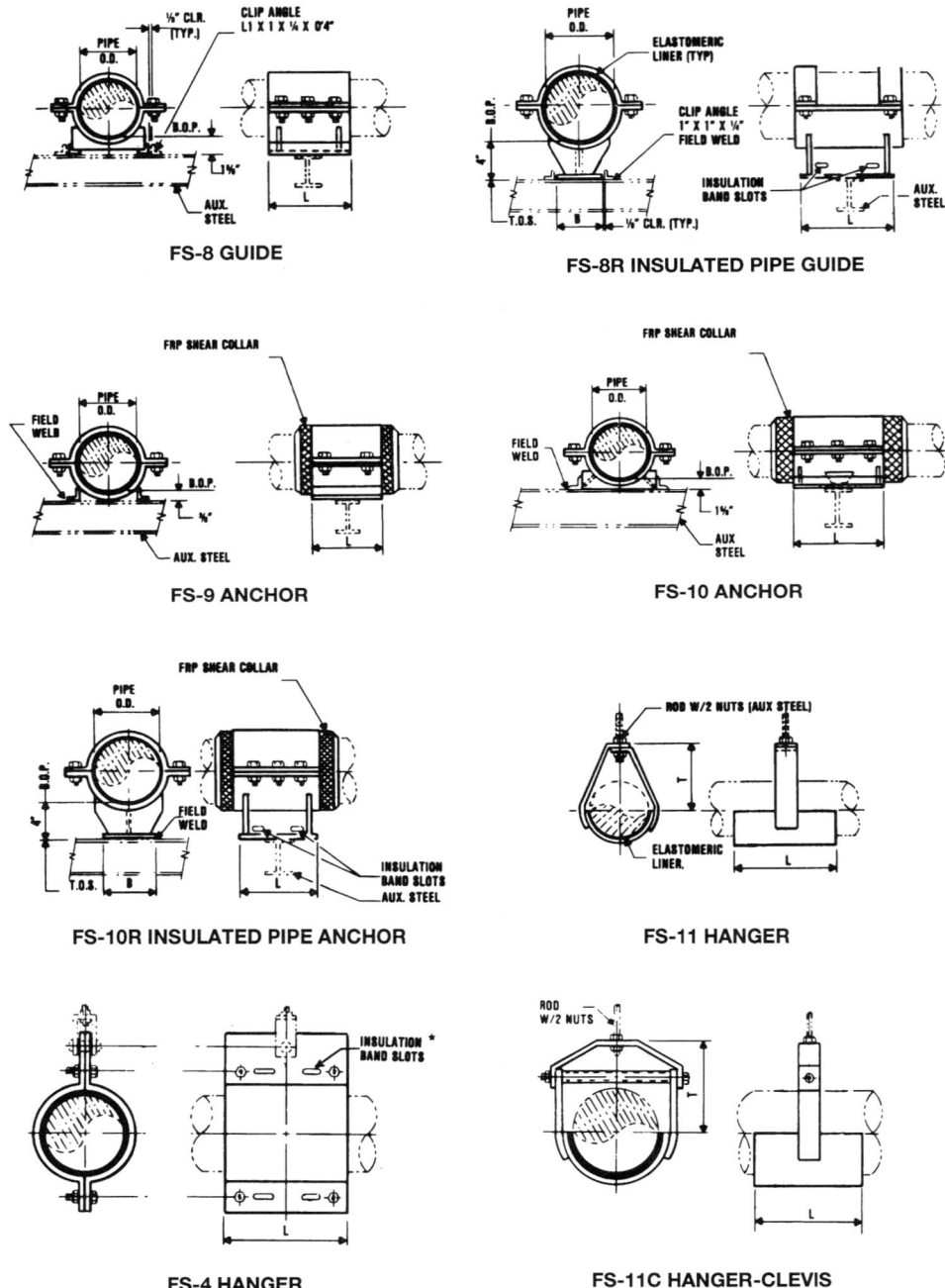

Figure 9.5. Examples of various types of supports, anchors and guides for fiberglass piping systems. (Source: Jove Manufacturing Company, a division of Jove Technologies, Inc.)

Structural Considerations

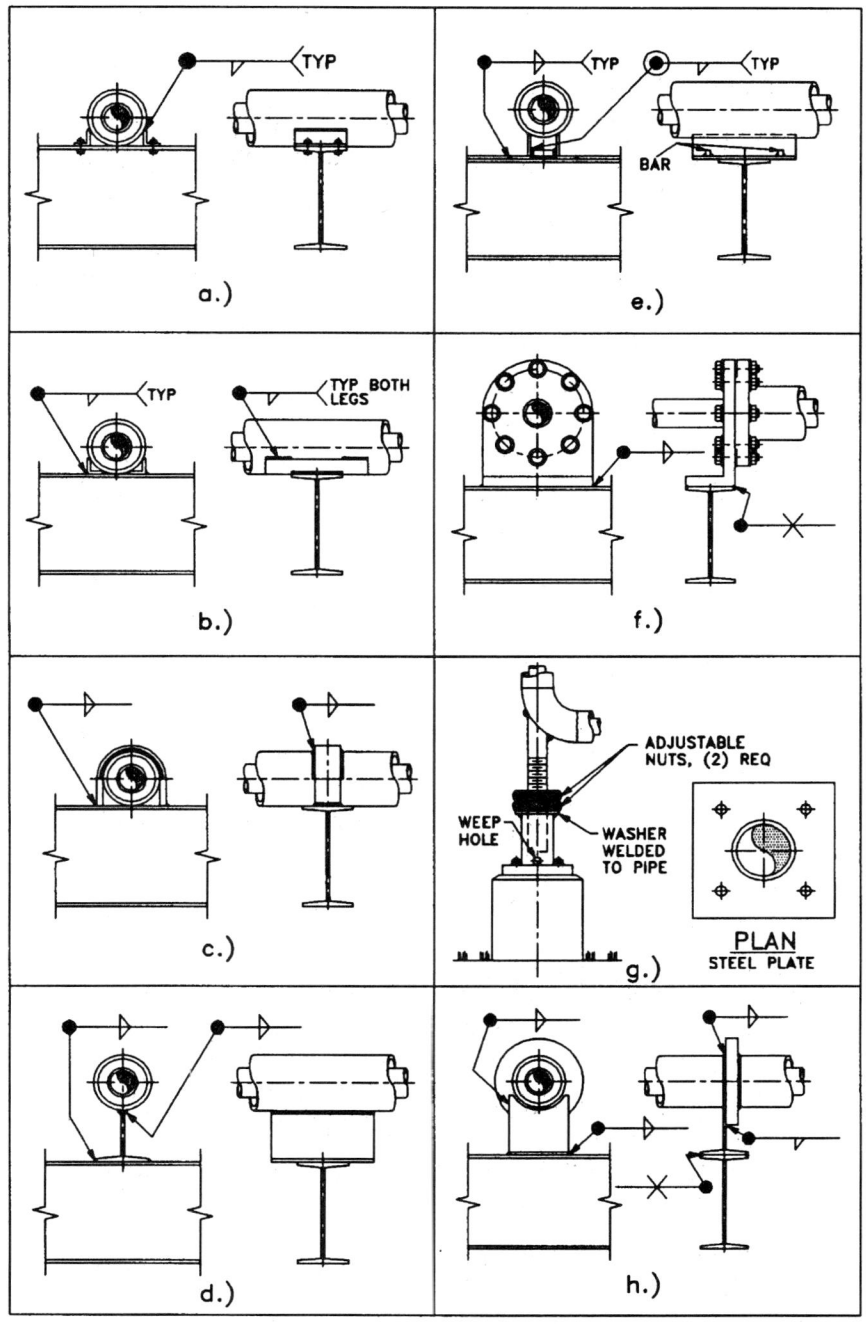

Figure 9.6. Examples of external anchor arrangements for double containment pipes with metallic outer piping. (Source: Becht Engineering.)

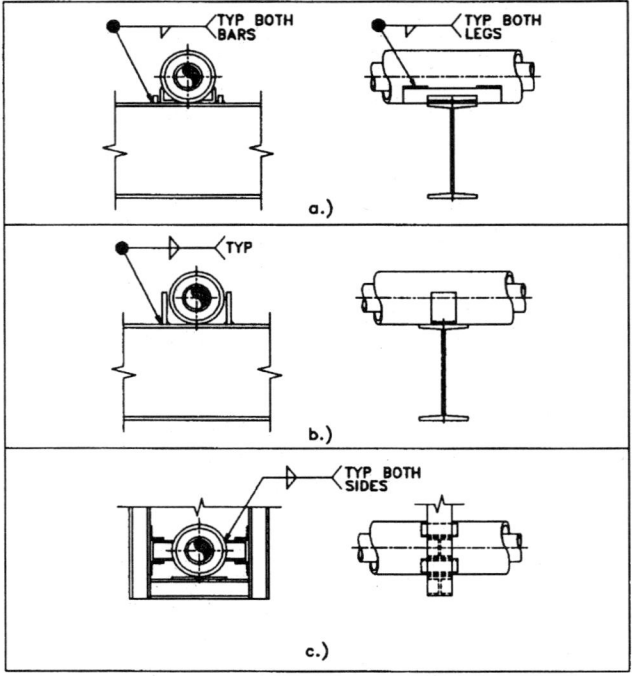

Figure 9.7. Examples of external guide arrangements for double containment pipes with metallic outer piping. (Source: Becht Engineering.)

9.3.1.2.2 Vane or Fin-Type Interstitial Supports

Vane or fin-type supports mainly consist of thin, short strips of sheet of the same material as that of the primary pipe. These are usually welded (bonded) directly to the surface of the primary pipe at several positions around the circumference. Common designs include three-, four-, five-, and six-vane arrangements in which the vanes are directly welded to the primary pipe, illustrated in Figure 9.8. Directly-welded vane types of supporting devices have been commonly used in jacketed-process piping systems, and are more readily acceptable for systems that use metallic components. Figure 9.9 presents other vane or fin-type supports, whereby the vanes are directly attached to a flexible base pad. In such a support the pad is then attached, or welded (bonded) directly to the primary piping. Thermoplastic piping that is extruded with continuous vanes along its entire external length, along the entire inner surface of a secondary containment pipe, or inner and outer pipes that are extruded along with longitudinal vanes as a homogeneous construction, all may be technically classified as having a vane or fin-type interstitial supporting structure.

9.3.1.2.3 Interstitial Support Pads

A support pad is a common type of device that is used in applications involving fiberglass and metallic primary pipes. A pad is essentially a wider vane which encompasses approximately 15°-30° of the annulus. Typically, three pads positioned at 120° on centers are used,

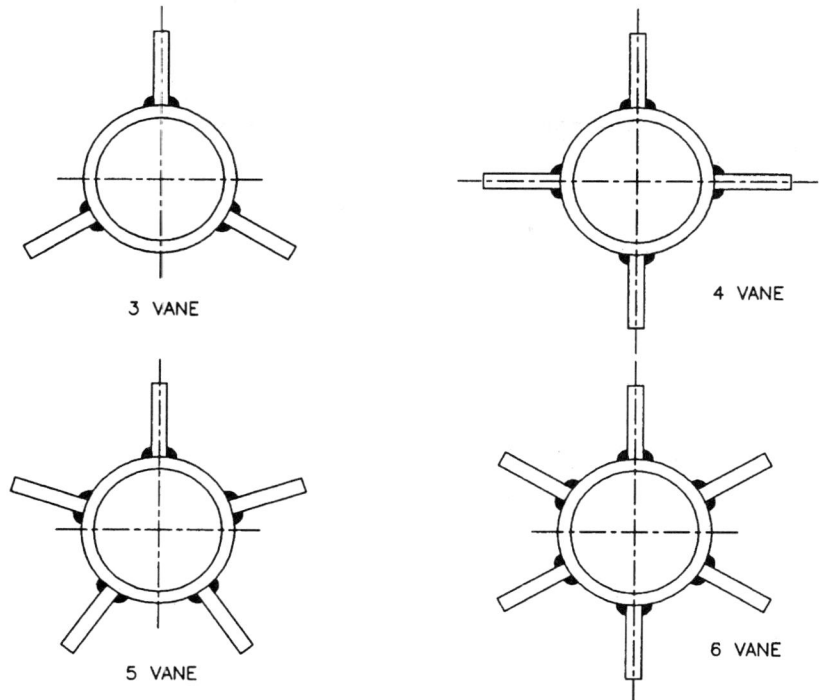

Figure 9.8. Illustration of three-, four-, five-, and six-vane fabricated vane-type interstitial supports, where the vanes or fins are welded directly to the primary pipe.

although as few as one at the 6 o'clock position and as many as four positioned at 90° on centers may be used. Typical three-pad arrangements are illustrated in Figure 9.10.

9.3.1.2.4 Welded or Bonded Interstitial Collar Attachments

Interstitial supporting devices that are of the collar style include any type of support that is slipped over the end of the primary pipe, and then slid into position. The requirement for slipping the support over the end of the pipe is not an absolute requirement. A more important defining characteristic is that the support encompasses a significant portion of the circumference of both the primary and secondary containment pipes once installed. These supports can be configured with a wide variety of cutout patterns and can have a wide variation in the amount of open surface area that it possesses. Examples of eight different types of collar-style interstitial supports are given in Figure 9.11.

A unique type of collar-style interstitial support is illustrated in Figure 9.12. This type of support may be constructed of two entirely different materials. This would allow the exterior portion of the support to be construction of a material that has a low amount of frictional resistance, such as PTFE or UHMW-HDPE. This is helpful in systems that require a substantial amount of unrestricted longitudinal movements of the primary piping. The support may be directly welded to the primary piping if the inside portion of the support is made of the same material as the primary pipe.

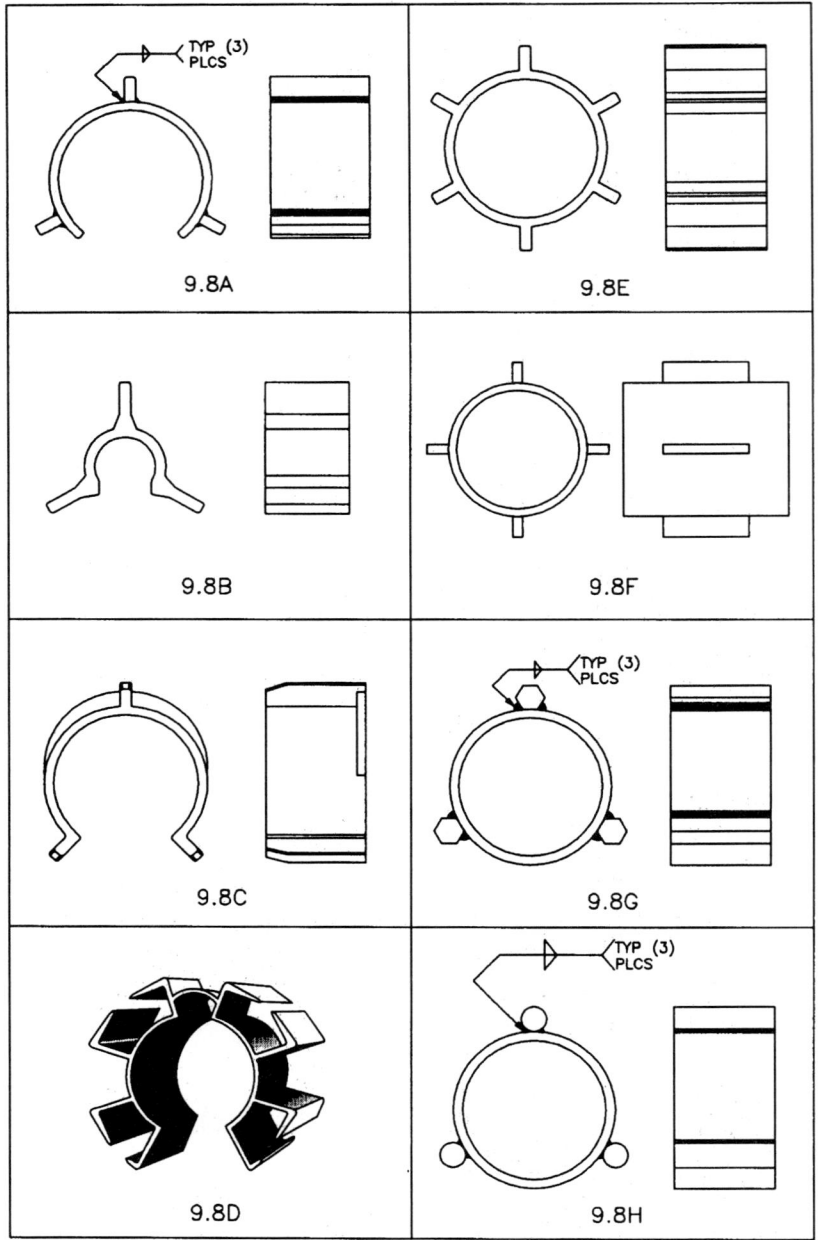

Figure 9.9. Examples of miscellaneous vane-type supports: (a) fabricated vane-type clip-on support; (b) molded or cast vane support with integrally reinforced vanes (Guardian); (c) molded vane support with lateral reinforcement of vanes (U.S. Patent No. 5,018,260)(Asahi/America and AGRU); (d) complex molded vane support with external pads (R&G Sloane); (e) collar-style support with vanes; (f) vane-type coupling; (g) welded hex-rod with 3 vanes; (h) welded round bar stock with 3 vanes.

Structural Considerations

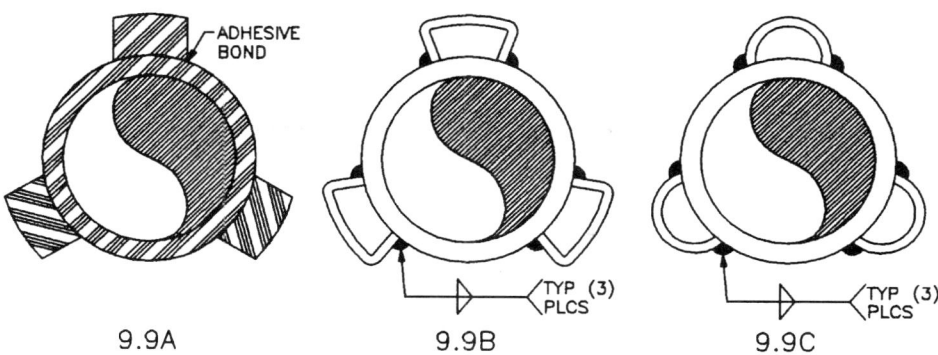

Figure 9.10. Examples of support pad-type internal supports: (a) Illustration of a typical bonded three-pad support arrangement for a fiberglass primary pipe; (b) Illustration of a metallic primary pipe having three pipe saddles welded to its surface; (c) use of welded half-pipe saddles on a primary pipe.

The support illustrated in Figure 9.12 is useful for another purpose. It may be constructed of two different thermoplastic materials, which may then be used in a hybrid system that uses the same combination of materials. Its inner portion may be welded (bonded) to a primary pipe, and its outer portion may be welded (bonded) to the associated secondary containment pipe. Inner and outer pipes, though they consist of dissimilar materials, may thus be rigidly interconnected by means of this support. The dissimilar inner-outer material assembly may then be welded (bonded) to a second assembly using simultaneous fusion techniques (see Section 12.3.3), a method that is otherwise limited to homogeneous combinations of materials. This allows an expensive material to be housed within a less expensive secondary containment housing, and still allow simultaneous fusion to be carried out as the primary sequence of joining.

An example would be a polypropylene primary pipe housed inside a high-density polyethylene secondary containment housing, instead of using a polypropylene-polypropylene combination, which would be much more expensive. One must be aware that the authority having local jurisdiction may not allow a dissimilar combination to be used for a double containment piping system. It should only be used where it is acceptable, and the combination of materials is right for the application, based on a risk analysis. Also, there are many technical drawbacks concerning the use of simultaneous fusion (see Sections 8.5.2, 12.3.3, 13.2.3.2, and 13.3.1). Simultaneous fusion is only acceptable in systems that experience little or no changes in temperature, low internal pressures, and where visual examination of primary pipes during initial pressure testing is not required.

9.3.1.2.5 Mechanically Joined Attachments for Interstitial Supporting of Pipes

There are a variety of multiple-piece boltable attachments that are available for use in supporting or centralizing a primary pipe inside of a secondary containment pipe. Boltable devices allow for more rapid attachment of the interstitial device in comparison to those types that would otherwise be fabricated and welded (or bonded) to the primary pipe. Thus, they provide labor savings in some applications. Such supporting devices can also be made to support a primary pipe uniformly over at least 120° of the bottom and top of the

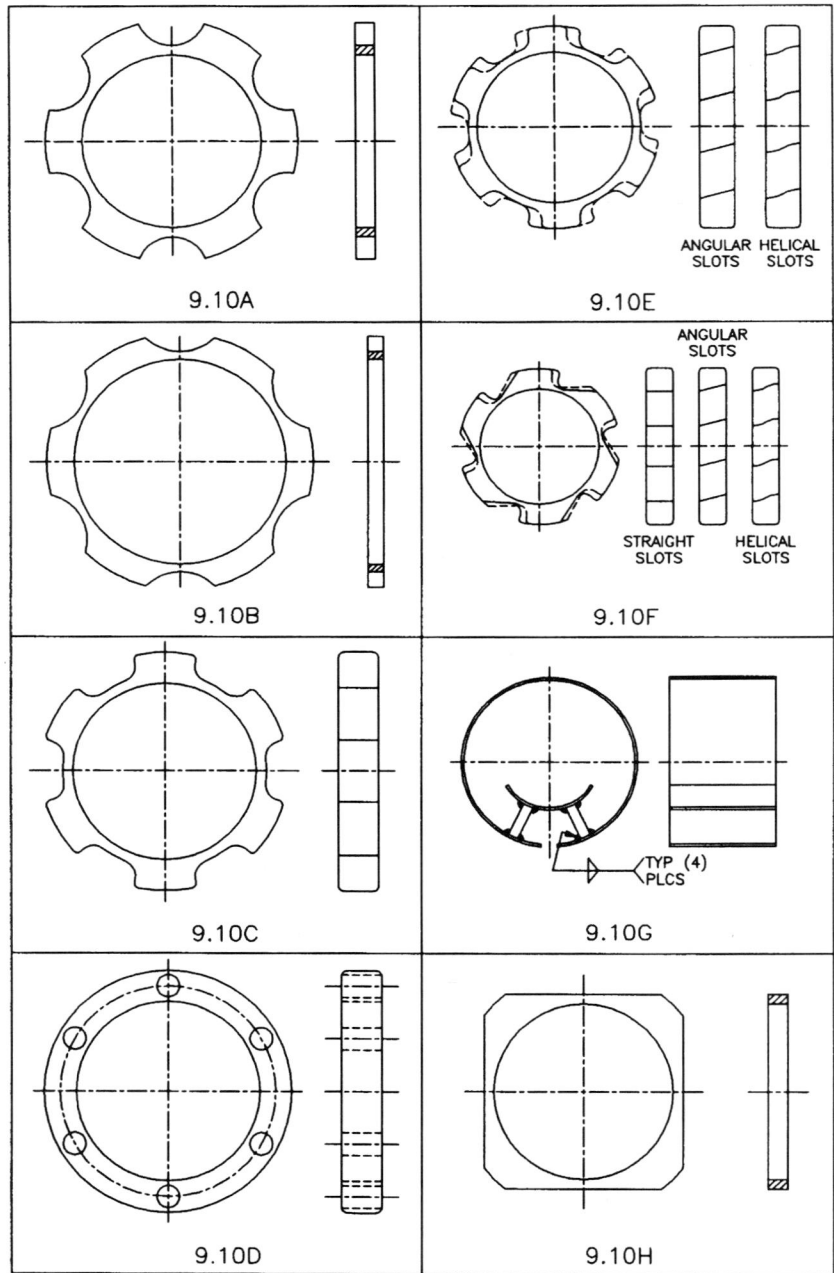

Figure 9.11. Various examples of collar-style interstitial supports: (a) basic support with half-moon cutouts; (b) with extended open area; (c) with annular slots; (d) with drilled holes; (e) with angular slots or curved helical vanes; (f) for improved annular flushing/drying (U.S. Patent #5,400,828); (g) fabricated saddle-type; (h) "square" shape for fabricated thermoplastic systems intended for simultaneous fusion.

Structural Considerations

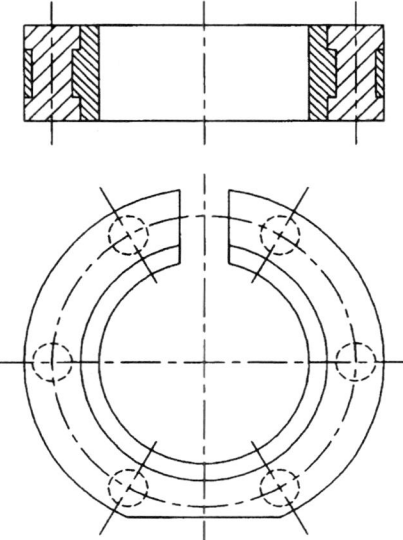

Figure 9.12. Interstitial collar support constructed of dissimilar materials on the inner and outer portions. (U.S. Patent No. 5,141,184)

pipe, allowing it to serve the equivalent function of a saddle-type support used in single-wall pipes. Used together with a cushioning rubber inner layer, these parts serve as effective support devices for use with fiberglass primary pipes. An example of a basic boltable support constructed of two halves is shown in Figure 9.13.

The support shown in Figure 9.13 provides the added benefit of being able to be used with either a singular primary pipe, or alternatively, with a bundle of one to four much smaller primary pipes all housed in a bundle inside of a given outer jacket size. This is commonly found in fuel oil or diesel fuel systems involving underground storage tanks housed outside a building. Note that the support in Figure 9.13(b) has curved surfaces which conform to the shape of a pipe which is one to two NPS sizes smaller than the corresponding outer jacket, thereby allowing a singular primary pipe to be housed inside of an outer jacket that is one or two sizes larger.

9.3.1.2.6 Flexibility Supports

Flexibility supports are designed to provide structural support to the primary pipe for dead weight, which allows both axial and radial movements to some degree. These supports are useful where thermal expansion is present, and pipe must be allowed to both grow axially and to bend and twist radially. A thorough explanation of why such features are desired is in close proximity to elbows is presented in Section 8.6.2, along with procedures for locating supports of this type. Basic flexibility supports are shown in Figures 9.14 and 9.15.

A new type of flexibility support that has added benefit is the internally-guided-ball flexibility support, and example of which is shown in Figure 9.16. This support provides the benefit of allowing angular movements, in addition to axial and limited bending and torsional movements.

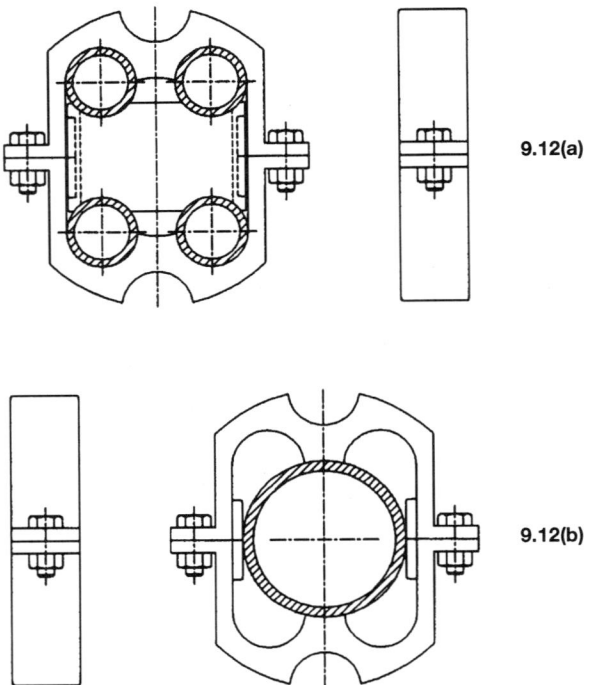

Figure 9.13. Example of a unique 2-part boltable centering support which is designed to support any one of a number of combinations of multiple primary pipes within a given outer jacket size as shown in (a), or a singular primary pipe for the given outer pipe size as shown in (b) (U.S. Patent #5,404,914).

9.3.1.2.7 Design Criteria and Overall Tolerances

Interstitial supports must be sized so that they fit snugly, yet allow the primary piping to be inserted into its secondary housing (or conversely, allow a secondary containment pipe to be slipped over its associated primary pipe), which has interstitial supports affixed at desired points along its external surface. They must also not be so snug that they restrict piping from being able to move in a desired manner.

The problem with setting a tolerance that seems to be reasonable is that piping and components are far from being precision-machined elements. They normally possess wide tolerances of their own, including tolerances of their diameter, wall thickness, and ovality. Thus, if both primary and secondary containment pipes are able to vary widely, it is difficult, if not impossible, to set general size and tolerance criteria for interstitial supports. It is more practical to set conditions for which the supports must be made, based on the actual dimensions of piping to be fabricated into each assembly. This means that different supports to be used on the same project may require different dimensions, which is one of the reasons that double containment piping is very conducive to prefabrication in sections, in a shop environment.

The tolerances that are desired depend on many factors, including the types of materials to be used, amount of longitudinal movement required, and the acceptable amount of vibra-

Structural Considerations

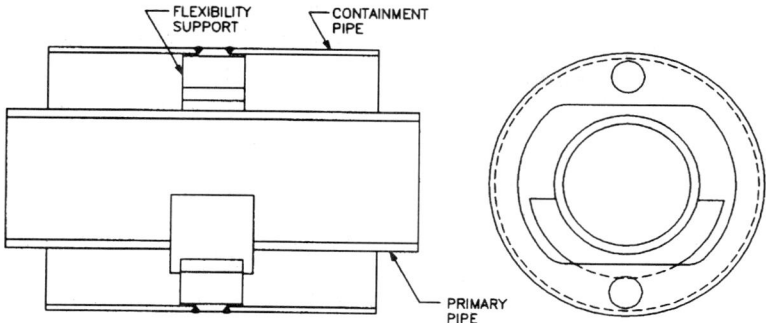

Figure 9.14. A basic flexibility support designed to be welded or bonded to the outer jacket. (U.S. Patent No. 5,197,518, additional patents pending)

tion to which the primary piping can be subjected. A looser fit will typically result in a large amount of vibration that a pipe will experience.

Supports must be wide enough such that concentrated point loads and discontinuity stresses are not created on the secondary containment or primary piping system. Loads must also be evenly distributed so that they are not excessive at any one point, in order to maintain a well-balanced system. A comprehensive study of these components, in their as-installed condition, is required.

9.3.1.2.8 Burial Design Criteria for Interstitial Supports

Most flexible pipes are required to be able to be diametrically deflected up to a maximum of 5% (some pipes can be deflected by more or less than this amount). In underground systems, attention must be paid to the design of the support, such that this allowance is made. Most spider-style supports allow a reasonable amount of deflection. However, problems can result with the design of many collar-style supports, since they are often tightly fitting.

An example of a collar-style interstitial support that does allow at least a 5% diametrical deflection is illustrated in Figure 9.17. The illustration shows the pipe deformed a full 5%;

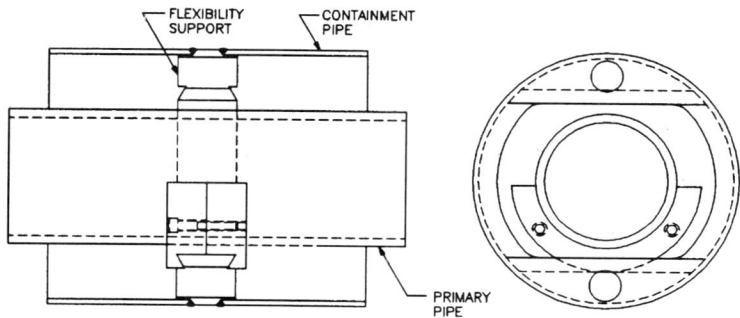

Figure 9.15. A flexibility support having a saddle that interlocks to the base via a dove-tailed arrangement. (U.S. Patent No. 5,197,518, additional patents pending.)

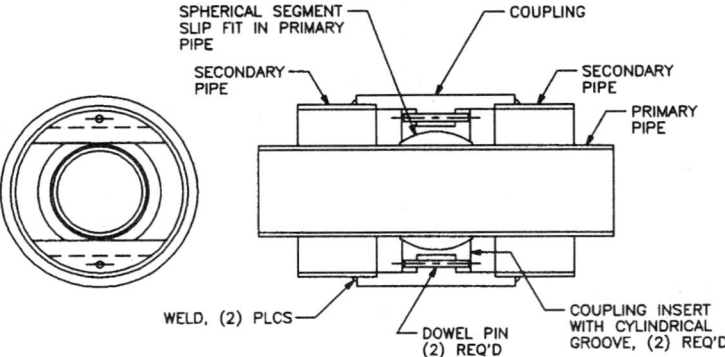

Figure 9.16. Illustration of an internally guided ball-type flexibility support. The version shown is designed for use in an all-welded system. Variations are also available that are intended for use in systems with a mechanically joined outer jacket. (U.S. Patent No. 5,197,518; additional patents pending.)

it further shows that there is still additional room for 25% more deflection (up to 6.25%), which is purposely allowed as an extra safety factor. Piping should not be buried in a condition that would allow more than 5% diametrical deflection, in most conditions. Therefore, a support that allows up to 6.25% deflection is considered to meet the minimum clearance requirement for burial.

9.3.2 Considerations for Double Containment Pipe Assemblies

9.3.2.1 Support of Homogeneous-Material Double Containment Assemblies

A double containment piping assembly can be supported by the use of external supports, spaced over a wide span, a span that is wider in comparison to either inner or outer pipes considered alone. The reason has to do with two major factors. The area moment of inertia of the assembly will be greater [refer to Eqs. (9.27) and (9.28)] than a single-wall pipe that has the same outside diameter. The other factor is an increase in rigidity that intersti-

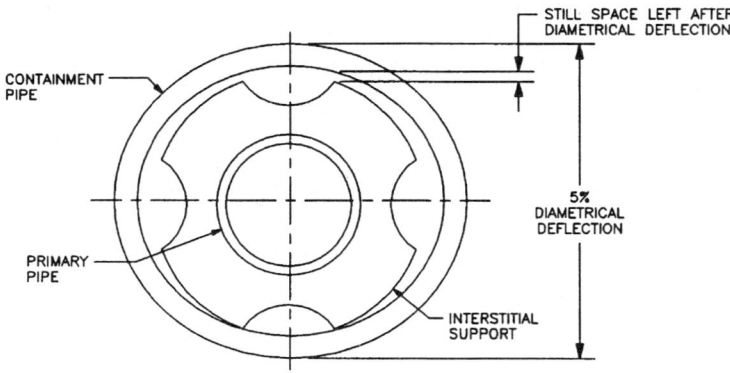

Figure 9.17. Illustration of an interstitial support that allows at least a 5% diametrical deflection.

Structural Considerations

tial supports and interconnecting parts provide, since they behave as structural members. The second factor tends to be the dominant factor in most situations. However, it has been rarely been accounted for in determining double containment support span length, since simplified equations have not yet been determined.

A conservative approach is to use the corrected moment of inertia of the two pipes in a double containment assembly shown in Eq. (9.27), ignoring any effects of additional stability that each pipe provides to the other via interstitial supports. This quantity will produce a conservative result, yet will allow somewhat wider spans than a single-wall pipe of equal outside diameter. If wider spans are desired, designers must conduct a comprehensive analysis, taking into account the added rigidity provided by the interstitial supporting devices.

In conducting simplified calculations described in the previous paragraph, it is essential that the unit weight of the double containment piping assembly take into account all contributing sources. These include the weight of secondary containment components, and the weight of the annular space, including the heaviest fluid that will be present during testing or operation of the system. It is best to determine the weight of the annular space based upon the need to contain the primary fluid safely (i.e., while fluid coexists in the associated primary system). The equation for corrected weight per unit length is:

$$W = \frac{(\pi D_2^2 - \pi d_2^2)}{4}\left(\frac{\rho_4}{1728}\right) + \frac{(\pi d_2^2 - \pi D_1^2)}{4}\left(\frac{\rho_3}{1728}\right) + \frac{(\pi D_1^2 - \pi d_1^2)}{4}\left(\frac{\rho_2}{1728}\right) + \frac{(\pi d_1^2)}{4}\left(\frac{\rho_1}{1728}\right) \quad (9.33)$$

where:
D_2 = outer diameter of the secondary containment pipe, in.
D_1 = outer diameter of the primary pipe, in.
d_2 = inner diameter of the secondary containment pipe, in.
d_1 = inner diameter of the primary pipe, in.
ρ_4 = density of the secondary containment pipe, lb/ft^3
ρ_3 = density of fluid in the annular space, lb/ft^3
ρ_2 = density of the primary pipe, lb/ft^3
ρ_1 = density of fluid in the primary pipe, lb/ft^3
W = unit weight per linear inch of double containment pipe, lb/linear in.

9.3.2.2 Support of Hybrid-Material Double Containment Assemblies

For dissimilar material assemblies, a mean average of the modulus of elasticity of the two materials may be used. However, in the case where the moduli are substantially different, as would be the case for a metal housed within a plastic, an average modulus is not suitable. Where a stronger/stiffer material is used for primary piping, the modulus of elasticity of the primary pipe material can be used as the modulus of the combined system, as in the example given. The relative stiffness of the metallic piping will prevent a nonmetallic material from sagging excessively at any one point, if a sufficient number of interstitial supports are used. On the other hand, if plastic components are housed within metallic components, very little stiffening will be added to the metallic material. Therefore, no credit should be taken, as a conservative measure.

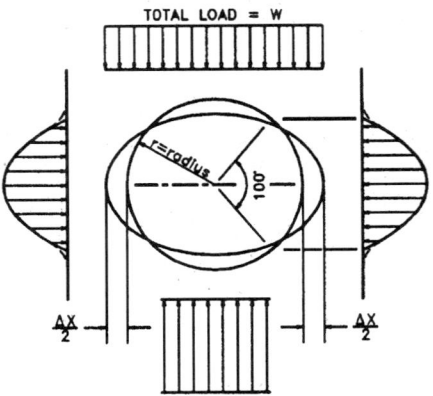

Figure 9.18. Illustration of the details involved in diametrical deflection of flexible pipes.

Appendix VII of ANSI/ASME B31.1 presents a method for analyzing buried single-walled pipes that are also subject to thermal expansion and contraction. This issue is discussed in more detail in Chapter 8 of this text (see Section 8.1.1). This method should be used for secondary containment portions of double containment pipes that are also subject to temperature changes.

9.4 Burial Considerations

Analysis of buried pipes requires that a wide range of possible burial factors and their resultant loads on the buried pipe be considered, along with reactions that a pipe subjected to the load may experience. Behavior of pipes under burial-type loading depends to some extent on whether a pipe is flexible or rigid. According to commonly accepted pipe burial theory, a flexible pipe is one that may be diametrically deflected by at least 2% (as measured by its inside diameter), without any long- or short-term harm being caused to the pipe. A pipe incapable of this degree of deformation is classified as rigid. Figure 9.18 illustrates what is meant by a pipe being diametrically deflected.

The suitability of a pipe for a specific burial conditions depends on the sum of all loads that are imposed upon it, and on the reaction of the pipe to those loads. The amount of load imposed on a pipe will typically be greater for a rigid pipe than it will be for a flexible conduit. Reactions to be considered include amount of deflection, creep, stress, and strain to which a buried pipe will be subjected. The rated maximum allowable deflection of the pipe, its allowable stress and strain, and the likelihood of its buckling, all vary greatly according to the type of material being used. Based on these concerns, simplified methods for both flexible and rigid single-wall pipes are presented in Sections 9.4.1.4 and 9.4.1.3, respectively. For many applications, these simplified means of analysis for single-wall pipes may also be adapted to judge double containment pipe assemblies in buried situations.

Modern-day simplified methods of burial analysis are largely the result of work performed by various researchers at the University of Iowa. The first significant work was developed by Professor Anston Marston; the original publication of his works dates back to 1913. A student of Professor Marston, M. G. Spangler, subsequently developed what is now referred

Table 9.2 Average Values of Modulus of Soil Reaction, E' (for Initial Flexible Pipe Deflection)

Soil type-pipe bedding material (Unified Classification System [a]) (1)	E' for degree of compaction of bedding, in lbs/sq. in.			
	Dumped (2)	Slight, <85% Proctor, <40% relative density (3)	Moderate 85-95% Proctor, 40-70% relative density (4)	High >95% Proctor, >70% relative density (5)
Fine-grained soils (LL>50)[b] Soils w/medium to high plasticity CH, MH, CH-MH	No data available; consult a competent soils engineer; otherwise use E'=0			
Fine-grained soils (LL<50) Soils w/medium to no plasticity CL, ML, ML-CL, w/<25% coarse-grained particles	50	200	400	1,000
Fine-grained soils (LL<50) Soils w/medium to no plasticity CL, ML, ML-CL, w/>25% coarse-grained particles Coarse-grained soils w/fines GM, GC, SM, SC [c] contains > 12% fines	100	400	1,000	2,000
Coarse-grained soils w/little or no fines GW, GP, SW, SP [c] contains < 12% fines	200	1,000	2,000	3,000
Crushed rock	1,000	3,000	3,000	3,000
Accuracy in terms of percentage deflation [d]				

[a] ASTM Designation D-2487, USBR Designation E-3.
[b] LL= Liquid limit.
[c] Or any borderline soil beginning w/one of these symbols (i.e., GM-GC, GC-SC).
[d] For + or - 1% accuracy and predicted deflection of 3% actual deflection would be between 2% and 4%.
Note: Values applicable only for fills < 50 ft. (15 m). Talbe does not include any safety factor. For use in predicting initial deflection only, appropriate deflection lag factor must be applied for long-term deflection. If bedding falls on the borderline between two compaction categories, select lower E' value or average the two values. Percentage Proctor based on laboratory maximum dry density from test standards using about 12,500 ft-lb/cu. ft. (598,000 J/m[3]) (ASTM D-698, AASHO T-99, USBR Designation E-11). 1 psi = 6.9 kN/m^2.

Source: "Soil Reaction for Buried Flexible Pipe," by Amster K. Howard, U.S. Bureau of Reclamation, Denver, Colorado. Reprinted with permission from American Society of Civil Engineers Journal of Geotechnical Engineering Division. January 1977. pp. 33-43.

to as the Iowa formula for analyzing flexible pipes under burial loads, based on the theory of ring deflection; a student of Spangler, Watkins, further developed the formula to its final form which is in use today. Spangler also developed other useful theories, including the soil strain formula, shown in Eq. 9.48. Other work of significance includes the work performed by Amster K. Howard of the United States Bureau of Reclamation. Howard developed values for the modulus of elasticity of soils, based on soil type and proctor densities. These are listed in Table 9.2. All of these various theories as they relate to the analysis of rigid, flexible, and double containment pipes are presented in this Section.

9.4.1 General Methods; Single-Wall Piping

There are two types of loads that a pipe can experience as a result of being buried. The two types of loads are dead-weight loads (sometimes referred to as earth loads) and live loads; they apply to pipes of all materials of construction. The magnitudes of the loads and the methods by which they are determined vary depending on the relative flexibility of the pipe under consideration. In analyzing a buried system using simplified methods, live loads and dead-weight loads are determined separately. However, it is their combined values that are compared against the maximum that the pipe can handle, since they will occur in concurrent fashion. All other forces that may act upon a pipe in concurrent fashion must be considered to determine the suitability of the piping system under the specified conditions.

9.4.1.1 Determination of Dead-Weight Load

The load exerted on a pipe by a column of soil superimposed upon it is modified by the response of that pipe to the load applied. In other words, a pipe will respond to a load imposed by buried soil to either lessen or increase the load. The total magnitude of the load will always change over time. The response can be either to reduce the load due to the weight of the column above it, or to magnify the weight above it. Typically for rigid pipes, the load is magnified beyond the total amount of weight of the column of soil above it. For flexible pipes, the total load is less than the initial load of the weight of the column of soil above the crown of the pipe.

The reason that a dead-weight load may be increased or decreased has to do with the dynamics of the column of soil above the crown of the pipe and the dynamics of the side fill. If the arch action in the soil transfers a portion of the weight of the column to the adjacent side prisms, the total amount of load on the conduit is reduced. This effect is normally found in flexible pipe applications. When some of the weight of the side prisms are transferred to the column of soil over the pipe, the effect is to increase the total amount of load imposed on the pipe. This effect is normally found in rigid pipe applications. The total magnitude of the modification depends upon the amount of relative movement, the unit weight of the soil, the height and depth of the side prisms, and the geometry of the installation.

The calculation for rigid pipes and conduits can be expressed as the rigid pipe form of the Marston equation[8]:

$$W_c = C_d w B_d^2 \tag{9.34}$$

where:
W_c = load on the conduit, lbs/linear ft
w = density of backfill, lbs/ft^3
B_d = horizontal width of the trench at the crown of the pipe, ft
C_d = load coefficient for pipe installed in trenches

The calculation for flexible pipes and conduits can be expressed as the flexible conduit form of the Marston equation:

$$W_c = C_d w B_d B_c \tag{9.35}$$

where:
W_c = load on the conduit, lbs/linear ft
B_c = outside diameter of the pipe, ft

For purposes of burial analysis, a flexible pipe is normally considered as any pipe that may be deflected by at least 2%, without any sign of detrimental structural effects.

To use Eqs. (9.34) and (9.35), the value of the constant C_d must be determined. C_d is a function of the ratio of the height of the backfill (H) to the width of the trench (B_d) and to the coefficient of friction between the backfill and the sides of the trench. The implicit formula for C_d is as follows:

$$C_d = \frac{1 - e^{(-2k\mu' H/B_d)}}{2k\mu'} \tag{9.36}$$

where:
k = Rankine's ratio of the lateral to vertical pressure for the backfill
μ' = coefficient of friction between the material of the backfill and the trench

Figure 9.19 presents a graphical solution of the values of C_d at a variety of values of ($k\mu'$) and H/B_d.

The ratio of the burial loads for a rigid pipe as compared to a flexible pipe depends entirely on the ratio of the width of the trench to the diameter of the pipe. Therefore, for a pipe of a given diameter, the load on a rigid pipe will be three times that of a flexible pipe if the trench is three times the outside diameter of the pipe being buried. The derivation of this relationship is shown as:

$$\frac{(W_c)_{rigid}}{(W_d)_{flexible}} = \frac{C_d w B_d^2}{C_d w B_d B_c} = \frac{B_d}{B_c} \tag{9.37}$$

The width of trench has more than a simple and direct affect on the total load on the pipe. The width of trench is included in the computation of C_d. Therefore, increasing trench width has an effect of increasing the load in a greater-than-linear fashion, even for a flexible pipe. However, as the width of a trench is increased for a given pipe at a given depth, the load eventually reaches a value called the transition width at which no additional load is imposed.

At the transition width and beyond, the embankment form of the Marston equation may be used. Embankment installation refers to the condition where the top of a pipe projects above the natural ground surface, or is buried in a trench at or beyond the transition width. The equation is the same for flexible and rigid pipes, and is shown as follows:

$$W_c = C_c w B_c^2 \tag{9.38}$$

where:
C_c = load coefficient for embankment installation

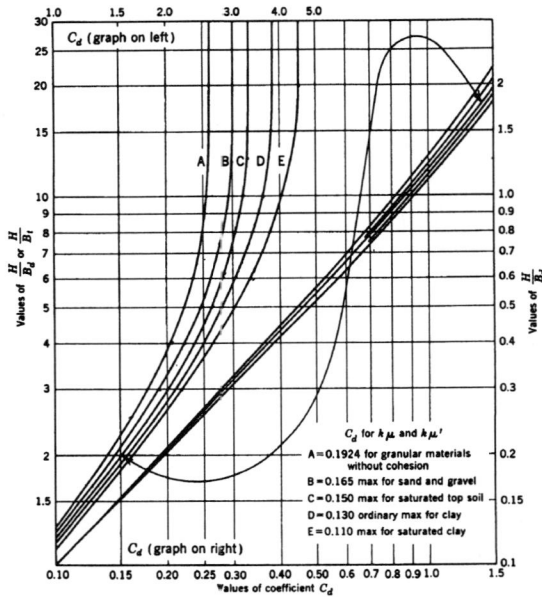

Figure 9.19. Graphical solution of the values of C_d based on various values of $K\mu'$ and H/B_d. (Source: Design & Construction of Sanitary & Storm Sewers. "Manuals & Reports on Engineering Practice No. 37," ASCE, and Manual of Practice No. 9, WPCF, 1969.)

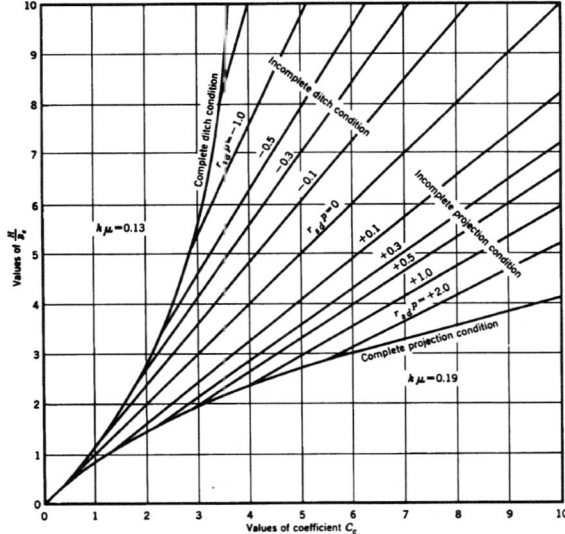

Figure 9.20. Graphical solution of C_c at various values of $K\mu'$, H/B_d, and r_{sd} for positive projecting conduits. (Source: Design & Construction of Sanitary & Storm Sewers. "Manuals & Reports on Engineering Practice No. 37," ASCE, and Manual of Practice No. 9, WPCF, 1969.)

Structural Considerations 343

The value of load coefficient C_c is dependent on the projection ratio (p), the settlement ratio (r_{sd}), and the ratio of the height of the fill (H) to the outside diameter of the pipe (B_c). Figure 9.20 presents a graphical solution of C_c at various values of ($k\mu'$) and H/B_d and also at a variety of r_{sd} values.

It can be seen from Figure 9.20 that as the value of the term r_{sd} approaches zero, the value of the coefficient C_c approaches the value of the ratio H/B_c. Most pipe installations have an rsd term that is approximately zero. When this is true, the value of C_c in Eq. (9.38) can be replaced by the ratio H/B_c. The value that is then determined is referred to as its prism load. Prism load is the maximum load that can be imposed by soil on a flexible pipe, except under very unusual circumstances. The prism load form of the Marston equation is shown as follows:

$$W_c = HwB_c \qquad (9.39)$$

where:
H = height of the fill above the pipe, ft

Expressed in terms of soil pressure, prism load can be determined as:

$$P = wH = \frac{W_c}{B_c} \qquad (9.40)$$

where:
P = pressure due to soil weight at depth H, lbs/ft^2

For flexible single-wall pipes, the long-term load imposed on a pipe may approach the prism load. This will happen if frost action or water action is expected, which is likely in many installations. These actions tend to dissipate the frictional forces of soil trenches, which otherwise reduce the load on a flexible pipe through the arching action of the column of soil above the pipe. Therefore, prism load must be used as the design basis for flexible single-wall pipes that will likely see frost or water action.

9.4.1.2 Determination of Live Loads

Underground pipes are also subjected occasionally to live loads that are the result of being buried underneath highways, roads, driveways, parking lots, railways, airport runways, and other traffic-bearing locations. A superimposed live load that is applied to a ground surface is transmitted through the soil, with a portion of the load potentially reaching a buried pipe.

The calculation of live loads is through the application of Boussinesq theory. Boussinesq theory simplifies an actual model to one where soil is assumed to be a semi-infinite elastic and isotropic medium through which stresses superimposed on its surface are distributed according to a curve. A Boussinesq curve for the distribution of live loads at a soil's surface for various depths of backfill is presented in Figure 9.21. The graphic shows that the live load diminishes rapidly with depth of burial. As the depth of burial reaches 6 ft of cover, the slope of the curve decreases to the point where the load no longer diminishes, and at this value the live load no longer appears to be significant. It should be pointed out that a Boussinesq solution to live loads gives only an estimation of the solution. Actual loads will vary slightly, although the theory tends to give good enough results.

Equation (9.41) represents the method of determining live loads resulting from concentrated live loads, produced from the equivalent of a truck wheel. When live loads are of the distributed type, Eq. (9.42) can be used to determine their value. The values of Cs in both equations can be determined by Table 9.3.

$$W_{sc} = C_s \frac{PF'}{L} \tag{9.41}$$

where:
W_{sc} = the load on the pipe, lb/unit length
P = the concentrated load, lbs
F' = impact factor (ranges from 1 to 1.75)
L = the effective length of the pipe (3 feet or less), ft
C_s = the load coefficient, which is a function of $B_c/(2H)$ and $L/(2H)$
H = the height of the fill above the pipe, ft

$$W_{sd} = C_s p F B_c \tag{9.42}$$

where:
p = intensity of the distributed load, lb/ft^2
C_s = the load coefficient, which is a function of $D/(2H)$ and $M/(2H)$
D = width of the area over which the distributed load acts, ft
M = length of the area over which the distributed load acts, ft

Figure 9.22 illustrates graphically the effect of H-20-type highway loading at varying depths of burial. H-20 highway loading assumes two 16,000 pound concentrated loads applied to two 18 in. by 20 in. areas. One of these areas is assumed to be located over the area being analyzed, the other being positioned a distance of 6 ft away. The total load of the truck in this question is assumed to be 20 tons. Figure 9.23 also illustrates graphically the effect of Cooper E-80 railway loads at varying depths of burial. Cooper E-80 loading assumes 80,000 pounds applied over three 2 ft by 8 ft areas spaced on 5 ft centers. This would be typical of a locomotive that has three 80,000 pound axle loads. Both graphs indicate that as depth of burial increases, the relative effect from each live load rapidly diminishes.

9.4.1.3 Rigid Piping Analysis

Simplified burial analysis for rigid pipes is historically based upon concrete pipe. Therefore, to conduct a simplified analysis of a rigid double containment pipe assembly, a concrete-pipe-based analysis would be used.

Most rigid pipes for underground use, including rigid double containment pipe assemblies, are installed in relatively narrow trenches, excavated in undisturbed soil, and back-

Structural Considerations

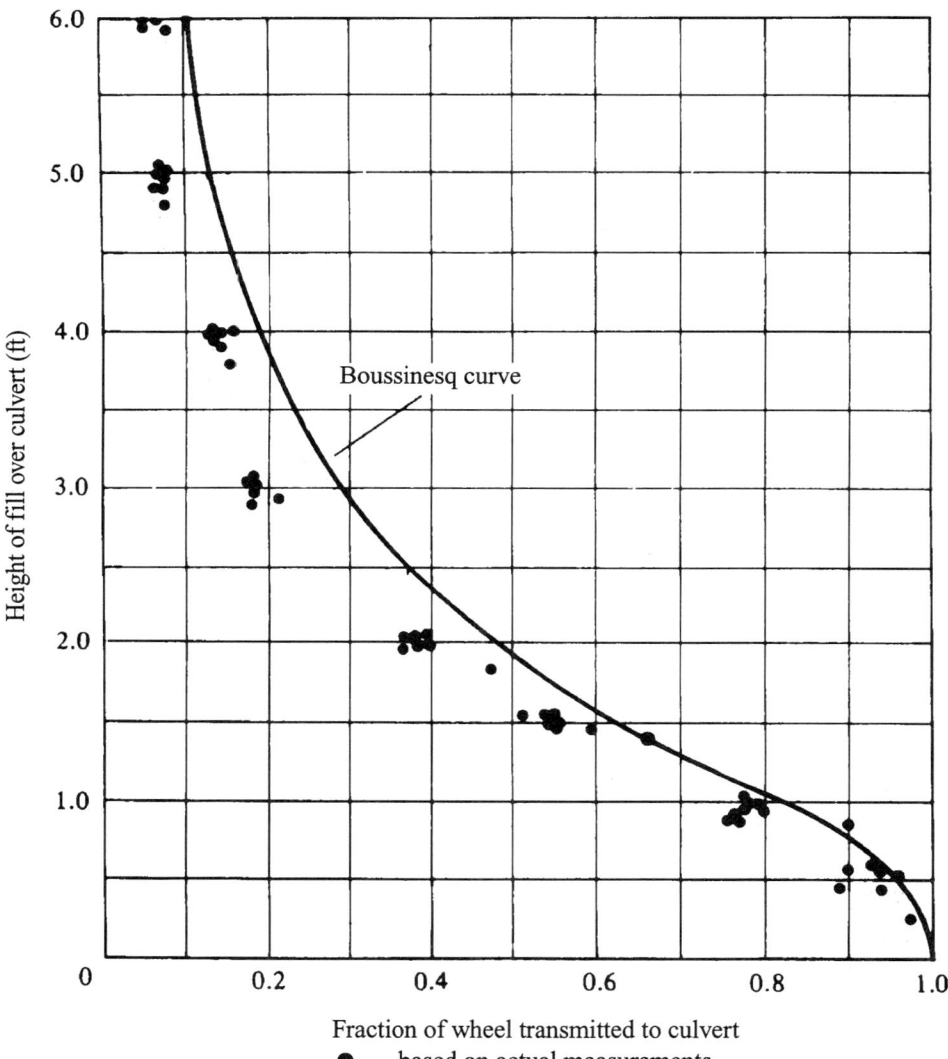

Figure 9.21. Boussinesq distribution of live loads at a soil's surface for various depths of backfill. (Source: Soil Engineering, 3rd Edition, Merlin G. Spangler and Richard L. Handy, 1973, Intext Press.)

filled to the original ground surface. When a rigid pipe assembly is installed in a slightly larger trench and backfilled, the backfill material will usually settle over a period of time.

The downward movement of backfill generates frictional forces along the walls of the soil trench, which tends to act upwards, thus helping to support the weight of the backfill material. The frictional force that is developed depends on such factors as the unit weight of the backfill material (w), the value of Rankine's lateral pressure ratio (K), and the coefficient of sliding friction between backfill materials and soil trench walls (μ').

Table 9.3 Values of Load Coefficients, C_s, for Concentrated and Distributed Superimposed Loads Vertically Centered Over Conduit*

D/2H or B[c]/2H	\multicolumn{13}{c}{M/2H or L/2H}
	0.1	0.2	0.3	0.4	0.5	0.6	0.7	0.8	0.9	1.0	1.2	1.5	2.0	5.0
0.1	0.019	0.037	0.053	0.067	0.079	0.089	0.097	0.103	0.108	0.112	0.117	0.121	0.124	0.128
0.2	0.037	0.072	0.103	0.131	0.155	0.174	0.189	0.202	0.211	0.219	0.229	0.238	0.244	0.248
0.3	0.053	0.103	0.149	0.190	0.224	0.252	0.274	0.292	0.306	0.318	0.333	0.345	0.355	0.360
0.4	0.067	0.131	0.190	0.241	0.284	0.320	0.349	0.373	0.391	0.405	0.425	0.440	0.454	0.460
0.5	0.079	0.155	0.224	0.284	0.336	0.379	0.414	0.441	0.463	0.481	0.505	0.525	0.540	0.548
0.6	0.089	0.174	0.252	0.320	0.379	0.428	0.467	0.499	0.524	0.544	0.572	0.596	0.613	0.624
0.7	0.097	0.189	0.274	0.349	0.414	0.467	0.511	0.546	0.584	0.597	0.628	0.650	0.674	0.688
0.8	0.103	0.202	0.292	0.373	0.441	0.499	0.546	0.584	0.615	0.639	0.674	0.703	0.725	0.740
0.9	0.108	0.211	0.306	0.391	0.463	0.524	0.574	0.615	0.647	0.673	0.711	0.742	0.766	0.784
1.0	0.112	0.219	0.318	0.405	0.481	0.544	0.597	0.639	0.673	0.701	0.740	0.774	0.800	0.816
1.2	0.117	0.229	0.333	0.425	0.505	0.572	0.628	0.674	0.711	0.740	0.783	0.820	0.849	0.868
1.5	0.121	0.238	0.345	0.440	0.525	0.596	0.650	0.703	0.742	0.774	0.820	0.861	0.894	0.916
2.0	0.124	0.244	0.355	0.454	0.540	0.613	0.674	0.725	0.766	0.800	0.849	0.894	0.930	0.956

* Influence coefficients for solution of Holl's and Newmark's integration of the Boussinesq equation for vertical stress.
Source: **Design & Construction of Sanitary & Storm Sewers.** "Manuals & Reports on Engineering Practice No. 37," American Society of Civil Engineers and "Manual of Practice No. 9," Water Pollution Control Federation. 1969.

Structural Considerations

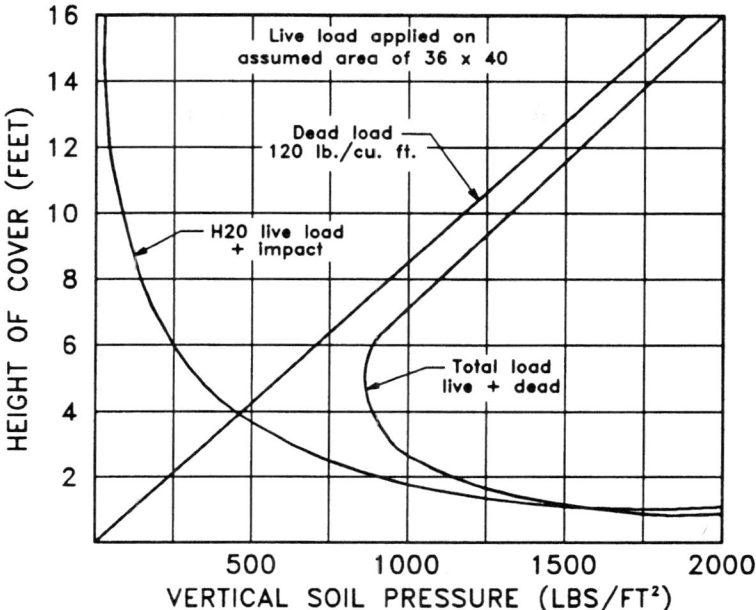

Figure 9.22. H-20 wheel loading at various depths of burial. (Source: American Iron and Steel Institute.)

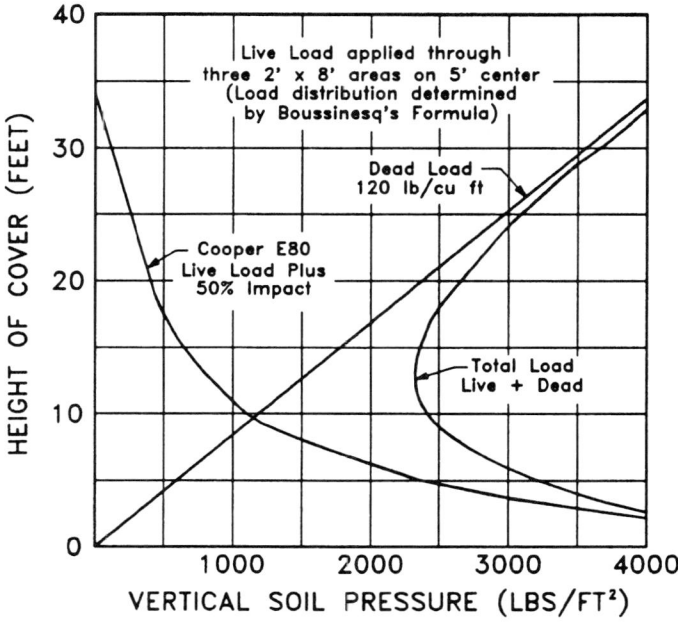

Figure 9.23. Cooper E-80 railway loads at varying depths of burial. (Source: American Iron and Steel Institute.)

The backfill load on a rigid double containment pipe assembly may be calculated by:

$$W_d = C_d w B_d^2 \tag{9.43}$$

To use Eq. (9.43), an appropriate value of the load coefficient must be selected. Load coefficients may be directly calculated for a specific condition if K and μ' are known. However, a series of nomographs are presented in Figures 9.24 through 9.27 that have loads already calculated for rigid pipes for positive- and negative-projecting pipes. For intermediate sizes that are not shown in Figures 9.24 through 9.27, designers may interpolate to determine the appropriate values.

The field supporting strength of a rigid double containment pipe assembly depends on the inherent strength of the pipe, the type of foundation on which it is installed, and the compaction of its adjacent side fill. For rigid pipes, a common method to establish its ultimate strength is a test referred to as a "three-edge bearing test." This test involves loads applied to its crown and invert. The load at which a pipe experiences failure is termed its "three-edge bearing strength," with units expressed in terms of pounds per linear foot (kilograms per meter).

The "three-edge bearing test" is a test that has been specifically designed for concrete pipe (ASTM C-14). However, it may be applied equally to a double containment pipe assembly that is considered to be a rigid assembly, as opposed to having a condition where its secondary containment pipe is free to behave as a flexible pipe.

Another way to express the strength of a rigid pipe is in terms of bearing strength per foot (meter) of inside diameter, referred to as the D load of a pipe, expressed in units of pounds per linear foot per foot of inside pipe diameter (kilograms per linear meter per meter of inside pipe diameter). The D load of a pipe is equal to the "three-edge bearing strength" divided by a pipe's inside diameter.

The actual strength of a rigid pipe in a field condition will be greater than its bearing strength (D load), to an extent that depends on the type of bedding that is selected. There are five general classes of trench bedding that have been established. Each of these is assigned a load factor based on concrete pipe buried under these conditions. Load factor is the ratio of the supporting strength of a rigid pipe in the field to its strength determined in a "three-edge-bearing-test." Figure 9.28 illustrates these five classes of bedding, and indicates the associated load factor. No work has been done as of this writing to understand if load factors accepted for concrete pipes are valid for analyzing rigid double containment pipe assemblies using a rigid pipe analysis. More work must be done in this area. Thus, readers are urged to use caution and employ good judgment before applying this type of analysis to "rigid" double containment pipe assemblies. A comprehensive analysis, based on finite element analysis, when conducted properly, will always produce an accurate prediction of performance (see Section 9.4.2.3).

Load factors for rigid pipes in three different nontrench conditions based on concrete pipes are shown in Figures 9.29 through 9.31. Required pipe strength for nontrench (embankment) conditions must be determined in accordance with a design procedure acceptable to the local jurisdiction.

To calculate the required strength of a pipe, based on backfill load, and the load factor, Eqs. (9.44) and (9.45) may be used.

Structural Considerations

$$W_d = \frac{TEB \times LF}{FS} \qquad (9.44)$$

where:
TEB = three-edge bearing load, lbs/linear ft
LF = load factor
FS = factor of safety

$$D_{load} = \frac{W_d(FS)}{L_f d} \qquad (9.45)$$

where:
D_{load} = D load to produce either a 0.01 in. (0.3 mm) crack or ultimate load in the three-edge bearing test, lb/ft/ft (kg/m/m)
W_d = backfill load, lb/ft (kg/m)
L_f = load factor
d = inside diameter of pipe, ft (m)
FS = factor of safety

9.4.1.4 Flexible Piping Analysis

Flexible pipes are defined in a burial context as any pipe that may be deflected at least 2%, as measured in terms of inside diameter, without any detrimental structural effects occurring to the pipe over its lifetime. Flexible pipes under soil loads tend to deflect, and when they are buried in a trench, develop passive soil support at the sides of the pipe. This inherent flexibility of a flexible pipe is what produces its soil load-carrying capacity. Ring deflection of a flexible pipe relieves a major portion of the vertical soil load by transferring this load to the surrounding soil. This is done through a mechanism of "arching action" of the soil over the pipe. For a flexible pipe with a reasonable stiffness, relatively large loads will produce only a moderate amount of deflection. Since most flexible pipes have their allowable deflection determined with a large safety factor, a moderate amount of deflection will not cause failure, even on a long-term basis. It takes a considerable load to cause a deflection of the magnitude that will cause a flexible pipe to fail in service.

Pipe stiffness for flexible nonmetallic pipes is determined by ASTM D 2412, "Standard Test for External Loading Properties of Plastic Pipe by Parallel-Plate Loading." The stiffness, or strength of the pipe under this test, is measured while the pipe is subjected to an arbitrary deflection value of 5% deflection. The equation for pipe stiffness is:

$$PS = \frac{F}{\Delta y} \geq \frac{EI}{0.149 r^3} = \frac{6.71 EI}{r^3} = \frac{6.71 E t^3}{12 r^3} \qquad (9.46)$$

$$= 0.559 E \left(\frac{t}{r}\right)^3$$

where:
PS = pipe stiffness, lbs/linear in.
F = force, lbs/linear in.
Δy = vertical deflection, in.
E = modulus of elasticity, psi
I = the area moment of inertia per unit length of pipe, in.^{4}/linear in. = in.^{3}
r = mean radius of pipe, in.
t = wall thickness, in.

For pipe that is manufactured based on controlled outside diameters, the equation can be written:

$$PS = 4.47 \frac{E}{(DR-1)^3} \tag{9.47}$$

where:
DR = ratio of outside diameter to the wall thickness; also called dimensional ratio

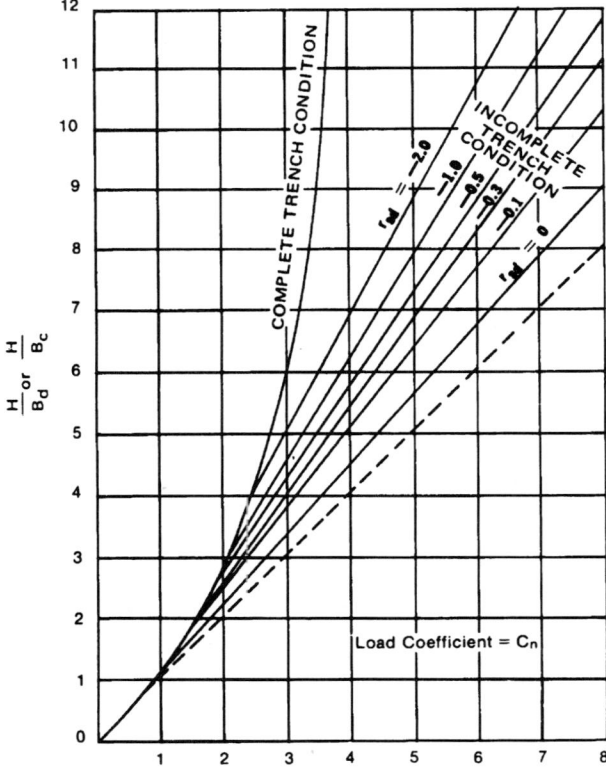

Figure 9.24. Load coefficient for negative projecting conduit at various values of r_{sd} for $p'=0.5$ and $k\mu=0.13$. (Source: Concrete Pressure Pipe, "Manual of Water Supply Practices," Manual M9, AWWA, 1979.)

Structural Considerations

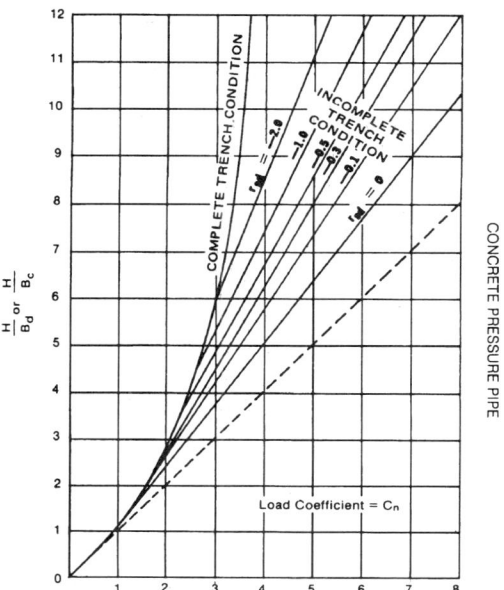

Figure 9.25. Load coefficient for negative projecting conduit at various values of r_{sd} for $p'=1.0$ and $k\mu=0.13$. (Source: Concrete Pressure Pipe, "Manual of Water Supply Practices," Manual M9, AWWA, 1979.)

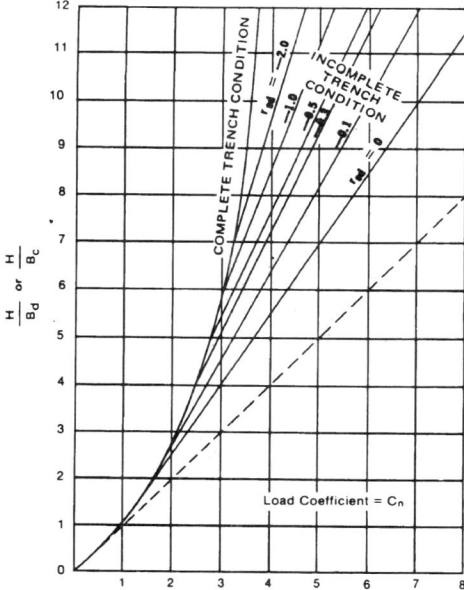

Figure 9.26. Load coefficient for negative projecting conduit at various values of r_{sd} for $p'=1.5$ and $k\mu=0.13$. (Source: Concrete Pressure Pipe, "Manual of Water Supply Practices," Manual M9, AWWA, 1979.)

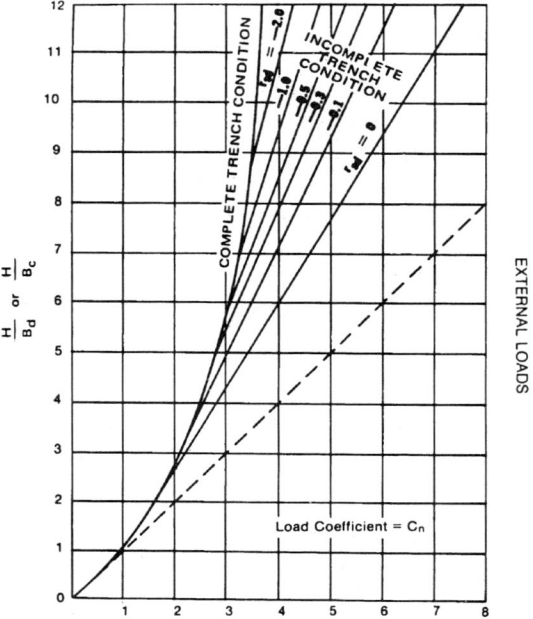

Figure 9.27. Load coefficient for negative projecting conduit at various values of r_sd for p'=2.0 and kμ=0.13. (Source: Concrete Pressure Pipe, "Manual of Water Supply Practices," Manual M9, AWWA, 1979.)

A flexible pipe interacts with the surrounding soil by deflecting and transferring a substantial portion of the vertical load to the side fill. For rigid pipes, bedding is seen as an important factor in minimizing soil pressure on the pipe. Likewise, for flexible pipes, the degree of soil compaction and the resulting soil density have a direct influence on the ability of a pipe to resist a soil load. The ability of a flexible pipe to resist a soil load is dependent on the relative flexibility of the pipe, the size of the pipe, its wall thickness, the type of soil, and the compaction of the soil and the side fill. This is shown in Spangler's Iowa formula[9,10], Eq. (9.48). Various terms used in the equation are illustrated in Figure 9.17.

$$\Delta X = D_L \frac{kW_c r^3}{EI + 0.061 er^4} \qquad (9.48)$$

where:
D_L = deflection lag factor
k = bedding constant
W_c = Marston's load per unit length of pipe, lb/linear in.
r = pipe radius, in.
E = modulus of elasticity of the pipe, psi
I = moment of inertia, in.^{4}
$e = 2h/\Delta X$
ΔX = horizontal deflection or change in diameter, in.

Structural Considerations

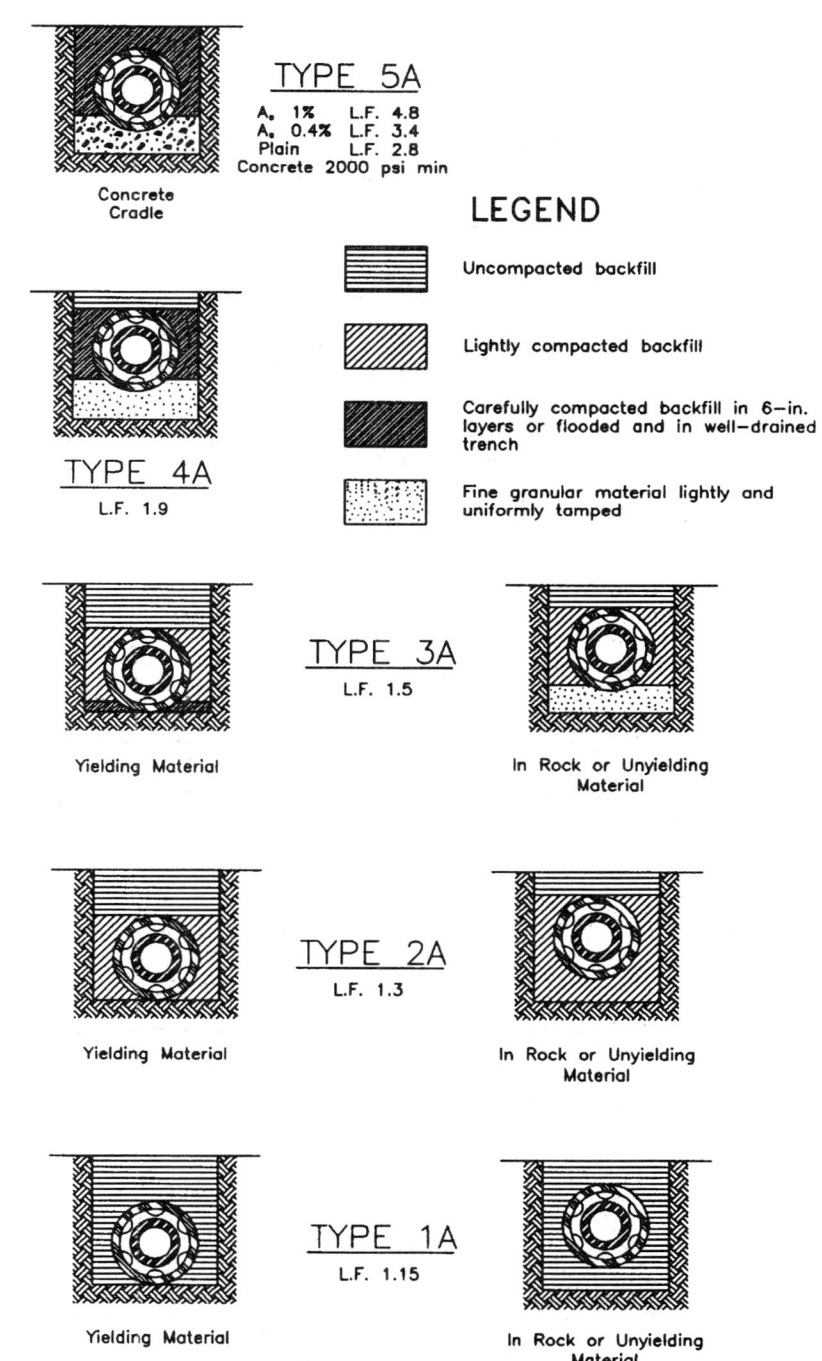

Figure 9.28. Five classes of bedding for rigid pipes based on concrete.

Concrete Cradle

Figure 9.29. Load factor for rigid pipe laid on a half crown.

A student of Spangler, Reynold K. Watkins, developed a modified form of the Iowa formula where the modulus of passive resistance of the side fill was replaced by a term called the modulus of soil reaction[11] (where $E' = er$):

$$\Delta X = D_L \frac{kW_c r^3}{EI + 0.061 E' r^3} \tag{9.49}$$

Table 9.2 presents average values of the modulus of soil reaction (E') for various types of soils as they relate to their relative proctor density. The work is based upon the study work of Amster K. Howard of the United States Bureau of Reclamation[12], and is accurate to within ±2% deflection, based upon the use of prism loads in the modified Iowa formula, Eq. (9.49). Table 9.4 lists values of bedding constants for a variety of bedding angles, based upon the theoretical work of Spangler. In most cases, a deflection lag factor of 1.5 can be used to produce conservative results.

The modified Iowa equation can also be written in other forms for use where values for pipe stiffness are normally used to specify the pipe, or the pipe is referred to by DR number:

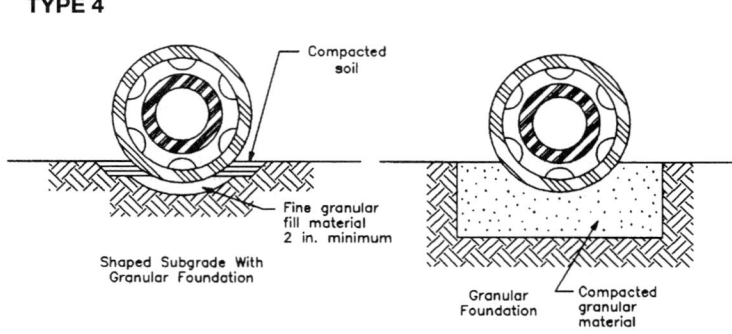

Figure 9.30. Load factor for rigid pipe laid on a full crown.

Structural Considerations

TYPE 3

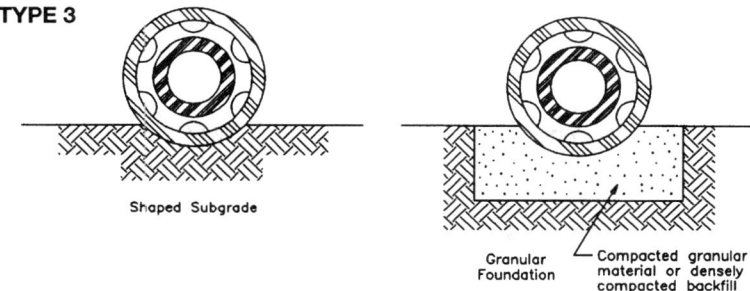

Figure 9.31. Load factor for rigid pipe buried directly on a shaped bottom.

$$\% \frac{\Delta}{D} = \frac{D_L k P(100)}{0.149\dfrac{F}{\Delta Y} + 0.061 E'} \tag{9.50}$$

where:
ΔY = vertical deflection, in.
P = prism load (soil pressure), psi
F = force, lbs/lin.

Figure 9.32 shows a photograph in which a flexible secondary containment pipe is used, and by virtue of the design, it is allowed to deflect diametrically by at least 2% as measured by its inside diameter. This allows it to be treated by the simplified equations of (9.48)-(9.52) developed for flexible pipes, since the vane-style supports [Figure 9.8(a)] allow for adequate clearance. It can be seen from the photograph however that perhaps too much clearance provided by the support is allowed in this example. The outer pipe is shown in its deflected position, whereby it can be noticed the primary pipe is sitting on the bottom invert of the secondary pipe, eliminating any bottom annulus. This is often a highly undesirable feature in double containment pipes, particularly where leak detection cable is used.

Figure 9.32. Photo of a section of a double containment pipe in which the secondary containment pipe behaves as a "flexible" buried pipe.

Table 9.4 Load Factors for Circular Pipe – Positive Projecting Embankment Installations

H/B_c	Type 3 bedding					Type 4 bedding				Type 5 bedding			
	\multicolumn{13}{c}{$p = 0.9$}
	$r_{sd}p=0$	0.1	0.3	0.5	1.0	$r_{sd}p=0$	0.1	0.3	0.5	$r_{sd}p=0$	0.1	0.3	0.5	1.0
0.5	3.01	2.82	2.82	2.82	2.82					11.26	8.87	8.87	8.87	8.87
1.0	2.55	2.35	2.35	2.35	2.35					6.61	5.37	5.37	5.37	5.37
1.5	2.42	2.26	2.16	2.16	2.16		Maximum			5.81	4.83	4.47	4.47	4.47
2.0	2.37	2.20	2.14	2.10	2.10		recommended projection			5.48	4.49	4.35	4.19	4.19
3.0	2.31	2.17	2.10	2.07	2.02		ratio of 0.7			5.18	4.50	4.21	4.06	3.88
5.0	2.27	2.14	2.08	2.04	2.00					4.97	4.37	4.11	3.97	3.81
10.0	2.24	2.12	2.06	2.03	1.99					4.82	4.28	4.04	3.90	3.76
15.0	2.23	2.10	2.05	2.02	1.98					4.77	4.25	4.01	3.88	3.74
	\multicolumn{13}{c}{$p = 0.7$}
	$r_{sd}p=0$	0.1	0.3	0.5	1.0	$r_{sd}p=0$	0.1	0.3	0.5	$r_{sd}p=0$	0.1	0.3	0.5	1.0
0.5	2.35	2.27	2.27	2.27	2.27	3.00	2.88	2.88	2.87	7.52	6.54	6.54	6.54	6.54
1.0	2.18	2.08	2.08	2.08	2.08	2.73	2.58	2.58	2.58	5.61	4.79	4.79	4.79	4.79
1.5	2.13	2.03	1.99	1.99	1.99	2.65	2.50	2.44	2.44	5.17	4.46	4.19	4.19	4.19
2.0	2.10	2.01	1.97	1.95	1.95	2.61	2.48	2.42	2.39	4.98	4.35	4.11	3.99	3.98
3.0	2.08	2.00	1.96	1.94	1.91	2.58	2.45	2.40	2.36	4.80	4.25	4.02	3.90	3.75
5.0	2.06	1.98	1.95	1.93	1.90	2.55	2.43	2.38	2.35	4.66	4.18	3.95	3.84	3.70
10.0	2.05	1.98	1.94	1.92	1.89	2.53	2.42	2.36	2.33	4.57	4.12	3.91	3.79	3.66
15.0	2.04	1.97	1.94	1.91	1.89	2.52	2.41	2.36	2.33	4.53	4.09	3.89	3.77	3.65
	\multicolumn{13}{c}{$p = 0.5$}
	$r_{sd}p=0$	0.1	0.3	0.5	1.0	$r_{sd}p=0$	0.1	0.3	0.5	$r_{sd}p=0$	0.1	0.3	0.5	1.0
0.5	1.94	1.92	1.92	19.2	---	2.37	2.33	2.33	2.33	4.84	4.54	4.55	4.55	---
1.0	1.90	1.86	1.86	1.86	---	2.31	2.25	2.25	2.25	4.33	3.97	3.97	3.97	---
1.5	1.88	1.85	1.83	1.83	---	2.28	2.23	2.20	2.20	4.18	3.83	3.68	3.68	---
2.0	1.88	1.84	1.83	1.82	---	2.27	2.22	2.20	2.19	4.11	3.79	3.65	3.58	---
3.0	1.87	1.84	1.82	1.81	---	2.26	2.22	2.19	2.18	4.04	3.75	3.62	3.54	---
5.0	1.86	1.83	1.82	1.81	---	2.26	2.21	2.19	2.17	3.99	3.72	3.58	3.51	---
10.0	1.86	1.83	1.81	1.80	---	2.25	2.20	2.18	2.17	3.95	3.69	3.56	3.49	---
15.0	1.86	1.83	1.81	1.80	---	2.25	2.20	2.18	2.17	3.94	3.68	3.56	3.48	---
	\multicolumn{13}{c}{$p = 0.3$}
	$r_{sd}p=0$	0.1	0.3	0.5	1.0	$r_{sd}p=0$	0.1	0.3	0.5	$r_{sd}p=0$	0.1	0.3	0.5	1.0
0.5	1.76	1.76	1.76	---	---	2.11	2.10	2.10	---	3.49	3.41	3.41	---	---
1.0	1.76	1.75	1.75	---	---	2.10	2.08	2.08	---	3.40	3.28	3.28	---	---
1.5	1.75	1.74	1.74	---	---	2.09	2.08	2.07	---	3.37	3.25	3.20	---	---
2.0	1.75	1.74	1.74	---	---	2.09	2.08	2.07	---	3.35	3.24	3.20	---	---
3.0	1.75	1.74	1.74	---	---	2.09	2.08	2.07	---	3.34	3.23	3.18	---	---
5.0	1.75	1.74	1.74	---	---	2.09	2.08	2.07	---	3.33	3.22	3.17	---	---
10.0	1.75	1.74	1.74	---	---	2.09	2.08	2.07	---	3.32	3.22	3.17	---	---
15.0	1.75	1.74	1.74	---	---	2.09	2.08	2.07	---	3.32	3.22	3.17	---	---
	\multicolumn{13}{c}{*Zero projecting*}
---	---	---	---	1.70	---	---	2.02	---	---	---	---	2.83	---	---

(Source: Concrete Pressure Pipe, "Manual of Water Supply Practices," Manual M9, AWWA, 1979.)

Structural Considerations

$$\%\frac{\Delta}{D} = \frac{D_L kP(100)}{\left[\dfrac{2E}{3(DR-1)^3}\right] + 0.061E'} \qquad (9.51)$$

In any application where live loads are involved, they must be included in the analysis:

$$\%\frac{\Delta}{D} = \frac{D_L kP + kW'}{\left[\dfrac{2E}{3(DR-1)^3}\right] + 0.061E'} \qquad (9.52)$$

Any buried flexible pipe will continue to deflect as long as its surrounding soil consolidates. This is true whether the flexible pipe is thermoplastic, reinforced-thermosetting-resin pipe, or even for some metallic pipes. The response of a pipe to the consolidation of its surrounding soil is referred to as the deflection lag of the soil. The amount of time during which deflection lag will occur is a function of soil density in the area of a pipe. The time during which deflection lag occurs and the total amount of deflection of a pipe are both inversely related to the density of soil at the sides of a pipe.

If final backfill is compacted to a high density, the full load on a buried pipe will be approximately reached in an immediate fashion. Otherwise, consolidation of the soil will occur over time along with continued deflection of the pipe. At burial temperatures above ambient, a pipe will have a lower stiffness and therefore will have a slightly larger equilibrium deflection. Also, a buried pipe that is at elevated temperatures will take a shorter time to reach equilibrium. This is due to the faster interaction between the soil and the pipe to achieve equilibrium, and thus a lower stiffness value of the material at the higher temperature.

A number of variations of the Spangler-Watkins Iowa formula have been suggested. All of them are based on the premise that the deflection of a pipe depends on the ratio of the soil load involved to the sum of the pipe stiffness and a constant times the soil stiffness. From this basic premise, Watkins derived his soil-strain theory.[13-15] This theory is expressed in dimensionless terms as follows:

$$\frac{Y}{D\varepsilon} = \frac{R_s}{AR_s + B} \qquad (9.53)$$

where:
Y = vertical deflection, in.
R_s = stiffness ratio. This is the ratio of soil stiffness E_s to pipe ring stiffness EI/D^3. This quantity includes all properties of materials, soil as well as pipe
D = pipe outside diameter, in.
ε = vertical soil strain, in.
A, B = empirical constants which include such terms as D_L and K of the Iowa formula

Table 9.5 Values of Bedding Constant, K

Bedding angle (degrees)	K
0	0.110
30	0.108
45	0.105
60	0.102
90	0.096
120	0.090
180	0.083

where R_S can be determined by:

$$R_s = \frac{12E_s(D)^3}{E(t)} \tag{9.54}$$

where:
$E_s = P/\varepsilon$ = vertical pressure at top of pipe / vertical soil strain, psi

Figure 9.33 presents ring deflection factor as a function of stiffness, which is based upon Eq. (9.55) and also on empirical data. For most flexible pipe applications, R_S is greater than 300. Therefore, the ring deflection factor is normally one, and ring deflection thereby becomes:

$$\frac{y}{D} = \varepsilon \tag{9.55}$$

The load on the flexible pipe can be calculated using prism load theory. If so, soil strain may be determined by the use of Figure 9.34. This figure has been shown by experience to provide adequate results for most soils. Since the results are adequate, the need for specific laboratory tests to generate these types of curves is not warranted in most applications.[16]

9.4.1.5 Performance Limits for Flexible Pipes

Pipes that are subjected to external loading can fail due to a number of different mechanisms.[17] These include situations where the primary cause of failure is due to: (1) stress; (2) fatigue; (3) deflection; (4) buckling; (5) wall crushing; (6) longitudinal bending; and (7) strain. Individual performance limits are described for each of these different mechanisms in the following paragraphs.

9.4.1.5.1 Stress Performance Limit

The stress performance limit refers to any application where an applied internal pressure results in hoop stresses that exceed the design strength of a pipe. This can be written simply as:

$$\sigma = \varepsilon E \tag{9.56}$$

where :
σ = wall stress, psi
ε = strain, in./in.
E = modulus of elasticity, psi

9.4.1.5.2 Fatigue Performance Limit

Fatigue failure can occur where materials are subjected to repeated stresses such as those that occur in "severe-cyclic" operation. Failure in the material can result at stress levels that are less than the actual design strength of the material. This type of failure is normally limited to soft materials, but can also occur in other types of materials. Live loading that is of a frequent nature for a flexible pipe buried to a shallow depth can produce this type of failure.

9.4.1.5.3 Deflection Performance Limit

Deflection performance limits have been historically set at levels of 5% for all materials. This is due to the fact that corrugated steel pipe was assigned a deflection performance limit of 5% based upon a 4 to 1 safety factor on its ultimate deflection limit of 20%. While 5% produces conservative results for most flexible materials, it may in fact be too conservative for some materials, such as high-density polyethylene; it is not enough for other materials such as polyvinylidene fluoride (PVDF), which usually is rated to 3% upper limits, depending on the grade of resin used (not true of newer flexible grades).

9.4.1.5.4 Buckling Performance Limit

The buckling performance limit governs the design of flexible pipes where the design conditions include internal vacuum, a water table above the depth of burial, loose soil burial, and any circumstance where external loads exceed the compressive strength of a pipe material. From Timoshenko,[18] the critical buckling pressure for a circular ring subjected to a uniform external pressure or internal vacuum is:

$$P_{cr} = \frac{3EI}{r^3} \tag{9.57}$$

where :
R_{cr} = critical buckling pressure, psi

For long pipes under combined stress, the equation can be rewritten as:

$$P_{cr} = \frac{E}{4(1-\mu^2)}\left(\frac{t}{r}\right)^3 \tag{9.58}$$

In most situations involving proper burial procedures, the surrounding soil will provide some resistance to buckling. Under these conditions, buckling pressure can be calculated as[7]:

$$P_b = 1.15\sqrt{P_{cr}E'} \tag{9.59}$$

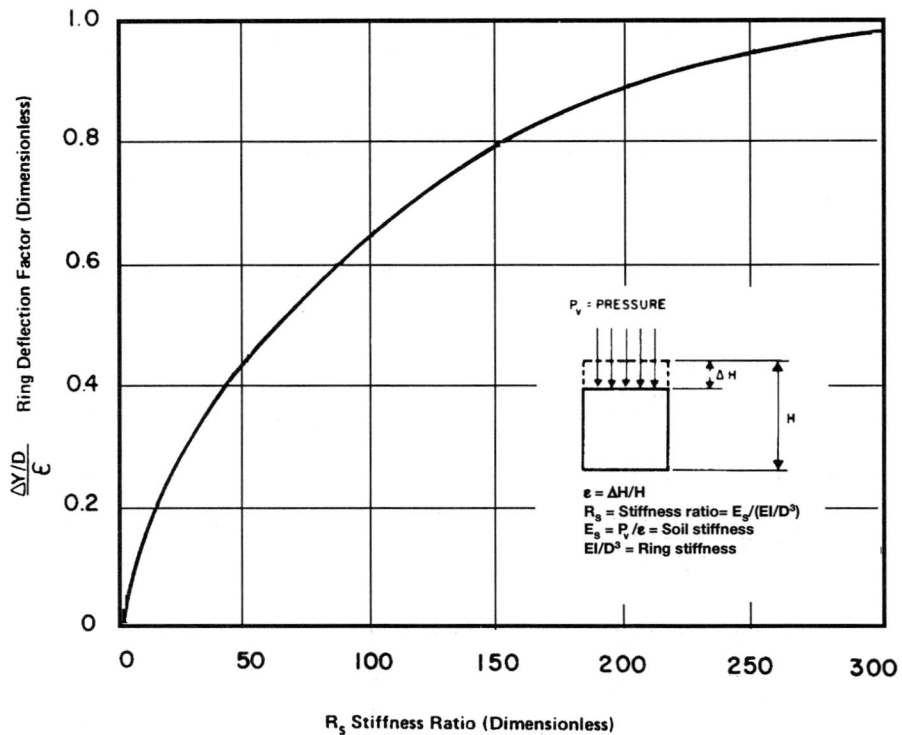

Figure 9.33. Ring deflection factor as a function of stiffness. (Source: Utah State University.)

where:
P_b = buckling pressure in a given soil, psi
E' = modulus of soil reaction, psi

The allowable buckling pressure for fiberglass-reinforced thermosetting-resin-plastic piping is stated in Appendix A of AWWA Standard C 950 as:[19]

$$q_a = \left(\frac{1}{FS}\right)\left[32 R_w B' E' \left(\frac{EI}{D^3}\right)\right]^{1/2} \qquad (9.60)$$

$$B' = \frac{1}{\left(1 + 4\theta^{-0.065H}\right)} \qquad (9.61)$$

$$R_w = 1 - 0.33\left(\frac{H_w}{H}\right) \qquad (9.62)$$

where:
$0 \leq H_w \leq H$

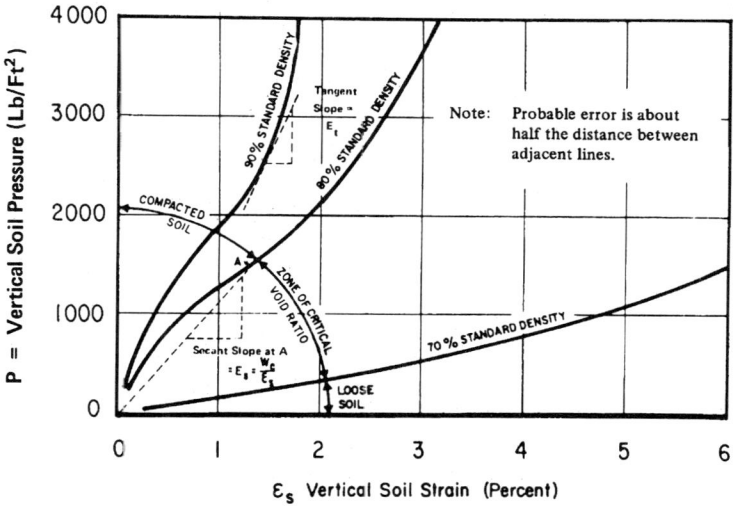

Figure 9.34. Nomograph of soil strain vs. soil pressure for typical trench backfill (except clay) from actual tests. (Source: Utah State University.)

Equation (9.61) is valid for the following conditions:

with internal vacuum: $4 \leq H \leq 80$ ft
without internal vacuum: $2 \leq H \leq 80$ ft

Where internal vacuum occurs with: $2 \leq H \leq 4$ ft, the allowable buckling pressure can also be determined by the von Mises formula as:

$$q_a = \frac{2Et_t}{\left[D(n^2-1)(1+k)^2\right]} + \left(n^2 - 1 + \frac{(2n^2 - 1 - \mu_{hl})}{(1+k)}\right)\left(\frac{8EI}{D^3(1 - \mu_{hl}\mu_{lh})}\right) \quad (9.63)$$

$$k = \frac{(2nl)}{\pi D} \quad (9.64)$$

where:
q_a = allowable buckling pressure, psi
t_t = combined thickness of the pipe reinforced wall, per ASTM D3567, and the liner (when used), in.
D = mean pipe diameter, in.
n = number of lobes formed at buckling ≥ 2 (the value of n must give the minimum value of q_a obtained by iterative solution)
μ_{hl} = Poisson's ratio, applied hoop stress
μ_{lh} = Poisson's ratio, applied longitudinal stress
E = ring flexural modulus of elasticity, psi
I = moment of inertia of pipe wall for ring bending, (in.^{4}/lineal in.)
l = distance between rigid ring stiffeners, in.

9.4.1.5.5 Wall-Crushing Performance Limit

When backfill is extremely compacted, soils become very rigid, and pipe buried under such conditions can experience large loads that cause it to fail via a wall-crushing mechanism. An example of where this can occur is where the pipe is completely encased in concrete or where it is concrete cradled. However, for flexible pipes, the load involved must be extremely large in order for this to occur; therefore, it is not possible to occur in practice, except in unusual circumstances. The performance limit due to wall crushing is expressed as follows:

$$\sigma_c = \frac{P_y D}{2A} \tag{9.65}$$

9.4.1.5.6 Longitudinal Bending Performance Limit

The longitudinal bending performance limit of a pipe and procedures for analyzing longitudinal bending are covered in an earlier part of this chapter (refer to Section 9.2). The usual response of a flexible pipe to longitudinal bending is that if it is bent to a radius that is smaller than its minimum bending radius, failure of the pipe can occur, usually as a result of buckling.

9.4.1.5.7 Strain Performance Limit

The strain performance limit for flexible pipes is not frequently used since it usually results in deflection that are far in excess of those allowed by deflection performance limits. Most flexible pipes are therefore not limited by definable strain limits for burial applications.

9.4.1.6 RTRP Piping Eight-Step Procedure

The American Water Works Association has outlined a procedure for the determination of the acceptability of fiberglass-reinforced thermosetting-resin-plastic piping in buried services, using an eight-part procedure. The procedure is outlined in Appendix A of AWWA Standard C 950-88. The procedure is described as follows:

1. Check that the pipe satisfies the pressure classification;
2. Check that the working pressure does not exceed the pressure classification;
3. Check that the working pressure plus surge pressure does not exceed the designated pressure class when functioning under a surge pressure;
4. Calculate the allowable deflection based on the pipe operating without internal pressure so long term (50 year) stress or strain is not exceeded;
5. Calculate the external dead and live loads that induce pipe deflection;
6. Calculate a predicted deflection and insure that it does not exceed the allowable;
7. Check the stress, or strain, for combined loading of external loads and internal pressure to assure that it does not exceed the long term (50 year) allowable;
8. Check that the capacity of the pipe to resist buckling exceeds the combined loading that can induce buckling.

The procedure has been developed for RTRP materials, but serves as a useful outline in the analysis of any flexible material for burial suitability.

9.4.2 General Methods; Double Containment Piping Assemblies

Double containment piping systems intended for underground service must be subjected to an analysis that is similar to that used for single-wall piping systems. A double containment pipe assembly is different from a single-wall pipe; thus its behavior with respect to burial loads is often different. The effect of rigid, nonyielding supports positioned in the interstice of the pipes is to stiffen the assembly at each support location. The same effect is produced by internal anchors and other nonyielding components. Where increased stiffening is possible, the usual result is to lessen the pipes resistance to soil loads, since a more flexible pipe has more inherent resistance (see Section 9.4.1.4). Localized discontinuity stresses are also created at each of the locations of the nonyielding components, since flexible pipe sections are immediately adjacent to the component location.

Not all interstitial supports are designed to be nonyielding or prevent secondary containment piping from a reasonable amount of diametrical deflection. There are many types of supports that may be used that allow the outer pipe to deflect, and thus allow it to behave in a truly "flexible" manner. To understand the difference between different styles of interstitial supports, refer to Subsection 9.3.1.2, "Interstitial Supporting Devices."

A substantial portion of the load imposed on a secondary containment pipe may also be transferred to its associated primary pipe through interstitial supports and interconnecting points. Thus, an analysis for this type of arrangement must also consider the effect on primary piping components.

When flexible secondary containment piping materials are used, the effect of interstitial supports and interconnecting parts is to add stiffness and rigidity to the secondary containment pipe in localized regions. Since flexible pipes gain their resistance to soil loads by their ability to deflect readily, the resistance characteristics of a secondary containment pipe may become completely modified. In some applications, the use of tightly fitting interstitial supports, or rigid interconnecting parts, may prevent what is an otherwise flexible secondary containment pipe from being able to be deflected by at least 2%, without structural harm being caused. An assembly of this type may have to be classified as a rigid pipe system, and its primary and secondary containment pipes analyzed as such, if a simplified analysis is desired. It is always more accurate to conduct a comprehensive analysis for a pipe system of this type, in order to study the effect of localized discontinuity stresses thoroughly.

9.4.2.1 Totally Rigid Assemblies

A double containment piping system that is subject to burial loads is often classified as a rigid pipe according to the same criteria used for single-wall pipes. A double containment pipe assembly may meet this classification if its secondary containment piping meets the same definition for a single-wall pipe. It may also be classified as such if the assembly makes use of tightly fitting interstitial supports and/or interconnecting components that prevent an otherwise flexible secondary containment pipe from being able to be deflected by at least 2% without structural harm being caused to the pipe.

When rigid secondary containment piping is used as part of the assembly, there can be two types of conditions present. The first condition is the easiest to analyze for. It consists of a primary pipe that is positioned within the secondary containment piping in a manner such that the loads imposed on the secondary containment pipe are not imposed on it as well. Under this design condition, only the secondary containment pipe needs to be analyzed, using the simplified analysis ordinarily applied to rigid single-wall piping described in Section 9.4.1.3.

The second rigid double containment piping assembly condition involves loads imposed on the secondary containment pipes being transferred to the primary pipe through points of interconnection. Pipes that have continuously welded (bonded) or extruded vanes along their annulus fall into this category, as do simultaneously welded assemblies, RTRP assemblies that make use of bonded collars at the end of straight sections, assemblies that make use of primary pipe anchoring arrangements, and others. This condition is possible when either rigid or flexible piping components (considered alone) are used for secondary containment pipes.

Under this second condition, both the secondary containment and primary piping need to be analyzed for suitability under the given load. A simplified means to analyze assemblies with this condition is to determine a three-edge-bearing load for the assembly, using the approach described in Section 9.4.1.3 for single-wall rigid pipes. A three-edge-bearing strength value can be established for the entire assembly, with supports in place during the test in order to simulate the actual installation closely. Three-edge-bearing strength values should be determined for individual inner and outer pipes as well. The smallest value determined can be selected as the controlling value for analyzing the complete assembly.

In the second condition mentioned, different sections of the pipe that have different types of interconnecting arrangements, different types of interstitial supports, or spacing of such supports should be tested separately. This should be done in case the three-edge-bearing strengths are different in the unique sections. If a double containment assembly makes use of interconnecting parts that are without "separation" (see Section 7.33), then the interconnecting part should be analyzed for burial as if it were a component of the secondary containment piping structure. This is in contrast to the analysis ordinarily used for internal pressure design (see Section 7.3), and thermal stress analysis (see Section 8.3), whereby a component of this nature must be considered as component of the primary piping system. For interconnecting parts that are designed with true "separation," their individual parts should be analyzed as parts of their respective systems. The three-edge-bearing load of all individual interconnecting components, regardless of whether "separation" exists, should be determined in order to conduct a proper simplified analysis. In addition to having an effect on the overall assembly of making it more rigid, these types of components also act as a localized area unto themselves that have their own unique deflection limitations. Therefore, their three-edge-bearing strengths should be considered and their unique values established.

When plastic pipes that would otherwise be flexible are designed with tightly fitted interstitial supports (allowing less than 2% deflection as measured by the inside diameter of the pipe), the piping will be subject to higher than normal burial loads. The loads for this type of system will be higher by the ratio shown in Eq. (9.37) and can exceed prism loading conditions as a result. The basic materials lack structural strength without their ability to deflect and reduce the load produced by a column of soil that is above it. If a pipe is buried sufficiently deep, and proper attention is not given to the design, premature failure of both primary and secondary containment pipes can result. Thus a "double failure" can occur, defeating the purpose of providing secondary containment. It may actually increase the likelihood of failure to its surrounding atmosphere, compared to a well-designed single-wall pipe. This type of situation can be addressed by the use of an appropriate bedding arrangement, as shown for rigid single-wall piping buried conditions in Figures 9.28 through 9.31, and discussed in Section 9.4.1.3.

9.4.2.2 Nonrigid (Flexible) Assemblies

A double containment piping system subject to burial loads is classified as a flexible pipe, according to the same criteria applied to single-wall pipes. A double containment pipe assembly may meet this classification if its secondary containment piping meets the definition for flexible pipes, which is being able to be deflected by at least 2% as measured by its inside diameter, without structural harm being caused to the pipe, if no other restrictions apply.

Thus, a flexible secondary containment pipe must be capable of resulting in a nonrigid arrangement, in order for a system to be classified as a flexible pipe. There are two basic ways that a nonrigid system can be obtained, which are both in effect very similar. The first method is the easiest to analyze for and has been used extensively in many landfill applications involving double containment polyethylene pipes. As is common practice in landfill piping, or in slip-lining applications, a flexible assembly will result from simply laying a primary pipe inside a secondary containment pipe without the use of any interstitial supports (provided there is sufficient annular clearance). A variation by which a nonrigid arrangement can be achieved is through the use of interstitial supports that provide vertical support, but do not significantly affect diametrical deflection of its secondary containment piping. In some cases with the use of vane or collar-type supports, vanes can be designed so that they allow the maximum amount of deflection required (5% in most cases), and prevent further deformation once the outside pipe becomes ovalized to the maximum allowable elliptical shape (refer to Figure 9.18). Under this design condition, only the secondary containment pipe needs to be analyzed, using the analysis outlined for flexible single-wall piping, described in Section 9.4.1.4.

A second method by which a double containment piping can be classified as a flexible conduit is where the complete assembly is allowed to flex simultaneously by at least 2%, as measured by the inside diameters of both inner and outer pipes. This situation can only result where both primary and secondary containment pipes are constructed of flexible materials, and interstitial supports and/or rigid interconnecting parts are used that permit this to be possible. When this type of design occurs, the analysis tends to be highly complex. The ordinary equations of flexible pipe analysis may be used with modification, but may not be valid. If the same inner and outer materials are used, a corrected moment of inertia can be substituted in the analysis, and the system treated as one single-wall pipe. Under this analysis, it is highly recommended that the pipe stiffness of the combined system be tested according to the parallel plate loading method (ASTM D 2412) in order to be sure that the combined system has adequate strength.

If different materials are used for primary and secondary containment pipes, the system can be analyzed by using Eq. (9.51) from the measured combined stiffness. In this equation, the modulus of elasticity and moment of inertia are not included. Thus, the result can be obtained by use of the measured stiffness. Systems of this type can also be analyzed by treating each pipe as a separate case.

9.4.2.3 Comprehensive Analysis

It is always more accurate to use some type of comprehensive analysis for an assembly that models inner and outer systems at the same time (i.e., finite element analysis), as opposed to using a simplified approach. This is particularly true if there are multiple forces acting concurrently on the assembly. An example would be a primary pipe system under internal pressure, subject to large changes in temperature, and with substantial dead-weight and live burial loads imposed. This is in actuality a very common condition. Comprehensive stress analysis is discussed in more detail in Section 8.3.3 of Chapter 8.

References

1. E. Reissner, "On Finite Bending of Pressurized Tubes," J. Appl. Mechanics Trans. ASME (Sept. 1959), pp. 386-392.
2. S. Timoshenko, Strength of Materials, Part I, Elementary Theory and Problems, 3rd. ed., Van Nostrand Company, Princeton, NJ, 1958.
3. S. Timoshenko, Strength of Materials, Part II, Advanced Theory and Problems, . ed., Van Nostrand Company, Princeton, NJ, 1968.
4. S. Timoshenko, and D. H. Young, Elements of Strength of Materials, 4th ed., Van Nostrand Company, Princeton, NJ.
5. S. Timoshenko, Theory of Elastic Stability, 2nd ed., McGraw-Hill, 1961.
6. S. Timoshenko, op. cit.
7. E. Reissner, op. cit.
8. A. Marston and A. O. Anderson, "The Theory of Loads on Pipes in Ditches and Tests of Cement and Clay Drain Tile and Sewer Pipe," Bul. 31, Iowa Engineering Experiment Station, Ames Iowa, 1913.
9. M. G. Spangler, "The Structural Design of Flexible Pipe Culverts," Bulletin 153, Iowa Engineering Experiment Station, Ames, Iowa, 1941.
10. M. G. Spangler and R. L. Handy, Soil Engineering, Intext Educational Publ., New York, NY, 1973.
11. R. K. Watkins and M. G. Spangler, "Some Characteristics of the Modulus of Passive Resistance of Soil – A Study in Similitude."
12. A. K. Howard, "Modulus of Soil Reaction (E´) Values for Buried Flexible Pipe," Journal of the Geotechnical Engineering Division, ASCE, Vol. 103, No. GT, Proceedings Paper 12700, Jan. 1977.
13. R. K. Watkins and A. P. Moser, "Response of Corrugated Steel Pipe to External Soil Pressures," Highway Research Record 373, 1971, pp. 88-112.
14. R. K. Watkins, A. P. Moser and R. R. Bishop, "Structural Response of Buried PVC Pipe," Modern Plastics, Nov. 1973, pp. 88-90.
15. R. K. Watkins and A. B. Smith, "Ring Deflection of Buried Pipe," Journal AWWA, Vol. 59, No. 3, March 1967.
16. "Handbook of PVC Pipe Design and Construction", Uni-Bell Plastic Pipe Association, 3rd printing, Dallas, TX, 1979, p. 154.
17. "Handbook of PVC Pipe Design and Construction", Uni-Bell Plastic Pipe Association, 3rd printing, Dallas, TX, 1979, p. 159.
18. S. Timoshenko, op. cit.
19. ANSI/AWWA C950-88, "AWWA Standard for Fiberglass Pressure Pipe," American Water Works Association, Denver, CO, 1989.

Notes

Notes

Chapter Ten

Heat Transfer In Double Containment Piping

10.1 General

The theory of heat transfer as applied to piping applications is well known, including applications involving a pipe within a pipe. The concept of a pipe within another pipe has a long history in heat transfer applications. Various heat transfer applications involving concentrically positioned pipes include: double pipe heat exchangers, insulated utility piping (equipped with external jacketing), jacketed process pipes, and others. Heat transfer coefficients are well known in these cases, as are the relevant insulating characteristics, and other such factors that pertain.

Applications that are considered relevant in terms of double-wall piping systems include: cooling water applications, refrigeration and air conditioning, heat exchangers, cogeneration, standard transport of hot or cold chemicals, and many others. A thorough understanding of heat transfer technology is important in order to develop computerized models for thermal stress analysis and for establishing a basis from which compensation must be provided. Heat transfer knowledge of pipes is also essential for selecting and sizing heat tracing for freeze protection purposes, establishing insulation requirements, and many other purposes.

Piping heat transfer theory can be divided into three general categories: conduction, convection, and radiation. These are discussed in Sections 10.1.1 through 10.1.3.

10.1.1 Conduction in Single-Wall Pipes and Double Containment Pipes

Conduction is the process whereby heat is transferred between molecules of substances that adjoin each other in a static arrangement. Conductivity for all materials generally increases with any increase in temperature. The conductivity of a substance is defined by a term known as the thermal conductivity of the substance. Thermal conductivity of a material is measured in terms of energy transfer units per cross-sectional area of a material per unit temperature gradient.

Thermal conductivities of nonmetals are typically lower than those of metallic materials. As an example, the thermal conductivity of pure copper is approximately 223 BTU in./ft h °F at 68°F, while the thermal conductivity of PVC is approximately 1.2 BTU in./ft h °F at

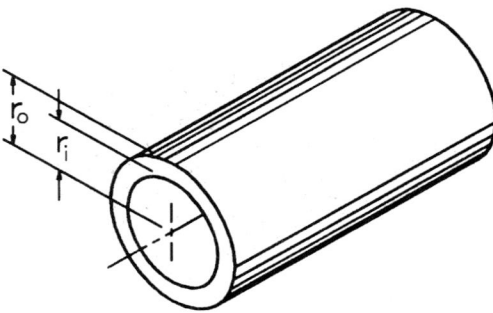

Figure 10.1. Illustration of an uninsulated single-wall pipe showing terms relating to heat transfer.

68°F. Materials that have thermal conductivities of less than 0.3 BTU in./ft h °F are often selected as insulating materials.

The discussion in this chapter is limited to steady-state conduction. Non-steady-state conduction is important in heat transfer during startups and cyclic operating conditions. The analysis of non-steady-state conduction involves the use of differential forms of the Fourier equation. For a comprehensive discussion of non-steady-state conductive heat transfer, there are many suggested readings.[1-4]

The flow of heat in a single-wall pipe is mainly limited to flow in the radial direction, with axial flow of heat in most cases being negligible. This energy transfer can typically be modeled as a one-dimensional steady-state conduction problem. The flow of heat across a pipe wall by conduction can be modeled in its simplest form by Eq. (10.1). Refer to Figure 10.1 for an illustration of the terms used for heat transfer in a single-wall pipe, without insulation.

$$q = \frac{\Delta T}{R} \tag{10.1}$$

where:
q = heat transfer = BTU/h ft^2/in.
ΔT = temperature difference = °F
R = conductive heat transfer coefficient = h ft^2 °F/BTU in.

For single-wall pipes, the resistance term, R, expressed in terms of its inner and outer radius is:

$$R = \frac{1}{2\pi k L} \ln\left(\frac{r_o}{r_i}\right) \tag{10.2}$$

where:
k = coefficient of thermal conductivity = $\left[\text{BTU per (ft}^2\text{)(h)(°F)(per inch)}\right]$
L = length of piping, in.
r_o = outside radius, in.
r_i = inside radius, in.

Heat Transfer in Double Containment Piping

The value of the resistance shown in Eq. (10.2) can be substituted into Eq. (10.1) to result in the following more common expression of steady-state conduction in a single-wall pipe not having insulation:

$$q = \frac{2\pi L \Delta T}{\ln\left(\dfrac{r_o}{r_i}\right)} \tag{10.3}$$

An alternative way of expressing Eq. (10.3) is in terms of the logarithmic mean area of the inner and outer surfaces of the pipe. Logarithmic mean area is expressed as:

$$A_L = \frac{A_o - A_i}{\ln\left(\dfrac{A_o}{A_i}\right)} = \frac{2\pi L(r_o - r_i)}{\ln\left(\dfrac{r_i}{r_o}\right)} \tag{10.4}$$

where:
$A_i = 2\pi r_i L$ = inner surface area of the pipe, in.^{2}
$A_o = 2\pi r_o L$ = outer surface area of the pipe, in.^{2}

Substitution in Eq. (10.3) results in a close approximation of the overall heat transfer that takes place:

$$q = \frac{k A_L \Delta T}{r_o - r_i} \tag{10.5}$$

This equation can also be written based on the arithmetic mean area A_a given in Eq. (10.6), or the geometric mean area A_g given in Eq. (10.7).

$$A_a = \frac{A_i + A_o}{2} \tag{10.6}$$

$$A_g = \sqrt{A_i A_o} \tag{10.7}$$

The rewritten expressions are found by substitution into Eq. (10.3), and are shown, respectively, in Eqs. (10.8) and (10.9). When using these two equations, only approximate results are obtained.

$$q = \frac{2\pi k A_a \Delta T}{(r_i + r_o) \ln\left(\dfrac{r_o}{r_i}\right)} \tag{10.8}$$

$$q = \frac{A_g k \Delta T}{\sqrt{r_o r_i} \ln\left(\frac{r_o}{r_i}\right)} \quad (10.9)$$

When a pipe is constructed of a composite material, such as a dual-laminate material, the value for conductive heat transfer must be found by the use of an overall heat transfer coefficient, U. An overall heat transfer coefficient may also be used when the problem involves a combination of conduction and convection. Convective heat transfer is described in more detail in Section 10.1.2. Most piping applications involve a combination of convection and conduction.

The use of an overall heat transfer coefficient to calculate conduction is expressed as:

$$q = UA\Delta T \quad (10.10)$$

where:

U = overall heat transfer coefficient = $\left[\text{BTU per } (\text{ft}^2)(\text{h})(°F)(\text{per inch})\right]$

A = surface area, ft^2

Since q is also defined by Eq. (10.3), the two equations can be solved in terms of U:

$$U = \frac{1}{A \Sigma R} \quad (10.11)$$

For a noncomposite pipe, not having insulation, and subjected to static internal and external fluid conditions, U is defined as:

$$U = \frac{1}{\dfrac{A \ln(r_o/r_i)}{2\pi k_o L}} \quad (10.12)$$

For a dual-laminate pipe, without external insulation and subjected to static internal and external fluid conditions, U is defined by Eq. (10.13). Figure 10.2 illustrates the terms for a dual-laminate pipe.

$$U = \frac{1}{A_s \left[\dfrac{\ln(r_3/r_2)}{2\pi k_2 L} + \dfrac{\ln(r_2/r_1)}{2\pi k_1 L}\right]} \quad (10.13)$$

where:

A_s = outside surface area = in.^2
r_1 = inside diameter of the liner, in.
r_2 = outside diameter of the liner, and also equals inside diameter of the structual layer, in.
r_3 = outside diameter of the structural layer, in.
k_1 = coefficient of thermal conductivity of the liner,
   $\left[\text{BTU per } (\text{ft}^2)(\text{h})(°F)(\text{per inch})\right]$
k_2 = coefficient of thermal conductivity of the structural layer,
   $\left[\text{BTU per } (\text{ft}^2)(\text{h})(°F)(\text{per inch})\right]$

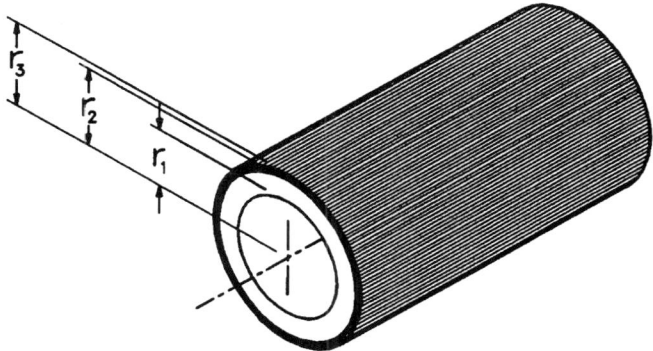

Figure 10.2. Illustration of a dual-laminate single-wall pipe showing resistance terms.

For the case of a noninsulated, noncomposite pipe, where convection is taken into account on the outside and inside surfaces, U is determined by:

$$U = \frac{1}{A_s \left[\dfrac{1}{A_i h_i} + \dfrac{\ln(r_o/r_i)}{2\pi k_o L} + \dfrac{1}{A_o h_o} \right]} \qquad (10.14)$$

where:
A_i = inside surface area = ft^2
$A_o = A_s$ = outside surface area = ft^2
h_i = inside contact coefficient due to convection,
  [BTU per (ft^2)(h)(°F)(per inch)]
h_o = outside contact coefficient due to convection,
  [BTU per (ft^2)(h)(°F)(per inch)]

For a dual-laminate pipe that is without insulation, where convective heat transfer is taken into account on the outside and inside surfaces:

$$U = \frac{1}{A_s \left[\dfrac{1}{A_i h_i} + \dfrac{\ln(r_3/r_2)}{2\pi k_2 L} + \dfrac{\ln(r_2/r_1)}{2\pi k_1 L} + \dfrac{1}{A_o h_o} \right]} \qquad (10.15)$$

For a single-walled pipe that contains external insulation (see Figure 10.3), and excluding convective effects, the term for U must contain a resistance term for the insulation, as well as for the inside air contact coefficient (air gap) between the insulation and the pipe wall:

$$U = \frac{1}{A_s \left[\dfrac{\ln(r_o/r_i)}{2\pi k_1 L} + \dfrac{1}{h_c A_o} + \dfrac{\ln(r_{ji}/r_o)}{2\pi k_{ins} L} \right]} \qquad (10.16)$$

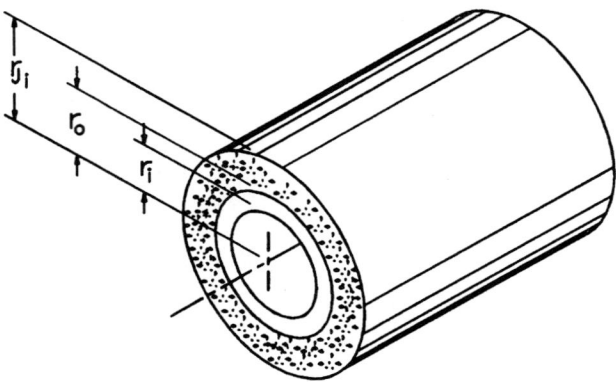

Figure 10.3. Illustration of a single-wall pipe with external insulation showing resistance terms.

where:
k_{ins} = coefficient of thermal conductivity of the insulation,
$$\left[\text{BTU per (ft}^2\text{)(h)(°F)(per inch)}\right]$$
r_o = inside diameter of the insulation; outside diameter of the pipe, in.
r_{ji} = outside diameter of the insulation, in.
h_c = inside air contact coefficient between the insulation and the pipe wall,
$$\left[\text{BTU per (ft}^2\text{)(h)(°F)(per inch)}\right]$$

In many cases, piping with insulation also is provided with a secondary protective jacketing to protect the insulation from weather effects; this is commonly referred to as a weather barrier, and is illustrated in Figure 10.4. Therefore, resistance terms for the inside air contact coefficient (the coefficient as the result of air gap) between the outside of the insulation and the inside diameter of the secondary jacket should be added to result in the expression

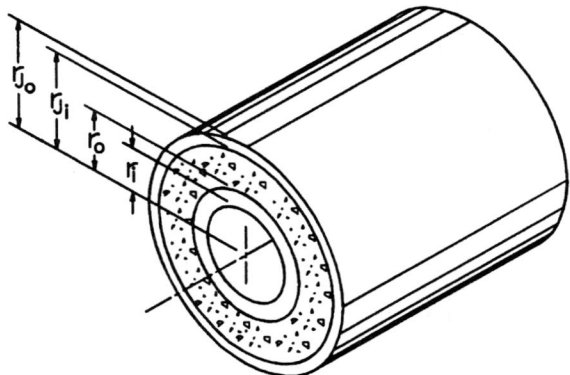

Figure 10.4. Illustration of a single-wall pipe with insulation and an external vapor barrier indicating resistance terms. This illustration also applies to a double containment pipe with insulation in its annulus.

shown in Eq. (10.17), when a weather barrier is used. The expression shown in Eq. (10.17) is also the equation used for double containment piping that contains insulation inside its annular space, when resistance terms for convection due to the inside and outside films are negligible. Thus, Figure 10.4 can also represent a section of double containment piping that also has insulation contained in its annulus.

$$U = \left[A_s \left(\frac{\ln(r_o/r_i)}{2\pi k_1 L} + \frac{1}{h_c A_o} + \frac{\ln(r_{ji}/r_o)}{2\pi k_g L} + \frac{1}{h_{c2} A_{ji}} + \frac{\ln(r_{jo}/r_{ji})}{2\pi k_2 L} \right) \right]^{-1} \quad (10.17)$$

For a double containment pipe that contains entrapped dead (stagnant) air or other stagnant gas, the nonmoving gas behaves as an effective means of insulation at moderate temperatures. Equation (10.17) is modified as shown in Eq. (10.18) to exclude the terms for the inside contact coefficients in this type of assembly. This is to be used for pipes where the secondary containment system is a closed system, and with entrapped air in the annulus, as illustrated in Figure 10.5. For a double containment pipe that contains vacuum in its annulus, the equation is not valid, as heat transfer between the inner and outer pipes will occur primarily through radiation. For a double containment pipe having an open annulus, convective effects will occur, as movement of air occurs as the result of natural gas circulation. These effects must be considered for these cases.

$$U = \frac{1}{A_s \left[\frac{\ln(r_o/r_i)}{2\pi k_1 L} + \frac{\ln(r_{ji}/r_o)}{2\pi k_g L} + \frac{\ln(r_{jo}/r_{ji})}{2\pi k_2 L} \right]} \quad (10.18)$$

where:
k_g = coefficient of thermal conductivity of the entrapped air or gas =
$\left[\text{BTU per (ft}^2\text{)(h)(°F)(per inch)} \right]$

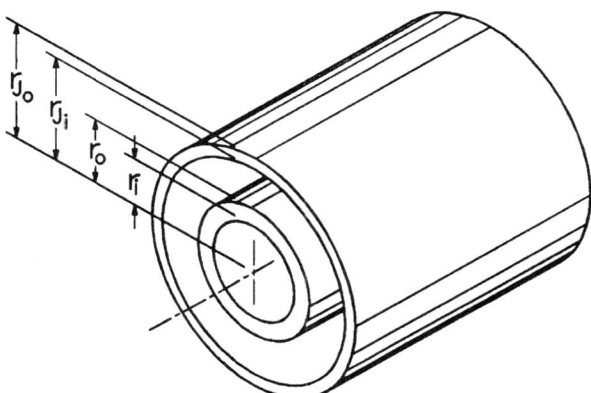

Figure 10.5. Illustration of a double containment pipe section having an enclosed air space or vacuum in its annulus, and indicating terms for resistance.

In double containment pipe that contains insulation in its annulus, the term for U must also contain a resistance term for the insulation. A term should also be included for the inside air contact coefficient (air gap) between the insulation and the outer pipe wall, if it is judged to be significant. The resulting expression is:

$$U = A_s \left(\frac{1}{A_i h_i} + \frac{\ln(r_o/r_i)}{2\pi k_1 L} + \frac{1}{h_c A_o} + \frac{\ln(r_{ji}/r_o)}{2\pi k_{ins} L} + \frac{1}{h_{c2} A_{ji}} + \frac{\ln(r_{jo}/r_{ji})}{2\pi k_2 L} + \frac{1}{A_{jo} h_{jo}} \right) \quad (10.19)$$

where:
r_{jo} = outside radius of the secondary containment jacket, in.
r_{ji} = inside radius of the secondary containment jacket, in.
h_{c2} = inside air contact coefficient between the outer surface of the insulation and the inside surface of the secondary containment pipe, $\left[\text{BTU per (ft}^2)(h)(°F)(\text{per inch})\right]$
A_{jo} = outside surface area of the secondary containment pipe = A_s = ft^2/ft
h_{jo} = outside film coefficient due to convection surrounding the secondary containment pipe, $\left[\text{BTU per (ft}^2)(h)(°F)(\text{per inch})\right]$

Equation (10.19) is the equation used for double containment piping that contains insulation inside its annular space, where resistance terms for inside surface layer convection and outside surface layer convection are taken into account. For a double containment pipe that uses entrapped dead air or other gas as its form of insulation, the equation is modified as shown in Eq. (10.20) to exclude the terms for the inside contact coefficients that would be expected due to the use of commercial insulation. For double pipe heat exchangers, or jacketed-process pipes, the calculation of the overall heat transfer coefficient can be even more complex due to the substantial convection that takes place in the annulus.

$$U = \frac{1}{A_s \left(\frac{1}{A_i h_i} + \frac{\ln(r_o/r_i)}{2\pi k L} + \frac{\ln(r_{ji}/r_o)}{2\pi k_{ins} L} + \frac{\ln(r_{jo}/r_{ji})}{2\pi k_2 L} + \frac{1}{A_{jo} h_{jo}} \right)} \quad (10.20)$$

In double containment piping applications where insulation is applied external to its secondary containment piping, the overall heat transfer coefficient will contain even more terms, as illustrated in Figures 10.6 and 10.7. The case for static conditions is shown in Eq. (10.21), and the case for convective heat transfer of the internal fluid and external environment is shown in Eq. (10.22). Note that the equations for the case with externally applied insulation include the protective weather barrier (external protective jacketing) of Figure 10.7. For a complex arrangement of this nature (shown in Figure 10.7), the piping is effectively provided with at least two layers of insulation. This is due to the entrapped air contained in the annulus, which functions as a layer of insulation. If a secondary containment pipe is constructed of non metallic materials, or other poorly heat conducting metal, the secondary containment pipe will also effectively function as a third layer of insulation.

Heat Transfer in Double Containment Piping

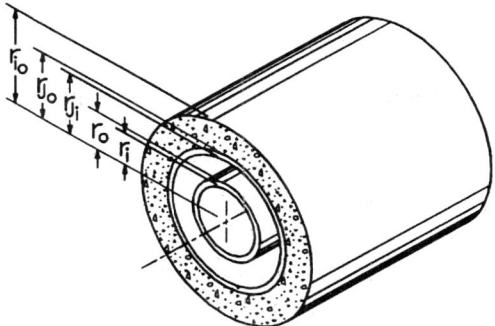

Figure 10.6. Illustration of a double containment pipe system with an enclosed air space or vacuum in its annulus, and equipped with externally applied insulation, showing terms for resistance.

$$U = A_s[x]^{-1} \tag{10.21}$$

where:

$$x = \begin{bmatrix} \dfrac{\ln(r_o/r_i)}{2\pi k_1 L} + \dfrac{\ln(r_{ji}/r_o)}{2\pi k_g L} + \dfrac{\ln(r_{jo}/r_{ji})}{2\pi k_2 L} + \dfrac{1}{h_c A_{jo}} + \dfrac{\ln(r_{io}/r_{jo})}{2\pi k_{ins} L} + \dfrac{1}{h_{io} A_{io}} \\ + \dfrac{\ln(r_{oo}/r_{io})}{2\pi k_j L} \end{bmatrix}$$

h_{io} = air contact coefficient between the insulation and the protective outer jacketing, $\left[\text{BTU per (ft}^2)(\text{h})(°\text{F})(\text{per inch})\right]$

A_{io} = inside surface area of the protective jacket, ft^2/ft

r_{jo} = outside radius of the jacket, in.

r_{ji} = inside radius of the protective jacket, in.

k_j = coefficient of thermal expansion of the protective jacket, $\left[\text{BTU per (ft}^2)(\text{h})(°\text{F})(\text{per inch})\right]$

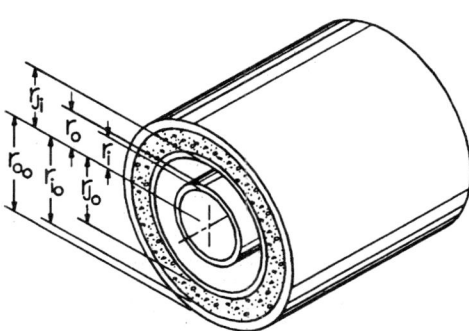

Figure 10.7. Illustration of the arrangement shown in Figure 10.6. which is provided with a tertiary protective jacketing, and indicating terms for resistance.

$$U = A_s \left(\frac{1}{A_i h_i} + x + \frac{1}{A_{oo} h_{oo}} \right) \quad (10.22)$$

The value of the overall heat transfer coefficient will be different depending on whether the inside or outside area is chosen as the design basis. However, the actual value of heat transfer calculated from either overall heat transfer coefficient will be the same, no matter which basis is used.

10.1.2 Convection in Single-Wall and Double Containment Pipes

Convection refers to molecular heat transfer that occurs between a flowing fluid and a stationary mass. In piping convection problems, the temperature difference, due to convection, varies as a function of length. There are equations that have been developed to calculate convective heat transfer coefficients for fluids flowing in a primary pipe, inside an annulus, or externally to a pipe. Each of these situations is described in this section.

10.1.2.1 Inside Convective Heat Transfer Coefficients

For turbulent gas flow in a long straight tube, the following correlation can be used to determine the inside heat transfer coefficient:

$$h = \frac{16.6 C_p (G')^{0.8}}{(D'_i)^{0.2}} \quad (10.23)$$

where:
h = convective heat transfer coefficient, $[\text{BTU per (ft}^2)(\text{h})(°F)(\text{per inch})]$
C_p = heat capacity, BTU per (h)(°F)
G' = mass velocity, lb/(h)(ft^2)
D'_i = inside diameter, ft

When gas velocities are very high, in the range of Mach numbers 0.2 to 1.0, h should be calculated from:

$$\left(\frac{h}{C_p G} \right) \left(\frac{C_p \mu}{k} \right)^{2/3} = \frac{0.027}{\left(\frac{DG}{\mu_m} \right)^{0.23}} \quad (10.24)$$

where:
G = mass velocity, lb/(h)(ft^2)
μ = viscosity (abs), lb/(ft)(h)
k = thermal conductivity, $[\text{BTU per (ft}^2)(\text{h})(°F)(\text{per inch})]$
μ_m = mean fluid viscosity (abs), lb/(ft)(h)

For liquids that are flowing under turbulent flow, the inside heat transfer coefficient can be calculated by:

$$\left(\frac{h}{C_p G}\right)\left(\frac{C_p \mu_f}{k}\right)^{2/3} = 0.023\left(\frac{DG}{\mu_f}\right)^{-0.2} \tag{10.25}$$

where:
μ_f = viscosity of the fluid, lb/(ft)(h)

One variation of these dimensionless groupings that are often used for calculating h for turbulent liquids is the Dittus-Boelter equation:

$$h = 0.023\frac{k}{d}\left(\frac{dG}{\mu}\right)^{0.8}\left(\frac{C_p \mu}{k}\right)^{n} \tag{10.26}$$

where:
d = inside diameter, in.

If the liquids are flowing under laminar flow conditions, as is often the case for high-viscosity fluids such as oils, the inside heat transfer coefficient can be calculated as follows:

$$\frac{hD}{k} = 2.0\left(\frac{WC_p}{kL}\right)^{1/3}\left(\frac{\mu'}{\mu_s}\right)^{0.14} \tag{10.27}$$

where:
μ_s = viscosity of the surface film (abs), lb/(ft)(h)
μ' = bulk viscosity (abs), lb/(ft)(h)

When using Eq. (10.27), the Grashof number should not be excessively large in comparison to the Reynolds number. If the Reynolds number is small in comparison to the Grashof number, the equation should be modified as follows:

$$\frac{hD}{k} = 2.0\left(\frac{WC_p}{kL}\right)^{1/3}\left(\frac{\mu'}{\mu_s}\right)^{0.14}\left[\frac{2.31 + 0.01 N_{Gr}^{1/3}}{\log N_{Re}}\right] \tag{10.28}$$

where:
$N_{Gr} = D^3 \rho^2 g L \Delta T / \mu^2$

For turbulent liquid flow in vertical pipes, h can be calculated from the following equation based on water flow:

$$h = 0.13\left(\frac{k_f^2 \rho_f^2 C_p g \beta(\Delta T)}{\mu_f}\right)^{1/3} \tag{10.29}$$

where:
k_f = thermal conductivity of the fluid, $\left[\text{BTU per (ft}^2\text{)(h)(°F)(per inch)}\right]$
ρ_f = density of the fluid, lb/ft^3
β = coefficient of thermal expansion, 1/°F
g = gravitational constant = 32.2 ft/s^2

10.1.2.2 Outside Convective Heat Transfer Coefficients

For a pipe that is subjected to free (natural) convection where condensation does not occur on the outside of its surface and the ratio of the length to the diameter is large, the outside convective heat transfer coefficient can be calculated from:

$$\frac{h_o D}{k_f} = \frac{1}{2.12} \left(\frac{D^3 \rho_f^2 g \beta \Delta T}{\mu_f^2} \right)^{0.27} \left(\frac{C_p \mu_f}{k_f} \right)^{0.25} \qquad (10.30)$$

where:
h_o = outside convective heat transfer coefficient,
$\left[\text{BTU per } (\text{ft}^2)(\text{h})(°F)(\text{per inch}) \right]$

For heat loss to the atmosphere whereby the heat transfer mechanism involves a combination of convection and conduction, the outside heat transfer coefficient can be determined by for horizontally positioned pipes and long vertical pipes:

$$h_c = 0.5 \left(\frac{\Delta T}{D_o} \right)^{0.25} \qquad (10.31)$$

where:
D_o = outside diameter, ft

10.1.2.3 Annular Heat Transfer Coefficients

For laminar flow in an annulus with a diameter range of d_2/d_1 = 1.09 to 2, the following correlation can be used[5]:

$$C_1 (N_{RE})^{n_1} (N_{PR})^{n_2} \left(\frac{d}{L} \right)^{n_3} \qquad (10.32)$$

$$= 1.02 (N_{RE})^{0.45} (N_{PR})^{0.5} \left(\frac{\mu_b}{\mu_1} \right)^{0.14} \left(\frac{d_e}{L} \right)^{0.4} \left(\frac{d_2}{d_1} \right)^{0.8} (N_{GR})^{0.05}$$

where:
C_1 = coefficient
d = inside diameter, ft
d_e = equivalent diameter, ft
L = length, ft
d_2 = inside diameter of the secondary containment pipe, in.
d_1 = outside diameter of the primary pipe, in.
N_{RE} = Reynolds number, dimensionless
N_{PR} = Prandtl number, dimensionless
N_{GR} = Grashof number, dimensionless
μ_b = viscosity at the arithmetic mean bulk temperature, lb/(ft)(h)
μ_1 = viscosity at temperature T_1 of the inner wall of the annulus, lb/(ft)(h)
n_i, n_2, n_3 = exponents

If fully developed turbulent flow ($N_{RE} > 10^4$) occurs in an annulus, the convective heat transfer coefficient can be determined by:

$$\left(\frac{hd}{k}\right) = 0.023\left(\frac{dG}{\mu}\right)^{0.8}\left(\frac{C_p\mu}{k}\right)^{0.4}\left(\frac{d_2}{d_1}\right)^{0.45} \tag{10.33}$$

10.1.3 Radiation in Pipes and Double Containment Pipes

Radiative heat transfer occurs between any two bodies and depends on the emissivity and absorptivity of the surfaces of the bodies. Radiation is determined by the Stefan-Boltzmann Law, which is given as:

$$q = \sigma\varepsilon A\left[(T_1 + 460)^4 - (T_2 + 460)^4\right] \tag{10.34}$$

where:
T_2 = temperature of the second surface, °F
T_1 = temperature of the first surface, °F
$\sigma = 0.173$
ε = emissivity, dimensionless

σ is the Stefan-Boltzmann constant and ε is equal to the emissivity of the surface. The emissivity of the radiating surface varies with wavelength as well as direction of emission.

The emissivity of a nonmetallic surface is generally inversely proportional to the temperature of emission. At most temperatures it is higher than for a comparable metallic surface. Emissivity for metallic surfaces generally increases with an increase in temperature. The formation of oxide layers on metallic surfaces generally increases the emissivity of metals. Conversely, the polishing of metallic surfaces generally results in a lowering of their emissivities.

The outside radiative heat transfer coefficient for a pipe can be calculated by:

$$h_r = 4\sigma\varepsilon T_m^3 \tag{10.35}$$

where:
h_r = radiative heat transfer coefficient = $\left[\text{BTU per (ft}^2\text{)(h)(°F)(per inch)}\right]$
T_m = mean temperature of the pipe wall, °R

10.1.4 Hot Pipe Design

When piping is designed to carry fluids that are hot, heat transfer that takes place from the pipe to the atmosphere normally consists of a combination of convection, conduction, and radiation. If the surrounding air is completely stagnant or still, there will be some flow that will occur due to air that is heated and replaced by cooler, more dense air. This process is referred to as natural convection. When natural convection occurs, the outside heat transfer coefficient may be determined by a combination of convection and radiation:

$$h = h_n + h_r \tag{10.36}$$

where:
h_n = natural convection heat transfer coefficient obtained from Eq. (10.30)
h_r = radiation heat transfer coefficient obtained from Eq. (10.35)

When the air surrounding the pipe is not stagnant, the situation is referred to as moving-air conditions. If air is being blown over the piping by a fan or compressor or by other forced action, it is referred to as forced convection. The heat transfer coefficient for forced convection is also found by a combination of convection and radiation, as given in Eq. (10.36). However, the term for the convective heat transfer coefficient can be determined by the following correlation for forced convection:

$$h_f = \frac{0.0266 k_f}{D}\left(\frac{\overline{n}D}{v_f}\right)^n \left(\frac{C_p \mu}{k_f}\right)^{1/3} \tag{10.37}$$

where:
h_f = heat transfer coefficient due to forced convection, $\left[\text{BTU per (ft}^2)(h)(°F)(\text{per inch})\right]$
k_f = thermal conductivity of the air evaluated at the average air-film temperature, $\left[\text{BTU per (ft}^2)(h)(°F)(\text{per inch})\right]$
D = outside diameter, ft
v_f = kinematic viscosity evaluated at the average air-film temp., sq. ft/s
$n = 0.805$
C_p = heat capacity, BTU/lb °F
μ = relative viscosity, centipoise
$\overline{V}$ = mean velocity, ft/s

10.1.5 Cold Pipe Design

Heat transfer for the condition of a cold fluid being conveyed on the inside of a pipe, with warmer ambient surroundings, is basically the same as that of the hot fluid case. However, the values used for the outside heat transfer coefficients used in calculations are typically different, as are the methods used for determining them. Also, one major consideration specific to piping that conveys cold fluids is that the transported fluids transported can produce surface temperatures below that of the dew point of the surrounding air. Therefore, condensation of the surrounding air or medium can result, and with it, damaging or corrosive effects to the insulation and/or piping materials. This atmospheric effect sets a limiting design parameter in the sizing of insulation and thus the determination of its adequacy.

When natural convection design conditions exist, the outside combined heat transfer coefficient is normally treated as a constant value independent of the outside diameter or surface area of the piping. Forced convection coefficients are treated similarly, except the magnitude of the constant is dependent on the velocity of the wind or air movement external to the pipe.

Whether the design condition is forced or natural convection, a limiting factor in the selection of insulation is that the minimum thickness must prevent excessive sweating from occurring. Sweating is the term commonly used for the formation of condensation on the

Heat Transfer in Double Containment Piping

outside surface of a solid, such as on the outside surface of thermal insulation. Condensation will occur when the temperature of the surface is below that of the dew point of the surrounding air. A small amount of infrequent sweating can be tolerated without harm caused to insulation, piping, or building components. However, sweating that is frequent or of a large magnitude can result in harmful damage to insulation, rendering it totally useless in some situations. It can also result in corrosion of metal piping, its supports, and damage to a building and its contents.

While insulation can be sized to prevent this occurrence for a certain percentage of the year (including the total prevention of sweating), selecting an insulation thickness based on minimum thickness to prevent sweating may produce larger than desired energy losses, unless a thicker than minimum required thickness is selected. Therefore, the selection of actual insulation thickness is usually a tradeoff between energy savings and the cost of purchasing, installing, and maintaining a pipe's insulation.

A solution to the prevention of sweating is to add a sufficient amount of insulation to a piping system until the temperature of the outer surface is raised above the design dew point. The design dew point selected may be based on the maximum condition ever anticipated in the region or design area, or it can be for a smaller value that represents a majority of the time (e.g., 98% of the time), since a limited amount of sweating may be tolerable.

At a given design dry-bulb temperature and design wet-bulb temperature (otherwise known as the dew point), the potential heat gain by the fluid at the minimum amount of insulation will be:

$$q = \frac{(T_d - T_w)}{R_s} \tag{10.38}$$

where;
q = heat flow rate, BTU/h ft
T_d = ambient still-air dry-bulb temperature, °F
T_w = wet-bulb temperature (dew point), °F
R_s = surface thermal resistance = $1/h_o = \left[(\text{ft}^2)(\text{h})(°\text{F}) \text{ per } (\text{BTU})(\text{inch})\right]$
where h_o = surface film coefficient of the air

When the minimum amount of insulation to prevent sweating at the design dew point and dry bulb is used, the following equality exists:

$$q = \frac{(T_d - T_w)}{R_s} = \frac{(T_w - T_o)}{R_i} \tag{10.39}$$

where :
T_o = operating temperature of the water
R_i = combined resistance of the insulation and piping = BTU in./ft² °F h

In most applications involving a nonviscous liquid flowing under fully turbulent conditions, a pipe's inside film resistance coefficient will be negligible. Therefore, this resistance term can be written as:

$$R_i = \frac{r_s \ln(r_s/r_i)}{k_{ins}} + \frac{r_i \ln(r_i/r_p)}{k_p} \tag{10.40}$$

where:
r_s = outer radius of insulation, in.
r_i = inner radius of insulation, in.
r_p = inner radius of pipe, in.
k_{ins} = mean thermal conductivity of the insulation,
$\quad$ [BTU per (ft^2)(h)(°F)(per inch)]
k_p = mean thermal conductivity of the pipe,
$\quad$ [BTU per (ft^2)(h)(°F)(per inch)]

Solution of the outer radius of insulation is implicit and must be solved either by computer program or by trial and error calculation. The American Society of Heating Refrigeration and Air Conditioning Engineers (ASHRAE),[6] in its ASHRAE Handbook, publishes nomographs for the solution of the insulation thickness for metallic piping systems. Other nomographs are available in the ASPE Databook,[7] and from the manufacturers of commercial insulation.

To determine the actual thickness of insulation to be used, based on a desired amount of heat gain, Eq. (10.41) can be used. The equation is based on a single-wall pipe, provided with insulation, and having a negligible inside film coefficient, and inside air contact coefficient (between the piping and its insulation). In some applications, additional energy savings may be desirable to reduce the size of chiller or downstream capital equipment (i.e., heat exchangers) that are part of the overall process.

$$q = \frac{\pi L \Delta T}{\frac{\ln(D/d)}{2k_p} + \frac{\ln(D_{ins}/d)}{2k_{ins}} + \frac{1}{h_o D_o}} \tag{10.41}$$

where:
ΔT = difference between the maximum ambient air temperature and the cold fluid temperature

To determine the outside surface temperature of insulation, the heat gain must first be determined by Eq. (10.41) or by some other analysis. Once the heat gain is known, the surface temperature of the insulation can be determined since the outside heat transfer coefficient is a constant. If no weather barrier is used, the temperature can be found by modeling the system as one of conductive heat transfer between the insulation and the air:

$$q = \frac{\pi(T_{air} - T_s)}{\frac{1}{HD_{ins}}} \tag{10.42}$$

where:
q = actual heat gain based on actual insulation thickness, BTU/°F ft
$\pi \cong 3.1416$
T_{air} = maximum possible ambient air temperature, °F
T_s = surface temperature of insulation based on the actual thickness, °F
H = outside heat transfer coefficient, $\left[\text{BTU per (ft}^2)(h)(°F)(\text{per inch})\right]$,
  h_n for still air (natural convection), h_r for moving air (forced convection), both of which are constant

Rearranging the equation yields the solution for the surface temperature:

$$T_s = T_{air} - \frac{qR_s}{\pi} \qquad (10.43)$$

where:
R_s = surface resistance = $1/(HD_{ins})$ from Eq. (10.42)

In many applications, it is desirable to add a protective vapor barrier (external jacketing) to the outside of the insulation to prevent against moisture damage, or against potential corrosive atmospheric gases. To calculate the rate of a mass of vapor diffusing through an insulation material, the following mass transfer equation is used:

$$W = \frac{uA\theta p}{7000L} \qquad (10.44)$$

where:
W = total weight of vapor transmitted, lb/ft
u = permeability of the insulation, ft/s
A = cross-sectional area of the flow path, ft^2
θ = time of transmissions, h
p = difference of the vapor pressure between ends of the flow path, psi
L = length of the flow path, ft

When double containment piping is provided with commercial insulation in its annulus, or has entrapped dead air/gas contained in its annulus, secondary containment piping acts as a natural vapor barrier. If insulation is placed externally to its secondary containment piping, as opposed to being positioned in its annulus, then the insulation will require an additional vapor barrier (depending on the insulation selected). A vapor barrier can consist of a mastic application, a rigid jacketing (which may in effect serve as tertiary containment), or a membrane jacketing.

If piping is exposed to cold temperatures for a prolonged period of time, its core fluid will freeze if the external temperature is at or below the freezing point of the fluid. This is true regardless of the type and amount of insulation, unless an infinite amount of insulation is provided. To calculate the time required for a fluid to freeze, one may use:

$$H = \frac{C(t_1 - 32)}{q_o\left(\frac{\pi}{6} \times r_3\right)} \qquad (10.45)$$

$$q_o = \frac{\dfrac{t_1 - 32}{2} - t_2}{\dfrac{r_3 \ln\left(\dfrac{r_3}{r_2}\right)}{K_{ins}} + \dfrac{r_2 \ln\dfrac{r_2}{r_1}}{K_p}} \qquad (10.46)$$

where :
C = capacity of pipe, lb. water/lin. ft = $\pi r_1^2 \times 62.4\,\text{lbs}/\text{ft}^3 / 144\,\text{in.}^2/\text{ft}^2$
H = time for water to cool to 32°F, h
q_o = heat loss from pipe, [BTU per (h)(ft² of insulation surface)]
r_1 = inner radius of pipe, in.
r_2 = outer radius of pipe or inner radius of insulation, in.
K_{ins} = thermal conductivity of insulation, [BTU per (ft²)(h)(°F)(per inch)]
t_1 = initial water temperature, °F
t_2 = temperature of outer surface of insulation, °F

10.1.6 Freeze Protection

Heat tracing of a double containment piping assembly requires special consideration, due to its complex design and the potential for exposing heat tracing components to corrosive chemicals. Double containment pipe assemblies may be heat traced with continuous heat tracing cables, or by steam tracing methods. Heat tracing may be positioned in annular spaces of the piping system, but it is more practical and more efficient in most applications to install heat tracers, or steam-carrying tubes, on the outside of the assembly (external to its secondary containment piping portion).

When electric heaters are to be positioned in the annulus of a system carrying corrosive chemicals, the type of heaters selected must be limited to those that are constructed with a corrosion-resistant fluoropolymer coating to withstand the corrosive attack of aggressive chemicals in the event of a leak. If steam tracing is used, the materials of the steam tracing tubes must also be constructed of a material that is resistant to the primary (core) fluids to which it may become exposed. If plastic piping is used for either the primary or secondary containment pipes, it is highly recommended to limit heat tracing to the use of self-regulating-type electrical heat tracing, with thermostatic controls also added.

10.1.6.1 Calculations and Sizing for Electrical Heat Tracing

Heat loss calculations for heat tracing are governed in the United States by the Institute of Electrical and Electronics Engineers (IEEE) standard 515-1983, Eq. (1). However, modification of this equation must be made in the case of single-wall plastic piping, or when providing heat tracing to the external portion of secondary containment pipes. The equation must be modified to include all of the additional resistance terms. To determine heat loss through an insulated pipe, the following procedure is recommended:

1. Determine all applicable design conditions. This includes the type of piping (including material, diameter, and wall thickness), minimum design temperature, desired maintenance temperature, outdoor and/or indoor environmental conditions, amount and type of insulation, and others.

2. The maximum design temperature difference is determined by subtracting the minimum design temperature from the desired maintenance temperature.
3. The total heat loss should be calculated by using the appropriate heat transfer equation, taking into account all of the applicable resistance terms, given the type of piping, insulation, and vapor barrier to be used. When using single-wall pipes with commonly used commercial insulation, the values can be found in tables published in a variety of well-known sources.[8] For purposes of sizing heat tracing, heat loss values are normally expressed in units of watts per linear foot in the United States.

10.1.6.2 Self-Regulating Electrical Heat Tracing Design

Most nonmetallic piping is subject to relatively low melting temperatures as compared to metallic piping materials. Therefore, it is critical that nonmetallic piping is prevented from being subjected to excessive temperatures. Plastic piping in particular is subject to charring and even combustion if high enough temperatures are achieved. Therefore, use of self-regulating types of electrical heat tracing is recommended for applications involving nonmetallic piping. Non-self-regulating-type electrical heat tracing may also be used for nonmetallic piping, but must always be used with thermostatic controls. Even self-regulating heat tracing should be equipped with backup thermostatic controls, as an extra conservative measure.

Self-regulating types of heaters are constructed of polymers and conductive carbon. Conductive carbon allows current to flow by the creation of electrical paths between parallel bus wires that run the length of each cable. As current flows between bus wires, the polymer expands, thus increasing resistance to the flow of electricity. This allows a heater to reduce its power output automatically. Heat-traced piping subsequently loses heat to its surrounding atmosphere. When it does, the polymer contracts, which in turn frees electrical paths between bus wires, which increases the flow of heat. By operation in this manner, a pipe's temperature can be evenly controlled along its surface, as positions that experience lower temperatures are subjected to more frequent heating. This is important so that parts of the piping that do not need heating do not become "hot spots," and will therefore be protected from overheating.

Figures 10.8 through 10.16 contain illustrations of commonly recommended installation details for self-regulating-style electrical heat tracing. Various details are provided for straight pipe sections in horizontal and vertical positions, changes in direction (elbows), branch connection fittings, valve installations, flange locations and at typical aboveground hanger locations.

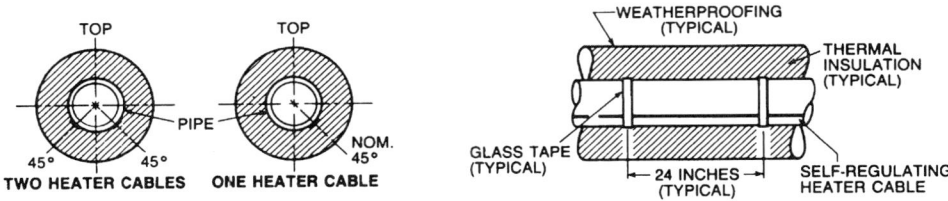

Figure 10.8. Installation of self-regulating heat tracing in a straight pipe section. (Source: Raychem.)

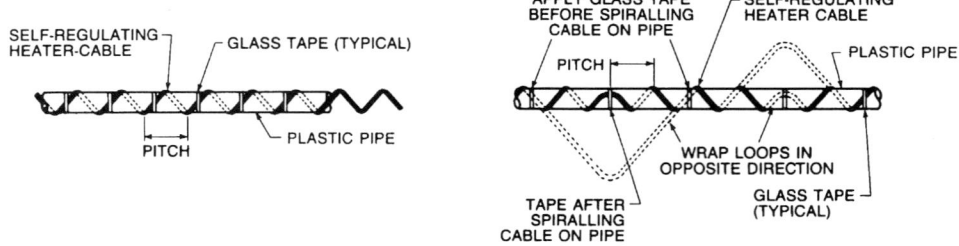

Figure 10.9. Spiralling method for self-regulating heat tracing cable. (Source: Raychem.)

10.1.6.2 Use of Steam or Hot-Oil Tracing

Both hot oil and steam can be run in tubes external to a primary and/or secondary containment piping system to heat trace the core fluid in the primary pipe. Hot oil is usually only used in the rare event that the core fluid is hotter than saturated steam at typical operating temperatures. Normally, a fluid of this type is only provided with heat tracing to maintain a certain viscosity, as opposed to prevent against freezing of the fluid. Therefore, it is rare to find applications of double containment piping where hot oil tracing is to be used.

Steam tracing on the other hand is common, particularly in applications where #6 fuel oil or caustic (sodium hydroxide) are to be pumped through the primary pipe in areas subject to cold climates. These are both examples of fluids which can solidify at relatively warm temperatures. Steam-tracing is more involved than hot oil tracing in that steam traps and condensate return lines are usually required. However, double containment piping does present the possibility of using the annulus for carrying the steam, which may

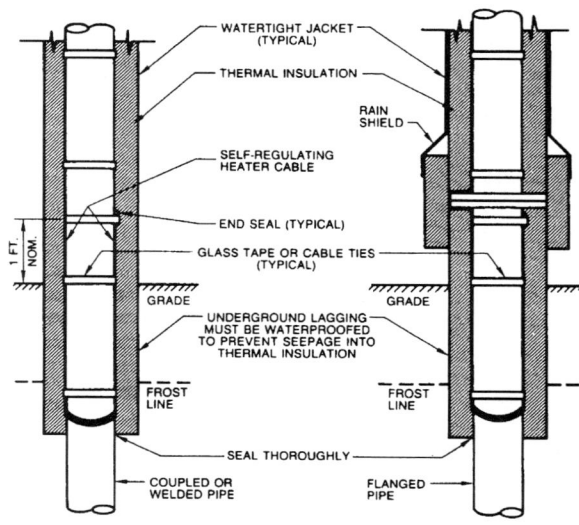

Figure 10.10. Installation of self-regulating heat tracing at a grade penetration. (Source: Raychem.)

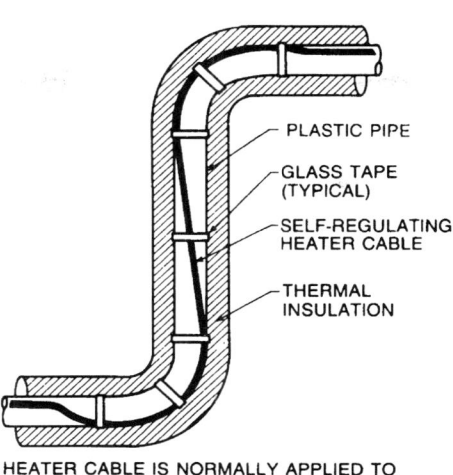

Figure 10.11. Typical installation of self-regulating heat tracing at elbows. (Source: Raychem.)

eliminate the need for traps and condensate return lines. When separate steam tubes are used, it is highly desirable to use heat transfer cement applied around the tracing tubes, and on the pipe to which the tubing is to be located adjacent to, because it provides better heat overall transfer.[9]

The following outlines procedures for designing steam or hot oil tracing systems, with and without the use of heat transfer cement. To design traced (core) piping without heat transfer cement, the following procedure is suggested:[10]

1. Assume a value of the temperature of the air space to be equal to or greater than the minimum temperature of the core process fluid.
2. Estimate the natural convection coefficient, h_c, from the tracing tube to the air space by using Eq. 10.31.
3. Determine the equivalent cylindrical insulation thickness, T_e, on a log mean basis.

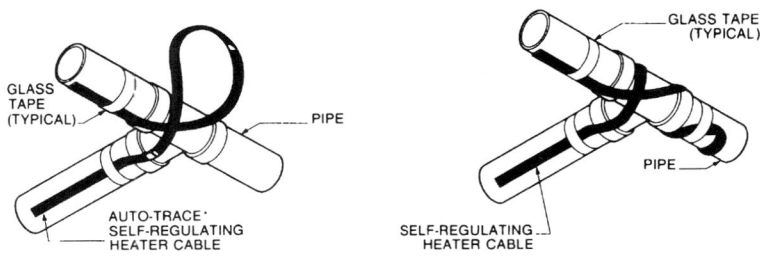

Figure 10.12. Example of installation of self-regulating heat tracing at a branch. (Source: Raychem.)

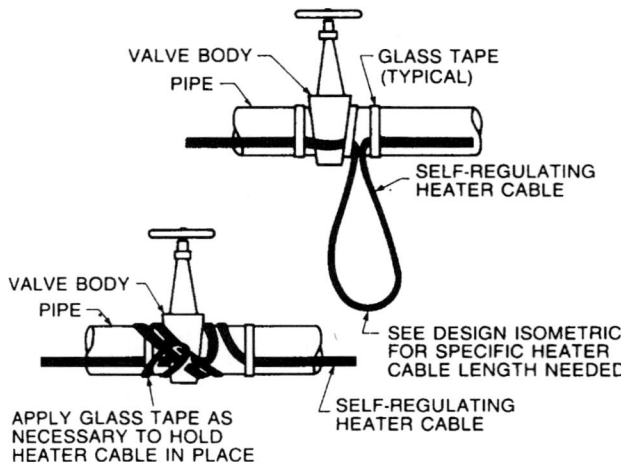

Figure 10.13. Example of installation of self-regulating heat tracing at a gate valve. (Source: Raychem.)

4. Calculate the outside film coefficient of the insulation to the atmosphere, h_o, by using Eq. 10.30, and the overall heat transfer coefficient U_o.
5. Conduct a heat balance around the air space and solve for the temperature of the air space, t_a.
6. If the temperature of the air space is greater than or equal to the temperature of the insulation, then the system is adequate.

To design traced (core) piping with heat transfer cement, the following procedure is used:[11]

1. Determine a starting number and spacing of heat tracing tubes to be used in the design.
2. Determine the metal wall area (equal to the wall thickness) A_m, the outside surface area A_{ji} of the insulation, the outside surface area of the core pipe, A_o, and the outside surface area of the heat transfer cement.
3. Assume a value of the minimum temperature, t_p, of the pipe wall that is equal to or greater than the minimum core fluid temperature.
4. Assume a value for the temperature of the air space, t_a.

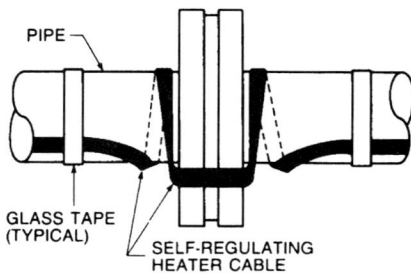

Figure 10.14. Example of a flange installation detail for self-regulating heat tracing. (Source: Raychem.)

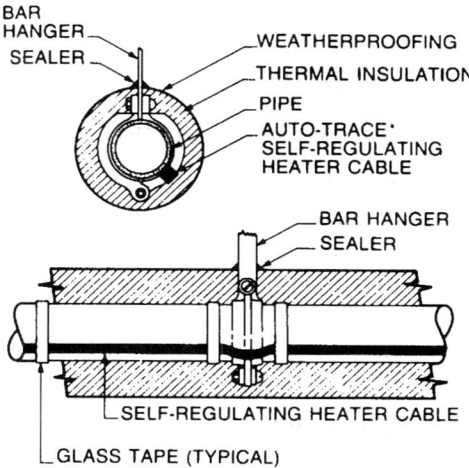

Figure 10.15. Example of a hanger installation detail for self-regulating heat tracing. (Source: Raychem.)

5. Estimate the natural convection coefficient, h_c, from the heat transfer cement to the air space using Eq. 10.31.
6. Determine the equivalent cylindrical thickness of the insulation.
7. Determine the outside film coefficient of the insulation to the atmosphere, h_o, from Eq. 10.30, and calculate the overall heat transfer coefficient U_o.
8. Calculate the approximate average pipe wall temperature, t_p, and estimate the natural convection coefficient from the pipe to the air space, h_p, using Eq. 10.31.
9. Conduct a heat balance around the pipe wall and air space and solve for the temperature of the air space and the temperature of the pipe wall by iteration.
10. If the temperature of the air space and the pipe wall is greater than or equal to that of the insulation, then the system is adequate.

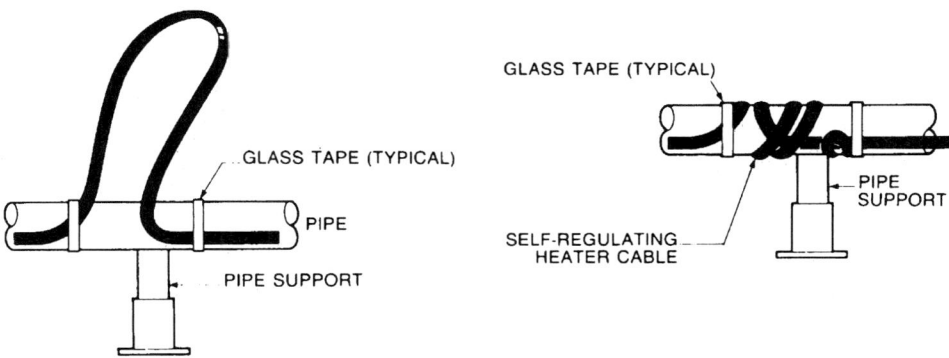

Figure 10.16. Example installation of self-regulating heat tracing at a pipe support. (Source: Raychem.)

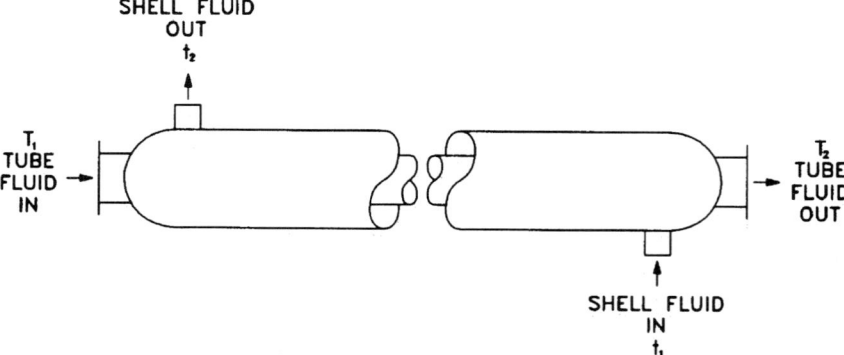

Figure 10.17. Illustration of a countercurrent, one-pass-style double-pipe heat exchanger.

10.2 Heat Transfer Theory Applied to Double Containment Pipe Assemblies

Heat transfer theory as applied to double containment piping assemblies has been generally defined in Eqs. (10.17) through (10.22). There are many additional resistances to consider in comparison to a comparable single-wall pipe, as evidenced by the number of resistance terms included in the equations.

When nonmetallic piping is used as the material of construction of secondary containment piping, a double containment pipe system becomes an efficiently insulated arrangement. Entrapped air in an annular space, or vacuum, functions as a highly efficient form of insulation at moderate temperatures, thus retarding the flow of heat. Entrapped dead air will also function as an insulator when metallic secondary containment piping is used. However, the overall heat transfer that occurs will be less when nonmetallic secondary containment piping is used, due to metallic materials having comparatively large thermal conductivities. When

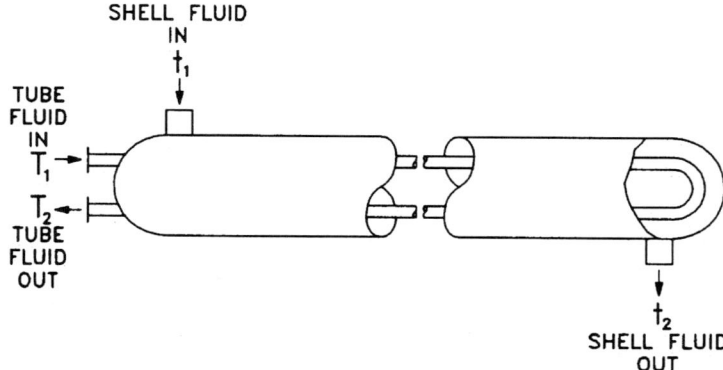

Figure 10.18. Illustration of a concurrent, two-pass-style double-pipe heat exchanger.

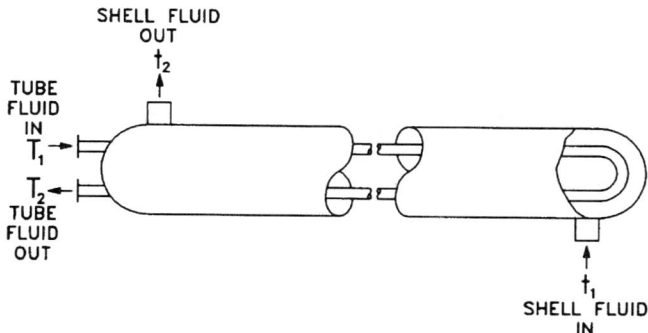

Figure 10.19. Illustration of a countercurrent, two-pass-style double-pipe heat exchanger.

piping is placed underground, the surrounding burial medium will enable either metallic or nonmetallic secondarily contained systems to function as highly energy-efficient systems.

In some applications involving double containment piping, there is a need to provide insulation to the arrangement in order to save energy further, to allow heat tracing to be added in order to warm the pipes, or for other purposes. It is usually inefficient to add another insulation material in lieu of entrapped air in the annulus, as the insulating value of entrapped, stagnant air is better than all types of commercially available insulation materials at close to ambient temperatures. This is true of situations where the annular space is closed in a pressure-tight manner. In systems where the secondary containment piping is open ended or nonsealed, it is more efficient to add another form of insulation. At high temperatures, stagnant air becomes a less effective insulating medium; in fact, it becomes less stagnant as the added energy excites the air molecules, resulting in more movement of the air. Therefore, for hot core fluid applications (> 150°F; > 66°C), the use of commercial insulation is also preferred.

10.2.1 Double Containment Assemblies with Externally Applied Insulation

As described in the previous paragraphs, it is often more efficient for the overall system to add insulation external to an assemblies' secondary containment piping. This may be done for reasons of supplementing the energy savings of the annular space, to minimize the diameter of the secondary containment piping needed, to provide an external means of heat tracing, to provide a clear path for fluid to drain in a system's annulus, etc. When external insulation is provided, there are many resistances to heat transfer to be considered in the model, as demonstrated in Eqs. (10.19) through (10.22). In most applications involving insulation external to secondary containment piping, a protective vapor barrier may be also be required.

Table 10.1 Terminal Temperature Differences

	Cold terminal	Hot terminal
Countercurrent	$T_2 - t_1$	$T_1 - t_2$
Concurrent	$T_2 - t_2$	$T_1 - t_1$

10.2.2 Tertiary Containment of Assemblies with Externally Applied Insulation

In a double containment pipe system that has insulation added to the outside diameter of its secondary containment piping, a protective vapor barrier is usually required. This vapor barrier may also be designed in a "pressure-tight" manner, like a primary and secondary containment piping system. When it is, a system may be considered as having tertiary containment; it can also be described as "triple-containment" piping. When this is the case, the tertiary containment piping should be designed in such a manner that it does not adversely affect the performance of the double containment system and its components. It will be necessary to model such a system as one complex system for stress and performance analysis purposes. However, it may not be necessary to apply all of the rules of Chapter 7 to the design of the vapor barrier system (tertiary containment piping). The designer or owner does have the option to decide whether this is required.

10.2.3 Double-Pipe Heat Exchangers

Double-pipe heat exchangers are a special case of double-walled piping that serves the primary purpose of providing heat transfer between fluids. These are among the oldest types of heat exchangers in use, and are among the simplest types of heat exchangers to design and analyze. All of the principles of primary (core) fluid flow, annular flow, thermal expansion, and pressure design follow the same principles as double containment piping. Additionally, fabrication and layout techniques that are used for double-pipe heat exchangers are similar to those used for double containment piping.

Most double containment heat exchangers are of the countercurrent, two-pass style, as shown in Figure 10.19. To design a heat exchanger of this type, it is necessary to determine its average design temperature difference, which is calculated as follows:

$$\Delta T_m = \frac{GTTD - LTTD}{\ln \frac{GTTD}{LTTD}} \tag{10.47}$$

where:
$GTTD$ = the greater terminal temperature difference
$LTTD$ = the least terminal temperature difference

Terminal temperature difference calculations for both concurrent- and countercurrent-type exchangers are shown in Table 10.1. In this table the temperatures of the hotter fluid in the tubes are indicated by an uppercase T, and the temperature of a colder fluid in the shell is indicated by a lowercase t, and is illustrated in Figures 10.17 through 10.19. The subscript 1 refers to the inlet condition, and the subscript 2 refers to the outlet condition. The lowest value for the LTTD for which a double-pipe exchanger should be designed is a 5°F (2.8°C) temperature difference for countercurrent flow, and a 10°F (5.6°C) difference for concurrent flow.

To determine the surface area of inside pipe required, Eq. (10.48) should be used with the overall heat transfer coefficient determined by Eq. (10.49). The temperature losses to a double-pipe exchanger's external environment are assumed to be negligible according to this calculation. However, they must be considered if they are determined to be significant.

$$A_o = \frac{q}{U_o \Delta T_m} \tag{10.48}$$

$$U_o = \frac{1}{\dfrac{1}{h_o} + R_{do} + \dfrac{LD_o}{KD_{avg}} + R_{di}\dfrac{D_o}{D_i} + \dfrac{1}{h_i}\dfrac{D_o}{D_i}} \tag{10.49}$$

where:

h_o = outside film coefficient, $\left[\text{BTU per (ft}^2)(\text{h})(°\text{F})(\text{per inch})\right]$
h_i = inside film coefficient, $\left[\text{BTU per (ft}^2)(\text{h})(°\text{F})(\text{per inch})\right]$
R_{do} = outside fouling resistance = $1/h_{do}$, $[(\text{h})(\text{ft})(°\text{F})\text{ per (BTU)(in.)}]$
R_{di} = inside fouling resistance = $1/h_{di}$, $[(\text{h})(\text{ft})(°\text{F})\text{ per (BTU)(in.)}]$
R_d = overall fouling resistance = $R_{di} + R_{do} = 1/h_f$,
  $[(\text{h})(\text{ft})(°\text{F})\text{ per (BTU)(in.)}]$
h_f = fouling resistance = $\left[\text{BTU per (ft}^2)(\text{h})(°\text{F})(\text{per inch})\right]$

To determine the diameter required from a known amount of surface area:

$$D_o = \sqrt{\frac{4A_o}{\pi L}} \tag{10.50}$$

where:
D_o = outside diameter, ft
A_o = value from equation 10.48, ft^3
L = length, ft
$\pi = 3.1428$

References

1. C. O. Bennet and J. E. Myers, Momentum, Heat and Mass Transfer, McGraw-Hill, NY (1982).

2. A. J. Chapman, Heat Transfer, Macmillan Co., NY (1967).

3. J. P. Holman, Heat Transfer, McGraw-Hill, NY (1976).

4. P. J. Schneider, Conduction Heat Transfer, Addison-Wesley, Reading, MA (1955).

5. C. Y. Chen, G. A. Hawkins, and H. L. Solberg, Trans. ASME, 68:99 (1946).

6. ASHRAE Handbook, American Society of Heating, Refrigeration and Air Conditioning Engineers, Philadelphia, PA.

7. ASPE Databook, American Society of Plumbing Engineers, Thousand Oaks, CA.

8. Institute of Electrical and Electronics Engineers (IEEE), standard 515-1983.

9. A.K. Escoe. Mechanical Design of Process Systems, Vol. 1. Gulf Publishing Co., Houston, TX, 1986. p. 106

10. A.K. Escoe. op. cit., p. 107

11. A.K. Escoe. op. cit., p. 109-110

Notes

Chapter Eleven

Layout Concepts for Double Containment Piping

A double containment piping system is only partially designed once its piping and components have been sized and their pressure ratings established. The layout of the system must then be determined, taking into account all of the requirements of the application. These include: flexibility requirements, installation requirements, inspection, examination and testing requirements, leak detection system requirements, and others. It also involves the detailed design of interconnecting parts, design and placement of centering devices (interstitial supports), and double containment fitting details. Space issues are of the utmost importance; potential interference that may exist between inner and outer components must be avoided in all systems. This includes interferences that may result from differential movements that occur when inner and outer systems are subjected to different amounts of thermal expansion. The allowable space can also have a profound effect on the fabrication and installation of a system.

System layout is interrelated to all other aspects of system design by virtue of performance criteria. For instance, the final pressure rating of a primary pipe system will be directly affected by the layout and the resulting frictional losses that are calculated. Each layout choice will also result in unique stress levels developed in the system components based on design temperatures and pressures. The distribution of these stresses will also change upon each change in layout detail. Layout choices will have an effect on other aspects of system design as well (i.e., structural, heat transfer, fluid dynamics, etc.). In each design, the layout process involves first selecting a layout and then determining its suitability for the given design conditions. Often this involves computing stresses or performing some other analysis to determine if the chosen layout will result in a safely working system over its design life; if it is determined that the system may fail under the layout that has been selected, aspects of the layout must be changed, and a new analysis performed. By its very nature, the layout process involves trial and error on the part of the engineering design team.

There are two layout issues to which the first time designer and customer of a double containment piping system must pay special attention at the start of a project. The first concerns the overall size of double containment piping and its components. By its nature, the overall diameter sizes are much greater than their corresponding single-wall primary

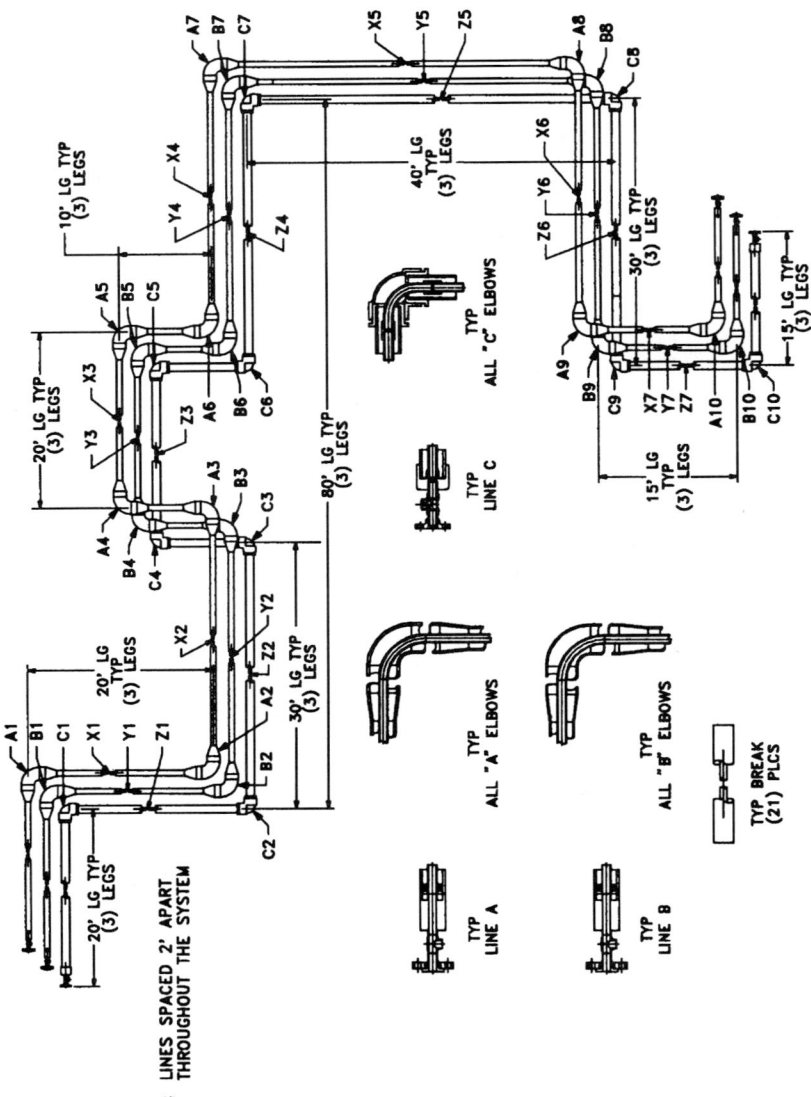

Figure 11.1. Plan view of three separate above-ground chemical transfer lines in a waste treatment facility, using thermoplastic materials. Internal anchor locations are represented by the letter "X" for Line A, "Y" for Line B and "Z" for Line C. Further details for elbow design and system layout are provided in Table 11-2 and Appendix F, based upon the design conditions shown in Table 11.1.

Layout Concepts for Double Containment Piping

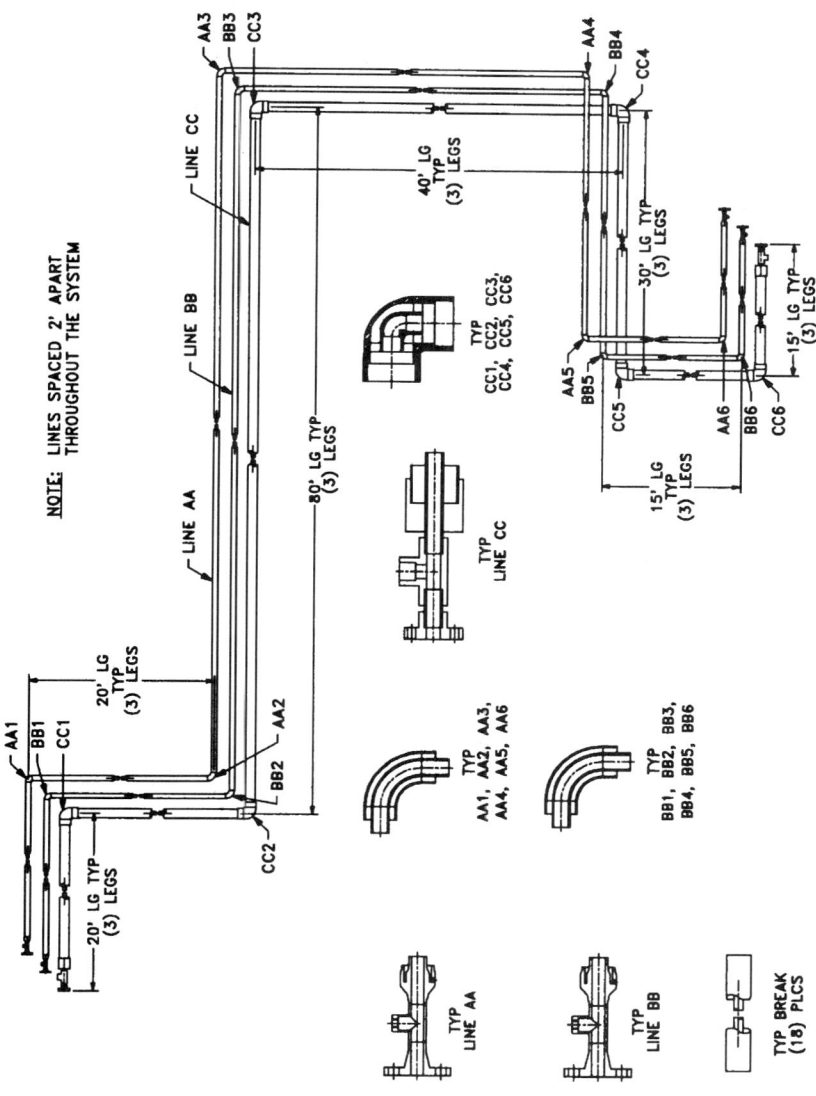

Figure 11.2. Plan view of three separate above-ground chemical transfer lines in a waste treatment facility, using metallic and RTRP materials. Internal anchors are assumed to be located at each approximate midpoint between consecutive elbows where breaks are shown [five (5) anchors provided on each line AA, BB and CC]. Further information for elbow design and system layout are provided in Table 11-3 and Appendix F, based on the design conditions shown in Table 11.1.

Table 11.1 Summary of Process Design Conditions, Used as a Basis for Figure 11.1 and 11.2 Examples

	Figure 11.1			Figure 11.2		
	Line A	Line B	Line C	Line AA	Line BB	Line CC
Fluid	Sodium Hydroxide NaOH	Sulfuric Acid H_2SO_4	Hydrochloric Acid HCl	Sodium Hydroxide NaOH	Sulfuric Acid H_2SO_4	Hydrochloric Acid HCl
Conc. (weight %)	50	98	37	50	98	37
Max. fluid temp. °F (°C)	130 (54.4)	130 (54.4)	130 (54.4)	130 (54.4)	130 (54.4)	130 (54.4)
Ambient temp. °F (°C)	60-80 (15.6-26.7)	60-80 (15.6-26.7)	60-80 (15.6-26.7)	60-80 (15.6-26.7)	60-80 (15.6-26.7)	60-80 (15.6-26.7)
Design pressure (psi)	100	100	100	100	100	100

piping systems. While this aspect of double containment piping sounds obvious, it often is a source of surprise and frustration for facility owners and piping designers. This aspect of double containment piping is a limiting factor, particularly when a system is being installed as a retrofit into an existing facility, or a specified slope has to be met.

A second major item of importance has to do with allowing sufficient clearance for primary fittings to be installed or fabricated within secondary containment fittings, in other words, providing adequate clearance between the two components or, stated differently, proving adequate "internal" clearance. This item is important for all systems, but it is highly important for those requiring an internally flexible design and layout. Components must allow for flexibility and expansion/contraction in many applications where thermal expansions will occur. Adequate internal flexibility is sometimes achieved by providing a secondary containment fitting that is larger in diameter than the adjacent secondary containment piping components. Where this layout technique is required, more space may be added; where there is, there may also be additional requirements for local venting and/or drainage, thus meaning higher costs.

It is difficult to present specific criteria for all layout details that may be encountered in a system design. There are an infinite set of conditions that may be encountered; no two piping systems are ever completely alike. However, considerations for many of the common aspects of double containment piping system layout can be described. Each of these commonly encountered major layout aspects is presented in this chapter.

Designers must understand that in almost every system the layout selected will have an effect on the overall system performance. The overall layout of any system, consisting of each individual detail acting together to form a complete system, must always be analyzed as to its suitability by a competent professional, to achieve a safe, working system.

11.1 Pressure Piping Systems

Pressure piping systems include both above-ground and underground systems that operate at 15 psig (1.1 bar) or greater internal pressure, or operate at vacuum conditions (below 0 psig). Secondary containment portions of these systems may also be pressure rated to the same extent as the primary, but not in all cases. The choice is normally up to the designer, although code requirements may dictate whether it is required. Systems that operate under vacuum conditions (less than 0 psig, 14.7 psia, 1 bar absolute) are also classified as "pressure piping systems," even if they will never be subjected to a positive internal pressure. Pressure piping systems represent the majority of industrial above-ground piping installations, although many underground piping systems are often pressure piping systems.

The layout of a pressure piping system is far from an exact science, and is not taught in schools. Most competent designers develop their techniques from experience and from working with other designers. The ASME B31 Pressure Piping Codes state many useful layout concepts based upon allowing adequate space and minimizing the stresses of pressure piping systems. To minimize stresses in a system layout, as described in the ASME Codes, one must provide a system whose stresses are well balanced during operation. A system whose stresses are evenly balanced will result in a system where detrimental stresses will not occur during operation. This criteria, which sets a worldwide accepted basis for determining the layout of a pressure piping system, also provides the designer with guidance to determine if a system can have a "restrained" or "flexible" layout.

11.1.1 Major Component Layout Considerations for Pressure Piping Systems

Throughout this section of the chapter, the example layout shown in the plan view illustrations, Figures 11.1 and 11.2, will be used to demonstrate certain aspects of layout. Both plan views illustrate 3 separate lines that are used to deliver neutralizing chemicals as part of a waste water treatment process. Figure 11.1 presents the plan view of the layout selected based on the three individual lines constructed of different thermoplastic materials. Figure 11.2 illustrates a redundant plan view, but using two different metallic materials for the services, and one line based on a typical restrained RTRP straight socket material system.

In Figure 11.1, line A consists of a schedule 80 copolymer-polypropylene piping used for 50% sodium hydroxide (NaOH) service; its nominal diameter consists of 1.0 in. placed inside 3.0 in. pipe (approximately 32 mm inside of 90 mm). Line B (Figure 11.1) consists of schedule 40 polyvinylidene fluoride (pvdf) piping that is used for 98% sulfuric acid (H_2SO_4) service, with nominal diameters 1.0 in. inside 3.0 in. (approximately. 32 mm inside 905 mm). Line C (Figure 11.1) consists of 1 inch inside of 3.0 inch schedule 80 CPVC (postchlorinated PVC) piping used for 37% Hydrochloric Acid (HCl) service.

The design temperature conditions for this example involve a maximum fluid service temperature of +130°F (+54.4°C) and ambient temperatures of +60°F (+15.5°C) to +80°F (+26.7°C). The piping is to be installed inside a climate-controlled process building, where the ambient temperatures will be controlled to these temperatures all year round. Each of the three lines is pressured by the use of a double-diaphragm pump. The design pressure is 100 psig (7 bar), at the maximum expected design temperature, for both the primary and secondary containment piping of all three lines. Table 11.1 summarizes the process conditions for all three services.

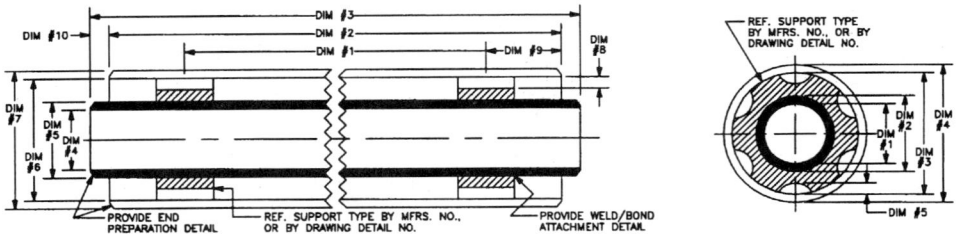

Figure 11.3. Typical layout details for a straight pipe section of double containment piping.

Under these conditions, the maximum differential temperature that can be expected is 70°F (38.9°C), which will occur when the system is installed at ambient conditions of 60°F (+15.5°C), and the maximum fluid temperature of 130°F (+54.4°C) is later pumped through the inner piping. Thus the maximum differential expansion that can occur between the two pipes occurs under these extreme conditions. This temperature difference also sets the basis for computing the stress range in the pipe and fittings, including calculations for localized discontinuity stresses at interconnecting locations due to the resulting differential thermal expansion that occurs in these elements.

The example layout in Figure 11.2 is presented for comparison to that shown in Figure 11.1. This drawing represents acceptable metallic and RTRP materials used in place of the equivalent nonmetallic materials in Figure 11.1 to transport the same three fluid services. The following materials are used for the services indicated: (1) Line AA, 50% NaOH: 316L SS; (2) Line BB 98% H_2SO_4: Alloy 20Cb3® stainless steel (a type of nickel-chromium intermediate alloy); and (3) Line CC, 37% HCl: Epoxy RTRP (Fibercast Dualcast®). Lines AA and BB are nominal 1.0 in. schedule 40 inside 2.0 in. Schedule 10 butt-weld pipe and fittings. Line CC is a nominal 1" in 3", based on a centrifugally cast Epoxy RTRP manufactured by Fibercast, and is laid-out as a restrained system as would normally be the case in these moderate design conditions. The layout indicated in Figures 11.1 and 11.2 is used throughout this section to compare example layout choices of both nonmetallic and metallic pressure pipe systems that operate under a given set of design conditions.

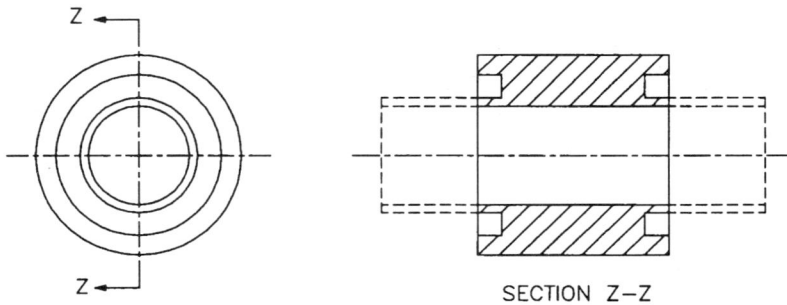

Figure 11.4. Illustration of a simple machined or molded block-style rigid termination fitting. (U.S. Patent # 4,930,544)

Layout Concepts for Double Containment Piping

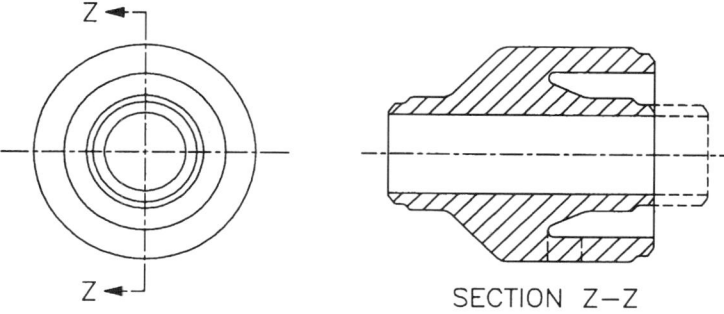

Figure 11.5. Illustration of a rigid termination fitting with contours, smooth radii and external reinforcement to withstand concurrent axial loads due to thermal expansion, radial differential thermal expansion, vibration and dead weight bending. (U.S. Patent #5,141,261, additional patents pending.)

11.1.1.1 Straight Piping Sections

Figure 11.3 illustrates required dimensional information needed for typical layout for straight piping sections for both non metallic and metallic piping material based systems, as may be used for instance in portions the sample layouts of Figures 11.1 and 11.2. On every project, each unique section of straight piping should have all details clearly indicated, except for highly fitting intensive projects where only short lengths of straight pipe exist.

Details that must be shown for a "typical" straight pipe section, or specified clearly, include: the materials of construction, with reference to the appropriate ASTM, DIN or AISI specifications, all inside and outside diameter sizes (including the annular space), wall thicknesses, applicable tolerances, location (spacing) and type of internal and external

Figure 11.6. Example of the type of termination fitting shown in Figure 11.5 made in carbon steel materials for a high pressure system transporting hot #6 fuel oil in a power plant. The fitting shown has a low point drain and pressure gauge connection. (Source: Process Engineers and Constructors, Inc.)

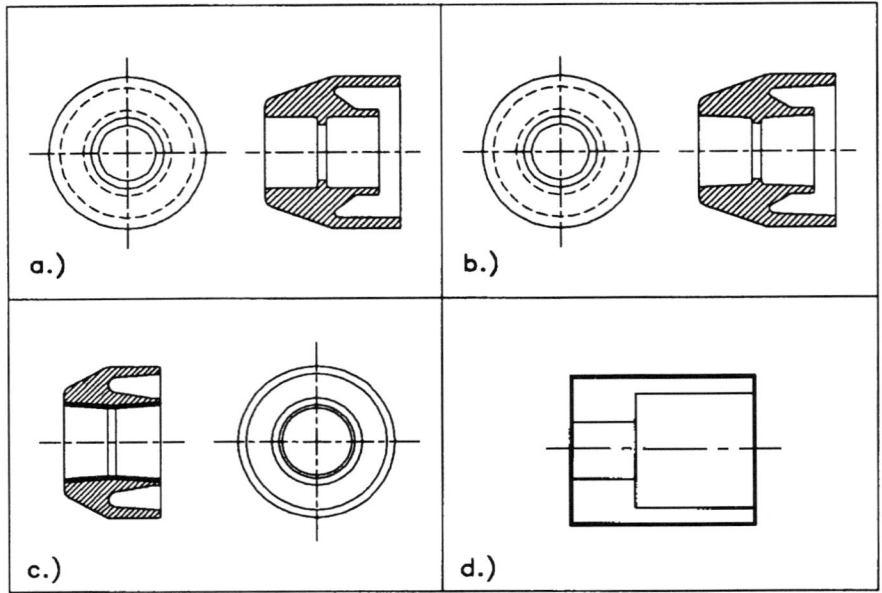

Figure 11.7. Example of socket style termination fittings: (a) Straight socket for solvent cement, adhesive or welded systems; (b) Tapered socket for solvent cement thermoplastic materials; (c) Tapered socket for adhesive bonded RTRP; (d) Straight socket for adhesive bonded RTRP. (Source for (d): Fibercast.)

supports and leak detection cable (if applicable), and references to detail drawings of internal and external supports. A section of piping can be considered as unique if any of these variables vary in any way. Both side and cross-sectional typical views should be shown. For intensive projects where separate spool drawings are to be prepared for each and every spool length of pipe, there is no need for "typical" straight pipe section details on drawings.

Figure 11.8. Illustration of a rubber termination fitting used to terminate a wide variety of inner and outer materials. This fitting finds is widely used in fiberglass lines for fueling facilities. (Source: R&G Sloane Manufacturing Company, Inc.)

Layout Concepts for Double Containment Piping

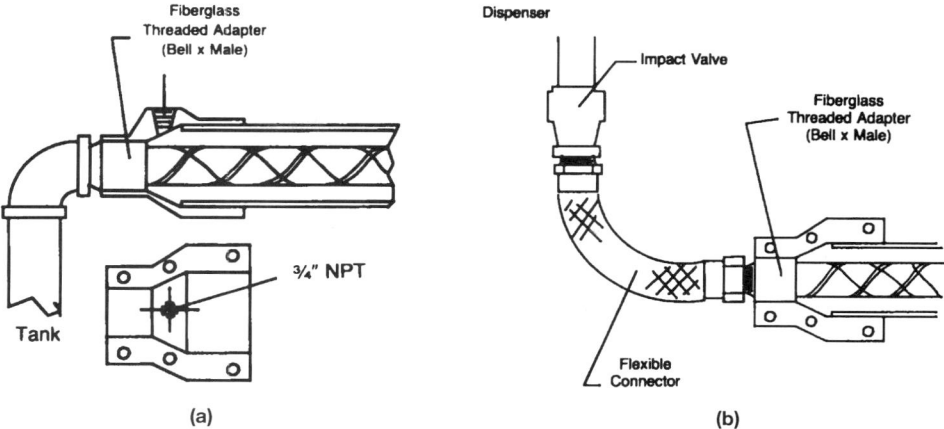

Figure 11.9. Examples of clam-shell fiberglass termination fittings: (a) Termination concentric reducer with 3/4" NPT; (b) Termination concentric reducer. (Source: Smith Fiberglass Products, Inc.)

11.1.1.2 Termination and Initiation Design

There are several options for terminating a double containment zone, thereby making a transition from double containment piping to ordinary single-wall piping. The most simple option is to keep the annulus open-ended, where it is permissible. For applications where closure of the ends is required, Figures 11.4 through 11.25 provide some examples.

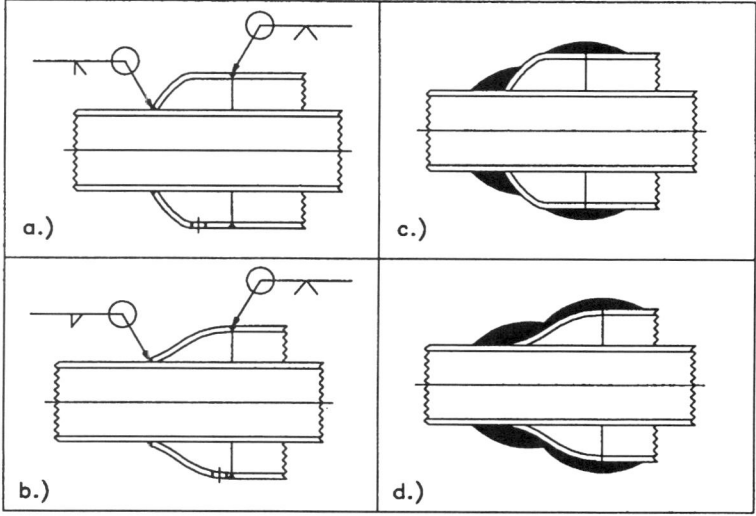

Figure 11.10. Examples of fabricated termination fittings: (a) Modified cap; (b) Modified concentric reducer; (c) Modified RTRP cap with overwrap; (d) Modified RTRP concentric reducer with overwrap.

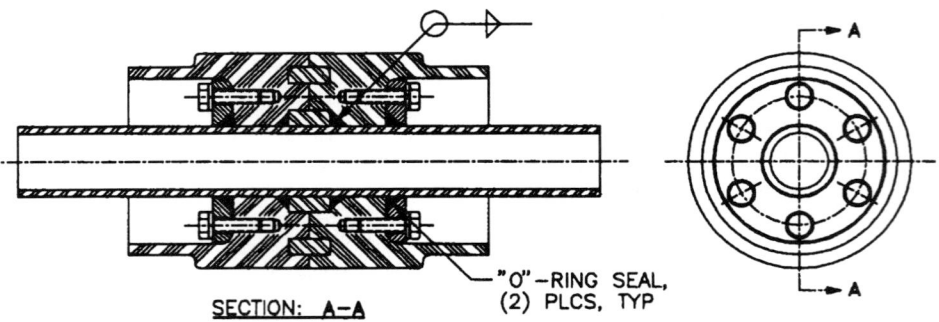

Figure 11.11. Illustration of a rigid termination fitting made for use in hybrid systems involving a metallic inner bulkhead housed inside a two-part heat-fused thermoplastic outer housing. (U.S. Patent #5,085,471.)

11.1.1.2.1 Termination and Initiation Transitional Fittings

Figures 11.4 through 11.11 illustrate a variety of termination fittings. Figures 11.5 and 11.6 illustrate a termination fitting designed to minimize the effects of concurrent internal pressures, axial loads, discontinuity stresses due to differential expansion, dead-weight bending, and vibration loads. However, when concurrent loads are great, a flexible termination concept is always recommended (e.g. Figs. 11.12, 11.14 through 11.16 and 11.25).

Various fabricated forms of termination fittings using modified standard concentric reducers and caps, are shown in Figure 11.10. Fittings of this type have been used for both systems that involve socket-joining techniques where the inside diameter of the small end of the reduction may be solvent cemented or adhesively bonded to the outside diameter of the primary pipe, and as a means of accomplishing transitions in welded-metallic systems. Figures 11.11 through 11.16 illustrate a variety of termination fittings that may be used in either homogeneous systems or in hybrid systems having dissimilar primary and secondary containment materials. Variations of the Figure 11.13 and 11.14 assemblies are also used as a means of internal anchoring (see Section 11.1.1.4).

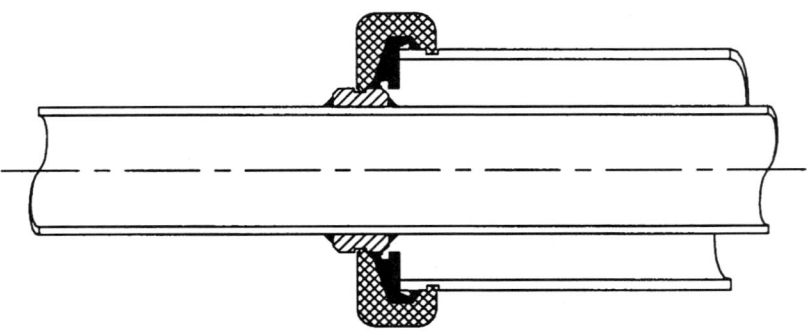

Figure 11.12. Illustration of a flexible-style termination arrangement using a reducing-style flexible grooved mechanical joint (shown is a Grinnell style 1010 Gruvlok® fitting), and a termination adapter.

Layout Concepts for Double Containment Piping

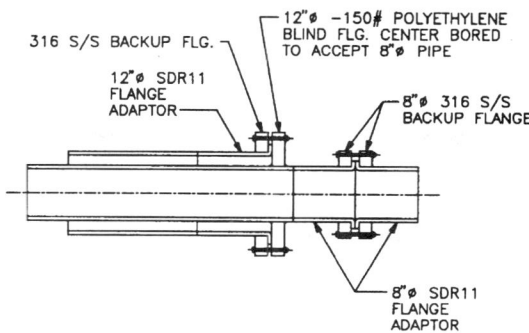

Figure 11.13. Example of a fabricated rigid termination arrangement for a heat-fusible primary pipe system using a stub-end and backing ring style secondary containment flange. Note that the welding method (e.g., hot gas welding or extrusion welding) used to bond the blind flange to the primary pipe will result in a relatively low weld quality factor of the fabricated part (0.05-0.8, depending on the method).

11.1.1.1.2.2 Termination/Transitional Flanges and Flexible Termination Assemblies

Another option for making single-wall to double containment zone transitions include the use of various fabricated termination flanges. Several varieties of rigid termination flanges have been used extensively in metallic jacketed process piping systems. Termination flanges consist of a primary pipe slip-on flange that are welded to both the primary and secondary containment pipes. Examples of two slip-on flange arrangements are illustrated in Figures 11.17 and 11.18. A variation of a termination flange is an insert flange (see Figures 11.19 and 11.20), wherein a groove is machined into the back side of the flange for the jacket pipe to be inserted into it. An insert or termination flange has the inherent problem of acting as a crevice former; thus, it should not be used where conditions of crevice corrosion are a possibility, once fluid is spilled into the annulus. A unique proprietary flange available for use with certain thermoplastic materials (polypropylene, PVDF) in systems joined by simultaneous welding is shown in Figure 11.21. A variety of flexible termination fitting arrangements are illustrated in Figures 11.14 through 11.16.

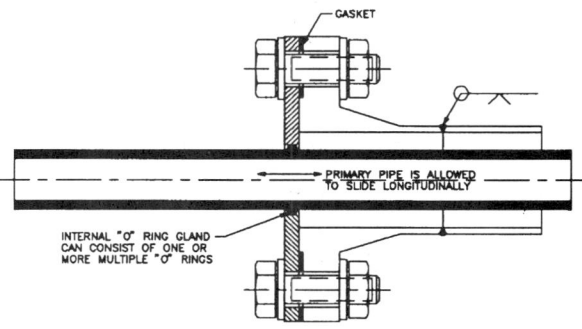

Figure 11.14. Example of a fabricated flexible termination arrangement for a metallic or thermoplastic primary pipe using a standard ANSI B16.5 weld-neck secondary containment flange.

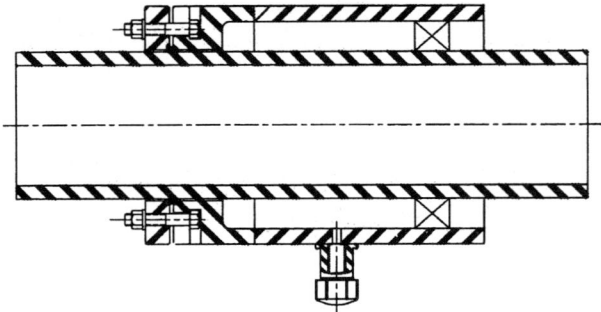

Figure 11.15. Illustration of a fabricated flexible termination fitting utilizing an internal "O"-ring gland seal (Source: Chevron Plexco® Division.)

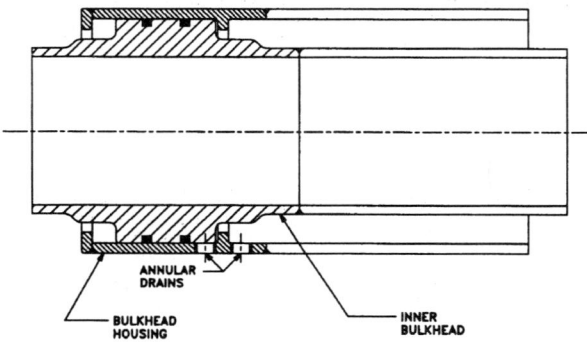

Figure 11.16. Illustration of a flexible termination fitting that functions as both a termination/transition and a piston-style expansionary joint. (U.S. Patent #5,141,261.)

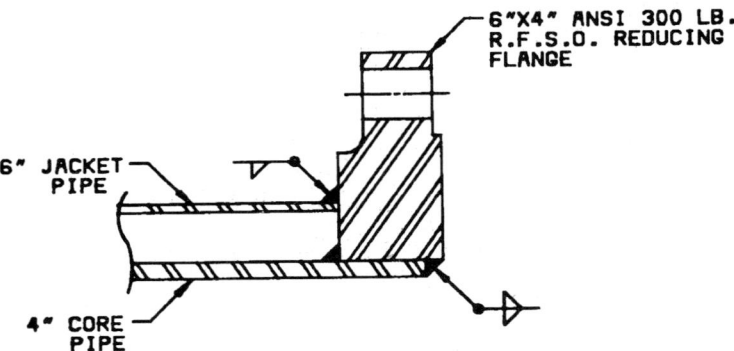

Figure 11.17. Example of a termination closure flange for metallic systems. (Source: "Failure Assessment and Redesign of a Jacketed Piping System, a Practical Approach," by W. E. Short II, Pressure Vessels and Piping Conference, 1988, ASME, PVP Vol. 139, # H00423.)

Layout Concepts for Double Containment Piping

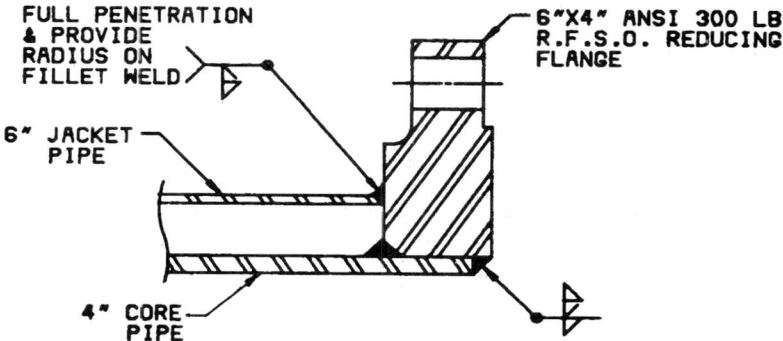

Figure 11.18. An improved termination closure flange for metallic systems using a counter bevel jacket weld. (Source: "Failure Assessment and Redesign of a Jacketed Piping System, a Practical Approach," by W. E. Short II, Pressure Vessels and Piping Conference, ASME, Vol. 139, # H00423.)

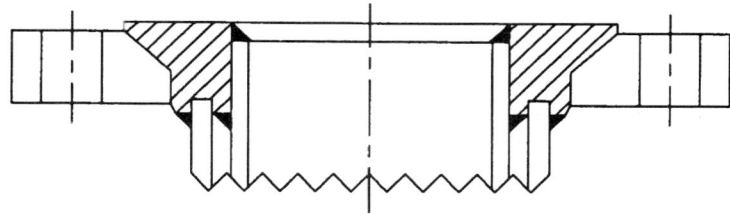

NOTE: LAP JOINT FLANGE O.D. THICKNESS AND HOLE PATTERN BASED ON THE SECONDARY PIPE SIZE FOR AN ANSI B16.5 150 PSI FLANGE.

Figure 11.19. Example of an insert closure flange for metallic systems. (Source: Bestweld, Inc.)

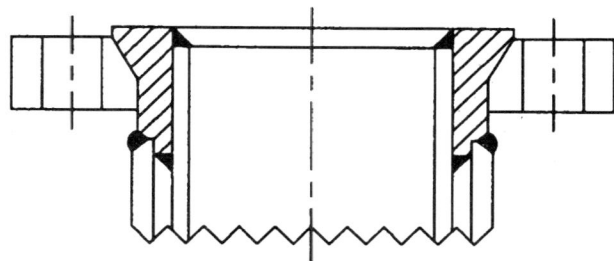

NOTE: LAP JOINT FLANGE O.D. THICKNESS AND HOLE PATTERN BASED ON THE PRIMARY PIPE SIZE FOR AN ANSI B16.5 150 PSI FLANGE.

Figure 11.20. Example of an insert closure flange for metallic systems. (Source: Bestweld, Inc.)

Figure 11.21. Example of a molded double-flange termination fitting using an "O"-ring double seal for use in heat-fused thermoplastic systems. (Source: PolyFlowLines Co.)

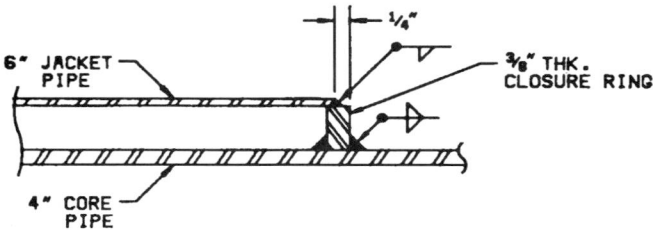

Figure 11.22. Example of a termination closure ring for metallic systems. (Source: "Failure Assessment and Redesign of a Jacketed Piping System, a Practical Approach," by W. E. Short II, Pressure Vessels and Piping Conference, 1988, ASME, PVP Vol. 139, # H00423.)

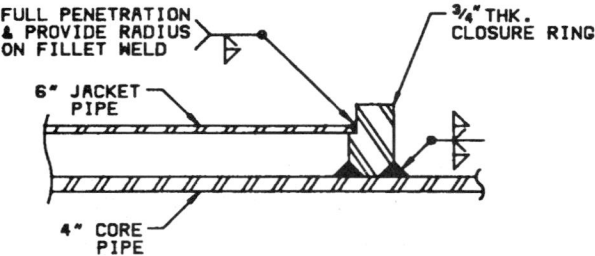

Figure 11.23. Example of a termination closure ring for metallic systems that has an insert step for the secondary containment pipe attachment, and that extends beyond the secondary containment pipe. Note the counter bevel weld details for the primary pipe attachment, and the thicker ring portion which together adds strength and resistance to cyclic loads of a differential thermal expansion nature. (Source: "Failure Assessment and Redesign of a Jacketed Piping System, a Practical Approach," by W. E. Short II, Pressure Vessels and Piping Conference, 1988, ASME, PVP Vol. 139, # H00423.)

Layout Concepts for Double Containment Piping

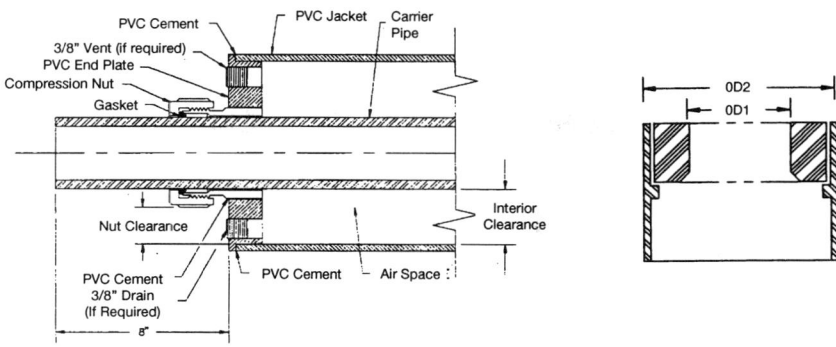

Figure 11.24. On the left, a PVC termination insert in a compression nut-type end seal. (Source: Rovanco Corp.) On the right, a termination insert ring for use in solvent-cementable thermoplastic or adhesive-bondable RTRP systems (Source: Guardian Division of Nibco, Inc.)

11.1.1.2.3 Closure Rings/Insert Rings

One of the more common termination arrangements for metallic systems that has been used extensively in jacketed pipe applications is the closure ring. Closure rings also find application as a termination device in double containment piping systems. Examples of closure rings are provided in Figures 11.22 and 11.23. A closure ring is a similar concept to the internal anchor/baffle (see Figures 11.44 and 11.45), except that an internal anchor/baffle is welded to the inner and outer pipes on both sides; a closure ring is welded only on one side to the secondary containment piping. A closure ring may be used on thermoplastic piping since it is weldable, although the resulting secondary containment pressure rating (pressure capability of the annulus) may be substantially less than that of the primary piping.

A variation of the closure ring is an insert ring (collar), which may be used in systems that use solvent cementing (PVC, CPVC, ABS) or adhesive bonding (all RTRP materials). A standard insert ring and a proprietary RTRP closure insert ring that contains an "O"-ring seal and locking ring groove are shown in Figures 11.24 and 11.25, respectively.

A further discussion of termination and transition components, such as closure rings, are discussed in Section 7.4.8.

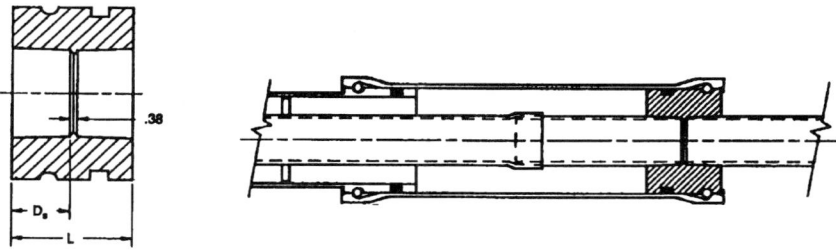

Figure 11.25. Example of a proprietary termination component that uses an external "O"-ring seal and a locking ring. (Source: Ameron Fiberglass Pipe Division.)

11.1.1.3 Instrumentation, Vent, and Drain Connections

Good design procedures for single-wall and double containment pressure piping allow the piping (primary and/or secondary) to be vented at all high points and drained at all low points. The venting and draining features are most important during system testing, initial filling, and start-up and during shutdown and maintenance procedures. Without venting capabilities, the systems will allow trapped air pockets that become compressed during hydrostatic testing, imposing the dangers of a sudden release of this form of stored energy during a failure.

Without drainage capability, fluids may collect at low points that exist in a piping system, which may then be left undrained. This is particularly dangerous when hazardous fluids are being transported and maintenance workers are to perform work on the piping system during shutdown or repair situations. In double containment piping, lack of drainage capability in the annulus will allow accumulation and extended storage of the hazardous chemicals at its low points, a condition whereby secondary containment of the fluid no longer exists. If the material of the secondary containment is not capable of withstanding the corrosive effects of the fluid for an extended period of time, failure to the external environment can occur, resulting in a "double failure." If drainage capability is not provided to the annulus, the only way that fluid can be removed is to cut into the pipe. This inevitably results in a spill, thus preventing the fluid from being safely removed from the piping system in a controlled manner. It is counterproductive to install secondary containment and then allow a fluid to spill readily during repair procedures. Provisions for venting and drainage are important to consider for all piping, however, the concept derives added importance for the annulus in a double containment piping system.

Pressure piping systems almost always require instrumentation of various sorts. Common types of instrumentation include such devices as pressure and temperature gauges, flow measuring instruments, pH probes, leak detection devices and connections, and many others. In a double containment piping system, instrumentation may be required for both the primary and/or secondary containment piping systems.

In order to facilitate drainage and venting and provide for instrumentation, connections must be designed into the primary piping and into the annulus of the double containment zone. Connections to the primary system may be kept outside the double containment zone, if the vents, drains, or instruments are positioned inside a building, sump, manhole, or diked area (outdoor above-ground systems). By doing so, the fabrication, installation, maintenance, and operation of these systems can be simplified. If it is required to maintain a primary piping vent, drain, or instrument under a pressurized form of secondary containment, the system design will be more complex and harder to get at when maintenance, operation, or observation of the item is required.

Figure 11.26 illustrates a possible arrangement for the initiation of a pressure double containment piping system, where the section is positioned vertically. This arrangement contains a tee with threaded-cap connection for both primary pipe and annulus pressure gauges, and a tee with threaded-cap valve connection to allow drainage of the secondary containment piping (annulus). Since the spool piece is vertically positioned, there is a local low point, requiring a drain connection. Since the need to commence the zone of containment is often downstream of such an initiation arrangement (i.e., after a penetration through a building wall), there is no need to maintain the area of the primary pipe pressure

Layout Concepts for Double Containment Piping

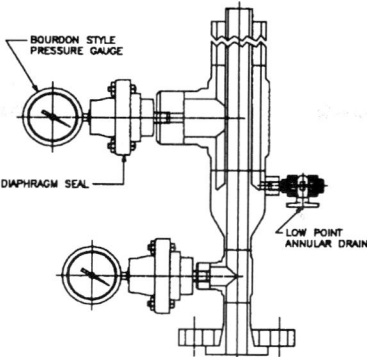

Figure 11.26. A possible initiation arrangement for a vertically positioned section. Note that in this arrangement, a "dead leg" does not exist in the annulus, due to an integral outlet at the extreme low point.

gauge connection under containment, depending on the area surrounding the pipe, and whether it constitutes secondary containment. The building where the section is to be located may function as a suitable means of containment according to most authorities having jurisdiction, if the flooring and walls are properly protected, or constructed of a corrosion-resistant material (i.e., concrete with an applied monolithic coating, acid brick, etc.).

The arrangement in the previous paragraph is provided with a drain at the extreme low-point of the annulus. This is done to solve the otherwise inherent problem created by an annular area below a secondary containment low-point drain not being at the extreme low-point (e.g. Figs. 11.28 and 11.29), thereby creating a "dead-leg" condition This undrained fluid can be dried out over a period of time, or alternatively, dried by vacuum means. A better arrangement is to provide a modified termination fitting with a connection (threaded, socket, or butt nipple) at the low point of the annulus. A termination fitting of this type, having a "reinforced opening," is illustrated in Figure 11.27. Termination fittings of this type may contain multiple connections (i.e., for drain, vent, or instruments; see Figure 11.6).

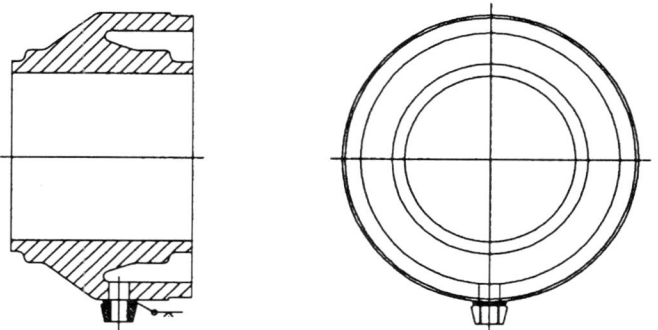

Figure 11.27. Example of a rigid termination fitting with an integral outlet at the extreme low point of the annulus (see also Fig. 11.6 for a photo of this fitting). (U.S. Patent #5,141,261, additional patents pending.)

It is difficult to drain/vent the extreme end of the annulus for most other types of termination arrangements (i.e., closure rings, termination flanges, etc.). In these cases, an adjacent tee may be positioned next to the termination means in the manner shown in Figure 11.28, although the same "dead-leg" problem results. Figure 11.29 provides a look at the similar situation for the typical termination flange without an adjacent tee. In metallic systems, a nipple can be tapped and welded directly into the wall as shown in Figures 11.29 and 11.30. A weld-o-let® or sock-o-let® can also be used. In RTRP and thermoplastic systems, saddle fittings can also be used. Care should be taken in the case of the termination flange that its bolting arrangement is not disturbed by closely positioning a drain or vent. Figure 11.30 and 11.31 illustrate possible arrangements for closure rings in vertical positions that eliminate any dead-leg problems. The version in Figure 11.31 may be used where adequate annular and external space exists.

Wherever possible, instrumentation connections for the primary piping systems should be provided outside the zone of double containment. This will allow for a simpler layout and will greatly add to the maintainability of such arrangements. Instrumentation connections in the secondary containment pipe may be added in a system through a tee branch connection with a cap wherever possible. Instrumentation connections may be grouped into more than one connection on a single branch header as well. Instrument connections into the primary pipe may be made by the means shown in Figures 11.26 and 11.32 through 39. Figure 11.34 is an example where all portions of the primary system are contained.

Figures 11.35 through 11.37 illustrate miscellaneous instrument connection header alternatives, in which gauge connections are made for both the primary and secondary containment systems. Some of these arrangements are unavoidably complex by their nature.

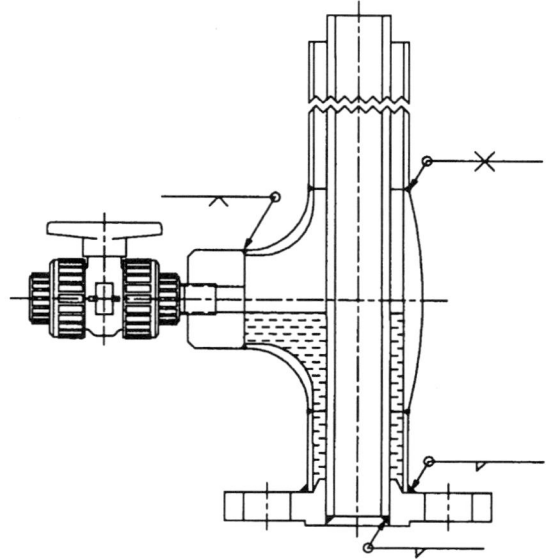

Figure 11.28. Example of a vertically positioned termination flange adjacent to a secondary containment tee with cap and threaded outlet, illustrating a "dead leg" in the annulus.

Layout Concepts for Double Containment Piping 415

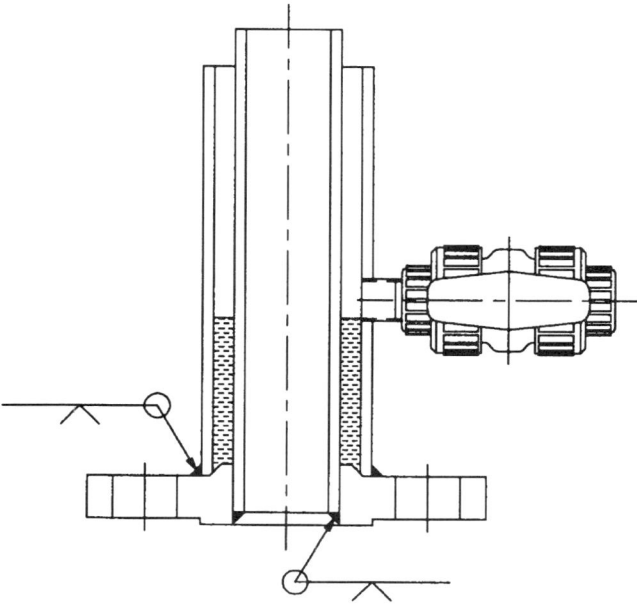

Figure 11.29. Example of a vertically positioned termination flange adjacent to a tapped outlet illustrating a "dead leg" in the annulus.

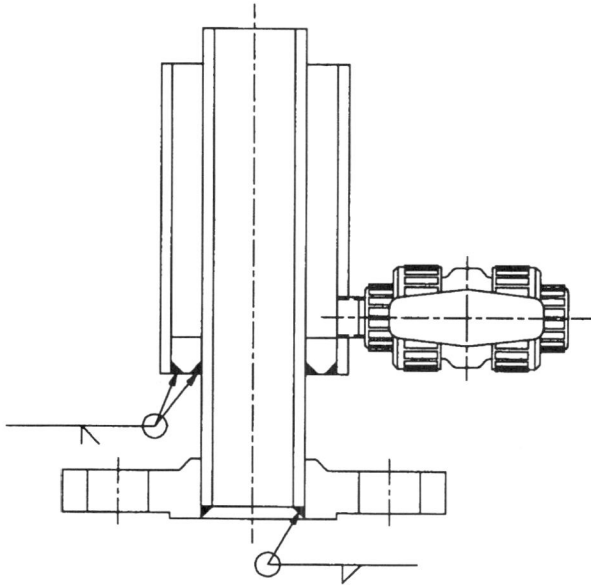

Figure 11.30. Illustration of a vertically positioned closure ring with a tapped outlet at the extreme low point of the annulus.

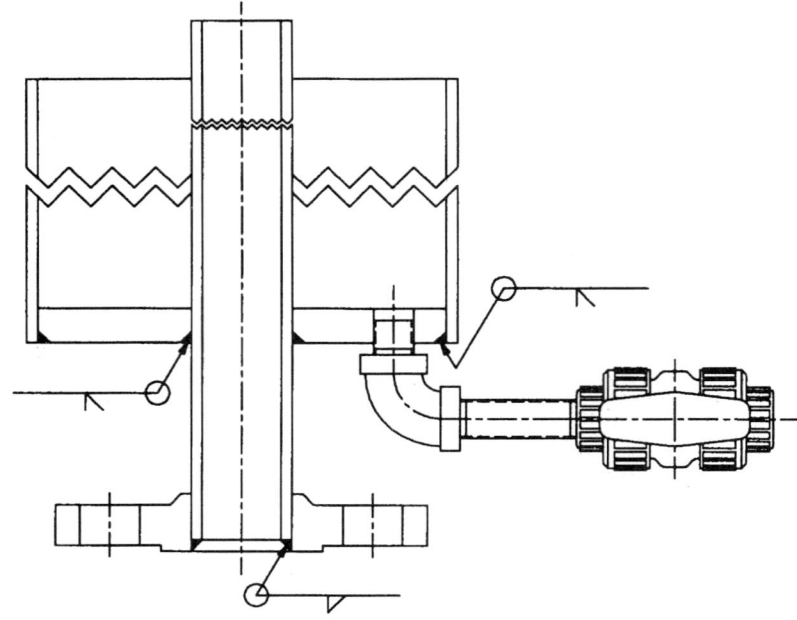

Figure 11.31. Illustration of a vertically positioned closure ring with a tapped outlet in the closure ring.

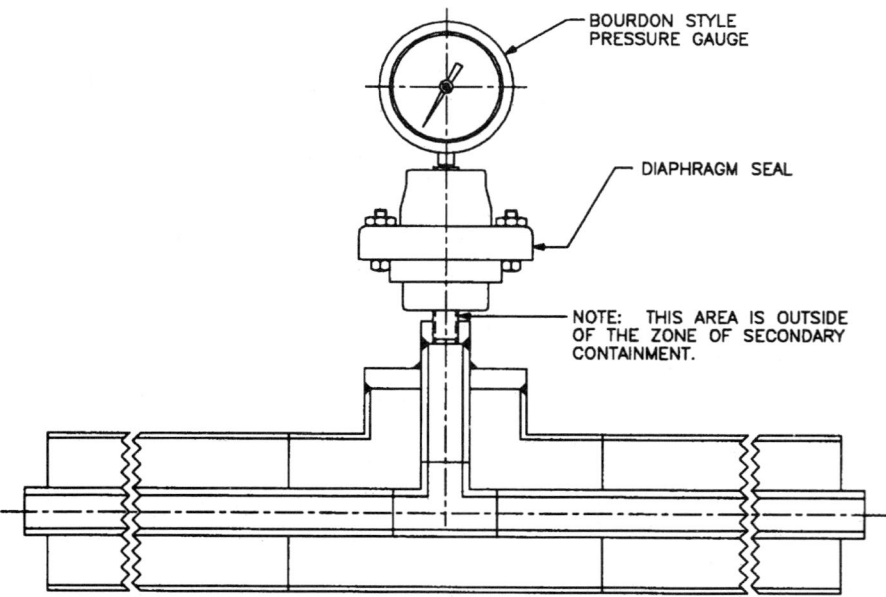

Figure 11.32. Example of a primary pipe pressure gauge connection.

Layout Concepts for Double Containment Piping 417

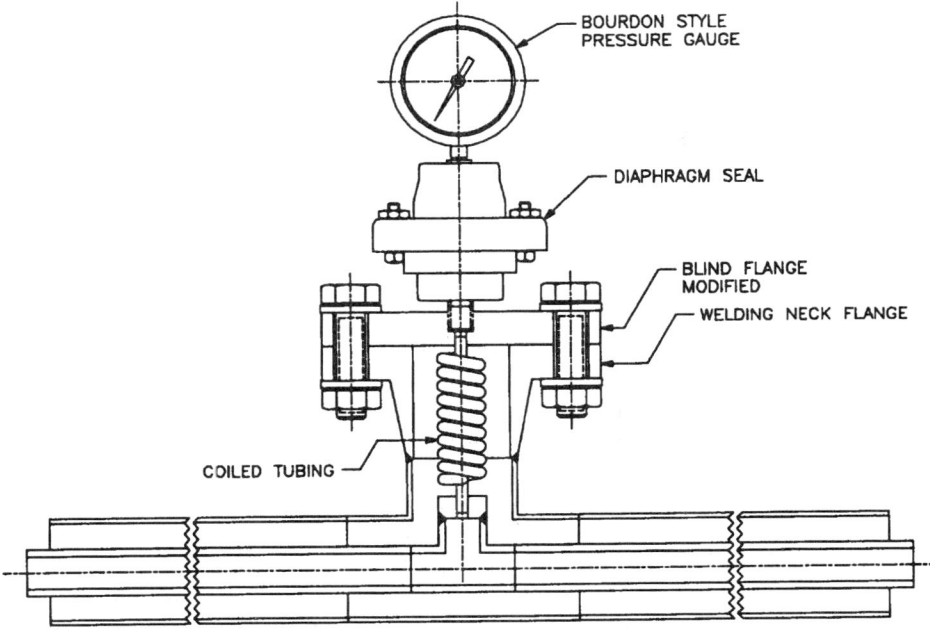

Figure 11.33. Example of a primary pipe pressure gauge connection.

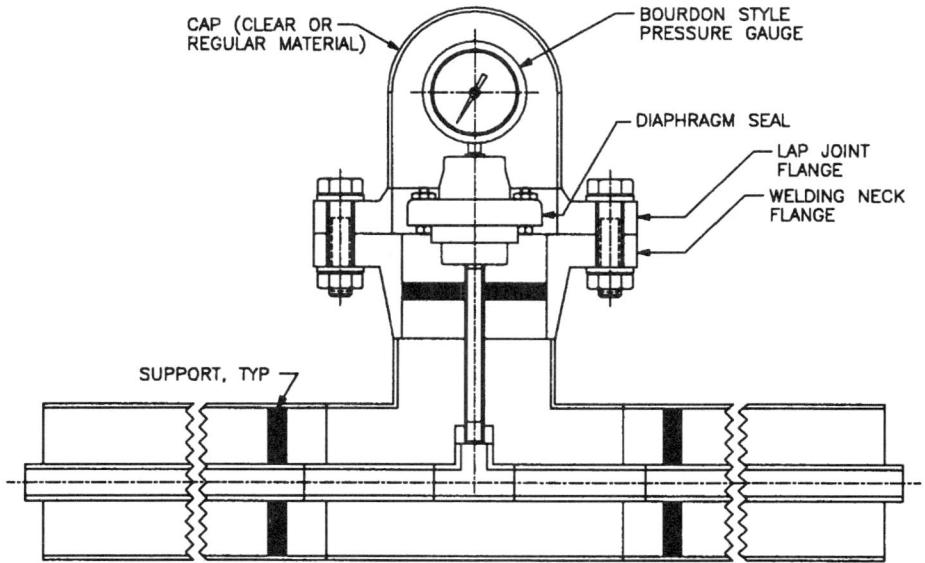

Figure 11.34. Example of a primary pipe pressure gauge connection.

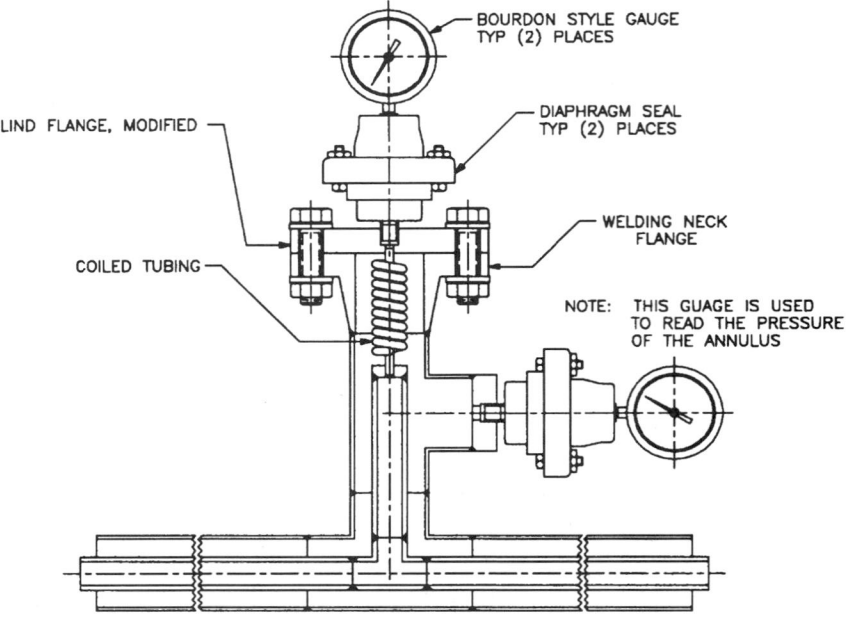

Figure 11.35. Example of a combined primary pipe and secondary containment pressure gauge connection header.

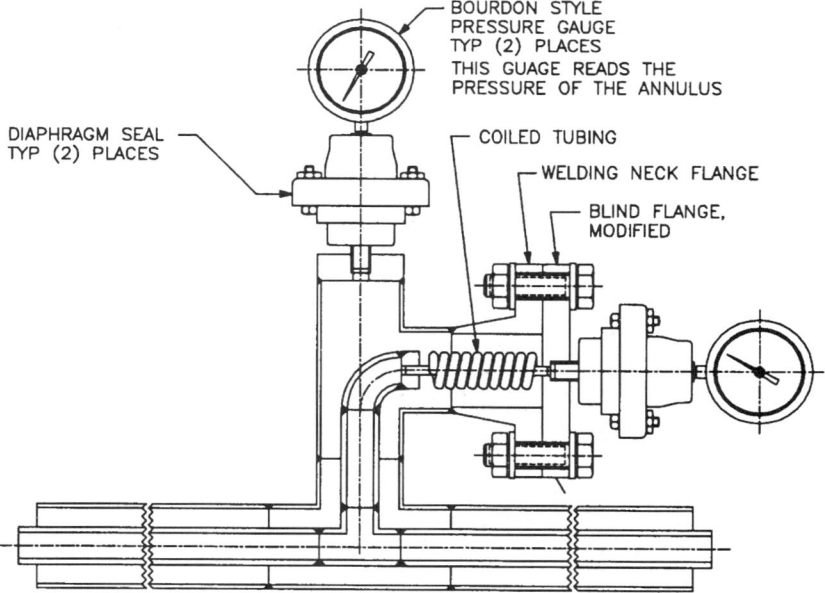

Figure 11.36. Example of a combined primary and secondary containment pressure gauge header.

Layout Concepts for Double Containment Piping

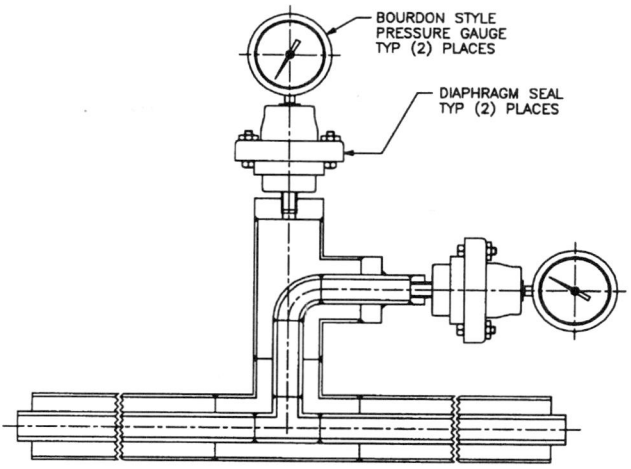

Figure 11.37. Example of a combined primary pipe and secondary containment pressure gauge connection header.

In jacketed process systems, it is often necessary to provide thermowell connections into the primary system. This is often done in 90° elbow arrangements, as shown in Figure 11.38.

If areas requiring instrumentation occur at the end of the line, the annulus may either have to be vented or drained at this point, depending on whether it is a high or low point. Thus, a combined arrangement may have to be used. Figure 11.39 illustrates a possible combined horizontally-positioned header arrangement.

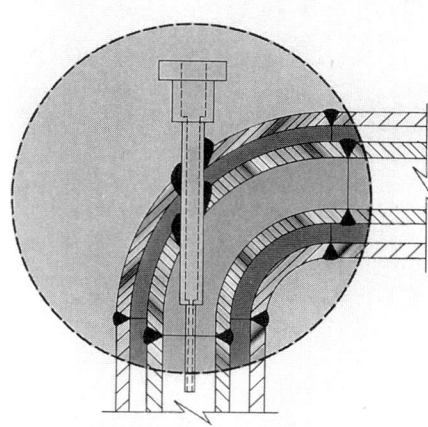

Figure 11.38. A typical fabricated thermowell connection in an elbow of a jacketed process pipe. (Source: Reprinted with permission from **CHEMICAL ENGINEERING** June 1993, Copyright 1993 by McGraw-Hill, Inc., with all rights reserved.)

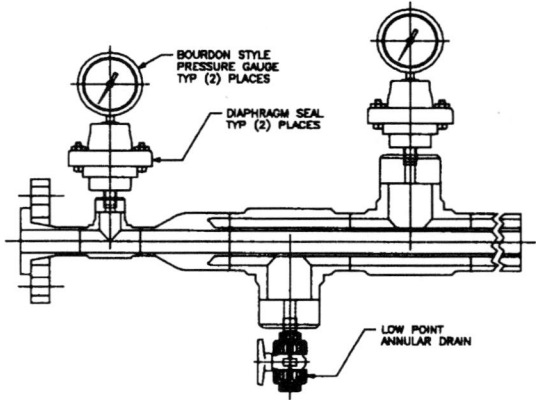

Figure 11.39. A possible combined horizontally positioned initiation arrangement having instrumentation (and leak detection connections) in the primary and secondary containment systems.

In some parts of the system, a localized high or low point may exist in the annulus. An example would be where the diameter of the secondary containment has been expanded in a section surrounding a primary pipe elbow to allow for adequate room for expansion/contraction. Consideration should be given to venting and/or draining this area. Figure 11.40 illustrates a typical vent connection and low point drain connection at the high and low points, respectively, of an expanded secondary containment elbow which is horizontally-positioned.

11.1.1.4 Anchoring Methods for Primary Pipe

It is occasionally necessary to restrain a piping system at a given point by securely anchoring it at that point. This may be for such reasons as controlling vibration, controlling, directing, and limiting the amount of thermal expansion in a pipe, and other reasons. Since

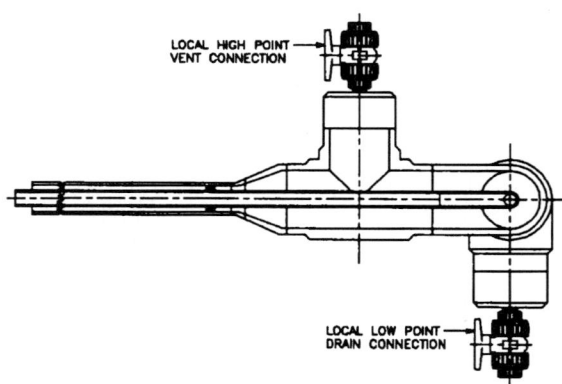

Figure 11.40. A typical vent connection positioned at the localized high point of a locally expanded annular section, along with a typical low point drain connection at the localized low point.

Layout Concepts for Double Containment Piping 421

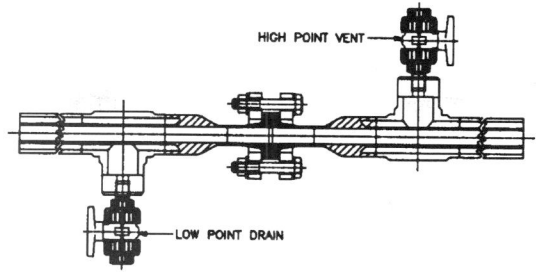

Figure 11.41. An example of a double-termination external anchoring arrangement.

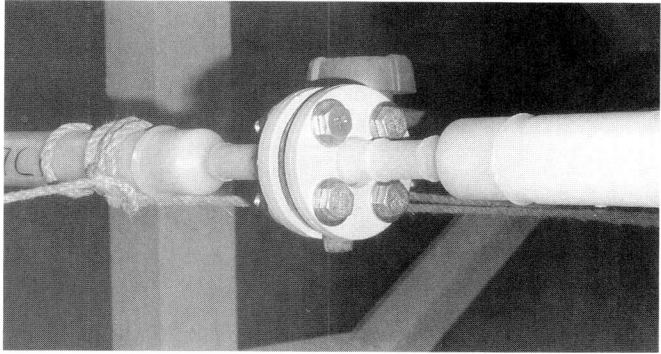

Figure 11.42. This photograph shows the arrangement of Figure 11.42 in an above-ground installation at a waste treatment facility.

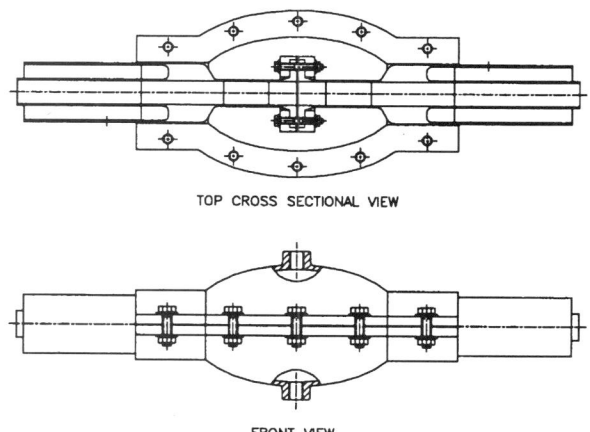

Figure 11.43. Illustration of a protective flange cover. (U.S. Patent #5,141,256.)

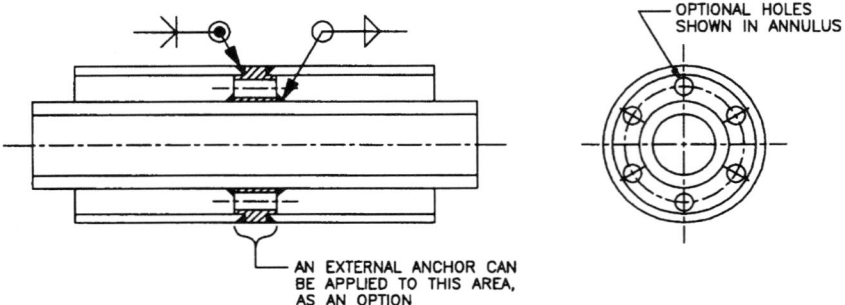

Figure 11.44. A basic internal-baffle style internal anchor, which is a structural attachment to the primary pipe. This type of device is limited to metallic pipes due to the limitations of welding/bonding nonmetallic materials to achieve adequate shear strength.

it is required in single-wall piping practice, anchoring is thus required for primary piping in double containment piping for the same reasons for its use in single-walled piping.

11.1.1.4.1 External-Anchoring Methods

A primary pipe in a double containment system may either be anchored "externally" (external to the annulus), or "internally" (internal to the annulus). An "external" arrangement involves interruption of the secondary containment feature, and is also referred to as a double-termination-flange arrangement. An example of this arrangement is illustrated in Figure 11.41 and shown in the accompanying photograph, Figure 11.42. In this arrangement, the inner and outer pipes are restrained on both sides since they are connected to a termination arrangement. Since termination arrangements are a natural point of interconnection, the systems are restrained in place. Further, since in this case they are bolted together via a flange, the flanges may be secured to a building part, resulting in a completely rigid arrangement. The same effect can be achieved if an outer pipe is anchored at points away from the flange on both sides and sufficient intermediate guiding is provided.

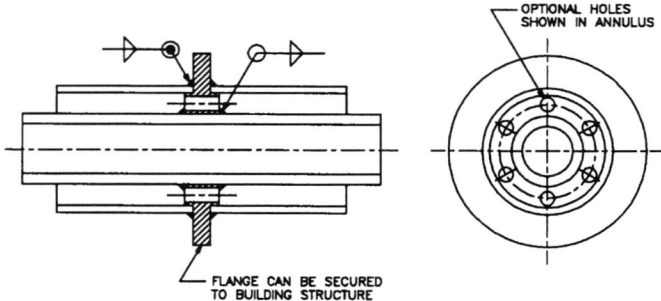

Figure 11.45. A variation of the internal-baffle concept shown in Figure 11.45, in which the baffle extends radially beyond the secondary containment pipe for the purpose of being welded/bolted to an external structural member.

Layout Concepts for Double Containment Piping

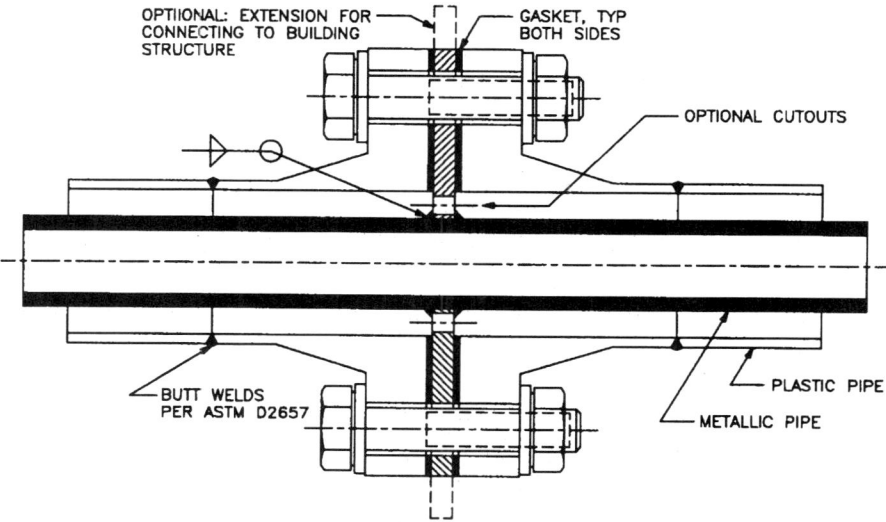

Figure 11.46. Internal anchoring method for metallic primary, nonmetallic secondary containment piping arrangements.

The main drawback with this type of arrangement is that this area is no longer secondarily contained. Some other means of secondary containment must be added. If secondary containment is required to be provided, then possible solutions for above-ground systems include either a dike structure to be constructed around it, or a protective flange cover provided. An example of a protective flange cover is illustrated in Figure 11.43. Another option in some above-ground applications is to leave the flange arrangement single contained, and provide daily visual inspection at this point. In underground systems, the arrangement will require either placement within a concrete sump or manhole, or the provision of a protective flange cover.

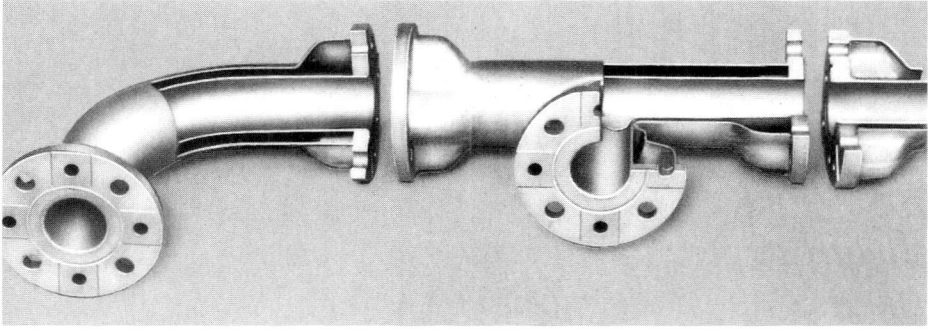

Figure 11.47. Example of the use of the patented Schulz Flange which is designed with a flow-through capability. (U.S. Patent # 4,121,858) (Source: Wilh. Schulz GmbH)

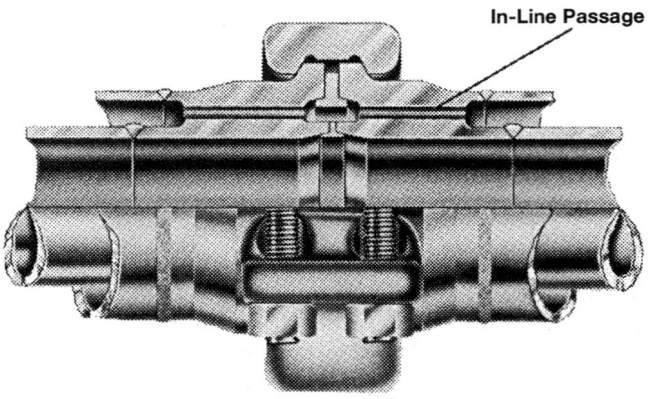

Figure 11.48. An example of a Grayloc® Connector having a "flow-through" annulus feature for use in jacketed process systems which have high pressures. (Source: ABB Vetco Gray, Inc.)

11.1.1.4.2 Internal-Anchoring Methods

There are options for anchoring the primary piping that eliminate the need to "interrupt" the containment casing. A variety of internal methods exist, some of which are fabricated, and others that involve specialized fittings.

A basic fabricated version, termed an "internal-baffle" arrangement, is illustrated in Figures 11.44 and 11.45. This method is a variation of the termination closure ring described earlier (see Section 11.1.1.2.3). It is suitable in metallic systems where the metal can be directly welded to the baffle.

A variation of the "internal-baffle" for use in systems that use a metallic primary piping and a nonmetallic outer system is illustrated in Figure 11.46. This type, like the version shown in Figure 11.45, may have the baffle extend radially, in order to allow it to be directly attached or secured to an external structure (e.g. Fig. 9.6h). These arrangements are suitable in above-ground applications; in underground applications the flanges would have to

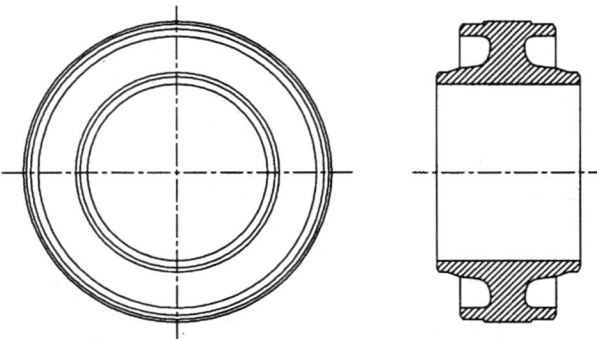

Figure 11.49. An example of a homogeneous internal anchor coupling having a solid annulus for compartmentalization purposes. (U.S. Patents #5,141,261 and 4,930,544.)

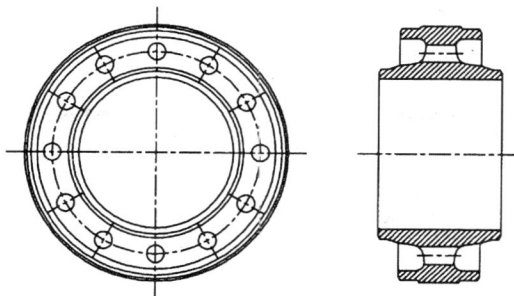

Figure 11.50. An example of a homogeneous internal anchor coupling having annular cutouts for flow and/or to allow for leak detection cable in the section. (U.S. Patents #5,141,261 and #4,930,544.)

be directly buried, and may therefore require extra protection (i.e., coating and cathodic protection of the bolts, in addition to protective shrink-wrap covers, etc.).

Figures 11.47 and 11.48 present viable layout alternatives to "external-anchoring" by way of some innovative "flow-through" mechanical connectors. Figure 11.46 illustrates the patented Schulz Flange, and Figure 11.48 illustrates a flow-through Grayloc® Connector.

A structural design of a homogeneous internal-anchor component is presented in Figures 11.49 and 11.50 that is capable of withstanding significant loads and is designed for use in metallic or thermoplastic material systems. Figure 11.49 illustrates a solid annulus version, while Figure 11.50 illustrates a version that is designed to allow annular flow (and maximize the flow through the component).

Any homogeneous-material internal-anchoring component is generally considered to possess single containment in its center portions (refer to Figures 7.2 through 7.4). Therefore, these components may require additional secondary containment around their exterior (i.e., a concrete sump, dike, etc.) to satisfy secondary containment requirements.

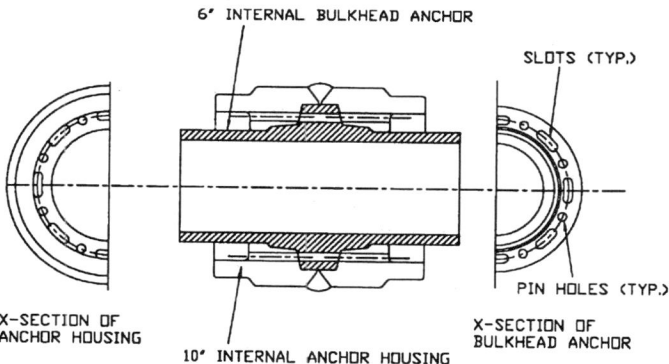

Figure 11.51. Example of a patented proprietary Internal Anchor Coupling for thermoplastic, or other material, primary piping with a separate secondary containment housing that can be constructed of the same, or a dissimilar plastic material. Shown is a copolymer polypropylene primary bulkhead housed inside an HDPE outer housing. (U.S. Patents #5,085,471 and #5,141,261)

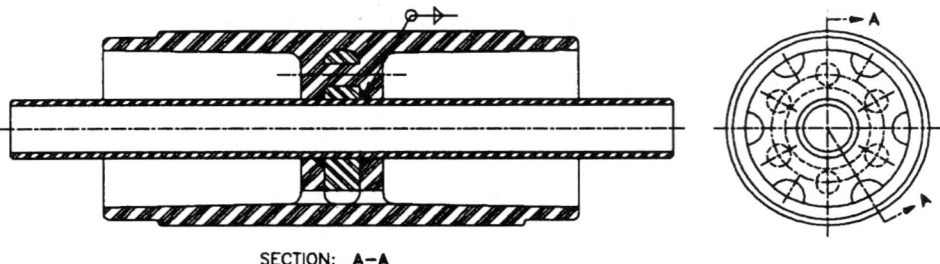

Figure 11.52. Example of a patented Internal anchor coupling involving a metallic primary bulkhead and an Epoxy RTRP resin-transfer-molded secondary containment housing. (U.S. Patent #5,085,471.)

A recently developed approach to internal anchoring involves the use of fitting components designed to anchor the primary pipe that also provides a means of secondary containment. Figures 11.51 and 11.52 illustrate types of "internal-anchor" fittings, designed to enable the primary piping to be anchored, while minimizing stresses on the component itself, lessening the chance of failure as compared to the homogeneous versions, yet maintaining 100% secondary containment, and also allowing any combination of materials.

These fittings shown in Figures 11.51 and 11.52 allow for annular flow capability. Alternatively, they may be provided with annular seal carriers to prevent flow. One would desire to prevent annular flow when designing a system in a compartmentalized fashion, or when using these fittings as termination fittings (see Figure 11.11).

Figures 11.51 and 11.52 are efficient methods for accomplishing internal anchoring in hybrid systems by eliminating the need for direct burial of flanges. For this reason and for reasons of their structural design (i.e., a secondary containment feature and low accumulation of concentrated localized stresses when high core fluid temperatures are present), the fittings illustrated in Figures 11.50 and 11.51 represent substantial improvements over previous methods of anchoring primary piping, particularly in thermoplastic systems.[1]

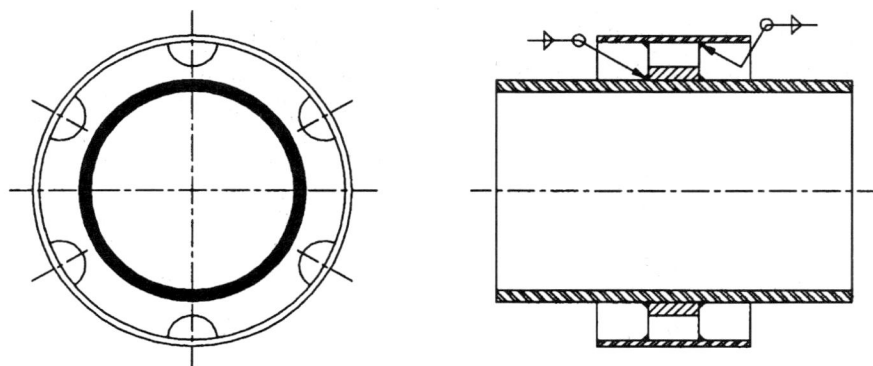

Figure 11.53. Example of a fabricated Internal Anchor Coupling involving a metallic inside a metallic combination and having annular cutouts for flow.

Layout Concepts for Double Containment Piping

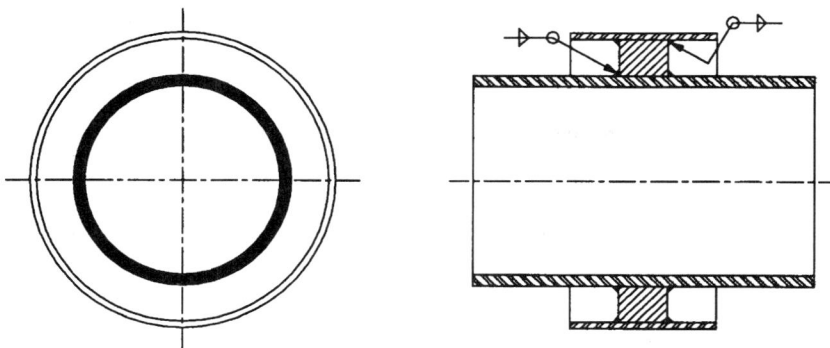

Figure 11.54. Example of a fabricated Internal Anchor Coupling involving a metallic inside a metallic combination and having a solid annular cross section for purposes of compartmentalization.

11.1.1.5 Systems Requiring Flexibility for Thermal Expansion

Chapter 8, "Thermal Expansion Considerations," describes various methods used to increase flexibility in systems that undergo a significant change in temperature. In a system containing elbows, it is often desirable to use the elbows to accommodate a portion or all of the expansion/contraction to which the system is subjected (see Section 8.6.2).

When adequate room to prevent contact due to differential expansion/contraction between primary and secondary containment elbows does not exist, the diameter of secondary containment elbows may be increased to allow primary elbows sufficient room to flex (refer to Section 8.6.3). In systems that have a significant amount of thermal expansion, expansion loops, offsets, or additional changes of direction might be required to achieve the desired level of flexibility (see Sections 8.6.2 through 8.6.12). Most nonmetallic materials have a linear coefficient of thermal expansion that is far greater than that of metals; additionally, they are subject to early failure due to creep strain accumulation when they are designed in a restrained fashion. RTRP materials do not yield, and as such, they must be laid out with sufficient flexibility to avoid premature failure. However, due to the design of certain components and the high strength of machine made RTRP components, it is usually best to lay out machine-made RTRP systems as restrained systems.

Consider the two piping schematic layouts illustrated in Figures 11.1 and 11.2. For the design conditions described earlier, the layouts have been decided upon to provide flexibility and balance out stresses in the system. In the RTRP system (Line CC of Fig. 11.2), the layout is a restrained design, as the temperature and pressure design conditions would not produce detrimental stresses in such a system, and the restrained approach is desirable as it is the more economical and better overall structural approach for the material. Careful review of the drawings reveal subtle differences between the chosen layout of the metallic and RTRP materials (Figure 11.2) and the nonmetallic materials (Figure 11.1).

Notice for instance that for the same service, the diameter sizes indicated for select secondary containment elbows in the nonmetallic system are increased in order to balance stresses, as are the number of elbows that are required. The standard size for most of the secondary containment elbows in the metallic systems possess adequate room for inner elbows to move and flex without having inner and outer elbows come in contact with each other. (The thermoplastic lines also require the use of specially fabricated long radius bends in order to achieve the required clearance. The metallic lines also make use of fabricated 3-D bends, but only so that a smaller and less expensive 2" diameter outer pipe can be used.) Also, the metallic systems of Figure 11.2 could optionally be designed in a restrained manner, since calculated stresses would all be within the allowable stress ranges of these materials, since the design conditions are relatively mild for metallic materials (a flexibility analysis would prove this to be true). However, the thermoplastic material systems must be treated in a different fashion, allowing for adequate flexibility in the system. A flexibility analysis would show that to use a restrained design/layout in each of the thermoplastic materials that detrimental stresses could occur under the design conditions.

The layout shown in Figure 11.1 is only one of many possible layout schemes that will result in sufficient flexibility for the materials shown under the services and conditions indicated. However, the example layout that was selected is such that total system cost is considered, while providing for the required flexibility of the pipe. There has also been an effort made to keep any one individual secondary containment elbow from becoming unreasonably large. Other alternative measures could have been taken to limit the differential growth that can occur. These include the use of expansion joints, loops or offsets, including the use of grooved mechanical joints, or other means. There are also many details which were not shown in the drawing that would need to be detailed in shop drawings. This includes items such as the length of straight 4" pipe on each side of expanded elbows, for instance, and the exact location and type of internal supports that are used.

Tables 11.2 and 11.3 tabulate the expansion of straight pipe sections between points of anchoring and the individual elbows for the Figure 11.1 and 11.2 layouts. The systems have been designed with the intention of installing a means of internal anchoring between points of alleviation, either consisting of the type-1 homogeneous type (see Figures 11.4, 11.49, and 11.50), the type-2 internal-anchoring components illustrated in Figures 11.51 through 11.54, or the type-3 fabricated versions illustrated in Figures 11.41 and 11.42. Tables 11.2 and 11.3 also show a tabulation of the diameter sizes and relative radii of the inner and outer elbows required of the secondary containment elbows in order to avoid any contact of primary and secondary containment elbows. It also lists the positioning of the closest interstitial support that can restrict radial bending of the primary pipe (i.e., the relative positions of axial guide supports), which along with the locations of flexibility supports is a key to achieving a successful internally-flexible design, (though often overlooked by designers).

It should be noted that a connection should be provided for a vent valve at every local annular high point and for draining fluids at every local low point in both Figures 11.1 and 11.2 systems. Both are important in order to operate the system properly and maintain a system free of accumulated fluids during actual operation, testing, and maintenance procedures. In the system, threaded valves can be typically provided for secondary containment drains, whereas welded (solvent-cemented) valves are recommended for systems that use pressurized annular nitrogen blankets. This is recommended in order to prevent against the slow, continual loss of the gas through the threads, which is typical of threaded joints.

An alternative layout detail to that shown would be to use a double termination flange arrangement at the midpoint of straight pipe sections (e.g., Figures 11.41, 11.42 and 11.47). If this design were to be used, protective flange covers (e.g., Figure 11.44) having both vent and drain connections for valves could also be added at every flanged spool connection to result in 100% secondary containment provided to the primary system at all points.

Table 11.2 Summary of Dimensional Analysis and Flexibility Requirements for Figure 11.1 Elbows

Elbow	Nominal elbow size (in. x in.)	Radius of primary elbow (in.)	Radius of secondary elbow (in.)	Inside diameter of outer elbow (in.)	Allowable space [1] (in.)	Ratio of upstream to downstream length	Thermal expansion of upstream side [2] (in.)	Thermal expansion of downstream side [2] (in.)	Min. dist. to rigid support of upstream side [3] (ft)	Min. Dist. to rigid support of downstream side [3]
A-1	1 x 4	6	4	3.83	1.26–2.08	2.0	1.76	0.88	3.0	4.3
A-2	1 x 4	6	4	3.83	1.26–2.08	0.67	0.88	1.32	3.7	3.0
A-3	1 x 4	6	4	3.83	1.26–2.08	3.0	1.32	0.44	2.1	3.7
A-4	1 x 4	6	4	3.83	1.26–2.08	0.5	0.44	0.88	3.0	2.1
A-5	1 x 4	6	4	3.83	1.26–2.08	2.0	0.88	0.44	2.1	3.0
A-6	1 x 4	6	4	3.83	1.26–2.08	0.33	0.44	1.32	3.7	2.1
A-7	1 x 4	6	4	3.83	1.26–2.08	0.75	1.32	1.76	4.3	3.7
A-8	1 x 4	6	4	3.83	1.26–2.08	1.33	1.76	1.32	3.7	4.3
A-9	1 x 4	6	4	3.83	1.26–2.08	2.0	1.32	0.66	2.6	3.7
A-10	1 x 4	6	4	3.83	1.26–2.08	0.5	0.66	1.32	3.7	2.6
B-1	1 x 4	6	4	4.03	1.36–2.18	2.0	1.11	0.55	3.0	4.3
B-2	1 x 4	6	4	4.03	1.36–2.18	0.67	0.55	0.83	3.7	3.0
B-3	1 x 4	6	4	4.03	1.36–2.18	3.0	0.83	0.28	2.2	3.7
B-4	1 x 4	6	4	4.03	1.36–2.18	0.5	0.28	0.55	3.0	2.2
B-5	1 x 4	6	4	4.03	1.36–2.18	2.0	0.55	0.28	2.2	3.0
B-6	1 x 4	6	4	4.03	1.36–2.18	0.33	0.28	0.83	3.7	2.2
B-7	1 x 4	6	4	4.03	1.36–2.18	0.75	0.83	1.11	4.3	3.7
B-8	1 x 4	6	4	4.03	1.36–2.18	1.33	1.11	0.83	3.7	4.3
B-9	1 x 4	6	4	4.03	1.36–2.18	2.0	0.83	0.42	2.6	3.7
B-10	1 x 4	6	4	4.03	1.36–2.18	0.5	0.42	0.83	3.7	2.6
C-1	1 x 4	4	2.32	3.83	0.98–1.93	2.0	0.64	0.32	2.5	3.5
C-2	1 x 4	4	2.32	3.83	0.98–1.93	0.67	0.32	0.48	3.0	2.5
C-3	1 x 4	4	2.32	3.83	0.98–1.93	3.0	0.48	0.16	1.8	3.0
C-4	1 x 4	4	2.32	3.83	0.98–1.93	0.5	0.16	0.32	2.5	1.8
C-5	1 x 4	4	2.32	3.83	0.98–1.93	2.0	0.32	0.16	1.8	2.5
C-6	1 x 4	4	2.32	3.83	0.98–1.93	0.33	0.16	0.48	3.0	1.8
C-7	1 x 4	4	2.32	3.83	0.98–1.93	0.75	0.48	0.64	3.5	3.0
C-8	1 x 4	4	2.32	3.83	0.98–1.93	1.33	0.64	0.48	3.0	3.5
C-9	1 x 4	4	2.32	3.83	0.98–1.93	2.0	0.48	0.24	2.2	3.0
C-10	1 x 4	4	2.32	3.83	0.98–1.93	0.5	0.24	0.48	3.0	2.2

Notes:
[1] Allowable space varies in Fig. 11.1 elbows depending on the direction of growth, as reflected by the ratio of upstream to downstream sides. A ratio of close to 1 provides maximum room, whereas an R approaching 0 or ∞ provides minimum. room.
[2] Based on α =1.05 x 10^{-4} (PP - Himont Profax® 7823 Copolymer), 6.6 x 10^{-5} (PVDF - Kynar® 460), 3.5 x 10^{-5} (CPVC 4120)
[3] Based on Eq. 16, using E = 150,000 psi (PP); 300,000 psi (PVDF); 420,000 psi (CPVC), and on opposite side expansion.

Table 11.3 Summary of Dimensional Analysis and Flexibility Requirements for Figure 11.2 Elbows

Elbow	Nominal elbow size (in. x in.)	Radius of primary elbow (in.)	Radius of secondary elbow (in.)	Inside diameter of outer elbow (in.)	Allowable space [1] (in.)	Thermal expansion of upstream side [2] (in.)	Thermal expansion of downstream side [2] (in.)	Min. dist. to rigid support of upstream side [3] (ft)	Min. dist. to rigid support of downstream side [3] (ft)
AA-1	1 x 2	3	3	2.16	.42	0.15	0.08	1.8	2.4
AA-2	1 x 2	3	3	2.16	.42	0.08	0.30	3.4	1.8
AA-3	1 x 2	3	3	2.16	.42	0.30	0.15	2.4	3.4
AA-4	1 x 2	3	3	2.16	.42	0.15	0.11	2.0	2.4
AA-5	1 x 2	3	3	2.16	.42	0.11	0.06	1.5	2.0
AA-6	1 x 2	3	3	2.16	.42	0.06	0.11	2.0	1.5
BB-1	1 x 2	3	3	2.16	.42	0.14	0.07	1.6	2.3
BB-2	1 x 2	3	3	2.16	.42	0.07	0.27	3.2	1.6
BB-3	1 x 2	3	3	2.16	.42	0.27	0.14	2.3	3.2
BB-4	1 x 2	3	3	2.16	.42	0.14	0.10	2.0	2.3
BB-5	1 x 2	3	3	2.16	.42	0.10	0.05	1.4	2.0
BB-6	1 x 2	3	3	2.16	.42	0.05	0.10	2.0	1.4
CC-1	1 x 3	NA	NA	NA	NA	0.14	0.07	NA	NA
CC-2	1 x 3	NA	NA	NA	NA	0.07	0.28	NA	NA
CC-3	1 x 3	NA	NA	NA	NA	0.28	0.14	NA	NA
CC-4	1 x 3	NA	NA	NA	NA	0.14	0.10	NA	NA
CC-5	1 x 3	NA	NA	NA	NA	0.10	0.05	NA	NA
CC-6	1 x 3	NA	NA	NA	NA	0.05	0.10	NA	NA

Notes:
[1] Allowable space is constant in the Line AA and BB elbows. Not applicable to Line CC elbow due to restrained design.
[2] Based on α =8.9 x 10^{-6} for 316L SS, 8.16 x 10^{-6} for Alloy 20 Cb-3®, and 8.2 x 10^{-6} for Fibercast Centricast Plus® RB-2530
[3] Based on Equation 8.16, using E of 28 x 10^6 psi for Lines AA & BB. Dist. based on opposite side expansion.
[4] Line CC is a restrained system design, and therefore the concepts of allowable space/min. rigid support dist. do not apply.

However, since protective flange covers of the type shown in Figure 11.43 may not always be designed for significant pressure, the outer system may lack fully pressurized containment at this point. This would represent a compromise solution as compared to designing a system to have a fully pressure-rated inner and outer system, at all points.

As an example of an alternative layout for nonmetallic or metallic piping that also provides for sufficient flexibility, yet is entirely different, is a layout which uses expansion devices on the outer jacket. In this layout scheme, flexible sleeve mechanical couplings could, for instance, be applied to the outer piping as both a means of joining, and to provide flexibility. In such a design, the inner and outer jackets are also anchored together at select-

Layout Concepts for Double Containment Piping

ed points between consecutive couplings. Although it is primary piping that is subject to greater amounts of thermal expansion, the addition of the flexible sleeve mechanical couplings to the secondary containment piping allows it to expand along with the inner piping in an accordion-like manner. Internal anchors are selectively positioned on the primary piping to anchor both systems to each other, so that they can grow in unison according to the growth of the primary piping, forcing the flexible sleeve couplings to open.

To illustrate this, Figure 11.55 illustrates one coupling of a system in its original (installed) condition; Figure 11.56 illustrates the coupling in its expanded condition. It should be noted that in such a system that each double containment elbow must be allowed to flex, since pipe will grow towards them. Therefore, attention needs to be directed to the position of the closest external hangers to each outer elbow so that bending at such a system's elbows is not restricted, which could otherwise result in overstrained behavior.

11.1.1.6 Systems Containing Branches

The example schematic illustrated in Figures 11.1 and 11.2 did not include any branch connections. However, many applications do involve branch connections of some type (i.e., tees, laterals, crosses, etc.). In pressure piping systems that contain branches, there exist several layout/installation options. One must first consider the manner in which the branch connection must be constructed, then decide upon the sequence by which it is to be installed into the rest of the system.

All forms of double containment piping branch design/construction fall within one of the following four types:

1. Standard (listed) primary branch fitting inside a standard (listed) secondary, containment branch fitting. This arrangement may be used if there is sufficient annular clearance between inner and outer systems, and also depends on the materials involved and if their fabrication methods allow for it. An example of such an arrangement is illustrated in Figure 11.57.
2. Standard (listed) primary branch fitting inside a fabricated secondary containment branch fitting. This a common arrangement whereby the secondary containment portion is fabricated from straight pipe around a standard primary containment branch fitting that has short sections of adjacent pipes attached. This type of arrangement is illustrated in Figure 11.58.
3. Standard (listed) primary branch fitting inside a split and rejoined secondary, containment branch fitting. This is an approach to branch construction for applications when the primary branch is welded (bonded) into the line first. The outer component serves as a type of midline secondary closure and is assembled into its position in the line sometime after the primary components have been assembled.

There are a variety of commercially available nonmetallic secondary containment products that have been designed in a clam-shell style to be reassembled in this fashion. These components vary greatly in pressure capability; most are capable of moderate internal pressure (< 5 to 85 psi). In order to result in a component having substantial pressure rating upon reassembly, a split standard fitting using welding (metals) or overwrapping (RTRP) techniques would have to be used. A commercially available clam-shell-style clear PVC branch connection is shown in Figure 11.59. A fabricated lateral is shown in Figure 11.60.

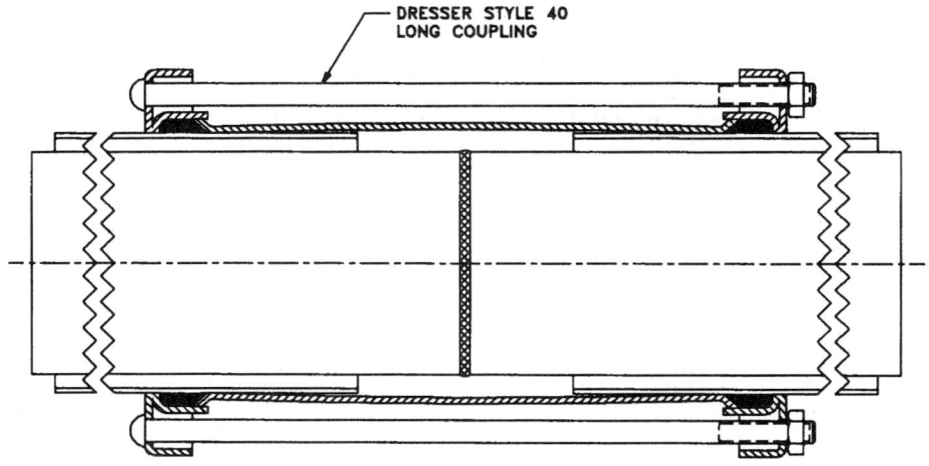

NOTE: THESE COUPLINGS MUST BE SECURED BY LUGS, SELECTIVE ANCHORING OF THE OUTER JACKET, OR BY OTHER MEANS TO PREVENT JOINT FAILURE DUE TO INTERNAL PRESSURE THRUST IN THE EVENT OF A GUILLOTINE FAILURE OF THE PRIMARY PIPE.

Figure 11.55. Illustration of the use of flexible sleve mechanical joints to increase flexibility and control expansion in above-ground systems. In a system that uses this type of fitting for flexibility, it is necessary to anchor the system between each consecutive coupling.

4. Fabricated primary and secondary containment branch fitting. In this approach both the primary and secondary containment fittings are prepared from their respective straight pipes.

   When a branch connection fitting is fabricated from pipe, reinforcement of the branch opening may be required, depending on the forces to be applied to the branch. This subject is covered for fabricated branches in the design chapters (Sections 5.2.1.3 and 6.3.4.1 for primary fittings, and Section 7.4.4, for secondary containment branches). When fabricating their double containment branch arrangements, some manufacturers include plate or fin-type supports positioned between the inner and outer fittings on each of the run legs and on branch legs, as an integral part of the fitting. This will not count towards reinforcement criteria (see Section 5.2.1.3) according to standard design rules (e.g., ANSI/ASME B31.3 Chemical Plant and Petroleum Refinery Piping Code). It also may not be desirable if system components must be free to flex (i.e., as a layout requirement due to thermal expansion).

Aside from design/construction of the branch fittings, there are options concerning installation sequence that will affect overall system layout. Alternative sequences for adding branch fittings into the rest of the piping system include:

1. In all-thermoplastic systems that use simultaneous fusion joining techniques, branch fittings must be prefabricated, usually in a shop environment, and subsequently simultaneously fused directly to its adjoining components.

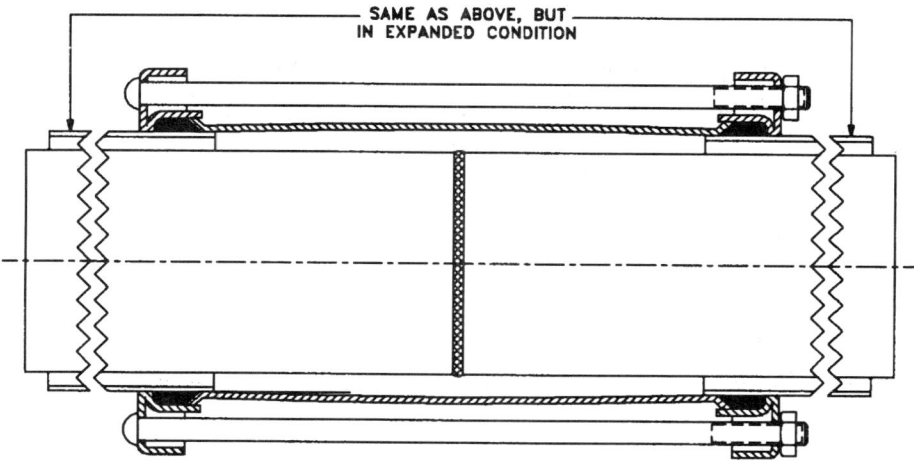

Figure 11.56. Illustration of the arrangement shown in Figure 11.55 in the expanded condition.

2. In systems that are field assembled in a staggered fusion, the branch connections may be preassembled with primary branches extending beyond the secondary branches. A branch of this type would require secondary closures to join the secondary containment branches to their adjoining secondary containment components, on one or more sides.
3. In systems that are field assembled in a staggered fusion, the primary branch fitting may be welded first to its adjoining primary piping components. The secondary containment branch fitting may then be installed as a split or clam-shell style and mechanically joined or welded (bonded/overwrapped) directly to its adjoining secondary containment piping components. In this approach, the secondary containment branch fitting serves as a secondary closure.
4. The double containment branch fitting is part of a complex fitting arrangement. In this instance it is best to prefabricate the entire complex assembly and join it to the rest of the system using one of the first three methods described for each individual branch connection (the most common choice would be alternatives 2 and 3).

The layout details for all unique branch connections should be indicated on project fabrication drawings. If multiple branches of the same type exist, then at least one "typical" detail drawing should be prepared. In order for a branch to be considered the same as another, all aspects of the branch design and installation, including its diameter, the sequence of joining, and the type of adjacent components to which it is to be joined, should all be identical. Otherwise, separate fabrication drawings should be prepared for each unique system branch connection.

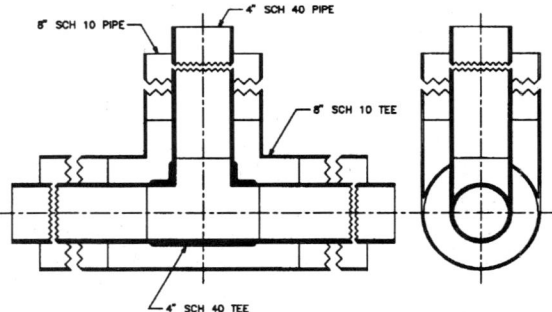

Figure 11.57. Standard primary fitting housed inside a standard secondary containment fitting (where enough space in the outer fitting exists).

11.1.1.7 Systems Requiring Easy Assembly and Frequent Disassembly; Use of Specialized Inner and Outer Flange Arrangements

One aspect of double containment piping systems that is often overlooked is whether a system may be readily installed in the field by laborers who possess only limited skills. The ease with which a system may be readily maintained, modified, or repaired in the event of an incident may also go unconsidered. These important features of systems must be considered as part of the design and layout; they may turn out to be the single most limiting factor of the installation. Without the ability to install readily, facilitate a repair, or maintain a system, a double containment piping system can later prove to be a major burden for a facility owner.

By designing a system to be easily installed and repaired, major costs and headaches can be avoided during the operating life of the system. The best way to achieve this level of repair is to design a system that is somewhat modular and highly prefabricated. Not only will the maintainability of the system be increased, but greater installation control can be gained through having complex arrangements fabricated under shop conditions. A system

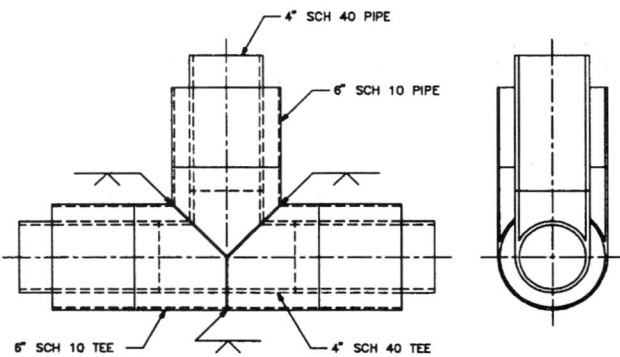

Figure 11.58. Standard primary tee inside a fabricated secondary containment tee, fabricated from straight pipe.

Figure 11.59. Example of a proprietary "clam-shell" style secondary containment fitting made of clear PVC, housing a standard PVC schedule 80 primary tee. (Source: R&G Sloane Manufacturing Co., Inc.)

that is designed in this fashion will require a completely different layout and will require that more detailed shop drawings be prepared than that of a field-constructed system.

Consider the layout of the systems shown in Figures 11.1 and 11.2. If this system were to be installed using a totally shop-prefabricated approach, one possible approach would be to have spool sections prepared using termination/flange arrangements at the end of each spool section. When separate closed spool sections are designed, a connection for high-point vents and low-point drains must be provided into each spool, since a unique, finite annular compartment is created.

Once the spool pieces are fabricated, they may be bolted together in the field with flanges or flexible mechanical joints, using a suitable gasket material. A proper selection of bolts, nuts, and washers must be made. The system must then be bolted in the proper sequence to a specified amount of torque using a torque wrench.

Drawings for typical flanged bolted connections are illustrated in Figures 11.41 and 11.42 for heat-element butt-fused thermoplastic systems (butt-welded metallic systems). An equivalent bolted arrangement could be used which uses companion-type PVC solvent cemented flanges or compression-molded RTRP flanges in PVC, CPVC or RTRP systems.

It should be emphasized that the arrangements illustrated in the three previous illustrations contain a section surrounding the flange that is not provided with secondary containment at that point. If secondary containment is required for 100% of the system, then a protective flange cover or open-air dike structure must be added. A protective flange cover used to provide secondary containment and thus act as a secondary closure for all three systems is illustrated in Figure 11.43.

In some cases, the local authority having jurisdiction, the plant owner, and the design engineer of record may all agree that the system may not require secondary containment at mid-line double-termination/flange locations like those in Figures 11.41, 11.42, and 11.60. Instead, these may become points that are visually inspected on a daily basis. All three of the parties mentioned must determine and agree that the arrangement selected is suitable.

Figure 11.60. Example of a clam-shell RTRP 45° lateral. (Source: Ershigs' Division of CBI, Inc.)

An alternative approach is to use protective flange access housings designed to accommodate an arrangement with both primary and secondary containment flanges. This would allow the system to be completely unbolted and disassembled in sections, without creating unique, finite annular spaces in the spool sections. This feature would eliminate the need to have separate high-point vents and drains for each spool section, although drains and vents are usually required in the housing itself. Also, housings of this design allow for a continuous annulus throughout the piping system. An example of a protective flange access housing of this type is illustrated in Figure 11.61. The Figure 11.61 protective flange access housings can be constructed and designed as a fully pressure-rated component (i.e., to ASME Boiler and Pressure Vessel Code Rules, Section 8, Division 1 for metals or Section 10 for RTRP).

With the Figure 11.61 protective flange housing, it is necessary to use split-backing rings, in conjunction with an ANSI B16.5-type stub end, for making inside lap-joint flange connections. Three examples of split backing rings are illustrated in Figures 11.62 through 11.64. These types of protective covers may be provided with point probes for leak detection, and may also be adapted to house valves, expansion joints, etc.

11.1.1.8 Automatic Flushing and Drying of Annuli

It has been stated repeatedly in this text that flow in an annulus tends to be less than ideal. There are methods that may be used to increase the efficiency of the distribution of flushing/decontamination fluid through the use of spray headers. The same methods also increase the efficiency of air distribution during a drying process. For example, Figure 11.65 illustrates a spray header system that is made possible by using an interstitial support system and associated perforated pipes, developed for this purpose by the author. The header system may further be controlled manually or may be controlled in an automatic fashion when a signal is sensed by a leak detection system. A system of this type would be justified in the event of the transport of an extremely hazardous substance such as nuclear irradiated fluids.

11.1.2 Installation-Related Layout Issues for Pressure Piping Systems

11.1.2.1 Joining Sequence for Inner and Outer Welds

Each sequence of joining (i.e., staggered versus simultaneous fusion) has a direct bearing on a system's layout. Shop fabrication and construction drawings are directly affected since they must indicate to the fabricator or contractor where welds are to be made (and in what sequence). Other chapters discuss this issue as well. For instance, Chapter 12, "Fabrication, Assembly, and Erection," discusses the details of installation sequence as it pertains to different materials. Chapter 8, "Thermal Expansion Considerations," also discusses this subject as it pertains to the flexibility requirements of various systems that are subject to temperature changes. The discussion in this section is limited to the effect of joining sequence on system layout.

Metallic materials may only be joined using staggered welding techniques. A method to join metallic materials by simultaneous welding does not exist, nor would it be advisable since it would prevent visual and other nondestructive examinations described in Chapter 13, "Inspection, Examination, and Testing," from being performed. Also, virtually all reinforced thermosetting resin piping (RTRP) systems and socket-fusion thermoplastic and solvent-cement-based thermoplastic systems are limited to being joined by use of staggered-joining techniques. Currently, simultaneous fusion techniques are applicable mostly to heat-element butt-fusion-based thermoplastic materials, including high-density polyethylene, polypropylene, and fluoropolymer materials. These techniques have been performed on a limited basis on some solvent-cemented-based thermoplastic systems.

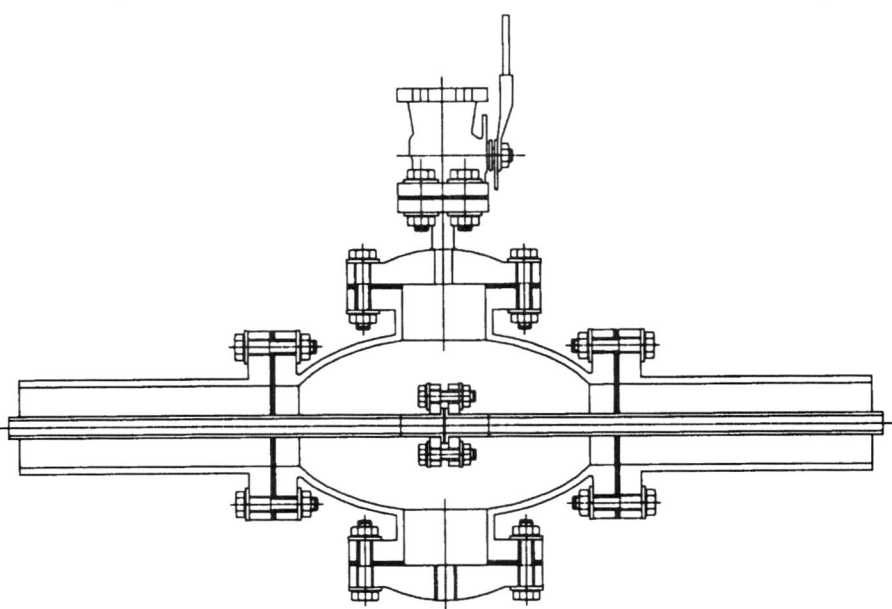

Figure 11.61. Example of an access housing designed for high-pressure systems. (U.S. Patent #5,141,256.)

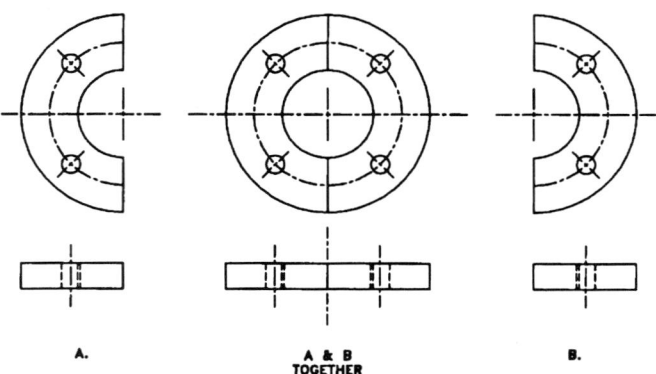

Figure 11.62. Plain split backing ring.

When a system is to be installed using some type of staggered method, designers have more than one option from which to choose in terms of how fittings are treated. These options, as they pertain to installation sequence of components (fittings), include:

1. Field, sequential joining of all components, no midline secondary closures used. This type of approach is selected when it is the preference of the designer or facility owner to avoid the use of any midline secondary closures. This tends to be the most difficult and labor-intensive of all the staggered welding approaches. Using this approach, all inner and outer system components must be assembled prior to testing either the inner or outer pipes (sections of the system could be assembled and tested in this fashion). This approach eliminates the opportunity for visual inspection of all primary pipe welds upon the initial pressure test. It also makes it very difficult and sometimes impractical to conduct required nondestructive examinations of primary pipe welds. Therefore, it may not be used when implementing a system per the ANSI/ASME B31.3 Chemical Plant & Petroleum Refining Code and other design codes. This technique is outlined in Section 12.3.2.1.5, specifically as it relates to elbow installation.

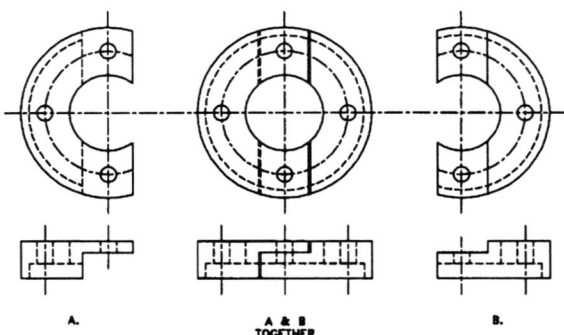

Figure 11.63. Offset stepped split backing ring.

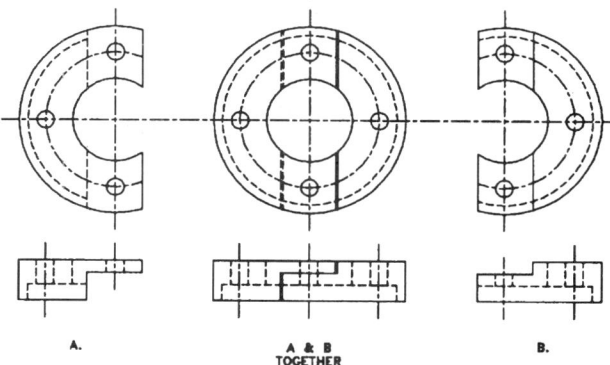

Figure 11.64. Stepped split backing ring.

2. Prefabrication in straight pipe sections using split secondary containment fittings as closures. Under this approach, straight sections of primary pipe are first welded, and interstitial supports are secured to their respective primary pipes according to their desired spacing. The primary pipes then have straight unwelded lengths of secondary containment pipe slipped "over" these supports and into position. Primary pipe fittings are then attached to their respective pipes. The primary pipe system may then be hydrostatically tested, and all welds (bonds) visually inspected by moving outer pipes back and forth. Once the primary pipe system is completely tested, and all inspections and nondestructive examinations performed, the secondary containment straight pipe welds (bonds) can be made. As a last step, split secondary containment fittings may be reassembled into their position to complete the containment casing. This approach is described and illustrated in Section 12.3.2.1.6 for systems containing elbows.
3. Prefabrication in "L," "T," "Y," or other complex-shaped sections with secondary closures in straight pipe sections. This approach is similar to #2, except that the fabricator/installer starts at fittings and works his/her way to the straight pipe sections. The last items to install with this approach are straight pipe secondary closures (i.e., slip couplings, weld wraps, split-secondary pipe, etc.). In applications where adequate flexibility is inherent in the system, the use of secondary closures can be avoided. In some all-thermoplastic systems where it is deemed acceptable, a simultaneous butt weld may be used as a means of secondary closure (and point of restraint).
4. Field fabrication using highly modified slip-lining techniques. This joining technique, initially developed by polyethylene pipe manufacturers for retrofitting old sewer systems, is adaptable to some double containment pipe systems. Under this approach, straight pipe sections are joined separately, and then inner pipes are "slipped" through "outer" pipes. This technique is illustrated in Chapter 16.

When simultaneous fusion is used as an option for the joining sequence of primary and secondary containment portions of thermoplastic systems, all ends of prefabricated piping and component must be parallel. This requires that an end-support baffle (disc) must be positioned an approximate distance of 1 in. (25 mm) from the end of the pipes and welded to both pipes using a hot-gas fillet weld with welding rod as filler material (see Fig. 12.96).

11.1.2.2 Field Sequential Joining

Double containment elbows (90°, 45°, 30°, etc.), mitered bends, and other elbow fittings may be assembled in a sequential fashion in the field to avoid the use of secondary closures. This sequence, along with the prefabricated approaches described in Section 11.1.2.3, will allow primary elbows to be constructed in an unrestrained manner within their respective secondary containment elbows. This is true so long as the closest interstitial supports are positioned far enough away that they do not restrict natural movements of the elbow. Additionally, they do not over direct bending of the elbows, otherwise caused by directed axial movements of the adjacent growing pipes. While the supports must be a minimum distance away to allow nominal bending and torsional movements, and to prevent over bending, they must also be close enough to provide structural support required of the primary pipe in the elbow area to prevent bending due to "dead-weight" load. This is discussed further in Sections 8.6.2 and 9.3.1.1.

If these conditions are met, the elbows may be considered as "freely floating" in the surrounding secondary containment elbows, with support provided by the adjacent interstitial supports. This is effectively the same approach used in single-wall pressure piping, which it is standard practice to provide support to elbows by the use of hangers located away from the elbows, and not directly at elbows themselves. Elbows can be provided by means of direct support (or even be restrained), although it is not typical for systems that experience significant thermal expansion/contraction. Direct support can be provided in the vertical direction, while also allowing for anticipated bending and torsional movements that will be concentrated at the bend. A support of this type is illustrated in Figures 9.14 through 9.16, and 8.37.

Sequential joining may be used as layout option for some prefabricated sections involving elbows. However, it is more common to start at the elbows themselves, except perhaps in expansion loop/offset construction. Sequential installation of elbows is therefore pre-

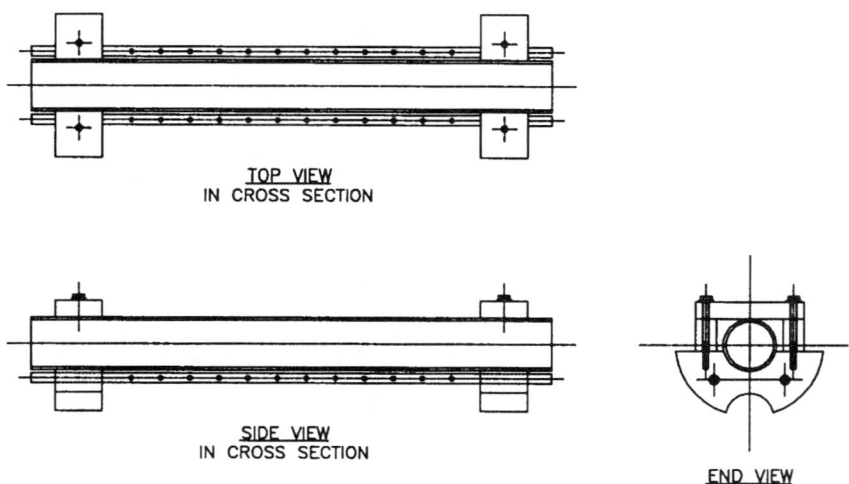

Figure 11.65. Illustration of a spray header system that is designed with perforated pipes to facilitate automatic flushing and drying of an annulus (patent pending).

Layout Concepts for Double Containment Piping

sented here as a field joining option, and only in a conceptual sense; the author does not endorse it. It must be remembered that existing design codes (i.e., ANSI/ASME B 31.3) may eliminate sequential joining as a possible joining/layout option due to inspection, examination, and testing requirements of the primary piping system. Also, sequential elbow layout techniques are not possible in systems using simultaneous joining methods; simultaneous joining methods represent a completely opposite approach.

The steps required for sequential joining of secondary containment elbows is illustrated in Chapter 12, Figures 12.70 through 12.78 for butt-style component systems and Figures 12.79 through 12.83 for socket-style component systems for a section having one 90° elbow. The same procedure is also discussed further in Chapter 12, "Fabrication, Assembly, and Erection" in Section 12.3.2.1.5.

For elbows other than 90°, the same sequence is used. It is a wise idea when planning to use this procedure to prepare at least one detailed set of drawings to illustrate to the fabricator the sequence of joining desired and the exact location and type of welds (bonds). The length of the small piece of pipe, sometimes termed a "pup," that is shown in Figure 12.73 as attached to the 90° elbow should be determined through dimensional analysis. If this information is not determined analytically, then it must be determined in the fabrication stage, using a trial and error approach.

11.1.2.3 Elbow Layout, Prefabricated Approaches

There are two major types of prefabrication joining/layout options that may be used when installing elbows into a system. They are both described in Section 12.3.2.1.6, "Systems Using Mostly Prefabricated Sections." The first describes a sequence where straight pipe sections are prefabricated, the second describes a sequence where the elbows are prefabricated. In this sense they are somewhat opposite approaches, although both tend to depend on the use of some form of secondary closure. The choice as to which approach is best for any one project depends on the preference for closure type that is favored by the designer/owner and the materials involved, sizes, etc.

The first approach is illustrated in Figures 12.85 through 12.87. In this approach, prefabricated straight primary pipe sections must have their ends extended a certain distance beyond their respective secondary containment pipes. This is for two reasons: (1) secondary containment elbows (components) have comparatively greater take-off dimensions than primary elbows; and (2) space must be allowed to join primary pipe elbows (i.e., by welding or bonding). The distance is usually a fixed amount and is equal to the difference between the "center line to end" dimensions of the inner and outer elbows (taking into account welding losses).

The second approach, entitled 'Prefabrication in "L," "V," "T," "Y," or Other Complex-Shaped Sections with Secondary Closures in Straight Pipe Sections,' is illustrated in Figures 12.88 through 12.91. In the basic version of this approach, prefabricated straight pipe sections must have ends of primary pipes extended a certain distance beyond the secondary containment pipes. This is to allow space to make the primary pipe welds (bonds). The distance required depends on the dimensions required by joining tools and is limited by the type of midline secondary closure selected.

A variation to this approach exists that eliminates the need to have primary pipe extended beyond the secondary containment pipe and thus eliminates the need to add a secondary

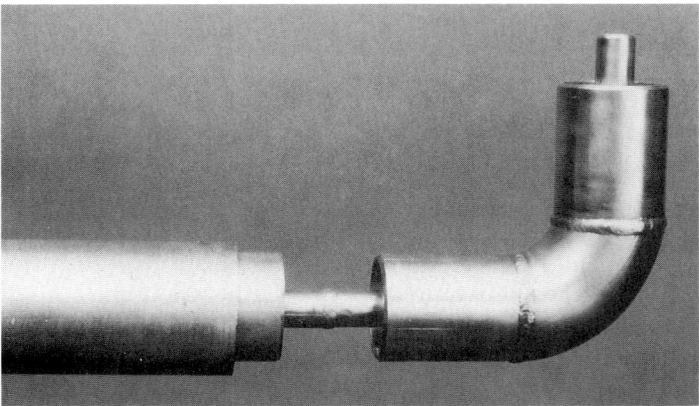

Figure 11.66. Illustration of a gap in a double containment pipe. (Source: Guardian division of Nibco, Inc.)

closure, if there is sufficient flexibility in a double containment piping system. This is illustrated for a system having consecutive elbows in Figures 12.92 through 12.94.

A second variation of this approach may be used in all-thermoplastic systems using simultaneous-welding (bonding, fusion) techniques at the midpoint between elbows, with all other joints being made in a staggered manner. The sequence, as applied in a system containing consecutively occurring elbows, is illustrated in Figures 12.95 and 12.96. Whenever a simultaneously welded (bonded) joint is introduced into a piping system, it will become a point of restraint by anchoring the two systems together at that point. Therefore, under this approach, this resulting section is anchored at the midpoint between the two elbows.

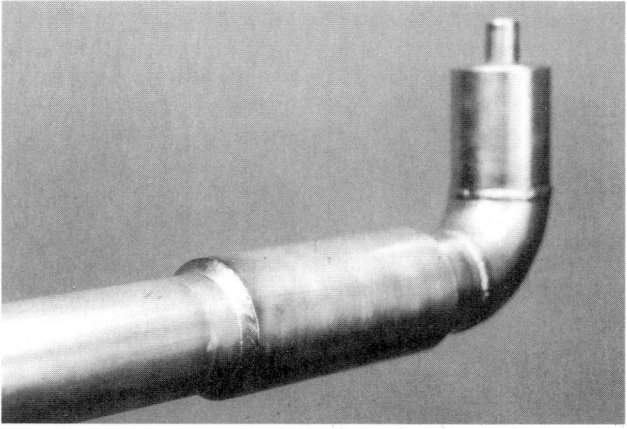

Figure 11.67. Illustration of a slip coupling. This component contains no splits and telescopes over the secondary containment piping and is welded as shown in Fig. 11.75. (Source: Guardian division of Nibco, Inc.)

Layout Concepts for Double Containment Piping 443

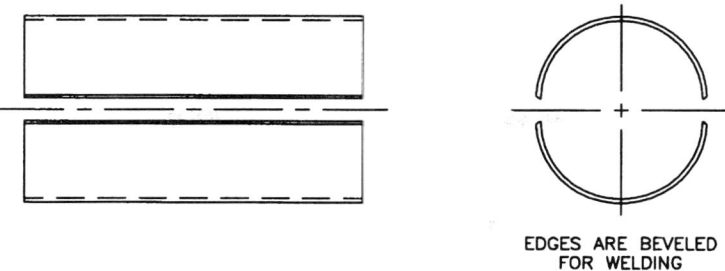

EDGES ARE BEVELED FOR WELDING

Figure 11.68. Illustration of a weld wrap and/or "clam-shell" coupling. This component contains two longitudinal splits at 180° apart. A weld wrap overlaps the secondary containment pipe on both sides, whereas a "clam-shell" coupling's diameter and wall thickness match the outer pipe and also has beveled edges on both sides for butt-welding to the outer pipe.

11.1.2.4 Types of Secondary Closures and Their Locations

Secondary closures consist of closures that are used to seal gaps remaining in a secondary containment piping system in order to complete a containment casing. Many secondary closure methods exist; however, they may be divided into two basic types: end-type closures (terminations) and midline closures.

End closures are described elsewhere in this chapter in Section 11.1.1.3; thus, midline-type secondary closures are discussed here. Midline closures are required whenever a gap exists either in a straight pipe section or in a secondary containment fitting. They are applicable mainly to systems where staggered welding is used as the sequence for joining and whenever sections are prefabricated. A gap in a straight pipe section of a secondary containment pipe requiring a midline secondary closure is illustrated in Figure 11.66. This type of gap exists for systems that use the joining sequence option illustrated in Figures 12.88 through 12.91.

There are a many options to complete containment casings where such a gap exists. Four basic welded (bonded) types include: (1) split-pipe sections, also referred to as "clam-shell" couplings; (2) weld-wraps, (3) sheet wraps; and (4) slip couplings. These

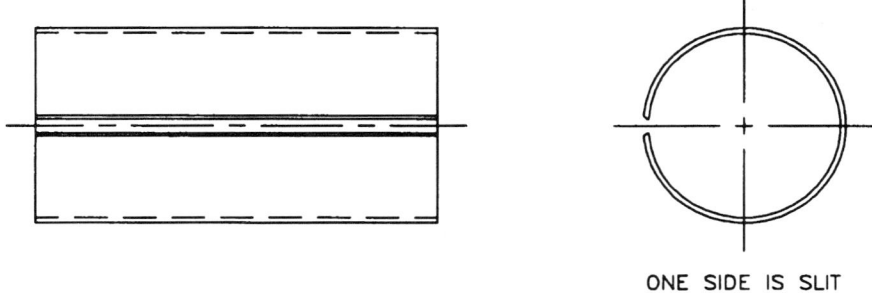

ONE SIDE IS SLIT

Figure 11.69. Illustration of a sheet wrap. This consists of a rolled metal sheet which overlaps the secondary containment pipe on both sides.

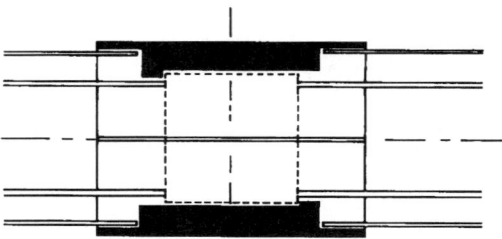

Figure 11.70. Illustration of a combination "clam-shell" closure coupling and internal anchor arrangement. This component contains two longitudinal splits at 180° apart, and is designed to entrap a primary pipe coupling. (Source: Conley Corp., patent pending.)

four types are illustrated in their as-installed condition, designed to close a midline gap, in Figures 11.67 through 11.69. Figure 11.70 shows a combination closure and anchor. Figure 11.71 is a photograph of a slip coupling being installed into an 8 in. (200 mm) nominal diameter secondary containment polyolefin pipe.

Cross-sectional views of these four closures after assembly are illustrated in Figures 11.72 ("clam-shell" coupling), 11.73 (weld wrap), 11.74 (sheet wrap), and 11.75 (slip coupling). In these side views it can be noticed that at the 6 o'clock position of the annulus only the "clam-shell" coupling (Figure 11.70 and 11.72) possesses a somewhat constant bottom profile. The other methods create a localized low point where fluids can collect and become difficult to flush/dry. The 12 o'clock position can have similar problems with entrapped air, which may affect any hydrostatic test of the annulus. This tends to favor the pipe window ("clam-shell" coupling) in deciding upon which method to use. However, there are other technical drawbacks that result from the use of pipe windows. They are more difficult to weld (bond) into position, and they possess the least structural strength of the three choices (true of thermoplastic materials). The slip coupling (Figures 11.67, 11.71 and 11.75) is the only method of the three that requires prepositioning on the secondary containment pipe prior to making the primary pipe connection.

Figure 11.71. Installation of a slip coupling in a polyolefin piping system, installed by means of conventional hot-gas welding techniques.

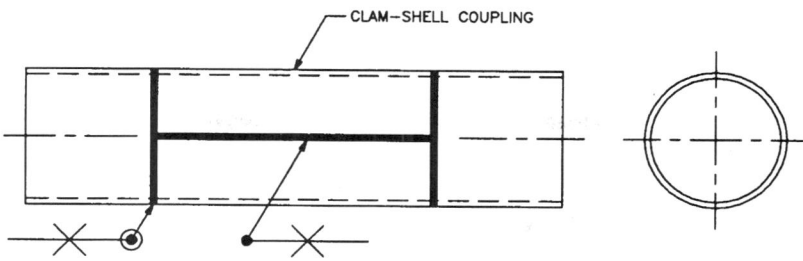

Figure 11.72. Cross-sectional view of an assembled "clam-shell" coupling.

When using midline closures, of the types illustrated in Figures 11.72 through 11.75, in RTRP systems, additional over wrapping is normally required to seal the closure and to add structural strength for the longitudinally split arrangements. The final appearance of an overwrapped component is illustrated in Figure 11.76. Figure 11.77 illustrates a proprietary modified PVC slip-coupling for PVC secondary containment systems. It is sometimes beneficial to overwrap thermoplastic closures with RTP materials to add structural strength as well. It is generally not recommended to use a fabricated "clam-shell" secondary closure (Figures 11.68 and 11.72) for use in RTRP or thermoplastic systems. There are a variety of machine-made secondary containment "clam-shell" closure couplings which various RTRP manufacturers have tooled up to make and which have high integrity.

Any time a secondary containment fitting is used as a means of secondary containment closure, it also can be classified as a midline secondary closure. This type of arrangement would be chosen when selecting a joining sequence option illustrated in Figures 12.88 through 12.90.

Other types of unique closures exist, including mechanical-style secondary closures exist, including clam-shell-style fittings and couplings, flange protective access housings, and other methods. There are also available a patented clear-PVC split piping and fittings that are joined by an injection- (adhesive-) bonding technique that may be used as a means of secondary closure. An "tee" example in this proprietary joining system is shown earlier in this chapter in Figure 11.59. A fabricated RTRP lateral is shown in Figure 11.60.

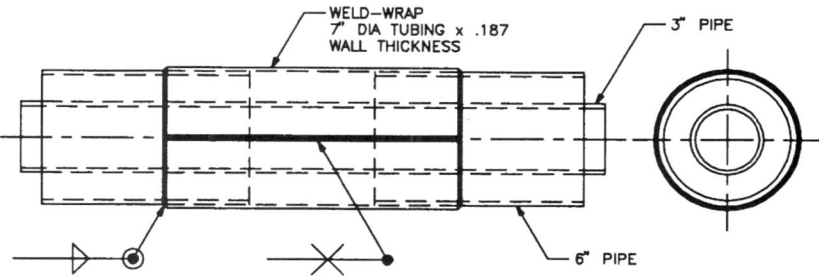

Figure 11.73. Cross-sectional view of a weld-wrap assembly.

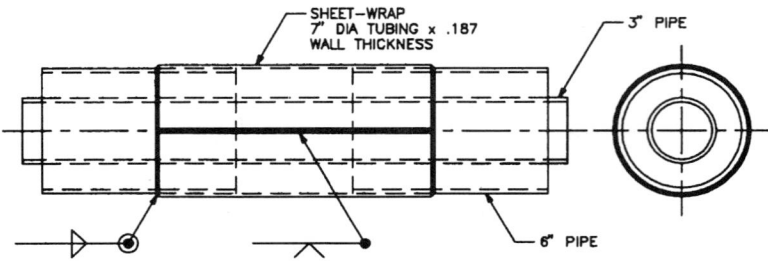

Figure 11.74. Cross-sectional view of an assembled sheet wrap.

11.2 Layout Details for Double Containment Pressure Fittings

Each double containment pressure inner-outer component arrangement in a system that is of a unique design must be clearly detailed in project drawings, in their intended as-installed condition. This includes any fittings that are totally prefabricated, or inner-outer fitting assemblies that are entirely field constructed. For inner-outer fittings that are partially or totally prefabricated, detail drawings of typical assemblies should be prepared for each unique assembly.

Fittings that are to be totally field constructed using sequential methods may have all details indicated as part of overall layout drawings or separate spool drawings. The easiest way to indicate details for these types of drawings is through the use of orthographic-view drawings (the same as those used in the illustrations throughout this chapter, for example). There should be at least one set of separate drawings prepared to indicate the sequence of joining for each field-assembled inner-outer component, as described in the section titled "Sequential Fitting Layout" of this chapter.

11.2.1 Dimensional and Material Information

Enough information should be clearly identified on project and shop drawings to ensure that the proper combination of parts is supplied and fabricated. Information that is pertinent includes construction materials, including the type and grade of the individual

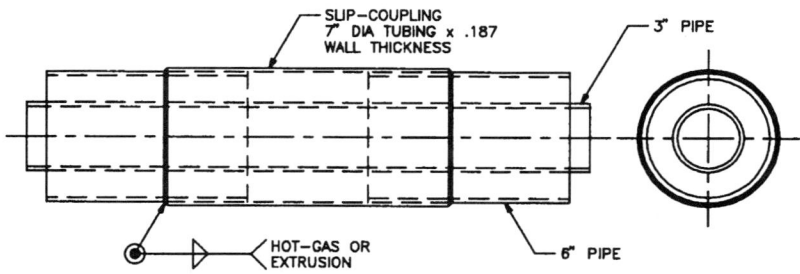

Figure 11.75. Cross-sectional view of an assembled slip coupling.

Layout Concepts for Double Containment Piping

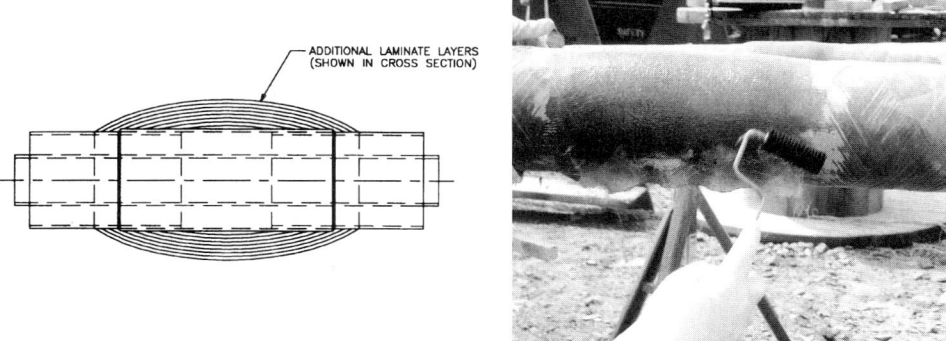

Figure 11.76. On the left is an illustration of a slip-coupling assembly provided with additional laminate layer overwrapping. On the right is an actual example using vinyl ester RTRP materials in a 3" by 6" system. (Source: Right Photo: Ershig's Division of Chicago Bridge and Iron, Inc.)

alloy/resin, and "manufactured brand," if it is important. Wherever possible, the grades of material should be related to an ASTM specification, UNS specification, or other consensus standard or appropriate European standard such as CEN, ISO, or DIN.

The dimensional information should include all applicable diameter length and thickness data. All applicable tolerances should be stated as well, with care taken not to state conflicting tolerance data. The temperature basis at which dimensional information should be measured should be stated as well. This is very important for nonmetallic materials, as they can experience substantial changes in dimensions for relatively small ambient temperature changes. Whenever possible, the dimensioning practices stated in a widely accepted consensus code or standard (i.e., ASME, ASTM, ANSI, ISO, etc.) should be used as a suitable basis for providing dimensions.

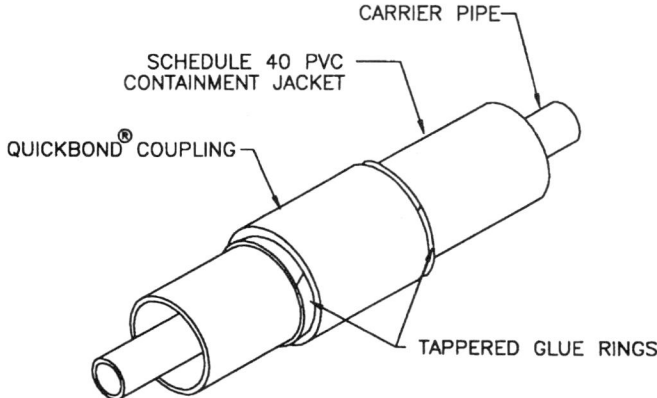

Figure 11.77. A proprietary modified PVC slip-coupling. (Source: Rovanco Corp.)

11.2.2 Layout Details for Fittings

The details of fittings in their as-installed condition should be clearly indicated on fabrication and/or construction drawings. It is the as-installed condition at approximate installation temperature that is of concern, since components may experience changes in relative position due to differential thermal expansion. Such details should be clearly indicated, regardless of whether the fitting is prefabricated or field assembled. Layout details must include enough information to determine the exact location of each fitting in the system, its relation to adjoining components and an indication of the type and locations of welds (bonds), or other methods of attachment. Any other information that is important to allow the fitting to be installed in its intended as-installed condition must be indicated as well.

11.2.3 Fitting Limitations

All limitations of fittings and their ability to be installed in their intended manner should be thoroughly investigated early in the project. Most items that limit the ability to install a fitting are either space related or material related. Space limitations may involve restrictions created by other adjacent lines, building parts, or other pieces of equipment. In some applications, the fittings themselves may be able to be installed, but the intended method of welding (bonding) may be restricted by the space involved. Where this occurs, an alternative method of joining must be considered. Alternatively, parts may be assembled by prefabrication. In the more drastic cases, the design may have to be modified to result in a totally different layout. Joining methods and their space requirements must always be considered early in a project for this reason.

If dissimilar materials are involved in a particular fitting design, the joining method selected must be given careful consideration. If welding or chemical joining techniques are not able to be used, then mechanical joining techniques may be required. If there is no possible way to join dissimilar products, or code limitations impose too many restrictions, then the design should be altered to involve compatible materials or changed in some other fashion.

11.2.4 Special Fittings and Elements for Double Containment

Double containment piping system design may occasionally require that a totally custom fitting or special component be provided as part of the system. When this is the case, it must be determined early in the project whether the part can be manufactured in the manner intended. Detailed machining or fabrication drawings should be prepared with sufficient information to manufacture the component. All pertinent dimensions should be clearly stated along with all applicable tolerances. Whenever machined parts are involved, the dimensioning practices stated in ANSI Y 14.5 can be used as a suitable basis for providing dimensions. Designers must be aware that if a system is to comply with the ANSI/ASME B31.3 Code, any unlisted components have to be qualified for use, as described in Chapters 5 through 7 of this text.

11.2.5 Complex Fitting Arrangements

Whenever more than one fitting is to be joined in a complex arrangement, or where headers are to be fabricated, it is recommended that assemblies should be prefabricated in a controlled shop environment, and a separate spool drawing should be prepared.

11.3 Nonpressure Double Containment Piping Layout and General Underground Double Containment Piping Considerations

Nonpressure systems include both above-ground and underground systems whose primary systems operate at a pressure between 0 and 15 psig (bar). The major type of nonpressure system is the gravity flow chemical drain waste and vent (DWV) system, which is designed to drain chemical waste to sewers by means of gravity to treatment or holding tanks, or to a treatment facility. The secondary containment portions of these systems are typically designed to the same performance specifications as the primary system (i.e., to a pressure rating of 10 ft of head, or 0.3 bar). In some instances, the secondary containment jackets of these systems may be designed in an "open-ended" fashion, whereby the secondary containment jacket flows "open ended" over a tank or treatment pond. When a system is allowed to be designed in such a manner, the pressure rating of the secondary containment should at least be capable of any head buildup due to developed back pressures. The secondary containment design normally is up to the designer, although code or regulatory requirements may dictate what is required.

While many underground waste systems are of the gravity-fed, nonpressure type, some double containment waste piping transfer lines can be pressurized due to the distance they must convey the fluids, or due to a lack of slope. Also, in some systems where a substantial change in elevation exists (> 33 ft, 10 m), the line may have to be reclassified as a pressure system, since some of its components will be subject to greater than 15 psi (1.1 bar). In many projects, the site does not allow for much change in elevation. Since elevation changes will be limited, the system may have both nonpressure and pressure characteristics. The usual design in such applications involves the nonpressure line (lines) draining into a sump, whereby the fluid is then moved within a force main by the use of a sump pump.

Underground piping may include pressure transfer lines that are routed underground at the choice of the owner. This may be due to reasons of surface aesthetics, unavailable surface space due to surface restrictions, the existence of underground tanks, or to control piping temperatures/safety concerns. In the case of petroleum marketing outlets, piping is placed underground for all four reasons. Some aspects of underground layout apply to nonpressure lines and pressure lines. General layout considerations for underground pressure and nonpressure double containment piping are discussed in this section.

Many aspects of nonpressure waste piping and other non-ASME Code underground piping layout practices are defined in many localities by building and/or fire codes. There are also many well-defined layout practices that have been developed by various professional disciplines and their organizations (e.g., American Society of Civil Engineers, American Water Works Association, American Society of Plumbing Engineers, etc.). However, most competent designers involved in nonpressure piping and underground pressure piping applications gain their layout knowledge from working on actual projects, and by following the practical methods of other accomplished designers.

Designers of nonpressure chemical DWV systems (e.g. from laboratories), should be aware of the differences between a plumbing code (e.g. BOCA, IAPMO, UBC, etc.) and a design code (e.g. ANSI/ASME B31.3, 31.9). Design codes cover the mechanical design integrity of the components and system, whereas a plumbing code typically does not. Though a design code may not be required for certain systems by local jurisdiction, designers should consider applying such codes to verify the mechanical integrity of the system.

11.3.1 Chemical Waste DWV Piping Systems (Chemical Sewers; Acid Waste DWV Piping Systems)

Chemical waste piping systems include those that serve as drain, waste, and vent piping for chemicals of all types. These types of systems are commonly referred to as "acid waste DWV systems" by plumbing engineers. Although the name implies that they are specific to the disposal of acids, they may be designed to convey acids, bases, organic chemicals, chlorinated solvents, or inorganic chemicals of all types. The name is designated as acid waste piping since acids are among the most commonly encountered chemicals in laboratory and plating applications, whereby such chemicals are commonly discharged into the waste piping after they have served their purpose. The plumbing and sanitary engineering profession has historically designated this piping with such a name to distinguish it from other sanitary drain, waste, and vent lines in building services.

Chemical waste materials should not be discharged into a regular sewer system without first being neutralized or treated in some fashion. Many chemical waste piping systems are designed to allow fluid to discharge into a neutralization basin or treatment pond where the chemicals are treated. The neutralized or treated product is then allowed to be discharged into the sewer system or waterways once the effluent is within prescribed purity limits.

Layout considerations for these systems include such items as determination of diameter sizes (capacity of the piping), slope determination, determination of component style, and venting requirements. Other system selection factors (such as material selection) can, in certain locations, be based on fire-related considerations such as flame and smoke spread ratings of the materials, and adherence to U.L. (Underwriter Laboratories) and F.M. (Factory Mutual) standards. Whenever a layout is to be determined for nonpressure waste systems or other type of underground system, the authority having local jurisdiction must be contacted as to what codes, standards, and permits apply.

Whereas the design and layout of most sanitary drain, waste, and vent and other plumbing-related piping is governed by strict building codes, chemical waste piping systems are not regulated in all areas. The reason that it may not be covered in certain local building codes is that in many municipalities it is classified as process piping. Process piping is ordinarily not covered by the general building codes. However, above-ground systems may be still subject to fire code regulations depending on the locality where it is to be installed. Although there may not be strict coverage under building codes, the design and layout of such systems should still follow the well-defined engineering practices of ordinary DWV piping. This includes sizing practices, slope determination, provision of proper venting, and other principles of design and layout.

One aspect of chemical waste piping design that tends to be common among systems of this type is that most systems have, at least part, if not all, of the system buried directly under a building slab, or behind walls, or between floors, which usually means very limited access to the piping system. If there is a leak detected, portions of the building usually have to be excavated to get at the pipes. It also means a disruption of ongoing activities within the building (and lost revenue) in the event of a repair. Therefore, the selection of material for the secondary containment pipe becomes critical in this situation. The fluid will have a tendency to remain in the annulus for a fairly long period of time, at least until a repair operation can be scheduled. For these reasons, there is added incentive to use homogeneous inner and outer materials.

Layout Concepts for Double Containment Piping 451

An example layout of an above-ground portion of a chemical waste double containment piping system is shown in Figure 11.78. An isometric view is used due to the complexity and fitting intensiveness that is characteristic of such systems. The design conditions for these types of systems typically involve only minor changes in temperature; the piping system is always close to that of the controlled-climate building in which it is installed. This is not true of every chemical DWV system, as occasionally hot-temperature fluids are discharged into them, or high temperatures due to heats of mixing can be experienced. When wide temperature swings are involved, a chemical DWV system layout must account for the anticipated expansion. Chapter 8 discusses the fact that in nonmetallic systems, even moderate temperature changes create trouble producing stresses even in low-pressure systems (see Section 8.4 and 8.5).

In Figure 11.78 there is no mention as to material used. The reason is that in an application of this type, the designer may select a suitable material from either the metallic, RTP, or thermoplastic groups. Layout concepts to be used would apply in principle, regardless of the material selected.

Figure 11.78 illustrates that providing secondary containment to an above-ground chemical DWV system presents many challenges to designers. A system of this type is highly fitting intensive, requiring close tolerance installation in some areas. There are many areas where restrictions exist due to building parts, minimum takeoff dimensions required by secondary containment fittings, and less-than-minimum space requirements for use of joining equipment. For these reasons, the use of above-ground double containment for DWV systems should be limited to areas where it is judged to be absolutely necessary. These might include areas where a piping leak could present a danger to people, or where damage to building parts or sensitive equipment might occur.

11.3.1.1 Floor Drains and Other Fixture Outlets

The example layout in Figure 11.78 is typical of an above-ground chemical DWV piping project. The purpose of a drainage system of this type is to collect waste from laboratory sinks, fume hoods, and floor drains. However, other applications exist for above-ground drain systems to allow drainage of fluids from process equipment, rinse-down areas, potential spill areas around tanks and vessels, etc., in order to transport the waste to a remote treatment area. These lab sinks, fume hoods, floor drains, etc., are the point of introduction of the waste fluids.

Typical diameters for outlet primary piping from fixtures range from 1-1/2 in. (40-50 mm) diameter to 4 in. (100-110 mm) diameter. These outlet pipes must be equipped with secondary containment piping ranging in sizes from 3 to 8 in. diameter sizes, respective of the aforementioned primary pipes. Outlets from floor drains are usually 2 in. (50-63 mm) in diameter to a maximum of 6 in. (160 mm).

Figure 11.79 details an ideal configuration for fixture outlets that must be equipped with secondary containment. In this example, the secondary containment jacket is shown as being directly welded (or bonded) to the underside of the sink, basin, or tank. This is only possible in systems involving homogeneous or highly compatible materials. It is something that is highly encouraged in the layout of such systems as it will lead to a transition involving high integrity. A reason why this is important is that it allows the transition from single containment to double containment to be made above the level of the first primary pipe

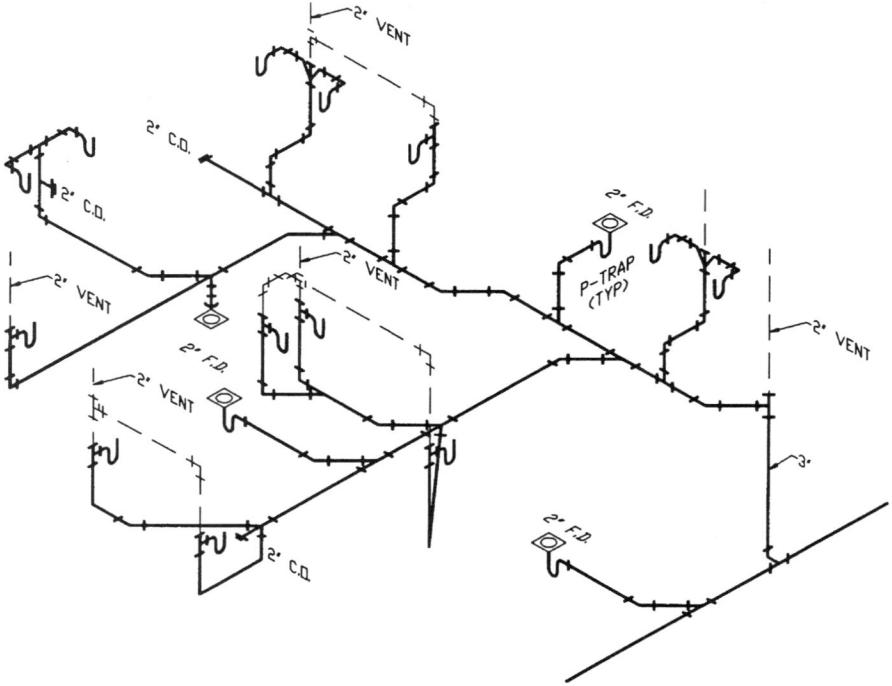

Figure 11.78. Isometric of an above-ground multi-floor acid DWV system. (Source: ALLCAD)

joint. In contrast, Figure 11.80 illustrates a transition from single containment to double containment where the transition is made below the level of the first joint. Either transition is acceptable if it occurs above ground in an area considered to provide adequate secondary containment. If the transition occurs underground, it is critical that it be made above the level of the first joint. Otherwise, a concrete sump equipped with a leak detection probe may have to be constructed around the area to prevent the possibility of leakage to the surrounding soil.

Floor drains also have many specific concerns that need to be addressed. The same transition concerns expressed in the previous paragraph apply. However, there are a number of additional concerns in comparison to fixture outlets. Figure 11.81 illustrates examples of typical double containment floor drains. These floor drains should be provided with a flashing in order to collect any leaks safely that occur where the brim is sealed into the floor, something that can be expected to occur eventually in many floor drains. There should be perforations provided in their basins that allow collected fluids that leak around the brim to drain into the "primary" portion of the drain, which will allow fluids to flow down their respective primary pipe outlets. It is important that fluid never be purposely introduced into the annulus of double containment piping since a leak will then be sensed and repair procedures commenced. Thus, collected leaks around a brim are discharged into the "primary" portion of the drain.

Layout Concepts for Double Containment Piping

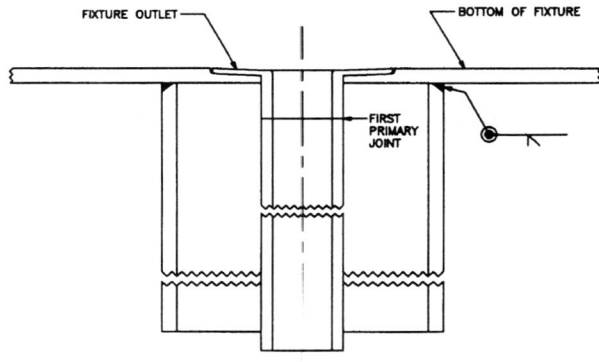

Figure 11.79. Ideal configuration for fixture outlets.

In Figure 11.81, the transition from single containment to double containment is made by directly attaching the secondary containment pipe directly to the underside of each drain's flashing by using a weld or bond. This is critical for floor drains, since they are typically directly imbedded in concrete flooring. This is readily possible if the drain, the primary pipe, and the secondary containment pipe are constructed of homogeneous materials; if it is not possible (i.e., in hybrid material systems), then a concrete sump has to be constructed around the area with some form of monitoring applied in most jurisdictions in the United States.

11.3.1.2 Use of Traps

Floor drains and fixtures need to be provided with traps at their outlets in chemical systems where there is a possibility of backup of gases from the system through the fixtures/drains. Chapter 4 presents further discussion concerning the requirements for traps in these systems. (See Section 4..4.2.3)

The most common type of trap in a chemical DWV system is the P trap. A typical double containment P trap is illustrated in Figure 11.82. A requirement of P-trap design is that a minimum of a 2 in. (50 mm) water seal be provided in the design. In a trap equipped with secondary containment, the annulus will always have the same theoretical seal dimension as the primary pipe, plus a small amount equal to the sum of the annulus available at the bottom of the bottom elbow and the wall thickness of the primary elbow. (One should note that the annulus is meant to remain normally empty, thus the seal dimension of the secondary pipe is described only in "theoretical terms.") Another feature of traps that must be considered is whether or not the primary P trap needs to be provided with a cleanout at its low point. Cleanouts are often provided at the bottom of traps in ordinary plumbing systems to allow the drain to be unclogged of collected aggregates, hardened foam, etc. Therefore, if the trap requires a cleanout to be provided, then a specialized arrangement is required.

Figure 11.83 illustrates an arrangement where the primary part of the trap has cleanout capability; the secondary containment portion is provided with a removable access cover that is further equipped with a digital liquid sensing probe. An arrangement of this type would

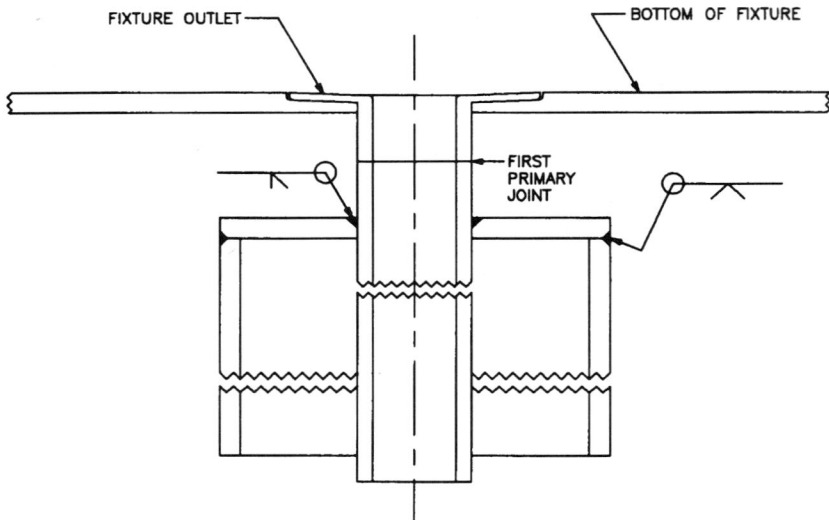

Figure 11.80. A configuration from a fixture outlet where the transition is made below the level of the first joint. This type of arrangement should be avoided.

need to be placed inside a sump that can safely house such an arrangement and allow the safe collection of fluids when they spill. An alternative arrangement is illustrated in Figure 11.84, where single-wall components are housed inside a secondary containment sump.

11.3.1.3 Above-Ground Horizontal Branches

Above-ground horizontal branch piping is similar in terms of layout requirements to other horizontal gravity-pressure drainage piping. It must be provided with a consistent slope,

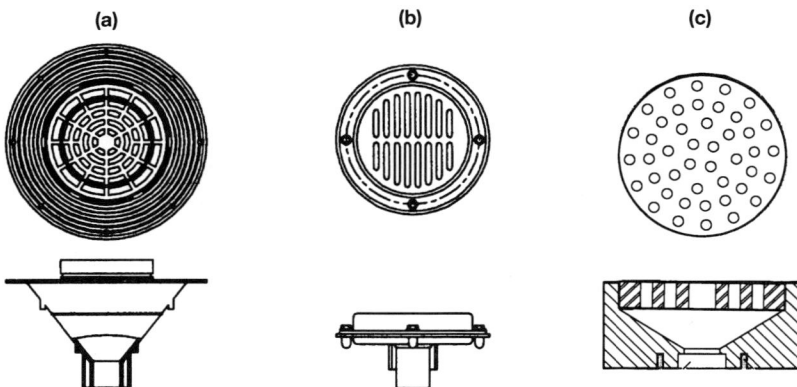

Figure 11.81. Illustration of three types of double containment floor drains: (a) Polypropylene (Enfield); (b) Stainless steel (J. R. Smith); (c) Epoxy or vinyl ester RTRP (Fibercast).

Layout Concepts for Double Containment Piping

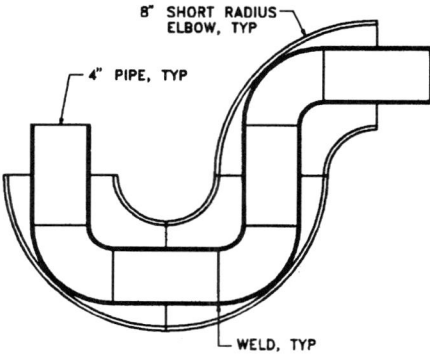

Figure 11.82. A typical double containment P trap.

usually either 1/8 in. (3 mm) per ft (0.3 m) or 1/4 in. (6 mm) per ft (0.3 m), depending on the desired rate of flow for the given diameter. Be sure to include all the appropriate information as to location and type of interstitial supports (centralizing devices), location of leak-detection devices or access ports, and any other pertinent data. For information on sizing of horizontal branch piping refer to Chapter 4, Section 4.4.2.6.

11.3.1.4 Vertical Risers

In a system that is provided with secondary containment above and underground, the vertical risers are usually required to be double containment below grade, and may be transitioned to single-wall pipe when they are above the floor slab. This is normally accomplished with the use of a termination transitional component, as shown in Figures

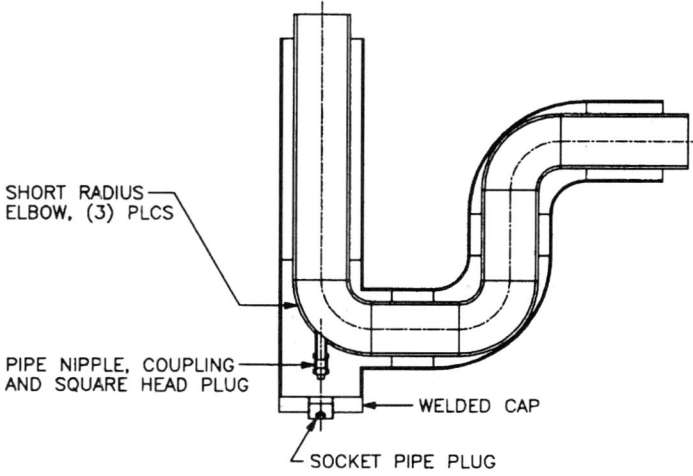

Figure 11.83. A double containment P trap with cleanout capability.

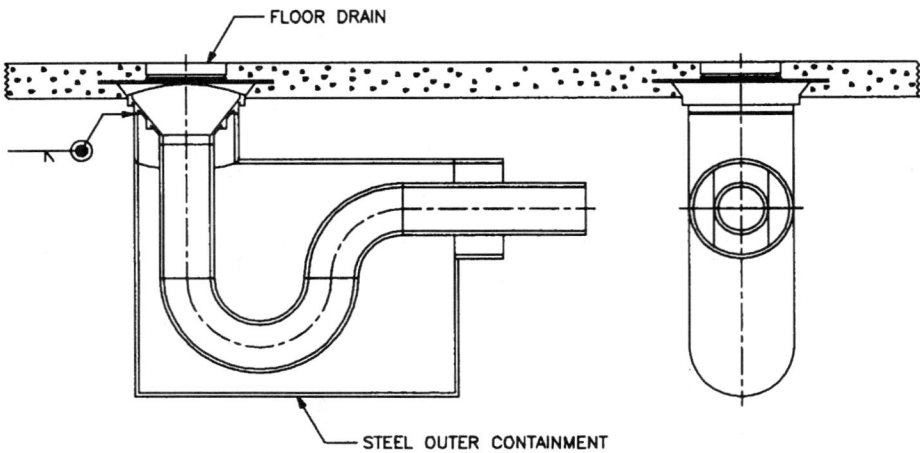

Figure 11.84. Alternative arrangement for P traps involving single-wall components housed inside a secondary containment sump.

11.4 through 11.25. It is important to consider that vertical risers usually connect into an underground header, which slopes at 1/8" to 1/4" per foot. A correction must be made to the angle at which the piping comes off of the fitting which connects to the horizontal header so that it is plumbed straight to the floor drain or vertical riser that it connects to, unless the lateral is designed with integral features to accommodate the angle. Installers often simply bend the vertical piping coming off of the sloping fitting, which induces stresses into the pipe and fitting, and should therefore be avoided.

11.3.1.5 Underground Horizontal Headers

Most systems involve one or more vertical waste stacks that collect waste from horizontal branches and/or floor drains and discharge waste vertically into a common header installed under the building. Each stack normally connects to a header through either a 90° long sweep elbow, or through a long-turn-tee-wye (45° lateral with a 1/8th bend). Floor drains that are used to collect emergency spills are typically connected to an underground P trap that is in turn connected to the header at the exit of the trap. Underground headers usually connect outside the building to an underground main, are fed into a common sump or basin, or are drained into a neutralization tank.

A typical underground header system used at a laboratory-type facility is illustrated in Figure 11.85 and shown in the accompanying photograph, Figure 11.86. The layout of the system is typical of a two-story laboratory building that collects wastes from various fixtures, floor drains, and vertical waste stacks. Underground header systems should have orthographic plan view drawings prepared, similar to that shown. An overall plan view showing all headers should be prepared, as well as individual drawings for each header. The individual headers should include details of all connections, including fitting types, location, and method of attachment (type of weld or bond).

An alternative design for underground headers that fall within a building's limits is to place a single-wall header within a lined or coated open trench. This can be done on all or

Layout Concepts for Double Containment Piping

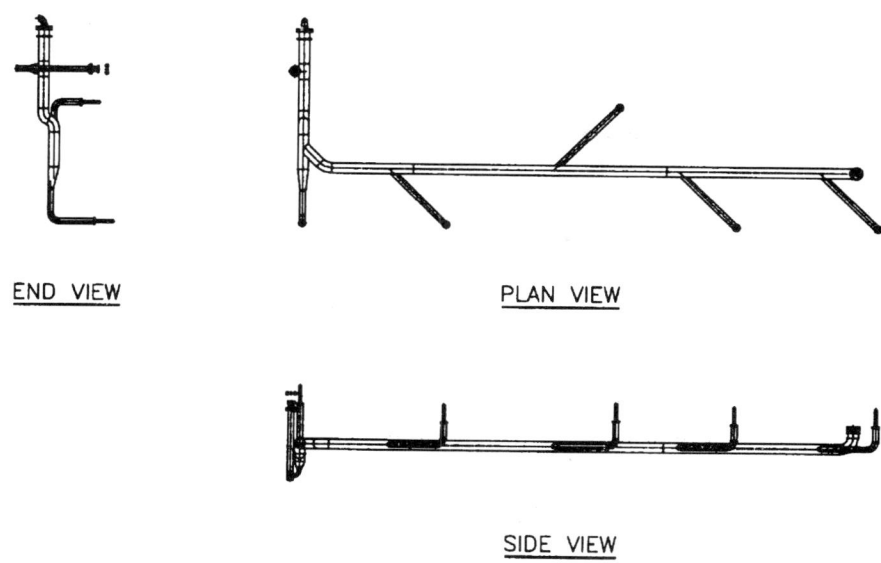

Figure 11.85. A typical underground chemical sewer header system.

a portion of an underground header that is within a building's limits. Figures 11.87 is an example of single-wall underground headers that is contained within a coated, open trench. In this example, perpendicular double containment pipes are terminated upon entry to the common trench, whereby multiple pipes are secondarily contained by the associated common open trench. Shown in Figure 11.88 is the structure designed to support the different pipes. While not indicated in the drawing, stainless steel or fiberglass pultruded grating is placed over the trench to prevent falling into the open trench.

Figure 11.86. Photograph of the header illustrated in Figure 11.85.

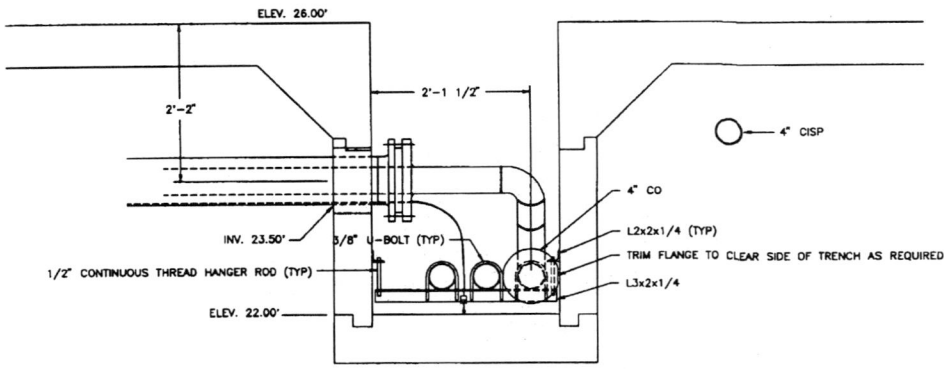

Figure 11.87. An example of single-wall underground chemical sewer headers housed in a coated secondary containment trench. (Source: Poole and Kent Co.)

In this example, there are also underground side headers feeding into the trench, each of which is individually double contained before it enters the common trench. When more than one header can run parallel to each other, it becomes economical to place them inside a common trench. Designers must consult with the authority having local jurisdiction to determine if a common means of secondary containment can be used for more than one primary pipe. Also, while open trenches are readily suitable inside a building, they lose some of their appeal outside a building. Outside building limits, rainwater will readily enter an open trench, meaning rainwater flow will have to be monitored and diverted to a waste treatment facility when contamination is detected. This can mean a very large increase in capacity for the waste treatment facility, making the use of open trenches less desirable outside a building's limits.

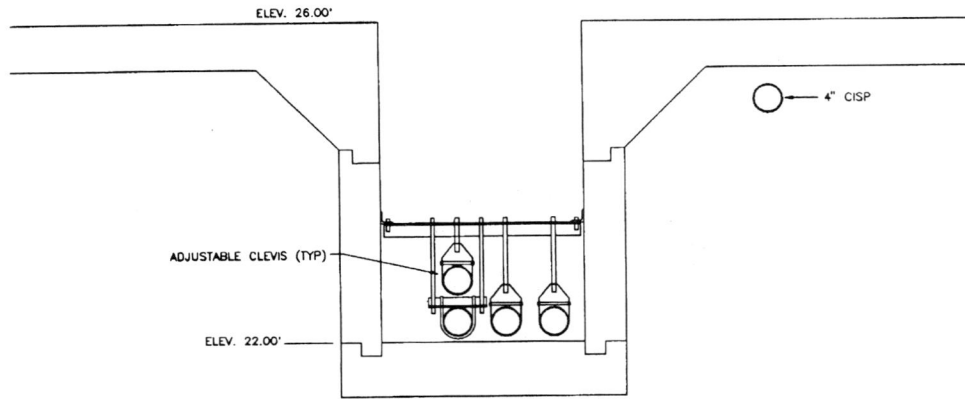

Figure 11.88. A structural supporting system used to support the pipes shown in Figure 11.87. (Source: Poole and Kent Co.)

11.3.1.6 Underground Horizontal Mains

When an underground header exits a building's limits, horizontal waste piping either continues as an underground main or ties in via branch connection to a main. Mains may collect wastes from one or more headers, possibly multiple headers from more than one building. An underground main typically carries waste to a remote site where it is processed, stored, or discharged into an industrial sewer. The main may either be a gravity-fed drain pipe, or it might operate as a pressure force main. The decision whether to move the chemical waste fluids by either gravity or pressure depends on the present and future capacity needs, the distance to be traveled and available slope.

Several drawings are needed to depict the layout of the underground mains accurately. The overall schematic of the mains, including their relation to buildings and other surface structures, should be shown on a plan view drawing, which should show all pipes, including branch connections and changes in direction of the piping. The location of fittings, manholes, and other major details should also be indicated. Any horizontal change in elevation should be shown by preparing a profile view drawing, indicating all changes in elevation of the pipe (or pipes), and the surface geography. The elevations should be indicated at various points along the drawing in order to determine burial depths, locations of manholes, etc., accurately. Whenever feasible, a combined drawing should be prepared so that it is easy to coordinate location and elevation. Both drawings should be drawn to scale in order to assist in preparing accurate takeoffs and estimates of project costs.

In addition to these drawings, detail drawings should be prepared for all unique aspects of underground mains. At least one typical detail drawing needs to be prepared for each unique section. A section is considered unique if there is any change in size, diameter (including annular space), wall thickness, materials of construction of the primary or secondary containment piping, type and spacing of interstitial supporting devices, or type of welds (bonds). If any of these variables changes, then a separate detail drawing needs to be prepared.

A typical drawing illustrating a combination plan view and profile view of an underground gravity-fed double containment main is shown in Figure 11.89.

11.3.1.7 Gradient of Internally-Supported Headers to Prevent Pocketing

If primary pipes are intermittently supported by means of internal centering supports (centralizers), they must not contain sag pockets in order to drain freely. To eliminate pockets, each downstream internal centralizing support must be lower than its upstream neighbor by an amount that depends on the sag of the pipe between them. A practical average gradient of internal support elevations to meet this requirement may be found by using the following formula.

$$G = \frac{4y}{L} \tag{11.1}$$

where:
G = gradient, in. per ft
L = span, ft
y = deflection, in.

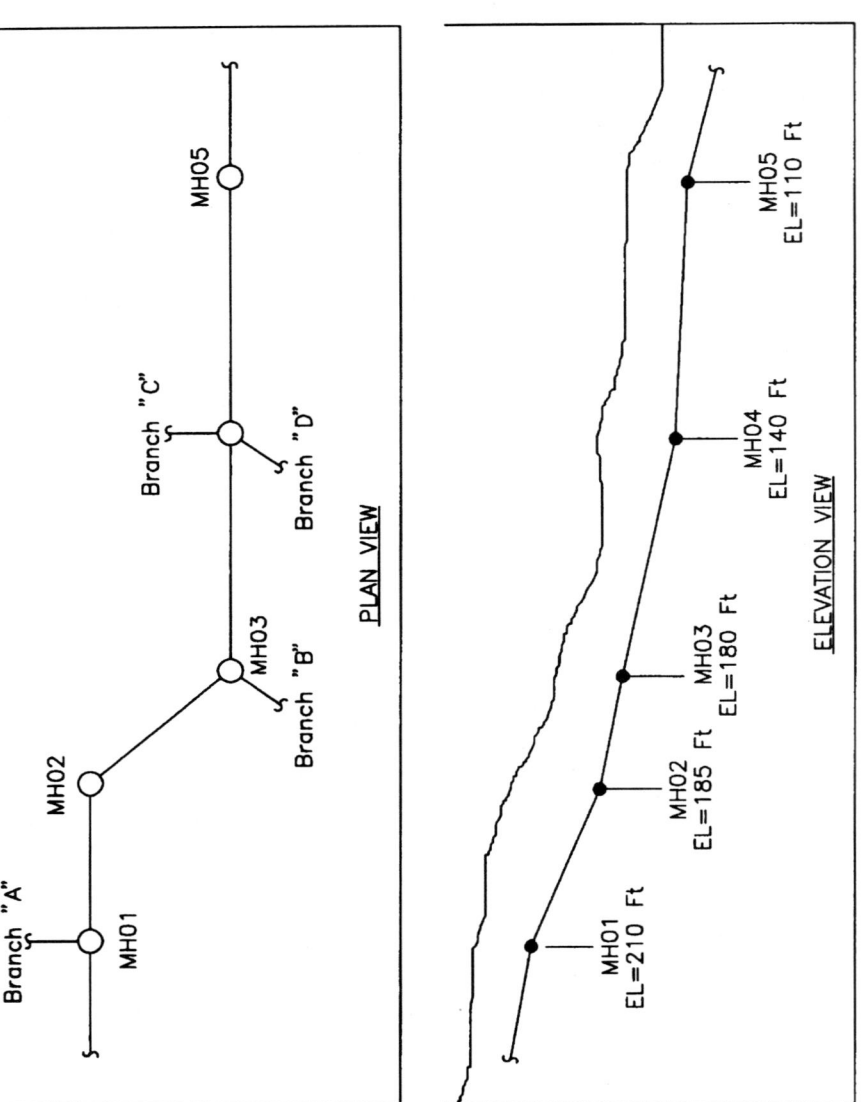

Figure 11.89. A typical combined plan and profile view drawing for an underground gravity-flow chemical sewer.

The difference in elevation between a downstream internal centering support and its upstream neighbor must be four times the theoretical deflection of the pipe between them, as determined by Equation 9.29 through 9.32 to establish the grade according to Equation 11.1. It has been suggested as a conservative measure to use twice the theoretical mid-span deflection[2] when determining the slope of the double containment pipe. If so, the elevation difference between successive supports would be eight times the theoretical mid-span. The elevation of the internal supports is equal to the invert of the secondary containment pipe.

11.3.1.8 Use of a Common Sump with Discharge into Pressure Force Mains

In Figure 11.89, the profile view indicates that the piping changes direction by 100 ft. What if a change in elevation is not available to allow the piping to experience the minimum change in elevation required to sustain the minimum required slope? The normal procedure in this instance is to drain all horizontal underground headers into a common sump and then pump the fluids involved by use of a sump pump through a pressure force main. When this method is used, the underground header should be designed as a pressure pipe, according to all of the rules of design for pressure pipes described throughout this book.

An example of a system of this type is illustrated in the plan view shown in Figure 11.90. When such a design is used, the sump may have to be provided with an integral liner and the interstitial space monitored for leaks, since it is in effect an underground storage tank. Local authorities having jurisdiction vary in their interpretation of this type of arrangement; thus, it is important to consult with them on any given project.

11.3.2 Spill-Collection Double Containment Piping Systems

Spill-collection systems are a special form of chemical drain, waste, and vent systems, the sole purpose of which is to collect spills from a tank overflow or failure, laboratory mishap, accidental hose discharge, or other event. They consist of floor drains feeding into an underground header, without any additional fixtures or stacks feeding into the system. However, some systems are designed as combination systems, where both fixtures and floor drains feed into the header, as described earlier (see Section 11.3.1.) A spill-collection header, which collects waste from various floor drains, discharges into either a holding tank or an underground main. An associated collection main collects waste from collection headers and discharges spilled fluids into a waste treatment system where chemicals await neutralization or processing.

Design criteria for sizing emergency spill systems are covered in Chapter 4 (see Section 4.4.5). The design of such systems is based upon an expected rate of collection of fluids and an estimate as to the frequency of mishap. Oftentimes local building codes govern the sizing of such systems, the spacing of drains, etc. The layout and drawing criteria follow the exact same guidelines as those of a typical chemical waste system. The details of spill collection header systems should be illustrated using orthographic drawings, with the position of all floor drain locations indicated.

In most applications, spill-collection floor drains are attached to a P trap before discharging into associated underground headers. This is required in order to prevent dangerous sewer gases from backing up into the above-ground environment whenever possible. The details of connecting riser pipes between P traps and floor drains should be indicated in a detail drawing. Figure 11.91 illustrates a typical arrangement for connecting a floor drain

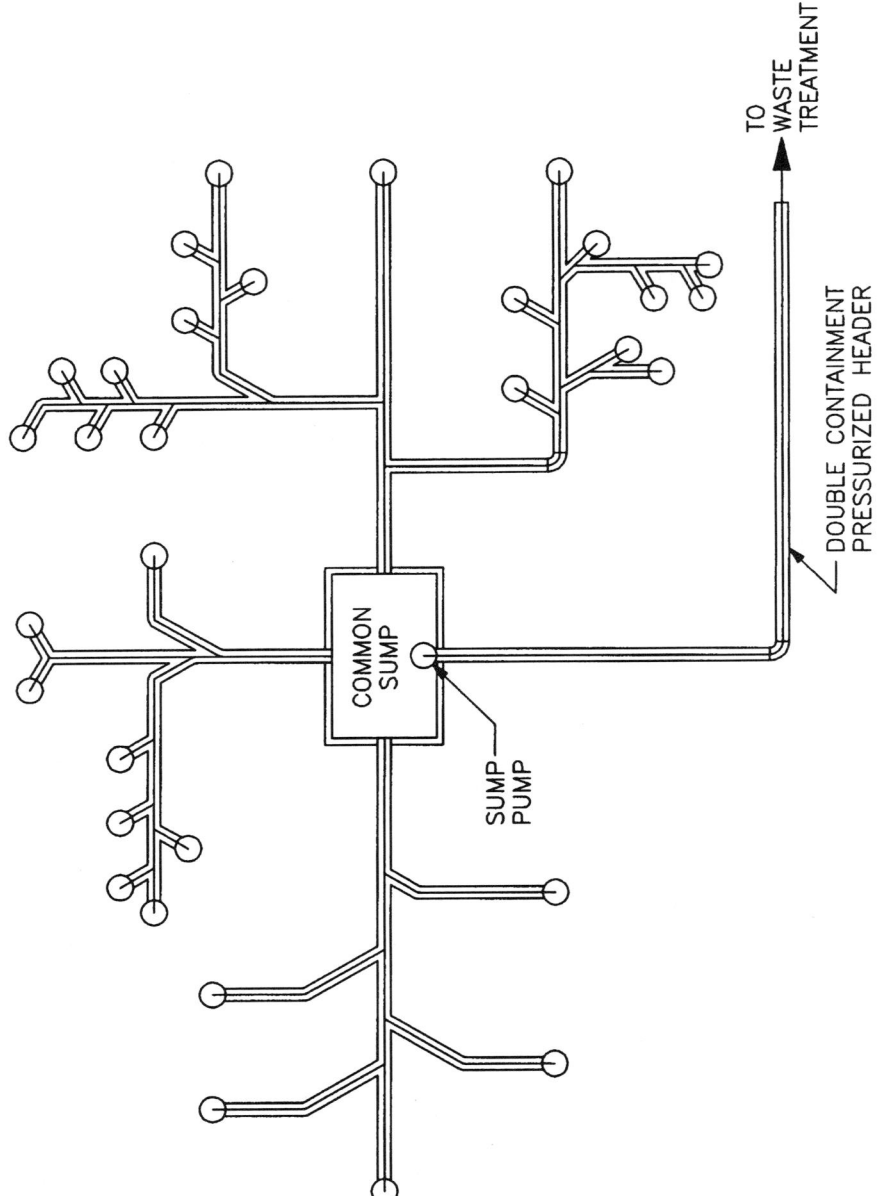

Figure 11.90. A typical common sump system with a pressurized discharge header.

Layout Concepts for Double Containment Piping

to an underground header by means of a riser/P trap/45° lateral with 1/8th bend combination. The corresponding photograph in Figure 11.92 shows the appearance of a similar arrangement after fabrication in a retrofit application.

Figure 11.93 illustrates an arrangement for connecting a floor drain to the underground header by means of a simple 90° bend, as may be used if backup of sewer gases is not a possibility. At each and every floor drain, there will be a set dimension required from the top of the drain to the centerline elevation of associated horizontal headers into which it ties. The reason that this dimension is critical at each and every manhole is that the accompanying horizontal header into which it ties must maintain a constant, fixed slope (i.e., a minimum of 1/8 in. per ft; 3 mm per 0.3 m).

Designers are cautioned that the minimum required space between the top of floor drain and the centerline of the underground header is set by the size of the secondary containment materials. Since the materials are much larger than corresponding single-wall component sizes, problems can develop due to a required dimension for a project that is less than the minimum needed actually to fabricate such an arrangement. In most applications, there is a greater likelihood of this occurring in the very beginning of the header since the line has not yet had a chance to slope to a significant depth.

One complication of using a double-contained P trap in conjunction with floor drains is that there are very limited means available to provide a cleanout. It is normal to have a cleanout positioned at the bottom of a P trap according to standard plumbing practice. This adds complexity to the design and fabrication of double containment P traps. It also would be very difficult to access in order to use the cleanout in a directly buried system, even if it were to be provided as part of the P trap. Therefore, consideration should be given to placing double-contained P traps with cleanouts (inner and outer cleanouts) in an accessible sump. One other viable alternative is to use single-wall P trap components and place the single-wall components in a lined and accessible sump. The piping secondary containment zone can then begin at the downstream exit to the P trap, prior to the piping exiting the lined secondary containment sump.

As a last point regarding emergency spill systems, remember that these systems will have limited access. Therefore, it becomes critical to make sure that materials of construction of secondary containment components can withstand leaked chemicals for an extended period of time (this is required under RCRA in the United States). The photograph in Figure 11.94 shows an example of how limiting a factor this can be. Although not shown in the photograph, the drain in this installation is located beneath a process reactor vessel. If a repair of a drain system like this must be made, oftentimes the entire process must be shut down for safety, meaning lost revenue. Once these systems are installed, they will not be readily accessible, as the photograph clearly shows.

11.4 Layout Details of Double Containment Drainage Fittings

Most nonpressure fittings are designed to handle relatively low internal pressures, and are instead designed to aid in smooth drainage of liquids. However, it is essential that all present and future requirements be considered in setting the design conditions. Many times, internal pressure ratings will need to be greater in the future as a system may require the addition of pumps to increase flow capacity, particularly in underground mains. Also, a system test pressure that is set may actually be a controlling factor in determining the

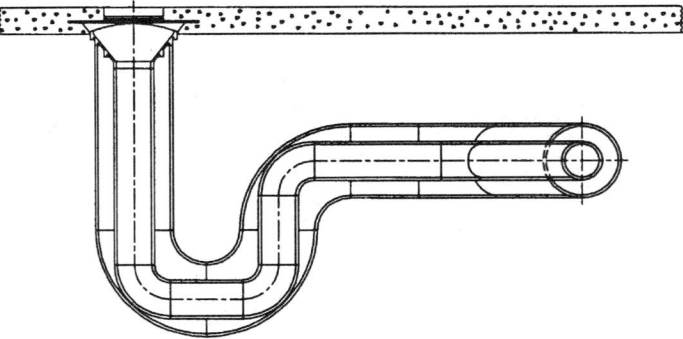

Figure 11.91. A typical detail from a floor drain to a header involving a riser to P trap to 45° lateral with a 1/8th bend combination.

design of fittings. While it is important to design a test that assures an owner of a quality system, care should be taken not to design a test that is overly conservative and unrepresentative of a system's intended use.

Laterals typically present unique problems due to the difficulty in fabricating double-contained branch connections. Fabrication of double-contained laterals from standard off-the-shelf parts is not possible for many sizes and types of materials. Therefore, double containment laterals are commonly fabricated from straight pipe (see Section 11.1.1.6). The design of lateral branch connections from pipe should involve careful consideration of details, following the reinforcement rules stated in Sections 5.2.1.3 and 6.3.4.1., if necessary. For nonmetallic parts and combination components involving multiple materials, the same basic philosophy should apply. However, the testing and analysis requirements stated in Chapters 5, 6, and 7 for unlisted components should be strictly adhered to. Many problems can result due to a lateral fitting design that is fabricated without following standard design rules, common methods of design verification, and good common sense.

Figure 11.92. Photograph of a floor drain to a header involving a riser to a P-trap to a 45° lateral.

Layout Concepts for Double Containment Piping

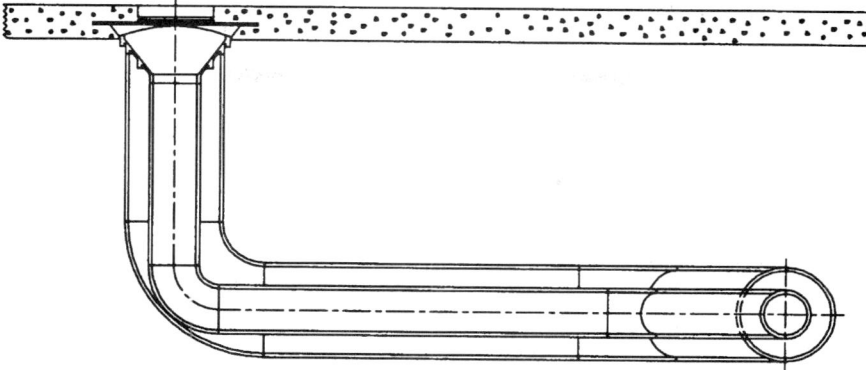

Figure 11.93. A typical detail from a floor drain to a header involving a riser to a 90° elbow to a 45° lateral with a 1/8th bend combination.

11.4.1 Above-Ground Fittings

Above-ground double containment chemical waste nonpressure fittings consist of typical drain, waste, and vent pattern fittings as shown in ASTM D3311, for nonmetallic fittings. These fittings consist of long and short sweep bends, laterals, laterals with 1/8th bends, sanitary tees, crosses, and other drainage fittings. These fittings are designed to allow a smoothly flowing fluid, which flows due to gravity.

Most systems involving above-ground chemical waste DWV fittings are subject to nearly constant temperatures. When this is true, fittings can be designed in a manner such that primary components are permanently and rigidly fixed to secondary containment fittings by means of supporting plates. However, if thermal expansion is a consideration, fittings

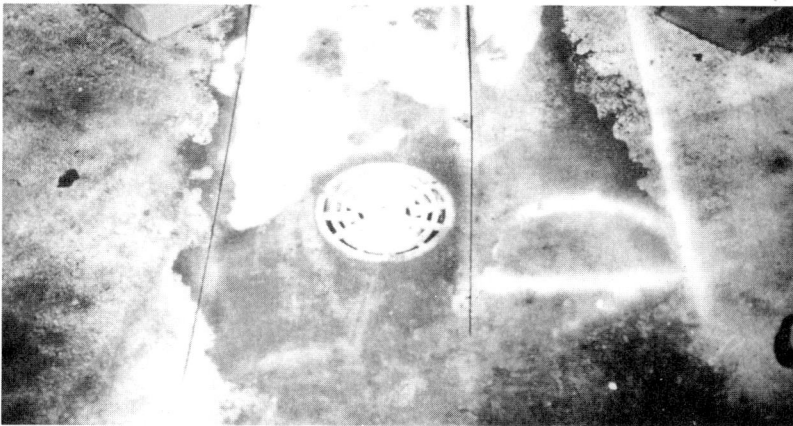

Figure 11.94. Photograph of an underground chemical sewer system in a pharmaceutical plant after installation. All that can be seen is the floor drain opening.

must be designed such that they have sufficient clearance and flexibility so that they are not overstressed. In some applications, systems and their components may be designed in a permanently restrained fashion, if the internal pressures are low, and the thermal expansion is not excessive. A simple or comprehensive stress analysis can be performed to determine if this is allowable. Where systems cannot be designed in a restrained manner due to extreme design conditions, the use of interconnecting plate or fin-type supports or discs as part of the fitting design must be avoided.

11.4.2 Underground Fitting-in-Manhole Arrangements

In nonpressure and pressure systems that are installed underground, it is common to avoid direct burial of double containment fittings. Most direct burial of double containment fittings is limited to bends and elbows. The direct burial of branch fittings is often avoided due to the difficulty in installing these fittings underground, leak-detection concerns, and other reasons. Where there is a substantial amount of thermal expansion and contraction expected, the installation of double containment fittings underground of all types is usually avoided.

A common design approach for underground system (pressure and nonpressure) layout of fittings is to place single-wall fitting connections inside secondary containment manholes. In most situations, it becomes more economical to terminate secondary containment piping in the entry to a secondary containment manhole, thereby making use of single-wall fittings. Manholes, if constructed of lined or protected concrete, effectively serve as a means of secondary containment. Any secondary containment piping penetrations can be made water tight by the use of mechanical seals, by welding secondary containment piping to manholes (if a compatible lining material is used), or by using a water-stop with grouting applied around the outer pipe circumference. Grouting can also be applied on the inside, outside, or both.

If manholes are designed to be readily accessible, primary piping fittings can be joined by using flanges. This greatly aids in the installation of such items, and makes maintenance easier. In addition, expansion joints can be incorporated in the layout, being positioned

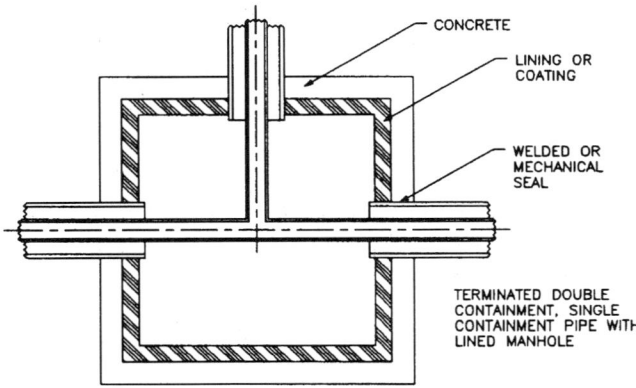

Figure 11.95. Plan details for manhole MH01 from Figure 11.89.

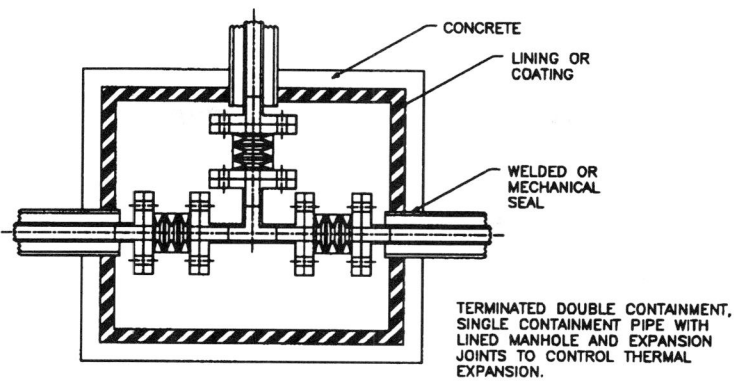

Figure 11.96. Plan details for manhole MH01, from Figure 11.89, showing optional installation of expansion joints.

between the fittings and the pipe flanges. This can serve as a means of alleviating thermal expansion in the straight pipe sections (if the primary piping is designed in an unrestrained manner) and serves as a means of isolating fittings and complex fitting arrangements. As examples of possible designs, details of underground fitting manhole arrangements, for manholes MH01 and MH02 that are referenced in Figure 11.89, are shown in Figures 11.95 through 11.99.

When using a lined or coated concrete structure as a means of secondary containment, it is necessary to incorporate some means of leak detection in such structures. This is due to the fact that single-wall fittings in such structures, or their flange connections, may be a source of leaks. Furthermore, double containment pipes that have a non-open-ended annulus at the entry to such structures may also have some means of leak detection added to the straight pipe sections as well. In the example illustrated in Figure 11.89, and Figures 11.95 through 11.99, the manholes could, for example, be equipped with both liquid level sensors and conductivity or pH probes as a means of detection. The associated double containment pipes

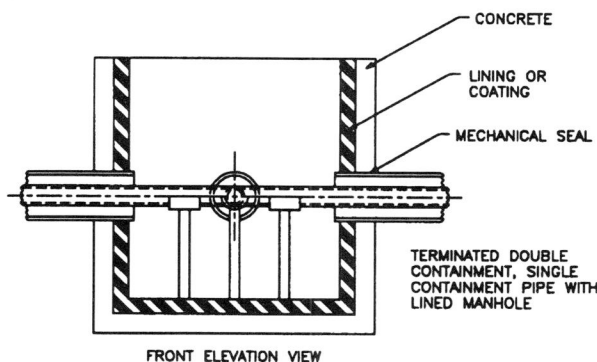

Figure 11.97. Front elevation details for MH01 from Figure 11.89.

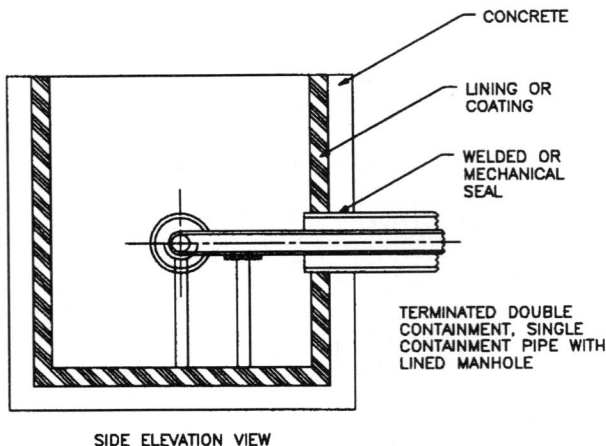

Figure 11.98. Side elevation details for MH01 from Figure 11.89.

are sealed by means of a gland-type termination expansion seal (see Figures 11.12, 11.15, and 11.16). They are also affixed prior to the termination with a drip leg and multiple leak detection capability (e.g. liquid level sensors, visual detection and manual backup/low-point drains, reference Section 15.2.2).

As an alternative design, the arrangement shown in Figure 11.100 may be used. In the alternative arrangement, the fitting area that is positioned in the manhole is part of an "uninterrupted" double containment piping zone. The arrangement may be positioned as such to allow for easier installation of the fitting or fittings, incorporation of leak-detection connections, future easy access, etc. When this type of arrangement is used, a manhole does not always have to be lined or coated, nor does it have to be provided with its own leak detection. The arrangement shown in Figure 11.100 is only partially constructed; to complete the arrangement, the pipe penetrations must be protected from ground water intrusion by sealing or grouting by any of a variety of means.

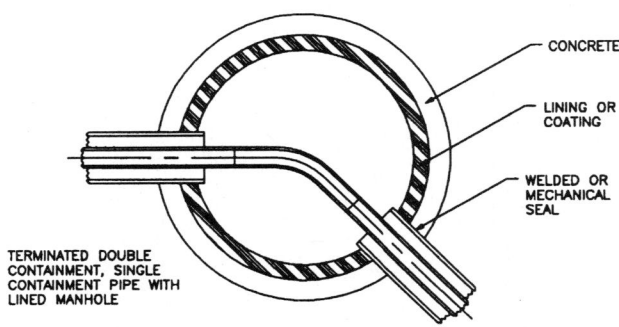

Figure 11.99. Plan details for manhole MH02 from Figure 11.89.

11.4.3 Dimensional and Material Information

Enough information should be clearly identified on fitting drawings to ensure that a proper combination of parts is supplied and fabricated. This information should include construction materials, including the type and grade of the individual product and manufactured brand, if important. Wherever possible, the grades of material should be related to an ASTM specification or applicable European standard such as ISO or DIN.

Dimensional information to be included must include all applicable diameters, lengths, and thickness specifications. All applicable tolerances should be stated as well, with care taken not to give redundant or conflicting tolerance data. The temperature basis at which dimensional information should be measured should be stated as well. This is particularly important for nonmetallic materials, as they can show substantial changes in dimensions upon wide ambient temperature changes. Whenever possible, the dimensioning practices stated in some consensus code or standard (i.e., ASME, ANSI, etc.) should be used as a suitable basis for providing dimensions.

11.5 Petroleum Station Double Containment Piping Layout

Petroleum station piping layout is controlled in many municipalities by strict building and fire codes. These codes often set minimum requirements for space allowances, isolation from structures, etc. Most underground systems consist of several tanks that have filling connections that extend above ground. These tanks also have individual discharge piping that is connected to above-ground service pumps, whereby the gasoline, gasohol, or diesel fuel is discharged into vehicles. Typical layouts for service station piping is illustrated in Figures 11.101 and 11.102.

One of the greatest concerns for rigid discharge piping that connects tanks to above-ground service pumps is that such piping often must involve a "crossover" of one or more lines. An example of crossover details is presented in Figure 11.103. Usually there is a minimum allowable space that is required by the materials in order to operate the pipes safely. Pipes that actually touch each other at crossover locations can lead to one pipe causing

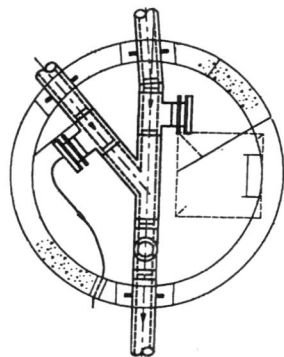

Figure 11.100. A plan view of a manhole detail whereby the piping components housed within feature continuous double containment.

abrasion on the other. Also, in the case of steel pipes, galvanic corrosion could occur due to the contact of the two pipes. In many municipalities, there are minimum crossover requirements established by the local codes for underground storage tank associated piping. Another requirement for the supply pipes from the storage tanks is that they be isolated from connections to the shear valve of the associated pumps to prevent damage due to vibration or cyclic usage. This is usually accomplished by means of a flexible braided-hose connection and swivel joint contained within a flexible secondary containment "boot" or corrugated hose. Figure 11.104 illustrates the more common arrangement consisting of the flexible braided hose within the corrugated hose.

An alternative layout for fueling facilities has been developed recently which uses innovative flexible primary hoses housed within flexible corrugated secondary containment piping systems. The hoses that have been developed consist of a multiple-composite material that is resistant to a variety of fuel grades, and as a result, these flexible hose systems, like their rigid RTRP counterparts, have gained U.L. listing (U.L. 971, "Nonmetallic Underground Piping for Petroleum Products"). An example of a flexible multi-composite hose suitable for use in fuels is shown in Figure 11.105. Flexible primary hose systems offer the installer several advantages over rigid piping alternatives. First, pipe sections can be completely installed in seamless lengths from start to finish, thereby minimizing field joints as compared to rigid piping. The end connections are typically made by means of mechanical couplings. Secondly, the secondary containment system can be completely installed and tested first prior to installing the primary pipe, in most cases. Most importantly of all, the primary pipe can be extracted and replaced in the event of a leak, without having to make a major excavation and repair, which is a major

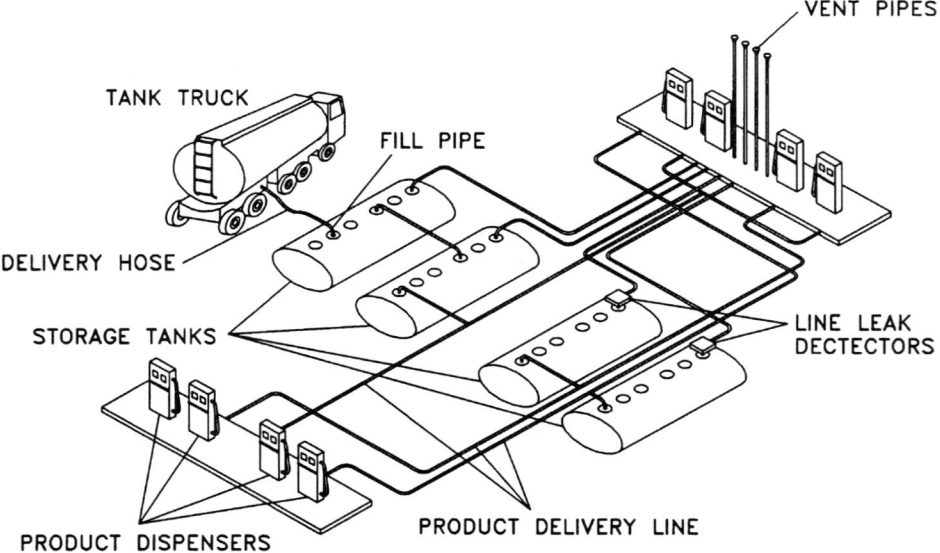

Figure 11.101. Typical four-tank service station layout. (Source: U.S. EPA – Musts for UST's.)

Layout Concepts for Double Containment Piping

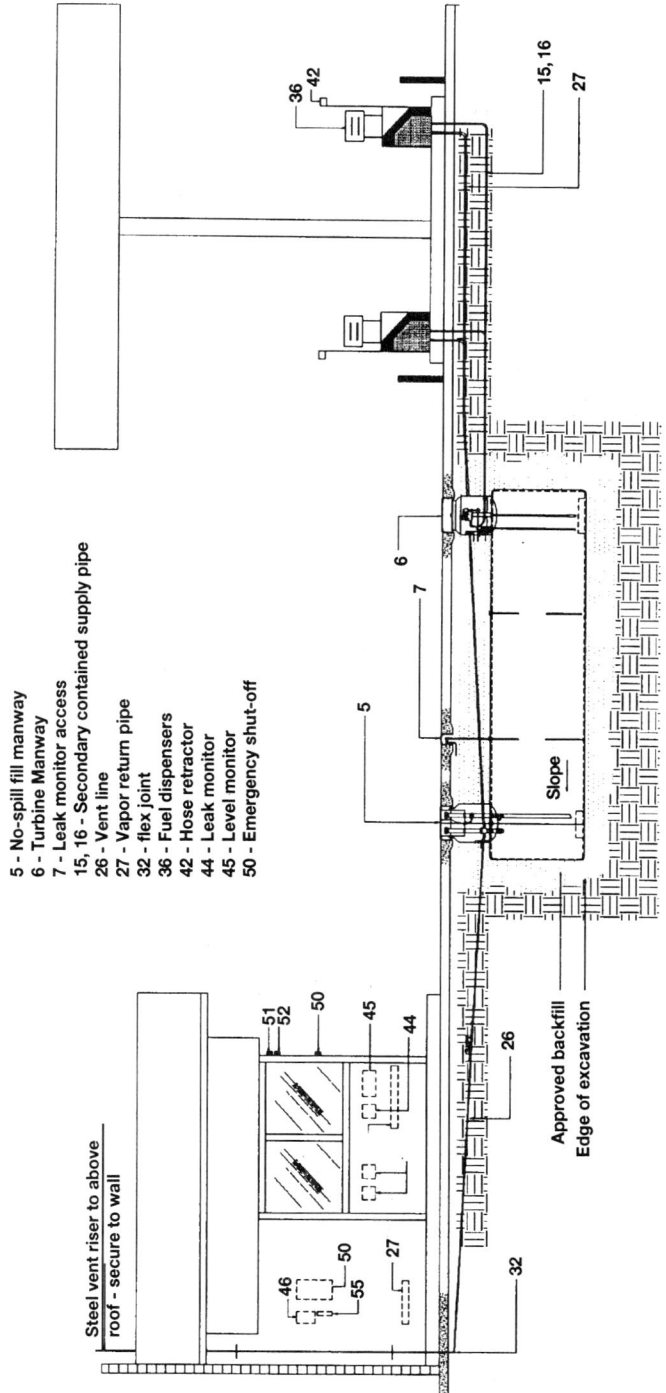

Figure 11.102. Typical service station details involving secondary containment and vapor return, meeting present day standards. (Source: Joor Manufacturing.)

Containment Pipe (in.)	A (in.)	B (in.)	C (in.)	D (in.)	E (in.)
3	7.5	10.6	6.5	10	14.5
4	8.5	12	7.5	12	16.5
6	10.65	15	8	14.75	22

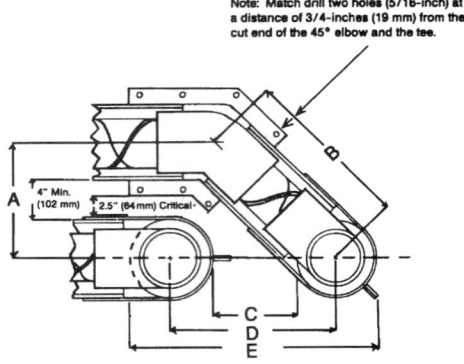

Figure 11.103. Typical crossover details for rigid piping layout. (Source: Smith Fiberglass Products Co.)

consideration for the future operation of a fueling facility. An example of inserting a flexible primary hose product into the secondary containment flexible corrugated hose, through the entryway in a tank sump, is shown in Figure 11.106. Flexible hose systems are laid out in series, as opposed to a traditional branched layout for rigid RTRP piping.

Other pertinent details for petroleum station piping layout are shown in Chapter 14 (see Section 14.9). This includes various details such as pipe connections to tank sumps, and other related details.

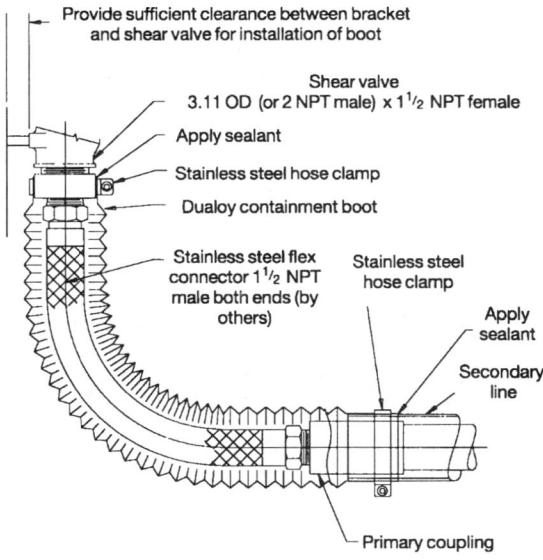

Figure 11.104. Flexible hose connection contained within a flexible secondary containment boot. (Source: Ameron Fiberglass Pipe Division.)

Layout Concepts for Double Containment Piping

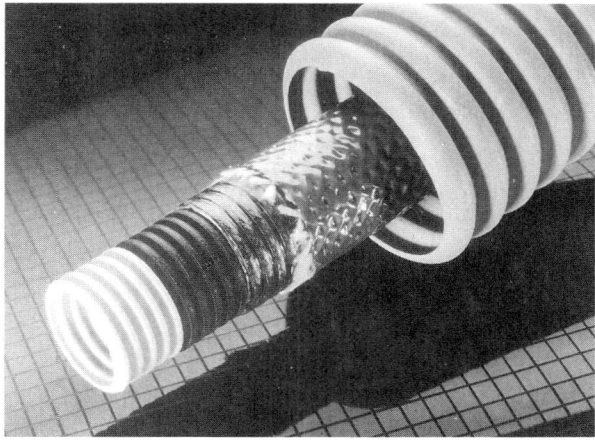

Figure 11.105. Example of a flexible multi-composite U.L.-Listed hose product suitable for use as a primary pipe to carry fuels and other fluids. (Source: Total Containment, Inc.)

11.6 Multiple-Primary-Pipe/Common Secondary Containment Pipe Systems

Multiple-primary-pipe systems (housed within a common secondary containment pipe) present additional layout complications, beyond the ordinary complications discussed in this chapter for a single-wall pipe within a single secondary containment pipe. Many of the complications arise from the increased importance of the orientation of the primary pipes within the common secondary containment pipe. In such systems, the following

Figure 11.106. Example of a flexible-primary pipe being inserted into the already-installed secondary containment hose system through a tank sump. (Source: Total Containment, Inc.)

Figure 11.107. Cross-sectional view of a two-pipe multiple primary pipe system housed inside a common secondary containment pipe.

factors have to be well thought out: (1) secondary containment sections must be initiated; (2) primary pipes have to follow bends; (3) in some systems they pick up branches along the way; and (4) secondary containment sections are eventually terminated. Throughout a system of this type, the pipes must maintain their relative positions within the cross-sectional plane. Figure 11.107 illustrates three possible cross-sectional arrangements for two-pipe, three-pipe, and four-pipe situations, respectively. Each of these arrangements is for applications where the primary piping is of the same size diameter. For those applications where the primary pipes vary in diameter, the arrangements become increasingly complex.

Referring to Figure 11.107, it can be seen that if one of the pipes must be added midstream, as opposed to all the pipes being initiated together, the position at which it enters the line through the branch is critical. Additionally, it should be pointed out that piping that enters a branch midstream will require special termination plates, the design of which depends on the number of pipes and their positioning. As an alternative to having special termination plates, the lines may enter the system through a rotated branch fitting. It may then enter the branch through the centerline of the branch, and turn 90° to initiate its point

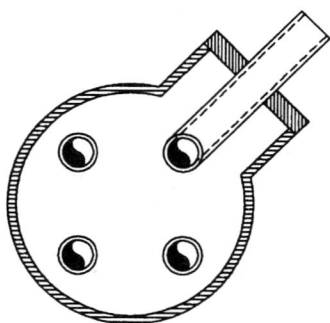

Figure 11.108. Example of the use of a rotated second containment tee fitting with concentric cap to allow for the concentric entry of the third line in a three-line primary piping system.

Layout Concepts for Double Containment Piping 475

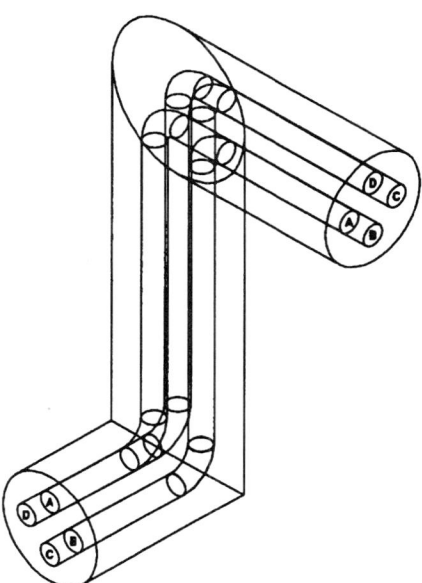

Figure 11.109. Example of multiple primary pipes that change their orientation in the cross section of the bundle after a change in direction. (Source: Reprinted with permission from **CHEMICAL ENGINEERING**, Sept. 1991, Copyright 1991 by McGraw-Hill, Inc., with all rights reserved.)

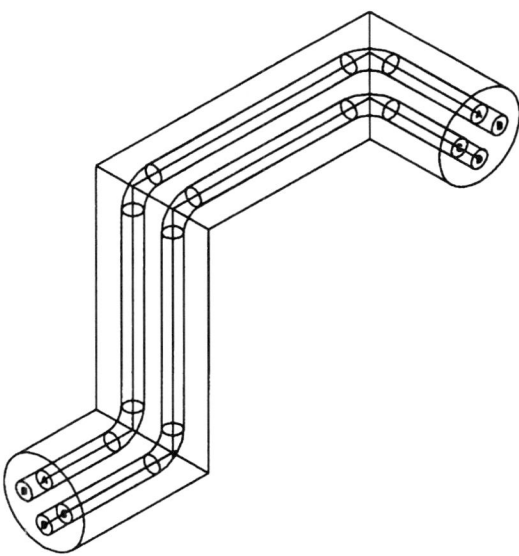

Figure 11.110. Example of correction for "rotation" by the addition of a third elbow. (Source: Reprinted with permission from **CHEMICAL ENGINEERING**, Sept. 1991, Copyright 1991 by McGraw-Hill, Inc., with all rights reserved.)

in the secondary containment zone. An example of a rotated tee secondary containment branch to allow the concentric entry of the third line in a three-line primary piping system is shown from the cross-sectional view in Figure 11.108.

Multiple-pipe systems by their nature require additional detailed drawings to be prepared as compared to single-primary-piping double containment systems. This is due to the fact that there are many more details to consider in order to create successfully installed system. The details should include enough information to determine the method and spacing of supporting devices, termination and initiation arrangements and parts, welds (bonds) at each of these areas, and any other instructions the fabricator might need. The additional costs of fabrication, engineering, and capital costs of enlarged secondary containment pipes make multiple-pipe systems more expensive than running a series of individually contained primary pipes, in many situations. They also tend to be difficult to repair or modify. Therefore, secondary containment of multiple pipes should only be considered where space is limited. Common secondary containment of underground multiple pipes may be better accomplished with the use of lined tunnels (manways or walkways) or lined trenches. An example of a common secondary containment trench which is designed to house multiple primary pipes is shown in Figures 11.87 and 11.88. The design and construction of lined concrete structures as a means for secondary containment is also discussed in Chapter 16.

11.6.1 "Rotation" in Multiple-Primary-Pipe/Common Secondary Containment Pipe Systems

An issue that arises in the design and layout of multiple-primary-pipe systems is called "rotation." In "rotation" of multiple-primary-pipe systems, the pipes change their relative orientation in the cross section of the bundle after a change in direction.[3] This will occur whenever a vertical pipe directional change occurs, followed by a perpendicular change in direction. This is illustrated in Figure 11.109. If each of the four lines is followed, it can be noticed that they all change orientation.

Rotation in and of itself is not a bad situation. However, it can be an undesirable feature if the carrier pipes are to maintain their original positions, due to equipment connections, maintaining positions for branch connections, etc.[3] To prevent rotation, a third elbow can be added into the system, as illustrated in Figure 11.110.

11.6.2 Types of Multiple-Primary-Pipe/Common Secondary Containment Pipe Systems

The complexity of a multiple-primary-pipe systems depends on the complexity of the primary pipe configuration and the makeup of the outer casing. The primary pipe configuration will fall into one of the following four categories: (1) systems with primary pipes of the same diameter and the same material; (2) systems with primary pipes that have the same diameter, but are constructed of different materials; (3) systems with primary pipes that have different diameters and are constructed of the same materials; (4) systems where the primary pipes vary in both diameter and materials of construction. The complexity of the system also depends upon the material used for the common secondary containment pipe (outer casing) in relation to that of the primary pipes.

11.6.2.1 Same Material/Same Size Primary Pipes

This type of system is the least complex of all possible arrangements in terms of fabrication and installation. At least one set of detail drawings needs to be prepared for each unique termination and initiation arrangement, supporting device, fitting assembly, and typical straight pipe section.

11.6.2.2 Combinations of Primary Pipe Material/Same Size

In systems where the materials vary, but they are all of the same actual dimensions (i.e., outside diameter, wall thickness), additional details must be added into fabrication drawings to explain clearly the required methods and filler materials for making dissimilar welds or bonds. Typical areas where this would be required are where the pipes are attached to supporting devices and where initiation and termination connections are made. The designer should be aware of the potential for galvanic corrosion any time dissimilar metals come in contact with each other in a system of this type.

11.6.2.3 Combinations of Primary Pipe Size/Same Material

In systems where the diameter sizes and/or wall thicknesses of the primary pipe vary, detail drawings need to be prepared as described in Section 11.6.2.1. Details must be added to specify the location and size of penetrations. Typical locations are at support plates and in initiation and termination assemblies. The orientation of the pipes in all straight pipe sections, initiation and termination areas, and fitting assemblies must be clearly shown. The design of support plates, baffles, and termination plates must reflect the required orientation.

11.6.2.4 Combinations of Materials and Sizes

The requirements of all three of the previous three paragraphs must be met. This situation typically requires many additional detail drawings to be prepared as compared to the other situations mentioned. It is the most complex type of multiple system to design, fabricate, and erect.

11.6.2.5 Variation of Secondary Containment Piping Material from One or More of the Primary Pipes

This situation can be present in any one of the four cases mentioned in the preceding four paragraphs. When this situation is applicable, welding (bonding) complications can arise depending on the choice of materials for the supporting devices and/or termination and initiation plates. All of the pertinent details of the types and location of welds (bonds) and the filler materials to be used must be clearly spelled out in the detail drawings. The designer should be aware of the potential for galvanic corrosion any time there are dissimilar metals that come in contact with each other in a system of this type.

11.6.3 Fitting Layout for Multiple-Primary-Pipe/Common Secondary Containment Pipe Systems

Fittings in a system of this type should be entirely prefabricated in a shop environment. If fitting assemblies are to be correctly fabricated and subsequently installed correctly into the remainder of the system, a set of detailed shop fabrication drawings needs to be prepared.

The drawings should include all necessary views and all pertinent dimensional and tolerance information. Additionally, all construction materials should be clearly indicated, including weld (bond) requirements (i.e., filler materials, type, and location). Every unique aspect of the assembly should be identified by a numbering system. A detailed takeoff of all materials involved in the fabrication should be described and totalled in an accompanying table. The proper orientation of primary pipe in all pipe sections and fittings assemblies should be clearly identified. Separate machining drawings should be prepared for any item of the fabricated assembly that requires machining. All pressure-retaining welds (bonds) should be completely examined and tested at the site of prefabrication, according to the required project specifications, as a means of qualifying the component for acceptance.

11.6.4 Termination and Supporting Components for Multiple-Primary-Pipe/Common Secondary Containment Pipe Systems

All termination components, assemblies, and internal supporting devices should have detail drawings prepared for their machining and/or fabrication. Detail drawings should include details of penetrations to be made for pipes that will penetrate the components/assemblies being drawn. The applicable orientation and tolerances of such penetrations must be clearly identified. All associated pipes must be identified, and details of the types of welds (bonds) to be used should also be indicated. Where it is appropriate, the type of filler material, solvents, or adhesives to be used to attach pipes through their penetrations should be identified.

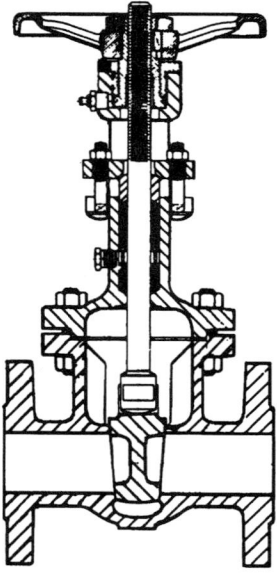

Figure 11.111. Typical stainless steel class 300 gate valve. (Source: Crane Co.)

11.7 Valves and Valve Layout for Double Containment Piping Systems

Valves present unique layout concerns for double containment piping systems. By their nature, they are designed to be operating mechanical devices. Their operation can be either by manual or some automatic means (i.e., pneumatic or electrical operation). Either way, they require a considerable degree of access, for either operation or maintenance purposes. Various valve types commonly encountered in chemical piping are discussed in this section, along with their associated layout alternatives.

Wherever possible, valve use should be maintained outside zones of pressurized secondary containment. By maintaining valves and other high-maintenance/operating mechanical items outside zones of secondary containment, added system complexity may be avoided. Whenever a valve must be added to a primary pipe system that is required to have pressurized secondary containment, there will be added expense. The added expense arises from the need to contain the valve in such a manner so as to enable ready access to its secondary containment housing, yet maintain its containment in a pressure-tight manner between times when access is required. Since valves are common sources of leaks and vapor emissions, additional and frequent maintenance of a double containment piping system will be required due to the increase in detected leaks. Thus it is much simpler and more cost effective to maintain a primary pipe system valve outside a zone of pressurized secondary containment.

Valves may still be secondarily contained; they cannot just be easily and inexpensively housed and still meet the necessary requirements. A common way to provide containment for such a valve is by the use of a lined or coated concrete dike, berm, or building floor in an above-ground application and by the use of a lined or coated concrete sump or accessible manhole in an underground application.

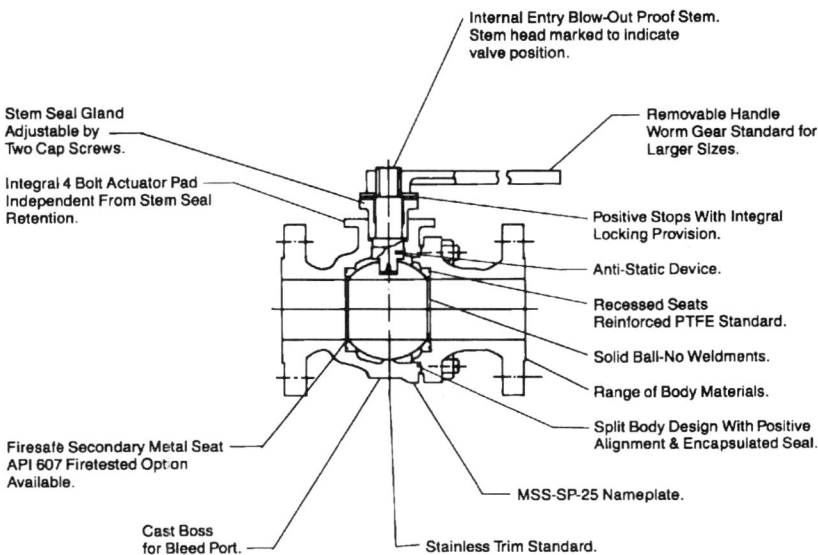

Figure 11.112. Typical multi-piece "floating" ball valve. (Source: Newco Valves.)

Valves may also be required as part of the secondary containment piping system (i.e., vent and drain valves, leak-detection and instrumentation isolation valves, etc.). These valves are usually small diameter (< 2 in.; 63 mm) and of the quarter-turn (on-off) variety. They are usually not required to be contained; however, as part of an underground system, they must be positioned inside a concrete housing (which does not normally have to be lined) in order to be accessed and operated, and to prevent their being directly buried.

11.7.1 Types of Valves

Valves can be grouped into five basic types, according to their function. These are: (1) valves for on-off service; (2) throttling valves; (3) backflow prevention valves; (4) mixing valves; and (5) other miscellaneous valves (including control valves, solenoid valves and pressure relief valves).[4] Each of these valve types is described in the following.

11.7.1.1 On-Off Valves

11.7.1.1.1 Gate Valves

The gate valve is among the most widely used type of valve in on-off service. Gate valves impose a smaller pressure drop than any other style valve due to their straight-through flow design. A typical gate valve is illustrated in Figure 11.111. Many variations of gate valves exist and can be divided into two general categories according to the type of fluid-control element used. These are the wedge-gate and the double-disk-gate styles. Wedge-gate styles include three basic variations: (1) plain; (2) solid-wedge style; and (3) flexible-solid-wedge style. Also included as a wedge-gate style is the split-wedge ball-and-socket type (sometimes classified as a double disk). The second major category is the double disk,

Figure 11.113. Typical thermoplastic ball valve. (Source: Nibco, Inc.)

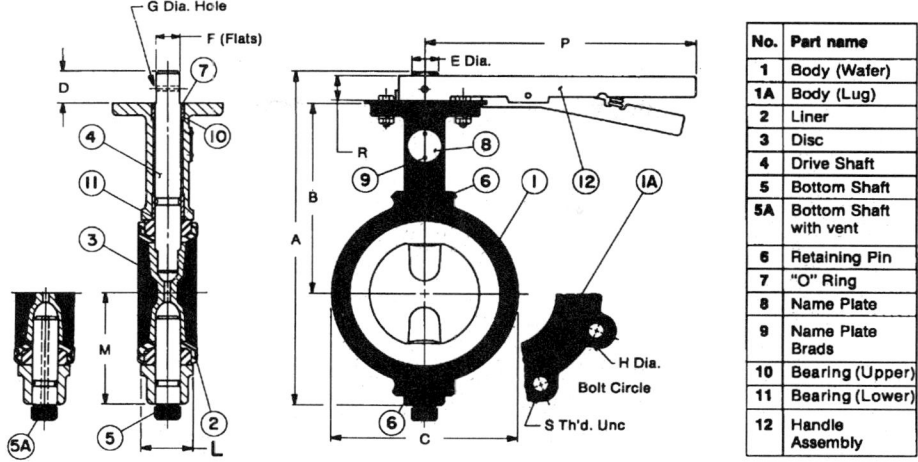

Figure 11.114. Typical stainless steel butterfly valve. (Source: Grinnell Sales Company)

whereby two disks are forced apart against parallel seats by a mechanical spreader device at the point of closure.

A gate valve must have seals located at four primary locations. These are: (1) at their end connections; (2) at their bonnet to body joint; (3) around their stem; and (4) at all wetted points of the wedge. The last item is to prevent flow leakage of the internal fluid when the valve is in its closed position. Gate valves typically require a lot of space for installation and must be well supported. Both of these items are a real handicap for installation within a secondary containment chamber.

11.7.1.1.2 Plug Valves

Plug valves are the oldest type of valve; they are a refinement of the simple cock. Plug valves consist of three basic parts: (1) plug; (2) body; and (3) cover. Plugs can be of either a tapered or a cylindrical shape and always contain a flow passage in at least one direction. In a single-port plug valve, the ball rotates through 90° to change from a fully open to a fully closed position. Multiport plug valves also exist, with a more complex flow opening geometry, and can rotate as much as 270°. Single-port plug valves are available in round-opening, regular, venturi, and short patterns.

Plug valves have sealing requirements that are very similar to those of gate valves. Stem seals are typically made with the use of "O" rings, by means of an injected fluid packing, or by mechanical seals. Other considerations for plug valves are whether or not it is of a "lubricated" or "nonlubricated" design. Plug valves take up very little physical space and weigh less than most valve types. This gives plug valves an advantage over other valve types in double containment applications.

11.7.1.1.3 Ball Valves

Ball valves began to find widespread acceptance in the chemical industry during the 1960s. They have continued to grow in popularity ever since, particularly in corrosive services.

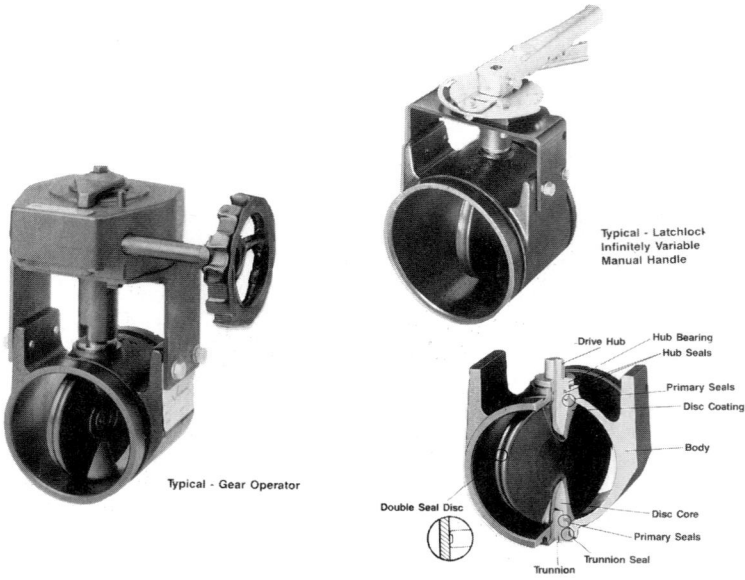

Figure 11.115. A unique grooved end butterfly valve (shown here is the Victaulic Vic-300). (Source: Victaulic Company of America.)

This is due to the manufacture of many valves that are constructed from corrosion-resistant materials, the development of all nonmetallic ball valves, and the development of linings of corrosion-resistant materials for valves having metallic bodies.

The ball valve is similar to a plug valve. Instead of a plug, it has a ball that contains at least one hole for fluid flow when it is in the open position. Normally, a ball valve is designed to rotate 90° from the fully open to the fully closed position. Ball valves can come in venturi, full port, and reduced port patterns, and are available in multiport style as well. In a single-port ball valve with full port design, the ball valve has approximately the same pressure drop as a gate valve of a similar size. A ball valve that has a "floating" design which allows the ball some freedom to move along the axis is shown in Figure 11.112.

Ball valves make use of either O rings or conventional packing to provide stem seals. The attachment of a ball valve to adjacent piping is similar to other piping or fitting components. Ball valves may be joined to piping by flanging, threading, or directly welding the valve to the pipe, with or without union connections. Ball valves normally have some type of seating device on at least the downstream side of the ball valve. On valves designed to operate at low differential pressures, the seats are normally constructed from precompressed plastic materials. Some nonmetallic valves also have elastomeric O rings behind the seats to reduce the operating torque.

A Typical thermoplastic ball valve configurations are shown in Figure 11.113. Ball valves are compact and lightweight compared to many other valve types, such as gate valves. Therefore, they are readily suitable for use in double containment services.

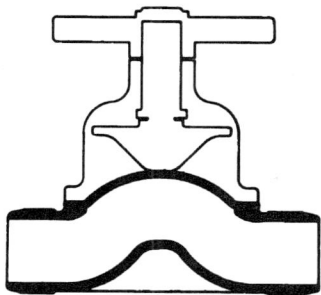

Figure 11.116. Typical weir-style diaphragm valve. (Source: Saunders Co.)

11.7.1.2 Throttling Valves

11.7.1.2.1 Butterfly Valves

Butterfly valves contain an internal flow control element, usually a disk, that is approximately the same diameter as the internal diameter of the piping. The disk, or vane, normally rotates 90° from the fully open to the fully closed position. Butterfly valves are either made in the "swing-through" style or they contain some type of seat or ring gland around the disk for shutoff. Some butterfly valves are completely lined with an elastomeric seal, or "boot." As the vane approaches its closed position, the disk completely compresses the lining around the entire 360° circumference to assure a positive seal. Some styles of butterfly valves contain an offset disk so that when the disk closes, it is not directly in line with its stem seal. A butterfly valve whose disk lines up with a projection of its centerline when it is in the closed position may experience problems maintaining the seal of its stem.

Butterfly valves vary widely in terms of pressure and temperature capability, depending on style and materials of construction. Most butterfly valves that have elastomeric seals are rated to a maximum of 150 psi, although some high-performance valves are available with metal-to-metal contact that are tight-shutoff valves. Butterfly valves have bodies that take up relatively little space, but they do require flanges on both the upstream and downstream sides of the pipe. Their handles are normally lever style. While their handles do not always need to be attached to their bodies, they still require that adequate space be provided during their operation. These requirements need to be taken into account to enable their use within a zone of secondary containment. A diagram of a typical butterfly valve is shown in Figure 11.114. A unique "grooved" design is shown in Figure 11.115.

11.7.1.2.2 Diaphragm Valves

Diaphragm valves are available in both weir styles and straight-through designs. The weir design (Figure 11.116) has become the most popular since it requires a relatively short stroke and a tighter shutoff. Diaphragm valves consist of three main parts: (1) body; (2) bonnet; and (3) diaphragm. The diaphragm valve gained popularity during the 1980s due to the development of many low-cost designs using corrosion-resistant materials. In the 1980s, the semiconductor and waste water treatment industries experienced rapid growth.

Both of these industries use considerable volumes of harsh corrosive chemicals, including concentrated acids, bases, and chlorinated chemicals. The diaphragm valve has significant advantages over many valves in that its primary seal can be effected with the use of a Teflon (PTFE) diaphragm and even a PVDF vapor barrier. PTFE is highly corrosion resistant to most chemicals and is rated to much higher temperatures than most nonmetallic materials. A PVDF vapor barrier will prevent permeation of small-molecule liquids or gases from escaping or corroding the valve.

The diaphragm valve is good for all types of throttling services, but is particularly good for viscous fluids and slurries. Diaphragm valves are most readily available with flanged ends, although some small-diameter styles exist with ends that can be directly welded to associated piping, with or without unions for body removal. Flanged diaphragm valves require significant space and therefore should be limited to outside zones of secondary containment wherever possible. Directly welded valves take up relatively little space and are therefore most suitable in applications requiring pressurized secondary containment.

11.7.1.2.3 Globe Valves

Globe valves are widely used for throttling services in the chemical industry. All globe valves have a disk or plug that moves within the valve body to mate against a seat of matching geometry. Many globe valves contain a flow path that changes as much as 90° in the direction of the fluid. In some, the flow path can take several changes in direction before continuing in the direction of the pipe. As a result, frictional losses through globe valves are often higher than in most valves. To correct for this, some manufacturers offer Y-pattern and angle-pattern globe valves that minimize such loss.

All of the same sealing concerns that are present in gate valves are also common to globe valves, with the exception of the seal to restrict leakage of the fluid downstream of the valve. Globe valves are all constructed with either a disk or plug that mates against a seat ring. The disk may either be of the same material as the body, or it may be constructed of a resilient material. Resilient seals are desirable where the pressure and temperatures are not too high and the fluid service contains solid particles. Metal disks are normally constructed of either a tapered or spherical seating surface that has line contact with a conical seat upon closure.

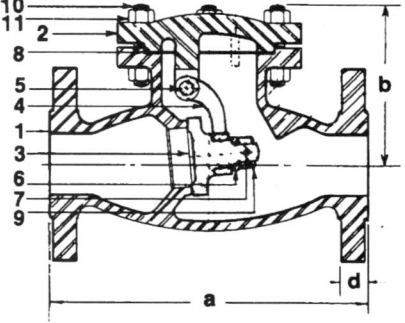

1. body
2. cover
3. disc
4. clapper arm
5. clapper arm shaft
6. disc nut washer
7. disc nut pin
8. gasket
9. disc nut
10. cover studbolt
11. cover studbolt nut

Figure 11.117. Typical stainless steel class 200 swing check valve. (Source: Crane Company)

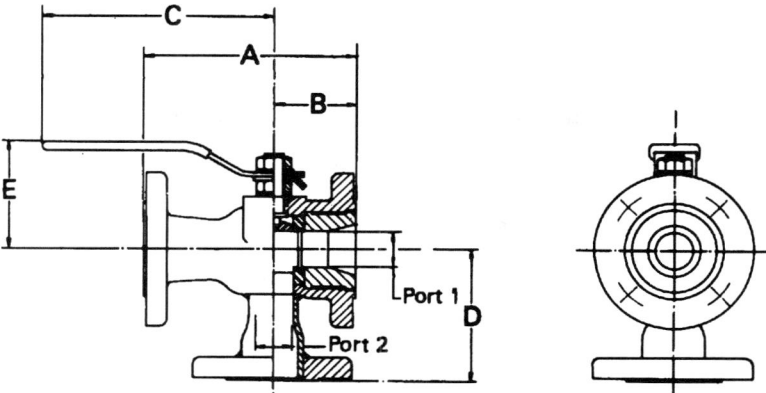

Figure 11.118. Typical multiport valve. (Source: SVF, Inc.)

Globe valves require essentially the same space requirements as gate valves. They must be well supported but are not as sensitive to loads from connected piping as the typical gate valve. Globe valves are all manufactured with flanges on either end to be connected to accompanying upstream and downstream pipe flanges. They take up considerable space and should therefore be used outside secondary containment zones wherever possible.

A needle valve is a variation of a globe valve that has tapered needle-like plugs that fit precisely in their seats. Needle valves are typically used where accurate throttling of small flows is required.

11.7.1.2.4 Pinch Valves

Pinch valves can be used either for on-off service or for throttling service where flow is to be throttled between 10 and 95% of rated flow capacity. Pinch valves are used for services where there are low pressure drops required, moderate temperatures, and slurry flow. The liquid is normally isolated from metal parts by elastomeric or synthetic tubes, thus enabling corrosive fluids to be controlled very well.

A pinch valve works by clamping a flexible tube with a pinching mechanism. Thus the two main parts consist of a flexible sleeve valve body and a pinching mechanism. The flexible sleeve is equipped either with flanged ends or with another mechanical arrangement. Pinching mechanisms are operated with hand wheels, chain wheels, or a hydraulically or electrically operated device.

11.7.1.3 Valves for Back-Flow Prevention

This category of valves includes a variety of check valves that are self-operating and prevent the reversal of flow in a piping system. The operation of these types of valves is facilitated by the weight of a check mechanism, by a spring or other mechanical device, or by the back-pressure of flow reversal.

Various check valve types that are available include swing check, tilting disk check (including two-flapper style), lift check (including disk, piston, and ball types), and stop-

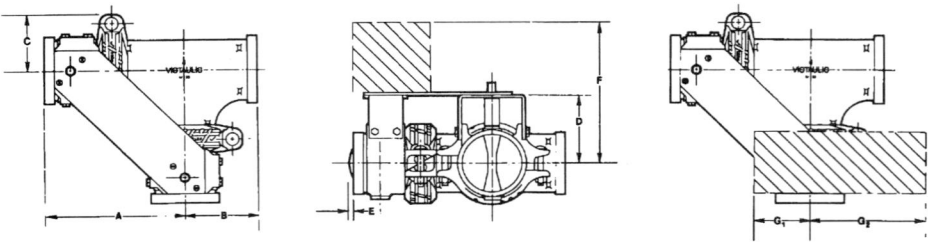

Figure 11.119. Typical tandem butterfly valve arrangement. (Source: Victaulic Company of America.)

check designs. Foot valves are a special type of check valve that are used on suction lines for horizontal pumps. They normally contain a strainer to keep solids from entering a pump, and also help to maintain the pump prime. The two most commonly used check valves in industrial services are the swing check and ball-check valve types.

11.7.1.3.1 Swing Check Valves

Swing check valves typically contain a flapper or vane that is hinged at the top. Their operation results in relatively little pressure drop during full line flow. When solid particles are present, when noise is objectionable, or when tight closure is required on shutoff, flappers may be equipped with a composition disk. An external lever, weight, or spring can be added to result in faster closing. When these accessories are used, there will be an added pressure drop during full line flow. A typical swing check valve is shown in Figure 11.117.

Figure 11.120. Illustration of a valve contained within a tee. (Source: Guardian division of Nibco, Inc.)

Layout Concepts for Double Containment Piping   487

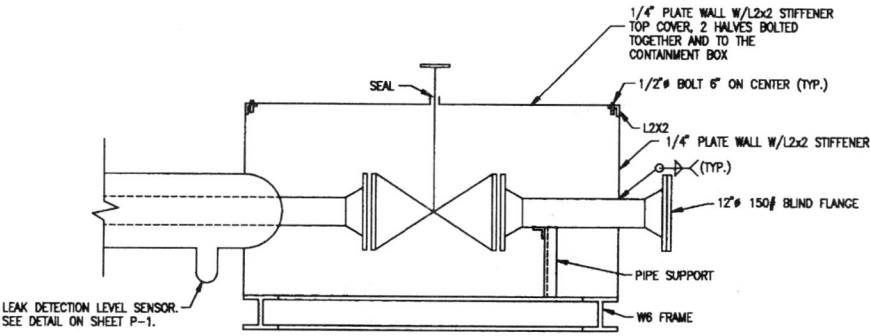

Figure 11.121. An example of a low-pressure valve housing. (Source: Becht Engineering.)

11.7.1.3.2 Ball Check Valves

A ball check valve operates by having its "ball" lifted by upward fluid-flow pressure. When flow reverses, its ball is forced to fall back onto its seat by the actions of gravity and back-flow. A ball check valve can be equipped with a spring to result in faster closure upon flow reversal. Ball check valve use should be limited to vertically positioned pipes.

11.7.1.4 Mixing Valves

11.7.1.4.1 Multiport Valves and Tandem Butterfly Arrangements

Multiport valves are modifications of single-port plug valves and single-port ball valves. A typical multiport ball valve is illustrated in Figure 11.118. A tandem butterfly valve arrangement is a fabricated arrangement of butterfly valves, pipe sections, and flanges that are equipped with a single lever or actuator to allow the valves to act in tandem. Tandem butterfly arrangements are most common in municipal applications, such as landfill facility designs. An example of a tandem butterfly valve is shown in Figure 11.119.

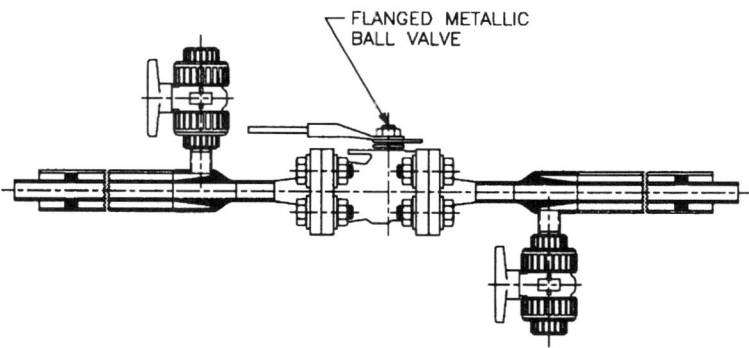

Figure 11.122. A single contained valve positioned outside the zone of secondary containment. The double containment piping arrangement is terminated on either side.

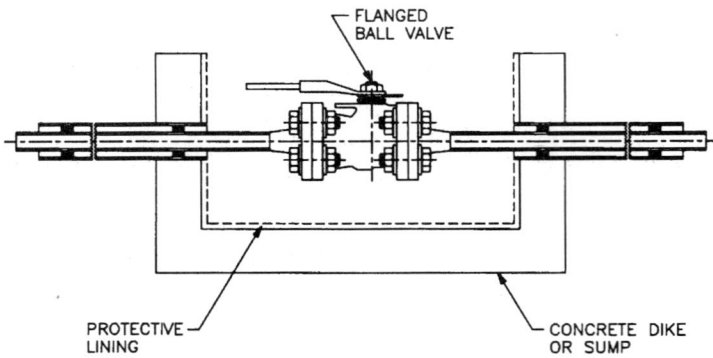

Figure 11.123. An example of a valve contained inside a secondary containment open sump.

11.7.2 Pressurized Secondary Containment of Valves; Layout and Fabrication Issues

There are two major layout options for situations where valves must be placed within a zone of pressurized secondary containment. A first option is to terminate the secondary containment pipe portion of the system with the use of a double termination arrangement and place the single containment valve inside of some nonpressurized means of secondary containment. A second option is to place the valve in some type of secondary containment vessel or component in a continuous pressurized secondary containment zone. Anytime a valve is to be housed in a pressure rated containment housing, it can be expected that the arrangement will involve significant expense. Wherever possible, it is always best to keep valves outside zones of secondary containment or provide a non-pressurized means..

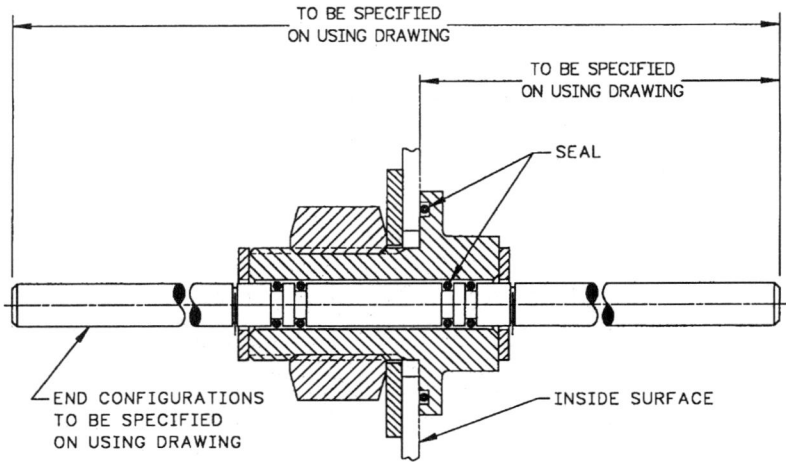

Figure 11.124. An example of a rotating stem shaft seal. (Source: Sudikatis, George, "Glovebox Shaft Seals for Rotating Equipment," 1989, American Glovebox Conference, Denver Colorado.)

Layout Concepts for Double Containment Piping

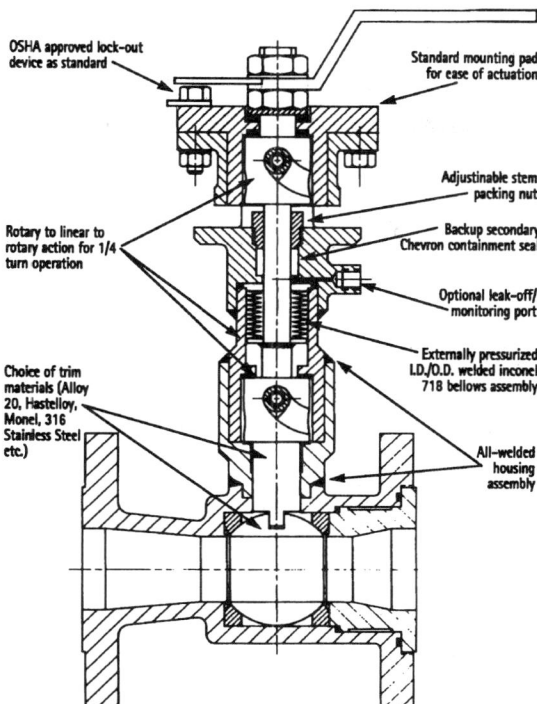

Figure 11.125. Illustration of a quarter turn ball valve which features a patented bellows seal to minimize fugitive emmisions from the valve. (Source: TBV, a division of the Victaulic. Company of America)

11.7.2.1 Valve/Secondary Containment Tee Arrangement

Small, compact primary piping valves such as ball valves and plug valves may be contained within a secondary containment tee. A typical arrangement is illustrated in Figure 11.120. The branch of the tee is equipped with a blind flange/handle assembly to allow access for operation of the valve.

11.7.2.2 Valves inside Access Vessels or Other Housings

Large valves for above-ground systems must be contained within a vessel that is specially designed for the application. Smaller valves can be contained inside of some type of manufactured or fabricated access housing. Figures 11.43 and 11.61 provide examples of two different types of access housings. An schematic illustration of a low-pressure access housing for a large valve is shown in Figure 11.121.

Underground valve assemblies must have housings placed inside underground manholes that provide further access to the surface. The valve may be directly contained inside a lined manhole; however, it is difficult to construct secondary containment manholes in a pressurized manner. Workers must be able to get into the manhole and gain access to the valve housing so they may readily operate the valve.

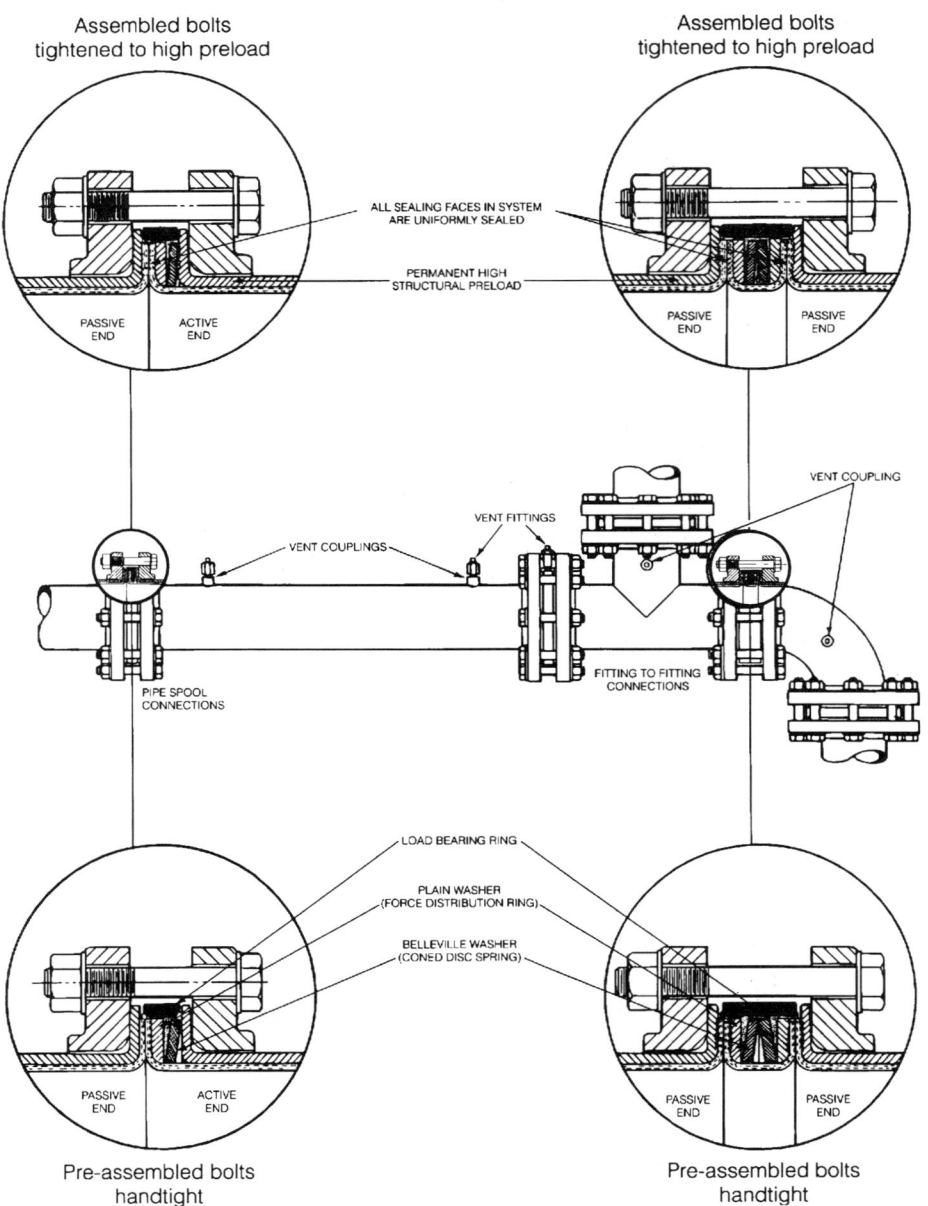

Figure 11.126. Overview of the Crane® Resistoflex High Integrity Flange (HIF) plastic-lined steel double containment piping system. (Source: Crane Resistoflex Company; U.S. Patent # 4,537,425.)

11.7.3 Nonpressurized Secondary Containment of Valves; Layout and Fabrication Issues

A variety of means exist to provide nonpressurized secondary containment to valves. There are options available for both above-ground and underground installations. Each of these options is described in the following.

11.7.3.1 Single Contained Valve with Double Termination/Flange Arrangement

An option for valves of all types is to terminate the zone of secondary containment and place the valve between termination arrangements on either side. If this is allowable, valve operability and maintainability will be greatly enhanced, and system costs will be minimized. A typical arrangement is illustrated in Figure 11.122. In most jurisdictions in the United States, if daily visual inspection is performed on the equivalent single containment portions in above-ground applications, then it is not necessary to provide secondary containment and leak detection. An example of above-ground piping systems that must either be provided with secondary containment and interstitial monitoring or given daily inspection include above-ground, outdoor chemical piping that convey a regulated chemical. However, it is usually a better design, and a safer approach to incorporate a secondary containment structure of some type around the valve. If the rest of the system is being provided with secondary containment, this will eliminate a weak link in the system.

11.7.3.2 Valve inside a Lined or Coated Concrete Dike or Berm

The valve arrangement referenced in Figure 11,122 may be provided with secondary containment if all of the area of the valve and its connections to the double containment piping system are placed inside an "open-air" lined or coated concrete dike area or sump. Figure 11.123 illustrates such an arrangement.

It is common for valves and other pipe connections that are associated with an above-ground tank that is provided with a secondary containment dike or berm to have the valves positioned within the dike or berm. If this is the arrangement used, the valve is usually said to be provided with adequate secondary containment.

11.7.3.3 Valve inside a Lined or Coated Concrete Sump or Manhole

In underground applications, a valve can readily be positioned inside a sump or manhole. The requirements for valves is similar to the enclosing of expansion joints and other maintenance-requiring devices of the primary piping system. Refer to Figures 11.95 through 11.99 for examples of manhole use as a means of secondary containment of similar items.

11.7.4 Actuation and Accessory Provisions in Secondary Containment Applications

Actuators and stem extensions may be used in conjunction with valves that are secondarily contained. However, they do present additional problems in facilitating an arrangement that maintains a pressure-tight seal. They should be limited to arrangements that depend on rotary movements, as opposed to piston-type movement to allow an entire actuator or handle to be placed outside of the secondary containment chamber. A mechanical seal can then be used to seal the rotating shaft as illustrated in Figure 11.124.

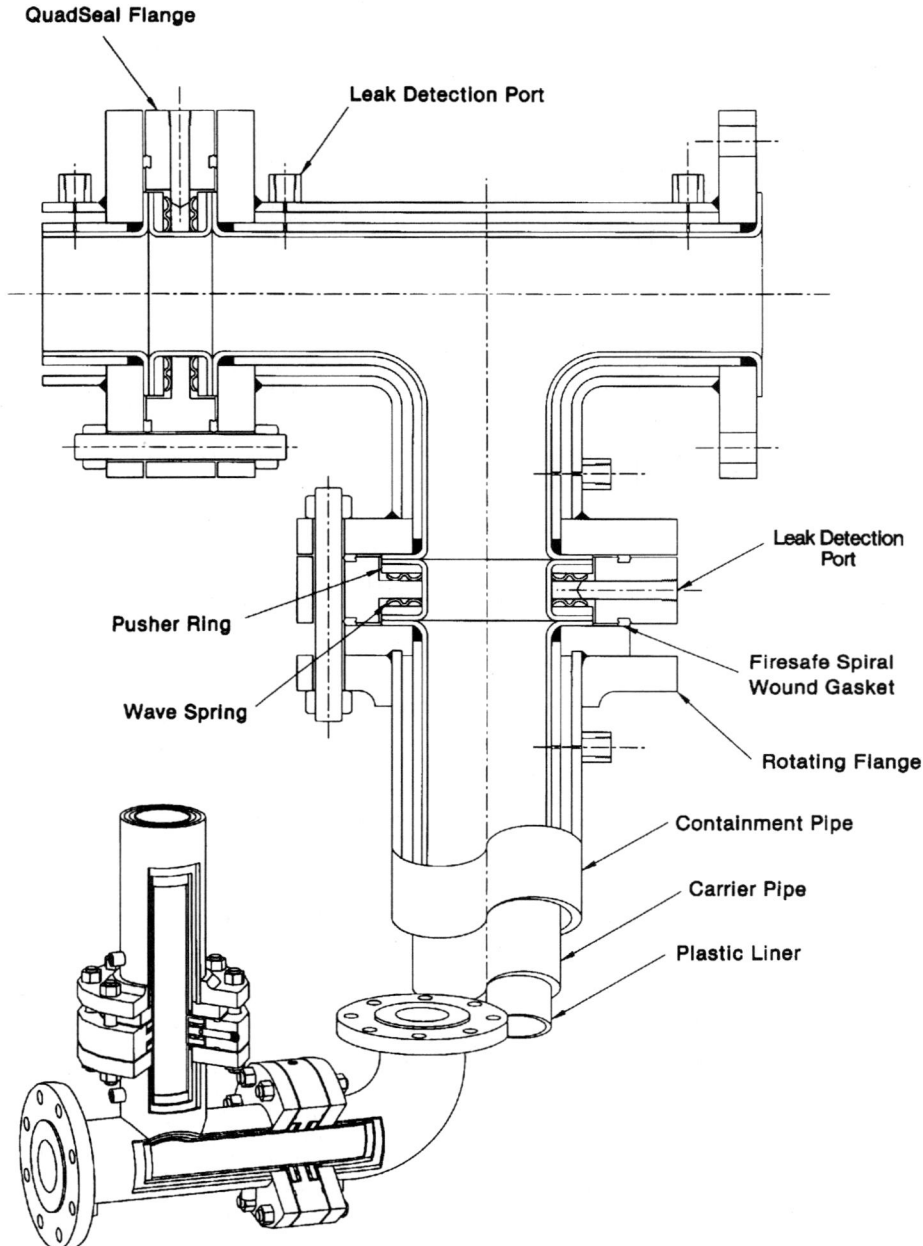

Figure 11.127. Overview of the Sentinel plastic-lined steel double containment piping system. (Source: Performance Plastic Products.)

11.7.5 Fugitive Emission Prevention and Monitoring Requirements for Valves

Clean air legislation in the United States, Canada, and throughout Europe has prompted pump seal and valve manufacturers to develop more sophisticated means of preventing fugitive losses of fluid vapors. There have been significant improvements in stem-sealing mechanisms and valve seals, as well as in mechanical seals for pumps. Many valves are not only being manufactured with an improved sealing means, but they are designed to allow ready monitoring of the seals by the use of probes. This is one of the most rapidly changing aspects of valve technology and one that is still developing at this time. An example of a ball valve with a specially designed shaft bellows seal, which prevents fugitive emissions, is shown in Figure 11.125. The valve incorporates a rotary-to-linear-to-rotary action for cost effective on-off operation. Its bellows operator is welded to its body to eliminate a potential leak path at the bonnet joint. The externally pressurized bellows assembly is also welded internally and backed by a secondary containment chevron seal to virtually eliminate stem leakage.

Other valve packing technology has been developed to incorporate antiextrusion rings and live loading in an integrated valve packing system to meet the challenges of fugitive emission reduction. The technology can be applied to polytetrafluoroethylene (PTFE) and graphite packing systems for sliding stem and rotary valves. It was developed for use in control valves, however, the concept can be applied to manually operated valves. A valve packing system that incorporates four major principles has been shown to result in a long, low maintenance service life, and as such can be expected to meet tough fugitive emmisions criteria. These are: (1) Prevent the packing from extruding out of the packing area by installing less pliable antiextrusion rings on either side of the packing, (2) Keep the valve stem aligned with stem bushings installed near the packing, (3) minimize the adverse effects of thermal cycling by using only the minimal amount of packing required to effect a seal, and (4) apply a constant and proper packing stress with live-load springs, which is dependent on the type of valve and packing system.

11.8 Plastic-Lined Metallic Systems

Recently, manufacturers of lined metallic piping systems have developed some innovative product changes to incorporate features of secondary containment and leak detection into their systems. In particular, the Crane/Resistoflex "High-Integrity-Flange" (HIF) product has received widespread attention due to its practicality, and relatively affordable approach to providing fugitive emission control and secondary containment of "spray-outs" at flange connections in aboveground systems. The main working feature of the system is the incorporation of Belleville Washers to maintain an optimum sealing pressure at the PTFE flare faces, independent of bolt torque. A load bearing ring provides the secondary containment feature which eliminates "spray-outs" common at flange connections when conveying acids and other highly corrosive media in lined metallic piping. Figure 11.126 provides an illustration of the system, which points out its major highlights.

Another product that has been developed is the "Sentinel System" by Performance Plastic Products. Though it is not as cost-effective for use as the Crane Resistoflex HIF approach, it is more of a true double containment piping system in that straight pipe portions do possess secondary containment at all points. The flanges of the system contain a pusher ring, and a wave spring to achieve a constant sealing pressure. A firesafe spiral wound gasket is used, which incorporates a leak detection port for the attachment of a leak detection probe. A schematic of the system is shown in Figure 11.127.

References

1. C. Becht IV and Y. Chen, "Double Containment Anchor Fitting Stress Evaluation," Becht Engineering, Liberty Corner, NJ, 1994.

2. "Steel Pipe – A Guide for Design and Installation," AWWA Manual M11, American Water Works Association, Denver, CO, 1989.

3. R. O. Couch, "A Guide to Multiple Pipe Containment," Chemical Engineering, Sept. 1991, pp. 158-164.

4. Pressure Vessels and Piping Design: Collected Papers, ASME, 1927-1959.

Notes

Notes

Chapter Twelve

Fabrication, Assembly, and Erection

In order to implement a double containment piping system, it must be successfully fabricated and installed. This chapter deals with the various fabrication and construction issues pertaining to double containment piping systems of all types. The information is presented in this chapter in three major parts. The first two parts discuss basic fabrication and joining practices for individual single-wall piping components. Most of the available fabrication and joining methods for materials considered throughout this book are presented in these first two parts. These include methods and techniques for metallic material components (Section 12.1) and nonmetallic material components (Section 12.2).

The reason for discussing joining practices for single-wall components in this text is that a double containment piping system consists of two separate single-wall piping systems, with one (a primary pipe system) being housed inside the other (its secondary containment piping system). In order to construct a double containment piping system, both the inner and outer pipe systems must be individually fabricated and installed as an integral part of a complex assembly. It is not possible to construct a double containment piping system without fabricating and assembling single-wall components. Thus, it is important to understand how to join single-wall pipes and components that are part of any double containment pipe system.

Section 12.3 of this Chapter describes fabrication and installation issues that are unique to double containment piping systems. Included in the discussion are issues pertaining to sequences of joining inner and outer pipes, methods for installing secondary closures, and other items that are unique to double-wall piping applications. Extensive discussion is also presented in Section 12.3 for burial of double containment piping systems, due to the applicability of such systems in underground service. The information in Section 12.3 is also applicable to double-wall piping systems for other underground double-walled pipe applications, including insulated and jacketed utility piping, and jacketed process piping systems.

12.1 Metallic Component Considerations

Metallic piping components may be joined and fabricated by a variety of methods. Welding is the predominant method used; however, other methods are available. The most common include: bending, forming, brazing, soldering, flanging, and a variety of mechanical and other proprietary joining techniques. Supplemental practices that are used in conjunction with welding, bending, and forming of metals include preheating and heat treatment. The various methods of joining and the requirements for assembly, erection, and fabrication of metallic pipe systems are discussed in this section.

12.1.1 Welding of Metallic Components

Welding is the most common method of joining steel, steel alloys, and many nonferrous metals. The process consists of creating a local coalescence of material at the intersection of two parts by heating the materials to suitable temperatures. The weld may be accomplished with or without the application of pressure, or by application of pressure alone, depending on the materials and components being welded. Also, a filler material sometimes may be used, depending on the materials, components, and thicknesses being welded. Unlike nonmetallic materials, many dissimilar metallic materials or dissimilar alloys can be welded together.

There is a wide variety of welding processes available, ranging from completely manual procedures to a variety of semiautomatic and automatic welding procedures. Specific welding procedures can be divided into general groups. One such group is termed gas welding. Gas welding refers to a group of welding processes where a gas flame (or flames) is used to generate heat, thereby producing coalescence. Gas welding may be performed with or without the use of pressure and/or filler materials.

The second major group of welding processes is arc welding (AW). Arc welding refers to a group of welding processes where a coalescence of the metals is produced by heating them with an arc or arcs. This process may be accomplished with or without the use of filler material. The major types of arc welding processes include: (1) gas-metal-arc welding (GMAW); (2) gas-tungsten-arc welding (GTAW); (3) shielded-metal-arc welding (SMAW); and (4) submerged-arc welding (SAW). Semiautomatic arc welding refers to arc welding processes whereby only the filler material is automatically controlled, but the actual welding process is controlled manually.

In gas-metal-arc welding (sometimes referred to as MIG or CO_2 welding), coalescence of the metals results by heating them with an arc between a single tungsten (nonconsumable) electrode and the work. Shielding results from the use of a gas or gas mixture. GMAW may be performed with or without the application of pressure and/or filler material. Figure 12.1 shows a typical metal-arc butt weld involving alloy 20Cb-3® parts made in a fixture with a grooved copper backing with filler materials. (20Cb-3 is a registered trademark of the Carpenter Technology Corporation.)

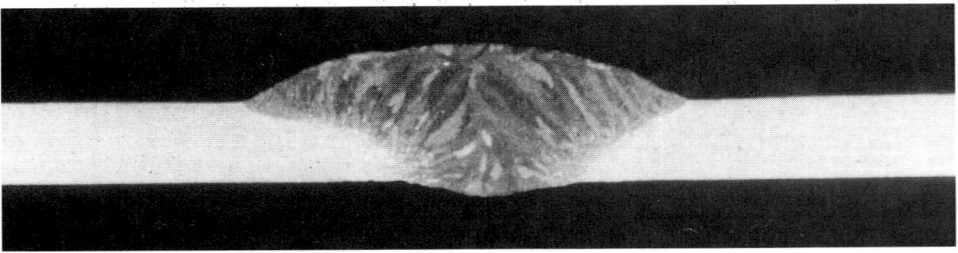

Figure 12.1. Typical metal-arc butt weld involving Alloy 20Cb-3® materials, sometimes referred to as "Carpenter 20" stainless steel. (Source: Carpenter Technology Corp.)

Fabrication, Assembly, and Erection

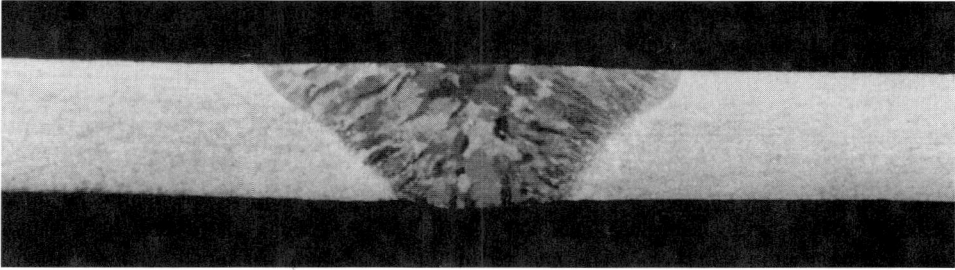

Figure 12.2. Example of a TIG weld with Alloy 20Cb-3® materials. (Source: Carpenter Technology Corp.)

Gas-tungsten-arc welding is performed by heating an arc between a single tungsten (nonconsumable) electrode and the work to produce coalescence of the metals. GTAW may be performed with or without the application of pressure and/or filler material. GTAW is sometimes referred to as TIG welding. Figure 12.2 shows a typical TIG weld involving alloy 20Cb-3® parts made in a fixture with grooved copper backing. Shielded-metal-arc welding is performed by heating the metals with an arc between a covered metal electrode and the work to result in their coalescence.

In submerged-arc welding, coalescence of metals occurs when the metals are heated by an arc, or arcs, between a bare metal electrode, or electrodes and the work. A blanket of granular, fusible material shields the arc. The process is performed without pressure. Filler material is obtained from the electrode, but is sometimes provided by the use of welding rod, flux, or metal granules.

12.1.1.1 Responsibility and Qualification for Welding

Whenever welding is to be performed, the responsibility for the qualification and quality control of welders, and welding procedures, rests with the organization whose personnel are performing the welding. The ASME Boiler and Pressure Vessel Code, Section IX, sets criteria for the qualification of welding procedures to be used on metallic materials. It also sets criteria for the performance qualification of welders and welding equipment operators. Supplemental requirements and exceptions to the ASME Boiler and Pressure Vessel Code, Section IX, requirements for chemical plant piping are stated in the current edition of the ANSI/ASME B 31.3 Code for Chemical Plant and Petroleum Refinery Piping, paragraph 328.2.1.

The most important aspect of welding involves developing a welding procedure specification (WPS) that is appropriate for the type of metals to be joined. A WPS is a document that lists the detailed methods and practices, along with all other parameters, to be used in the construction of weldments. Since there are potentially thousands of possible metal combinations that can be welded, P numbers, S numbers and Group numbers are assigned to groupings of metals, generally based on composition, weldability, and mechanical properties. By using the same specification for metals having the same P number or S number, the need for developing a large number of individual WPSs is substantially reduced. The P numbers for many common metallic piping materials are listed in Table 12.1.

Table 12.1 P Numbers and A Numbers for Common Metal Groups

Base metal group	Base metal P No. [1]	Weld metal analysis [2]
Carbon steel	1	1
Alloy steels, Cr ≤1/2%	3	2,11
Alloy steels, 1/2% < Cr ≤ 2%	4	3
Alloy steels, 2 1/4% < Cr ≤ 10%	5	4,5
High alloy steels martensitic	6	6
High alloy steels ferritic	7	7
High alloy steels austenitic	8	8,9
Nickel alloy steels	9A, 9B	10
Cr-Cu steel	10	---
Mn-V steel	10A	---
27Cr steel	10E	---
8Ni, 9Ni steel	11A SG 1	---
5Ni steel	11A SG 2	---
---	21-52	---

Notes:
[1] P Number from BPV Code, Section IX, Table QW-422, Special P Numbers (SP-1, SP-2, SP-3, SP-4, and SP-5) require special consideration. The required thermal treatment for Special P Numbers shall be established by the engineering design and demonstrated by the welding procedure qualification.
[2] A Number from BPV Code, Section IX, QW-422.

Also critical to the success of any welding program is the adequate maintenance of records relating to procedures, performance qualification tests, and welder personnel identification symbols that are assigned, etc. The proper maintenance of records can help reduce the need for duplication of effort, and can also prove to be critical when trouble shooting or analyzing failed joints. A duplication of effort will produce added costs to an employer; therefore it is a direct incentive to an employer to keep and maintain adequate records related to welding.

There are several types of supplementary materials that are used occasionally in the welding process. The most important types include filler materials, weld-backing materials, and consumable inserts. Filler materials include welding rod, flux, and metal granules. Acceptable filler materials are listed in Section IX of the ASME Boiler and Pressure Vessel Code, although any type of filler material can be used if a procedure qualification test is developed and passed, and the owner gives it approval. Typical backing rings (solid, split, and nonmetallic removable types) for the purpose of functioning as a weld-backing material, and various consumable insert geometries are shown in Figure 12.3.

Nonferrous or nonmetallic backing rings may be used if the designer approves their use, and the welding procedure that involves their use is qualified in the normal manner. A consumable insert is a replaced filler metal that is completely fused into the root of the joint and becomes part of the weld. The consumable insert normally matches the geometrical shape of the weld, and should be constructed of the same nominal composition as the filler

Fabrication, Assembly, and Erection

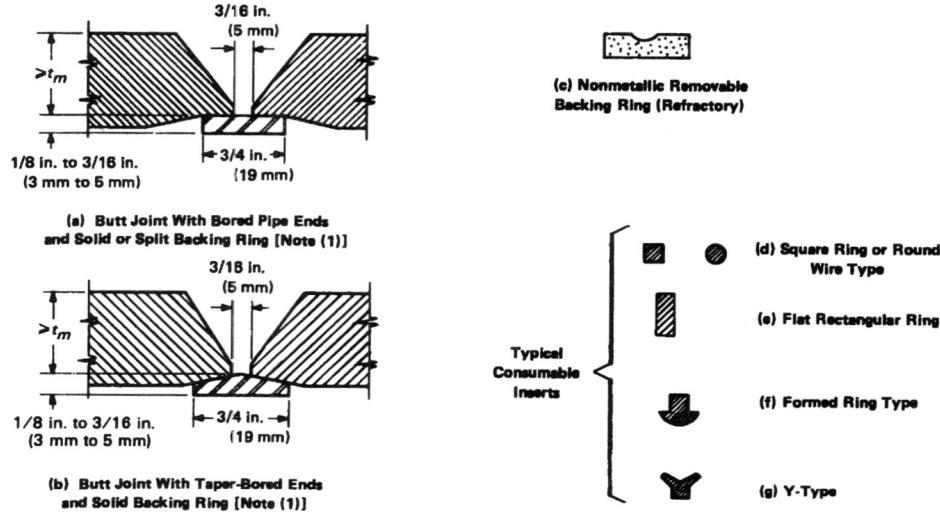

Figure 12.3. Typical backing rings used as weld-backing materials. (Source: ANSI/ASME B31.3 Chemical Plant and Petroleum Refinery Piping Code, 1993 Edition.)

material. Additionally, a consumable insert should not cause detrimental alloying of the weld metal. Typical consumable insert cross sections are shown in Figure 12.3(d-g).

12.1.1.2 Preliminary Welding Steps

Prior to welding or thermal cutting, the metal surfaces must first be cleaned and prepared. Attention must be given to metallic piping materials that contain paint, oil, rust, scale, or other material that can contaminate a metal once heat is applied. The ends must be prepared until the surface is already reasonably smooth and true. Also, any slag that results from oxygen or arc cutting should be thoroughly removed. However, discoloration that results from proper thermal cutting procedures is not a defect, and does not have to be removed.

When groove welds are to be used, the procedures outlined in ASME B16.25 may be used, or other acceptable means that may be outlined in a WPS. Some examples of basic bevel angles for butt welds are shown in Figure 12.4(a) and (b). Figure 12.5(a) and (b) show some examples of trimming and permitted misalignment for pipes. When pipes are prepared in this fashion, or as shown in Figures 12.3(a) and (b), the trimming should not result in a finished wall thickness below the required minimum wall thickness, t_m, from Chapter 5, for the component. The ends of components may be bored to provide for a recessed backing ring only if the resulting thickness of the finished ends is a minimum of t_m.

It is common to weld components having the same nominal diameter, but with different actual dimensions. When this is the case, the component ends must be prepared to improve alignment so that they are nearly matching, if the wall thickness requirements are still able to be met after such modification. When a girth or miter groove weld is intended for components of which one has a 50% greater thickness, the rules of ASME B16.25 should be followed. Also, when required, weld metal can be deposited inside or outside a component to improve alignment or provide for machining so that rings or inserts can be seated properly.

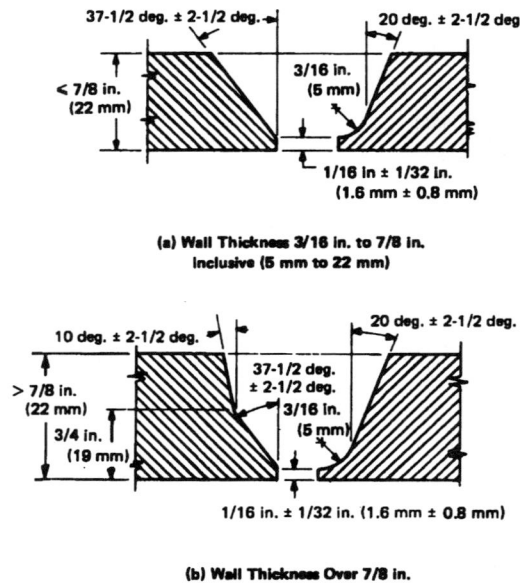

Figure 12.4. Examples of basic bevel angles for butt weld ends. (Source: ANSI/ASME B31.3 Chemical Plant and Petroleum Refinery Piping Code, 1993 Edition.)

Metallic piping components must also be aligned properly prior to the performance of any welding work. The requirements depend on whether the weld is of the circumferential, longitudinal, or branch connection variety. For circumferential welds, the inside surfaces of components at the ends to be joined in girth or miter groove welds should be aligned. The tolerance for this alignment should be spelled out in the WPS and in the engineering design. The external surfaces of these types of welds do not also need to be aligned. However, unaligned external surfaces must be tapered evenly to prevent the formation of stress risers.

The alignment criteria for longitudinal groove welds follow the same basic one as those described for circumferential welds in the preceding paragraph. The inside surfaces must be aligned to the tolerances set forth in the WPS and engineering design, but the external surfaces need not be aligned (although it is good practice that they be evenly tapered).

For branch connections, there are several weld details to consider. If a fabricated branch is prepared so that it butts directly to the outside surface of the run pipe, the branch connection should be contoured for groove welds in accordance with the WPS requirements. This is illustrated in Figure 12.6(a) and (b). If a fabricated branch connection is designed so that its branch pipe is to be inserted into the run opening, the same basic rules apply. However, the branch in this case should be inserted at least as far as the inside surface of the run pipe at all points. This is illustrated in Figure 12.6(c). For both types of branch connections, the dimension m sets a minimum condition for the deviation from the required contour.

Care should be taken also to check that the deviations of the shape of a branch opening do not cause the root spacing to exceed tolerance limits set in the engineering design and

Fabrication, Assembly, and Erection

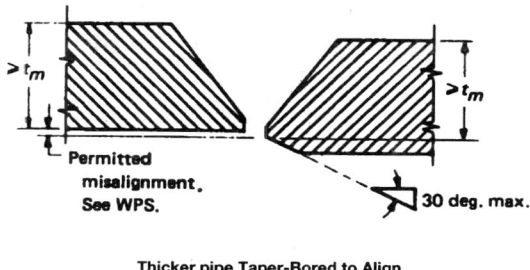

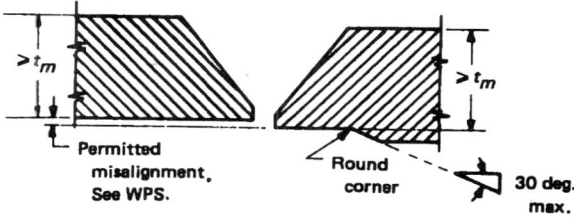

Figure 12.5. Trimming and permitted misalignment. (Source: ANSI/ASME B31.3 Chemical Plant and Petroleum Refinery Piping Code, 1993 Edition.)

the WPS. If this is exceeded, filler material followed by refinishing can be used as a correctional measure. In any case, the WPS shall set the tolerance limits for the root opening of the branch connection.

12.1.1.3 The Welding Process

Once piping components are cleaned, prepared, and aligned, welding can proceed in accordance with the welding procedure specification. Each person to perform welding should be individually subjected to a qualification test. Each qualified welder must be assigned an identification symbol, and their symbol recorded in some permanent fashion. Each com-

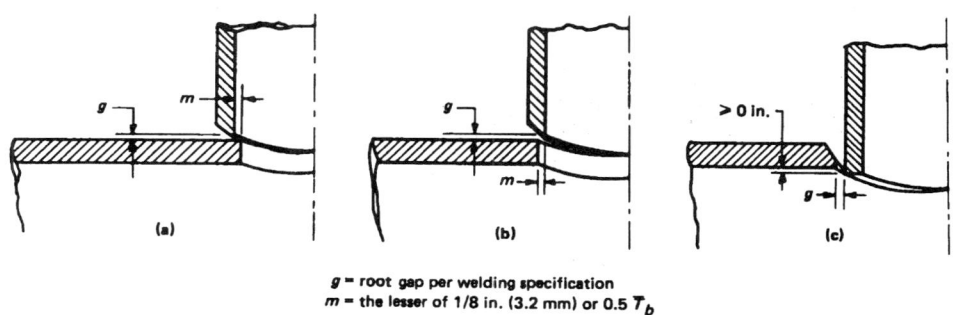

g = root gap per welding specification
m = the lesser of 1/8 in. (3.2 mm) or 0.5 T_b

Figure 12.6. Weld details for fabricated branch connections. (Source: ANSI/ASME B31.3 Chemical Plant and Petroleum Refinery Piping Code, 1993 Edition.)

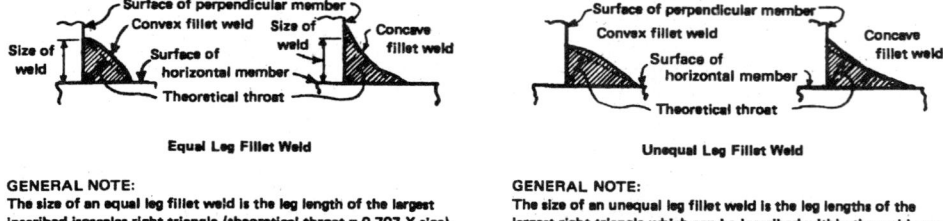

Figure 12.7. Determination of fillet weld size. (Source: ANSI/ASME B31.3 Chemical Plant and Petroleum Refinery Piping Code, 1993 Edition.)

ponent should either be marked with the symbol of the individual welder, or it should be recorded somewhere as to which welder completed each individual weldment. These records should be maintained in a permanent record-keeping system that has been established ahead of time.

Tack welds may be used initially to hold parts of a weldment in proper alignment until the final welds are made. The tack welds at the root of the joint should always be made with the same material to be used in the root pass. Tack welds should only be performed by those qualified to perform the actual welding. If a tack weld cracks prior to performing the root pass weld, it should be removed prior to welding. Otherwise, the tack weld can be left and fused to the root pass weld.

When the root pass and final pass of a weld are performed, peening should be completely avoided. Welding should not be allowed to proceed if adverse weather conditions exist such as rain, snow, sleet, or excessive wind. Welding can proceed under these circumstances only if the entire area of the joint is fully protected from the environmental conditions.

Fillet weld geometries can vary from concave to convex, and can vary from equal to unequal. Figure 12.7 illustrates how to determine the size of a fillet weld. Similar details for slip-on flanges and socket welding flanges are shown in Figure 12.8, and for socket welds in Figure 12.9. If slip-on flanges are single welded, the weld should be at the hub of the slip-on flange.

Figures 12.10, and Figure 12.11(nos. 1-5) show examples and details for typical welded branch connections and branch attachment welds. Other types of branch connections are possible, and are also valid where they pass all the requirements stated in Chapter 5. Welds

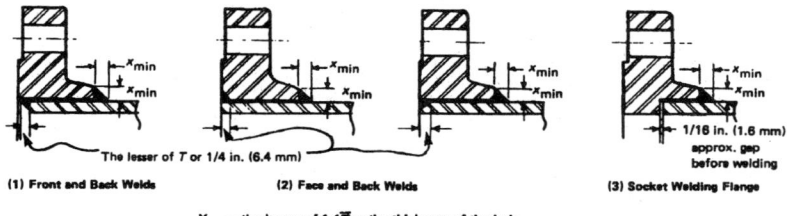

Figure 12.8. Typical details for double-welded slip-on and socket welding flange attachment welds. (Source: ANSI/ASME B31.3 Chemical Plant and Petroleum Refinery Piping Code, 1993 Edition.)

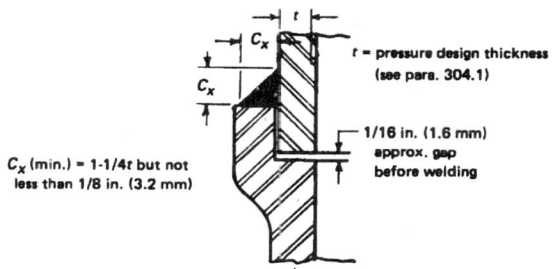

Figure 12.9. Minimum welding dimensions for socket welding components other than flanges. (Source: ANSI/ASME B31.3 Chemical Plant and Petroleum Refinery Piping Code, 1993 Edition.)

shown in Figure 12.11 should be verified in accordance with the methods stated in Chapter 5 for branch connections, but should be at a minimum equal to the sizes shown in Figure 12.11. The nomenclature and symbols used in Figure 12.11 are as follows:

t_c = lesser of $0.7\overline{T}_b$ or 1/4 in. (6.4 mm)
$\overline{T}_b$ = nominal thickness of branch, in.
$\overline{T}_h$ = nominal thickness of header, in.
$\overline{T}_r$ = nominal thickness of reinforcing pad or saddle, in.
t_{min} = lesser of $\overline{T}_b$ or $\overline{T}_r$

The welds used to attach branch connections should be full-penetration groove welds. These welds should be finished with cover fillet welds having a throat dimension at least equal to t_c from Figure 12.11(a) and (b). If a reinforcement pad, or saddle, is to be welded to the branch pipe, it may be welded in one of two ways. It may be attached by the use of a full-penetration groove weld as described for attachment of the branch. An alternative is the use of a fillet weld that has a throat dimension equal to at least 70% of the t_m. The second alternative is illustrated in Figure 12.11(e). As illustrated in Figure 12.11(c)-(e), the outer edge of the reinforcing pad or saddle should be attached to the run pipe by a fillet weld that has a throat dimension of at least 50% of T_r.

Vent holes should be provided at the side of a reinforcing pad or saddle to allow venting during welding and heat treatment, and to reveal leakage in a weld between the branch and

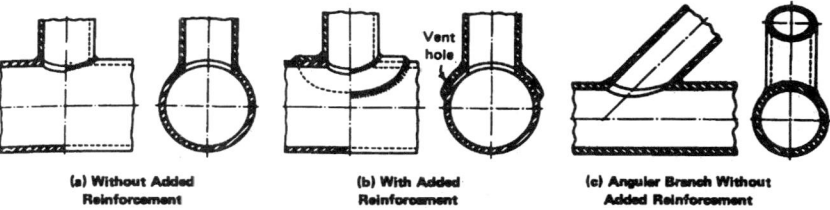

Figure 12.10. Typical welded branch connections. (Source: ANSI/ASME B31.3 Chemical Plant and Petroleum Refinery Piping Code, 1993 Edition.)

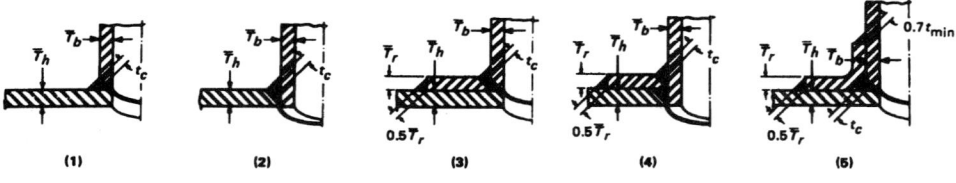

GENERAL NOTE:
These sketches show minimum acceptable welds. Welds may be larger than those shown here.

Figure 12.11. Acceptable details for branch attachment welds. (Source: ANSI/ASME B31.3 Chemical Plant and Petroleum Refinery Piping Code, 1993 Edition.)

the run. The pad or saddle can be constructed multipiece in a split fashion if the joints between pieces have equivalent strength to the base pad or saddle material. Multipiece pads or saddles should be provided with a vent hole for each individual piece. Prior to welding the reinforcing pad or saddle, defects in the weld of the branch should be repaired.

Figure 12.12 shows an illustration of branch attachments that are suitable for 100% radiography as a means of their qualification.

Examples of typical fabricated laps are illustrated in Figure 12.13(a)-(e). The requirements for welding lap joints follow the same requirements as those for attaching branch connections in the preceding paragraphs.

A special welding procedure specification should be prepared for services that are subject to "severe-cyclic" conditions. In general, welding for joints subject to this type of condition should result in a finished part that has a smooth, regular inner surface. In addition, the welds should fully penetrate the joint, down to the inner surface. Chapter 5 states many specific requirements, and limitations, as to joint types intended for "severe-cyclic" services.

12.1.1.4 Preheating Considerations

Preheating of metallic piping components is a process that is used to reduce the adverse effects of high-temperature exposure, and subsequent residual stresses that result from welding. The situations where preheating is required apply to all types of welds, including tack welds. The requirements relating to preheating should be spelled out in detail in the WPS, and if it is not required, it should be stated. Table 12.2 presents various recommend-

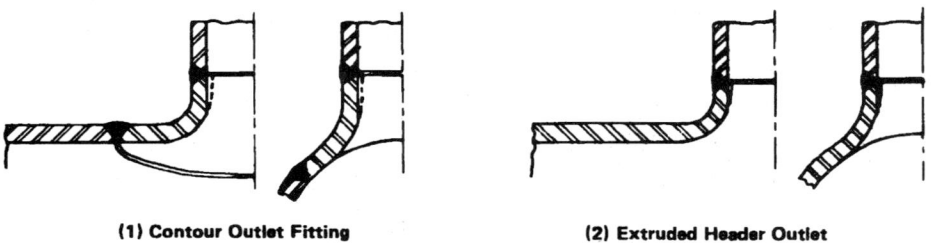

(1) Contour Outlet Fitting   (2) Extruded Header Outlet

Figure 12.12. Acceptable details for attachments suitable for 100% radiography. (Source: ANSI/ASME B31.3 Chemical Plant and Petroleum Refinery Piping Code, 1993 Edition.)

Fabrication, Assembly, and Erection

ed and required preheating temperatures for a variety of steel and steel alloy products. If the ambient temperature is below 32°F (0°C), preheating recommendations listed in Table 12.2 become mandatory, according to ANSI/ASME B31.3 Code. If materials of two different thicknesses are to be welded, the thickness of the thicker component, measured at the location of the joint, should be used as the basis for determining preheat requirements. The area to be preheated is referred to as the preheat zone, and extends a distance of at least 1 in. (25 mm) beyond the extreme edge of the weld.

The temperature of preheating should be tested by the use of a suitable measuring device. Suitable devices include measuring crayons, pyrometers (thermocouple type), or other device having a reasonable accuracy. A thermocouple may be attached directly to pressure containing parts temporarily, using the capacitor discharge method. Once an attached thermocouple is removed, the component should be checked to see if defects have been created. If so, they should be repaired immediately by a suitable means.

In some cases, metals that are dissimilar are to be welded together. The recommended preheat temperatures vary from metal to metal. Therefore, the higher value is normally selected. If welding must be interrupted, care should be taken to control the rate of cooling to the greatest extent possible. When welding is resumed, the interrupted weld should be subjected to preheating in the area of the preheat zone.

12.1.1.5 Postwelding Heat Treatment

When metals are subjected to high heat such as in welding, bending, and forming operations, large residual stresses can be created. These stresses can be substantially reduced by subjecting the fabricated part to heat-treatment techniques. The following paragraphs outline the basic requirements for heat treatment that are appropriate for most services.

Table 12.3 presents heat-treatment data for various steel and steel alloy groupings at a variety of thicknesses. Like preheating requirements for materials of two different thicknesses that are to be welded, the thickness of the thicker component should set the basis for heat-treatment requirements, measured at the location of the joint.

For branch connections, metal that serves as a means of reinforcement should not be considered as part of the thickness in determining heat-treatment requirements, whether it is attached or an integral part of the fitting. However, when the thickness through a plane exceeds twice the minimum thickness of that which requires heat treatment, heat treatment

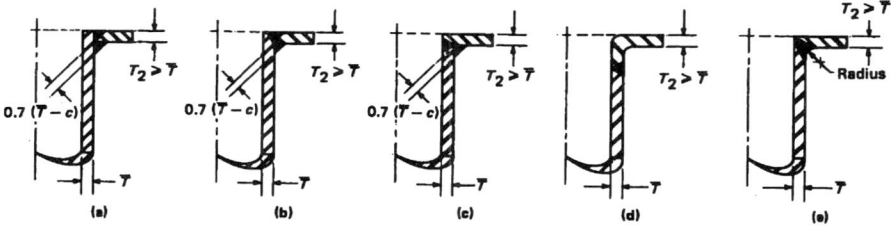

GENERAL NOTE: Laps shall be machined (front and back) or trued after welding. Plate flanges per para. 304.5 or lap joint flanges per ASME B16.5 may be used. Welds may be machined to radius, as in sketch (e), if necessary to match ASME B16.5 lap joint flanges.

Figure 12.13. Typical fabricated laps. (Source: ANSI/ASME B31.3 Chemical Plant and Petroleum Refinery Piping Code, 1993 Edition.)

Table 12.2 Preheat Temperatures

Base Metal Group	Nominal wall thickness		Specified min. tensile strength, base metal		Min. temperature			
					Required		Recommended	
	(in.)	(mm)	(ksi)	(MPa)	(°F)	(°C)	(°F)	(°C)
Carbon steel	<1	<25.4	≤71	≤490	---	---	50	10
	≥1	≥25.4	All	All	---	---	175	79
	All	All	>71	>490	---	---	175	79
Alloy steels, Cr ≤ 1/2%	<1/2	<12.7	<71	≤490	---	---	50	10
	≥1/2	≥12.7	All	All	---	---	175	79
	All	All	>71	>490	---	---	175	79
Alloy steels, 1/2% < CR ≤ 2%	All	All	All	All	300	149	---	---
Alloy steels, 2 1/4% < Cr ≤ 10%	All	All	All	All	350	177	---	---
High alloy steels martensitic	All	All	All	All	---	---	300	149
High alloy steels ferritic	All	All	All	All	---	---	50	10
High alloy steels austenitic	All	All	All	All	---	---	50	10
Nickel alloy steels	All	All	All	All	---	---	200	93
Cr-Cu steel	All	All	All	All	300-400	149-204	---	---
Mn-V steel	All	All	All	All	---	---	175	79
27Cr steel	All	All	All	All	300	149	---	---
8Ni, 9Ni steel	All	All	All	All	---	---	50	10
5Ni steel	All	All	All	All	50	10	---	---
---	All	All	All	All	---	---	50	10

Source: ANSI/ASME B31.3 Code – 1993 Edition.

Fabrication, Assembly, and Erection

Table 12.3 Requirements for Heat Treatment

Base Metal Group	Nominal wall thickness		Specified min. tensile strength, base metal		Metal temperature range		Holding time	
	(in.)	(mm)	(ksi)	(MPa)	(°F)	(°C)	nominal wall (h/in.)	min. time, (h)
Carbon steel	3/4	19	All	All	None	None	---	---
	>3/4	>19	All	All	1100-1200	593-649	1	1
Alloy steels, Cr ≤ 1/2%	3/4	19	71	490	None	None	---	---
	>3/4	>19	All	All	1100-1325	593-718	1	1
	All	All	>71	>490	1100-1325	593-718	1	1
Alloy steels, 1/2% < CR ≤ 2%	1/2	12.7	71	490	None	None	---	---
	>1/2	>12.7	All	All	1300-1375	704-746	1	2
	All	All	>71	>490	1300-1375	704-746	1	2
Alloy steels, 2 1/4% < Cr ≤ 10% (≤3%Cr, ≤.15%C) (>3%Cr or >.15%Cor)	1/2	12.7	All	All	None	None	---	---
	>1/2	>12.7	All	All	1300-1400	704-760	1	2
High alloy steels martensitic	All	All	All	All	1350-1450	732-788	1	2
A 240 Gr. 429	All	All	All	All	1150-1225	621-663	1	2
High alloy steels ferritic	All	All	All	All	None	None	---	---
High alloy steels austenitic	All	All	All	All	None	None	---	---
Nickel alloy steels	3/4	19	All	All	None	None	---	---
	>3/4	>19	All	All	1100-1175	593-635	1/2	1
Cr-Cu steel	All	All	All	All	1400-1500	760-816	1/2	1/2
Mn-V steel	3/4	19	71	490	None	None	---	---
	>3/4	>19	All	All	1100-1300	593-704	1	1
	All	All	>71	>490	1100-1300	593-704	1	1
27Cr steel	All	All	All	All	1225-1300	663-704	1	1
Cr-Ni-Mo steel	All	All	All	All			1/2	1/2
8Ni, 9Ni steel 1	2	51	All	All	None	None	---	---
	>2	>51	All	All	1025-1085	552-585	1	1
5Ni steel	>2	>51	All	All	1025-1085	552-585	1	1

Source: ANSI/ASME B31.3 Code – 1993 Edition.

is also required. This is so even if the thickness at the joint is less than the minimum required thickness. The thickness at the weld can be computed for the details shown in Figure 12.11 as follows:

$$\text{sketch (1) from Fig. 12.11} = \overline{T}_b + t_c$$

$$\text{sketch (2) from Fig. 12.11} = \overline{T}_h + t_c$$

$$\text{sketch (3) from Fig. 12.11} = \text{greater of } \overline{T}_b + t_c \text{ or } \overline{T}_r + t_c$$

Heat-treatment requirements for fillet welds at slip-on flanges, socket welding flanges, nominal pipe sizes 2 in. (approximately 60.3 mm outside diameter) and less, and attachment of external nonpressure parts are determined in the same manner as for branch connections. If the thickness through any one plane is more than twice the minimum thickness, heat treatment is required. For the fillet welds described, there are certain exceptions. It is suggested to check the paragraphs for details for governing thicknesses stated in the ANSI/ASME B 31.3 Code (paragraph 331.1.3 of the 1993 Code) or other applicable code including Section 111 (Nuclear) or B31.3 (Power).

Various methods that are used for heat treatment include enclosed furnaces, local flame heating, electrical resistance, electrical induction, and induced exothermic-type chemical reaction. Cooling may be accomplished to the desired cooling conditions by the use of local insulation, local heat, air cooling, and cooling in a furnace. Heat treatment should be checked by the application of a thermocouple pyrometer, or any other method that is deemed feasible and reasonably accurate. A thermocouple may be attached directly to pressure-containing parts temporarily, using the capacitor discharge method. Once the thermocouple is removed, the component should be checked to see if defects have been created. If so, they should be repaired immediately by a suitable means.

Once heat treatment is performed, it may be verified by the performance of a hardness test. Hardness limits described in Table 12.4 apply to the heat-affected zone. The heat-affected zone (HAZ) is that portion of the base material that has been affected by the heat-treatment process. If dissimilar materials are welded and subsequently treated by heat treatment, the base and welding hardness limits should be met for both materials.

Alternative heat-treatment methods may be used where it is determined to be necessary or feasible based on experience or service conditions. Alternative methods include normalizing, tempering, annealing, and also such treatments as delayed heat treatments, partial heat treatments, and local heat treatments. Normalizing, tempering, and annealing may be used if the material continues to meet specifications after treatment by such processes. When performing partial heat treatments, the treated sections should overlap by at least 1 ft (0.3 m) between successive heats. For local heat treatments, the heat affected zone consists of a circumferential band that extends at least 1 in. (25.4 mm) beyond the extreme edge of the weld. If the heat-treatment process is delayed once the weld is performed, the rate of cooling should be controlled to the greatest extent possible in order to minimize the buildup of residual stresses.

Table 12.4 Requirements for Heat Treatment

Base metal group	Brinell hardness, [1] max.
Carbon steel	---
Alloy steels, Cr ≤1/2%	225
Alloy steels, 1/2% < Cr ≤ 2%	225
Alloy steels, 2 1/4% < Cr ≤ 10%	241
High alloy steels martensitic A 240 Gr. 429	241
High alloy steels ferritic	---
High alloy steels austenitic	---
Nickel ally steels	---
Cr-Cu steel	---
Mn-V steel	225
27Cr steel	---
Cr-Ni-Mo steel	---
8Ni, 9Ni steel	---
5Ni steel	---

Notes:
[1] See 331.1.7 of the ASME B31.3 Code – 1990 Edition. Source: ANSI/ASME B31.3 Code – 1993 Edition.

12.1.2 Bending and Forming of Metallic Materials

Metallic piping can be bent and formed into a variety of configurations and components. The applications are limited only by the performance limits of the material, and the bending and forming process. The finished surfaces should be free from cracks, and show no signs of buckling damage. Also, a component that has been formed or bent should have a final thickness at all points that meets the minimum required design thickness.

When piping is fabricated using bending techniques, the resulting thickness of the bent pipe should result in a maximum flattening of 8% for internal pressure applications and 3% for external pressure applications. Hot bending of piping should be performed at a temperature above the transformation range of the metal. Cold bending of ferritic stainless steel materials should be done at a temperature below the transformation range of the material. Bends created from bending corrugated piping should be qualified for pressure design by the same rules as described in Chapter 5 for unlisted components. (See Section 5.1.10) The temperature range for forming varies according to the type of material used and the specifics of the application.

For applications of hot and cold bending and forming of various piping, specific applications that require heat treatment are stated in the current edition of the ASME B 31.3 Chemical Plant and Petroleum Refinery Piping Code, para. 332.4.

12.1.3 Brazing and Soldering

Brazing and soldering are two methods that are used to join nonferrous materials such as copper and its alloys. They are similar processes, but each is distinguishable by subtle differences.

The brazing process produces coalescence of the metal materials by the use of a nonferrous filler material that has a melting point above 800°F (427°C) and below that of the base metal being brazed. In this process, the filler material distributes itself by capillary action between the closely fitted surfaces of the joint. A similar process is braze welding, with the difference being that the filler material is not distributed by capillary action in the braze welding process.

Soldering is a process whereby coalescence is produced by the use of a nonferrous alloy, fusible at temperatures below 800°F (427°C), and below that of the base metal being soldered. In the soldering process, the filler material distributes itself by capillary action between the closely fitted surfaces of the joint. Most solders are lead-tin alloys that may contain one or more other constituents such as antimony, bismuth, and other elements.

The ASME Boiler and Pressure Vessel Code, Section IX, Part QB outlines procedures for the qualification of brazing procedures and individuals performing brazing operations. Care should be taken that the appropriate type of filler material is selected for the particular application. Normally, a flux is also employed to eliminate oxidation of the filler material and the surfaces to be joined, and to allow the alloy or solder to flow freely. The procedures that are most commonly used for soldering are those described in the Copper Tube Handbook of the Copper Development Association.

For brazing and soldering, the surfaces of the materials to be joined must be first cleaned and prepared for joining. Any impurities on the surface will affect the final quality of the joint. For brazing and soldering, the parts to be joined should be provided with adequate clearance to allow complete capillary flow of the filler material. Oxidation can be minimized by bringing the joint to the brazing or soldering temperature in as short a time as possible. The development of localized underheating or overheating should be avoided as well, in order to keep oxidation to a minimum. Excess flux must be removed in order to maintain final joint quality.

12.1.4 Assembly and Erection of Single-Wall Metallic Piping Components

The most important consideration in the assembly and erection of metallic piping materials is that they be fitted without developing detrimental distortions. Oftentimes it is possible that piping be distorted slightly in order to bring it into alignment for the connection of flanges, threaded joints, etc. This practice should be avoided, and in any case should not result in the development of detrimental strains in piping components and connected equipment, or excessive movement of their supports.

When cold spring is used, the dimensions and positions of all guides, supports, and anchors should be checked to see that they are in their right location. In no case should an undesired gap be corrected by the use of heating, as the creation of residual stresses will tend to offset the benefits of cold springing.

Special attention should be paid to the tolerance of flange face matchups. All flange faces should align to within 1/16 in./ft (1.5 mm/0.3 m) (0.5%) measured across any diameter, and bolt holes should align to within 1/8 in. (3 mm) maximum offset. Gaskets should

Fabrication, Assembly, and Erection

be checked to be sure that their seating surface are not damaged. In no case should multiple gaskets be used between a single set of contact faces. Bolts that are selected should ideally extend all the way through their nuts, but can be acceptably sized to be within one thread of complete engagement. The flanges should be tightened to a predetermined torque whenever possible. This is very important when assembling flanges that have widely differing mechanical properties. In every application, flange gaskets must be uniformly compressed to their proper design loading in order to maintain flanged joint integrity.

When assembling threaded joints, thread compound or lubricant should be selected that does not react unfavorably with a service fluid or attack piping materials. A thread compound should not be used if the joint is to be seal welded after threading. However, a threaded joint that is to be seal welded to correct a leak may already contain a thread compound. It is acceptable to seal weld under this circumstance. However, thread compound should be removed from any exposed threads to the greatest extent possible. Typical straight threaded joints are shown in Figure 12.14. When installing straight threaded joints where the seal is not at the threads, care should be taken so that gaskets, "O" rings, or seats do not become damaged.

Other types of joints such as flared tubing joints, flareless, and compression tubing joints, caulked joints, expanded joints, packed joints, and proprietary joints should be installed with careful practices according to the instructions of the manufacturer of the product being used.

After installation, if cleaning is necessary, it should proceed with consideration of the materials involved and the service fluids to be used. The cleaning compound or procedure should not adversely affect the piping material or contaminate the service fluid. Proper protective clothing should be provided to the installers where cleaning poses a safety hazard. Cleaning fluids of this type should be disposed of properly and in a responsible manner.

12.2 Nonmetallic Component Considerations

12.2.1 Bonding of Nonmetallic Piping Components

Nonmetallic piping components may be joined by a variety of methods. The methods differ considerably when comparing thermoplastic, reinforced-thermosetting-resin plastics, lined metallic materials, borosilicate glass, porcelain, and other materials (such as dual laminates, concrete, etc.). Thermoplastics are joined by either heat welding processes (butt

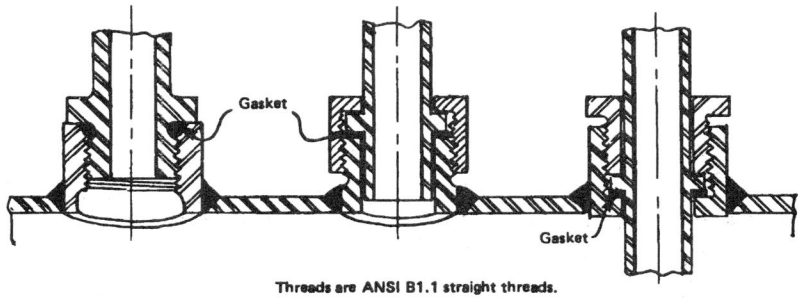

Figure 12.14. Typical details for straight threaded joints. (Source: ANSI/ASME B31.3 Chemical Plant and Petroleum Refinery Piping Code, 1993 Edition.)

fusion, socket fusion, extrusion welding, hot gas welding, saddle fusion, sidewall fusion, electrical-resistance fusion), solvent cementing, or by mechanical means. RTRP materials are joined by either adhesive bonding, butt-and-wrap techniques (sometimes referred to as butt-and-strap), or by mechanical means. Most other nonmetallic piping is joined by some type of mechanical joining technique, or by methods that use elastomers to affix a seal. Dual laminates often use combinations of many methods, particularly when the dual laminate consists of a thermoplastic material that is laminated with a reinforced-thermosetting-resin material. In the following section, the various methods of joining thermoplastics and RTRP materials and their requirements as they pertain to assembly, erection, and fabrication are presented.

12.2.1.1 Responsibility and Qualification for Bonding

The joining of plastic piping (thermoplastic and RTRP) is generally referred to as "bonding," which is an all-encompassing term that includes all methods for joining nonmetallic piping components. Whenever bonding of plastic piping components is to be performed, the responsibility for the qualification and quality control of bonders and bonding procedures rests on the organization whose personnel are performing the bonding.

The most important aspect of bonding is to develop a bonding procedure specification (BPS) that is appropriate for the materials to be joined. A BPS is a document that lists the detailed methods and practices, along with all other parameters, to be used in the construction of plastic weldments. In order to qualify a BPS, all tests and examinations that it specifies must be completed, along with any additional requirements for analysis and testing that may be stated in Chapter 6. The bonding procedure specification should go beyond the description of the actual steps in performing the bond. It should also describe in detail the following considerations:

1. All materials and supplies, including their proper care, storage, and handling.
2. Tools and fixtures, and their proper care, storage, and handling.
3. Environmental requirements such as the temperature and humidity required for successful completion of the bond. Also included should be requirements for the measurements of these conditions, including the types of instruments and their accuracy.
4. Joint preparation.
5. Dimensional requirements and tolerances (it is important for nonmetallic materials to state a reference temperature for this item).
6. Cure time for the bond.
7. Protection of the work.
8. Tests and examinations from Chapter 13.
9. Acceptance criteria for the completed test assembly.

A critical part of the success of any bonding program involves adequate maintenance of records relating to procedures, performance qualification tests, bonder personnel identification symbols that are assigned, etc. The proper maintenance of records can help to reduce the need for duplication of efforts later on, and can also prove to be critical when trouble shooting or analyzing failed joints. A duplication of effort will produce added costs for an employer. Therefore, there is a direct incentive to an employer to keep and maintain adequate records relating to bonding plastic weldments.

Fabrication, Assembly, and Erection

The ANSI/ASME B 31.3 Code states requirements for sizes of assemblies, number and types of joints, etc. for qualification tests. The requirement stated in paragraph A328.2.5 of the 1993 edition of the Code is:

a. A test assembly shall be fabricated in accordance with the BPS and at least one of each different type of joint identified in the BPS. More than one test assembly may be prepared if necessary to accommodate all of the joint types. Design of the test assembly shall be in conformance with the rules of para. A304 of the B31.3 Code. The size of the pipe used for the test assembly shall be as follows:

    1. When the largest size to be joined is NPS 4 or smaller, the test assembly shall be the largest size to be joined.

    2. When the largest size to be joined is greater than NPS 4, the size of the test assembly shall be either NPS 4 or 25% of the largest piping size to be joined, whichever is greater.

b. When the test assembly has been cured, it shall be subjected to a hydrostatic test pressure of four times the design pressure (see para. A301.2 of the Code), for not less than 1 h with no leakage or separation of the joints. The test shall be conducted so that the joint is loaded in both the circumferential and longitudinal directions.

The requirement of four times the design pressure often creates confusion. The design pressure of the test assembly is not to be confused with the design stress of the base piping material or the pressure rating of the component. If this were the case, a thermoplastic component tested in this fashion would very likely burst before the hour expires in most cases. The use of the term design pressure in this statement refers to the design pressure of the piping system, which is a number selected for the project by the engineer of record.

Under the wording of the 1993 Edition of the ANSI/ASME B31.3 Code, if a thermoplastic pipe's wall thickness has been sized to meet a certain pressure rating, or a thermoplastic pipe is being used right up to its design limit, the bonding qualification test will result in failure of the piping assembly within the required one hour duration. In order to arrive at a cure for this, a new approach to determining a test pressure for thermoplastic piping that will adequately stress the bonds of the assembly, yet prevent outright failure of the assembly, has been proposed by Bradshaw in the following equation[2]:

$$P_T = 0.80 \left(\frac{(S_S + S_H)\overline{T}}{D - \overline{T}} \right) \qquad (12.1)$$

where:
D = outside diameter of pipe, in.
$\overline{T}$ = total wall thickness of pipe, in.
S_S = mean short-term burst stress in accordance with ASTM D-1599, from Table B-1 of the ANSI/ASME B31.3 Code, or from manufacturer's data if not printed in Table B-1 of the Code, psi
S_H = mean long-term hydrostatic strength (LTHS) in accordance with ASTM D-2837. Use twice the HDB design stress at 73°F, from Table B-1 of the ANSI/ASME B31.3 Code, or from manufacturer's data if not printed in Table B-1 of the Code, psi

This new equation is the subject matter of a letter ballot to revise para. A328.2.5 of the ANSI/ASME B31.3 Code, which has been approved by the B31 Main Committee, and has been incorporated into para. A328.2.5 of the Code in a recent addenda as a requirement for thermoplastic assemblies intended for a bonder qualification test.

If the design temperature is greater than the test temperature, consideration should be given to increase the test pressure by the use of Eq. (12.5). It should be pointed out that the ASME B 31 Codes do not require this to be done for performance qualification tests of nonmetallic piping components. However, there are some circumstances where the test pressure described would result in a stress that is not truly indicative of that which the piping could see under actual service, even though it may be tested at four times the design pressure. This is due to the often dramatic decrease in pressure capability of thermoplastic materials as a service temperature is substantially increased. Thus, the test pressure may not accurately reflect the ratio of the actual stress the part will see to its allowable stress under actual operating temperatures. Therefore, the author suggests the conservative approach of upwardly adjusting the test pressure in the manner described using Eq. 12.2.

$$P_T = \frac{1.5 P S_T}{S} \quad (12.2)$$

where:
P_T = minimum test gage pressure, psi
P = internal design gage pressure, psi
S_T = stress value at test temperature, psi
S = stress value at design temperature, psi

It is suggested that a bonder that has not performed any bonding of the type needed for more than three months be requalified according to the same BPS. Also, any persons performing bonding whose work is suspect should be requalified, if there are substantial or specific reasons to question such an individual's work.

12.2.1.2 Preliminary Bonding Steps

Prior to bonding, materials must be cut to size and their surfaces cleaned and prepared. Particular attention must be given to nonmetallic piping materials that contain paint, oil, rust, scale, or other materials that can contaminate the material or the joint once heat, solvent cement, or adhesive is applied. The ends must be prepared until the surface is already reasonably smooth and true. Also, any grease or dirt that results from cutting should be cleaned until it is thoroughly removed. However, care should be taken in the selection of cleaning agents in order that the material is not swollen, softened, or otherwise damaged. It is very important when using thermoplastics that have been oxidized by exposure to sunlight to remove the oxidized layer by sanding prior to welding or cementing. The material should still meet the minimum thickness requirements after sanding; if it has less than the required thickness, it should not be installed into the system. The specific preparation requirements for bonding should be spelled out in detail in the bonding procedure specification.

All bonding materials, including solvent cements, adhesives, welding rod, or other supplementary materials should be inspected prior to use. If they have been adversely affected

Fabrication, Assembly, and Erection

by exposure to environmental conditions, they should be discarded. All fixtures and tools should be inspected and tested for alignment, calibration, etc. to ensure that they are in satisfactory working condition prior to being used.

12.2.1.3 The Bonding Process

Once the piping components are cleaned, prepared, and aligned, bonding can proceed in accordance with the bonding procedure specification. Each person to perform bonding should be individually qualified by being subjected to a qualification test as described in Section 12.2.1.1. Each qualified bonder should be assigned with an identification symbol, and the symbol recorded in some permanent fashion. Each component should either be marked with the symbol of the individual bonder, or it should be recorded as to which bonder completed each individual plastic weldment. These records should be maintained in a permanent record-keeping system that is established ahead of time. Qualification in one BPS should never be used as a basis for qualifying a bonder or bond operator for another bonding procedure or process.

12.2.2 Thermoplastic Materials

One of the major differences between bonding of thermoplastic materials and welding of metallic materials is that unlike thermoplastic materials, dissimilar thermoplastics cannot be directly bonded together. There are some materials that can be broadly considered to be exceptions to this rule, such as copolymers of one type of resin to a homopolymer of the same major constituent. An example would be a copolymer of polypropylene/ethylene to a homopolymer of polypropylene.

In order for dissimilar thermoplastics to be directly welded (bonded), the copolymer normally has to have a 90% or greater content of the major constituent. A copolymer and homopolymer of this type are different thermoplastics only in a broad sense (comparable to a closely matched low alloy to a parent metal material); some would classify two thermoplastic materials of this type as being the "same" and not "dissimilar."

Having greater than 90% base resin content does not automatically mean that a copolymer and homopolymer of the same thermoplastics may be directly welded (bonded) together. Some may be completely incapable of being joined.

12.2.2.1 Hot-Gas-Welded Joints and Extrusion-Welded Joints in Thermoplastic Piping

Tack welds may be used initially to hold parts of a thermoplastic weldment in proper alignment until the final welds are made. The tack welds at the root of the joint should always be made with the same welding rod used in the root pass. Tack welds should only be performed by those qualified to perform the actual bonding (welding). If a tack weld cracks prior to performing the root pass weld, it should be removed prior to bonding (welding). Otherwise, the tack weld should be left and fused to the root pass weld.

In order to hot-gas weld or extrusion weld a thermoplastic material, it is necessary to use welding rod from the same type of resin. As an example, "homopolymer polypropylene welding rod" for "homopolymer polypropylene" pipe must be used. When the root pass and final pass of the weld are performed, peening should be completely avoided. Additionally, welding should not be allowed to proceed if adverse weather conditions exist

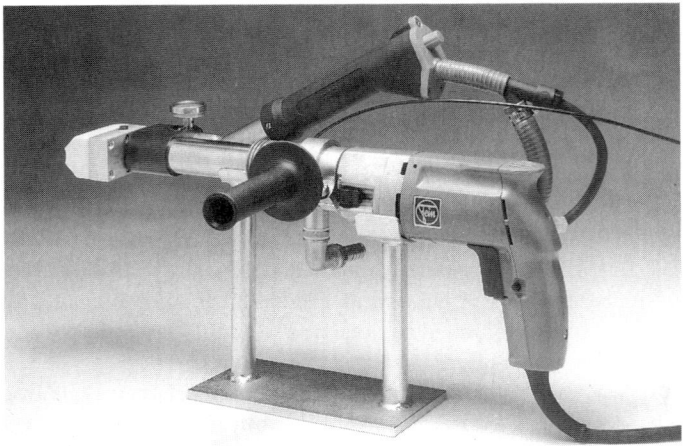

Figure 12.15. Example of an extrusion welding tool. (Source: Wegener North America.)

such as rain, snow, sleet, or excessive wind. Welding can proceed under these circumstances if the entire area of the joint is fully protected from the environmental condition.

Examples of extrusion-welding equipment are shown in Figure 12.15 and 12.16, respectively. The equipment is much larger than that for hot-gas welding, since the nozzle must be large enough to hold an extrusion screw, and a motor must be attached to drive the screw. As a result, extrusion welding requires plenty of space to conduct the welding; approximately 3 ft of clearance in all directions is the usual requirement. The advantage extrusion welding has over hot-gas welding is the application of a thick weld bead in a single pass (usually a five-bead to twelve-bead equivalent, depending on the size of the tool). The result is a substantially lower amount of residual stresses in the heat-affected zone, since only one application of heat is required.

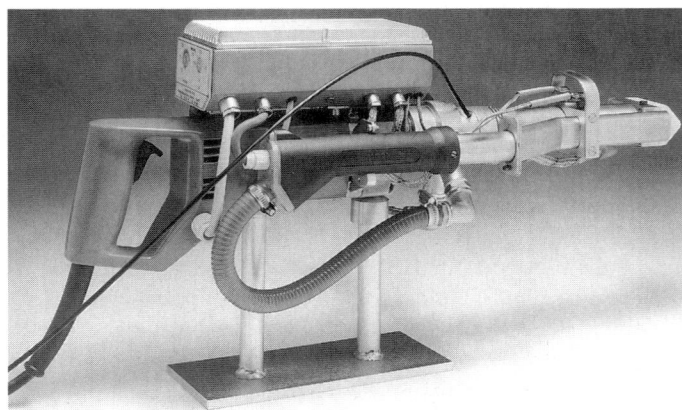

Figure 12.16. Example of an extrusion welding tool for large welds. (Source: Wegener North America.)

12.2.2.2 Solvent-Cemented Joints in Thermoplastic Piping

Polyvinyl chloride (PVC), chlorinated polyvinyl chloride (CPVC), and acrylonitrile butadiene styrene copolymer (ABS) are three types of thermoplastic piping materials that are joined by solvent cementing. Joints that are prepared using solvent cementing techniques may use ASTM D 2855 as a suitable basis for formulating a BPS. Solvent cements to be used should conform to ASTM D 2564 for PVC, ASTM D 2846 for CPVC, and ASTM D 2235 for ABS.

There are several important considerations involved in solvent cementing thermoplastic materials. First, joints should always be suitably prepared and cleaned by wiping with an appropriate solvent. Suitable solvents for PVC and CPVC include acetone or methyl ethyl ketone. Suitable cleaning procedures for ABS are discussed in ASTM D 2235. Second, joints should always be prepared using a suitable primer material prior to applying the actual solvent cement. Suitable primers are described in an ASTM specification that applies to the material being solvent cemented; the primer selected should always match the base solvent of the cement being used.

Another consideration is that there should always be a slight interference fit between the pipe and the fitting socket. The diametrical clearance between the pipe and entrance of the fitting socket should never exceed 0.04 in. (1 mm). For pipe sizes where a relatively large gap is possible, the use of a heavy-bodied cement is preferred. The manufacturers of solvent cements state specific size recommendations for their products and have printed instructions available. Primer and cement should be applied to both of the surfaces being joined, according to recommended procedures. When inserting the fully wetted male surface into the fully wetted female socket, a continuous bond must be produced, with a small bead of cement appearing at the outer limits of the joint.

Branch connections that are fabricated using solvent cementing procedures should be achieved with the use of a manufactured full-reinforcement saddle having an integral branch socket. The reinforcement saddle must be solvent cemented to the run pipe by applying primer and solvent cement over the entire contact surface of both the saddle and the run piping. The rules of Chapter 6 should always be applied in establishing the design of a fabricated branch fitting and in determining its adequacy (see Section 6.3.4).

12.2.2.3 Heat-Fusion Joints in Thermoplastic Piping

Many thermoplastic piping materials may be joined by heat-fusion techniques. This group of methods include heat-element butt-fusion, heat-element socket-fusion, electrical-resistance-fusion, and branch fabrication techniques that include saddle fusion and sidewall fusion. Heat-element butt-fusion is also adaptable to mitered parts to produce fabricated elbows and branch connections. Heat-fusion techniques serve as the primary joining methods for all polyolefins (polyethylene, polypropylene, and polybutylene), the fluoropolymers (PVDF, ECTFE, ETFE, and PFA), and the polyketones (PEEK). The majority of commercially available materials use heat-element butt fusion or heat-element socket fusion. Electrical-resistance fusion has been limited thus far to high-density polyethylene (HDPE), mainly for natural gas services, and polypropylene for use in acid-waste systems (nonpressure). It has also seen limited application in some polypropylene pressure piping.

ASTM D 2657 may serve as a suitable basis for the development of a BPS for butt fusion, socket fusion, and saddle fusion. This standard has been developed based on polyolefin materials, but can be used as a suitable basis for other thermoplastic materials. The specification covers three different procedures: technique I - socket fusion, technique II - butt fusion, and technique III - saddle fusion. When performing these procedures, it is critical to have all surfaces cleaned of foreign matter prior to the application of heat. The use of fixtures to align the components is required in nearly all situations, except for very small diameters where a hand joint can be successfully produced.

The basic procedures for heat-element butt fusion are illustrated in Figure 12.17. When performing butt fusion or socket fusion, the surfaces should be uniformly heated at a controlled temperature. As the heated parts are assembled, a properly performed joint will have a small bead of material appear uniformly around the entire circumference of the outer limits of the joint. A variety of heat-element butt-fusion equipment is shown in Figures 12.18 through 12.22. In Figure 12.18, the equipment is used for pipes having nominal diameters of 1/2 - 2 in. (20-63 mm); Figure 12.19 is used for 1/2 - 4 in. (20-110 mm); Figure 12.20 is used for pipes 1-1/2 - 10 inch (50-250 mm); Figure 12.21 is a benchtop-style fabrication machine designed to be used for pipes of 3-12 in. (90-300 mm), including mitered joints; and Figure 12.22 is a trench-style machine used to join pipes of 3-12 in. (90-300 mm).

Molded branch connections should be used in lieu of fabricated branch connections whenever they are available. Where they are not available, they may be produced with saddle fusion, by mitered heat-element butt-fusion, or by sidewall-fusion techniques. The rules of Chapter 6 should always be applied in establishing the design of a fabricated branch fitting, and in determining its adequacy (see Section 6.3.4).

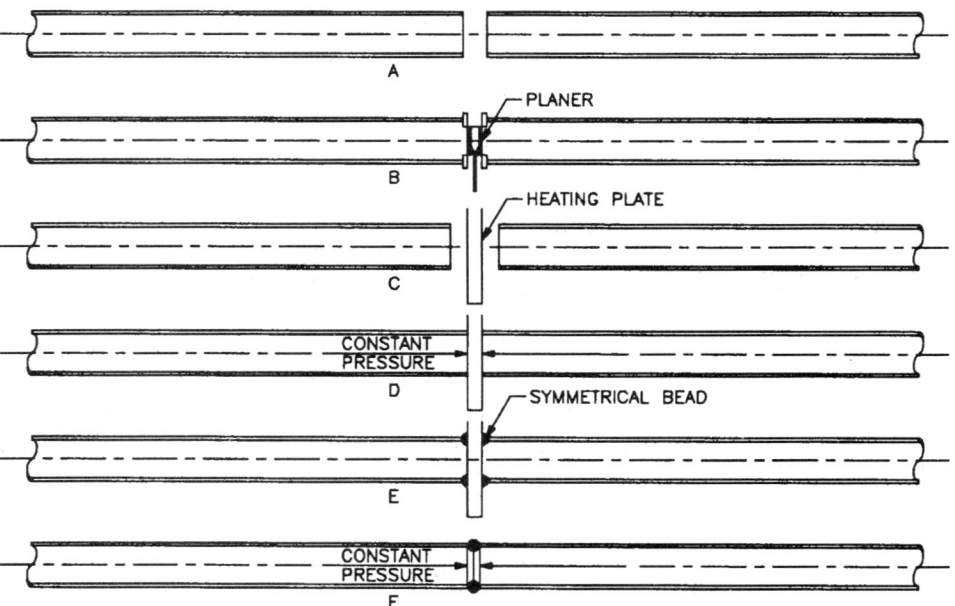

Figure 12.17. Outline of the basic procedures for heat-element butt fusion according to ASTM D 2657.

Fabrication, Assembly, and Erection

Figure 12.18. Example of a butt-fusion tool for sizes 1/2–2 in. (20–63 mm). (Source: Wegener North America.)

Figure 12.19. Example of a butt-fusion tool for sizes 1/2–4 in. (20–110 mm). (Source: Wegener North America.)

Figure 12.20. Example of a bench-style butt-fusion tool for sizes 1 1/2–10 in. (50–250 mm). (Source: Wegener North America.)

Figure 12.21. Example of a bench fabrication butt-fusion tool for sizes 3–12 in. (90–300 mm). (Source: Wegener North America.)

12.2.3 Reinforced-Thermosetting-Resin-Plastic Materials

12.2.3.1 Adhesive Joints in RTR and RTP Piping

All reinforced-thermosetting-plastic piping materials may be joined by adhesive bonding techniques. Adhesive joints may be used with either tapered or straight socket designs, and are applicable to pipes from 1–72 in. (30-1800 mm) piping. Depending on the materials and specific make of piping, the joint may require curing by one of many methods. This may include chemical heat wraps, application of heat through some other means, or by allowing the joint to cure naturally. An example of an adhesive joint of the bell-and-spigot type is illustrated in Figure 12.23.

For every application, a bonding procedure specification must be prepared; the BPS is usually prepared on the basis of the manufacturer's recommendations. Basic recommendations for the consistent production of high-quality joints include[2]:

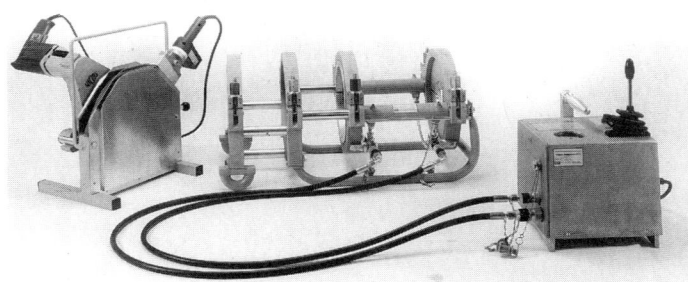

Figure 12.22. Example of a trench butt-fusion tool for sizes 3–12 in. (90–300 mm). (Source: Wegener North America.)

Fabrication, Assembly, and Erection

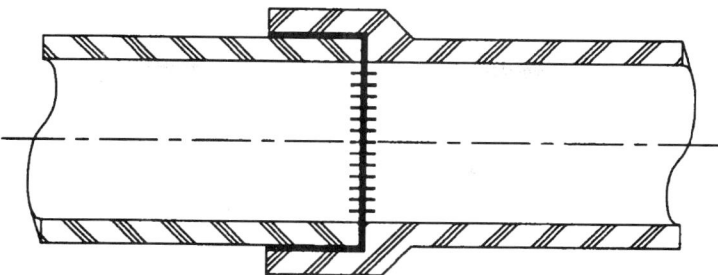

Figure 12.23. Illustration of a bell-and-spigot adhesive joint.

1. Clean and roughen (sand) boding surfaces just prior to bonding.
2. Wet the interfaces well with properly mixed adhesive.
3. Do not allow any voids or pockets in the bond line.
4. Keep the adhesive bond line as thin as possible.
5. Hold the joint together during the entire curing process.

The application of the adhesive to the surfaces to be joined and subsequent assembly of the surfaces should result in a continuous bond between them. Special attention should be paid to any cuts so that they do not result in a lack of complete bonding at their location.

Fabricated branch connections may be made using manufactured full-reinforcement saddles that have a socket or integral length of branch pipe suitable for a nozzle or coupling. The hole in the run pipe should be made with a hole saw. The cut edges of the hole should be sealed with adhesive at the time the saddle is bonded to the run pipe. The rules of Chapter 6 should always be applied in establishing the design of a fabricated branch fitting of this type and in determining its adequacy (see Section 6.3.4).

12.2.3.2 Butt-and-Wrapped Joints in RTR and RPM Piping

As an alternative to adhesive bonding, RTR and RTM piping may be joined using butt-and-wrap techniques. This technique may be facilitated by either directly butting pipes together, or with a bell-and-spigot arrangement. In either method, the area of the joint is overwrapped with layers of resin-impregnated glass fiber cloths. The cloths are applied in layers that become increasingly wider, resulting in a joint capable of withstanding internal pressure and longitudinal forces. Figures 12.24 and 12.25 illustrate the before and after appearance of butt-and-wrap joints involving plain-ended directly butted pipes. Figure 12.26 shows the appearance of a finished joint as a result of a variation of this method when both pipes have tapered ends. Tapering the ends reduces the need for the number and buildup of layers for the butt-and-wrap joint. Figure 12.27 illustrates a bell-and-spigot joint that uses a butt-and-wrap overlay.

For each installation involving a butt-and-wrap technique, a BPS should be prepared. The procedure may be based upon the procedures outlined in NBS PS 15-69, which details a sound basis for butt-and-wrap installation of RTR and RTM piping. Wrapping a joint with plies of reinforcement (glass cloth) saturated with a catalyzed resin (using the same resin as the base pipe) to the surfaces to be joined should result in a continuous structure in

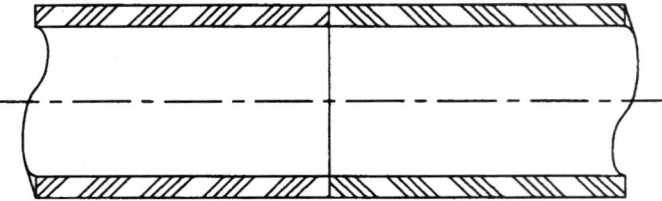

Figure 12.24. Butted RTRP pipes before overwrapping.

every case. Cuts should always be sealed so that the glass-fiber reinforcement can be protected from contacting the service fluid.

Branch fittings may be fabricated using similar hand layup techniques. The hole in the run pipe should be made with a hole saw, and the joint then prepared using the same techniques described for butt-and-wrap joints. The rules of Chapter 6 should always be applied in establishing the design of a fabricated branch fitting of this type, and in determining its adequacy (see Section 6.3.4).

12.2.4 Seal Bonds and Repair Work

Threaded joints may be seal bonded in certain circumstances; Chapter 6 describes the conditions where this practice is permitted (see Section 6.5.1). When threaded joints are seal bonded, the bonder should be a qualified bonder according to the rules mentioned in the beginning of this section. When a threaded joint is seal bonded, all exposed threads should be covered by the seal bond.

If any joints, materials, or workmanship appear to be defective, or fail under testing, it should be replaced or repaired by qualified bonders using known and acceptable techniques. The work should be repaired to the satisfaction of the job-site inspector.

12.2.5 Fabrication of Metallic Piping Lined with Nonmetals

Metallic piping components lined with nonmetals are normally delivered to a job site with linings in place, and with flanges attached to their ends. However, many applications require that subassemblies have field preparation. When subassemblies have to be field

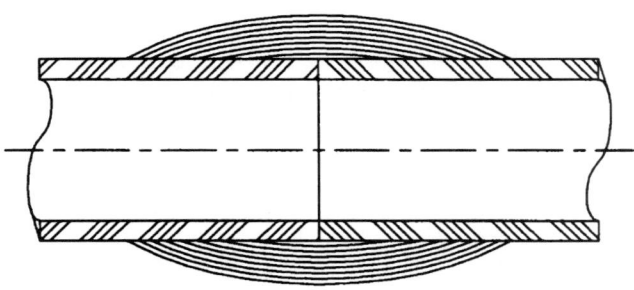

Figure 12.25. Illustration of a butt-and-wrap joint.

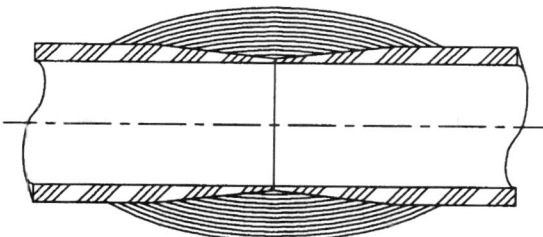

Figure 12.26. Example of a butt-and-wrap joint with tapered pipe ends.

prepared, the welds performed on the metallic external piping should be limited to those areas where the serviceability of the lining will not be affected.

The welding should conform to the ordinary rules of welding metallic materials described in the first part of this chapter. However, for nonmetallic-lined piping, there are additional welding considerations. First, many manufacturers have recommendations that suggest modifications to be made in preparation for welding. These modifications should be addressed in the engineering design. Special attention should be paid to the lining so it is not damaged. If the lining is damaged during the welding process, it should be completely repaired or replaced. Each different type of lining material or manufactured product will require a separate WPS. Therefore, the welder should be qualified for each specific product and type of liner that has a separate WPS.

In many cases, a shorter spool section must be prepared in the field by flaring a pipe that has been previously lined with a nonmetal. Many times this involves first cutting a piece of pipe that has been previously lined to the desired size. When field flaring needs to be performed, a flaring procedure specification (FPS) should be prepared based upon the specific recommendations of the manufacturer of the lined pipe materials. Only operators that are specifically qualified to the FPS should be allowed to perform flaring procedures.

12.2.6 Bending and Forming of Thermoplastic Piping

Thermoplastic piping can be bent and formed into a variety of configurations and components. The applications are limited only by the performance limits of the material and the bending and forming process. The finished surfaces should be free from cracks, and must show no signs of buckling damage. Also, a component that has been formed or bent

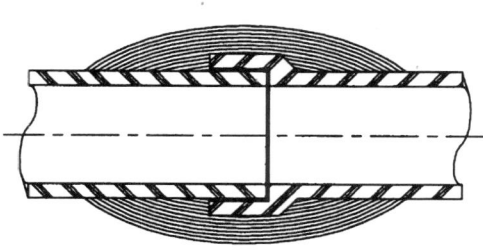

Figure 12.27. Illustration of a bell-and-spigot joint with a butt-and-wrap overlay.

should have a final thickness at all points that meets the minimum required design thickness described in Chapter 6 (see Section 6.3.3).

When thermoplastic piping is fabricated using bending techniques, the resulting thickness of the bent pipe should have a maximum flattening of 5% for both internal and external pressure applications. Hot bending of piping should be performed at a temperature above the glass transition temperature of the plastic. Bends created from bending corrugated piping should be qualified for pressure design by the same rules as described in Chapter 6 for "unlisted" components (see Section 6.2.2). The temperature range for forming varies according to the type of material and the specific application.

12.2.7 Threaded Joints

When assembling threaded joints, thread compounds or lubricants should be selected that do not react unfavorably with a service fluid or attack the nonmetallic piping materials. A thread compound should not be used if the joint is to be seal-welded after threading. However, a threaded joint that is to be seal welded to correct a leak may already contain a thread compound. It is acceptable to seal weld under this circumstance, but the thread compound should be removed from the exposed threads to the greatest extent possible. When installing straight-threaded joints where the seal is not at the threads, care should be taken so that the gasket, O-ring, or seat is not damaged. For thermoplastic threaded joints, either strap wrenches or other full circumferential wrenches should be used to tighten threaded pipe joints. Tools that score or scratch thermoplastic piping significantly should be completely avoided. For RTP pipe threads, care should be taken to prevent service fluids from being able to attack the reinforcing material. If necessary, apply a sufficient amount of resin over the threads to fill the clearance between the pipe and the fitting completely.

Other types of joints such as flared tubing joints, flareless and compression tubing joints, caulked joints, expanded joints, packed joints, and proprietary joints should be installed with careful practices according to the instructions of the manufacturer. Flared thermoplastic joints should be prepared in accordance with ASTM D 3140, "Flared Joints for Polyolefins," or other appropriate method.

12.2.8 Flexible Elastomeric Sealed Joints

A variety of flexible elastomeric sealed joints are available for use in many materials, primarily for underground service. The assembly of these types of joints should follow closely with the specific manufacturer's instructions for the product being used. Seal and bearing surfaces of these joints should be inspected first to be sure that there are no injurious imperfections. If a lubricant is recommended for these types of services, the lubricant should be compatible with the service fluid. Also, if expansion is likely to occur due to thermal or pressure effects, proper restraint should be provided to the pipe to ensure that joint separation will not occur.

12.2.9 Joining Nonplastic Nonmetallic Piping

12.2.9.1 Borosilicate Glass Piping

Most nonplastic, nonmetallic piping is joined using mechanical techniques, and arrive at the job site as spool pieces that are ready to be joined. In some cases, however, there are

Fabrication, Assembly, and Erection

short pieces required to be made up to correct for differences between fabrication drawings and field dimensions. Pieces like this may be cut to length and joined in the field, following the recommended practice of the manufacturer. With borosilicate glass piping, it is important that defects be detected, repaired, and replaced prior to being joined in the system and placed into service.

12.2.10 Assembly and Erection of Single-Wall Nonmetallic Piping

The most important consideration in the assembly and erection of nonmetallic piping materials is that they be fitted without the development of detrimental distortions. Oftentimes it is possible that piping need to be distorted slightly in order to bring it into alignment for the connection of flanges, threaded joints, etc. This practice should be avoided, and in any case should not result in the development of detrimental strains in piping components and connected equipment, or excessive movement of their supports. When cold spring is to be employed, the dimensions and positions of all guides, supports, and anchors should be checked to see that they are in their right location. In no case should an undesired gap be corrected by the use of heating, as the creation of residual stresses will tend to offset the benefits of cold springing.

Special attention should be paid to the tolerance of the flange face matchups. All flange faces should align to within 1/16 in./ft (1.5 mm/ 0.3 m) (0.5%) measured across any diameter and bolt holes should align to within 1/8 in. (3 mm) maximum offset. Gaskets should be checked to be sure that its seating surface is not damaged. In no case should multiple gaskets be employed between a single set of contact faces. Bolts that are selected should ideally extend all the way through their nuts, but can be acceptably sized to be within one thread of complete engagement. All flanges should be tightened to a predetermined torque whenever possible. This is especially important when assembling flanges that have widely differing mechanical properties. In every application, gaskets must be uniformly compressed to the proper design loading in order to maintain the integrity of each flanged joint.

When other than flat face flanges and full face gaskets are used, consideration should be given to the strength of the flanges, and to sustained loads, displacement strains, and other occasional loads described in Chapter 6. Also, an appropriate boltup sequence should be specified in all cases. Appropriate limits for boltup torque should be established, and a torque wrench used to ensure that these limits are never exceeded. Flat washers should be used under all bolt heads and nuts so that the nonmetallic materials are not damaged when tightening metallic bolts and nuts.

In assembling nonmetallic lined metallic piping, there should be means provided to ensure that electrical continuity of the pipe sections exist where this is a potential problem. Otherwise static sparking could occur to result in the ignition of flammable vapors.

12.2.10.1 Assembly of Borosilicate Glass Piping

Borosilicate glass piping and other brittle materials should be handled carefully to avoid scratching and notching during all procedures involving handling and during storage. If there are any scratches, notches, or chips in the surface of a borosilicate glass component, the part should be discarded and replaced. Borosilicate glass piping should also be protected from weld splatter. If it is damaged in such a manner, the part should be discarded and replaced. Care should be used in the joining of flanges and mechanical joints. The gaskets used should be limited to the type recommended by the manufacturer. The same is true of the sequence for torquing of bolts for flanges.

For other types of brittle piping, such as glass-lined steel and cement-lined steel, care should be taken to protect against impact damage. Oftentimes the material can sustain an external blow that is nondamaging to itself, but harmful to the interior lining. If such an impact does occur, lining materials should be inspected for damage prior to use, and any damaged material discarded and replaced.

12.3 Double Containment Piping Installation Issues

The first two sections of this chapter discuss installation and fabrication issues for single-wall piping materials that make up the inner and outer portions of a double containment system. This section concentrates on those issues that relate only to combining the individual single-wall pipe systems to form a complex double containment system. Of primary importance in double containment piping assembly and installation is the sequence selected for assembling and joining the inner and outer pipes. The choice of sequence depends on code limitations (i.e., required visual and nondestructive examinations), system economics, available joining technologies based on the materials to be used, and any limitations that the design places on the system. These design limitations may include such items as required flexibility (degree of nonrestraint), support requirements, and other issues. Flexibility requirements may arise from a number of reasons, due to thermal forces, external loads imposed by burial forces, and others. Therefore, the selection of joining process and sequence requires careful consideration, as it has a direct effect on both the cost of the system and on the system performance.

12.3.1 Joining Sequence

There are two basic joining sequences to be considered for assembling double containment pipe systems. These are referred to as: (1) staggered welding and (2) simultaneous fusion. Staggered welding is performed either by joining inner materials first, followed by outer materials, or by joining outer components first, followed by joining and inserting the inner components. A staggered welding method allows for complete inspection of each and every weld in both inner and outer systems, during and after the welding process. Staggered methods can be used with any combination of materials, and is the sequence of choice whenever a hybrid material system is involved. A staggered welding method is also chosen if visual inspection and nondestructive examinations are required for primary components.

Simultaneous fusion is a process that is limited to plastic piping materials. In the simultaneous fusion process, inner and outer components are permanently affixed to each other by welding (bonding) or mechanical means. Once inner and outer components are fixed to each other, both inner and outer welds (bonds) can be formed at the same time. Simultaneous fusion (see Section 12.3.3) may be used for hybrid material systems that use the same type of heat joining method (i.e., butt fusion and butt fusion, socket welding and socket welding, etc.), or the same type of chemical bonding method (i.e., solvent cementing and solvent cementing, adhesive bonding and adhesive bonding, etc.). Simultaneous fusion does not allow for any inspection of inner pipe welds during the bonding process, or while conducting a hydrostatic test. It typically results in a permanently restrained system; a system that lacks sufficient flexibility is not desirable for a thermoplastic system (see Section

8.5 and 6.6.1.1). The burial resistance of such systems may also be lessened, since rigid interconnecting welding plates are often used to affix inner and outer systems together.

Simultaneous joining techniques have many inherent drawbacks that often preclude their use. The reason that simultaneous joining techniques have any appeal is that overall labor costs can be reduced. However, the catch is that labor costs can only be reduced if primary pipe welds may be made without initial pipe leaks. Every primary pipe leak detected in an initial pressure test tends to counteract much of the potential labor savings. If enough primary piping leaks are encountered, this method will result in a more expensive installation.

12.3.2 Staggered Welding (Bonding) Techniques

Staggered welding (bonding) can be broadly divided into two basic types: "inner-first" and "outer-first" sequences. The majority of installations involve joining inner (primary) components first, followed by the fabrication of the associated secondary containment components at some later time. This process is referred to as an inner-first staggered joining sequence. There are occasions, however, where the secondary containment pipes are welded (bonded) first, or separately from associated inner pipes, followed by welding (bonding) and insertion of the primary piping components. Such an "outer-first" approach is used extensively in landfill double containment piping applications involving the assembly of long straight runs of HDPE-HDPE double containment pipes. This sequence of assembling a pipe within a pipe has another common name; when used for retrofitting existing concrete sewer and gas pipes it is termed slip-lining. Each of these staggered-joining sequences is discussed in the following sections.

12.3.2.1 Assembly Using an "Inner-First" Sequence

The approach most commonly used in constructing a double containment piping assembly is to weld (bond) inner pipe joints first, and outer pipe welds at a later time. There are several reasons why this sequence is the most common approach. First, it allows all welds to be visually inspected during the welding process, as well as during hydrostatic testing of the primary pipe. These visual inspections are mandatory for many systems that are designed and installed in accordance with common codes such as the ANSI/ASME B31 Pressure Piping Codes.[3] In addition, all welds of the inner pipe system can be examined nondestructively prior to completing the secondary containment casing. Installing a pipe in this fashion has the added advantage of allowing the installed primary piping system to move freely in longitudinal directions within its secondary containment casing, allowing components to undergo desired movements. This feature is desirable where the installation of a freely flexible, unrestrained system is important due to thermal expansion design conditions (see Section 8.3).

The "inner-first" staggered welding (bonding) assembly sequence of straight piping segments is illustrated in Figures 12.28 through 12.52 for various materials. Figures 12.28 through 12.33 and 12.34 through 12.39 illustrate the two sequences used for assembling straight sections of metallic pipes via some type of butt-welding method, as would be done on a long, straight run. Figures 12.40 through 12.46 illustrate a similar sequence for joining thermoplastic pipes that use thermal-butt-fusion joining methods. This sequence illustrates the use of split-and-hinged heating elements used to make the secondary containment joints Figures 12.47 through 12.52 illustrate an "inner-first" joining sequence for materials that use socket-joining techniques (i.e., adhesively bonded RTRP materials, solvent-cemented thermoplastics). For RTRP systems that use butt-and-strap techniques for joining outer pipes, the details shown in Figures 12.24 and 12.25 can be substituted.

The reader must be aware that these are simplified representations of three possible sequences. There are many combinations of inner and outer materials and joining methods that may be used; thus, there can be many different possible representations of "inner-first" sequences, other than those shown here. Actual construction of a double containment piping assembly is complex; the actual sequence of joining and the specific techniques used depend on all of the site factors involved in the application, the extent of prefabrication, etc.

Chapter 11 discusses sequential methods and options for assembling components other than straight pipe, such as branch fittings and elbows, as they pertain to system layout and preparation of shop drawings (see Sections 11.1.1.6 and 11.1.2.3).

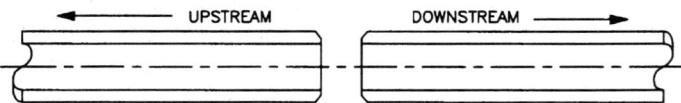

Figure 12.28. In the first step of the "inner-first" staggered assembly of small-diameter metallic pipe straight sections where the outer jacket is capable of being telescoped, and involving a butt-welding procedure for inner and outer pipes, the ends of the primary pipes are prepared for welding.

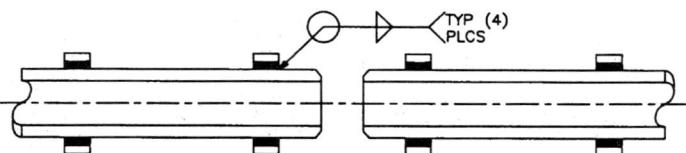

Figure 12.29. In the second step, interstitial supports are attached to the primary pipes.

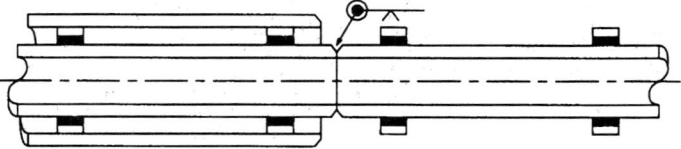

Figure 12.30. In the third step, the outer jacket is positioned into place on the upstream side (with ends prebeveled), and the primary pipes are welded. Note that the secondary pipe must fall short of the weld on the upstream side.

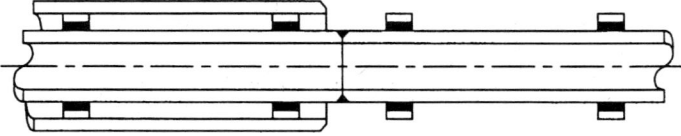

Figure 12.31. Illustration of the assembly section after the primary pipe weld is completed.

Fabrication, Assembly, and Erection

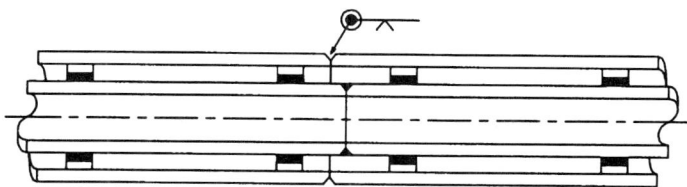

Figure 12.32. The secondary pipe is slipped into position for welding. After the primary pipe has been tested and the weld visually inspected, the secondary jacket is telescoped back into position, and the outer joint is then welded.

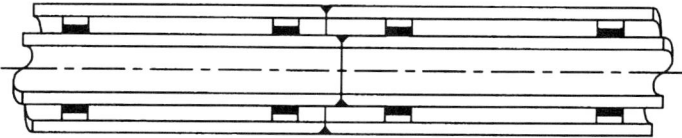

Figure 12.33. Illustration of the section of the double containment piping system after the secondary containment jacket is finished being welded.

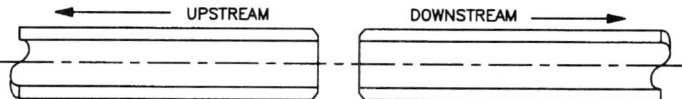

Figure 12.34. In the first step of the "inner-first" staggered assembly of metallic pipe straight sections using a nontelescoping method, the ends of the primary pipes are prepared for welding.

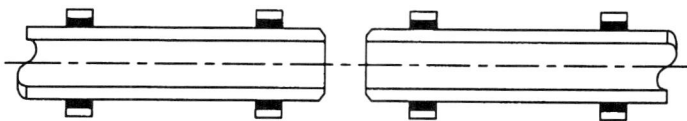

Figure 12.35. In the second step, interstitial supports are attached to the primary pipes.

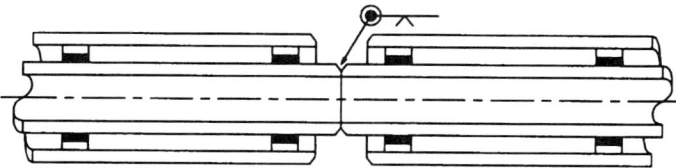

Figure 12.36. In the third step, the secondary containment pipes are prepositioned into place on both sides of the primary joint, and the primary joints are then welded. Note that the secondary containment pipes must fall short of the weld on either side, usually by 3 to 6 in. on each side.

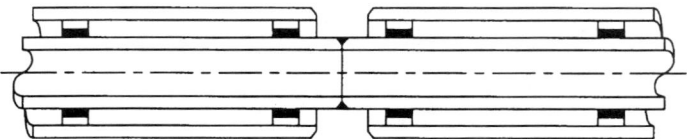

Figure 12.37. Illustration of the appearance of the assembly section after the primary pipe weld is completed. The primary pipe weld can be visually inspected during an initial pressure test, and/or nondestructively examined, prior to coverup.

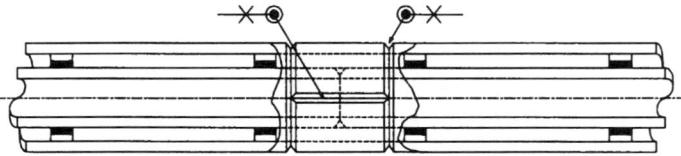

Figure 12.38. Once the inspection and nondestructive exams of the primary pipe joint are completed, the closure component can be positioned into place, tack welds made, and then the root pass weld completed.

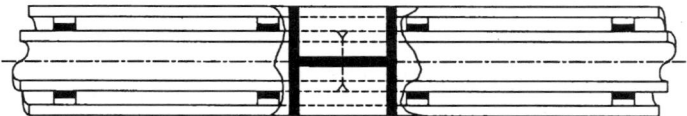

Figure 12.39. Illustration of the section of the double containment piping system after the secondary jacket is finished being welded.

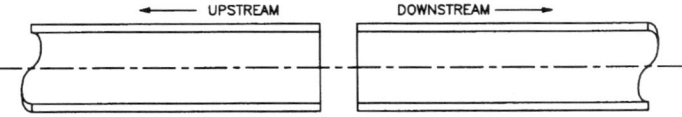

Figure 12.40. In the first step of the "inner-first" staggered assembly of thermoplastic piping where the outer jacket is capable of being telescoped, and involving a heat element butt-fusion procedure, the primary pipes are prepared for joining.

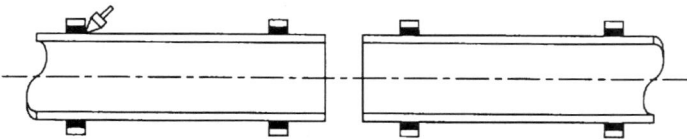

Figure 12.41. In the second step, interstitial supports are secured to the primary pipes (in the example shown, by means of hot gas welding).

Fabrication, Assembly, and Erection

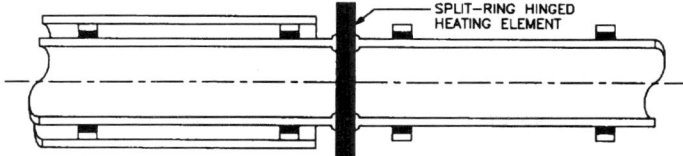

Figure 12.42. Next, the secondary containment piping is slipped over the primary pipe on at least the upstream side. A buttweld is then made on the primary pipe using a heating element. Note that the secondary containment pipe must fall short of the weld, usually by 3 to 12 in..

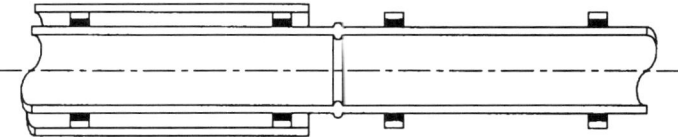

Figure 12.43. Illustration of the assembly section after the primary pipe weld is completed.

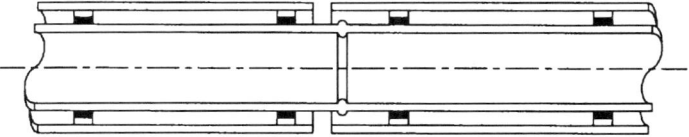

Figure 12.44. In the next step, the secondary containment pipe is slipped over the interstitial supports on the downstream side. The primary pipe weld can be visually inspected during the pressure test of the primary pipe by telescoping the secondary containment jacket, prior to completing the secondary containment weld.

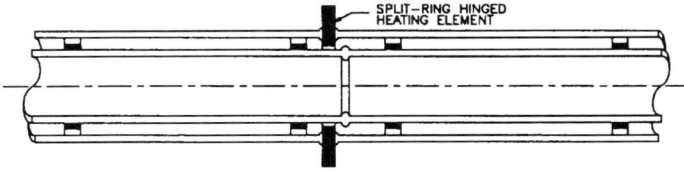

Figure 12.45. Once the primary pipe weld visual inspection is completed, the secondary containment pipe sections can then be welded using a split-ring annular heating element (see Figures 3.1 and 3.2).

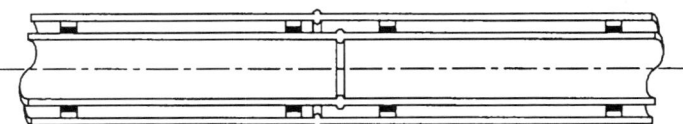

Figure 12.46. Illustration of the section of the double containment piping system after the secondary containment jacket is finished being welded.

12.3.2.1.1 "Inner-First" Assembly in All-Metallic Straight Pipe Sections

Figures 12.28 through 12.39 illustrate common sequences of assembling multiple lengths of straight metallic pipes, as in a long straight run. This sequence can be used in certain systems involving relatively small-diameter metallic pipes, whereby the secondary containment piping segments can be telescoped back and forth over the interstitial supports. Telescoping is required in order to enable the primary pipe joints to be visually inspected during a pressure test, or nondestructively examined in accordance with Code criteria (i.e., the ANSI/ASME B31.3 and B31.1 Codes). By telescoping secondary containment piping segments, one can avoid having to use a closure coupling (e.g., a "clam-shell" coupling as shown in Figures 12.38 and 12.39) at every joint location on the secondary containment shell. However, it should be pointed out that at least one gap must exist in the secondary containment pipe in order to allow for telescoping of the other segments. The gap must be filled with a "clam-shell" coupling, or other method as shown in Figures 12.38 and 12.39. In larger pipes, the method illustrated in Figures 12.28 through 12.33 becomes impractical, and instead the method illustrated in Figures 12.34 through 12.39 must be used.

The step-by-step procedure presented in Figures 12.34 through 12.39 illustrate the traditional means of assembly all-welded double-walled piping systems utilizing a "clam-shell" secondary coupling. This type of joint can be used at every joint location, as is the norm in systems having primary piping sizes that are greater than nominal 4 in. However, the sequence shown is also the detailed procedure used when installing one "clam-shell" coupling in a long straight run whereby all other joints use the procedures illustrated in Figures 12.28 through 12.33.

12.3.2.1.2 "Inner-First" Assembly for All-Thermoplastic Straight Pipe Sections (Butt Fusion Inner and Outer)

This procedure can be used on relatively small systems that allow for the secondary containment piping sections to be telescoped back and forth over the interstitial support. Telescoping is required in order to enable the primary pipe joints to be visually inspected during a pressure test, or nondestructively examined in accordance with Code criteria (i.e. ,the ANSI/ASME B31.3 and B31.1 Codes). By telescoping secondary containment piping segments, one can avoid having to use a closure coupling (e.g., a slip coupling as shown in Figures 11.67 and 11.70 at every joint location on the secondary containment. However, it should be pointed out that at least one gap must exist in the secondary containment pipe in order to allow for telescoping of the other segments. The gap must be filled with a slip coupling, or other method as shown in Figures 11.67 and 11.71. In larger pipes, the method illustrated in Figures 12.28 through 12.33 becomes impractical, and a method comparable to that illustrated in Figures 12.34 through 12.39 for metallic piping must be used.

12.3.2.1.3 "Inner-First" Assembly for Socket-Based Inner and Outer Straight Pipe Sections

Figures 12.47 through 12.52 illustrate a sequence of joining when socket-type joints are used for joining straight sections of pipes, as in a long straight run. In Figure 12.47, straight primary piping lengths are prepared for joining. In Figure 12.48, the pipes are attached to a coupling, by means of socket welding (metallic materials), adhesive bonding (RTRP materials), or solvent cementing (thermoplastic materials). In Figure 12.49, intersti-

Fabrication, Assembly, and Erection   535

tial supports (centering devices) are attached to the primary pipes (these can alternatively be attached to the pipes prior to the first step). In Figure 12.50, the secondary containment pipes, and a coupling, are all slipped over the interstitial supports and positioned for bonding. Figure 12.51 illustrates the secondary containment pipes being bonded into a coupling by either adhesive bonding (RTRP) or solvent cementing (thermoplastics). The final step, Figure 12.52, illustrates a completed primary/secondary containment pipe section.

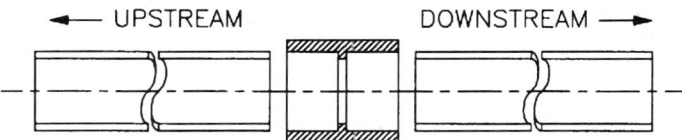

Figure 12.47. In the first step of an "inner-first" staggered assembly of metallic or nonmetallic straight pipe section where the outer pipe can be telescoped, and a "socket" joining method is used, the primary pipe sections, and the coupling, are prepared for joining.

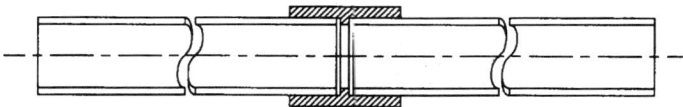

Figure 12.48. The primary pipes are joined to the coupling by the prescribed method (e.g., socket-welding-metals, solvent cementing or socket fusion-thermoplastics, adhesive bonding RTRP materials).

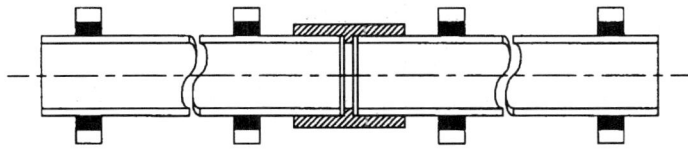

Figure 12.49. Once the coupling is joined, interstitial supports are attached to the primary pipe sections, either mechanically, or by welding, cements, or adhesives. Alternatively, these interstitial supports can be attached prior to the first step.

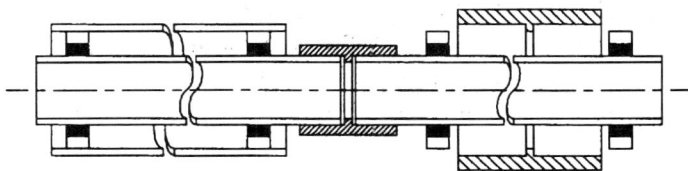

Figure 12.50. Next the secondary containment pipe must be slipped over the interstitial supports. The primary system must then be pressure tested and the joints visually inspected.

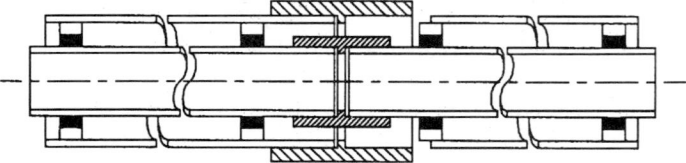

Figure 12.51. The secondary containment coupling is joined to the adjacent secondary containment pipes.

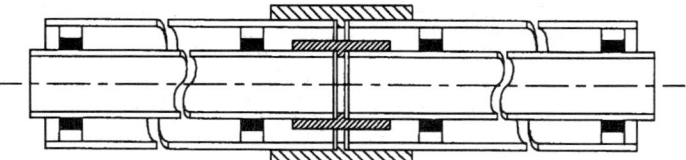

Figure 12.52. Illustration of the section of the double containment piping system after the secondary containment jacket is finished being assembled.

12.3.2.1.4 Sequential Installation of Fixed-Point Components

A basic sequence of assembly of fixed-point components (i.e., internal anchors and termination/transition components) is illustrated in Figures 12.53 through 12.69. In systems that are welded using a staggered "inner-first" sequence, certain components are not capable of being consecutively installed in sequence. These components are termed "fixed-point" components and include fittings such as internal anchors (see Section 11.1.1.4.2). The installation sequence must normally proceed in one of two ways for these components. They can serve as one of the starting points for installation of a double containment assembly; where this is the sequence used, either midline closures or end closures will be required at a point away from the fixed-point component. If a fixed-point component is installed as it is happened upon in a sequential installation, then a secondary (midline) closure will be required to connect the outer portion of the component to the secondary containment pipe on its "upstream" side. The downstream side can continue a normal staggering sequence of "inner-first" where component geometry permits. On some installations involving plastic materials, the need for secondary closures can be avoided with the use of a simultaneous weld on the upstream side of the component; the simultaneous fusion replaces a primary pipe weld where a secondary closure is otherwise required. (Also refer to the discussion on simultaneous fusion, Section 12.3.3.)

Figures 12.53 through 12.59 illustrate the joining of a midline internal-anchor component of homogeneous construction (e.g., Type-1 internal anchor, shown in Figure 11.50), and using a slip-coupling midline secondary closure (shown in Figures 11.71, 11.75), on the upstream side of the component. Figures 12.60 through 12.64 illustrate the sequence of using Type-1 termination/transition component as one of the starting points in the assembly of a system. An example involving the use of a simultaneous weld on the consecutive upstream side of a "fixed-point" component in an all-thermoplastic system is presented in Figures 12.65 through 12.69, which is discussed further in Section 12.3.3.

Fabrication, Assembly, and Erection

For situations involving dissimilar primary-secondary containment materials, internal anchors of the types shown in Figures 11.11 and 11.51 have many important advantages (see Section 11.1.1.4.2). If a homogeneous-material component is used in a system having a hybrid combination of primary-secondary containment pipes, installation will require a flanged secondary containment connection on both sides of the solid anchor. Use of this type of system is less than ideal for underground installations, due to a need to provide cathodic protection and/or covering to the flange bolts, nuts, and washers. Installing a Type-2 internal anchor makes the final arrangement more efficient and thus more economical. This takes five circumferential field welds, as opposed to eight welds and two bolted-up joints, and is more suited to direct burial. In all-metallic hybrid systems, a Type-3 internal anchor may be used of the types shown in Figures 11.44 and 11.45, which can further reduce the number of field welds to four.

Unless a double containment branch fitting assembly is constructed using a "clam-shell"-type secondary midline closure, it will have the same assembly criteria as other "fixed-point" components. The same sequences described in the preceding paragraphs apply to branch-fitting installation in systems using a staggered-type, "inner-first" approach to welding and assembly.

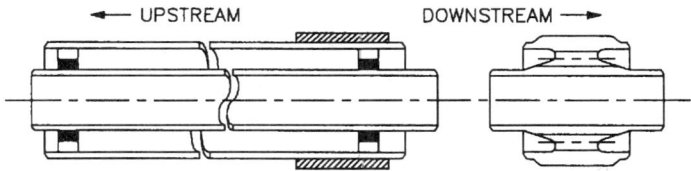

Figure 12.53. Illustration of the first step when installing a type-1 internal anchor in a metallic system using a butt-welding method, and using a slip-coupling closure method on the upstream side of the secondary containment pipe. Note that the slip coupling is pretelescoped back on the secondary contain-

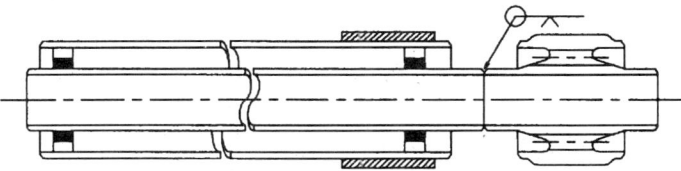

Figure 12.54. In the second step, the primary pipe weld is performed.

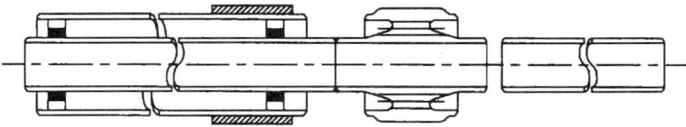

Figure 12.55. Illustration of the appearance of the fitting after the primary pipe weld is made. The next sequential downstream primary pipe is brought into position.

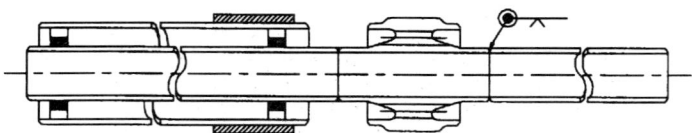

Figure 12.56. The next primary pipe segment on the downstream side is welded.

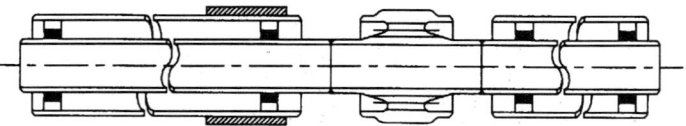

Figure 12.57. The interstitial supports are attached to the downstream secondary containment pipe.

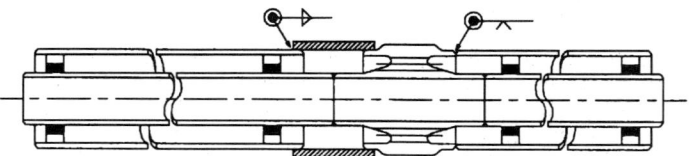

Figure 12.58. After the primary pipe is tested and the primary joints are visually inspected and nondestructive exams are performed, the secondary containment jacket is ready for closure. The slip coupling is positioned in place, the downstream pipes are telescoped into position, and the various welds are performed.

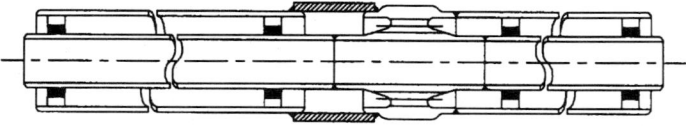

Figure 12.59. Illustration of the completed assembly after the secondary containment piping is welded.

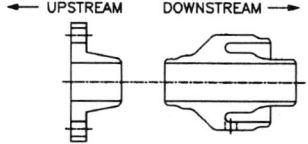

Figure 12.60. In the first step, the type-1 termination component and flange are positioned in place for welding. Note that this version of termination fitting has the primary pipe on the double containment system extending beyond the secondary portion. This allows the primary pipe system to be staggered beyond its outer shell. Alternatively, the inner and outer portions could end in a parallel fashion, depending on the needs of the system.

Fabrication, Assembly, and Erection

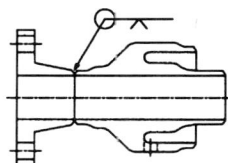

Figure 12.61. In the second step, the single-wall flange is welded to the termination fitting.

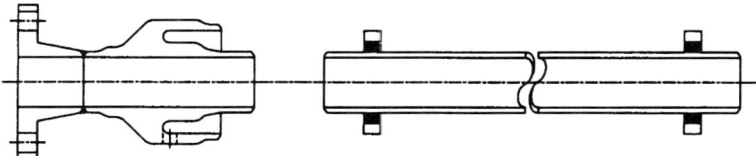

Figure 12.62. In step 3 the welded termination fitting and flange combo is positioned next to the first primary pipe section.

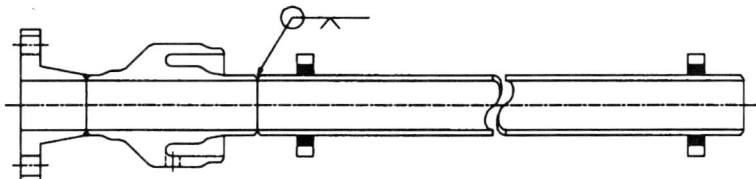

Figure 12.63. In step four, the primary pipe is welded to the termination/transition arrangement.

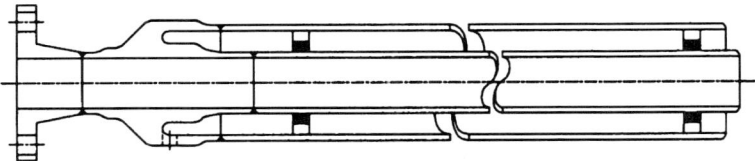

Figure 12.64. In the last step, the secondary containment pipe is welded to the outer part of the termination fitting, after the primary weld passes the required visual and other nondestructive exams.

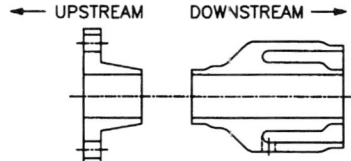

Figure 12.65. In the first step, the type-1 termination fitting and flange are positioned into place for heat-element butt welding.

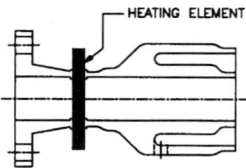

Figure 12.66. In the second step, the single-wall flange is welded to the termination fitting.

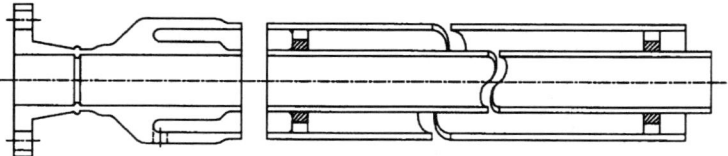

Figure 12.67. In the third step, a prefabricated double containment pipe section is positioned into place for welding.

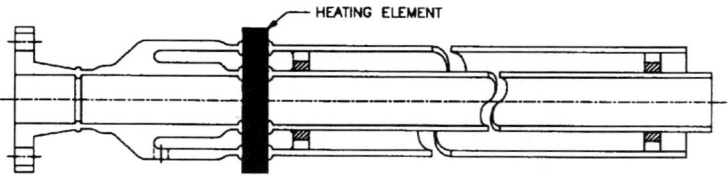

Figure 12.68. In the fourth step, the termination/transition arrangement is simultaneously fused to the prefabricated double containment pipe section using the heat-element butt-fusion method.

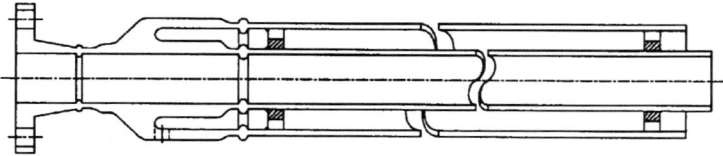

Figure 12.69. Illustration of the final appearance of the assembly.

12.3.2.1.5 Completely Field-Fabricated Systems

Systems that are joined by staggered fusion can be completely fabricated in the field in sequential fashion (see also Section 11.1.2.2). The use of these techniques can minimize the need to make field secondary closures, which are otherwise required to complete a double containment casing when extensive shop prefabrication is used. Use of these methods can result in a quality installation, but they do require the use of detailed planning, careful measurements, and extra setup time on the part of contractor personnel/installers. It is helpful to identify the location and sequence of every weld in order that unanticipated problems do not arise in the field. Fabrication in this manner does involve cutting spool lengths to accurate dimensions in order for the centerline of fittings to match correctly and the pipes to match the desired layout.

Complete field sequential installation of a double containment assembly will mean that the outer casing is completely sealed prior to conducting a pressure test of the primary piping system. This prevents the primary pipe welds from being visually inspected during the pressure test, which may not allow the system to be in compliance with ANSI/ASME B31 Pressure Piping Codes.[3] It also goes against well-established piping practices.

One must be aware that systems may also be field fabricated without having to fabricate in a sequential manner. A system that is fabricated in a nonsequential manner will require additional secondary closures. However, it will allow a system to be field fabricated and still meet code requirements for inspection, examination, and testing.

In general, the use of pre-fabrication will tend to increase efficiency overall by minimizing lengthy and complex setups in the field.

In performing field sequential installations, elbows, branch fittings, and other components beyond pipe gain added importance. Double containment elbows (90°, 45°, 30°, etc.), mitered bends, and other elbow fittings may be assembled in a sequential fashion in the field to avoid the use of secondary closures. However, branch fittings must be assembled in a manner which is similar to fixed-point components (see Section 12.3.2.1.4).

Sequential joining may be used as layout option for some prefabricated sections involving elbows. However, it is more common to start at the elbows themselves, as described in Section 11.1.2.2.3, except perhaps in expansion loop/offset construction. Sequential installation of elbows is therefore presented here as a field joining option, and only in a conceptual sense. It must be remembered that existing design codes (e.g., ANSI/ASME B31.3) may eliminate sequential joining as a possible joining/layout option due to inspection, examination, and testing requirements of the primary piping system. Also, sequential elbow layout techniques are not possible in systems using simultaneous joining methods; simultaneous joining methods represent a completely opposite approach.

The steps required for sequential joining of sections of double containment piping involving elbows in order to avoid secondary closures is illustrated in Figures 12.70 through 12.78 for butt-style component systems and Figures 12.79 through 12.83 for socket-style component systems. Socket-type systems that allow this procedure include solvent-cemented thermoplastics (e.g., PVC, CPVC, ABS), and adhesively bonded "commodity" RTRP systems (e.g., epoxy FRP and vinyl ester FRP). It is not suggested here that secondary closures should not be used or should be avoided. However, these sequences are shown here for those situations where such a procedure is appropriate and is the preference of the owner.

For elbows other than 90°, the same sequence is used. It is a wise idea when planning to use this procedure to prepare at least one detailed set of drawings to illustrate to the installing contractor the sequence of joining desired and the exact location and type of welds (bonds). The length of the small piece of pipe used in the procedure involving butt welding, sometimes termed a "pup," which is shown in Figure 12.73, as attached to the downstream side of the 90° elbow should be determined through dimensional analysis, if this information is not determined in the installation stage, using a trial and error approach.

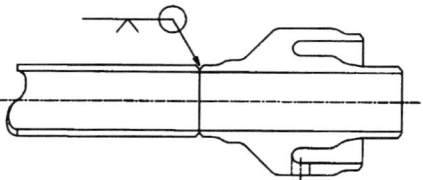

Figure 12.70. Starting with a termination fitting, a sequence is illustrated that allows for a complete field installation involving an elbow section downstream without having to use a secondary closure fitting. The sequence shown here involves a butt-style system as is used in metallic piping. Note that the piping on the double containment side extends beyond the outer jacket. This is a key part of the procedure.

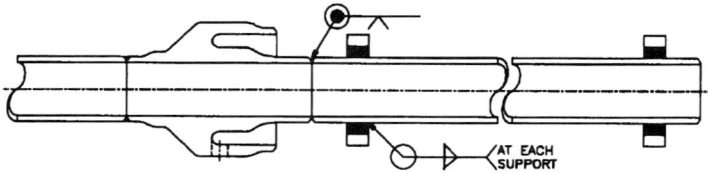

Figure 12.71. In the second step, the next downstream primary pipe is attached.

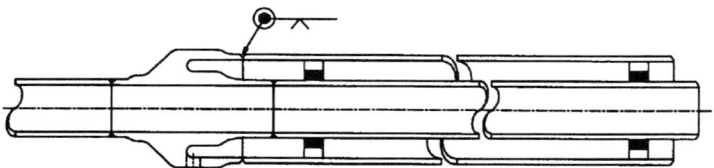

Figure 12.72. In the third step, the secondary containment jacket is joined to the termination/transition fitting. Note that if the lengths of the primary and secondary containment piping sections are equal, the primary pipe will extend beyond the secondary containment pipe by the same amount provided on the termination fitting. The second and third steps are repeated for each straight piping segment until the first downstream elbow is reached.

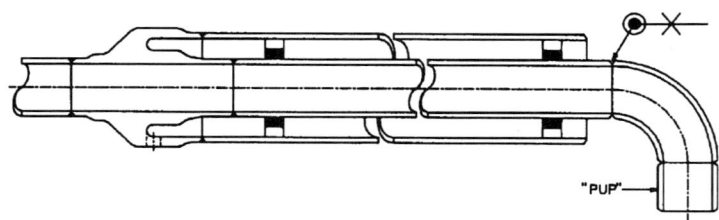

Figure 12.73. Next, the primary elbow is joined to the primary pipe. Note that a straight piece, termed a "pup," must be joined to the downstream side of the primary elbow. The length of the pup must be short enough to allow the secondary containment elbow to slip around the primary containment elbow to allow for continuation of the sequence downstream of the elbow location.

Fabrication, Assembly, and Erection

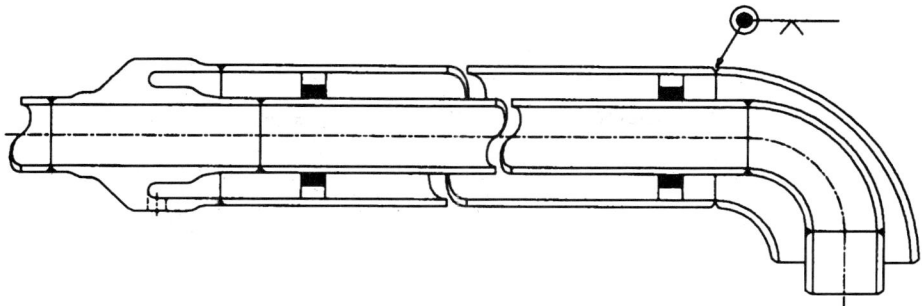

Figure 12.74. Next, the secondary containment elbow is slipped over the primary elbow and welded. Note that the secondary containment elbow may require a "pup" as shown in the illustration; this depends on the relative length of "pups" specified for the primary pipe. If the right dimension is determined, then no "pup" is required on the upstream side of the secondary containment elbow.

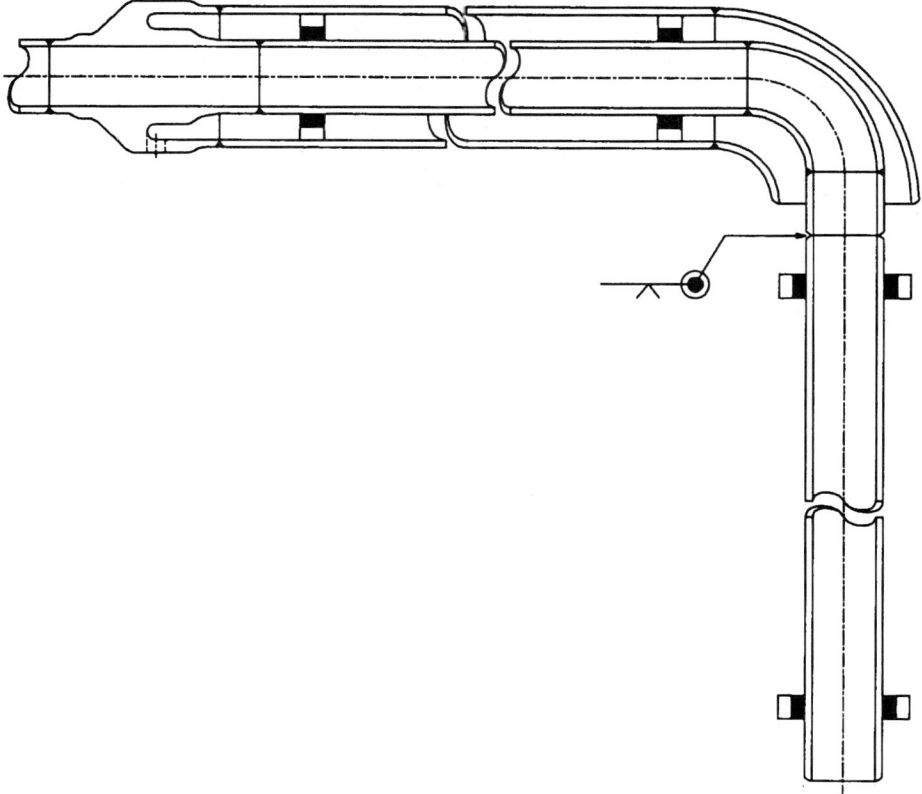

Figure 12.75. The next downstream primary pipe is then welded to the downstream "pup" of the primary elbow.

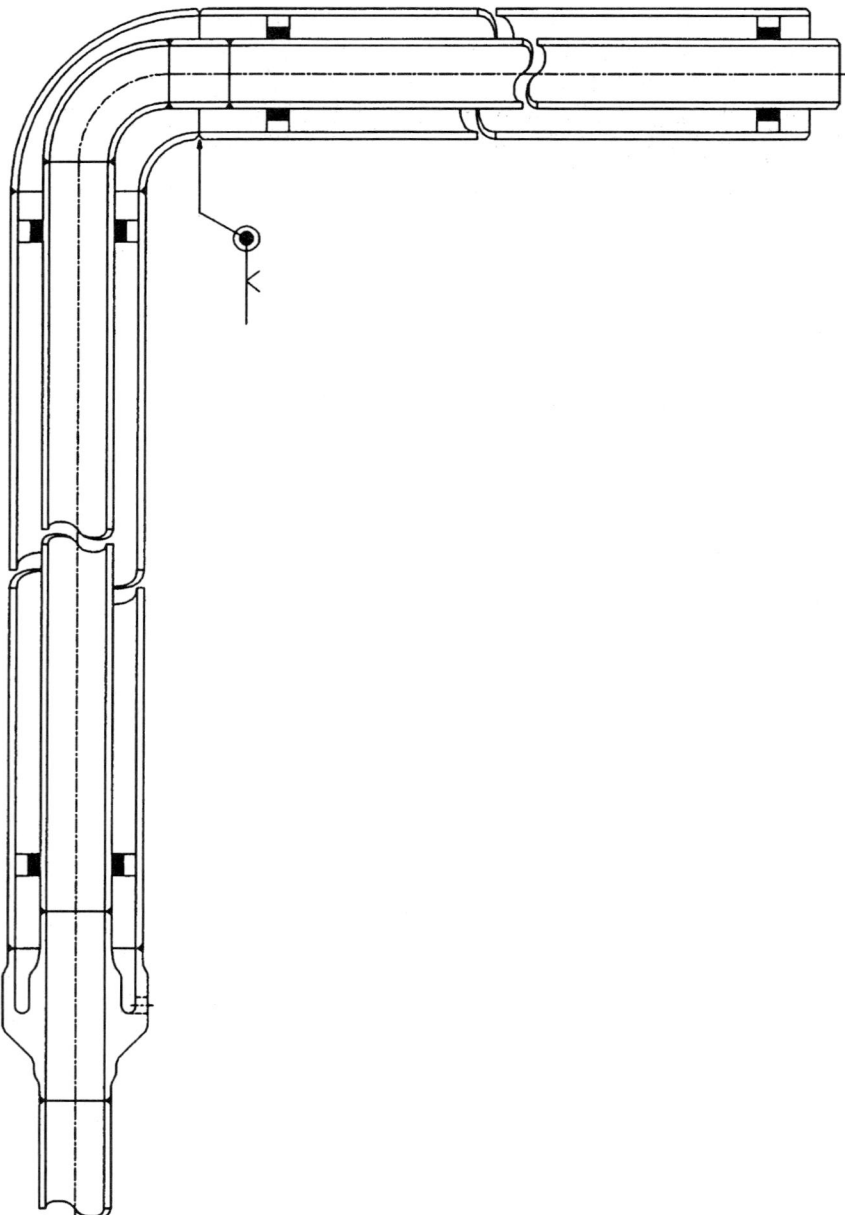

Figure 12.76. In the next step, the secondary containment pipe is joined to the secondary containment elbow. Note that provision is once again made for extension of the primary pipe beyond the secondary containment jacket. Inner and outer pipes are joined in sequence until the next downstream elbow is encountered.

Fabrication, Assembly, and Erection 545

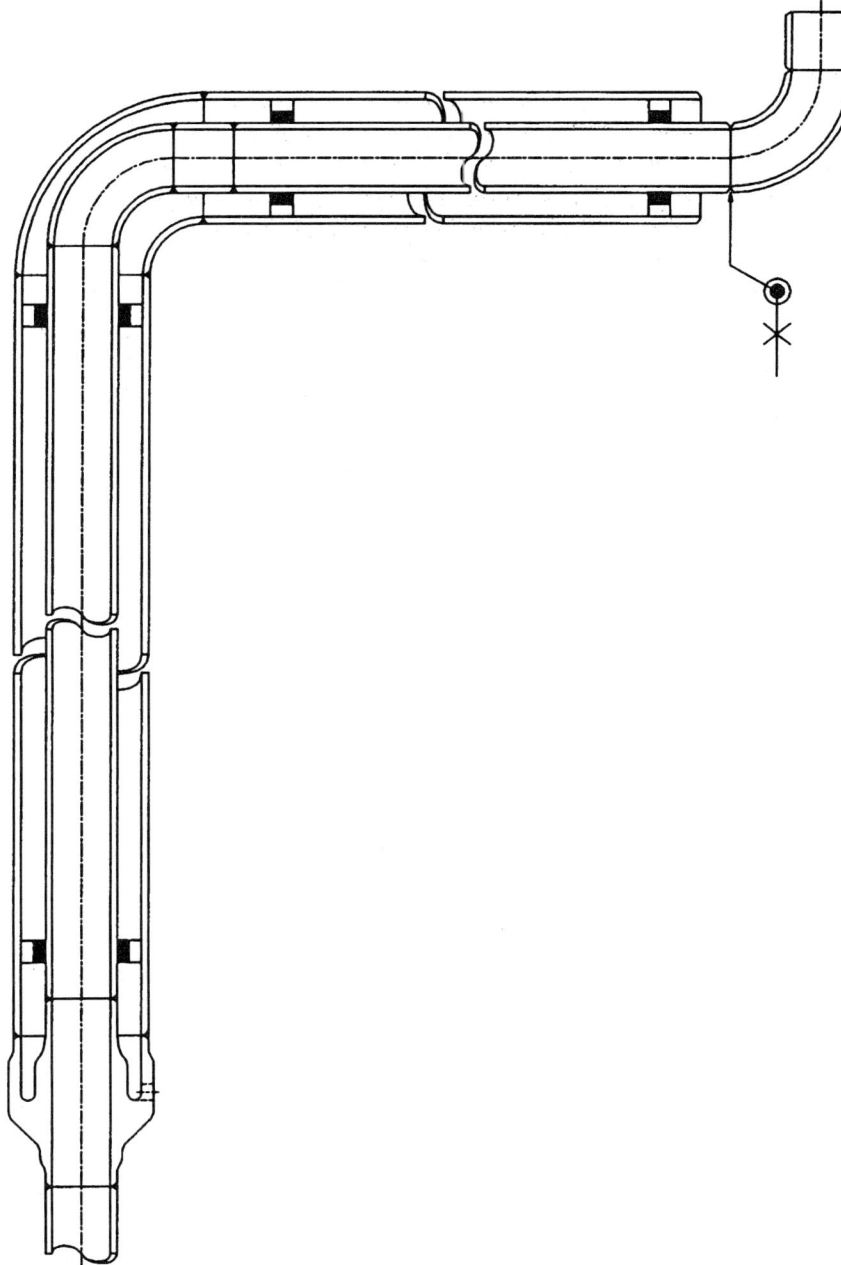

Figure 12.77. To install the next downstream elbow, the procedure described in Figure 12.73 is followed.

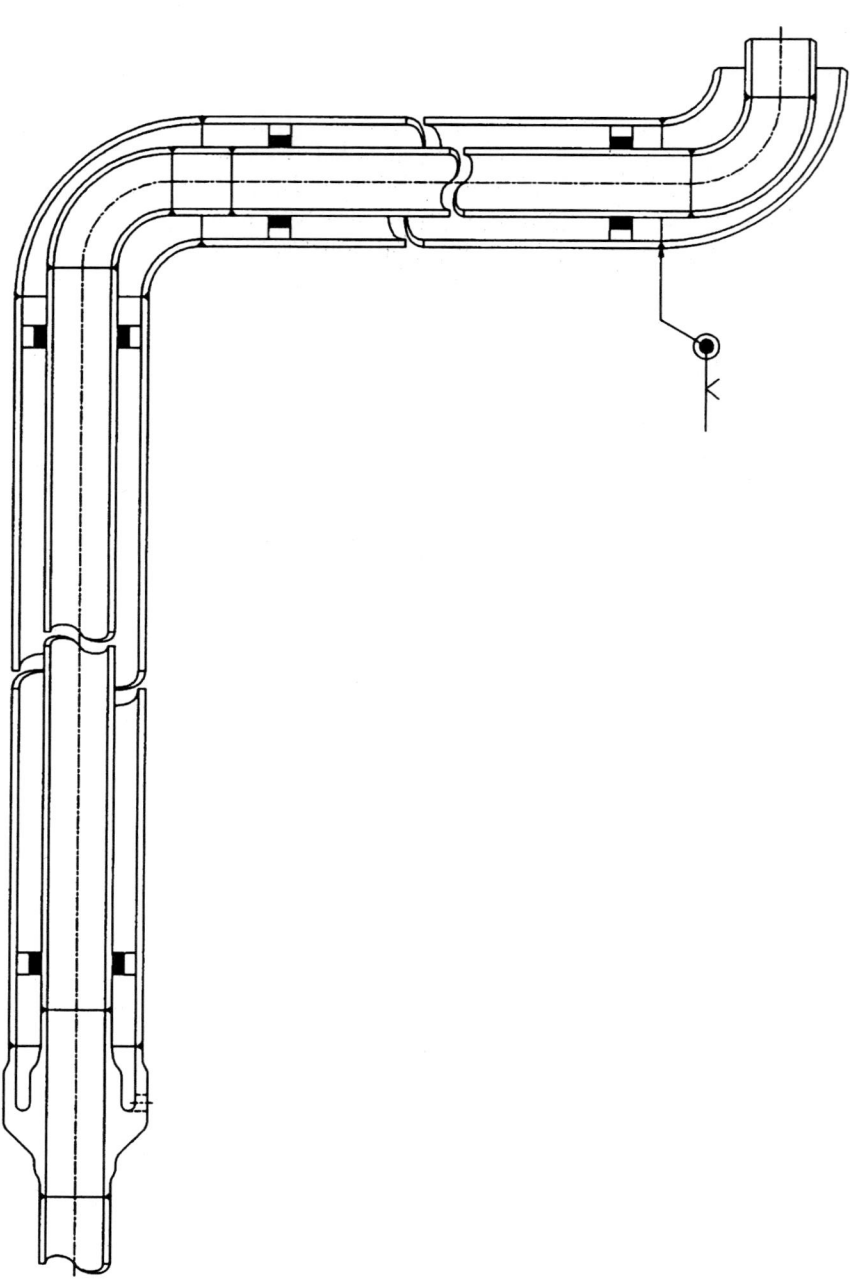

Figure 12.78. The secondary containment elbow is then installed as described in Figure 12.74. To install further sections of pipe, the procedures described in Figures 12.75 and 12.76 are followed, etc., until the system is completed.

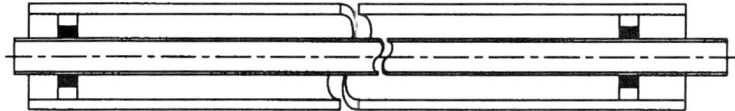

Figure 12.79. In the first step of a field sequential layout involving socket joining systems, start with primary (inner) piping that extends beyond the secondary containment (outer) pipe. The distance that it extends should be the same as the difference between the centerline to end dimensions of the inner and outer elbows.

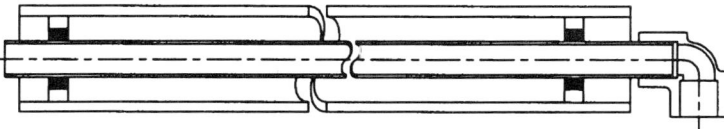

Figure 12.80. In the second step, the primary (inner) elbow is joined to the primary (inner) pipe extension.

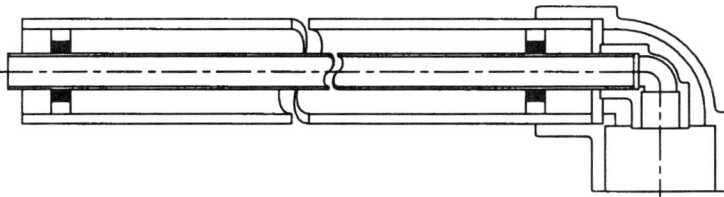

Figure 12.81. Next, the secondary containment (outer) elbow is joined to the secondary containment (outer) piping, over the already joined primary (inner) elbow.

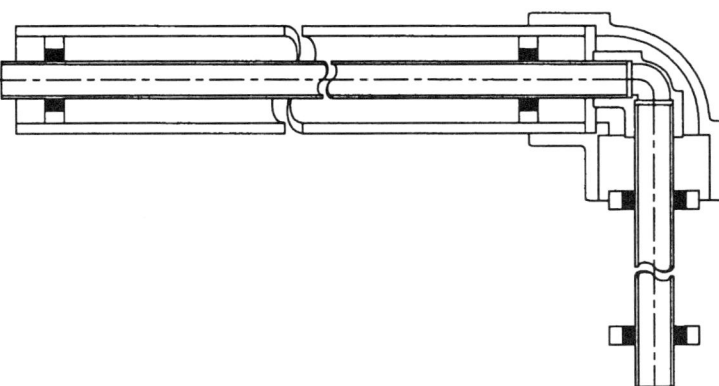

Figure 12.82. The next straight length of primary (inner) piping must be joined to the outlet side of the primary (inner) elbow. In this drawing the interstitial supports are shown as already having been attached to the primary (inner) pipe.

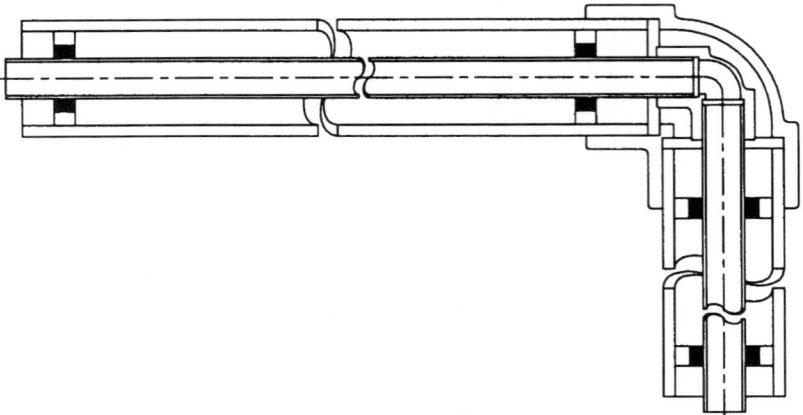

Figure 12.83. The next straight length of secondary containment (outer) piping can then be joined to the outlet side of the secondary containment (outer) elbow, after it has been slipped over the primary (inner) piping/interstitial supports. The primary (inner) piping must again extend beyond the secondary containment piping, in order to be assembled to the next coupling or elbow.

12.3.2.1.6 Systems Using Mostly Prefabricated Sections

In many systems, the greatest overall installation efficiency can be achieved by having substantial portions of a system prefabricated in a controlled environment, and subsequently delivered to a job site for final assembly. In a shop environment, difficult welds and assemblies can be made under a controlled set of conditions, thereby maximizing the chances for a leak-free final assembly. In the field, conditions will always be less efficient due to weather, lack of needed equipment, lack of joining expertise, and other reasons. Higher efficiencies gained by using shop fabrication will mean overall installation costs are likely to be reduced.

On the down side, the more prefabrication that is used, the less the opportunity to minimize the need for labor-intensive midline secondary closures that the system will require. Secondary closures such as "clam-shell" couplings may require longer weld times compared to other types of welds, and tend to be more difficult welds to make. The cost of using them may sometimes offset some of the savings gained through the use of prefabrication. The number of midline closures depends on several aspects of a system. This includes the layout of the system, the materials and joining systems used, and the preferences of each designer and owner. Since their number can very, a good compromise is to use prefabrication and minimize the number of possible midline secondary closure locations.

There are four basic types of secondary closures that are utilized for straight pipe sections. These are: clam-shell couplings (illustrated in Figure 11.72), weld wraps (Figure 11.73), sheet wraps (Figure 11.74), and slip couplings (Figure 11.75). Examples of a slip-coupling-type secondary-closure in thermoplastic pipes is shown in Figure 12.84. The method used in Figure 12.84 to join the closure is hot-gas welding, using filler rod of the same resin as the closure and secondary containment pipe sections.

In Chapter 11, two major types of prefabrication layout and installation sequence options were discussed in relation to installing elbow sections into a system. They are both described in Section 11.1.2.1, "Joining Sequence for Inner and Outer Welds." One describes a sequence where straight pipe sections are prefabricated, the other describes a sequence where the elbows are prefabricated. In this sense they are somewhat opposite approaches, although both tend to depend on the use of some form of secondary closure. The choice as to which approach is best for any one project depends on the preference for closure type that is favored by the designer/owner and the materials involved sizes, etc.

The first approach, entitled "Prefabrication in Straight Pipe Sections with Split Secondary Containment Fittings as Secondary Closures" in Section 11.1.2.1, is illustrated in Figures 12.85 through 12.87. Further discussion to the layout details regarding dimensional aspects of using this procedure is described in Section 11.1.2.1.

The second approach, entitled "Prefabrication in "L," "V," "T," "Y," or Other Complex-Shaped Sections with Secondary Closures in Straight Pipe Sections" is illustrated in Figures 12.88 through 12.91. In the basic version of this approach, prefabricated straight pipe sections must have ends of primary pipes extended a certain distance beyond the secondary containment pipes. This is to allow space to make the primary pipe welds (bonds). The distance required depends on the dimensions required by joining tools and is limited by the type of midline secondary closure selected.

There are two variations to this second approach that eliminate the need to have the primary pipe extended beyond the secondary containment pipe and thus eliminates the need to add a secondary closure. The first variation of this second approach may be used for any joining method, if sufficient flexibility exists in a double containment piping system. This is illustrated for a system having consecutive elbows in Figures 12.92 through 12.94.

The second variation may be used in all-thermoplastic systems where simultaneous-welding (bonding, fusion) techniques can be applied. The sequence, as applied in a system containing consecutively occurring elbows, is illustrated in Figures 12.95 and 12.96. Whenever a simultaneously welded (bonded) joint is introduced into a piping system, it will become a point of restraint by anchoring the two systems together at that point. Therefore, under this approach, this resulting section is anchored at the midpoint between the two elbows.

Figure 12.84. A slip coupling is joined to the secondary containment piping in a polypropylene double containment piping system. (Source: Grewe Plastics.)

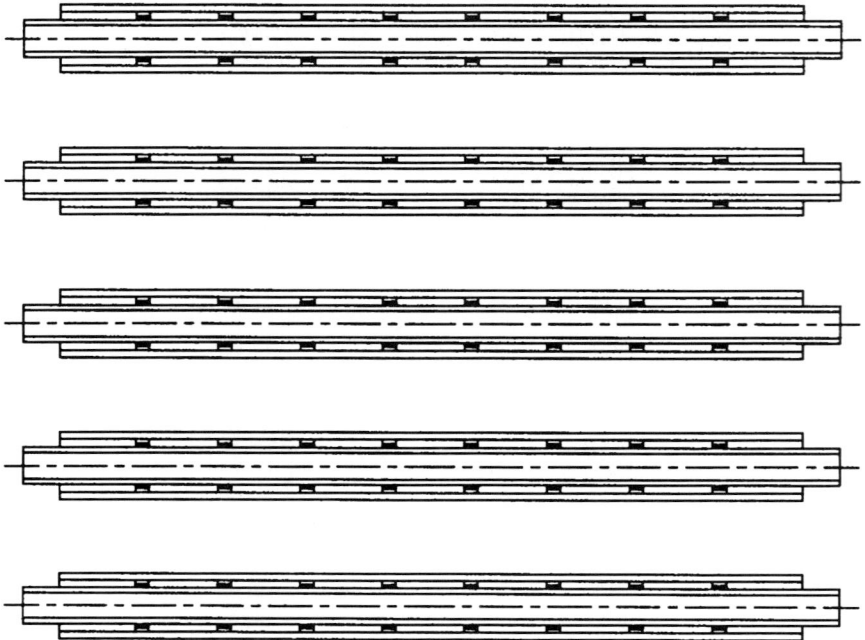

Figure 12.85. In the first step of prefabrication of straight pipe sections using split "clam-shell" fitting secondary closures, individual straight lengths of piping between components are prefabricated by attaching supports to the primary pipe and then sliding the inner pipe in place. The inner pipe must be longer than the outer by a fixed amount, depending on the joining procedure.

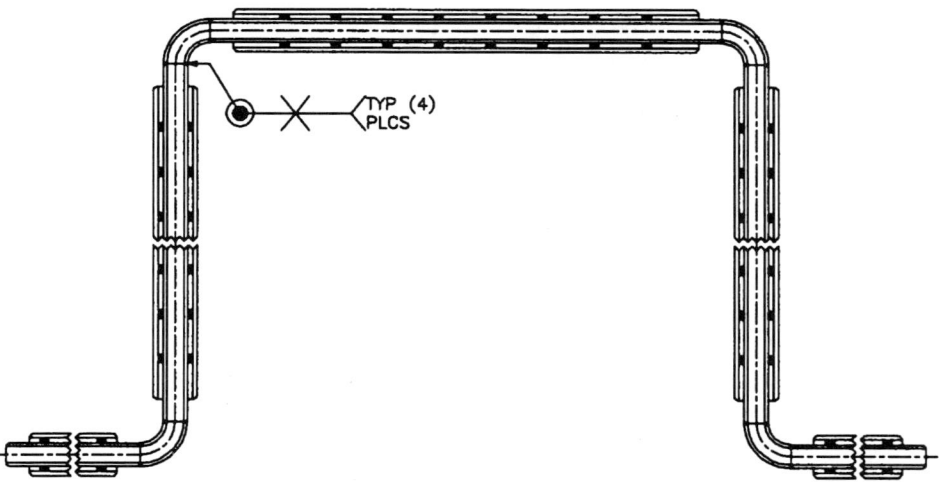

Figure 12.86. Next, the primary 90° elbows are attached to adjacent primary pipe extensions.

Fabrication, Assembly, and Erection

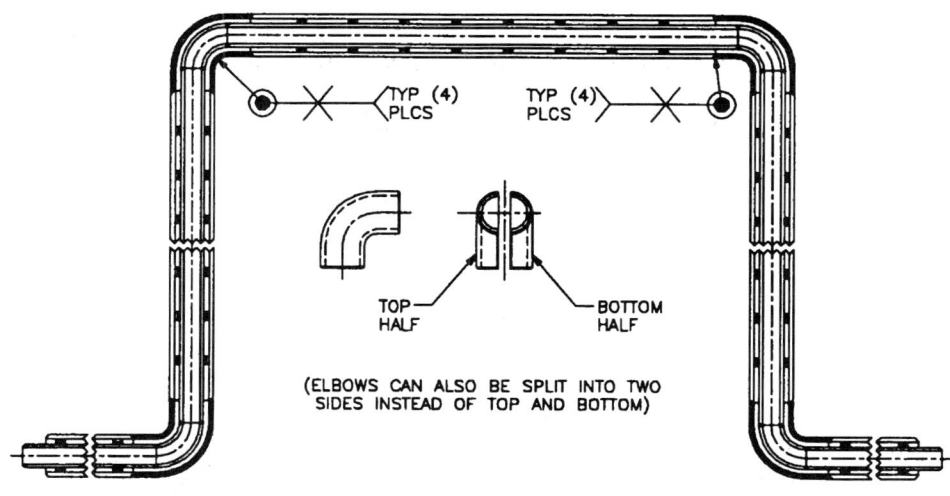

Figure 12.87. The split outer elbows can be fit-up in place and welded to complete the containment casing.

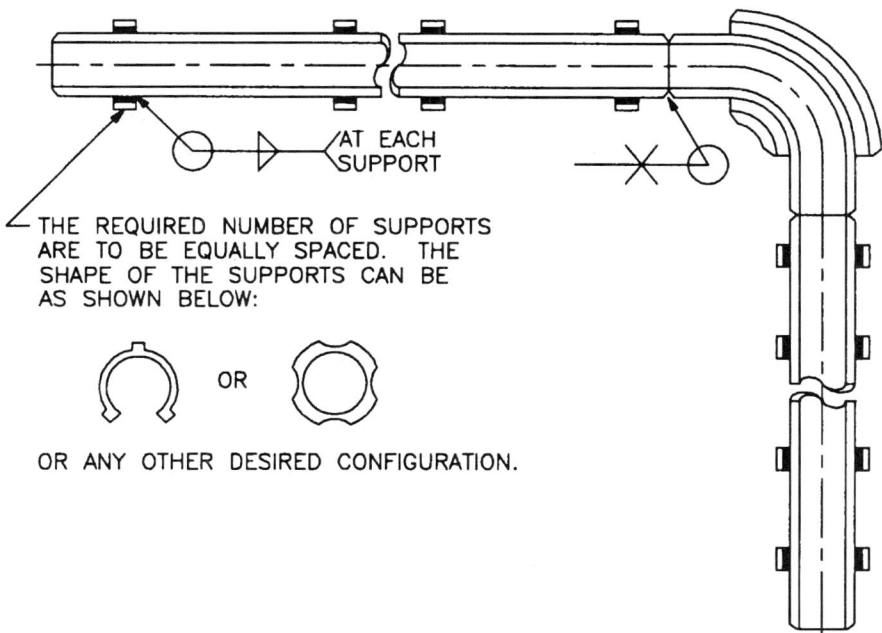

Figure 12.88. In the prefabrication "L" or "V" sections, fabrication starts at the elbows. Note that the secondary containment elbow must be prepositioned in place prior to welding the adjacent primary pipe sections.

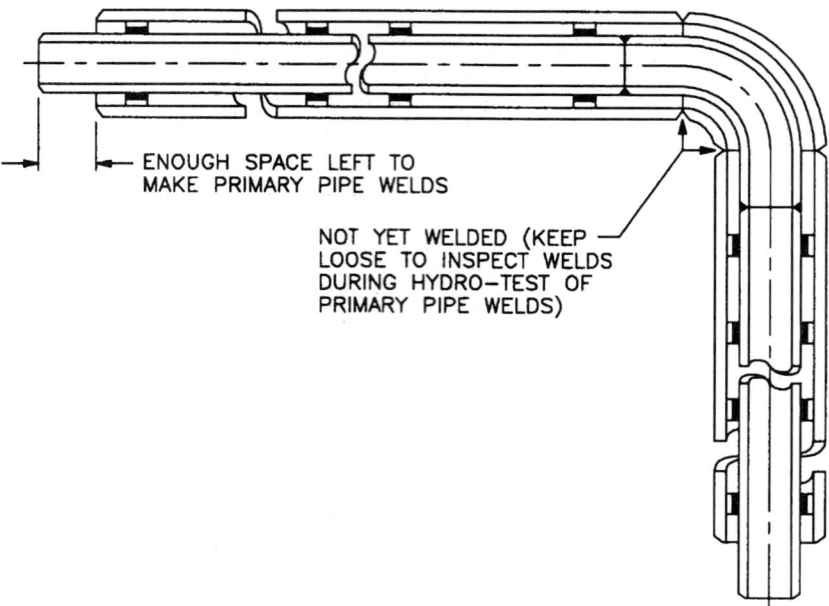

Figure 12.89. Next, secondary containment pipes are slipped over the interstitial supports. The primary welds are not made, however; also note that the secondary containment pipe sections allow the primary pipes to extend beyond the secondary containment pipe on both ends.

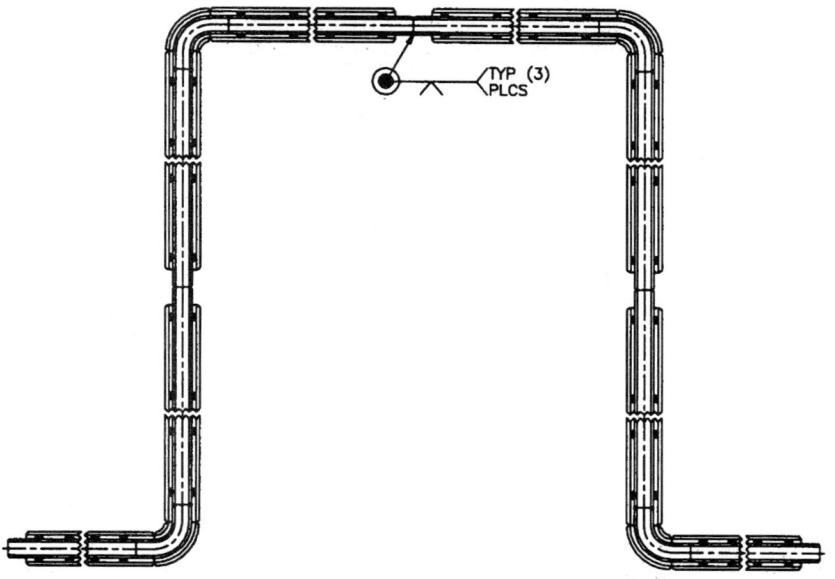

Figure 12.90. The prefabricated "L" sections are fit-up in the field and welded together.

Fabrication, Assembly, and Erection

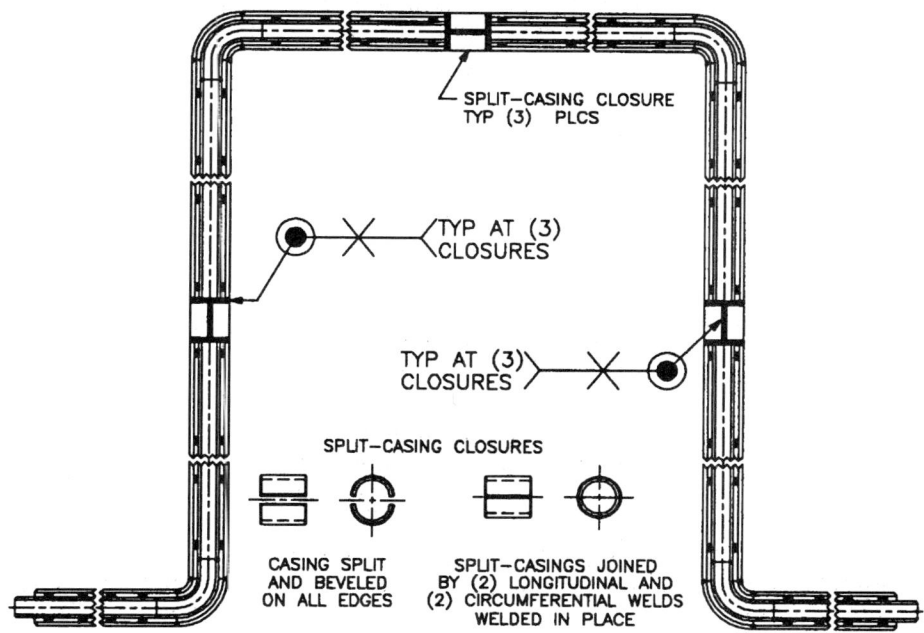

Figure 12.91. After the primary piping system is tested and all welds are visually inspected, and all nondestructive examinations completed, the "clam-shell" couplings are fit-up and welded, thereby completing the secondary containment casing.

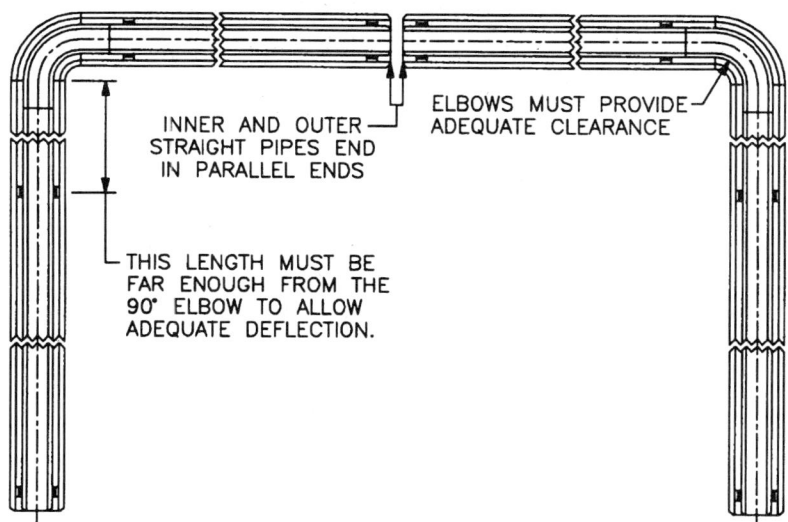

Figure 12.92. In order to eliminate secondary closures in performing field assembly of prefabricated elbow sections, sufficient tolerance must exist in the elbows to allow the primary pipes to be pulled out.

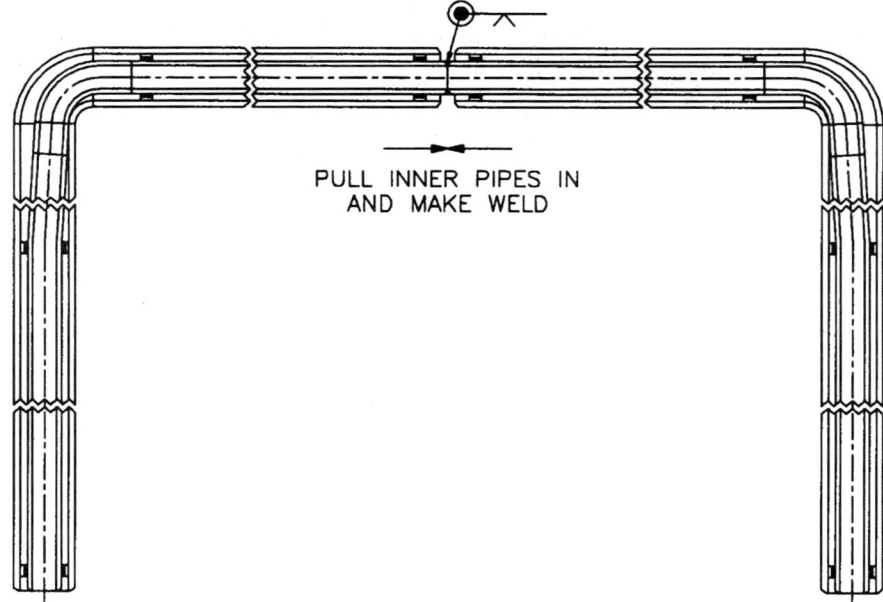

Figure 12.93. The primary pipes must be pulled out, clamped, and then welded.

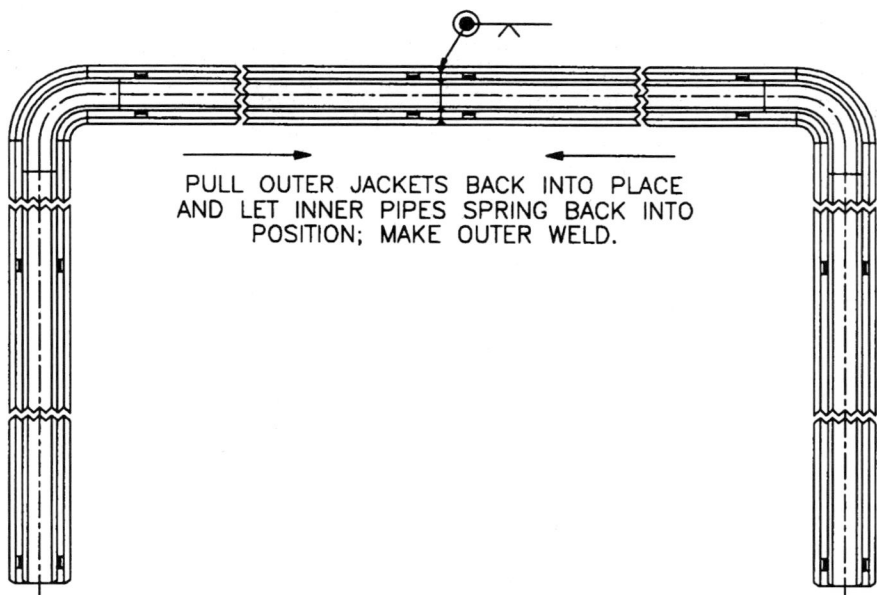

Figure 12.94. After the weld is performed, the outer jackets are pulled back into position, allowing the primary pipes to spring back into the proper position.

Fabrication, Assembly, and Erection

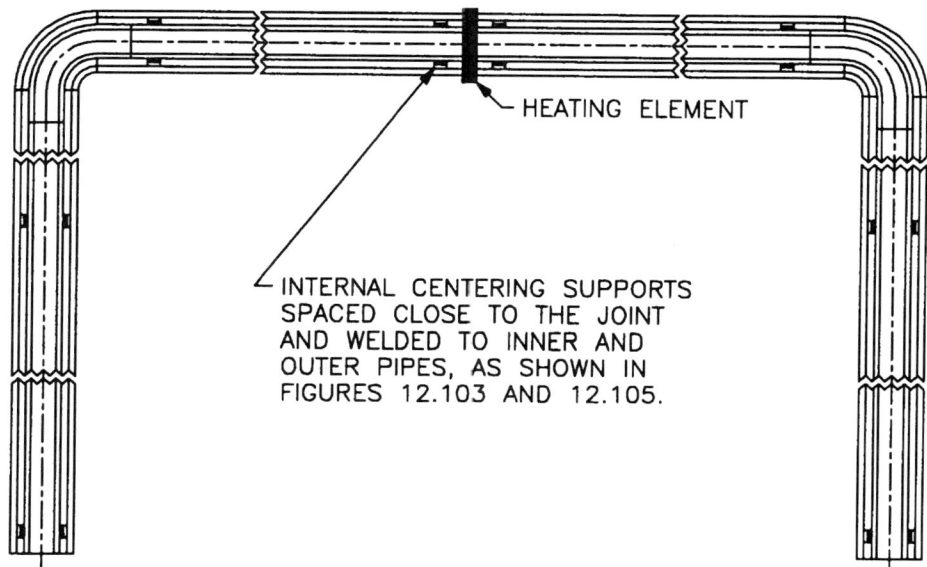

Figure 12.95. Example of the use of simultaneous fusion as applied to prefabricated elbow sections. In the illustration shown, the two sections are heated by the heating element in the alignment jig.

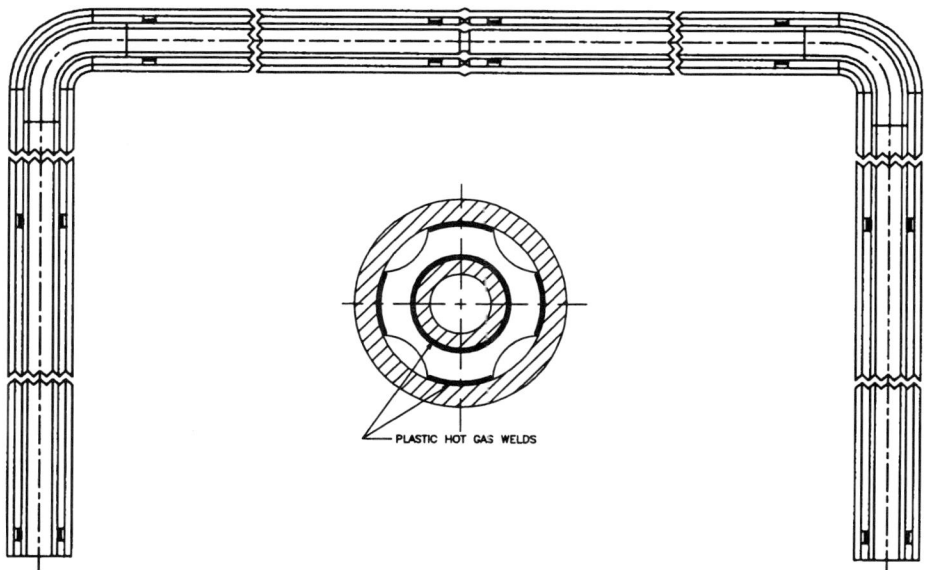

Figure 12.96. Illustration after the simultaneously fused joint is complete, along with a cross section of the internal support. Simultaneous fusion procedures are discussed further in Section 12.3.3.

12.3.2.2 "Outer-First" Assembly

For installations involving long straight runs of pipe, an "outer-first" staggered welding sequence can be used. This type of installation involves making outer pipe joints prior to making inner pipe joints. The secondary containment pipes do no necessarily have to be joined first. The distinction in using this method is that they be joined at a separate time than the primary pipes. Once separate joining and pretesting is completed, straight primary pipe sections are inserted into their associated welded (bonded) secondary containment pipes, after the secondary containment piping is laid in place in its trench.

The primary pipes may have interstitial supports (centering devices) attached, or they may be laid in the secondary containment pipes without any centering devices, as is often the practice in large-diameter HDPE double containment piping installations. If there is any degree of expansion/contraction involved, or if the application is for pressure transfer of fluids subject to cyclic conditions, the use of centering devices to center and support primary components is recommended. If the installation does involve the use of centering/supporting devices that are attached to primary pipes, associated secondary containment pipes should be free of any internal restrictions that will prevent insertion ("slip lining") of its primary piping. If butt-welded thermoplastic secondary containment piping is used, internal-weld beads must be removed prior to insertion of primary piping having attached interstitial supports (centering devices). Care must be used not to cause damage in removing the internal weld beads.

The practice of "slip lining" is generally described and illustrated in Section 16.1.1, Figures 16.1 through 16.3.

In applications involving an "outer-first" staggered joining sequence, it is common practice to have secondary containment pipes terminated at manholes, with branch connections and changes in direction designed as single-wall fittings positioned inside a secondary containment manhole. Normally, a manhole must be made of lined and coated concrete or other corrosion-resistant material, and further equipped with leak detection. This allows straight double containment pipe sections to be readily assembled, and attachment of fittings into a double containment assembly to be easier. It also enhances future access and maintenance.

Figure 12.97 illustrates a proprietary system using specially equipped patented manhole assemblies that permit future removal and reinsertion of the primary pipe system, without having to perform an excavation. This type of system is installed by "outer-first" techniques.

For branch connections in a system that uses "outer-first" insertion techniques, the branch connection fittings must either be prefabricated or split-secondary containment (clam-shell) components must be used. In either case, a secondary closure of some type is required to join the branch into the rest of the system. In some systems simultaneous fusion techniques may be used as an option to replace the requirement for using at least one secondary closure. Only all-plastic double containment systems may use simultaneous welding techniques where design conditions and Codes permit this practice. However, it should be pointed out that ANSI/ASME Codes typically do not permit the use of simultaneous welds (bonds) due to visual inspection and testing criteria.

Fabrication, Assembly, and Erection

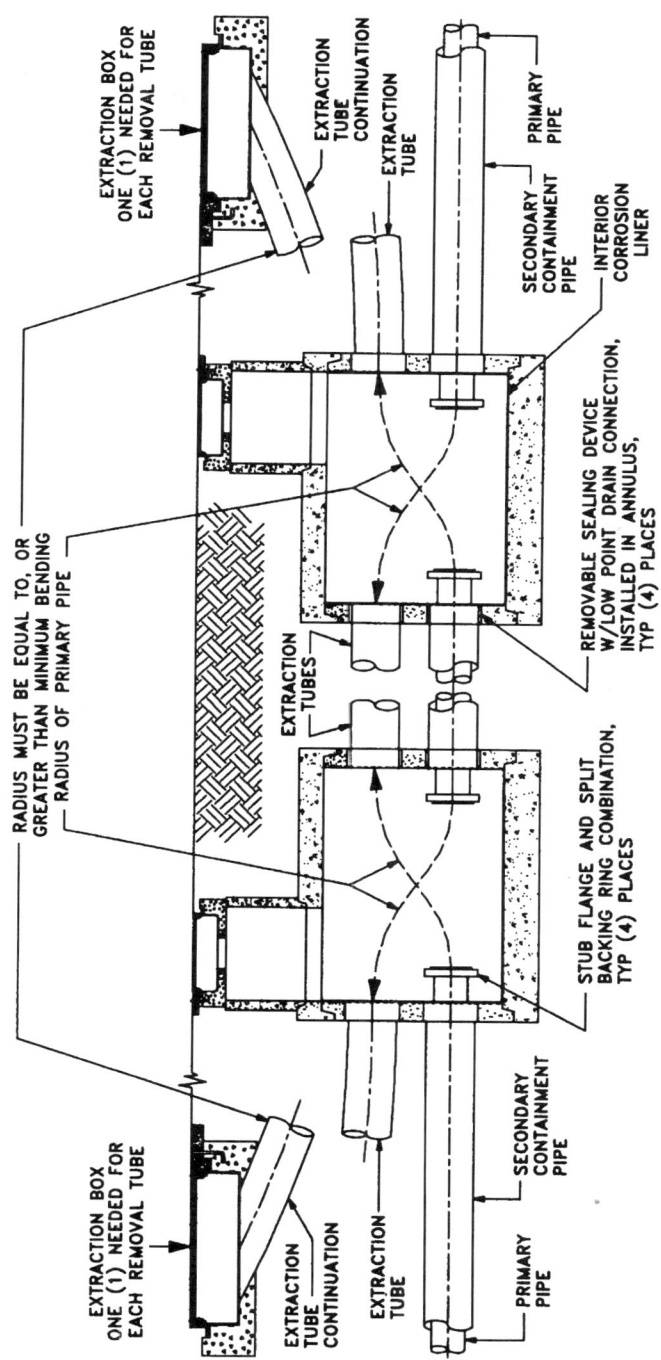

Figure 12.97. Illustration of a system that allows for insertion and removal of the primary pipe system, without having to excavate, in a chemical sewer system that uses HDPE as the primary pipe material. This system can be installed using "outer-first" techniques. (U.S. Patent #5,152,635, Other Patents Pending)

12.3.2.3 All-Metallic Systems

All systems that use metallic primary and secondary containment materials are joined using staggered welding techniques. In most applications, the use of the "inner-first" sequence is more practical due to the weight and rigidity of the materials involved. However, "outer-first" techniques may be used in applications involving long straight runs of pipe, as well as in applications where inspection requirements make it more feasible.

In some applications, an economical selection of materials dictates that a high stainless or nickel alloy be contained within a lower alloy or carbon steel secondary containment piping. Due to the service fluid involved, an expensive metal is often required for a primary material, but in many applications the contact time expected for the secondary containment system is short. Therefore, based upon the statistical likelihood of a leak and the expected maximum duration of contact, a relatively inexpensive secondary containment metal material can be selected, if it has sufficient chemical resistance for the anticipated maximum duration of contact. (Please note that galvanic corrosion concerns must be taken into account; see Section 2.1.3).

In these systems, dissimilar metals are may be welded together, where permitted and where dielectric isolation is not required, at initiation and termination areas, and possibly where interstitial supports attach. In this case, special attention needs to be paid to the welding metallurgy in selecting appropriate filler materials, in order to obtain the desired properties of the finished weld.

Stainless steel, heat-resisting steel, and other high-alloy-type welding rod and electrodes should be used for joining carbon and low-alloy steels in applications such as these.[4] The need for a high-alloy filler metal is required where carbon or low-alloy steel is joined to stainless steel or a heat-resisting steel.

A high-alloy filler material may be desired even where carbon or low-alloy steel is used for primary and secondary containment components, if the properties of a high-alloy weld deposit, or a surfacing weld, are desired. These properties include greater toughness, corrosion resistance, low magnetic permeability, and crack resistance, among others. It may also be desired if cracking in the base metal heat-affected zone must be prevented, as at interconnecting locations that are subject to stresses, and where a weld will serve to seal a starting or termination location of the annulus. Also, it may be desired where a base metal is coated with a protective metal, like aluminum, and a weld metal is needed that is immune to the metallurgical effects of picking up the material of the coating in the weld deposit. The compositions of standardized high-alloy steel covered welding electrodes and of bare welding rods or electrodes of the corrosion-resistant chromium and chromium-nickel types are listed in Tables 12.5 and 12.6, respectively.

If an improper selection of welding electrodes is made in the welding of dissimilar metals, the result can be cracking during fabrication or, even worse, cracking while in service. Many times in the joining of carbon and low-alloy steels to stainless or heat-resisting steels, Type 308, 347, 316, and 318 filler materials have been chosen. In virtually every case, the choice of filler material of these types can be said to be incorrect or illogical.[5] Whenever carbon or low-alloy steels are welded with higher-alloy filler materials, dilution of the high-alloy weld deposit occurs by entry of melted carbon or alloy-steel base metal. The amount of any dilution must be taken into consideration, and a determination or estimate of the composition of the final weld should be made. In some

Table 12.5 Chromium and Chromium-Nickel Types Covered Welding Electrodes

Chemical composition requirements in percents for all-weld-metal deposit, AWS A5.4 (ASTM A298) [1]

Sulfur is limited to 0.03 max percent in all cases; phosphorous is limited to 0.04 max percent in all cases except E310, E310Cb, E310Mo and E16-8-2 for which the limit is 0.03 max percent.

AWS-ASTM Classification	C	Cr	Ni	Mo	Cb+Ta	Mn	Si
E308	0.08	18.0-21.0	9.0-11.0	---	---	2.50	0.90
E308L	0.04	18.0-21.0	9.0-11.0	---	---	2.50	0.90
E309	0.15	22.0-25.0	12.0-14.0	---	---	2.50	0.90
E309Cb	0.12	22.0-25.0	12.0-14.0	---	0.70-1.00	2.50	0.90
E309Mo	0.12	22.0-25.0	12.0-140	2.00-3.00	---	2.50	0.90
E310	0.20	25.0-28.0	20.0-22.5	---	---	2.5	0.75
E310Cb	0.12	25.0-28.0	20.0-22.0	---	0.70-1.00	2.50	0.75
E310Mo	0.12	25.0-28.0	20.0-22.0	2.00-3.00	---	2.50	0.75
E312	0.15	28.0-32.0	8.0-10.5	---	---	2.50	0.90
E16-8-2	0.10	14.5-16.5	7.5-9.5	1.00-2.00	---	2.50	0.50
E316	0.08	17.0-20.0	11.0-14.0	2.00-2.50	---	2.50	0.90
E316L	0.04	17.0-20.0	11.0-14.0	2.00-2.50	---	2.50	0.90
E317	0.08	18.0-21.0	12.0-14.0	3.00-4.00	---	2.50	0.90
E318	0.08	17.0-20.0	11.0-14.0	2.00-2.50	6xC min, 1.00 max	2.50	0.90
E330	0.25	14.0-17.0	33.0-37.0	---	---	2.50	0.90
E347	0.08	18.0-21.0[2]	9.0-11.0	---	8xC min, 1.00 max	2.50	0.90
E349 [3]	0.13	18.0-21.0	8.0-10.0	0.35-0.65	0.75-1.2	2.50	0.90
E410	0.12	11.0-13.5	0.60	---	---	1.00	0.90
E430	0.10	15.0-18.0	0.60	---	---	1.00	0.90
E502	0.10	4.0-6.0	0.40	0.45-0.65	---	1.00	0.90
E505	0.10	8.0-10.5	0.40	0.85-1.20	---	1.00	0.90
E7Cr	0.10	6.0-8.0	0.40	0.45-0.65	---	1.00	0.90

Notes:
[1] Single values shown are maximum percentages except where otherwise indicated.
[2] Chromium may be specified as 1.9 X Ni min.
[3] Also titanium 0.15 and tungsten 1.25 to 1.75

Table 12.6 Chromium and Chromium-Nickel Welding Rods & Bare Electrodes

Chemical composition requirements in percents for the rod or electrode AWS AS5.9 (ASTM A371) Phosphorus and sulfur are each limited to 0.03 max percent. Single values shown are maximum percentages except where otherwise specified.

AWS-ASTM Classification	C	Cr	Ni	Mo	Cb+Ta	Mn	Si
ER308	0.08	19.5-22.0 [1]	9.0-11.0	---	---	1.0-2.5	0.25-0.60
ER308L	0.03	19.5-22.0 [1]	9.0-11.0	---	---	1.0-2.5	0.25-0.60
ER309	0.12	23.0-25.0	12.0-14.0	---	---	1.0-2.5	0.25-0.60
ER310	0.08-0.15	25.0-28.0	20.0-22.5	---	---	1.0-2.5	0.25-0.60
ER312	0.08-0.15	28.0-32.0	8.0-10.5	---	---	1.0-2.5	0.25-0.60
ER316	0.08	18.0-20.0	11.0-14.0	2.0-3.0	---	1.0-2.5	0.25-0.60
ER316L	0.03	18.0-20.0	11.0-14.0	2.0-3.0	---	1.0-2.5	0.25-0.60
ER317	0.08	18.5-20.5	13.0-15.0	3.0-4.0	---	1.0-2.5	0.25-0.60
ER318	0.08	18.0-20.0	11.0-14.0	2.0-3.0	8xC min, 1.0 max	1.0-2.5	0.25-0.60
ER321 [2]	0.08	18.5-20.5	9.0-10.5	0.5	---	1.0-2.5	0.25-0.60
ER330 [4]	0.15-0.25	15.0-17.0	34.0 min.	---	---	1.0-2.5	0.25-0.60
ER347	0.08	19.0-21.5 [1]	9.0-11.0	---	10xC min, 1.0 max	1.0-2.5	0.25-0.60
ER348	0.08	19.0-21.5 [1]	9.0-11.0	---	10xC min, 1.0 max	1.0-2.5	0.25-0.60
ER349 [3]	0.07-0.13	19.0-21.5	8.0-9.5	0.35-0.65	1.0-4.0	1.0-2.5	0.25-0.60
ER410	0.12	11.5-13.5	0.6	0.6	---	0.6	0.50
ER420	0.25-0.40	12.0-14.0	0.6	---	---	0.6	0.50
ER430	0.10	15.5-17.0	0.6	---	---	0.6	0.50
ER502	0.10	4.5-6.0	0.6	0.45-0.65	---	0.6	0.25-0.60

Notes:
[1] Chromium, min = 1.9 x nickel, when so specified.
[2] Also titanium 9 x C min. to 1.0 max.
[3] Also titanium 0.10 to 0.30 and tungsten 1.25 to 1.75.
[4] Not a recognized classification in the AWS-ASTM specification.

Fabrication, Assembly, and Erection

cases, the final metal composition of the weld could result in a weld of lesser qualities than that desired, or required, for the application.

As an example of dilution in a weld, consider an application where two sheets of 3/32-in. (2.4 mm) thick carbon steel are welded with a 1/8 in. (3.2 mm) electrode of the classification AWS E308. In this case, the weld metal may become diluted with as much as 35% or more of the base metal. The final weld composition could therefore have a final composition with as low as 13% chromium and 6.5% nickel. A weld with this final composition would transform on cooling to have a hard, relatively brittle martensitic type of structure. This is in contrast to the tough, austenitic-type structure normally expected of an AWS-type E308 weld metal. However, if an AWS-type E310 electrode is used, the final weld metal for a 35% dilution would consist of approximately 17.5% chromium and 13% nickel, assuring an austenitic weld-metal structure.[6]

If the extent of dilution in the above example increases, an AWS-type E310 electrode may still be too lean, requiring an electrode with an even higher chromium and nickel content. Much depends on the amount of dilution achieved in the welding process. In many arc-welding procedures involving the use of filler materials, a relatively high level of dilution is achieved.

In some applications, it may be desirable to use a nickel or nickel-alloy filler material to join a carbon or low-alloy steel to a dissimilar, highly alloyed steel. The selection process is very difficult and complex. However, the welding should never be done with a carbon steel or low-alloy filler metal, because it is likely that the weld deposit, upon picking up the alloying agents from the high-alloy base metal, would have an unpredictable hardening propensity. The joint's mechanical properties would therefore be unpredictable as well. Table 12.7 indicates various nickel and nickel-alloy filler metals that may be used to weld dissimilar metals. Covered electrode ENiCrFe-3 and its related bare filler metal, ERNiCR-3, are good choices for making satisfactory dissimilar metal joints between carbon and low-alloy steels and highly alloyed materials and for depositing a surface weld of nickel-chromium alloy on carbon steel.

As an example of a highly alloyed substance to be joined to a dissimilar metal, consider alloy 20Cb-3® stainless steel. (20Cb-3 is a registered trademark of the Carpenter Technology Corporation.) Where corrosion resistance of the joint is not critical, AWS (A5.4) type E309 or types E320 or E320LR electrodes may be used. Where attachments are made on the outside of a vessel or pipe, such as on a stiffening frame, or annular carbon steel centering/supporting device, where the shell or primary pipe material of alloy 20Cb-3® stainless steel is greater than 1/8 in. (3 mm), the attachments may be made direct using E320, E320LR, or E309 electrodes. Where it is applicable, it is important to control the weld penetration so that no weld fusion appears on the inside of the container.

Where attachments are made to gauges thinner than 1/8 in. (3 mm), it is necessary to attach a pad of alloy 20Cb-3 stainless plate to the shell with E320 or E320LR electrodes. Then, the dissimilar metal attachment can be welded with less expensive E309 or with E320, E320LR electrodes. The purpose of such a pad is to avoid dilution of the metal in contact with the corrodent by structural material fused into the weld bead. For all of the dissimilar joints mentioned in this example, nickel-base material of the type AWS ENiCrFe-2 (AWS A5.11) may also be used.[7]

Figure 12.98 shows a welder attaching a carbon steel ring stiffener to the outside of an alloy 20Cb-3® stainless steel vessel using an electrode that is among the choices

Table 12.7 Guide to Identification of Nickel and Nickel-Alloy Welding Rods & Electrodes

Present classification AWS 5.11–64T (ASTM B295–64T)	Previous classification AWS 5.11–54T (ASTM B295–54T)	Frequently used trade designation	Welding process and applications [1]
	Covered electrodes		
ENi-1	E4N11	Nickel	SMAW of nickel, and nickel to steel.
ENiCu-1	E4N10	Monel	SMAW where dilution of weld with steel is expected.
ENiCu-2	---	Monel	SMAW for nickel-copper alloys.
ENiCu-3	E3N14	"K" Monel	SMAW precipitation hardening nickel-copper-aluminum alloy.
ENiCu-4	E3N10	Monel	SMAW for application where presence of Cb or Ta is undesirable.
ENiCr-1	E4N12	80Ni-20Cr	SMAW of nickel-chromium alloys and other dissimilar alloys.
ENiCrFe-1	E3N12	Inconel	SMAW of nickel-chromium-iron alloys.
ENiCrFe-2	---	Incoweld "A"	SMAW of nickel-chromium-iron alloys and other dissimilar alloys.
ENiCrFe-3	---	Inconel-182	SMAW of nickel-chromium-iron alloys & other dissimilar metals. Highly resistant to fissuring.
ENiMo-1	E3N1B, E4N1B	Hastelloy "B"	SMAW of nickel-molybdenum alloys.
ENiMo-2	E3N1C, E4N1C	Hastelloy "C"	SMAW of nickel-molybdenum-chromium alloys.
ENiMo-3	---	Hastelloy "W"	SMAW of dissimilar metal combinations of nickel-base, cobalt base, and iron alloys.
----	---	Inconel "X"	SMAW of nickel-chromium-iron-aluminum-titanium precipitation hardening alloys.
	Bare welding rods and electrodes		
RNi-2	RN41	Nickel	OAW of nickel.
ERNi-3	ERN61	Nickel	GTAW & GMAW of nickel.
RNiCu-5	RN40	Monel	OAW of nickel-copper alloys.
RNiCu-6	RN43	Monel	OAW of nickel-copper alloys.
ERNiCu-7	ERN60	Monel	GTAW, GMAW & SAW of nickel-copper alloys.
ERNiCu-8	ERN64	"K" Monel	GTAW and GMAW of precipitation hardening nickel-copper-aluminum alloys.
ERNiCr-2	ERN6N	80Ni-20Cr	GTAW, GMAW, & AHW of 80Ni-20Cr alloy.
ERNiCr-3	---	Inconel-82	GTAW, GMAW, SAW, & AHW of nickel-chromium-iron alloys & other dissimilar metals. Highly resistant to fissuring.
RNiCrFe-4	RN42	Inconel	OAW of nickel-chromium-iron alloys.
ERNiCrFe-5	ERN62	Inconel	GTAW, GMAW, SAW, & AHW of nickel-chromium-iron alloys.
ERNiCrFe-6	---	Inconel-92	GTAW, GMAW, SAW, & AHW of nickel-chromium-iron alloys and other dissimilar metals.
ERNiCrFe-7	ERN69	Inconel"X"	GTAW, GMAW, & AHW of nickel-chromium-iron-titanium precipitation hardening alloys.
ERNiMo-4	ERN7B	Hastelloy "B"	GTAW, GMAW, & AHW of nickel-molybdenum alloys.
ERNiMo-5	ERN7C	Hastelloy "C"	GTAW, GMAW, & AHW of nickel-molybdenum-chromium alloys.
ERNiMo-6	ERN7W	Hastelloy "W"	GTAW, GMAW, & AHW of dissimilar high-alloy materials.

Notes:
[1] AWS Designations for Welding Processes:
SMAW – Shielded metal-arc welding
GTAW – Gas tungsten-arc welding
GMAW – Gas metal-arc welding
SAW – Submerged-arc welding
AHW – Atomic hydrogen welding
OAW – Oxyacetylene welding

Figure 12.98. A welder attaching a carbon steel structural attachment to the exterior of an Alloy 20 Cb-3® Vessel. (Source: Carpenter Technology Corp.)

described in the previous paragraph. In making attachments of this type, it is important to control the weld penetration so that no weld fusion appears on the inside of the vessel, as described earlier.

When using nickel and nickel alloys as filler metals, the problem of microfissuring or hot cracking must be considered. Hot cracking is intergranular in nature, and is greatly aggravated by certain residuals in a weld metal. The severity of cracking can vary significantly. It can produce subtle results such as internal microfissures, detectable only by bend testing or by a searching metallographic examination at approximately 500X. It can also result in surface hot cracking, detectable by the human eye. Resistance to hot cracking has been obtained by the use of manganese-modified ENiCrFe-2, ENiCrFe-3, ERNiCr-3, and ERNiCrFe-6.[8] Tables 12.8 and 12.9 list nickel and nickel-alloy covered electrodes and nickel and nickel-alloy bare welding rods and electrodes, respectively.

Figures 12.99 and 12.100 show a comparison of bend test specimens for two different GMAW welds on alloy 20Cb-3® materials. The specimens are bent 60° around a 1.5 in. (38.1 mm) pin. In Figure 12.99, an ER320 filler material was used. The weld in Figure 12.81 was produced using Carpenter 20Cb-3LR filler material, which is a special formulation that aids in the resistance to hot cracking of alloy 20Cb-3® stainless steel.

12.3.3 Simultaneous Fusion

An alternative sequence of constructing a double containment piping assembly is termed simultaneous fusion, and pertains only to systems composed of plastic piping inner and outer components. Simultaneous fusion is most readily applied to heat-element-based butt-fusion-based thermoplastic systems due to the ability to combine primary and secondary

Table 12.8 Nickel and Nickel-Alloy Covered Electrodes

Typical chemical composition of deposited metal (for details of AWS-ASTM classifications see AWS 5.11-64T or ASTM B295-64T)

AWS-ASTM Classification	Trade name or example	Typical Chemical Composition					
		Ni	Cu	Cr	Si	Fe	Others
ENi-1	Nickel (131&141)	94	---	---	1.0	---	Al 1, Ti 2.5
ENiCu-1	Monel (140)	66	22	---	1.0	2	Cb + Ta 2
ENiCu-2	Monel	64	22	---	1.0	2	Cb + Ta 2
ENiCu-3	"K" Monel (134)	64	24	---	1.0	2	Al 3, Ti 1, Mn 4
ENiCu-4	Monel (130)	67	23	---	0.5	2	Al 1, Ti 1
ENiCr-1	80Ni-20Cr (142)	77	---	18	0.5	4	Cb + Ta 2
ENiCrFe-1	Inconel (132)	68	---	15	0.5	10	Cb + Ta 3
ENiCrFe-2	Incoweld "A"	70	---	16	0.5	6	Mn 2, Mo 2, Cb + Ta 2
ENiCrFe-3	Inconel (182)	67	---	15	0.5	6	Mn 7, Cb + Ta 2, Ti 0.5
ENiMo-1	Hastelloy B	60	---	---	1.0	6	Mo 29
ENiMo-2	Hastelloy C	54	---	15	1.0	6	Mo 16, W 4
ENiMo-3	Hastelloy W	62	---	5	1.0	6	Mo 25
---	Inconel "X" (139)	66	---	15	1.0	10	Al 0.7, Ti 2, Cb + Ta = 4xSi min.

Table 12.9 Nickel and Nickel-Alloy Bare Welding Rods and Electrodes

Typical chemical composition of bare filler metal. For details of AWS-ASTM classification see AWS 5.14-64T (ASTM B304-64T)

AWS-ASTM Classification	Trade name or example	Typical chemical composition, %					
		Ni	Cu	Cr	Si	Fe	Others
RNi-2	Nickel (41)	99	---	---	0.5	---	Ti 0.5
ERNi-3	Nickel (61)	95	---	---	0.5	1	Al 1, Ti 2
RNiCu-5	Monel (40)	66	23	---	0.5	2	---
RNiCu-6	Monel (43)	58	38	---	1.0	1	---
ERNiCu-7	Monel (60)	66	24	---	1.0	2	Al 1, Ti 2
ERNiCu-8	"K" Monel (64)	67	27	---	1.0	2	Al 3, Ti 0.5, Mn 1.5
ERNiCr-2	80Ni-20Cr	75	---	20	0.3	2	Al 0.4, Ti 0.3
ERNiCr-3	Inconel (82)	72	---	20	0.3	2	Mn 3, Cb + Ta 2.5, Ti 0.3
RNiCrFe-4	Incoweld (42)	74	---	16	0.5	8	---
ERNiCrFe-5	Incoweld (62)	72	---	16	0.5	8	Cb + Ta 2
ERNiCrFe-6	Inconel (92)	70	---	16	0.2	6	Ti 3, Mn 2.5
ERNiCrFe-7	Inconel "X" (69)	70	---	16	0.5	7	Al 0.7, Ti 2, Cb + Ta 1
ERNiMo-4	Hastelloy B	60	---	---	1.0	6	Mo 29, V 0.4
ERNiMo-5	Hastelloy C	54	---	15	1.0	6	Mo 16, W 4
ERNiMo-6	Hastelloy W	62	---	5	1.0	6	Mo 25

Figure 12.99. Bend test specimen on a GMAW weld of an Alloy 20 Cb-3® material, made using an ER320 filler material. Notice the presence of the cracks in the specimen. (Source: Carpenter Technology Corp.)

containment pipe sizes that are relatively close in diameter. The method can also be applied to small-diameter heat-element-based socket-fusion systems and solvent-cement based socket systems, but not without difficulty.

The object of any simultaneous fusion system is to prepare both primary and secondary containment components so that they are permanently fixed to each other, and can thus be joined to a mating set of components. In some systems, simultaneous fusion can be used to join 100% of all components involved in the piping system. It can also be selectively used for situations where staggered welding is the primary method, in lieu of using secondary closures. Simultaneous fusion can substantially reduce overall labor involved on a project, unless primary piping joints leak upon initial pressure test. However, the inspection of primary welds is substantially limited, and the method may only be used where a flexible (unrestrained) layout is not required. A simultaneously welded system will result in a structurally more rigid assembly.

Simultaneous fusion is more easily applied to systems that have identical primary and secondary containment materials. If a heat-element butt-fusion-based system is used, the materials should have a welding temperature range that overlaps and specific welding pressures that are nearly the same.

Figure 12.100. Bend test specimen on a GMAW weld of an Alloy 20 Cb-3® material, made using a Carpenter 20 CB-3LR filler material, which is specially formulated to resist hot cracking. (Source: Carpenter Technology Corp.)

When homogeneous materials are used for the primary and secondary containment components in a butt-fusion thermoplastic system, the welding forces required for the various steps can be easily calculated using the distributive law of arithmetic. Each thermoplastic resin has its own unique specific welding pressure for butt welding, which represents the optimum amount of force to be applied to a given cross section of the plastic during welding. This is a function of the density and crystallinity of the resin structure. The welding pressures at the various stages of the thermoplastic butt-welding process are determined by multiplying this specific welding pressure of the thermoplastic resin times the cross-sectional area. Therefore, if a primary and secondary containment pipe combination (of identical resin makeup) is to be welded simultaneously, it follows from the distributive law of mathematics that the cross-sectional areas can be added together, and the sum of the two multiplied by the specific welding pressures to obtain the combined welding forces that are required.

The procedure for simultaneous fusion is illustrated in Figures 12.101 through 12.112. The single most important consideration in performing simultaneous fusion is that the components should be prefabricated so that they align in the same plane at their ends; the alignment must occur within very close tolerances (< 5% of diameter). The standards according to DIN, and accepted worldwide, requires that thermoplastic butt-fusion pipes be aligned to within 10% of their wall thicknesses in terms of their diametrical alignment. Therefore, the use of thick-wall pipes (SDR 17.6 or thicker) as the material of the primary pipes for primary pipes 8 in. nominal diameter (approximately 200 mm ISO 161/1 outside diameter) and below is highly recommended. The procedure for simultaneous fusion, based on a heat-element butt-fused-based thermoplastic component system, is described as follows:

Primary piping materials are first prepared for fusion as depicted in Figure 12.101. It is not necessary to have primary pipe ends fully planed prior to this step, since both primary and secondary containment pipes will be planed (shaved) together. Next, insert simultaneous-weld baffles (discs) or other centering devices approximately 1 in. (25 mm) from the end of both primary pipes as shown in Figure 12.102. Each of the simultaneous weld baffles is then welded to the primary pipes (Figure 12.103), using a single bead pass of 1/8 in. (3 mm) or 5/32 in. (4 mm) welding rod of the same material as the baffle and pipe. Assuming all other midline centering devices are installed, secondary containment pipes are then positioned (Figure 12.104). The outer vanes of both simultaneous weld baffles are then welded to the inside diameter of their respective secondary containment pipes using a single bead pass of 1/8 in. (3 mm) or 5/32 in. (4 mm) welding rod of the same material as the baffle and the pipe (Figure 12.105). Both of the welded-together double containment pipe sections are then simultaneously planed (shaven) using a motor-driven planing unit, simply illustrated in Figure 12.106. The alignment of the pipes after planing should be checked with an alignment device, as shown in Figure 12.107, or other device before proceeding.

A normal, flat, Teflon-coated aluminum heating element is then lowered into place in the butt-fusion equipment (Figure 12.108), and the ordinary butt-fusion procedure followed. (It is not shown in these drawings, but it is assumed that the pipes are held in place in all the drawings using a butt-fusion alignment jig, otherwise called a "butt-fusion machine," as referenced in ASTM D 2657, part 2.) After the welding process has been completed, the finished appearance of the simultaneous-fusion butt welds will appear as shown in Figures 12.112.

The same general process can be used in principle for thermally socket fused systems, as well as for solvent-cemented systems, and for adhesively bonded RTRP systems. It is usually best if the systems are permanently attached to each other, by welding (in the case of thermoplastics only) or bonding, although the procedure can be used if the components are mechanically attached or held together by other means.

The simultaneous-butt-fusion process, or any other simultaneous fusion process, is not covered or described in specific terms by any consensus standards, such as ASTM D 2657. However, in the ASTM standards, there is no specific reference to the number of concentrically positioned pipes that may be joined by the stated procedures for butt and/or socket fusion. Nor are there any limitations stated concerning the use of such practice. Therefore, one interpretation is that as long as the procedures as described in the specifications are followed precisely for both the primary and the secondary containment pipes, each complies with the standard, although they are fused at the same time. Another interpretation is that any variation from a standard, stated or unstated in the standard, means that the standard is not being fully met. The way most consensus standards work, it is up to the user of the standard to interpret its meaning.

Simultaneous fusion is a relatively new method, and like all new methods, there must be some period of use and demand established prior to the development of any new standard. All practices and methods described in ASTM and other consensus standards were unlisted at some point prior to the writing of the first standard describing their use. Therefore, just because a new practice does not have a consensus standard written does not mean that it is unacceptable by default. (In the case of simultaneous fusion, there are many other drawbacks that may preclude its use, such the inability to meet code inspection and examination requirements.)

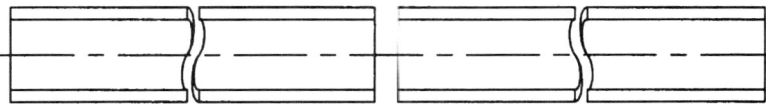

Figure 12.101. Position the primary pipes in place.

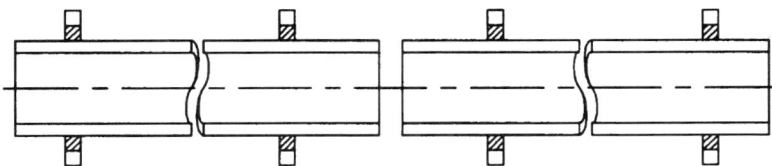

Figure 12.102. Position the interstitial centering supports in place approximately one inch from the end of the primary pipe.

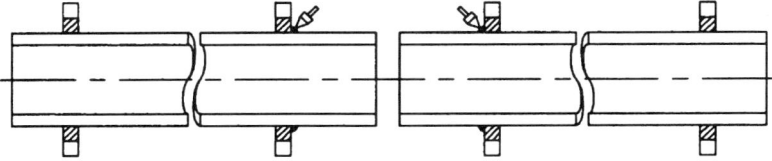

Figure 12.103. Weld the interstitial supports to the exterior of the primary pipe.

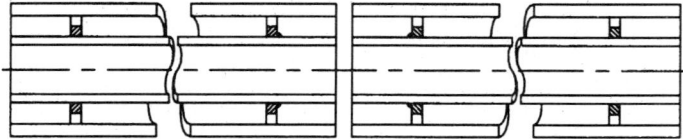

Figure 12.104. Slip the secondary pipe over the primary pipe interstitial supports.

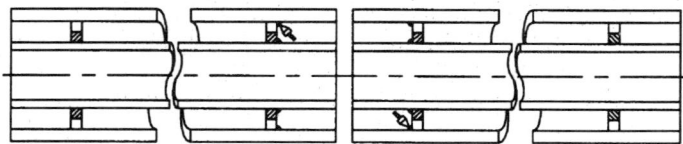

Figure 12.105. Weld the interstitial supports to the I.D. of the secondary containment pipe.

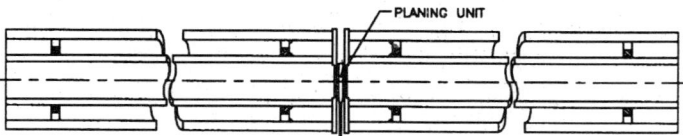

Figure 12.106. Plane the ends of the pipes to be sure that they are parallel and to expose a clean surface for welding.

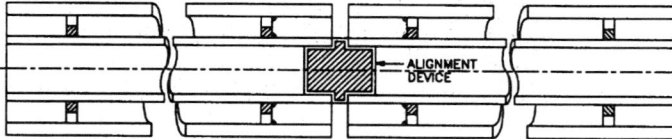

Figure 12.107. Check the alignment of the primary pipe after planing.

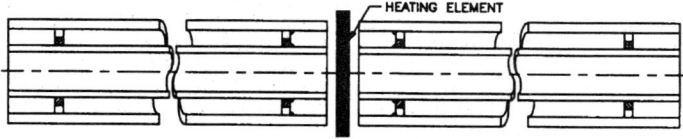

Figure 12.108. Position the heating element in place.

Figure 12.109. Bring the pipes up to the heating element under the specified initial weld pressure.

Fabrication, Assembly, and Erection

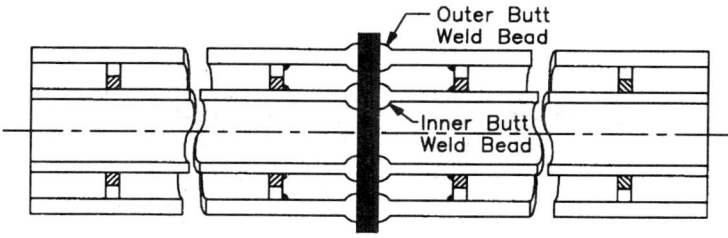

Figure 12.110. Allow the pipes to heat up for the required heating duration.

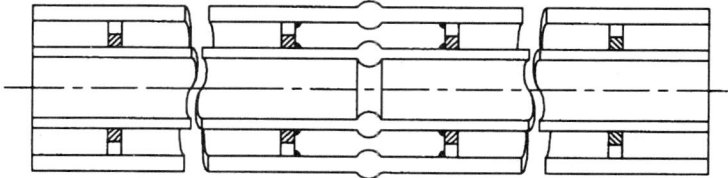

Figure 12.111. Remove the heater plate, and bring the pipes together using the specified final weld pressure.

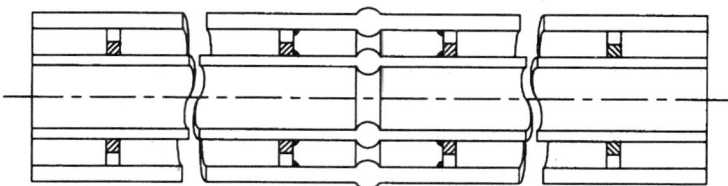

Figure 12.112. The illustration shows the final appearance of the joint section.

12.3.4 Multiple-Primary-Pipe/Common Secondary Containment Pipe Systems

All multiple-primary-pipe applications are joined by some type of staggered-welding procedure. Most are joined using "inner-first" procedures. However, some applications involve situations where the primary pipes may prefabricated and inserted as a bundle into the secondary containment jacket. This may be done for short runs, or where inspection requirements mandate that "outer-first" procedures be followed. All multiple-primary-pipe situations are unique; thus, the details of their joining must be thoroughly analyzed before any fabrication can proceed. The exact sequences depend upon the number of pipes, their sizes, and the materials of construction. They also depend on changes of direction, branch frequency and location, starting and termination arrangements, etc.

12.3.5 Installation of Above-Ground Double Containment Piping

There are three types of above-ground installations involving double containment piping systems. The most common type of installation involves changes of direction, branch connections, diameter changes, connections to equipment, etc.; piping in a system of this type is suspended in some manner. This type of piping system is typical of those in chemical plants, refineries, and manufacturing operations. These piping systems are normally suspended, anchored, and guided at various points along a trestle, or supporting by hangers, guides, or other devices from portions of the building structure or other permanent structures.

The second type of above-ground installation involves transfer pipes and other pipelines that are mostly long straight runs of pipe. In this type of installation, the pipes are supported, anchored, and guided on above-ground saddle supports or ring girders. The third type of installation consists of pipelines that are the same as the type just mentioned, but instead of being intermittently supported, are laid directly on the ground. This third type of installation mentioned is not a common installation type for double containment piping, but is frequently the method used for above-ground oil-patch piping transfer lines.

12.3.5.1 Support, Guiding, and Anchoring of Above-Ground Double Containment Piping

In general, the objective in installing above-ground piping is to keep the deflection of the piping system at all points within acceptable short- and long-term limits, by providing properly positioned and properly constructed supports, guides, and anchors. Double containment piping has an advantage over single-wall piping in that it is a structurally more rigid assembly. Chapter 9, "Structural Design" presents design theory necessary to determine frequency of supports for above-ground double containment piping systems. Additionally, the details of supporting primary piping within a double containment piping assembly are also presented in Chapter 9. In this section, the physical aspects of how to carry out the support requirements of double containment piping assemblies are presented.

Figure 9.4 illustrates various types of external supporting devices that are commonly used for the installation of above-ground piping systems. These supports must be attached or suspended from structural steel or other building parts. In recent years, a variety of nonmetallic corrosion-resistant materials have been introduced for supporting piping systems. Various corrosion-resistant alloy materials are also available. The construction material of the supports must be kept in mind if atmospheric corrosion problems exist, or in applications where contact with corrosive service fluids from another pipe, tank, or vessel is a possibility.

Figure 9.5 illustrates several common methods of anchoring, hanging and guiding for RTRP secondary containment piping systems. Figure 9.6 and 9.7 illustrate various types of guides and methods of anchoring secondary containment piping systems.

Other considerations for the selection and design of supports include the relative bearing surface of contact. All supports should have a wide enough contact surface such that a narrow bearing load is not imposed on the inner walls of secondary containment components. For notch-sensitive nonmetallic pipes, this is very important. It is usually good piping practice to position a compressible material between the piping material and the material of the support, in order to reduce the possibility of point loads. Standard sling-type supports, clevis hangers, shoe supports, and trapezes are among the most common types of supports used.

Small-diameter secondary containment piping (less than 12 in.; 315 mm) must be provided with a type of support that provides a wide underpinning of at least 120° of contact with the secondary containment pipe. For small-diameter nonmetallic secondary containment pipe that is subject to relatively high temperatures, consideration should be given to providing continuous support with a v channel, u channel, or rectangular channel.

For secondary containment piping larger than 12 in. (315 mm) in diameter, thin-wall pipes and secondary containment pipes with relatively low stiffness, it is recommended to provide an underpinning of at least 150° to 180° of contact. The width of the accompanying bands should be at least 1/3 the nominal pipe diameter, and bearing stresses should be kept within reasonable limits. Support diameters should always match as closely as possible the outside diameter of the secondary containment pipe. A compressible material is always recommended as the elastomer will tend to the reduce bearing stress and offer a more intimate contact between the pipe and its support.

Valves and other heavy equipment and assemblies should be provided with direct support, as these types of components can impose large bending moments on a piping system if the system is undersupported.

If flexibility is required in the layout of the secondary containment system due to thermal or other stresses, the installation must follow the design. Rigid attachment of secondary containment to supports and placement of anchors should be limited to those areas where it is necessary to ensure the proper flexibility of the system.

Care should be taken to isolate double containment piping systems from vibrating equipment such as pumps and compressors. This may be accomplished by providing expansion joints prior to the start of the zone of secondary containment. As an alternative, a vibration damper may be provided prior to the start of the secondary containment zone. Since double containment piping systems are structurally rigid and contain internal supports, they should not be subjected to vibrating forces.

12.3.5.2 Weatherability Considerations

Many types of secondary containment piping materials are subject to corrosion or damage due to the effects of ultraviolet radiation in above-ground applications. Several of the metallic piping materials are damaged by corrosion of atmospheric moisture, or airborne chemical corrosives in chemical facilities. Unless metallic secondary containment piping is constructed of corrosion-resistant alloys, it must be protected. There are many types of coatings and paints available to protect metallic piping. These range from ordinary solvent-based paints to coal tar or epoxy coatings and thermoplastic-backed tape products. The type of protection required depends on the application, the type of metallic material of the secondary containment piping, and whether or not external insulation is used.

Thermoplastic and RTRP secondary containment piping are also subject to weathering damage, but of a completely different nature than that of metallic materials. The damage is due to the tendency of plastic materials to be affected by ultraviolet radiation. Many thermosetting plastics and thermoplastic products are affected, although the fluoropolymers and polyketones tend to be completely unaffected. It should be pointed out that in some natural, unpigmented transparent materials, ultraviolet penetration can affect and alter a service fluid contained within a primary pipe. This is particularly important when the service fluid involved is chlorine or a chlorine-related compound, as active free-radical chlorine molecules can be generated, which may subsequently chlorinate or attack a plastic piping material.

Thermoplastic and thermosetting resins that are susceptible to ultraviolet light damage should be protected by painting or wrapping the piping to block out the light. RTP piping is easily protected with a solvent-based paint, or by using an outer liner of ultraviolet-resistant coating materials. Paint adhesion of RTP piping is often enhanced by delaying the application until some weathering has occurred. PVC and CPVC are easily protected by painting with a water-based paint. However, most other thermoplastics are not easily painted without some type of harsh primer being applied first. Such materials are more easily protected by taping or wrapping the piping. In installations where external insulation is applied, the piping will be completely protected from the effects of ultraviolet light. In any event, these effects, while readily detectable by visual means, do not always significantly alter a component's performance capability. In most cases, ultraviolet effects will act to reduce the ductility of the plastic materials involved.

12.3.6 Installation of Underground Double Containment Piping

While the method of analysis may differ for rigid and flexible pipes, the methods of installing all types of piping underground are very similar. The main areas of concern in underground installation are the type of bedding or trenching that the design calls for, the type of backfill materials, and the degree to which the backfill is to be compacted. In Chapter 9, the methods for analyzing underground pipes of all types are presented. In this section, the concentration is on implementing the design. A majority of single-wall pipes fall into the flexible piping category in terms of burial analysis (with the exception of concrete pipes, thick-wall steel pipes, etc.). However, double containment piping that is constructed of flexible pipes may behave more like a rigid pipe if frequent, rigid supports are positioned in its annulus. For this reason, double containment piping occasionally should be installed in many applications using considerations that ordinarily apply to rigid single-wall pipes such as concrete piping (see Section 9.4.2.1). The concepts and techniques for proper burial of both rigid and flexible pipes are presented in this section.

12.3.6.1 Rigid Double Containment Piping Considerations

Various trench beddings and embankment bedding details are shown in Figures 9.27 through 9.30 (details ordinarily used for single-wall concrete pipes). For each type of bedding, there is an accompanying load factor recommended for rigid pipe. Load factors for the five types of trench beddings are shown next to each individual type in Figure 9.27. Load factors for the three types of embankment beddings are shown in Figures 9.28 through 9.30. The load factor is defined as the ratio of the strength of a pipe under the installed condition of loading and bedding to its strength in the three-edge-bearing test. It should be pointed out that a three-edge-bearing test is normally only performed on concrete pipe. For double containment piping constructed of other materials, a suitable value must be found by conducting an appropriate test on the double containment assembly. The applicability of concrete load factors shown in Figures 9.27 through 9.30 should be determined for a double containment pipe assembly in the equivalent condition.

The Type 3 bedding condition shown in Figures 9.27 and 9.30 is the most frequently specified type of bedding and backfilling condition used for rigid piping in a trench condition and embankment condition, respectively. For moderate covers (6 ft; 1.8 m or less), the Type 3 bedding condition may represent a more rigorous bedding design than is actually required to support concrete pipe.[9] For more severe external loading situations (>6 ft; 1.8 m), Types 4 and 5 details represent the beddings that provide the needed additional pile support. In any given situation, the design should be governed by the most economical method that will provide the needed structural support and result in a safe design.

Other considerations for rigid piping installation include attention as to the type of surface upon which a pipe will be laid. Avoid laying the pipe on an unyielding or hard surface such as shale, hard clay, or rock. Where these conditions are present, fines from the excavation can be placed at the bottom of the trench or embankment to provide sufficient support. If fines are not available from the installation, they should be brought in to the job site from an outside source.

12.3.6.2 Buried "Flexible" Double Containment Piping Considerations

As discussed in Chapter 9, "Structural Considerations," flexible pipes obtain their ability to resist soil loads primarily from their ability to deflect and transfer a portion of the load to their sidefill. Therefore, flexible pipes are always better able to resist loads if they are installed in a trench condition. Recommendations as to appropriate widths of a trench for various materials of construction vary widely. Individual manufacturers must be consulted as to specific recommendations for their product. Alternatively, if a consensus standard exists for the product being used, the standard can be followed. In general, a conservative value for the minimum width of a trench, sheeted or unsheeted, measured at the spring line

Table 12.10 Comparison of Standard Density Tests*

Test	Compactive energy (ft-lb/cu ft)
Standard AASHTO (Standard Proctor) AASHTO T99-74, [1] ASTM D698-78 [2]	12, 400
Modified AASHTO (Modified Proctor) AASHTO T180-74, [3] ASTM D1557-78 [4]	56, 250

Notes:
* Natural in-place deposits of soils have densities from 60% to 100% of maximum obtained by the standard AASHTO compaction method. The designer should be sure that the E value used in design is consistent with this specified degree of compaction and method of testing that will be used during construction.
[1] Design and Construction of Sanitary & Storm Sewers. ASCE Manual No. 37. ASCE, New York (1969).
[2] Tests for Moisture-Density Relations of Soils & Soils-Aggregate Mixtures, Using 5.5-lb. (2.5-kg) Rammer & 12-in. (304.8-mm) Drop. ASTM, Philadelphia, PA. (1978).
[3] Moisture-Density Relations of Soils Using a 10-lb. (4.54-kg.) Rammer & an 18-in. (457 mm) Drop. AASHTO Standard T180-74. AASHTO, Washington, D.C. (1974).
[4] Test methods for Moisture-Density Relations of Soils & Soil-Aggregate Mixtures Using 10-lb. (4.54-kg.) Rammer & 18-in. (457-mm) Drop. ASTM Standard D1557-78. ASTM, Philadelphia, PA. (1978).

should be 1 ft. (305 mm) greater than the outside diameter of the pipe. The maximum clear width of a trench at the top of a pipe should not exceed a width equal to the pipe outside diameter plus 2 ft (610 mm). Where these widths must be exceeded, or for applications where it is necessary to install double containment piping in a compact embankment, the soil on each side of the secondary containment pipe should be compacted to a distance of at least 2-1/2 diameters on each side of the pipe, or to the trench walls, if that value is less.

Figure 12.113 illustrates a typical cross section of a trench for flexible pipe installation, which details all of the applicable terms. An equivalent diagram is used by the Composites Institute of the Society of the Plastics Industry. The terminology is slightly different, but all of the same considerations apply.[10] For buried metal piping, the American Society of Civil Engineers, and the American Water Works Association has devoted much attention to this subject.

The performance of a flexible double containment pipe combination is affected directly by a trench's width, depth of backfill, and the soil types. It also depends upon the preparation of a trench's bottom, the imbedment of the pipe, and the type of final backfill to be used. Table 12.10 presents a comparison of standard soil density tests. Proper procedures need to be carefully followed to place the completed pipe assembly in its trench. The backfill procedure is also critical to the success of any installation. Care must be taken in the compaction procedures to avoid vibrational damage to welded internal supports. It should always be remembered that loads imposed on the outside of a secondary containment pipe will be transmitted partially to its associated primary pipe.

All trenches must be de-watered using appropriate de-watering techniques. Water will interfere with the ability to follow recommended burial procedures. In addition, unstable subgrades should be excavated and refilled with a suitable foundation material. Items such as ledge rock, boulders, and large stones should be removed to allow at least a 4 in. (100 mm) soil cushion to be provided on all sides of the double containment piping assembly.

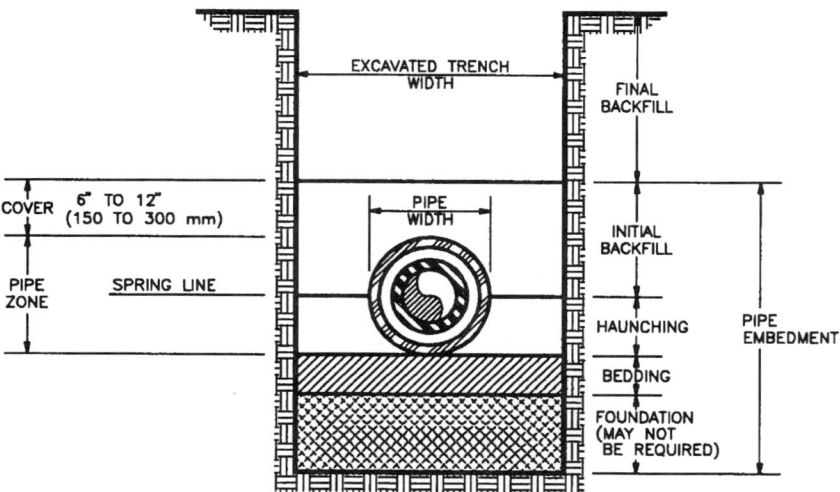

Figure 12.113. Trench cross section showing terminology for thermoplastic and other flexible pipes.

Fabrication, Assembly, and Erection

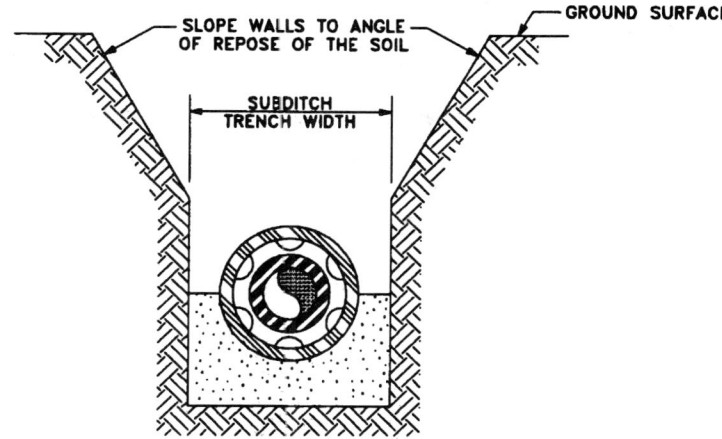

Figure 12.114. Example of an unsupported subditch trench with sloped side walls.

An alternative in some applications is the unsupported subditch type of trench. These are illustrated in Figures 12.114 and 12.115. The only limitation to using this type of design is where there is interference expected from pre-existing subsurface structures, property, buildings, trees, and pavements. If one of the aforementioned subsurface structures does present itself, then perhaps a narrow trench that extends all the way to the surface should be used. The recommended minimum trench widths for PVC single-wall pressure pipe are shown in Table 12.11. The minimum values for single-wall reinforced-thermosetting-resin piping is shown in Table 12.12. The method of compaction of the sidefill requires close attention to ensure that the sidefill can be compacted to their desired values in these narrow trenches.

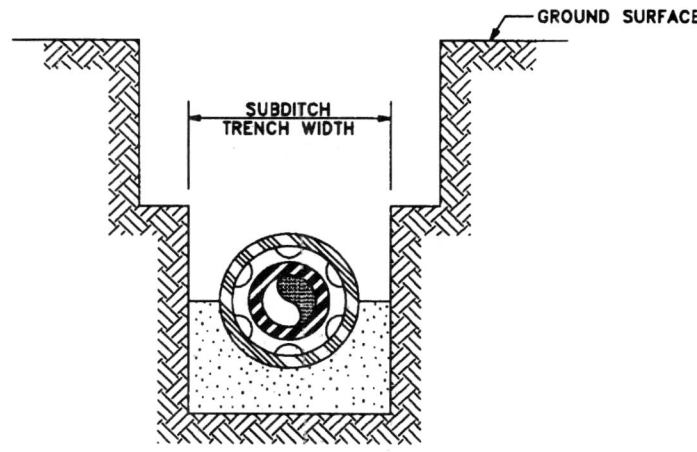

Figure 12.115. Example of an unsupported subditch trench with straight side walls.

Table 12.11 Recommended Minimum Trench Widths for Plastic Single-Wall Pressure Pipe

Nominal pipe diameter, in.	Minimum trench width width, in.
< 12	12 plus one pipe diameter
14 to 20	16 plus one pipe diameter
21 to 36	24 plus one pipe diameter
39 to 60	38 plus one pipe diameter
66 to 96	48 plus one pipe diameter

12.3.6.3 Installation by Trenching for Rigid and Flexible Double Containment Piping Systems

Specific details for trenching, laying, and backfilling should be drawn up once the design details of the project are known. The detailed requirements should be firmly stated in the project plans, drawings, and specifications. The installation methods chosen, and the actual equipment to be used, should be based on the requirements and conditions of the project. A checklist, recommended by the American Water Works Association for concrete pipe installations, is shown in Table 12.13. This list, indicating all of the direct and indirect factors that affect installation, serves as a suitable basis for preparing specifications for underground double containment piping systems. Installation of double containment piping systems by tunneling methods is not recommended.

Trenching is an item that requires considerable coordination with various trades, double containment piping system suppliers, and other parts of the project construction team. The final details can vary significantly due to findings uncovered at a field site. In almost every situation, changes will be necessitated due to field conditions discovered as construction

Table 12.12 Minimum Trench Widths for Fiberglass Piping

Nominal pipe diameter, in.	Minimum trench width width, in.
< 12	12 plus one pipe diameter
14 to 20	16 plus one pipe diameter
21 to 36	24 plus one pipe diameter
39 to 60	38 plus one pipe diameter
66 to 96	48 plus one pipe diameter

Table 12.13 Checklist and Guide for Installation Planning of Underground Systems

Site Preparation	Installation and Backfill
1. Location and protection of existing structures and plantings.	1. Pipe weight and length.
2. Existing utilities, protection, and access for use.	2. Selection of equipment.
3. Pavement removal.	3. Pipe jointing.
4. Storage locations – pipe and equipment	4. Control of grade and alignment.
5. Rights-of-way.	5. Service connections.
6. Construction of detours or traffic controls.	6. Connections to existing pipe.
7. Access roads.	7. Preparation of trench bottom.
8. Barricades.	8. Pipe bedding material – imported or native.
	9. Backfill placement and compaction.
Excavation	10. Backfill of appurtenances or structures.
1. Selection of equipment.	11. Pavement or sidewalk replacement.
2. Classification of excavation (type of material).	12. Removal of excess soil.
3. Limitations of trench width.	13. Surface restoration.
4. Spoil placement.	14. Acceptance tests.
5. Excavation for appurtenances.	15. Measurement and payment.
6. Sheeting or bracing.	
7. Tunnels and/or jacketing.	**Appurtenances and Miscellaneous Materials**
8. Groundwater.	1. Air relief outlets and valves.
9. Moving existing utilities.	2. Drain outlets, valves, and piping.
10. Installation of temporary utilities.	3. Manhole outlets and structures.
	4. In-line valves.
	5. Structure construction.
	6. Encasement or cradles.
	7. Concrete and reinforcing steel.
	8. Cathodic protection systems.
	9. Measurement and payment.

Source: AWWA Manual M9, "Concrete Pressure Pipe," Denver, CO, 1979, p. 74.

progresses. These should be kept to an absolute minimum. The double containment system supplier should be involved in the project early on, and should be immediately informed of any changes. At least one meeting is necessary to coordinate manufacture, prefabrication, and delivery with the construction schedule. Double containment piping systems normally involve a considerable amount of fabrication, in addition to the procurement of nonstandard items. Consideration should be given to the inherently long lead times of these items in planning project schedules.

12.3.6.4 Trench Safety Considerations

In the United States, the rules and regulations of the Occupational Safety and Health Administration and the state and local authorities having jurisdiction must be followed. There should be an individual on every project who is authorized to assume responsibility for safety and should be aware of all the rules, regulations, and requirements of the agencies having jurisdiction.

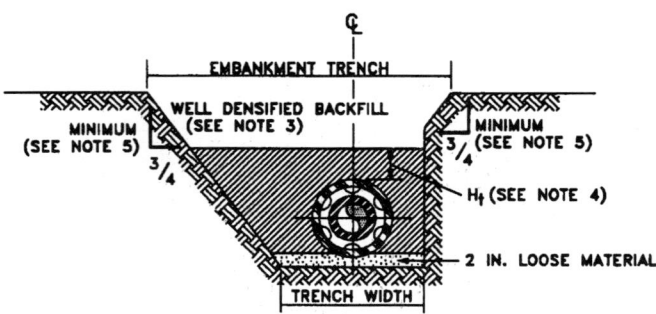

Figure 12.116. Densified pipe zone bedding and backfill for class C bedding conditions. (Source: "Steel Pipe – A Guide for Design and Installation," AWWA Manual M11, Figure 12.1, p. 124.)

12.3.6.5 Construction Survey and Controls

Each project should have the complete right of way of the project to be reviewed through a comprehensive reconnaissance survey. All existing utilities and substructures should be identified through pre-existing utility company records. The types of information that should be gathered include: (1) the location of streets and their intensity; (2) trees; (3) above-ground utilities; and (4) space requirements for equipment. Surveys should be made on the entire route and checked with existing information to establish survey ties. For vertical control, benchmarks should be checked. For horizontal control, traverses should be closed. Variances of prefabricated double containment piping from the plan and profile should be avoided to the greatest extent possible to avoid interfering with utilities or existing structures. There is also unwanted expense and delay involved when materials require additional reworking to meet such variations.

A contractor normally assumes responsibility for the transfer of line and grade to the excavation work from control points. It is very important that the control points be preserved. The line and grade can be transferred to the bottom of the trench by a number of methods including tape and plumb-bob units, use of a tape and level, or batter boards. An alternative method for accurate alignment is the use of a double string line and grade rod technique. Other more sophisticated means include setting the line and grade by means of a transit and level, or even the use of lasers.

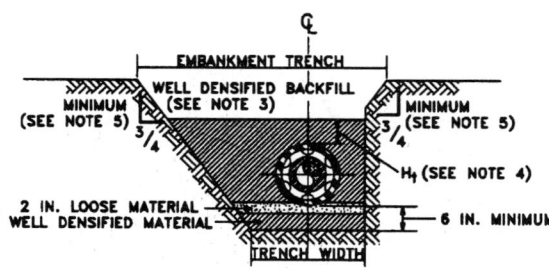

Figure 12.117. Special subgrade densification for class C bedding conditions. (Source: "Steel Pipe – A Guide for Design and Installation," AWWA Manual M11, Figure 12.2, p. 124.)

12.3.6.6 Site Preparation

On some projects, the site may require preliminary work before trenching procedures can be started. This includes work such as clearing of trees and brush, removal of rocks and unstable soils, access road construction, and protection and relocation of existing utilities. If this preliminary type of work needs to be done, it should be coordinated with other phases of the project.

12.3.6.7 Size of the Installation Trench

Based on the project design conditions, each trench width should be specified. It is always desirable to have a trench as small as possible to allow for proper laying, backfilling, and compacting of the soil. A smaller trench not only is less expensive to excavate in most applications, but it normally aids in a pipe's ability to resist soil loads. Where feasible, the sides of a trench above the top of a pipe are often sloped, as shown in Figures 12.116 and 12.117, for economic reasons. A trench with a sloped side will save on the need to reinforce the sides of the trench with sheeting and bracing, except where safety standards still require it.

12.3.6.8 Excavation

Depending on the application, excavation of a trench may be accomplished in one or more steps. A trench should always be excavated as close to grade as possible. Sufficient width should be left to allow backfill to be properly placed around a pipe. The type of equipment to be used varies widely based on conditions of the job, the local availability of equipment, and contractor preferences. Common types of equipment used to excavate trenches include backhoes, draglines, clamshells, power shovels, front end loaders (for large and deep trenches), and rotary trenching machines.

12.3.6.9 Sheeting and Bracing of Trench Walls

The design of sheeting and bracing for reinforcing of trench walls should follow sound rules of engineering, and must comply with existing safety standards. The responsibility for trenching and bracing is normally delegated to the installing contractor. This includes any damage sustained by piping, due to inadequate supporting of the trench wall or during removal of sheeting and bracing. Various types of reinforcing methods include use of: (1) skeleton sheeting; (2) continuous sheeting; (3) trench shields; and (4) steel sheet piling. For most applications, skeleton sheeting or continuous sheeting is sufficient. Both of these types involve the use of vertical members supported by spreaders or trench jacks. The difference between the two is that in skeleton sheeting vertical members are wooden planks, and in continuous sheeting they are panels prefabricated with plank stiffeners. In noncohesive soils with ground water, it becomes necessary to use steel sheet piling.

12.3.6.10 De-watering of Installation Trenches

If ground water exists, trenches must be completely de-watered prior to laying any pipe. If the ground water causes conditions involving unstable soils to exist, then trenches should be overexcavated, and unstable soil removed and replaced with steady or stable soil. If ground water is expected to be an ongoing problem, then unstable soil can be replaced with crushed stone or gravel to assist in trench drainage. In extreme cases, a well-point system must be provided. This type of system consists of a series of perforated pipes connected to a header pipe and pump.

12.3.6.11 Pipe Foundations

Double containment pipes should always be provided with a uniform foundation (bedding). A firm, cohesive soil is an adequate foundation if it is undisturbed. If secondary containment pipe fittings have bells, the foundation should be shaped to accommodate bells and other features. Normally, it is more practical to overexcavate a foundation and provide backfilling to grade with a uniform bedding material. In some applications where a double containment piping system is classified as a rigid assembly, an appropriate bedding must be provided, as in Figures 9.27 through 9.30, to result in an improved load-bearing capacity.

12.3.6.12 Pipe Delivery

Double containment piping components is normally prefabricated to desired lengths from single-wall piping components, and as a result are usually not regularly stocked items. Fittings are almost without exception fabricated items. Therefore, their schedules should be closely coordinated between all parties involved. If there are any changes to project requirements or schedules, the double containment piping system manufacturer should be notified right away, so their production schedule can be modified.

12.3.6.13 Job-Site Storage

Since double containment piping is expensive and is used in hazardous or corrosive-fluid applications, it should be stored in a careful, protected manner. Avoid stacking the pipe where possible. If space requirements dictate that it must be stacked, then keep the stacks to within the recommended stacking heights for the ordinary single-wall pipe requirements of the secondary containment pipe. Piping should be kept in enclosed, protected buildings, containers, or shelters, and the ends covered and protected from moisture. Rain water and other liquids should be prevented from entering a system's annular space, so that leak-detection systems are not affected at a later date by false readings. Fittings and assemblies should also be kept thoroughly protected in some type of enclosure.

12.3.6.14 Pipe Handling

Double containment piping should be handled very carefully, with care taken so that no impacts, vibrations, or other forces are imparted to the piping. It should be remembered that piping normally contains interstitial supports and interconnecting parts. Thus, a force imparted to a secondary containment pipe will be partially transmitted to the associated primary pipe components as well.

12.3.6.15 Laying of Double Containment Piping

Double containment piping should be laid in the trench as gently as possible, with the use of proper equipment. Additionally, the piping should not be subjected to excessive bending forces during the laying process (see Section 9.2). The best type of equipment to use is a crane that has horizontal extensions, thus allowing the piping to be supported over a wide

Table 12.14 Classification of Native Soils[11]

Group 1 Very stable soils	Group 2 Stable soils	Group 3 Soil mixtures	Group 4 Cohesive soils
1A Dense [1] gravels or sands according to ASTM-D2487 GW, GP, SW & SP containing less than 5% fines.	2A Medium [1] slightly silty or clayey gravels or sands according to ASTM GM, GC, SM & SC containing less than 15% fines.	Typically medium cohesive or loose granular [1] soils. According to ASTM ML or CL with liquid limit less than 50 and GM, GC, SM & SC.	Soft and very loose [1] soils per ASTM MH, CH, OL & OH.
1B Hard or very stiff cohesive soils [1]	2B Stiff cohesive soils [1].		

Notes: [1] See Table 12.15.

area. A crane of this type is shown in Figure 12.106. Other types of equipment that can be used are backhoes having cable slings, or those equipped with a strong-back for lowering pipes into a trench. Also, for small piping, a side-boom-type tractor can be used.

12.3.6.16 Field Joints

Most field joints for double containment piping should be out of the trench whenever possible. If an in-trench joint is required, a double containment pipe should be supported above its foundation, so that the foundation is not disturbed.

12.3.6.17 Backfilling

The quality and workmanship used during backfill procedures are critical to the success of any double containment project. Double containment piping is unique in that it normally contains internal supports, some of which may be welded. Therefore, the compaction procedures used are critical. It is also important that compaction loads be distributed evenly, so that the compaction is performed to the desired specifications.

Table 12.15 Soil Group Classifications[12]

Soil group	1	2	3	4
Cohesive [2] (fine grained)	Hard and very stiff	Stiff	Medium	Soft
Granular [3] (coarse grained)	Very dense and dense	Medium	Loose	Loose
Blow count [1]	>30	16-30	6-15	3-5

Notes:
[1] Blows as measured with a 2" OD 1.375" ID sampler driven 1' by a 140 lb. hammer falling 30". ASTM D1586, "Standard Method for Penetration Test and Split Barrel Sampling of Soils."
[2] Soils containing a large portion of fine particles (clay & colloidal size). Shear strength is largely, or entirely, derived from cohesion (natural attraction of particles). Includes clays, silty clays, and clays mixed sand or gravel.
[3] Soils not exhibiting a natural attraction. Shear strength is derived from the compactness (density) of the confined grains.

Table 12.16 Classification of Soils[13]

Cohesive soils [2]

Consistency	Blows [1]/ft (N)	qu [4] (kN/m2)	Identification characteristic
Very soft	0-2	0-25	Sample tends to lose shape under own weight.
Soft	3-5	26-50	Molded with slight finger pressure.
Medium	6-15	51-150	Molded with moderate finger pressure.
Stiff	16-30	151-300	Molded with substantial finger pressure.
Very stiff	31-50	301-500	Not moldable by finger pressure. Requires picking to remove.
Hard	>50	>500	Difficult to remove by picking.

Granular soils [3]

Compactness	Blows [1]/ft (N)	Relative density (%)	Moist unit weight (kN/m3)
Very loose	0-5	0-15	11-16
Loose	6-15	16-40	14-18
Medium	16-30	41-65	17-20
Dense	31-50	66-85	18-22
Very dense	>50	>85	20-23

Notes:
[1] Blows as measured with a 2" OD 1.375" ID sampler driven 1' by a 140 lb. hammer falling 30". ASTM D1586, "Standard Method for Penetration Test and Split Barrel Sampling of Soils."
[2] Soils containing a large portion of fine particles (clay & Colloidal size). Shear strength is largely, or entirely, derived from cohesion (natural attraction of particles). Includes clays, silty clays and clays mixed sand or gravel.
[3] Soils not exhibiting a natural attraction. Shear strength is derived from the compactness (density) of the confined grains.
[4] Unconfined compressive strength.

12.3.6.17.1 Backfill Materials

Backfill that is specified should be based on the conditions of the project and the pipe combination to be buried. Backfill materials should be free from rocks, stumps, pavement, sharp rocks, or other potentially damaging materials. If the native material does not meet the project specifications, then material must be imported. The best type of backfill material for most projects is sand or a combination of sand and gravel.

Various types of native soils are classified in Table 12.14. Additionally, soil group classifications are shown in Table 12.15. Other miscellaneous characteristics of soils are shown in Table 12.16. A detailed classification of soils is shown in Table 12.17.

Table 12.17 Detailed Classification of Soil[14]

	Criteria for assigning group symbols and group names using laboratory tests.				
Class	Category	Test		Symbol	Group
Coarse grain soils more than 50% retained on No. 200 sieve	**Gravels** More than 50% of coarse fraction retained on No. 4 sieve	Clean gravels less than 5% fines	$Cu > 4$ and $1 < Cc < 3$	GW	Well graded gravel
			$Cu < 4$ and/or $1 > Cc > 3$	GP	Poorly graded gravel
		Gravels with fines over 12% fines	Fines classify as ML or MH	GM	Silty gravel
			Fines classify as CL or CH	GC	Clayey gravel
	Sands 50% or more of coarse fraction passes No 4 sieve	Clean sands less than 5% fines	$Cu > 6$ and $1 < Cc < L\ 3$	SW	Well graded sand
			$Cu < 6$ and/or $1 > Cc > 3$	SP	Poorly graded sand
		Sands with fines over 12% fines	Fines classify as ML or MH	SM	Silty sand
			Fines classify as CL or CH	SC	Clayey sand
	Silts and clays Liquid limit less than 50%	Inorganic	Pl > 7, plots on/above A-line	CL	Lean clay
			Pl < 4 or plots below A-line	ML	Silt
		Organic	LL-OD/Ll-ND < 0.75	OL	Organic clay or silt
Fine grained soils 50% or more passes the No. 200 sieve	**Silts and clays** Liquid limit less 50% or more	Inorganic	Pl plots on/above A-line	CH	Fat clay
			Pl plots below A-line	MH	Elastic silt
		Organic	LL-OD/Ll-ND < 0.75	OH	Organic clay or silt
Highly organic soils	Primarily organic matter	Dark in color and organic odor		PT	Peat

Notes:
a. Based on material passing the 3-in. sieve.
b. If field sample contained cobbles or boulders, or both, add "with cobbles or boulders, or both" to group name.
c. Gravels with 5 to 12% fines require dual symbols: GW-GM well graded gravel with silt; GW-GC well graded gravel with clay; GP-GM poorly graded gravel with clay.
d. Sands with 5 to 12% fines require dual symbols: SW-SM well graded sand with silt; SW-SC well graded sand with clay; SP-SM poorly graded sand with silt; SP-SC poorly graded sand with clay.
e. $Cu = D60/D10 * (D30)^2/d10 * D60$
f. If soil contains > 15% sand add "with sand" to group name.
g. If fines classify as CL-ML use dual symbol GC-GM or SC-SM.
h. If fines are organic, add "with organic fines" to group name.
i. If soil contains > 15% gravel, add "with gravel" to group name.
j. If Atterberg limits plot in hatched area, soil is a CL-ML silty clay.
k. If soil contains 15-29% plus No. 200, add "with sand" or "with gravel" whichever is dominant.
l. If soil contains > 30% plus No. 200, predominately sand, add "sandy" to group name.
m. If soil contains > 30% plus No. 200, predominately gravel, add "gravelly" to group name.
n. Pl 4 and plots on or above "A" line.
o. Pl < 4 or plots below the "A" line.
p. Pl plots on or above the "A" line.
q. Pl plots below the "A" line.

12.3.6.17.2 Placing the Backfill

Once the pipe assembly has been backfilled to the level of the pipe zone height, the rest of the fill can be put in place with bulldozers and other equipment. Care should be taken to place the fill materials uniformly on both sides of the pipe to prevent shifting or lateral movement of the pipe assembly. Care should also be taken that excessive live loads not be imparted to the pipe by the bulldozer. Bracing should not be removed until the backfilling procedure is complete.

12.3.6.17.3 Backfill Densification

If densification is required, there are two methods that can be used. These are consolidation (using water in a flooding or jetting operation) and compaction (tamping the soil with vibrating equipment, air tools, or other mechanical means). Compaction is the method most commonly used, although consolidation may be more practical for soils that are porous and self-draining (i.e., sand and gravel). Backfill should be compacted in layers 6-12 in. (150-300 mm) in depth. Extreme care should be used where a brittle pipe material (such as PVC or PVDF at cold ambient temperatures) is used as the primary piping material, so that impact damage does not occur. Caution should also be observed where hot-gas welds or extrusion welds are used with thermoplastic secondary containment piping, due to the inherent lack of ductility and toughness of those types of welds.

12.3.6.17.4 Special Considerations for Thermoplastic Pipes

When installing a system that uses PVC secondary containment piping, extra precautions should be taken depending on the joining method used. Very often these systems use straight socket solvent cement or adhesive slip-couplings in the secondary jacket. These types of joints are inherently weak in comparison to the pipe, and as such represent a weak link in the secondary containment piping system. When these types of systems are installed in the heat of day, they will later cool to the temperature of the surrounding soil. The pipe will want to therefore contract, but due to the friction of the soil acting on the pipe's surface, and due to the restraint imposed by elbows (see Figure 8.1) the pipe will be restrained from contracting, and will instead be placed under a longitudinal tensile stress. Since the slip-coupling to pipe straight socket joint is much weaker than the corresponding pipe, the bond will typically shear resulting in secondary containment joint failure at one or more joint locations. Groundwater will then usually work its way into the failed joint, if the hydrostatic head is great enough in the surrounding soil. Depending on the type of leak detection system used, the owner of the system may or may not be alerted that such a failure has taken place. A similar effect can be achieved in systems that use hot-gas-welded slip-couplings or clam-shell couplings (See Figures 11.67 through 11.77). It is usually best to wait until early morning to backfill systems of this type, when the systems will be at the closest temperature to the surrounding soil.

12.3.6.17.5 Restoration of the Surface

Upon completion of backfilling, the surface should be restored to a condition that is agreed to and stated in the project specifications. Normally, the conditions that existed prior to construction are required to be restored after the construction is complete. Before any permanent pavement or driveway is replaced, the subgrade should be restored and compacted until it is smooth and unyielding.

12.3.6.17.6 Marine Installations

There are inherent complexities involved with double containment piping in marine subsurface installations. If it is used for these purposes, the extra buoyancy provided by the lack of fluid in the annular space must be taken into account. In general, there is an inherent difficulty in facilitating the repair of such a system; normally, a system of this type has to be brought above the surface to make a repair or modification.

References

1. Code for Chemical Plant and Petroleum Refinery Piping, ANSI/ASME B31.3-1990 Edition, Paragraph A328.2.5, American Society of Mechanical Engineers, New York, NY (1990).

2. Bradshaw, Roger D., "A Proposed Bonder Qualification Test for ASME B31 Plastic Piping Systems," PVP Conference, 1993.

3. Code for Chemical Plant and Petroleum Refinery Piping, ANSI/ASME B-31.3-1990 Edition, Paragraph 345.2.2, 345.2.5, and 345.3.1, American Society of Mechanical Engineers, New York, NY (1990).

4. George E. Linnert, Welding Metallurgy, Vol. 2, 3rd ed., American Welding Society, New York, 1967, pp. 82-93.

5. Ibid. p. 85.

6. Ibid. p. 87.

7. Example taken from the brochure: "Carpenter 20Cb-3 Stainless Steel," Carpenter Technology Corporation, 1980.

8. Linnert et al., op. cit., p. 93.

9. "Concrete Pressure Pipe," AWWA Manual of Water Supply Practices, AWWA M9, 1979, Denver, CO, p. 51.

10. Fiberglass Pipe Handbook, Fiberglass Pipe Institute, The Composites Institute of the Society of the Plastics Industry, Inc., New York, NY, 1989, p. 119.

11. ibid., p. 124.

12. ibid., p. 125.

13. ibid., p. 126.

14. ibid., p. 127.

Notes

Chapter Thirteen

Inspection, Examination, and Testing

Double containment piping systems are used in applications involving hazardous or corrosive fluids. If these fluids are allowed to escape to the surroundings, they can harm the environment, or pose a threat to living organisms. To design and implement a double containment system to handle a dangerous fluid means that a considerable investment is being made to safeguard against such incidents. Therefore, it is imperative that installed and welded (bonded) components of such systems be inspected and examined for defects and quality of workmanship prior to allowing the system to start up.

This chapter covers three distinct ways to qualify the integrity of a double containment piping system: inspection, examination, and testing. Inspection is a function that is performed by an owner or an owner-appointed inspector and is a process that is performed visually. Examination, on the other hand, refers to quality-control functions performed by any organization that carries out welding or bonding of system joints. They include double containment piping system suppliers, single-wall piping component manufacturers, fabricators of double containment weldments, mechanical contractors, and owner personnel. There is a variety of nondestructive examination techniques that may be used to examine a system and its components. Testing usually refers to some means of placing the system components under an internal pressure that is greater than the design pressure of the component for a set period of time. In a double containment system, there are testing issues to consider for both the individual pipes and the double containment assembly. On any successful project, a system must be subjected to all three forms of qualification. They are discussed in the following.

13.1 Metallic Component Considerations

13.1.1 Inspection Requirements

The ultimate responsibility for the quality control of a double containment pipe project is left to the owner. For this reason, an owner should provide an inspector to inspect the work on every project of this type. One of the basic duties of an owner-supplied inspector is to verify that all required examinations and tests are performed and passed with satisfactory results. An inspector's other major duty is to inspect the piping system until there is reasonable assurance that the installation satisfies the requirements of the engineering design and is in conformance with applicable codes, standards, and laws.

The ANSI/ASME B31.3 Code for Chemical Plant and Petroleum Refinery Piping suggests that qualifications to serve as an inspector of the owner include at least 10 years in the design, fabrication, or inspection of industrial pressure piping. Of these 10 years, 5 may consist of schooling in an accredited engineering program. (The program should be accredited by the "Accreditation Board for Engineering and Technology" located at 345 E. 47th Street, New York, NY 10017.) These are the same experience qualifications necessary to apply for a Professional Engineering license in most of the individual U.S. states and territories. If the inspector delegates any portion of their responsibility to another, care should be taken in the selection so that only a qualified person is chosen.

13.1.2 Examination Requirements

Although the owner has responsibility for inspecting the work, double containment system suppliers, fabricators of double containment weldments, and each contractor involved in a project are all responsible for completing the examination requirements specified. These suppliers and contractors not only must complete all required examinations, but they must also prepare and maintain suitable records for the inspector and for other future purposes. The suppliers, including manufacturers of single-wall components that wind up in double containment parts, are responsible for providing quality components that meet applicable codes and standards and the engineering design.

Prior to the startup of a double containment piping system, each weld in the piping system must either be subjected to examination or be qualified by a pressure test. The intent of the examination, where it is required, is to provide the examiner and the inspector with reasonable assurance that the engineering design and applicable codes have been met. On some materials that require heat treatment, the examination should take place only after heat treatment is completed (for instance P-Nos. 3, 4, and 5 type materials). On welded branch connections, the pressure-retaining weld should be examined prior to adding any reinforcing pad or saddle for the primary branch connection (and secondary containment branch connection if it is pressure rated).

Most examination performed in the field relates to welding (bonding) aspects of piping components. The acceptance criteria for each of the different types of welds to be performed on a project should be clearly spelled out in project specifications. Tables 13.1 through 13.4 list acceptance criteria for various types of welds subject to a variety of conditions, according to the 1993 edition of the ANSI/ASME B 31.3 Code. Additionally, Table 13.5 lists types of weld imperfections versus the kinds of examinations that may be used to detect them. Figure 13.1 shows a variety of imperfections in metallic welds.

Figure 13.2 shows an example of a cracked weld crater in a weld bead involving alloy 20Cb-3® materials. The weld was made with 85 A and 1/8 in. (3.2 mm) shielded metal-arc electrodes; the arc was abruptly broken, resulting in the crack. Figure 13.3 shows how crater cracking can be eliminated in the same materials by an accelerated travel speed before breaking the arc. Alternatively, a "crater-eliminator" attachment could be used, by which the current could be decreased before the arc is broken. (20Cb-3 is a registered trademark of Carpenter Technology Corporation.)

Table 13.1 Acceptance Criteria for Welds According to ANSI/ASME B31.3

Kind of imperfection	Criteria (A-M) for types of welds and for required examination methods [1]					
	Normal fluid service					
	Methods		Types of Welds			
	Visual	Spot or random radiography	Girth and miter groove	Longitudinal groove [2]	Fillet [3]	Branch connection [4]
Crack	X	X	A	A	A	A
Lack of fusion	X	X	A	A	A	A
Incomplete penetration	X	X	B	A	NA	B
Internal porosity	---	X	E	E	NA	E
(a) Slag inclusion, tungsten inclusion, or elongated indication	---	X	G	G	NA	G
(a) Undercutting	X	---	H	A	H	H
Surface porosity or exposed slag inclusion [5]	X	---	A	A	A	A
Surface finish	---	---	---	---	---	---
(a) Concave root surface (suck-up)	X	X	K	K	NA	K
Reinforcement or internal protrusion	X	---	L	L	L	L

Notes:
[1] Criteria given are for required examination. More stringent criteria may be specified in the engineering design. See also paras. 341.5 and 341.5.3 of the ASME B31.3 Code-1993 Edition.
[2] Longitudinal groove weld includes straight and spiral seam. Criteria are not intended to apply to welds made in accordance with a standard listed in Table A-1 of Appendix A of this text or Table 326.1 of the ASME B31.3 Code-1993 Edition.
[3] Fillet weld includes socket & seal welds, & attachment for slip-ons & branch reinforcement.
[4] Branch connection weld includes pressure containing welds in branches & fabricated laps.
[5] These imperfections are evaluated only for welds ≤ 3/16" (5 mm) in nominal thickness.

Chapter 5 lists acceptance criteria for such items as castings and the like, which should also be subject to examination. If at any time defects in products or workmanship are detected, they should be repaired or replaced. All new work should be re-examined by the same criteria until the work satisfies the acceptance criteria.

If random examination is used on a project, a progressive examination should be used. A progressive examination is one where the unveiling of a defect results in having to go back and inspect more items of the same kind. If the additional items that are examined reveal another defect, then further items have to be examined. Usually, after the third such iteration, any items that show up as faulty require replacement and re-examination or a complete examination with repair and re-examination of defects.

Table 13.2 Acceptance Criteria for Welds According to ANSI/ASME B31.3

Kind of imperfection	Criteria (A-M) for types of welds and for required examination methods [1]							
	Severe cyclic conditions							
	Methods [1]				Types of welds [2]			
	V. [5]	100% R. [6]	M.P. [7]	L.P. [8]	G.& M.G. [9]	L.G. [2] [10]	Fillet [3]	B.C. [4] [11]
Crack	X	X	X	X	A	A	A	A
Lack of fusion	X	X	---	---	A	A	A	A
Incomplete penetration	X	X	---	---	A	A	NA	A
Internal porosity	---	X	---	---	D	D	NA	D
(a) Slag inclusion, tungsten inclusion, or elongated indication	---	X	---	---	F	F	NA	F
(a) Undercutting	X	X	---	---	A	A	A	A
Surface porosity or exposed slag inclusion [12]	X	---	---	---	A	A	A	A
Surface finish	X	---	---	---	J	J	J	J
(a) Concave root surface (suck-up)	X	X	---	---	K	K	NA	K
Reinforcement or internal protrusion	X	---	---	---	L	L	L	L

Notes:
[1] Criteria given are for required examination. More stringent criteria may be specified in the engineering design. See also paras. 341.5 and 341.5.3 of the ASME B31.3 Code-1993 Edition.
[2] Longitudinal groove weld includes straight and spiral seam. Criteria are not intended to apply to welds made in accordance with a standard listed in Table A-1 of Appendix A of this text or Table 326.1 of the ASME B31.3 Code-1993 Edition.
[3] Fillet weld includes socket & seal welds, & attachment for slip-ons & branch reinforcement.
[4] Branch connection weld includes pressure containing welds in branches & fabricated laps.
[5] V. = Visual
[6] 100% R. = 100% Radiography
[7] M.P. = Magnetic Particle
[8] L.P. = Liquid Penetrant
[9] G.&M.G. = Girth & Miter Groove
[10] L.G. = Longitudinal Groove
[11] B.C. = Branch Connection
[12] These imperfections are evaluated only for welds ≤ 3/16" (5 mm) in nominal thickness.

Table 13.3 Acceptance Criteria for Welds According to ANSI/ASME B31.3

Kind of imperfection	Criteria (A-M) for types of welds and for required examination methods [1]				
	Category D fluid service				
	Methods		Types of welds		
	Visual	Girth and miter groove	Longitudinal groove [2]	Fillet [3]	Branch connection [4]
Crack	X	A	A	A	A
Lack of fusion	X	A	A	NA	A
Incomplete penetration	X	C	A	NA	B
Internal porosity	---	---	---	---	---
(a) Slag inclusion, tungsten inclusion, or elongated indication	---	---	---	---	---
(a) Undercutting	X	I	A	H	H
Surface porosity or exposed slag inclusion [5]	X	A	A	A	A
Surface finish	---	---	---	---	---
(a) Concave root surface (suck-up)	X	K	K	NA	K
Reinforcement or internal protrusion	X	M	M	M	M

Notes:
[1] Criteria given are for required examination. More stringent criteria may be specified in the engineering design. See also paras. 341.5 and 341.5.3 of the ASME B31.3 Code-1993 Edition.
[2] Longitudinal groove weld includes straight and spiral seam. Criteria are not intended to apply to welds made in accordance with a standard listed in Table A-1 of Appendix A of this text or Table 326.1 of the ASME B31.3 Code-1993 Edition.
[3] Fillet weld includes socket & seal welds, & attachment for slip-ons & branch reinforcement.
[4] Branch connection weld includes pressure containing welds in branches & fabricated laps.
[5] These imperfections are evaluated only for welds ≤ 3/16" (5 mm) in nominal thickness.

There are many specific types of examination that are required for every installation involving metallic components intended for chemical services. These include visual examination of randomly selected materials and components. Visual inspection of these components is needed to ensure that they comply with specifications and have been made with good workmanship. At least 5% of all fabrications should be visually inspected, with the work of each fabricator and welder represented. Flanged joints and other bolted joints should be randomly examined to be sure that they comply with the design requirements. However, if pneumatic testing is used, 100% of these joints should be examined. The piping erection should be randomly examined to be sure that alignment, supports, and cold spring are being followed correctly and for any other irregularities. If longitudinal welds

Table 13.4 Criterion Value Notes for Tables 13.1, 13.2, and 13.3

Criterion		Acceptable value limits [1]
Symbol	Measure	
A	Extent of imperfection	Zero (no evident imperfection)
B	Depth of incomplete penetration	≤ 1/32" 90.8 mm) and ≤ 0.2T_w
	Cumulative length of incomplete penetration	≤ 1.5" (38 mm) in any 6" (150 mm) weld length
C	Depth of lack of fusion & incomplete penetration	≤ 0.2T_w
	Cumulative length of lack of fusion & incomplete penetration [2]	≤ 1.5" (38 mm) in any 6" (150 mm) weld length
D	Size & distribution of internal porosity	See BPV Code, Section VIII, Division 1, Appendix 4
E	Size & distribution of internal porosity	T_w ≤ 1/4" 96.4 mm), limit is same as D For T_w > 1/4" 96.4 mm), limit is 1.5 X D
(a) F	Slag inclusion, tungsten inclusion, or elongated indication	
	Individual length	≤T_w/3
	Individual width	≤3.32" (2.4 mm) and ≤ T_w/3
	Cumulative length	≤T_w in any 12T_w weld length
(a) G	Slag inclusion, tungsten inclusion, or elongated indication	
	Individual length	≤ 2T_w
	Individual width	≤ 1/8" (3.2 mm) and ≤ T_w/2
	Cumulative length	≤ 4T_w in any 6" (150 mm) weld length
H	Depth of undercut	≤ 1/32" (0.8 mm) and ≤ T_w/4
I	Depth of undercut	≤ 1/16" (1.6 mm) and ≤ [T_w/4 or 1/32" (0.8 mm)]
J	Surface roughness	≤ 500 min. Ra per ASME B46.1
K	Depth of root surface concavity	Total joint thickness incl. weld reinforcement, ≥ T_w
L	Height of reinforcement or internal protrusion [3] in any plane through the weld shall be within limits of the applicable height value in the tabulation. Weld metal shall merge smoothly into the component surfaces.	For T[w], in. (mm) Height, in. (mm) ≤ 1/4 (6.4) ≤ 1/16 (1.6) > 1/4 (6.4), 1/2 (12.7) ≤1/8 (3.2) > 1/2 (12.7), 1 (25.4) ≤ 5/32 (4.0) >1 (25.4) ≤ 3/16 (4.8)
M	Height of reinforcement or internal protrusion [3] as described in L	Limit is 2L

X = required examination   NA = not applicable   --- = not required

Notes:
[1] Where two limiting values are separated by "and," the lesser of the values determines acceptance. Where two sets of values are separated by "or," the larger value is acceptable. T_w is the nominal wall thickness of the thinner of 2 components joined by a butt weld.
[2] Tightly butted unfused root faces are unacceptable.
[3] For groove welds, height is the lesser of the measurements made from the surfaces of the adjacent components. For fillet welds, height is measured form the theoretical throat, Fig. 328.5.2A of the ASME B31.3 Code-1993 Edition; internal protrusion does not apply.

Table 13.5 Types of Examination for Evaluating Imperfections [1]

Kind of Imperfection	Type of examination			
	Visual	Liquid penetrant or magnetic particle	Ultrasonic or radiographic	
			Random	100%
Crack	X	X	X	X
Incomplete penetration	X	---	X	X
Lack of fusion	X	---	X	X
Weld undercutting	X	---	---	---
Weld reinforcement	X	---	---	---
Internal porosity	---	---	X	X
External porosity	X	---	---	---
Internal slag inclusions	---	---	X	X
External slag inclusions	X	---	---	---
Concave root surface	X	---	X	X

Notes:
[1] Evaluation, any necessary repair of imperfections, and examination of additional items shall be limited to the requirements of paragraph 341.3.3 of the ANSI/ASME B31.3 Code; unless more stringent requirements are specified by the engineering design.

are performed, they should be completely examined along their entire length. These are commonly used for secondary closures, split secondary containment fittings, and other welds in double containment piping systems.

Other types of examinations that are required as a minimum relate to butt-and-miter groove welds and brazed joints. At least 5% of all circumferential butt-and-miter groove welds should be fully examined by random radiography or by random ultrasonic examination. As an option to these two methods, at least 5% of the welds mentioned can be inspected by in-process examination. At least 5% of all brazed joints should be inspected by in-process examination. For both brazed joints and the types of butt welds described, the work of each different welder or brazer on a project should be represented.

On every project, it is very important that an appropriate record-keeping system be established and maintained. Additionally, a certification process should be established to certify quality control requirements are being followed. The inspector and examiner should have complete access to all records and certifications.

For piping systems that are to be subjected to "severe-cyclic" conditions, the examination should be more widespread. One-hundred percent of all fabrications and threaded or bolted joints should be visually examined. All aspects of the piping erection should be examined as well to check for alignment, support placements, guides, and implementation of cold spring. Also, all circumferential butt-and-miter groove welds should be subjected to radiographic or ultrasonic examination. Socket welds and branch connection welds should be examined by magnetic particle or liquid penetrant methods. Any of the commonly used nondestructive tests may be used as a supplementary test.

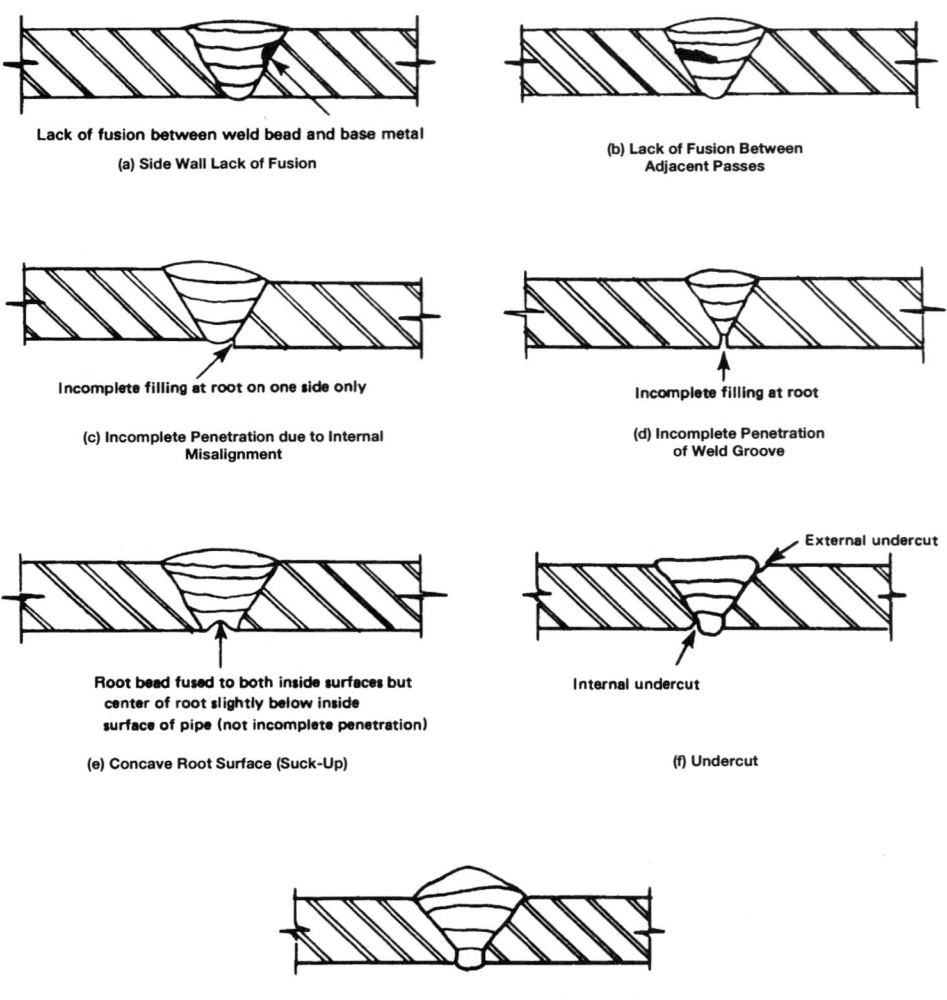

Figure 13.1. Typical weld imperfections. (Source: 1993 ANSI/ASME B31.3 Chemical Plant and Petroleum Refinery Piping Code.)

For welds that require heat treatment, hardness testing should be performed to verify that the part has been satisfactorily heat treated. The weld and the heat affected zone (HAZ) should both be tested as close to the edge of the weld as possible. For dissimilar metals that are welded together, the hardness limits for each of the base materials and the welding metals should be met. At least 10% of all welds should be tested for those metals where a hardness limit is specified in Table 12.4.

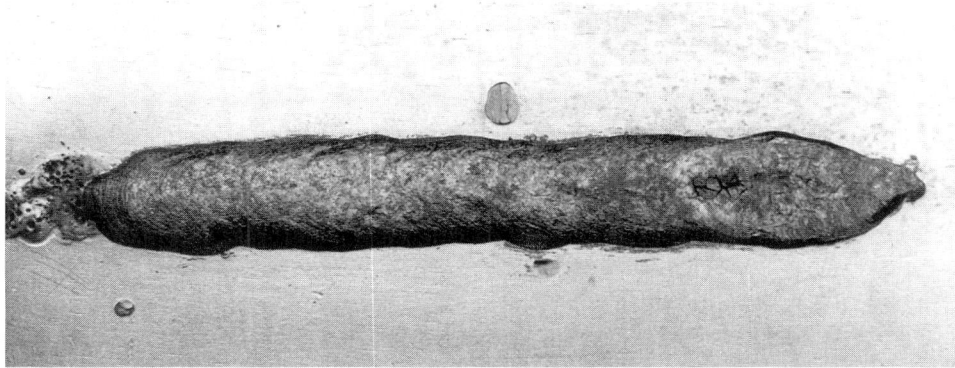

Figure 13.2. Example of a cracked weld crater in a weld bead of Alloy 20Cb-3® materials. (Source: Carpenter Technology Co.)

13.1.2.1 Examination Procedures

A suitable outline for the preparation of examination procedures is stated in the ASME Boiler and Pressure Vessel Code, Section V, Article 1, T-150. Each employer of welding personnel who perform work for projects should maintain his/her own records of the examination procedures used. These records should keep track of dates and results of procedure qualifications. All of these records should be made available to the owner's inspector.

13.1.2.2 Types of Examinations

The various types of examinations that are used on metallic components include: (1) visual examinations; (2) magnetic particle examination; (3) liquid penetrant examination; (4) radiographic examination; (5) ultrasonic examination; and (6) in-process examination. Other methods may be used, but they should be described in sufficient detail to allow qualifica-

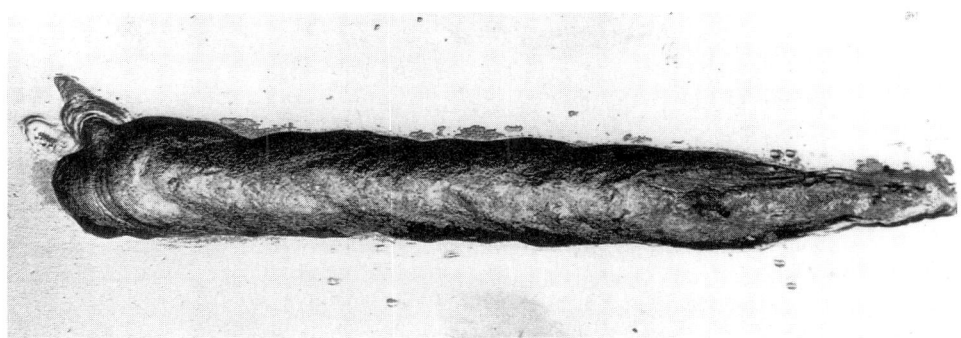

Figure 13.3. Example of a satisfactory weld (no cracking) in Alloy 20Cb-3® materials, using an improved welding procedure. (Source: Carpenter Technology Co.)

tion of the necessary procedures and examiners. The criteria for acceptance for any alternative method should also be described in detail.

13.1.2.2.1 Visual Examination

The ASME Boiler and Pressure Vessel Code, Section V, Article 9, outlines a suitable method of practice of visual examination. The visual examination of a piping system involves observation of those portions that are exposed to view at any time in the manufacturing, fabrication, and assembly process. Materials, components, dimensions, joint preparation, alignment, welding, bonding, brazing, bolting, threading, other joining methods, supports (interstitial or external), and the assembly and erection of the double containment piping system may all be observed for verification of the engineering design and applicable code compliance.

13.1.2.2.2 Magnetic Particle Examination

The ASME BPV Code, Section V, Article 7, outlines a suitable method for the examination of welds and other components by magnetic particle examination. Castings may also be examined by magnetic particle examination, as discussed in Chapter 5.

13.1.2.2.3 Liquid Penetrant Examination

The ASME BPV Code, Section V, Article 6, outlines a suitable method for the examination of welds and other components by liquid penetrant examination. Castings may also be examined by liquid penetrant examination, as discussed in Chapter 5.

13.1.2.2.4 Radiographic Examination

The ASME BPV Code, Section V, Article 2, outlines a suitable method for the examination of welds and other components by radiographic examination. Radiography can be performed to a varying degree, including 100% of all welds, randomly (girth and miter groove welds), or on a spot basis to a specified percentage of welds. For a longitudinal weld, the weld should be examined at least 6 in. along its length.

13.1.2.2.5 Ultrasonic Examination

The ASME BPV Code, Section V, Article 5, outlines a suitable method for the examination of welds and other components by ultrasonic examination. Examination using ultrasonic techniques is a very involved and complex science. An excellent reference is "Non-Destructive Testing of Large-Diameter Pipe for Oil and Gas Transmission Lines."[1] Each separate size, wall thickness, and type of material should be included in the examination. Reference levels for monitoring discontinuities should be modified to reflect the transfer correction when the transfer method is used.

13.1.2.2.6 In-Process Examination

An in-process examination may be conducted for many aspects of double containment piping installation. This includes joint preparation and cleanliness, preheating requirements of materials, the fit-up, clearance and internal alignment of joints (prior to welding), variables specified for welding and brazing, the condition of the root pass after cleaning, slag condition, and weld removal between passes and appearance of the finished joint. The condition of the root pass after cleaning may also be inspected on an internal basis where it is accessible. The internal examination can be aided with the use of liquid penetrant or magnetic particle techniques. An in-process examination is usually of the visual type, unless otherwise specified in the engineering design. Appropriate records should be established and maintained when using in-process examination techniques.

13.1.3 Testing

All piping must be tested to ensure tightness of both the primary and secondary containment piping systems prior to their use. In most situations, a hydrostatic leak test is required for the primary piping. If the fluid service is other than Category D or Category M (i.e. normal fluid service; see the current edition of the ANSI/ASME B31.3 Code), an initial service leak test can be conducted of the primary piping system at the owner's option.

In some instances, a hydrostatic pressure test may be impractical if linings, insulation, or leak detection will be damaged. Also, some fluid services may be contaminated when contacted with water or test fluids. Under these conditions, a pneumatic pressure test can be used, at the owner's option. In making such a choice, the hazards of the energy stored in compressed gases should be thoroughly considered. A combination hydrostatic-pneumatic test can also be used; in a combination test, the pipes are filled with water (or test fluid), and the pressure raised by pneumatic action.

In some rare instances, hydrostatic or pneumatic testing of the primary piping systems may be impracticable; in this case, an alternative leak test can be used. Alternative leak testing is described in Section 13.1.3.4.5.

A hydrostatic test is always the preferable method when it can be implemented, since it is usually safe and highly reliable. If a hydrostatic test cannot be used, a pneumatic test should be used unless an undue hazard is created, or there is a possibility of a brittle fracture. Only when these two tests are ruled out should an alternative leak test be used.

13.1.3.1 Limitations of Pressure/Leak Testing

When applying test pressures to a piping system, the test pressure selected should never result in a stress that exceeds the yield strength of the piping material at the test temperature. Instead, the test pressure should be reduced to a level so that the yield strength of the material will not be exceeded. If the testing fluid is subject to thermal expansion, precautions should be taken if the test duration is long enough such that a problem may result. If the hydrostatic pressure test is to be performed for an extended period of time (> 10 min), or the test water is to be left in the pipes after the test is completed, the possibility of microbiologically induced corrosion (MIC) should be considered (see Section 2.1.9) and if necessary, treatment chemicals added.

13.1.3.2 General Requirements for Pressure/Leak Testing

Prior to conducting a full-scale test of a system, it may be wise to pretest the system using air at a pressure of 25 psig (170 kPa). This is done in order to locate any major leaks in the system and eliminate any expense that may result from large losses of water.

Pressure tests of primary metallic piping systems should be conducted for at least 10 min. During this time, all joints and other connections must be left exposed to enable them to be examined for any sign of leaks (unless previously tested and examined), according to ANSI/ASME B31.3 Code.[2] If the piping requires any heat treatment, the system should be tested only after the heat treatment is complete. Special precautions should be taken if the test must be conducted near the ductile-brittle transition temperature of the material.

Subassemblies that are flanged at both ends only need to be tested once. Therefore, if they are tested prior to assembly, it is necessary to test only the remainder of the system. If a blank is inserted to isolate a certain section of a piping system during the hydrostatic test, the flanged joint does not have to be retested. If there are any repairs made to the system, the repaired sections must be retested. This is required, unless the repairs are extremely minor, and the owner decides that retesting is not required.

13.1.3.3 Preparation for Pressure/Leak Testing

All joints should be left uninsulated and fully exposed for visual examination during leak testing, whenever possible. This may not always be practical or feasible for primary piping; it is still required by Code.[2] This problem is addressed in Section 13.3. If the welds have already been tested, they do not need to be exposed. This is very important for primary piping, since once it is tested as part of a prefabricated section, it can be subsequently covered. For primary piping that is intended for vapor or gas, temporary supports should be provided to the secondary containment piping if it is needed. Primary piping that is not designed to withstand the weight of a liquid should not be tested with liquid. In Chapter 9, it is suggested that systems be designed to withstand the weight of actual fluids under the most extreme condition, including test fluids or service fluids, whichever fluid is heaviest (see Section 9.3.1.1). For piping that is not going to be pressure tested, it should be isolated by providing blinds, or other methods, during the test. Expansion joints should be tested in accordance with the ANSI/ASME B31.3 Code (in the United States and Canada) or other applicable design criteria (e.g., Expansion Joint Manufacturer Association rules).

13.1.3.4 Details of Testing

Details are presented in this section for hydrostatic testing, pneumatic testing, combined hydrostatic-pneumatic testing, initial service leak testing, and alternative leak testing.

13.1.3.4.1 Hydrostatic Pressure Test

Hydrostatic tests should always be conducted with water as the test medium, unless there is a possibility of freezing. Also, some metals may be damaged or corroded by water, or by microbiologically induced corrosion as a result of microbes in the water (see Section 2.1.9). Under these conditions, an alternative fluid can be used, so long as the fluid is safe and nontoxic. Flammable fluids should be limited to those that have a flash point greater than 120°F (49°C).

A system should be tested at a pressure of 1-1/2 times the design pressure. However, the test pressure selected should never result in a stress that exceeds the yield strength of the piping material at the test temperature. Instead, the test pressure should be reduced to a level so that the yield strength of the material will not be exceeded.

If the design temperature of the system is above the testing temperature, the minimum test pressure should be increased to reflect the ratio of the stress to the component strength at that temperature. The following equation can be used:

Inspection, Examination, and Testing

$$P_T = \frac{1.5 P S_T}{S} \qquad (13.1)$$

where:
P_T = minimum test gage pressure, psi
P = internal design gage pressure, psi
S_T = stress value at test temperature, psi
S = stress value at design temperature, psi

The ratio of S_T/S should not exceed a value of 6.5 when using this equation. If it does, the test pressure can be determined at this maximum ratio.

If the piping is attached to a pressure vessel that has a test pressure the same, or greater than the piping, then the piping and the vessel can be tested at the same time. However, if the test pressure of the pressure vessel is less, the piping should be isolated from the vessel and tested separately. Where this is considered to be impracticable, the piping can be tested at the vessel test pressure so long as the test pressure of the vessel is at least 77% of the piping test pressure.[3] The owner has the final decision as to whether to allow this reduction in piping test pressure. Testing of tanks is discussed further in Chapter 14.

If the piping is designed for external pressure service, it should be tested to an internal gauge pressure of 1.5 times the differential pressure. The minimum pressure test required for external pressure service is 15 psig (103 kPa).

13.1.3.4.2 Pneumatic Leak Test

Pneumatic testing is a procedure that involves compressing air, or some other gas, inside the pipe to a predetermined test pressure. Since air and other gases are compressible fluids, this process results in the storing of large amounts of potential energy. Upon the sudden release of this stored energy, significant damage can result, posing a threat of injury to those nearby. The ductility of the piping at the test temperature is a primary consideration, and the chance of brittle fracture to occur should be minimized during such a procedure.

Pneumatic tests should be conducted to a test pressure of 110% of the design pressure of the piping system. The pressure should be slowly raised to the desired test pressure. First, an initial test of 25 psi (1.8 bar) should be conducted, unless 25 psi (1.8 bar) is more than half of the desired test pressure. In this case, the initial test should be performed at one-half of the desired test pressure. At this point, all joints and connections should be visually examined for leaks. If the piping passes the initial test, the pressure should be raised gradually. At each step, the pressure should be held until the piping strains are somewhat equalized. Once the test pressure is reached, the pressure should be reduced to the design pressure, and the system examined for leaks. A pressure-relief device should be provided, set to a pressure of the test pressure plus 50 psi (340 kPa), or 110% of the test pressure, whichever is less.

13.1.3.4.3 Combined Hydrostatic-Pneumatic Leak Test

A combination test consists of filling the system with water (or other fluid) completely, thoroughly venting all air, and then using compressed gas as a means of raising the pressure of the system. In this type of test, the test pressure should be that of a hydrostatic pressure test. Caution should be exercised that the pressure is not raised too rapidly, and the testing pressure thereby exceeded.

13.1.3.4.4 Initial Service Leak Testing

An initial service leak test is one that is performed with the actual fluids upon initial start-up. It is limited to services other than Category D and M fluids (see definitions in para. 300.1.C of the ANSI/ASME B31.3 Code), and is performed instead of a pressure test. When a fluid other than Category D or M fluids is used, an initial service test can be used at the option of the owner. The key to such a test is to raise the pressure gradually in small steps. At each step, the pressure should be held until the piping strains are somewhat equalized. At all stages, the system should be thoroughly examined for leaks. This test is not feasible, nor recommended, for double containment piping systems.

13.1.3.4.5 Alternative Leak Testing

An alternative leak test consists of 100% radiography of all welds, combined with complete liquid penetrant inspection. Even structural welds, such as those used to attach piping centering supports should be examined using liquid penetrant. For systems using magnetic materials, the magnetic particle method may be used in place of the liquid penetrant method. Additionally, a sensitive leak test as described in the ASME Boiler and Pressure Vessel Code, Section V, Article 10, or other equivalent method must be provided. The sensitivity for such a test should be at least 10-3 atm ml/sec (100 Pa ml/s) under test conditions. Any piping system subjected to an alternative leak test must also be subjected to a flexibility analysis, as an extra precaution.

13.1.3.4.6 Testing Records

Records should be kept of the testing program for each piping system. The information should include at least: (1) date of test; (2) identification of the piping system tested; (3) the type of test fluid or fluids used; (4) testing pressure; and (5) certification of the results (by the examiner). Records of the examination procedures and the examination personnel qualifications should both be retained for a minimum period of at least 5 years.

13.2 Nonmetallic Component Considerations

13.2.1 Inspection Requirements

As stated in Section 13.1.1, the ultimate responsibility for the quality control of a double containment pipe project is left to the owner. For this reason, the owner should provide an inspector to check inspect the work on every project of this type. This applies equally to projects that use nonmetallic materials. The basic duties of an owner-supplied inspector remain the same as those described in Section 13.1.1: to verify that all required examinations and tests are performed and passed with satisfactory results and to inspect the piping system until there is reasonable assurance that the installation satisfies the requirements of the engineering design and is in conformance with applicable codes, standards, and laws.

The ASME B 31.3 Code suggests that qualifications to serve as an inspector on the part of the owner include at least 10 years in the design, fabrication, or inspection of industrial pressure piping. Of these 10 years, 5 may consist of schooling in an accredited engineering program. (The program should be accredited by the "Accreditation Board for Engineering Board for Engineering and Technology" located at 345 E. 47th Street, New York, NY 10017.) These are the same experience qualifications necessary to apply for a Professional

Engineering license in most of the individual U.S. states and territories. If an inspector delegates any portion of his/her responsibility to another, care should be taken so that only a qualified person is selected.

13.2.2 Examination Requirements

Although the owner has responsibility for inspecting the work, double containment system suppliers, fabricators of double containment weldments, and each contractor involved in a project are still responsible for completing all examination requirements specified. These suppliers and contractors not only must complete all required examinations, but they must also prepare and maintain suitable records for the inspector, and for other future purposes. The suppliers, including manufacturers of individual single-wall components that wind up in double contained parts, are responsible for providing quality nonmetallic components that meet applicable codes and standards and the engineering design.

Prior to the startup of the double containment piping system, each nonmetallic bond in the piping system must either be subjected to examination or be qualified by pressure testing prior to use. The intent of the examination, where it is required, is to provide the examiner, and the inspector, with reasonable assurance that the engineering design and applicable codes have been met.

Most examination performed in the field relates to bonding aspects of nonmetallic piping components. The acceptance criteria for each of the different types of bonds to be performed on a project should be clearly spelled out in the project specifications. Table 13.6 lists acceptance criteria for various types of bonds verses types of defects, according to the 1993 edition of the ASME B 31.3 Code. Figures 13.4 and 13.5 illustrate a variety of imperfections in thermoplastic hot-gas fillet welds and thermal butt fusion welds, respectively, taken from German DIN standards and the DVS handbook (Deutscher Verlag fur SchweiSStechnik, Dusseldorf).

Table 13.6 Acceptance Criteria for Various Types of Plastic Bonds

Kind of imperfection	Thermoplastic			RTR & RPM
	Hot gas weld	Solvent cemented	Heat fusion	Adhesive cemented
Cracks	None permitted	Not applicable	Not applicable	Not applicable
Unfilled areas in joint	None permitted	None permitted	None permitted	None permitted
Unbonded areas in joint	Not applicable	None permitted	None permitted	None permitted
Inclusions of charred material	None permitted	Not applicable	Not applicable	Not applicable
Unfused filler material inclusions	None permitted	Not applicable	Not applicable	Not applicable
Protrusion of material into pipe bore, % of pipe wall thickness	Not applicable	Cement, 50%	Fused material, 25%	Adhesive, 25%

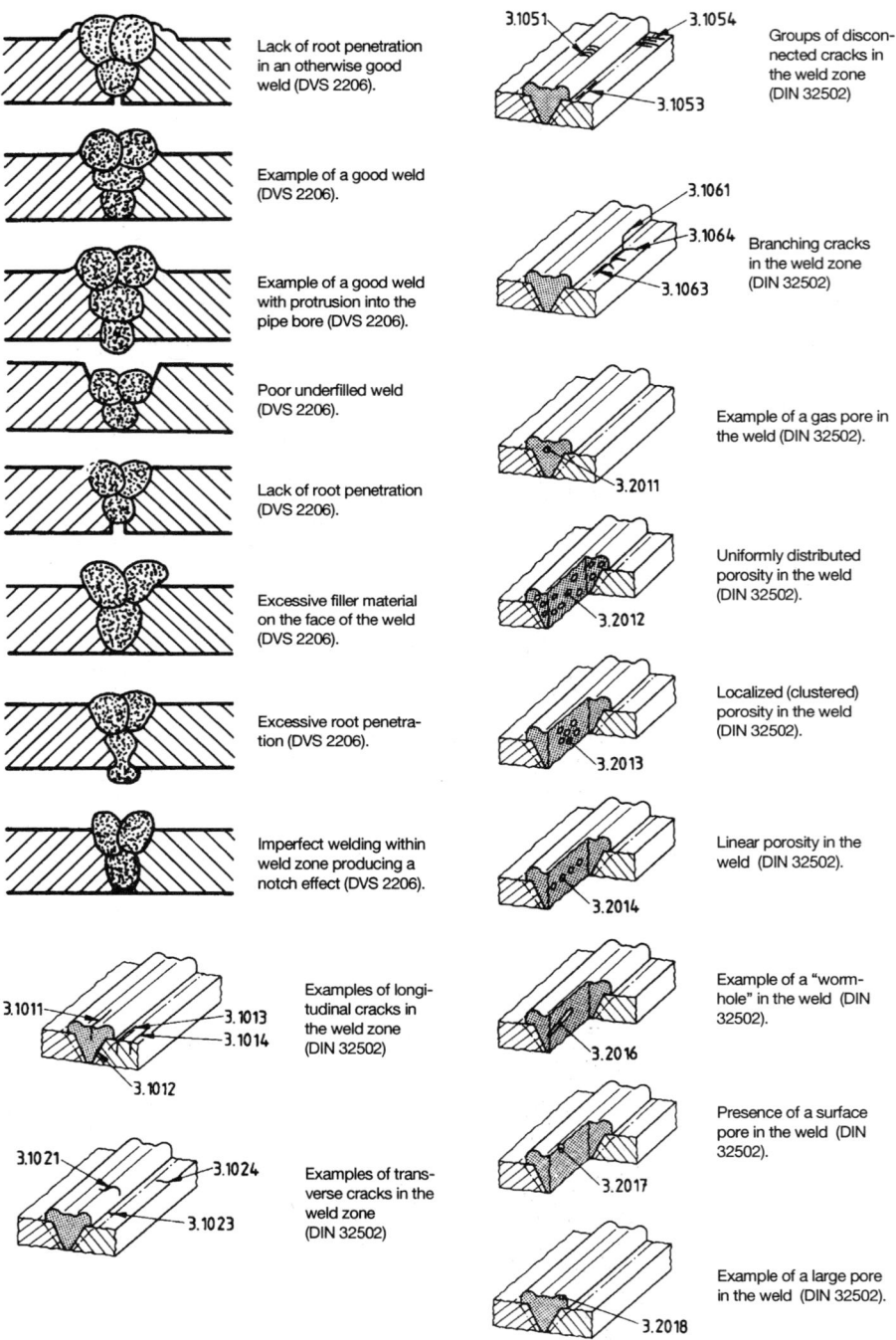

Figure 13.4. Examples of imperfections in hot gas fillet welds in thermoplastic materials. (Source: DVS 2206 and DIN 32502.)

Inspection, Examination, and Testing

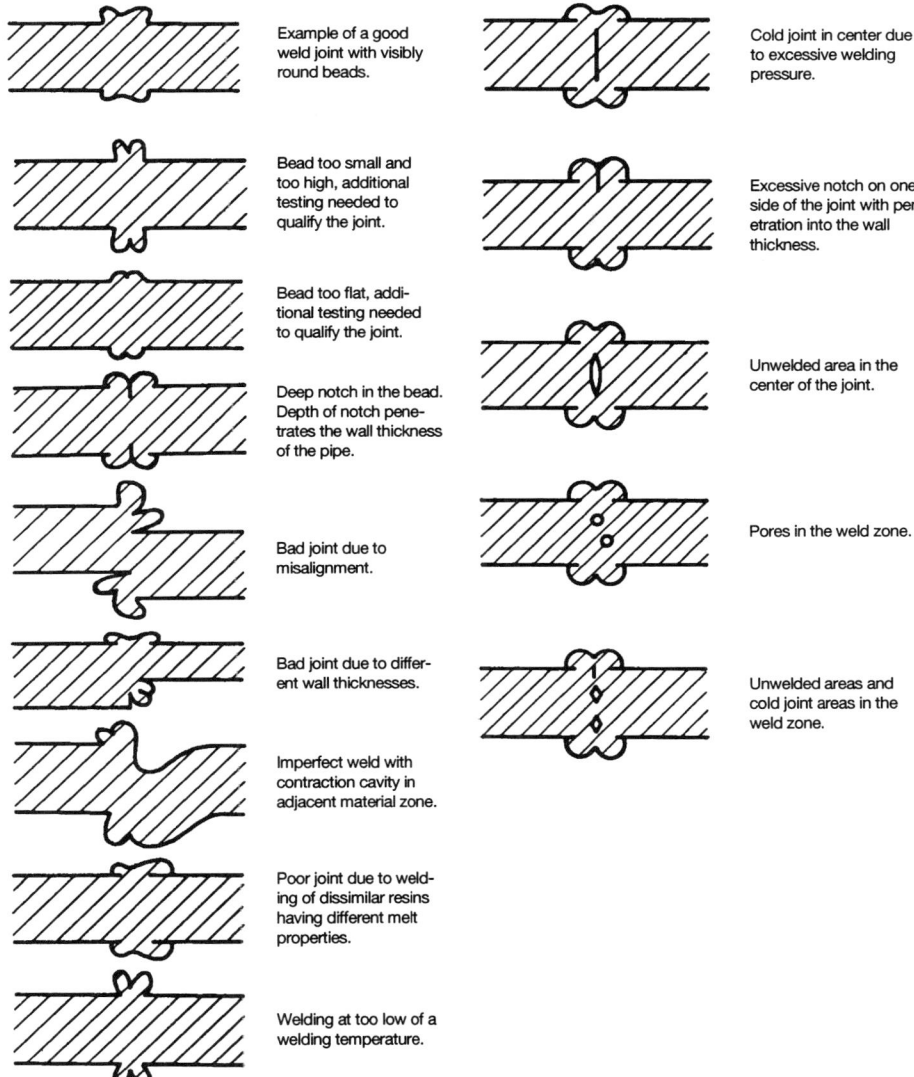

Figure 13.5 Examples of imperfections in heat-element butt-fusion joints in thermoplastic materials. (Source: DVS 2206.)

If at any time defects in products or workmanship are detected, they should be repaired or replaced. All new work should be re-examined by the same criteria until the work satisfies the acceptance criteria.

If random examination is used on a project, it should be a progressive examination, in which the unveiling of a defect results in having to go back and inspect more items of the same kind. If the additional items reveal another defect, then further items have to be

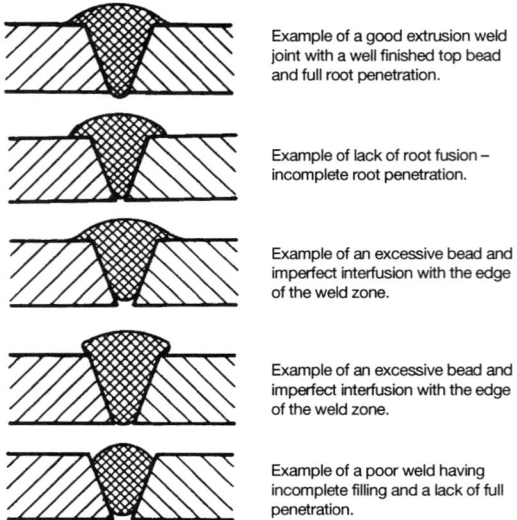

Figure 13.6 Examples of imperfections in extrusion welds in thermoplastic materials. (Source: DVS 2206.)

examined. Usually, after the third such iteration, any items that show up as faulty require replacement and re-examination, or complete examination with repair and re-examination of defects.

There are many types of examination that should be required for every installation involving nonmetallic components. These include visual examination of randomly selected materials and components. The visual inspection of these components should be done to ensure that they comply with specifications and have been made with good workmanship. At least 5% of all fabrications should be visually inspected, with the work of each fabricator and welder represented.

Flanged joints and other bolted joints should be randomly examined to be sure that they comply with the design requirements. However, if pneumatic testing is used, 100% of these joints should be soaped and visually inspected for bubbles.

The piping erection should be randomly examined to be sure that alignment, supports, and cold spring are being followed correctly and for any other irregularities. If bonds other than circumferential types are used (i.e., longitudinal hot-gas welds, adhesively bonded longitudinal seams, RTP-overwrapped longitudinal seams, etc.); they should be completely examined along their entire length. These are commonly used for secondary closures, split secondary containment fittings, and other bonds in nonmetallic secondary containment portions of a double containment assembly.

On every project, it is very important that an appropriate record-keeping system be established and maintained. Additionally, a process should be established to certify whether quality-control requirements are being followed. The inspector and examiner should have complete access to all records and certifications.

13.2.2.1 Examination Procedures

An outline that may serve as a suitable basis for the preparation of examination procedures is stated in the ASME Boiler and Pressure Vessel Code, Section V, Article 1, T-150; this outline can be adapted to any nonmetallic material. Each employer of bonding personnel who perform bonding for projects should maintain his/her own records of the examination procedures used. These records should keep track of dates and results of procedure qualifications. All of these records should be made available to the owner's inspector.

13.2.2.2 Types of Examinations

The various types of examinations that may be applied to nonmetallic piping components include: (1) visual examination; (2) ultrasonic examination; (3) holiday (spark) testing; (4) acoustic emission testing; and (5) in-process examination. Other methods may be used, but they should be described in sufficient detail to allow qualification of the necessary procedures and examiners. The criteria for acceptance for any alternative method should also be described in detail.

13.2.2.2.1 Visual Examination

The ASME Boiler and Pressure Vessel Code, Section V, Article 9, outlines a suitable method of practice of visual examination that may be used as a basis for examining nonmetallic piping systems. The visual examination of piping systems involves observation of those portions that are exposed to view at any time in the manufacturing, fabrication, and assembly process. Materials, components, dimensions, joint preparation, alignment, bonding, bolting, threading, other joining methods, supports (interstitial or external), and the assembly and erection of the double containment piping system may all be observed for verification of the engineering design and applicable code compliance.

13.2.2.2.2 Ultrasonic Examination

The ASME BPV Code, Section V, Article 5, outlines a suitable method for the examination of bonds and other components by ultrasonic examination. Examination using ultrasonic techniques is a very involved and complex science. An excellent reference concerning the subject of ultrasonic examination is "Non-Destructive Testing of Large-Diameter Pipe for Oil and Gas Transmission Lines."[4]

13.2.2.2.3 In-Process Examination

An in-process examination may be conducted for many aspects of double containment piping component installation. This includes joint preparation and cleanliness, the fit-up, clearance and internal alignment of joints (prior to bonding), and other variables specified for bonding. An in-process examination is usually of the visual type, unless otherwise specified in the engineering design. Appropriate records should be established and maintained when using in-process examination techniques.

13.2.3 Testing of Nonmetallic Double Containment Piping Systems

All systems must be tested to ensure tightness of both the primary and secondary containment piping systems prior to their use. In most situations, a hydrostatic leak test is required for the primary piping. If the fluid service is other than Category D or M (according to ANSI/ASME B31.3), an initial service leak test can be conducted of the primary piping system at the user's option, although it is usually not practical for systems of this type.

In some instances, a hydrostatic leak test may be impractical if linings, insulation, or leak detection will be damaged. Also, some fluid services may be contaminated when contacted with water or test fluids. Under these conditions, a pneumatic pressure test can be used at the owner's option. In making such a choice, the hazards of the stored energy contained in compressed gases should be thoroughly considered. A combination hydrostatic-pneumatic test can be used as an option (see Section 13.2.3.3.3).

The hydrostatic pressure test is the best method when it can be implemented since it is safer and more reliable than other methods. If a hydrostatic test cannot be used, a pneumatic test should be used unless an undue hazard is created, or there is a possibility of a brittle fracture. Only when these two tests are ruled out should an alternative leak test be used.

13.2.3.1 General Requirements for Pressure/Leak Testing

Prior to conducting a full scale test of a system, it is usually a wise idea to pretest using air at a maximum pressure of 15 psig (170 kPa), recognizing the potential hazards of stored energy in compressed air. This is done in order to locate any major leaks in a system and eliminate any expense that may result from major water spills.

Many nonmetallic materials have the additional concern of their light-weight characteristics. Upon the release of stored energy in a compressed gas, nonmetallic piping and components may be subject to a severe whipping effect, thereby imposing grave danger to those in the test area. Therefore, care should be taken that nonmetallic components are adequately restrained from moving during any pneumatic testing procedure. It is strongly recommended that the use of pneumatic testing for PVC or CPVC materials be completely avoided due to the likelihood of a brittle fracture of the materials if they should experience a failure. The same is true of many other nonmetallic materials at low ambient temperatures, since many experience a decrease in ductility at low temperatures.

All pressure/leak testing of nonmetallic primary piping systems should be conducted for at least 10 min. For plastic piping, it is strongly recommended that the hydrostatic test be conducted for at least 1 hour due to the significant tensile creep inherent in plastics. During this time, all joints and other connections should be examined for any sign of leaks. Special precautions should be taken if the test must be conducted near the ductile-brittle transition temperature of the nonmetallic material.

Subassemblies that are flanged at both ends only need to be tested once. Therefore, if they are tested as part of a prefabricated subassembly, it is not necessary to retest the subassembly once it has been added into the system. Only untested and uninspected bonds need to be tested and examined during a system pressure test. If a blank is inserted to isolate a certain section of a piping system during the hydrostatic test, the flanged joint does not have to be retested. If there are any repairs made to the system, the repaired sections should be retested. This is required unless the repairs are extremely minor and the owner decides that retesting is not required.

In some systems, it may be wise to conduct an initial low pressure (<5 psi) air test to avoid experiencing a substantial loss of water in the event of gross initial leaks. Also, temporary annular seals (e.g. a Link-Seal®) may be installed to prevent annular water intrusion.

13.2.3.2 Preparation for Pressure/Leak Testing

All joints should be left uninsulated and fully exposed for examination during leak testing, whenever possible. (This may not always be practical or feasible for primary piping, though it is required by code.[5] This problem is further addressed in Section 13.3 of this chapter.) If the bonds have already been tested, they need not be exposed. This is very important for primary piping, since once it is tested, it can be subsequently covered up. For primary piping that is intended for vapor or gas, temporary supports should be provided to the secondary containment piping in order to withstand the weight of the test liquid during the pressure test. Primary piping that is not designed to withstand the weight of a liquid should not be tested with liquid. In Chapter 9, it is suggested that the system be designed to withstand the weight of actual fluids under the most extreme conditions, including test fluids or service fluids, whichever fluid is heaviest. (see Section 9.3.1.1). For piping that is not to be pressure tested, it should be isolated by providing blinds or other methods during the test. Expansion joints should be tested in accordance with the ASME B31.3 Code or other applicable code (i.e., Expansion Joint Manufacturer Association rules).

13.2.3.3 Details of Testing

Details are presented in this section for hydrostatic testing, pneumatic testing, combined hydrostatic-pneumatic testing, initial service leak testing, and alternative leak testing for nonmetallic systems.

13.2.3.3.1 Hydrostatic Leak Test

Hydrostatic tests should always be conducted with water as the test medium, unless there is a possibility of freezing. Under these conditions, an alternative fluid can be used so long as the fluid is safe, nontoxic, and does not harm the nonmetallic material. The use of flammable fluids should be limited to those that have a flash point greater than 120°F (49°C).

The system should be tested at a pressure of 1-1/2 times its design pressure. If the design temperature of the system is above the testing temperature, the minimum test pressure should be increased to reflect the ratio of the stress to the pipe's strength at that temperature. Equation (13.1) is used for this purpose.

If the piping is attached to a pressure vessel that has a test pressure the same as or greater than the associated piping, then the piping and the vessel can be tested at the same time. However, if the test pressure of the pressure vessel is less, the piping should be isolated from the vessel and tested separately. Where this is considered to be impracticable, the piping can be tested at the vessel test pressure so long as the test pressure of the vessel is at least 77% of the piping test pressure.[6] The owner has the final decision as to whether to allow this reduction in piping test pressure.

If the piping is designed for external pressure service, it should be subjected to an internal gauge pressure of 1.5 times the differential pressure. The minimum pressure test required for external pressure service is 15 psig (103 kPa).

13.2.3.3.2 Pneumatic Leak Test

Pneumatic testing is a procedure that involves compressing air or some other gas inside the pipe to a predetermined test pressure. Since air and other gases are compressible fluids, this process results in the storage of large amounts of potential energy. Upon the sudden release of this stored energy, significant damage can result, posing a threat of injury to those nearby. The ductility of the piping at the test temperature is a primary consideration, and the chance of brittle fracture should be minimized during such a procedure. Pneumatic tests should only be conducted if the owner decides that it is all right to do so after careful evaluation of the potential hazards.

Pneumatic tests should be conducted to a test pressure of 110% of the design pressure of the piping system. The pressure should be slowly raised to the desired test pressure. First, an initial test of 25 psi (1.8 bar) should be conducted, unless 25 psi (1.8 bar) is more than half of the desired test pressure. In this case, the initial test should be performed at one-half of the desired test pressure. At this point, all joints and connections should be examined for leaks. If the piping passes the initial test, the pressure can be raised gradually. At each step, the pressure must be held until the piping strains are somewhat equalized. Once the test pressure is reached, the pressure should be reduced to the design pressure, and the system examined for leaks. A pressure-relief device should be provided and set to a pressure of the test pressure plus 50 psi (340 kPa) or 110% of the test pressure, whichever is less.

13.2.3.3.3 Combined Hydrostatic-Pneumatic Leak Test

A combination test consists of filling the system with water (or other fluid), completely and thoroughly venting out all of the air, and then using compressed gas as a means of raising the pressure of the system. In this type of test, the test pressure should be that of a hydrostatic pressure test (1.5 times the design pressure). Caution should be observed that the pressure is not raised too rapidly, and the testing pressure exceeded.

13.2.3.3.4 Initial Service Leak Testing

An initial service leak test is one that is performed with the actual fluids upon initial startup. The key to such a test is to raise the pressure gradually in small steps. At each step, the pressure should be held until the piping strains are somewhat equalized; this may take longer for a nonmetallic material than it would for a metallic material. At all stages, the system should be thoroughly examined for leaks. This procedure is generally not recommended for double containment piping systems, regardless of which materials of construction are selected.

13.2.3.3.5 Alternative Leak Testing

If a hydrostatic, pneumatic, or combination test cannot be used, a sensitive leak test can be conducted, as described in the ASME Boiler and Pressure Vessel Code, Section V, Article 10, or other equivalent method. The sensitivity for such a test should be at least 10^{-3} atm ml/sec (100 Pa ml/s) under test conditions. Any piping system that is subjected to an alternative leak test must also be subjected to a flexibility analysis, as an extra precaution.

13.2.3.4 Testing Records

Records should be kept of the testing program for each separate nonmetallic system. The information should include at least: (1) date of the tests; (2) identification of the piping system tested; (3) the type of test fluid or fluids used; (4) testing pressure; and (5) certification of the results (by the examiner). The examination procedures and related personnel qualifications should be retained for a minimum period of at least 5 years.

13.3 Considerations for Double Containment Assemblies

There are unique inspection, examination, and testing issues for double containment piping assemblies. With respect to inspection, complications arise due to a lessened ability to inspect or examine primary piping system welds (bonds) and installed components, whenever a portion or most of the secondary containment housing is in place and/or already welded (bonded). Inspection and examination of secondary containment components is relatively straightforward; however, it is more difficult and time consuming for the associated primary components. With respect to testing, there are a number of issues that are unique. These relate to external pressure and buoyancy effects that are imposed on primary piping as a result of pressure testing an annulus between inner and outer pipes.

Some of the other effects are of a more subtle nature. An example is the effect of a rupture of a primary pipe component, within a closed annulus that is completely filled with an incompressible fluid. Many inspection and examination issues for double containment pipe systems can be readily addressed. Others require additional analysis and planning in order to avert potential problems.

13.3.1 Inspection and Examination Requirements

Inspection and examination requirements for double containment piping assemblies follow the same general guidelines as for single-wall installations. The primary, or inner, system follows the same basic requirements as that of the basic single-wall pipe in piping codes. The requirements for outer piping depend on one's interpretation of applicable piping codes. Most of the difficulties involve the inspection and examination of elements of the primary portion of the double containment assembly. These include inspection and examination of the following aspects of a primary system: (1) pipe connections; (2) fittings; (3) weldments; (4) assemblies; (5) interstitial support connections; (6) internal alignment; (7) internal cold spring; and other aspects of primary piping that can be hidden by its associated secondary containment housing.

If inspection and examination of every aspect of a primary system is required, it is best to have the system prefabricated to the greatest extent possible under shop conditions. A greater number of secondary closures are required on a project where inspection of internal (primary) components is mandatory; it is usually best to limit the number of secondary closures to be joined in the field.

On projects where extensive inspection and examination is required due the fluid service conditions, there may be certain limitations in terms of joining methods. For instance, simultaneous fusion involving thermoplastic materials is not recommended for applications having requirements for inspection and examination of primary piping welds (see Section 12.3.3, and paragraphs 345.3.1, 345.2.5, and A345.2.2 of the ANSI/ASME B31.3 Code).

There are some aspects of double containment piping other than welds that may be not be possible to inspect visually. These include the as-installed positioning of interstitial supports, and the structural welds used to attach interstitial supports to primary and/or secondary containment piping. Similarly, the final alignment of primary components may not be readily inspected. Radiography may be used in some applications, although it is usually an expensive approach. One possible solution to this dilemma, although not always feasible, is the use of a clear or transparent material for all, or a portion, of the secondary containment piping.

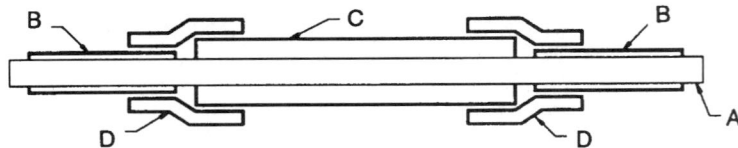

Figure 13.7 Illustration of a telescoping system that uses alternating sizes of secondary pipes for 100% visual inspection of primary pipe. (Source: Smith Fiberglass Products, Inc., U.S. Patent #4,805,444)

Primary piping may be 100% visually inspected by means of using a patented telescoping method developed for RTRP petroleum service station piping. 100% visual inspection during initial pressure testing via soap testing of the primary pipe's exterior is required by some municipalities for underground fuel piping. It is not required for any other double containment piping services, and is not required in any of the ANSI/ASME B31 Codes (visual inspection of joints during testing is the usual requirement.) This type of system is illustrated in Figure 13.7.

It is much easier to inspect and examine secondary containment piping components. Since these piping components are directly exposed, they can be readily inspected and examined. It should be remembered that where examination and inspection is required, that priming, painting, coating, or insulation must not yet be added to the secondary containment piping until the examination and inspection requirements are satisfied. This requirements is mandatory when a sensitive leak test, or similar requirement, is specified by the engineering design for qualification of the integrity of the outer jacket.

Radiography of secondary containment welds is difficult due to shadows that may be created by the presence of the primary components. For this reason, methods other than radiography are more readily suitable as a means of examining secondary containment components.

13.3.2 Testing Requirements

Testing of double containment piping systems requires careful consideration of the sequence by which the tests are conducted. One must consider all possible effects and loads that occur as a result of the testing process. Pressure/leak testing is a must for every double containment piping system; it is definitely required for the primary piping system; secondary containment piping does not have to be pressure tested, although it must always be at least subjected to a leak test. A carefully planned and executed testing protocol will save time and cost by avoiding problems created by the testing process.

One simple way to avoid problems is to specify a test pressure that is within safe working limits of the components. It is a common error to require a nonpressure piping system to be tested to the maximum rating of the basic pipe, based on its wall thickness, even though the design pressure is significantly below the rating of other components. A thick wall may be used in straight pipe sections of a nonpressure system for reasons other than pressure rating. These may include availability of materials, corrosion resistance, fabrication issues, resistance to soil loads, or other reasons. It is not a recommended practice to require system components to be tested above their ultimate capability; if a pressure test level greater than the capabilities of the components is viewed as necessary, then the components should be resized with greater thickness and/or reinforcements (see Sections 5.1, 6.2, and 7.2).

It is possible also to specify a test pressure that is too low for the system. This is one of the worst mistakes to make in system design, since defects may not be identified during initial pressure testing, leading to possible failure during actual operation.

One of the most important concerns involves external pressures that are exerted upon primary piping and its components by virtue of a pressure test that is conducted on the secondary containment piping (annulus). Most pipes are less capable of withstanding external pressures than they are of equivalent internal pressures. Therefore, secondary containment piping must not be tested at a pressure that creates a differential pressure in excess of that for which the primary piping is rated. An easy solution is to fill a primary pipe system with an incompressible fluid and maintain it under a fixed internal pressure while testing the secondary containment piping. In most applications, it makes sense to maintain primary piping at the same pressure to which its associated secondary containment (annulus) is to be tested. If this is feasible, then the primary piping will never be subjected to a differential pressure that will cause it to collapse, nor will it allow the possibility of overpressurizing the outer pipe should the inner pipe burst.

A problem can arise in systems having a lower pressure-rated outer jacket if both pipes are filled with incompressible fluids during a simultaneously conducted pressure test. If the primary piping is being tested at a pressure that is in excess of the rating of the secondary containment pipe system, a rupture of the primary system can produce a subsequent failure of the secondary containment piping. The effect of the pressure rise will be virtually instantaneous. Since both incompressible fluid systems are closed, a failure in a component of the primary piping will result in the two systems being equalized to the higher pressure, almost instantaneously. For this reason, two pipes with substantially different ratings should never be tested in this manner. In many double containment piping systems, the secondary containment piping is designed for a much lower pressure rating than the primary for economic reasons. The design of the pressure testing program should not allow for a double failure due to secondary overpressurization.

A testing problem that is also related to a primary pipe being unfilled during a test of the secondary containment (annulus) involves empty primary pipes becoming buoyant during the test. This can also occur when the primary pipes are filled, but interstitial supports are used that do not address buoyant effects. It is also more of a problem when a material that has a density less than the test fluid is used for the primary piping (i.e., polyethylene or polypropylene, for example). These problems due to buoyancy can be minimized by maintaining the primary piping full of incompressible fluid. If the appropriate type of annular supports is used, then primary piping components can be prevented from become buoyant altogether.

Where buoyancy forces will create an adverse bending stress on the components, a pneumatic test should be carried out in the annulus, in lieu of using an incompressible fluid to test secondary containment piping. During a pneumatic test of secondary containment piping, the adverse effect of a sudden release of the stored energy must be considered, including the effect on internal piping and its components. Weighing the system down by maintaining a fluid in a primary portion of a double containment piping system will help to minimize the effects of a sudden release of the stored energy involved.

The type of leak detection to be used plays an important part in the determination of whether a hydrostatic or pneumatic test is to be conducted on the secondary containment piping. If a very sensitive leak detection system is to be used, it is better to test the secondary containment (annulus) pneumatically, taking into account all of the hazards of stored energy inherent in compressed gases. Liquid normally becomes entrapped behind interstitial supports during a hydrostatic pressure test of the secondary containment piping (annulus). The annulus can be totally dried after pressure testing with water, but this can only be thoroughly accomplished by blowing dry air, warm dry air, or gas through the annulus for an extended period of time. False alarms may result in a sensitive leak detection system after startup, unless all crevices in the annulus are completely dry. If false alarms occur after startup, the likely result is the loss of confidence in the system and a mistrust developed by the plant operators. This could result in plant operators ignoring an alarm during the event of a real operating leak.

13.3.3 Sequence of Testing

The sequence of testing inner and outer pipes is important in order to conduct a safe and reliable test program. For double containment piping assemblies, it is important to consider the exact step-by-step sequence to be used. The sequence recommended depends to an extent on the types of test to be used for the inner and outer pipes, and whether the outer jacket has been fully closed. Step-by-step sequences are presented for six major testing protocols in the following paragraphs. On each project, it is a wise idea to prepare a step-by-step written checklist procedure that is even more detailed; every individual valve opening and closing should be described. The testing inspector should check off each item as it is physically performed.

13.3.3.1 Hydrostatic/Hydrostatic Test; Secondary Jacket Fully Assembled

This sequence of testing applies to those systems where a hydrostatic pressure test is to be applied to both the inside (primary) pipe system and in the annular space (to the outer pipe system; secondary containment pipe system). It is also the procedure used for systems that will have their secondary containment system fully completed prior to conducting the initial pressure test of the primary piping system. This procedure will not allow most primary pipe joints, or other sections of the primary pipe, to be inspected visually during its pressure test. This type of system may not be in compliance with various codes for this reason (i.e., ASME B31.3 Code for Chemical Plant and Petroleum Refinery Piping).[7]

1. Open all high-point vent valves of the primary pipe system.
2. Attach a fill hose to the overall low-point drain valve of the primary pipe system. In some systems, the valve which serves as the fill point is one that is attached to a blind flange that is connected to the single-wall pipe flange of an initiation assembly (see Figure 13.8). It is important to have a pressure gauge connected to the "upstream" side of the low-point drain valve so that the pressure in the primary piping system can be determined at all times during the test. The pressure gauge may alternatively be connected at another point in the system; however, it must be in full view of the testing operator when the low-point valve is turned on and off.
3. Turn on the water source and open the low-point drain valve in the primary pipe system, allowing water to flow in. Allow water to enter slowly: less than 5 ft/s (1.5 m/s) velocity.

Inspection, Examination, and Testing

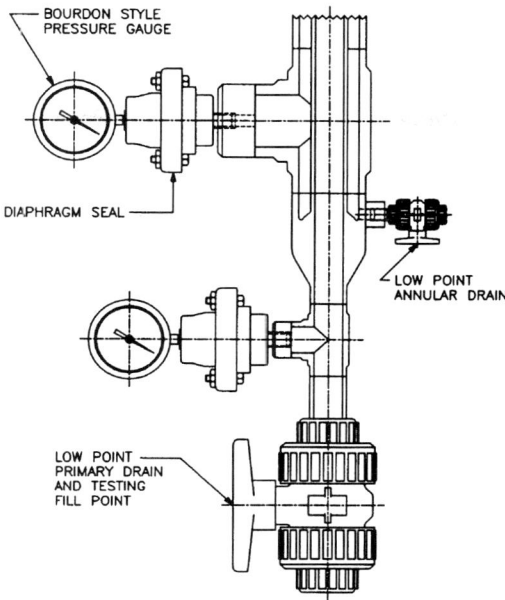

Figure 13.8 Illustration of an overall low-point fill valve for testing purposes in a primary pipe of a double containment piping system.

4. Allow the primary pipe system to fill completely until water spills out of the high-point vent valves of the primary pipe system.
5. Make sure that all air is completely vented from the high-point vent valves by shaking the pipes gently. When there is a high degree of certainty that all air is out of the system, shut off the low-point drain valve into which water is being introduced.
6. Close the primary pipe low-point drain valve, shut off the water source, and disconnect the fill hose.
7. Turn off all primary pipe high-point vent valves to seal off the primary pipe system.
8. Attach a source of pressure that can introduce pressure in a gradual, stepwise manner (i.e., hand pump, air compressor with pressure regulator valve, etc.) to the low-point drain valve of the primary pipe system.
9. At this point, the primary system is a closed system, filled with incompressible fluid, and at atmospheric pressure. Open the low-point drain valve to the source of pressure.
10. Open any low-point drains in the outer pipe system so they may be used as visual leak-detection points if there is a failure of the primary system during the test.
11. Raise the pressure in increments of 25 psi in a slow, gradual manner.
12. When the final test pressure of the primary system is reached, it must be held at least 10 min. The typical test pressure used to conduct a hydrostatic pressure test of the primary pipe system is 150% of the design pressure of the primary system, adjusted for operating temperature [see Section 13.1.3.4.1 and Eq. (13.1)]. The primary pipe low-point drain valve can be closed at this point and the source of pressure removed. This is an optional procedure.

If a plastic pipe is being tested, the pressure must be applied for at least 1 h due to the substantial creep inherent in plastics. For thermoplastic pipes, a modified procedure as described in Section 13.3.4.4 should be used to correct for the anticipated initial drop in pressure due to expansion and creep of the plastic pipe.

13. If the internal gauge pressure of the primary system does not drop by more than an amount specified as an "unacceptable pressure drop" for the system, after the specified duration of the test, the system can be described as having "passed" the pressure/leak test. If leaks or defects are observed during the tests, then specified corrective action procedures must be followed.
14. For a primary piping system that passes its pressure test, the low-point drain valve of the system should be turned off (if it has not already been turned off in step 12).
15. Disconnect the source of pressure from the primary pipe low-point drain valve (if it has not already been disconnected in step twelve). If the test pressure of the secondary containment pipe is less than that of the primary pipe pressure test, the procedures in steps 16-18 should be followed to avoid overpressurizing the secondary containment pipe. If the outer pipe system is to be tested to the same pressure of the inner pipe system, then steps 16-18 can be skipped.
16. Connect a drain hose to the primary pipe low-point drain valve. The hose should be drained to a storm sewer, storm drain, or other appropriate point in the facility. If a test fluid other than water is used, it may have to be collected and treated as chemical waste.
17. Open the primary pipe low-point drain valve and allow the pressure in the primary pipe system to be reduced to the test pressure of the secondary containment piping system.
18. When the pressure in the primary system drops to the test pressure of the secondary containment piping system, the primary piping drain valve should be closed. The drain hose may remain connected for draining of the primary pipe system at the conclusion of testing. The secondary containment piping system may now be tested.
19. Open all high point vent valves of the secondary containment system.
20. Attach a fill hose to the overall low-point drain valve of the secondary containment system (annulus). Remember, it is important to have a pressure gauge connected to the "upstream" side of the low-point drain valve so that the pressure in the secondary containment piping system can be determined at all times during the test. The pressure gauge must be in full view of the testing operator when the low-point valve is turned on and off.
21. Turn on the water source and open the low-point drain valve in the secondary containment piping system, allowing water to flow into the annular space. Allow water to enter slowly less than 5 ft/s (1.5 m/s) velocity.
22. Allow the secondary containment pipe system (annulus) to fill completely until water spills out of the high-point vent valves of the secondary containment pipe system.
23. Make sure that all air is completely vented from the high-point vent valves by shaking the pipes gently. When there is a high degree of certainty that all air is out of the system, shut off the low-point drain valve into which water is being introduced.
24. Close the secondary containment low-point drain valve, shut off the water source and disconnect the fill hose.
25. Turn off all secondary containment high-point vent valves to seal off the annulus.
26. Attach a source of pressure that can introduce pressure in a gradual, stepwise manner (i.e., hand pump, air compressor with pressure regulator valve, etc.) to the low-point

Inspection, Examination, and Testing

drain valve of the annulus. At this point, the annulus is a closed system, filled with incompressible fluid, and at atmospheric pressure.

27. Open the secondary containment low-point drain valve to the source of pressure.
28. Raise the pressure in increments of 10-25 psi in a slow, gradual manner. Be sure to inspect visually all secondary containment pipe welds (bonds) when each increment of 10-25 psi is reached. Randomly inspect other points in the system, including straight pipe sections, for the presence of water.
29. When the final test pressure of the secondary containment system is reached, it must be held at least 10 min, or at least long enough to be able to inspect all necessary points visually. (Refer to step 31 for a description of typical secondary containment hydrostatic test pressure.) The secondary containment low-point drain valve can be closed at this point and the source of pressure removed. This is an optional procedure. If a plastic pipe is being tested, the pressure must be applied for at least 1 h due to the substantial creep inherent in plastics. For thermoplastic pipes, a modified procedure as described in Section 13.3.4.4 should be used to correct for the anticipated initial drop in pressure due to expansion and creep of the plastic pipe.
30. If the internal gauge pressure of the secondary containment system does not drop by more than an amount specified as an "unacceptable pressure drop" after the specified duration of the test, the secondary containment system can be described as having "passed" the pressure/leak test. If leaks or defects are observed during the test, then specified corrective action procedures must be followed.
31. For a secondary containment system that passes the pressure test, its low-point drain valve should be turned off (if it has not already been turned off in step 29).
32. Disconnect the source of pressure from the low-point drain valve (if it has not already been disconnected in step 29).
33. Connect a drain hose to the low-point drain valves of both the primary and secondary containment piping system. The hoses should be drained to a storm sewer, storm drain, or other appropriate point in the facility. If a test fluid other than water is used, it may have to be collected and treated as chemical waste. If more than one local low point exists in the system, connect drain hoses from these points as well.
34. Open the low-point drain valve of the secondary containment pipe system and allow the pressure in the annulus to be reduced to atmospheric pressure.
35. When the pressure in the annulus drops to atmospheric pressure, at least one secondary containment high-point vent valve should be opened to allow the system to drain smoothly. If there is more than one low point in the annulus, the additional valve(s) should also be opened.
36. When all of the water (test fluid) has drained from the annulus, the high- and low-point valves should be kept open to allow the procedures described in steps 55-57 to be carried out.
37. Open the primary pipe low-point drain valve and allow the pressure in the primary system to be reduced to atmospheric pressure.
38. When the pressure in the primary system drops to atmospheric, at least one primary pipe high-point vent valve should be opened to allow the primary system to drain smoothly. If there is more than one low point in the primary pipe system, the additional valve(s) should also be opened.
39. When all of the water (test fluid) has drained from the primary pipe, the high- and

low-point valves should be kept open to allow the procedures in steps 58-60 to be carried out.
40. Disconnect the drain hose(s).
41. Connect a dry air source to the secondary containment pipe, either at the high or low point of the annulus.
42. Blow air through the annulus until the desired level of drying is reached. If necessary, use warm, dry air to dry the annular space thoroughly. It is critical to have an annulus that is thoroughly dry to prevent false leak alarms or corrosion to take place.
43. Turn off the air source and disconnect it from the secondary containment pipe system. Turn all secondary containment high- and low-point valves to the off position.
44. Connect a dry air source to the primary pipe, either at its high or low point.
45. Blow air through the primary pipe until the desired level of drying is reached.
46. Turn off air source and disconnect it from the primary pipe system. Turn all primary high- and low-point valves to their off position.

13.3.3.2 Hydrostatic/Hydrostatic Test; Secondary Jacket Not Fully Assembled

This sequence of testing applies to those systems where a hydrostatic pressure test is to be applied to both the inside (primary) pipe system and in the annular space (to the outer pipe system; secondary containment pipe system). It is also the procedure used for systems that will have primary pipe sections tested prior to completing secondary closures of the secondary containment system. This allows previously untested primary pipe joints to be inspected visually during the pressure test. Primary pipe welds that have been previously tested and visually examined during a test of a prefabricated section do not have to be reinspected during a test of the assembled system (see Section 13.1.3.3). This sequence, if followed, will not allow the primary pipe to collapse while pressure testing the outer pipe system.

1. Open all high-point vent valves of the primary pipe system.
2. Attach a fill hose to the overall low-point drain valve of the primary pipe system. In some systems, the valve that serves as the fill point is one that is attached to a blind flange, which in turn is connected to the single-wall pipe flange of an initiation assembly (see Figure 13.8). It is important to have a pressure gauge connected to the "upstream" side of the primary pipe low-point drain valve so that the pressure in the primary piping system can be determined at all times during the test. The pressure gauge may alternatively be connected at another point in the system; however, it must be in full view of the testing operator when the low-point valve is turned on and off.
3. Turn on the water source and open the low-point drain valve in the primary pipe system, allowing water to flow in. Allow water to enter slowly: less than 5 ft/s (1.5 m/s) velocity.
4. Allow the primary pipe system to fill completely until water spills out of the high-point vent valves of the primary pipe system.
5. Make sure that all air is completely vented from the high-point vent valves by shaking the pipes gently. When there is a high degree of certainty that all air is out of the system, shut off the low-point drain valve into which water is being introduced.
6. Close the primary piping system low-point drain valve, shut off the water source, and disconnect the fill hose.
7. Turn off all primary piping high-point vent valves to seal off the primary pipe system.

Inspection, Examination, and Testing

8. Attach a source of pressure that can introduce pressure in a gradual, stepwise manner (i.e., hand pump, air compressor with pressure regulator valve, etc.) to the low-point drain valve of the primary pipe system.
9. At this point, the primary system is a closed system, filled with incompressible fluid, and at atmospheric pressure. Open the low-point drain valve to the source of pressure.
10. Open any low-point drains in the outer pipe system so they may be used as visual leak-detection points if there is a failure of the primary system during the test. There will also be many open gaps in the secondary system since midline secondary closures are not installed at this point; these will also function as visual inspection points.
11. Raise the pressure in increments of 25 psi in a slow, gradual manner. Be sure to inspect all primary pipe welds visually when each increment of 25 psi is reached. Randomly inspect other points in the primary pipe system and check all visual leak-detection points for the presence of water.
12. When the final test pressure of the primary system is reached, it must be held at least 10 min, or at least long enough to be able to inspect all necessary points visually. The typical test pressure used to conduct a hydrostatic pressure test of the primary pipe system is 150% of the design pressure of the primary system, adjusted for operating temperature [see Section 13.1.3.4.1 and Eq. (13.1)]. The primary pipe drain valve can be closed at this point and the source of pressure removed. This is an optional procedure. If a plastic pipe is being tested, the pressure must be applied for at least 1 h due to the substantial creep inherent in plastics. For thermoplastic pipes, a modified procedure as described in Section 13.3.4.4 should be used to correct for the anticipated initial drop in pressure due to expansion and creep of the plastic pipe.
13. If the internal gauge pressure of the primary system does not drop by more than an amount specified as an "unacceptable pressure drop" for the system, after the specified duration of the test, the system can be described as having "passed" the pressure/leak test. If leaks or defects are observed during the tests, then specified corrective action procedures must be followed.
14. For a primary piping system passes its pressure test, the low-point drain valve of the primary system should be turned off (if it has not already been turned off in step 12).
15. Disconnect the source of pressure from the primary pipe low-point drain valve (if it has not already been disconnected in step 12).
16. Connect a drain hose to the primary pipe low-point drain valve. The hose should be drained to a storm sewer, storm drain, or other appropriate point in the facility. If a test fluid other than water is used, it may have to be collected and treated as chemical waste. If there is more than one local low point in the primary pipe system, connect drain hoses from these points as well.
17. Open the primary pipe low-point drain valve and allow the pressure in the primary pipe system to be reduced to atmospheric pressure.
18. When the pressure in the primary system drops to atmospheric pressure, at least one high-point vent valve should be opened to allow the system to drain smoothly. If there is more than one low point in the system, the additional valve(s) should also be opened.
19. When all of the water (test fluid) has drained, the high- and low-point valves of the primary piping system should be closed. Consideration should be given to drying the primary pipe if the extended presence of moisture will present a corrosion problem. (If this is so, the same drying sequence presented in steps 58–60 should be followed.)

20. Disconnect the drain hose(s).
21. Finish constructing the secondary containment system; make all secondary midline closure welds (bonds).
22. When the secondary containment system is finished being constructed, point valves of the primary pipe system should be opened. The primary pipe system must be filled with incompressible fluid and raised to the test pressure as the secondary containment system prior to introducing fluid to the secondary containment pipe. This will prevent the primary pipe from collapsing while testing the outer system.
23. Attach a fill hose to the overall low-point drain valve of the primary pipe system. The same valve used in step 2 should be used for this purpose.
24. Turn on the water source and open the low-point drain valve in the primary pipe system, allowing water to flow in. Allow water to enter slowly; less than 5 ft/s (1.5 m/s) velocity.
25. Allow the primary pipe system to fill completely until water spills out of the high-point vent valves of the primary pipe system.
26. Make sure that all air is completely vented from the high-point vent valves by shaking the pipes gently. When there is a high degree of certainty that all air is out of the system, shut off the low-point drain valve into which water is being introduced.
27. Close the primary pipe low-point drain valve, shut off the water source, and disconnect the fill hose.
28. Turn off all primary pipe high-point vent valves to seal off the primary pipe system.
29. Attach a source of pressure that can introduce pressure in a gradual, stepwise manner (i.e., hand pump, air compressor with pressure regulator valve, etc.) to the low-point drain valve of the primary piping system. At this point, the system is a closed system, filled with incompressible fluid, and at atmospheric pressure.
30. Open the primary pipe low-point drain valve to the source of pressure.
31. Raise the pressure of the primary piping in a slow, gradual manner to the test pressure of the secondary containment piping system. The typical test pressure at which to conduct a hydrostatic pressure test of the secondary containment system is 150% of its design pressure, adjusted for operating temperature [see Section 13.1.3.4.1 and Eq. (13.1)].
32. Turn off the low-point drain valve of the primary pipe system and disconnect the source of pressure.
33. Open all high-point vent valves of the secondary containment system.
34. Attach a fill hose to the overall low-point drain valve of the secondary containment system (annulus). Remember, it is important to have a pressure gauge connected to the "upstream" side of the low-point drain valve so that the pressure in the secondary containment piping system can be determined at all times during the test. The pressure gauge must be in full view of the testing operator when the low-point valve is turned on and off.
35. Turn on the water source and open the low-point drain valve in the secondary containment piping system, allowing water to flow into the annular space. Allow water to enter slowly; less than 5 ft/s (1.5 m/s) velocity.
36. Allow the secondary containment pipe system (annulus) to fill completely until water spills out of the high-point vent valves of the secondary containment pipe system.
37. Make sure that all air is completely vented from the high-point vent valves by shaking

Inspection, Examination, and Testing

the pipes gently. When there is a high degree of certainty that all air is out of the system, shut off the low-point drain valve into which water is being introduced.

38. Close the secondary containment low-point drain valve, shut off the water source, and disconnect the fill hose.
39. Turn off all high-point vent valves to seal off the annulus.
40. Attach a source of pressure that can introduce pressure in a gradual, stepwise manner (i.e., hand pump, air compressor with pressure regulator valve, etc.) to the low-point drain valve of the annulus. At this point, the annulus is a closed system, filled with incompressible fluid, and at atmospheric pressure.
41. Open the secondary containment low-point drain valve to the source of pressure.
42. Raise the pressure in increments of 10-25 psi in a slow, gradual manner. Be sure to inspect all secondary containment pipe welds (bonds) visually when each increment of 10-25 psi is reached. Randomly inspect other points in the system, including straight pipe sections, for the presence of water.
43. When the final test pressure of the secondary containment system is reached, it must be held at least 10 min, or at least long enough to be able to inspect all necessary points visually. (Refer to step 31 for a description of typical secondary containment hydrostatic test pressure.) The secondary containment drain valve can be closed at this point and the source of pressure removed. This is an optional procedure. If a plastic pipe is being tested, the pressure must be applied for at least 1 h due to the substantial creep inherent in plastics. For thermoplastic pipes, a modified procedure as described in Section 13.3.4.4 should be used to correct for the anticipated initial drop in pressure due to expansion and creep of the plastic pipe.
44. If the internal gauge pressure of the secondary containment system does not drop by more than an amount specified as an "unacceptable pressure drop" after the specified duration of the test, the secondary containment system can be described as having "passed" the pressure/leak test. If leaks or defects are observed during the test, then specified corrective action procedures must be followed.
45. For a secondary containment system that passes the pressure test, its low-point drain valve should be turned off (if it has not already been turned off in step 43).
46. Disconnect the source of pressure from the secondary containment low-point drain valve (if it has not already been disconnected in step 43).
47. Connect drain hoses to the low-point drain valves of both the primary and secondary containment piping system. The hoses should be drained to a storm sewer, storm drain, or other appropriate point in the facility. If a test fluid other than water is used, it may have to be collected and treated as chemical waste. If more than one local low point exists in the inner and/or outer systems, connect drain hoses from these points as well.
48. Open the low-point drain valve of the secondary containment pipe system and allow the pressure in the annulus to be reduced to atmospheric pressure.
49. When the pressure in the annulus drops to atmospheric pressure, at least one secondary containment high-point vent valve should be opened to allow the system to drain smoothly. If there is more than one low-point in the annulus, the additional valve(s) should also be opened.
50. When all of the water (test fluid) has drained from the annulus, the high- and low-point valves should be kept open to allow the procedures described in steps 55-57 to be carried out.

51. Open the primary pipe low-point drain valve and allow the pressure in the primary system to be reduced to atmospheric pressure.
52. When the pressure in the primary system drops to atmospheric, at least one primary pipe high-point vent valve should be opened to allow the primary system to drain smoothly. If there is more than one low point in the primary pipe system, the additional valve(s) should also be opened.
53. When all of the water (test fluid) has drained from the primary pipe, the high- and low-point valves should be kept open to allow the procedures in steps 58-60 to be carried out.
54. Disconnect the drain hose(s).
55. Connect a dry air source to the secondary containment pipe, either at the high or low point of the annulus.
56. Blow air through the annulus until the desired level of drying is reached. If necessary, use warm, dry air to dry the annular space thoroughly. It is critical to have an annulus that is thoroughly dry to prevent false leak alarms or corrosion to take place.
57. Turn off the air source and disconnect it from the secondary containment pipe system. Turn all secondary containment high- and low-point valves to the off position.
58. Connect a dry air source to the primary pipe, either at its high or low point.
59. Blow air through the primary pipe until the desired level of drying is reached.
60. Turn off air source and disconnect it from the primary pipe system. Turn all primary high- and low-point valves to their off position.

13.3.3.3 Hydrostatic/Pneumatic Test; Secondary Jacket Fully Assembled

This sequence of testing applies to those systems where a hydrostatic pressure test is to be applied to the inside (primary) pipe system and a pneumatic pressure test is to be applied in the annulus (to the outer pipe system; secondary containment pipe system). It also applies to systems that will have their secondary containment system fully completed prior to conducting the initial pressure test of the primary piping system. This procedure will not allow many primary pipe welds, or other sections of the primary pipe, to be inspected visually during the pressure test. For this reason, this type of system may not be in compliance with various codes (i.e., ANSI/ASME B31.3 Code for Chemical Plant and Petroleum Refinery Piping).[8]

In some applications where a pneumatic test is to be applied to the secondary containment system, it is important that the annulus be kept as moisture-free as possible. Therefore, if the annulus becomes wetted during the primary pipe pressure test, be sure to dry the annulus thoroughly prior to conducting the test of the secondary containment jacket.

Since a pneumatic pressure test is being conducted, be sure to consider all of the potential danger inherent in such a test, particularly if a material is used that is subject to brittle-type failure. Do not conduct a pneumatic pressure test with a material such as PVC (or other plastic when it is in a cold nonductile state) since there is too much danger inherent in such a test. While the inside pipe will weigh the system down somewhat since it will be filled with water/test fluid, it is still a good idea to restrain the system properly so that it does not move excessively in the event of a failure, or when the pressure is decreased in step 28.

1-18. Follow steps 1-18 as described in Section 13.3.3.1, "Hydrostatic/Hydrostatic Test; Secondary Jacket Fully Assembled." Continue on to step 19 described in the follow-

Inspection, Examination, and Testing

ing. A typical pneumatic test pressure for the secondary containment piping system is 110% of its design pressure, adjusted for operating temperature [see Section 13.1.3.4.1 and Eq. (13.1)].

19. Check all high-point vent valves and low-point drain valves of the secondary containment piping system to be sure that they are closed.
20. Attach a source of pneumatic pressure (i.e., air compressor with pressure regulator valve, nitrogen bottle with pressure regulator valve, etc.) to the overall low-point drain valve of the secondary containment system (annulus). Remember, it is important to have a pressure gauge connected to the "upstream" side of the low-point drain valve so that the pressure in the secondary containment piping system can be determined at all times during the test. The pressure gauge must be in full view of the testing operator when the low-point valve is turned on and off. At this point, the annulus is a closed system, filled with a compressible gas under atmospheric pressure.
21. Turn on the pneumatic pressure source and open the low-point drain valve in the secondary containment piping system to the source of pressure, allowing the pressure to rise in a gradual, stepwise manner.
22. Raise the pressure in increments of 10-25 psi in a slow, gradual manner. Completely and thoroughly cover each secondary containment weld (bond) with a soapy solution in order to detect bubbles if there are leaks. Be sure to inspect all secondary containment pipe welds (bonds) visually when each increment of 10-25 psi is reached. Randomly inspect other points in the system, including straight pipe sections, for the presence of water.
23. When the final test pressure of the secondary containment system is reached, it must be held at least long enough to allow piping strains to become somewhat equalized, or at least long enough to be able to inspect all necessary points visually. The secondary containment low-point drain valve can be closed at this point and the source of pressure removed while the system is being visually inspected. This is an optional procedure. If is a plastic pipe is being tested, the pressure must be applied for at least 1 h due to the substantial creep inherent in plastics. For thermoplastic pipes, a modified procedure as described in Section 13.3.4.4 should be used to correct for the anticipated initial drop in pressure due to expansion and creep of the plastic pipe.
24. If the internal gauge pressure of the secondary containment system does not drop by more than an amount specified as an "unacceptable pressure drop" after the specified duration of the test, and no "soap bubbles" are detected at the joints, the secondary containment system can be described as having "passed" the pressure/leak test. If leaks or defects are observed during the test, then specified corrective action procedures must be followed.
25. For a secondary containment system that passes the pressure test, its low-point drain valve should be turned off (if it has not already been turned off in step 23).
26. Disconnect the source of pressure from the low-point drain valve (if it has not already been disconnected in step 23).
27. Connect a drain hose to the low-point drain valve(s) of the primary piping system. The hose(s) should be drained to a storm sewer, storm drain, or other appropriate point in the facility. If a test fluid other than water is used, it may have to be collected and treated as chemical waste. If more than one local low point exists in the system, connect drain hoses from these points as well.

28. Open a secondary containment pipe system high-point vent valve and carefully allow the pressure in the annulus to be reduced to atmospheric pressure. Other valves can be opened to accelerate this process.
29. When the pressure in the annulus drops to atmospheric pressure, close off the high-point vent valve that was opened, and any other valves that were opened for this purpose.
30. Open the primary pipe low-point drain valve and allow the pressure in the primary system to be reduced to atmospheric pressure.
31. When the pressure in the primary system drops to atmospheric, at least one primary pipe high-point vent valve should be opened to allow the primary system to drain smoothly. If there is more than one low point in the primary pipe system, the additional valve(s) should also be opened.
32. When all of the water (test fluid) has drained from the primary pipe, the high- and low-point valves should be kept open to allow drying procedures to be carried out.
33. Disconnect the drain hose(s).
34. Connect a dry air source to the primary pipe, either at its high or low point.
35. Blow air through the primary pipe until the desired level of drying is reached.
36. Turn off air source and disconnect it from the primary pipe system. Turn all primary high- and low-point valves to their off position.

13.3.3.4 Hydrostatic/Pneumatic Test; Secondary Jacket Not Fully Assembled

This sequence of testing applies to those systems where a hydrostatic pressure test is to be applied to the inside (primary) pipe system and a pneumatic pressure test is to be applied in the annulus (to the outer pipe system; secondary containment pipe system). It also applies to systems that will have primary pipe sections tested prior to completing secondary closures of the secondary containment system. This allows previously untested primary pipe joints to be inspected visually during the pressure test. Primary pipe welds that have been previously tested and visually examined during a test of a prefabricated section do not have to be reinspected during a test of the assembled system (see also Section 13.1.3.2). This sequence, if followed, will not allow the primary pipe to collapse while pressure testing the outer pipe system.

In some applications where a pneumatic test is to be applied to the secondary containment system, it is important that the annulus be kept as moisture-free as possible. Therefore, if the annulus becomes wetted during the primary pipe pressure test, be sure to dry the annulus thoroughly prior to conducting the test of the secondary containment jacket.

Since a pneumatic pressure test is being conducted, be sure to consider all of the potential danger inherent in such a test, particularly if a material is used that is subject to brittle-type failure. Do not conduct a pneumatic pressure test with a material such as PVC (or other plastic when it is in a cold nonductile state) since there is too much danger inherent in such a test. While the inside pipe will weigh the system down somewhat since it will be filled with water/test fluid, it is still a good idea to restrain the system properly so that it does not move excessively in the event of a failure, or when the pressure is decreased in step 42.

1-30. Follow steps 1-30 in Section 13.3.3.2, "Hydrostatic/Hydrostatic Test; Secondary Jacket Not Fully Assembled." Continue on to step 31 described in the following.

31. Raise the pressure of the primary piping in a slow, gradual manner to the test pressure of the secondary containment piping system. A typical pneumatic test pressure for the secondary containment piping system is 110% of its design pressure, adjusted for operating temperature [see Section 13.1.3.4.1 and Eq. (13.1)].
32. Turn off the low-point drain valve of the primary pipe system and disconnect the source of pressure.
33. Check all high-point vent valves and low-point drain valves of the secondary containment piping system to be sure that they are closed.
34. Attach a source of pneumatic pressure (i.e., air compressor with pressure regulator valve, nitrogen bottle with pressure regulator valve, etc.) to the overall low-point drain valve of the secondary containment system (annulus). Remember, it is important to have a pressure gauge connected to the "upstream" side of the secondary containment low-point drain valve so that the pressure in the secondary containment piping system can be determined at all times during the test. The pressure gauge must be in full view of the testing operator when the low-point valve is turned on and off. At this point, the annulus is a closed system, filled with a compressible gas under atmospheric pressure.
35. Turn on the pneumatic pressure source and open the low-point drain valve in the secondary containment piping system to the source of pressure, allowing the pressure to rise in a gradual, stepwise manner.
36. Raise the pressure in increments of 10-25 psi in a slow, gradual manner. Completely and thoroughly cover each secondary containment weld (bond) with a soapy solution in order to detect bubbles if there are leaks. Be sure to inspect all secondary containment pipe welds (bonds) visually when each increment of 10-25 psi is reached. Randomly inspect other points in the system, including straight pipe sections, for the presence of water.
37. When the final test pressure of the secondary containment system is reached, it must be held at least long enough to allow piping strains to become somewhat equalized, or at least long enough to be able to inspect all necessary points visually. The secondary containment low-point drain valve can be closed at this point and the source of pressure removed while the system is being visually inspected. This is an optional procedure. If a plastic pipe is being tested, the pressure must be applied for at least 1 h due to the substantial creep inherent in plastics. For thermoplastic pipes, a modified procedure as described in Section 13.3.4.4 should be used to correct for the anticipated initial drop in pressure due to expansion and creep of the plastic pipe.
38. If the internal gauge pressure of the secondary containment system does not drop by more than an amount specified as an "unacceptable pressure drop" after the specified duration of the test, and no "soap bubbles" are detected at the joints, the secondary containment system can be described as having "passed" the pressure/leak test. If leaks or defects are observed during the test, then specified corrective action procedures must be followed.
39. For a secondary containment system that passes the pressure test, its low-point drain valve should be turned off (if it has not already been turned off in step 37).
40. Disconnect the source of pressure from the low-point drain valve (if it has not already been disconnected in step 37).
41. Connect a drain hose to the low-point drain valve(s) of the primary piping system. The hose(s) should be drained to a storm sewer, storm drain, or other appropriate point in

the facility. If a test fluid other than water is used, it may have to be collected and treated as chemical waste. If more than one local low point exists in the inner and/or outer systems, connect drain hoses from these points as well.

42. Open a secondary containment pipe system high-point vent valve and carefully allow the pressure in the annulus to be reduced to atmospheric pressure. Other valves can be opened to accelerate this process.
43. When the pressure in the annulus drops to atmospheric pressure, close off the high-point vent valve that was opened, and any other valves that were opened for this purpose.
44. Open the primary pipe low-point drain valve and allow the pressure in the primary system to be reduced to atmospheric pressure.
45. When the pressure in the primary system drops to atmospheric, at least one primary pipe high-point vent valve should be opened to allow the primary system to drain smoothly. If there is more than one low point in the primary pipe system, the additional valve(s) should also be opened.
46. When all of the water (test fluid) has drained from the primary pipe, the high- and low-point valves should be kept open to allow drying procedures to be carried out.
47. Disconnect the drain hose(s).
48. Connect a dry air source to the primary pipe, either at its high or low point.
49. Blow air through the primary pipe until the desired level of drying is reached.
50. Turn off air source and disconnect it from the primary pipe system. Turn all primary high- and low-point valves to their off position.

13.3.3.5 Pneumatic/Pneumatic Test; Secondary Jacket Fully Assembled

This sequence of testing applies to those systems where a pneumatic pressure test is to be applied to both the inside (primary) pipe system and in the annular space (to the outer pipe system; secondary containment pipe system). It is also the procedure used for systems that will have their secondary containment system fully completed prior to conducting the initial pressure test of the primary piping system. This procedure will not allow many primary pipe joints, or other sections of the primary pipe, to be visually inspected during its pressure test. This type of system may not be in compliance with various codes for this reason (i.e., ANSI/ASME B31.3 Code for Chemical Plant and Petroleum Refinery Piping).[9]

Since a pneumatic pressure test is being conducted, be sure to consider all of the potential danger inherent in such a test, particularly if a material is used that is subject to brittle-type failure. Do not conduct a pneumatic pressure test with a material such as PVC (or other plastic when it is in a cold nonductile state) since there is too much danger inherent in such a test. Since the inside pipe will not weigh the system down, as it too will be filled with air or gas, it is highly important to restrain the system properly so that it does not move excessively or violently in the event of a failure, or when the pressure of either system is decreased (see steps 8, 18, and 20).

1. Check all high-point vent valves and low-point drain valves of both the primary and secondary containment piping systems to be sure they are closed.
2. Attach a source of pneumatic pressure (i.e., air compressor with pressure regulator valve, nitrogen bottle with pressure regulator valve, etc.) to the overall low-point drain

valve of the primary piping system. Remember, it is important to have a pressure gauge connected to the "upstream" side of the low-point drain valve so that the pressure in the primary piping system can be determined at all times during the test. The pressure gauge must be in full view of the testing operator when the low-point valve is turned on and off. At this point, the primary piping is a closed system, filled with a compressible gas under atmospheric pressure.

3. Turn on the pneumatic pressure source and open the primary pipe low-point drain valve to the source of pressure, allowing the pressure to rise in a gradual, stepwise manner.
4. Raise the pressure in increments of 10-25 psi in a slow, gradual manner. At each increment of 10-25 psi, pause to allow the piping strains to become somewhat equalized.
5. When the final test pressure of the primary piping system is reached, it must be held for at least 10 min, or at least long enough to allow piping strains to become somewhat equalized. The primary pipe low-point drain valve can be closed at this point and the source of pressure removed while the primary piping test is being conducted. This is an optional procedure. If a plastic pipe is being tested, the pressure must be applied for at least 1 h due to the substantial creep inherent in plastics. For thermoplastic pipes, a modified procedure as described in Section 13.3.4.4 should be used to correct for the anticipated initial drop in pressure due to expansion and creep of the plastic pipe.
6. If the internal gauge pressure of the primary piping system does not drop by more than an amount specified as an "unacceptable pressure drop" after the specified duration of the test, the primary piping system can be described as having "passed" the pressure/leak test. If leaks or defects are observed during the test, then specified corrective action procedures must be followed.
7. For a primary piping system that passes this pressure test, its pressure must be reduced to the pressure of the secondary containment piping pneumatic pressure test (if the secondary containment system is being tested to a lower pressure than that of the primary piping system).
8. Open a primary pipe high-point vent valve and carefully allow the pressure of the primary system to be reduced to that of the secondary containment test pressure by allowing gas to escape (if the secondary containment system is being tested to a lower pressure than that of the primary piping system). Close the primary pipe high-point vent valve once the pressure is reduced to the desired point.
9. The primary pipe system low-point drain valve may now be closed, and the source of pressure to the low-point drain valve removed. The primary pipe system is at this point a closed system, filled with a compressible gas, under the same pressure to which the primary pipe system will be pneumatically tested.
10. Recheck all high-point vent valves and low-point drain valves of the secondary containment piping system to be sure that they are closed.
11. Attach a source of pneumatic pressure (i.e., air compressor with pressure regulator valve, nitrogen bottle with pressure regulator valve, etc.) to the overall low-point drain valve of the secondary containment system (annulus). Remember, it is important to have a pressure gauge connected to the "upstream" side of the low-point drain valve so that the pressure in the secondary containment piping system can be determined at all times during the test. The pressure gauge must be in full view of the testing operator when the low-point valve is turned on and off. At this point, the annulus is a closed system, filled with a compressible gas under atmospheric pressure.

12. Turn on the pneumatic pressure source and open the low-point drain valve in the secondary containment piping system to the source of pressure, allowing the pressure to rise in a gradual, stepwise manner.
13. Raise the pressure in increments of 10-25 psi in a slow, gradual manner. Completely and thoroughly cover each secondary containment weld (bond) with a soapy solution in order to detect bubbles if there are leaks. Be sure to inspect all secondary containment pipe welds (bonds) visually when each increment of 10-25 psi is reached. Randomly inspect other points in the system, including straight pipe sections, for the presence of water.
14. When the final test pressure of the secondary containment system is reached, it must be held at least long enough to allow piping strains to become somewhat equalized, or at least long enough to be able to inspect all necessary points visually. The secondary containment low-point drain valve can be closed at this point and the source of pressure removed while the secondary containment system is being visually inspected. This is an optional procedure. If a plastic pipe is being tested, the pressure must be applied for at least 1 h due to the substantial creep inherent in plastics. For thermoplastic pipes, a modified procedure as described in Section 13.3.4.4 should be used to correct for the anticipated initial drop in pressure due to expansion and creep of the plastic pipe.
15. If the internal gauge pressure of the secondary containment system does not drop by more than an amount specified as an "unacceptable pressure drop" after the specified duration of the test, and no "soap bubbles" are detected at the joints, the secondary containment system can be described as having "passed" the pressure/leak test. If leaks or defects are observed during the test, then specified corrective action procedures must be followed.
16. For a secondary containment system that passes the pressure test, its low-point drain valve should be turned off (if it has not already been turned off in step 14).
17. Disconnect the source of pressure from the low-point drain valve (if it has not already been disconnected in step 14).
18. Open a secondary containment pipe system high-point vent valve and carefully allow the pressure in the annulus to be reduced to atmospheric pressure. Other valves can be opened to accelerate this process.
19. When the pressure in the annulus drops to atmospheric pressure, close off the high-point vent valve that was opened, and any other valves that were opened for this purpose.
20. Open the primary pipe high-point vent valve and carefully allow the pressure in the primary system to be reduced to atmospheric pressure. Other valves can be opened to accelerate this process.
21. When the pressure in the primary system drops to atmospheric, close off the high-point vent valve that was opened, and any other valves that were opened for this purpose.

13.3.3.6 Pneumatic/Pneumatic Test; Secondary Jacket Not Fully Assembled

This sequence of testing applies to those systems where a pneumatic pressure test is to be applied to both the inside (primary) pipe system and in the annular space (to the outer pipe system; secondary containment pipe system). It is also the procedure used for systems that will have primary pipe sections tested prior to completing secondary closures of the secondary containment system. This allows previously untested primary pipe joints to be

Inspection, Examination, and Testing

inspected visually during the pressure test. Primary pipe welds that have been previously tested and visually examined during a test of a prefabricated section do not have to be reinspected during a test of the assembled system (see Section 13.1.3.2). This sequence, if followed, will not allow the primary pipe to collapse while pressure testing the outer pipe system.

Since a pneumatic pressure test is being conducted, be sure to consider all of the potential danger inherent in such a test, particularly if a material is used that is subject to brittle-type failure. Do not conduct a pneumatic pressure test with a material such as PVC (or other plastic when it is in a cold nonductile state) since there is too much danger inherent in such a test. Since the inside pipe will not weigh the system down, as it too will be filled with air or gas, it is highly important to restrain the system properly so that it does not move excessively or violently in the event of a failure, or when the pressure of either system is decreased (see steps 7, 22, and 24).

1. Check all high-point vent valves and low-point drain valves of the primary piping system to be sure they are closed.
2. Attach a source of pneumatic pressure (i.e., air compressor with pressure regulator valve, nitrogen bottle with pressure regulator valve, etc.) to the overall low-point drain valve of the primary piping system. Remember, it is important to have a pressure gauge connected to the "upstream" side of the low-point drain valve so that the pressure in the primary piping system can be determined at all times during the test. The pressure gauge must be in full view of the testing operator when the low-point valve is turned on and off. At this point, the primary piping is a closed system, filled with a compressible gas under atmospheric pressure.
3. Turn on the pneumatic pressure source and open the primary pipe low-point drain valve to the source of pressure, allowing the pressure to rise in a gradual, stepwise manner.
4. Raise the pressure in increments of 10-25 psi in a slow, gradual manner. At each increment of 10-25 psi, pause to allow the piping strains to become somewhat equalized.
5. When the final test pressure of the primary piping system is reached, it must be held for at least 10 min, or at least long enough to allow piping strains to become somewhat equalized. Do not use a soap solution to inspect the primary pipe welds unless it is considered acceptable for the project; if the joints are soaped, they must be thoroughly cleaned and dried prior to completing construction of the secondary containment jacket. If a plastic pipe is being tested, the pressure must be applied for at least 1 h due to the substantial creep inherent in plastics. For thermoplastic pipes, a modified procedure as described in Section 13.3.4.4 should be used to correct for the anticipated initial drop in pressure due to expansion and creep of the plastic pipe.
6. The primary pipe low-point drain valve can be closed at this point and the source of pressure removed while the primary piping system test is being conducted. This is an optional procedure.
7. If the internal gauge pressure of the primary piping system does not drop by more than an amount specified as an "unacceptable pressure drop" after the specified duration of the test, the primary piping system can be described as having "passed" the pressure/leak test. If leaks or defects are observed during the test, then specified corrective action procedures must be followed.
8. For a primary piping system that passes this pressure test, its pressure must be reduced to atmospheric pressure prior to commencing work on the outer jacket. Open the primary

pipe high-point vent valve and carefully allow the pressure in the primary system to be reduced to atmospheric pressure. Other valves can be opened to accelerate this process.
9. When the pressure in the primary system drops to atmospheric, close off the high-point vent valve that was opened, and any other valves that were opened for this purpose.
10. Finish constructing the secondary containment system; make all secondary midline closure welds (bonds).
11. When the secondary containment system is finished being constructed, check all high-point vent valves and low-point drain valves of the primary piping system to be sure that they are closed.
12. Reconnect the source of pneumatic pressure (i.e., air compressor with pressure regulator valve, nitrogen bottle with pressure regulator valve, etc.) to the overall low-point drain valve of the primary piping system. At this point, the primary piping is a closed system, filled with a compressible gas under atmospheric pressure.
13. Turn on the pneumatic pressure source and open the primary pipe low-point drain valve to the source of pressure, allowing the pressure to rise in a gradual, stepwise manner.
14. Raise the pressure slowly to the pneumatic test pressure of the secondary containment system.
15. Recheck all high-point vent valves and low-point drain valves of the secondary containment piping system to be sure they are closed.
16. Attach a source of pneumatic pressure (i.e., air compressor with pressure regulator valve, nitrogen bottle with pressure regulator valve, etc.) to the overall low-point drain valve of the secondary containment system (annulus). Remember, it is important to have a pressure gauge connected to the "upstream" side of the low-point drain valve so that the pressure in the secondary containment piping system can be determined at all times during the test. The pressure gauge must be in full view of the testing operator when the low-point valve is turned on and off. At this point, the annulus is a closed system, filled with a compressible gas under atmospheric pressure.
17. Turn on the pneumatic pressure source and open the low-point drain valve in the secondary containment piping system to the source of pressure, allowing the pressure to rise in a gradual, stepwise manner.
18. Raise the pressure in increments of 10-25 psi in a slow, gradual manner. Completely and thoroughly cover each secondary containment weld (bond) with a soapy solution in order to detect bubbles if there are leaks. Be sure to inspect all secondary containment pipe welds (bonds) visually when each increment of 10-25 psi is reached. Randomly inspect other points in the system, including straight pipe sections, for the presence of water.
19. When the final test pressure of the secondary containment system is reached, it must be held at least long enough to allow piping strains to become somewhat equalized, or at least long enough to be able to inspect all necessary points visually. The secondary containment low point drain valve can be closed at this point and the source of pressure removed while the secondary containment system is being visually inspected. This is an optional procedure. If a plastic pipe is being tested, the pressure must be applied for at least 1 h due to the substantial creep inherent in plastics. For thermoplastic pipes, a modified procedure as described in Section 13.3.4.4 should be used to correct for the anticipated initial drop in pressure due to expansion and creep of the plastic pipe.

20. If the internal gauge pressure of the secondary containment system does not drop by more than an amount specified as an "unacceptable pressure drop" after the specified duration of the test, and no "soap bubbles" are detected at the joints, the secondary containment system can be described as having "passed" the pressure/leak test. If leaks or defects are observed during the test, then specified corrective action procedures must be followed.
21. For a secondary containment system that passes the pressure test, its low-point drain valve should be turned off (if it has not already been turned off in step 19).
22. Disconnect the source of pressure from the low-point drain valve (if it has not already been disconnected in step 19).
23. Open a secondary containment pipe system high-point vent valve and carefully allow the pressure in the annulus to be reduced to atmospheric pressure. Other valves can be opened to accelerate this process.
24. When the pressure in the annulus drops to atmospheric pressure, close off the secondary containment high-point vent valve that was opened, and any other valves that were opened for this purpose.
25. Open the primary pipe high-point vent valve and carefully allow the pressure in the primary system to be reduced to atmospheric pressure. Other valves can be opened to accelerate this process.
26. When the pressure in the primary system drops to atmospheric, close off the primary pipe high-point vent valve that was opened, and any other valves that were opened for this purpose.

13.3.4 Other Testing Considerations

There are many variations of testing procedures that may be used to test a double containment piping system. For every given application, a designer must take into account the design of the piping system, its materials of construction, and any environmental factors that may affect testing parameters. The procedures described in Section 13.3.3 outline steps for testing piping systems to relatively high test pressures. The basic procedures have to be modified slightly if a low-pressure test is to be conducted, or if other factors apply. These modifications, and other testing factors, are described in the following paragraphs.

13.3.4.1 10-ft (3-m) Standing Water Column Test for Drainage Pipes

The most common test applied to single-wall drain, waste, and vent (nonpressure) piping is a 10-ft (3-m) standing water test. In this test, a 10 ft (3-m) riser is attached to the high point in the system. Plugs are positioned in other open points in the system (floor drains, etc.). The system is then filled with water until it spills out of the riser. When the riser is filled, every point in the system is subjected to a minimum pressure of 10 ft (3-m) of water (4.33 psi; 0.3 bar). At the low point in the system, the components are subjected to a pressure equal to 10 ft (3-m) of water (4.33 psi; 0.3 Bar), plus the overall change in elevation. This test is sufficient for systems that are intended for gravity-induced drainage under atmospheric pressure. Systems that will be subjected to a change in elevation of more than 33 ft (10 m) (14.7 psi; 1 bar) or will convey pumped waste should not be subjected to this type of test. Instead, a hydrostatic test should be conducted.

13.3.4.2 Low-Pressure Pneumatic Leak Testing of Secondary Containment

This procedure is a variation of the tests described in Sections 13.3.3.1 through 13.3.3.6. It may be applied to those systems that have a secondary containment design pressure rating of 0 to 15 psig (0 to 1 bar gauge). Instead of testing a secondary containment jacket to a pneumatic pressure of 110% of the design pressure, a low-pressure test is conducted (usually in the range of 5 to 15 psig; 0.4 to 1.0 bar). The remainder of the procedures described in the typical procedures still apply. However, the procedures for soaping and visually inspecting the welds take on added importance.

13.3.4.3 Testing of Open-Ended Secondary Containment Jacketing

All of the procedures described in Sections 13.3.3.1 through 13.3.3.6 are for systems that have a tightly sealed annular space. Circumstances may require that a system be designed with an open-ended annulus. If this is the case, then temporary end-type secondary closures must be welded (bonded), mechanically joined, or joined with the use of compressed elastomeric seals. Once the sections are sealed, the testing procedures described in Sections 13.3.3.1 through 13.3.3.6 may be followed as written. After testing is concluded, the temporary end-type secondary closures may be removed.

13.3.4.4 Modified Test to Account for Pressure Decrease in Thermoplastic Piping

Thermoplastic piping presents unique problems in terms of conducting a hydrostatic or pneumatic test. All thermoplastic materials undergo a linear and radial expansion when an internal pressure is initially applied. In some plastics, application of 150 psi (10 bar) internal pressure can result in 1% or more linear growth in the piping. This, in addition to a certain amount of radial expansion, will translate into a substantial drop in the internal pressure of the pipe, often as much as 15 to 20 percent, even though no leaking has taken place. The plastic will tend to deform permanently to a partial extent due to the substantial tensile creep that occurs in a thermoplastic material. This effect presents a dilemma in establishing a reasonable correlation between pressure drop and what constitutes "passing" of a system's pressure test. After prolonged exposure to initial test pressure, by raising the pressure back to the specified test pressure, the plastic will no longer experience significant expansion. Therefore, the following modifications are suggested when testing a thermoplastic material:

Upon initial introduction of pressure, wait for 15 min and check the pressure of the piping. If it has dropped by more than 2 psi (0.15 bar), increase the pressure in the system to the desired level. Wait for 15 min and again check the pressure of the system. If it has dropped by more than 2 psi (0.15 bar) after the second application of pressure, increase the pressure in the system to the desired level. Repeat the process for a third time. If the pressure decreases by more than 2 psi (0.15 bar) after the third application of pressure, a leak must be present in the system, unless other ambient effects have contributed to this decrease. (See Section 13.3.4.5.)

13.3.4.5 Ambient Effects

Temperature will have some effect on testing since materials may be affected by temperature changes (i.e., thermal expansions/contractions, change in ductility, etc.).

Thermoplastic materials are notorious for being affected, since they undergo greater dimensional changes as compared to metallic or RTRP materials over normal changes in ambient temperature. Any time there is the chance for materials to change temperatures by more than 5 - 10°F (3 to 5°C), there may be changes in system pressure. In some applications, a test that is started in the cool temperatures of an early morning may have materials that can change by almost 100°F (55°C) due to radiative effects of the sun on the piping surface. This not only causes a significant drop in pressure due to thermal expansion, but it can also rapidly diminish the pressure-retaining capability of a thermoplastic material (i.e., PVC at 140°F/60°C surface temperature). These factors must be taken into account in designing a testing program.

Equally important to the testing of a pipe temperature decreases. There are two problems that may result from a decrease in temperature. First, a pipe may increase in pressure if it is filled with a testing fluid (compressible or incompressible fluid). However, it will also cause a material to lose ductility rapidly. Therefore, it is important to consider a drop in temperatures if a pneumatic testing program is to be completed (particularly for a secondary containment jacket).

References

1. F. J. Weisweiler and G. N. Sergeev, "Non-Destructive Testing of Large-Diameter Pipe for Oil and Gas Transmission Lines," VCH Verlagsgesellschaft mbH, Weinheim (1987).

2. Code for Chemical Plant and Petroleum Refinery Piping, ANSI/ASME B31.3-1993 Edition, Paragraph 345.2.2, 345.2.5, and 345.3.1, American Society of Mechanical Engineers, New York, NY (1990).

3. Code for Chemical Plant and Petroleum Refinery Piping, op. cit.

4. Weisweiler et. al., op. cit.

5. Code for Chemical Plant and Petroleum Refinery Piping, op. cit., Paragraph 345.3.1.

6. Code for Chemical Plant and Petroleum Refinery Piping, op. cit., Paragraph 345.4.3(b).

7. Code for Chemical Plant and Petroleum Refinery Piping, op. cit., Paragraph. 345.2.2, 345.2.5 and 345.3.1.

8. Code for Chemical Plant and Petroleum Refinery Piping, op. cit., Paragraph. 345.2.2, 345.2.5 and 345.3.1.

9. Code for Chemical Plant and Petroleum Refinery Piping, op. cit., Paragraph. 345.2.2, 345.2.5 and 345.3.1.

Notes

Chapter Fourteen

Associated Storage Tanks and Pressure Vessels

14.1 Overview

Many double containment piping systems either terminate or originate from a tank or pressure vessel. If a double containment piping system is designed adequately, but associated tanks/vessels or connections of the pipe to the associated tanks/vessels leak, there is little gained by the use of double containment piping. Therefore, a double containment piping system, any associated tanks/vessels, and connections to the associated tanks/vessels should be viewed as an integral system. The reason for including information on tanks and vessels is to allow a wider scope of facilities to be adequately constructed in accordance with modern methods. Also, most government regulations that mandate the use of secondary containment for piping have similar regulations regarding tanks. Thus, it is very useful to a designer to gain a basic understanding of both.

This chapter presents some of the basic requirements for construction and installation of above-ground and underground single- and double-wall storage tanks. Rules are also presented for pressure vessel construction, connections to pressure vessels, and other issues involving double containment pipes associated with above-ground pressure vessels.

Certain secondary containment housing components that are part of a double containment assembly may in some cases be technically required to be designed as a pressure vessel (e.g., Figure 11.61). For components meeting this criterion, the general rules concerning design, construction, and operation of pressure vessels presented in this chapter can be followed.

14.1.1 Source

Information and data available from many professional organizations have contributed to the information presented in this chapter. These include: the Steel Tank Institute, the Fiberglass Petroleum Tank and Pipe Institute, the Petroleum Equipment Institute, the American Petroleum Institute, the American Society of Mechanical Engineers (Boiler and Pressure Vessel Code, Section VIII (Divisions 1 and 2), Section X and Section III; Standard RTP-1; B31.3 Code), the American Society for Testing and Materials (Volume 8.04), the Underwriters Laboratories, the National Fire Protection Association, and the National Association of Corrosion Engineers. Technical information supplied by OPW, Smith

Figure 14.1. Example of a single-wall storage tank which has 50% of its volume above ground and 50% underground. This tank would be classified as a UST (underground storage tank) according to RCRA standards. (Source: Steel Tank Institute.)

Fiberglass Products, Inc., OC Tanks Corporation (subsidiary of Owens-Corning Fiberglass), and Xerxes Corporation also contributed to the development of this chapter.

There are many government agencies involved in regulating the construction and installation of underground storage tanks, including local governments and fire departments. Requirements may vary considerably from location to location due to differences in the fire and building codes of the state or local municipality. Therefore, designers, purchasers, and installers of underground storage tanks and pressure vessels are encouraged to check with the requirements of local authorities and to consult with the manufacturers selected for specific handling requirements prior to procuring and installing an underground or aboveground storage tank or pressure vessel.

14.2 Storage Tanks (Nonpressure Vessels)

The underground storage of petroleum or other organic liquids is common for reasons of space savings, safety concerns, and sometimes for improving the visual esthetics of the surrounding environment. At petroleum marketing locations and tank terminals, all of these conditions tend to be underlying factors for placing tanks underground. By placing storage tanks underground, stored fluid temperatures can be controlled. Fugitive losses of vapors tend to be low as a result. The worldwide trend is towards minimizing evaporative losses as a means of lessening hydrocarbon emissions. Therefore, underground storage is environmentally desirable since it lessens emissions at surface pumps and connections.

If the tank to be buried is a double containment tank equipped with interstitial monitoring, and is designed and installed properly, there will be further environmental benefit. Locating tanks underground also substantially lowers the risk of fire and explosions, thus also providing safety benefits.

A tank does not have to be physically underground in order to be technically classified as an underground storage tank, according to U.S. federal regulations. RCRA, Subtitle I, (40

CFR Part 280) defines a storage tank to be an underground storage tank if 10% of the volume of a tank, or its associated piping, is underground. Therefore, a tank may itself be physically above ground, but if its associated piping is located underground, the entire system may be classified as an underground storage tank system. Figure 14.1 shows a tank that is 50% above ground and 50% underground. Since more than 10% of its overall volume is underground, the entire tank and the associated connected piping is technically classified as an underground storage tank system (UST) in the United States.

By calling a vessel a "storage tank," it is inferred that a tank will contain less than 5 psi internal pressure (a typical pressure is 1-2 psi max).

A metallic tank that contains a pressure of greater than 5 psi, but less than 15 psi, would usually be designed in the same manner as a pressure vessel according to Section VIII, Division 1 rules (although 15 psi is the lower pressure limit for classification as a pressure vessel according to the ASME Boiler and Pressure Vessel Code). However, it would most likely not be given an ASME stamp, in order to save on administrative costs. An RTRP vessel with an internal design pressure of between 0 and 15 psig may be designed and constructed according to the rules of ASME Standard RTP-1.

14.2.1 Types of Storage Tanks

Nonpressure tanks may be fabricated from a variety of materials, including metallic and nonmetallic. Tank types used in fluid services that would typically be associated with double containment piping systems include: cathodically protected single-wall metallic tanks (coated and noncoated), single-wall metallic tanks "clad" with corrosion-resistant materials (e.g., metallic with epoxy RTRP cladding), double containment tanks (metallic, FRP, and small-diameter thermoplastic), and single-wall tanks provided with secondary containment protection (such as impervious membrane liners or plastic-lined or coated concrete).

Most underground storage tanks for chemical applications are either of single-wall or double containment metallic or RTRP construction. This is due to the fact that these materials possess the structural strength needed for underground burial and also have the needed corrosion resistance for most petroleum and petroleum-related products. In applications where a harsh corrosive fluid is involved, storage tanks may be internally lined with a thermoplastic material. Most metallic tanks gain their corrosion resistance to the external environment by the addition of an external coating (e.g., epoxy) and cathodic protection.

Underwriters Laboratories has developed standards for the safe underground storage of petroleum products based on storage systems constructed of steel and FRP materials. These standards are UL 58, "Steel Underground Storage Tanks for Petroleum Products" and UL 1316, "Glass-Fiber-Reinforced Plastic Underground Storage Tanks for Petroleum Products." UL 58 for steel tanks is a manufacturing standard only. UL 1316 for fiberglass tanks covers installation recommendations in addition to outlining manufacturing requirements.

14.3 General Considerations for Underground Storage Tanks

Nonpressure tanks intended for underground use require special attention be paid to their design and installation. Particular attention must be given to the procedures by which they are tested and installed, and also the manner of their transportation, loading, and unloading. This subsection discusses each of these issues concerning underground storage tanks.

14.3.1 Plans, Drawings, and Specifications

A written plan should be prepared for all underground tank projects in order to assist in obtaining permits, to solicit bids, and to provide precise guidance for installers. The plans should describe the property and specify the size of all tanks and their location. In addition to new tanks, the location of all existing tanks should be identified. For service stations, plans should identify the location of product dispensers and associated piping. Plans must detail the location of electrical service components, as well as vent locations, inspection wells, vapor recovery systems, and gauge monitoring systems.

Tank project plans should include a set of detailed specifications that describe all materials of construction and other pertinent details for procurement. They also should describe all installation and quality control requirements. Where applicable, tank project plans should show any cathodic protection component locations, the configuration of electronic-release monitoring devices, manway positions, tank-hole lining details, hold-down pads or other anchoring devices, and other factors needed to procure, install, or test the tanks properly.

14.3.2 Underground Storage Tank Handling Criteria

The protective coatings of steel tanks and the shells of fiberglass-reinforced plastic (FRP) tanks are designed to withstand normal handling. However, they can be damaged during transportation or by improper handling procedures. Tanks must not be allowed to fall, be dragged, contacted with sharp objects, or rolled (although small rolling movements may have to be used during tank inspection and testing). If tank shells or their coatings become damaged, they must be repaired or replaced, according to the requirements of the tank manufacturer.

14.3.2.1 Unloading Requirements

Tanks should always be constructed with lifter lugs to allow each tank to be lifted (Figure 14.2). Cables or chains having adequate length should be used to lift and lower each tank. The recommended included angle of the cables varies depending on whether the tank is of metallic or fiberglass construction. In some situations, a spreader bar may also be needed. Guide wires should be attached to each end of the tank to allow for complete control during maneuvering. A tank should never be rolled to reposition it prior to its installation. The proper way to reposition a tank is by lifting it using its lifter lugs. Figure 14.3 is a photograph showing proper techniques. Hoisting equipment must have sufficient capacity and reach to lift and lower a tank (or tanks) without having to drop or drag it. The adequacy of such hoisting equipment should be evaluated before any attempt is made to move tanks. Chains and slugs should never be positioned around a tank shell on any project.

14.3.2.2 Storing Tanks Prior to Installation

Tanks should be protected from accidental damage or vandalism by keeping them in a secure area prior to installation. Indoor or fenced-in secure areas are the best choices for storage of tanks. The area should in all cases be free from rocks, sharp objects, and high traffic. If high winds are expected, the tank (tanks) should be tied down with at least a 1/2 in. (13 mm) diameter nylon or hemp rope that is secured to stakes. The size and number of the stakes required must be determined for each given application.

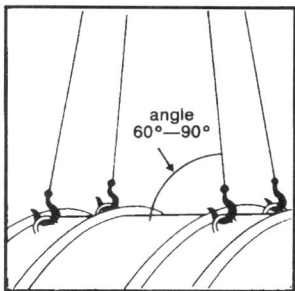

Figure 14.2. Illustration of the proper technique for lifting tanks by means of lifter lugs that are integral to the tank. (Source: OC Tanks.)

14.3.2.3 Inspection Requirements Prior to Installation

All construction materials must be physically inspected before being installed on any project, including tanks, piping, and associated equipment. The materials should be inspected to determine that they meet the requirements of the engineering design and specifications. If defects or damage is detected, repairs should be made in a manner approved by the manufacturer. Alternatively, a new tank may be provided if an existing tank cannot be repaired to the original design specifications.

14.3.2.4 Testing of Tanks Prior to Installation

Tanks should always be provided with factory-installed plugs for shipping. Upon arrival at the site, they should be removed, doped, and reinstalled, taking care not to cross-thread

Figure 14.3. Example of placing steel tanks by means of lifter lugs into an excavation. (Source: Steel Tank Institute.)

them. If tanks have been equipped with metal or plastic thread protectors, they should be replaced with liquid-tight cast iron plugs. Some tank manufacturers do not tighten tank fittings that are delivered as part of a tank. This is done to allow for temperature changes during shipping and storage.

Single-wall, nonpressure storage tanks must be subjected to a 3 to 5 psig air pressure test. All surfaces should be soaped and 100% of the seams and fittings inspected visually for leaks by looking for bubbles (see Figure 14.4). Testing with compressed air that is over 5 psig (3 psig for 12 ft diameter FRP tanks) is not recommended due to the hazards involved. It may also cause damage to a tank in the event of a leak, by action of the sudden release of stored energy that would result. If a tank has previously been used to store flammable or combustible liquids, it should not be subjected to a compressed air test, regardless of test pressure used. Prior to commencing a test, warning barricades should be placed at the ends of tanks to be tested. While the tank is under pressure, avoid the tank ends, manways, and fitting connections. However, it is always necessary to get near these locations when applying soap solution and inspecting for bubbles. While a tank pressure test is being conducted, an inspector should be present for the duration of the test.

Individual tank manufacturers have their own specific instructions for installation and testing. The specific recommendations of an individual manufacturer should be followed, in addition to the rules of the authority having local jurisdiction. Air tests are inconclusive without soaping and careful inspection for bubbles. Soap solution should be applied uniformly by spraying or by using a mop. Be sure to select a proper gauge to conduct the air test. The type of gauge selected should be capable of detecting small changes in pressure. A gauge with a maximum limit of 10 to 15 psig is recommended, since the best accuracy for detecting a change in pressure is at the midrange of a gauge. Gauges should be calibrated and checked for operation before use. The use of vacuum-type gauges poses a serious accident threat, and should be avoided.

A pressure relief device is recommended to be provided to the tank during the pressure test to prevent overpressurization. A pressure relief device should have sufficient capacity to relieve the total capacity of air. The set pressure of the relief device should be not more than 6 psig. These points are illustrated in Figure 14.5.

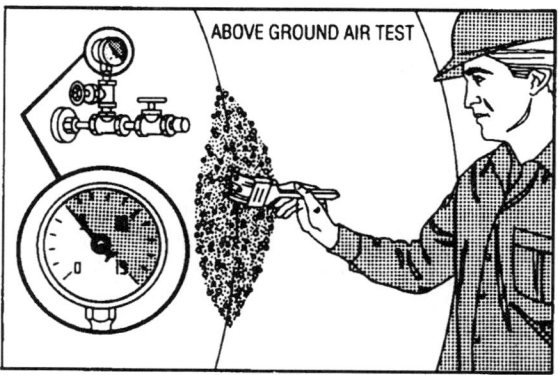

Figure 14.4. Illustration of a soap test of a tank seam. (Source: Steel Tank Institute.)

Associated Storage Tanks and Pressure Vessels

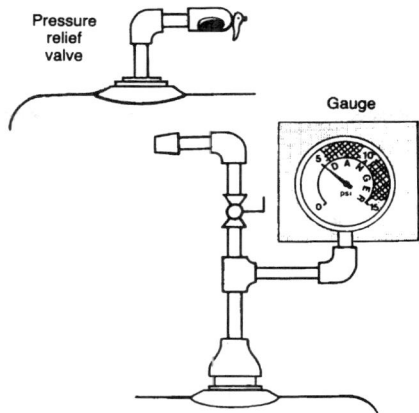

Figure 14.5. Selection of proper gauges and the provision of adequate pressure relief valves, set to no more than 6 psig, are essential to a safe pressure test. (Source: Petroleum Equipment Institute.)

It is normal to expect a slight decrease in pressure in the primary tank when a double-wall tank's interstice is pressurized with gas from the primary pipe. The pressure drop to be expected is around 0.3 psig (0.02 bar) or less. Double-wall tank interstitial volumes may vary from product to product. The individual manufacturer should be consulted for any specific requirements for their tanks.

14.3.3 Overview of Underground Storage Tank Installation

The installation of a tank involves many separate phases of work. These major phases of work include:

1. Initial preparation and layout.
2. Excavation and paving removal (if required).
3. Base preparation for the tanks.
4. Setting the tanks, anchors, and other component systems.
5. Backfilling and compaction.
6. Connecting of piping and electrical and equipment installation.
7. Final backfilling.
8. Finishing (repaving).

14.3.4 Excavation Requirements

A large enough excavation must be planned so that tanks, their associated piping, and related equipment can all be properly installed. It should allow for the backfill materials to be placed in the specified manner and properly compacted. General requirements of the Steel Tank Institute are illustrated in Figures 14.6 and 14.7.

Special attention should be paid to the areas under the circumference of the tank shell and its ends. The degree of slope to be provided for the excavation walls should not be arbitrarily decided. There are many factors to consider before selecting an appropriate slope. These included: soil conditions, the depth of the excavation, weather conditions,

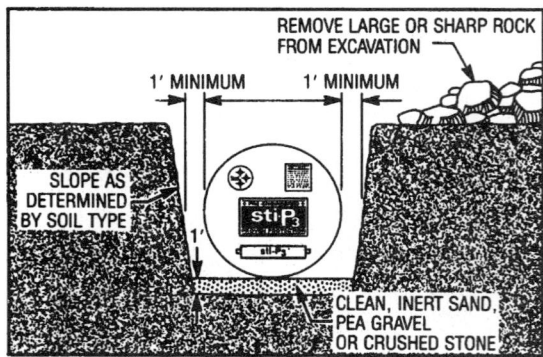

Figure 14.6. Excavation requirements for steel storage tanks. (Source: Steel Tank Institute.)

and other factors. The safety of those working in and around the area must be taken into consideration (there are OSHA standards to consider for projects in the United States). Any safety rules of the local authority having jurisdiction must be implemented.

Some important factors in determining the size, shape, and depth of the excavation required for a project are:

1. The recommendations of the tank manufacturer.
2. Soil stability.
3. Requirements for compacting, bedding, and backfill.
4. Space requirements for piping and associated equipment.
5. Overall burial depth.

There are many factors that can adversely affect the installation. These include such items as the presence of surface water and ground water, unstable soil conditions, the presence of other backfill areas that are in close proximity, frost or frozen soil conditions, mechanically induced vibration, and contaminated soil conditions. Figure 14.8 illustrates this point.

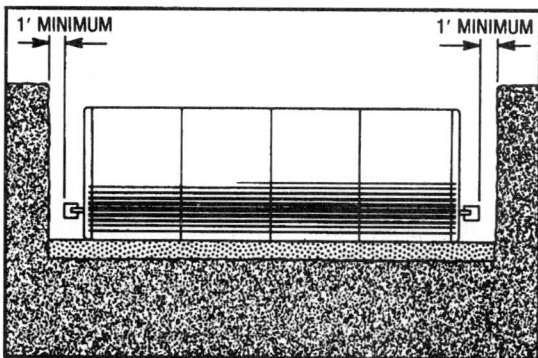

Figure 14.7. Excavation requirements for steel storage tanks. (Source: Steel Tank Institute.)

Associated Storage Tanks and Pressure Vessels

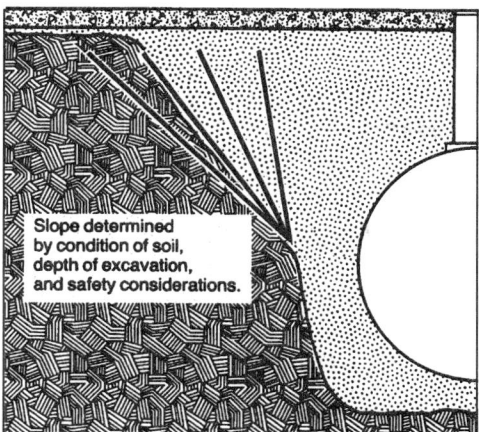

Figure 14.8. Typical excavation considerations include the stability of the soil, bedding, and backfill requirements, space for components, manufacturer's recommendations, and safety. Cave-ins require more backfill, since fallen materials cannot usually be reused as backfill.[1] (Source: Petroleum Equipment Institute.)

If unstable soil is present or water infiltration occurs, tanks that are already set have to be cleared of any fallen materials. If movement of a tank occurs as a result, it will normally require repositioning. If a cave-in occurs, backfill materials may need to be replaced, since fallen materials should not be used as backfill.

Surface water may be prevented from entering the excavation by building a dike around the excavation, but only if it is determined to be a potential problem.

14.3.4.1 Burial, Bedding, and Backfill Details

The relative position of underground tanks with respect to the foundations of existing structures must be investigated thoroughly, and with careful consideration, to decide upon where to excavate and locate any new tank. A general rule that may be used is that each tank must be located a minimum distance of 5 ft (1.5 m) from the base of existing adjacent structures or property lines, as shown in Figure 14.9. This rule should be used in the absence of local building code rules or regulations, regarding the same.[2]

Consideration should also be given to potential downward forces from loads that are carried by other foundations and supports, to ensure that such loads cannot be transmitted to the tanks.

Each tank must be buried to a depth that depends on several factors. These include: its diameter, the thickness of bedding to be used, whether or not a holddown pad is to be used, the minimum depth of cover required, and the desired slope of the associated piping. Since compacted backfill materials provide most of a tank's support, downward forces will be dissipated uniformly, and over a relatively wide area. A minimum of 1 ft (0.3 m) of backfill material should be added beyond a tank's boundary, in order to provide a firm foundation.

In any tank application, a suitably graded, leveled, and compacted backfill material must be provided to a depth of at least 1 ft (0.3 m), at the bottom of its excavation. If a hold-down pad is required, a minimum bedding of 6 in. (150 mm) may be used (metallic tanks) (see

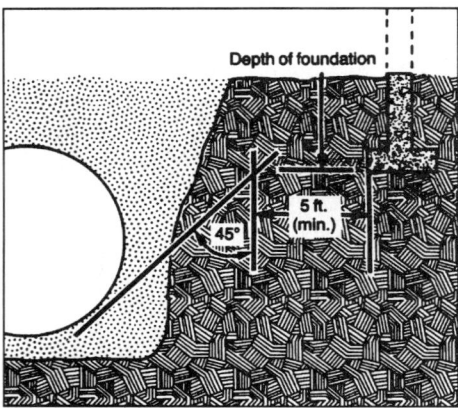

Figure 14.9. Care must be exercised to avoid undermining nearby structures during construction or afterwards, permitting transfer of foundation loads onto the tank. The 45° factor illustrated will accomplish this in most cases. (Source: Petroleum Equipment Institute.)

Figure 14.10). For nonmetallic (RTRP or thermoplastic) tank applications that require a bottom hold-down pad to be installed, a minimum of 1 ft (0.3 m) of suitably graded, leveled, and compacted backfill material should be placed over the top of the pad, to cushion each tank.

To ensure that the bottom quadrant of a nonmetallic tank is fully and evenly supported, backfill materials must be carefully placed along its bottom, sides, and end caps by hand shoveling and tamping. For metallic tanks, a minimum of 6 in. (150 mm) of compacted backfill materials is a requirement of most manufacturers. In order to prevent damage to a tank and its connections or coatings, backfill must also be placed carefully around and over the top of each tank, to a depth of at least 1 ft (0.3 m).

A tank should never be directly placed on a hold-down pad, or be provided with a pad that has a smaller area than the total area of the tank bottom. Unlike above-ground piping, the use of intermediate supports (saddles) must never be applied for tanks, due to the resulting uneven distribution of loads. Each of these items may contribute to structural failure of a tank, and therefore should be avoided in all applications.

The external burial load imposed on an underground storage tank should never exceed 5 psig (0.3 bar). Reinforcement of a tank may be required if the burial depth is greater than the tank diameter. The manufacturer of each tank should be consulted to determine if reinforcement is required. The maximum recommended burial depth for reinforced thermosetting resin plastic (RTRP) tanks is 7 ft (approx. 2 m), measured from the top of the tanks.

14.3.4.2 Cover Requirements

Cover requirements differ as to whether an area is considered to be dry or nondry, and whether or not the area under consideration is subject to vehicular traffic. A dry area is characterized as an area where the water table is not likely to reach the level of a tank bottom. In dry areas that are subject to vehicular traffic, a minimum cover of at least 30 in. (762 mm) of compacted backfill must be provided, in addition to 4-8 in. (102-203 mm) of reinforced concrete, depending on the manufacturer consulted. In traffic areas where

Associated Storage Tanks and Pressure Vessels

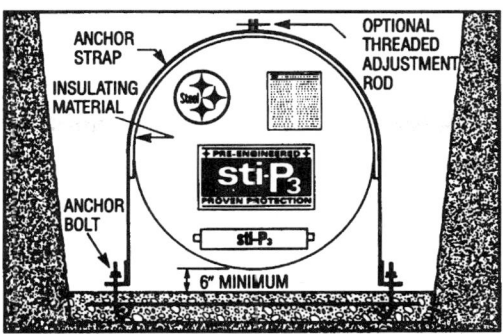

Figure 14.10. When anchoring by means of a concrete slab is required, the tank must not be placed directly on the pad. A layer of fine or pea gravel, sand, or No. 8 crushed stone (No. 8 coarse aggregate ASTM C 448) at least 6 in. deep, spread evenly over the concrete pad, is required for steel tanks. (Source: Steel Tank Institute.)

paving will be placed over the tanks, the paving should extend at least 1 ft (0.3 m) beyond the perimeter of each tank.

At most gasoline stations and other petroleum storage and distribution facilities, fully loaded transports may be required to pass over the tank area. Recommendations by various authorities for reinforced concrete paving vary from 6-8 in. (150-200 mm) in such applications. The requirements of NFPA 30 and 31 should be consulted where applicable, as well as any local codes.

In dry areas that are not subjected to above-ground traffic, the cover should consist of at least 2 ft (0.6 m), with a minimum of 1 ft (0.3 m) of backfill. Backfill should be covered by filter fabric in order to prevent the possibility of soil migration. A minimum of 1 ft (0.3 m) of earth should be provided over backfill, together with filter fabric. A minimum of at least 4 in. (100 mm) of reinforced concrete may be used as an alternative, instead of the usual 1 ft (0.3 m) of top earth cover.

As an example of recommended cover requirements/depth of excavation, the recommendations of Xerxes Corporation are presented for their tanks in Figures 14.11 through 14.14. The recommendations described in those illustrations are typical for RTRP storage tanks.

14.3.5 Ballasting Procedures

After backfill is placed up to the top of the tank, there is some danger that a tank may be lifted due to buoyant forces, caused from ground-water infiltration. Either the product to be stored, or water, should be added to fill the tank as a ballast until its piping is in place and the remainder of the backfilling and paving is completed. (Also see Section 14.3.11 concerning anchorage requirements.)

All tank spaces must be provided with adequate venting during the installation process to allow ballast to be added readily. Tank manufacturers may have recommendations for ballasting of their own specific tank models, based upon each tank's unique design. If the actual service fluid to be stored is to be used for ballasting purposes, special care and consideration must be used. Some of the important details to consider include: (1) safeguarding against fire; (2) safeguarding against product spills; (3) protection against accident or

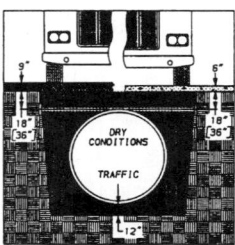

Figure 14.11. Recommended minimum depth for excavation for a typical fiberglass tank in traffic areas and dry conditions. (Source: Xerxes Corp.)

Figure 14.12. Recommended minimum depth for excavation for a typical fiberglass tank in traffic areas and wet soil conditions. (Source: Xerxes Corp.)

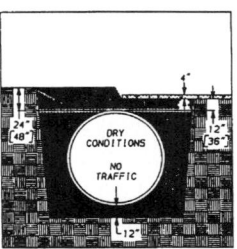

Figure 14.13. Recommended minimum depth for excavation for a typical fiberglass tank in dry conditions and nontraffic areas. (Source: Xerxes Corp.)

Figure 14.14. Recommended minimum depth for excavation for a typical fiberglass tank in wet soil conditions and nontraffic areas. (Source: Xerxes Corp.)

Associated Storage Tanks and Pressure Vessels

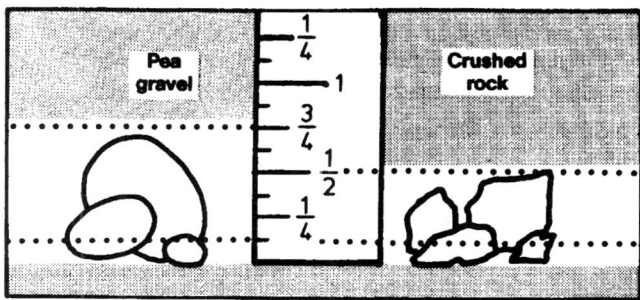

Figure 14.15. Recommended types of backfill include sand, pea gravel, and crushed rock in the sizes shown. Pea gravel and crushed rock are self-compacting, eliminating the need for further compaction. (Source: Petroleum Equipment Institute.)

thefts; and (4) general inventory control as a means of leak detection. All fill caps and pumps must be secured during periods of time when they are left unattended.

14.3.6 Backfill Materials for Metallic or Fiberglass Clad Metallic Tanks

Normally recommended backfill materials consists of a clean, washed, well-granulated, free-flowing, noncorrosive, inert material. These include sand, crushed rock, or pea gravel, the largest particle of which is no larger than 3/4 in. (18 mm) in diameter. Figure 14.15 summarizes typical recommendations for backfill material. Backfill materials must be free of debris, rock, ice, snow, or organic matter. The presence of any of these materials could damage the tank or its coating and adversely affect compaction.

At a majority of project sites, excavated material may not be immediately removed from the site. Excavated materials have to be stockpiled away from the edge of tank excavations. Additionally, excavated materials must be kept separate from approved backfill materials and removed as soon as practical, except where it has been approved for reuse as backfill.

14.3.7 Backfill Materials for Nonmetallic Tanks

Materials used for backfilling nonmetallic tanks should be free of ice and snow; not more than 3% by weight should pass through a #8 sieve. The material to be used should also conform to ASTM C-33, Paragraph 9.1 requirements. Approved materials are:

1. Pea gravel: Naturally rounded particles with a minimum diameter of 1/8 in. (3 mm) and a maximum size of 3/4 in. (18 mm).
2. Crushed rock or gravel: Washed and free-flowing angular particles between 1/8 in. (3 mm) and 1/2 in. (13 mm) in size.

14.3.8 Compaction Options

14.3.8.1 Compaction Techniques

The compaction of bedding and backfill materials must be adequate in order to provide proper support to a tank (tanks). However, it also serves the purpose of preventing move-

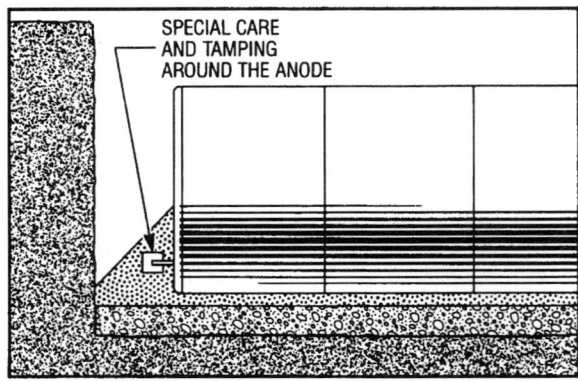

Figure 14.16. Special care should be exercised when compacting around anodes used for cathodic protection purposes. (Source: Steel Tank Institute.)

ment or settlement of a tank (tanks). Many manufacturers require a specified minimum compaction density, when sand is used as its backfill material. Mechanical compaction is the most commonly used means of achieving compaction. When using a mechanical means to compact backfill, a tank must be protected from being damaged.

In applications involving pea gravel or crushed rock, they may not require compaction after backfilling. These materials are relatively self-compacting. However, all materials require a minimum amount of compaction to force material under the lower quadrant of the tank, thereby preventing voids and ensuring the highest degree of compaction required. Special care should be used when compacting around sacrificial anodes, if they are used in conjunction with a metal tank (see Figure 14.16).

14.3.8.2 Sand-Slurry Backfilling Techniques

The sand-slurry method uses a controlled mixture of sand and water to provide compaction to a tank's backfill materials. A sand-slurry method may be used to provide compaction for underground storage tanks if performed by personnel trained by the manufacturer of the tank and only with the approval of the tank manufacturer. Any water that is introduced with the backfill must be removed in order to achieve a satisfactory degree of compaction.

14.3.9 Tank Deflection after Burial

If a buried tank is deflected by the soil load imposed upon it, the deflection can create a distortion in tank dimensions. This effect is more pronounced when a tank is constructed of an RTRP material. Diametrical deflection can result in structural damage to storage tanks. However, it may also result in penetration of the tank bottom by tank internal parts such as suction stubs or submersible pumps. Each individual tank will have its own unique limits for deflection, as determined by its manufacturer. If the measured deflection of a tank exceeds its rated maximum, the manufacturer should be consulted.

The quality of the backfilling and compaction can actually be measured by measuring the total amount of diametrical deflection of an installed tank. Excessive deflection indicates that residual stresses are present in the tank's walls.

Associated Storage Tanks and Pressure Vessels

Deflection in the vertical diameter of a tank may also occur. Vertical deflection may be caused by improper bedding, by voids under the tank bottom, or due to poor compaction of backfill material on a tank's sides.

14.3.10 Migration of Backfill

If unstable soils, bogs, swampy areas, or landfills exist adjacent to a tank installation, backfill may be subject to migration. This would lessen the support required for a tank or paving. A filter fabric must be installed between the backfill and the adjacent source of the migrating water to prevent the possibility of migration from occurring. When sand and pea gravel are used as backfilling materials, they must be separated with filter fabric to prevent sand fines from migrating into voids between pea gravel particles. Use of filter fabrics may make the early detection of cave-ins during construction difficult.

Filter fabrics are geotextile materials designed to permit water to pass through, while preventing the movement of backfill materials. Depending on the material, it might be capable of resisting the deterioration caused by soil corrosion or by the escape of liquids from the underground tank system. All geotextiles are suitable for basic, direct-burial service. Careful consideration in the selection of a suitable fabric is important, since the composition, construction, and mechanical properties of commercially available materials vary widely.

14.3.11 Anchorage and Support of Underground Storage Tanks

Tanks must be prevented from floating in the event of a rise in water table. This possibility exists when potential installations are located in areas subject to high water tables or flooding. Tanks must be prevented from floating during a rise in the water level, up to the established maximum flood stage, whether or not they are full or empty. Flotation might result from the infiltration of surface water in areas with impervious soil, sometimes termed a "bathtub effect." If the tank has vents or other connections that are not liquid-tight, they should be extended above the maximum flood-stage water level until its associated piping is complete.

The water level in the excavation must be reduced to the lowest level that can be practically achieved during construction. If a water ballast must be used to sink a tank in a wet hole, the maximum ballast level to be used is equal to the level of the water in the hole. Care must be used in the addition of ballast. Tanks have to be free to roll slightly, and lifting equipment should only be used to steady a tank in position during the process. Lifting cables must be carefully tended to during the process in order to minimize the possibility of tank damage. The use of cradles, beams, or timbers must be completely avoided during the excavation and during the addition of ballast.

Supplemental restraints are unnecessary if a tank's burial depth is at least 60% of the tanks' vertical diameter, as measured from the top of the tank. The most common method of restraint in areas subject to flooding is to increase the burial depth, without exceeding the 60% rule. In making this determination, each additional inch (25 mm) of reinforced concrete above the tank can be considered equal to inches (50 mm) of compacted backfill. The determination is based on worst-case conditions of an empty tank and a water level at the finished grade.

The diameter of a tank is the primary factor in determining its buoyancy. The weight of its backfill material and the paving over it are the main factors in offsetting buoyancy.

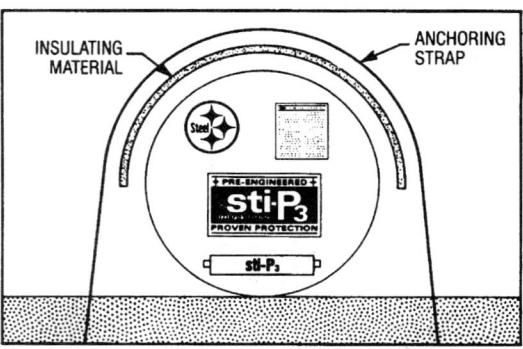

Figure 14.17. Metallic tanks must be separated from hold-down straps by means of a dielectric material. (Source: Steel Tank Institute.)

The weight of an empty tank and its attached equipment, friction between a tank and its backfill, and voids in backfill due to manway openings over a tank must also be given thorough consideration.

Requirements for anchorage are determined by the conditions of each installation. However, all nonmetallic (RTRP) tanks require that tanks 12 ft (3.7 m) or larger in diameter must be anchored, if the water level can be expected to reach the tank bottom. If soil conditions and the depth of water table are unknown at the time of installation, it is best that anchoring be added to the layout requirements. Plans must include having pumps, hoses, straps, cables, and various anchoring materials available for such an installation.

The weight of backfill and paving on top of a tank is an important component of any method used to offset buoyancy forces. Various mechanical anchoring systems will increase the amount of backfill ballast for a tank. If the burial depth is at least 60% of a tank diameter, placement of normal backfill and paving on top of a tank will usually result in an adequate amount of restraint. However, a deep burial may not be appropriate for projects involving unstable soils, bedrock, or extremely high ground-water condi-

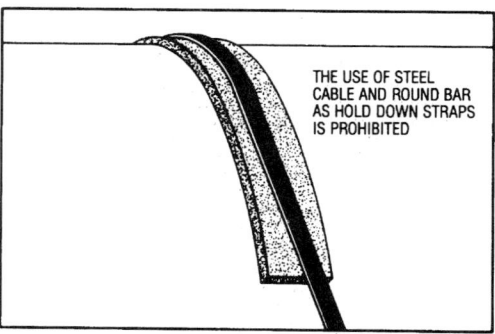

Figure 14.18. Closeup view of the use of an insulating material to separate a metallic tank from its hold-down strap. (Source: Steel Tank Institute.)

Associated Storage Tanks and Pressure Vessels

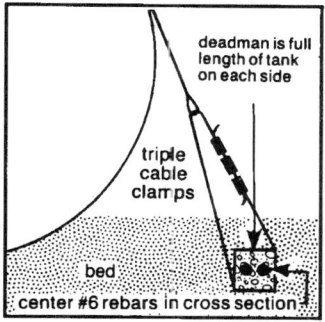

Figure 14.19. Illustration of a fiberglass tank that is anchored by means of a deadman anchor. (Source: OC Tanks.)

tions. For shallow burial depths, a supplemental method of restraint may be used. Some of the methods to be considered include: (1) the use of a slab at grade; (2) additional backfill; (3) the use of deadman anchors; (4) the installation of a bottom hold-down pad; and (5) application of midanchoring.

Tanks must be firmly secured with properly sized straps and/or cables prior to placement in an excavation. Special consideration must be given to securing RTRP tanks with straps; their manufacturer should be consulted on every project. For metallic tanks, straps should be a nonconductive material or should be electrically isolated from the tank structure (see Figures 14.17 and 14.18). A minimum of three cable clamps must be used at each connection, if cables are to be used in the installation. Cables and straps must always be secured before backfilling. Overtightening must be avoided to prevent damage to the tank.

A main concern is always that coatings can become damaged during the installation process, if the tank is of the coated type. If damage to a tank's coating is detected, the coating must be repaired before backfilling.

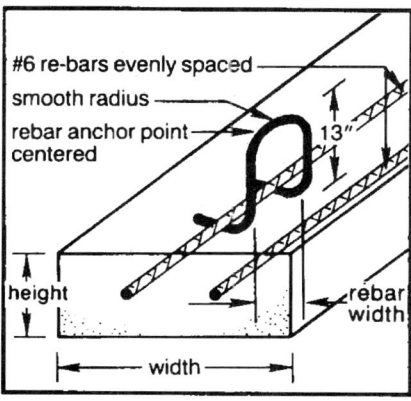

Figure 14.20. Sample details for deadman anchors. (Source: OC Tanks.)

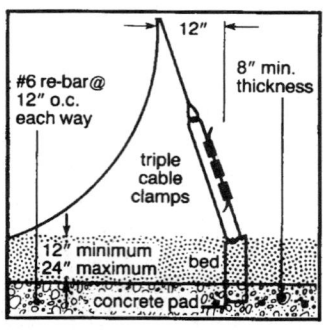

Figure 14.21. Illustration of a concrete bottom hold-down pad. (Source: OC Tanks.)

14.3.11.1 Use of a Slab at Grade/Greater Burial Depth

Ballast weight can be added on top of a tank to offset its buoyancy, by increasing pavement thickness or the depth of its burial. The weight differential between the submerged weight of concrete and gravel will set a limit to additional weight that may be added.

14.3.11.2 Use of Deadman Anchors

A typical deadman anchor is constructed of a beam of reinforced concrete that is lowered into position with cables or straps attached, and sometimes with both. Deadman anchors are an effective means of offsetting ballast. A typical deadman anchor arrangement is illustrated in Figures 14.19 and 14.20.

14.3.11.3 Use of Bottom Hold-Down Pads

Typical bottom hold-down pad construction involves 8 in. (200 mm) of reinforced concrete extending at least 18 in. (450 mm) beyond each end of a tank. They provide a firm foundation in addition to offsetting buoyancy. Each installation requires calculation of the number of pads to be used, as well as the thickness required. An adequate amount of backfill material must be used as a bedding to separate a tank and its bottom hold-down pad. An example of bottom hold-down pad construction is illustrated in Figure 14.21.

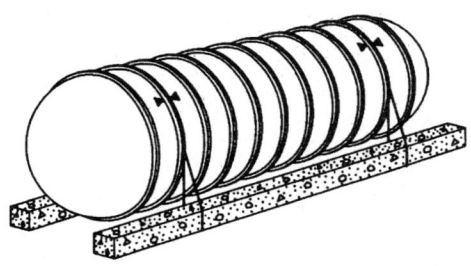

Figure 14.22. A bottom hold-down pad or deadman anchor should never be shorter than the tank it is anchoring. (Source: Xerxes Corp.)

Associated Storage Tanks and Pressure Vessels

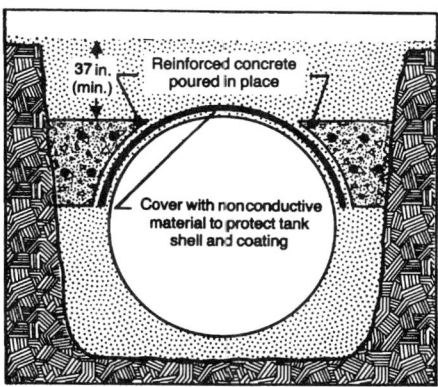

Figure 14.23. If midanchoring is applied, care must be taken to protect the tank shell or coating from damage. (Source: Petroleum Equipment Institute.)

A tank must never be placed directly on concrete itself. If a bottom hold-down pad is constructed shorter than the full length of a tank, uneven stresses can develop, which can result in structural failure of the tank. Therefore, a bottom hold-down pad must never be shorter than the full length of a tank, as illustrated in Figure 14.22.

14.3.11.4 Use of Midanchoring

Midanchoring is used on a limited basis for single-wall tanks, and is only acceptable for certain types of tanks. An example of midanchoring is illustrated in Figure 14.23. Tank manufacturers have to be consulted as to whether it is appropriate for their tanks. The concept of midanchoring involves the placement of reinforced concrete over the top one-third of the tank diameter. A layer of nonconductive material should be placed over the tank and between the concrete as a means of protecting the shell and coating of a tank. For cathodically protected metallic tanks, a nonconductive cover helps to maintain its electrical isolation.

14.3.12 Tank Installation Safety Requirements

The topic of safety in tank applications is covered in documents prepared by the U.S. Department of Labor, Occupational Safety and Health Administration, including: "Construction Industry Standards and Interpretations," 1983 (29 CFR 1926/1910), and OSHA Publication 2079 (revised, June 1985). Individuals involved in the installation of tanks must thoroughly review these and other documents that apply within their local jurisdiction.

Some general rules for safety to every tank application: At all project sites, work areas should be kept clear of stockpiled materials. In areas with unstable soil, excavation walls must be sloped and/or shored, if people are required to enter the excavation. Many underground storage tank projects pose a risk to both installation personnel and to the general public. Therefore, all work areas must be barricaded, particularly at night, to protect both the public and installation personnel, and to prevent accidental damage to and from vehicles and equipment. Fire extinguishers and first-aid supplies should be kept on hand. Members of the installation crew should always be provided with personal protective gear. They should be required to wear them if the project conditions demand it.

Many tank projects involve the removal and replacement of used or existing underground storage tanks. Liquids may remain in existing tanks and in some applications there may be an uncontained release of such liquids. The uncontained product must be recovered, removed, and disposed of in an approved manner. Used underground tanks can then be made safe prior to their removal. During the removal process, previously released liquids can re-enter the tank. Therefore, tanks should be monitored during their removal process. Further detailed procedures are covered in "Removal and Disposal of Used Underground Petroleum Storage Tanks," API Recommended Practice, 1604 (1987).

14.3.13 Common Backfilling Mistakes

Careful placement and compaction of approved backfill materials is essential to protect underground tanks during the installation process. Common deficiencies that may adversely affect tank structural integrity and tank coatings include:

1. Use of incorrect backfill material.
2. Inadequate or improper placement and/or compaction.

Table 14.1 Standard Test Requirements for Underground Storage Tank Systems

Before placing components in the excavation		Before backfilling (after assembly but before connection to the tank or dispenser)		Before placing the system in service	
Condition	Test	Condition	Test	Condition	Test
Single-walled tank	Inspection & 5 psig air test & soaping	New piping (double or single wall)	See Chapter 13	Tank shell deflection	Compare measurement to that taken before backfilling
Double-walled tank	Same as above, avoiding over-pressurization of the interstice. Soaping & inspection are applied to the secondary containment portion of the tank only	Piping has contained flammable or combustible liquids & has not been purged.	110% hydrostatic test (according to PEI, unless ASME B31.3 applies, otherwise see Chapter 13).	Internal & external monitoring & gauging systems	Manufacturers' recommendations
Piping	Careful inspection			Cathodic protection systems	Electrical measurements
Cathodic protection	Test continuity/isolation			Tank system	Precision test (NFPA 329)

Source: Modified from PEI Recommended Practices 100-87, p.20.

3. Rocks or debris left in the excavation.
4. Voids under the perimeter of a tank.
5. Failure to prevent migration of backfill materials.

14.4 Testing of Storage Tanks

All tanks must be tested to ensure that they are correctly built and installed. A tank's accompanying monitoring system must also be proved before a completed system is placed into operation after it is installed. All components of its secondary containment system, overfill prevention devices, impact valves, and leak detection equipment/cable must be tested to verify that they all are capable of satisfactory operation.

Testing is required to verify that installation of the tank system is satisfactory. Standard tests that are applied to storage tank systems and components are summarized in Table 14.1.

14.5 Double-Wall Storage Tanks and Other Secondary Containment Alternatives for Tanks

A variety of secondary containment options exist to allow for safe containment of fluid spills from tanks. In addition to containing the released fluid, secondary containment methods help to facilitate the detection of any release of stored fluid, while providing access for the safe recovery of released fluid. Secondary containment methods for underground installations include the use of double containment tanks (double-walled tanks), common secondary-containment tanks that house multiple-primary-tank compartments, lined-concrete structures, impervious geotextile membrane liners, or thermoplastic liners. Overfill prevention is not a form of secondary containment, but does represent a means of safeguarding against leaks. Therefore, it is discussed, along with the other options, in Section 14.5.

Additional methods for above-ground tanks include the use of dikes and containment berms. The design and installation of tank-associated double containment piping is covered in Chapters 3 through 13 of this text. Discussion in this chapter with respect to piping issues is therefore limited to double containment pipe connections to single- and double-wall tanks.

Liners and concrete structures provide useful alternatives as a means of secondary containment of tank structures. An overview of these alternatives is presented in this chapter. For further discussion on secondary containment uses of liners (geotextile, thermoplastic, etc.) and the use of concrete structures to protect tanks and piping, refer to Chapter 16, "Trenchless Reconstruction and Alternatives to Secondary Containment Piping."

14.5.1 Double-Wall Underground Storage Tanks

Double containment tanks are typically constructed with a thin-gauge material serving as the secondary containment tank (outer shell), and are normally designed to have as small an annulus (interstice) as possible. Many double containment tanks are designed with manways that permit access to the primary tank and its associated piping. Double containment tanks must be installed and tested in accordance with the specific recommendations of each tank manufacturer. However, general recommendations are outlined in this section.

Underwriters Laboratories (UL) currently lists two specific double-wall tank designs, Types I and II. The Type I design has the exterior shell designed so that it is wrapped in direct contact with its primary vessel. Additionally, the exterior shell need not wrap the full

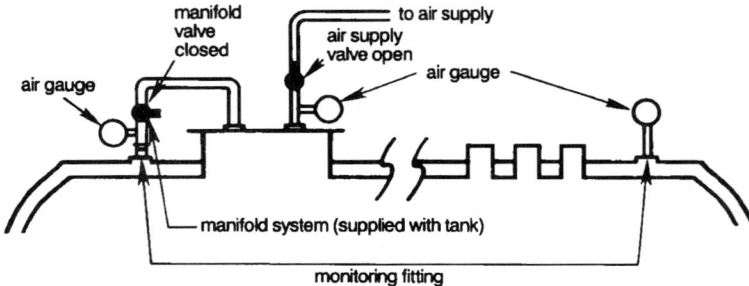

Figure 14.24. Illustration of a typical testing schematic for double-wall tanks. In the first step, pressurize the inner tank to 5 psi maximum. (Source: OC Tanks.)

360 degrees of a primary tank's circumference. A Type-II double-wall tank consists of the outer tank being physically separated from its primary vessel. A Type-II outer tank is separated from its primary tank by the use of spacers or standoffs, and each shell is a full 360 degrees. A Type-II tank lacks the rigidity of a Type-I design in buried situations, as backfill provides limited support to this type of tank. An inner tank must be designed and fabricated in much the same manner as a single-wall above-ground tank, since it is to be supported on spacers or standoffs. Inner tanks for petroleum service must be constructed in the United States according to UL 42 specifications, titled "Steel Aboveground Tanks for Flammable and Combustible Liquids."

Detailed design considerations for double-wall tanks parallel those of double containment piping. Material selection may be the same, or it may be constructed of hybrid materials, due to economy or to external corrosion considerations. Very often metallic single- and double-wall tanks are provided with a corrosion-resistant outer lamination of fiberglass reinforced plastic material, or with an organic coating, for protection against soil corrosion. In terms of pressure ratings, most double-wall tanks are designed with a thin-gauge wall thickness for the secondary tank; thus they are not intended to carry any appreciable internal pressures.

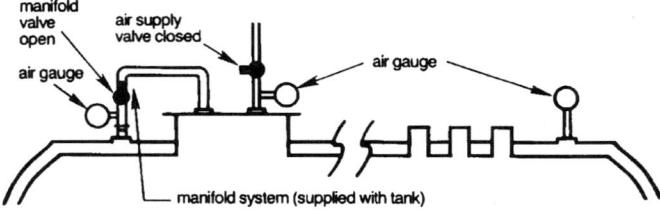

Figure 14.25. Using a manifold from the inner tank, pressurize the tank cavity to 5 psi maximum. (Source: OC Tanks.)

Associated Storage Tanks and Pressure Vessels

Double containment tanks weigh significantly more than single-wall tanks, due to the outside jacket and interstitial supports (baffles). Double containment tanks must never be rolled due to the lack of structural integrity of the secondary containment shell. Instead, they should only be moved by lifting with equipment that has sufficient capacity for handling the weight of the tank assembly.

Most considerations that are important to single-wall tanks also apply in principle to the design and installation of double containment tanks. These include such design and installation issues as inspection, determination of burial depth, backfill requirements, amount and type of compaction, whether or not cathodic protection must be applied, the type and design of the cathodic protection system, and others.

14.5.1.1 Testing of Double-Wall Tanks Prior to Installation

Double-wall tanks require unique testing requirements, as compared to those of a single-wall tank (refer to Table 14.1). Considerations must be given to the integrity of both primary tanks, as well as their secondary walls. The following general procedures are recommended:

1. Inner tanks should be pressurized to a maximum of 5 psig, as when testing a normal, single-wall tank. Once a primary (inner) tank is pressurized, it may be completely sealed and its external air supply disconnected. Refer to Figure 14.24.
2. The pressure applied to a primary (inner) tank should be monitored for a period of 1 h. This will allow a very large leak or defect to be detected. Air tests generally require soaping and careful inspection for bubbles, in order for a test to be effective. However, double-wall tank design has the obvious limitation of not allowing visual inspection of the primary tank to be performed.
3. The interstice of a secondary containment tank should be pressurized using air from the primary tank. A second gauge should be used to measure pressure in the interstice. Refer to Figure 14.25.
4. The outer wall of secondary containment tanks must be soaped and subjected to 100% visual inspection, by observing for bubbles. Both gauges should be monitored throughout the test to detect any pressure drop of either the primary and secondary containment piping.
5. After testing has been successfully completed for both primary tank and its interstice, the test pressure should be released and both the primary tank and the interstice vented. The pressure in the tank interstice should be released first, so that the primary tank is not subjected to damaging external pressures.
6. An interstice should never be directly pressurized from an outside air source, unless its outer tank shell is of a pressure-rated design (i.e., to ASME Boiler and Pressure Vessel Code rules). Pressurization from an outside air source is dangerous and should be avoided.
7. An inner tank should never be entered while its interstice is under pressure, as a safety precaution. The capacity of an interstice of a typical double-wall tank is small compared to the capacity of a primary tank. The use of a compressor for testing an interstice could result in overpressurization of an outer tank wall in seconds, resulting in serious damage of the tank.

8. High-point vents should be installed in manways and at the high end of sloped tanks to vent trapped vapors. This will help facilitate tightness testing.
9. It is normal to expect a slight decrease in pressure in a primary tank when the interstice of a double wall is pressurized from the primary tank. The pressure drop to be expected is of the magnitude 0.3 psig or less. Double-wall tank manufacturers produce tanks that have unique interstitial volume specifications, by virtue of different designs. Interstitial volumes therefore vary according to tank brand.

14.5.2 Common Secondary Containment Outer Shell for Multiple Primary Tanks

Figure 14.27 and 14.28 illustrate an arrangement whereby one outer tank shell is designed to house more than one primary tank compartment. The tank shown is typical of a petroleum marketing station installation, and is part of the typical overall system shown in Figure 11.102. Each of the different compartments is separated as shown in the detail illustration, Figure 14.30. The interstitial space of the common outer shell may be monitored by means of a dip-tube leak-monitoring riser. Such a tank arrangement may provide secondary containment for several different services (i.e., leaded gasoline, unleaded gasoline, diesel oil, gasohol, etc.). Further details for the associated turbine sump enclosures and optional access manway are provided in Figures 14.26 and 14.29 through 14.31. The legend for the parts in Figure 11.102 and Figures 14.26 through 14.31 is shown in Table 14.2; notes to the drawings are shown in Table 14.3.

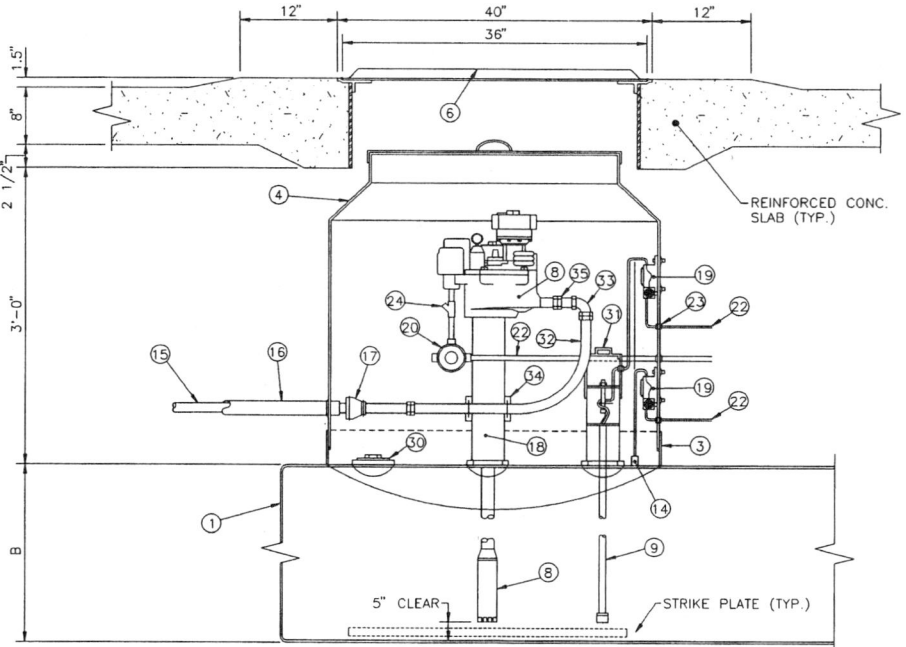

Figure 14.26 Illustration of a containment sump which serves as a turbine enclosure. Refer to Tables 14.2 and 14.3 for the legend and notes pertaining to this illustration. (Source: Joor Manufacturing.)

Associated Storage Tanks and Pressure Vessels

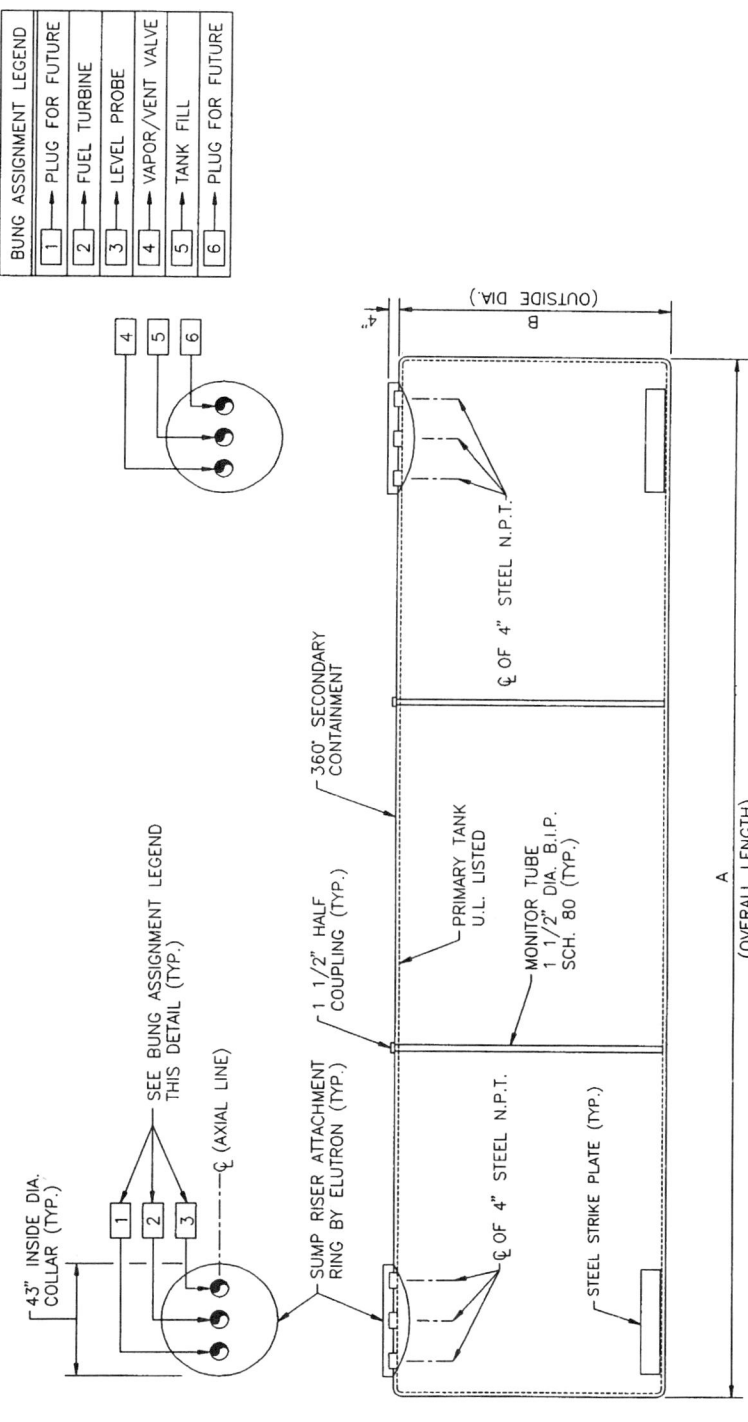

Figure 14.27. Illustration of a multiple-primary compartment steel tank with a common secondary containment shell. For the legend and notes pertaining to this figure, see Tables 14.2 and 14.3. (Source: Joor Manufacturing.)

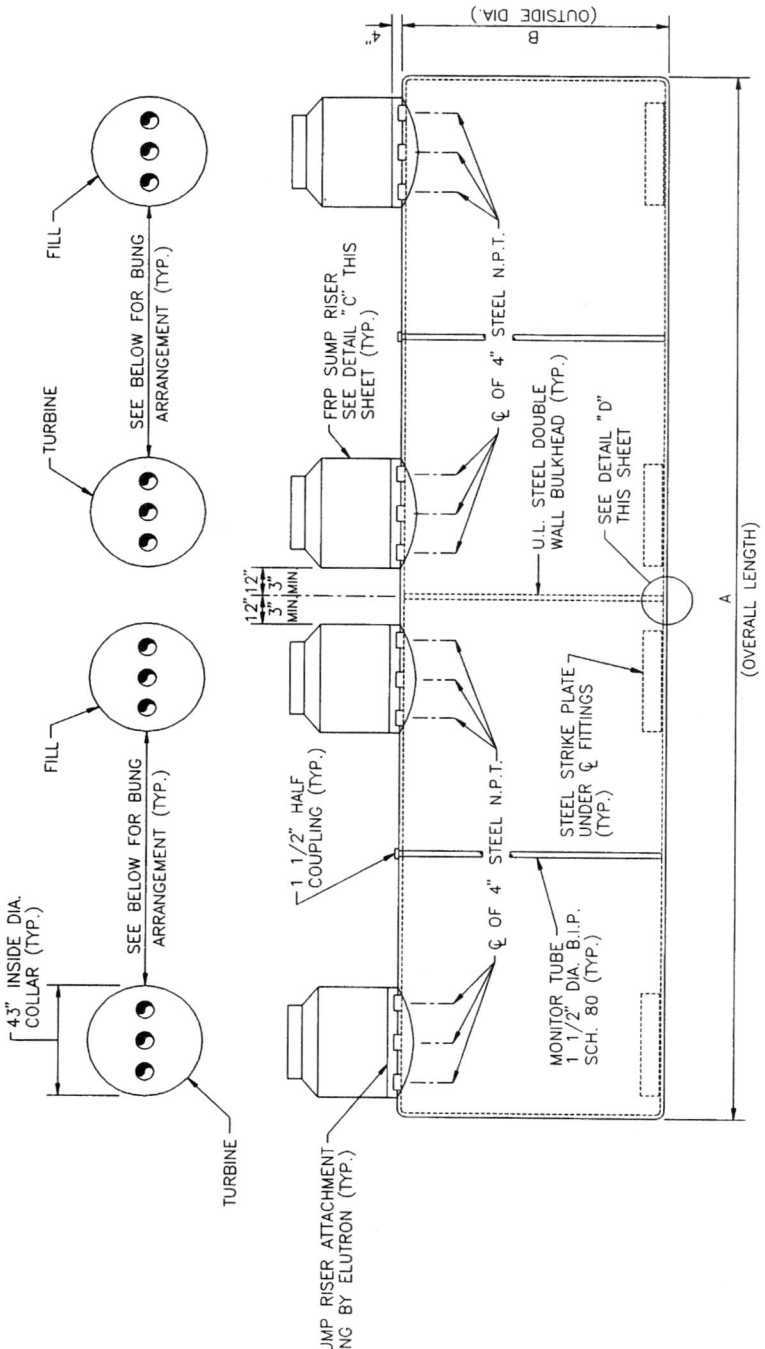

Figure 14.28. Detail of the multiple primary compartment steel tank. Note the use of a 1-1/2 in. monitor tube that provides for leak detection by a variety of means. For legend and notes pertaining to this figure, see Tables 14.2 and 14.3. (Source: Joor Manufacturing.)

Associated Storage Tanks and Pressure Vessels

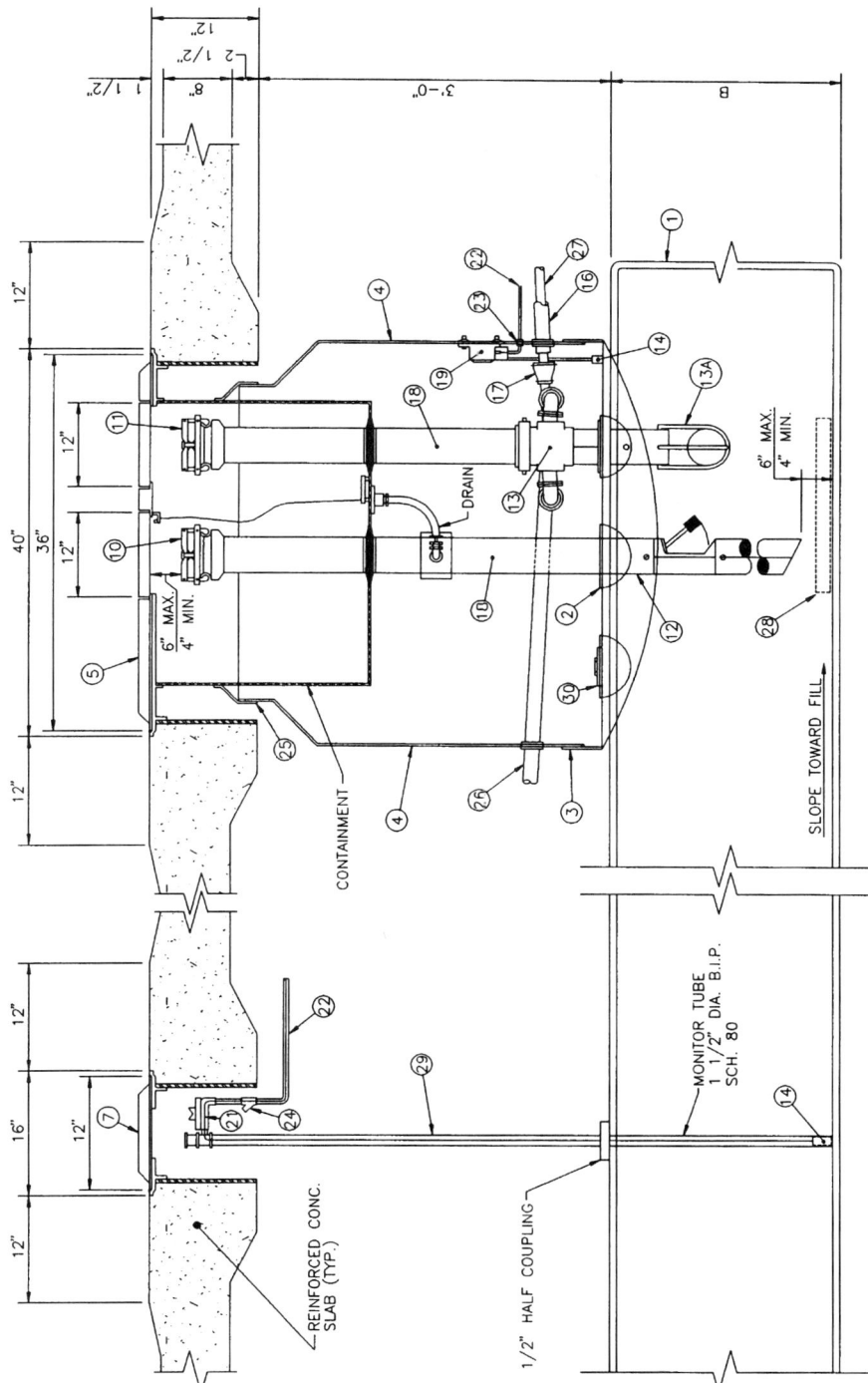

Figure 14.29. Details of the no-spill sump and monitor system. For the legend and notes pertaining to this figure, see Tables 14.2 and 14.3. (Source: Joor

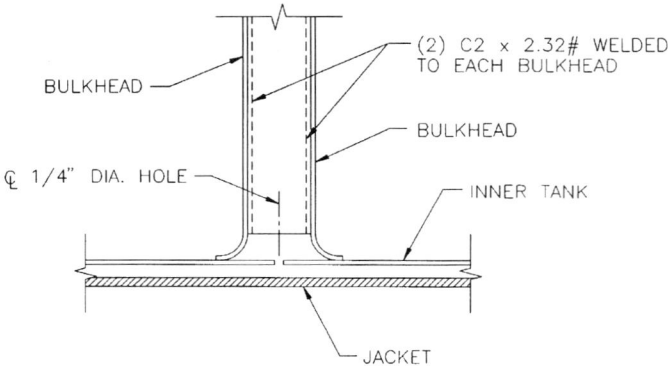

Figure 14.30. Fabrication details for the multiple-primary compartment double-wall tank. (Source: Joor Manufacturing.)

14.5.3 Impervious Geotextile Membrane Liners

There are two types of geotextiles: woven and nonwoven. The purpose of using an impervious geotextile material is to line excavations and pipe trenches in order to contain releases from single-wall or double containment tanks and piping. They also facilitate detection, but recovery of released fluids may require treatment and disposal of a finite amount of contaminated backfill as hazardous materials. Liners tend to stabilize surrounding soil and prevent migration of backfill materials. Installation techniques for liners vary widely depending on the manufactured product. Therefore, manufacturer's instructions should be followed carefully, and carried out by personnel trained and certified in accordance with standards set by the individual manufacturers.

Training and certification for field-fabrication, hot-gas-welding, extrusion-welding, and adhesive-bonding procedures should be required for all persons who will perform these procedures on a given project. A welding/bonding procedure specification must be drafted to maintain minimum standards. Training should address joining methods under all actual

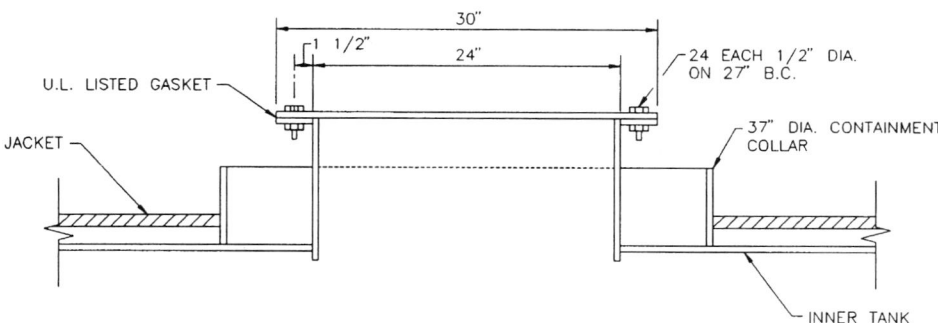

Figure 14.31. Details of an optional manway opening for the multiple primary compartment steel tank. (Source: Joor Manufacturing.)

Associated Storage Tanks and Pressure Vessels

Table 14.2 Legend for Figures 14.26 through 14.31 and Figure 11.102

I.D. Number		Equipment
1.	Elutron Storage Tank	Elutron Jacketed Tank
2.	NPT Tank Fittings	Elutron (6) 4" and (2) 1 1/2"
3.	Containment Collar	Elutron 13" to 43"
4.	Sump Riser System	Elutron – Match Collar I.D.
5.	36" Fill Manway	CNI
6.	36" Turbine Manway	CNI
7.	Monitor Access	CNI
8.	Submersible Turbine	Red Jacket – 1.5 HP, 3/4 HP, 1/2 HP
9.	Tank Level Probe	EMCO Wheaton – Std.; EMCO Wheaton – Stls.
10.	Fill Adapter & Cap	EMCO Wheaton
11.	Vapor Adapter & Cap	EMCO Wheaton
12.	Alum. Drop Tube	OPW
13.	Extractor Vent Valve	Universal
13A.	Ball Check Valve	
14.	Liquid Leak Sensor	EMCO Wheaton
15.	2" or 3" Primary	UL Red Thread 2"; CIBA Dualoy 3"
16.	Secondary Pipe	UL Red Thread 2"; CIBA Dualoy 3"
17.	Test Clamp Fitting	
18.	4" Riser Pipe	American Made Sch. 40; B.P. or Stainless Stl.
19.	Class-Div1 J-Box	OZ/ Gedney at Sensors & Level Space
20.	Class-Div1 J-Box	OZ – Seal Unused Openings
21.	Class1-Div1 J-Box	OZ – 6" Max. Below Manway Lid
22.	Conduit to Panels	PVC Coated Rigid, Install Conductors per '90 NEC
23.	Conduit Expansion	OZ (Optional)
24.	Conduit Seals	OZ – Install per '90 NEC
25.	Sheet Metal Trim	Field Fabricate & Attach to Sump Riser
26.	2" Tank Vent Line	2" Fiberglass Reinforced Plastic Pipe
27.	2" Vapor Return	2" Fiberglass Reinforced Plastic Pipe – Dbl. Cnthd.
28.	Strike Plate	Steel Plate
29.	Leak Monitor Riser	2" Sch. 40 Steel Pipe Riser w/ 2" x 2" x 3/4" T & Plug
30.	4" Plug	Plug Center NPT Fitting for Future
31.	Probe Cap	Provide 4" Cap if Probe is Not Installed
32.	Flex Connector	UL. Teleflex; UL Flexever
33.	2" – 9° Elbow	Sch. 40 Galvanized Steel
34.	Pipe Brace	W/Bolts, Nuts, Stl. Plate, Vibration Isolators
35.	Supply Exit	Sch. 40 Steel Pipe with 2" Steel Union
Equipment at Dispenser Island		
36.	Fuel Dispensers	Wayne
37.	Vapor Valves	EMCO
38.	Shear Valve	EMCO
39.	Spill Containment	Bravo Containment Box; Model to Match Dispenser Base
40.	Nozzles	EMCO
41.	Fuel Hoses	Goodyear Maxim; Coaxial
42.	Overhead Retractor	CNI
43.	Junction Box.	OZ – Explosion Proof
Equipment at Control Building		
44.	Leak Alarm Panel	EMCO Wheaton
45.	Inventory Panel	EMCO Wheaton
46.	Voltage Regulator	SOLA
47.	Elec. Wireways	Square D
48.	Motor Starter	Red Jacket
49.	Lighting Contractor	Square D
50.	Emergency Shut-Off	Square D
51.	Remote Leak Alarm	Leak Alert – Field Locate
52.	Remote Level Alarm	EMCO – Field Locate
53.	Conduit Seal	OZ
54.	Electrical Conduit	PVC Coated Steel per '90 NEC
55.	Elec. Sub Panel	Square D – Size for Load, Voltage 7 Breakers
56.	Fuel Management	Gasboy; 2 thru 8 Hose; 1 thru 4 Product
57.	Computerized Cash – Credit Dispenser	Southwest; 2 Prod. – 4 Hose > 2PD; 3 Prod.– 6 Hose > 3PD
58.	Retail Cash – Credit; 8;16;24	Southwest; ECS – 1 Console; ECS – 1 Console; ECS – 1 Console

Table 14.3 Notes for Figures 14.26 through 14.31 and Figure 11.102

1. The underground tank shall be an Underwriters Laboratories incorporated jacketed tank listed for the underground storage of all flammable and combustible motor fuels. Listing must include ability to withstand a vacuum test of 13.8 in. of Hg.

2. Underground tank installation and testing shall be per UL listed "Manufacturer's Installation Instructions," a copy of which shall accompany each delivered tank. Installation shall conform to NFPA 30 and all applicable local and state regulations.

3. Monitor systems for the interstice shall be: by a liquid sensor, by visual determination, or by manual determination, in accordance with applicable local and state regulations.

4. Backfill may be sand, pea gravel, 3/4" rounded aggregate or native soil in accordance with manufacturer's current installation instructions.

5. See drawing for number of fittings, configuration, accessories, and sump riser details.

6. The design, assembly, and testing of the piping system shall be in conformance with the applicable section of ANSI-B31. American National Standard Code for Pressure Piping and NFPA 30. Flammable and Combustible Liquids Code.

7. All associated piping shall be protected from corrosion.

8. Tank slabs shall extend a minimum of 2' beyond all edges of the tank.

9. Vents for tank interstice are optional. If provided they must:
   A. Be installed with a minimum 1/4" slope to 1' of run back to tank.
   B. Discharge to a point not subject to possible liquid or vapor intrusion.
   C. Interstitial space tank vents may be manifolded.

10. Standard fitting arrangement shall be on 12" centers on top center line. Other fitting arrangements available.

---

field conditions including wet, dusty, cold, and/or windy conditions. Installation personnel should be required to provide proof of adequate training and have documents available at each job site for an inspector's review. A permanent record of the training of each operator should be maintained by the manufacturer or persons responsible to supervise such training.

Liners must be pitched to sumps, in order to allow gravity collection of released liquids. Methods to test the effectiveness of containment, concentration, and any detection systems must be considered, and a testing procedure drafted. A liner system should not inhibit operation of associated cathodic protection systems or other aspects of tank/piping systems. Impervious liners and barriers are discussed further in Section 16.

14.5.4 Tank Hole Liners

Tank hole liners can be used to line an entire tank excavation area, or, alternatively, they can be used to line an area above tanks. In either case, geotextile fabric (or thermoplastic sheet material) requires protection from damage due to puncture or abrasion caused by placement of tanks. After installation, they must continue to be protected from puncture due to future work, from placement of observation wells, piping, and other equipment, etc. Puncture-resistant filter fabric may also be used during installation, in conjunction with a liner, to protect it from the edges of paving, rocks, and other sharp or hard objects. If penetrations for piping and conduit are needed, or overliners are used, liner system interfaces must be provided with liquid-tight penetrations and seals. This may be accomplished by welding or adhesive bonding (depends on material), or by the use of a mechanical seal, applied at the point of penetration. Examples of tank hole liners for use in providing secondary containment for tanks are presented in Section 16.2.1.

14.5.5 Trench Liners

Trench liners may be used to enclose the areas around piping that is connected to its associated tanks, around tank drains, and beneath dispensers. They are applied in such areas to contain any released liquid and to prevent liquid from infiltration. Trench liners may be used in combination with any method of tank area containment. Liner components are often mechanically fastened together and can be mechanically fastened to flanges that are connected to tanks and associated equipment. Trench liners, like membrane liners, should be used in conjunction with sumps and sloped to the sump so that release fluids can be more readily collected. Tank liners are discussed further in Section 16.2.1.

14.5.6 Overfill Protection

Accidental releases resulting from overfilling tanks cannot be distinguished from leak releases from a tank or associated piping. Devices to prevent a release from accidental overfilling of a tank may be incorporated into an underground tank system. Fill pipe enclosures capture small quantities released when delivery hoses are uncoupled. Vent float valves slow deliveries by restricting venting when a tank's maximum desirable capacity is reached. A high-level sensor, frequently incorporated into a gauging system, may sound an alarm or interrupt delivery when the liquid level reaches a predetermined upper level. Each system has unique installation requirements that should be carefully followed to ensure protection.

Protection against overfills may also be accomplished with equipment that may be installed on delivery vehicles. Devices may interrupt delivery when a predetermined volume has been delivered, or when the liquid level in the tank reaches a predetermined level.

Operating and inventory control procedures should include measuring the level of the liquid in a tank before attempting a delivery. "Outage charts" (tank charts from which 150 to 200 gallons of capacity have been deducted) may be provided to operators for determining allowable delivery quantities.

If a restriction in venting is used to prevent overfilling, care must be exercised to prevent overpressurizing such a tank.

14.5.7 Above-Ground Secondary Containment Structures

14.5.7.1 Concrete Dikes and Berms

The most common method to provide secondary containment to an above-ground storage tank system is to place a tank inside a dike structure. Dikes have been required for many

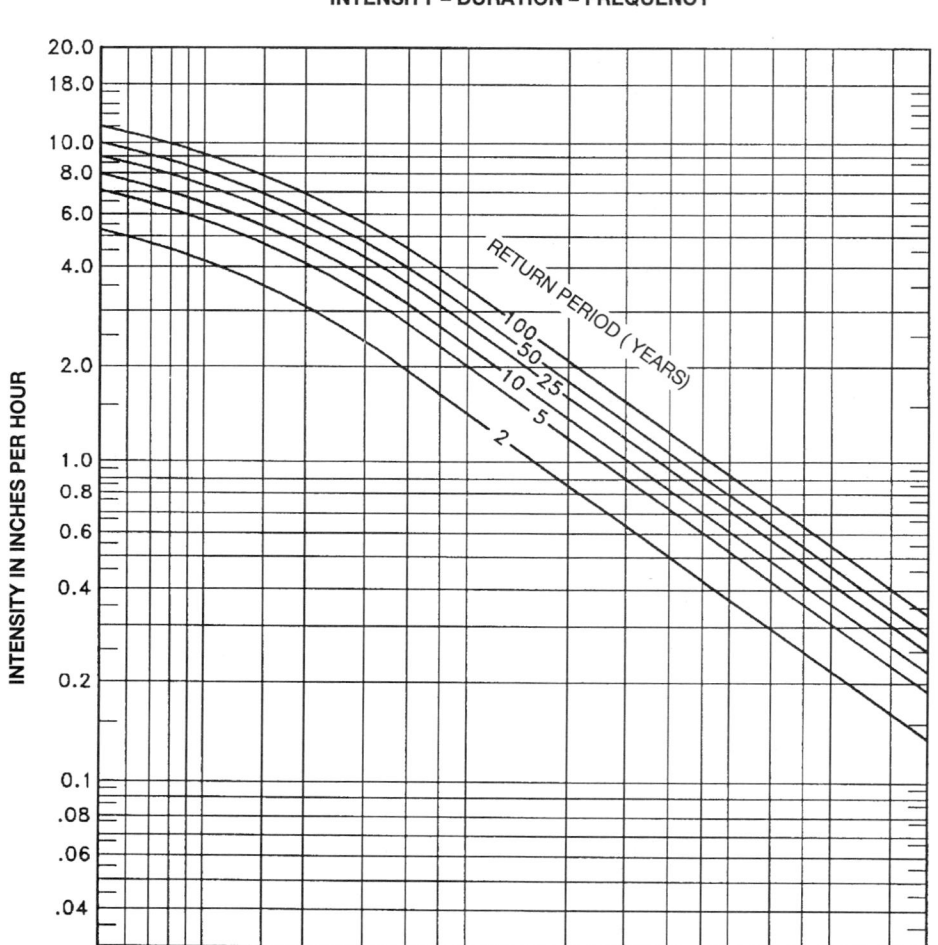

Figure 14.32. Example of a frequency nomograph for typical rainstorms. For a given application, the user must use data which is specific to the geographic area, provided by the local governmental weather bureau. It is usually up to the owner, but sometimes up to the authority having local jurisdiction or the applicable fire code as to which return period is to be used. (Source: A. Steele, "Review Manual for the CIPE Exam," American Society of Plumbing Engineers, Sherman Oaks, CA, 1984.)

Figure 14.33. Example of an above-ground storage tank protected by means of a secondary containment dike.

years in new tank installations for both environmental and fire safety reasons. Modern-day dikes and berms must be constructed of a lined or coated reinforced concrete material, or alternatively a metallic construction specifically designed for such purposes. Piping that is connected to tanks that are protected by dikes may be considered as having secondary containment if they are within the dike area and the fluid they are carrying is a liquid or nonvolatile gas.

The typical standard for dike construction is that it must be capable of handling 110% of the volume of the tank it is protecting, plus some accumulation of rain that can reasonably be expected. The usual basis consists of a 10 to 25 year rainstorm, depending on the rules of the local municipality, according to most local fire code rules, or the preference of the owner. A frequency chart for rainstorms is presented in Figure 14.32.

Figure 14.34. Example of an above-ground storage tank protected by means of a secondary containment dike.

Examples of above-ground storage tanks protected by means of secondary containment dikes are shown in the photographs in Figures 14.33 and 14.34.

Dikes must possess an integrity such that they can withstand the effects of a sudden release of fluid, typically under maximum seismic conditions. It is reasonable to anticipate that one of the most likely times for a single-wall tank to fail catastrophically is during a severe earthquake. Dikes may be constructed of the same materials as the basic tank. However, very often steel tanks are contained within a coated or lined concrete dike; sometimes the wall of the dike is constructed of steel and the floor of the dike of lined or coated concrete.

A berm is similar to a dike, except that a dike is designed to protect a single tank, whereas a berm is designed to house fluid spills from more than one tank and possibly other associated process equipment. Berms must be designed to handle 110% of the maximum liquid that may be spilled by the units that the berm is protecting and the amount of rain produced by the most severe rainstorm to be expected.

14.5.7.2 Double-Wall Above-Ground Tanks

The use of double-wall above-ground tanks has not been as popular as in underground service to date. The main reason has to do with the comparative expense of using a double-wall tank versus the construction of a concrete dike. Also, many double-wall tanks present special structural concerns in terms of providing a supporting arrangement that does not damage the outer shell. This is particularly true of vertical tanks. Most vertically positioned above-ground double-wall tanks are of Type-I design (Underwriters Laboratories). For horizontally positioned tanks, both Type-I and -II double-wall tanks can be readily supported. In burial applications, it is much easier to achieve a uniform distribution of structural loads by use of proper burial and bedding techniques.

In above-ground applications, ordinary means of providing support to a vertical Type-II tank are not adequate, since the outer shell may become damaged. A specialized supporting arrangement has to be created that provides uniform support over the entire tank bottom. Special attention also must be paid to the design of interstitial standoffs/spacers such that they will not result in structural harm on the primary tank.

14.5.7.3 Double-Bottom Tanks

Another common way of providing that above-ground tanks with some means of secondary containment is by the use of a double-bottom tank. A tank of this type usually has two bottoms that house a normally empty space (annulus) in between. The space is normally also equipped with some means of monitoring for leaks. Most metallic tanks of this type have both their outer and inside bottom shell provided with cathodic protection, although considerable debate exists as to whether this is necessary. Since there is no spill protection from a failure of the tank's sides, a vessel of this type must be housed inside a secondary containment dike, if secondary containment of the service is mandatory.

14.5.8 Safety Consideration of Released Liquids

Hazardous or petroleum-based volatile liquids may escape from unsealed manhole access covers of tanks or their containment sumps in the form of vapors. High concentrations of flammable or combustible liquids and their vapors may collect in these areas, presenting a severe hazard to employees and others in the immediate project area. Therefore, monitoring equipment must be put in place to monitor concentrations of vapors continuously, if they are flammable or otherwise hazardous. Appropriate safety standards should be completely followed, including federal rules [in the United States, the Occupational, Safety and Health Administration (OSHA)] and all rules and regulations of authorities having local jurisdiction.

14.6 Pressure Vessels

A pressure vessel is a container that is of limited length and has closed ends that enable it to operate in a closed, pressure-tight manner. A pressure vessel is distinguishable from piping in that, by comparison, piping is usually of indefinite length. The largest dimension of associated piping is usually considerably smaller than the smallest dimension of a vessel.

Pressure vessels are usually designed to handle pressures above 15 psi (kPa) or designed to contain a vacuum, which distinguishes them from storage tanks. However, vessels can be designed to handle internal pressures as low as 1 to 2 psi, and still be classified as a pressure vessel. A boiler is different from a pressure vessel in its intended purpose, which in most cases involves the generation of steam for use external to itself.

14.6.1 Use of the ASME Boiler and Pressure Vessel Code

The ASME Boiler and Pressure Vessel Code contains rules for the design, fabrication, and inspection of boilers and pressure vessels. The ASME Boiler and Pressure Vessel Code is prepared by the American Society of Mechanical Engineers (345 E. 47th Street, New York, NY 10017). The ASME BPV Code has been adopted as an American National Standard and is law in most of the United States, and in each of the Provinces of Canada. It consists of eleven separate sections, which are listed in Table 14.4.

Every 3 years, a new code is issued, which incorporates alterations and revisions from the semiannual addenda and contains alterations and revisions that have been voted on. These alterations and revisions to the code are published in semiannual addenda, which may be purchased by subscription. The sections of the ASME BPV Code that pertain to projects that have fluid services of the type where a double containment piping system is likely to be used include: Sections II, III, V, VIII, IX, X, and XI.

Table 14.4 Sections of the ASME BPV Code

I.	Power boilers
II.	Material specifications (three parts)
III.	Nuclear power plant components
IV.	Heating boilers
V.	Nondestructive examination
VI.	Recommended rules for care and operation of heating boilers
VII.	Recommended rules for care of power boilers
VIII.	Pressure vessels, Division 1
	Pressure vessels, Division 2, alternative rules
IX.	Welding and brazing qualifications
X.	Fiberglass-reinforced plastic pressure vessels
XI.	Rules for inservice inspection of nuclear power plant components

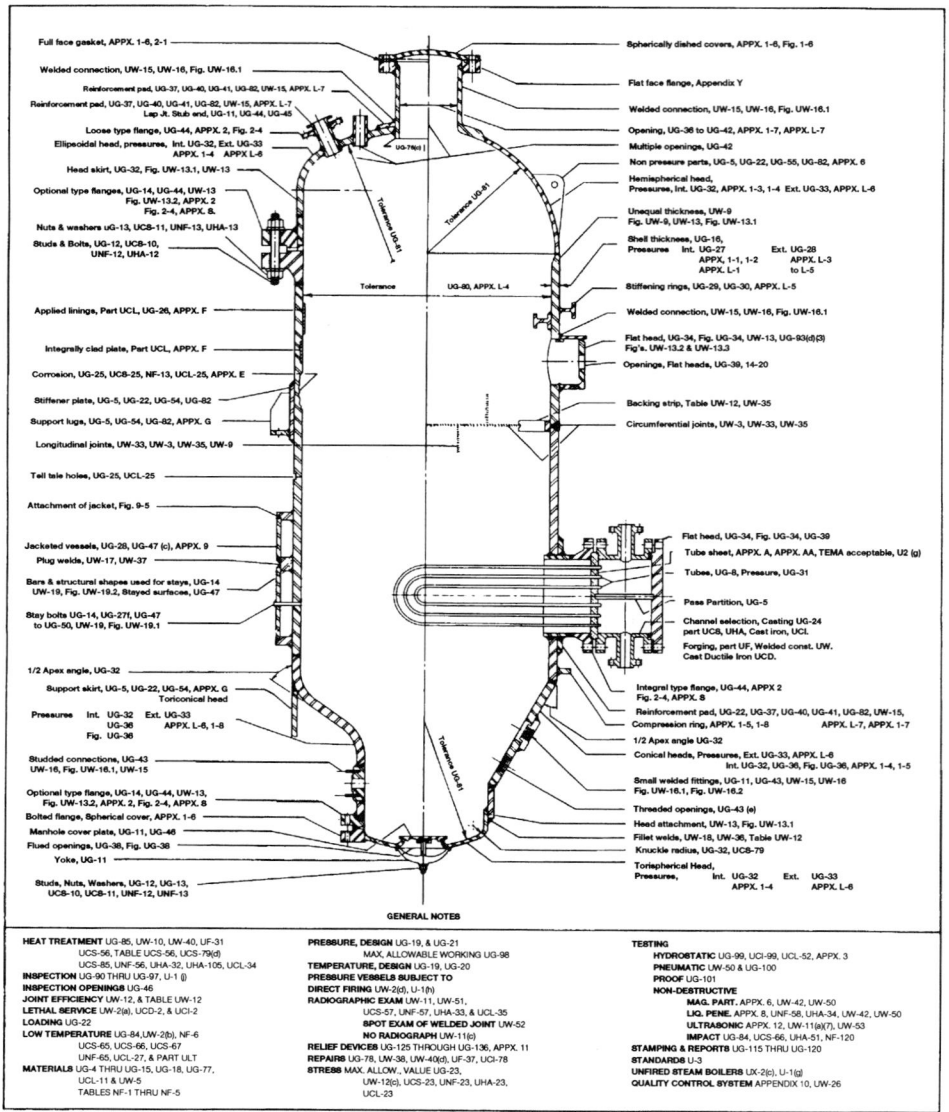

Figure 14.35. ASME Code Sec. VIII, Div. 1 applicable paragraphs for design and construction details based on the 1983 Code. (Source: Missouri Boiler and Tank Company, division of Nooter Corporation.)

14.6.2 Overview of ASME Code Section VIII, Division 1

Section VIII, Division 1, covers vessels that operate with an internal pressure between 15 psi (103 kPa) and 3,000 psi (20,670 kPa), and those that contain a vacuum. Most process pressure vessels are designed and constructed in accordance with Section VIII, Division 1, for applications in the United States and Canada. If higher pressures are to be encountered, it is usually necessary to use rules outside the scope of this Division. A vessel that is designed, fabricated, and tested in accordance with all the rules of Division 1 may be stamped with the code U symbol.

Division 1 of Section VIII contains three subsections (A, B, and C) and a series of mandatory and nonmandatory appendixes. The highlights of each of its major sections are described in Sections 14.6.2.1 through 14.6.2.5. Figure 14.35 illustrates an example of a pressure vessel. In the figure, Section VIII, Division 1, code paragraphs that apply to individual elements of the vessel are noted. Other code paragraphs pertaining to the example vessel are noted at the bottom.

14.6.2.1 Section VIII, Div. 1, Subsection A

Subsection A covers general requirements concerning materials and methods of construction, provides definitions for design temperature and pressure, and specifies the various loadings that must be considered in the vessel design. Subsection A uses the maximum-stress theory of failure as its basis for stress failure and yielding.

Appendix P defines the basis for the establishment of allowable stresses. However, depending on the material used, this subsection states that the safety factors used for internal-pressure loads are either 1.6 or 1.5 on yield strength. Based on ultimate strength, the safety factors for internal loads is 4. For cylindrical shells that are subjected to external-pressure loads, they must be designed using a safety factor of 3 for both elastic buckling and plastic collapse. In terms of external forces acting on vessels of other shapes, a safety factor of 4 must be used based on either elastic buckling or plastic collapse. The same is true for longitudinal shell compression. Cylindrical elements that are subjected to longitudinal compressive stress must be analyzed for both stress failure or buckling failure, and the lower value used as the limiting basis for design. Tables elsewhere in Division 1 in which the maximum allowable tensile-stress values are tabulated are referred to in this subsection.

Various design rules and formulas for internal-pressure design of cylindrical and spherical shells and for ellipsoidal, hemispherical, conical, and ASME (torispherical) heads, based on membrane-stress failure, are presented in subsection A. Rules stated for heads include consideration for buckling failure in the knuckle area (the area of transition from cylinder to head). Also, the minimum allowable thickness is usually specified in the knuckle area since there is usually some thinning from the original plate thickness. The code also takes into account the fact that longitudinal joints in cylinders are more highly stressed than circumferential joints.

Subsection A provides rules to determine design based on external pressures imposed on most shells through the use of different charts. These charts allow determination of the allowable stress for cylinders, spheres, and hemispherical, ellipsoidal, conical, and ASME (torispherical) heads.

Rules and formulas are also presented in this subsection for the design of unstayed flat heads and covers. These formulas have an allowance for additional edge moments imposed on heads, covers, or blind flanges when they are attached by bolts. Since there is always a risk of incomplete attachment or opening of a closure while a vessel is under pressure, rules for designing quick-opening closures are provided.

Rules for designing braced and stayed surfaces are also stated in this subsection. Rules for designing and locating stiffeners allow a reduction in the effective length-to-diameter ratio. By reducing the effective L to D ratio, a reduction in shell thickness may be accomplished, which translates into a cost savings for a completed cylindrical shell.

Pressure vessels always require openings of various sorts in their shells and/or their heads. Stress intensification is created by placing a hole in what is otherwise a symmetrical section. To compensate for the stress intensification, an area-replacement method is required in this subsection. A cross section must be taken through each opening, whereby the area of the metal of the required shell that is removed is measured. It is then replaced in the cross section by added material. The material may be added in the form of shell wall, nozzle wall, reinforcing plate, or weld, and must be placed within a certain distance of the centerline of the opening. The intensification of stresses is thereby effectively compensated.

Rules for tube-hole patterns and their associated strength reductions for the cylindrical shell are presented here. These rules must be used if a cylindrical shell must be drilled for multiple tube insertion as part of the design of the vessel.

Tolerances, inspection, and testing requirements are covered in detail. Also discussed are repair requirements for defects identified during fabrication, material identification, heat treatment, and impact testing.

The standard test that is applied to pressure vessels is a hydrostatic test, at a pressure equal to 1-1/2 times the maximum allowable working pressure, with water the most common medium. In unusual instances, a vessel is permitted under Subsection A to be hydrostatically tested at pressures lower than 1-1/2 times its maximum allowable working pressure. This is true if the vessel has a fragile internal lining (i.e., glass).

Pneumatic tests are also permitted to be carried to at least 1-1/4 times the maximum allowable working pressure, in lieu of a hydrostatic test. There is also an allowance for proof testing if the strength of the vessel or any of its parts cannot be computed with satisfactory assurance of accuracy.

All of the details related to the safe relief of overpressurization are provided in Subsection A. Various factors that are defined include: (1) set point; (2) maximum pressure during relief; (3) the cause of overpressure; and (4) the number of relief devices required. The rules for pressure relief devices stated here apply to pressure piping. They are referenced in the ASME B 31.3 Code for Chemical Plant and Petroleum Refinery Piping and in Chapter 5 of this text (see Section 5.6.)

14.6.2.2 Section VIII, Division 1, Subsection B

Subsection B covers rules concerning the fabrication of pressure vessels, including welded vessels (Part UW). Subsection B defines all services restrictions and further defines those services that must be classified as "lethal service." Lethal services include those poisonous/toxic gases or liquids that upon a very small release of the gas or vapor of a liquid (mixed or unmixed with air) pose a threat to life when inhaled. The responsibility for advising the designer or manufacturer that a service is lethal belongs to each owner/user.

Temperature restrictions are also defined, including "low-temperature" service. It is defined as operating temperatures below -20°F (-29°C) whereby impact testing of many materials is a requirement, and the type of welds to be used are restricted. In Subsection B, efficiencies for welded joints are defined, and vary for joints of varying quality. The efficiencies for welded-joint types are used elsewhere in the code for determining vessel thicknesses (i.e., in Subsection A design formulas).

Subsection B provides rules for the designs of welded joints and specifies the required extent of radiography for each. Types of welds that are covered include: (1) head-to-shell welds; (2) tube-sheet-to-shell welds; and, (3) nozzle-to-shell welds. For staying plates, the subsection also covers allowable welded-stay-bolt arrangements and plug-and-slot welds.

The fabrication items that are covered in Subsection B include: (1) welding processes; (2) record-keeping requirements for welding procedures (by manufacturer); (3) qualification of welders; (4) cleaning of weld areas; (5) alignment tolerances; (6) repair requirements for weld defects; (7) detailed requirements for postweld heat treatments; and (8) verification of procedures, welders, and nondestructive examinations of welded joints.

Some vessels are fabricated using forging techniques; rules for these techniques are covered in Part UF. Attention must be paid to stress risers, fabrication, heat treatment, repair of defects, and inspection. Part UB covers vessels that are fabricated from brazing.

All vessels fabricated of carbon or low-alloy steel must be treated by postweld heat treatment. Vessels that are fabricated for lethal service should have all butt-welded joints, which must be subjected to 100% radiography. Direct-fired pressure vessels have specific requirements relative to types of welded joints and postweld heat treatment.

14.6.2.3 Section VIII, Division 1, Subsection C

Subsection C includes allowable stresses and safety factors over a range of temperatures for all materials covered by Section VIII, Division 1. Classes of materials/vessel construction are separated into different sections, including: (1) cast iron (Part UCI); (2) ductile iron (Part UCD); (3) carbon and low-alloy steels (Part UCS); (4) nonferrous materials (Part UNF); (5) high-alloy steels (Part UHA); (6) steels with heat-treatment-enhanced tensile properties (Part UHT); (7) those using clad-plate materials (Part UCL); and (8) vessels fabricated by layered construction (Part ULW). For vessels from one of these classes of materials or construction methods, rules are stated in the specific Subsection C part references concerning their application, fabrication, and heat treatment.

A vessel constructed by means of layering is usually intended for use where the service pressure is in excess of 2,000 psi (13,800 kPa). There are three major methods of layering that are used to construct a Part ULW-type vessel: (1) thick layers that are wrapped and then shrunk to each other; (2) thin layers, where one is wrapped over the other and longitudinally seam welded, with the prior layer used as a backup; and (3) thin layers that are spirally wrapped. The code rules are written for use of both thick or thin layers. Rules and details are provided in Part ULW for nozzle reinforcement and welded-joint design.

In any layered vessel, only the outer layer of the vessel can be considered to contribute to its support. Therefore, any supporting arrangement that is designed for the vessel must take this into account. Also, to prevent the buildup of pressure between layers in the event that a leak develops in the inner shell, vent holes are typically added through each layer, all the way to the inner shell to prevent the buildup of pressure. For this and other reasons, radiography is not practical for examination of these vessels; however, magnetic particle and ultrasonic inspection may be readily applied. If radiography is used, the results must be given careful interpretation. What would appear to be an indication sufficient to reject a single-layer vessel may be a false reading for an acceptable layered vessel.

14.6.2.4 Section VIII, Division 1, Mandatory Appendixes

There are several different appendixes included among the "mandatory" heading. These include coverage of: (1) supplementary design formulas for those shells and heads that are not covered in Subsection A (i.e., for thick shells, heads, and dished covers); (2) specific rules, formulas, and charts for bolted-flange design, which is referenced in the ASME B 31.3 Code for Chemical Plant & Petroleum Refinery Piping, and in Chapter 5 of this text (see Section 5.2.1.6); (3) determination of external pressure of shells by the use of charts; (4) an Appendix that specifically covers the design of jacketed vessels; and (5) other Appendixes that deal specifically with issues pertaining to inspection and quality control.

14.6.2.5 Section VIII, Division 1, Nonmandatory Appendixes

There are several nonmandatory appendixes, covering a variety of subjects. These suggest recommended practices and offer useful suggestions as to understanding the code and designing in accordance with the code.

14.6.3 Overview of ASME Code Section VIII, Division 2

The rules of Section VIII, Division 2, permit higher design stress intensity values to be used in the range of temperatures over which the design stress intensity value is controlled by the ultimate strength or the yield strength. However, it is more restrictive in the choice of materials that may be used, and more precise design procedures are required. In addition, certain common design details are prohibited, fabrication procedures permitted in Division 1 are further delineated, and additional testing and inspection requirements are imposed. Division 2 rules are often applied to large vessels or those that are intended for high pressure. It is usually an economic tradeoff between material and labor savings resulting from smaller safety factors provided under Division 2, versus the additional engineering, administrative, and inspection costs that Division 2 imposes.

Division 2 has a different organization than that of Division 1. Its major parts consist of: (1) Part A; (2) Part AM; and (3) Part AD. The highlights of each of its major sections are described in Section 14.6.3.1 through 14.6.3.3.

14.6.3.1 Section VIII, Division 2, Part A

Part A describes the scope of Division 2 and describes the responsibilities of all parties involved in a Section VIII, Division 2, vessel. Division 2 does not specify any upper limitation in pressure; it is the user's design specification that is required. In Division 2, a lot of emphasis is placed on users or their agents. They must provide enough information to form a suitable basis to enable material selection, and to design, fabricate, and inspect a vessel in accordance with code rules. The design specifications of users must also take into account how a vessel is to be supported. Requirements for a fatigue analysis must also be described, and if it is required, information must be provided in sufficient detail to include cyclic conditions as part of the analysis.

14.6.3.2 Section VIII, Division 2, Part AM

Part AM includes materials that are permitted to be used. This information includes design stress-intensity values, their applicable specifications, miscellaneous property data, and any special requirements. Toughness characteristics of materials are presented, along with ultrasonic-testing requirements. Design stress-intensity values are based on a safety factor of 3 on the ultimate strength at metal temperature, or 1.5 on yield strength at metal temperature.

14.6.3.3 Section VIII, Division 2, Part AD

Part AD is the subsection that describes the design of Division 2 vessels, based on the maximum-shear theory of failure for stress failure and yielding. Under Part AD, higher stresses are permitted if seismic or wind loads are involved. Rules are presented for determining which applications require fatigue analysis to be performed.

Division 1 and Division 2 rules differ in their approach to the design of shells of revolution that are to contain a fluid under internal pressure. This includes rules to be used for designing formed heads, if the basis for failure is plastic deformation in a vessel's knuckle area. Using the same basis, shells of revolution that are to be subjected to external pressure may be designed, using safety factors given in Division 1. The area-replacement method, as described in Division 1, may be used to design reinforcement requirements for openings. Actual placement of reinforcement metal must be closer to the opening centerline in a Division 2 vessel, as compared to that designed by application of Division 1 rules.

Division 1 rules for the design of flat heads, bolted and studded connections, quick-acting closures, and layered vessels are the same as those used for vessels designed under Division 2. The approach to support skirt design differs slightly in Division 2.

14.6.4 Overview of ASME Code Section X: Fiberglass-Reinforced-Plastic Pressure Vessels

Section X covers the design of four types of fiberglass-reinforced-plastic pressure vessels. These include: (1) bag molded (up to 150 psi; 1,000 kPa); (2) centrifugally cast (up to 150 psi; 1,000 kPa); (3) filament-wound with cut filaments (1,500 psi; 10,000 kPa); and (4) filament wound with uncut filaments (3,000 psi; 21,000 kPa). Operating temperatures for each of these four types are limited to -65°F (-54°C) to +150°F (66°C).

There are many differences between the properties of fiberglass-reinforced-plastic materials and metallic materials used in the design of vessels. For instance, the plastic materials considered in Section X typically have much lower moduli of elasticity than those of a typical metallic material listed in Section XIII, Division 1 and 2. As a result, the procedures and rules for design of plastic vessels are different from those for code sections that describe metallic vessel design and construction. The properties of a plastic material are such that it often makes the accurate prediction of vessel performance a difficult task. Therefore, Section X requirements used to state that at least one vessel of a particular design and fabrication must be tested to destruction. (Now this is allowed as an option; a vessel may also be qualified by analysis.) When conducting a destructive test of this type, results from the combined fatigue and burst test must allow a safety factor of 6 (the burst pressure must be at least six times the design pressure of the vessel). This requirement previously steered users away from implementing the section rules for FRP pressure vessel design and construction.

14.6.5 Overview of ASME Code Section III: Nuclear Power Plant Components

Section III of the Code includes information and rules for the design of vessels, storage tanks, and components for use in nuclear applications. Also covered in Section III are rules for the design of concrete containment vessels and items other than vessels.

Concrete pressure vessels and tanks are used for large-volume nuclear plant requirements. Post-tensioned (prestressed) concrete is usually used to field fabricate vessels and tanks of this type. The post-tensioning process involves the placement of reinforcing steel bars in tubes or plastic covers, which are prepositioned to have concrete cast around them. After the concrete has acquired most of its strength, tension is applied to stress the steel.

Concrete is often used to construct large-diameter nuclear containment vessels, as large as 50 ft (15 m) in diameter and length. These vessels are normally designed with inner linings of steel, which function as the primary means of retaining pressure. Tubes for housing high-strength steel reinforcing cables or wires are put in place, and concrete is poured, after the metallic primary liner has been fabricated. Thousands of reinforcing tendons are often used. The result is a vessel that has a high degree of safety factor designed into the construction.

14.6.6 Vessels of Unusual Construction

Vessels that are intended for high pressures have unique design concerns and present additional challenges to a designer. ASME Code Section VIII, Division 1, covers vessels that are designed up to 3,000 psi (20,670 kPa). ASME Code Section VIII, Division 2, does not place an upper limitation on the pressure rating of a vessel. At pressures beyond 3,000 psi (20,670 kPa), the design of a vessel may be outside the normal coverage of the Code. The most commonly used designs consist of: (1) a vessel made from high-strength steel plate; (2) vessels constructed from a solid forging; and (3) vessels made from multilayer construction (discussed in Section 14.6.2.3). At high pressures, large-diameter vessels designed from ordinary low-carbon-steel plate would become excessively thick to fabricate, or would become extraordinarily expensive.

14.6.6.1 Vessels for High-Pressure Use Constructed of High-Strength Steels

High-strength steels [tensile strengths over 200,000 psi (1380 MPa)] may be used to construct vessels for high-pressure use. They are useful where overall weight is a concern; however, they are brittle over most temperature ranges. Vessels of this type must be designed, fabricated, and operated to prevent brittle fracture. Careful attention must be paid to shape and design details, and careful welding procedures must be followed. Nondestructive examination in these vessels gains added importance, as the size of flaws must be carefully controlled.

14.6.7 Miscellaneous Pressure Vessel Issues

14.6.7.1 Vessel Load Calculations Not Provided in ASME BPV Codes

The ASME Boiler and Pressure Vessel Code, in each section, instructs designers to consider loads due to: (1) impact; (2) vessel weight under all anticipated conditions; (3) loads from other equipment and piping that are superimposed on a vessel; (4) seismic

Associated Storage Tanks and Pressure Vessels 675

and wind loads; (5) stresses due to differential temperatures; and (6) localized stresses that arise from internal and external supports. Like the ASME B31 Piping Code, the Boiler and Pressure Vessel Code does not actually present specific methods for determining these loads, or the stresses that result. Representative values are also not given. It is up to the designer to solve these problems by application of the principles of applied mechanics and strength of materials. The designer must be prepared to demonstrate that the method that he or she has used is accurate for the application. There are many good references on the subject of vessel design that do present formulas and methods to solve for these types of loads.[3-5] The Pressure Vessel Research Committee (PVRC) of the Welding Research Council (WRC) in New York City also often publishes important papers on these and other topics.

14.6.7.2 Special Hazards Affecting Pressure Vessels

The ASME Boiler and Pressure Vessel Code presents a set of conservative rules for the design of pressure vessels. However, it does not present the designer with rules for every conceivable set of conditions to which a vessel might be subjected. There are many unusual conditions to which pressure vessels may be subjected, including: (1) extreme low temperatures; (2) differential thermal stresses; (3) concentrated localized stresses; (4) thermal cycling that results in stress ratcheting; and (5) vibration caused by several different means (i.e., wind, extremely high pressures, runaway chemical reactions, localized and repeated overheating, explosions, exposure to fire, rapid corrosion attack, and extreme size requirements for a vessel).[5] The designer must recognize these, and any other conditions that apply, and design vessels to accommodate them.

Other concerns that designers must address include: (1) the possibility of localized overheating and weakening of a vessel due to fire; (2) failure of a pressure-reducing valve in connected piping; (3) operating cycles involving the solidification and remelting of solids, leading to excessive pressures once the solid melts; (4) runaway chemical reactions and/or explosions; (5) condensation of the contents of a vessel leading to internal vacuum conditions; (6) exceeding vessel design temperature due to lack of operational controls; (7) excessively low temperatures that pose a threat of brittle fracture; and (8) corrosion of all types.[6] The design and operation of a vessel must take into account all of these factors.

14.6.7.3 Vessel Failure Modes

Two common failure mechanisms by which pressure vessels may fail in a catastrophic manner include metal fatigue and brittle fracture. Both of these failure mechanisms can result in a catastrophic failure of a vessel, thereby presenting a danger to plant personnel and to the surrounding environment. Each of these failure mechanisms is discussed as follows.

14.6.7.3.1 Metal Fatigue

Metal fatigue may be caused by rapidly fluctuating pressures, or other cyclic loadings. In some applications, process fluids may contribute to the endurance limit of the vessel material, which is better known as corrosion fatigue. Corrosion fatigue is similar to corrosion-stress cracking (see Section 2.1.8), except that in corrosion fatigue, cyclic loading of a material is usually also involved. Typical Section VIII materials can experience an endurance limit loss of up to 50% under extreme conditions of corrosion fatigue. By comparison, some high-strength heat-treated steels can lose 75% of their endurance due to corrosion fatigue. Some steels do not even have an endurance limit when corrosion fatigue is involved.

14.6.7.3.2 Brittle Fracture

The reason that materials to be used for Section VIII vessel construction are often required to undergo impact testing is to determine their transition temperatures [Division 1 vessels under -20°F (-29°C) and extensively for Division 2 vessels]. The transition temperature is that temperature at which a material no longer fails in a ductile mode, and instead fails in a brittle mode with a flat fracture surface and almost no elongation. The transition temperature is usually stated as a single temperature point, although the transition in reality takes place over a range of temperatures.

A common way to determine the transition temperature of a material is by means of Charpy impact testing,[7] which is usually correlated with service experience on full-size plates. A more precise but elaborate method of dealing with the ductile-brittle transition is the fracture-analysis diagram, which has been described by Pellinin and Puzak.[8] In this approach, a transition referred to as the nil-ductility temperature (NDT) of the material is determined by the drop-weight test[9] or the drop weight tear test.[10]

14.6.7.4 Nondestructive Testing of Pressure Vessels

Chapter 13 discusses the use of nondestructive examination methods as they are applied to piping systems. It is also important to examine pressure vessels for the presence of flaws and voids. The most widely used methods of nondestructive examination for pressure vessel construction are: (1) radiographic examination; (2) magnetic-particle examination; (3) liquid-penetrant examination; and (4) ultrasonic examination. Sections 14.6.6.2.1 through 14.6.6.2.6 discuss the use of these and other methods as they pertain to pressure vessels.

Each of these most commonly used methods may be used in conjunction with one of the other methods. Magnetic particle and liquid penetrant examinations will only indicate surface cracks. Radiographic examination or ultrasonic testing must be used if the concern is to detect other types of flaws. There is no known method of testing that may be performed with 100% accuracy.

14.6.7.4.1 Radiographic Examination

Radiography is performed on a pressure vessel by means of either x rays or gamma radiation. The advantage of x rays is that they can penetrate thicker surfaces. However, gamma-radiation radiographic equipment is more lightweight and easier to handle. Radiography is usually limited to vessels with wall thicknesses of 12 in. (300 mm) and under.

14.6.7.4.2 Ultrasonic Examination

Ultrasonic examination involves the measurement of fine vibrations. Vibrations are transmitted to a metallic (or FRP) material by a transducer that puts out a signal with a frequency between 0.5 and 20 MHz. The instrument sends out a series of pulses that show on a cathode-ray screen as they are sent out and again when they return after being reflected from the opposite side of the member being examined. If there is a flaw in a material, part of the signal will be reflected, thus signalling that a flaw exists. The magnitude of a flaw and its relative position in the material can be determined. The technology can be used to examine a material of virtually any thickness; its only limitation is that it is not applicable to complex or irregularly shaped surfaces.

14.6.7.4.3 Magnetic-Particle Examination

Materials that have inherent magnetic characteristics may be examined by magnetic-particle examination. In this type of examination, a magnetic flux is passed through a part, parallel to its surface. If there is a crack or void in the surface of a material, fine magnetic particles will collect near the surface of the defect, after being dusted over it. A surface area must be examined in two directions in order to pick up all cracks and voids. This is because the sensitivity of magnetic-particle examination is dependent upon the sine of the angle between the direction of the magnetic flux and the direction of the crack.

14.6.7.4.4 Liquid-Penetrant Examination

Liquid-penetrant examination consists of coating the surface of a metal with a special penetrant fluid (usually a dye). Once excess penetrant fluid is removed, any liquid that has penetrated the cracks will reveal locations where surface cracks exist. Some liquid penetrants require an additional fluid ("whitewash") to be added, which becomes stained by the dye. Some methods involve the use of fluorescent liquids.

14.6.7.4.5 Other Nondestructive Examinations

Other methods that are occasionally used to examine vessels nondestructively include: (1) eddy-current; (2) electrical-resistance; (3) acoustic-emission; and (4) infrared (thermal) testing. The Nondestructive Testing Handbook [Robert C. McMaster (ed.), Ronald, New York (1959)] gives information on many testing techniques.

14.6.7.4.6 Hydrostatic Testing of Pressure Vessels

Hydrostatic testing is required for most pressure vessels according to ASME BPV Code rules (unless a pneumatic test or other test is permitted). A hydrostatic test will often reveal major flaws, inadequate fabrication, improper design, and leaks that may exist at bolted connections. Vessels that are subject to and pass a hydrostatic test can still fail in service, a fact that is often not realized. Failure can, in fact, occur on the very first application of internal pressure after a hydrostatic test is conducted. This is why it is important to have a vessel designed and fabricated using careful techniques, and to subject a vessel to other nondestructive forms of examination.

Hydrostatic tests should always be carried out at a temperature above a vessel material's nil-ductility temperature. This is something strongly recommended in the ASME BPV Codes. By doing so, vessel materials will in effect have been subjected to a form of pressure-temperature treatment, which may be termed "notch nullification" (if cracks exist in the vessel material). This is because testing under this condition will cause the material to yield at crack tips and at other points where high residual stresses exist (e.g., weld areas). This form of heat treatment acts to reduce residual stresses by means of redistributing them, resulting in safer operation.

If a vessel is tested at a temperature that is at or below its minimum operating temperature, its hydrostatic test can sometimes be considered as a proof test of the vessel. Vessels that are intended for cryogenic services, having been constructed of ultrahigh-strength steel, are often subjected to a proof test of this type.

14.6.7.5 Inspection, Care, and Maintenance of Pressure Vessels

Vessels can be damaged in a number of ways during installation and/or operation. Damage to a vessel can sometimes be repaired. However, it is usually safest to cut out a damaged area and replace it, since it may become strained or overly stressed in localized portions where damage is inflicted. Repairs must be carried out in accordance with the rules of the ASME BPV Code.

Pressure vessels should be inspected as frequently as necessary, depending on the specifics of the application. The major thrust of such an inspection is to gauge the extent of corrosion to which the vessel has been subjected. Common inspection intervals range from once a year to once every 5 years. Measurement of corrosion is an important inspection item. Any type of nondestructive testing may be used for examining vessels during an inspection. However, ultrasonic examination is the most common method used for ongoing inspections. The National Board of Boiler and Pressure Vessel Inspectors (Columbus, Ohio) develops many guidelines for the inspection of pressure vessels.

After inspection or maintenance has been performed, gaskets should be located properly and carefully put back into their position. Bolts should be tightened in their specified sequence using a torque wrench or such devices as micrometers, proprietary bolt-tightening devices, or heating bolts. After a vessel has been reassembled, it must be subjected to another hydrostatic test to ensure bolt tightness and overall integrity of the vessel.

14.6.7.6 Other Vessel Regulations and Standards

14.6.7.6.1 Pressure Vessels Carried on Board Ships

The U.S. Coast Guard has developed rules for pressure vessels that are carried aboard United States-registered ships. Title 46 of the Code of Federal Regulations, Subchapter F (marine engineering applications), Parts 50 through 61 and 98, describe the rules for pressure vessels carried on board ships. The rules tend to be similar to those in the ASME Code.

Insurance underwriters usually require that pressure vessels that are a permanent part of a ship must be designed and constructed in accordance with the rules of the American Bureau of Shipping (ABS). Included in these rules are pressure cargo tanks that are supported at several points, but are independent of a ship's structure. These are distinguishable from tankers, which have internal cargo tanks that are part of a ship's main structure.

14.6.7.6.2 Tubular Heat Exchangers

The Tubular Equipment Manufacturers Association (TEMA) develops rules for the design and construction of tubular heat exchangers. TEMA is not a regulatory body. Therefore, there is no legal requirement for the use of its standards. However, it is generally accepted that their standards serve as a good basis for design and construction of such process units. TEMA has developed formulas to determine the required thickness of tube sheets, which are not covered in the ASME BPV Codes.

14.6.7.6.3 Foreign Codes and Standards

There are several countries besides the United States that have developed their own Pressure Vessel Design Codes. The most commonly accepted codes include those from: (1) Great Britain (British Standards); (2) West Germany (A. D. Merkblatter); (3) Netherlands (Grondslagen); (4) Sweden; and (5) France. Most codes differ in such items as factors of safety, whether or not ultimate strength of materials is considered in design calculations, and in construction and inspection rules. In general, a vessel that is designed to the ASME Boiler and Pressure Vessel Code, Section VIII, Division 1, will have a heavier wall thickness than a vessel built to one of the European codes. This is not necessarily true at high temperatures, however.

There are other differences between codes, including such items as: (1) the design and amount of reinforcement of openings; (2) flange design; (3) head design; (4) design of heat exchanger tube sheets; (5) calculation of wind loads; and (6) many others. The designer must be aware of which code or codes apply to a given project, and is always responsible to hire an appropriate expert(s) concerning the code or codes to be dealt with, where such expertise is required.

14.7 Internally Lined Tanks and Vessels

Pressure vessels and storage tanks may be internally lined to provide added corrosion protection from the service fluid. The ASME Boiler and Pressure Vessel Code allows a vessel to be fitted with any liner that is compatible with the base metal and the service fluid. Suitable internal linings for both pressure vessels and tanks include: (1) clad metals; (2) rubbers; (3) plastics; (4) rare metals; (5) ceramics; and (6) others. If a metal material is used, the lining may be installed in the form of clad plate. For nonmetallic inner linings, it must be installed separately after the fact. Cladding on plate may in some cases be classified as a stress-carrying component of the vessel. However, nonmetallic linings must not be considered as contributing to the pressure-retaining capability of a vessel. Nonmetallic materials are added strictly as a means of providing additional corrosion protection for the base material.

Ceramic linings will behave as an insulator in high-temperature applications. When high temperatures are present, a ceramic lining's inside surface temperature may be high, but its steel outer shell remains at moderate temperatures. Ceramic linings may be applied in the form of unstressed, or prestressed brick. It may also be cast in place.

To protect the outer shell of a metallic pressure vessel, a prestressed-brick lining can be used. To install a prestressed brick lining, a special thermosetting-resin mortar must be used to create an adhesion between the bricks and the metal. The vessel must be subjected to heat and internal pressure after the external lining is applied to expand its shell, and the mortar allowed to expand and take up the space. By reducing the temperature and pressure after the mortar is set, a vessel will contract and place bricks in compression. A vessel constructed as such can be used up to 375°F (190°C). This metal/prestressed-brick vessel combination can be construed as a specialized form of an internally lined vessel (the base metal is actually the primary lining; the prestressed brick outer shell acts as secondary containment).

Metallic pressure vessels may also be constructed with graphite or ceramic vessels functioning as an inner liner. A vessel construction of this type is referred to as one where the inner liner is "fully armored."

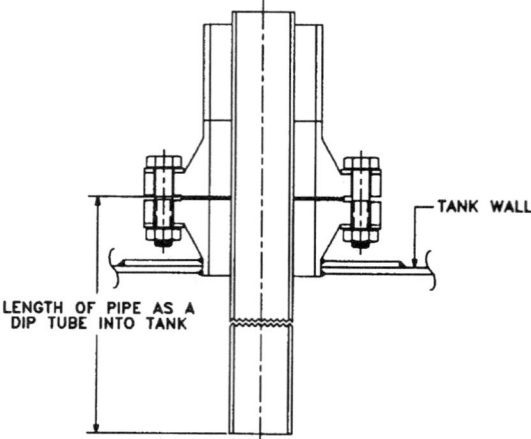

Figure 14.36. Example of a double containment pipe tank termination, whereby the secondary containment pipe terminates at a single-wall tank to a matching flange nozzle, the annulus is left open-ended, and the primary pipe extends as a dip tube. This type of arrangement should be avoided due to the difficulty in detecting leaks.

14.8 Associated Double Containment Piping

Piping materials must be compatible with tanks and pressure vessels, at least in terms of their ability to be joined with leak-free or pressure-tight connections. Piping should be installed so as to facilitate testing, resist corrosion, and prevent damage from movement of system components, as described in Chapters 3 through 13. It is essential that local building codes, consensus standards and codes, the project design (and its detailed plans, drawings, and specifications), and component installation instructions provided by each manufacturer be followed in their entirety. Where conflicts between requirements of different sources are identified, use the one that is the most conservative and provides the greatest margin of safety. The following are key considerations that apply to all components that might be involved in a combined tank/piping installation.

1. The details of double containment piping installation are covered in Chapter 12, "Fabrication, Assembly, and Erection," and in Chapter 13, "Inspection, Examination, and Testing."
2. All components should be inspected to insure compliance with the engineering design and with the plans and specifications. They should also be inspected for signs of damage, and defective materials reported and immediately repaired or replaced.
3. Adequate clearance is required between piping and other components of an underground storage tank installation. Particular attention must be paid to clearance between piping and tanks, trench walls, monitoring wells, conduit and other utilities, existing structures, and other system components.
4. Crossing of associated pipes should be avoided whenever feasible. Where pipes must cross each other, keep pipes within their minimum crossover requirements, according to the manufacturer. Refer to Section 11.5 for further discussion, and Figure 11.103.

Associated Storage Tanks and Pressure Vessels

5. Piping materials, their coatings, insulation, and insulation jacketing (where applicable) may become damaged during installation. Damage may also occur during transportation, handling, testing, and backfilling. Care should be taken to eliminate the possibility of damage to these components.
6. It is important to protect piping and piping components from direct contact with rain water, moisture, dirt, debris, and other corrosive fluids during the installation process.
7. In order to avoid damage to a tank or vessel's bottom, careful procedures must be followed when installing a tank or vessel's internal components. Damage to a tank's bottom may be avoided by installing suction and fill lines, submersible pumps, monitor probes, and gauge probes to be well within the bottom of the tank. A good general rule for suction and fill lines is to keep them at least 4 in. (103 mm) from the bottom of a tank.
8. Piping system accessories such as pumps, valves (manual and actuated), check valves, impact valves, expansion joints and flexible connectors, overfill devices, instrumentation, and leak-detection components should all be examined to be sure they are in accordance with the engineering design and all recommendations of the manufacturer.
9. In applications involving submersible pumps, the pumps and associated piping must be free to move with the tank. A street box may be used to keep them separated from the traffic slab in order to prevent damage from settling.

14.8.1 Piping Tightness Testing

In applications involving underground storage tanks, all piping should be isolated from the tanks (and product dispensers, if they are involved) before testing. In retrofit applications where previously existing piping is to be reused, lines should not be pneumatically tested if they have previously contained hazardous, flammable, or combustible liquids. If such pipes for reuse must be tested using a pneumatic test, the lines should be completely

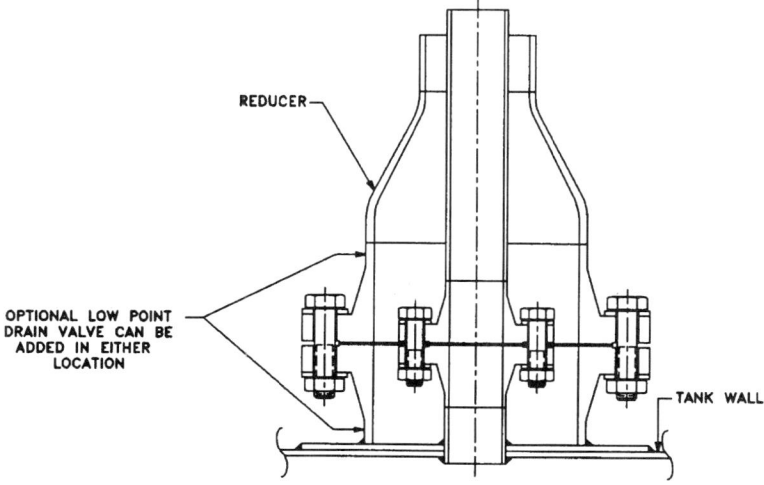

Figure 14.37. Example of a double containment pipe tank termination to a single-walled tank, whereby both inner and outer pipes terminate to a flange on the tank.

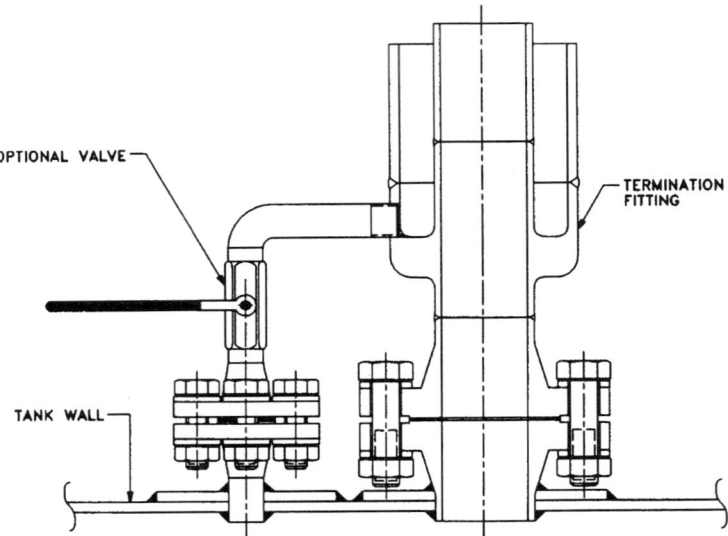

Figure 14.38. Illustration of a double containment pipe termination to a single-walled tank, whereby the secondary containment pipe is terminated by means of a termination fitting, the primary pipe is joined to the tank by means of a single-wall flange, and a low-point drain line connects the annulus to the primary tank by means of a flanged nozzle. Note the optional valve which is highly recommended.

purged and made thoroughly safe before the test is commenced. For further discussion regarding testing of double containment piping systems, refer to Chapter 13, Section 13.3.

14.9 Double Containment Piping Connections to Associated Tanks and Vessels

14.9.1 Double Containment Piping Termination Connections to Single-Wall and Double-Wall Tanks and Pressure Vessels

When a double containment piping system terminates at a tank or vessel, it must be provided with a connection for either its primary pipe, or for both of its primary and secondary containment pipe systems, to the tank or vessel. There are several ways to make such a transition. The choice typically depends on the preference of the facility owner, and technical issues having to do with the fluids involved. Several common options for making such connections are illustrated in this section.

A simple way to terminate a double containment piping system (for tanks and not pressure vessels) is to provide a flange connection on the tank to mate with the secondary containment piping system. In such an arrangement, secondary containment is not terminated upon entry to the tank, and instead is maintained in an open-ended fashion. Primary piping is then extended into the tank, which would in effect become a discharge dip tube. This type of arrangement is illustrated in Figure 14.36. Figure 14.37 illustrates a similar arrangement, but shows how to prevent the annulus from automatically emptying into the primary tank. However, note that no provision is made to drain leaks at the

Associated Storage Tanks and Pressure Vessels

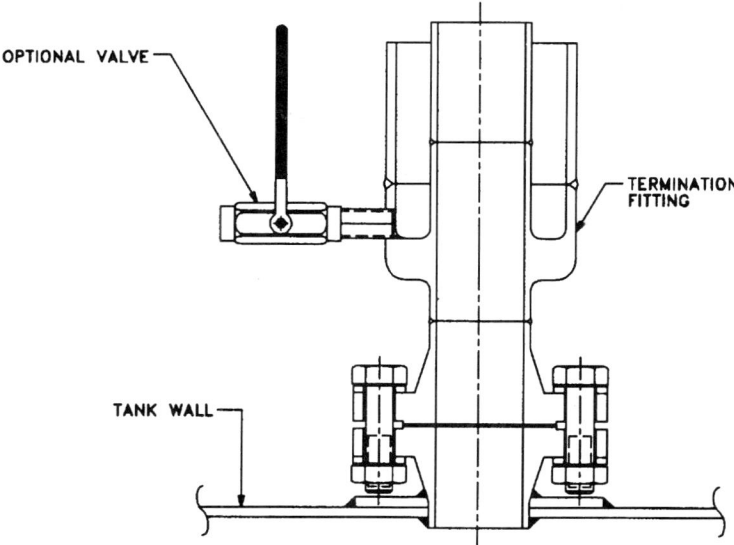

Figure 14.39. Illustration of an arrangement similar to Figure 14.39, except that the low-point tank annular drain attachment is not connected to the tank at all. It is not always desirable to allow primary pipe leaks to be drained into the primary tank, especially where contamination of fluids is a concern.

low point of the annulus (see Figures 14.38 and 14.39 for examples of how to provide for draining the annulus).

One problem the arrangement in Figure 14.36 often presents is that if a storage tank contains any volatile liquids, vapors will be generated, which, if they are light enough, will work their way into the annulus of the double containment piping system. As this vapor travels up through the annulus, it may eventually be subjected to a colder temperature condition that may cause it subsequently to condense. This will then fill the annulus with liquid contaminate, which any leak-detection system will interpret as a leak. This arrangement is therefore unsuitable for any applications involving gasses that may evolve from a tank, and must never be used if a vapor-probe detection system is used for the piping.

Another problem the arrangement in Figure 14.36 presents is that any leaks in the secondary containment piping will automatically be discharged into the tank. This alone is not enough to cause a problem. However, if any flushing water is introduced, it may present a problem since it may contaminate or dilute fluids in the tank. The same effect may be produced if there is any possibility of condensation.

Figure 14.38 illustrates another option for terminating a double containment piping system at a tank (or vessel). In this variation, a secondary containment pipe system is terminated, and instead the primary pipe system is joined to a connection on the tank. A low-point drain is used to allow a separate connection for the annulus to be connected to a mating flange on the tank. In this arrangement, both the primary system and the annulus may allow fluids to flow directly into the tank, if a valve is not provided on the low-point annular drain. However, without an annular low point drain valve the same problems described in the Figure 14.36 arrangement (preceding paragraph) would be present.

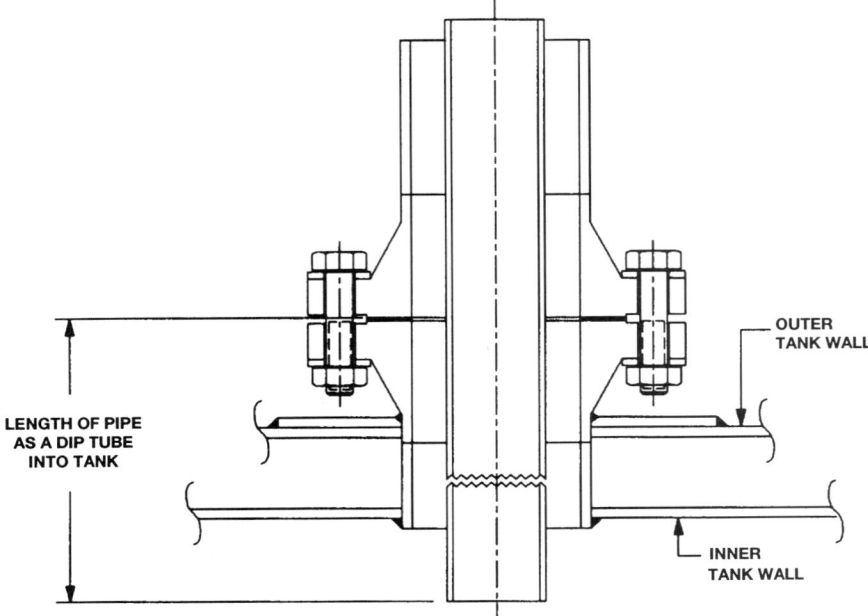

Figure 14.40. Example of a double containment pipe tank termination to a double-walled tank whereby the annulus is left open-ended, and the primary pipe extends as a dip tube into the primary tank.

Figure 14.41. This example shows three separate double containment pipes connected to a double-walled FRP tank by the method shown in Figure 14.40.

Associated Storage Tanks and Pressure Vessels

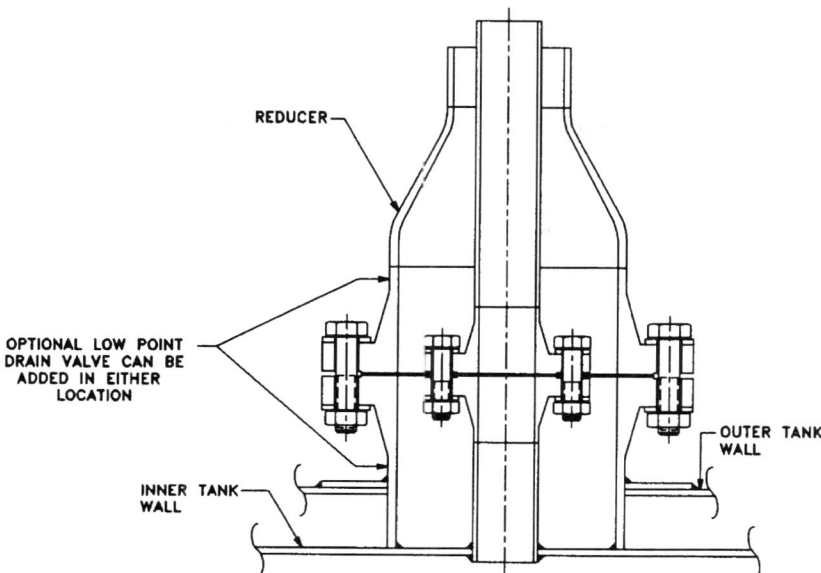

Figure 14.42. Example of a double containment pipe tank termination to a double-walled tank whereby both inner and outer pipes terminate to a flange on the primary tank. In this type of arrangement, the annulus of the pipe must not be interfaced in any way with the interstice of the the double-walled tank.

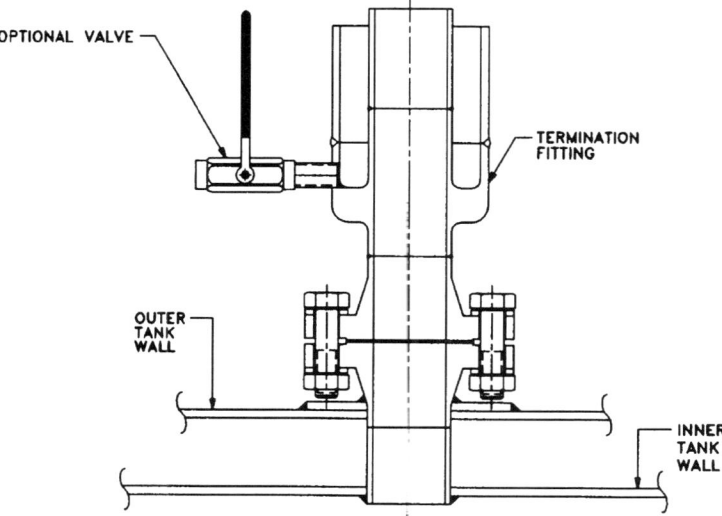

Figure 14.43. Example of a double containment pipe tank termination to a double-walled tank whereby the secondary containment pipe is terminated by means of a termination fitting, the primary pipe is joined to the primary tank by means of a single-wall flange, and a low-point annular drain is provided on the pipe. The low-point annular drain may or may not be interfaced to the primary tank at the option of the designer (and never to a tank's interstice!). However, a valve should always be provided on the low point of the drain.

Figure 14.44. An example photograph of a PVC single-wall suction pipe connection to a double-walled vertical storage tank. The tank and the single-wall portion of the piping system in this example are provided with secondary containment by means of an epoxy-coated concrete dike.

The arrangement shown in Figure 14.36 may not be connected to a pressure vessel, since a pressure vessel operates completely full, and under pressure.

To solve the dilemma of preventing vapors from going up into the annulus and preventing fluid from automatically entering the tank, an on/off valve may be added into the low-point drain line of the annulus. This is illustrated in Figures 14.38 and 14.39. The valve may also be actuated to allow for automatic opening once a leak is sensed by a leak-detection system. This arrangement is also only suitable for tank connections, as an annulus of a double containment piping should not be directly exposed to the contents of a pressure vessel.

Figure 14.38 offers a solution to the dilemma of vapors automatically entering an annulus of a double containment piping system. However, there are many applications for which it is never desirable to have fluids drained into the contents of a tank once a leak occurs in a double containment piping system. When this is the case, the arrangement illustrated in Figure 14.39 is recommended. In this arrangement, a valve is provided on the low-point drain, which is not connected to the tank (or vessel) in any way. Instead, a connection (threaded, hose connection, or flange connection) is provided to one end of the valve to allow the contents of the annular space of the double containment piping system to be removed safely for treatment and/or disposal. The arrangements in Figures 14.38 and 14.39 are both suitable for terminating a double containment piping system by direct connection to a pressure vessel.

Figure 14.40 illustrates an example of a dip-tube assembly termination connection to a double-wall tank. Notice in the illustration that the secondary containment pipe mates to a flange that originates from the inside tank. The outer wall of the double-wall tank is welded (bonded) directly to the flange. Thus, flow cannot occur between the annulus of the double containment pipe and the interstice of the double-wall tank. An example of an

Associated Storage Tanks and Pressure Vessels

Figure 14.45. An example photograph of a Polypropylene single-wall suction pipe connected to a double-walled vertical storage tank. The tank and the single-wall portion of the piping system in this example are provided with secondary containment by means of an epoxy-coated concrete dike.

arrangement of this type is shown in Figure 14.41, involving two 3 in. (90 mm) inside 6 in. (160 mm) reinforced vinyl ester double containment piping systems, and a 2 in. inside 4 in. PVC double containment pipe. In the photograph, the three secondary containment pipes are shown to mate to corresponding flanges provided on the tank. Since the tank in the photograph is a double-wall tank constructed of reinforced polyester, the three flanges arise from the inner tank wall, and in no way allow any flow to the interstice of the tank. This issue is discussed further in Section 14.9.2. The couplings shown on the 6 in. (160 mm) secondary containment pipe just above the flanges are examples of midline RTRP "slip-coupling" secondary closure (see Figure 11.76).

Figures 14.42 and 14.43 illustrate alternate termination arrangements for double containment piping connected to a double-wall tank. In the examples, all connections, whether they are primary pipe connections or connections from the secondary containment piping system (annulus), are made into the inside tank and not into its interstice.

14.9.2 Double Containment Piping Initiation Arrangements to Single-Wall and Double-Wall Tanks and Pressure Vessels

In many tank applications, a double containment piping system that is initiated from a tank does so by means of single-wall piping that acts as suction piping, upstream of a pump. The suction single-wall piping, tank, and pump are normally contained within a secondary containment berm or dike, and thus are provided with secondary containment in this manner. The connection to a tank involves an ordinary single-wall pipe flange connection mated to a single-wall flange connection on the tank. The double containment piping system usually is initiated after discharge from its pump, prior to exiting the berm/dike area.

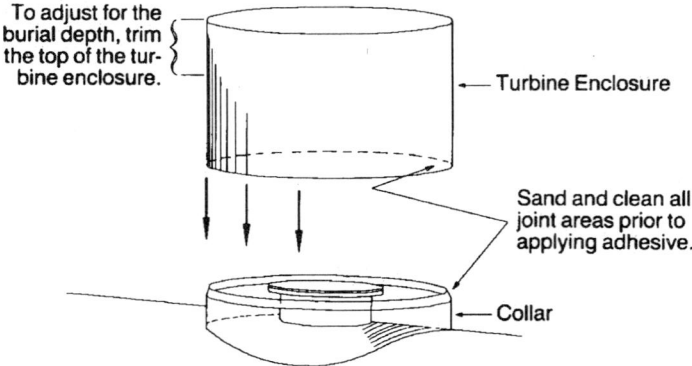

Figure 14.46. Mounting details for the bottom half of a turbine enclosure sump to a fiberglass tank. (Source: OC Tanks.)

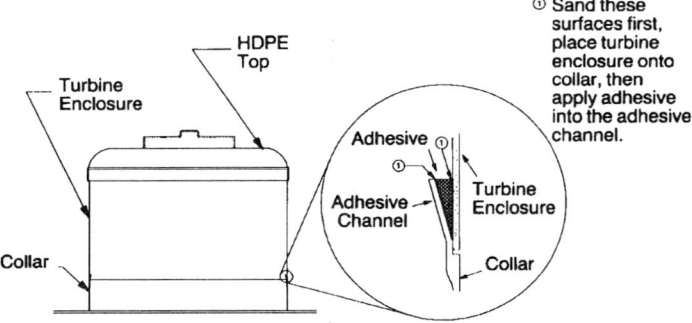

Figure 14.47. Mounting details for the top half of a turbine enclosure sump to its respective bottom half. (Source: OC Tanks.)

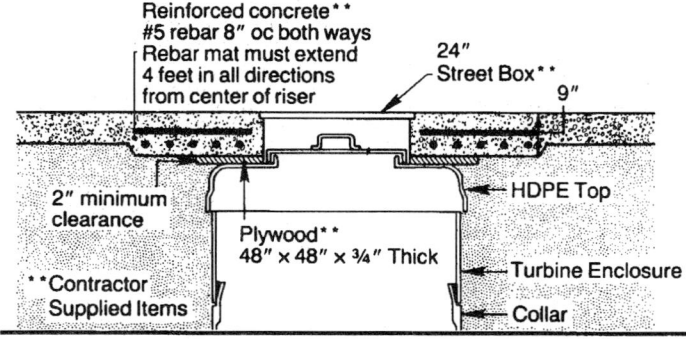

Figure 14.48. Burial considerations for a turbine sump enclosure in traffic areas. (Source: OC Tanks.)

Associated Storage Tanks and Pressure Vessels

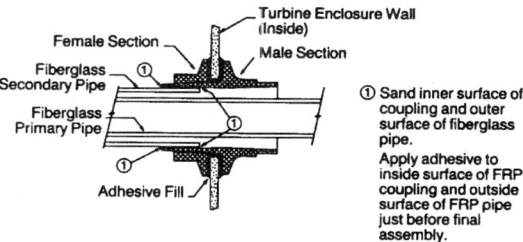

Figure 14.49. Illustration of a fiberglass double containment pipe termination arrangement using a special coupling at a tank sump enclosure. (Source: OC Tanks.)

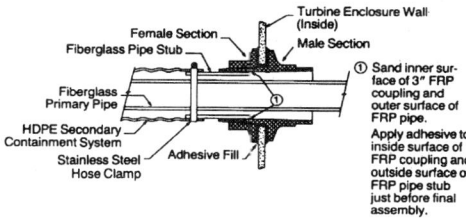

Figure 14.50. Illustration of a fiberglass in a corrugated HDPE double containment pipe termination arrangement using a special coupling at a tank sump enclosure. (Source: OC Tanks.)

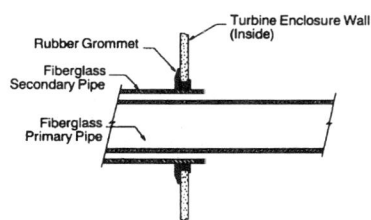

Figure 14.51. Illustration of a fiberglass double containment pipe termination arrangement using a rubber grommet at a tank sump enclosure. (Source: OC Tanks.)

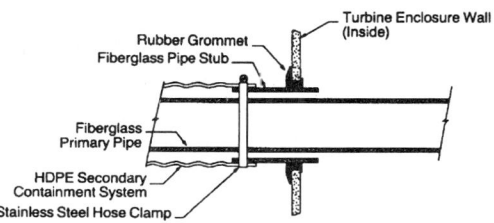

Figure 14.52. Illustration of a fiberglass in a corrugated HDPE double containment pipe termination arrangement using a rubber grommet at a tank sump enclosure. (OC Tanks.)

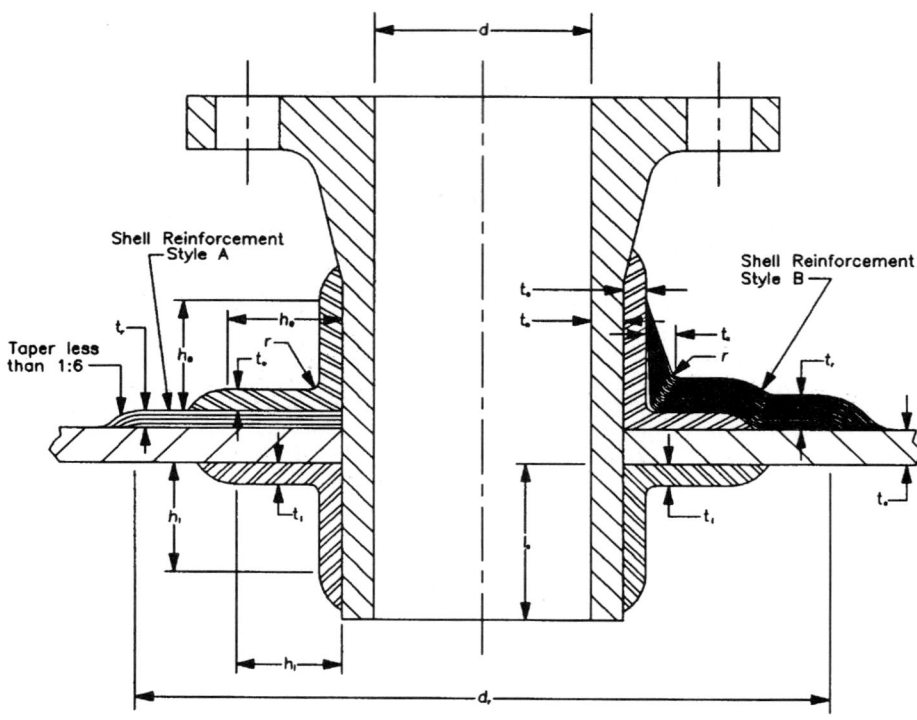

d = nozzle diameter
d_r = cutout reinforcement diameter, in., measured on the vessel surface (greater of $2d$ or $d + 6$)
h_i = inside shear bond length
h_o = outside shear bond length
h_s = total shear bond length
    = $h_i + h_o$
l_p = penetration length (2 in. min.)
r = fillet radius (3/8 in. min.)
t_i = inside installation laminate thickness
    = $t_v - t_o$ [minimum t_i = 0.14 in. (3M,V)]
t_n = minimum nozzle stub thickness
t_o = outside installation laminate thickness
    = $t_v - t_i$ [minimum t_o = 0.26 in. (6M)]
t_r = cutout reinforcement laminate thickness
t_s = shell thickness
t_v = total installation laminate thickness (minimum t_v = 0.40 in.)
    = greater of t_r or $2t_s$ for Style A
    = lesser of t_r or $2t_s$ for Style B
t_x = reinforcement base thickness
    = $t_r/2$ (Style B reinforcement)

Figure 14.53. Illustration of a fiberglass penetration nozzle in a fiberglass tank or pressure vessel showing pertinent reinforcement criteria. (Source: ASME Boiler and Pressure Vessel Code, Section X.)

Many of the issues that apply to connecting a double containment piping system to a tank apply in principle when a connection must be made to a double-wall tank. However, there are some additional considerations. There should be no direct or indirect connection made from the piping system to the interstice of a double-wall tank. That includes connections from the annulus of the double containment pipe to the interstice of a double-wall tank. Whenever liquid finds its way into the interstice of a double-wall tank, the tank must be taken out of service, decontaminated, and completely dried out. Its leak-detection system will also require some amount of servicing. In most double-wall tank designs, these procedures are not easy to carry out, and will require considerable downtime. Thus, the interstice of a double-wall tank and the annulus of a double-containment piping system must be considered as entirely separate and unrelated systems and must never be cross-connected.

Figures 14.44 (3/4" schedule 80 PVC piping) and 14.45 (25mm SDR 11 Polypropylene piping) show examples of single-wall suction piping connections to a double-wall tank. Each of these suction pipe systems are part of a double containment piping system that initiates after discharge from a centrifugal pump. The piping is single wall since it is contained in a dike/berm structure. The single-wall piping is joined by a flange connection to a nozzle on the primary tank. The secondary containment flange connection surrounding the primary nozzle connection only serves the purpose of allowing the outer pipe jacket to be bonded to it, to allow for unrestricted access to the primary tank nozzle.

14.9.3 Double Containment Piping Connections to Storage Tank Sumps

In petroleum marketing station storage tank applications, it is common to use sumps connected to the top of tanks (e.g., see Figures 14.26 and 14.29). A mounting detail for a turbine sump to a tank is shown in Figures 14.46 and 14.47. Burial details are shown in Figure 14.48.

A variety of typical penetrations of various combinations of double containment pipes to turbine enclosure sumps are illustrated in Figures 14.49 through 14.52.

14.9.4 Reinforcement of Piping Connections to Tanks and Vessels

Rules for the reinforcement of nozzle connections to vessels are provided in Section VIII, Division 1, Subsection A and Section VIII, Division 2, Part AD of the ASME Boiler and Pressure Vessel Code for metallic materials. Section X of the Code describes rules for providing reinforcement to fiberglass-reinforced plastic pressure vessels. In the rare event that a pressure vessel is designed as a double-wall vessel for safety purposes, it would most likely have to be designed as a jacketed vessel, per Section VIII, Division 1, Appendix 9 rules of the ASME Boiler and Pressure Vessel Code. The rules presented there for piping connection reinforcement would have to be followed.

Other rules for designing reinforcement of nozzle attachments for metallic pressure vessels are described in various publications of the American Petroleum Institute.

For reinforced fiberglass single-wall tanks, there are several documents that describe rules for providing reinforcement to flange and nozzle connections. These include Section X of the ASME BPV Code, ASME Standard RTP-1, and ASTM Standard D 3299. A typical reinforcement design by these standards is illustrated in Figure 14.53.

There are no specific standards described for the reinforcement of piping connections to double-wall tanks. If the piping penetration must penetrate both walls of a double-wall pipe, the piping connection penetration may be provided with reinforcements to the inside of the inner wall of the tank and the outside of the outer wall of the tank. In some cases, the inside tank penetration may be provided with reinforcement on its inner and outer wall.

References

1. PEI/RP100-87, "Recommended Practices for Installation of Underground Liquid Storage Systems," Petroleum Equipment Institute, 1987, p. 5.

2. PEI/RP 100-87, op. cit., p. 5.

3. Process Equipment Design, Wiley, New York (1959).

4. Pressure Vessels and Piping Design: Collected Papers, ASME, 1927-1959.

5. Pressure Vessels and Piping Design and Analysis, four volumes, and International Conference: Pressure Vessel Technology, published annually, ASME.

6. Perry and Green, Perry's Chemical Engineering Handbook, 6th Edition, McGraw-Hill, New York, 1984.

7. ASTM E 23.

8. Pellinin and Puzak (Trans. Am. Soc. Mech. Eng., 429 (October 1964); Welding Res. Counc. Bull. 88, 1963.

9. ASTM Standard E208.

10. ASTM Standard E436.

Notes

Notes

Chapter Fifteen

Leak Detection

15.1 Introduction

15.1.1 Overview

Leak detection is one of the most important aspects of double containment piping and double-wall tank systems. Without an effective and reliable means to sense that a leak has occurred, the additional protection otherwise added by means of providing secondary containment may be compromised. Once a fluid leaks into an annulus of a pipe or the interstice of a tank, there is no longer any means to contain the leaked fluid secondarily; only a primary means of containment exists at that point, that is, unless tertiary containment has been incorporated into a system design.

Left undetected, a fluid may have a chance to corrode through its secondary containment casing, or may find its way through defects in the containment to the surrounding environment. It is important that a leak be detected as early as possible, so that any chance of fluid escaping to the surrounding environment is minimized. A system whose leaks are detected early can be repaired right away, and quickly returned to its original safe working condition.

Both single-wall and double containment piping systems may be monitored for leaks by a wide variety of methods. There are likewise numerous methods to detect leaks that originate from single-wall or double-wall tanks and vessels. Both double containment piping and any associated tanks/vessels can be equipped with automatic sensing mechanisms to monitor continuously and receive alarms as soon as system upset occurs. Alternatively, manual and visual leak detection methods may be incorporated by adding the necessary components into the design of a system. In addition to interstitial monitoring, there are both internal methods and external methods that exist to monitor for leaks, although most of these methods work best for tanks (as opposed to pipes). More than one method of leak detection may be used as a means of redundancy.

The successful implementation of any leak-detection system is related to many different aspects of the design and installation of a piping/tank system. On any given project, it is essential that leak detection be considered from the conceptual phase of the project. By doing so, any needed design details or installation procedures may be taken into account from the start, and carefully coordinated. When a project follows this practice, costly design changes and construction delays/problems can be avoided. The best overall result is achieved when leak-detection decisions are considered from the very start of a project.

15.1.2 General Considerations of Leak-Detection Systems

There are many considerations that need to be examined prior to making a determination of which leak-detection method is best for a given application. The expectations or goals of the system should be evaluated prior to deciding on the method(s) to be used. Based upon knowledge of what the specified performance criteria of the system are the choices for system types can be narrowed. Then, a specific system can be selected based on its compatibility with the proposed piping or tank system. The overall design of the system, the materials of construction involved, the fluids to be detected, the location and type of installation, and the intended operation practices for the system are all among the factors that will determine what leak-detection system is compatible with a piping/tank system. Several major performance criteria that must be considered prior to selecting a leak-detection system are described in Sections 15.1.2.1 through 15.1.2.5.

15.1.2.1 Ability to Locate Leaks

For piping systems of a considerable length, it is important to consider how it is to be determined where a leak is originating from along its length. This is a critical consideration for underground systems, since lack of knowledge of a leak location may require that an entire pipe be excavated in order to find and correct a defect. For double containment pipes, leak location may be determined by means of compartmentalizing an annulus, thereby creating zones of secondary containment. Each secondary containment zone may then be monitored for leaks. Alternatively, leak-detection cable may be used to detect leaks. For single-wall piping systems, multiple testing stations may have to be installed to monitor soil or ground-water contamination. All methods for detecting for the location of leaks vary in their degree of accuracy.

All methods of double containment piping annular sensing provide information that may be used to determine the location of leaks; this feature is not just limited to leak-detection cable. However, while each method provides data that can be used to determine the location of a leak "theoretically," the accuracy of locating leaks varies considerably according to the method used.

15.1.2.2 Ability to Distinguish Fluids/Source of Leak

Not all leak-detection systems are capable of distinguishing between fluids. Many methods can only sense that some fluid has leaked (i.e., periodic visual inspection methods). In double containment pipe systems, or double-wall tank systems, the integrity of both the primary and secondary containment systems need to be monitored. A breach of primary containment will usually result in process fluid leaking into the annulus (unless a greater pressure exists in the annular space). In services having multiple primary pipes with a common secondary containment housing, there may be several different process fluids involved, each of which may spill into an annulus. A breach of secondary containment, on the other hand, may result in an influx of water or groundwater, rainwater, or moist air (which may subsequently condense). A system that has warm, moist air in its annulus may be subject to future condensation forming in its annular space, which may trip a leak-detection system. Each of these circumstances may dictate that more than one leak-detection method may be needed.

15.1.2.3 Time of Response and Sensitivity

All leak-detection systems vary in terms of the time of response required to detect a leak, which is often is described as the sensitivity of a system. For many methods, time of response and sensitivity can be quantified in terms of an amount of fluid that must be leaked prior to detection. For leak-detection cable, sensitivity and time of response can usually be quantified in terms of an amount of cable that must be wetted, when it is fully immersed in a unit volume of fluid.

Some leak-detection systems may have their sensitivities quantified in terms of a percentage change in some physical measurement. However, this may not be indicative of the time it takes for the system to respond and generate an alarm. Measurement of flow in a primary pipe is an example of a method where this is so. A system that has a slow, gradual leak may never generate an alarm condition in a leak-detection method using flow measurement as its basis.

15.1.2.4 System Reliability and Backup

All leak-detection systems may fail once they are put into operation. It is important to know whether a system is operational at all times, if it is of the continuous type. Thus, a means has to be implemented into a system, or an ongoing inspection procedure put in place, to determine if a system is working.

If it is determined that a system is no longer operational, a backup means must be used to detect leaks once the primary system fails. In order for this to be possible, a redundant means must be made available as part of the initial design. Redundancy is a very important feature, since a leak could occur during the time a leak-detection system is not operational (e.g., during an earthquake or power failure).

In addition to the worry concerning a missed reporting of a leak, there is an equal concern that false leaks may be reported. This has to be known if possible, since frequent false alarms may lead plant operators to ignore a leak-detection system, a dangerous practice indeed. Environmental regulations in the United States which require leak-detection systems have standards for allowable accuracy for both determination of leaks and the frequency of false alarms. There must be a 98% probability of reporting a leak, and no more than a 5% probability of false alarms, according to the provisions stated in 40 CFR Part 280.

15.1.2.5 Ease of Installation, Operation and Maintenance

The overall difficulty in installing, operating, and maintaining a leak-detection system should be fully assessed prior to finalizing a selection of the type of system to be implemented. If facility or contractor personnel lack the available expertise to install, operate, or maintain a specific leak-detection system, it will do very little good for a piping or tank system, and for the facility owner. In some cases, a less sophisticated means of leak detection might be a more logical choice if plant personnel will have trouble with a more complicated system. This factor needs to be evaluated for every project.

15.2 Interstitial (Annular) Monitoring of Double Containment Piping

There are two broad categories of leak-detection systems that are available to monitor the annular space of double containment piping and/or the interstice of double-wall tanks. These are continuous and noncontinuous types. Continuous types include any means that continually sense for leaks. There are specific methods that feature the direct ability to locate leaks (e.g., leak-detection cable, also called line sensing and locating systems) and others that do not provide direct information as to the location of leaks (e.g., probe sensing systems). Noncontinuous methods include both manual and visual means of sensing. Noncontinuous means of inspection may be combined with continuous means of sensing, for redundancy and other reasons. Continuous monitoring methods may be combined with other continuous and/or noncontinuous methods as well.

Methods that provide a general alarm (e.g., probe systems, manual, visual detection, etc.) can be used in a system having a compartmentalized annulus, to provide information as to leak location. It is not completely necessary to compartmentalize the annulus. Multiple probe stations in a continuous annulus may sometimes effectively serve this purpose. The smaller the compartments used, or the closer the probe stations, the greater the accuracy in determining leak location by these methods. The larger the compartment used, or the greater the distance between probe stations, the less the accuracy in determining leak location. In other words, a probe system installed in a compartment that is 200 ft (68 m) in length can only determine a leak to be "somewhere" in that 200 ft (68 m) compartment. Thus "within 200 ft" ("within 68 m") is the accuracy of leak location of this type of system. For a compartment that is 20 ft (6.8 m) in length, a much greater accuracy can be achieved.

Leak-detection components and systems that are intended for annular sensing vary widely in their degrees of accuracy, reliability, ease of installation, and operation. It is important that one understand the capabilities and limitations of leak-detection components and systems so that the system being designed will function in a desired manner.

15.2.1 Leak-Detection Cable Systems

Continuous line leak sensing and locating systems, commonly referred to as "leak-detection cable," have been installed in many double containment piping systems. The appeal of leak-detection cable is that approximate locations of leaks can be sensed early after a leak occurs. Several types exist, each based upon the continuous measurement of an energy source, thereby monitoring for changes in the energy source behavior. The two most common methods include: (1) conductance (resistance) -based cable systems, whereby the conductance (resistance) of an electrical signal is continuously measured; and (2) impedance-based systems, also referred to as TDR (time-domain-reflectometry) systems, whereby the impedance of an energy pulse wavelength is continuously monitored against a set pattern to detect changes. Each has subtle differences and capabilities; both are capable of performing their desired functions, namely, determining the approximate location of a leak, and alarming a user as to its existence.

In one type of conductance-based cable, which is intended to sense conductive fluids, an electrical current is short circuited when a fluid bridges a gap between sensor and signal wires. Resistance monitoring is a straightforward concept that applies Ohm's Law to determine the location of a short. It is a method that can immediately sense small leaks, even over relatively long lengths of cable. Thus it has relatively "sensitive" reporting capabilities, and is a well- proved technology. Unless condensation is eliminated or controlled, versions of conductivity-based cable which is designed to detect conductive fluids will report false leak signals due to its sensitivity.

Impedance-based (TDR) leak-detection cable technology is based upon the measurement of the impedance of an energy pulse, and is similar in concept to radar technology. It has the subtle advantage of locating multiple leaks, if a prior leak is of small magnitude and confined to a local area. Impedance-based systems compare favorably with conductance-based systems, although impedance methods lack some of the precision inherent in resistance-based methods. This is particularly true if a leak occurs at the far end of a long length of cable.

Thus, both types of cable technologies have subtle differences that could prove useful for a given design condition. For instance, it was mentioned that conductance-based cables have the ability to detect very small leaks (some are capable of signaling a leak with < 0.25 in. of cable length wetted, when fully immersed), thereby aiding in early detection of a leak. The ability to detect a leak early can help to limit the total quantity of a leak (release). Accordingly, the amount of work to repair a piping system could be kept to a smaller scope. If a leak were to remain undetected for a longer time, there is greater potential for more work to be involved in the repair of the system.

A unique feature of TDR-based technology is the ability to detect multiple leaks. This feature is interesting, but there are limited applications where it is of benefit. To continue to operate a double containment piping system once a leak has been detected is a somewhat self-defeating practice for the concept of double containment. Good operating practices suggest that a system be immediately repaired upon the detection of a leak. In fact, RCRA requirements in the United States would not allow a user to continue to operate a regulated system once an initial leak has been detected.[1]

In order for a TDR-based cable to detect a second leak once the first has occurred, the system must map over the first leak. When this occurs, the cable sensing system is subject to a signal-attenuation effect, which will lead to a reduced sensitivity. Figure 15.1 graphically shows this effect, based upon a first leak at 150 ft down a TDR cable's length. The sensitivity of the TDR cable, as measured by an amount of wetted cable required to create an alarm condition, depends upon the magnitude of the mapped-over leak, as shown in Figure 15.1. However, it also depends on the distance of the first leak from the source of the signal. Sensitivity values will improve for TDR-based sensor cables exposed to fluids at less than the effective range[2]; this also means that sensitivity will lessen as cables are exposed to fluids at greater than the effective range.

Leaks that occur in a near simultaneous manner could be detected using mapping-over techniques in a TDR-based system. However, it is very rare for multiple leaks to occur simultaneously once hydrostatic testing has been concluded. The time when multiple leaks may be reasonably expected to occur is during an initial primary pipe hydrostatic test. Thus, having a cable of this type in place during initial pressure testing might be desirable. However, cable systems are not recommended to be installed prior to, or during, a hydrostatic-pressure test, according to manufacturers of both resistance-based and TDR-based sensing cables. Pull cable should always be installed first, and left in place during pressure testing (see Section 15.2.1.6). Only upon completion of all testing should actual cable be installed. Thus, the ability to detect multiple leaks is of little real benefit.

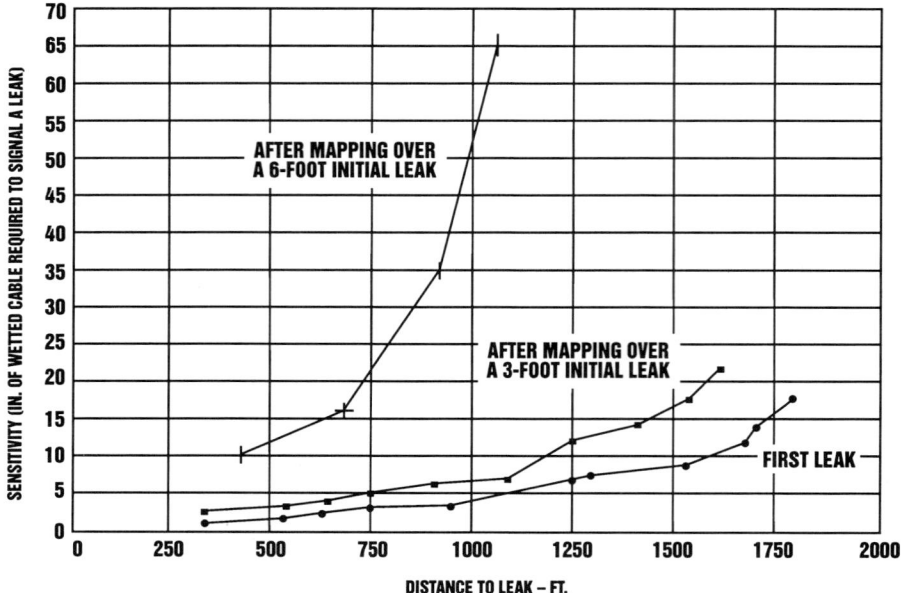

Figure 15.1. Response of a typical TDR-based cable to a second leak, based upon a first leak that occurs 150 ft away. Data is shown for a leak that spreads 3 ft, and for one that spreads 6 ft. First leak data is shown for comparison. (Source: Based on test data supplied by Raychem Corp.)

Another difference between the two cable technologies involves their sensitivity in detecting a first leak, in relation to their overall length. For resistance-based systems, sensitivity is nearly constant over the full length of a cable, even for very long lengths (> 1,000 ft). Sensitivities of TDR-based cables do vary according to length, however. The further down the length of cable, the greater the wetted length of cable required to create an alarm condition, due to signal attenuation. For a long pipeline (>1000 ft), this effect can be significant, as illustrated by Figure 15.2.

Both resistance-based and TDR-based cable sensing systems claim an accuracy in detecting leak location to within plus or minus 1% of the total cable length, or 5 feet, whichever is greater. Accuracy of leak location is functionally different from cable sensitivity. In assessing the performance capability of a leak detection cable system, one needs to consider both the claimed accuracy, and the sensitivity of the cable to obtain a complete picture.

Installation of leak-detection cable requires carefully planned coordination with other aspects of the piping system installation or associated tanks. It also requires that close inspection and control practices be followed during an installation. These functions can either be provided by outside inspectors or by the installers of the cable systems, provided they receive adequate training. Each detail must be carefully considered, both in the engineering design and during the installation process. Detailed requirements should be clearly described in project plans and specifications. Most of the major considerations for designing a system that has cable and installing such systems are described in Sections 15.2.1.1 through 15.2.1.11.

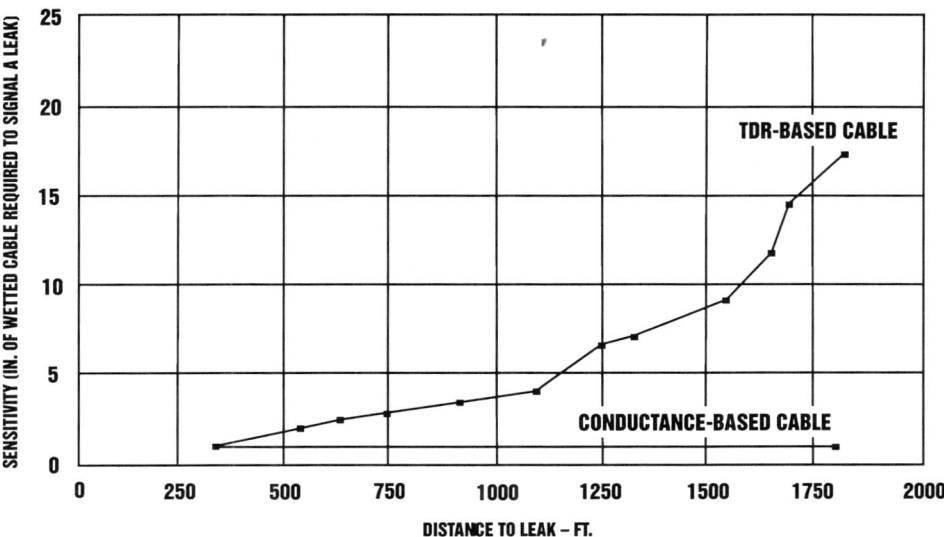

Figure 15.2. Sensitivity of a typical TDR-based cable based on the length down the cable where a leak occurs. (Source: Based on test data supplied by Raychem Corp.)

15.2.1.1 Design Considerations for Leak-Detection Cable

Engineering a double containment piping system that incorporates leak-detection cable demands careful consideration of the piping layout, access requirements of the cable, and the design of access fittings. It also requires careful consideration of allowable space details inherent in all centralizing devices interstitial supports, interconnecting components (e.g., baffles, internal anchors), and the 6 o'clock position of the annulus for all components. (In elbows, the allowable space at the 3 o'clock, 9 o'clock, or 12 o'clock position may also be important, depending on the rotation/orientation of the elbow.)

The design of each component must address the ability to install cable readily. It must also allow for future ability to extract and/or replace leak-detection cable with relative ease. Leak-detection cable must be designed to be installed and operated under all conditions to which it will be subjected. Therefore, any aspect of the design that affects cable installation or operation must be considered in the design of each element of the piping system. This will involve components, and their annular clearances, movements of the piping system and supports (external and interstitial) due to thermal expansion, vibration, and other effects.

15.2.1.2 Leak-Detection Cable System Layout

The layout of a sensing-cable leak-detection system should take into account the overall length of the associated piping system, the number and types of branches, the number of circuits required, and the desired location of its alarm module. Each leak-detection cable system is limited by the maximum length specified by its specific cable manufacturer. Different manufactured brands have different recommendations for most layout details,

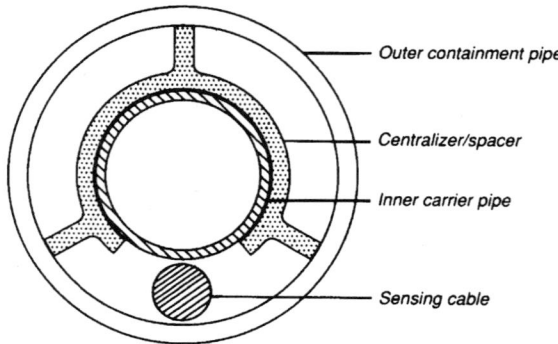

Figure 15.3. Illustration of the cross section of a double containment pipe spacer that provides a minimum amount of required clearance for leak-detection cable. (Source: Raychem Corp.)

such as the means to accomplish cable branch connections. For instance, some systems utilize branch connectors, others use individual cables in a continuous fashion in branched systems, and others utilize field splicing. Each unique type of cable connection will require separate detail drawings to be made.

Depending on the piping layout, a multiple circuit or zone system might be desirable. A zone system is feasible if double containment piping travels through different portions of a facility, through different parts of a building, or through multiple buildings. Also, if a double containment pipe annulus is separated into different compartments, a zone system might be applicable. If a multiple circuit or zone system is to be used, leak-detection layout drawings must indicate all wiring and connection details.

One common layout rule that applies to all sensing-cable systems is that sensing cable should always enter a pipe at the closest point to its alarm module, unless there are restrictions that require it to enter the system at another location. Manufacturer's specifications, and other layout data provided with leak-detection cable, should be thoroughly reviewed to be sure that all layout requirements are being satisfied.

15.2.1.3 Piping System Requirements for Leak Detection Cable Use

A double containment piping system and its components must contain sufficient annular (interstitial) clearance in order to incorporate leak-detection cable. All types of leak-detection cable require a minimum annular clearance for installation and removal. The minimum amount of required clearance does vary from brand to brand. Figure 15.3 illustrates a system that contains an acceptable amount of annular clearance. Figure 15.4 illustrates a theoretical arrangement involving a system that does not contain sufficient annular clearance to install leak-detection cable of the size shown. For resistance-based leak-detection cable products, typical minimum-clearance requirements are 0.75 in. (18 mm). This clearance is usually required to be maintained in the vertical direction at the 6 o'clock position in the annulus. TDR-based leak-detection cables typically require greater annular clearance, usually in the range of 1-1/2 to 2 in. (38 to 50 mm).

There must be horizontal clearance at the 6 o'clock position as well. As a general rule, the same value assigned for minimum vertical clearance should be used to determine the

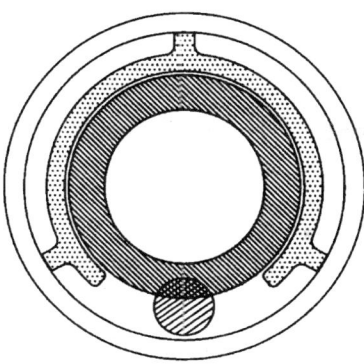

Figure 15.4. In the illustration shown, the cross section of the double containment pipe annulus does not provide sufficient clearance required for leak-detection cable. (Source: Raychem Corp.)

minimum required dimension at the position shown in Figure 15.5. In determining annular clearance, the thickness of any interstitial supporting/centralizing devices that extend into the 6 o'clock area should be taken into account. Aside from the basic diameters of the inner and outer pipes, the minimum vertical and horizontal required clearances also set minimum requirements for design annular clearances in interstitial supports, annular baffles, internal anchors, and other double containment piping system components.

Figure 15.6 illustrates the 6 o'clock annular position requirements for a variety of four centralizer types. These examples of centralizing devices are shown for illustrative purposes. Designers must always determine the appropriate details of the supporting/centralizing device to be used in each specific application. Interstitial supporting devices must also be designed to support the primary piping properly, and prevent excessive loads being transmitted between either pipe, as is described elsewhere in this book. A support could be designed "properly" to accommodate a leak-detection system, yet it might be "improperly" designed for its other mechanical and structural needs. Refer to Section 9.3.1.2 for a discussion on the design of interstitial supporting/centralizing

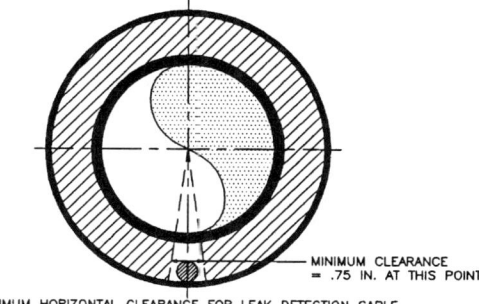

Figure 15.5. Sufficient minimum clearance must be provided in both the vertical and horizontal directions at the 6 o'clock position of components positioned in the annulus such as internal anchors.

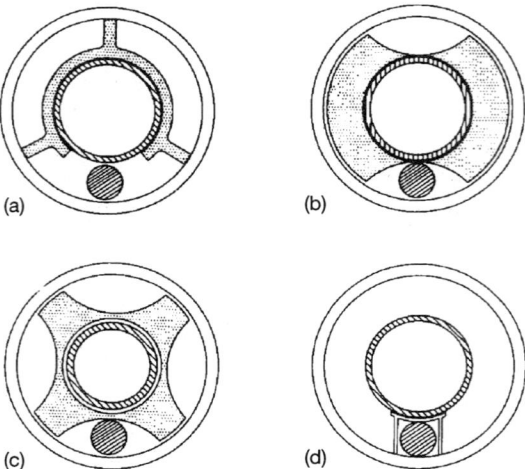

Figure 15.6. Cross-section detail for various interstitial supporting devices that provide the minimum required clearance for leak-detection cable: (a) vane-style support; (b) two-hole collar style; (c) four-hole collar-style; (d) channel-profile support. (Source: Raychem Corp.)

devices. Care should be taken to ensure that all interstitial supporting devices (centralizing devices) are installed properly. Figure 15.7 illustrates potential problems created by improperly installed interstitial supporting devices. Figure 15.8 is a photo showing a cable positioned at the 6 o'clock position of an annulus.

Joining and welding practices for certain materials can result in a buildup of internal restrictions on the inside diameters of welded (bonded) pipe components. These types of restrictions must be taken into account when determining the amount of clearance available for cable installation. Internal restrictions created by application of heat-element butt-welding procedures for thermoplastic piping, and through the use of socket-based systems, are illustrated in Figure 15.9.

Internal anchor components that are intended to be installed "midline" in a double containment pipe system must also allow for adequate cable clearance. Designers should be

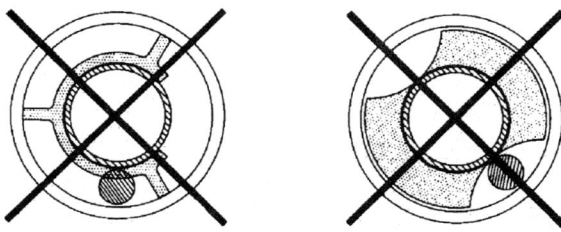

Figure 15.7. Examples of improperly positioned interstitial supports that prevent clearance at the 6 o'clock position of the annulus. (Source: Raychem Corp.)

Leak Detection

Figure 15.8. Example of a cable located at the 6 o'clock position of an annulus at a gap in a straight section of pipe. Shown in the view is a spliced connection. (Source: Permalert ESP, Inc.)

aware of the mechanical effect of creating such an opening in an interconnecting component, and must determine its suitability through analysis (see Section 8.4). Figure 15.10 illustrates an example of a cable opening in an interconnecting component. Where an interconnecting part contains a solid annulus for purposes of creating separate annular compartments, leak-detection cable must be terminated at these locations, and a "zone-type" leak-detection cable system created. A continuous system can be provided if the system uses jumper cables to transfer from one compartment to another. It can also be accomplished by incorporating a feedthrough assembly into the annular cross section of the annular baffle/interconnecting part.

The minimum clearance required in straight piping is also the same for double containment fittings as well. A minimum clearance required at bends and elbows must be provided at the crotch of each elbow, when its orientation is as shown in Figure 15.11. There must also be sufficient clearance at branch connections to allow cable to be installed, and leak-detection cable branch connections or splices to be made. It is usually a requirement to position an access port next to a branch connection, regardless of whether a leak-detection cable system requires connectors or splices at branch locations.

In applications where installation is predicted to be difficult, a channel profile can be incorporated into the annulus as part of the design, as shown in Figure 15.6(d). In applications where any of the components would cause undue restriction of the leak-detection cable, serious consideration must be given to altering the design of the components or selecting a secondary containment piping that has a larger inside diameter and thus more generous annular space.

15.2.1.4 Access Requirements

Regardless of the type of cable selected, sufficient access must be provided in the layout of a double containment piping system. Without sufficient access, a leak-detection system cannot be installed easily, or be maintained and/or serviced readily. Specific requirements for access vary according to manufactured brand, and whether or not the leak-detection cable system uses connectors. Several general requirements that apply to all commercially available products are discussed in the following paragraphs.

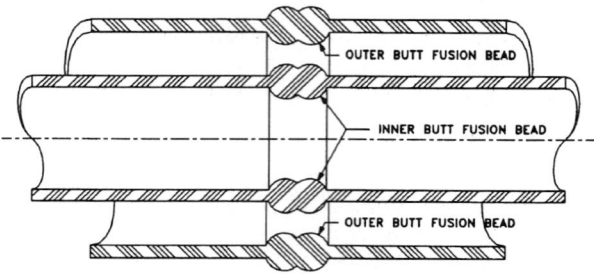

Figure 15.9. Example of internal and external joint protrusions. The height of these restrictions must be counted as occupying part of the annulus. The specified minimum required clearance must be provided in addition to these internal and external protrusions.

Access must be provided at each entry point of leak-detection cable to allow for the installation and removal of the cable. Additionally, access must also be provided at every cable termination location, including branches and sub-branches (if cable is terminated at such locations). This allows for a pull rope/cable to be pulled through a double containment piping annulus during the process of cable installation. Access is also required at each piping branching connection, whether or not the cable product uses branch connectors. Consideration should be given to providing access at all branch connections, regardless of the product used, as it greatly assists in the installation and maintenance of leak-detection cable. Table 15.1 describes the access that must be provided at all other intermediate points in a double containment piping system.

Access ports normally consist of standard secondary containment pipe tees that have a blind flange/feed-through assembly connected to its branch. The use of long-turn-tee-wye's (45° laterals with 1/8th bends) or 90° elbow/cleanout access ports should be kept to a minimum, as they add difficulty in installing cables when used as for this purpose.

An access port should be equipped with a riser that extends it to grade level for buried double containment pipes, as shown in Figure 15.12. The cable should be provided with a service loop, like that in Figure 15.13, which consists of an extra loop of sensing cable that reaches to the top of the access port. Access ports must have at least a 4 in. (100 mm) clearance in order to provide room for manual access to the cable.

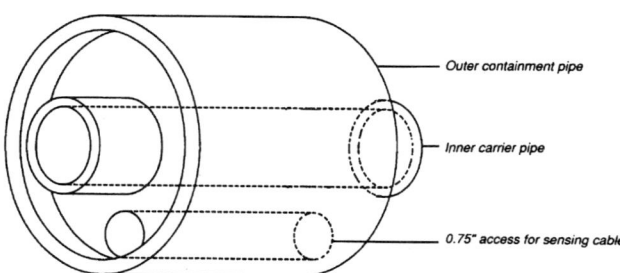

Figure 15.10. Illustration of required cable openings in internal anchors. (Source: Raychem Corp.)

Table 15.1 Typical Containment Access Requirements for Conductance-Based Cable

At the beginning and end of each pipe run.
At each branch or lateral that is sensed.
At intervals specified below.

Annular clearance	Required location of access points
< 1 in.	After 180° of pipe bend or every 250 feet of straight pipe
> 1 in.	After every 360° of pipe bend or every 400 feet of straight pipe

(Source: Raychem Corp.)

Access can be provided with a number of alternative arrangements. Some of the more common arrangements include: (1) vertical access; (2) horizontal access; (3) pipe-thread access; (4) intermediate access (including branch connection access); and (5) initiation/termination access. Two examples of typical vertical access are illustrated in Figures 15.13 and 15.14.

Horizontal access is illustrated in Figure 15.15, shown in conjunction with a termination flange. Pipe-thread access arrangements are a very economical means in situations where a threaded nonpressure joint is acceptable for use in a secondary containment piping system. An illustration of a typical pipe thread-access arrangement is shown in Figure 15.16. Intermediate access is illustrated in Figure 15.17 for straight piping and in Figure 15.18 for branch connections. Typical termination access, for a design that uses termination flanges, is illustrated in Figure 15.19.

15.2.1.5 Leak-Detection Cable System Component Detail

There are many individual components that collectively make up a continuous sensing cable system. The exact design of components varies by manufacturer, and not all systems require the same type of components. For instance, some cable manufacturers do not use connectors to install straight lengths of cable. Instead, they use continuous cable spliced together at the ends. Examples of a variety of cable constructions are illustrated in Figures 15.20 through 15.24.

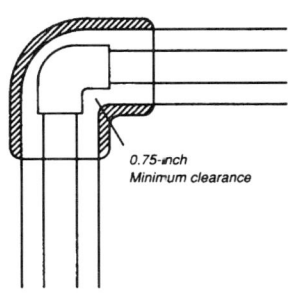

Figure 15.11. Minimum required clearance at elbows. The specified minimum clearance must be in addition to any internal or external weld beads or socket hubs. (Source: Raychem Corp.)

Figure 15.12. Examples of access port risers provided in a double containment straight pipe section.

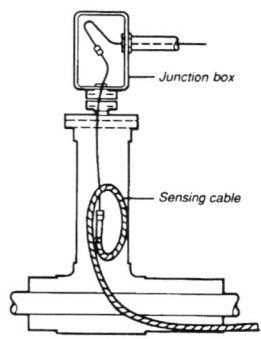

Figure 15.13. Illustration of a service loop provided in a double containment pipe vertical access riser. (Source: Raychem Corp.)

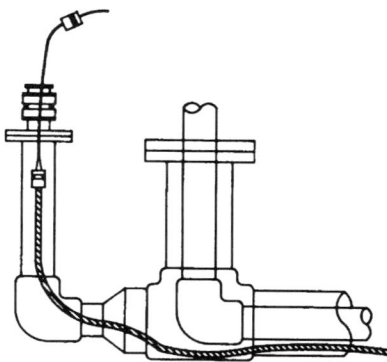

Figure 15.14. Illustration of a vertical access port provided by means of a 90° elbow located at the end of a straight run. (Source: Raychem Corp.)

Leak Detection

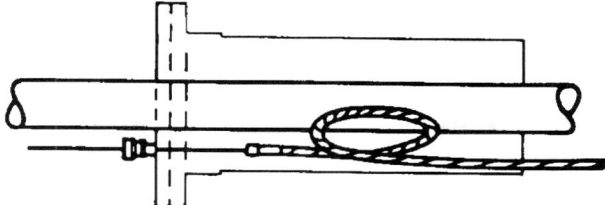

Figure 15.15. Illustration of a horizontally positioned access port provided in a termination flange arrangement. (Source: Raychem Corp.)

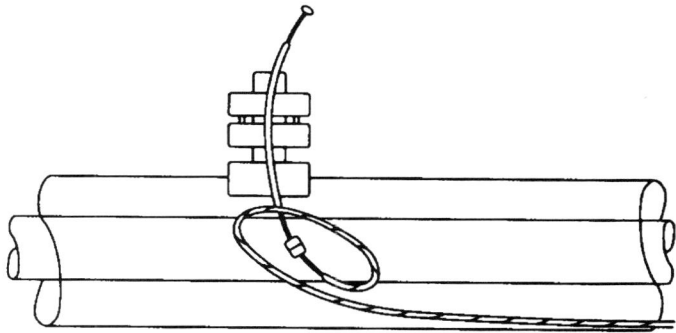

Figure 15.16. Illustration of a pipe thread access detail provided in a horizontally positioned double containment pipe. (Source: Raychem Corp.)

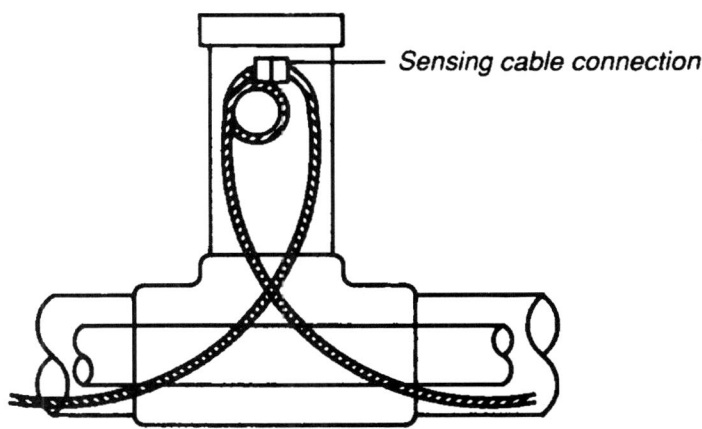

Figure 15.17. Illustration of an intermediate access port riser in a horizontally positioned section of a double containment pipe. (Source: Raychem Corp.)

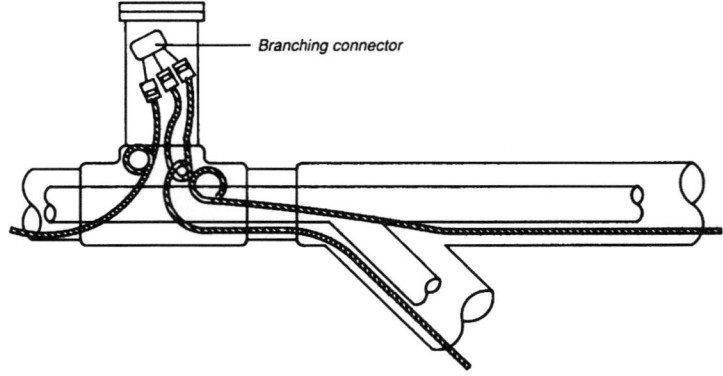

Figure 15.18. Illustration of an intermediate access port riser adjacent to a branch connection. (Source: Raychem Corp.)

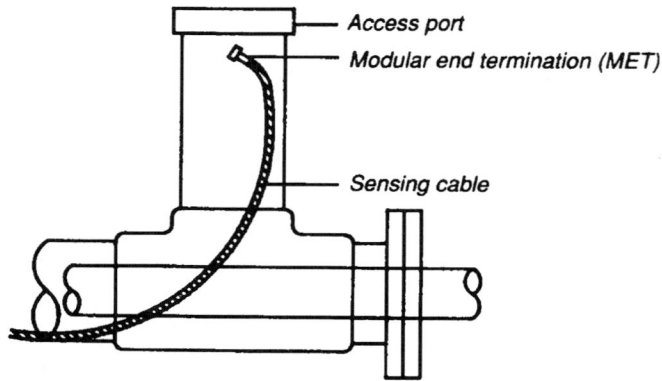

Figure 15.19. Illustration of a typical termination access port adjacent to a termination flange. (Source: Raychem Corp.)

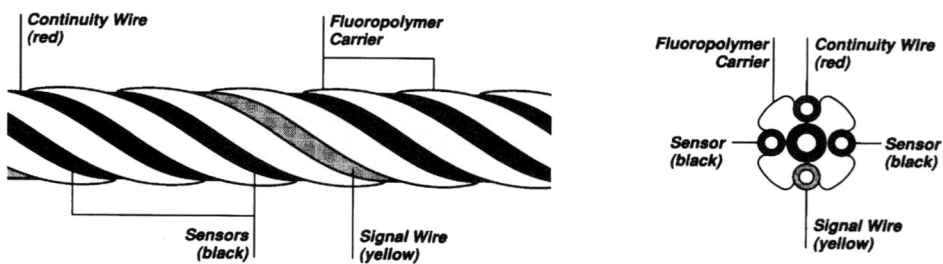

Figure 15.20. Example of the construction of a typical conductance-based leak-detection cable, which is specific to conductive fluids. (Source: Raychem Corp.)

Leak Detection

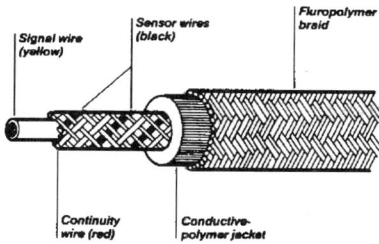

Figure 15.21. Example of the construction of a typical conductance-based leak detection cable that is specific to nonconductive fluids. (Source: Raychem Corp.)

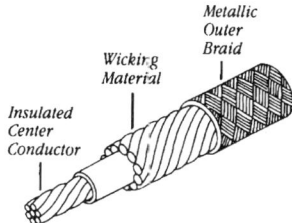

Figure 15.22. Example of the construction of a TDR-based cable designed to detect corrosive chemicals. (Source: Midwesco.)

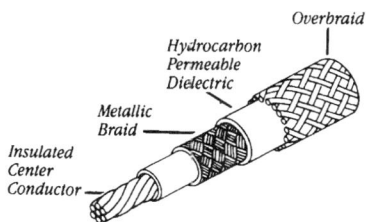

Figure 15.23. Example of the construction of a TDR-based cable designed to detect hydrocarbons. (Source: Midwesco.)

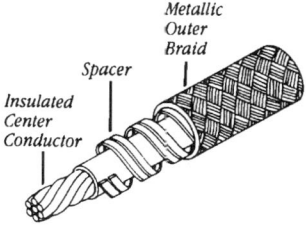

Figure 15.24. Example of the construction of a TDR-based cable designed to detect water and hydrocarbon fluids, in addition to steam and hot water up to 400°F. (Source: Midwesco.)

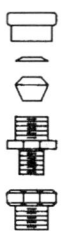

Figure 15.25. Blowup details of a jumper cable feedthrough fitting. (Source: Raychem Corp.)

The major components that may be encountered in any leak-detection cable system, aside from the alarm panel, consist of: (1) cable; (2) jumper cable; (3) jumper feedthrough fitting; (4) feedthrough assembly; (5) branch connector; and (6) termination device. All cable systems utilize cable, jumper cable, and some type of jumper feedthrough fitting. However, the actual design of the components varies considerably from manufacturer to manufacturer.

A typical jumper feedthrough fitting is illustrated in Figure 15.25, and a typical feedthrough assembly is shown in Figure 15.26. Jumper feed through assemblies are used whenever a cable system exits a pipe. Jumper cable is run outside of a double containment piping system through conduit, to connect the cable to the alarm module.

15.2.1.6 Installing Leak-Detection Cable; Requirements of the Pipe Installation

The installation and welding of piping is critical to the successful installation of a continuous-sensing-cable system. The sequence of joining, types of welds to be performed, and the amount of annular clearance, all have a direct bearing on its installation. Therefore, installation of double containment piping must be closely coordinated with the installation of leak-detection cable. By incorporating cable into the system, certain additional installation needs are imposed on the piping system. This section describes the major aspects of the piping installation that require thorough review or change, in order to incorporate a leak-detection cable system successfully.

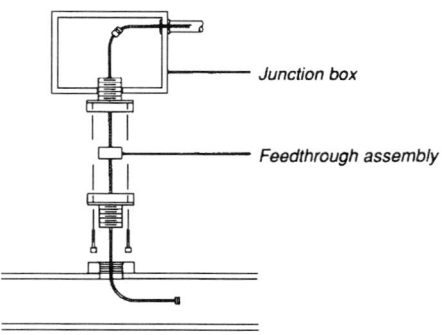

Figure 15.26. Illustration of a jumper cable feedthrough assembly. (Source: Raychem Corp.)

The annulus between primary and secondary containment piping must be maintained in a clean and dry manner. If the annulus of a double containment piping system does become contaminated during any of the stages described, it must be cleaned and dried prior to final assembly. All water, oily residues, and solvent traces from manufacture of components or welding procedures must be removed. Manufacturers who supply double containment piping assemblies must be consulted as to their specific recommendations for cleaning and drying of a piping annulus, based on the materials, components, and layout involved.

The minimum annular clearance specified by a leak-detection product manufacturer must be maintained at the 6 o'clock position of the annulus in all components. Care should be taken in installing centering devices so that they are securely attached in their proper position and orientation in the annulus. They should be restricted from moving out of position so that they do not interfere with leak-detection cable installation and with the maintenance of cable components.

When constructing pipe subassemblies, a pull rope must already be in place at the bottom of the pipe. The purpose of pull rope is to pull sensing cable into place, after subassemblies have been completely assembled, tested, cleared, and dried. Pull rope should consist of a 1/4 in. (6 mm) minimum-diameter hollow-braid polypropylene rope (same quality as that for water skiing). Each length of pull rope used between access points should be one continuous piece, where possible. The practice of splicing together more than one section of pull rope should be kept to a minimum due to the adverse consequences that can result, should the knot break. The length of each pull rope should be a minimum of the distance between access points, plus 6 ft (2 m).

One of the keys to a successful leak-detection installation is the ability to have a freely moving pull rope in place prior to pulling in leak-detection cable. Therefore, it is essential that the pull rope be free to move once it is installed into an annulus. In most cases, to install a pull rope successfully, it must be installed at the same time as a double containment piping system.

Prior to installing pull rope, an electrician's "fish tape" may be used to insert or pull a pull rope into place through a full length of pipe. Systems that use staggered-welding (bonding) assembly techniques (see Sections 12.3.2) must have pull ropes installed as the piping installation progresses. Systems that use simultaneous-fusion techniques (see Section 12.3.3) may require that pull rope be "fished" through with a "fish tape" after assembly of each pipe section (weld or bond). Fish tape should be 1/4 in. (6 mm) fiberglass electrician's "fish tape," in order that it may be easily "fished" through a double containment pipe system's annulus to each weld joint (bonded joint). After completing a simultaneous weld (bond), pull rope should be snagged with a "fish tape" and immediately pulled through to the other side. Lubricants should not be applied to the pull rope to assist in its installation, due to the fact that it might contaminate an annulus and thus the leak-detection cable as well. It is always more difficult to incorporate leak-detection cable into a system that uses a simultaneous-fusion joining sequence. However, recently split heating elements with holes have been developed that allow the pull-wire to be in place during the simultaneous fusion process.

When installing pull rope, it may be necessary to secure it to primary piping to be able to make certain types of secondary containment pipe welds (bonds). This is common during staggered-welding (bonding) procedures to avoid damaging the pull rope for reasons of heat, spark, solvent-cement, or adhesive-bonding-agent damage. When this needs to be

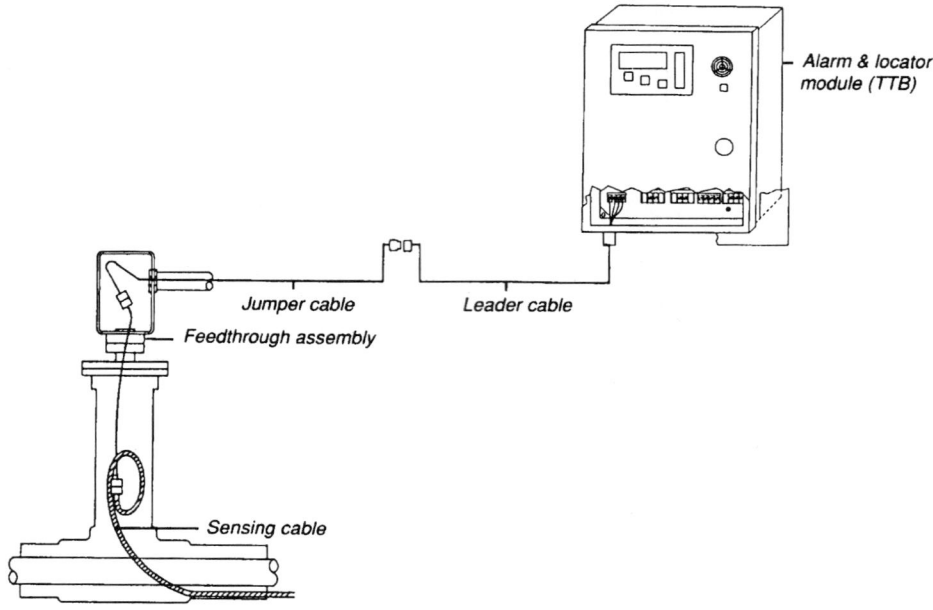

Figure 15.27. Schematic of pertinent details of a leak-detection cable system. (Source: Raychem Corp.)

done, the pull rope must be secured to a primary pipe by means of ordinary masking tape. The use of duct, electrician's, friction, or glass-reinforced tape should be avoided.

It is also very important to pressure test both primary and secondary containment systems prior to installation of leak-detection cable. The sudden release of pressurized gas caused by the failure of a pipe during a pneumatic test could cause damage to leak-detection cable components. Contact with water in a hydrostatic test would require that a cable be extracted, dried, and reinstalled after such contact. Therefore, the pull rope should be left in place for the duration of pipe testing. Leak-detection cable should be installed only after the successful completion of primary and secondary containment piping tests.

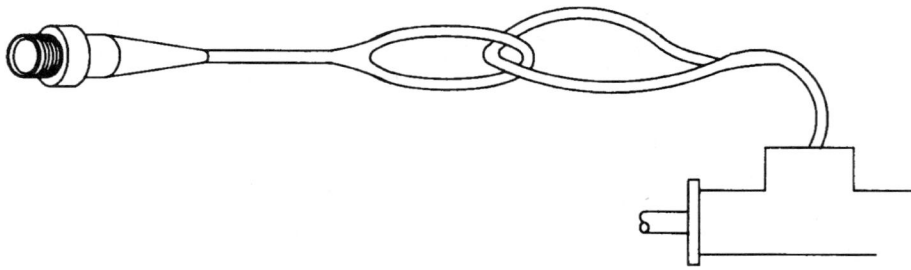

Figure 15.28. Illustration of a pulling tool preparation detail. (Source: Raychem Corp.)

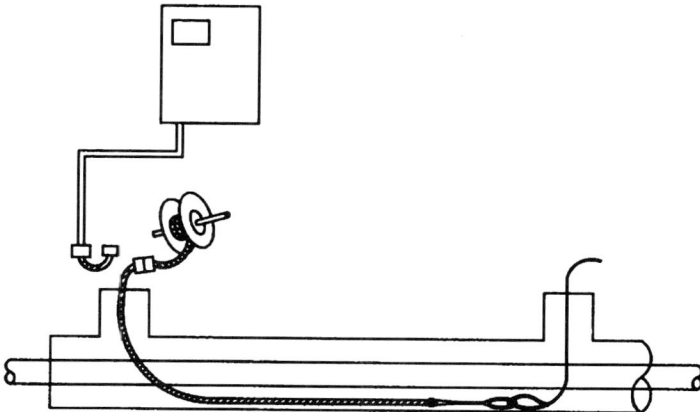

Figure 15.29. Illustration of leak-detection cable being pulled through an annulus between consecutive access ports. (Source: Raychem Corp.)

15.2.1.7 Installing Leak-Detection Cable; General Requirements

Once a double containment piping system and pull rope have been successfully installed and tested, leak-detection cable has to be installed. Leak-detection cable installation involves many considerations beyond those of the piping, and the procedure can be divided into five sequential steps, described as follows:

1. The alarm module must be installed/mounted and the power turned on. The alarm module should be tested at this point.
2. The leader cable (where used) should be connected to the jumper cable, and the jumper cable connected to its electrical junction box at the cable-entry access point provided to the annulus of the double containment piping system. The completed arrangement is illustrated in Figure 15.27.
3. A last check should be made on the system to ensure that the annulus is clean and dry, and that the pull rope is free and loose. The pressure testing records must also be checked at this point.
4. Leak-detection cable may now be installed. A pulling tool is usually required to be connected to the pull rope. This may be done by looping the pull rope through a pulling tool and securing it with electrician's tape. It is important to keep the diameter of any taped area within the minimum required dimensions of the annulus. This is illustrated in Figure 15.28. The other end of a pulling tool can be threaded if the leak-detection cable is equipped with a connector having a female-threaded end. Leak-detection cable can then be pulled through the annulus from access point to access point, as illustrated in Figure 15.29.
5. After each cable length is installed, it must be tested. If the system is installed with modular lengths and connectors, the first length installed should be connected and tested. Other lengths can then be installed, connected, and tested in sequence. If connectors are used, they should be covered with a heat-shrunk polyethylene tube prior to insertion.

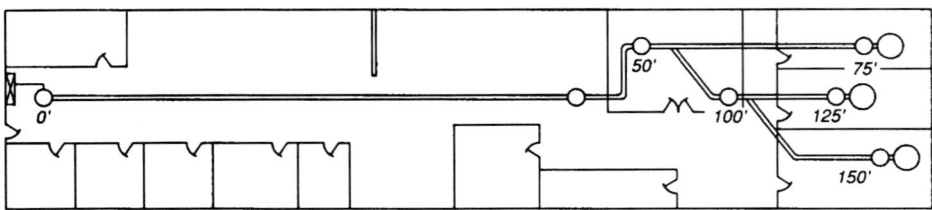

Figure 15.30. Example of a typical leak-detection system "map." (Source: Raychem Corp.)

15.2.1.8 Purging Requirements

In any double containment system, it is always recommended to purge the annular space with an inert gas. Purging refers to the replacement of ambient air in the annular space with a dry gas that is inert to both piping and leak-detection system components. For nonmetallic systems, nitrogen is normally the inert gas used. However, clean, dry air can also be used. For metallic systems, argon is sometimes used due to the potential of nitrogen embrittlement of some metals. Purging is a good idea in any system, but it is particularly important for systems that use leak-detection cable. For additional information regarding purging of a double containment pipe system annulus, refer to Section 15.2.4, "Condensation Effects in Annuli."

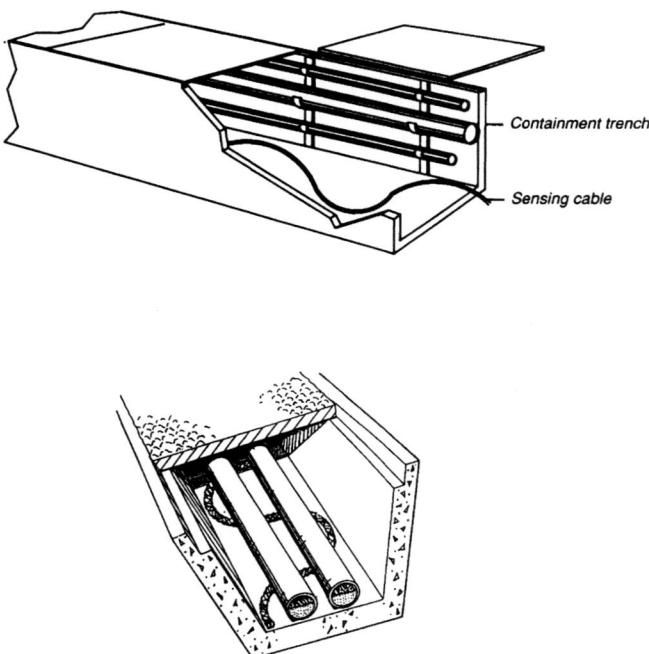

Figure 15.31. Two examples of a leak-detection cable installation in piping contained within secondary containment trenches. (Source: Raychem Corp.)

15.2.1.9 Mapping Requirements

After the leak-detection cable system is completely installed, a system map must be created. This is a graphic display map that must accurately depict the layout of leak-detection cable on an as-built basis. The system map must indicate the location of the alarm module, cable connectors, and all access points of the cable. The map should indicate actual cable distance readings, related to easily identifiable building landmarks. Recognizable landmarks should be related to such cable system components as entry points, access points, and end seals. The graphic display map should be constructed of a moisture-resistant, durable material. It should be mounted near an alarm module in a conspicuous area. Backup copies must be created and kept on file. Without the ability to access an accurately created system map readily when a leak occurs, the location of a leak will not be able to be determined readily. A typical system map for a chemical waste double containment piping system is illustrated in Figure 15.30.

15.2.1.10 Inspection, Examination, and Testing of Leak-Detection Cable

After installing any leak-detection cable, it must be inspected to verify that it is in compliance with the engineering design, and with the plans and specifications. The cable and modules should be thoroughly tested to see that there is continuity, and that the system is functioning properly. Every connection in the system should be subject to 100% visual inspection to ensure that it is installed properly and appropriate heat-shrink tubing added as well. The system map should be inspected to verify that it accurately reflects the layout details of the installation. The individuals responsible for the operation of the system should be provided with sufficient training. An operation manual should be provided and kept on file in a readily accessible place. The inspected items should be recorded in a permanent inspection report, and the records maintained in a permanent file.

15.2.1.11 Installation of Leak-Detection Cable in Trenches

The installation of leak-detection cable in secondary containment trenches requires a layout in which it is snaked in a sine-wave pattern on the bottom of the trench. The period of the sine wave should touch the sides of the trench, and should be approximately 4 to 6 ft (1.2 to 1.8 m) in length. At this value, the ratio of available cable to trench is at its maximized economic value. Typical trench leak-detection cable arrangements are illustrated in Figure 15.31.

When installed in trenches, leak-detection cable must be secured with the use of hold-down clips that are secured with plastic adhesive, as shown in Figure 15.32. Pipe must be positioned on structural supports such that a minimum 0.75 in. (18 mm) clearance (for

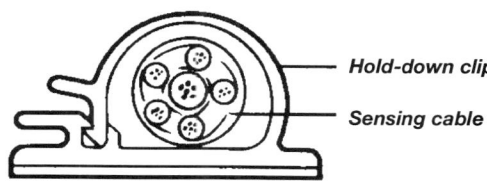

Figure 15.32. Illustration of two types of hold-down clips. (Source: Raychem Corp.)

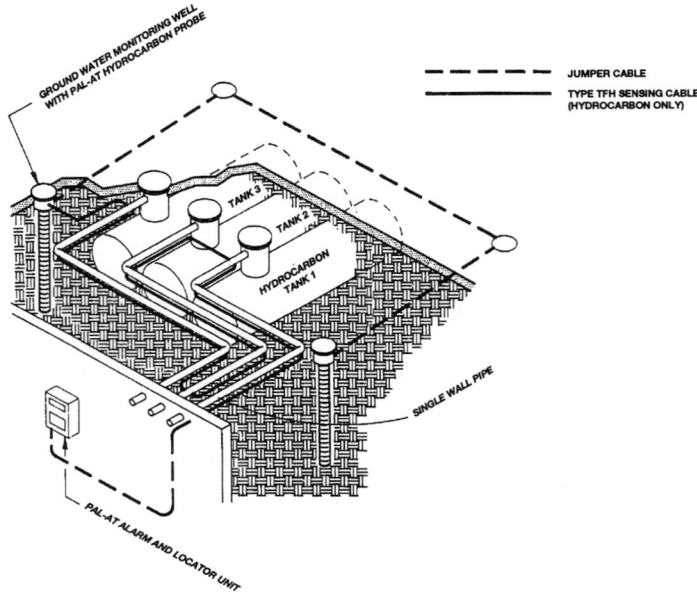

Figure 15.33. Typical installation for leak-detection cable in an application involving single-wall piping and tanks for hydrocarbon liquids. (Source: Midwesco.)

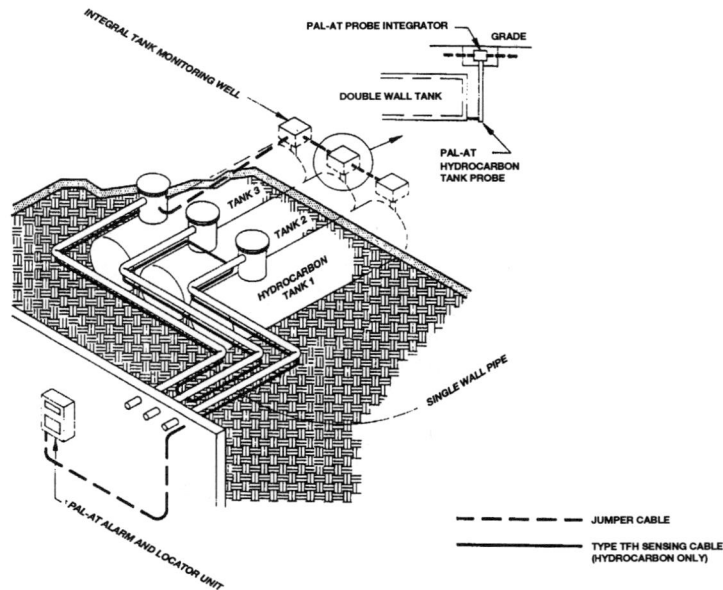

Figure 15.34. Typical installation for leak-detection cable in an application involving single-wall piping and double-wall tanks for hydrocarbon liquids. (Source: Midwesco.)

Leak Detection

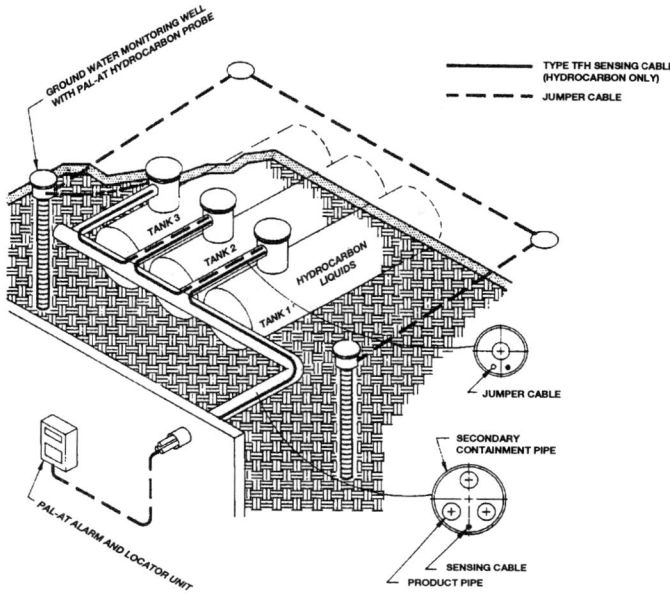

Figure 15.35. Typical installation for leak-detection cable in an application involving double-wall piping and single-wall tanks for hydrocarbon liquids. (Source: Midwesco.)

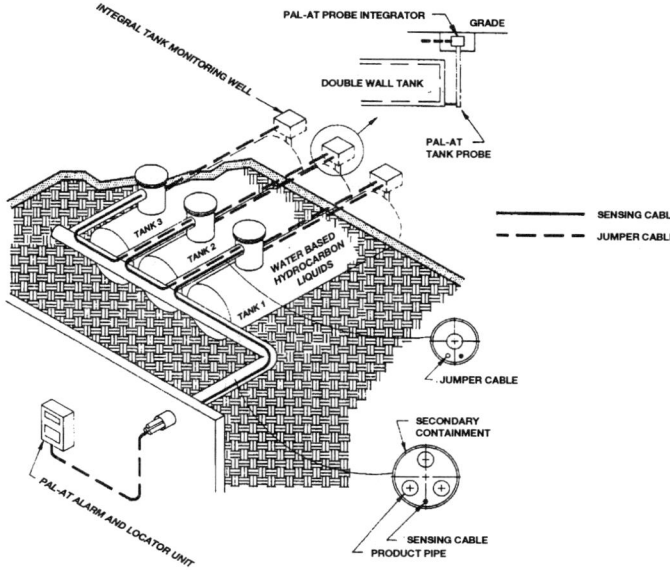

Figure 15.36. Typical installation for leak-detection cable in an application involving double-wall piping and double-wall tanks. (Source: Midwesco.)

resistance-based cables only; TDR cables require greater clearance) is maintained between the bottom of the pipe and the bottom of the trench. Other considerations for installing leak-detection cable in a double containment piping annulus remain the same for installation in a secondary containment trench, including testing and mapping requirements. See also Section 15.4.6, "Monitoring of Secondary Containment Structures."

15.2.1.12 Leak-Detection Cable Used in Combined Piping and Tank Systems

Figures 15.33 through 15.36 show layout details for leak-detection cable systems used in combined piping and tank systems. Figure 15.33 illustrates the general layout for a single-wall piping and tank application. Figure 15.34 illustrates the general layout for an application involving single-wall piping and double-wall tanks. Figure 15.35 illustrates the use of cable in conjunction with double containment piping and single-wall tanks. Figure 15.36 illustrates the use of cable where double containment piping is associated with double-wall tanks.

15.2.2 Point-Probe Systems

Leak detection may be effectively provided by dividing a double containment piping system's annulus into separate, isolated leak-detection compartments. Each compartment may then be monitored with some form of probe measuring device. The types of probes typically used include: (1) liquid level sensing; (2) moisture detection; (3) vapor detection; (4) conductivity/resistivity; (5) pH measurement; (6) pressure sensing; (7) flow measurement (primary piping); (8) density measurement; (9) wavelengths (light, radar, and sonar); (10) motion detection; and (11) others. While providing compartments in an annulus may be the most efficient means of incorporating a point-probe-based detection system, it is not the only approach. Point-probe systems may also be used in a system designed with a continuous annulus, although the ability to locate a leak reasonably is significantly less.

Normally, probes are housed within a branch outlet, or reducing-branch outlet, which is provided in the secondary containment pipe. Examples of downward and upward positioned arrangements are shown in Figures 15.37 and 15.38, respectively. Such arrangements often function as a means to house a probe, in addition to functioning as a combination low-point drain (high-point vent) and manual detection point. Normally, probe systems are more readily applied to above-ground systems. In underground systems, probes can be attached at manhole locations using the branch arrangement illustrated in Figure 15.39, immediately after piping/manhole penetration.

15.2.2.1 Liquid Level Sensing

Detection of a fixed level of liquid is among the simplest probe systems to design and implement in a double containment piping system. A liquid level system will function in a very reliable manner to detect leaks. However, it will not distinguish between water (including ground water or condensate) and service fluids. In this type of system it is important to have a pressure-tight annulus and to eliminate condensation by providing a moisture-free inert gas in the annulus (both items are important for most methods of sensing leaks). A liquid level detection system can be implemented by installing a liquid level measuring device in a vertically positioned drop-leg branch, as shown in Figure 15.37 and 15.39, or in the low-point sumps shown in the photographs in Figures 15.43 and 15.44.

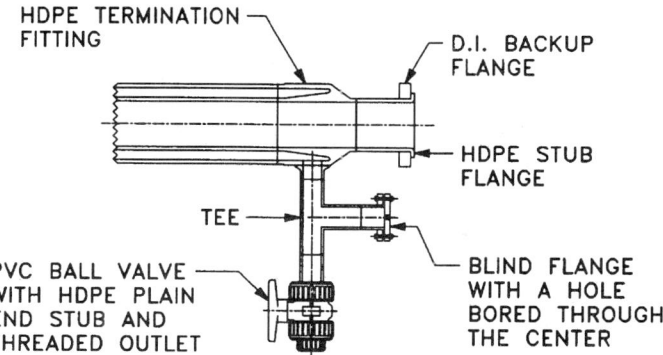

Figure 15.37. Photo of a downward positioned probe attachment for a horizontally mounted liquid level sensor future attachment.

A liquid level probe (or probes) may be connected to an alarm panel that signals an alarm when a specified level of liquid is detected. The system is good for establishing that a leak in a certain zone of an annulus. It will not allow the source of a leak to be specifically identified within a leak zone, however. Also, it is not designed to detect small leaks quickly caused by a spraying, but will rapidly respond to major leaks. A liquid level sensing system may be coupled with other types of probes to distinguish between types of fluids, resulting in a very effective overall approach to detecting leaks.

15.2.2.2 Moisture Detection

Moisture probes are useful for detecting the presence of ground water or service fluids that have a high water content. There are a wide variety of types of probes available; the sensitivities of available probes vary widely. The wide range of moisture probe styles available allows a system designer to select a customized system designed for a specific application. Probes can be positioned as shown in Figure 15.37.

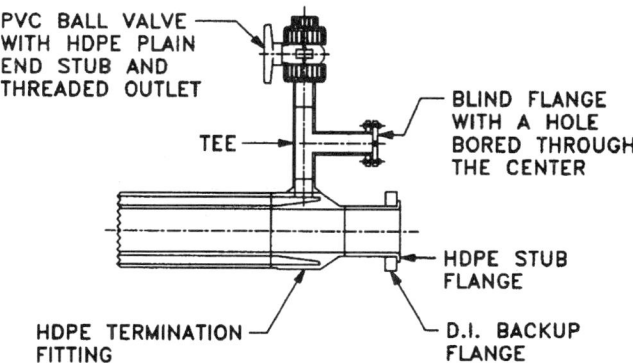

Figure 15.38. Photo of an upward positioned probe attachment for a vertically mounted pressure sensor.

A moisture probe system can be implemented by installing a moisture measuring device in a vertically positioned drop-leg branch. This arrangement is similar to the liquid level sensing arrangement described in the Section 15.2.2.1, although a branch for a probe of this type need not be at the 6 o'clock position of an annulus. Most sensitive probes can be positioned at the 12 o'clock position of an annulus, since they detect fine vapors. Therefore, an arrangement such as that illustrated in Figure 15.38 may be used.

A moisture probe (or probes) may be connected to an alarm panel that signals when a change in moisture content of the annular space is detected. Such a system is good for verifying that a leak of a primary service fluid has occurred in a certain zone of the annulus. It will not allow the position of a leak to be specifically located within a leak zone. However, it can be good for detecting small leaks if the probe used is highly sensitive.

15.2.2.3 Vapor Detection

Vapor-detection systems are very similar to moisture-detection systems, except that vapor probes detect volatile chemicals other than water. This type of probe is normally designed to detect fluids that have high vapor pressures, including chemicals, halogens, and other chemicals that are relatively volatile. There are many types of vapor probes available with varying values of sensitivity. The type of organic chemical being carried in a primary piping system will determine the type of probe that should be selected for any given application. The sensitivity of the probe to be selected is based upon the leak desired to be detected (e.g., small leak versus major leak), although most vapor probes tend to be highly sensitive compared to other forms of leak detection. When an earlier response is desired, a probe that has a greater sensitivity can be specified.

15.2.2.4 Conductivity (Resistivity) Measurement

The conductivity or resistivity of an annular space can be measured to detect whether a leak has occurred. This type of probe measurement system is most effectively used in conjunction with a liquid level or pressure sensing system as a means of classifying the type of fluid being sensed. In such an arrangement, liquid level or pressure monitoring may be used as the primary system to detect the presence of fluids. The conductivity (resistivity) probe will then help to distinguish between water from ground water or condensation and the service fluid (which must be of the conductive type), based upon a change in conductivity. Conductive fluids include such fluids as acids, bases, salts, other organic chemicals, and a host of polar and semipolar organic solvents. Conductivity probes have also been used as the sole means of detection for double containment piping systems.

Conductivity probe systems can be incorporated into a double containment pipe annulus by installing a conductivity measuring device in a vertically positioned drop-leg branch (see Figures 15.37 and 15.39). The probe (or probes) may then be connected to a computer or recording device to record the conductivity continually, or may be directly connected to an audio alarm. Monitoring conductivity is good for identifying the fluid that has leaked, but can also be used as the sole means to sense for leaks. Conductivity probes can detect small leaks in addition to major leaks. In a very general sense, leak-detection cable is a form of conductivity (resistivity) probe. This is not the primary purpose of leak-detection cable, but it has been used in this manner. Therefore, a short section of leak-detection cable could theoretically be used as part of a point sensing approach.

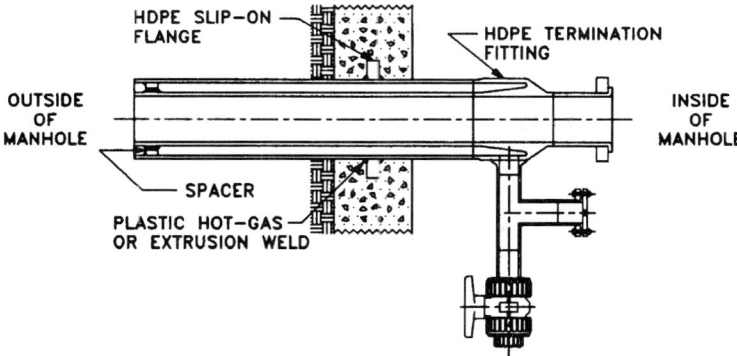

Figure 15.39. Illustration of a probe attachment in a manhole location.

15.2.2.5 pH Probes

A pH measurement probe functions in much the same way as a conductivity (resistivity) measurement system. However, pH measurement is limited to detection of acids or bases since only these types of chemicals can be measured in terms of pH. Since pH measuring devices are mostly designed to function in a wetted capacity, they normally are used as a secondary means of detection, in combination with some other type of probe. A pH probe system can be implemented by installing a pH measuring device in a vertically positioned drop-leg branch (see Figures 15.37 and 15.38). The probe (or probes) may be connected to an alarm panel that signals when a change in pH is detected.

15.2.2.6 Flow Measurement

A change in flow of primary piping may be related to the possibility that a leak has occurred. The means for accomplishing flow monitoring is to place a flow measuring device at the beginning and end of the piping system. If the flow of the primary pipe changes by more than a set amount, an alarm can be sounded. This method is good for detecting large leaks. However, it may not detect a fine spraying of fluid, such as would initially occur in a ductile material that is subject to stress cracking (e.g., polypropylene). Flow measurement can be very effective when used in conjunction with other methods.

15.2.2.7 Pressure Sensing

The pressure of an annular space can be measured to determine changes in pressure. A pressure change can be used to determine whether primary and/or secondary containment has been breached, making it a very useful form of monitoring. The value and sign of the pressure change depend on the relative pressures of the annular space, the primary (core) pipe, and the pressure of the surrounding local atmosphere. For pressurized primary piping systems, a failure of the primary piping systems can be detected by a pressure rise in its annular space (assuming the annular space is maintained under a lower pressure than its primary pipe). In pressure piping systems where a positive pressure is maintained in its annular space, at an intermediate pressure between the primary pressure and the external environment, a drop in annular pressure will reflect a failure (breach) of the secondary containment system.

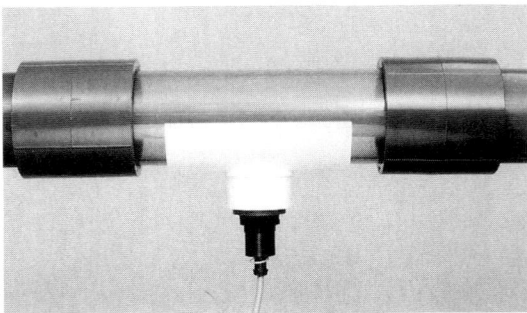

Figure 15.40. Example of an externally attached density probe. (Source: Guardian div. of Nibco, Inc.)

For drainage systems that maintain an annular space under a positive pressure, a pressure drop can signal a leak in the primary piping. However, a pressure drop in this type of system could also mean that the secondary containment has been breached. Therefore, a further investigation has to be made, unless a separate probe is added to provide additional information as to the origin of the leak.

Gas that is maintained under a low pressure tends to find its way out of threads, valve seals, etc. over a period of time. Therefore, some loss of gas could occasionally be expected from secondary containment threaded joints, and resulting false (low-point) alarms triggered. This must be taken into account in operational training procedures so that false alarms do not lead to a leak-detection system's being ignored.

The pressure of a primary piping system can also be monitored. A substantial drop in primary piping pressure will often mean that a leak has occurred somewhere over its length. This type of approach is useful for detecting large leaks.

The pressure of a double-wall tank's interstitial space can also be measured to determine if changes in pressure of a contained gas or liquid (e.g., brine) occur. The value and sign of the pressure change depend on the relative pressures of the interstitial space, the primary tank, and the pressure of the external atmosphere. Most double-wall tanks are for low-pressure service. Therefore, it may be difficult to distinguish between a breach of primary or secondary containment of a double-wall tank, in most applications.

15.2.2.8 External Density Sensing Probes

A proprietary probe method has been developed and patented in the recent past which allows the use of an externally attached probe sensing device to detect for the presence of annular fluids. The method, developed by the Guardian division of Nibco, Inc., uses density measuring devices, attached to the external surface of the secondary containment piping, as shown in Figure 15.40. The devices are activated when a change in density is detected.

15.2.3 Noncontinuous Sensing of Annuli

Noncontinuous sensing of double containment piping systems includes those methods that do not involve the incorporation of any type of continuous measuring device, or instrument, to the secondary containment piping system. The three main categories of noncontinuous sensing include manual detection, visual detection, and periodic annular pressure monitoring.

15.2.3.1 Manual Detection Methods

Manual detection typically is accomplished by adding valves into a secondary containment piping system at various low points in its annulus. The low-point valves can be occasionally opened to detect the presence of fluids. Such valves are normally positioned on drip legs, usually at each low point and other intermediate points of the double containment piping system annulus. Typical low-point drip legs incorporating the use of manual testing valves for above-ground double containment piping systems are illustrated in Figures 15.37 and 15.39. For underground systems, access is required. The same arrangements may be used, but they must be positioned in a manhole, as illustrated in Figure 15.39. Common methods of manual leak detection for underground nonpressure chemical waste systems are illustrated in Figures 15.41 and 15.42. These low-point sumps may be accessed by means of a removable top, and some manual or visual means of leak detection (e.g., manual gauge stick) used to determine if a leak has occurred. Automatic liquid level sensing systems can also be provided on these types of arrangements. Figures 15.43 and 15.44 are photographs of manual sensing sumps in plastic and metallic systems, respectively. When using a system of this type, more than one sensing point is highly recommended.

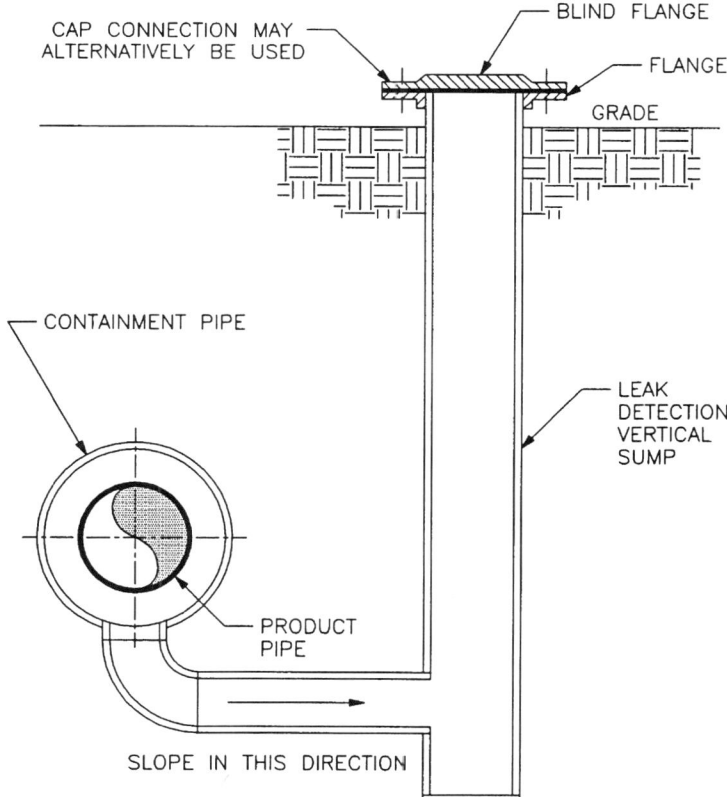

Figure 15.41. Example of a low-point sump.

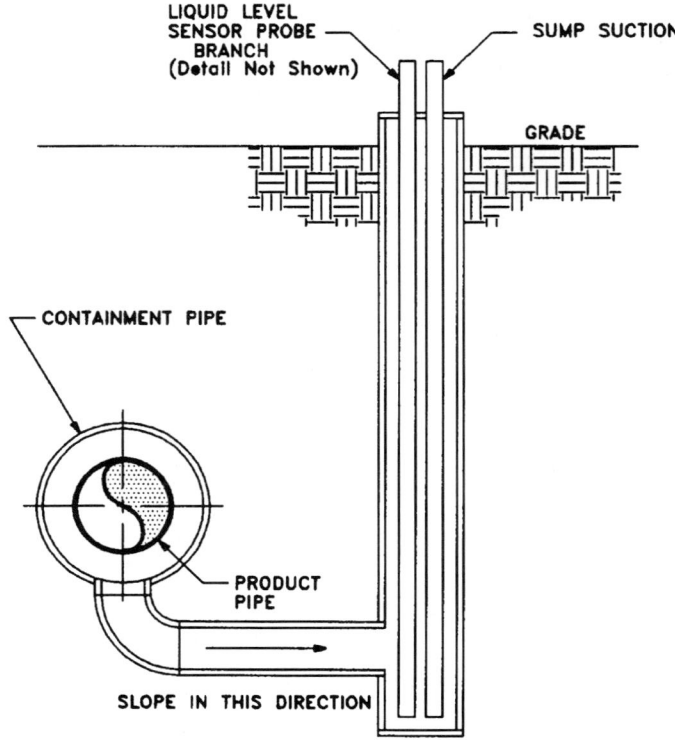

Figure 15.42. Example of a low-point sump showing sensor probe branch attachment and a sump suction tube.

Manual leak-detection methods are only effective when used as part of a regular inspection program, where the system is routinely inspected. Manual systems serve as a very effective secondary means of leak detection, used in conjunction with any other method or methods. It is usually a built-in feature to most well-designed systems, where appropriate low-point drains are provided (see Section 11.1.1.3).

15.2.3.2 Visual Detection Methods

Visual detection may be provided in a number of ways. It can be provided for the entire secondary containment piping system or any partial portion. Visual leak detection can be provided by using a transparent material (borosilicate glass, clear PVC, etc.) as the outside piping (and fittings where applicable). A more practical application involves flanging in a short section of clear piping, or providing clear piping or sight glass as part of a secondary containment drip-leg branch. Visual containment can also be provided where an "open-ended" secondary containment piping system is used, by observing the end of the pipe for fluid leakage. Visual detection is most effectively applied as part of a regular inspection program where the system is routinely inspected. Visual leak-detection systems also serve as a very effective secondary means of leak detection, used in combination with any other method.

Figure 15.43. Example of a low-point leak-detection sump in a plastic gravity drain system.

15.2.3.3 Periodic Pressure Monitoring

The pressure of an annulus can be occasionally monitored to detect if a leak has occurred. To do so, a pressure gauge (instead of a transducer) must be provided in an annulus in at least one position. A typical connection for providing a pressure gauge in an annulus is illustrated in Figure 15.45.

It may be desirable for some installations to maintain an inert gas under pressure and monitor the pressure of the gas (refer to discussion in Section 15.2.2.7). In applications involving pressure-rated primary piping, the annulus does not have to be maintained under a fixed pressure. A rise in pressure will signify that a breach of primary containment has occurred. However, it is not possible to detect a breach of secondary containment in such a system, unless a fixed pressure is maintained in the annulus.

Figure 15.44. Example of a low-point leak detection sump in a metallic fuel oil system.

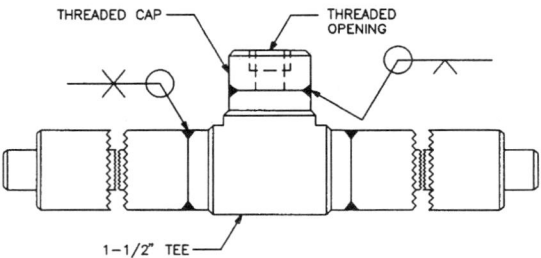

Figure 15.45. Illustration of a pressure tap in a secondary containment pipe for attachment of a pressure gauge.

15.2.4 Condensation Effects in Annuli

If air that is warm and relatively humid is introduced into the annulus of a double containment piping system prior to final closure, moisture will condense when the system is subjected to a colder temperature. This often occurs when a system is constructed during the summer season and later subjected to cold winter operating temperatures.

Condensation is a common concern among designers and owners-to-be of double containment piping systems, and for very good reasons. The principal concern rests with the fact that a leak-detection system may interpret this action as a false alarm. It does depend to some extent on the sensitivity of the leak-detection system selected. However, no leak-detection method is immune to this effect, including visual and manual means. Frequent false alarms may subsequently lead to ignoring actual alarms on the part of plant operations personnel, a very dangerous practice, and one that is self-defeating to the system. Also, metal materials could be subjected to various corrosion effects produced by the water (galvanic, pitting, microbiologically induced corrosion, etc.). Therefore, there are many reasons to avoid the possibility of condensation in an annulus.

The solution to the condensation dilemma is a simple and inexpensive one. Replace the original air with a dry or inert atmosphere. Alternatively, a vacuum may be pulled in the annular space, if inner and outer pipe components are mechanically strong enough to withstand vacuum conditions. It has been assumed by many that this is a very involved and expensive activity. However, if an annulus is designed in a leak-tight manner, and has been provided with adequate high-point vents and drains, this task will actually be one of the least expensive aspects of the entire project. The tables provided in Section 15.1.4.2 show that, for a majority of systems, it takes a relatively small amount of nitrogen to decrease the moisture content of a system to the equivalent of the saturation level (dew point) at -20°F, a level at which no condensation could be expected in most applications. Since bottled nitrogen currently (1994) costs approximately $0.14 - 0.15 (U.S.) per cubic foot when supplied in bottles that hold the gas at 2200 psi, the cost for nitrogen will typically be a fraction of the overall cost of the project. The data assume 100% mixing in all cases, which is an extremely conservative assumption.

The discussion in the preceding paragraph is valid for systems where the annular space is closed in a leak-tight manner. However, in a system that terminates in an open end (including at a manhole entry), natural draft circulation will occur due to temperature differences between the gas in the annulus and the outside ambient air. This will eventually mean that

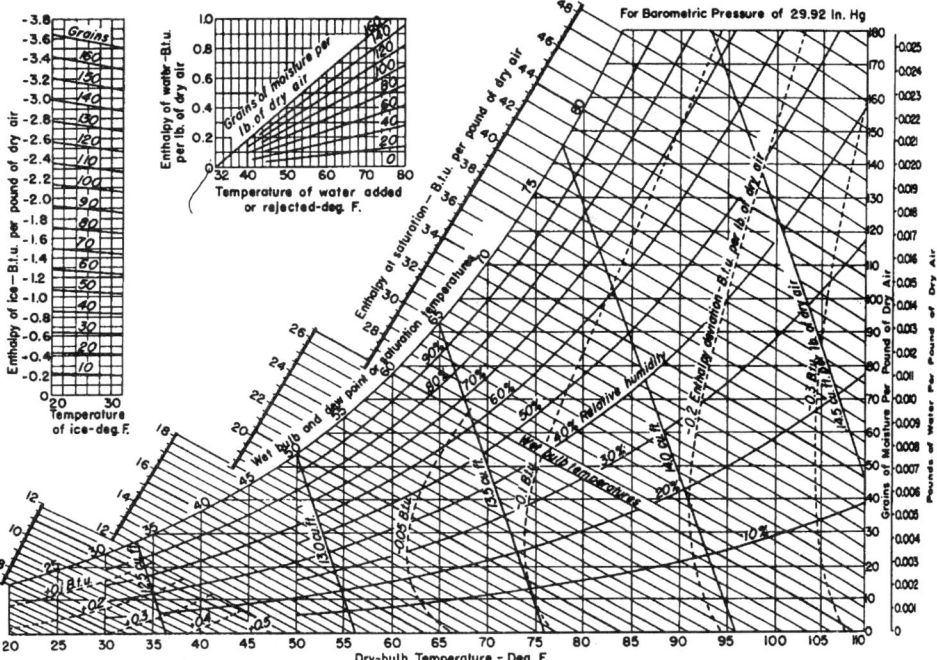

Figure 15.46. Psychometric chart for low to medium temperature ranges.

moist air will return to the annular space through this process. The answer to this dilemma is that the atmosphere should be replaced initially, and then periodically recharged with the same dry gas to displace the moist air that periodically returns. The period of time for replacing the air will depend on the specifics of the climate, the season of the year, and the layout of the system. Another alternative is to install an air line in the beginning of the annulus, from a source that continuously puts out relatively dry air at a pressure slightly higher than atmospheric, thereby continually displacing the volume of air in the system. This would most likely be an overly conservative and expensive approach for most systems, but an alternative none the less. In general, close-ended systems are preferable to open-ended systems for the reason that recharging is not needed on a frequent basis.

15.2.4.1 Condensation Tables/Psychometric Chart

Air has a greater capacity to hold moisture at higher temperatures. The total amount of moisture that air can hold at any given temperature is termed its saturation level, values of which are well known. Any moisture that is in excess of this saturation level will condense; as air increases in temperature, the drops of liquid that have condensed will tend to evaporate until the air becomes saturated.

Figures 15.46 and 15.47 are psychometric charts containing the properties of air at most temperatures. The total moisture at saturation at various temperatures can be found from the saturation curve.

Table 15.2 Moisture Content of Annuli (in pounds) for 1,000 ft Long Double Containment Straight Pipe Sections of Common Size Combinations

Size	Starting condition of air in the annulus		
	90°F (32°C)/90%R.H.	80°F (27°C)/80%R.H.	70°F (21°C)/70%R.H.
1" IPS/2"Sch.10	0.03	0.02	0.01
1" IPS/3"Sch.40	0.09	0.06	0.03
2" IPS/3"Sch.10	0.06	0.04	0.02
2" IPS/4"Sch.40	0.12	0.08	0.05
3" IPS/4"Sch.10	0.07	0.04	0.03
3" IPS/6"Sch.40	0.28	0.18	0.11
4" IPS/6"Sch.10	0.23	0.15	0.09
4" IPS/8"Sch.40	0.49	0.31	0.20
6" IPS/8"Sch.10	0.29	0.18	0.11
6" IPS/10"Sch.40	0.63	0.41	0.25
8" IPS/10"Sch.10	0.38	0.25	0.15
8" IPS/12"Sch.40	0.78	0.50	0.31
10" IPS/12"Sch.10	0.43	0.27	0.17
10" IPS/14"Sch.40	0.64	0.41	0.25
12" IPS/16"Sch.40	0.70	0.45	0.28
12" IPS/18"Sch.40	1.37	0.88	0.55

Note: The chart assumes an annular pressure of 0 psig. For compressed air, a higher moisture content would be present.

Table 15.2 contains information on the total amount of moisture, in pounds (kilograms), that will condense for several 1,000 foot systems, based on starting air conditions of 90°F (32°C) and 90% relative humidity, 80°F (27°C) and 80% relative humidity, and 70°F (21°C) and 70% relative humidity, to achieve a final dew point temperature of -20°F.

15.2.4.2 Purge/Blanket Gas Tables

Table 15.3 contains information on the total amount of dry, clean nitrogen, in cubic feet, that must be added to a 1,000 foot length of straight double containment pipe size combinations, to achieve a dew point in the annulus of -20°F (-29°C), based on an initial starting annulus conditions of air at 90°F (32°C) and 90% relative humidity, 80°F (27°C) and 80% relative humidity, and 70°F (21°C) and 70% relative humidity, respectively.

15.3 Internal Leak-Detection Methods

Leak detection for piping and tank systems can be facilitated by a variety of internal means. Available methods include monitoring tank and piping inventories, periodic tightness testing, in-tank methods, and a variety of external measurements that detect leaks to the soil surrounding the system. Table 15.4 summarizes various internal leak-detection options for tanks, pressure piping, and suction piping, along with other available means.

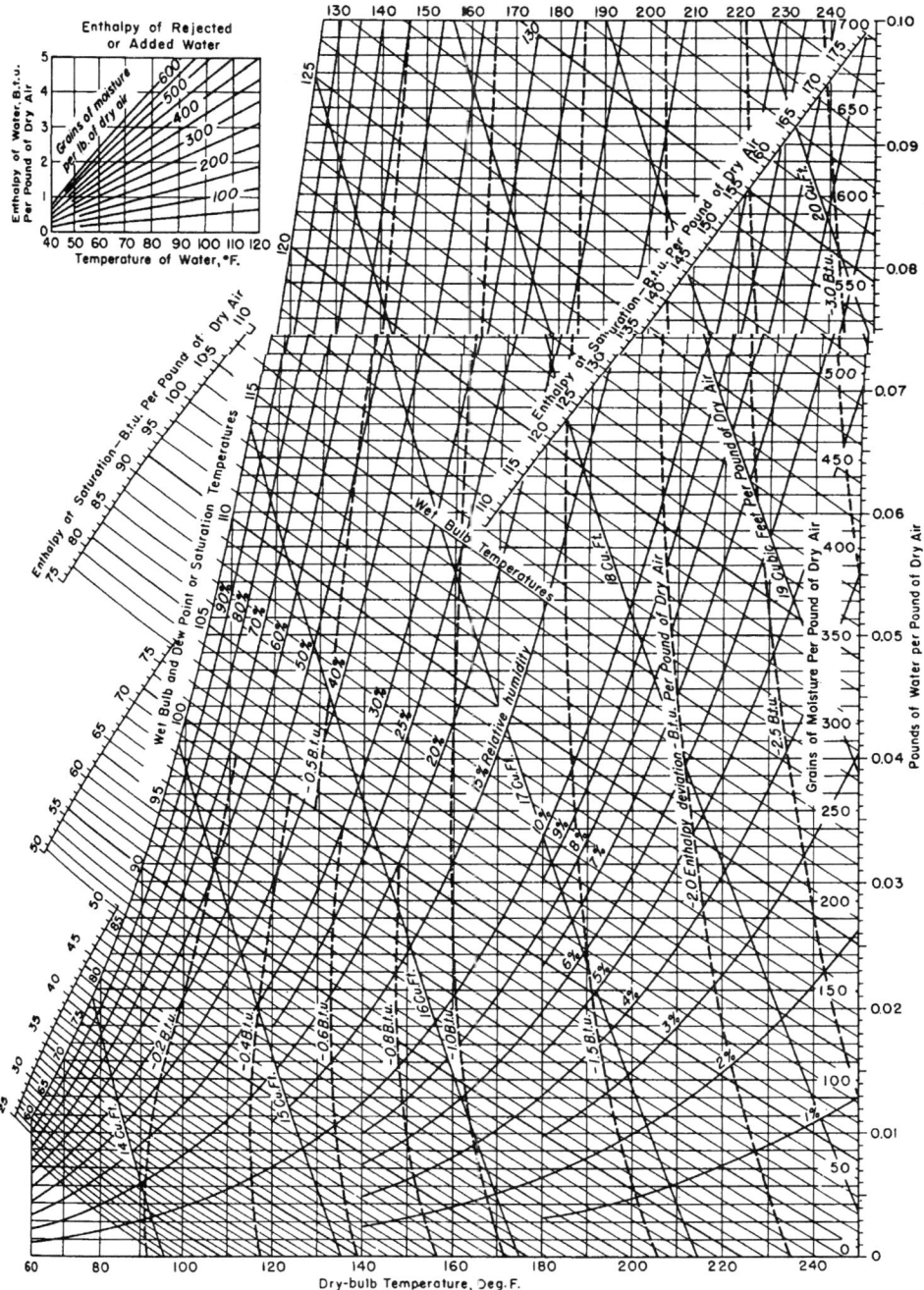

Figure 15.47. Psychometric chart for medium temperature ranges.

Table 15.3 Minimum Amount of ft^3 of N$_2$ that Must Be Added to Purge Various 1,000 ft Lengths of Double Containment Piping to Achieve a Dew Point of -20°F (-29°C)

Size	Starting condition		
	90°F (32°C)/90%R.H.	80°F (27°C)/80%R.H.	70°F (21°C)/70%R.H.
1" IPS/2"Sch.10	1,380	890	550
1" IPS/3"Sch.40	3,630	2,340	1,450
2" IPS/3"Sch.10	2,360	1,520	950
2" IPS/4"Sch.40	4,990	3,220	2,000
3" IPS/4"Sch.10	5,790	3,730	2,320
3" IPS/6"Sch.40	11,590	7,460	4,650
4" IPS/6"Sch.10	9,530	6,130	3,820
4" IPS/8"Sch.40	20,530	13,220	8,230
6" IPS/8"Sch.10	12,040	7,750	4,830
6" IPS/10"Sch.40	26,690	17,190	10,700
8" IPS/10"Sch.10	16,160	10,400	6,480
8" IPS/12"Sch.40	32,190	20,730	12,910
10" IPS/12"Sch.10	17,930	11,540	7,190
10" IPS/14"Sch.40	26,770	17,240	10,730
12" IPS/16"Sch.40	29,500	18,990	11,830
12" IPS/18"Sch.40	57,750	37,180	23,160

Notes:
1. Table assumes nitrogen to be completely dry at atmospheric pressure and 70°F, and assumes 100% mixing.
2. In many applications, a dew point of -20°F is lower than is necessary. A higher design dew point will require less nitrogen.
3. This table assumes 100% mixing, which yields results that are highly conservative. Actual practice should consume less nitrogen.

15.3.1 Inventory Monitoring

Inventory control techniques are commonly used for piping and associated storage tanks, in those systems that operate with noncontinuous flow. A typical example would be petroleum storage facilities where product is added and removed intermittently. Inventory control techniques involve monitoring the volume of a liquid/gas that is delivered to and removed (dispensed) from its storage system. The volumes are tracked on at least a monthly basis, and often with additional frequency. When using these methods, the volume of stored fluid is often also required to be checked on a daily basis with the use of a gauge stick. When this is the case, monthly checks are more of a reconciliation of fluid inventories. If accounts do not balance, such inventories are a possible indication that a leak has occurred.

The daily measurements of tank volume must occur at least in the morning and evening, and in some cases at the end of each work shift. The level on the gauge stick can be translated to a volume of product using a calibration chart designed specifically for each tank.

Leak Detection

Table 15.4 Summary of Leak-Detection Options for Piping and Tanks

Methods	Automatic method		Single-wall tanks		Double-wall tanks		Single-wall pressure pipes		Double contain. pressure pipes		Single-wall suction pipes		Double-wall nonpressure pipes	
	Yes	No	Yes	No	Yes	No	Yes	No	Yes	No	Yes	No	Yes	No
Internal methods														
Inventory monitoring/ volumetric testing														
Manual reconciliation		√	√		√				√		√		√	
Metered reconciliation	√		√		√				√		√		√	
Temp. compensated reconciliation		√	√		√				√		√		√	
Automatic gauging*	√		√		√				√		√		√	
Nonvolumetric testing														
Helium leak detection		√	√		√				√		√		√	
Tracer leak detection		√	√		√				√		√		√	
Nondestructive Examination**		√	√A		√A		√		√		√		√	
Interstitial methods														
Leak-detection cable (line leak sensing and locating systems)	√			√	√				√			√	√	
Probe systems***	√			√	√				√			√	√	
Visual detection		√	√A		√A		√A		√		√A		√	
Manual valve systems		√		√	√				√	√			√	√
Integrity testing		√	√				√	√		√		√		√
External methods														
Groundwater detection techniques														
Detection wells		√	√				√	√		√		√		√
Soil sampling		√	√				√	√		√		√		√
Dyes and tracers		√	√				√	√		√		√		√
Surface geophysics		√	√				√	√		√		√		√
Vapor-detection techniques														
Grab sampling of soil cores		√	√		√				√		√		√	
Surface flux chambers		√	√		√				√		√		√	
Downhole flux chambers		√	√		√				√		√		√	
Accumulator systems		√	√		√				√		√		√	
Ground probe testing		√	√		√				√		√		√	
Observation wells		√	√		√				√		√		√	
Vapor wells	√		√		√				√		√		√	
U-tubes	√		√		√				√		√		√	
Visual inspection		√	√A		√A						√A		√A	
Integrity testing		√	√A		√A		√A		√		√A			√

Notes:
* Automatic gauging techniques include: capacitance devices, ultrasound devices, resistance sensor devices, magneto strictive device, servo-level gauges.
** Nondestructive examination includes: radiography, ultrasonic examination, liquid penetrant examination, magnetic-flux examination, and acoustic emission.
*** Probe systems include: capacitance probes, moisture sensors, vapor sensors, pH sensors, pressure sensors, liquid level sensors (dry and wetted systems).
A= Above-ground applications only.

An alternative to performing manual gauging involves automatic gauging through the use of liquid level sensing probes and temperature measuring devices. Level and temperature data may be analyzed and recorded by the use of a computer.

An industry standard that is used is that if a shortage or overage (determined at the monthly reconciliation) is greater than or equal to 1.0 percent of the tank's flow through volume plus 130 gallons (liters), it must be assumed that the tank under consideration has experienced a leak.[3]

15.3.1.1 Regulatory Requirements for Inventory Monitoring

In the United States, RCRA Subtitle I regulations require that inventory control systems be used in conjunction with periodic tank-tightness tests. Gauge sticks are required to be long enough to reach a tank's bottom. They must also be marked so that the fluid level can be determined to the nearest 1/8 in. (3 mm). Monthly measurements must be made to determine if there is any water at the bottom of each tank. Deliveries are required to be made through a drop tube that extends to within 1 ft (0.3 m) of the tank bottom. Product dispensers must be calibrated to the local weights and measures standards.[4]

15.3.1.2 Accuracy of Inventory Monitoring

Various considerations that affect the accuracy of volumetric measurements include: (1) the coefficients of expansion of the tank materials and of the contained fluids; (2) temperature effects; (3) vapor pockets; (4) water table effects; (5) tank distortion (due to burial loads, etc.); (6) vibrations; (7) evaporation and condensation; and (8) head pressure effects. These factors can cause wide variations in volume at any given time.

Other factors that have an effect upon inventory monitoring include: (1) meter accuracy; (2) evaporation; (3) gauging accuracy; and (4) tank geometry. Temperature can have an effect on available tank volume due to expansion and contraction. It also causes a stored fluid to expand and contract. Both of these factors must be taken into account in interpreting the results of daily gauging and monthly reconciling.

15.3.2 Tightness Testing

Internal tightness leak-detection methods include both volumetric testing and nonvolumetric methods. Volumetric testing is described in Section 15.2.1, and involves the use of either manual or automatic gauging systems. Both methods are more commonly applied to tanks.

Leak detection can be facilitated, in a theoretical sense, for pressure-piping systems by testing the primary and/or secondary containment piping system for pressure tightness at regularly scheduled times. This method is best applied as a backup means of leak detection; it is not recommended as the primary means of leak detection for chemical piping systems. Pressure tests of piping systems are usually expensive to conduct, and can involve a significant amount of down time. Therefore, it is only practical to conduct a pressure test on an infrequent basis. In most pressure piping systems, pressure testing is only conducted prior to initial startup as a means of verifying the system has been installed properly. The U.S. EPA has recommended pressure tightness testing as an option for existing systems that contain petroleum-based fluids, but has mandated other means for systems that convey hazardous chemicals.

When performing line tightness testing, a line must be taken out of service and pressurized with a test fluid. A pressure drop over the duration of the test suggests a possible leak. Suction and drain lines are pressurized only to the extent needed (10 ft of head for a drainage pipe, 1.5 times the suction pressure required for suction piping.) Tightness tests must be conducted at least every 3 years for existing systems containing petroleum-based fluids, according to U.S. RCRA Subtitle I (Title 40 of the Code of Federal Regulations, part 280), in lieu of other mandated optional methods.

Wherever possible, it is always recommended to perform piping tightness tests in conjunction with the tightness testing of associated tanks. Performing separate tests for the piping and tanks means added testing expense and extra revenue losses due to additional down time.

15.3.2.1 Internal Integrity Testing

Nonvolumetric forms of internal integrity testing can be performed on piping/tank systems using many methods. These include: (1) helium leak detection; (2) tracer leak detection; (3) use of continuous in-tank monitors; and (4) conducting an internal physical inspection.

In associated tanks and vessels, inspection of the internal primary wall may be conducted if the tank (vessel) contains manholes. When internal inspection is to be performed, safety procedures must be followed. The contents of each tank/vessel must be completely removed, and all vapors must be evacuated. The inside primary wall must be completely decontaminated prior to entrance by an inspector. Appropriate breathing apparatus and fire-resistant clothing are standard requirements for many installations, according to federal and local health and safety rules. A detailed, written set of procedures should be prepared, and adequate training for these rules given to all inspectors. Safety rules should be posted near the vessel if internal inspections are to be conducted. Additional information concerning the inspection and maintenance of pressure vessels is given in Section 14.

15.3.3 Continuous Pressure Monitoring of Piping and Pressure Vessels

Use of continuous pressure sensing to monitor double containment piping systems and double-wall tanks has been described in Section 15.2.2.7. In principle, pressure monitoring may also be used to determine if a leak has occurred in a single-wall pressure piping system. The same is true of pressure vessels. In most above-ground, single-wall piping systems where leak detection is required, the monitoring of pump output pressure is the standard means of facilitating leak detection.

Gas that is maintained under low pressure in a tank interstice tends to find its way out of threads, flanges, valve seals, etc. over a period of time. Therefore, some loss of gas could occasionally be expected from threaded joints, meaning false (low-point) alarms may be triggered. Some double-wall tanks have been monitored by storing a brine in its interstice under a slight pressure, to prevent gas loss, and subsequent false leak alarms. It must be remembered whenever pressurizing a tank interstice that the internal tank will be subjected to a slight external pressure if it is not under an internal pressure equal to at least the level of the interstitial pressure. This may require that the storage tank be designed according to the rules of the ASME Boiler and Pressure Vessel Code. The vessel would be required to be designed in the same fashion as a jacketed pressure vessel, with the outer wall acting as the outer jacket (according to the rules of ASME BPV Code, Section 9, Division 1, Appendix 9). Designers must be aware of this, and avoid this situation if it is not desired.

15.4 External Leak-Detection Techniques

There is a host of external-based methods that can be used to detect leaks from single-wall piping and tanks, double containment piping, and double-wall tanks. These methods involve the monitoring of the area outside of a tank (or pipe) to detect the presence of a fluid that has escaped the piping or tank. Depending on where the tank is installed, the method may consist of monitoring surrounding soil, ground water, or air (in the case of above-ground piping). The most common fluids that may be monitored by such methods include hydrocarbons and other volatile organic chemicals. Some inorganic chemicals may be detected by these methods, although it is much more difficult to detect such a chemical externally.

Many external methods can be applied to piping systems; however, in many applications, the monitoring of piping by means of external methods proves to be expensive. External monitoring methods tend also to be unreliable when applied to piping; therefore, they are not widely used as a leak-detection means for piping applications. It is always more efficient to monitor a tank externally by placing it inside a concrete secondary containment housing and monitoring the free space for the presence of the fluid in question.

External leak detection may be accomplished by taking samples of the surrounding environment (soil, air, or ground water) on a periodic basis. Some methods also function on a continuous basis or in a permanent fashion. One simple way to provide external leak detection for above-ground systems or underground tanks installed in an accessible vault is to inspect the tank visually on a periodic basis. Various external integrity testing methods also exist to check tanks that are readily accessible.

15.4.1 Ground-Water-Based Systems

In applications where the ground-water table is close to the surface (< 50 ft; 15 m), various methods may be used to test for leaks. The most common methods include: (1) detection wells; (2) direct sampling of soils; (3) use of dyes and tracers; and (4) surface geophysical techniques. These methods are typically applied on a periodic basis; many are not capable of being applied on a continuous basis by their very nature. Ground-water-based systems are best applied to associated tanks, and are impractical for use in detecting piping leaks.

15.4.1.1 Observation and Detection Wells

Observation and detection wells have been used extensively due to their relative low cost and an established performance record. They usually provide a visual confirmation of subsurface conditions in an accurate manner.

To construct a well, the depth of the ground-water table must first be determined. The depth of the well, or monitoring point, can then be established. Once this is done, suitable drilling equipment must be selected and positioned at the site.

Next, a hole is bored, and casings and liners installed into the hole. Well screens and fittings are also installed. It is necessary to pack the annular space between the bore hole and the screen with a suitable gravel. The annular space (of the well) can then be sealed against ground-water penetration. Once these steps are concluded, the well can be developed, and either a manual or automatic system put into place. There are three common types of wells[5]: (1) screened or open over a single vertical interval, as shown in Figure 15.48; (2) well clusters, an example of which is shown in Figure 15.49; and (3) nested wells (single well with multiple-sample points). Those having a single vertical interval can provide composite ground samples if the screen covers a saturated portion of the ground-water table. Both well clusters and nested wells are good for providing information on the vertical migration of a chemical that has leaked from a tank or pipe.

There is a wide range of sampling techniques and products for collecting subsurface ground-water samples from wells. These include the following: (1) downhole collection devices (e.g., bailers, mechanical depth-specific samplers, and pneumatic depth-specific bailers); (2) suction-lift methods; (3) positive-displacement methods (e.g., submersible centrifugal pumps, submersible piston pumps, gas-squeeze pumps, gas-lift methods, gas-drive methods, and jet pumps); and (4) destructive methods (mostly for reconnaissance surveys).

Downhole collection systems are not driven by a differential. Thus, they tend to maintain the integrity of samples beyond that of other methods. Many are inexpensive to purchase, fabricate, and construct, and therefore are a good first choice. Suction-lift methods are also inexpensive, but are limited to water table depths of less than 23 ft (7 m) below the ground surface. Positive-displacement methods are costly, large, and can be difficult to clean. However, they do reduce the possibility of degassing as samples taken do not contact the atmosphere. Once liquid is removed, detection of a contaminant is normally carried out via laboratory destructive testing (e.g., infrared spectroscopy, etc.). Laboratory methods, some of which can be field applied, can be used to obtain preliminary data, and can also be used as a source of data for both the liquid and solid phases.

Other nondestructive methods for detecting for the presence of leaks that may be used in conjunction with observation and detection wells include: (1) use of hydrocarbon-sensitive pastes; (2) interference probes; (3) thermal conductivity probes; (4) electrical resistivity probes; (5) use of hydrocarbon-soluble devices; and (6) the use of hydrocarbon-permeable devices. Each probe-type method may be used to conduct measurements on a continuous or periodic basis, whereas the other methods may be conducted on a noncontinuous basis only. For more information regarding sampling and sensing methods to be used in conjunction with observation and detection wells, many suggested readings exist.[6-8]

Section 15.4.3, "Design and Construction of Wells," provides further discussion on the construction of wells of all types, including observation and detection wells and vapor wells (see Section 15.4.2.6). For additional information concerning the advantages and disadvantages of ground-water detection methods in general and the suitability of various well sampling methods, the American Petroleum Institute publishes information on the topic.[9]

15.4.1.2 Direct Soil Sampling

When soil core sampling is performed during the installation of a leak-detection well, meaningful detection data can be obtained, particularly if the data are interpreted along with information gathered by detection well sampling. Soil samples may be collected at various times during the drilling of a well, and results visually classified and permanently recorded. Laboratory analysis may also be conducted on a portion of the samples collected. Some field analysis tests can be performed as well. Soil core sampling is most commonly used to measure the presence of vapors, although ground-water measurements may also be made on collected samples, where a high ground-water table exists.

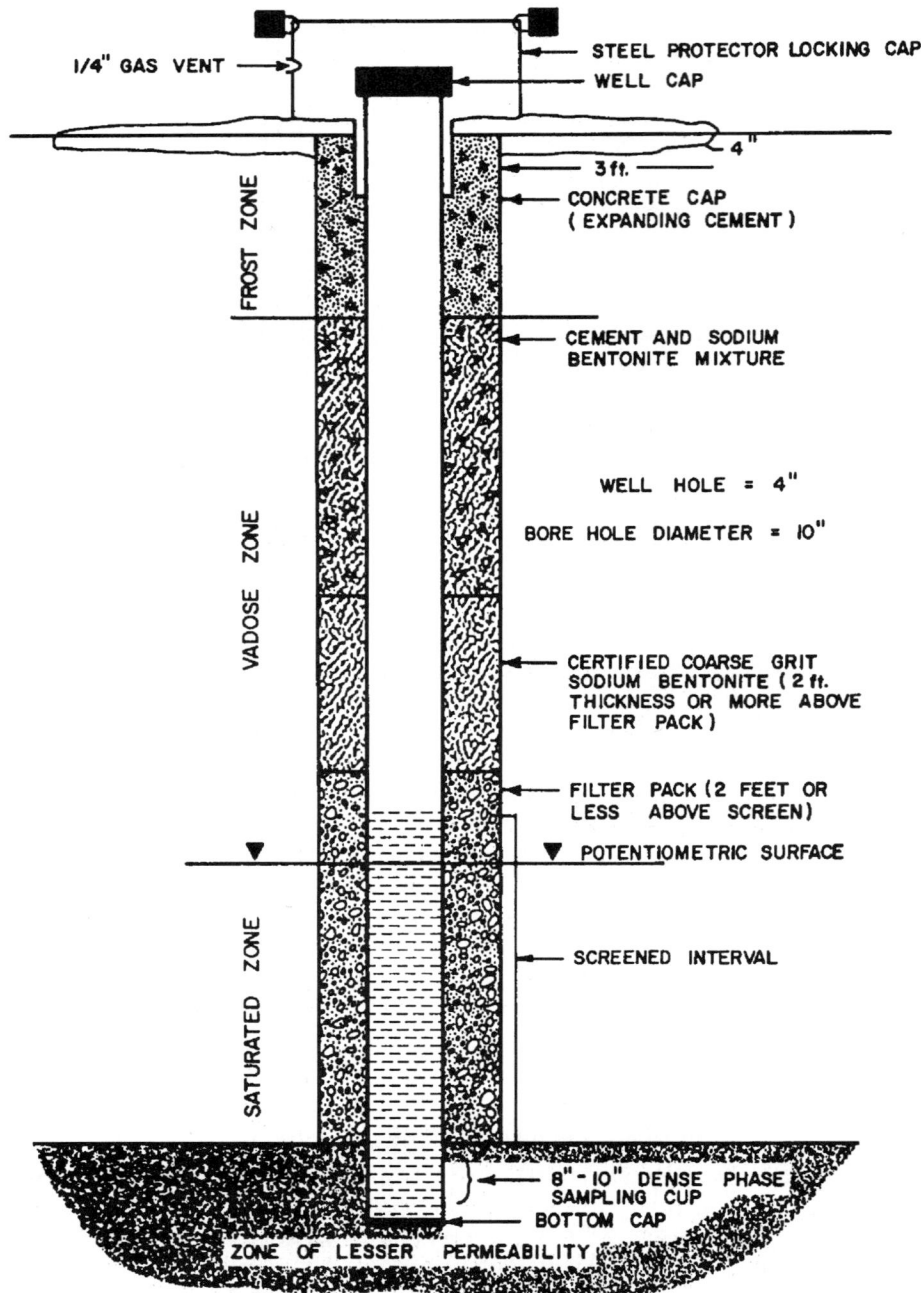

Figure 15.48. Typical detection well. (Source: EPA, "Draft RCRA Ground Water Monitoring Technical Enforcement Guidance Document," 1985.)

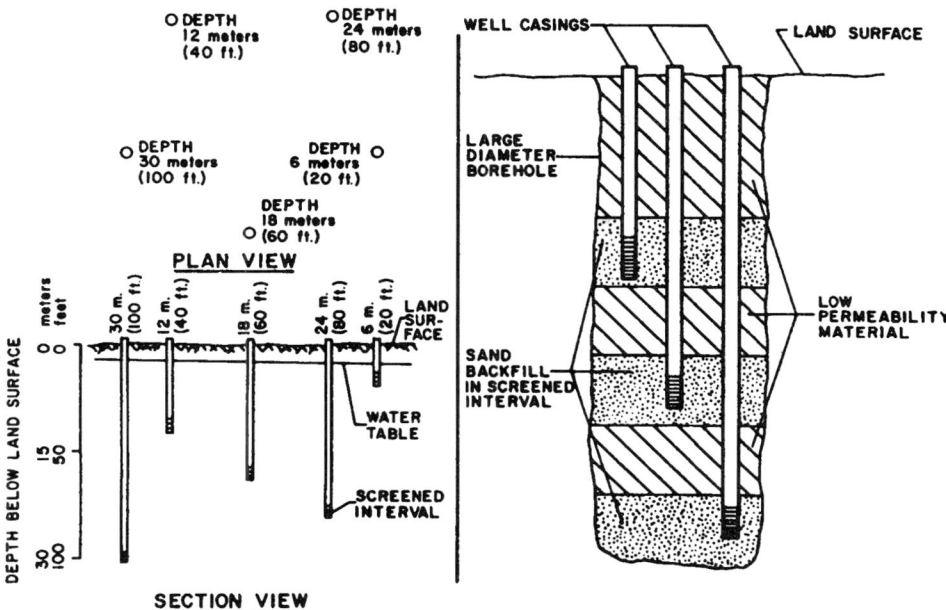

Figure 15.49. Typical detection well cluster. (Source: New York State Department of Environmental Conservation, "Technology for the Storage of Hazardous Liquids," 1983.)

15.4.1.3 Use of Dyes and Tracers

When a storage system is suspected of leaking, dyes and tracers may be used to verify that a leak exists. Typical dyes and tracer materials include: organic dyes, fluorescent dyes, and metallic tracers.[10] When geologic conditions involve high silt, clay, and organic content, dye compounds will be adsorbed onto the formation, and thus are ineffective. Monitoring is usually conducted at the point at which the release is first detected.

One must always be concerned that dyes or tracer chemicals can possibly contaminate the ground water being tested. Aside from obvious environmental side effects, there are a number of other subtle technical drawbacks to the use of dyes and tracers. For many reasons, dyes and tracers are rarely an effective means for detecting leaks from piping and associated tanks, and thus are rarely used for such purposes.

15.4.1.4 Use of Surface Geophysical Techniques

A number of surface geophysical methods may be used to identify that a leak exists. Some are used to gauge how much leakage of a specific fluid has taken place. The methods available include: (1) resistivity techniques; (2) ground-penetrating radar; (3) electromagnetic induction; (4) metal detection; (5) seismic refraction; and (6) magnetometry. Of these, seismic refraction and electrical resistivity are the most commonly used methods.

Surface geophysical methods are best applied to the detection of inorganic contaminants. There are a number of factors that affect the results obtained by these methods.

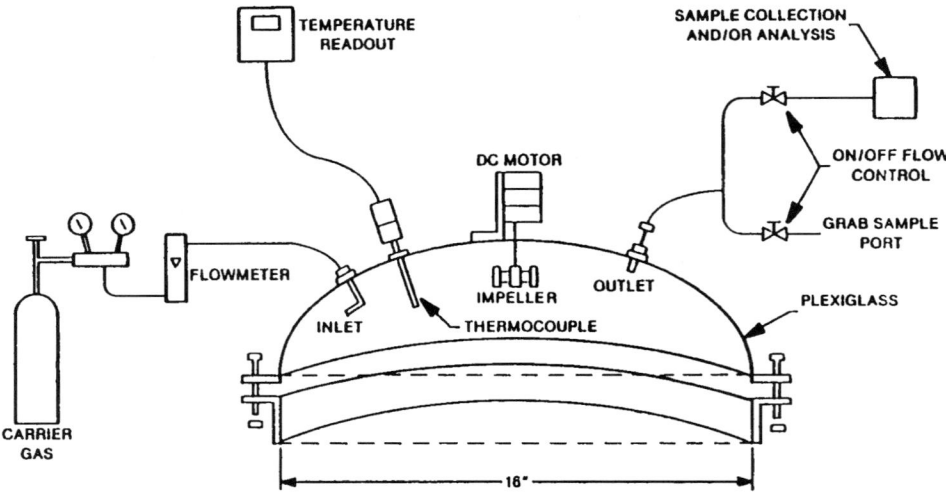

Figure 15.50. Schematic of a surface flux chamber arrangement. (Source: API Publication 4394.)

These include: (1) depth to be measured; (2) background conductivity of the soil and ground water being studied; and (3) whether or not aquifers or other interferences are present. Surface geophysical methods, when they are applicable, may help to reduce the number of test wells drilled, if test wells are the primary method used to determine the extent of a leak.

15.4.2 Soil-Vapor-Based Methods

In applications involving the underground storage or transportation of volatile organic chemicals, soil vapor techniques may be used as a means of detection. Above-ground ambient air monitoring could be used, as organic chemicals can migrate to the surface by molecular diffusion and convection. However, soil vapor techniques tend to produce more reliable results than above-ground air monitoring.[8,11] Soil core sampling techniques may be used to monitor the presence of vapors and facilitate further qualitative testing. (Also refer to Section 15.4.1.2, "Direct Soil Sampling.") Various common soil-vapor-based methods can be grouped as: (1) grab sampling of cores; (2) surface flux chambers; (3) downhole flux chambers; (4) accumulator systems; (5) ground probe testing; and (6) vapor wells.

15.4.2.1 Grab Sampling of Soil Cores

To conduct grab sampling, an undisturbed solid core must be collected from the portion of the soil that contains vapors. This is typically accomplished by driving a hollow tube into the ground, and sealing the resulting core in a sample container. Alternatively, an auger may be used in place of a hollow tube. In all cases, the container should have as small a head space as possible to reduce the possibility of outgassing, and thus loss of the vapor when the container is opened. This is easier to prevent if an organic chemical is adsorbed into the soil, or if organic chemicals exist in the soil particles, instead of existing in soil pores. Analysis may be conducted on "grabbed" samples using gas chromatography, or any other suitable technique.

15.4.2.2 Surface Flux Chambers

A typical surface flux chamber is shown in Figure 15.50. This approach uses a flux chamber, which is an enclosed system, for sampling vapors that collect within a given surface area. Clean, dry sweep air is introduced at one end of the chamber. Concentration of a specific organic vapor is measured at the opposite end of the chamber as the gas exits. Many types of analytical techniques may be used; however, gas chromatography is usually the preferred method of analysis.

15.4.2.3 Downhole Flux Chambers

A schematic illustration of the construction of a typical downhole flux chamber is shown in Figure 15.51. The main difference between a surface flux chamber and a downhole flux chamber is that the latter is used to collect and provide measurements beneath the surface of the ground. When using a surface flux chamber, it is best to sample vapors as soon as a hole has been bored. This will minimize outgassing of vapors and help maintain the accuracy of the results obtained.

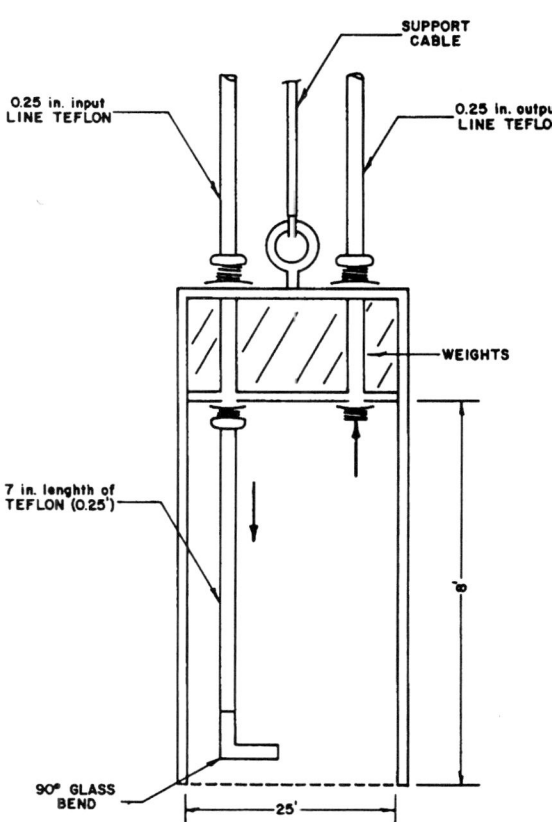

Figure 15.51. Schematic of a downhole isolation flux chamber. (Source: API Publication 4394.)

Table 15.5 Comparison of Ground Probe and Soil Vapor Testing Techniques

Technique	Disturbance of soil gas equilibrium		Suitable for rocky soil		Sampling time		On-site analysis practical		Sampling at greater depths if necessary		Required sampling equipment		Required analytical equipment		Estimated sampling manpower	
	Sml.	Lrg.	Yes	No	Hrs.	Days	Yes	No	Yes	No	Simple	Complex	Variable	Complex	1	2
1. Grab sampling of soil cores																
Auger		√	√		√		√		√		√			√	√	
Driven sleeve	√	√	√		√		√			√	√			√	√	
2. Surface flux chamber	√		√		√		√			√	√	√				√
3. Subsurface flux chamber																
Auger/enclosure		√	√		√		√		√			√	√			√
Ground probe type	√	√	√		√		√		√			√	√			√
4. Accumulator device																
Curie-point wire	√	√				√	√		√		√		√	√		
Absorbent/pump		√	√		√		√		—	—	√		√	√		
5) Ground probe																
Passively emplaced		√	√		√		√		√		√		√			√
Driven	√	√	√		√		√		√		√		√	√		
Driven, sm. volume	√	√	√		√		√		√		√		√	√		

Source: API Publication 4394.

15.4.2.4 Accumulator Systems

Accumulator systems can be used in a continuous or periodic manner. Either way, an averaged sample may be obtained that corrects for any short-term fluctuations in contaminant level. Accumulator systems are somewhat immune to the effects of ambient conditions. However, they tend to be expensive compared to other methods. Sample times may also be lengthy, and the accuracy of the results obtained depend on the quality of the collection method used. In some cases, accuracy can prove to be unreliable. An example of a Curie-point accumulator device is illustrated in Figure 15.52.

15.4.2.5 Ground-Probe Testing

To sample vapors by the use of ground probes, a tube must first be placed into the ground at a desired depth. As the vapors enter the tube, they may be withdrawn by the use of a syringe positioned within the tube. The vapors are extracted by the syringe and injected

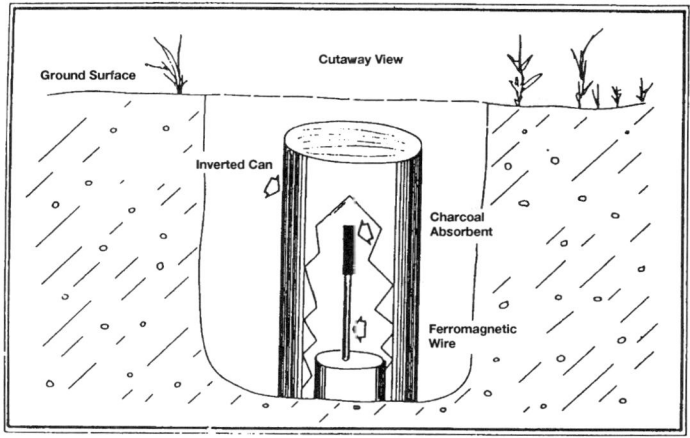

Figure 15.52. Illustration of a Curie-point accumulator device. (Source: API Publication 4394.)

into a sample bottle. The concentration of the vapor can then be established by the use of either gas chromatography or by passing sampled vapors through a packed bed that changes color to indicate the presence of a specific organic chemical.

A dual-tracer method has been recently developed that can distinguish a vapor leak from liquid.[12] This technique may have significant potential as a future means of determining contamination in older, previously existing underground system installations.[13]

Ground probe techniques are very useful, except where conditions involving high moisture and low permeability are present. If near-surface rock strata exist at the site, detection by probes may produce inaccurate results. Ground probe techniques and other soil vapor techniques are compared in Table 15.5.

15.4.2.6 Vapor Wells

In applications involving deep water tables (> 40 ft below ground surface elevation), detection may be performed by monitoring samples collected from isolated vapor wells. This method is possible only where the content of piping and associated tanks involves volatile chemicals, whose vapors can readily move upward through the soil. Background levels of the same or similar chemicals can lead to erroneous results and possibly false alarms. Many systems that are in use can be adjusted to account for background levels of contamination.

A typical vapor well installation is illustrated in Figure 15.53. Vapor wells vary in diameter from 3/4 in. (25 mm, nominal diameter) to 2 in. (63 mm, nominal diameter), depending on the material of construction of the well. A vapor well is normally installed to a depth, depending on site conditions, varying from just below the ground surface to the lowest elevation of the excavation. The well is provided with slots or perforations for air entry, which are usually located at the lower end of the well. A variety of sampling and detection means exist to be used in conjunction with vapor wells, including: (1) detector tubes; (2) combustible gas indicators; (3) catalytic detectors; (4) metal oxide semiconductors; (5) photoionization detectors; and (6) flame ionization detectors.

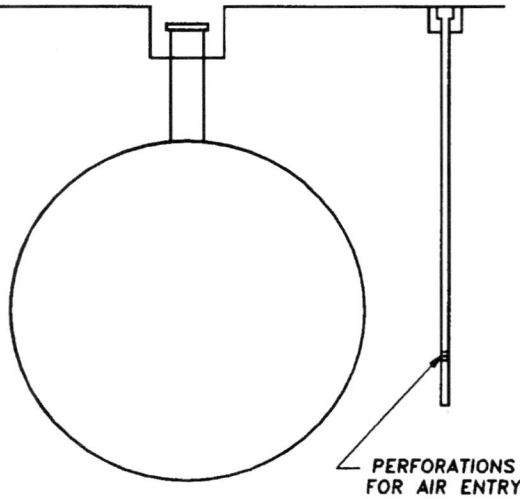

Figure 15.53. Schematic of a vapor well monitoring system. (Source: Thompson, 1985.)

15.4.3 Design and Construction of Wells

The design and construction of observation and leak-detection wells, or vapor wells, and the number of wells to be used in a given application depend on several factors. The depth of the wells, the diameters selected, their materials of construction, chosen location, and the number and size of perforations represent key variables of observation/detection and vapor well construction. Variables are selected based on the geologic conditions present, the nature of the stored substance, the details of the system layout to be protected, and the preferred leak-detection method to be used.

For observation and detection wells, the deeper the ground water is located from the surface, the more expensive the installation. This is due to the need to drill and install more wells, and the need to use more expensive drilling methods. Also, a ground water that has a lower concentration gradient will mean a lower driving force, and will also require more wells. If the level of the water table varies greatly, most leak-detection systems will vary in accuracy during their operational lives.

For piping, significantly more wells are required, as compared to applications involving only tanks or vessels. This is true for both vapor wells and observation/detection wells. A multiple-well system always means higher costs. For this reason, the use of observation/detection wells and vapor wells is not very common for piping systems. This is true for both single-wall and double containment piping. Conversely, tanks may be monitored with relatively few wells. A single tank may only require one or two wells. Multiple tank installations may be provided with as few as two to four wells.[8] Depending on the details of the installation, multiple tank projects may require more than four wells. Secondarily contained tanks need only be provided with one well.

Leak Detection

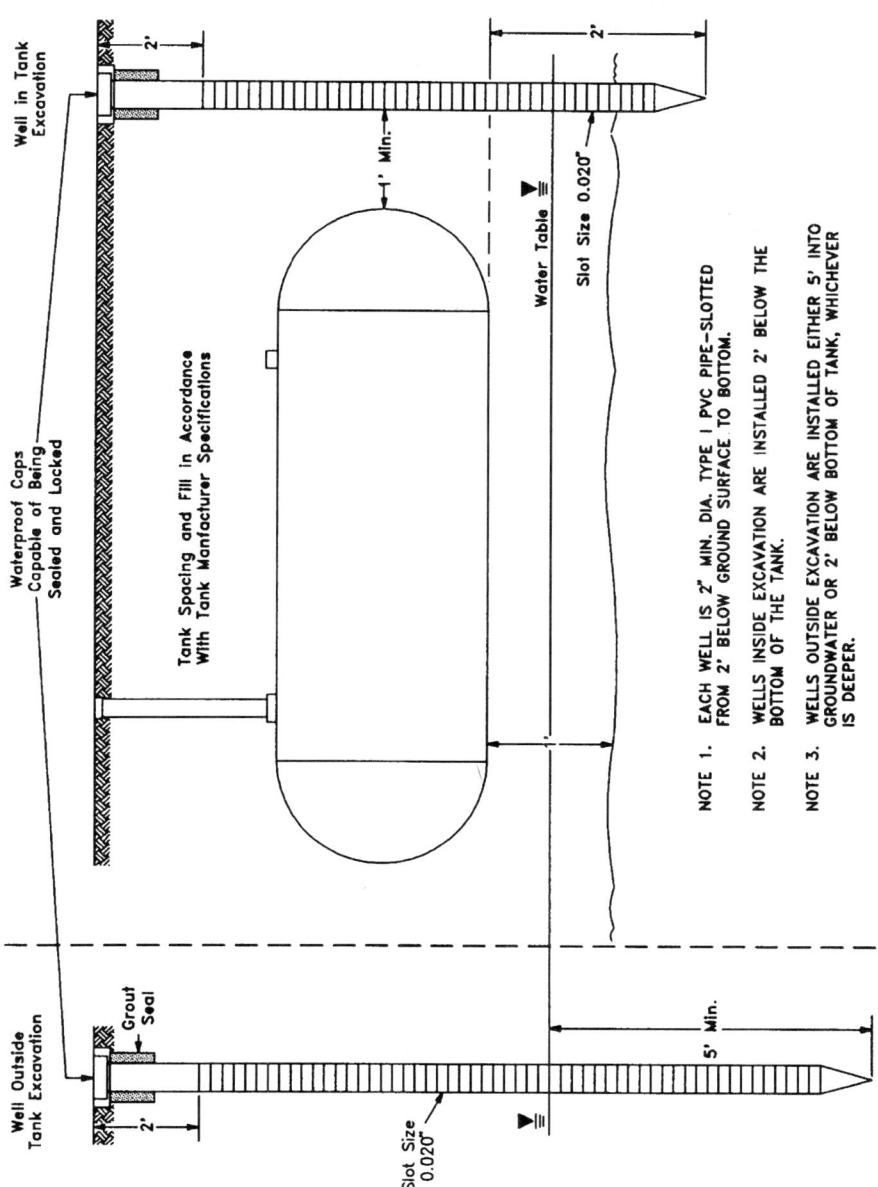

Figure 15.54. Schematic of a typical observation well installation. (Source: API Draft, "Observation Wells as a Release Detection Technique," 1986.)

In many applications, sloping the excavation can reduce the number of wells required (for the monitoring of piping and/or associated tanks), depending on other factors.

Most wells are constructed of at least 2 in. (50 - 63 mm) schedule 40 (approx. SDR 17.6) pipe, although vapor wells may be constructed from pipes as small as 3/4 in. (25 mm, nominal diameter). PVC is a common material of construction due to its soil corrosion resistance, although any material that is determined to be suitable for a given application may be used. The well should have a casing that is provided with 0.020 in. (0.5 mm) slots. Observation and detection wells located in a tank hole excavation should have their slots extend from 2 ft (0.6 m) below grade to the well bottom. Vapor wells do not have any set requirements for how far slots should extend.

For observation/detection wells installed outside an excavation zone, their slots should extend into the water table. The amount that it extends should be the deeper of 5 ft (1.5 m) below the water table, or 2 ft (0.6 m) below the tank bottom. These details for observation/detection wells are shown in Figure 15.54 for a typical petroleum underground storage tank installation. Vapor wells should always terminate above the level of the ground water, or the results may provide an inaccurate measure of soil contamination.

15.4.4 Associated Tank Monitoring by Use of U Tubes

U tubes may be used to monitor leaks in new underground tank applications, where a low water table exists. If a high or rising water table exists, inaccurate results can occur. Most U tubes are positioned directly underneath a tank, and depend upon downward travel of contaminants. Most U tubes are susceptible to flooding by rain water or a high or rising water table.

A typical U-tube installation is illustrated in Figure 15.55. A U tube commonly consists of a 4 in. (110 mm, nominal diameter) plastic tube, having 0.06 in. (1.5 mm) slots at its 12 o'clock installed position. The horizontal piping is pitched/sloped at 1/4 in. (6 mm) per foot (0.3 meter) towards a low-point-sump. Low-point sumps usually extend 2 ft (0.6 m) below the bottom liner, as illustrated in Figure 15.55. A bottom liner is usually required to prevent tank leaks from escaping around the tube, thereby assisting in its collection by the tube. U-tube samples may be sampled and analyzed in the same manner as an observation well (see Section 15.4.1.1). U tubes are not practical as a means of detecting external releases from piping, although technically they could be used.

15.4.5 Physical Inspection and Integrity Testing

Piping, and any associated tanks or vessels, may be inspected and/or subjected to nondestructive examination methods, to determine whether defects are present. Underground piping systems must be made accessible by placement in an accessible trench or manway; associated tanks and piping connections may be placed in an accessible vault.

15.4.5.1 Nondestructive Examination

Procedures that are used for the inspection and examination of single-wall and double containment piping systems are described in Chapter 13, "Inspection, Examination, and Testing." Many of the procedures described there are similar to, or the same as, nondestructive examinations that are applied to tanks and vessels, as described in Chapter 14, "Associated Storage Tanks and Pressure Vessels." The methods described in Chapters 13 and 14 are used to verify the integrity of components and welds prior to conducting a

Leak Detection

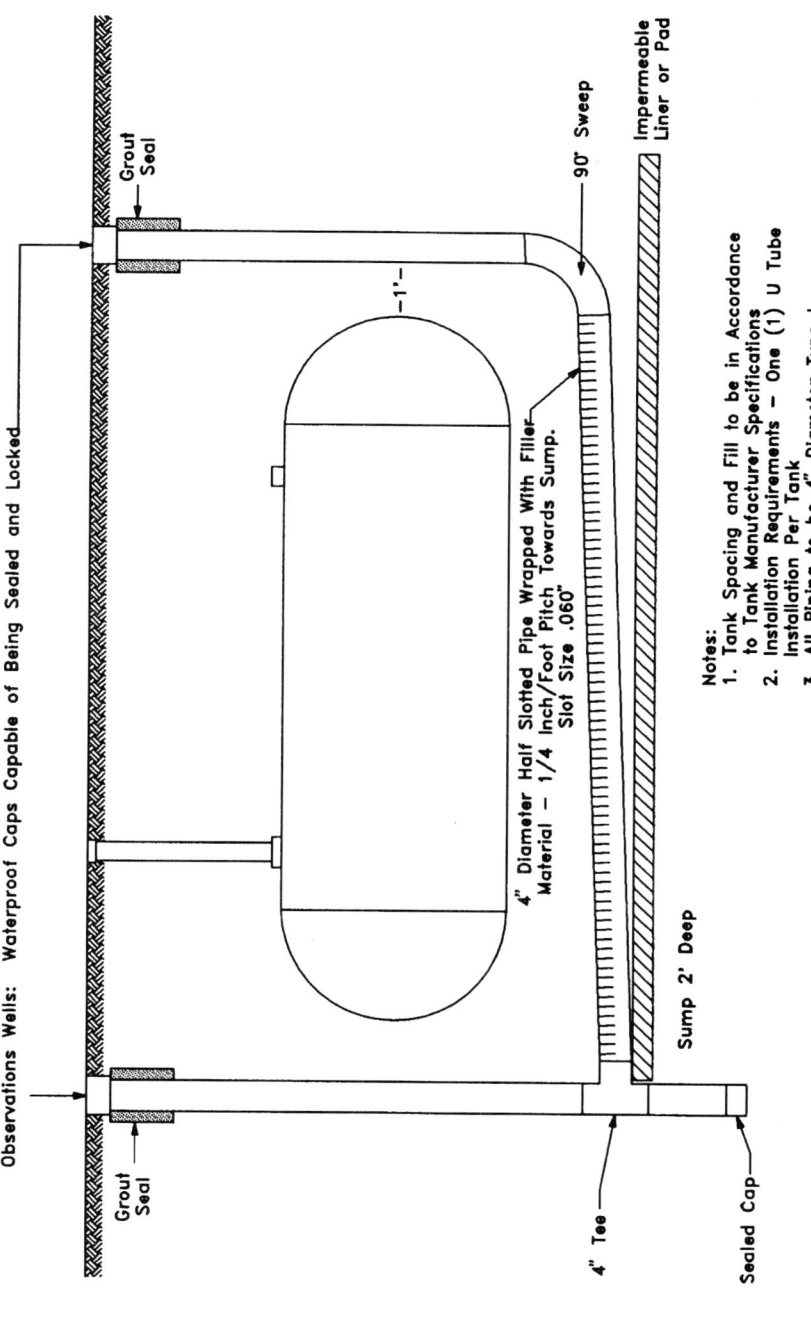

Figure 15.55. An overall schematic of a U-tube monitoring installation being used to externally monitor a tank for leaks. (Source: API Draft, Observation Wells as a Release Detection Technique," Washington, D.C., American Petroleum Institute, 1986.)

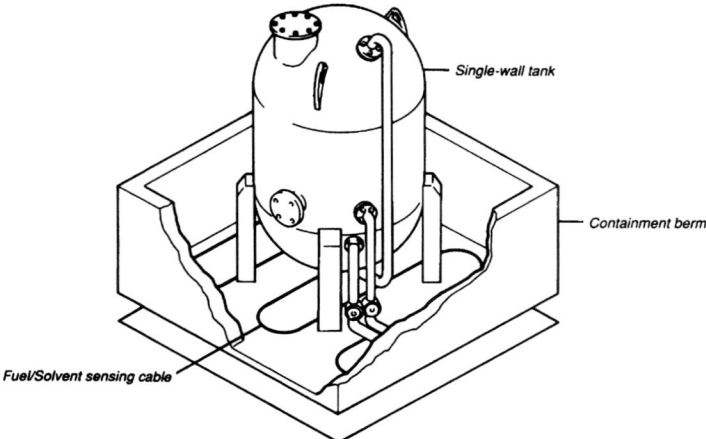

Figure 15.56. Illustration of leak detection cable layout for a typical above-ground storage tank having a secondary containment dike. (Source: Raychem Corp.)

hydrostatic test and starting up a system. However, piping and tank systems may also be inspected and/or examined periodically after startup using these same methods. Evaluations of this type will help determine if a system has maintained its integrity, or if it has been corroded or otherwise adversely affected by its service fluid.

The two most common methods applied to piping and its associated tanks are ultrasonic testing and radiography. (Radiography is common only for metallic materials.) However, these methods, as applied to double containment piping and tanks, tend to produce results that are more difficult to interpret. This is particularly true of radiography, as shadow images of primary components may appear. For more information concerning the use of ultrasonic testing and radiography, refer to Chapters 13 and 14.

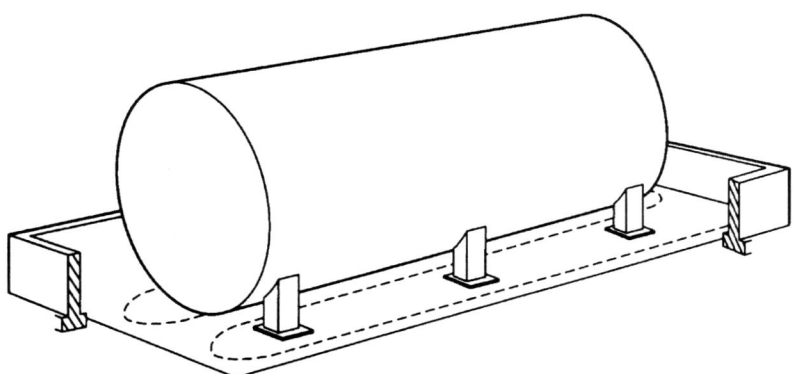

Figure 15.57. Illustration of leak-detection cable layout for a typical underground storage tank provided with a secondary containment housing. (Source: Raychem Corp.)

15.4.6 Monitoring of Secondary Containment Structures

Leak detection can be provided to above-ground and underground storage tanks, vessels, and their associated piping connections that are placed within a secondary containment area, or that have some other secondary containment means. In above-ground tanks, secondary containment normally consists of a lined or coated concrete containment dike or the construction of a lined or coated concrete containment berm. Leak detection may be either by means of probes (e.g., liquid level, vapor sensing probes) or by the use of leak-detection cable. The method used should be capable of ignoring any collection of rain water.

The layout of leak-detection cable for an above-ground tank provided with a containment dike is illustrated in Figure 15.56.

The layout of leak-detection cable for a typical underground storage tank having a secondary containment housing is illustrated in Figure 15.57.

References

1. 40 CFR, part 280.

2. Permalert ESP Inc. Data Sheet, Niles, IL.

3. Underground Storage Tank Guide, Thompson Publishing. Washington, DC, 1989..

4. Underground Storage Tank Guide, op. cit.

5. Todd G. Schwendeman and Kendall H. Wilcox, "Underground Storage Systems – Leak Dectection and Monitoring," Lewis Publishers, Chelsea, MI, 1987.

6. R. A. Scheinfeld, J. B. Robertson, and T. G. Schwendeman, "Underground Storage Tank Monitoring: Observation Well Based Systems," Ground Water Monitoring Rev. 6, No. 4 (Fall 1986).

7. American Petroleum Institute, "Observation Wells as a Release Detection Technique: Draft Research Report," Washington, D.C., April 1986.

8. Todd G. Schwendeman, et. al., op. cit..

9. American Petroleum Institute Publication 4367.

10. New York Department of Environmental Conservation. "Technology for the Storage of Hazardous Liquids: A State-of-the-Art Review," Albany, NY: Bureau of Water Resources, Division of Water, New York State Department of Environmental Conservation, January 1983.

11. W. D. Balfour and C. E. Schmidt, "Sampling Approaches for Measuring Emission Rates From Hazardous Waste Disposal Facilities," San Francisco, CA: Seventy-Seventh Annual Meeting of the Air Pollution Control Association, June 24-29, 1984.

12. G. E. Thompson, "Tracer Leak Detection Technology," Tucson, AZ; Tracer Research Corporation, 1985.

13. Schwendeman and Wilcox, et al., op. cit.

Notes

Chapter Sixteen

Trenchless Reconstruction and Alternatives to Secondary Containment Piping

16.1 Trenchless Reconstruction

In general, trenchless reconstruction as it applies to underground chemical sewer systems involves a variety of techniques to fix or repair existing underground systems while minimizing or eliminating the need for excavating the old system. These methods may be used to retrofit the old system to result in either a rehabilitated single-walled system or a functional double containment system. Most of the methods produce some loss in cross-sectional area due to the smaller diameter of the newly installed pipe or liner. However, there may not be much, if any, of a loss of flow capacity as the new pipe or lining typically exhibits a smooth bore which reduces frictional losses as compared to the host pipe. These methods are typically used in large diameter underground chemical (process) sewers.

The reason that these methods have been developed, and are being used is due largely to economics. Excavating can be very costly in old facilities due to contaminated soils, the cost to obtain environmental permits, and other related issues. Also, in many older facilities, the underground piping and utilities are not very well documented, leading to many delays which can be expected as unknown lines are encountered. Excavating very often also causes a disruption in facility operations, presents dangers to laborers and plant employees, may damage the roadbed or surface, and requires a costly restoration of the open-cut section to its original condition.[1]

In many facilities, these older chemical sewer lines have simply been abandoned, being replaced in favor of the installation of an above-ground system. However, above-ground systems require above-ground supporting structures and require the fluid to be pumped. Such highly mechanized systems involve substantial maintenance costs and upkeep over time. The piping of lines above-ground also takes away valuable free space and further clutters a facility with piping and pipe trestles. Given the choice, it is better to retrofit the existing system, and to make use of gravity to convey fluids, thereby eliminating the need for an above-ground mechanized piping system.

This section discusses six major methods for rehabilitating an older pipe to result in a new, functioning double containment pipe. These include: Slip-lining with HDPE; pipe bursting; diameter reduction/reversion techniques; modified slip-lining using the Double-Butt® technique; soft-lining techniques using cured-in-place pipe (CIPP); and folded and formed liners.

Figure 16.1. Example of an existing brick-lined concrete sewer pipe being "slip-lined" with flexible HDPE piping. (Source: Phillips Driscopipe.)

16.1.1 Slip-Lining with HDPE

Conventional slip-lining is the process of installing a new pipe into an older host pipe by pulling or pushing the new pipe into the host pipe. Slip-lining is sometimes referred to as pipe insertion. While virtually any type of piping material may be used for slip-lining, most slip-lining is performed with either HDPE or fiberglass pipes. HDPE is the most common overall due to its combined properties of ductility, toughness, durability, flexibility, handability, and acceptable economic cost, particularly in large diameters.[2] An example of an HDPE pipe being slip-lined into an existing concrete sewer is shown in Figure 16.1.

The two basic techniques that exist involve the pulling method, which is illustrated in Figure 16.2, and the pushing method, which is illustrated in Figure 16.3. In the pulling method, a pulling head is securely attached to the pipe end and connected through a freely rotating shackle to a winch line.[3] The winch line is then fed up and out of an existing manhole to a pulling mechanism located on the surface, as illustrated in Figure 16.2. By contrast, the push technique involves pushing the pipes into position by using a jacking device, such as a backhoe, which is illustrated in Figure 16.3.

Where short length sections or small diameter sections are to be relined, insertion can be accomplished from existing manhole to existing manhole. However, if existing manholes cannot be used to accomplish the insertion, some limited amount of excavation must be tolerated, as both Figures 16.2 and 16.3 indicate. ASTM Standard F 585, "Standard Practice for Insertion of Flexible Polyolefin Pipe into Existing Sewers," covers procedures and basic requirements for conventional slip-lining.

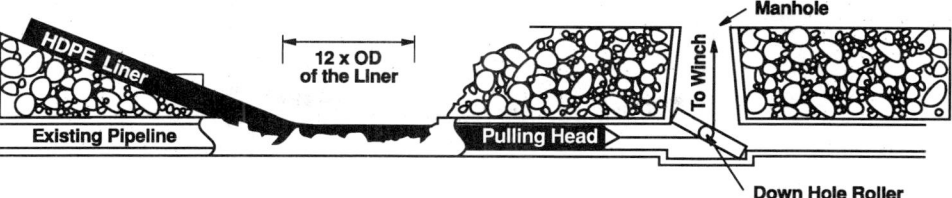

Figure 6.2. Insertion of flexible HDPE pipes into an existing pipeline using the "pull" technique. (Source: Phillips Driscopipe.)

When slip-lining is to be performed, it is highly important to identify whether or not the host pipe has collapsed, and if so, where the collapses are located. Branch connections present another complication; it is usually best to install a new manhole in locations where existing branches are located. In some applications where a substantial reduction in diameter is to be achieved, the inserted liner pipe can be suspended within the outer host pipe by the use of a grout between the liner and host pipe. Care has to be taken that the pressures achieved in applying the grout do not collapse the liner.

Polyethylene pipe is relatively flexible in comparison to other pipes, and as such, it is ideally suited for use as a slip-lining material. However, while it is a relatively flexible material, it too has its limitations. The piping cannot be bent beyond an amount where the strain would exceed the maximum longitudinal strain recommended by the pipe manufacturer. The following formula is useful in determining the minimum recommended radius of curvature to be used in the insertion process:[4]

$$R_c = \frac{D}{2\varepsilon_a} \qquad (16.1)$$

where:
R_c = minimum radius of curvature, in.
D = diameter of the inserted pipe, in.
ε_a = allowable axial strain, %

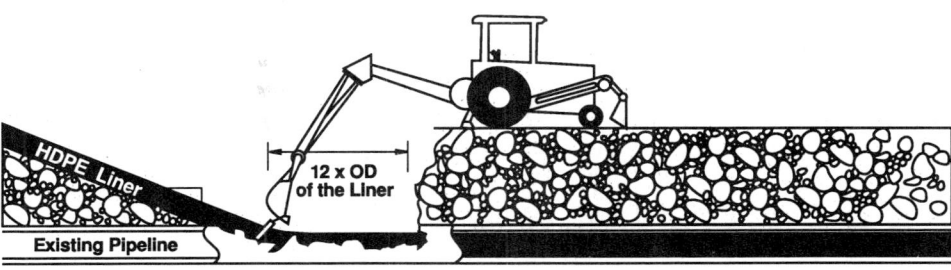

Figure 6.3. Insertion of flexible HDPE pipes into an existing pipeline using the "push" technique. (Source: Phillips Driscopipe.)

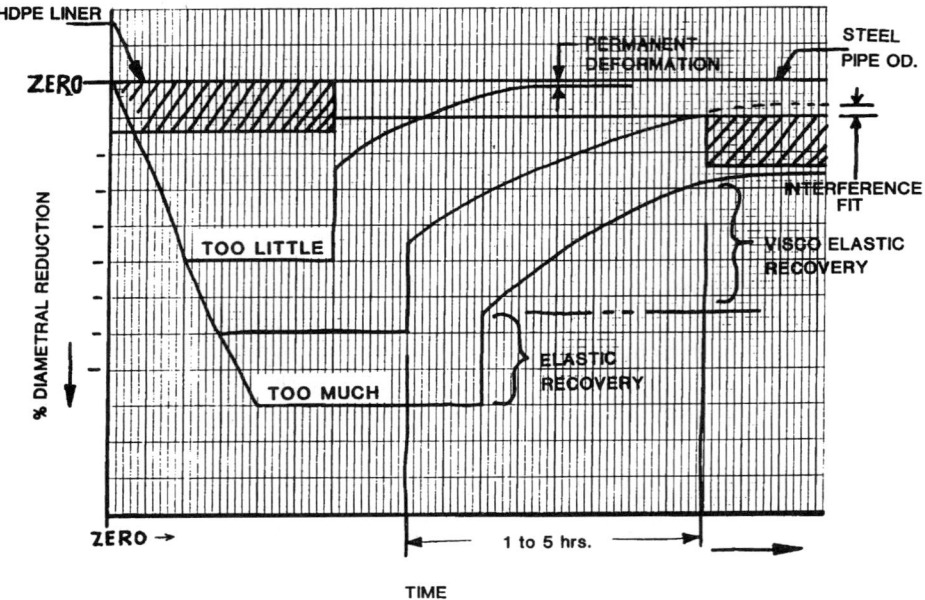

Figure 16.4. Diametrical reduction versus the working time to achieve an interference fit for HDPE materials. (Source: Svetlik, H. E., "Reduction, Renovation, Reversion: R3," Buried Plastic Pipe Technology, ASTM STP 1093, George S. Buczala and Michael J. Cassady, Eds. ASTM, Philadelphia, 1990, presented on Sept. 10-13, 1990 in Dallas, TX.)

For example, if the allowable axial strain is 1.5% for the pipe material, then the minimum radius of curvature would be $R_c = D/2(0.015) = 33D$.

If a double containment pipe is to be installed into the host pipe using slip-lining techniques, the secondary liner is first installed using the above-described techniques. The new primary pipe is then installed into the re-lined "secondary containment" pipe, typically using a pull technique. If the pull technique is used to install the first liner, this does present the option of pulling in both new lines simultaneously using a special termination pulling head. In any event, the insertion of a new double containment arrangement into an existing host pipe will result in a substantial lessening of cross-sectional flow area.

In Chapter 12, Figure 12.97 illustrates a patented system that allows removal of a primary pipe by means of specially adapted manholes. This system works well in conjunction with re-lining techniques such as slip-lining, thereby providing the added benefit of future maintenance without having to excavate in the future. Using the Figure 12.97 method, new manholes would be installed in locations whereby excavations are required to slip-line the new liner into the existing host pipe. For short sections where existing manholes are used, the existing manholes can be modified to allow the features illustrated in Figure 12.97.

16.1.2 Use of Slip-Lining and Pipe Bursting Techniques

A variety of techniques and equipment have been developed to allow an existing host pipe to be burst and expanded. This allows re-lining by means of a pipe sized so as to not have

Alternatives to Secondary Containment Piping

any reduction in cross-sectional area or flow capacity. In fact, these usually can be a net gain in flow capacity since the insertion of a new smoother pipe of equal diameter to the host pipe will reduce frictional losses, and will thus allow more flow according to the flow equations [see Equations (4.20, 4.29, 4.87 and 4.89)].

The practice of pipe bursting is also known by some as pipe splitting. In this process, the host pipe is broken and enlarged by means of radial forces which act outward to force the fragments into the soil. As the newer larger bore is formed by the bursting tool, a new liner pipe is pulled into the bore. Procedures were originally developed around 1960,[5] and were further developed in the 1970's in Britain with the introduction of the "pipeline insertion machine" or "PIM." The PIM technology was introduced in the U.S. in the mid-1980's.[6]

The replacement pipe is almost always HDPE, however, the bursting tool (which is a specially adapted pneumatic mole with a unique pointed mole and hydraulic breaker arms) often pulls in a PVC protective sleeve to protect the new liner pipe from sharp fragments created in the bursting process. A typical bursting tool can work at the pace of 60 to 120 ft. (18 to 37 meters) per hour. In applications where a heavy wall sewer pipe is being burst, there is a reduced risk of damage from pipe fragments, and as a result, the PVC liner is often omitted.

To install a new double-walled HDPE system using pipe bursting, the techniques and options are the same as that of conventional slip-lining. The purpose of the bursting operation is simply to provide a larger bore so that the replacement liner and new primary pipe do not reduce by much, if any, the cross-sectional flow area that was previously afforded by the existing host pipe.

16.1.3 Use of Slip-Lining and Diameter Reduction/Reversion Techniques

In the mid-1980's, a process was developed to diametrically reduce relatively thick-walled HDPE pipes to an O.D. sufficiently small enough to allow insertion of the liner into the existing host pipe.[7] The process, which involves axial tension being applied to the HDPE pipe as it is pulled through, if controlled properly, allows the liner to maintain its reduced diameter, until the tension is lessened. The liner then recovers and expands outward to achieve a snug fit against the host pipe. This process, which uses semi-elliptical, dual roller machines, is also commonly referred to as "liner rolldown reduction."

Two variations of the liner rolldown reduction technique include "hot swage reduction" and "viscoelastic reduction." In hot swaging, the process consists of introducing a prefused continuous HDPE liner (with external fusion beads removed) into a pre-heating chamber. Residence time in the heating zone depends on pipe wall thickness, ambient conditions and starting pipe temperature.[8] Pipe is typically heated above 200°F for a prescribed time, until the pipe's inside diameter reaches approximately 140°F, resulting in a thermal gradient in the pipe's wall. This process can be carried out at a speed of approximately 8 to 10 ft./min., or 480 to 600 ft./hr. After insertion is completed, the hot swage liner will diametrically recover, but can also be expected to axial contract to a degree.

The process of "viscoelastic reduction" reduces the diameter of standard HDPE sizes to temporarily undersize the liner as in the standard liner rolldown reduction procedure. However, after a quickly-forced diametrical reduction, the process holds the size reduction (compressive strain) for an appropriate time to allow the viscoelastic component of the HDPE material to creep under fixed compressive strains. This reduces the severity of the diametrical reduction required to result in the undersizing lasting an extended duration and provides for a greater degree or percentage of recovery. It also preserves the effective

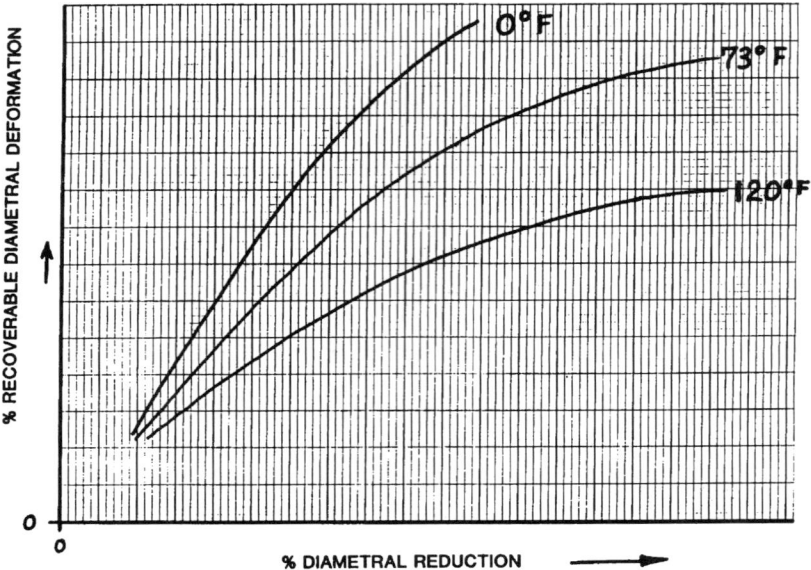

Figure 16.5. Recoverable deformation versus diametrical reduction for HDPE pipes used in rolldown or hot swaging processes. (Source: Svetlik, H. E., "Reduction, Renovation, Reversion: R3," Buried Plastic Pipe Technology, ASTM STP 1093, George S. Buczala and Michael J. Cassady, Eds. ASTM, Philadelphia, 1990, presented on Sept. 10-13, 1990 in Dallas, TX.)

diameter to thickness ratio of the liner. The net result of all of these considerations is a smaller amount of permanent deformation.[9]

Figure 16.4 illustrates a comparison of how different percentages and rates of reduction and recovery affect the final fit of the inserted liner. The optimum reduction is achieved by applying the right amount of deformation in a sufficiently slow manner (1 to 5 hrs.) to allow installation with subsequent diametrical contact having an interference fit. In such a scenario, the liner will wind up in both radial compression and axial tension to a degree.

Figure 16.5 illustrates the relationship of the recoverable creep that can be expected versus the percent of diametrical reduction at given temperatures. While specific values are not shown, they would depend on the resin used, its size and the percent reduction required.

To install a new double-walled HDPE system using liner reduction/reversion techniques, the existing host pipe is first relined by means of using one of the three major methods described in this section.. The existing host pipe, once relined with the snug, dimensionally-recovered liner then is prepared to receive a new primary pipe using conventional slip-lining methods. There is no need to use a reduction/reversion technique to install the new primary pipe. In fact, it is undesirable to do so in that the primary pipe should not fit snugly into the liner so as to provide a true annulus which will allow for leak detection to be performed. The new primary pipe can be optionally equipped with a boltable internal centering support, such as that illustrated in Figure 9.13, as it is being inserted into the liner, using the recommended support spacing of the primary pipe. It is not recommended to slip-line both primary and secondary containment pipes at once using this method.

Alternatives to Secondary Containment Piping 757

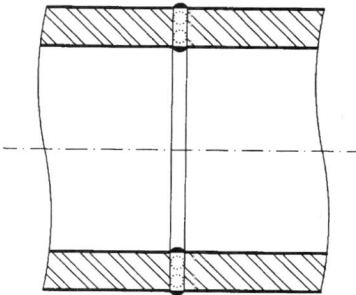

Figure 16.6. Sectional view of a joint section of thermoplastic pipe which uses a butt-electrofusion method with a Unicore® cable. (Source: Evanco Environmental.)

16.1.4 Modified Slip-Lining Using the Double-Butt® System

The Double-Butt® system is a proprietary method that was developed by Evanco Environmental of Bracebridge, Ontario, as a means to install an HDPE double containment pipe into an existing host chemical sewer pipe. The Evanco system uses a unique butt electrofusion joint that features a stainless steel wire embedded in a polyethylene resin (refer to Figure 16.6 for an illustration of the joint). The assembly, called a Unicore™ welding rod is spot welded to an HDPE ring, termed a Fusebutt™ ring, which is custom made to fit the end of the pipe. Alternatively, the Unicore™ rod can be attached directly to the pipe ends. After butting together ends of the double pipe arrangement, electric power is applied from a remote transformer through a connector on the Fusebutt™ ring. A packer is inserted into the pipe sections, with air valves, inflating bladders and screw clamps for compressing the pipe ends together during the fusion process. A major advantage of this system is that joints can be made and air tested (inner and outer joints) prior to winching the completed section into the host pipe.[10] The use of the inflated center bladder of the packer during joint cooldown minimizes the formation of a weld bead on the pipe interior, which is an advantage in maintaining maximum flow capability of the new double containment pipe. The entire new double containment pipe is fairly low profile, resulting in a reduction of flow capacity which is not too excessive, particularly for larger host pipe retrofits. Figure 16.7 is an overview of the Double-Butt™ system which highlights its important points.

16.1.5 Soft-Lining Techniques Using Cured-in-Place (CIPP) Pipe

Soft-lining involves the use of a soft, permeable tube or liner that is impregnated with resins, and is inserted via an inversion process into a host pipe by hydraulic or pneumatic action, and then cured in place by hot water to form a new pipe. The process was pioneered by Insituform and was first introduced into the U.S. in 1977.[11]

Insituform is a process for reconstructing damaged pipeline systems in municipal and industrial applications without digging. A flexible, resin absorbent, fabric Insitutube®, coated on eh outside with an elastomeric material, is custom engineered and manufactured to fit the cross-section, length and required design thickness for the damaged pipe. The Insitutube® is saturated (wet-out) with a liquid thermosetting resin. Resin selection is determined by the effluent characteristics and type of service (gravity versus pressure). Insituform

758   Handbook of Double Containment Piping Systems

Figure 16.7. Overview of the patented Evanco Double Butt™ system. (Source: Evanco Environmental.)

Alternatives to Secondary Containment Piping

is installed using an inversion technique which allows the material to negotiate bends, accommodate changes in pipe alignment and size, and span missing pipe segments. A new cured-in-place pipe (CIPP), or Insitupipe®, is formed inside of the existing conduit as shown in Figures 16.8 through 16.11. Other inversion techniques have been developed for use in other applications. An example of an actual installation is shown in the photo in Figure 16.12.

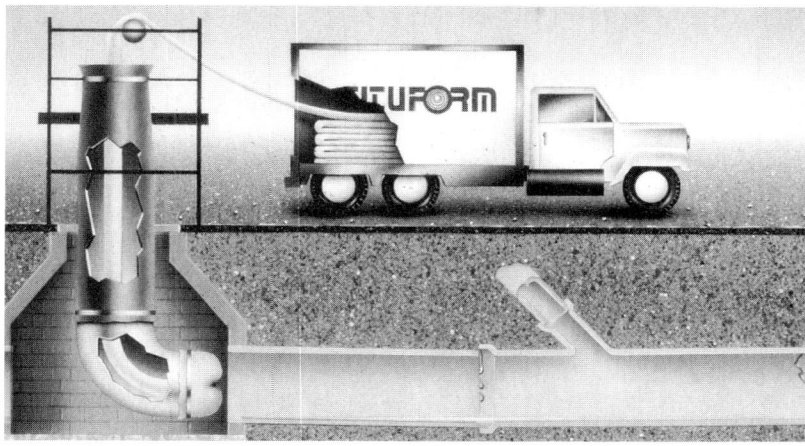

Figure 16.8. Stage 1: The resin saturated material is installed in the existing pipe through a manhole or other access point via an inversion standpipe and inversion elbow. The Insitutube® is cuffed back and banded to the inversion elbow, creating a closed system that allows the water inversion process to take place. (Source: Insituform Technologies, Inc.)

Figure 16.9. Stage 2: Water from nearby hydrants, or other convenient source, is used to fill the inversion standpipe. The force of the column of water turns the wet-out Insitutube® inside-out and into the pipe being reconstructed. As the Insitutube® travels through the pipe, water is continually added to maintain a constant pressure. The water pressure keeps the Insitutube® pressed tightly against the walls of the old pipe. (Source: Insituform Technologies, Inc.)

Figure 16.10. Stage 3: After the Insitutube® reaches the termination point, the water in the line is circulated through a heat exchanger where it is heated and returned to the Insitutube®. The hot water cures the thermosetting resin, causing it to harden into a structurally sound, jointless "pipe-within-a-pipe," an Insitupipe®. (Source: Insituform Technologies, Inc.)

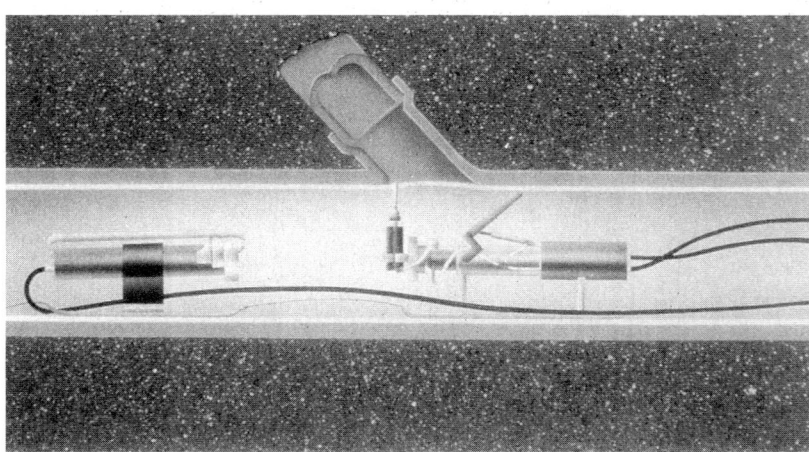

Figure 16.11. Stage 4: Once the Insitupipe® has hardened and cooled, the water pressure is released and the ends are trimmed. Service connections are reinstated internally with a remote control cutting device or by man-entry techniques. The Insituform® operation is then completed, and the newly installed pipe is ready for immediate use. All this is accomplished without excavation. (Source: Insituform Technologies, Inc.)

The Insituform process can be used to install a double containment pipe system. This is accomplished by essentially repeating the process to install a primary pipe once the host pipe is re-lined with a first liner. Insituform has developed a liner that used DuPont Kevlar® aramid fiber, which restricts the inflation to a predetermined diameter.[12] The diameter of the primary pipe is thus restricted to ensure an annular space to facilitate leak

Alternatives to Secondary Containment Piping

Figure 16.12. An example of installing cured-in-place pipe (CIPP) during the inversion process, using an existing manhole in an industrial facility. (Source: Insituform Technologies, Inc.)

detection, an important consideration for meeting the criteria of a true double containment piping system.

16.1.6 Use of Folded and Formed Liners

The use of folded and formed liners, also referred to as collapsed liners, is a simple and straightforward concept that uses PVC or HDPE liners. As shown in Figure 16.13, a liner is extruded into a folded shape. The liner is then inserted into the host pipe, and by means of hot, air or steam, is then reverted to its circular form to make a close fit within the host pipe. To install a double containment pipe system, using this technology, the process is simply repeated twice.[13]

16.1.7 Other Trenchless Reconstruction Techniques

Other methods do exist, but have not yet been used in practice to install a double containment pipe system. Examples include the use of a spiral winding method using an interlocking PVC strip developed in Australia, and the use of spraying techniques using a thin mortar coating or resin coating. While these methods have not been used to install a double containment pipe, it is possible that they can be adapted to at least retrofit an existing host pipe so that it may serve as a secondary containment pipe, and thereby allowing insertion of a new primary pipe system by means of one of the other methods described in this section.

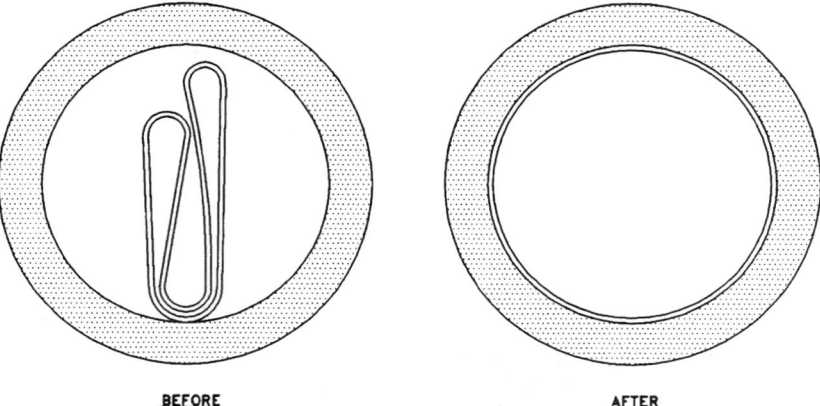

Figure 16.13. An illustration of the use of folded and formed liners. On the left (before) the folded liner is inserted into the existing host pipe. The liner is inflated by the use of steam, hot air, or hot water to take its final shape, as shown on the right (after).

16.2 Impervious Barriers

Impervious barriers include the use of thermoplastic or geotextile membrane liners and low-permeability soils as a means of providing secondary containment for piping and associated tank systems. These technologies have found widespread use as means of providing secondary containment for underground tank installations. However, these methods are limited as means of providing secondary containment for extensive lengths of underground piping installations. Tank installations having limited amounts of associated piping have been protected by these methods for years.

Thermoplastic and geotextile liners are constructed from a variety of materials. The most common include high-density polyethylene, medium-density polyethylene, polyvinyl chloride, polypropylene, polyester elastomer, and several types of synthetic rubbers. Most sheets are plain extruded sheets, ranging in thickness from 8 mils (0.2 mm) to 100 mils (2.5 mm).

Low-permeability soils (soils having a permeability less than 10^{-6} cm/sec) have also been used as a form of secondary containment for underground tank and piping systems. Typically, the use of low-permeability soils has been restricted to clay soils that occur naturally on site. Only on rare occasions have low-permeability soils been transported to a site and applied to piping trenches.

Lining a piping trench with low-permeability soils is difficult. The material is usually hard to handle, is expensive to transport, requires compacting in a pipe trench, and further requires extensive quality control testing during the installation. Once an installation is completed, the final cost is usually greater than other forms of piping secondary containment.

16.2.1 Impervious Barriers for Secondary Containment of Piping and Tanks

When impervious barriers are used as a means of secondary containment for piping and tank applications, the idea is to encompass a finite area of the soil surrounding the piping

Alternatives to Secondary Containment Piping

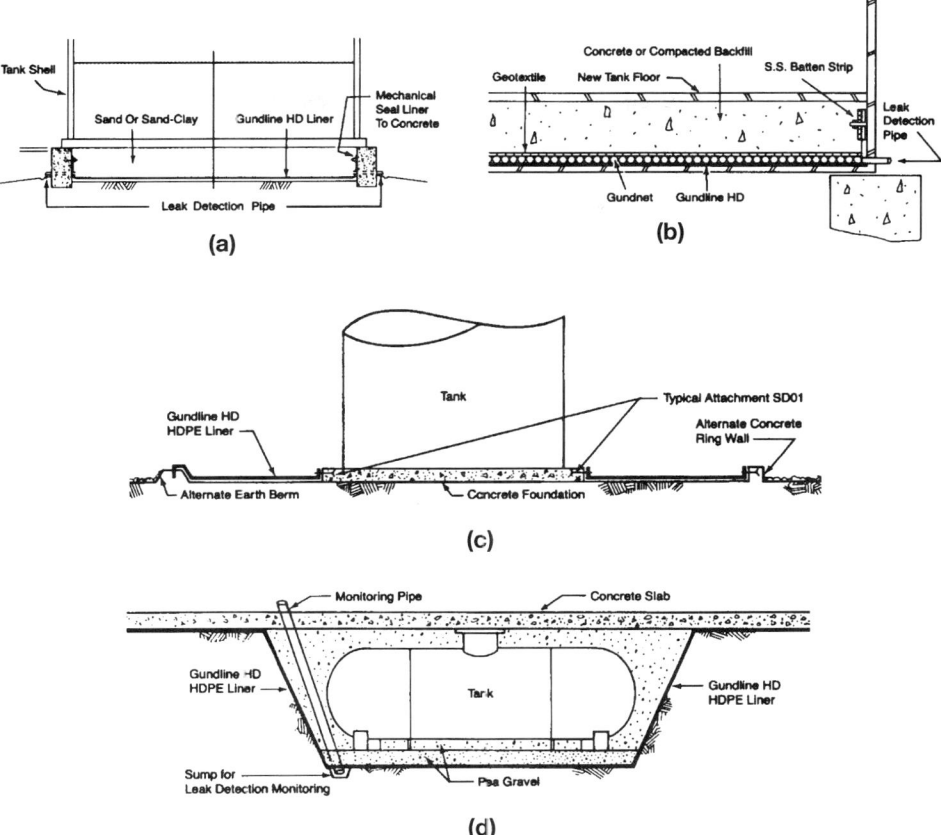

Figure 16.14. (a) Typical leak detection under an above-ground storage tank; (b) Retrofit of a new leak detection system with a new tank floor; (c) Typical spill containment liner in a dike area of an above-ground vertical tank; (d) Underground tank leak containment liner. (Source: Gundle Lining Systems, Inc.)

and associated tank. In the event of a release of the fluid contained within a single-wall piping or tank system, there is a limited amount of soil within the secondary containment zone that becomes contaminated.

If a release occurs in a system that uses a membrane or geotextile liner, or a low-permeability soil, there is a limited amount of soil that becomes contaminated. In most applications of this type, contaminated soil will subsequently become classified as a hazardous material according to federal or local regulations. It will have to be disposed of by one of various allowable regulated methods for hazardous waste, which vary by country, state, and, in some cases, by local county or municipality.

In the United States, Canada, most parts of western Europe, and other developed nations, there are very definite rules as to how contaminated soil must be handled and disposed of. In the United States, this practice is defined by either the Resource, Conservation and Recovery Act (RCRA), or by the Comprehensive Emergency and Response Clean Land Act (CER-

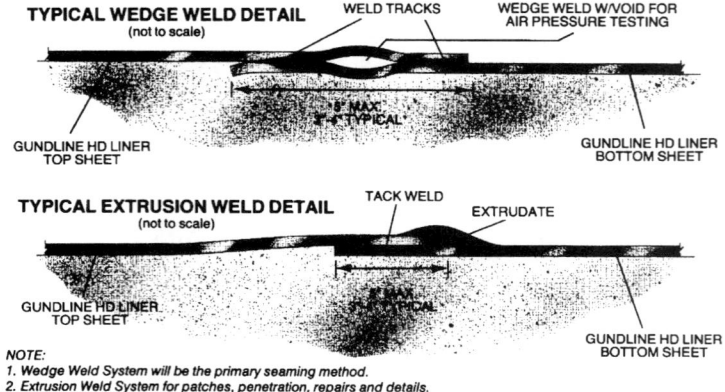

Figure 16.15. Examples of hot wedge weld details and extrusion weld details for HDPE liner joints. Wedge welds are used as a primary sealing method whereas extrusion welds are used for patches, penetration, repairs and miscellaneous attachments. (Source: Gundle Lining Systems, Inc.)

CLA), including parts of CERCLA that are commonly referred to as the Superfund Act. Any contaminated soil is required to be replaced with fresh soil after the contaminated soil has been removed, in order to place the piping/tank system back into service. Figure 16.14 illustrates typical installation details for above-ground and underground piping and tanks.

Typical weld details for HDPE secondary containment liners are shown in Figure 16.15. To test a field weld involving the hot wedge weld shown in Figure 16.15, a seam air pressure test can be used. Normally, it is best to install same type of soil-monitoring or vapor-monitoring system of the types described in Section 15.4, at least within the zone of secondary containment. In the event of a release of fluid, a soil-monitoring device serves as a means of leak detection, signalling that there is a leak. In most installations of this type, it would be wise to monitor the soil outside the zone of secondary containment as well, to detect for a containment failure of the liner, if and when a leak occurs.

Above-ground, vertically positioned storage tanks may be protected by means of the arrangement shown in Figure 16.16. Various details of the piping penetrations equipped with secondary containment pipe sleeves are illustrated in Figures 16.17 through 16.20.

The use of a polyester elastomeric flexible sleeve method represents a unique method of protecting pipe sections for certain underground piping applications involving petrochemicals. Close-up details showing how this type of sleeve method may be used are illustrated in Figures 16.19 through 16.20. A horizontally positioned above-ground storage tank illustration protected by means of a flexible liner is illustrated in Figure 16.21. Details of the liner connections to the concrete footings are shown in Figures 16.22 and 16.23.

Another recent development in secondary containment includes sheet materials that are extruded with a textured profile on one or both sides. The capability of having a textured profile is beneficial when installing plastic liners on embankments, sloped trench walls, and other nonflat surfaces. Perhaps the most unique sheet material developed for soil lining purposes is extruded with one side containing a spiked profile. A uniform spiked profile, produced during the extrusion process, allows the spiked sheet to resist movement on even the steepest of slopes. The profile of a spiked-profile sheet is illustrated in Figure 16.24.

Alternatives to Secondary Containment Piping 765

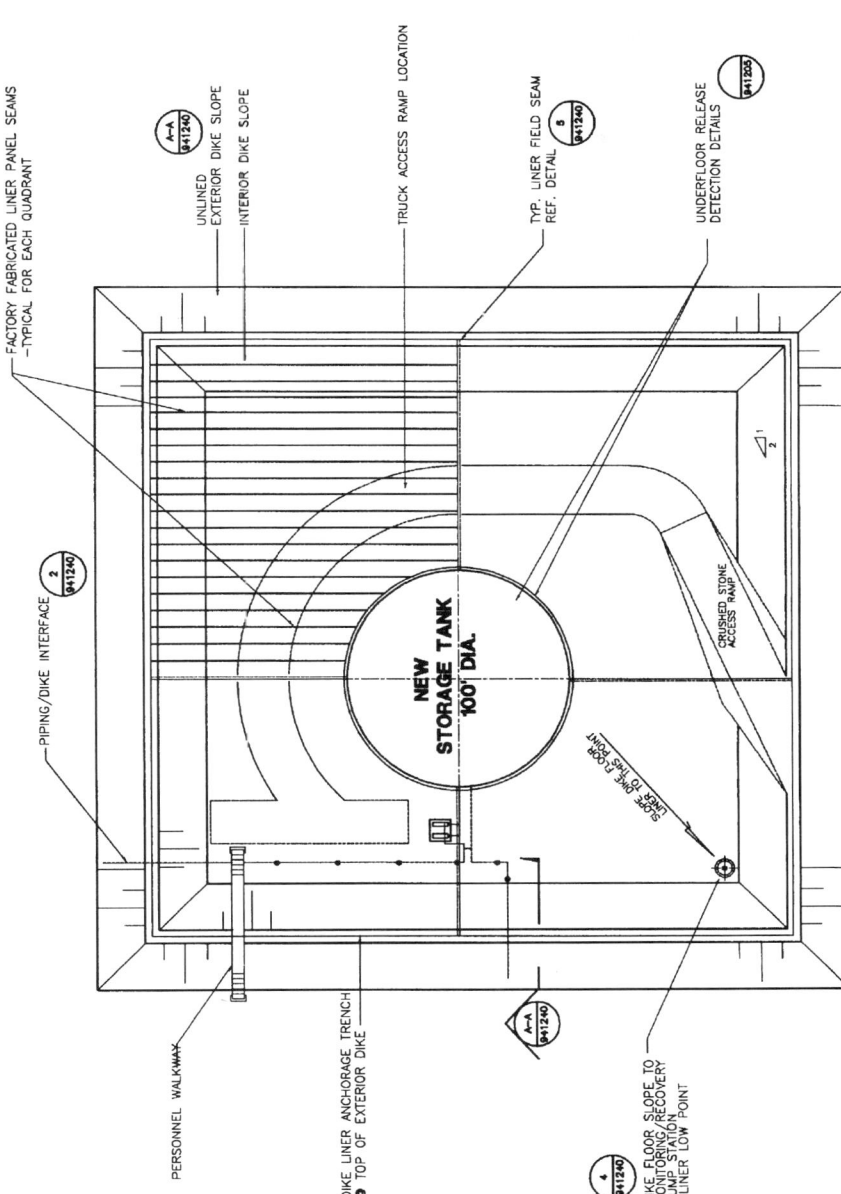

Figure 16.16. Schematic representation of a large diameter vertically-positioned storage tank which is provided with secondary containment by means of a lined dike. (Source: MPC Containment Systems, Inc.)

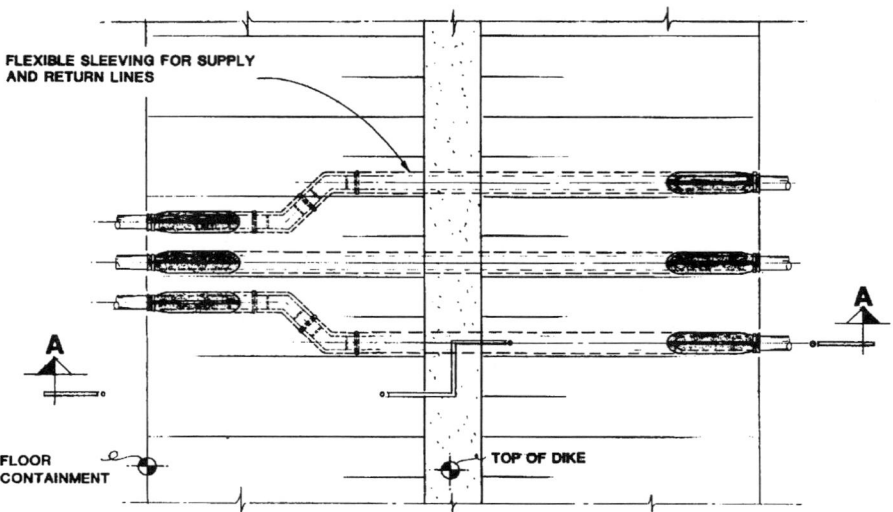

Figure 16.17. Typical site plan of an underground section of piping from an above-ground storage tank which passes through a dike section, and which is provided with secondary containment by means of a flexible membrane liner. (Source: MPC Containment Systems, Inc.)

16.2.2 Impervious Barriers in Landfill Applications

By far the largest use of impervious barrier technologies is in the design of modern-day municipal landfills and hazardous waste dump sites. In many of these applications throughout the United States, Canada, and Europe, the design consists of multiple layers of these barriers, with a primary layer and a secondary layer designed to provide secondary containment for hazardous landfill liquids that drain down to the liners, and in which there is a subsequent failure of the first (primary) liner.

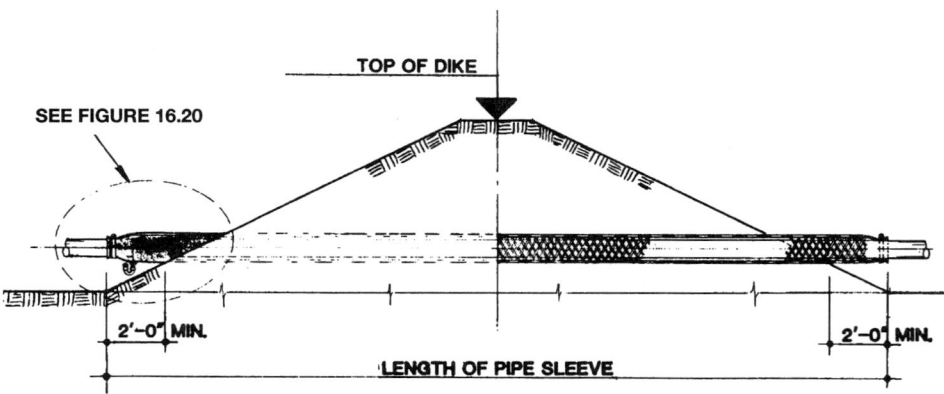

Figure 16.18. Side view of the illustration shown in Figure 16.17. This detail would also be typical of that indicated as section A-A on Figure 16.16. (Source: MPC Containment Systems, Inc.)

Alternatives to Secondary Containment Piping 767

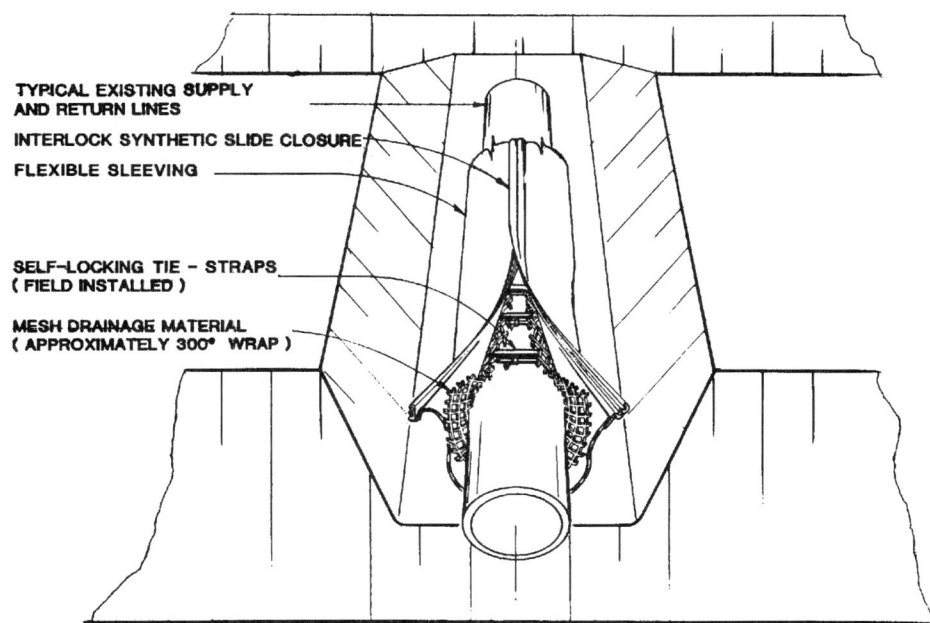

Figure 16.19. A schematic illustration of a flexible membrane liner secondary containment sleeve system which uses a special "zipper" closure means. This is the joining method typically used for the pipe sleeves shown in Figures 16.17 and 16.18. (Source: MPC Containment Systems, Inc.)

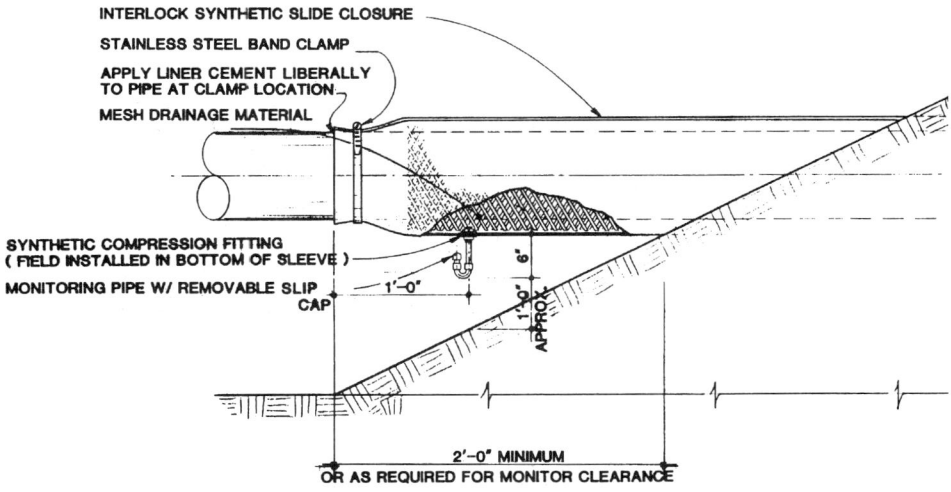

Figure 16.20. Illustration of a pipe sleeve monitor detail for the type of sleeve shown in Figure 16.18. (Source: MPC Containment Systems, Inc.)

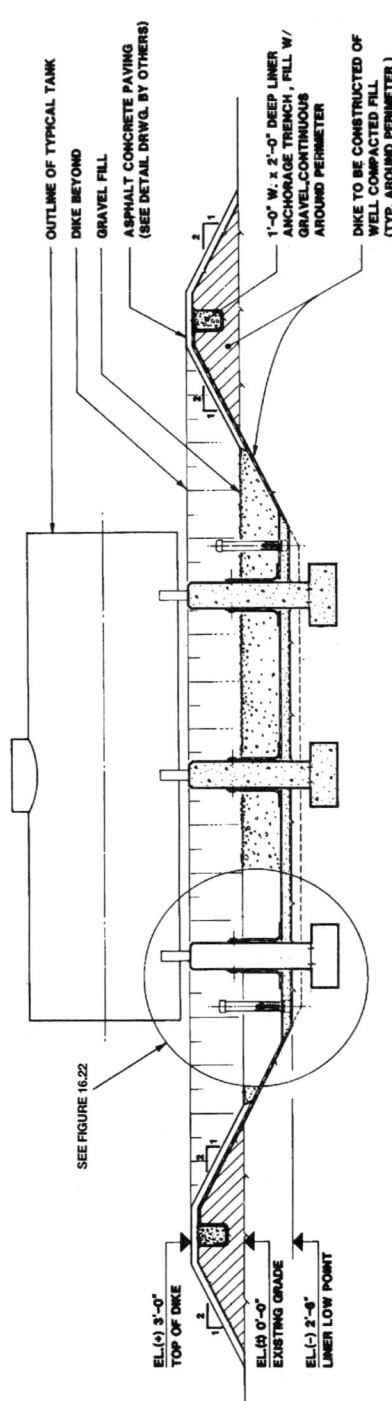

Figure 16.21. Side view of an above-ground horizontally-positioned storage tank which is protected by means of a lined dike. (Source: MPC Containment Systems, Inc.)

Alternatives to Secondary Containment Piping

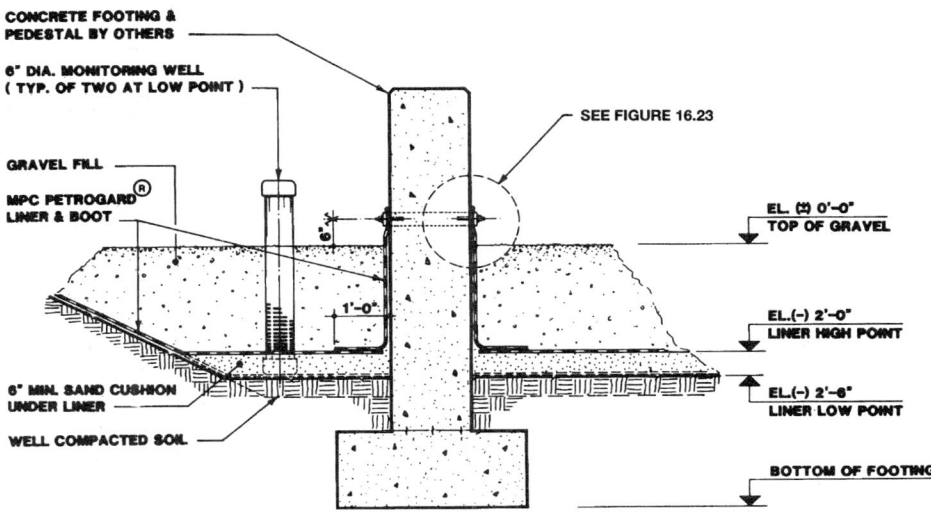

Figure 16.22. Details of foundation connections for the tank footings shown in Figure 16.21. (Source: MPC Containment Systems, Inc.)

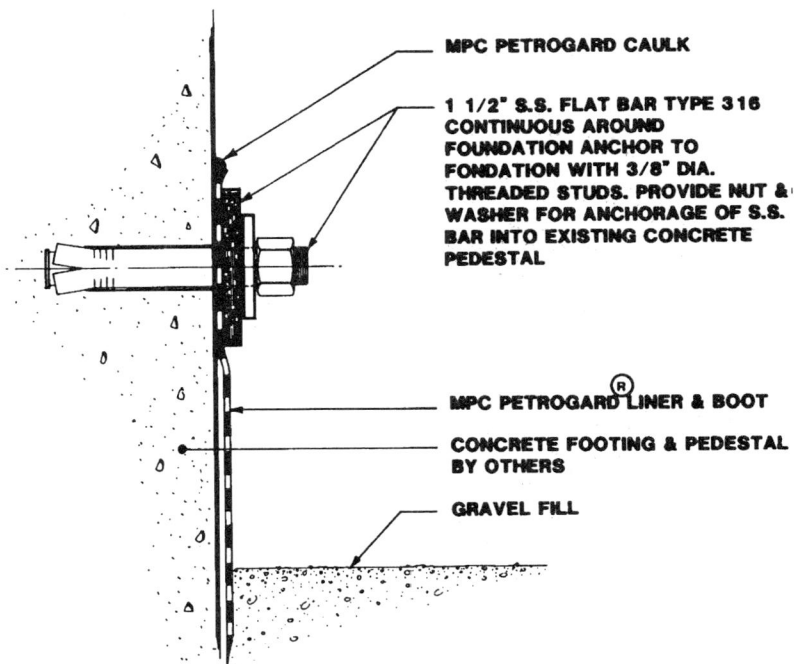

Figure 16.23. Typical details for liner boot anchorage for the foundation connections illustrated in Figure 16.22. (Source: MPC Containment Systems, Inc.)

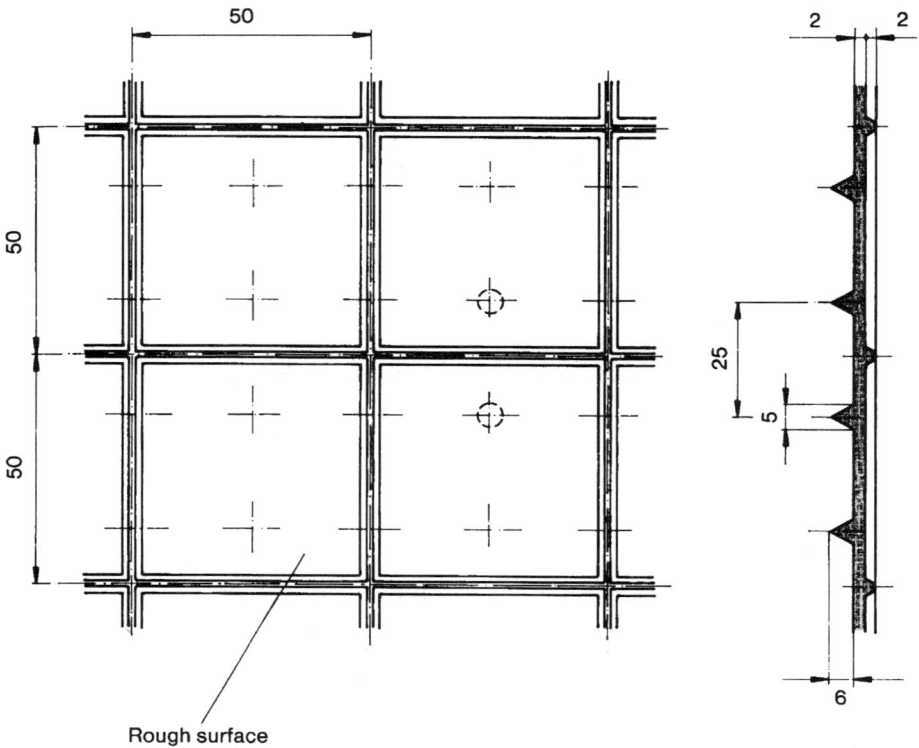

Figure 16.24. Top and side profile views of a spike thermoplastic liner which is used for lining sloped soil embankments to act as an impervious barrier. (Source: AGRU.)

Figures 16.25 through 16.27 are illustrations of modern-day landfill designs, typical of European installations. These types of arrangements are required for all new landfill facilities in many parts of western Europe; in the United States, similar requirements have been adopted as a result of the revisions enacted to the Resource Conservation and Recovery Act (RCRA), as of September 1991.

Typical design of a landfill facility includes two membrane liners, one being a primary and the other a secondary containment liner. Above the liner is a perforated gas collection main to collect methane gas that evolves as a result of decomposing landfill substances. The entire system is sloped so that liquid wastes collect in a sump at the low point of the system. Liquid landfill wastes are then pumped under the liner system. In the majority of applications, an HDPE/HDPE double containment piping system is used for this line, unless the liquid contains less than trace levels of corrosive or toxic substances commonly encountered in municipal landfill applications (referred to by the U.S. EPA as a de minimis quantity). A double containment piping system (which is typically located under the liners) transports collected waste to a treatment system. The waste treatment system is typically located at a distance of 1,000 ft. or more from the collection sumps. Liquid wastes are then neutralized, combusted, or remediated, thereby reducing them to a less harmful state.

Alternatives to Secondary Containment Piping

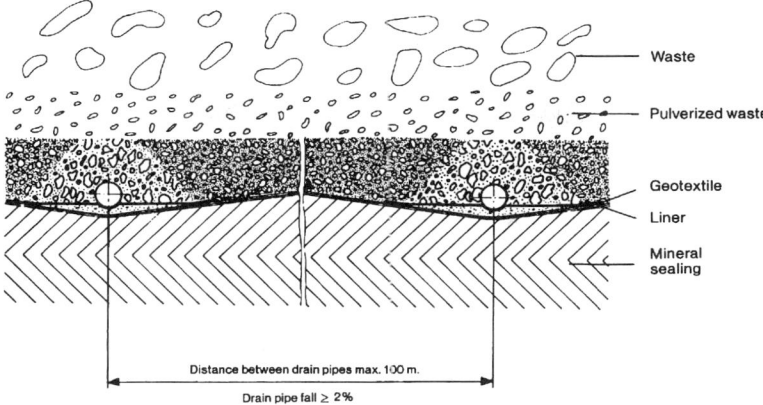

Figure 16.25. Normal profile of a landfill bottom construction according to modern-day standards. (Source: AGRU.)

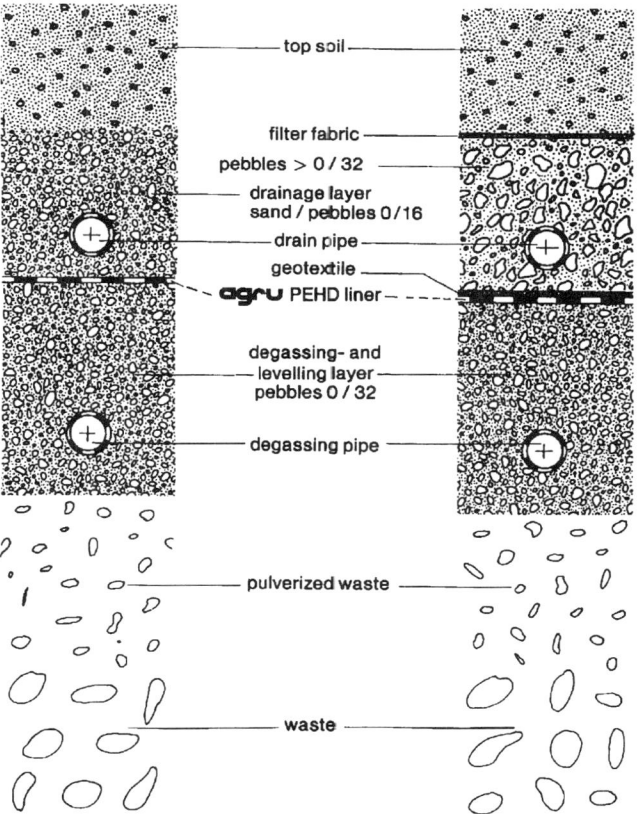

Figure 16.26. Profile of landfill cover materials using HDPE sheet liners. (Source: AGRU.)

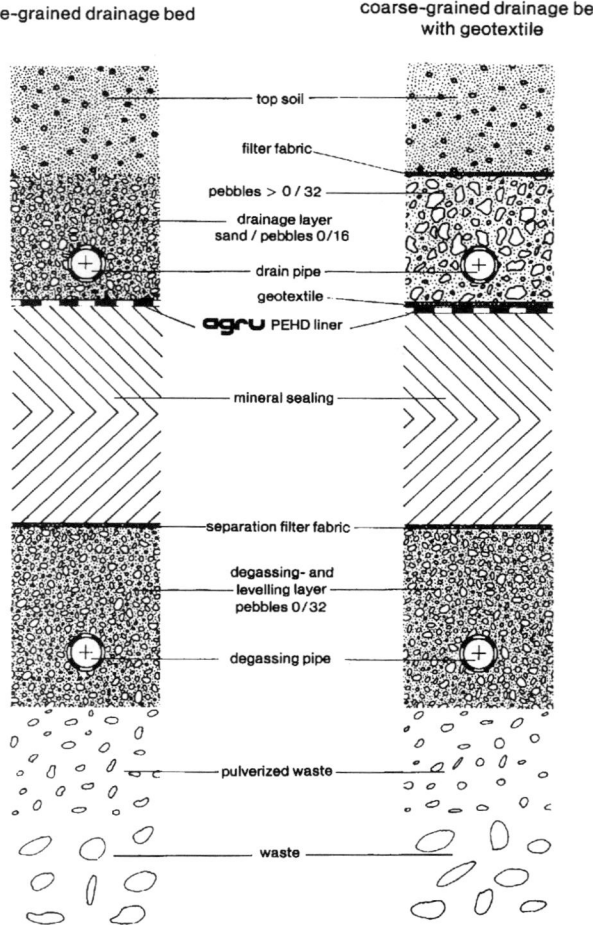

Figure 16.27. Profile of landfill cover materials using a combination of lining methods. (Source: AGRU.)

16.3 Concrete Encasement

Perhaps the earliest and simplest method of providing secondary containment for a piping system involves completely encasing a pipe with concrete. This can be done by pouring concrete directly over a single-wall piping system so that it is thoroughly covered with concrete. When the concrete hardens, the pipe is imbedded within the concrete, forming an opening between the pipe and hardened concrete that is in the shape of a pipe of circular cross section, much in the same way that a mold is formed. The cross-sectional view in Figure 16.28 illustrates this point. The method, as crude as it may be, does represent a way to protect pipes, and has been used for many years in the protection of domestic plumbing pipes and other such systems. It has also been used as a method of protecting pipes of all types from dynamic, structural live loads for pipes that are buried to a shallow depth.

Alternatives to Secondary Containment Piping

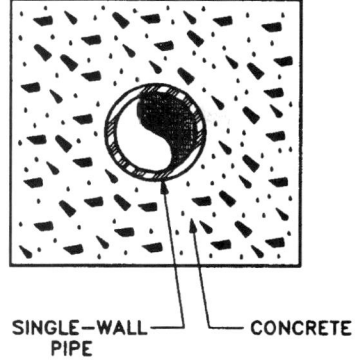

SINGLE-WALL PIPE — CONCRETE

THE CONCRETE FORMS A HOLE AROUND THE PIPE ESSENTIALLY FORMING A SECONDARY CONTAINMENT PIPE

Figure 16.28. Concrete that is directly poured around a pipe essentially forms a surrounding secondary pipe shape. However, there is not an annulus form which makes detection of leaks very difficult. Also, though low-porosity concretes have been developed, many authorities having jurisdiction do not accept concrete as a viable means of secondary containment, unless it is lined or coated.

There are several drawbacks to this simplified approach to providing secondary containment for the protection of chemical pipes. While concrete does provide structural protection, concrete itself is typically not resistant to highly corrosive chemicals. Also, concrete itself is usually very porous, which may allow chemicals to seek their way out of the material and into surrounding soil. Since concrete is porous by nature, it is also subject to degradation by means of being alkylated once ground water permeates it, unless it is further protected. Therefore, unprotected concrete is not considered an effective material for secondary containment purposes. Another drawback to direct encasement in concrete is that concrete readily forms cracks, which could represent a potential leak path for chemicals that must be contained. A third reason for avoiding this practice is the difficulty in facilitating a repair once it has been determined that a leak has occurred. It is very difficult to remove concrete without subjecting the encased piping to some type of further damage in the removal process. Yet another reason for avoiding this practice is the lack of a "true" annular space, and thus the difficulty in allowing leaking fluids to drain safely, and to incorporate some type of leak-detection method. While the method lacks sophistication to a large extent, it does represent some interesting possibilities as a means of protecting piping systems, especially from structural damage.

16.4 Concrete, Masonry and RTRP Structures

Concrete may be used in a variety of ways as a means of housing piping components and tanks. In some applications, it may even be used as a means of providing tertiary containment to double containment piping components and/or double-wall tanks. While concrete is itself not considered to be a corrosion-resistant material in most jurisdictions (and by most corrosion experts), there are many forms of linings and protective coatings available to increase its impermeability and its corrosion resistance. Protective linings and coating systems that are available to protect concrete in immersion service applications are discussed further in Section 16.4.2.

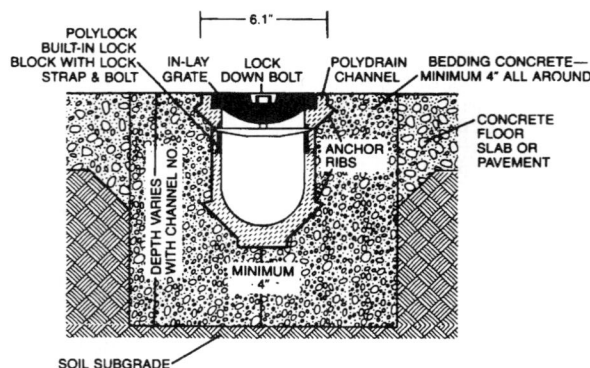

Figure 16.29. Cross-sectional view of a precast polymer concrete trench with in-lay type grate in a concrete floor slab or pavement. (Source: Polydrain Co.)

Polymer concrete also exists as a possible material of construction, representing a mix of the corrosion-resistant properties of plastic materials with the broad range of construction possibilities that concrete offers. Another structural material that may be used is acid brick, which also allows for a broad range of construction possibilities and has had a long history of successful application in corrosive environments.

Concrete structures may either be precast into their final form, or they can be cast in the field to form the needed structural arrangement. Depending on the structural requirements of the project, concrete may either be of the reinforced or unreinforced type. Concrete is normally reinforced by means of adding steel wires or steel bars. The percentage of steel added is determined by the amount of reinforcement needed.

There are many types of concrete structures that are useful in piping and tank system secondary containment design. These include buried structures such as trenches, tanks, sumps, pits, vaults, and manholes. These may be constructed in a wide variety of geometric configurations.

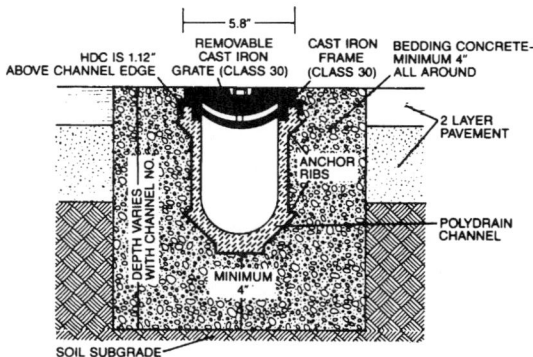

Figure 16.30. Cross-sectional view of a precast polymer concrete trench with HDC frame and grate in a two-layer flexible pavement. (Source: Polydrain Co.)

Alternatives to Secondary Containment Piping

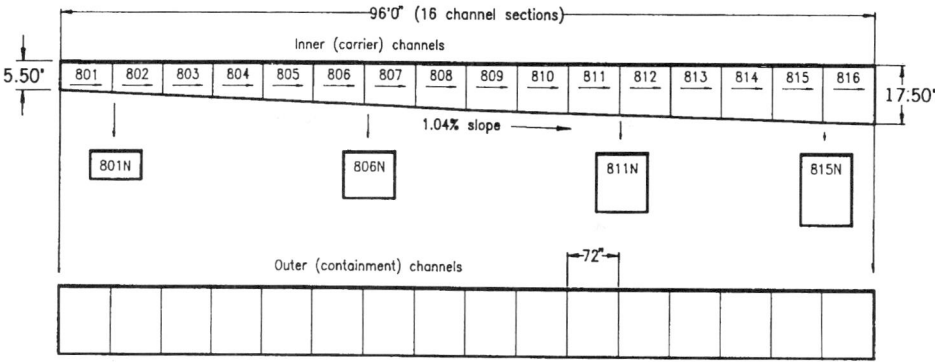

Figure 16.31. Overview of a double containment sloping FRP trench system. (Source: Aquaduct, Inc.)

Concrete may also be used as a means of secondary containment in above-ground applications. An example of a common structure used to provide secondary containment for an above-ground tank is a concrete dike. Another common above-ground secondary containment structure is a concrete berm.

When concrete is used as a means of secondary containment in corrosive or regulated services, it is almost always lined, coated, or protected in some fashion to add to its corrosion resistance (unless polymer concrete is used).

Polymer concrete is a unique material that consists of a thermosetting plastic resin that is reinforced with the addition of 80% or more aggregate. Aggregate provides a base polymer with added structural rigidity; polymer provides the mixture with corrosion-resistance properties, which vary according to the base resin selected. Polymer concrete may be cast into a number of forms, either precast in sections or cast in the field. The two most common uses for polymer concrete thus far have been for the protection of floors and in precast open trench systems used for drainage in factory floors, airports, and other locations. Polymer concrete that is used for protecting floors is normally cast in a several inch layer at the site for areas where there is the possibility of spills of corrosive fluids.

Figures 16.29 and 16.30 illustrate trench cross-sections for typical precast polymer concrete trench arrangements involving a sloping trench. The trench is precast in sections wherein each section is designed to be progressively deeper in order to maintain a designed slope. Figures 16.31 and 16.32 illustrates a unique product involving a vinylester fiberglass concrete "double containment" trench for applications involving drainage of a hazardous fluid. The primary trench is designed as the primary conduit, replacing the need for a primary piping system. The outside trench thereby acts as a means of secondary containment, should there be a leak of fluid from the primary trench. This type of arrangement could also be described as a trench on top of a trench, rather than a double containment trench. Figure 16.33 illustrates a custom RTRP trench system which also can be made in a double containment configuration, similar to Figures 16.31 and 16.32.

Virtually any type of trench or structure that can be constructed out of concrete can be made from acid brick materials. These materials are discussed further in Chapter 2. Both polymer concrete and acid brick tend to be expensive relative to concrete, which is provided

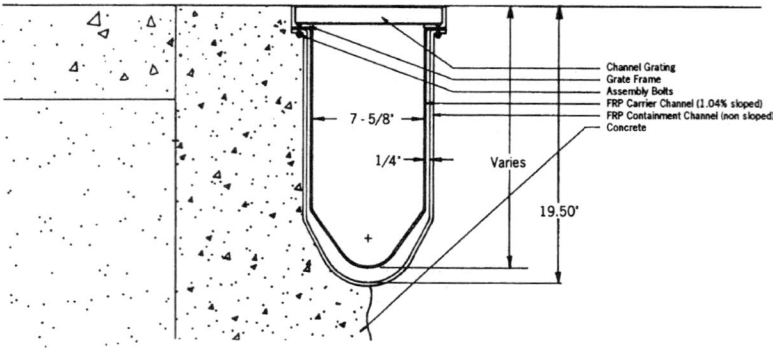

Figure 16.32. Cross-sectional view of a precast polymer "double containment" trench. (Source: Aquaduct, Inc.)

with added protection. Cost is usually a limiting factor in the use of polymer concrete and acid brick when considering the construction of a large structure, and usually precludes its use. Table 16.1 lists the relative costs of various protective means for concrete, including polymer concrete, versus the use of acid brick.

16.4.1 Protective Linings and Coatings for Concrete

There are many varieties and methods for lining and protecting concrete from the effects of corrosive chemicals. These include a variety of thermosetting plastic and organic coatings that may be applied by spraying or applied with the use of a trowel directly on the concrete

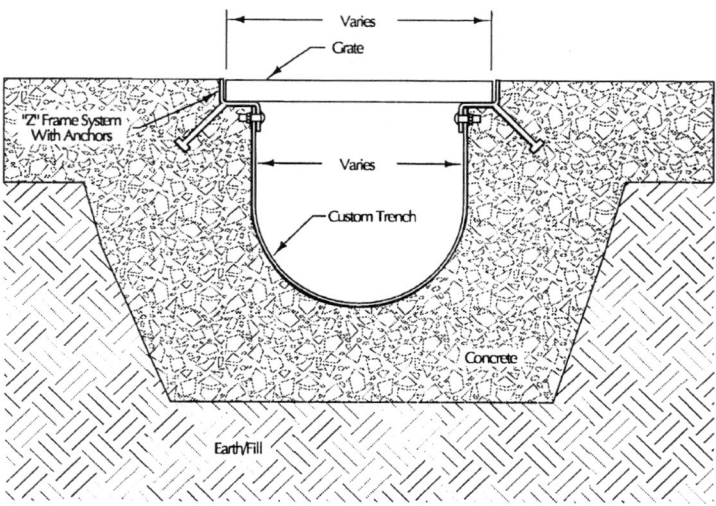

Figure 16.33. Illustration of a custom FRP trench system to be installed in a concrete slab or pavement. (Source: Aquaduct, Inc.)

Table 16.1 Relative Installed Unit Costs (Labor and Materials) of Floor Systems

Conditions	Floor type and application method	Cost range ($/ft^2/yr)*
A	Acidproof brick floors with membrane and chemical resistant mortars	18-25
B	Reinforced 1/8 in. floor system, trowel application	8-10
C	Unreinforced 1/4 in. floor toppings, trowel application	4-6
D	High build floor toppings, broadcast seed application, 60-100 mils	2.50-3
E	High build floor coatings, brush, roll, or squeegee application, 15-40 mils	1.50-2.50

Source: Materials Performance, "Monolithic Linings and Coatings for Secondary Containment Structures," Hazen, Fred E., August, 1991, Vol. 30, Number 8, p. 36-41.
* This compares the relative costs. The A-B range represents heavy duty or critical process floor service zones. The C-D range represents moderate floor service zones. E represents light floor service zones.

substrate, to a thickness of 30 mils (0.8 mm) or more. A lining for immersion service (or secondary containment service) is usually a mixture of liquid thermosetting resins and inert fillers, with reinforcement provided by flake fillers, glass, or synthetic fibers.

Some of the more common monolithic reinforced-thermosetting-plastic linings and coatings used for concrete protection are listed in Table 16.2. The three most common thermosetting resins used in the forms described in Table 16.2 include epoxy, polyester, and vinyl ester. The chemical resistance of these coatings varies greatly, as do the methods and difficulties in applying them. The specific lining should be selected based on the chemical or chemicals to be encountered and the resistance of the lining to these chemicals. The available expertise to apply the coatings and the available quality control to inspect such linings should be considered in making such a selection.

In order to coat or line concrete successfully, attention must be paid to the surface preparation of the concrete structure. The condition of the concrete surface and the degree to which it is prepared will determine to a great extent how well a lining or coating system will perform in service. A high-quality concrete with a well-prepared surface will enable a high-quality bond between a protective liner/coating and a concrete substrate it is protecting. Proper procedures must also be followed during the application of a lining/coating. A rigorous inspection program must be implemented to ensure that proper procedures are being followed.

16.4.2 Lining Concrete with Studded Structured Thermoplastic Liners

An efficient method of providing corrosion protection to concrete structures is by means of studded structured thermoplastic liners. Thermoplastic liners designed specifically for concrete protection are manufactured with profiled studs on one side to allow a liner to become mechanically imbedded into concrete, once it is cast. Liner sections may be welded at their seams, with or without the use of profiled strips, depending on the struc-

Table 16.2 Major Coating and Lining Classifications

	Heavy-duty reinforced linings	Light-duty linings/heavy duty coatings	Thin-film protective coatings
Type of construction: reinforcement and resin	Glass cloth or mat-reinforced polyester, vinyl ester, or epoxy resin	Flake-filled thick-film coatings with or without mat-reinforced polyester, vinyl ester, or epoxy resin	Unreinforced epoxy, vinyl, urethane, and other thin-film coatings
Methods of application	Trowel	Spray, brush, roller	Spray, brush, roller
Thickness range	1/8 to 3/16 in.	30 to 60 mils DFT without mat-reinforcement layer	10 to 15+ mils DFT
Reason for selection	(1) Aggressive chemistry (2) Abrasion (3) Temperature	(1) Moderate to light chemistry (2) Temperature	(1) Very mild chemistry (2) Atmospheric fumes
Typical uses	(1) Interior vessel lining (2) Harsh spillage (3) Heavy-duty traffic or abrasion (4) Secondary chemical containment	(1) Interior coating of vessels-mild immersion (2) MIld to aggressive spillage (3) Light-duty floor traffic (4) Secondary chemical containment	(1) Exterior of tanks/vessels (2) Wall protection above splash/spillage zone (3) Structural steel coating in corrosive fume service (4) Secondary containment-diesel fuel and other non-aggressive materials

Source: Materials Performance, "Monolithic Linings and Coatings for Secondary Containment Structures," Hazen, Fred E., August, 1991, Vol. 30, Number 8, p. 36-41.
Notes:
(1) Carbon-filled and graphite flake-filled formulations are available in the above heavy-duty and light-duty systems for applications requiring resistance to hydrofluoric acid, fluoride salts, and strong alkalis.
(2) Heavy-duty reinforced lining products are characterized by resistance to hostile environments and the capacity to withstand varying physical impositions.
(3) Combinations of the above products/systems can be used for protection of secondary chemical containment structures, i.e., use of heavy-duty reinforced systems for lining floor, trench, collection sump, and splash zone of walls combined with the use of light-duty linings/thick-film or thin-film coatings for upper walls.

tural integrity needed. The three most common profiles for studded structured thermoplastic sheets products are illustrated in Figure 16.34(a,b,c).

Using the profile geometry illustrated in Figure 16.34(c), installation details for lining a concrete structure with a studded thermoplastic liner is described in Section 16.4.2.1.

16.4.2.1 Installation Details for Lining Concrete Structures Using Structured Thermoplastic Linings

To line concrete with a structured thermoplastic liner, standard wooden forms must be erected in their normal manner. Structured thermoplastic concrete protective liners are fitted to such forms, with or without the use of profiled strips, depending on the structural integrity required of the lining. An illustration of typical "H"-profiled connector strips (couplings), as well as buffer strips and floor strips are illustrated in Figure 16.35.

Liners are normally fastened to wooden forms with the use of finishing nails. Nails may be placed next to the joints (where profiled strips are to be placed). At the joints, liners may be secured together by means of wire loops at every sixth row of studs, and thereby held together inside the boarding. In many applications, protective liners are prefabricated

Alternatives to Secondary Containment Piping

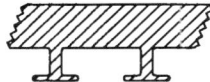
a.) CONTINUALLY-EXTRUDED "TEE" PROFILED STUDS

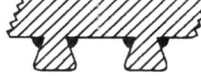
b.) WELDED-ON CONICAL STUDS IN SEPARATELY EXTRUDED SHEET

c.) VACUUM-FORMED STUDS IN EXTRUDED SHEET

Figure 16.34. Cross-sectional profile views of "studded" structural thermoplastic liners that are to be imbedded in and are used to line concrete for secondary containment purposes. (Source: (a) Ameron, (b) Atlas; (c) AGRU.)

in large sections to their respective wooden forms. If concrete is to be reinforced with steel, steel reinforcement bars or rods must be put in place after the structured thermoplastic concrete protective liners have been secured to the forms. Outer boarding can then be installed. Once the boarding has been removed, drill holes and nail holes created by securing structured liners to the forms must be sealed with sealing caps, or by directly welding over the holes using hot-gas-welding techniques.

Figure 16.36 illustrates basic details of liner joints when using H-profile strips. The left side of figure 16.37 illustrates a detail of a liner joint area where nail holes have been sealed by means of thermoplastic extrusion welding or thermoplastic hot-gas welding. This technique is used to seal the holes left by the finishing nails once the wooden forms have been stripped away, and the finishing nails have been extracted.

The right side of figure 16.37 illustrates a detail of a liner joint area where liners have been directly welded without the use of H-profile strips. This detail is typical of tank bottoms where raised channels are first cast with non-H-profile bottom strips imbedded in place. This practice is feasible in small areas only and in areas that do not require high structural integrity (e.g., in applications involving negligible temperature fluctuations). However, this practice can be used in additional applications if the thermoplastic material is chemically etched and provided with an overlay of reinforced-thermosetting-plastic material. It should be avoided where there can be significant thermal expansion/contraction.

Examples of details at inside and outside ceiling corners are illustrated in Figure 16.38. The left and right sides of figure 16.39 illustrates liner installation details at a bottom corner joint seam and a typical vertical offset using corner profiles, respectively.

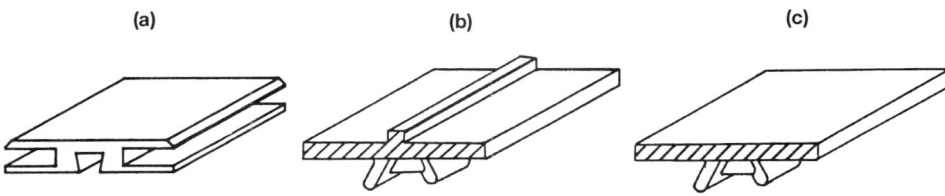

Figure 16.35. Illustration of: (a) typical H-profile strip; (b) a buffer strip; and (c) a floor strip. Profiles are also available for several angle corner joints and for making terminations. (Source: AGRU; A/A Manufacturing, Inc.)

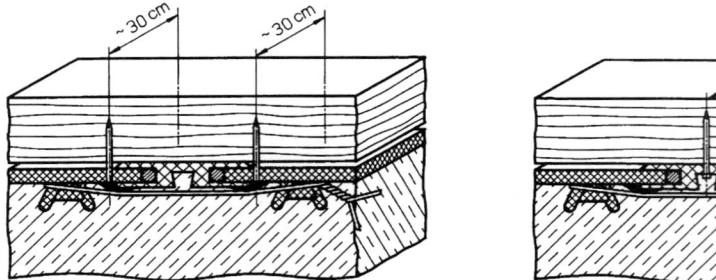

Figure 16.36. Joint details when using an H-profile strip. (Source: AGRU; A/A Manufacturing, Inc.)

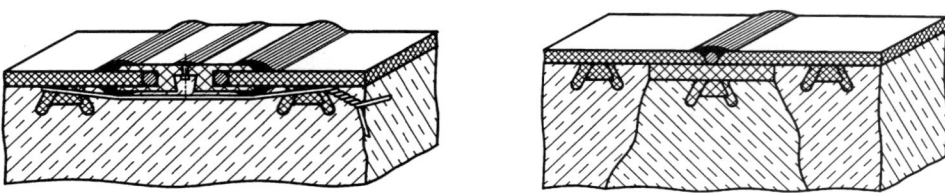

Figure 16.37. Detail of joining of liners at joints using H-profile strips showing extrusion weld seams applied after the wooden forms are stripped away (left). Example of a bottom detail showing the connection of floor plates to floor strips (right). (Source: AGRU; A/A Manufacturing, Inc.)

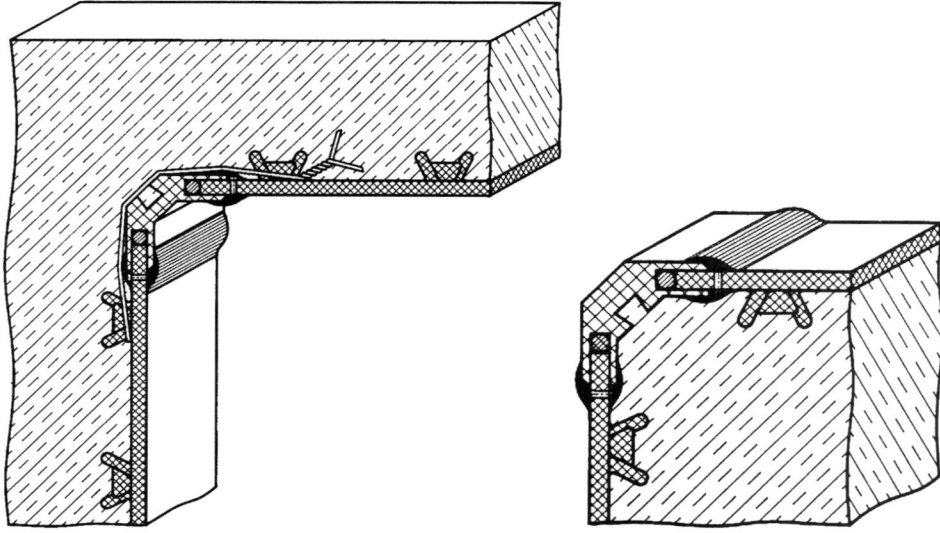

Figure 16.38. Connection details of a liner at a 90° inside corner using a 90° corner profile (left). Example of connection details of a liner at a 90° outside corner using a 90° corner profile (right). (Source: AGRU; A/A Manufacturing, Inc.)

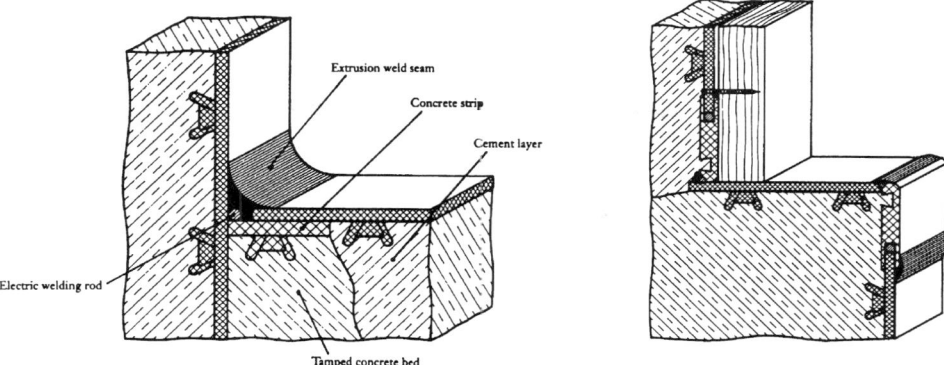

Figure 16.39. Connection details of a liner at a bottom corner connection between a floor and a wall (left). Connection detail of a liner at a vertical offset area (right). (Source: AGRU; A/A Manufacturing, Inc.)

If hollow spots occur in a concrete structure that is to be lined, the top and bottom of Figure 16.40 (a & b) provide a before-and-after graphical depiction of the steps necessary to fill such a void. A filling tube and overflow tube must be used to enable concrete to be poured in any void areas. Once a void has been filled, filling tubes and overflow tubes must be removed and the resulting drill holes closed with stoppers and sealed by thermoplastic hot-gas-welding techniques.

In piping applications using lined concrete as a means of secondary containment, such as in a double containment fitting manhole area, pipe penetrations may be readily sealed, particularly where a compatible thermoplastic piping material is used. Details of a typical pipe penetration are illustrated in Figure 16.41. If a dissimilar-thermoplastic or nonthermoplastic piping material is used, some type of mechanical sealing device (e.g. a flanged joint) must be used. An example of a pipe penetration in a secondary containment vault that is protected with this type of liner is illustrated in Figure 16.41b.

16.4.3 Unique Concrete Structures

Using protective means, such as structured thermoplastic concrete liners and other coatings, a number of concrete structures can be constructed as an excellent means of providing secondary containment for one or more primary pipes, or pipe components, in lieu of using a secondary containment piping encasement. The advantage of using protected concrete secondary containment structures is that it greatly increases the maintainability and access to a primary system. Ordinary buried secondary containment piping, on the other hand, is difficult to maintain, and has only limited access. These concrete structures can be constructed so that they are water tight, and some may even be constructed so that they are rated for pressure (> 15 psi).

Some of the various structures that can be constructed include: trenches and trenching systems, underground tunnels and walkways (for multiple pipe networks), manholes, fitting chambers, vaults, and many others, which are shown throughout this text.

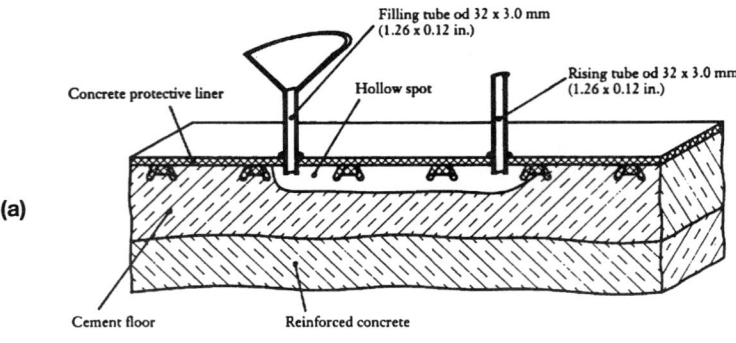

(a)

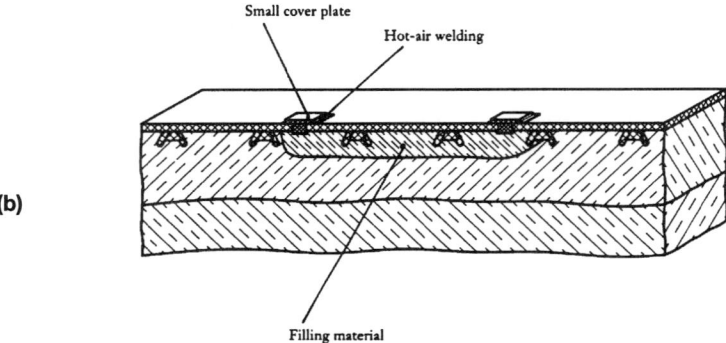

(b)

Figure 16.40. a: Illustration of the repair of a hollow spot in concrete in a structured studded liner application. b: Illustration of the appearance of the hollow spot shown in "a" after the concrete has been poured and is hardened. (Source: AGRU; A/A Manufacturing, Inc.)

16.5 Cathodic Protection for Coated Single-Walled Piping and Tanks

Cathodic protection, applied together with external coatings, is allowable in some jurisdictions as a means of providing adequate protection for coated, single-wall metallic systems that carry nonhazardous chemicals (e.g., certain petroleum byproducts). In these applications, metallic piping components, together with applied coating and cathodic protection, may be used in lieu of providing a double containment piping/tank system. This practice is considered acceptable according to RCRA regulations (Title 40, Code of Federal Regulations, Part 280) in the United States, if the fluid is one of certain petroleum compounds listed in the regulation. Hazardous chemicals listed in the RCRA (Title 40, Code of Federal Regulations, Part 281, Subpart D) do not allow this practice as an option for underground systems (see Title 40, Code of Federal Regulations, Part 280).

Metallic piping and metallic underground storage tanks must be provided with coating and cathodic protection, whenever an electrolyte that will allow corrosion to take place is present (e.g., wetted, active soil as an electrolyte that allows soil corrosion to occur). Tanks constructed of corrosion-resistant materials do not require additional coating and cathodic

Alternatives to Secondary Containment Piping

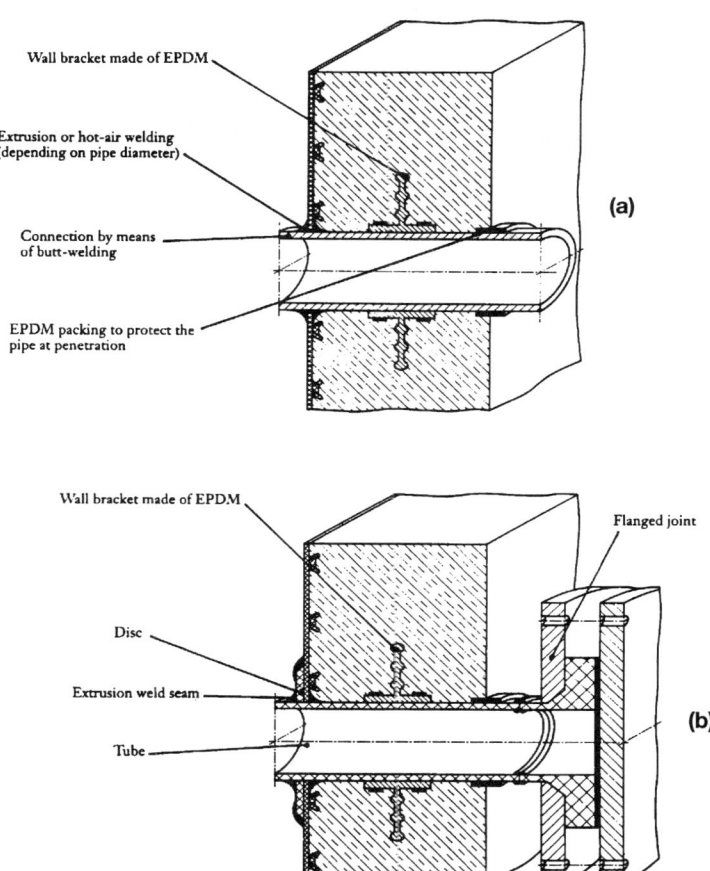

Figure 16.41. (a) Pipe penetration detail in a manhole or sump that is lined by means of structured studded thermoplastic liners, (b) Pipe penetration detail whereby the pipe shown in "a" is provided with a flanged connection outside the concrete structure. (Source: AGRU; A/A Manufacturing, Inc.)

protection where a relatively inactive soil condition is present. Where cathodic protection is to be added, the installation of a cathodic protection system must be carried out in accordance with the installation instructions of the manufacturer of the cathodic protection system, and in accordance with the rules of the authority having local jurisdiction.

For a double containment piping or tank system constructed of a metallic secondary containment material, coating and cathodic protection must also be applied in the majority of the applications where an active soil is present. The majority of metallic tanks to be protected with cathodic protection are supplied "pre-equipped" with a factory-installed cathodic protection system that is provided as part of the tank.

Field-installed cathodic protection systems are also common, but are usually limited to piping and associated nontank components, where the ability to preinstall a cathodic protection system is not possible or feasible. Piping and other components should be electrically isolated from cathodically protected metallic tanks by using insulated flanges and bushings. In some installations, a common cathodic protection system may be used to provide protection to both piping systems and their associated tanks.

Most manufacturers (suppliers) of cathodic protection equipment also supply plans and specifications as part of their equipment package. Where specific plans are not provided, Sections 16.5.2 through 16.5.12 serve as a set of general recommendations for the design and installation of cathodic protection systems.

16.5.1 Overview of Cathodic Protection

The process of cathodic protection involves replacing current that flows out of a metal as a result of the natural soil corrosion process, by the use of a sacrificial electrode or by supplying an impressed current. In other words, cathodic protection creates a reversal of electric current flow within the defined corrosion cell.[14] This can be physically accomplished by attaching a more galvanically active metal, termed a sacrificial anode to a metal requiring protection. The same effect can also be achieved by supplying a direct current (impressed current) to the metal, through a comparatively inert anode. Whether sacrificial anodes or impressed currents are employed, an electrolyte must be present to enable an electrical current to flow within the defined cell. In most applications, the use of cathodic protection is used in conjunction with a protective coating. When this is the case, the total amount of current supplied should not exceed the energy that will create disbonding of a dielectric coating from a metal.

The choice over the type of anode to be used depends on the substrate to be protected, the conditions of the electrolyte (soil condition or solution condition), and the cost of power that is available. The most commonly used material for galvanic anodes is magnesium, although zinc is quite prevalent as well.

When the resistivity of the soil condition is high (greater than 3000 ohms/cm), the use of impressed current represents a better choice. When impressed current is the form of protection selected, an inert material such as platinum, aluminum, scrap iron, high-silicon cast iron, or graphite is used for the anode. In areas where the cost of electricity is high, the choice often favors the selection of galvanic anodes as the method of cathodic protection.

Whenever cathodic protection systems are used, they must be periodically inspected to ensure a properly functioning system. In addition, considerations must be given to the design of the system to ensure proper functioning. This includes considerations to ensure electrical continuity necessary for functioning of the cathodic protection system, and whether the system should be designed with isolated sections to obtain better control over the whole system. In long pipelines, for instance, certain sections may be separated by the use of insulated couplings or mechanical joints. Also, it is common to employ protective coatings in addition to cathodic protection. Therefore, the choice of coating and its possible reaction to the cathodic protection method chosen should be thoroughly analyzed during the design stages. When proper material selection is coupled with adequate protective measures, prolonged life of metallic (or concrete) structures can be achieved.

16.5.2 General Comments

Sections 16.5.3 through 16.5.12 include general recommendations that apply to many typical installations of cathodic protection systems. They will not take the place of a specifically engineered plan for cathodic protection or a detailed corrosion survey. A competent cathodic engineer must always be consulted, regardless of whether the application is typical or unusual, and no matter where the components are to be installed.

Aside from protecting piping and tanks, systems that are to receive cathodic protection must also have protection provided for gauges, monitoring devices, and other systems that are subject to corrosion. If these systems are not protected, their failure could cause the release of a process fluid. The failure of such systems could also disrupt or prevent the operation of monitoring systems. In order to provide protection to these peripheral components, additional anodes and bonding will most likely be needed.

Also, gauging and monitoring devices that must be positioned in the interstice of a double-wall tank, or inside a single-wall tank, must be separated from the tank with the use of dielectric bushings to prevent electrical continuity.[15] Separate cathodic protection systems must be provided to a (double-wall) tank and its associated peripheral components, which are dielectrically separated. Any component that is not made of nonmetallic or other corrosion-resistant material should also be provided with cathodic protection.

Project drawings and plans, particularly "as-built" drawings, must have the location of cathodic protection system components clearly indicated. Included in drawing notations should be the method and location of cathodic protection connections, the location of test stations, and references to buildings and piping/tank system positions.[16] Specific details and requirements for implementing a cathodic protection system are provided in Sections 16.5.3 through 16.5.12.

16.5.3 Factory-Installed Cathodic Protection

All aspects of the cathodic protection system (anodes, dielectric bushings, and coatings) must be inspected for possible damage that may have been incurred during shipping and handling, immediately after the system has been delivered to a project location. Verification must be provided that electrical continuity has been maintained between the anodes and tanks (and/or piping, where cathodic protection components are preinstalled).[16] If damage to anode connections or dielectric coatings is detected, damaged or defective components must be repaired in strict accordance with the manufacturer's instructions. Once repair has been carried out, the components must be subjected to reinspection. During the installation and backfilling of a piping and tank system where preinstalled cathodic protection systems are provided, there is a high probability of damage if care is not taken. Extra careful procedures must be followed to ensure that system units are not damaged during the installation and backfilling process.

If tank parts have been supplied with plastic or metal thread protectors equipped with cast iron plugs, the cast iron plugs must be removed, redoped, and reinstalled prior to setting the tank.[17] Dielectric bushings must not be removed from unused openings, however.

16.5.4 Dielectric Coatings

The purpose of a dielectric coating is to ensure that external surfaces of metallic piping and associated metallic tanks may be functionally separated from the external environment. By isolating the piping and associated tanks from surrounding corrosive soil, the current demand on a cathodic protection system may be substantially reduced.[18,19] If a coating remains intact, any reduction in current demand that can be achieved can be maintained throughout the life of the piping/tank system. A coating may be selected based on several characteristics, including: (1) an ability to maintain high dielectric properties over its life; (2) to prevent or minimize moisture absorption and moisture transfer rates; (3) chemical resistance to a highly corrosive soil condition; and (4) chemical resistance to any contained liquids. The most commonly used dielectric coatings are epoxy, enamel, urethane, and mastics, applied in liquid form, as tape, or by extrusion.[20] Most metallic pipes that require coatings have such coatings applied in a controlled environment at a piping system manufacturer's (fabricator's) location. The preapplication of a coating tends to improve its quality control, thereby minimizing any defects. Field coatings should be limited to those areas where a preapplication is not readily possible. These include: (1) piping field weld areas; (2) exposed threads on tank or connections; (3) tank fittings; and (4) areas that become damaged from handling and fabrication procedures.

A defect in an applied dielectric coating results in severe and rapid galvanic corrosion at a defect site. Corrosion will occur rapidly where a defect occurs, due to a high concentration effect generated by the electrochemical process. This type of corrosion occurs by means of concentration-cell-type corrosion or pit-type corrosion, discussed further in Sections 2.1.5 and 2.1.4, respectively. Cathodic protection, when properly applied in conjunction with dielectric coatings, will minimize or eliminate concentration-cell corrosion and pitting corrosion due to holidays (coating pinhole-sized defects).

There are many important steps in the successful implementation of a protective dielectric coating. The first step is always the selection of a proper coating, based on metallic materials to be protected and the corrosivity of the electrolyte and/or process fluid to which it will be exposed. The preparation of the surface of a metal to be coated is also highly important and should be done under controlled conditions. Once a metal surface has been properly prepared, it is essential that its coating be thoroughly and evenly applied.

16.5.5 Field-Applied Dielectric Coatings

For coatings, or portions of coatings, that must be applied outdoors, their application must be limited to personnel who have been provided with special training in the use of the materials and installation techniques. Any coating to be field applied must be capable of outdoor application.

After a piping and associated tank and leak-detection system has been fabricated and tested in accordance with Chapters 12 through 15, all of its exposed metal surfaces must be cleaned and coated. Surfaces of the metallic substrate must be carefully prepared to ensure adequate cohesion and to prevent holidays (pinhole-sized flaws). Where bonding and anode wire connections exist, they must also be coated with the same dielectric coating material. Any damage to a system's dielectric coating that occurs during its application must be repaired thoroughly by wrapping or coating with a material designed for this purpose.

16.5.6 Electrical Isolation

In order for a cathodic protection system to function properly, there must be electrical isolation between piping and tanks and other metallic structures. A minimum physical separation of at least 12 in. (305 mm) between structures is desirable.[21] However, it is not always feasible to provide this type of physical separation. When it is not possible, isolation may be achieved by providing an insulating material between points of contact. Insulating material that is used in this fashion is referred to as a dielectric bushing. In some cases, an entire fitting is constructed of a dielectric material to achieve isolation between a metallic piping system and an underground storage tank. Any dielectric bushing or fitting must be constructed of a material that is chemically resistant to the contained fluid, rated to the design pressure of the system, and in accordance with applicable design codes (i.e., ANSI/ASME B31.3 Code for Chemical Plant and Petroleum Refinery Piping).

If electrical isolation is not provided, there is a distinct possibility that a given cathodic protection system will not function properly. An example would be where a chemical waste pipe system vertical riser penetrates a metal building, providing contact with other structures. In such an arrangement, part of the current supplied by its cathodic protection system would flow to the building foundation, and to other connected underground structural or mechanical systems. This would most likely result in an insufficient level of cathodic protection for the underground chemical waste piping system described in the example.

In a piping and tank system that has a combined cathodic protection system by design, there is no need to isolate piping and tank parts electrically. As described earlier, it is not common to have a combined piping and tank cathodic protection system.

16.5.7 Impressed-Current Systems

In an impressed-current system, an inert anode is placed in the area of the substrate to be protected. A current is then supplied to the anode, which then forms a blanket of protection around the metal substance to be protected. The amount of current to be protected depends on the expected loss of current from the metallic substrate in the corrosion process. A schematic of the process is shown in Figure 16.42.

The most common materials from which anodes used for impressed current systems are constructed include: (1) high-silicon-alloy iron; (2) graphite, (3) platinum, (4) scrap iron, (5) aluminum, and (6) titanium. Materials such as these are subject to damage during shipping and handling. Thus, they must undergo a thorough visual inspection and a rigorous electrical continuity testing program prior to their installation. Visual inspection of impressed-current anodes must include observation for damaged or broken lead wires and inspection for any damage to insulation. Anodes must be installed as indicated in the engineering design and according to project plans and specifications. Visual inspection of the system must confirm that this has been done. Backfilling must proceed carefully to ensure that there are no voids present in backfill, which could otherwise prevent current from being properly distributed over the surface area of piping (or tanks) to be protected. The rectifier negative terminal must be connected to the structure to be protected. The positive terminal is to be connected to the anodes. A rectifier system may be used to protect a combined system, including a piping system and any accompanying tanks, or other associated equipment. Electrical continuity must be provided to a rectifier system that is designed to protect a combined system. Electrical power must be maintained at all times to an impressed current system. Special means must be taken to ensure that power is not allowed to be turned off by switches that might be mistakenly turned off by plant operators.[22]

16.5.8 Sacrificial (Galvanic) Anodes

In a sacrificial anode cathodic protection system, an anode becomes "sacrificed" by corroding, thus replacing electrons that are otherwise lost in the natural corrosion process of the structure being protected. In order for this to occur, an anode with a greater negative

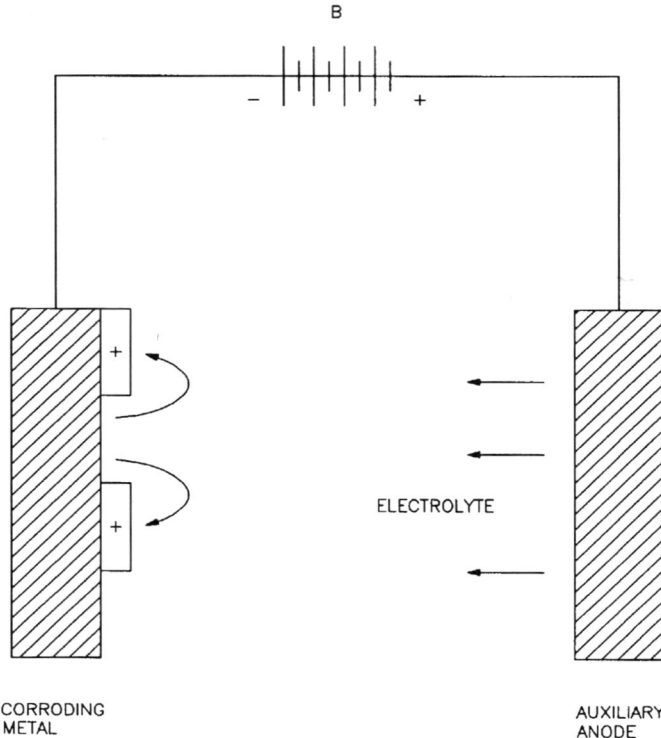

Figure 16.42. Overview of cathodic protection using impressed current on a local action cell (Source: Corrosion and Corrosion Protection Handbook, Schweitzer, Philip A., Marcel Dekker, New York, 1983.)

potential must be selected for use. Table 2.1 in Section 2.1.3 presents a partial list of the galvanic series of metals, which shows metals ranked in terms of their electrical potential. In Table 2.1, materials are listed in order, from lowest potential to highest potential (from top to bottom in the table).

The most commonly used types of sacrificial anodes are constructed of magnesium or zinc. These anodes function by galvanic action to supply a flow of current that replaces the flow of current that otherwise flows from the metallic structure to the surrounding electrolyte. Moist soil or ground water serves as the electrolyte in any underground application. The typical form of anode that is commercially available is packaged in low-resistivity backfill, with a copper lead wire attached for connection to the part to be protected.

Sacrificial galvanic anodes are subject to damage during shipping and handling, and must be subjected to thorough visual inspection, and electrical continuity testing, prior to their installation. Most anodes arrive in a waterproof packaging, with the anode centered in prepackaged backfill. Anodes and their backfill must be installed with compacted native soil that is thoroughly soaked with water (3-5 gal is the normal recommendation). When installing anodes, lead wires must not be used as a means to hold the anode, due to the fragile nature of the connection, and the importance of maintaining electrical continuity.

Table 16.3 Maximum Lengths of Protection for Well-Coated Carbon Steel Piping*

Nominal pipe diameter	Two 17 lb. magnesium anodes	Two 32 lb. magnesium anodes
2 in. steel pipe	1000 ft	1200 ft
3 in. steel pipe	680 ft	800 ft
4 in. steel pipe	530 ft	620 ft
Theoretical anode life	23 yrs	37 yrs

* Mildly corrosive soil; period of protection = 20-30 yrs.

16.5.9 Design of Sacrificial Anode Systems

Table 16.3 provides general recommendations for coated carbon steel piping systems that are to be cathodically protected by the use of sacrificial anodes.[23] Actual conditions always vary from site to site. Therefore, it is suggested that a corrosion expert, knowledgeable in the science of sacrificial-anode cathodic protection, must be consulted for specific recommendations.

As a general rule, protection may be provided for most well-coated metallic piping systems by installing at least two magnesium anodes, vertically positioned, 5 ft from the center of a pipe trench and below the level of the piping to be protected. Protected piping must be electrically continuous and well coated. Piping that is uncoated may be bonded with a #12 TW or THHN stranded or solid wire. Piping must not be ground to other piping systems, tanks, conduit, or other building structures.

These figures are based upon a soil resistivity of 50,000 ohms/cm, a current output of the anode of 1.5 mA/ft^2 of exposed steel, and a well-coated carbon steel piping (less than 5% exposed surface).

In many applications, a single anode will provide a sufficient amount of protection. However, it is best to design a degree of redundancy in the system, in order to reduce the possibilities of loss should the lead wire become damaged or a single anode malfunctions. If two or more anodes are used, they must be equally spaced in order to distribute the current equally. If piping is longer than 1000 ft (305 m), adequate protection may be achieved by the addition of multiple anodes. The use of high-potential magnesium anodes also may be used to provide adequate protection to a single, long-run pipe.

16.5.10 Wiring and Electrical Connections

Electrical wiring for cathodic protection systems is like any other electrical wiring used on a project. Therefore, it should be designed and installed in conformance with applicable national and local electrical codes. The authority having local jurisdiction should be consulted as to local rules and building regulations regarding the installation of electrical components. Wiring that is to be directly buried must be buried at least 20 in. deep, to prevent damage that might occur during construction or during normal operation. Wiring must be provided with sufficient slack to allow it to have its required flexibility. All wires must be clean, dry, and free of foreign matter and must be properly prepared before making connections.

Recommended practices for connecting electrical wiring to anodes and surfaces to be protected typically include the use of thermite welding, pressure-type grounding clamps, or other devices. Bare wires and connections must be coated with an insulating material that is compatible with the wire itself, insulation, and other structural coatings. If piping or tanks previously contained flammable or combustible liquids, the piping/tank system and its surrounding area must be made safe before connecting any anodes by thermite welding.

Wiring components must be listed by a recognized testing laboratory. They must always be installed according to manufacturer's installation instructions and the rules of the authority having local jurisdiction. The practice of directly burying splices must be disallowed under all circumstances. All wiring must be thoroughly tested for electrical continuity before backfilling.

16.5.11 Test Stations

It is wise to include test stations as part of the overall system design, in order to monitor the functioning of the cathodic protection system. Various measurements that may be taken on a continuous or manual basis to test a system include: (1) piping-to-soil potential; (2) tank-to-soil potential; and (3) piping-to-tank (or other structure) isolation.[24] It is only important to measure item #3 prior to system startup. Anode current output also provides a measurement of the protection being provided. Therefore, the current output of an anode may be used to measure its projected life. It is good practice to have all lead wires and terminals color coded and clearly marked. If color coding is used, that selected must be specified in the engineering specifications and clearly marked on the "as-built" drawings, for easy future reference.

16.5.12 Inspection and Testing of Cathodic Protection Systems

Components and connections must always be carefully inspected before backfilling. The effectiveness of cathodic protection must be tested by a competent technician. Electrical continuity between a piping system and any associated tanks must also be tested. If dielectric bushings, flanges, or unions are installed, there should not be any continuity between the tank and piping, except in the rare circumstance that a combined cathodic protection is used. Other inspection and testing practices are stated elsewhere in this section. The rules of the authority having local jurisdiction must always be followed.

When testing galvanic anode or impressed current cathodic protection systems, a negative voltage of at least -0.85 V is usually sufficient to provide corrosion protection.[25] Such a measurement is usually provided by a copper-copper sulfate reference electrode. Other means are also available.

References

1. S. Kramer, W. McDonald and J. Thomson, "An Introduction to Trenchless Technology," Van Nostrand Reinhold, New York, NY, 1992, p. 3.

2. H. Svetlik, "Reduction, Reversion, Renovation: R^3," Buried Plastic Pipe Technology, ASTM STP 1093, G. Buczala and M Cassady, Eds., American Society for Testing and Materials, Philadelphia, PA 1990, p. 313.

3. S. Kramer, et. al., op. cit., p. 133.

4. ASTM Standard F 585, "Standard Practice for Insertion of Flexible Polyolefin Pipe into Existing Sewers," ASTM Annual Book of Standards, Vol. 8.04, Philadelphia, PA, 1986.

5. S. Kramer, et. al., op. cit., p. 124.

6. S. Kramer, et. al., op. cit., p. 124.

7. H. Svetlick, op. cit., p. 316.

8. H. Svetlick, op. cit., p. 319.

9. H. Svetlick, op. cit., p. 320.

10. C. Faller, "In-Place Rehabilitation of Existing Process Sewers," Proceedings of the American Power Conference, Illinois Institute of Technology, Chicago, IL, 1994, pp. 184-87.

11. S. Kramer, et. al., op. cit., p. 138.

12. C. Faller, op. cit., p. 185.

13. C. Faller, op. cit., pp. 185-86.

14. P. Schweitzer, "Corrosion and Corrosion Protection Handbook," Marcel Dekker, New York, NY, 1983, p. 14.

15. PEI/RP 100-87, "Recommended Practices for the Installation of Underground Liquid Systems," Petroleum Equipment Institute, Tulsa, OK, 1987, p. 19.

16. PEI/RP 100-87, op. cit., p. 17.

17. PEI/RP 100-87, op. cit., p. 17.

18. P. Schweitzer, op. cit., p. 15.

19. PEI/RP 100-87, op. cit., p. 17.

20. PEI/RP 100-87, op. cit., p. 18.

21. PEI/RP 100-87, op. cit., p. 18.

22. PEI/RP 100-87, op. cit., p. 18.

23. PEI/RP 100-87, op. cit., p. 19.

24. PEI/RP 100-87, op. cit., p. 18.

25. PEI/RP 100-87, op. cit., p. 20.

Notes

Appendix A

Table A-1 Annulus and Annular Area Dimensional Data for IPS Primary Pipe Inside of ASTM F-17 IPS-Size SDR-Rated Secondary Containment Pipe

Nominal primary size (in.)	Nominal sec. cont. size (in.)	IPS size, schedule no., and/or SS schedule no.	Nominal annulus	Nominal annular area (in.^{2})	Nominal annular area (ft^2)	Nominal primary size (in.)	Nominal sec. cont. size (in.)	IPS size, schedule no., and/or SS schedule no.	Nominal annulus	Nominal annular area (in.^{2})	Nominal annular area (ft^2)
						(continued)					
1/2	1-1/4	5S	0.35	1.28	0.009	3/4	1-1/4	10S	0.20	0.77	0.005
1/2	1-1/4	10S	.030	1.08	0.007	3/4	1-1/4	STD, 40, 40S	0.17	0.63	0.004
1/2	1-1/4	STD, 40, 40S	0.27	0.94	0.007	3/4	1-1/2	5S	0.36	1.10	0.008
1/2	1-1/4	XS, 80, 80S	0.22	0.73	0.005	3/4	1-1/2	10S	0.32	0.86	0.006
1/2	1-1/4	160	0.16	0.50	0.003	3/4	1-1/2	STD, 40, 40S	0.28	0.68	0.005
1/2	1-1/4	XXS	0.03	0.08	0.001	3/4	1-1/2	XS, 80, 80S	0.23	0.41	0.003
1/2	1-1/2	5S	0.47	1.60	0.011	3/4	2	5S	0.60	1.79	0.012
1/2	1-1/2	10S	0.42	1.36	0.009	3/4	2	10S	0.55	1.49	0.010
1/2	1-1/2	STD, 40, 40S	0.39	1.17	0.008	3/4	2	STD, 40, 40S	0.51	1.19	0.008
1/2	1-1/2	XS, 80, 80S	0.33	0.90	0.006	3/4	2	XS, 80, 80S	0.44	0.79	0.005
1/2	1-1/2	160	0.25	0.54	0.004	3/4	2	160	0.32	0.07	0.000
1/2	1-1/2	XXS	0.13	0.08	0.001	3/4	2	XXS	0.23	-0.39	-0.003
1/2	2	5S	0.70	2.60	0.018	3/4	2-1/2	5S	0.83	2.93	0.020
1/2	2	10S	0.66	2.30	0.016	3/4	2-1/2	10S	0.79	2.62	0.018
1/2	2	STD, 40, 40S	0.61	2.00	0.014	3/4	2-1/2	STD, 40, 40S	0.71	1.95	0.014
1/2	2	XS, 80, 80S	0.55	1.60	0.011	3/4	2-1/2	XS, 80, 80S	0.64	1.40	0.010
1/2	2	160	0.42	0.88	0.006	3/4	2-1/2	160	0.54	0.71	0.005
1/2	2	XXS	0.33	0.42	0.003	3/4	2-1/2	XXS	0.36	-0.37	0.003
1/2	2-1/2	5S	0.93	3.60	0.025	3/4	3	5S	1.14	4.30	0.030
1/2	2-1/2	10S	0.90	3.29	0.023	3/4	3	10S	1.11	3.92	0.027
1/2	2-1/2	STD, 40, 40S	0.81	2.62	0.018	3/4	3	STD, 40, 40S	1.01	2.96	0.021
1/2	2-1/2	XS, 80, 80S	0.74	2.07	0.014	3/4	3	XS, 80, 80S	0.93	2.18	0.015
1/2	2-1/2	160	0.64	1.38	0.010	3/4	3	160	0.79	0.98	0.007
1/2	2-1/2	XXS	0.47	0.03	0.002	3/4	3	XXS	0.63	-0.28	-0.002
1/2	3	5S	1.25	5.90	0.041	1	1-1/2	5S	0.23	1.10	0.008
1/2	3	10S	1.21	5.51	0.038	1	1-1/2	10S	0.18	0.86	0.006
1/2	3	STD, 40, 40S	1.11	4.56	0.032	1	1-1/2	STD, 40, 40S	0.15	0.68	0.005
1/2	3	XS, 80, 80S	1.03	3.77	0.026	1	2	5S	0.46	2.60	0.018
1/2	3	160	0.89	2.57	0.018	1	2	10S	0.42	2.30	0.016
1/2	3	XXS	0.73	1.32	0.009	1	2	STD, 40, 40S	0.38	2.00	0.014
3/4	1-1/4	5S	0.24	0.97	0.007	1	2	XS, 80, 80S	0.31	1.60	0.011

(continued)

Table A-1 (cont.) Annulus and Annular Area Dimensional Data for IPS Primary Pipe Inside of ASTM F-17 IPS-Size SDR-Rated Secondary Containment Pipe

Nominal primary size (in.)	Nominal sec. cont. size (in.)	IPS size, schedule no., and/or SS schedule no.	Nominal annulus	Nominal annular area (in.²)	Nominal annular area (ft²)	Nominal primary size (in.)	Nominal sec. cont. size (in.)	IPS size, schedule no., and/or SS schedule no.	Nominal annulus	Nominal annular area (in.²)	Nominal annular area (ft²)
						(continued)					
1	2	160	0.19	0.88	0.006	1-1/2	3	STD, 40, 40S	0.58	4.56	0.032
1	2	XXS	0.09	0.42	0.003	1-1/2	3	XS, 80, 80S	0.50	3.77	0.026
1	2-1/2	5S	0.70	4.41	0.031	1-1/2	3	160	0.36	2.57	0.018
1	2-1/2	10S	0.66	4.10	0.028	1-1/2	3	XXS	0.20	1.32	0.009
1	2-1/2	STD, 40, 40S	0.58	3.43	0.024	1-1/2	4	5S	1.22	11.92	0.083
1	2-1/2	XS, 80, 80S	0.50	2.88	0.020	1-1/2	4	10S	1.18	11.42	0.079
1	2-1/2	160	0.41	2.19	0.015	1-1/2	4	STD, 40, 40S	1.06	9.90	0.069
1	2-1/2	XXS	0.23	1.11	0.008	1-1/2	4	XS, 80, 80S	0.96	8.67	0.060
1	3	5S	1.01	7.37	0.051	1-1/2	4	120	0.86	7.48	0.052
1	3	10S	0.97	6.99	0.049	1-1/2	4	160	0.77	6.45	0.045
1	3	STD, 40, 40S	0.88	6.04	0.042	1-1/2	4	XXS	0.63	4.97	0.035
1	3	XS, 80, 80S	0.79	5.25	0.036	2	3	5S	0.48	4.30	0.030
1	3	160	0.65	4.05	0.028	2	3	10S	0.44	3.92	0.027
1	3	XXS	0.49	2.80	0.019	2	3	STD, 40, 40S	0.35	2.96	0.021
1	4	5S	1.51	13.40	0.093	2	3	XS, 80, 80S	0.26	2.18	0.015
1	4	10S	1.47	12.90	0.090	2	4	5S	0.98	10.33	0.072
1	4	STD, 40, 40S	1.36	11.38	0.079	2	4	10S	0.94	9.83	0.068
1	4	XS, 80, 80S	1.26	10.14	0.070	2	4	STD, 40, 40S	0.83	8.30	0.058
1	4	120	1.15	8.96	0.062	2	4	XS, 80, 80S	0.73	7.07	0.049
1	4	160	1.06	7.93	0.055	2	4	120	0.62	5.89	0.041
1	4	XXS	0.92	6.45	0.045	2	4	160	0.53	4.86	0.034
1-1/4	2	5S	0.29	1.79	0.012	2	4	XXS	0.39	3.37	0.023
1-1/4	2	10S	0.25	1.49	0.010	2	5	5S	1.49	18.02	0.125
1-1/4	2	STD, 40, 40S	0.20	1.19	0.008	2	5	10S	1.46	17.60	0.122
1-1/4	2-1/2	5S	0.52	3.60	0.025	2	5	STD, 40, 40S	1.34	15.58	0.108
1-1/4	2-1/2	10S	0.49	3.29	0.023	2	5	XS, 80, 80S	1.22	13.77	0.096
1-1/4	2-1/2	STD, 40, 40S	0.40	2.62	0.018	2	5	120	1.09	11.93	0.083
1-1/4	2-1/2	XS, 80, 80S	0.33	2.07	0.014	2	5	160	0.97	10.18	0.071
1-1/4	2-1/2	160	0.23	1.38	0.010	2	5	XXS	0.84	8.54	0.059
1-1/4	3	5S	0.84	6.57	0.046	2	6	5S	2.02	27.82	0.193
1-1/4	3	10S	0.80	6.19	0.043	2	6	10S	1.99	27.32	0.190
1-1/4	3	STD, 40, 40S	0.70	5.23	0.036	2	6	STD, 40, 40S	1.85	24.47	0.170
1-1/4	3	XS, 80, 80S	0.62	4.44	0.031	2	6	XS, 80, 80S	1.69	21.65	0.150
1-1/4	3	160	0.48	3.24	0.023	2	6	120	1.56	19.34	0.134
1-1/4	3	XXS	0.32	1.99	0.014	2	6	160	1.41	16.71	0.116
1-1/4	4	5S	1.34	12.59	0.087	2	6	XXS	1.26	14.41	0.100
1-1/4	4	10S	1.30	12.09	0.084	2-1/2	3	5S	0.23	2.24	0.016
1-1/4	4	STD, 40, 40S	1.18	10.57	0.073	2-1/2	3	10S	0.19	1.86	0.013
1-1/4	4	XS, 80, 80S	1.08	9.34	0.065	2-1/2	4	5S	0.73	8.26	0.057
1-1/4	4	120	.98	8.15	0.057	2-1/2	4	10S	0.69	7.76	0.054
1-1/4	4	160	0.89	7.12	0.049	2-1/2	4	STD, 40, 40S	0.58	6.24	0.043
1-1/4	4	XXS	0.75	5.64	0.039	2-1/2	4	XS, 80, 80S	0.48	5.01	0.035
1-1/2	2	5S	0.17	1.12	0.008	2-1/2	4	120	0.37	3.82	0.027
1-1/2	2-1/2	5S	0.40	2.93	0.020	2-1/2	4	160	0.28	2.79	0.019
1-1/2	2-1/2	10S	0.37	2.62	0.018	2-1/2	4	XXS	0.14	1.31	0.009
1-1/2	2-1/2	STD, 40, 40S	0.28	1.95	0.014	2-1/2	5	5S	1.24	15.95	0.111
1-1/2	2-1/2	XS, 80, 80S	0.21	1.40	0.010	2-1/2	5	10S	1.21	15.53	0.108
1-1/2	3	5S	0.72	5.90	0.041	2-1/2	5	STD, 40, 40S	1.09	13.52	0.094
1-1/2	3	10S	0.68	5.51	0.038	2-1/2	5	XS, 80, 80S	0.97	11.71	0.081

(continued)

Appendix A

Table A-1 (cont.) Annulus and Annular Area Dimensional Data for IPS Primary Pipe Inside of ASTM F-17 IPS-Size SDR-Rated Secondary Containment Pipe

Nominal primary size (in.)	Nominal sec. cont. size (in.)	IPS size, schedule no., and/or SS schedule no.	Nominal annulus	Nominal annular area (in.²)	Nominal annular area (ft²)	Nominal primary size (in.)	Nominal sec. cont. size (in.)	IPS size, schedule no., and/or SS schedule no.	Nominal annulus	Nominal annular area (in.²)	Nominal annular area (ft²)
						(continued)					
2-1/2	5	120	0.84	9.86	0.069	4	6	160	0.34	5.23	0.036
2-1/2	5	160	0.72	8.12	0.056	4	6	XXS	0.20	2.93	0.020
2-1/2	5	XXS	0.59	6.48	0.045	4	8	5S	1.95	39.62	0.275
2-1/2	6	5S	1.77	25.76	0.179	4	8	10S	1.91	38.60	0.268
2-1/2	6	10S	1.74	25.26	0.175	4	8	20	1.81	35.96	0.250
2-1/2	6	STD, 40, 40S	1.60	22.41	0.156	4	8	30	1.79	35.27	0.245
2-1/2	6	XS, 80, 80S	1.44	19.58	0.136	4	8	STD, 40, 40S	1.74	34.14	0.237
2-1/2	6	120	1.31	17.28	0.120	4	8	60	1.66	32.05	0.223
2-1/2	6	160	1.16	14.65	0.102	4	8	XS, 80, 80S	1.56	29.77	0.207
2-1/2	6	XXS	1.01	12.35	0.086	4	8	100	1.47	27.55	0.191
3	4	5S	0.42	5.13	0.036	4	8	120	1.34	24.67	0.171
3	4	10S	0.38	4.63	0.032	4	8	140	1.25	22.60	0.157
3	4	STD, 40, 40S	0.26	3.11	0.022	4	8	XXS	1.19	21.23	0.147
3	4	XS, 80, 80S	0.16	1.88	0.013	4	8	160	1.16	20.56	0.143
3	5	5S	0.92	12.82	0.089	4	10	5S	2.99	70.42	0.489
3	5	10S	0.90	12.40	0.086	4	10	10S	2.96	69.40	0.482
3	5	STD, 40, 40S	0.77	10.39	0.072	4	10	20	2.88	66.64	0.463
3	5	XS, 80, 80S	0.66	8.58	0.060	4	10	30	2.82	64.81	0.450
3	5	120	0.53	6.73	0.047	4	10	STD, 40, 40S	2.76	62.98	0.437
3	5	160	0.41	4.99	0.035	4	10	XS, 80, 80S	2.63	58.78	0.408
3	5	XXS	0.28	3.35	0.023	4	10	80	2.53	55.93	0.388
3	6	5S	1.45	22.63	0.157	4	10	100	2.41	52.22	0.363
3	6	10S	1.43	22.13	0.154	4	10	120	2.28	48.61	0.338
3	6	STD, 40, 40S	1.28	19.28	0.134	4	10	XXS, 140	2.13	44.25	0.307
3	6	XS, 80, 80S	1.13	16.45	0.114	4	10	160	2.00	40.86	0.284
3	6	120	1.00	14.15	0.098	5	6	5S	0.42	7.94	0.055
3	6	160	0.84	11.51	0.080	5	6	10S	0.40	7.44	0.052
3	6	XXS	0.70	9.22	0.064	5	6	STD, 40, 40S	0.25	4.59	0.032
3	8	5S	2.45	45.91	0.319	5	8	5S	1.42	31.22	0.217
3	8	10S	2.41	44.88	0.312	5	8	10S	1.38	30.19	0.210
3	8	20	2.31	42.24	0.293	5	8	20	1.28	27.55	0.191
3	8	30	2.29	41.56	0.289	5	8	30	1.25	26.87	0.187
3	8	STD, 40, 40S	2.24	40.42	0.281	5	8	STD, 40, 40S	1.21	25.73	0.179
3	8	60	2.16	38.34	0.266	5	8	60	1.13	23.65	0.164
3	8	XS, 80, 80S	2.06	36.06	0.250	5	8	XS, 80, 80S	1.03	21.37	0.148
3	8	100	1.97	33.83	0.235	5	8	100	0.94	19.14	0.133
3	8	120	1.84	30.96	0.215	5	8	120	0.81	16.27	0.113
3	8	140	1.75	28.89	0.201	5	8	140	0.72	14.20	0.099
3	8	XXS	1.69	27.51	0.191	5	8	XXS	0.66	12.82	0.089
3	8	160	1.66	26.85	0.186	5	8	160	0.63	12.16	0.084
4	5	5S	0.42	6.54	0.045	5	10	5S	2.46	62.01	0.431
4	5	10S	0.40	6.12	0.042	5	10	10S	2.43	60.99	0.424
4	5	STD, 40, 40S	0.27	4.10	0.028	5	10	20	2.34	58.23	0.404
4	5	XS, 80, 80S	0.16	2.29	0.016	5	10	30	2.29	56.41	0.392
4	6	5S	0.95	16.34	0.113	5	10	STD, 40, 40S	2.23	54.57	0.379
4	6	10S	0.93	15.84	0.110	5	10	XS, 80, 80S	2.09	50.38	0.350
4	6	STD, 40, 40S	0.78	12.99	0.090	5	10	80	2.00	47.52	0.330
4	6	XS, 80, 80S	0.63	10.17	0.071	5	10	100	1.87	43.82	0.304
4	6	120	0.50	7.87	0.055	5	10	120	1.75	40.21	0.279

(continued)

Table A-1 (cont.) Annulus and Annular Area Dimensional Data for IPS Primary Pipe Inside of ASTM F-17 IPS-Size SDR-Rated Secondary Containment Pipe

Nominal primary size (in.)	Nominal sec. cont. size (in.)	IPS size, schedule no., and/or SS schedule no.	Nominal annulus	Nominal annular area (in.²)	Nominal annular area (ft²)	Nominal primary size (in.)	Nominal sec. cont. size (in.)	IPS size, schedule no., and/or SS schedule no.	Nominal annulus	Nominal annular area (in.²)	Nominal annular area (ft²)
						(continued)					
5	10	XXS, 140	1.59	35.84	0.249	10	12	STD, 40S	0.63	22.34	0.155
5	10	160	1.47	32.45	0.225	10	12	40	0.59	21.18	0.147
6	8	5S	0.89	21.05	0.146	10	12	XS, 80S	0.50	17.68	0.123
6	8	10S	0.85	20.02	0.139	10	12	60	0.44	15.40	0.107
6	8	20	0.75	17.38	0.121	10	12	80	0.31	10.85	0.075
6	8	30	0.72	16.70	0.116	10	12	100	0.16	5.35	0.037
6	8	STD, 40, 40S	0.68	15.56	0.108	10	14	5S	1.47	56.41	0.392
6	8	60	0.59	13.48	0.094	10	14	10S	1.44	55.04	0.382
6	8	XS, 80, 80S	0.50	11.20	0.078	10	14	10	1.38	52.40	0.364
6	8	100	0.41	8.97	0.062	10	14	20	1.31	49.78	0.346
6	8	120	0.28	6.10	0.042	10	14	STD, 30	1.25	47.14	0.327
6	8	140	0.19	4.03	0.028	10	14	40	1.19	44.53	0.309
6	10	5S	1.93	51.84	0.360	10	14	XS	1.13	41.99	0.292
6	10	10S	1.90	50.82	0.353	10	14	60	1.03	38.17	0.265
6	10	20	1.81	48.06	0.334	10	14	80	0.88	31.97	0.222
6	10	30	1.76	46.24	0.321	10	14	100	0.69	24.69	0.171
6	10	STD, 40, 40S	1.70	44.40	0.308	10	14	120	0.53	18.83	0.131
6	10	XS, 60, 80S	1.56	40.21	0.279	10	14	140	0.38	13.11	0.091
6	10	80	1.47	37.35	0.259	10	14	160	0.22	7.55	0.052
6	10	100	1.34	33.65	0.234	10	16	5S	2.46	102.13	0.709
6	10	120	1.22	30.04	0.209	10	16	10S	2.44	101.00	0.701
6	10	XXS, 140	1.06	25.67	0.178	10	16	10	2.38	97.97	0.680
6	10	160	0.94	22.28	0.155	10	16	20	2.31	94.96	0.659
8	10	5S	0.93	27.88	0.194	10	16	STD, 30	2.25	91.93	0.638
8	10	10S	0.90	26.86	0.187	10	16	XS, 40	2.13	85.99	0.597
8	10	20	0.81	24.10	0.167	10	16	60	1.97	78.71	0.547
8	10	30	0.76	22.27	0.155	10	16	80	1.78	70.14	0.487
8	10	STD, 40, 40S	0.70	20.44	0.142	10	16	100	1.59	61.84	0.429
8	10	XS, 60, 80S	0.56	16.24	0.113	10	16	120	1.14	53.72	0.373
8	10	80	0.47	13.39	0.093	10	16	140	1.19	44.53	0.309
8	10	100	0.34	9.68	0.067	10	16	160	1.03	38.17	0.265
8	10	120	0.22	6.07	0.042	12	14	5S	0.47	19.48	0.135
8	12	5S	1.91	63.10	0.438	12	14	10S	0.44	18.11	0.126
8	12	10S	1.88	62.17	0.432	12	14	10	0.38	15.47	0.107
8	12	20	1.81	59.46	0.413	12	14	20	0.31	12.85	0.089
8	12	30	1.73	56.40	0.392	12	14	STD, 30	0.25	10.21	0.071
8	12	STD, 40S	1.69	54.69	0.380	12	16	5S	1.46	65.20	0.453
8	12	40	1.66	53.53	0.372	12	16	10S	1.44	64.07	0.445
8	12	XS, 80S	1.56	50.03	0.347	12	16	10	1.38	61.04	0.424
8	12	60	1.50	47.75	0.332	12	16	20	1.31	58.03	0.403
8	12	80	1.37	43.20	0.300	12	16	STD, 30	1.25	55.00	0.382
8	12	100	1.22	37.70	0.262	12	16	XS, 40	1.13	49.06	0.341
8	12	XXS, 120	1.06	32.35	0.225	12	16	60	0.97	41.78	0.290
8	12	140	0.94	28.18	0.196	12	16	80	0.78	33.21	0.231
8	12	160	0.75	22.11	0.154	12	16	100	0.59	24.91	0.173
10	12	5S	0.84	30.75	0.214	12	16	120	0.41	16.79	0.117
10	12	10S	0.82	29.82	0.207	12	18	5S	2.46	117.60	0.817
10	12	20	0.75	27.11	0.188	12	18	10S	2.44	116.32	0.808
10	12	30	0.67	24.05	0.167	12	18	10	2.38	112.90	0.784

(continued)

Appendix A

Table A-1 (cont.) Annulus and Annular Area Dimensional Data for IPS Primary Pipe Inside of ASTM F-17 IPS-Size SDR-Rated Secondary Containment Pipe

Nominal primary size (in.)	Nominal sec. cont. size (in.)	IPS size, schedule no., and/or SS schedule no.	Nominal annulus	Nominal annular area (in.²)	Nominal annular area (ft²)	Nominal primary size (in.)	Nominal sec. cont. size (in.)	IPS size, schedule no., and/or SS schedule no.	Nominal annulus	Nominal annular area (in.²)	Nominal annular area (ft²)
						(continued)					
12	18	20	2.31	109.50	0.760	16	18	XS	0.50	25.93	0.180
12	18	STD	2.25	106.07	0.737	16	18	40	0.44	22.63	0.157
12	18	30	2.19	102.67	0.713	16	18	60	0.25	12.77	0.089
12	18	XS	2.13	99.34	0.690	16	20	5S	1.81	101.44	0.074
12	18	40	2.06	96.04	0.667	16	20	10S	1.78	99.59	0.692
12	18	60	1.88	86.18	0.598	16	20	10	1.75	97.62	0.678
12	18	80	1.69	76.54	0.532	16	20	STD, 20	1.63	90.01	0.625
12	18	100	1.47	65.65	0.456	16	20	XS, 30	1.50	82.50	0.573
12	18	120	1.25	55.00	0.382	16	20	40	1.41	76.91	0.534
12	18	140	1.06	46.15	0.320	16	20	60	1.19	64.18	0.446
12	18	160	0.84	36.06	0.250	16	20	80	0.97	51.68	0.359
14	16	5S	0.84	38.93	0.270	16	20	100	0.72	37.78	0.262
14	16	10S	0.81	37.80	0.263	16	20	120	0.50	25.93	0.180
14	16	10	0.75	34.77	0.241	16	20	140	0.25	12.77	0.089
14	16	20	0.69	31.76	0.221	16	20	160	0.03	1.56	0.011
14	16	STD, 30	0.63	28.73	0.199	16	22	5S	2.81	166.26	1.155
14	16	XS, 40	0.50	22.79	0.158	16	22	10S	2.78	164.22	1.140
14	16	60	0.34	15.51	0.108	16	22	10	2.75	162.05	1.125
14	18	5S	1.84	91.32	0.634	16	22	STD, 20	2.63	153.66	1.067
14	18	10S	1.81	90.05	0.625	16	22	XS, 30	2.50	145.36	1.009
14	18	10	1.75	86.63	0.602	16	22	60	2.13	121.05	0.841
14	18	20	1.69	83.23	0.578	16	22	80	1.88	105.33	0.731
14	18	STD	1.63	79.80	0.554	16	22	100	1.63	90.01	0.625
14	18	30	1.56	76.40	0.531	16	22	120	1.38	75.08	0.521
14	18	XS	1.50	73.07	0.507	16	22	140	1.13	60.55	0.420
14	18	40	1.44	69.77	0.485	16	22	160	0.88	46.41	0.322
14	18	60	1.25	59.91	0.416	18	20	5S	0.81	48.01	0.333
14	18	80	1.06	50.27	0.349	18	20	10S	0.78	46.16	0.321
14	18	100	0.84	39.37	0.273	18	20	10	0.75	44.20	0.307
14	18	120	0.63	28.73	0.199	18	20	STD, 20	0.63	36.58	0.254
14	18	140	0.44	19.87	0.138	18	20	XS, 30	0.50	29.07	0.202
14	20	5S	2.81	148.58	1.032	18	20	40	0.41	23.49	0.163
14	20	10S	2.78	146.73	1.019	18	22	5S	1.81	112.83	0.784
14	20	10	2.75	144.77	1.005	18	22	10S	1.78	110.79	0.769
14	20	20	2.63	137.16	0.952	18	22	10	1.75	108.63	0.754
14	20	30	2.50	129.64	0.900	18	22	STD, 20	1.63	100.23	0.696
14	20	40	2.41	124.06	0.862	18	22	XS, 30	1.50	91.93	0.638
14	20	60	2.19	111.32	0.773	18	22	60	1.13	67.62	0.470
14	20	80	1.97	98.82	0.686	18	22	80	0.88	51.91	0.360
14	20	100	1.72	84.92	0.590	18	22	100	0.63	36.58	0.254
14	20	120	1.50	73.07	0.507	18	22	120	0.38	21.66	0.150
14	20	140	1.25	59.91	0.416	18	24	5S	2.78	181.71	1.262
14	20	160	1.03	48.70	0.338	18	24	10, 10S	2.75	179.34	1.245
16	18	5S	0.84	44.18	0.307	18	24	STD, 20	2.63	170.16	1.182
16	18	10S	0.81	42.90	0.298	18	24	XS	2.50	161.07	1.119
16	18	10	0.75	39.48	0.274	18	24	30	2.44	156.60	1.088
16	18	20	0.69	36.08	0.251	18	24	40	2.31	147.59	1.025
16	18	STD	0.63	32.66	0.227	18	24	60	2.03	127.86	0.888
16	18	30	0.56	29.25	0.203	18	24	80	1.78	110.72	0.769

(continued)

Table A-1 (cont.) Annulus and Annular Area Dimensional Data for IPS Primary Pipe Inside of ASTM F-17 IPS-Size SDR-Rated Secondary Containment Pipe

Nominal primary size (in.)	Nominal sec. cont. size (in.)	IPS size, schedule no., and/or SS schedule no.	Nominal annulus	Nominal annular area (in.²)	Nominal annular area (ft²)	Nominal primary size (in.)	Nominal sec. cont. size (in.)	IPS size, schedule no., and/or SS schedule no.	Nominal annulus	Nominal annular area (in.²)	Nominal annular area (ft²)
						(continued)					
18	24	100	1.47	89.89	0.624	26	28	30	0.38	31.08	0.216
18	24	120	1.19	71.64	0.498	26	30	5S	1.75	152.63	1.060
18	24	140	0.94	55.83	0.388	26	30	10, 10S	1.69	146.89	1.020
18	24	160	0.66	38.46	0.267	26	30	STD	1.63	141.08	0.980
20	22	5S	0.81	53.11	0.369	26	30	XS, 20	1.50	129.64	0.900
20	22	10S	0.78	51.08	0.355	26	30	30	1.38	118.30	0.822
20	22	10	0.75	48.91	0.340	26	32	10	2.69	242.36	1.683
20	22	STD, 20	0.63	40.51	0.281	26	32	STD	2.63	236.16	1.640
20	22	XS, 30	0.50	32.21	0.224	26	32	XS, 20	2.50	223.93	1.555
20	24	5S	1.78	121.99	0.847	26	32	30	2.38	211.80	1.471
20	24	10, 10S	1.75	119.63	0.831	26	32	40	2.31	205.72	1.429
20	24	STD, 20	1.63	110.44	0.767	28	30	5S	0.75	67.77	0.471
20	24	XS	1.50	101.36	0.704	28	30	10, 10S	0.69	62.03	0.431
20	24	30	1.44	96.89	0.673	28	30	STD	0.63	56.23	0.390
20	24	40	1.31	87.88	0.610	28	30	XS, 20	0.50	44.79	0.311
20	24	60	1.03	68.15	0.473	28	30	30	0.38	33.44	0.232
20	24	80	0.78	51.01	0.354	28	32	10	1.69	157.50	1.094
20	24	100	0.47	30.17	0.210	28	32	STD	1.63	151.30	1.051
20	26	10	2.69	191.67	1.331	28	32	XS, 20	1.50	139.07	0.966
20	26	STD	2.63	186.66	1.296	28	32	30	1.38	126.94	0.882
20	26	XS, 20	2.50	176.79	1.228	28	32	40	1.31	120.87	0.839
22	24	5S	0.78	55.99	0.389	28	34	10	2.66	255.90	1.777
22	24	STD, 20	0.75	53.63	0.372	28	34	STD	2.63	252.66	1.755
22	24	10, 10S	0.63	44.44	0.309	28	34	XS, 20	2.50	239.64	1.664
22	24	XS	0.50	35.36	0.246	28	34	30	2.38	226.73	1.574
22	24	30	0.44	30.89	0.214	28	34	40	2.31	220.26	1.530
22	24	40	0.31	21.88	0.152	30	32	10	0.69	66.36	0.461
22	26	10	1.69	125.67	0.873	30	32	STD	0.63	60.16	0.418
22	26	STD	1.63	120.66	0.838	30	32	XS, 20	0.50	47.93	0.333
22	26	XS, 20	1.50	110.79	0.769	30	32	30	0.38	35.80	0.249
22	28	10	2.69	208.56	1.448	30	32	40	0.31	29.72	0.206
22	28	STD	2.63	203.16	1.411	30	34	10	1.66	164.76	1.144
22	28	XS, 20	2.50	192.50	1.337	30	34	STD	1.63	161.51	1.122
22	28	30	2.38	181.94	1.263	30	34	XS, 20	1.50	148.50	1.031
24	26	10	0.69	53.38	0.371	30	34	30	1.38	135.58	0.942
24	26	STD	0.63	48.37	0.336	30	34	40	1.31	129.11	0.897
24	26	XS, 20	0.50	38.50	0.267	30	36	10	2.69	276.15	1.918
24	28	10	1.69	136.28	0.946	30	36	STD	2.63	269.16	1.869
24	28	STD	1.63	130.87	0.909	30	36	XS, 20	2.50	255.36	1.773
24	28	XS, 20	1.50	120.21	0.835	30	36	30	2.38	241.66	1.678
24	28	30	1.38	109.66	0.762	30	36	40	2.25	228.05	1.584
24	30	5S	2.75	231.20	1.606	32	34	10	0.66	67.33	0.468
24	30	10, 10S	2.69	225.46	1.566	32	34	STD	0.63	64.08	0.445
24	30	STD	2.63	219.66	1.525	32	34	XS,20	0.50	51.07	0.355
24	30	XS, 20	2.50	208.21	1.446	32	34	30	0.38	38.16	0.265
24	30	30	2.38	196.87	1.367	32	34	40	0.31	31.68	0.220
26	28	10	0.69	57.71	0.401	32	36	10	1.69	178.72	1.241
26	28	STD	0.63	52.30	0.363	32	36	STD	1.63	171.73	1.193
26	28	XS, 20	0.50	41.64	0.289	32	36	XS,20	1.50	157.93	1.097

(continued)

Appendix A

Table A-1 Annulus and Annular Area Dimensional Data for IPS Primary Pipe Inside of ASTM F-17 IPS-Size SDR-Rated Secondary Containment Pipe

Nominal primary size (in.)	Nominal sec. cont. size (in.)	IPS size, schedule no., and/or SS schedule no.	Nominal annulus	Nominal annular area (in.^{2})	Nominal annular area (ft^2)
32	36	30	1.38	144.23	1.002
32	36	40	1.25	130.63	0.907
34	36	10	0.69	75.01	0.521
34	36	STD	0.63	68.01	0.472
34	36	XS, 20	0.50	54.21	0.376
34	36	30	0.38	40.51	0.281

Table A-2 Annulus and Annular Area Dimensional Data for IPS Primary Pipe Inside of ASTM F-17 IPS-Size SDR-Rated Secondary Containment Pipe

Nominal primary size (in.)	Nominal sec. cont. size (in.)	SDR No.	Nominal annulus	Nominal annular area (in.^{2})	Nominal annular area (ft^2)	Nominal primary size (in.)	Nominal sec. cont. size (in.)	SDR No.	Nominal annulus	Nominal annular area (in.^{2})	Nominal annular area (ft^2)
						(continued)					
1/2	1-1/4	32.5	0.36	1.35	0.009	1/2	3	13.5	1.07	6.43	0.045
1/2	1-1/4	26	0.35	1.29	0.009	1/2	3	11	1.01	5.89	0.041
1/2	1-1/4	21	0.33	1.22	0.008	1/2	3	9.3	0.95	5.38	0.037
1/2	1-1/4	17	0.31	1.13	0.008	1/2	3	9	0.94	5.27	0.037
1/2	1-1/4	15.5	0.30	1.09	0.008	1/2	3	8.3	0.91	4.99	0.035
1/2	1-1/4	13.5	0.29	1.02	0.007	1/2	3	7.3	0.85	4.52	0.031
1/2	1-1/4	11	0.26	0.89	0.006	3/4	1-1/2	32.5	0.37	1.63	0.011
1/2	1-1/4	9.3	0.23	0.78	0.005	3/4	1-1/2	26	0.35	1.55	0.011
1/2	1-1/4	9	0.23	0.76	0.005	3/4	1-1/2	21	0.33	1.46	0.010
1/2	1-1/4	8.3	0.21	0.69	0.005	3/4	1-1/2	17	0.31	1.34	0.009
1/2	1-1/4	7.3	0.18	0.59	0.004	3/4	1-1/2	15.5	0.30	1.29	0.009
1/2	1-1/2	32.5	0.47	1.94	0.013	3/4	1-1/2	13.5	0.28	1.19	0.008
1/2	1-1/2	26	0.46	1.86	0.013	3/4	1-1/2	11	0.25	1.03	0.007
1/2	1-1/2	21	0.44	1.77	0.012	3/4	1-1/2	9.3	0.22	0.88	0.006
1/2	1-1/2	17	0.42	1.65	0.011	3/4	1-1/2	9	0.21	0.85	0.006
1/2	1-1/2	15.5	0.41	1.60	0.011	3/4	1-1/2	8.3	0.20	0.77	0.005
1/2	1-1/2	13.5	0.39	1.50	0.010	3/4	1-1/2	7.3	0.16	0.63	0.004
1/2	1-1/2	11	0.36	1.34	0.009	3/4	2	32.5	0.59	3.04	0.021
1/2	1-1/2	9.3	0.33	1.19	0.008	3/4	2	26	0.57	2.91	0.020
1/2	1-1/2	9	0.32	1.16	0.008	3/4	2	21	0.55	2.76	0.019
1/2	1-1/2	8.3	0.30	1.08	0.007	3/4	2	17	0.52	2.58	0.018
1/2	1-1/2	7.3	0.27	0.94	0.007	3/4	2	15.5	0.51	2.50	0.017
1/2	2	32.5	0.69	3.35	0.023	3/4	2	13.5	0.49	2.35	0.016
1/2	2	26	0.68	3.22	0.022	3/4	2	11	0.45	2.10	0.015
1/2	2	21	0.65	3.07	0.021	3/4	2	9.3	0.41	1.86	0.013
1/2	2	17	0.63	2.90	0.020	3/4	2	9	0.40	1.81	0.013
1/2	2	15.5	0.61	2.81	0.019	3/4	2	8.3	0.38	1.69	0.012
1/2	2	13.5	0.59	2.66	0.018	3/4	2	7.3	0.34	1.47	0.010
1/2	2	11	0.55	2.41	0.017	3/4	2-1/2	32.5	0.82	4.85	0.034
1/2	2	9.3	0.51	2.18	0.015	3/4	2-1/2	26	0.80	4.67	0.032
1/2	2	9	0.50	2.13	0.015	3/4	2-1/2	21	0.78	4.45	0.031
1/2	2	8.3	0.48	2.00	0.014	3/4	2-1/2	17	0.74	4.19	0.029
1/2	2	7.3	0.44	1.78	0.012	3/4	2-1/2	15.5	0.73	4.06	0.028
1/2	2-1/2	32.5	0.93	5.17	0.036	3/4	2-1/2	13.5	0.70	3.85	0.027
1/2	2-1/2	26	0.91	4.98	0.035	3/4	2-1/2	11	0.65	3.48	0.024
1/2	2-1/2	21	0.88	4.76	0.033	3/4	2-1/2	9.3	0.60	3.14	0.022
1/2	2-1/2	17	0.85	4.50	0.031	3/4	2-1/2	9	0.59	3.06	0.021
1/2	2-1/2	15.5	0.83	4.37	0.030	3/4	2-1/2	8.3	0.57	2.88	0.020
1/2	2-1/2	13.5	0.80	4.16	0.029	3/4	2-1/2	7.3	0.52	2.56	0.018
1/2	2-1/2	11	0.76	3.79	0.026	3/4	3	32.5	1.12	7.61	0.053
1/2	2-1/2	9.3	0.71	3.45	0.024	3/4	3	26	1.09	7.33	0.051
1/2	2-1/2	9	0.70	3.37	0.023	3/4	3	21	1.06	7.01	0.049
1/2	2-1/2	8.3	0.67	3.19	0.022	3/4	3	17	1.02	6.63	0.046
1/2	2-1/2	7.3	0.62	2.87	0.020	3/4	3	15.5	1.00	6.44	0.045
1/2	3	32.5	1.22	7.92	0.055	3/4	3	13.5	0.97	6.12	0.042
1/2	3	26	1.20	7.65	0.053	3/4	3	11	0.91	5.58	0.039
1/2	3	21	1.16	7.32	0.051	3/4	3	9.3	0.85	5.06	0.035
1/2	3	17	1.12	6.94	0.048	3/4	3	9	0.84	4.96	0.034
1/2	3	15.5	1.10	6.75	0.047	3/4	3	8.3	0.80	4.68	0.032

(continued)

Appendix A

Table A-2 (cont.) Annulus and Annular Area Dimensional Data for IPS Primary Pipe Inside of ASTM F-17 IPS-Size SDR-Rated Secondary Containment Pipe

Nominal primary size (in.)	Nominal sec. cont. size (in.)	SDR No.	Nominal annulus	Nominal annular area (in.²)	Nominal annular area (ft²)	Nominal primary size (in.)	Nominal sec. cont. size (in.)	SDR No.	Nominal annulus	Nominal annular area (in.²)	Nominal annular area (ft²)
						(continued)					
3/4	3	7.3	0.75	4.21	0.029	1-1/4	2	21	0.24	1.46	0.010
1.0	2	32.5	0.46	2.54	0.018	1-1/4	2	17	0.22	1.29	0.009
1.0	2	26	0.44	2.42	0.017	1-1/4	2	15.5	0.20	1.20	0.008
1.0	2	21	0.42	2.27	0.016	1-1/4	2	13.5	0.18	1.05	0.007
1.0	2	17	0.39	2.09	0.015	1-1/4	2	11	0.14	0.80	0.006
1.0	2	15.5	0.38	2.00	0.014	1-1/4	2-1/2	32.5	0.52	3.55	0.025
1.0	2	13.5	0.35	1.86	0.013	1-1/4	2-1/2	26	0.50	3.37	0.023
1.0	2	11	0.31	1.61	0.011	1-1/4	2-1/2	21	0.47	3.15	0.022
1.0	2	9.3	0.27	1.37	0.010	1-1/4	2-1/2	17	0.44	2.89	0.020
1.0	2	9	0.27	1.32	0.009	1-1/4	2-1/2	15.5	0.42	2.76	0.019
1.0	2	8.3	0.24	1.19	0.008	1-1/4	2-1/2	13.5	0.39	2.55	0.018
1.0	2	7.3	0.20	0.98	0.007	1-1/4	2-1/2	11	0.35	2.18	0.015
1.0	2-1/2	32.5	0.69	4.36	0.030	1-1/4	2-1/2	9.3	0.30	1.84	0.013
1.0	2-1/2	26	0.67	4.18	0.029	1-1/4	2-1/2	9	0.29	1.76	0.012
1.0	2-1/2	21	0.64	3.96	0.027	1-1/4	2-1/2	8.3	0.26	1.58	0.011
1.0	2-1/2	17	0.61	3.70	0.026	1-1/4	2-1/2	7.3	0.21	1.26	0.009
1.0	2-1/2	15.5	0.59	3.57	0.025	1-1/4	3	32.5	0.81	6.31	0.044
1.0	2-1/2	13.5	0.57	3.35	0.023	1-1/4	3	26	0.79	6.04	0.042
1.0	2-1/2	11	0.52	2.99	0.021	1-1/4	3	21	0.75	5.71	0.040
1.0	2-1/2	9.3	0.47	2.64	0.018	1-1/4	3	17	0.71	5.33	0.037
1.0	2-1/2	9	0.46	2.57	0.018	1-1/4	3	15.5	0.69	5.14	0.036
1.0	2-1/2	8.3	0.43	2.38	0.017	1-1/4	3	13.5	0.66	4.82	0.033
1.0	2-1/2	7.3	0.39	2.06	0.014	1-1/4	3	11	0.60	4.28	0.030
1.0	3	32.5	0.98	7.12	0.049	1-1/4	3	9.3	0.54	3.77	0.026
1.0	3	26	0.96	6.84	0.048	1-1/4	3	9	0.53	3.66	0.025
1.0	3	21	0.93	6.52	0.045	1-1/4	3	8.3	0.50	3.38	0.023
1.0	3	17	0.89	6.13	0.043	1-1/4	3	7.3	0.44	2.91	0.020
1.0	3	15.5	0.87	5.94	0.041	1-1/4	4	32.5	1.28	11.85	0.082
1.0	3	13.5	0.83	5.63	0.039	1-1/4	4	26	1.25	11.39	0.079
1.0	3	11	0.77	5.08	0.035	1-1/4	4	21	1.21	10.86	0.075
1.0	3	9.3	0.72	4.57	0.032	1-1/4	4	17	1.16	10.22	0.071
1.0	3	9	0.70	4.46	0.031	1-1/4	4	15.5	1.13	9.90	0.069
1.0	3	8.3	0.67	4.19	0.029	1-1/4	4	13.5	1.09	9.38	0.065
1.0	3	7.3	0.61	3.71	0.026	1-1/4	4	11	1.01	8.49	0.059
1.0	4	32.5	1.45	12.65	0.088	1-1/4	4	9.3	0.94	7.64	0.053
1.0	4	26	1.42	12.20	0.085	1-1/4	4	9	0.92	7.46	0.052
1.0	4	21	1.38	11.67	0.081	1-1/4	4	8.3	0.88	7.00	0.049
1.0	4	17	1.33	11.03	0.077	1-1/4	4	7.3	0.80	6.22	0.043
1.0	4	15.5	1.30	10.71	0.074	1-1/2	2-1/2	32.5	0.40	2.88	0.020
1.0	4	13.5	1.26	10.19	0.071	1-1/2	2-1/2	26	0.38	2.70	0.019
1.0	4	11	1.18	9.29	0.065	1-1/2	2-1/2	21	0.35	2.48	0.017
1.0	4	9.3	1.11	8.44	0.059	1-1/2	2-1/2	17	0.32	2.22	0.015
1.0	4	9	1.09	8.27	0.057	1-1/2	2-1/2	15.5	0.30	2.09	0.015
1.0	4	8.3	1.05	7.81	0.054	1-1/2	2-1/2	13.5	0.27	1.88	0.013
1.0	4	7.3	0.98	7.03	0.049	1-1/2	2-1/2	11	0.23	1.51	0.010
1-1/4	2	32.5	0.28	1.74	0.012	1-1/2	2-1/2	9.3	0.18	1.17	0.008
1-1/4	2	26	0.27	1.61	0.011	1-1/2	2-1/2	9	0.17	1.09	0.008

(continued)

Table A-2 (cont.) Annulus and Annular Area Dimensional Data for IPS Primary Pipe Inside of ASTM F-17 IPS-Size SDR-Rated Secondary Containment Pipe

Nominal primary size (in.)	Nominal sec. cont. size (in.)	SDR No.	Nominal annulus	Nominal annular area (in.²)	Nominal annular area (ft²)	Nominal primary size (in.)	Nominal sec. cont. size (in.)	SDR No.	Nominal annulus	Nominal annular area (in.²)	Nominal annular area (ft²)
						(continued)					
1-1/2	3	32.5	0.69	5.64	0.039	2	5	9.3	1.00	10.55	0.073
1-1/2	3	26	0.67	5.36	0.037	2	5	9	0.98	10.28	0.071
1-1/2	3	21	0.63	5.04	0.035	2	5	8.3	0.92	9.58	0.067
1-1/2	3	17	0.59	4.66	0.032	2	5	7.3	0.83	8.39	0.058
1-1/2	3	15.5	0.57	4.46	0.031	2	6	32.5	1.92	25.94	0.180
1-1/2	3	13.5	0.54	4.15	0.029	2	6	26	1.87	24.95	0.173
1-1/2	3	11	0.48	3.61	0.025	2	6	21	1.81	23.80	0.165
1-1/2	3	9.3	0.42	3.09	0.021	2	6	17	1.74	22.42	0.156
1-1/2	3	9	0.41	2.99	0.021	2	6	15.5	1.70	21.73	0.151
1-1/2	3	8.3	0.38	2.71	0.019	2	6	13.5	1.63	20.59	0.143
1-1/2	3	7.3	0.32	2.24	0.016	2	6	11	1.52	18.65	0.130
1-1/2	4	32.5	1.16	11.18	0.078	2	6	9.3	1.41	16.82	0.117
1-1/2	4	26	1.13	10.72	0.074	2	6	9	1.39	16.43	0.114
1-1/2	4	21	1.09	10.19	0.071	2	6	8.3	1.33	15.44	0.107
1-1/2	4	17	1.04	9.55	0.066	2	6	7.3	1.22	13.75	0.095
1-1/2	4	15.5	1.01	9.23	0.064	2-1/2	3	32.5	0.20	1.98	0.014
1-1/2	4	13.5	0.97	8.71	0.060	2-1/2	3	26	0.18	1.71	0.012
1-1/2	4	11	0.89	7.81	0.054	2-1/2	3	21	0.15	1.38	0.010
1-1/2	4	9.3	0.82	6.97	0.048	2-1/2	4	32.5	0.67	7.52	0.052
1-1/2	4	9	0.80	6.79	0.047	2-1/2	4	26	0.64	7.06	0.049
1-1/2	4	8.3	0.76	6.33	0.044	2-1/2	4	21	0.60	6.53	0.045
1-1/2	4	7.3	0.68	5.55	0.039	2-1/2	4	17	0.55	5.89	0.041
2	3	32.5	0.45	4.04	0.028	2-1/2	4	15.5	0.52	5.58	0.039
2	3	26	0.43	3.77	0.026	2-1/2	4	13.5	0.48	5.05	0.035
2	3	21	0.40	3.45	0.024	2-1/2	4	11	0.40	4.16	0.029
2	3	17	0.36	3.06	0.021	2-1/2	4	9.3	0.33	3.31	0.023
2	3	15.5	0.34	2.87	0.020	2-1/2	4	9	0.31	3.13	0.022
2	3	13.5	0.30	2.55	0.018	2-1/2	4	8.3	0.27	2.67	0.019
2	3	11	0.24	2.01	0.014	2-1/2	4	7.3	0.20	1.89	0.013
2	3	9.3	0.19	1.50	0.010	2-1/2	5	32.5	1.17	14.92	0.104
2	3	9	0.17	1.39	0.010	2-1/2	5	26	1.13	14.22	0.099
2	4	32.5	0.92	9.58	0.067	2-1/2	5	21	1.08	13.41	0.093
2	4	26	0.89	9.13	0.063	2-1/2	5	17	1.02	12.44	0.086
2	4	21	0.85	8.59	0.060	2-1/2	5	15.5	0.99	11.95	0.083
2	4	17	0.80	7.96	0.055	2-1/2	5	13.5	0.93	11.15	0.077
2	4	15.5	0.77	7.64	0.053	2-1/2	5	11	0.84	9.78	0.068
2	4	13.5	0.73	7.11	0.049	2-1/2	5	9.3	0.75	8.49	0.059
2	4	11	0.65	6.22	0.043	2-1/2	5	9	0.73	8.21	0.057
2	4	9.3	0.58	5.37	0.037	2-1/2	5	8.3	0.67	7.51	0.052
2	4	9	0.56	5.19	0.036	2-1/2	5	7.3	0.58	6.32	0.044
2	4	8.3	0.52	4.73	0.033	2-1/2	6	32.5	1.67	23.88	0.166
2	4	7.3	0.45	3.95	0.027	2-1/2	6	26	1.62	22.89	0.159
2	5	32.5	1.42	16.98	0.118	2-1/2	6	21	1.56	21.74	0.151
2	5	26	1.38	16.29	0.113	2-1/2	6	17	1.49	20.35	0.141
2	5	21	1.33	15.47	0.107	2-1/2	6	15.5	1.45	19.67	0.137
2	5	17	1.27	14.50	0.101	2-1/2	6	13.5	1.38	18.53	0.129
2	5	15.5	1.24	14.01	0.097	2-1/2	6	11	1.27	16.59	0.115
2	5	13.5	1.18	13.21	0.092	2-1/2	6	9.3	1.16	14.75	0.102
2	5	11	1.09	11.85	0.082	2-1/2	6	9	1.14	14.37	0.100

(continued)

Appendix A

Table A-2 (cont.) Annulus and Annular Area Dimensional Data for IPS Primary Pipe Inside of ASTM F-17 IPS-Size SDR-Rated Secondary Containment Pipe

Nominal primary size (in.)	Nominal sec. cont. size (in.)	SDR No.	Nominal annulus	Nominal annular area (in.^{2})	Nominal annular area (ft^2)	Nominal primary size (in.)	Nominal sec. cont. size (in.)	SDR No.	Nominal annulus	Nominal annular area (in.^{2})	Nominal annular area (ft^2)
						(continued)					
2-1/2	6	8.3	1.08	13.37	0.093	4	6	17	0.67	10.94	0.076
2-1/2	6	7.3	0.97	11.68	0.081	4	6	15.5	0.64	10.25	0.071
3	4	32.5	0.36	4.39	0.030	4	6	13.5	0.57	9.11	0.063
3	4	26	0.33	3.93	0.027	4	6	11	0.46	7.17	0.050
3	4	21	0.29	3.40	0.024	4	6	9.3	0.35	5.34	0.037
3	4	17	0.24	2.76	0.019	4	6	9	0.33	4.95	0.034
3	4	15.5	0.21	2.44	0.017	4	6	8.3	0.26	3.96	0.027
3	4	13.5	0.17	1.92	0.013	4	6	7.3	0.15	2.27	0.016
3	5	32.5	0.86	11.79	0.082	4	8	32.5	1.80	35.57	0.247
3	5	26	0.82	11.09	0.077	4	8	26	1.73	33.89	0.235
3	5	21	0.77	10.28	0.071	4	8	21	1.65	31.94	0.222
3	5	17	0.70	9.31	0.065	4	8	17	1.56	29.60	0.206
3	5	15.5	0.67	8.82	0.061	4	8	15.5	1.51	28.43	0.197
3	5	13.5	0.62	8.02	0.056	4	8	13.5	1.42	26.50	0.184
3	5	11	0.53	6.65	0.046	4	8	11	1.28	23.22	0.161
3	5	9.3	0.43	5.36	0.037	4	8	9.3	1.14	20.10	0.140
3	5	9	0.41	5.08	0.035	4	8	9	1.10	19.45	0.135
3	5	8.3	0.36	4.38	0.030	4	8	8.3	1.02	17.76	0.123
3	5	7.3	0.27	3.19	0.022	4	8	7.3	0.88	14.90	0.103
3	6	32.5	1.36	20.75	0.144	4	10	32.5	2.79	64.06	0.445
3	6	26	1.31	19.76	0.137	4	10	26	2.71	61.46	0.427
3	6	21	1.25	18.60	0.129	4	10	21	2.61	58.42	0.406
3	6	17	1.17	17.22	0.120	4	10	17	2.49	54.78	0.380
3	6	15.5	1.14	16.54	0.115	4	10	15.5	2.43	52.97	0.368
3	6	13.5	1.07	15.40	0.107	4	10	13.5	2.33	49.98	0.347
3	6	11	0.96	13.46	0.093	4	10	11	2.15	44.87	0.312
3	6	9.3	0.85	11.62	0.081	4	10	9.3	1.97	40.03	0.278
3	6	9	0.83	11.24	0.078	4	10	9	1.93	39.02	0.271
3	6	8.3	0.76	10.24	0.071	4	10	8.3	1.83	36.40	0.253
3	6	7.3	0.65	8.55	0.059	4	10	7.3	1.65	31.95	0.222
3	8	32.5	2.30	41.85	0.291	5	6	32.5	0.33	6.06	0.042
3	8	26	2.23	40.18	0.279	5	6	26	0.28	5.07	0.035
3	8	21	2.15	38.22	0.265	5	6	21	0.22	3.91	0.027
3	8	17	2.06	35.88	0.249	5	8	32.5	1.27	27.16	0.189
3	8	15.5	2.01	34.71	0.241	5	8	26	1.20	25.49	0.177
3	8	13.5	1.92	32.79	0.228	5	8	21	1.12	23.53	0.163
3	8	11	1.78	29.50	0.205	5	8	17	1.02	21.19	0.147
3	8	9.3	1.64	26.39	0.183	5	8	15.5	0.97	20.02	0.139
3	8	9	1.60	25.73	0.179	5	8	13.5	0.89	18.10	0.126
3	8	8.3	1.52	24.05	0.167	5	8	11	0.75	14.81	0.103
3	8	7.3	1.38	21.18	0.147	5	8	9.3	0.60	11.70	0.081
4	5	32.5	0.36	5.50	0.038	5	8	9	0.57	11.04	0.077
4	5	26	0.32	4.81	0.033	5	8	8.3	0.49	9.36	0.065
4	5	21	0.27	3.99	0.028	5	8	7.3	0.35	6.49	0.045
4	5	17	0.20	3.02	0.021	5	10	32.5	2.26	55.65	0.386
4	5	15.5	0.17	2.53	0.018	5	10	26	2.18	53.05	0.368
4	5	32.5	0.86	14.46	0.100	5	10	21	2.08	50.01	0.347
4	6	26	0.81	13.47	0.094	5	10	17	1.96	46.38	0.322
4	6	21	0.75	12.32	0.086	5	10	15.5	1.90	44.56	0.309

(continued)

Table A-2 Annulus and Annular Area Dimensional Data for IPS Primary Pipe Inside (cont.) of ASTM F-17 IPS-Size SDR-Rated Secondary Containment Pipe

Nominal primary size (in.)	Nominal sec. cont. size (in.)	SDR No.	Nominal annulus	Nominal annular area (in.²)	Nominal annular area (ft²)	Nominal primary size (in.)	Nominal sec. cont. size (in.)	SDR No.	Nominal annulus	Nominal annular area (in.²)	Nominal annular area (ft²)
						(continued)					
5	10	13.5	1.80	41.57	0.289	8	12	9	0.65	18.82	0.131
5	10	11	1.62	36.47	0.253	8	12	8.3	0.53	15.14	0.105
5	10	9.3	1.44	31.63	0.220	8	12	7.3	0.32	8.88	0.062
5	10	9	1.40	30.61	0.213	8	14	32.5	2.26	77.18	0.536
5	10	8.3	1.30	28.00	0.194	8	14	26	2.15	72.77	0.505
5	10	7.3	1.12	23.55	0.164	8	14	21	2.02	67.61	0.470
6	8	32.5	0.73	16.99	0.118	8	14	17	1.86	61.45	0.427
6	8	26	0.67	15.32	0.106	8	14	15.5	1.78	58.37	0.405
6	8	21	0.59	13.36	0.093	8	14	13.5	1.65	53.30	0.370
6	8	17	0.49	11.02	0.077	8	14	11	1.41	44.64	0.310
6	8	15.5	0.44	9.85	0.068	8	14	9.3	1.18	36.44	0.253
6	8	13.5	0.36	7.93	0.055	8	14	9	1.13	34.71	0.241
6	8	11	0.22	4.64	0.032	8	14	8.3	1.00	30.28	0.210
6	10	32.5	1.73	45.48	0.316	8	14	7.3	0.77	22.73	0.158
6	10	26	1.65	42.88	0.298	10	12	32.5	0.61	21.69	0.151
6	10	21	1.55	39.84	0.277	10	12	26	0.51	18.03	0.125
6	10	17	1.43	36.21	0.251	10	12	21	0.39	13.76	0.096
6	10	15.5	1.37	34.39	0.239	10	12	17	0.25	8.64	0.060
6	10	13.5	1.27	31.40	0.218	10	12	15.5	0.18	6.09	0.042
6	10	11	1.09	26.30	0.183	10	14	32.5	1.19	44.83	0.311
6	10	9.3	0.91	21.46	0.149	10	14	26	1.09	40.42	0.281
6	10	9	0.87	20.44	0.142	10	14	21	0.96	35.26	0.245
6	10	8.3	0.77	17.83	0.124	10	14	17	0.80	29.10	0.202
6	10	7.3	0.59	13.38	0.093	10	14	15.5	0.72	26.02	0.181
6	12	32.5	2.67	78.01	0.542	10	14	13.5	0.59	20.95	0.145
6	12	26	2.57	74.35	0.516	10	14	11	0.35	12.29	0.085
6	12	21	2.46	70.07	0.487	10	16	32.5	2.13	86.35	0.600
6	12	17	2.31	64.96	0.451	10	16	26	2.01	80.59	0.560
6	12	15.5	2.24	62.41	0.433	10	16	21	1.86	73.86	0.513
6	12	13.5	2.12	58.20	0.404	10	16	17	1.68	65.80	0.457
6	12	11	1.90	51.02	0.354	10	16	15.5	1.59	61.78	0.429
6	12	9.3	1.69	44.21	0.307	10	16	13.5	1.44	55.16	0.383
6	12	9	1.65	42.78	0.297	10	16	11	1.17	43.85	0.305
6	12	8.3	1.53	39.10	0.272	10	16	9.3	0.90	33.13	0.230
6	12	7.3	1.32	32.84	0.228	10	16	9	0.85	30.88	0.214
8	10	32.5	0.73	21.52	0.149	10	16	8.3	0.70	25.09	0.174
8	10	26	0.65	18.92	0.131	10	16	7.3	0.43	15.23	0.106
8	10	21	0.55	15.88	0.110	12	16	32.5	1.13	49.42	0.343
8	10	17	0.43	12.24	0.085	12	16	26	1.01	43.66	0.303
8	10	15.5	0.37	10.43	0.072	12	16	21	0.86	36.93	0.256
8	10	13.5	0.27	7.44	0.052	12	16	17	0.68	28.87	0.200
8	12	32.5	1.67	54.04	0.375	12	16	15.5	0.59	24.86	0.173
8	12	26	1.57	50.38	0.350	12	16	13.5	0.44	18.23	0.127
8	12	21	1.46	46.11	0.320	12	18	32.5	2.07	96.48	0.670
8	12	17	1.31	40.99	0.285	12	18	26	1.93	89.19	0.619
8	12	15.5	1.24	38.44	0.267	12	18	21	1.77	80.66	0.560
8	12	13.5	1.12	34.24	0.238	12	18	17	1.57	70.47	0.489
8	12	11	0.90	27.05	0.188	12	18	15.5	1.46	65.39	0.454
8	12	9.3	0.69	20.25	0.141	12	18	13.5	1.29	57.00	0.396

(continued)

Table A-2 (cont.) Annulus and Annular Area Dimensional Data for IPS Primary Pipe Inside of ASTM F-17 IPS-Size SDR-Rated Secondary Containment Pipe

Nominal primary size (in.)	Nominal sec. cont. size (in.)	SDR No.	Nominal annulus	Nominal annular area (in.²)	Nominal annular area (ft²)	Nominal primary size (in.)	Nominal sec. cont. size (in.)	SDR No.	Nominal annulus	Nominal annular area (in.²)	Nominal annular area (ft²)
						(continued)					
12	18	11	0.99	42.69	0.296	18	22	13.5	0.37	21.38	0.148
12	18	9.3	0.69	29.12		18	24	32.5	2.26	144.01	1.000
12	18	9	0.63	26.27	0.202	18	24	26	2.08	131.05	0.910
12	18	8.3	0.46	18.94	0.182	18	24	21	1.86	115.90	0.805
12	18	7.3	0.16	6.46	0.132	18	24	17	1.59	97.78	0.679
14	16	32.5	0.51	23.15	0.045	18	24	15.5	1.45	88.74	0.616
14	16	26	0.38	17.39	0.161	18	24	13.5	1.22	73.84	0.513
14	16	21	0.24	10.65	0.121	18	24	11	0.82	48.39	0.336
14	18	32.5	1.45	70.20	0.074	18	24	9.3	0.42	24.28	0.169
14	18	26	1.31	62.91	0.488	18	24	9	0.33	19.21	0.133
14	18	21	1.14	54.39	0.437	20	22	32.5	0.32	20.64	0.143
14	18	17	0.94	44.20	0.378	20	24	32.5	1.26	84.30	0.585
14	18	15.5	0.84	39.11	0.307	20	24	26	1.08	71.34	0.495
14	18	13.5	0.67	30.73	0.272	20	24	21	0.86	56.19	0.390
14	18	11	0.36	16.42	0.213	20	24	17	0.59	38.06	0.264
14	20	32.5	2.38	122.79	0.114	20	24	15.5	0.45	29.03	0.202
14	20	26	2.23	113.79	0.853	20	24	13.5	0.22	14.12	0.098
14	20	21	2.05	103.27	0.790	20	26	32.5	2.20	153.50	1.066
14	20	17	1.82	90.69	0.717	20	26	26	2.00	138.29	0.960
14	20	15.5	1.71	84.41	0.630	20	26	21	1.76	120.50	0.837
14	20	13.5	1.52	74.06	0.586	20	26	17	1.47	99.23	0.689
14	20	11	1.18	56.39	0.514	20	26	15.5	1.32	88.63	0.615
14	20	9.3	0.85	39.64	0.392	20	26	13.5	1.07	71.14	0.494
14	20	9	0.78	36.12	0.275	20	26	11	0.64	41.27	0.287
14	20	8.3	0.59	27.07	0.251	20	26	9.3	0.20	12.97	0.9090.
16	18	32.5	0.45	23.06	0.188	22	24	32.5	0.26	18.30	127
16	18	26	0.31	15.77	0.160	22	26	32.5	1.20	87.50	0.608
16	20	32.5	1.38	76.65	0.110	22	26	26	1.00	72.29	0.502
16	20	26	1.23	66.65	0.525	22	26	21	0.76	54.50	0.379
16	20	21	1.05	56.13	0.463	22	26	17	0.47	33.23	0.231
16	20	17	0.82	43.54	0.390	22	26	15.5	0.32	22.63	0.157
16	20	15.5	0.71	37.27	0.302	22	28	32.5	1.20	87.50	0.608
16	20	13.5	0.52	26.92	0.259	22	28	26	1.92	144.59	1.004
16	22	32.5	2.32	133.78	0.187	22	28	21	1.67	123.97	0.861
16	22	26	2.15	122.89	0.929	22	28	17	1.35	99.30	0.690
16	22	21	1.95	110.16	0.853	22	28	15.5	1.19	87.00	0.604
16	22	17	1.71	94.93	0.765	22	28	13.5	0.93	66.72	0.463
16	22	15.5	1.58	87.34	0.659	22	28	11	0.45	32.08	0.223
16	22	13.5	1.37	74.81	0.607	24	28	32.5	1.14	89.95	0.625
16	22	11	1.00	53.43	0.520	24	28	26	0.92	72.30	0.502
16	22	9.3	0.63	33.17	0.371	24	28	21	0.67	51.68	0.359
16	22	9	0.56	28.91	0.230	24	28	17	0.35	27.01	0.188
18	20	32.5	0.38	22.22	0.201	24	30	32.5	2.08	170.22	1.182
18	20	26	0.23	13.22	0.154	24	30	26	1.85	149.96	1.041
18	22	32.5	1.32	80.35	0.092	24	30	21	1.57	126.29	0.877
18	22	26	1.15	69.46	0.558	24	30	17	1.24	97.97	0.680
18	22	21	0.95	56.73	0.482	24	30	15.5	1.06	83.86	0.582
18	22	17	0.71	41.50	0.394	24	30	13.5	0.78	60.57	0.421
18	22	15.5	0.58	33.91	0.235	24	30	11	0.27	20.81	0.144

(continued)

Table A-2 Annulus and Annular Area Dimensional Data for IPS Primary Pipe Inside of ASTM F-17 IPS-Size SDR-Rated Secondary Containment Pipe

Nominal primary size (in.)	Nominal sec. cont. size (in.)	SDR No.	Nominal annulus	Nominal annular area (in.^{2})	Nominal annular area (ft^2)
26	30	32.5	1.08	91.64	0.636
26	30	26	0.85	71.39	0.496
26	30	21	0.57	47.72	0.331
26	30	17	0.24	19.40	0.135
26	32	32.5	2.02	177.45	1.232
26	32	26	1.77	154.41	1.072
26	32	21	1.48	127.47	0.885
26	32	17	1.12	95.25	0.661
26	32	15.5	0.94	79.19	0.550
26	32	13.5	0.63	52.70	0.366
28	32	32.5	1.02	92.59	0.643
28	32	26	0.77	69.55	0.483
28	32	21	0.48	42.62	0.296
28	34	32.5	1.95	183.94	1.277
28	34	26	1.69	157.92	1.097
28	34	21	1.38	127.52	0.886
28	34	17	1.00	91.14	0.633
28	34	15.5	0.81	73.01	0.507
28	34	13.5	0.48	43.10	0.299
30	34	32.5	0.95	92.79	0.644
30	34	26	0.69	66.78	0.464
30	34	21	0.38	36.37	0.253
30	36	32.5	1.89	189.67	1.317
30	36	26	1.62	160.51	1.115
30	36	21	1.29	126.42	0.878
30	36	17	0.88	85.64	0.595
30	36	15.5	0.68	65.31	0.454
30	36	13.5	0.33	31.78	0.221
32	36	32.5	0.89	92.24	0.641
32	36	26	0.62	63.08	0.438
32	36	21	0.29	28.99	0.201
32	42	32.5	3.71	416.09	2.890
32	42	26	3.38	376.40	2.614
32	42	21	3.00	330.00	2.292
32	42	17	2.53	274.49	1.906
32	42	15.5	2.29	246.83	1.714
34	42	32.5	2.71	312.38	2.169
34	42	26	2.38	272.68	1.894
34	42	21	2.00	226.29	1.571
34	42	17	1.53	170.78	1.186
34	42	15.5	1.29	143.11	0.994
36	42	32.5	1.71	202.38	1.405
36	42	26	1.38	162.68	1.130
36	42	21	1.00	116.29	0.808
36	42	17	0.53	60.78	0.422
36	42	15.5	0.29	33.11	0.230
42	48	32.5	1.52	208.34	1.447
42	48	26	1.15	156.49	1.087
42	48	21	0.71	95.89	0.666

(continued)

Table A-3 Annulus and Annular Area Dimensional Data for IPS Primary Pipe Inside of ISO 160/1 Metric-Size Plastic SDR-Rated Secondary Containment Pipe

Nominal primary size (in.)	Nominal sec. cont. size (in.)	SDR No.	Nominal annulus	Nominal annular area (in.^{2})	Nominal annular area (ft^2)	Nominal primary size (in.)	Nominal sec. cont. size (in.)	SDR No.	Nominal annulus	Nominal annular area (in.^{2})	Nominal annular area (ft^2)
						(continued)					
1/2	40	32.5	0.32	1.16	0.008	1	50	32.5	0.27	1.32	0.009
1/2	40	26	0.31	1.11	0.008	1	50	26	0.25	1.24	0.009
1/2	40	21	0.29	1.04	0.007	1	50	21	0.23	1.13	0.008
1/2	40	17	0.27	0.96	0.007	1	50	17	0.21	1.01	0.007
1/2	40	11	0.22	0.75	0.005	1	50	11	0.15	0.68	0.005
1/2	50	32.5	0.50	2.13	0.015	1	63	32.5	0.51	2.90	0.020
1/2	50	26	0.49	2.04	0.014	1	63	26	0.49	2.76	0.019
1/2	50	21	0.47	1.94	0.013	1	63	21	0.46	2.60	0.018
1/2	50	17	0.45	1.82	0.013	1	63	17	0.44	2.40	0.017
1/2	50	11	0.39	1.48	0.010	1	63	11	0.36	1.88	0.013
1/2	63	32.5	0.74	3.70	0.026	1	75	32.5	0.73	4.67	0.032
1/2	63	26	0.72	3.56	0.025	1	75	26	0.71	4.48	0.031
1/2	63	21	0.70	3.40	0.024	1	75	21	0.68	4.25	0.030
1/2	63	17	0.67	3.21	0.022	1	75	17	0.65	3.97	0.028
1/2	63	11	0.59	2.68	0.019	1	75	11	0.55	3.23	0.022
1/2	75	32.5	0.97	5.48	0.038	1	90	32.5	1.01	7.33	0.051
1/2	75	26	0.94	5.28	0.037	1	90	26	0.98	7.05	0.049
1/2	75	21	0.92	5.05	0.035	1	90	21	0.95	6.72	0.047
1/2	75	17	0.88	4.78	0.033	1	90	17	0.91	6.32	0.044
1/2	75	11	0.79	4.03	0.028	1	90	11	0.79	5.24	0.036
1/2	90	32.5	1.24	8.13	0.056	1	110	32.5	1.37	11.62	0.081
1/2	90	26	1.22	7.85	0.055	1	110	26	1.34	11.20	0.078
1/2	90	21	1.18	7.52	0.052	1	110	21	1.30	10.70	0.074
1/2	90	17	1.14	7.13	0.049	1	110	17	1.25	10.11	0.070
1/2	90	11	1.03	6.05	0.042	1	110	11	1.11	8.51	0.059
3/4	40	32.5	0.21	0.85	0.006	1-1/4	63	32.5	0.33	2.09	0.015
3/4	40	26	0.20	0.79	0.006	1-1/4	63	26	0.31	1.95	0.014
3/4	40	21	0.19	0.73	0.005	1-1/4	63	21	0.29	1.79	0.012
3/4	40	17	0.17	0.65	0.005	1-1/4	63	17	0.26	1.60	0.011
3/4	50	32.5	0.40	1.82	0.013	1-1/4	63	11	0.18	1.07	0.007
3/4	50	26	0.38	1.73	0.012	1-1/4	75	32.5	0.56	3.87	0.027
3/4	50	21	0.37	1.63	0.011	1-1/4	75	26	0.53	3.67	0.025
3/4	50	17	0.34	1.50	0.010	1-1/4	75	21	0.51	3.44	0.024
3/4	50	11	0.28	1.17	0.008	1-1/4	75	17	0.47	3.17	0.022
3/4	63	32.5	0.64	3.39	0.024	1-1/4	75	11	0.38	2.42	0.017
3/4	63	26	0.62	3.25	0.023	1-1/4	90	32.5	0.83	6.52	0.045
3/4	63	21	0.60	3.09	0.021	1-1/4	90	26	0.81	6.24	0.043
3/4	63	17	0.57	2.90	0.020	1-1/4	90	21	0.77	5.91	0.041
3/4	63	11	0.49	2.37	0.016	1-1/4	90	17	0.73	5.51	0.038
3/4	75	32.5	0.86	5.17	0.036	1-1/4	90	11	0.62	4.44	0.031
3/4	75	26	0.84	4.97	0.035	1-1/4	110	32.5	1.20	10.81	0.075
3/4	75	21	0.81	4.74	0.033	1-1/4	110	26	1.17	10.39	0.072
3/4	75	17	0.78	4.47	0.031	1-1/4	110	21	1.13	9.90	0.069
3/4	75	11	0.68	3.72	0.026	1-1/4	110	17	1.08	9.31	0.065
3/4	90	32.5	1.14	7.82	0.054	1-1/4	110	11	0.94	7.70	0.053
3/4	90	26	1.11	7.54	0.052	1-1/4	125	32.5	1.48	14.59	0.101
3/4	90	21	1.08	7.21	0.050	1-1/4	125	26	1.44	14.05	0.098
3/4	90	17	1.04	6.81	0.047	1-1/4	125	21	1.40	13.41	0.093
3/4	90	11	0.92	5.74	0.040	1-1/4	125	17	1.34	12.65	0.088

(continued)

Table A-3 (cont.) Annulus and Annular Area Dimensional Data for IPS Primary Pipe Inside of ISO 160/1 Metric-Size Plastic SDR-Rated Secondary Containment Pipe

Nominal primary size (in.)	Nominal sec. cont. size (in.)	SDR No.	Nominal annulus	Nominal annular area (in.^{2})	Nominal annular area (ft^2)	Nominal primary size (in.)	Nominal sec. cont. size (in.)	SDR No.	Nominal annulus	Nominal annular area (in.^{2})	Nominal annular area (ft^2)
						(continued)					
1-1/4	125	11	1.18	10.57	0.073	2-1/2	110	17	0.47	4.98	0.035
1-1/2	75	32.5	0.44	3.20	0.022	2-1/2	110	11	0.33	3.37	0.023
1-1/2	75	26	0.41	3.00	0.021	2-1/2	125	32.5	0.87	10.26	0.071
1-1/2	75	21	0.39	2.77	0.019	2-1/2	125	26	0.83	9.72	0.067
1-1/2	75	17	0.35	2.50	0.017	2-1/2	125	21	0.79	9.08	0.063
1-1/2	75	11	0.26	1.75	0.012	2-1/2	125	17	0.73	8.32	0.058
1-1/2	90	32.5	0.71	5.85	0.041	2-1/2	125	11	0.58	6.24	0.043
1-1/2	90	26	0.69	5.57	0.039	2-1/2	140	32.5	1.15	14.53	0.101
1-1/2	90	21	0.65	5.24	0.036	2-1/2	140	26	1.11	13.84	0.096
1-1/2	90	17	0.61	4.84	0.034	2-1/2	140	21	1.06	13.05	0.091
1-1/2	90	11	0.50	3.77	0.026	2-1/2	140	17	0.99	12.09	0.084
1-1/2	110	32.5	1.08	10.14	0.070	2-1/2	140	11	0.82	9.48	0.066
1-1/2	110	26	1.05	9.72	0.067	2-1/2	160	32.5	1.52	20.96	0.146
1-1/2	110	21	1.01	9.23	0.064	2-1/2	160	26	1.47	20.07	0.139
1-1/2	110	17	0.96	8.64	0.060	2-1/2	160	21	1.14	19.03	0.132
1-1/2	110	11	0.82	7.03	0.049	2-1/2	160	17	1.34	17.78	0.123
1-1/2	125	32.5	1.36	13.92	0.097	2-1/2	160	11	1.14	14.38	0.100
1-1/2	125	26	1.32	13.38	0.093	3	110	32.5	0.28	3.35	0.023
1-1/2	125	21	1.28	12.74	0.088	3	110	26	0.25	2.93	0.020
1-1/2	125	17	1.22	11.98	0.083	3	110	21	0.21	2.44	0.017
1-1/2	125	11	1.06	9.90	0.069	3	110	17	0.16	1.85	0.013
2	90	32.5	0.48	4.26	0.030	3	125	32.5	0.56	7.13	0.050
2	90	26	0.45	3.97	0.028	3	125	26	0.52	6.59	0.046
2	90	21	0.42	3.64	0.025	3	125	21	0.48	5.95	0.041
2	90	17	0.38	3.25	0.023	3	125	17	0.42	5.19	0.036
2	90	11	0.26	2.17	0.015	3	125	11	0.26	3.11	0.022
2	110	32.5	0.84	8.55	0.059	3	140	32.5	0.84	11.40	0.079
2	110	26	0.81	8.12	0.056	3	140	26	0.79	10.71	0.074
2	110	21	0.77	7.63	0.053	3	140	21	0.74	9.91	0.069
2	110	17	0.72	7.04	0.049	3	140	17	0.68	8.96	0.062
2	110	11	0.58	5.43	0.038	3	140	11	0.50	6.35	0.044
2	125	32.5	1.12	12.33	0.086	3	160	32.5	1.21	17.83	0.124
2	125	26	1.08	11.78	0.082	3	160	26	1.16	16.94	0.118
2	125	21	1.04	11.15	0.077	3	160	21	1.10	15.90	0.110
2	125	17	0.98	10.38	0.072	3	160	17	1.03	14.65	0.102
2	125	11	0.83	8.31	0.058	3	160	11	0.83	11.25	0.078
2	140	32.5	1.40	16.59	0.115	3	180	32.5	1.58	25.13	0.174
2	140	26	1.36	15.91	0.110	3	180	26	1.52	24.00	0.167
2	140	21	1.31	15.11	0.105	3	180	21	1.46	22.68	0.157
2	140	17	1.24	14.15	0.098	3	180	17	1.38	21.10	0.146
2	140	11	1.07	11.55	0.080	3	180	11	1.15	16.79	0.117
2	160	32.5	1.77	23.03	0.160	3	200	32.5	1.94	33.28	0.231
2	160	26	1.72	22.13	0.154	3	200	26	1.88	31.88	0.221
2	160	21	1.66	21.09	0.146	3	200	21	1.81	30.25	0.210
2	160	17	1.59	19.84	0.138	3	200	17	1.72	28.30	0.197
2	160	11	1.39	16.44	0.114	3	200	11	1.47	22.99	0.160
2-1/2	110	32.5	0.59	6.48	0.045	4	140	32.5	0.34	5.11	0.035
2-1/2	110	26	0.56	6.06	0.042	4	140	26	0.29	4.43	0.031
2-1/2	110	21	0.52	5.57	0.039	4	140	21	0.24	3.63	0.025

(continued)

Table A-3 (cont.) Annulus and Annular Area Dimensional Data for IPS Primary Pipe Inside of ISO 160/1 Metric-Size Plastic SDR-Rated Secondary Containment Pipe

Nominal primary size (in.)	Nominal sec. cont. size (in.)	SDR No.	Nominal annulus	Nominal annular area (in.²)	Nominal annular area (ft²)	Nominal primary size (in.)	Nominal sec. cont. size (in.)	SDR No.	Nominal annulus	Nominal annular area (in.²)	Nominal annular area (ft²)
						(continued)					
4	140	17	0.18	2.67	0.019	6	225	11	0.31	6.79	0.047
4	160	32.5	0.71	11.55	0.080	6	250	32.5	1.31	32.55	0.226
4	160	26	0.66	10.65	0.074	6	250	26	1.23	30.37	0.211
4	160	21	0.60	9.61	0.067	6	250	21	1.14	27.82	0.193
4	160	17	0.53	8.36	0.058	6	250	17	1.03	24.77	0.172
4	160	11	0.33	4.96	0.034	6	250	11	0.71	16.47	0.114
4	180	32.5	1.08	18.84	0.131	6	280	32.5	1.86	49.60	0.344
4	180	26	1.02	17.71	0.123	6	280	26	1.78	46.87	0.325
4	180	21	0.96	16.39	0.114	6	280	21	1.67	43.67	0.303
4	180	17	0.88	14.81	0.103	6	280	17	1.55	39.85	0.277
4	180	11	0.65	10.50	0.073	6	280	11	1.20	29.43	0.204
4	200	32.5	1.44	26.99	0.187	8	250	32.5	0.31	8.59	0.060
4	200	26	1.38	25.60	0.178	8	250	26	0.23	6.41	0.044
4	200	21	1.31	23.97	0.166	8	280	32.5	0.86	25.64	0.178
4	200	17	1.22	22.02	0.153	8	280	26	0.78	22.91	0.159
4	200	11	0.97	16.70	0.116	8	280	21	0.67	19.71	0.137
4	225	32.5	1.91	38.39	0.267	8	280	17	0.55	15.89	0.110
4	225	26	1.84	36.62	0.254	8	280	11	0.20	5.47	0.038
4	225	21	1.76	34.56	0.240	8	315	32.5	1.51	47.98	0.333
4	225	17	1.66	32.09	0.223	8	315	26	1.41	44.52	0.309
4	225	11	1.37	25.36	0.176	8	315	21	1.30	40.47	0.281
5	180	32.5	0.54	10.44	0.072	8	315	17	1.16	35.63	0.247
5	180	26	0.49	9.31	0.065	8	315	11	0.76	22.44	0.156
5	180	21	0.42	7.99	0.055	8	355	32.5	2.25	76.72	0.533
5	180	17	0.34	6.40	0.044	8	355	26	2.14	72.33	0.502
5	180	11	0.12	2.10	0.015	8	355	21	2.01	67.19	0.467
5	200	32.5	0.91	18.59	0.129	8	355	17	1.85	61.04	0.424
5	200	26	0.85	17.19	0.119	8	355	11	1.41	44.29	0.308
5	200	21	0.78	15.56	0.108	10	315	32.5	0.44	15.63	0.109
5	200	17	0.69	13.61	0.095	10	315	26	0.35	12.17	0.084
5	200	11	0.44	8.29	0.058	10	315	21	0.24	8.12	0.056
5	225	32.5	1.38	29.98	0.208	10	355	32.5	1.18	44.37	0.308
5	225	26	1.31	28.22	0.196	10	355	26	1.08	39.98	0.278
5	225	21	1.23	26.15	0.182	10	355	21	0.95	34.84	0.242
5	225	17	1.13	23.69	0.164	10	355	17	0.79	28.69	0.199
5	225	11	0.84	16.96	0.118	10	355	11	0.34	11.94	0.083
5	250	32.5	1.84	42.72	0.297	10	400	32.5	2.01	80.81	0.561
5	250	26	1.76	40.54	0.282	10	400	26	1.89	75.23	0.522
5	250	21	1.67	37.99	0.264	10	400	21	1.75	68.71	0.477
5	250	17	1.56	34.94	0.243	10	400	17	1.57	60.91	0.423
5	250	11	1.24	26.64	0.185	10	400	11	1.07	39.64	0.275
6	200	32.5	0.38	8.42	0.058	12	400	32.5	1.01	43.89	0.305
6	200	26	0.32	7.02	0.049	12	400	26	0.89	38.30	0.266
6	200	21	0.25	5.39	0.037	12	400	21	0.75	31.78	0.221
6	200	17	0.16	3.44	0.024	12	400	17	0.57	23.98	0.167
6	225	32.5	0.84	19.81	0.138	12	450	32.5	1.94	89.47	0.621
6	225	26	0.78	18.05	0.125	12	450	26	1.80	82.41	0.572
6	225	21	0.69	15.98	0.111	12	450	21	1.64	74.15	0.515
6	225	17	0.60	13.52	0.094	12	450	17	1.44	64.27	0.446

(continued)

Table A-3 Annulus and Annular Area Dimensional Data for IPS Primary Pipe Inside of ISO 160/1 Metric-Size Plastic SDR-Rated Secondary Containment Pipe

Nominal primary size (in.)	Nominal sec. cont. size (in.)	SDR No.	Nominal annulus	Nominal annular area (in.²)	Nominal annular area (ft²)	Nominal primary size (in.)	Nominal sec. cont. size (in.)	SDR No.	Nominal annulus	Nominal annular area (in.²)	Nominal annular area (ft²)
12	450	11	0.87	37.36	0.259	24	710	17	0.33	25.40	0.176
14	400	32.5	.039	17.61	0.122	24	800	32.5	2.78	233.88	1.624
14	400	26	0.27	12.03	0.084	24	800	26	2.54	211.56	1.469
14	450	32.5	1.31	63.20	0.439	24	800	21	2.25	185.47	1.288
14	450	26	1.18	56.13	0.390	24	800	17	1.90	154.25	1.071
14	450	21	1.01	47.88	0.332	26	800	32.5	1.78	155.31	1.079
14	450	17	0.82	38.00	0.264	26	800	26	1.54	132.99	0.924
14	450	11	0.25	11.09	0.077	26	800	21	1.25	106.89	0.742
14	500	32.5	2.24	114.15	0.793	26	800	17	0.90	75.68	0.526
14	500	26	2.09	105.43	0.732	26	900	32.5	3.63	337.65	2.345
14	500	21	1.91	95.23	0.661	26	900	26	3.35	309.40	2.149
14	500	17	1.68	83.04	0.577	26	900	21	3.03	276.37	1.919
16	450	32.5	0.31	16.05	0.111	28	800	32.5	0.78	70.45	0.489
16	500	32.5	1.24	67.00	0.465	28	800	26	0.54	48.13	0.334
16	500	26	1.09	58.28	0.405	28	900	32.5	2.63	252.79	1.755
16	500	21	0.91	48.09	0.334	28	900	26	2.35	224.54	1.559
16	500	17	0.68	35.90	0.249	28	900	21	2.03	191.52	1.330
16	560	32.5	2.35	135.22	0.939	30	900	32.5	1.63	161.65	1.123
16	560	26	2.18	124.28	0.863	30	900	26	1.35	133.40	0.926
16	560	21	1.97	111.50	0.774	30	900	21	1.03	100.37	0.697
16	560	17	1.73	96.20	0.668	30	1000	41	3.72	394.80	2.742
16	560	11	1.02	54.52	0.379	30	1000	32.5	3.47	365.44	2.538
18	500	32.5	0.24	13.57	0.094	30	1000	26	3.17	330.56	2.296
18	560	32.5	1.35	81.79	0.568	30	1000	21	2.81	289.79	2.012
18	560	26	1.18	70.85	0.492	32	1000	41	2.72	297.37	2.065
18	560	21	0.97	58.07	0.403	32	1000	32.5	2.47	268.01	1.861
18	560	17	0.73	42.77	0.297	32	1000	26	2.17	233.13	1.619
18	630	32.5	2.64	171.14	1.188	32	1000	21	1.81	192.36	1.336
18	630	26	2.45	157.29	1.092	34	1000	41	1.72	193.66	1.345
18	630	21	2.22	141.11	0.980	34	1000	32.5	1.47	164.30	1.141
18	630	17	1.94	121.75	0.846	34	1000	26	1.17	129.42	0.899
20	560	32.5	0.35	22.08	0.153	34	1000	21	0.81	88.65	0.616
20	630	32.5	1.64	111.42	0.774	36	1200	41	4.47	568.51	3.948
20	630	26	1.45	97.58	0.678	36	1200	32.5	4.17	526.23	3.654
20	630	21	1.22	81.40	0.565	36	1200	26	3.80	476.01	3.306
20	630	17	0.94	62.04	0.431	42	1200	41	1.47	200.80	1.394
20	710	32.5	3.12	226.40	1.572	42	1200	32.5	1.17	158.52	1.101
20	710	26	2.90	208.82	1.450	42	1400	41	5.21	773.81	5.374
20	710	21	2.65	188.27	1.307	42	1400	32.5	4.86	716.26	4.974
20	710	17	2.33	163.68	1.137	48	1600	41	5.96	1010.6	7.019
22	630	32.5	0.64	45.42	0.315	48	1600	32.5	5.56	935.52	6.497
22	630	26	0.45	31.58	0.219						
22	710	32.5	2.12	160.40	1.114						
22	710	26	1.90	142.82	0.992						
22	710	21	1.65	122.27	0.849						
22	710	17	1.33	97.68	0.678						
24	710	32.5	1.12	88.12	0.612						
24	710	26	0.90	70.53	0.490						
24	710	21	0.65	49.98	0.347						

(continued)

Appendix A

Table A-4 Annulus and Annular Area Dimensional Data for IPS Primary Pipe Inside of ISO 160/1 Metric-Size Plastic SDR-Rated Secondary Containment Pipe

Nominal primary size (in.)	Nominal sec. cont. size (in.)	SDR No.	Nominal annulus	Nominal annular area (in.²)	Nominal annular area (ft²)	Nominal primary size (in.)	Nominal sec. cont. size (in.)	SDR No.	Nominal annulus	Nominal annular area (in.²)	Nominal annular area (ft²)
						(continued)					
20	40	32.5	8.8	793	0.000793	32	50	32.5	7.5	925	0.000925
20	40	26	8.5	757	0.000757	32	50	26	7.1	869	0.000869
20	40	21	8.1	715	0.000715	32	50	21	6.6	803	0.000803
20	40	17	7.6	664	0.000664	32	50	17	6.1	725	0.000725
20	40	11	6.4	527	0.000527	32	63	32.5	13.6	1942	0.001942
20	50	32.5	13.5	1416	0.001416	32	63	26	13.1	1853	0.001853
20	50	26	13.1	1359	0.001359	32	63	21	12.5	1748	0.001748
20	50	21	12.6	1294	0.001294	32	63	17	11.8	1623	0.001623
20	50	17	12.1	1215	0.001215	32	63	11	9.8	1283	0.001283
20	50	11	10.5	1001	0.001001	32	75	32.5	19.2	3088	0.003088
20	63	32.5	19.6	2432	0.002432	32	75	26	18.6	2961	0.002961
20	63	26	19.1	2343	0.002343	32	75	21	17.9	2813	0.002813
20	63	21	18.5	2239	0.002239	32	75	17	17.1	2636	0.002636
20	63	17	17.8	2114	0.002114	32	75	11	14.7	2154	0.002154
20	63	11	15.8	1773	0.001773	32	90	32.5	26.2	4801	0.004801
20	75	32.5	25.2	3578	0.003578	32	90	26	25.5	4618	0.004618
20	75	26	24.6	3452	0.003452	32	90	21	24.7	4405	0.004405
20	75	21	23.9	3304	0.003304	32	90	17	23.7	4150	0.004150
20	75	17	23.1	3127	0.003127	32	90	11	20.8	3456	0.003456
20	75	11	20.7	2644	0.002644	32	110	32.5	35.6	7568	0.007568
20	90	32.5	32.2	5291	0.005291	32	110	26	34.8	7296	0.007296
20	90	26	31.5	5109	0.005109	32	110	21	33.8	6978	0.006978
20	90	21	30.7	4895	0.004895	32	110	17	32.5	6597	0.006597
20	90	17	29.7	4641	0.004641	32	110	11	29.0	5560	0.005560
20	90	11	26.3	3946	0.003946	40	63	32.5	9.6	1489	0.001489
25	40	32.5	6.3	616	0.000616	40	63	26	9.1	1400	0.001400
25	40	26	6.0	580	0.000580	40	63	21	8.5	1296	0.001296
25	40	21	5.6	538	0.000538	40	63	17	7.8	1171	0.001171
25	40	17	5.1	488	0.000488	40	63	11	5.8	830	0.000830
25	50	32.5	11.0	1239	0.001239	40	75	32.5	15.2	2635	0.002635
25	50	26	10.6	1183	0.001183	40	75	26	14.6	2509	0.002509
25	50	21	10.1	1117	0.001117	40	75	21	13.9	2361	0.002361
25	50	17	9.6	1038	0.001038	40	75	17	13.1	2184	0.002184
25	50	11	8.0	824	0.000824	40	75	11	10.7	1701	0.001701
25	63	32.5	17.1	2255	0.002255	40	90	32.5	22.2	4348	0.004348
25	63	26	16.6	2166	0.002166	40	90	26	21.5	4166	0.004166
25	63	21	16.0	2062	0.002062	40	90	21	20.7	3953	0.003953
25	63	17	15.3	1937	0.001937	40	90	17	19.7	3698	0.003698
25	63	11	13.3	1597	0.001597	40	90	11	16.8	3003	0.003003
25	75	32.5	22.7	3401	0.003401	40	110	32.5	31.6	7116	0.007116
25	75	26	22.1	3275	0.003275	40	110	26	30.8	6844	0.006844
25	75	21	21.4	3127	0.003127	40	110	21	29.8	6525	0.006525
25	75	17	20.6	2950	0.002950	40	110	17	28.5	6145	0.006145
25	75	11	18.2	2468	0.002468	40	110	11	25.0	5107	0.005107
25	90	32.5	29.7	5114	0.005114	40	125	32.5	38.7	9555	0.009555
25	90	26	29.0	4932	0.004932	40	125	26	37.7	9204	0.009204
25	90	21	28.2	4719	0.004719	40	125	21	36.5	8793	0.008793
25	90	17	27.2	4464	0.004464	40	125	17	35.1	8301	0.008301
25	90	11	24.3	3769	0.003769	40	125	11	31.1	6961	0.006961

(continued)

Table A-4 (cont.) Annulus and Annular Area Dimensional Data for IPS Primary Pipe Inside of ISO 160/1 Metric-Size Plastic SDR-Rated Secondary Containment Pipe

Nominal primary size (in.)	Nominal sec. cont. size (in.)	SDR No.	Nominal annulus	Nominal annular area (in.²)	Nominal annular area (ft²)	Nominal primary size (in.)	Nominal sec. cont. size (in.)	SDR No.	Nominal annulus	Nominal annular area (in.²)	Nominal annular area (ft²)
						(continued)					
50	75	32.5	10.2	1928	0.001928	75	110	11	7.5	1945	0.001945
50	75	26	9.6	1802	0.001802	75	125	32.5	21.2	6393	0.006393
50	75	21	8.9	1654	0.001654	75	125	26	20.2	6041	0.006041
50	75	17	8.1	1477	0.001477	75	125	21	19.0	5630	0.005630
50	75	11	5.7	994	0.000994	75	125	17	17.6	5138	0.005138
50	90	32.5	17.2	3641	0.003641	75	125	11	13.6	3799	0.003799
50	90	26	16.5	3459	0.003459	75	140	32.5	28.2	9143	0.009143
50	90	21	15.7	3245	0.003245	75	140	26	27.1	8702	0.008702
50	90	17	14.7	2991	0.002991	75	140	21	25.8	8187	0.008187
50	90	11	11.8	2296	0.002296	75	140	17	24.3	7570	0.007570
50	110	32.5	26.6	6409	0.006409	75	140	11	19.8	5889	0.005889
50	110	26	25.8	6136	0.006136	75	160	32.5	37.6	13295	0.013295
50	110	21	24.8	5818	0.005818	75	160	26	36.3	12719	0.12719
50	110	17	23.5	5437	0.005437	75	160	21	34.9	12046	0.012046
50	110	11	20.0	4400	0.004400	75	160	17	33.1	11240	0.011240
50	125	32.5	33.7	8848	0.008848	75	160	11	28.0	9045	0.009045
50	125	26	32.7	8496	0.008496	90	110	32.5	6.6	2009	0.002009
50	125	21	31.5	8085	0.008085	90	110	26	5.8	1736	0.001736
50	125	17	30.1	7594	0.007594	90	125	32.5	13.7	4448	0.004448
50	125	11	26.1	6254	0.006254	90	125	26	12.7	4096	0.004096
63	90	32.5	10.7	2487	0.002487	90	125	21	11.5	3685	00.3685
63	90	26	10.0	2304	0.002304	90	125	17	10.1	3194	0.003194
63	90	21	9.2	2091	0.002091	90	125	11	6.1	1854	0.001854
63	90	17	8.2	1836	0.001836	90	140	32.5	20.7	7199	0.007199
63	90	11	5.3	1142	0.001142	90	140	26	19.6	6758	0.006758
63	110	32.5	20.1	5255	0.005255	90	140	21	18.3	6242	0.006242
63	110	26	19.3	4982	0.004982	90	140	17	16.8	5625	0.005625
63	110	21	18.3	4664	0.004664	90	140	11	12.3	3945	0.003945
63	110	17	17.0	4283	0.004283	90	160	32.5	30.1	11351	0.011351
63	110	11	13.5	3246	0.003246	90	160	26	28.8	10775	0.010775
63	125	32.5	27.2	7694	0.007694	90	160	21	27.4	10101	0.010101
63	125	26	26.2	7342	0.007342	90	160	17	25.6	9296	0.009296
63	125	21	25.0	6931	0.006931	90	160	11	20.5	7101	0.007101
63	125	17	23.6	6440	0.006440	90	180	32.5	39.5	16056	0.016056
63	125	11	19.6	5100	0.005100	90	180	26	38.1	15327	0.015327
63	140	32.5	34.2	10444	0.010444	90	180	21	36.4	14475	0.014475
63	140	26	33.1	10003	0.010003	90	180	17	34.4	13455	0.013455
63	140	21	31.8	9488	0.009488	90	180	11	28.6	10677	0.010677
63	140	17	30.3	8871	0.008871	90	200	32.5	48.8	21315	0.021315
63	140	11	25.8	7191	0.007191	90	200	26	47.3	20415	0.020415
63	160	32.5	43.6	14596	0.014596	90	200	21	45.5	19363	0.019363
63	160	26	42.3	14020	0.014020	90	200	17	43.2	18104	0.018104
63	160	21	40.9	13347	0.013347	90	200	11	36.8	14675	0.014675
63	160	17	39.1	12541	0.012541	110	140	32.5	10.7	4056	0.004056
63	160	11	34.0	10346	0.010346	110	140	26	9.6	3615	0.003615
75	110	32.5	14.1	3953	0.003953	110	140	21	8.3	3099	0.003099
75	110	26	13.3	3681	0.003681	110	140	17	6.8	2482	0.002482
75	110	21	12.3	3363	0.003363	110	160	32.5	20.1	8208	0.008208
75	110	17	11.0	2982	0.002982	110	160	26	18.8	7632	0.007632
(continued)											

Table A-4 (cont.) Annulus and Annular Area Dimensional Data for IPS Primary Pipe Inside of ISO 160/1 Metric-Size Plastic SDR-Rated Secondary Containment Pipe

Nominal primary size (in.)	Nominal sec. cont. size (in.)	SDR No.	Nominal annulus	Nominal annular area (in.²)	Nominal annular area (ft²)	Nominal primary size (in.)	Nominal sec. cont. size (in.)	SDR No.	Nominal annulus	Nominal annular area (in.²)	Nominal annular area (ft²)
						(continued)					
110	160	21	17.4	6958	0.006958	140	200	21	20.5	10327	0.010327
110	160	17	15.6	6153	0.006153	140	200	17	18.2	9069	0.009069
110	160	11	10.5	3958	0.003958	140	200	11	11.8	5639	0.005639
110	180	32.5	29.5	12913	0.012913	140	225	32.5	35.6	19632	0.019632
110	180	26	28.1	12184	0.012184	140	225	26	33.8	18493	0.018493
110	180	21	26.4	11332	0.011332	140	225	21	31.8	17161	0.017161
110	180	17	24.4	10312	0.010312	140	225	17	29.3	15568	0.015568
110	180	11	18.6	7534	0.007534	140	225	11	22.0	11227	0.011227
110	200	32.5	38.8	18172	0.018172	140	250	32.5	47.3	27849	0.027849
110	200	26	37.3	17272	0.017272	140	250	26	45.4	26443	0.026443
110	200	21	35.5	16220	0.016220	140	250	21	43.1	24799	0.024799
110	200	17	33.2	14961	0.014961	140	250	17	40.3	22832	0.022832
110	200	11	26.8	11532	0.011532	140	250	11	32.3	17473	0.017473
110	225	32.5	50.6	25525	0.025525	160	200	32.5	13.8	7565	0.007565
110	225	26	48.8	24386	0.024386	160	200	26	12.3	6665	0.006665
110	225	21	46.8	23054	0.023054	160	200	21	10.5	5613	0.005613
110	225	17	44.3	21461	0.021461	160	200	17	8.2	4354	0.004354
110	225	11	37.0	17120	0.017120	160	225	32.5	25.6	14918	0.014918
125	160	32.5	12.6	5438	0.005438	160	225	26	23.8	13778	0.013778
125	160	26	11.3	4862	0.004862	160	225	21	21.8	12447	0.012447
125	160	21	9.9	4189	0.004189	160	225	17	19.3	10854	0.010854
125	160	17	8.1	3383	0.003383	160	225	11	12.0	6513	0.006513
125	180	32.5	22.0	10144	0.010144	160	250	32.5	37.3	23135	0.023135
125	180	26	20.6	9415	0.009415	160	250	26	35.4	21728	0.021728
125	180	21	18.9	8562	0.008562	160	250	21	33.1	20085	0.020085
125	180	17	16.9	7543	0.007543	160	250	17	30.3	18118	0.018118
125	180	11	11.1	4765	0.004765	160	250	11	22.3	12759	0.012759
125	200	32.5	31.3	15403	0.015403	160	280	32.5	51.4	34137	0.034137
125	200	26	29.8	14503	0.014503	160	280	26	49.2	32373	0.032373
125	200	21	28.0	13450	0.013450	160	280	21	46.7	30311	0.030311
125	200	17	25.7	12192	0.012192	160	280	17	43.5	27844	0.027844
125	200	11	19.3	8762	0.008762	160	280	11	34.5	21122	0.021122
125	225	32.5	43.1	22755	0.022755	180	225	32.5	15.6	9575	0.009575
125	225	26	41.3	21616	0.021616	180	225	26	13.8	8436	0.008436
125	225	21	39.3	20284	0.020284	180	225	21	11.8	7104	0.007104
125	225	17	36.8	18691	0.018691	180	225	17	9.3	5511	0.005511
125	225	11	29.5	14351	0.014351	180	250	32.5	27.3	17792	0.017792
125	250	32.5	54.8	30972	0.030972	180	250	26	25.4	16386	0.016386
125	250	26	52.9	29566	0.029566	180	250	21	23.1	14742	0.014742
125	250	21	50.6	27922	0.027922	180	250	17	20.3	12775	0.012775
125	250	17	47.8	25955	0.025955	180	250	11	12.3	7416	0.007416
125	250	11	39.8	20597	0.020597	180	280	32.5	41.4	28795	0.028795
140	160	32.5	5.1	2315	0.002315	180	280	26	39.2	27030	0.027030
140	180	32.5	14.5	7020	0.007020	180	280	21	36.7	24968	0.024968
140	180	26	13.1	6291	0.006291	180	280	17	33.5	22501	0.022501
140	180	21	11.4	5439	0.005439	180	280	11	24.5	15779	0.015779
140	180	17	9.4	4420	0.004420	180	315	32.5	57.8	43205	0.043205
140	200	32.5	23.8	12279	0.012279	180	315	26	55.4	40972	0.040972
140	200	26	22.3	11379	0.011379	180	315	21	52.5	38363	0.038363

(continued)

Table A-4 (cont.) Annulus and Annular Area Dimensional Data for IPS Primary Pipe Inside of ISO 160/1 Metric-Size Plastic SDR-Rated Secondary Containment Pipe

Nominal primary size (in.)	Nominal sec. cont. size (in.)	SDR No.	Nominal annulus	Nominal annular area (in.²)	Nominal annular area (ft²)	Nominal primary size (in.)	Nominal sec. cont. size (in.)	SDR No.	Nominal annulus	Nominal annular area (in.²)	Nominal annular area (ft²)
						(continued)					
180	315	17	49.0	35240	0.035240	280	355	17	16.6	15491	0.015491
180	315	11	38.9	26733	0.026733	280	400	32.5	47.7	49118	0.049118
180	355	32.5	76.6	61750	0.061750	280	400	26	44.6	45517	0.045517
180	355	26	73.8	58915	0.058915	280	400	21	41.0	41309	0.041309
180	355	21	70.6	55600	0.055600	280	400	17	36.5	36274	0.036274
180	355	17	66.6	51634	0.051634	280	400	11	23.6	22556	0.022556
180	355	11	55.2	40829	0.040829	315	400	32.5	30.2	32755	0.032755
200	280	32.5	31.4	22823	0.022823	315	400	26	27.1	29155	0.029155
200	280	26	29.2	21059	0.021059	315	400	21	23.5	24946	0.024946
200	280	21	26.7	18997	0.018997	315	400	17	19.0	19912	0.019912
200	280	17	23.5	16530	0.016530	315	400	11	6.1	6193	0.006193
200	280	11	14.5	9808	0.009808	315	450	32.5	53.7	62165	0.062165
200	280	32.5	31.4	22823	0.022823	315	450	26	50.2	57608	0.057608
200	280	26	29.2	21059	0.021059	315	450	21	46.1	52282	0.052282
200	280	21	26.7	18997	0.018997	315	450	17	41.0	45910	0.045910
200	280	17	23.5	16530	0.016530	315	450	11	26.6	28547	0.028547
200	280	11	14.5	9808	0.009808	355	400	32.5	10.2	11698	0.011698
200	315	32.5	47.8	37234	0.037234	355	450	32.5	33.7	41108	0.041108
200	315	26	45.4	35001	0.035001	355	450	26	30.2	36551	0.036551
200	315	21	42.5	32391	0.032391	355	450	21	26.1	31225	0.031225
200	315	17	39.0	29269	0.029269	355	450	17	21.0	24853	0.024853
200	315	11	28.9	20761	0.020761	355	500	32.5	57.1	73977	0.073977
200	355	32.5	66.6	55779	0.055779	355	500	26	53.3	68351	0.068351
200	355	26	63.8	52943	0.052943	355	500	21	48.7	61776	0.061776
200	355	21	60.6	49628	0.049628	355	500	17	43.1	53909	0.053909
200	355	17	56.6	45663	0.045663	400	450	32.5	11.2	14413	0.014413
200	355	11	45.2	34857	0.034857	400	500	32.5	34.6	47282	0.047282
250	315	32.5	22.8	19555	0.019555	400	500	26	30.8	41657	0.041657
250	315	26	20.4	17322	0.017322	400	500	21	26.2	35081	0.035081
250	315	21	17.5	14713	0.014713	400	500	17	20.6	27215	0.027215
250	315	17	14.0	11590	0.011590	400	560	32.5	62.8	91293	0.091293
250	355	32.5	41.6	38100	0.038100	400	560	26	58.5	84236	0.084236
250	355	26	38.8	35265	0.035265	400	560	21	53.3	75987	0.075987
250	355	21	35.6	31950	0.031950	400	560	17	47.1	66120	0.066120
250	355	17	31.6	27984	0.027984	450	560	32.5	37.8	57900	0.057900
250	355	11	20.2	17179	0.017179	450	560	26	33.5	50843	0.050843
250	400	32.5	62.7	61611	0.061611	450	560	21	28.3	42594	0.042594
250	400	26	59.6	58010	0.058010	450	560	17	22.1	32727	0.032727
250	400	21	56.0	53802	0.053802	450	630	32.5	70.6	115542	0.115542
250	400	17	51.5	48767	0.048767	450	630	26	65.8	106611	0.106611
250	400	11	38.6	35049	0.035049	450	630	21	60.0	96171	0.096171
250	400	32.5	62.7	61611	0.061611	450	630	17	52.9	83683	0.083683
250	400	26	59.6	58010	0.058010	500	560	32.5	12.8	20578	0.020578
250	400	21	56.0	53802	0.053802	500	630	32.5	45.6	78221	0.078221
250	400	17	51.5	48767	0.048767	500	630	26	40.8	69290	0.069290
250	400	11	38.6	35049	0.035049	500	630	21	35.0	58850	0.058850
280	355	32.5	26.6	25608	0.025608	500	630	17	27.9	46361	0.046361
280	355	26	23.8	22772	0.022772	500	710	32.5	83.2	152402	0.152402
280	355	21	20.6	19457	0.019457	500	710	26	77.7	141058	0.141058

(continued)

Table A-4 Annulus and Annular Area Dimensional Data for IPS Primary Pipe Inside of ISO 160/1 Metric-Size Plastic SDR-Rated Secondary Containment Pipe

Nominal primary size (in.)	Nominal sec. cont. size (in.)	SDR No.	Nominal annulus	Nominal annular area (in.^{2})	Nominal annular area (ft^2)
500	710	21	71.2	127799	0.127799
500	710	17	63.2	111937	0.111937
560	630	32.5	15.6	28249	0.028249
560	630	26	10.8	19318	0.019318
560	710	32.5	53.2	102430	0.102430
560	710	26	47.7	91087	0.091087
560	710	21	41.2	77828	0.077828
560	710	17	33.2	61966	0.061966
630	710	32.5	18.2	36980	0.036980
630	710	26	12.7	25637	0.025637
630	800	32.5	60.4	131021	0.131021
630	800	26	54.2	116620	0.116620
630	800	21	46.9	99786	0.099786
710	800	41	25.5	58916	0.058916
710	800	32.5	20.4	46793	0.046793
710	800	26	14.2	32391	0.032391
710	900	32.5	67.3	164430	0.164430
710	900	26	60.4	146204	0.146204
710	900	21	52.1	124898	0.124898
800	900	41	28.0	72995	0.072995
800	900	32.5	22.3	57652	0.057652
800	900	26	15.4	39425	0.039425
800	900	41	28.0	72995	0.072995
800	1000	32.5	69.2	189129	0.189129
800	1000	26	61.5	166627	0.166627
800	1000	21	52.4	140324	0.140324
900	1000	41	25.6	74500	0.074500
900	1000	32.5	19.2	55558	0.055558
900	1200	41	120.7	387309	0.387309
900	1200	32.5	113.1	360032	0.360032
900	1200	26	103.8	327629	0.327629
1000	1200	41	70.7	238023	0.238023
1000	1200	32.5	63.1	210746	0.210746
1000	1200	26	53.8	178343	0.178343
1000	1400	41	165.9	607706	0.607706
1000	1400	32.5	156.9	570579	0.570579
1200	1400	41	65.9	261992	0.261992
1200	1400	32.5	56.9	224865	0.224865
1200	1600	41	161.0	688549	0.688549
1200	1600	32.5	150.8	640057	0.640057
1400	1600	41	61.0	279978	0.279978
1400	1600	32.5	50.8	231485	0.231485

Table A-5 Minimum Outside Diameter Which Provides for a 1/4" (6.3mm) Annular Space (ANSI B36.19 Schedule Pipe Inside of ANSI B36.19 Schedule Pipe)

Inner pipe		Outer pipe									
Nom. size (in.)	Actual O.D. (in.)	Sched. 5S		Sched. 10S		Sched. 20S		Sched. 40S & XH		Sched. 80S	
		Size	A	Size	A	Size	A	Size	A	Size	A
3/8	0.675	1	0.26	1-1/4	.39	1-1/4	.38	1-1/4	.36	1-1/4	.30
1/2	0.840	1-1/4	.34	1-1/4	.30	1-1/4	.29	1-1/4	.27	1-1/2	.33
3/4	1.050	1.5	.36	1.5	.32	1.5	.29	1.5	.28	2	.44
1	1.315	2	.47	2	.42	2	.40	2	.38	2	.312
1-1/4	1.660	2	.29	2	.25	2.5	.44	2.5	.40	2.5	.33
1-1/2	1.900	2.5	.40	2.5	.37	2.5	.32	2.5	.28	3	.50
2	2.375	3	.47	3	.44	3	.40	3	.35	3	.26
2-1/2	2.875	3.5	.48	3.5	.44	3.5	.40	3.5	.34	4	.48
3	3.500	4	.42	4	.38	4	.34	4	.26	5	.66
3-1/2	4.000	5	.67	5	.65	5	.58	5	.52	5	.41
4	4.500	5	.42	5	.40	5	.33	5	.27	6	.63
5	5.563	6	.42	6	.40	6	.33	6	.25	8	1.03
6	6.625	8	.89	8	.85	8	.75	8	.68	8	.50
8	8.625	10	.93	10	.90	10	.81	10	.70	10	.56
10	10.750	12	.84	12	.82	12	.75	12	.63	12	.50
12	12.750	14	.47	14	.44	14	.31	14	.25	16	1.125
14	14.000	16	.84	16	.81	16	.69	16	.63	16	.50
16	16.000	18	.84	18	.81	18	.69	18	.63	18	.50
18	18.000	20	.81	20	.78	20	.63	20	.63	20	.50
20	20.000	22	.81	22	.78	22	.63	22	.63	22	.50
22	22.000	24	.78	24	.75	24	.63	24	.63	24	.50
24	24.000	30	2.75	30	2.69	30	2.63	30	2.63	30	2.50

Table A-6 Minimum Outside Diameter Which Provides for a 1/2" (12.7mm) Annular Space (ANSI B36.19 Schedule Pipe Inside of ANSI B36.19 Schedule Pipe)

Inner pipe size (in.)	Nom. O.D. (in.)	Outer pipe									
		Sched. 5S		Sched. 10S		Sched. 20S		Sched. 40S		Sched. 80S	
		Size	A	Size	A	Size	A	Size	A	Size	A
3/8	0.675	1-1/2	.55	2	.74	2	.72	2	.70	2	.63
1/2	0.840	2	.71	2	.66	2	.63	2	.61	2	.55
3/4	1.050	2	.60	2	.55	2	.53	2	.51	2.5	.64
1	1.315	2.5	.70	2.5	.66	2.5	.62	2.5	.58	2.5	.50
1-1/4	1.660	2.5	.52	3	.80	3	.76	3	.70	3	.62
1-1/2	1.900	3	.72	3	.68	3	.64	3	.58	3	.50
2	2.375	3.5	.73	3.5	.69	3.5	.65	3.5	.59	4	.73
2-1/2	2.875	4	.73	4	.69	4	.65	4	.58	5	.97
3	3.500	5	.92	5	.90	5	.83	5	.77	5	.66
3-1/2	4.000	5	.67	5	.65	5	.58	5	.52	6	.88
4	4.500	6	.95	6	.93	6	.86	6	.78	6	.63
5	5.563	8	1.42	8	1.38	8	1.28	8	1.21	8	1.03
6	6.625	8	.89	8	.85	8	.75	8	.68	8	.50
8	8.625	10	.93	10	.90	10	.81	10	.70	10	.56
10	10.750	12	.84	12	.82	12	.75	12	.63	12	.50
12	12.750	16	1.46	16	1.44	16	1.31	16	1.25	16	1.125
14	14.000	16	.84	16	.81	16	.69	16	.63	16	.50
16	16.000	18	.84	18	.81	18	.69	18	.63	18	.50
18	18.000	20	.81	20	.78	20	.63	20	.63	20	.50
20	20.000	22	.81	22	.78	22	.63	22	.63	22	.50
22	22.000	24	.78	24	.75	24	.63	24	.63	24	.50
24	24.000	30	2.75	30	2.69	30	2.63	30	2.63	30	2.50

Table A-7 Minimum Outside Diameter Which Provides for a 3/4" (19.1mm) Annular Space (ANSI B36.19 Schedule Pipe Inside of ANSI B36.19 Schedule Pipe)

Inner pipe size (in.)	Nom. O.D. (in.)	Outer pipe									
		Sched. 5S		Sched. 10S		Sched. 20S		Sched. 40S		Sched. 80S	
		Size	A	Size	A	Size	A	Size	A	Size	A
3/8	0.675	2	.79	2.5	.98	2.5	.94	2.5	.90	2.5	.82
1/2	0.840	2.5	.93	2.5	.90	2.5	.85	2.5	.81	3	1.03
3/4	1.050	2.5	.83	2.5	.79	2.5	.75	3	1.01	3	.93
1	1.315	3	1.01	3	.98	3	.963	3	.88	3	.79
1-1/4	1.660	3	.84	3	.80	3	.76	3.5	.93	3.5	.83
1-1/2	1.900	3.5	.97	3.5	.93	3.5	.89	3.5	.82	4	.96
2	2.375	4	.98	4	.94	4	.90	4	.83	5	1.22
2-1/2	2.875	5	1.24	5	1.21	5	1.14	5	1.09	5	.97
3	3.500	5	.92	5	.90	5	.83	5	.77	6	1.13
3-1/2	4.000	6	1.20	6	1.18	6	1.11	6	1.03	6	.88
4	4.500	6	.95	6	.93	6	.86	6	.78	8	1.56
5	5.563	8	1.42	8	1.38	8	1.28	8	1.21	8	1.03
6	6.625	8	.89	8	.85	8	.75	10	1.70	10	1.56
8	8.625	10	.93	10	.90	10	.81	12	1.69	12	1.56
10	10.750	12	.84	12	.82	12	.75	14	1.25	14	1.13
12	12.750	16	1.46	16	1.44	16	1.31	16	1.25	16	1.125
14	14.000	16	.84	16	.81	18	1.69	18	1.63	18	1.50
16	16.000	18	.84	18	.81	20	1.69	20	1.63	20	1.5
18	18.000	20	.81	20	.78	22	1.63	22	1.63	22	1.5
20	20.000	22	.81	22	.78	24	1.63	24	1.63	24	1.5
22	22.000	24	.78	24	.75	30	3.63	30	3.63	30	3.5
24	24.000	30	2.75	30	2.69	30	2.63	30	2.63	30	2.50

Table A-8 Minimum Outside Diameter Which Provides for a 1" (25.4mm) Annular Space (ANSI B36.19 Schedule Pipe Inside of ANSI B36.19 Schedule Pipe)

Inner pipe size (in.)	Nom. O.D. (in.)	Outer pipe									
		Sched. 5S		Sched. 10S		Sched. 20S		Sched. 40S		Sched. 80S	
		Size	A	Size	A	Size	A	Size	A	Size	A
3/8	0.675	2.5	1.02	3	1.29	3	1.25	3	1.20	3	1.11
1/2	0.840	3	1.25	3	1.21	3	1.17	3	1.11	3	1.03
3/4	1.050	3	1.14	3	1.11	3	1.06	3.5	1.24	3.5	1.14
1	1.315	3	1.01	3.5	1.22	3.5	1.18	3.5	1.12	3.5	1.02
1-1/4	1.660	3.5	1.09	3.5	1.05	3.5	1.01	4	1.18	4	1.08
1-1/2	1.900	4	1.22	4	1.18	4	.89	4	1.06	4	1.45
2	2.375	5	1.49	5	1.46	5	1.39	5	1.34	5	1.22
2-1/2	2.875	5	1.24	5	1.21	5	1.14	5	1.09	6	1.44
3	3.500	6	1.45	6	1.43	6	1.36	6	1.28	6	1.13
3-1/2	4.000	6	1.20	6	1.18	6	1.11	6	1.03	8	1.81
4	4.500	8	1.95	8	1.91	8	1.81	8	1.74	8	1.56
5	5.563	8	1.42	8	1.38	8	1.28	8	1.21	8	1.03
6	6.625	10	1.93	10	1.90	10	1.81	10	1.70	10	1.56
8	8.625	12	1.91	12	1.88	12	1.81	12	1.69	12	1.56
10	10.750	14	1.47	14	1.44	14	1.31	14	1.25	14	1.13
12	12.750	16	1.46	16	1.44	16	1.31	16	1.25	16	1.125
14	14.000	18	1.84	18	1.81	18	1.69	18	1.63	18	1.50
16	16.000	20	1.81	20	1.78	20	1.63	20	1.63	20	1.50
18	18.000	22	1.81	22	1.78	22	1.63	22	1.63	22	1.50
20	20.000	24	1.78	24	1.75	24	1.63	24	1.63	24	1.50
22	22.000	30	3.75	30	3.69	30	3.63	30	3.63	30	3.50
24	24.000	30	2.75	30	2.69	30	2.63	30	2.63	30	2.50

Table A-9 Minimum Outside Diameter Which Provides for a 1/4" (6.4mm) Annular Space (ANSI B36.10 or B36.19 Schedule Pipe Inside of ASTM F 714 IPS Sizing System OD Plastic Pipe)

Inner pipe size (in.)	Nom. O.D. (in.)	Outer pipe									
		SDR 32.5		SDR 21		SDR 17		SDR 13.5		SDR 11	
		Size	A	Size	A	Size	A	Size	A	Size	A
1	1.315	3	.9845	3	.9255	3	.8865	3	.8335	3	.7745
1-1/4	1.660	3	.812	3	.753	3	.714	3	.661	3	.602
1-1/2	1.900	3	.692	3	.633	3	.594	3	.541	3	.482
2	2.375	3	.4545	3	.3955	3	.3565	3	.3035	4	.6535
2-1/2	2.875	4	.6745	4	.5985	4	.5475	4	.4795	4	.4035
3	3.500	4	.362	4	.286	5	.7045	5	.6195	5	.5255
3-1/2	4.000	5	.6105	5	.5165	5	.4545	5	.3695	5	.2755
4	4.500	5	.3605	5	.2665	6	.6725	6	.5715	6	.4605
5	5.563	6	.327	7	.441	7	.361	7	.253	8	.747
6	6.625	8	.735	8	.589	8	.493	8	.361	10	1.0855
8	8.625	10	.7315	10	.5505	10	.4305	10	.2665	12	.9035
10	10.750	12	.608	12	.393	12	.25	13	.3215	14	.352
12	12.750	16	1.133	16	.863	16	.684	16	.44	18	.989
14	14.000	16	.508	18	1.143	18	.941	18	.667	18	.364
16	16.000	18	.446	20	1.048	20	.824	20	.519	22	1.00
18	18.000	20	.385	22	.952	22	.706	22	.37	24	.818
20	20.000	22	.323	24	.857	24	.588	26	1.074	26	.636
22	22.000	24	.262	26	.762	26	.471	28	.926	28	.455
24	24.000	28	1.138	28	.667	28	.353	30	.778	30	.273

Table A-10 Minimum Outside Diameter Which Provides for a 1/2" (12.7mm) Annular Space (ANSI B36.10 or B36.19 Schedule Pipe Inside of ASTM F 714 IPS Sizing System OD Plastic Pipe)

Inner pipe size (in.)	Nom. O.D. (in.)	Outer pipe									
		SDR 32.5		SDR 21		SDR 17		SDR 13.5		SDR 11	
		Size	A	Size	A	Size	A	Size	A	Size	A
1	1.315	3	.9845	3	.9255	3	.8865	3	.8835	3	.7745
1-1/4	1.660	3	.812	3	.753	3	.714	3	.661	3	.602
1-1/2	1.900	3	.692	3	.633	3	.594	3	.541	4	.891
2	2.375	4	.9245	4	.8485	4	.7975	4	.7295	4	.6535
2-1/2	2.875	4	.6745	4	.5985	4	.5475	5	.932	5	.838
3	3.500	5	.8605	5	.7665	5	.7045	5	.6195	5	.5255
3-1/2	4.000	5	.6105	5	.5165	6	.9225	6	.8215	6	.7105
4	4.500	6	.8585	6	.7475	6	.6725	6	.5715	7	.6645
5	5.563	7	.562	8	1.12	8	1.024	8	.892	8	.747
6	6.625	8	.735	8	.589	10	1.4305	10	1.2665	10	1.0855
8	8.625	10	.7315	10	.5505	12	1.3125	12	1.1185	12	.9035
10	10.750	12	.608	13	.6755	13	.5255	14	.588	16	1.17
12	12.750	16	1.133	16	.863	16	.684	18	1.292	18	.989
14	14.000	16	.508	18	1.143	18	.941	18	.667	20	1.182
16	16.000	20	1.385	20	1.048	20	.824	20	.519	22	1.000
18	18.000	22	1.323	22	.952	22	.706	24	1.222	24	.818
20	20.000	24	1.262	24	.857	24	.588	26	1.074	26	.636
22	22.000	26	1.2	26	.762	28	1.353	28	.926	30	1.273
24	24.000	28	1.138	28	.667	30	1.235	30	.778	32	1.091

Table A-11 Minimum Outside Diameter Which Provides for a 3/4" (19.1mm) Annular Space (ANSI B36.10 or B36.19 Schedule Pipe Inside of ASTM F 714 IPS Sizing System OD Plastic Pipe)

Inner pipe size (in.)	Nom. O.D. (in.)	Outer pipe									
		SDR 32.5		SDR 21		SDR 17		SDR 13.5		SDR 11	
		Size	A	Size	A	Size	A	Size	A	Size	A
1	1.315	3	.9845	3	.9255	3	.8865	3	.8835	3	.7745
1-1/4	1.660	3	.812	3	.753	4	1.155	4	1.087	4	1.011
1-1/2	1.900	4	1.162	4	1.086	4	1.035	4	.967	4	.891
2	2.375	4	.9245	4	.8485	4	.7975	5	1.182	5	1.088
2-1/2	2.875	5	1.173	5	1.079	5	1.017	5	.932	5	.838
3	3.500	5	.8605	5	.7665	6	1.1725	6	1.0715	6	.9605
3-1/2	4.000	6	1.1085	6	.9975	6	.9225	6	.8215	7	.9145
4	4.500	6	.8585	7	.9725	7	.8925	7	.7845	8	1.2785
5	5.563	8	1.266	8	1.12	8	1.024	8	.892	10	1.6165
6	6.625	10	1.7315	10	1.5505	10	1.4305	10	1.2665	10	1.0855
8	8.625	12	1.6705	12	1.4555	12	1.3125	12	1.1185	12	.9035
10	10.750	13	.9005	14	.958	14	.801	16	1.44	16	1.17
12	12.750	16	1.133	16	.863	18	1.566	18	1.292	18	.989
14	14.000	18	1.446	18	1.143	18	.941	20	1.519	20	1.182
16	16.000	20	1.385	20	1.048	20	.824	22	1.37	22	1.00
18	18.000	22	1.323	22	.952	24	1.588	24	1.222	24	.818
20	20.000	24	1.262	24	.857	26	1.471	26	1.074	28	1.455
22	22.000	26	1.2	26	.762	28	1.353	28	.926	30	1.273
24	24.000	28	1.138	30	1.571	30	1.235	30	.778	32	1.091

Table A-12 Minimum Outside Diameter Which Provides for a 1" (25.4mm) Annular Space (ANSI B36.10 or B36.19 Schedule Pipe Inside of ASTM F 714 IPS Sizing System OD Plastic Pipe)

Inner pipe size (in.)	Nom. O.D. (in.)	Outer pipe									
		SDR 32.5		SDR 21		SDR 17		SDR 13.5		SDR 11	
		Size	A	Size	A	Size	A	Size	A	Size	A
1	1.315	4	1.4545	4	1.3785	4	1.3275	4	1.2595	4	1.1835
1-1/4	1.660	4	1.282	4	1.206	4	1.155	4	1.087	4	1.011
1-1/2	1.900	4	1.162	4	1.086	4	1.035	5	1.4195	5	1.3255
2	2.375	5	1.423	5	1.329	5	1.267	5	1.182	5	1.088
2-1/2	2.875	5	1.173	5	1.079	5	1.017	6	1.384	6	1.273
3	3.500	6	1.3585	6	1.2475	6	1.1725	6	1.0715	7	1.1645
3-1/2	4.000	6	1.1085	7	1.2225	7	1.1425	7	1.0345	8	1.5285
4	4.500	8	1.7975	8	1.6515	8	1.555	8	1.4235	8	1.2785
5	5.563	8	1.266	8	1.12	8	1.024	10	1.7975	10	1.6165
6	6.625	10	1.7315	10	1.5505	10	1.4305	10	1.2665	10	1.0855
8	8.625	12	1.6705	12	1.4555	12	1.3125	12	1.1185	13	1.159
10	10.750	14	1.194	16	1.863	16	1.684	16	1.44	16	1.17
12	12.750	16	1.133	18	1.768	18	1.566	18	1.292	20	1.807
14	14.000	18	1.446	18	1.143	20	1.824	20	1.519	20	1.182
16	16.000	20	1.385	20	1.048	22	1.706	22	1.37	22	1.000
18	18.000	22	1.323	24	1.857	24	1.588	24	1.222	26	1.636
20	20.000	24	1.262	26	1.762	26	1.471	26	1.074	28	1.455
22	22.000	26	1.2	28	1.667	28	1.353	30	1.778	30	1.273
24	24.000	28	1.138	30	1.571	30	1.235	32	1.63	32	1.091

Table A-13 Minimum Outside Diameter Which Provides for a 1/4" (6.4mm) Annular Space (ANSI B36.19 Schedule Pipe Inside of ISO 161/1 SDR Plastic Pipe)

Inner pipe size (in.)	Nom. O.D. (in.)	Conversion to mm	Outer pipe							
			SDR 32.5		SDR 21		SDR 17		SDR 11	
			Size (mm)	A (mm)	Size (mm)	A (mm)	Size (mm)	A (mm)	Size (mm)	A (mm)
1	1.315	33.401	N/A	N/A	90	23.99	90	22.99	90	20.096
1-1/4	1.660	42.164	N/A	N/A	90	19.62	90	18.62	90	15.718
1-1/2	1.900	48.26	N/A	N/A	90	16.58	90	15.58	90	12.67
2	2.375	60.325	N/A	N/A	90	10.538	90	9.548	90	6.64
2-1/2	2.875	73.025	110	15.086	110	13.288	110	11.988	110	8.488
3	3.500	88.9	110	7.15	160	27.95	160	26.15	160	21.05
3-1/2	4.000	101.6	160	24.3	160	21.6	160	19.8	160	14.7
4	4.500	114.3	160	17.95	160	15.25	160	13.45	160	8.35
5	5.563	141.3002	200	23.15	200	19.85	200	17.55	200	11.15
6	6.625	168.275	200	9.663	250	28.963	250	26.16	250	18.163
8	8.625	219.075	250	7.763	280	17.163	280	13.97	315	38.725
10	10.750	273.05	315	11.275	355	24.075	355	20.08	355	2.675
12	12.750	323.85	400	25.775	400	19.075	400	14.58	N/A	N/A
14	14.000	355.6	400	9.9	450	25.8	450	20.7	N/A	N/A
16	16.000	406.4	450	8.0	500	23	500	17.4	N/A	N/A
18	18.000	457.2	560	34.2	560	24.7	560	18.5	N/A	N/A
20	20.000	508	560	8.8	630	31	630	25.4	N/A	N/A
22	22.000	558.8	630	16.2	710	41.8	710	33.8	N/A	N/A
24	24.000	609.6	710	28.4	710	16.4	710	8.4	N/A	N/A

Table A-14 Minimum Outside Diameter Which Provides for a 1/2" (12.7mm) Annular Space (ANSI B36.19 Schedule Pipe Inside of ISO 161/1 SDR Plastic Pipe)

Inner pipe size (in.)	Nom. O.D. (in.)	Conversion to mm	Outer pipe							
			SDR 32.5		SDR 21		SDR 17		SDR 11	
			Size	A	Size	A	Size	A	Size	A
1	1.315	33.401	N/A	N/A	90	23.995	90	22.995	90	20.09
1-1/4	1.660	42.164	N/A	N/A	90	19.618	90	18.618	90	15.718
1-1/2	1.900	48.26	N/A	N/A	90	16.57	90	15.57	110	20.87
2	2.375	60.325	110	21.436	110	19.6375	110	18.338	110	14.8375
2-1/2	2.875	73.025	110	15.088	110	13.2875	160	34.088	160	28.9875
3	3.500	88.9	160	30.65	160	27.95	160	26.15	160	21.05
3-1/2	4.000	101.6	160	24.3	160	21.6	160	19.8	160	14.7
4	4.500	114.3	160	17.95	160	15.25	160	13.45	200	24.65
5	5.563	141.3	200	23.15	200	19.849	200	17.5499	250	31.6499
6	6.625	168.28	250	33.163	250	28.9625	250	26.1625	250	18.1625
8	8.625	219.075	280	21.863	280	17.1625	280	13.9625	315	38.725
10	10.750	273.05	355	30.075	355	24.075	355	20.075	400	27.075
12	12.750	323.85	400	25.775	400	19.075	400	14.575	N/A	N/A
14	14.000	355.6	450	33.4	450	25.8	450	20.7	N/A	N/A
16	16.000	406.4	500	23	500	23	500	17.4	N/A	N/A
18	18.000	457.2	560	34.2	560	24.7	560	18.5	N/A	N/A
20	20.000	508	630	41.6	630	31	630	25.4	N/A	N/A
22	22.000	558.8	630	16.2	710	41.8	710	33.8	N/A	N/A
24	24.000	609.6	710	28.4	710	16.4	800	48.1	N/A	N/A

Table A-15 Minimum Outside Diameter Which Provides for a 3/4" (19.1mm) Annular Space (ANSI B36.19 Schedule Pipe Inside of ISO 161/1 SDR Plastic Pipe)

Inner pipe size (in.)	Nom. O.D. (in.)	Conversion to mm	Outer pipe							
			SDR 32.5		SDR 21		SDR 17		SDR 11	
			Size	A	Size	A	Size	A	Size	A
1	1.315	33.401	N/A	N/A	90	23.99	90	22.99	90	20.09
1-1/4	1.660	42.164	N/A	N/A	90	19.62	110	27.418	110	23.918
1-1/2	1.900	48.26	110	27.47	110	25.67	110	24.37	110	20.87
2	2.375	60.325	110	21.438	110	19.64	160	40.438	160	35.338
2-1/2	2.875	73.025	160	38.59	160	35.89	160	34.088	160	28.99
3	3.500	88.9	160	30.65	160	27.95	160	26.15	160	21.05
3-1/2	4.000	101.6	160	24.3	160	21.6	160	19.8	200	31
4	4.500	114.3	200	36.65	200	33.35	200	31.05	200	24.65
5	5.563	141.3	200	23.15	200	19.85	250	39.65	250	31.65
6	6.625	168.26	250	33.163	250	28.96	250	26.163	280	30.363
8	8.625	219.06	280	21.863	315	32.97	315	29.463	315	19.363
10	10.750	273.05	355	30.08	355	24.08	355	20.075	400	27.075
12	12.750	323.85	400	25.76	450	41.68	450	36.575	N/A	N/A
14	14.000	355.6	450	33.4	450	25.8	450	20.7	N/A	N/A
16	16.000	406.4	500	31.4	500	23	560	43.9	N/A	N/A
18	18.000	457.2	560	34.2	560	24.7	630	50.8	N/A	N/A
20	20.000	508	630	41.6	630	31	630	25.4	N/A	N/A
22	22.000	558.8	710	53.8	710	41.8	710	33.8	N/A	N/A
24	24.000	609.6	710	28.4	800	57.1	800	48.1	N/A	N/A

Appendix A

Table A-16 Minimum Outside Diameter Which Provides for a 1" (25.4mm) Annular Space (ANSI B36.19 Schedule Pipe Inside of ISO 161/1 SDR Plastic Pipe)

Inner pipe size (in.)	Nom. O.D. (in.)	Conversion to mm	Outer pipe							
			SDR 32.5		SDR 21		SDR 17		SDR 11	
			Size	A	Size	A	Size	A	Size	A
1	1.315	33.401	N/A	N/A	110	33.099	110	31.799	110	28.300
1-1/4	1.660	42.164	110	30.518	110	28.718	110	27.418	160	44.418
1-1/2	1.900	48.26	110	27.47	160	48.27	160	46.47	160	41.37
2	2.375	60.325	160	44.938	160	42.238	160	40.438	160	35.338
2-1/2	2.875	73.025	160	38.588	160	35.888	160	34.088	160	28.988
3	3.500	88.9	160	30.65	160	27.95	160	26.15	200	37.35
3-1/2	4.000	101.6	200	43	200	39.7	200	37.4	200	31
4	4.500	114.3	200	36.65	200	33.35	200	31.05	250	45.15
5	5.563	141.3002	250	46.650	250	42.45	250	39.65	250	31.649
6	6.625	168.275	250	33.163	250	28.97	250	26.163	280	30.363
8	8.625	219.075	280	43.725	315	32.97	315	29.463	355	35.663
10	10.750	273.05	355	30.075	400	44.475	400	39.98	400	27.075
12	12.750	323.85	400	25.775	450	41.675	450	36.58	N/A	N/A
14	14.000	355.6	450	33.4	450	25.8	500	42.8	N/A	N/A
16	16.000	406.4	500	31.4	560	50.1	560	43.9	N/A	N/A
18	18.000	457.2	560	34.2	630	56.4	630	50.8	N/A	N/A
20	20.000	508	630	41.6	630	31	630	25.4	N/A	N/A
22	22.000	558.8	710	53.8	710	41.8	710	33.8	N/A	N/A
24	24.000	609.6	710	28.4	800	57.1	800	48.1	N/A	N/A

Table A-17 Minimum Outside Diameter Which Provides for a 1/4" (6.4mm) Annular Space (ISO 161/1 Sizing System SDR Plastic "Metric" Pipe Inside of ISO 161/1 Sizing System SDR Plastic "Metric" Pipe)

Inner pipe size		Nom. O.D.		Outer pipe size									
				SDR 32.5		SDR 26		SDR 21		SDR 17		SDR 11	
(mm)	(in)	(mm)	(in)	Size (mm)	A (mm)	Size (mm)	A (mm)	Size (mm)	A (mm)	Size (mm)	A (mm)	Size (mm)	A (mm)
20	0.79	20	0.79	110*	41.6	90*	31.5	90*	30.7	63	17.8	50	10.5
25	0.98	25	0.98	110*	39.1	90*	29.0	90*	28.2	63	15.3	50	8.0
32	1.26	32	1.26	110*	35.6	90*	25.5	90*	24.7	63	11.8	63	9.8
40	1.57	40	1.57	110*	31.6	90*	21.5	90*	20.7	63**	7.8	75	10.7
50	1.97	50	1.97	110*	26.6	90*	16.5	90*	15.7	75**	8.1	90	11.8
63	2.48	63	2.48	110*	20.1	90	10.0	90	9.2	90**	8.2	110	13.5
75	2.95	75	2.95	110*	14.1	110	13.3	110	12.3	110	11.0	110*	7.5
90	3.54	90	3.54	110**	6.6	125	12.7	125	11.6	125	10.1	140	12.3
110	4.33	110	4.33	140	10.7	140*	9.6	140*	8.3	140**	6.7	160	10.5
125	4.92	125	4.92	160	12.6	160	11.3	160	9.9	160**	8.1	180	11.1
140	5.51	140	5.51	180	14.5	180	13.1	180	11.4	180	9.4	200	11.8
160	6.30	160	6.30	200	13.8	200	12.3	200	10.5	200	8.2	225	12.1
180	7.09	180	7.09	225	15.6	225	13.9	225	11.8	225**	9.3	250	12.3
200	7.87	200	7.87	250	17.3	250	15.4	250	13.1	250	10.3	280	14.5
225	8.86	225	8.86	280	18.9	280	16.7	280	14.2	280	11.0	315	16.4
250	9.84	250	9.84	315	22.8	315	20.4	315	17.5	315	14.0	355	20.2
280	11.02	280	11.02	315**	7.8	355	23.8	355	20.6	355	16.6	400	23.6
315	12.4	315	12.4	355	9.1	400	27.1	400	23.5	400	19.0	450	26.6
355	13.98	355	13.98	400	10.2	450	30.2	450	26.1	450	21.0	---	---
400	15.75	400	15.75	450	11.2	500	30.8	500	26.2	500	20.6	---	---
450	17.72	450	17.72	500	9.6	560	33.5	560	28.3	560	22.1	---	---
500	19.69	500	19.69	560	12.8	560**	8.5	630	35.0	630	27.9	---	---
560	22.05	560	22.05	630	15.6	630	10.8	710	41.2	710	33.2	---	---
630	24.80	630	24.80	710	18.2	710	12.7	800	46.9	800	37.9	---	---

Table A-18 Minimum Outside Diameter Which Provides for a 1/2" (12.7mm) Annular Space (ISO 161/1 Sizing System SDR Pipe Inside of ISO 161/1 Sizing System SDR Plastic Piping)

Inner pipe size		Nom. O.D.		Outer pipe size									
				SDR 32.5		SDR 26		SDR 21		SDR 17		SDR 11	
(mm)	(in)	(mm)	(in)	Size (mm)	A (mm)	Size (mm)	A (mm)	Size (mm)	A (mm)	Size (mm)	A (mm)	Size (mm)	A (mm)
20	0.79	20	0.79	110*	41.6	90*	31.5	90*	30.7	63	17.8	63	15.8
25	0.98	25	0.98	110*	39.1	90*	29.0	90*	28.2	63	15.3	63	13.3
32	1.26	32	1.26	110*	35.6	90*	25.5	90*	24.7	75	17.1	75	14.7
40	1.57	40	1.57	110*	31.6	90*	21.5	90*	20.7	75	13.1	90	16.8
50	1.97	50	1.97	110*	26.6	90	16.5	90	15.7	90	14.7	110	20.0
63	2.48	63	2.48	110	20.1	110	19.3	110	18.3	110	17.0	110	13.5
75	2.95	75	2.95	110	14.1	125	20.2	125	19.1	125	17.7	125	13.6
90	3.54	90	3.54	140	20.7	140	19.6	140	18.3	140	16.8	160	20.5
110	4.33	110	4.33	160	20.1	160	18.8	160	17.4	160	15.6	180	18.6
125	4.92	125	4.92	180	22.0	180	20.6	180	18.9	180	16.9	200	19.3
140	5.51	140	5.51	180	14.5	200	22.3	200	20.5	200	18.2	225	22.1
160	6.30	160	6.30	200	13.8	225	23.9	225	21.8	225	19.3	250	22.3
180	7.09	180	7.09	250	27.3	250	25.4	250	23.1	250	20.3	280	21.4
200	7.87	200	7.87	250	17.3	250	15.4	250	13.2	280	23.5	280	14.5
225	8.86	225	8.86	280	18.9	280	16.7	280	14.2	315	26.5	315	16.4
250	9.84	250	9.84	315	22.8	315	20.4	315	17.5	315	14.0	355	20.2
280	11.02	280	11.02	355	26.6	355	23.8	355	20.6	355	16.6	400	23.6
315	12.4	315	12.4	400	30.2	400	27.1	400	23.5	400	19.0	450	26.6
355	13.98	355	13.98	450	33.7	450	30.2	450	26.1	450	21.0	---	---
400	15.75	400	15.75	500	34.6	500	30.8	500	26.2	500	20.6	---	---
450	17.72	450	17.72	560	37.8	560	33.5	560	28.3	560	22.1	---	---
500	19.69	500	19.69	630	45.6	630	40.8	630	35.0	630	27.9	---	---
560	22.05	560	22.05	630	15.6	710	47.7	710	41.2	710	33.2	---	---
630	24.80	630	24.80	710	18.2	710	12.7	800	46.9	800	37.9	---	---

Table A-19 Minimum Outside Diameter Which Provides for a 3/4" (19.1mm) Annular Space (ISO 161/1 Sizing System SDR Plastic "Metric" Pipe Inside of ISO 161/1 Sizing System SDR Plastic "Metric" Pipe)

Inner pipe size		Nom. O.D.		Outer pipe size									
				SDR 32.5		SDR 26		SDR 21		SDR 17		SDR 11	
(mm)	(in)	(mm)	(in)	Size (mm)	A (mm)	Size (mm)	A (mm)	Size (mm)	A (mm)	Size (mm)	A (mm)	Size (mm)	A (mm)
20	0.79	20	0.79	110*	41.6	90*	31.5	90	30.7	75	23.1	75	20.7
25	0.98	25	0.98	110*	39.1	90*	29.0	90	28.2	75	20.6	90	24.3
32	1.26	32	1.26	110*	35.6	90	25.5	90	24.7	90	23.7	90	20.8
40	1.57	40	1.57	110*	31.6	90	21.5	90	20.7	90	19.7	110	26.8
50	1.97	50	1.97	110	26.6	110	25.8	110	24.8	110	23.5	110**	20.0**
63	2.48	63	2.48	110**	20.1**	110**	19.3**	125	25.1	125	23.7	125**	19.6**
75	2.95	75	2.95	125**	21.2**	125	20.2	125**	19.1**	140	24.3	140	19.8
90	3.54	90	3.54	140**	20.7**	140**	19.6**	160	27.4	160	25.6	160	20.5
110	4.33	110	4.33	160**	20.1**	180	28.1	180	26.4	180	24.4	200	26.8
125	4.92	125	4.92	180	22.0	180	20.6	200	28.0	200	25.7	200	19.3
140	5.51	140	5.51	200	23.8	200**	22.3**	200**	20.5**	225	29.3	225	22.1
160	6.30	160	6.30	225	25.6	225	23.9	225**	21.8**	225**	19.3*	250	22.3
180	7.09	180	7.09	250	27.3	250	25.4	250	23.1	250**	20.3**	280	24.5
200	7.87	200	7.87	280	31.4	280	29.2	280	26.7	280	23.5	315	28.9
225	8.86	225	8.86	315	35.3	315	32.9	315	30.0	315	26.5	355	32.7
250	9.84	250	9.84	315	22.8	315	20.4	355	35.6	355	31.6	355**	20.2**
280	11.02	280	11.02	355	26.6	355	23.8	355**	20.6**	400	36.5	280	23.6
315	12.4	315	12.4	400	30.2	400	27.1	400	23.5	450	41.0	---	---
355	13.98	355	13.98	450	33.7	450	30.2	450	26.1	450**	21.0**	---	---
400	15.75	400	15.75	500	34.6	500	30.8	500	26.2	500**	20.6**	---	---
450	17.72	450	17.72	560	37.8	560	33.5	560	28.3	560**	22.1**	---	---
500	19.69	500	19.69	630	45.6	630	40.8	630	35.0	630	27.9	---	---
560	22.05	560	22.05	710	53.2	710	47.7	710	41.2	710	33.2	---	---
630	24.80	630	24.80	800	60.4	800	54.2	800	46.9	800	37.9	---	---

Table A-20 Minimum Outside Diameter Which Provides for a 1" (25.4mm) Annular Space (ISO 161/1 Sizing System SDR Pipe Inside of ISO 161/1 Sizing System SDR Plastic Piping)

Inner pipe size		Nom. O.D.		Outer pipe size									
				SDR 32.5		SDR 26		SDR 21		SDR 17		SDR 11	
(mm)	(in)	(mm)	(in)	Size (mm)	A (mm)	Size (mm)	A (mm)	Size (mm)	A (mm)	Size (mm)	A (mm)	Size (mm)	A (mm)
20	0.79	20	0.79	110*	41.6	90	31.5	90	30.7	90	29.7	90	26.8
25	0.98	25	0.98	110*	39.1	90	29.0	90	28.2	90	27.2	110	32.5
32	1.26	32	1.26	110*	35.6	90	25.5	110	33.8	110	32.5	110	29.0
40	1.57	40	1.57	110	31.6	110	30.8	110	29.8	110	28.5	125A	31.1
50	1.97	50	1.97	110	26.6	110	25.8	125	31.6	125	30.2	125	26.1
63	2.48	63	2.48	125	27.2	125	26.2	140A	31.8	140	30.3	140	25.8
75	2.95	75	2.95	140	28.2	140	27.1	140	25.8	160	33.1	160	28.0
90	3.54	90	3.54	160	30.1	160	28.8	160	27.4	160	25.6	180	28.6
110	4.33	110	4.33	180	29.5	180	28.1	180	26.4	200	33.2	200	26.8
125	4.92	125	4.92	200	31.3	200	29.8	200	28.0	200	25.7	225	29.6
140	5.51	140	5.51	225	35.6	225	33.9	225	31.8	225	29.3	250	32.3
160	6.30	160	6.30	225	25.6	250	35.4	250	33.1	250	30.3	280	34.5
180	7.09	180	7.09	250	27.3	250	25.4	280	36.7	280	33.5	315	38.9
200	7.87	200	7.87	280	31.4	280	29.2	280	26.7	315	39.0	315	28.9
225	8.86	225	8.86	315	35.3	315	32.9	315	30.0	315	26.5	355	32.7
250	9.84	250	9.84	355	41.6	355	38.8	355	35.6	355	31.6	400	38.6
280	11.02	280	11.02	355	26.6	400	44.6	400	41.0	400	36.5	450	44.1
315	12.4	315	12.4	400	30.2	400	27.1	450	46.1	450	41.0	450	26.6
355	13.98	355	13.98	450	33.7	450	30.2	450	26.1	500	43.1	---	---
400	15.75	400	15.75	500	34.6	500	30.8	500	26.2	560	47.1	---	---
450	17.72	450	17.72	560	37.8	560	33.5	560	28.3	630	52.9	---	---
500	19.69	500	19.69	630	45.6	630	40.8	630	35.0	630	27.9	---	---
560	22.05	560	22.05	710	53.2	710	47.7	710	41.2	710	33.2	---	---
630	24.80	630	24.80	800	60.4	800	54.2	800	46.9	800	37.9	---	---

Table A-21 Pipe Data – Carbon and Alloy Steel – Stainless Steel

Nominal pipe size in.	Outside diameter in.	Steel Iron pipe size	Steel Schedule number	Stainless steel schedule number	Wall thickness (t) in.	Inside diameter (d) in.	Area of metal sq. in.	Transverse internal area (a) sq. in.	Transverse internal area (A) sq. ft	Moment of inertia (I) in.^{4}	Weight pipe lbs. per ft	Weight water lbs. per ft of pipe	External surface sq. ft per ft of pipe	Section modulus $\left(2\frac{1}{O.D.}\right)$
1/8	0.405	---	---	10S	.049	.307	.0548	.0740	.0051	.00088	.19	.032	.106	.00437
		STD	40	40S	.068	.269	.0720	.0568	.00040	.00106	.24	.025	.106	.00523
		XS	80	80S	.095	.215	.0925	.0364	.00025	.00122	.31	.016	.106	.00602
1/4	0.540	---	---	10S	.065	.410	.0970	.1320	.00091	.00279	.33	.057	.141	.01032
		STD	40	40S	.088	.364	.1250	.1041	.00072	.00331	.42	.045	.141	.01227
		XS	80	80S	.119	.302	.1574	.0716	.00050	.00377	.54	.031	.141	.01395
3/8	0.675	---	---	10S	.065	.545	.1246	.2333	.00162	.00586	.42	.101	.178	.01736
		STD	40	40S	.091	.493	.1670	.1910	.00133	.00729	.57	.083	.178	.02160
		XS	80	80S	.126	.423	.2173	.1405	.00098	.00862	.74	.061	.178	.02554
1/2	0.840	---	---	5S	.065	.710	.1583	.3959	.00275	.01197	.54	.172	.220	.02849
		---	---	10S	.083	.674	.1974	.3568	.00248	.01431	.67	.155	.220	.03407
		STD	40	40S	.109	.622	.2503	.3040	.00211	.01709	.85	.132	.220	.04069
		XS	80	80S	.147	.546	.3200	.2340	.00163	.02008	1.09	.102	.220	.04780
		---	160	---	.187	.466	.3836	.1706	.00118	.02212	1.31	.074	.220	.05267
		XXS	---	---	.294	.252	.5043	.050	.00035	.02424	1.71	.022	.220	.05772
3/4	1.050	---	---	5S	.065	.920	.2011	.6648	.00462	.02450	.69	.288	.275	.04667
		---	---	10S	.083	.884	.2521	.6138	.00426	.02969	.86	.266	.275	.05655
		STD	40	40S	.113	.824	.3326	.5330	.00371	.03704	1.13	.231	.275	.07055
		XS	80	80S	.154	.742	.4335	.4330	.00300	.04479	1.47	.188	.275	.08531
		---	160	---	.219	.612	.5698	.2961	.00206	.05269	1.94	.128	.275	.10036
		XXS	---	---	.308	.434	.7180	.148	.00103	.05792	2.44	.064	.275	.11032
1	1.315	---	---	5S	.065	1.185	.2553	1.1029	.00766	.04999	.87	.478	.344	.07603
		---	---	10S	.109	1.097	.4130	.9452	.00656	.07569	1.40	.409	.344	.11512
		STD	40	40S	.133	1.049	.4939	.8640	.00600	.08734	1.68	.375	.344	.1328
		XS	80	80S	.179	.957	.6388	.7190	.00499	.1056	2.17	.312	.344	.1606
		---	160	---	.250	.815	.8365	.5217	.00362	.1251	2.84	.230	.344	.1903
		XXS	---	---	.358	.599	1.0760	.282	.00196	.1405	3.66	.122	.344	.2136
1-1/4	1.660	---	---	5S	.065	1.530	.3257	1.839	.01277	.1038	1.11	.797	.435	.1250
		---	---	10S	.109	1.442	.4717	1.633	.01134	.1605	1.81	.708	.435	.1934
		STD	40	40S	.140	1.380	.6685	1.495	.01040	.1947	2.27	.649	.435	.2346

Appendix A

Table A-21 Pipe Data – Carbon and Alloy Steel – Stainless Steel, cont.

Nominal pipe size in.	Outside diameter in.	Identification			Wall thickness (t) in.	Inside diameter (d) in.	Area of metal sq. in.	Transverse internal area		Moment of inertia (I) in.⁴	Weight pipe lbs. per ft	Weight water lbs. per ft of pipe	External surface sq. ft per ft of pipe	Section modulus $\left(2\frac{1}{O.D.}\right)$
		Steel		Stainless steel schedule number				(a) sq. in.	(A) sq. ft					
		Iron pipe size	Schedule number											
1-1/4	1.660	XS --- XXS	80 160 ---	80S --- ---	.191 .250 .382	1.278 1.160 .896	.8815 1.1070 1.534	1.283 1.057 .630	.00891 .00734 .00438	.2418 .2839 .3411	3.00 3.76 5.21	.555 .458 .273	.435 .435 .435	.2913 .3421 .4110
1-1/2	1.900	--- --- STD XS --- XXS	--- --- 40 80 160 ---	5S 10S 40S 80S --- ---	.065 .109 .145 .200 .281 .400	1.770 1.682 1.610 1.500 1.338 1.100	.3747 .6133 .7995 1.068 1.429 1.885	2.461 2.222 2.036 1.767 1.406 .950	.01709 .01543 .01414 .01225 .00976 .00660	.1579 .2468 .3099 .3912 .4824 .5678	1.28 2.09 2.72 3.63 4.86 6.41	1.066 .963 .882 .765 .608 .42	.497 .497 .497 .497 .497 .497	.1662 .2598 .3262 .4118 .5078 .5977
2	2.375	--- --- STD XS --- XXS	--- --- 40 80 160 ---	5S 10S 40S 80S --- ---	.065 .109 .154 .218 .344 .436	2.245 2.157 2.067 1.939 1.687 1.503	.4717 .7760 1.075 1.477 2.190 2.656	3.958 3.654 3.355 2.953 2.241 1.774	.02749 .02538 .02330 .02050 .01556 .01232	.3149 .4992 .6657 .8679 1.162 1.311	1.61 2.64 3.65 5.02 7.46 9.03	1.72 1.58 1.45 1.28 .97 .77	.622 .622 .622 .622 .622 .622	.2652 .4204 .5606 .7309 .979 1.104
2-1/2	2.875	--- --- STD XS --- XXS	--- --- 40 80 160 ---	5S 10S 40S 80S --- ---	.083 .120 .203 .276 .375 .552	2.709 2.635 2.469 2.323 2.125 1.771	.7280 1.039 1.704 2.254 2.945 4.028	5.764 5.453 4.788 4.238 3.546 2.464	.04002 .03787 .03322 .02942 .02463 .01710	.7100 .9873 1.530 1.924 2.353 2.871	2.48 3.53 5.79 7.66 10.01 13.69	2.50 2.36 2.07 1.87 1.54 1.07	.753 .753 .753 .753 .753 .753	.4939 .6868 1.064 1.339 1.638 1.997
3	3.500	--- --- STD XS --- XXS	--- --- 40 80 160 ---	5S 10S 40S 80S --- ---	.083 .120 .216 .300 .438 .600	3.334 3.260 3.068 2.900 2.624 2.300	.8910 1.274 2.228 3.016 4.205 5.466	8.730 8.347 7.393 6.605 5.408 4.155	.06063 .05796 .05130 .04587 .03755 .02885	1.301 1.822 3.017 3.894 5.032 5.993	3.03 4.33 7.58 10.25 14.32 18.58	3.78 3.62 3.20 2.86 2.35 1.80	.916 .916 .916 .916 .916 .916	.7435 1.041 1.724 2.225 2.876 3.424
3-1/2	4.000	--- --- STD XS	--- --- 40 80	5S 10S 40S 80S	.083 .120 .226 .318	3.834 3.760 3.548 3.364	1.021 1.463 2.680 3.678	11.545 11.104 9.886 8.888	.08017 .07711 .06870 .06170	1.960 2.755 4.788 6.280	3.48 4.97 9.11 12.50	5.00 4.81 4.29 3.84	1.047 1.047 1.047 1.047	.9799 1.378 2.394 3.140

Table A-21 Pipe Data – Carbon and Alloy Steel – Stainless Steel, cont.

Nominal pipe size in.	Outside diameter in.	Identification Steel Iron pipe size	Identification Steel Schedule number	Stainless steel schedule number	Wall thickness (t) in.	Inside diameter (d) in.	Area of metal sq. in.	Transverse internal area (a) sq. in.	Transverse internal area (A) sq. ft	Moment of inertia (I) in.⁴	Weight pipe lbs. per ft	Weight water lbs. per ft of pipe	External surface sq. ft per ft of pipe	Section modulus $\left(2\frac{1}{O.D.}\right)$
4	4.500	---	---	5S	.083	4.334	1.152	14.75	.10245	2.810	3.92	6.39	1.178	1.249
		---	---	10S	.120	4.260	1.651	14.25	.09898	3.963	5.61	6.18	1.178	1.761
		STD	40	40S	.237	4.026	3.174	12.73	.08840	7.233	10.79	5.50	1.178	3.214
		XS	80	80S	.337	3.826	4.407	11.50	.07986	9.610	14.98	4.98	1.178	4.271
		---	120	---	.438	3.624	5.595	10.31	.0716	11.65	19.00	4.47	1.178	5.178
		---	160	---	.531	3.438	6.621	9.28	.0645	13.27	22.51	4.02	1.178	5.898
		XXS	---	---	.674	3.152	8.101	7.80	.0542	15.28	27.54	3.38	1.178	6.791
5	5.563	---	---	5S	.109	5.345	1.868	22.44	.1558	6.947	6.36	9.72	1.456	2.498
		---	---	10S	.134	5.295	2.285	22.02	.1529	8.425	7.77	9.54	1.456	3.029
		STD	40	40S	.258	5.047	4.300	20.01	.1390	15.16	14.62	8.67	1.456	5.451
		XS	80	80S	.375	4.813	6.112	18.19	.1263	20.67	20.78	7.88	1.456	7.431
		---	120	---	.500	4.563	7.953	16.35	.1136	25.73	27.04	7.09	1.456	9.250
		---	160	---	.625	4.313	9.696	14.61	.1015	30.03	32.96	6.33	1.456	10.796
		XXS	---	---	.750	4.063	11.340	12.97	.0901	33.63	38.55	5.61	1.456	12.090
6	6.625	---	---	5S	.109	6.407	2.231	32.24	.2239	11.85	7.60	13.97	1.734	3.576
		---	---	10S	.134	6.357	2.733	31.74	.2204	14.40	9.29	13.75	1.734	4.346
		STD	40	40S	.280	6.065	5.581	28.89	.2006	28.14	18.97	12.51	1.734	8.496
		XS	80	80S	.432	5.761	8.405	26.07	.1810	40.49	28.57	11.29	1.734	12.22
		---	120	---	.562	5.501	10.70	23.77	.1650	49.61	36.39	10.30	1.734	14.98
		---	160	---	.719	5.187	13.32	21.15	.1469	58.97	45.35	9.16	1.734	17.81
		XXS	---	---	.864	4.897	15.64	18.84	.1308	66.33	53.16	8.16	1.734	20.02
8	8.625	---	---	5S	.109	8.407	2.916	55.51	.3855	26.44	9.93	24.06	2.258	6.131
		---	---	10S	.148	8.329	3.941	54.48	.3784	35.41	13.40	23.61	2.258	8.212
		---	20	---	.250	8.125	6.57	51.85	.3601	57.72	22.36	22.47	2.258	13.39
		---	30	---	.277	8.071	7.26	51.16	.3553	63.35	24.70	22.17	2.258	14.69
		STD	40	40S	.322	7.981	8.40	50.03	.3474	72.49	28.55	21.70	2.258	16.81
		---	60	---	.406	7.813	10.48	47.94	.3329	88.73	35.64	20.77	2.258	20.58
		XS	80	80S	.500	7.625	12.76	45.66	.3171	105.7	43.39	19.78	2.258	24.51
		---	100	---	.594	7.437	14.96	43.46	.3018	121.3	50.95	18.83	2.258	28.14
		---	120	---	.719	7.187	17.84	40.59	.2819	140.5	60.71	17.59	2.258	32.58
		---	140	---	.812	7.001	19.93	38.50	.2673	153.7	67.76	16.68	2.258	35.65
		XXS	---	---	.875	6.875	21.30	37.12	.2578	162.0	72.42	16.10	2.258	37.56
		---	160	---	.906	6.813	31.97	36.46	.2532	165.9	74.69	15.80	2.258	38.48

Appendix A

Table A-21 Pipe Data – Carbon and Alloy Steel – Stainless Steel, cont.

Nominal pipe size in.	Outside diameter in.	Identification Steel Iron pipe size	Identification Schedule number	Identification Stainless steel schedule number	Wall thickness (t) in.	Inside diameter (d) in.	Area of metal sq. in.	Transverse internal area (a) sq. in.	Transverse internal area (A) sq. ft	Moment of inertia (I) in.^{4}	Weight pipe lbs. per ft	Weight water lbs. per ft of pipe	External surface sq. ft per ft of pipe	Section modulus $\left(2\dfrac{1}{O.D.}\right)$
10	10.750	---	---	5S	.134	10.482	4.36	86.29	.5992	63.0	15.19	37.39	2.814	11.71
		---	---	10S	.165	10.420	5.49	85.28	.5922	76.9	18.65	36.95	2.814	14.30
		---	20	---	.250	10.250	8.24	82.52	.5731	113.7	28.04	35.76	2.814	21.15
		---	30	---	.307	10.136	10.07	80.69	.5603	137.4	34.24	34.96	2.814	25.57
		STD	40	40S	.365	10.020	11.90	78.86	.5475	160.7	40.48	34.20	2.814	29.90
		XS	60	80S	.500	9.750	16.10	74.66	.5185	212.0	54.74	32.35	2.814	39.43
		---	80	---	.594	9.562	18.92	71.84	.4989	244.8	64.43	31.13	2.814	45.54
		---	100	---	.719	9.312	22.63	68.13	.4732	286.1	77.03	29.53	2.814	53.22
		---	120	---	.844	9.062	26.24	64.53	.4481	324.2	89.29	27.96	2.814	60.32
		XXS	140	---	1.000	8.750	30.63	60.13	.4176	367.8	104.13	26.06	2.814	68.43
		---	160	---	1.125	8.500	34.02	56.75	.3941	399.3	115.64	24.59	2.814	74.29
12	12.75	---	---	5S	.156	12.438	6.17	121.50	.8438	122.4	20.98	52.65	3.338	19.2
		---	---	10S	.180	12.390	7.11	120.57	.8373	140.4	24.17	52.25	3.338	22.0
		---	20	---	.250	12.250	9.82	117.86	.8185	191.8	33.38	51.07	3.338	30.2
		---	30	---	.330	12.090	12.87	114.80	.7972	248.4	43.77	49.74	3.338	39.0
		STD	---	40S	.375	12.000	14.58	113.10	.7854	279.3	49.56	49.00	3.338	43.8
		---	40	---	.406	11.938	15.77	111.93	.7773	300.3	53.52	48.50	3.338	47.1
		XS	---	80S	.500	11.750	19.24	108.43	.7528	361.5	65.42	46.92	3.338	56.7
		---	60	---	.562	11.626	21.52	106.16	.7372	400.4	73.15	46.00	3.338	62.8
		---	80	---	.688	11.374	26.03	101.64	.7058	475.1	88.63	44.04	3.338	74.6
		---	100	---	.844	11.062	31.53	96.14	.6677	561.6	107.32	41.66	3.338	88.1
		XXS	120	---	1.000	10.750	36.91	90.76	.6303	641.6	125.49	39.33	3.338	100.7
		---	140	---	1.125	10.500	41.08	86.59	.6013	700.5	139.67	37.52	3.338	109.9
		---	160	---	1.312	10.126	47.14	80.53	.5592	781.1	160.27	34.89	3.338	122.6
14	14.00	---	---	5S	.156	13.688	6.78	147.15	1.0219	162.6	23.07	63.77	3.665	23.2
		---	---	10S	.188	13.624	8.16	145.78	1.0124	194.6	27.73	63.17	3.665	27.8
		---	10	---	.250	13.500	10.80	143.14	.9940	255.3	36.71	62.03	3.665	36.6
		---	20	---	.312	13.376	13.42	140.52	.9758	314.4	45.61	60.89	3.665	45.0
		STD	30	---	.375	13.250	16.05	137.88	.9575	372.8	54.57	59.75	3.665	53.2
		---	40	---	.438	13.124	18.66	135.28	.9394	429.1	63.44	58.64	3.665	61.3
		XS	---	---	.500	13.000	21.21	132.73	.9217	483.8	72.09	57.46	3.665	69.1
		---	60	---	.594	12.812	24.98	128.96	.8956	562.3	85.05	55.86	3.665	80.3
		---	80	---	.750	12.500	31.22	122.72	.8522	678.3	106.13	53.18	3.665	98.2
		---	100	---	.938	12.124	38.45	115.49	.8020	824.4	130.85	50.04	3.665	117.8

Table A-21 Pipe Data – Carbon and Alloy Steel – Stainless Steel, cont.

Nominal pipe size in.	Outside diameter in.	Iron pipe size	Schedule number	Stainless steel schedule number	Wall thickness (t) in.	Inside diameter (d) in.	Area of metal sq. in.	Transverse internal area (a) sq. in.	Transverse internal area (A) sq. ft	Moment of inertia (I) in.⁴	Weight pipe lbs. per ft	Weight water lbs. per ft of pipe	External surface sq. ft per ft of pipe	Section modulus $\left(2\dfrac{1}{O.D.}\right)$
14	14.00	---	120	---	1.094	11.812	44.32	109.62	.7612	929.6	150.79	47.45	3.665	132.8
		---	140	---	1.250	11.500	50.07	103.87	.7213	1027.0	170.28	45.01	3.665	146.8
		---	160	---	1.406	11.188	55.63	98.31	.6827	1117.0	189.11	42.60	3.665	159.6
16	16.00	---	---	5S	.165	15.670	8.21	192.85	1.3393	257.3	27.90	83.57	4.189	32.2
		---	---	10S	.188	15.624	9.34	191.72	1.3314	291.9	31.75	83.08	4.189	36.5
		---	10	---	.250	15.500	12.37	188.69	1.3103	383.7	42.05	81.74	4.189	48.0
		---	20	---	.312	15.376	15.38	185.69	1.2895	473.2	52.27	80.50	4.189	59.2
		STD	30	---	.375	15.250	18.41	182.65	1.2684	562.1	62.58	79.12	4.189	70.3
		XS	40	---	.500	15.000	24.35	176.72	1.2272	731.9	82.77	76.58	4.189	91.5
		---	60	---	.656	14.688	31.62	169.44	1.1766	932.4	107.50	73.42	4.189	116.6
		---	80	---	.844	14.312	40.14	160.92	1.1175	1155.8	136.61	69.73	4.189	144.5
		---	100	---	1.031	13.938	48.48	152.58	1.0596	1364.5	164.82	66.12	4.189	170.5
		---	120	---	1.219	13.562	56.56	144.50	1.0035	1555.8	192.43	62.62	4.189	194.5
		---	140	---	1.438	13.124	65.78	135.28	.9394	1760.3	223.64	58.64	4.189	220.0
		---	160	---	1.594	12.812	72.10	128.96	.8956	1893.5	245.25	55.83	4.189	236.7
18	18.00	---	---	5S	.165	17.670	9.25	245.22	1.7029	367.6	31.43	106.26	4.712	40.8
		---	---	10S	.188	17.624	10.52	243.95	1.6941	417.3	35.76	105.71	4.712	46.4
		---	10	---	.250	17.500	13.94	240.53	1.6703	549.1	47.39	104.21	4.712	61.1
		---	20	---	.312	17.376	17.34	237.13	1.6467	678.2	58.94	102.77	4.712	75.5
		STD	---	---	.375	17.250	20.76	233.71	1.6230	806.7	70.59	101.18	4.712	89.6
		XS	30	---	.438	17.124	24.17	230.30	1.5990	930.3	82.15	99.84	4.712	103.4
		---	40	---	.500	17.000	27.49	226.98	1.5763	1053.2	93.45	98.27	4.712	117.0
		---	60	---	.562	16.876	30.79	223.68	1.5533	1171.5	104.67	96.93	4.712	130.1
		---	80	---	.750	16.500	40.64	213.83	1.4849	1514.7	137.17	92.57	4.712	168.3
		---	100	---	.938	16.124	50.23	204.24	1.4183	1833.0	170.92	88.50	4.712	203.8
		---	120	---	1.156	15.688	61.17	193.30	1.3423	2180.0	207.96	83.76	4.712	242.3
		---	140	---	1.375	15.250	71.81	182.66	1.2684	2498.1	244.14	79.07	4.712	277.6
		---	160	---	1.562	14.876	80.66	173.80	1.2070	2749.0	274.22	75.32	4.712	305.5
		---	---	---	1.781	14.438	90.75	163.72	1.1369	3020.0	308.50	70.88	4.712	335.6
20	20.00	---	---	5S	.188	19.624	11.70	302.46	2.1004	574.2	39.78	131.06	5.236	57.4
		---	---	10S	.218	19.564	13.55	300.61	2.0876	662.8	46.06	130.27	5.236	66.3
		---	10	---	.250	19.500	15.51	298.65	2.0740	765.4	52.73	129.42	5.236	75.6

Appendix A

Table A-21 Pipe Data – Carbon and Alloy Steel – Stainless Steel, cont.

Nominal pipe size in.	Outside diameter in.	Identification – Steel Iron pipe size	Identification Schedule number	Identification Stainless steel schedule number	Wall thickness (t) in.	Inside diameter (d) in.	Area of metal sq. in.	Transverse internal area (a) sq. in.	Transverse internal area (A) sq. ft	Moment of inertia (I) in.^{4}	Weight pipe lbs. per ft	Weight water lbs. per ft of pipe	External surface sq. ft per ft of pipe	Section modulus $\left(2\frac{1}{O.D.}\right)$
20	20.00	STD	20	---	.375	19.250	23.12	290.04	2.0142	1113.0	78.60	125.67	5.236	111.3
		XS	30	---	.500	19.000	30.63	283.53	1.9690	1457.0	104.13	122.87	5.236	145.7
		---	40	---	.594	18.812	36.15	278.00	1.9305	1703.0	123.11	120.46	5.236	170.4
		---	60	---	.812	18.376	48.95	265.21	1.8417	2257.0	166.40	114.92	5.236	225.7
		---	80	---	1.031	17.938	61.44	252.72	1.7550	2772.0	208.87	109.51	5.236	277.1
		---	100	---	1.281	17.438	75.33	238.83	1.6585	3315.2	256.10	103.39	5.236	331.5
		---	120	---	1.500	17.000	87.18	226.98	1.5762	3754.0	296.37	98.35	5.236	375.5
		---	140	---	1.750	16.500	100.33	213.82	1.4849	4216.0	341.09	92.66	5.236	421.7
		---	160	---	1.969	16.062	111.49	202.67	1.4074	4585.5	379.17	87.74	5.236	458.5
22	22.00	---	---	5S	.188	21.624	12.88	367.25	2.5503	766.2	43.80	159.14	5.760	69.7
		---	---	10S	.218	21.564	14.92	365.21	2.5362	884.8	50.71	158.26	5.760	80.4
		---	10	---	.250	21.500	17.08	363.05	2.5212	1010.3	58.07	157.32	5.760	91.8
		STD	20	---	.375	21.250	25.48	354.66	2.4629	1489.7	86.61	153.68	5.760	135.4
		XS	30	---	.500	21.000	33.77	346.36	2.4053	1952.5	114.81	150.09	5.760	117.5
		---	60	---	.875	20.250	58.07	322.06	2.2365	3244.9	197.41	139.56	5.760	295.0
		---	80	---	1.125	19.75	73.78	306.35	2.1275	4030.4	250.81	132.76	5.760	366.4
		---	100	---	1.375	19.25	89.09	291.04	2.0211	4758.5	302.88	126.12	5.760	432.6
		---	120	---	1.625	18.75	104.02	276.12	1.9175	5432.0	353.61	119.65	5.760	493.8
		---	140	---	1.875	18.25	118.55	261.59	1.8166	6053.7	403.00	113.36	5.760	550.3
		---	160	---	2.125	17.75	132.68	247.45	1.7184	6626.4	451.06	107.23	5.760	602.4
24	24.00	---	---	5S	.218	23.564	16.29	436.10	3.0285	1151.6	55.37	188.98	6.283	96.0
		---	10	10S	.250	23.500	18.65	433.74	3.0121	1315.4	63.41	187.95	6.283	109.6
		STD	20	---	.375	23.250	27.83	424.56	2.9483	1942.0	94.62	183.95	6.283	161.9
		XS	---	---	.500	23.000	36.91	415.48	2.8853	2549.5	125.49	179.87	6.283	212.5
		---	30	---	.562	22.876	41.39	411.00	2.8542	2843.0	140.68	178.09	6.283	237.0
		---	40	---	.688	22.624	50.31	402.07	2.7921	3421.3	171.29	174.23	6.283	285.1
		---	60	---	.969	22.062	70.04	382.35	2.6552	4652.8	238.35	165.52	6.283	387.7
		---	80	---	1.219	21.562	87.17	365.22	2.5362	5672.0	296.58	158.26	6.283	472.8
		---	100	---	1.531	20.938	108.07	344.32	2.3911	6849.9	367.39	149.06	6.283	570.8
		---	120	---	1.812	20.376	126.31	326.08	2.2645	7825.0	429.39	141.17	6.283	652.1
		---	140	---	2.062	19.876	142.11	310.28	2.1547	8625.0	483.12	134.45	6.283	718.9
		---	160	---	2.344	19.312	159.41	292.98	2.0346	9455.9	542.13	126.84	6.283	787.9

Table A-21 Pipe Data – Carbon and Alloy Steel – Stainless Steel

Nominal pipe size in.	Outside diameter in.	Identification – Steel – Iron pipe size	Identification – Steel – Schedule number	Identification – Stainless steel schedule number	Wall thickness (t) in.	Inside diameter (d) in.	Area of metal sq. in.	Transverse internal area (a) sq. in.	Transverse internal area (A) sq. ft	Moment of inertia (I) in.^{4}	Weight pipe lbs. per ft	Weight water lbs. per ft of pipe	External surface sq. ft per ft of pipe	Section modulus $\left(2\frac{1}{O.D.}\right)$
26	26.00	---	10	---	.312	25.376	25.18	505.75	3.5122	2077.2	85.60	219.16	6.806	159.8
		STD	---	---	.375	25.250	30.19	500.74	3.4774	2478.4	102.63	216.99	6.806	190.6
		XS	20	---	.500	25.000	40.06	490.87	3.4088	3257.0	136.17	212.71	6.806	250.5
28	28.00	---	10	---	.312	27.376	27.14	588.61	4.0876	2601.0	92.26	255.07	7.330	185.8
		STD	---	---	.375	27.250	32.54	583.21	4.0501	3105.1	110.64	252.73	7.330	221.8
		XS	20	---	.500	27.000	43.20	572.56	3.9761	4084.8	146.85	248.11	7.330	291.8
		---	30	---	.625	26.750	53.75	562.00	3.9028	5037.7	182.73	243.53	7.330	359.8
30	30.00	---	---	5S	.250	29.500	23.37	683.49	4.7465	2585.2	79.43	296.18	7.854	172.3
		---	10	10S	.312	29.376	29.10	677.76	4.7067	3206.3	98.93	293.70	7.854	213.8
		STD	---	---	.375	29.250	34.90	671.96	4.6664	3829.4	118.65	291.18	7.854	255.3
		XS	20	---	.500	29.000	46.34	660.52	4.5869	5042.2	157.53	286.22	7.854	336.1
		---	30	---	.625	28.750	57.68	649.18	4.5082	6224.0	196.08	281.31	7.854	414.9
30	30.00	---	---	5S	.250	29.500	23.37	683.49	4.7465	2585.2	79.43	296.18	7.854	172.3
		---	10	10S	.312	29.376	29.10	677.76	4.7067	3206.3	98.93	293.70	7.854	213.8
		STD	---	---	.375	29.250	34.90	671.96	4.6664	3829.4	118.65	291.18	7.854	255.3
		XS	20	---	.500	29.000	46.34	660.52	4.5869	5042.2	157.53	286.22	7.854	336.1
		---	30	---	.625	28.750	57.68	649.18	4.5082	6224.0	196.08	281.31	7.854	414.9
32	32.00	---	10	---	.312	31.376	31.06	773.19	5.3694	3898.9	105.59	335.05	8.378	243.7
		STD	---	---	.375	31.250	37.26	766.99	5.3263	4658.5	126.66	332.36	8.378	291.2
		XS	20	---	.500	31.000	49.48	754.77	5.2414	6138.6	168.21	327.06	8.378	383.7
		---	30	---	.625	30.750	61.60	742.64	5.1572	7583.4	209.43	321.81	8.378	474.0
		---	40	---	.688	30.624	67.68	736.57	5.1151	8298.3	230.08	319.18	8.378	518.6
34	34.00	---	10	---	.344	33.312	36.37	871.55	6.0524	5150.5	123.65	377.67	8.901	303.0
		STD	---	---	.375	33.250	39.61	868.31	6.0299	5599.3	134.67	376.27	8.901	329.4
		XS	20	---	.500	33.000	52.62	855.30	5.9396	7383.5	178.89	370.63	8.901	434.3
		---	30	---	.625	32.750	65.53	842.39	5.8499	9127.6	222.78	365.03	8.901	536.9
		---	40	---	.688	32.624	72.00	835.92	5.8050	9991.6	244.77	362.23	8.901	587.7
36	36.00	---	10	---	.312	35.376	34.98	982.90	6.8257	5569.5	118.92	425.92	9.425	309.4
		STD	---	---	.375	35.250	41.97	975.91	6.7771	6658.9	142.68	422.89	9.425	369.9
		XS	20	---	.500	35.000	55.76	962.11	6.6813	8786.2	189.57	416.91	9.425	488.1
		---	30	---	.625	34.750	69.46	948.42	6.5862	10868.4	236.13	417.22	9.425	603.8
		---	40	---	.750	34.500	83.06	934.82	6.4918	12906.1	282.35	405.09	9.425	717.0

Notes

Notes

Appendix B

Table B-1 Maximum Permissable Fixture Unit Loads for Horizontal Branches

Branch Diameter, inches	Total Load, FU
1.5	3
2	6
2.5	12
3	20
4	160
5	360
6	620
8	1400

Table B-2 Maximum Permissable Fixture Unit Loads for Vertical Stacks

Stack Diameter inches	Stack Three Stories or Less in Height FU	Stacks More than Three Stories in Height FU	Total Discharge into One Branch Interval FU
2	10	24	6
2.5	20	42	9
3	30*	60*	16**
4	240	500	90
5	540	1100	200
6	960	1900	350
8	2200	3600	600
10	3800	5600	1000
12	6000	8400	1500

* Not more than six water closets permitted.
** Not more than two water closets permitted.

Table B-3 Maximum Permissable Fixture Unit Loads for Horizontal Mains

Pipe Diameter, inches	Building Drain and Runouts Slope (inches per foot)		
	1/8	1/4	1/2
2	---	21	26
2.5	---	24	31
3	20	27	36
4	180	216	250
5	390	480	575
6	700	840	1000
8	1600	1920	2300
10	2900	3500	4200
12	4600	5600	6700
15	8300	10000	12000

Appendix B

Table B-4 Vent Header Sizing

Size of Soil or Waste Stack (in)	Fixture Units Connected	Diameter of Vent Required (in) Maximum length of vent (ft.)								
		1.25	1.5	2	2.5	3	4	5	6	8
1.25	2	30								
1.5	8	50	150							
2	10	30	100							
2	12	30	75	200						
2	20	26	50	150						
2.5	42	--	30	100	300					
3	10	--	30	100	200	600				
3	30	--	--	60	200	500				
3	60	--	--	50	80	400				
4	100	--	--	35	100	260	100			
4	200	--	--	30	90	250	900			
4	500	--	--	20	70	180	700			
5	200	--	--	--	35	80	350	1000		
5	500	--	--	--	30	70	300	900		
5	1100	--	--	--	20	50	200	700		
6	350	--	--	--	25	50	200	400	1300	
6	620	--	--	--	15	30	125	300	1100	
6	960	--	--	--	--	24	100	250	1000	
6	1900	--	--	--	--	20	70	200	700	
8	600	--	--	--	--	--	50	150	500	1300
8	1400	--	--	--	--	--	40	100	400	1200
8	2200	--	--	--	--	--	30	80	350	1100
8	3600	--	--	--	--	--	25	60	250	800
10	1000	--	--	--	--	--	--	75	125	1000
10	2500	--	--	--	--	--	--	50	100	500
10	3800	--	--	--	--	--	--	30	80	350
10	5600	--	--	--	--	--	--	25	60	250

General Note: 20% of the total shown may be installed in a horizontal position.

Table B-5   Values of R, $R^{2/3}$ & Cross-Sectional Areas for Circular Pipes, Full Flow

Pipe Size inches	R= D/4 feet	$R^{2/3}$	A Cross-Sectional area of flow sq. feet
1.5	0.0335	0.1040	0.01412
2	0.0417	0.1200	0.02180
2.5	0.0521	0.1396	0.03408
3	0.0625	0.1570	0.04910
4	0.0833	0.1910	0.08730
5	0.1040	0.2210	0.13640
6	0.1250	0.2500	0.19640
8	0.1670	0.3030	0.34920
10	0.2080	0.3510	0.54540
12	0.2500	0.3970	0.78540
15	0.3125	0.4610	1.22700

Table B-6   Values of R, $R^{2/3}$ & Cross-Sectional Areas for Circular Pipes, Half Flow

Pipe Size inches	R= D/4 feet	$R^{2/3}$	A Cross-Sectional area of flow sq. feet
1.5	0.0335	0.1040	0.00706
2	0.0417	0.1200	0.01090
2.5	0.0521	0.1396	0.01704
3	0.0625	0.1570	0.022455
4	0.0833	0.1910	0.04365
5	0.1040	0.2210	0.06820
6	0.1250	0.2500	0.09820
8	0.1670	0.3030	0.17460
10	0.2080	0.3510	0.27270
12	0.2500	0.3970	0.39270
15	0.3125	0.4610	1.61350

Table B-7 Flow Rate for Horizontal Mains at .25" Slope

Pipe Size inches	Full or Half-Full Flow Velocity V, fps	Half-Full Capacity q, gpm	Full Flow Capacity q, gpm
1.5	1.85	5.85	11.7
2	1.98	9.70	19.4
2.5	2.3	17.60	35.2
3	2.59	28.60	57.2
4	2.91	57.00	114.0
5	3.15	96.50	193.0
6	3.58	157.50	315.0
8	4.07	318.50	637.0
10	4.69	574.00	1148.0
12	5.31	936.00	1872.0
15	6.15	1690.00	3380.0

Table B-8 Flow Rate for Horizontal Storm Drains at .25" Slope and Full Flow

Pipe Size inches	Velocity V, fps	Capacity q, gpm
2	1.72	17.4
2.5	1.99	31.5
3	2.25	51.3
4	2.74	111.0
5	3.15	201.0
6	3.58	327.0
8	4.35	705.0
10	5.04	1268.0
12	5.67	2070.0
15	6.58	3730.0

Notes

Appendix C

Table C-1 Metallic Component Standards Listed in the ANSI/ASME B31.3 Code for Chemical Plant & Petroleum Refinery Piping (1993 Edition)

Standard or Specification	Designation [2]
BOLTING	
Square & hex bolts & screws, inch series, incl. hex cap screws & lag screws	*ASME B18.2.1
Square & hex nuts	ANSI B18.2.2
METALLIC FITTINGS, VALVES & FLANGES	
Ductile iron fittings, 3"-24", for gas	ANSI A21.14
Cast iron pipe flanges & flanged fittings, Class 25,125,250 & 800	ANSI B16.1
Malleable iron threaded fittings, Class 150 & 300	*ASME B16.3
Cast iron threaded fittings, Class 150 & 250	*ASME B16.4
Pipe flanges & flanged fittings	ANSI B16.5
Factory-made wrought steel buttwelding fittings	*ASME B16.9
Face-to-face & end-to-end dimensions of ferrous valves	*ASME B16.10
Forged steel fittings, socket welding & threaded	ANSI B16.11
Ferrous pipe plugs, bushings & locknuts with pipe threads	ANSI B16.14
(a) Cast bronze threaded fittings, Class 125 & 250 [5][6]	*ASME B16.15
Cast copper alloy solder joint pressure fittings	ANSI B16.18
Wrought copper & copper alloy solder joint pressure fittings	ANSI B16.22
Bronze pipe flanges & flanged fittings, Class 150 & 300	ANSI B16.24
Cast copper alloy fittings for flared copper tubes	ANSI B16.26
Wrought steel buttwelding short radius elbows & returns [3]	*ASME B16.28
Valves, flanged & buttwelding end	ANSI
Steel orifice flanges, Class 300,600,900,1500 & 2500	B16.34
Malleable iron threaded pipe unions, Class 150,250 & 300	ANSI B16.36
Ductile iron pipe flanges & flanged fittings, Class 150 & 300	*ASME B16.39

Table C-1 (cont.) Metallic Component Standards Listed in the ANSI/ASME B31.3 Code for Chemical Plant & Petroleum Refinery Piping (1993 Edition)

Standard or Specification	Designation [2]
Flanged steel safety relief valves	*ASME B16.42
Ductile iron plug valves, flanged ends	API 526
Wafer check valves	API 593
Steel plug valves, flanged or buttwelding ends	API 594
Steel gate valves, flanged, & buttwelding ends	API 600
Compact carbon steel gate valves	API 602
Class 150, Corrosion-resistant gate valves	API 603
Large-diameter carbon steel flanges	API 605
Compact carbon steel gate valves (extended body)	API 606
Butterfly valves, lug-type & wafer-type	API 609
Ductile-iron & gray-iron fittings, 3"-48", for water & other liquids	*AWWA C110
Rubber gasket joints for ductile-iron & gray-iron pressure pipe & fittings	*AWWA C111
Flanged ductile-iron & gray-iron pipe with threaded flanges	*AWWA C115
Steel pipe flanges for water works service, 4"-144"	*AWWA C207
Dimensions for steel water pipe fittings	*AWWA C208
Gate valves, 3"-48", for water & sewage systems	*AWWA C500
Rubber-seated butterfly valves	*AWWA C504
Standard finishes for contact faces of pipe flanges & connecting-end flanges of valves & fittings	MSS SP-6
Spot facing for bronze, iron & steel flanges	MSS SP-9
Standard marking system for valves, fittings, flanges & unions	MSS SP-25
Corrosion-resistant gate, globe, angle & check valves with flanged & buttwelded ends	MSS SP-42
Wrought stainless steel buttwelding fittings	MSS SP-43
Steel pipe line flanges	MSS SP-44
Bypass & drain connection standard	MSS SP-45
Class 150LW Corrosion resistant cast flanges & flanged fittings	MSS SP-51
High pressure chemical industry flanges & threaded stubs for use with Lens Gaskets	MSS SP-65
Cast iron gate valves, flanged & threaded ends	MSS SP-70
Cast iron swing check valves, flanged & threaded ends	MSS SP-71
Ball valves with flanged or buttwelding ends for general service	MSS SP-72
Specifications for high test wrought buttwelding fittings	MSS SP-75
Socket-welding reducer inserts	MSS SP-79
Bronze gate, globe, angle & check valves	MSS SP-80
Stainless steel, bonnetless, flanged, knife gate valves	MSS SP-81
Carbon steel pipe unions, socket welding & threaded	MSS SP-83
Cast iron globe and angle valves, flanged & threaded ends	MSS SP-85

Appendix C

Table C-1 (cont.) Metallic Component Standards Listed in the ANSI/ASME B31.3 Code for Chemical Plant & Petroleum Refinery Piping (1993 Edition)

Standard or Specification	Designation [2]
Diaphragm type valves	MSS SP-88
Swage (d) nipples & bull plugs	MSS SP-95 (a)
Forges carbon steel branch outlet fittings-socket welding, threaded, & buttwelding ends	MSS SP-97 (a)
Refrigeration flare type fittings	*SAE J513
Hydraulic tube fittings	*SAE J514
Hydraulic flanged tube, pipe & hose connections, 4-bolt split flanged type	*SAE J518
METALLIC PIPE & TUBES [4]	
Ductile-iron pipe, centrifugally cast in metal molds or sand-lined molds for gas	ANSI A21.52
Welded & seamless wrought steel pipe	*ASME B36.10M
Stainless steel pipe	*ASME B36.19M
Seamless low carbon steel hydraulic line tubing	ANSI B93.11
Flanged cast-iron & ductile-iron pipe with threaded flanges	*AWWA C115
Thickness design of ductile-iron pipe	*AWWA C150
Ductile-iron pipe, centrifugally cast in metal molds or sand-lined molds, for water & other liquids	*AWWA C151
Steel water pipe 6" & larger	AWWA C200
MISCELLANEOUS	
Unified inch screw threads (UN & UNR thread form)	ANSI B1.1
Pipe threads (except Dryseal)	*ASME B1.20.1
Dryseal pipe threads (inch)	ANSI B1.20.3
Hose coupling screw threads	*ASME B1.20.7
Ring-joint gaskets & grooves for steel pipe flanges	ANSI B16.20
Nonmetallic flat gaskets for pipe flanges	ANSI B16.21
Buttwelding ends	*ASME B16.25
Surface texture	*ASME B46.1
Threading, gaging & thread inspection of casing, tubing & line pipe threads	API 5B
Metallic gaskets for piping, double-jacketed corrugated & spiral wound	API 601
Rubber gasket joints for cast-iron & ductile-iron pressure pipe & fittings	*AWWA C111
Pipe hangers & supports – Material, design, & manufacture	*MSS SP-58
Silver brazing joints for wrought & cast solder joint fittings	MSS SP-73
Screw threads & gaskets for fire hose connections	NFPA

Notes:
[1] It is not practical to refer to a specific edition of each standard throughout the Code text. Instead, the approved edition references, along with the names and addresses of the sponsoring organizations, are shown in Appendix E of the ANSI/ASME B31.3 Code.
[2] An asterisk (*) preceding the designation indicates that the standard has been approved an an American National Standard by the American National Standards Institute.
[3] Cautionary Note: See this Standard for special provisions concerning pressure ratings.
[4] See also Appendix A of the ASME b31.3 Code- 1990 edition.
[5] This standard allows the use of unlisted materials; see para. 323.1.2 of the ASME B31.3 Code- 1990 edition.
[6] This standard allows straight pipe threads in sizes ≤ NPS 1/2; see para. 314.2.1(d) of the ASME B31.3 Code- 1990 edition.

Table C-2 Casting Quality Factors

Spec. No.	Description	Ec[1]
Iron		
A 47[3]	Malleable iron castings	1.00
A 48[3]	Gray iron castings	1.00
A126[3]	Gray iron castings	1.00
A 197[3]	Cupola malleable iron castings	1.00
A 278[3]	Gray iron castings	1.00
A 395[3][4]	Ductile and ferritic ductile iron castings	0.80
A 571[3][4]	Austenitic ductile iron castings	0.80
Carbon Steel		
A 216[3][4]	Carbon steel castings	0.80
A 352[3][4]	Ferritic steel castings	0.80
Low and Intermediate Alloy Steel		
A426[5]	Centrifugally cast pipe	1.00
A 217[3][4]	Martensitic stainless and alloy castings	0.80
A 352[3][4]	Ferritic steel castings	0.80
Stainless Steel		
A451[[4][5]	Centrifugally cast pipe	0.90
A 452[4]	Centrifugally cast pipe	0.85
A 351[3][4]	Austensitic steel castings	0.80
Copper and Copper Alloy		
B 61[3][4]	Steam bronze castings	0.80
B 62[3][4]	Composition bronze castings	0.80
B 148[3][4]	Al-Bronze and Si-Al-Bronze castings	0.80
B 584[3][4]	Copper alloy castings	0.80
Nickel and Nickel Alloy		
A 494[3][4]	Nickel and nickel alloy castings	0.80
Aluminum Alloy		
B 26, Temper F[3][5]	Aluminum alloy castings	1.00
B 26, Temper T6, T71[3][4]	Aluminum alloy castings	0.80

Notes:
[1] These notes are Code requirements. Those marked with an asterisk(*) restate requirements found in the text of the Code. Others are special requirements which apply to particular materials.
[2] *The quality factors for castings Ec in this table are basic factors in accordance with para. 302.3.3(b) of the ANSI/ASME B31.3 Code.
[3] *Pressure-temp. ratings of cast and forged parts as published in standards referenced in this Code section may be used for parts meeting requirements of these standards. Allowable stresses for castings and forgings, where listed, are for use in design of special components not furnished in accordance with such standards.
[4] *This casting quality factor can be enhanced by supplementary examination in accordance with para. 302.3.3(c) and Table 302.3.3(c) of the ANSI/ASME B31.3 Code. The higher factor form Table 302.3.3(c) may be substituted for this factor in pressure design equations.
[5] *This casting quality factor is applicable only when proper supplementary examination has been performed (see para. 302.3.3) of the ANSI/ASME B31.3 Code.

Table C-3 Weld Joint Quality Factors

Spec. No.	Class (Type)	Description	Ec[1]
Carbon Steel			
API 5L	---	Seamless pipe	1.00
	---	Electric resistance welded pipe	0.85
	---	Electric fusion welded pipe, double butt, straight or spiral seam	0.95
	---	Furnace butt welded	0.60
A 53	Type S	Seamless pipe	1.00
	Type E	Electric resistance welded pipe	0.85
	Type F	Furnace butt welded pipe	0.60
A 105[2]	---	Forgings and fittings	1.00
A 106	---	Seamless pipe	1.00
A 120	---	Seamless pipe	1.00
	---	Electric resistance welded pipe	0.85
	---	Furnace butt welded pipe	0.60
A 134	---	Electric fusion welded pipe, single butt, straight or spiral seam	0.80
A 135	---	Electric resistance welded pipe	0.85
A 139	---	Electric fusion welded pipe, straight or spiral seam	0.80
A 179	---	Seamless tube	1.00
A 181[2]	---	Forgings and fittings	1.00
A 211	---	Spiral welded pipe	0.75
A 234[3]	---	Seamless and welded fittings	1.00
A 333	---	Seamless pipe	1.00
	---	Electric resistance welded pipe	0.85
A 334	---	Seamless pipe	1.00
A 350[2]	---	Forgings and fittings	1.00
A 369	---	Seamless pipe	1.00
A 381[4]	---	Electric fusion welded pipe, 100% radiographed	1.00
[5]	---	Electric fusion welded pipe, spot radiographed	0.90
	---	Electric fusion welded pipe, as manufactured	0.85
A 420[3]	---	Welded fittings, 100% radiographed	1.00
A 524	---	Seamless pipe	1.00
A 587	---	Electric resistance welded pipe	0.85
A 671	12,22	Electric fusion welded pipe, 100% radiographed	1.00
	13,23	Electric fusion welded pipe, double butt seam	0.85
A 672	12,22	Electric fusion welded pipe, 100% radiographed	
	13,23	Electric fusion welded pipe, double butt seam	
A 691	12,22	Electric fusion welded pipe, 100% radiographed	1.00
	13,23	Electric fusion welded pipe, double butt seam	0.85
Low and Intermediate Alloy Steel			
A 182[2]	---	Forgings and fittings	1.00
A 234[3]	---	Seamless and welded fittings	1.00
A 333	---	Seamless pipe	1.00
	---	Electric resistance welded pipe	0.85

Table C-3 (cont.) Weld Joint Quality Factors

Spec. No.	Class (Type)	Description	Ec[1]
A 334	---	Seamless tube	1.00
A 335	---	Seamless pipe	1.00
A 350	---	Forgings and fittings	1.00
A 369	---	Seamless pipe	1.00
A 420[3]	---	Welded fittings, 100% radiographed	1.00
A 671	12,22	Electric fusion welded pipe, 100% radiographed	1.00
	13,23	Electric fusion welded pipe, double butt seam	0.85
A 672	12,22	Electric fusion welded pipe, 100% radiographed	1.00
	13,23	Electric fusion welded pipe, double butt seam	0.85
A 691	12,22	Electric fusion welded pipe, 100% radiographed	1.00
	13,23	Electric fusion welded pipe, double butt seam	0.85
Stainless Steel			
A 182	---	Forgings and fittings	1.00
A 268	---	Seamless tube	1.00
	---	Electric fusion welded tube, double butt seam	0.85
	---	Electric fusion welded tube, single butt seam	0.80
A 269	---	Seamless tube	1.00
	---	Electric fusion welded tube, double butt seam	0.85
	---	Electric fusion welded tube, single butt seam	0.80
A 312	---	Seamless pipe	1.00
	---	Electric fusion welded pipe, double butt seam	0.85
	---	Electric fusion welded pipe, single butt seam	0.80
A 358	1,3,4	Electric fusion welded pipe, 100% radiographed	1.00
	5	Electric fusion welded pipe, spot radiographed	0.90
	2	Electric fusion welded pipe, double butt seam	0.85
A 376	---	Seamless pipe	1.00
A 403	---	Seamless fittings	1.00
[3]	---	Welded fitting, 100% radiographed	1.00
	---	Welded fitting, double butt seam	0.85
	---	Welded fitting, single butt seam	0.80
A 409	---	Electric fusion welded pipe, double butt seam	0.85
	---	Electric fusion welded pipe, single butt seam	0.80
A 430	---	Seamless pipe	1.00
A 789	---	Seamless	1.00
	---	Electric fusion welded, 100% radiographed	1.00
	---	Electric fusion welded, double butt	0.85
	---	Electric fusion welded, single butt	0.80
A 790	---	Seamless	1.00
	---	Electric fusion welded, 100% radiographed	1.00
	---	Electric fusion welded, double butt	0.85
	---	Electric fusion welded, single butt	0.80
Copper and Copper Alloy			
B 42	---	Seamless pipe	1.00
B 43	---	Seamless pipe	1.00

Appendix C

Table C-3 (cont.) Weld Joint Quality Factors

Spec. No.	Class (Type)	Description	Ec[1]
B 68	---	Seamless tube	1.00
B 75	---	Seamless tube	1.00
B 88	---	Seamless water tube	1.00
B 280	---	Seamless tube	1.00
B 466	---	Seamless pipe and tube	1.00
B 467	---	Electric resistance welded pipe	0.85
	---	Electric fusion welded pipe, double butt seam	0.85
	---	Electric fusion welded pipe, single butt seam	0.85
Nickel and Nickel Alloy			
B 160[2]	---	Forgings and fittings	1.00
B 161	---	Seamless pipe and tube	1.00
B 164[2]	---	Forgings and fittings	1.00
B 165	---	Seamless pipe and tube	1.00
B 166[2]	---	Forgings and fittings	1.00
B 167	---	Seamless pipe and tube	1.00
B 366[3]	---	Seamless and welded fittings	1.00
B 407	---	Seamless pipe and tube	1.00
B 444	---	Seamless pipe and tube	1.00
B 464	---	Welded pipe	0.80
B 619	---	Electric resistance welded pipe	0.85
	---	Electric fusion welded pipe, double butt seam	0.85
	---	Electric fusion welded pipe, single butt seam	0.80
B 622	---	Seamless pipe and tube	1.00
Titanium and Titanium Alloy			
B 337	---	Seamless pipe	1.00
	---	Electric fusion welded pipe, double butt seam	0.85
Aluminum Alloy			
B 210	---	Seamless tube	1.00
B 241	---	Seamless pipe and tube	1.00
B 247[2]	---	Forgings and fittings	1.00
B 345	---	Seamless pipe and tube	1.00
B 361	---	Seamless fittings	1.00
[4][6]	---	Welded fittings, 100% radiograph	1.00
[6]	---	Welded fittings, double butt	0.85
[6]	---	Welded fittings, single butt	0.80

Notes:
[a] These notes are requirements cf the ANSI/ASME B31.3 Code. Those marked with an asterisk (*) restate requirements found in the text of the Code. Others are special requirements which apply to particular materials.

Notes Continued:
[1] *The quality factors for longitudinal weld joints Ej in this table are basic factors in accordance with para. 302.3.4(b) of the ANSI/ASME B31.3 Code.
[2] *Pressure-temp. ratings of cast and forged parts as published in standards referenced in this Code section may be used for parts meeting requirements of these standards. Allowable stresses for castings and forgings, where listed, are for use in design of special components not furnished in accordance with such standards.

Table C-3 Weld Joint Quality Factors

Spec. No.	Class (Type)	Description	Ec[1]

[3] An Ej factor of 1.00 may be applied only if all welds, incl. welds in the base material, have passed 100% radiographic examination. Substitution of ultrasonic examination for radiography is not permitted for the purpose of obtaining an Ej of 1.00.

[4] *This specification does not incl. requirements for 100% radiographic inspection. If this higher joint factor is to be used, the material shall be purchased to the special requirements of Table 341.3.2A (of the ANSI/ASME B31.3 Code) for longitudinal butt welds with 100% radiography in accordance with Table 302.3.4 of the ANSI/ASME B31.3 Code.

[5] *This specification includes requirements for random radiographic inspection for mill quality control. If the 0.90 joint factor is to be used, the welds shall meet the requirements of Table 341.3.2A (of the ANSI/ASME B31.3 Code) for longitudinal butt welds with spot radiography in accordance with Table 302.3.4 of the ANSI/ASME B31.3 Code. This shall be a matter of special agreement between purchaser and manufacturer.

[6] Light-weight aluminum alloy welded fittings conforming to dimensions in MSS SP-43 shall have full penetration welds.

Notes

Notes

Appendix D

Table D-1 Nonmetallic Component Standards Listed in the ANSI/ASME B31.3 Code for Chemical Plant & Petroleum Refinery Piping (1993 Edition)

Standard or Specification	Designation [2]
NONMETALLIC FITTINGS	
Process glass pipe and fittings	ASTM C 599
Threaded PVC plastic pipe fittings, Sch 80	ASTM D 2464
PVC plastic pipe fittings. Sch 40	ASTM D 2466
Socket-type PVC plastic pipe fittings, Sch 80	ASTM D 2467
Socket-type ABS plastic pipe fittings, Sch 40	ASTM D 2468
Thermoplastic gas pressure pipe, tubing, and fittings	ASTM D 2513
Reinforced epoxy resin gas pressure pipe and fittings	ASTM D 2517
Plastic insert fittings for PE plastic pipe	ASTM D 2609
Socket-type PE fittings for outside diameter-controlled PE pipe and tubing	ASTM D 2683
CPVC plastic hot and cold water distribution systems	ASTM D 2846
Butt heat fusion PE plastic fittings for PE plastic pipe and tubing	ASTM D 3261
PB plastic hot-water distribution systems	ASTM D 3309
Fiberglass RTR pipe fittings for nonpressure applications [3]	ASTM D 3840
RTR flanges	ASTM D 4024
PTFE plastic-lined ferrous metal pipe and fittings	ASTM F 423
Threaded CPVC plastic pipe fittings, Sch 80	ASTM F 437
Socket-type CPVC plastic pipe fittings, Sch 40	ASTM F 438
Socket-type CPVC plastic pipe fittings, Sch 80	ASTM F 439
PVDF plastic-lined ferrous metal pipe and fittings	ASTM F 491
Propylene and PP plastic-lined ferrous metal pipe and fittings	ASTM F 492
FEP plastic-lined ferrous metal pipe and fittings	ASTM F 546
PVDC plastic-lined ferrous metal pipe and fittings	ASTM F 599
PFA plastic-lined ferrous metal pipe and fittings	ASTM F 781
Custom contact-molded reinforced-polyester chemical resistant process equipment	NBS PS15
NONMETALLIC PIPES AND TUBES	
PE line pipe	API 15LE
Thermoplastic line pipe (PVC and CPVC)	API 15LP
Low pressure fiberglass line pipe	API 15LR

Table D-1 Nonmetallic Component Standards Listed in the ANSI/ASME B31.3 Code
(cont.) for Chemical Plant & Petroleum Refinery Piping (1993 Edition)

Standard or Specification	Designation [2]
Reinforced concrete low-head pressure pipe	ASTM C 361
Process glass pipe and fittings	ASTM C 599
ABS plastic pipe, Sch 40 and 80	ASTM D 1527
PVC plastic pipe, Sch 40, 80 and 120	ASTM D 1785
PE plastic pipe, Sch 40	ASTM D 2104
PE plastic pipe (SIDR-PR) based on controlled inside diameter	ASTM D 2239
PVC plastic pressure-rated pipe (SDR series)	ASTM D 2241
ABS plastic pipe (SDR-PR)	ASTM D 2282
Classification for machine-made RTR pipe	ASTM D 2310
PE plastic pipe, Sch 40 and 80, based on outside diameter	ASTM D 2447
Thermoplastic gas pressure pipe, tubing, and fittings	ASTM D 2513
Reinforced epoxy resin gas pressure pipe and fittings	ASTM D 2517
PB plastic pipe (SDR-PR)	ASTM D 2662
PB plastic tubing	ASTM D 2666
Bell end PVC plastic pipe	ASTM D 2672
PE plastic tubing	ASTM D 2737
CPVC plastic hot and cold water distribution system	ASTM D 2846
Filament-wound fiberglass RTR pipe [3]	ASTM D 2996
Centrifugally cast RTR pipe	ASTM D 2997
PB plastic pipe (SDR-PR) based on outside diameter	ASTM D 3000
PE plastic pipe (SDR-PR) based on controlled outside diameter	ASTM D 3035
PB plastic hot-water distribution systems	ASTM D 3309
Fiberglass RTR pressure pipe [3]	ASTM D 3517
Fiberglass RTR sewer and industrial pressure pipe [3]	ASTM D 3754
PTFE plastic-lined ferrous metal pipe and fittings	ASTM F 423
CPVC plastic pipe	ASTM F 441
CPVC plastic pipe (SDR-PR)	ASTM F 442
PVDF plastic-lined ferrous metal pipe and fittings	ASTM F 491
PP plastic-lined ferrous metal pipe and fittings	ASTM F 492
FEP plastic-lined ferrous metal pipe and fittings	ASTM F 546
PVDC plastic-lined ferrous metal pipe and fittings	ASTM F 599
PFA plastic-lined ferrous metal pipe and fittings	ASTM F 781
Reinforced concrete pressure pipe, steel cylinder type, for water and other liquids	AWWA C300
Prestressed concrete pressure pipe, steel cylinder type, for water and other liquids	AWWA C301
Reinforced concrete pressure pipe, noncylinder type, for water and other liquids	AWWA C302
PVC pressure pipe, 4' through 12", for water	AWWA C900
Glass-fiber-reinforced thermosetting resin pressure pipe	*AWWA C950
Custom contact-molded reinforced-polyester chemical resistant process equipment	NBS PS15

MISCELLANEOUS

Contact-molded reinforced thermosetting plastic (RTP) laminates for corrosion resistant equipment	ASTM C 582
Threads for fiberglass RTR pipe (60° stub) [3]	ASTM D 1694
Solvent cements for ABS plastic pipe and fittings	ASTM D 2235
Solvent cements for PVC plastic pipe and fittings	ASTM D 2564
Bell end PVC plastic pipe	ASTM D 2672
Joints for plastic pressure pipes using flexible elastomeric seals	ASTM D 3139
Fiberglass RTR pipe joints using flexible elastomeric seals [3]	ASTM D 4161

Appendix D

Table D-1 Nonmetallic Component Standards Listed in the ANSI/ASME B31.3 Code
(cont.) for Chemical Plant & Petroleum Refinery Piping (1993 Edition)

Standard or Specification	Designation [2]
Design and construction of nonmetallic enveloped gaskets for corrosive service	ASTM F 336
Solvent cements for CPVC plastic pipe and fittings	ASTM F 493

Notes:
[1] It is not practical to refer to a specific edition of each standard throughout the Code text. Instead, the approved edition references, along with the names and addresses of the sponsoring organizations, are shown in Appendix E of the ASME B31.3 Code.
[2] An asterisk (*) preceding the designation indicates that the standard has been approved as an American National Standard by the American National Standards Institute.
[3] The term fiberglass RTR takes the place of the ASTM designation: "fiberglass" (glass-fiber-reinforced thermosetting resin).

Notes

Appendix E

Table E-1 Expansion Factor, c

Temp. T. °F	Carbon steel C ≤ .30%	Carbon steel C > .30%	C-Moly and low Cr.-Moly Cr ≤ 3%	Cr.-Moly 5% ≤ Cr. Mo ≤ 9%	Austenitic Stainless Steels	Cr. Stainless Steels 12 Cr., 17 Cr. & 27 Cr.	25 Cr.- 20 Ni	Wrought iron
70	0	0	0	0	0	0	0	0
100	37	40	40	35	54	34	47	44
150	98	106	106	92	143	90	125	120
200	160	171	171	149	232	145	204	195
250	228	244	244	212	323	204	287	273
300	294	315	315	271	414	264	368	352
350	365	391	391	335	509	326	455	434
400	436	467	467	396	603	389	541	514
450	510	547	547	465	699	455	629	598
500	584	626	626	531	794	520	716	681
550	664	711	711	603	893	590	809	768
600	743	796	796	672	989	659	901	855
650	827	886	886	714	1089	730	995	946
700	909	974	974	815	1189	799	1088	1035
750	996	1068	1068	891	1292	874	1186	1125
775	1038	1113	1113	929	1344	909	1235	1171
800			1159	967	1395	946	1284	1216
825			1208	1005	1448	983	1335	
850			1256	1043	1500	1022	1384	
875			1303	1081	1552	1061	1435	
900			1351	1121	1605	1097	1484	
925			1398	1161	1659	1134	1533	
950			1445	1200	1713	1174	1585	
975			1492	1240	1766	1212	1634	
1000			1538	1278	1820	1250	1681	
1050			1639	1357	1928	1328	1781	
1100			1737	1435	2036	1404	1879	
1150				1511	2144	1480	1980	

(1) Expansion Factor, c = (Expansion in inches per 100 ft. x E_c)/(1728 x 100).
(2) Data taken from "Piping Design and Engineering," ITT Grinell Industrial Piping, 5th Edition, Providence, RI, 1976.

Table E-2 Expansion Loops with Equal Tangents (Ginnell Method; ref. Fig. 8.31)

$\dfrac{L_C+L_B+L_D}{L_B}$	2		3		4		5	
$\dfrac{L_C+L_B+L_D}{L_A}$	k_x	k_b	k_x	k_b	k_x	k_b	k_x	k_b
1.0	2.40	7.20	2.46	8.2	2.52	8.82	2.58	9.29
1.2	3.70	9.25	4.46	10.9	4.65	12.0	4.78	12.8
1.4	5.31	11.37	6.46	13.6	6.79	15.2	6.98	16.3
1.6	7.22	13.53	8.46	16.3	8.93	18.4	9.20	19.8
1.8	9.45	15.75	10.48	19.0	11.08	21.6	11.42	23.4
2.0	12.0	18.0	12.5	21.8	13.24	24.8	13.87	27.1
2.2	14.85	20.25	15.8	24.9	16.6	28.5	16.9	31.0
2.4	18.0	22.5	19.6	28.0	20.4	32.2	20.8	35.3
2.6	21.52	24.83	23.4	31.1	24.4	35.9	25.5	39.7
2.8	25.32	27.1	27.3	34.2	28.9	39.7	30.6	44.0
3.0	29.45	29.45	31.2	37.4	33.6	43.7	35.8	48.7
3.2	33.9	31.8	35.6	40.6	39.0	47.6	41.2	53.3
3.4	38.7	34.1	40.0	43.8	44.5	51.6	46.9	58.0
3.6	43.7	36.5	46.1	47.0	50.3	55.6	53.0	62.8
3.8	49.1	38.8	52.3	50.2	57.0	59.8	60.2	67.6
4.0	54.9	41.1	58.5	53.6	64.0	64.0	69.1	72.5
4.2	60.8	43.4	64.7	57.0	71.1	68.2	78.1	77.5
4.4	67.3	45.9	71.0	60.4	78.9	72.4	87.2	82.5
4.6	73.9	48.2	79.1	63.8	87.0	76.6	96.3	87.5
4.8	81.0	50.6	87.2	67.3	95.8	80.8	105.4	92.5
5.0	88.2	52.9	95.3	70.8	104.6	85.2	114.7	97.8
5.2	95.9	55.3	104.4	74.3	114.0	89.5	125.0	103.0
5.4	103.8	57.7	113.5	77.8	123.6	93.9	136.3	108.3
5.6	112.1	60.1	122.6	81.3	134.0	98.3	147.6	113.5
5.8	120.7	62.4	132.0	84.8	144.6	102.7	159.0	118.8
6.0	129.6	64.8	141.6	88.4	155.8	107.0	171.3	124.1
6.2	138.8	67.2	152.4	91.9	167.2	111.5	184.0	129.5
6.4	148.4	69.6	163.3	95.4	179.1	116.0	198.0	134.9
6.6	158.2	71.9	174.2	98.9	191.0	120.5	212.2	140.3
6.8	168.4	74.3	185.2	102.4	204.0	125.0	226.4	145.7
7.0	178.9	76.7	196.3	106.0	217.0	129.4	240.7	151.1
7.2	189.8	79.1	209.1	109.5	230.5	133.9	256.0	156.6
7.4	200.9	81.5	221.9	113.0	244.2	138.4	271.5	162.1
7.6	212.4	83.8	234.7	116.5	259.2	142.9	287.5	167.6
7.8	224.2	86.2	247.6	120.0	274.5	147.4	304.3	173.1
8.0	236.2	88.6	260.7	123.5	289.8	152.0	322	178.6
8.2	248.7	91.0	275.0	127.0	305	156.6	340	184.1
8.4	261.5	93.4	289.3	130.5	320	161.2	358	189.7
8.6	274.6	95.8	304	134.0	336	165.8	377	195.3
8.8	287.9	98.2	318	137.5	351	170.4	395	200.9
9.0	302	100.5	332	141.0	367	175.0	416	206.6
9.2	316	102.9	348	144.5	384	179.6	437	212.4
9.4	330	105.4	365	148.0	402	184.2	458	218.2
9.6	345	107.7	381	151.5	422	188.8	480	224.0
9.8	360	110.1	397	155.0	443	193.4	502	229.8
10.0	375	112.5	414	158.7	466	198.1	525	236.1

Appendix E

Table E-2 (cont.) Expansion Loops with Equal Tangents (Grinnell Method)

$\dfrac{L_C+L_B+L_D}{L_B}$	6		7		8		9		10	
$\dfrac{L_C+L_B+L_D}{L_A}$	k_x	k_b	k_x	k_b	k_x	k_b	k_x	k_b	k_x	k_b
1.0	2.64	9.69	2.67	9.92	2.70	10.1	2.73	10.3	2.75	10.45
1.2	4.84	13.3	5.0	13.9	5.2	14.0	5.29	14.4	5.35	14.9
1.4	7.1	17.0	7.4	17.9	7.7	18.1	7.85	18.6	7.95	19.4
1.6	9.5	20.8	9.8	22.0	10.2	22.3	10.41	22.9	10.55	23.9
1.8	11.9	24.7	12.3	26.1	12.7	26.7	12.97	27.4	13.15	28.5
2.0	14.4	28.8	14.9	30.2	15.3	31.2	15.53	32.3	15.79	33.2
2.2	17.5	33.4	18.0	34.8	18.6	36.2	20.0	38.0	21.0	38.6
2.4	21.3	38.0	22.5	40.0	23.8	41.7	25.2	43.7	26.3	44.4
2.6	26.2	42.7	27.5	45.3	29.0	47.3	30.7	49.5	31.7	50.5
2.8	31.7	47.5	33.0	50.7	34.7	53.0	36.3	55.4	37.2	56.9
3.0	37.7	52.7	39.3	56.2	40.7	59.0	41.9	61.5	43.0	63.6
3.2	43.7	58.0	45.7	61.7	48.0	65.1	50.0	67.8	50.8	70.6
3.4	49.5	63.3	52.2	67.3	55.5	71.3	58.3	74.2	59.2	77.7
3.6	57.5	68.7	59.5	73.0	63.2	77.7	66.7	80.7	68.0	84.9
3.8	65.5	74.1	68.5	79.4	71.5	84.2	75.2	87.3	77.1	92.2
4.0	73.6	79.7	77.5	85.8	80.9	91.0	84.2	94.4	86.6	99.5
4.2	82.0	85.2	87.0	92.2	90.4	97.8	95.0	102.0	97.0	107.0
4.4	91.0	90.8	96.5	98.6	100.5	104.7	106.2	109.6	108.0	114.7
4.6	101.7	96.3	106.5	105.0	112.0	111.7	117.8	117.2	120.0	122.8
4.8	112.4	101.9	118.0	111.4	124.2	118.7	129.8	125.0	133.3	131.0
5.0	122.5	107.5	130.0	117.8	136.7	125.9	142.5	133.0	147.9	139.4
5.2	134.0	113.7	142.0	124.5	149.4	133.2	157.5	141.0	163.0	147.9
5.4	146.0	120.0	155.0	131.3	162.4	140.6	172.6	149.0	178.5	156.5
5.6	159.0	126.2	169.0	138.1	177.0	148.1	187.0	157.1	194.5	165.2
5.8	172.0	132.5	183.0	144.9	192.6	155.6	202.7	165.2	211.0	173.9
6.0	185.2	138.8	197.8	151.81	209.0	163.3	219.2	173.5	228.3	182.6
6.2	199.0	145.1	213.0	58.8	225.0	171.0	236.0	181.8	145.8	191.4
6.4	213.0	151.4	228.5	165.8	241.8	178.8	253.0	190.2	163.8	200.2
6.6	228.0	157.7	245.0	172.9	259.5	186.6	271.0	198.7	282.8	209.1
6.8	244.2	164.0	262.5	180.0	279.0	194.5	292.0	207.2	305	218.0
7.0	261.8	170.3	280.7	187.1	298.7	202.5	314	216.0	328	227.1
7.2	279.8	176.7	299.5	194.3	319	210.5	336	224.7	351	236.6
7.4	297.8	183.1	319.0	201.6	339	218.5	358	233.5	374	246.3
7.6	316	189.5	339.0	208.9	359	226.5	381	242.3	398	256.0
7.8	334	195.9	359.0	216.2	381	234.6	404	251.1	422	265.8
8.0	352	202.4	379	223.5	405	242.7	427	260.0	448	275.6
8.2	372	208.9	400	231.0	429	250.9	451	268.9	475	285.5
8.4	392	215.4	422	238.5	455	259.2	476	277.8	502	295.4
8.6	413	221.9	445	246.0	480	267.5	502	286.7	530	305
8.8	434	228.4	470	253.5	505	275.8	529	295.7	560	315
9.0	456	235.0	495	261.0	530	284.1	559	305	590	325
9.2	479	241.6	520	268.6	556	292.4	589	314	620	335
9.4	503	248.2	545	276.2	584	301	618	323	651	345
9.6	527	254.8	570	283.8	611	309	648	332	684	355
9.8	551	261.4	596	291.4	639	317	680	342	717	365
10.0	575	268.2	624	299.0	666	326	711	351	750	375

Table E-3 Table for Loops with Tangents $L_1/L_2 = 2$ (Grinnell Method; ref. Fig. 8.31)

$\frac{L_C+L_B+L_D}{L_B}$	2			4			6			8			10		
$\frac{L_C+L_B+L_D}{L_A}$	k_x	k_y	k_b	k_x	k_y	k_b	k_x	k_y	k_b	k_x	k_y	k_b	k_x	k_y	k_b
1.0	2.5	0.5	8.7	2.7	1.1	9.9	2.8	1.4	10.7	2.9	1.5	10.9	3.0	1.6	11.4
1.2	4.4	0.7	11.7	5.3	1.6	12.5	5.0	1.8	14.6	5.5	2.0	15.5	5.5	2.1	16.2
1.4	6.4	1.0	14.6	8.0	2.1	15.1	7.6	2.2	18.8	8.2	2.5	20.1	8.2	2.6	21.0
1.6	8.4	1.2	17.5	10.8	2.6	17.7	10.2	2.7	23.0	10.9	2.9	24.8	11.0	3.0	25.8
1.8	10.4	1.5	20.4	13.6	3.1	20.4	12.9	3.1	27.2	13.6	3.4	29.5	14.0	3.5	30.6
2.0	12.4	1.7	23.3	16.5	3.6	23.3	15.6	3.6	31.5	16.5	3.8	34.2	17.2	4.0	35.5
2.2	16.0	2.0	26.6	20.3	3.9	28.3	20.6	4.0	36.6	20.0	4.2	39.0	21.5	4.5	41.8
2.4	19.7	2.3	29.9	24.1	4.3	33.4	25.6	4.5	41.7	24.0	4.7	44.0	27.5	4.9	48.2
2.6	23.4	2.7	33.2	28.0	4.6	38.5	30.6	4.9	46.8	28.0	5.1	49.1	33.5	5.5	54.6
2.8	27.1	3.0	36.5	32.0	4.9	43.6	35.6	5.4	51.9	33.0	5.6	54.6	39.5	6.0	61.0
3.0	30.9	3.3	40.1	36.1	5.3	48.7	40.7	5.9	57.0	40.8	6.0	59.6	46.5	6.5	67.0
3.2	36.3	3.7	43.7	41.0	5.7	53.2	48.0	6.3	62.8	47.0	6.5	68.0	54.0	6.9	74.0
3.4	41.7	4.0	47.3	47.0	6.2	57.7	56.0	6.8	69.0	55.0	7.0	75.0	62.0	7.4	81.0
3.6	46.1	4.4	50.9	53.0	6.6	62.0	64.0	7.3	74.0	65.0	7.4	83.0	71.0	7.8	88.0
3.8	51.5	4.8	54.5	60.0	7.1	67.0	72.0	7.8	80.0	76.0	7.9	91.0	81.0	8.3	95.0
4.0	58.0	5.2	58.0	69.0	7.5	71.0	80.0	8.2	86.0	87.0	8.4	100	92.0	8.8	103
4.2	65.0	5.5	62.0	77.0	8.0	76.0	90.0	8.7	92.0	98.0	8.9	107	105	9.3	111
4.4	72.0	5.9	65.0	86.0	8.4	80.0	100	9.1	99.0	109	9.4	115	118	9.8	120
4.6	79.0	6.3	69.0	95.0	8.8	85.0	111	9.5	105	120	9.8	122	132	10.3	128
4.8	86.0	6.7	72.0	104	9.0	89.0	122	9.9	111	133	10.3	130	145	10.8	137
5.0	94.0	7.0	76.0	113	9.6	94.0	132	10.4	116	147	10.8	138	159	11.3	146
5.2	103	7.5	80.0	124	10.0	99.0	145	10.9	122	158	11.3	146	172	11.7	154
5.4	112	7.9	84.0	135	10.5	104	158	11.3	128	173	11.7	154	188	12.1	163
5.6	121	8.3	88.0	146	10.9	108	172	11.8	134	189	12.2	162	206	12.6	171
5.8	131	8.7	92.0	157	11.4	113	185	12.3	140	206	12.6	171	223	13.0	180
6.0	140	9.2	96.0	168	11.8	118	199	12.7	146	224	13.1	179	242	13.5	188
6.2	150	9.5	99.0	180	12.3	123	216	13.2	153	241	13.6	186	261	14.0	198
6.4	161	9.9	103	193	12.7	127	232	13.6	160	258	14.1	193	283	14.5	208
6.6	171	10.3	107	206	13.1	132	248	14.1	166	275	14.5	201	306	15.0	218
6.8	182	10.7	110	220	13.6	137	264	14.5	173	292	15.0	208	329	15.5	228
7.0	192	11.1	114	234	14.0	142	280	14.9	180	310	15.5	215	352	16.0	238
7.2	204	11.5	117	249	14.4	147	299	15.4	186	329	15.9	222	375	16.4	247
7.4	216	11.9	121	264	14.8	152	319	15.8	192	348	16.3	229	399	16.8	256
7.6	228	12.3	125	279	15.2	157	338	16.2	198	367	16.7	236	425	17.2	265
7.8	240	12.7	128	295	15.6	162	357	16.6	204	386	17.1	243	450	17.7	274
8.0	252	13.1	132	312	16.1	167	377	17.1	211	408	17.5	251	476	18.2	283
8.2	266	13.5	136	330	16.6	172	398	17.5	217	430	17.9	259	502	18.6	293
8.4	280	13.9	140	348	17.0	176	420	18.0	224	453	18.4	268	530	19.0	303
8.6	295	14.3	143	366	17.5	181	442	18.4	231	476	18.9	277	561	19.5	313
8.8	309	14.7	147	384	18.0	186	465	18.9	237	503	19.3	286	592	20.0	324
9.0	323	15.2	151	402	18.4	191	487	19.4	244	531	19.8	295	624	20.6	335
9.2	339	15.6	156	421	18.8	196	512	19.7	251	560	20.2	304	657	21.0	345
9.4	355	16.0	160	440	19.2	201	537	20.1	257	588	20.7	312	689	21.4	354
9.6	371	16.4	165	460	19.6	205	562	20.5	264	618	21.1	320	722	21.8	364
9.8	387	16.8	169	480	20.0	210	587	20.9	270	646	21.6	329	754	22.2	374
10.0	403	17.3	175	500	20.4	215	613	21.3	277	675	22.0	337	787	22.7	384

Table E-4 Table for Loops with Tangents $L_1/L_2 = 3$ (Grinnell Method; ref. Fig. 8.31)

$\frac{L_C+L_B+L_D}{L_B}$	2			4			6			8			10		
$\frac{L_C+L_B+L_D}{L_A}$	k_x	k_y	k_b	k_x	k_y	k_b	k_x	k_y	k_b	k_x	k_y	k_b	k_x	k_y	k_b
1.0	2.5	0.7	9.3	2.8	1.4	10.2	3.1	2.0	11.4	3.2	2.1	11.8	3.3	2.2	12
1.2	4.0	1.1	12.7	4.0	1.9	15	4.0	2.6	15.6	5.8	2.8	16.6	6.0	2.9	17
1.4	6.0	1.5	16.1	6.0	2.4	20	7.0	3.3	20	8.4	3.5	21.4	9.0	3.6	22
1.6	8.0	1.8	19.5	9.0	2.9	25	10	3.9	25	11	4.2	26	11	4.3	27
1.8	10	2.2	23	12	3.5	30	13	4.6	29	14	4.9	31	14	5.0	32
2.0	13	2.6	26	15	4.0	35	17	5.3	33	18	5.6	36	19	5.8	37
2.2	16	3.1	30	18	4.6	38	21	5.6	39	22	6.3	40	22	6.6	44
2.4	20	3.6	34	21	5.2	41	26	5.9	44	26	7.1	46	28	7.4	50
2.6	24	4.2	39	26	5.8	44	31	6.2	49	31	7.8	52	34	8.2	57
2.8	28	4.7	43	31	6.4	47	37	6.5	54	39	8.6	59	43	9.0	64
3.0	33	5.2	47	38	7.0	50	45	7.0	60	48	9.3	66	52	9.8	71
3.2	39	5.8	51	44	7.6	55	51	7.9	66	57	10.0	72	60	10.5	78
3.4	44	6.3	55	51	8.2	59	59	9.0	72	65	10.7	79	70	11.1	86
3.6	50	6.9	59	59	8.8	64	63	10.0	78	74	11.5	86	80	11.8	93
3.8	56	7.5	64	67	9.4	69	77	11.2	83	84	12.2	93	90	12.4	100
4.0	62	8.1	68	73	10.0	73	87	12.4	89	95	12.8	100	101	13.1	109
4.2	69	8.7	73	82	10.6	78	96	13.1	95	107	13.6	108	115	13.8	117
4.4	76	9.3	78	92	11.3	83	107	13.8	101	119	14.3	115	129	14.5	125
4.6	84	9.9	82	102	11.9	88	118	14.5	108	132	15.1	122	143	15.3	134
4.8	92	10.6	87	111	12.6	93	130	15.2	115	145	15.8	130	157	16.0	142
5.0	101	11.2	92	120	13.2	97	144	16.0	122	160	16.5	138	172	16.7	152
5.2	110	11.8	96	131	13.9	102	157	16.7	128	174	17.2	145	188	17.4	161
5.4	119	12.4	100	142	14.5	107	170	17.4	134	190	17.9	153	205	18.1	170
5.6	129	13.0	105	154	15.2	112	184	18.1	141	208	18.7	161	222	18.8	178
5.8	139	13.6	109	166	15.8	117	199	18.8	148	226	19.4	169	241	19.6	187
6.0	148	14.2	114	179	16.5	121	216	19.7	155	244	20.1	178	262	20.3	196
6.2	159	14.9	119	191	17.1	127	231	20.3	161	262	20.7	185	283	21.0	205
6.4	170	15.7	124	204	17.8	132	249	21.0	167	280	21.4	193	305	21.8	215
6.6	181	16.4	129	219	18.5	137	267	21.6	174	300	22.1	201	330	22.6	224
6.8	194	17.1	134	234	19.1	142	285	22.3	181	321	22.9	209	353	23.4	233
7.0	208	17.8	139	250	19.8	147	304	23.0	187	344	23.6	219	376	24.2	243
7.2	220	18.5	144	266	20.5	152	323	23.7	194	367	24.2	227	403	24.9	253
7.4	232	19.1	149	282	21.2	157	344	24.4	201	390	24.8	235	430	25.6	263
7.6	246	19.8	155	299	21.8	162	365	25.1	208	413	25.5	242	457	26.2	273
7.8	260	20.4	159	317	22.5	167	387	25.8	215	438	26.1	251	485	26.9	283
8.0	276	21.1	163	335	23.2	172	410	26.5	222	464	26.7	260	513	27.6	294
8.2	290	21.8	168	354	23.9	177	432	27.2	230	490	27.5	268	542	28.3	303
8.4	305	22.5	173	373	24.6	182	455	27.9	237	518	28.2	277	573	29.0	313
8.6	321	23.2	178	392	25.2	187	480	28.6	244	546	29.0	286	605	29.7	323
8.8	338	23.9	183	411	25.9	192	506	29.3	251	574	29.7	294	637	30.4	333
9.0	355	24.6	188	430	26.6	197	532	30.2	258	604	30.5	303	669	31.1	343
9.2	372	25.3	192	450	27.3	203	560	31.0	265	634	31.2	312	703	31.7	353
9.4	389	25.9	197	470	27.9	209	588	31.8	273	665	32.0	321	738	32.4	363
9.6	406	26.6	202	491	28.6	214	619	32.6	281	697	32.7	330	774	33.0	373
9.8	423	27.2	207	513	29.3	219	649	33.4	289	730	33.4	339	810	33.7	383
10.0	440	27.9	212	538	30.0	224	680	34.2	298	765	34.2	348	845	34.3	395

Table E-5 Table for Loops with Tangents $L_1/L_2 = 4$ (Grinnell Method; ref. Fig. 8.31)

$\frac{L_C+L_B+L_D}{L_B}$	2			4			6			8			10		
$\frac{L_C+L_B+L_D}{L_A}$	k_x	k_y	k_b	k_x	k_y	k_b	k_x	k_y	k_b	k_x	k_y	k_b	k_x	k_y	k_b
1.0	2.6	0.8	9.7	2.9	1.85	10.8	3.2	2.3	11.8	3.4	2.4	12.2	3.5	2.5	12.5
1.2	4.5	1.3	11.0	5.0	2.5	15	5.2	3.1	16.4	6	3.2	17.1	6.2	3.4	17.6
1.4	6.7	1.8	15.0	7.7	3.3	19	8	3.9	20.9	9	4.0	22.1	9	4.2	22.8
1.6	8.9	2.3	19.0	10.5	4.0	23	11	4.7	25.5	12	4.8	27.1	12	5.0	28.0
1.8	11.2	2.7	23.0	13.3	4.7	27	14	5.4	30	15	5.6	32	16	5.8	33.2
2.0	13.5	3.2	28.5	16.3	5.4	31	18	6.2	35	19	6.5	37	20	6.7	38.6
2.2	17.5	3.8	33.0	20	6.3	35	22	7.0	40	23	7.4	43	24	7.6	45.2
2.4	21.5	4.5	37.5	25	7.2	40	27	7.9	46	29	8.3	49	30	8.4	52
2.6	25.5	5.1	42.0	30	8.0	45	33	8.8	51	36	9.1	56	37	9.3	58
2.8	29.7	5.8	46.5	36	8.9	50	40	9.6	57	43	10.0	62	45	10.2	65
3.0	34.2	6.4	51	4.3	9.8	55	48	10.5	63	52	10.9	68	54	11.0	72
3.2	40.4	7.1	56	49	10.6	59	56	11.4	69	60	11.8	75	63	11.9	80
3.4	47	7.8	61	56	11.5	64	65	12.3	75	70	12.7	82	72	12.8	88
3.6	52	8.5	66	64	12.3	69	74	13.1	81	80	13.6	89	82	13.7	96
3.8	59	9.2	71	72	13.2	74	83	14.0	87	90	14.5	96	94	14.6	104
4.0	66	10.0	76	81	14.1	79	93	15.0	93	102	15.5	104	108	15.7	112
4.2	72	10.8	81	90	15.0	84	104	15.8	99	113	16.4	111	123	16.5	121
4.4	80	11.6	86	100	15.9	89	116	16.7	106	126	17.2	118	138	17.4	130
4.6	88	12.4	91	111	16.8	94	128	17.6	112	140	18.1	126	153	18.3	138
4.8	97	13.2	96	122	17.7	99	141	18.5	118	155	18.9	133	168	19.2	147
5.0	107	14.0	102	133	18.5	104	154	19.4	125	170	19.8	142	183	20.1	155
5.2	117	14.8	107	144	19.4	109	168	20.3	132	185	20.7	150	200	21.0	164
5.4	127	15.6	112	157	20.3	114	183	21.2	139	201	21.6	158	218	21.9	173
5.6	137	16.4	117	171	21.1	119	200	22.2	146	219	22.4	166	236	22.8	182
5.8	148	17.2	123	185	22.0	125	217	23.1	153	238	23.3	174	257	23.7	191
6.0	159	18.0	129	199	22.9	131	234	24.0	160	258	24.2	182	278	24.5	200
6.2	170	18.9	135	215	23.8	136	251	24.9	167	277	25.1	191	299	25.5	209
6.4	181	19.7	140	230	24.7	141	269	25.9	174	296	26.0	199	321	26.4	219
6.6	194	20.6	146	245	25.5	146	288	26.8	181	316	26.9	208	346	27.3	229
6.8	208	21.5	152	260	26.4	151	309	27.8	188	340	27.9	216	371	28.2	239
7.0	222	22.4	158	276	27.4	156	330	28.7	195	367	28.8	225	399	29.2	249
7.2	236	23.2	163	293	28.3	161	351	29.4	202	389	29.6	233	427	30.0	259
7.4	250	24.2	169	311	29.2	166	372	30.2	208	412	30.4	241	455	30.9	269
7.6	264	25.1	175	330	30.1	171	393	31.0	215	436	31.2	249	483	31.7	279
7.8	279	26.0	181	350	31.0	177	415	31.8	221	463	32.0	257	511	32.6	289
8.0	296	27.0	187	370	32.0	184	437	32.7	228	490	32.8	265	540	33.4	299
8.2	313	27.9	192	390	32.8	189	460	33.4	235	519	33.7	274	570	34.3	309
8.4	330	28.8	198	410	33.7	194	484	34.3	241	548	34.6	283	603	35.2	319
8.6	347	29.7	204	430	34.6	199	510	35.1	248	578	35.5	292	637	36.0	330
8.8	364	30.6	210	451	35.5	204	536	36.0	255	608	36.4	301	673	36.9	341
9.0	382	31.4	216	473	36.3	209	562	37.0	262	640	37.4	310	710	37.8	352
9.2	400	32.3	222	496	37.2	214	590	37.8	268	672	38.1	318	747	38.6	362
9.4	419	33.2	228	519	38.1	219	615	38.7	275	704	38.9	326	784	39.5	372
9.6	438	34.1	234	542	39.0	224	645	39.6	282	736	39.7	334	821	40.3	382
9.8	457	35.0	240	566	40.0	229	680	40.4	288	768	40.5	342	858	41.2	392
10.0	476	36.0	246	590	41.0	235	710	41.3	296	800	41.3	351	896	42.0	403

Table E-6 Table for Expansion Offset (Grinnell Method; ref. Fig. 8.40)

$\frac{L_1}{L_3}$	1			1.5			2			3			4		
$\frac{L_1+L_3}{L_2}$	k_x	k_y	k_b	k_x	k_y	k_b	k_x	k_y	k_b	k_x	k_y	k_b	k_x	k_y	k_b
0.6	9.25	43.0	83.8	8.5	38	91	7.3	32	85	6.5	25	73	6.0	22	66
0.8	12.8	39.0	69.0	11.8	35	76	10.5	29	71	9.2	23	62	8.5	20	56
1.0	17.2	37.9	61.9	15.9	34	69	14.4	29	66	12.6	22	52	11.8	19	50
1.2	22.5	37.8	57.8	21.0	35	69	18	29	66	16.0	23	53	14	20	51
1.4	28.3	37.7	60.6	27	36	69	22	30	67	20	24	55	19	21	52
1.6	35.4	42.1	66.3	34	37	71	30	32	69	27	25	57	24	21	55
1.8	43.0	43.2	72.0	41	39	75	38	33	71	34	26	60	30	22	58
2.0	52.8	45.7	79.3	50	41	81	46	35	76	42	27	67	40	24	63
2.2	63.0	48.0	86.5	60	43	88	57	38	83	51	29	73	48	25	68
2.4	76.0	51.0	93.8	71	46	96	68	40	90	61	31	80	58	27	74
2.6	89.0	54.5	101.2	83	49	102	79	43	97	71	33	86	69	29	80
2.8	102	58.2	109.0	96	53	110	91	46	105	82	35	92	80	30	87
3.0	116	62.2	116.1	110	56	118	104	49	115	92	37	99	90	32	93
3.2	132	66.0	124.5	124	59	126	118	51	121	106	39	107	104	34	99
3.4	149	70.0	133.0	140	63	134	133	54	128	121	41	114	118	36	105
3.6	168	74.0	141.0	168	66	143	149	57	135	136	44	120	132	38	111
3.8	188	78.0	149.0	177	70	151	165	60	142	151	46	127	147	40	117
4.0	210	82.0	157.8	197	73	159	181	63	150	166	49	133	163	42	124
4.2	235	86.4	166.0	219	77	168	201	66	157	185	51	140	182	44	130
4.4	260	90.6	174.5	241	81	177	221	69	164	204	53	147	201	46	137
4.6	285	94.8	183.0	263	85	186	241	72	173	223	56	154	220	48	143
4.8	310	99.0	192.0	287	88	194	263	75	182	243	58	161	239	50	150
5.0	336	103.2	201.4	314	92	203	288	78	190	264	61	168	260	52	156
5.2	364	107.6	210.0	341	96	212	313	81	198	286	63	175	281	54	163
5.4	393	111.8	219.5	370	100	221	339	85	206	310	66	182	304	56	169
5.6	425	116.2	228.0	399	104	230	365	88	214	335	68	189	329	58	176
5.8	457	120.5	237.5	430	108	239	392	92	223	360	71	197	355	61	182
6.0	491	124.8	245.5	461	112	248	422	95	232	386	73	205	381	63	190
6.2	526	129.4	254.5	493	116	258	450	98	240	414	76	212	408	65	196
6.4	562	133.8	263.5	526	120	267	478	102	248	443	79	219	436	67	203
6.6	598	138.2	273.0	561	124	276	506	105	256	472	81	227	465	69	210
6.8	633	142.6	282.0	598	128	285	535	108	265	502	84	234	495	72	217
7.0	670	145.0	287.0	636	132	294	565	111	274	533	86	242	526	74	224
7.2	715	152.0	300.0	674	136	303	601	115	282	565	89	248	557	76	230
7.4	758	156.5	309.0	714	140	312	639	118	290	599	92	256	588	78	237
7.6	803	161.0	319.0	756	143	321	680	122	299	633	94	263	620	81	244
7.8	850	165.5	328.0	798	148	330	724	125	308	668	97	270	655	83	250
8.0	898	170.0	337.0	840	152	340	770	129	317	703	99	279	694	85	257

Note:
[1] Data in tables E-2 through E-6 taken from "Piping Design and Engineering," ITT Grinnell Industrial Piping, 5th Edition, Providence, RI, 1976.

Notes

Appendix F

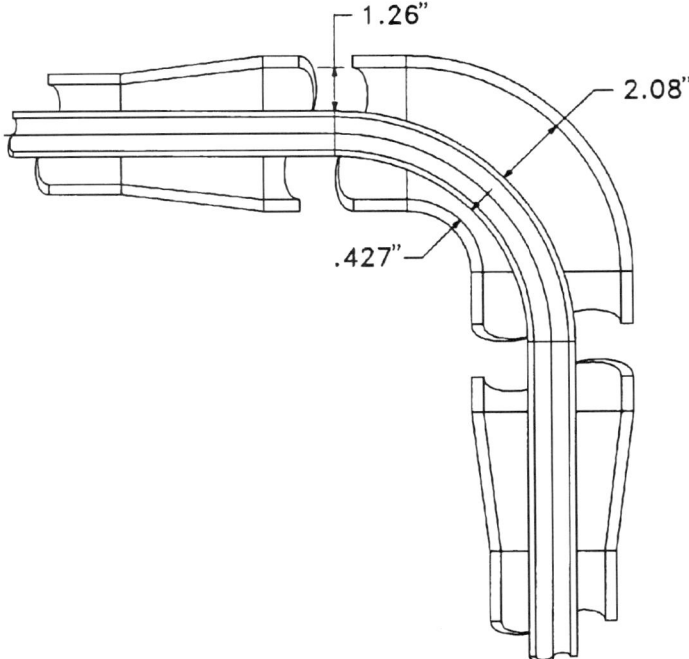

Figure F.1. Dimensional analysis of elbow section used in Line "A" from figure 11.1. The primary section consists of 1" schedule 80 PP pipe with a 6" radius primary elbow (fabricated bend.) The secondary section consists of a 3" schedule 80 PP pipe expanded to 4" schedule 80 in the area of the elbow. The secondary containment elbow is a short radius elbow (4" radius.) The maximum room for growth is directly in the center, and the least amount of room is in the straight pipe section between the 1" and 3" pipes.. The direction of movement of the primary relative to the secondary is determined by the relative amounts of straight pipe on either side of the elbow. (Patent Pending)

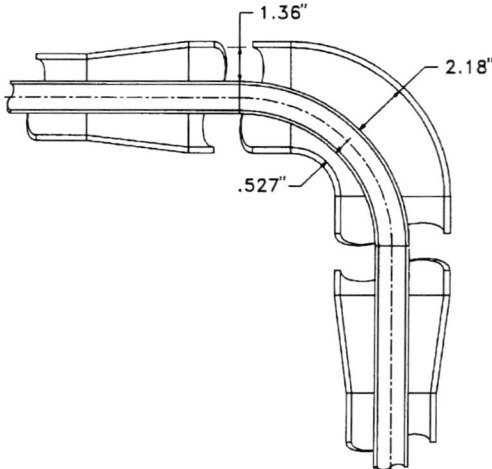

Figure F.2. Dimensional analysis of elbow section used in Line "B" from figure 11.1. The primary section consists of 1" schedule 40 PVDF pipe with a 6" radius primary elbow (fabricated bend.) The secondary section consists of a 3" schedule 40 PVDF pipe expanded to 4" schedule 40 in the area of the elbow. The secondary containment elbow is a short radius elbow (4" radius.) The maximum room for growth is directly in the center, and the least amount of room is in the straight pipe section between the 1" and 3" pipes.. The direction of movement of the primary relative to the secondary is determined by the relative amounts of straight pipe on either side of the elbow.

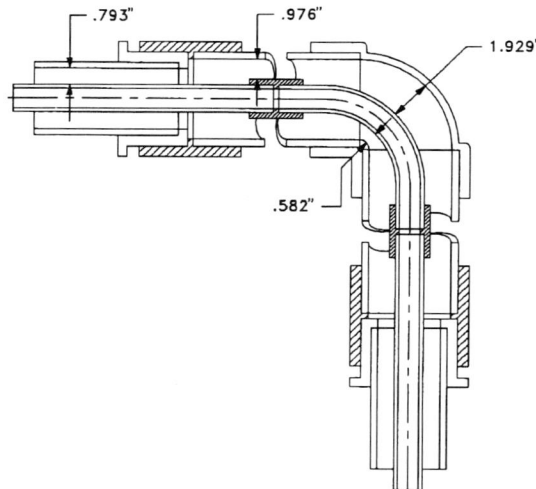

Figure F.3. Dimensional analysis of elbow section used in Line "C" from figure 11.1. The primary section consists of 1" schedule 80 CPVC pipe with a 4" radius primary elbow. The secondary section consists of a 3" schedule 80 CPVC pipe expanded to 4" schedule 80 in the area of the elbow. The secondary containment elbow is a standard socket elbow (ASTM F 439.) The maximum room for growth is directly in the center, and the least amount of room is in the straight pipe section between the 1" and 3" pipes.. The direction of movement of the primary relative to the secondary is determined by the relative amounts of straight pipe on either side of the elbow.

Appendix F

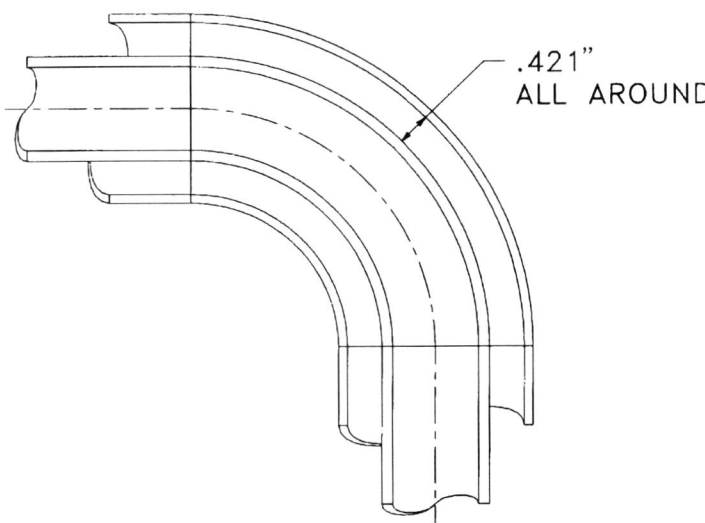

Figure F.4. Dimensional analysis of elbow section used in Lines "AA" and "BB" from figure 11.2. The primary section consists of schedule 40S pipe with a 3" radius primary elbow. The secondary section consists of a 2" schedule 10S pipe with a long radius secondary containment elbow (3" radius.) The maximum room for growth is directly in the center, and the least amount of room is in the straight pipe section between the 1" and 2" pipes.. The direction of movement of the primary relative to the secondary is determined by the relative amounts of straight pipe on either side of the elbow.

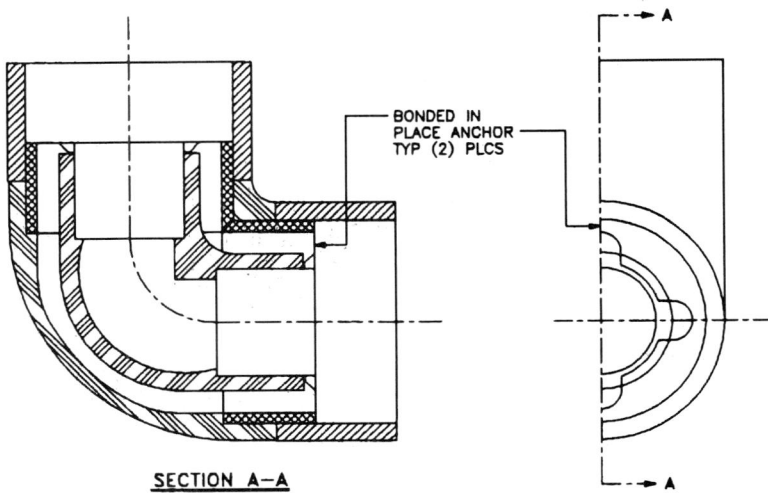

Figure F.5. Illustration of the restrained RTRP elbow shown used in the layout of line "CC" from figure 11.2. A dimensional analysis is not required as the elbow is manufactured in a fully restrained manner. The primary elbow is a compression molded straight socket elbow (1" dia.) to match the 1" pipe. The secondary containment elbow is a compression molded straight socket elbow (3" dia.) to match the 3" secondary containment pipe. (U.S. Patent #4,886,305)

Notes

Index

—A—

Allowable stress range, 181
Aluminum, 34
Aluminum alloys, 34, 76
American Society of Civil Engineers, 11
American Society of Plumbing Engineers, 11
American Water Works Association, 576
Anchors,
    deadman, 650
    midanchoring, 651
ANSI/ASME B31 Code, 12-13, 72, 157, 159, 186, 256, 401, 541, 675
ANSI/ASME B31.1 Code, 251, 272, 314, 338, 534
ANSI/ASME B31.3 Code, 13-14, 72, 85, 126, 151, 157-158, 162, 164, 166, 173-175, 179-180, 184-186, 191, 195-196, 198, 200, 215, 229, 244, 263, 272, 278, 314, 438, 441, 448-449, 499, 507, 510-511, 515-516, 534, 541, 588, 597-598, 600-601, 606-607, 609, 612, 620, 624, 670, 672, 787
ANSI/ASME Standard, 13
ANSI/AWWA C950, 202, 360-362
ASME Boiler and Pressure Vessel Code, 9, 13, 157, 162, 165, 171-173, 177-178, 191, 209, 229, 263, 266, 436, 499-500, 510, 512, 595-596, 600, 608, 633, 635, 655, 667-679, 691
ASTM Annual Book of Standards, 38, 65, 79
Artesian, 5-6
Attendent vent stacks, 132
Austenitic stainless steel, 27, 39, 40
AWWA, 576
Axial stress, 256-258

—B—

Backfilling, 581-584
    tanks, 641-647
Bending, 511
Bending stress, 316, 358-360
Bernoulli's Theorem, 92-93, 99, 112
Bond,
    adhesive bonding, 522, 523, 534-535
    procedure specification, 514-515, 523-524
    seal bonds, 524
Borosilicate glass, 62, 80, 83, 201, 214, 526-528
BPS, 514-515, 523-524
Branch connection(s),
    design,
        of metallic, 165-171
        of nonmetallic, 207-208
        of secondary containment, 238-239
    fittings, 431-434
Brass, 34
Bronze, 34
Buckling,
    maximum allowable, 201
Burial analysis,
    flexible pipes, 349-352, 354-355, 357-358, 365, 573-575
Butt-and-wrapped joints, 523-525

—C—

Cantilever beam, 284, 292
Carbon steel, 38-39, 82
Cast alloys, 46-47
Cathodic protection, 38, 537, 662, 782-790
    impressed-current systems, 787
    sacrificial (galvanic) anode systems, 787-789

testing, cathodic protection systems, 790
Cavitation, 112, 117, 151, 194
Cement, 214
Centroid, 313
CERCLA, 763-764
Cleanouts, 453
Closure ring, 410-411
Closures,
    design, of metallic, 171-172
    design, of nonmetallic, 208
Cold spring, 303
Concentration cell corrosion, 20, 24-25
Concrete, 63, 80, 83
Condensation effects, 716, 728-730
Conduction, 369-378
Conductivity probes, 722
Convection, 378-381
Copolymer, 517
Copper, 33, 82
Copper alloys, 33-34, 76, 82
Corrosion,
    concentration cell, 20, 25
    crevice, 20, 26
    electrochemical, 20
    erosion, 20, 33-34, 43
    galvanic, 19, 23-24, 26, 35-36
    general, 19, 21-22
    intergranular, 19, 27
    microbiologically induced, 20, 30, 32, 597
    pitting, 20, 24-25, 33, 35-36
    uniform, 19, 21, 22
Coupling,
    clam-shell, 537, 548, 553, 556
    slip, 537-538, 548-549
Crevice corrosion, 20, 26
Critical stress, 53, 67-68
Cupro-nickel, 34
Cured-in-place pipe, (CIPP), 751, 757-761

— D —

Darcy formula, 99-100, 105, 112
Dead-leg, 413-415
Deadman anchors, 650
Delamination, 50
Density sensing systems, 724
Design pressure, 152-153, 194-195, 222-223
Design temperature, 153, 195
Detection,
    moisture, 721-722
    pH, 723
    vapor, 722
Diameter,
    minimum, 132, 135
    reduction/reversion techniques, 755-756
Diametrical deflection, 319, 335-336, 338, 349, 355, 357, 359, 363-365
    tanks, 646-647
Dielectric coatings, 785-786
Differential thermal expansion, 227, 250-256
Dimensional analysis, 284, 286-289
Displacement strains, 182, 226
Displacement stresses, 182-183, 226
Double-Butt® System, 757-758
Double failure, 9, 221, 225, 228, 230, 254, 412
Double-walled tanks, 653-660
Drag, 119-120
Duplex stainless steel, 40
DWV, 449, 451, 465

— E —

ECTFE, 57-58, 519
Elastic behavior, 156, 215, 269
Elastic follow-up, 183-184
Elastic plastic follow-up, 271-274
Elbows,
    design,
        of metallic, 163-164
        of nonmetallic, 206-207
        of secondary containment, 234-237
Electrochemical corrosion, 20
Entrance and exit loss, 111
Environmental stress cracking, 30, 50-51, 53-58, 65-68
Epoxy, 401
Equivalent diameter, 118-119
Equivalent length, 107-109, 111
Erosion corrosion, 20, 33-34, 43
ETFE, 57-58, 519
Examination,
    in-progress, 596, 605
    liquid penetrant, 596
    magnetic particle, 596
    radiographic, 596
    ultrasonic, 596, 605
Exfoliation, 20, 37
Expansion joints, 275-278
External supports, 323-327

— F —

Factory Mutual, 450
Fanning Friction Factor, 100, 102-103
FEP, 57, 59
Ferritic stainless steel, 39

Fixture outlet, 129
Fixture unit, 129, 140
Flanges,
    design, of metallic, 173-174
    design, of nonmetallic, 209
    protective flange cover, 421, 423, 428, 435-436
Flaring procedure specification (FPS), 525
Flexible hose systems, 470, 472, 473
Flexibility supports, 278, 333, 335-336, 428-430
Floor drains, 451-453, 461-465
Flow sensing systems, 723
Fluoropolymers, 57, 78, 80, 105, 437
F.M., 450
FPS, 525
Freely floating, 278
Fugitive emissions, 493, 634
Furan, 62

— G —

Galvanic corrosion, 19, 23-24, 26, 35-36, 477
General corrosion, 19, 21-22
Getter, 29
Geysering, 151, 194
GMAW, 498
Grinell Method, 301, 294-295
Grooved mechanical joints, 406, 428
Ground water, 4-6
GTAW, 498-499
GTTD, 394

— H —

Harmonic frequency, 123
Hazardous fluid, 2
Hazen and Williams, 105, 112
HDPE, 56, 78, 82-83, 85, 141, 259, 273-274, 329, 407, 425, 431, 519, 529, 556, 689, 751-757, 762, 764, 770
Heat element butt-welding, 54, 520-522, 566
Heat tracing, 386-391
    electrical heat tracing, 386-391
    electrical heat tracing, self-regulating, 387-391
    hot oil tracing, 388-391
    steam tracing, 388-391
Heat treatment,
    preheating, 506-507
    postwelding, 507-511
High density polyethylene, 56, 78, 82-83, 85, 141, 259, 273-274, 329, 407, 425, 431, 519, 529, 556, 689, 751-757, 762, 764, 770
Hold-down pads, 650-651

Homogeneous, 75, 255, 536-537
Homopolymer, 517
Hybrid, 79-81, 426, 453, 537
Hydraulic radius, 118-119, 135, 140
Hydrologic cycle, 4
Hydrolic jump, 128-129

— I —

Initiation and termination components, 241-244
In-progress examination, 596, 605
Insert ring, 410-411
Interconnecting components, 151, 193, 220, 224, 228, 230, 233, 254-255, 260, 264-266, 306
Intergranular corrosion, 19, 27
Internal anchors, 241, 424-426
Internally-flexible, 279-286, 292-303, 400
Interstitial supporting devices, 323, 328-336
Iowa Formula, 339, 354-355, 357

— J —

Joints,
    butt-and-wrapped joints, 523-525

— K —

Kinematic viscosity, 90

— L —

Laminar flow, 96, 98, 120-121, 123
Lap-joint flange, 153-154
Leak detection,
    manual, 725-726
    visual, 726
Leak detection cable, 323, 355, 404, 698-720
Liners,
    folded and formed, 761-762
    impervious barriers, 762-772
    profiled thermoplastic sheet, 764, 777-781
Liquid level sensing, 720-721
Liquid penetrant examination, 596
Listed components, 157, 161, 198-199, 209-210, 219, 229-230
Live load(s), 154, 343-344
Longitudinal bending, 313-321, 362
Lowest stressed state (LSS), 279
LTTD, 394

— M —

Magnetic particle examination, 596

Manholes, 466-469
Manning Equation, 135, 144-145
Mapping, 716-717
Marston equation, 154, 196, 338-341, 343
Martensitic stainless steel, 39
Microbiologically induced corrosion, 20, 30, 32, 597
Midanchoring, 651
MIG, 498
Minimum diameters, 114-116
Moisture detection, 721-722
Moment of inertia, 313, 320-321
Momentum transfer, 93, 95
Moody Friction Factors, 100-103
Multiple-primary pipes, 473-478, 569-570

— N —

"n"-span, 321-322
Negative surges, 117
Newtonian liquids, 89
NFPA 30, 643
Nickel, 41, 76, 82, 561-564
Nickel alloys, 41, 76, 82, 561-564
Non-differential thermal expansion, 227, 250-254
Non-Newtonian liquids, 89
NPSH, 112

— O —

Overfill protection, 663
Oxidation-type reaction, 20, 50, 52

— P —

P number, 499
Parallel flow, 112
PEEK, 59, 519
Periodic pressure testing, annuli, 727
PFA, 57, 59, 78, 519
pH detection, 723
Phenolic, 62
Pipe bursting, 754-755
Piping flexibility, 181-189, 213-216, 260-264, 274-304, 400, 427-431
Pitting, 20, 24-25, 33, 35-36
Plastic-lined metallic systems, 490, 492-494
Plasticization, 50, 53
Point probe sensing systems,
    conductivity probes, 722
    density sensing, 724
    flow sensing, 723
    liquid level, 720-721

    moisture detection, 721-722
    pH detection, 723
    pressure, 723-724
    vapor detection, 722
Polybutylene, 57
Polyester, 61
Polyethylene, 56, 78, 82-83, 85, 141, 259, 273-274, 329, 407, 425, 431, 519, 529, 556, 689, 751-757, 762, 764, 770
Polyketones, 105
Polypropylene, 50, 54-55, 57, 78, 83, 85, 141, 304, 401, 407, 437, 517, 691
Polyolefin, 56, 78, 80, 444
Polytetrafluoroethylene, 57, 59, 329 484
Polyvinyl chloride, 55, 78, 304, 401, 411, 435, 445, 519, 541, 572, 575, 584, 606, 620, 622, 624, 631, 687, 691, 746, 755, 761
Polyvinylidene fluoride, 54, 57-58, 78, 401, 407, 484, 519, 584
Precipitation hardening stainless steel, 40
Pre-engineered, 71
Preliminary pipe sizing, 114
Pressure measurement, 91
Pressure relief, 190, 217
Pressure sensing systems, 723-724
Protective flange cover, 421, 423, 428, 435-436
PTFE, 57, 59, 329 484
P-trap, 130, 131, 453-456, 461, 463-464
Pup, 541-543, 545-546
Purging requirements, 716, 728-730
PVC, 55-56, 78, 85-86, 304, 401, 411, 435, 445, 519, 541, 572, 575, 584, 606, 620, 622, 624, 631, 687, 691, 746, 755, 761
PVDF, 54, 78, 83, 274, 401, 407, 484, 519, 584

— R —

Radiation, 381-382RCRA, 2, 76, 219-220
Radiographic examination, 596
RCRA , 2, 76, 219-220, 463, 634, 763, 782
Reducers, design,
    of metallic, 172-173
    of secondary containment, 240
Reduction, 20, 50, 52
Reinforced thermosetting resin plastics, 19, 60-62, 74, 76-77, 79-81, 84-85, 193, 201, 213-214, 259, 265, 269, 271-273, 279, 364, 401, 404-405, 411, 427, 435-437, 445, 470, 472, 513-517, 522-524, 526, 529-530, 534-535, 541, 566, 571-572, 604, 610, 631, 635, 642-643, 646, 648, 654, 676, 687, 690, 752, 775-777
Relative viscosity, 90
Resource Conservation and Recovery Act, 2, 76,

219-220, 463, 634, 763, 782
Restrained systems, 269, 271
Reynold's Number, 96, 111, 119-121, 123
RTRP, 19, 60-62, 74, 76- 77, 79-81, 84-85, 193, 201, 213-214, 259, 265, 269, 271-273, 279, 364, 401, 404-405, 411, 427, 435-437, 445, 470, 472, 513-517, 522-524, 526, 529-530, 534-535, 541, 566, 571-572, 604, 610, 631, 635, 642-643, 646, 648, 654, 676, 687, 690, 752, 775-777

— S —

S number, 499
Safeguarding, 85
Seal bonds, 524
Seal welds, 513
Secondary closure(s), 73
Secondary containment,
    structures, 663-666, 773-781
    trenches, 458, 717, 749
Seismic loading, 155
Sensitization, 27
Sensitizing temperature range, 27
Separation, 75, 225, 229-230, 242, 255, 260
Severe cyclic service, 181, 193, 228, 506, 593
Simultaneous fusion, 437, 442, 528-529, 536, 549 555, 563, 565-569
Slip-coupling, 443-445, 447
Slip-lining, 556, 752-757
SMAW, 498
Snaking, 303-304, 313
Soil loads, 313, 340-344
Solvation, 50, 53
Specific gravity, 89-90
Spill containment, 137-140
Stainless steel, 39, 75, 82, 561
    duplex, 40
    ferritic, 39
    martensitic, 39
    precipitation hardening, 40
Steel, 38-39, 75
Straight pipe,
    design,
        of metallic, 161-163
        of secondary containment, 233
        under external pressure, 162, 204-206
    filament wound RTR, RPM, 203
    laminated RTR, thermoplastic, 203
    sections, 401-403
Stress corrosion cracking, 19, 29-30, 35-37, 42
Subsurface water, 4

Support spacing, 321-322
Surcharging, 136-137, 145-146

— T —

Tanks,
    backfilling, 641-647
    diametrical deflection, 646-647
    double-walled, 653-660
    inventory monitoring, 732-734
    monitoring,
        continuous pressure, 735
        downhole flux chambers, 741
        inventory, 732-734
        observation/detection wells, 736-737
        soil vapor, 740
        surface flux chambers, 741
        surface geophysical techniques, 739
        tightness testing, 734-735
        U-tubes, 746
        vapor wells, 743-746
    multi-compartment, 656-660
    testing, double-walled tanks, 655-656
    testing, tanks, 652-653
Tantalum, 44-45, 76, 82
TDR, 10, 698-702, 720
Terminal length, 126, 140, 142
Terminal velocity, 126, 129
Termination and initiation components, 241-244, 402-411, 413-416
Testing, 597, 600, 605-608, 610-631, 681-682
    cathodic protection systems, 790
    double-walled tanks, 655-656
    periodic pressure testing, annuli, 727
    tanks, 652-653
        tightness testing of tanks, 734-735
Thermoplastic, 19, 55-60, 74, 77-78, 80, 82, 86, 193, 200, 213-214, 404-406, 428, 435, 437, 444-445, 513-522, 535, 548, 563, 565-569, 571-572, 628, 630-631, 642
TIG, 499
Time domain reflectometry, 10
Timoshenko's Equation, 316
Titanium, 42-44, 76, 82
Titanium alloys, 42-43
Trench,
    secondary containment, 458
Trenchless reconstruction, 751-761
Turbulent flow, 96, 98, 121, 123

— U —

U.L., 9, 450, 470, 473, 635, 653-654, 666

Ultrasonic examination, 596
Underwriters Laboratories (U.L.), 9, 450, 470, 473, 635, 653-654, 666
Uniform corrosion, 19, 21-22
Uniform flow, 96-98, 135
Unlisted components, 157, 161, 198-199, 209-211, 219, 229-230, 236-237, 239-240, 242-243, 264
UST, 84

— V —

Valves, 478-493
Vapor,
   detection, 722
   pressure, 90-91
Velocity,
   maximum, 114-115
Vent extension, 127
Vent stack, 127
Venting, 73, 131-134, 142-143
Vertical stacks, 126-128, 140-141, 455-457
Vinylester, 62
Vinylidene chloride, 56
Viscosity, 90
Vortex shedding, 122-123

— W —

Water hammer, 116-117, 151, 194
Weld-wrap, 443-444
Welding,
   extrusion, 517-518
   hot-gas, 517-518
   procedure specification, 499, 501-503, 506, 525
   seal welds, 513
   simultaneous, 364
   staggered, 437-445, 528-554, 556-557
WPS, 499, 501-503, 506, 525

— Z —

Zirconium, 45-46, 76, 82

ABOUT THE AUTHOR

Christopher G. Ziu is a principal in Herzog-Hart Corporation, and holds 14 U.S. patents in double containment piping systems design. He previously served as president of Double Containment Systems, Inc., as vice president of engineering for Asahi/America, as a process engineer for Fluor-Daniels and John Brown Engineers, and as a field engineer for R&G Sloane Manufacturing Company. He is an active member of various ASME technical committees and other societies.